CITY COLLEGE OF SAN FRANCISCO
ROSENBERG LIBRARY
50 PHELAN AVENUE
SAN FRANCISCO, CA 94112

Y0-BZA-967

Ref TJ 151 .M395 2014 v.2

Mechanical engineers'
 handbook

FOR REFERENCE

Do Not Take From This Room

Mechanical Engineers' Handbook

Mechanical Engineers' Handbook
Fourth Edition

Design, Instrumentation, and Controls

Edited by
Myer Kutz

CITY COLLEGE OF SAN FRANCISCO
ROSENBERG LIBRARY
50 PHELAN AVENUE
SAN FRANCISCO, CA 94112

WILEY

Ref TJ 151 .M395 2014 v.2

Mechanical engineers'
handbook

APR 2 3 2015

Cover image: © denisovd / Thinkstock
Cover design: Wiley

This book is printed on acid-free paper.

Copyright © 2014 by John Wiley & Sons, Inc. All rights reserved

Published by John Wiley & Sons, Inc., Hoboken, New Jersey
Published simultaneously in Canada

No part of this publication may be reproduced, stored in a retrieval system, or transmitted in any form or by any means, electronic, mechanical, photocopying, recording, scanning, or otherwise, except as permitted under Section 107 or 108 of the 1976 United States Copyright Act, without either the prior written permission of the Publisher, or authorization through payment of the appropriate per-copy fee to the Copyright Clearance Center, 222 Rosewood Drive, Danvers, MA 01923, (978) 750–8400, fax (978) 646–8600, or on the web at www.copyright.com. Requests to the Publisher for permission should be addressed to the Permissions Department, John Wiley & Sons, Inc., 111 River Street, Hoboken, NJ 07030, (201) 748–6011, fax (201) 748–6008, or online at www.wiley.com/go/permissions.

Limit of Liability/Disclaimer of Warranty: While the publisher and author have used their best efforts in preparing this book, they make no representations or warranties with the respect to the accuracy or completeness of the contents of this book and specifically disclaim any implied warranties of merchantability or fitness for a particular purpose. No warranty may be created or extended by sales representatives or written sales materials. The advice and strategies contained herein may not be suitable for your situation. You should consult with a professional where appropriate. Neither the publisher nor the author shall be liable for damages arising herefrom.

For general information about our other products and services, please contact our Customer Care Department within the United States at (800) 762 2974, outside the United States at (317) 572–3993 or fax (317) 572–4002.

Wiley publishes in a variety of print and electronic formats and by print-on-demand. Some material included with standard print versions of this book may not be included in e-books or in print-on-demand. If this book refers to media such as a CD or DVD that is not included in the version you purchased, you may download this material at http://booksupport.wiley.com. For more information about Wiley products, visit www.wiley.com.

Library of Congress Cataloging-in-Publication Data:

Mechanical engineers handbook : design, instrumentation, and controls / edited by Myer Kutz. – Fourth edition.
    1 online resource.
  Includes index.
  Description based on print version record and CIP data provided by publisher; resource not viewed.
  ISBN 978-1-118-93080-9 (ePub) – ISBN 978-1-118-93083-0 (Adobe PDF) – ISBN 978-1-118-11899-3 (4-volume set) – ISBN 978-1-118-11283-0 (cloth : volume 2 : acid-free paper) 1. Mechanical engineering–Handbooks, manuals, etc. I. Kutz, Myer, editor of compilation.
  TJ151
  621–dc23   2014005952

Printed in the United States of America

10 9 8 7 6 5 4 3 2 1

To Arlene, Bill, Merrilyn, and Jayden

Contents

Preface

The second volume of the fourth edition of the *Mechanical Engineers' Handbook* is comprised of two parts: Part 1, Mechanical Design, with 14 chapters, and Part 2, Instrumentation, Systems, Controls and MEMS, with 11 chapters. The mechanical design chapters were in Volume I in the third edition. Given the introduction of 6 new chapters, mostly on measurements, in Volume I in this edition, it made sense to move the mechanical design chapters to Volume II and to cull chapters on instrumentation to make way for the measurements chapters, which are of greater use to readers of this handbook. Moreover, the mechanical design chapters have been augmented with 4 chapters (updated as needed) from my book, *Environmentally Conscious Mechanical Design*, thereby putting greater emphasis on sustainability. The 4 chapters are Design for Environment, Life-Cycle Design, Design for Maintainability, and Design for Remanufacturing Processes. They flesh out sustainability issues that were covered in the third edition by only one chapter, Product Design and Manufacturing Processes for Sustainability. The other 9 mechanical design chapters all appeared in the third edition. Six of them have been updated.

In the second part of Volume 2, Instrumentation, Systems, Controls and MEMS, 5 of the 11 chapters were new to the third edition of the handbook, including the 3 chapters I labeled as "new departures": Neural Networks in Control Systems, Mechatronics, and Introduction to Microelectromechanical Systems (MEMS): Design and Application. These topics have become increasingly important to mechanical engineers in recent years and they are included again. Overall, 3 chapters have been updated for this edition. In addition, I brought over the Electric Circuits chapter from the fifth edition of *Eshbach's Handbook of Engineering Fundamentals*. Readers of this part of Volume 2 will also find a general discussion of systems engineering; fundamentals of control system design, analysis, and performance modification; and detailed information about the design of servo actuators, controllers, and general-purpose control devices.

All Volume 2 contributors are from North America. I would like to thank all of them for the considerable time and effort they put into preparing their chapters.

Vision for the Fourth Edition

Basic engineering disciplines are not static, no matter how old and well established they are. The field of mechanical engineering is no exception. Movement within this broadly based discipline is multidimensional. Even the classic subjects, on which the discipline was founded, such as mechanics of materials and heat transfer, keep evolving. Mechanical engineers continue to be heavily involved with disciplines allied to mechanical engineering, such as industrial and manufacturing engineering, which are also constantly evolving. Advances in other major disciplines, such as electrical and electronics engineering, have significant impact on the work of mechanical engineers. New subject areas, such as neural networks, suddenly become all the rage.

In response to this exciting, dynamic atmosphere, the Mechanical Engineers' Handbook expanded dramatically, from one to four volumes for the third edition, published in November 2005. It not only incorporated updates and revisions to chapters in the second edition, published seven years earlier, but also added 24 chapters on entirely new subjects, with updates and revisions to chapters in the Handbook of Materials Selection, published in 2002, as well as to chapters in Instrumentation and Control, edited by Chester Nachtigal and published in 1990, but never updated by him.

The fourth edition retains the four-volume format, but there are several additional major changes. The second part of Volume I is now devoted entirely to topics in engineering mechanics, with the addition of five practical chapters on measurements from the Handbook of Measurement in Science and Engineering, published in 2013, and a chapter from the fifth edition of Eshbach's Handbook of Engineering Fundamentals, published in 2009. Chapters on mechanical design have been moved from Volume I to Volumes II and III. They have been augmented with four chapters (updated as needed) from Environmentally Conscious Mechanical Design, published in 2007. These chapters, together with five chapters (updated as needed, three from Environmentally Conscious Manufacturing, published in 2007, and two from Environmentally Conscious Materials Handling, published in 2009) in the beefed-up manufacturing section of Volume III, give the handbook greater and practical emphasis on the vital issue of sustainability.

Prefaces to the handbook's individual volumes provide further details on chapter additions, updates and replacements. The four volumes of the fourth edition are arranged as follows:

Volume 1: Materials and Engineering Mechanics—27 chapters

   Part 1. Materials—15 chapters

   Part 2. Engineering Mechanics—12 chapters

Volume 2: Design, Instrumentation and Controls—25 chapters

   Part 1. Mechanical Design—14 chapters

   Part 2. Instrumentation, Systems, Controls and MEMS —11 chapters

Volume 3: Manufacturing and Management—28 chapters

   Part 1. Manufacturing—16 chapters

   Part 2. Management, Finance, Quality, Law, and Research—12 chapters

Volume 4. Energy and Power—35 chapters

Part 1: Energy—16 chapters

Part 2: Power—19 chapters

The mechanical engineering literature is extensive and has been so for a considerable period of time. Many textbooks, reference works, and manuals as well as a substantial number of journals exist. Numerous commercial publishers and professional societies, particularly in the United States and Europe, distribute these materials. The literature grows continuously, as applied mechanical engineering research finds new ways of designing, controlling, measuring, making, and maintaining things, as well as monitoring and evaluating technologies, infrastructures, and systems.

Most professional-level mechanical engineering publications tend to be specialized, directed to the specific needs of particular groups of practitioners. Overall, however, the mechanical engineering audience is broad and multidisciplinary. Practitioners work in a variety of organizations, including institutions of higher learning, design, manufacturing, and consulting firms, as well as federal, state, and local government agencies. A rationale for a general mechanical engineering handbook is that every practitioner, researcher, and bureaucrat cannot be an expert on every topic, especially in so broad and multidisciplinary a field, and may need an authoritative professional summary of a subject with which he or she is not intimately familiar.

Starting with the first edition, published in 1986, my intention has always been that the Mechanical Engineers' Handbook stand at the intersection of textbooks, research papers, and design manuals. For example, I want the handbook to help young engineers move from the college classroom to the professional office and laboratory where they may have to deal with issues and problems in areas they have not studied extensively in school.

With this fourth edition, I have continued to produce a practical reference for the mechanical engineer who is seeking to answer a question, solve a problem, reduce a cost, or improve a system or facility. The handbook is not a research monograph. Its chapters offer design techniques, illustrate successful applications, or provide guidelines to improving performance, life expectancy, effectiveness, or usefulness of parts, assemblies, and systems. The purpose is to show readers what options are available in a particular situation and which option they might choose to solve problems at hand.

The aim of this handbook is to serve as a source of practical advice to readers. I hope that the handbook will be the first information resource a practicing engineer consults when faced with a new problem or opportunity—even before turning to other print sources, even officially sanctioned ones, or to sites on the Internet. In each chapter, the reader should feel that he or she is in the hands of an experienced consultant who is providing sensible advice that can lead to beneficial action and results.

Can a single handbook, even spread out over four volumes, cover this broad, interdisciplinary field? I have designed the Mechanical Engineers' Handbook as if it were serving as a core for an Internet-based information source. Many chapters in the handbook point readers to information sources on the Web dealing with the subjects addressed. Furthermore, where appropriate, enough analytical techniques and data are provided to allow the reader to employ a preliminary approach to solving problems.

The contributors have written, to the extent their backgrounds and capabilities make possible, in a style that reflects practical discussion informed by real-world experience. I would like readers to feel that they are in the presence of experienced teachers and consultants who know about the multiplicity of technical issues that impinge on any topic within mechanical engineering. At the same time, the level is such that students and recent graduates can find the handbook as accessible as experienced engineers.

Contributors

H. Barry Bebb
ASI
San Diego, California

Bert Bras
Georgia Institute of Technology
Atlanta, Georgia

Sujeet Chand
Rockwell Automation
Milwaukee, Wisconsin

James H. Christensen
Holobloc, Inc.
Cleveland Heights, Ohio

Abigail Clarke
Michigan Technological University
Houghton, Michigan

B. S. Dhillon
University of Ottawa
Ottawa, Ontario, Canada

Shane Farritor
University of Nebraska–Lincoln
Lincoln, Nebraska

Daniel P. Fitzgerald
Stanley Black & Decker
Towson, Maryland

Shuzhi Sam Ge
University of Electronic Science and
Technology of China
Chendu, China
and
National University of Singapore
Singapore

John K. Gershenson
Michigan Technological University
Houghton, Michigan

Thornton H. Gogoll
Stanley Black and Decker
Towson, Maryland

James A. Harvey
Under the Bridge Consulting, Inc.
Corvallis, Oregon

Jeff Hawks
University of Nebraska–Lincoln
Lincoln, Nebraska

Jeffrey W. Herrmann
University of Maryland
College Park, Maryland

E. L. Hixson
University of Texas
Austin, Texas

I. S. Jawahir
University of Kentucky
Lexington, Kentucky

Robert J. Kretschmann
Rockwell Automation
Mayfield Heights, Ohio

Ming C. Leu
Missouri University of Science and
Technology
Rolla, Missouri

Gordon Lewis
Digital Equipment Corporation
Maynard, Massachusetts

F.L. Lewis
The University of Texas at Arlington
Fort Worth, Texas

Charalambos A. Marangos
Lehigh University
Bethlehem, Pennsylvania

James E. McMunigal
MCM Associates
Long Beach, California

Philip C. Milliman
Weyerhaeuser Company
Federal Way, Washington

Maury A. Nussbaum
Virginia Tech,
Blacksburg, Virginia

O. Geoffrey Okogbaa
University of South Florida,
Tampa, Florida

Wilkistar Otieno
University of South Florida,
Tampa, Florida

William J. Palm III
University of Rhode Island
Kingston, Rhode Island

Xiaobo Peng
Missouri University of Science and
Technology
Rolla, Missouri

A. Ravi Ravindran
The Pennsylvania State University
University Park, Pennsylvania

G. V. Reklaitis
Purdue University
West Lafayette, Indiana

E. A. Ripperger
University of Texas
Austin, Texas

Albert J. Rosa
University of Denver
Denver, Colorado

Andrew P. Sage
George Mason University
Fairfax, Virginia

Peter A. Sandborn
University of Maryland
College Park, Maryland

Linda C. Schmidt
University of Maryland
College Park, Maryland

Sekar Sundararajan
Lehigh University
Bethlehem, Pennsylvania

John Turnbull
Case Western Reserve University
Cleveland, Ohio

K. G. Vamvoudakis
University of California
Santa Barbara, California

Jaap H. van Dieën
Free University,
Amsterdam, The Netherlands

X. Wang
University of Kentucky
Lexington, Kentucky

P. C. Wanigarathne
University of Kentucky
Lexington, Kentucky

K. Preston White, Jr.
University of Virginia
Charlottesville, Virginia

Kazuhiko Yokoyama
Yaskawa Electric Corporation
Tokyo, Japan

M. E. Zaghloul
George Washington University
Washington, D.C.

Wenjuan Zhu
Missouri University of Science and
Technology
Rolla, Missouri

Emory W. Zimmers, Jr., and Technical Staff
Lehigh University
Bethlehem, Pennsylvania

PART 1
DESIGN

CHAPTER 1

COMPUTER-AIDED DESIGN

**Emory W. Zimmers Jr., Charalambos A. Marangos,
Sekar Sundararajan, and Technical Staff**
Lehigh University
Bethlehem, Pennsylvania

1 INTRODUCTION TO CAD

Computers have a prominent, often controlling role throughout the life cycle of engineering products and manufacturing processes. Their role is vital as global competitive pressures call for improvements in product performance and quality coupled with significant reductions in product design, development, and manufacturing timetables. Design engineers vastly improve their work productivity using computers. For example, performance of a product or process can be evaluated prior to fabricating a prototype using appropriate simulation software.

Computer-aided design (CAD) uses the mathematical and graphic processing power of the computer to assist the engineer in the creation, modification, analysis, and display of designs. Many factors have contributed to CAD technology being a necessary tool in the engineering

3

world for applications including shipbuilding, automotive, aerospace, medical, industrial, and architectural design, such as the computer's speed in processing complex equations and managing technical databases. CAD at one time was thought of simply as computer-aided drafting, and its use as an electronic drawing board is still a powerful tool in itself. Geometric modeling, engineering analysis, simulation, and the communication of the design information can also be performed using a CAD system. However, the functions of a CAD system are evolving far beyond its ability to represent and manipulate graphics. The CAD system is being integrated into the overall product life cycle as part of collaborative product design, sustainability impact analysis, product life-cycle management, and product data management.

1.1   Historical Perspective on CAD

Graphical representation of data, in many ways, forms the basis of CAD. An early application of computer graphics was used in the SAGE (Semi-Automatic Ground Environment) Air Defense Command and Control System in the 1950s. SAGE converted radar information into computer-generated images on a cathode ray tube (CRT) display. It also used an input device, the light pen, to select information directly from the CRT screen.

Another significant advancement in computer graphics technology occurred in 1963, when Ivan Sutherland, in his doctoral thesis at MIT, described the SKETCHPAD (Fig. 1) system. A Lincoln TX-2 computer drove the SKETCHPAD system. SKETCHPAD is a graphic user interface that enables a design to be input into a computer using a light pen on the CRT monitor. With SKETCHPAD, images could be created and manipulated using the light pen. Graphical manipulations such as translation, rotation, and scaling could all be accomplished on-screen using SKETCHPAD. Computer applications based on Sutherland's approach have become known as interactive computer graphics (ICG), which are the foundation of CAD design processes. The graphical capabilities of SKETCHPAD showed the potential for computerized drawing in design.

During his time as a professor of electrical engineering at the University of Utah, Sutherland continued his research on head-mounted displays (HMDs), the precursor to virtual reality head displays. The field of computer graphics (Fig. 2), as we know it today, was born from among the many new ideas and innovations created by the researchers who made the University a hub for this kind of research. Together with Dave Evans, the founder of the University's

Figure 1   Ivan E. Sutherland and the SKETCHPAD system.

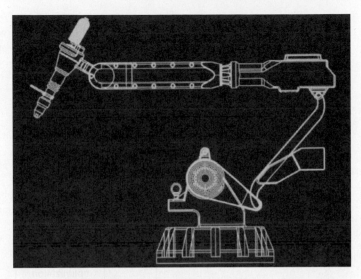

Figure 2 Image on a line drawing graphics display.

Computer Science Department, Sutherland co-founded Evans and Sutherland in 1968, which later went on to pioneer computer modeling systems and software.

While at the California Institute of Technology, Sutherland served as the chairman of the Computer Science Department from 1976 to 1980. While he was there, he helped to introduce the integrated circuit design to academia. Together with Professor Carver Mead, they developed the science of combining the mathematics of computing with the physics of real transistors and real wires and subsequently went on to make integrated circuit design a proper field of academic study. In 1980, Sutherland left Caltech and launched the company Sutherland, Sproull, and Associates. Bought by Sun Labs in 1990, the acquisition formed the basis for Sun Microsystems Laboratories.

The high cost of computer hardware in the 1960s limited the use of ICG systems to large corporations such as those in the automotive and aerospace industries, which could justify the initial investment. With the rapid development of computer technology, computers became more powerful, with faster processors and greater data storage capabilities. As computer cost decreased, systems became more affordable to smaller companies allowing entrepreneurs to innovate using CAD tools and technologies.

In more recent times, increased impact of computer-aided design has been facilitated by advances in Web-based technologies and standards, use of mobile computing platforms and devices, cloud-based storage, software as a service, and functional integration into enterprise-wide systems. Additionally, the proliferation of CAD systems running on a wide variety of platforms has promoted global collaboration as well as concurrent design and manufacturing approaches. In the view of many, CAD has become a necessary business tool for any engineering, design, or architectural firm.

1.2 Design Process

Before any discussion of computer-aided design, it is necessary to understand the design process in general. What is the series of events that leads to the beginning of a design project? How does the engineer go about the process of designing something? How does one arrive at

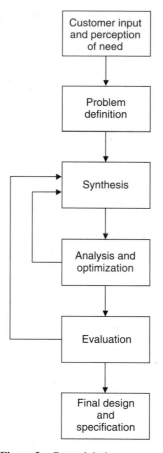

Figure 3 General design process.

the conclusion that the design has been completed? We address these questions by defining the process in terms of six distinct stages (Fig. 3):

1. Customer or sales field input and perception of need
2. Problem definition
3. Preliminary design
4. Analysis and optimization
5. Testing/evaluation
6. Final design and specification

A need is usually perceived in one of two ways. Someone from sales field reports or customer feedback must recognize either a problem in an existing design or a customer-driven opportunity in the marketplace for a new product. In either case, a need exists which can be addressed by modifying an existing design or developing an entirely new design. Because the need for change may only be indicated by subtle circumstances—such as noise, increased sustainability concerns, marginal performance characteristics, or deviations from quality standards—the design engineer who identifies the need has taken a first step in correcting the problem. That step sets in motion processes that may allow others to see the need more readily and possibly enroll them in the solution process.

Once the decision has been made to take corrective action to the need at hand, the problem must be defined as a particular problem to be solved such that all significant parameters in the problem are defined. These parameters often include cost limits, quality standards, size and weight characteristics, and functional characteristics. Often, specifications may be defined by the capabilities of the manufacturing process. Anything that will influence the engineer in choosing design features must be included in the definition of the problem. Careful planning in this stage can lead to fewer iterations in subsequent stages of design.

Once the problem has been fully defined in this way, the designer moves on to the preliminary design stage where knowledge and creativity can be applied to conceptualize an initial design. Teamwork can make the design more successful and effective at this stage. That design is then subjected to various forms of analysis, which may reveal specific problems in the initial design. The designer then takes the analytical results and applies them in an iteration of the preliminary design stage. These iterations may continue through several cycles of preliminary design and analysis until the design is optimized.

The design is then tested/evaluated according to the parameters set forth in the problem definition. A scale prototype is often fabricated to perform further analysis and testing to assess operating performance, quality, reliability, and other criteria. If a design flaw is revealed during this stage, the design moves back to the preliminary design/analysis stages for redesign, and the process moves in this circular manner until the design clears the testing stage and is ready for presentation.

Final design and specification represent the last stage of the design process. Communicating the design to others in such a way that its manufacture and marketing are seen as vital to the organization is essential. When the design has been fully approved, detailed engineering drawings are produced, complete with specifications for components, subassemblies, and the tools and fixtures required to manufacture the product and the associated costs of production. These can then be transferred manually or digitally using the CAD data to the various departments responsible for manufacture.

In every branch of engineering, prior to the implementation of CAD, design had traditionally been accomplished manually on the drawing board. The resulting drawing, complete with significant details, was then subjected to analysis using complex mathematical formulas and then sent back to the drawing board with suggestions for improving the design. The same iterative procedure was followed, and because of the manual nature of the drawing and the subsequent analysis, the whole procedure was time consuming and labor intensive. CAD has allowed the designer to bypass much of the manual drafting and analysis previously required, making the design process flow smoothly and efficiently.

It is helpful to understand the general product development process as a stepwise process. However, in today's engineering environment, the steps outlined above have become consolidated into a more streamlined approach called "concurrent engineering" or "simultaneous engineering." This approach enables teams to work concurrently by providing common ground for interrelated product development tasks. Product information can be easily communicated among all development processes: design, manufacturing, marketing, management, and supplier networks. Concurrent engineering recognizes that fewer iterations result in less time and money spent in moving from concept to manufacture and from manufacturing to market. The related processes of design for manufacturing (DFM) and design for assembly (DFA) have become integral parts of the concurrent engineering approach.

Design for manufacturing and DFA use cross-disciplinary input from a variety of sources (e.g., design engineers, manufacturing engineers, suppliers, and shop-floor representatives) to facilitate the efficient design of a product that can be manufactured, assembled, and marketed in the shortest possible period of time. Often, products designed using DFM and DFA are simpler, cost less, and reach the marketplace in far less time than traditionally designed products. DFM focuses on determining what materials and manufacturing techniques will result in the most efficient use of available resources in order to integrate this information early in the design process. The DFA methodology strives to consolidate the number of parts, uses gravity-assisted assembly techniques wherever possible, and calls for careful review and consensus approval of designs early in the process. By facilitating the free exchange of information, DFM and DFA methods allow engineering companies to avoid the costly rework often associated with repeated iterations of the design process.

1.3   Applying Computers to Design

Many of the individual tasks within the overall design process can be performed using a computer. As each of these tasks is made more efficient, the efficiency of the overall process increases as well. The computer is especially well suited to design in four areas, which correspond to the latter four stages of the general design process (Fig. 4). Computers function in

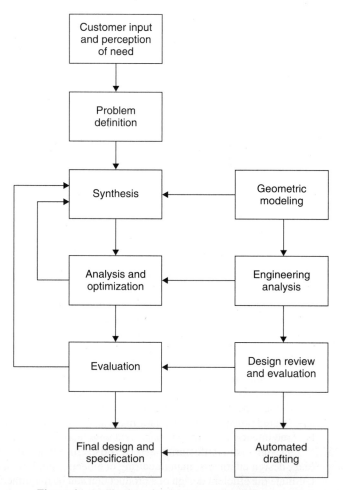

Figure 4   Application of computers to the design process.

the design process through geometric modeling capabilities, engineering analysis calculations, automated evaluative procedures, and automated drafting.

Geometric Modeling

Geometric modeling is one of the keystones of CAD systems. It uses mathematical descriptions of geometric elements to facilitate the representation and manipulation of graphical images on a computer display screen. While the computer central processing unit (CPU) and the graphics processing unit (GPU) provide the ability to quickly make the calculations specific to the element, the software provides the instructions necessary for efficient transfer of information between the user and the CPU and the GPU.

Three types of commands are used by the designer in computerized geometric modeling:

1. Input commands allow the user to input the variables needed by the computer to represent basic geometric elements such as points, lines, arcs, circles, splines, and ellipses.
2. Transformation commands are used to transform these elements. Commonly performed transformations in CAD include scaling, rotation, and translation.
3. Solid commands allow the various elements previously created by the first two commands to be joined into a desired shape.

During the entire geometric modeling process, mathematical operations are at work which can be easily stored as computerized data and retrieved as needed for review, analysis, and modification. There are different ways of displaying the same data on the computer monitor, depending on the needs or preferences of the designer.

One method is to display the design as a two-dimensional (2D) representation of a flat object formed by interconnecting lines.

Another method displays the design as a three-dimensional (3D) representation of objects. In 3D representations, there are four types of modeling approaches:

- Wire frame modeling
- Surface modeling
- Solid modeling
- Hybrid solid modeling

Wire Frame Model

A wire frame model is a skeletal description of a 3D object. It consists only of points, lines, and curves that describe the boundaries of the object. There are no surfaces in a wire frame model. 3D wire frame representations can cause the viewer some confusion because all of the lines defining the object appear on the 2D display screen. This makes it hard for the viewer to tell whether the model is being viewed from above or below, from inside the object or looking from outside.

Surface modeling defines not only the edge of the 3D object but also its surface. Two different types of surfaces can be generated: faceted surfaces using a polygon mesh and true curve surfaces.

A polygonal mesh is a surface approximated by polygons such as squares, rectangles, and hexagons. The surface is created as if a mosaic of fine polygons. Depending on the detail required by the designer, very fine surfaces cannot be created this way. Instead, polygonal meshes allow for faster rendering of shapes, as opposed to using curves.

The nonuniform rational basis spline (NURBS) is a B-spline curve or surface defined by a series of weighted control points and one or more knot vectors. It can exactly represent a wide range of curves such as arcs and conics. The greater flexibility for controlling continuity is one advantage of NURBS. NURBS can precisely model nearly all kinds of surfaces more robustly than the polynomial-based curves that were used in earlier surface models. Surface modeling is more sophisticated than wire frame modeling. Here, the computer still defines the object in terms of a wire frame but generates a surface "skin" to cover the frame, thus giving the illusion of a "real" object. However, because the computer has the image stored in its data as a wire frame representation having no mass, physical properties cannot be calculated directly from the image data. Surface models are very advantageous due to point-to-point data collections usually required for numerical control (NC) programs in computer-aided manufacturing (CAM) applications. Most surface modeling systems also produce the stereolithographic data required for rapid prototyping systems.

Solid Modeling

Solid modeling defines the surfaces of an object, with the added attributes of volume and mass. This allows data to be used in calculating the physical properties of the final product. Solid modeling software uses one of two methods to represent solid objects in a computer: constructive solid geometry (CSG) or boundary representation (B-rep).

The CSG method uses Boolean operations such as union, subtraction, and intersection on two sets of objects to define composite solid models. For example, to create a hole in a cube, a small cylinder can be subtracted from a large cube. See Fig. 5.

B-rep is a representation of a solid model that defines an object in terms of its surface boundaries: faces, edges, and vertices. In the case of the cube with a hole, a square surface could be created with a hole (as two mirrored surfaces) and then extruded to create the model. See Fig. 6.

Hybrid Solid Modeling

Hybrid solid modeling allows the user to represent a part with a mixture of wire frame, surface modeling, and solid geometry. The Siemens product lifecycle management (PLM) program offers this representation feature.

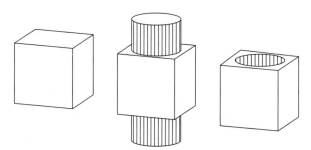

Figure 5   Solid subtraction.

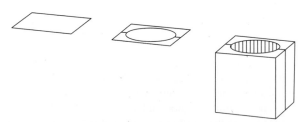

Figure 6   Surface solid subtraction.

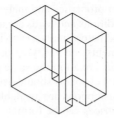

Figure 7 Wire frame model.

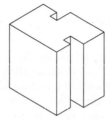

Figure 8 Wire frame model with hidden lines removed.

In CAD software, certain features have been developed to minimize the ambiguity of wire frame representations (Fig. 7). These features include using dashed lines to represent the background of a view or removing those background lines altogether (Fig. 8). The latter method is appropriately referred to as "hidden-line removal." The hidden-line removal feature makes it easier to visualize the model because the back faces are not displayed. Shading removes hidden lines and assigns flat colors to visible surfaces. Rendering is the process by which light is added and adjusted and textures are applied to the surfaces in order to produce realistic effects. Shading and rendering can greatly enhance the realism of the 3D image.

Engineering analysis can be performed using one of two approaches: analytical or experimental. Using the analytical method, the design is subjected to simulated conditions using any number of analytical formulas. By contrast, the experimental approach to analysis requires that a prototype be constructed and subsequently subjected to various experiments to yield data that might not be available through purely analytical methods.

There are various analytical methods available to the designer using a CAD system, such as finite-element analysis (FEA), static and dynamic analysis, and kinematic analysis.

Finite-Element Analysis

Finite-element analysis is a computer numerical analysis program used to solve complex problems in many engineering and scientific fields, such as structural analysis as it relates to stress, deflection, vibration, thermal analysis (steady state and transient), and fluid dynamics analysis (laminar and turbulent flow).

The finite-element method (FEM) divides a given physical or mathematical model into smaller and simpler elements, performs analysis on each individual element using required mathematics, and then assembles the individual solutions of the elements to reach a global solution for the model. FEA software programs usually consist of three parts: the preprocessor, the solver, and the postprocessor.

The program inputs are prepared in the preprocessor. Model geometry can be defined or imported from CAD software. Meshes are generated on a surface or solid model to form the elements. Element properties and material descriptions can be assigned to the model. Finally, the boundary conditions and loads are applied to the elements and their nodes. Certain checks must be completed before analysis solving is executed. These include checking for duplication of nodes and elements and verifying the element connectivity of the surface elements so that the surface normals are all in the same direction. In order to optimize disk space and running time, the nodes and elements should usually be renumbered and sequenced.

Many analysis options are available in the analysis solver to execute the model. The element stiffness matrices can be formulated and solved to form a global stiffness value for the model solution. The results of the analysis data are then interpreted by the postprocessor. The postprocessor in most FEA applications offers graphical output and animation displays. Vendors of CAD software are developing pre- and postprocessors that allow the user to graphically visualize their input and output. FEA is a powerful tool in effectively developing a design to achieve a superior product.

Kinematic Analysis and Synthesis

Kinematic analysis and synthesis allow for the study of the motion or position of a set of rigid bodies in a system without reference to the forces causing that motion or the mass of the bodies. It allows engineers to see how the mechanisms they design will function and interact in motion. This kinematic modeler enables the designer to avoid a faulty design and to apply a variety of scenarios to the model without constructing a physical prototype. A superior design may be developed after analyzing the data extracted from kinematic analysis after numerous motion iterations. The behavior of the resulting model mechanism may be understood prior to production.

Static Analysis

Static analysis determines reaction forces at the joint positions of resting mechanisms when a constant load is applied. As long as zero or constant velocity of the entire system under study is assumed, static analysis can also be performed on mechanisms at different points of their range of motion. Static analysis allows the designer to determine the reaction forces on mechanical systems as well as interconnection forces transmitted to individual joints. Data extracted from static analysis can be useful in determining compatibility with the various criteria set out in the problem definition. These criteria may include reliability, fatigue, and performance considerations to be analyzed through stress analysis methods.

Dynamic Analysis

Dynamic analysis combines motion with forces in a mechanical system to calculate positions, velocities, accelerations, and reaction forces on parts in the system. The analysis is performed stepwise within a given interval of time. Each degree of freedom is associated with a specific coordinate for which initial position and velocity must be supplied. Defining the system in various ways creates the computer model from which the design is analyzed. The user must supply joints, forces, and overall system coordination either directly or through a manipulation of data within the software.

Experimental Analysis

Experimental analysis involves fabricating a prototype and subjecting it to various experimental methods. Although this usually takes place in the later stages of design, CAD systems enable the designer to make more effective use of experimental data, especially where analytical methods are thought to be unreliable for the given model. CAD also provides the platform for incorporating experimental results into the design process.

Design review can be easily accomplished using CAD. The accuracy of the design can be checked using automated routines for tolerancing and dimensioning to reduce the possibility of error. Layering is a technique that allows the designer to superimpose images on one another. This can be quite useful during the evaluative stage of the design process by allowing the designer to visually check the dimensions of a final design against the dimensions of stages of the design's proposed manufacture, ensuring that sufficient material is present in preliminary stages for correct manufacture. Interference checking can also be performed using CAD. This procedure checks the models and identifies when two parts of a design occupy the same space at the same time.

Automated Drafting

Automated drafting capabilities in CAD systems facilitate presentation, which is the final stage of the design process. CAD data, stored in computer memory, can be sent to a plotter or other hard-copy device to produce a detailed drawing printout. In the early days of CAD, this feature was the primary rationale for investing in a CAD system. Drafting conventions, including but not limited to dimensioning, crosshatching, scaling of the design, and enlarged views of parts or other design areas, can be included automatically in nearly all CAD systems. Detail and assembly drawings, bills of materials (BOM), and cross-sectioned views of design parts are also automated and simplified through CAD parts databases. In addition, most systems are capable of presenting as many as six views of the design automatically (front, side left, side right, top, bottom, rear). Drafting standards defined by a company can be programmed into the system such that all final drafts will comply with the company standards.

Product Data Management

Product data management (PDM) is an important application associated with CAD. PDM allows companies to make CAD data available across the enterprise on computer networks. For example, PDM software may operate in conjunction with CAD software and word processing software. This approach holds significant advantages over conventional data management. PDM is not simply a database holding CAD data as a library for interested users. PDM systems offer increased data management efficiency, for example, through a client-server environment. Benefits of implementing a PDM system include faster retrieval of CAD files through keyword searches and other search features such as model parameters like color or serial number, automated distribution of designs to management, manufacturing engineers, and shop-floor workers for design review, record-keeping functions that provide a history of design changes, and data security functions limiting access levels to design files. PDM facilitates the exchange of information characteristic of the agile workplace. As companies face increased pressure to provide clients with customized solutions to their individual needs, PDM systems allow an augmented level of teamwork among personnel at all levels of product design and manufacturing, cutting down on costs often associated with information lag and rework.

Although CAD has made the design process less tedious and more efficient than traditional methods, the fundamental design process remains unchanged. It still requires the human input and ingenuity to initiate and proceed through the many iterations of the design process. CAD is a powerful, time-saving design tool that competing in the engineering world without it is difficult if not impossible. The CAD system will now be examined in terms of its components: the hardware and software of a computer.

2 HARDWARE

Just as a draftsman traditionally required pen and ink to bring creativity to bear on the page, there are certain essential components to any working CAD system. The use of computers for interactive graphics applications can be traced back to the early 1960s, when Ivan Sutherland developed the SKETCHPAD system. The prohibitively high cost of hardware made general use of interactive computer graphics uneconomical until the 1970s. With the development and subsequent popularity of personal computers, interactive graphics applications became an integral part of the design process.

CAD systems are available for many hardware configurations. CAD systems have been developed for computer systems ranging from mainframes to workstations, desktops, and laptop computers. A difference between a desktop and a workstation is the power and cost of the components and peripherals, a workstation being more expensive. Turnkey CAD systems come with all of the hardware and software required to run a particular CAD application and are supplied by specialized vendors. The use of CAD on smaller computer devices such as smart phones and other mobile or wireless devices is currently limited to viewing and annotating.

Hardware is the tangible element of the computer system. Any physical component of a computer system whether internal to the computer or an external such as printers, machinery, or other equipment is considered to be hardware. In a personal computer (PC) or in a workstation, common external hardware is the display monitor, a keyboard, network cables, and a mouse, while the CPU, memory chips, graphic cards, and hard drives are internal. The term hardware also extends beyond PCs to include any information technology (IT) devices such as routers and hubs.

2.1 Central Processing Unit

The CPU is a computer microprocessor component with specialized circuitry that interprets and processes programming instructions. The processing speed is called the clock rate and it is measured in hertz, that is, number of cycles per second. Recent CPU speeds are 3.6 GHz, or 3.6×10^9 cycles per second. In contrast, the early 1980s 8086 Intel chips would run at 12 MHz, or 12×10^6 cycles per second.

A CAD system requires a computer with a fast CPU. CAD software typically consists of millions of lines of computer code that require a high degree of computer processing power. This processing power is provided by the CPU. Because of the demands that CAD software has on the CPUs, it is written by the CAD vendors to utilize specific CPU capabilities, such as the rendering of geometry to the screen, or multiprocessing on CPU processors, such as the Intel Core i7 CPU or the AMD 990FX.

2.2 Operating System

Windows is the predominant operating system (OS), with support for UNIX and other systems waning with respect to high-end CAD support. Windows XP, 7, 8, the 32- and 64-bit versions are the ones supported by CAD software manufacturers such as AutoCAD and Creo. However, the faster computations on 64-bit OSs may see a decline in support of the slower 32-bit OSs in favor of 64-bit ones, e.g., Windows 7 and 8, as well as the Mac® OS X® v10.8 or later.

2.3 Bus

Communication between the CPU and the CAD program passes through a hardware system called the bus. Other peripherals communicate with each other via the bus. For the hardware within a computer, the bus is called the local bus. External to computer devices such as USB cameras use the universal serial bus (USB). PC bus sizes range from 64 bits to 32 bits, 16 bits, and 8 bits. As the bus gets wider (increase in the bit count), more data can pass through, which means more of the programming code can get from the hard drive to the CPU and back, and CAD software executes commands even faster. Note: 8 bits = 1 byte. One byte is used to describe 256 different values.

Devices such as CD/DVD-ROM drives, flash drives, printers, and LANs (local-area networks) and graphics cards are all part of the data exchange with the CPU. This input and output of data is known as I/O. Data in the I/O process are carried via specialized buses such as FireWire, PCIe (Peripheral Component Interconnect Express), IDE (Integrated Drive Electronics), and SCSI (Small Computer System Interface).

2.4 Memory

The CPU is the traffic controller of all the requests from the computer hardware and software. As the hardware and software flood the CPU with requests, such as to move the cursor across the screen while computing the distance between two points in 3D space, the local bus has only a limited capacity to process the requests. So, data must be stored in a temporary location so that hardware and software do not try to access the bus at the same time. This location is called the random access memory (RAM), or just plain "computer memory."

The size of the computer memory chip (chip storage capacity) is measured in gigabytes (GB), that is, 10^9 bytes. It is not uncommon for a CAD turnkey system to have 4 G of RAM. Workstations with 16 GB are also becoming common. The limiting function of RAM is the width of the motherboard local bus and the capacity of the OS to access this space.

In considering a CAD system, one should consider whether the RAM is dedicated to the CPU or shared among other hardware components such as the video card. Lower end systems use shared RAM which for CAD applications may cause a reduction in computer performance. Opting for dedicated RAM ensures that the RAM size listed for a computer system matches or exceeds the CAD software manufacturer specifications.

2.5 Video Cards/Graphics

Computer graphics cards are essential in any CAD computer system design. The cards allow the display of 2D and 3D graphics while performing many of the calculations that in early PCs were tasked to the CPU. Graphics chips have been designed to process floating-point arithmetic needed to display and render 3D models and simulations such as the calculations required for FEA. These graphics cards often require their own cooling devices, sometimes with extra fans and other times with liquid cooling devices, and extra power connections. Some cards may require 150 W or more power supply (such as the AMD FirePro W7000), which means that the computer power supply to the motherboard must be able to handle the card's power requirements.

A graphics card has to be compatible with the motherboard it will connect to [AGP (Accelerated Graphics Port, IDE, PCI (Peripheral Component Interconnect), and the current standard PCIe] as well as the operating system and the CAD software to be considered for a CAD computer system. A graphics card's characteristics and performance are defined, among others, by the following:

- *GPU*. The graphics processing unit is a chip like the CPU but designed specifically for graphics processing. The speed of the GPU is measured in megahertz with high-end cards clocking in at the high 800s per core. A dual-core GPU is a single chip designed with two cores so that it can perform in theory double the instructions, making it a faster processing and higher performing card when given the appropriate instructions to each core.
- *Video Memory*. Supporting the GPU is the video memory, which is dedicated to the graphics processing. Higher end cards have between 1 and 2.5 GHz speed and up to 4 GB size. Memory bus size exceeds 320 bits.
- *Open Graphics Library* (*OpenGL*). Version 3.1 is a minimum for CAD applications such as Creo 2.0 by PTC Corp. OpenGL is a graphics application programming interface

(API) standard that allows 2D and 3D graphics programmers of CAD software to utilize high-performance computation capabilities of the standards to interact with the GPU and CPU. A GPU that complies with the OpenGL standard means that the hardware was designed to utilize the standards. A GPU that does not comply will likely cause a degradation of performance or even not run the CAD software.

2.6   Hard-Drive/Flash Drive/Cloud Storage

Another type of memory in a computer is the primary storage called a hard drive. A hard drive stores the operating system which is how the hardware communicates. It also stores software, that is, lines of code that are commands and instructions that serve specific functions. CAD software manufacturers provide the minimum system specification for their software to run as well as the optimum or recommended configuration. For example, AutoCAD LT 2013 32-bit requires 1.4 GB of free space. Another example is that a 7-min high-definition animation video requires approximately 1 GB of storage. It is not uncommon to see computers with hard drives 350–500 GB in size, with some 2 or 3 TB in size (1 TB = 10^{12} bytes).

Hard drives can be internal to the computer system or external, with internal hard drives being typically faster to store and access data and external slower. USB or Firewire connections are typical for external hard drives, but serial ATA (Serial Advanced Technology Attachment or SATA), SCSI, serial attached SCSI (SAS), and IDE are also possible.

Hard-drive performance is established based on, among others, a combination of spindle rotations per minute (RPM) and connectivity. The higher the RPM, the faster typically is the access time, that is, the time to access data on the hard drive. The seek time is the average time the heads of the hard drive get to the data anywhere on the drive. Such times are measured in nanoseconds with fast drives at 15,000 RPM having a seek time of 2 ns and off-the-shelve internal hard drive with 7200 RPM at 4-5 ns. External drives, such as USB connected drives, are slower to access data. The type of connection, whether IDE, USB, SATA, or SAS, also has an effect on the drive performance with SATA and SCSI being faster than the others.

Hard drives without rotating plates are called solid-state drives (SSDs). They use similar technology as the memory chips, but they are larger in capacity, fast, and sold at a premium for systems where speed and size are more important than cost. SSDs are usually not meant to be removed like USB flash drives.

Hard drives fail and many times the data on them are lost. For that reason, mission-critical CAD applications require further system design considerations. Warranties on new hard drives vary by manufacturer and drive model, with 1 year being standard and up to 10 years or more on parts for some slower drives mostly used for backups. Multiple hard drives may be configured in a way that data are simultaneously duplicated on two or more hard drives (e.g., RAID 1). Alternatively or concurrently, daily data backups on devices like external hard drives or USB flash drives might give the needed confidence that work will be easily recoverable and design process delays are kept at a minimum. Cloud storage should also be considered as a possible backup choice, in addition to being a common design collaboration tool as mentioned later in the chapter. Retrieval from the cloud requires an Internet connection to use and restore the data. Autodesk, for example, has introduced products that take advantage of cloud computing; AutoCAD WS is a cloud-based editor for .dwg files and runs on Web browsers as well as mobile platforms.

2.7   Classes of Computers

There are four basic classes of computers that usually define a computer's size, power, and purpose. They are:

- *Supercomputers*. These are the largest, fastest, and most expensive computers available. Although they technically fall under the mainframe class, their difference lies in the fact that supercomputers are designed to handle relatively few extremely complicated tasks in a short amount of time. Some applications of supercomputers lie in calculations involving intensive research and sophisticated applications such as theoretical physics, turbulence calculations, weather forecasting, and advanced animated graphics. These applications are characterized by the need for high precision and repetitive performance of floating-point arithmetic operations on large arrays of numbers.

- *Mainframes*. Designed to run as many simultaneous applications as possible, these computers are typically large and fast enough to handle many users when connected to multiple terminals. They are most commonly used by research facilities, large businesses, and the military.
- *Desktops and Servers*. Similar to mainframes in their function, desktop computers are smaller in size and power, though they are actually midrange computers. Serving multiple users, servers are often networked together with other desktops and servers.
- *Laptops*. Used almost synonymously with personal computers, laptops have foldable screens; some have touch screens as a mode of interface beyond a mouse or a keypad, are portable, and run on batteries in addition to direct current. Large-screen laptop computers (greater than 17 in. diagonal) are powerful enough to run 2D and 3D graphics applications, assuming they have equally powerful hard drives, memory, CPU, and GPU. Laptops, essentially portable versions of personal desktop computers, can be equipped with equivalent components to a desktop computer. The primary difference between a laptop and desktop is that the devices and components used in a laptop are selected for their reduced energy consumption compared to a desktop. The size of the display monitor is also limited to the size of the lid used in the laptop.
- *Tablets*. Tablets are becoming popular due to their small size, light weight, and touch screen interface. Their relative small screen size, lightweight processors, and limited memory limit their usage to special applications, such as drawing reviews and presentations.

One must be careful not to take these computer classes as an absolute breakdown because there is some overlap, as well as other classes that fall between the four mentioned above. For example, a mini-supercomputer is simply one that falls between a supercomputer and a mainframe.

2.8 Engineering PCs

Computer-aided design projects often range from simple 2D drawings to graphics-intensive engineering design applications. Computationally intensive number crunching in 3D surface- and solid-modeling, photorealistic rendering, and FEA applications demands a great deal from a personal computer. Careful selection of a PC for these applications requires an examination of the capabilities of the CPU, RAM capacity, disk space, operating system, network features, and graphics capabilities. The computer industry advances quickly, especially in microprocessor capabilities. The following PC configurations list minimum requirements for various CAD applications. It should be noted that, because of rapid advances in the industry, this listing may be dated by the time of publication.

For 2D drafting applications, an Intel Pentium Duo Core or AMD Athlon dual-core processor, 4 GB RAM, 1024×768 resolution display, and an Internet browser are typical requirements for most PC installations. The size of the files is such that if used for CAD alone, a 10-GB hard drive would be adequate. However, since the engineering PC is usually used for multiple applications, gigabyte and terabyte hard drives are usually provided.

Additional requirements for 3D modeling include additional CPU memory, typically 6 GB RAM, a video display adaptor with graphics processor unit and onboard memory, and video accelerators and converters that vary depending on the software package being used. The actual requirements vary with the software being utilized.

2.9 Engineering Workstations

The Intel Pentium CPU microprocessor reduced the performance gap between PCs and workstations. A current trend is the merging of PCs and workstations into "personal workstations." Pentium Duo Core and i3, i5, and i7 are all powerful personal workstations with high-performance graphics accelerators.

Until the early 2000s, operating systems were the main distinction between a low-end workstation and a high-end PC. The UNIX operating system, which supports multitasking and networking, usually ran on a workstation. DOS, Windows, and Macintosh operating systems, which were limited to single tasks, usually ran on PCs. That distinction is beginning to disappear because of the birth of the Intel Duo Core and i3, i5, and i7 multicore processor. Microsoft responded with the Windows 7 64-bit operating system. This powerful combination allowed the PC to perform multithreaded, multitasking operations that were previously the domain of the dedicated mainframe computer.

Many CAD and FEA software applications were traditionally UNIX-based applications. Since there are significant differences in price between UNIX and Windows 7, more and more CAD and FEA vendors have released versions of their software for Windows. RAM capacity for a personal computer workstation can be 64 GB with disk space in the terabyte range, such as those from Dell, Hewlett-Packard, and Apple. Thanks to 64-bit technology and a scalable modular platform, high-end workstation performance can now boast supercomputer-like performance at a fraction of the cost of a mainframe or supercomputer. The noted performance gain is a result of using more powerful 64-bit word addressing and higher clock speed. The term "workstation" has colloquially become context dependent. However, technically speaking, a workstation is any networked computer that can be used to access or input data to the system.

2.10 Parallel Processing

To reach higher levels of productivity in the analysis of complex structures with thousands of components, such as in combustion engines or crash simulations, the application of parallel processing was introduced. Two parallel processing methods used in engineering applications are:

- Symmetric multiprocessing (SMP)
- Massively parallel processing (MPP)

Parallel processing is the use of multiple processors to complete complex tasks and calculations by dividing up the work. The ultimate goal of parallel processing is to constantly increase the power of the systems. This takes place in a parallel computer or supercomputer and is similar to that of having multiple workers work together to complete a single or many tasks in a shorter amount of time. SMP is the shared memory while MPP is the distributed memory used in these processes (Fig. 9).

3 INPUT AND OUTPUT DEVICES

Commonly used input devices in CAD systems include the alphanumeric keyboard, the mouse, and the graphics tablet. All of these allow information transfer from the device to the CPU. The information being transferred can be alphanumeric, functional (in order to use command paths in the software), or graphic in nature. In each case, the devices allow an interface between the designer's thoughts and the machine which will assist in the design process.

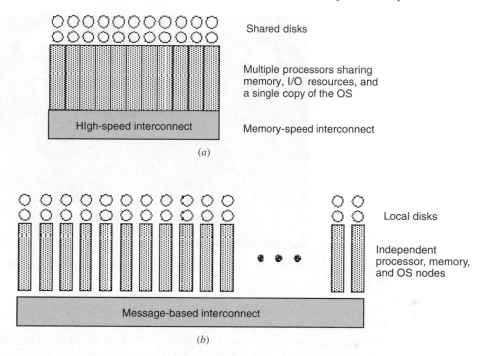

Figure 9   (*a*) SMP and (*b*) MPP parallel processor layouts.

3.1 Keyboard

The alphanumeric keyboard is one of the most recognizable computer input devices, as well as the principal method of text input in most systems. Rows of letters and numbers (typically laid out like a typewriter keyboard) with other functional keys (such as CTRL, ALT, and ESC), either dedicated to tasks such as control of cursor placement on a display screen or definable by the user, transfer bits of information to the CPU in one of several ways. The layout of the keyboard is dictated by international standards organizations, for example, the American National Standards Institute (ANSI) for the United States, International Organization for Standardization (ISO) for worldwide standardization, and Japanese Industrial Standards (JIS) for Japan. The most common keyboard layout used in the United States is the QWERTY, called so because of the layout of the keys at the top row. Other keyboard layouts exist, for hobbyists and enthusiasts, but QWERTY is the most commonly used one.

Keyboards connect to computers through PS/2 or USB wired interfaces. Wireless keyboards are also available and they may use radio transmitters/receivers or Bluetooth technology. For smaller devices like tablets that do not have their own external keyboard, projected keyboards like the one in Fig. 10 may provide alternative input methods.

In CAD systems, as with most software, time-saving alternatives to using the computer's mouse to carry out menu functions are keyboard shortcuts. For example, in most applications CTRL-I is the shortcut to italicize text in the Format menu under Font and CTRL-A is the shortcut to select an entire document under the Edit menu. There are also keyboards with programmable buttons that can be used in conjunction with a CAD system; tasks that may require multiple keystrokes can then be cut down to just one or two keystrokes.

Projected keyboards are virtual keyboards that can be projected and touched on any surface. These work using electronic perception technology (EPT) and can detect up to

Figure 10 Projected keyboard by Canesta Corp. (Microsoft Corp.).

350 characters per minute. An optical sensor is paired with an infrared layer to determine the exact position of the user's fingers while typing, and those positions are then translated into characters on the software/electronic device. These are compatible with the latest desktop and mobile platforms and connect to the platform wirelessly using Bluetooth technology or through USB. With the rise in mobile computing and the availability of CAD services on these devices, projected keyboards are becoming prevalent.

3.2 Mouse

The mouse is used for graphical cursor control and conveys cursor placement information in the X–Y coordinate plane to the CPU. Low-end mice are mechanical devices that use a spherical roller housed within the mouse such that the roller touches the plane upon which the mouse is resting. When the mouse is moved along a flat surface, the spherical roller simultaneously contacts two orthogonal potentiometers, each of which is connected to an analog-to-digital converter. The orthogonal potentiometers send X- and Y-axis vector information via a connecting wire to the CPU, which performs the necessary vector additions to allow cursor control in any direction on a 2D display screen. Often a mouse will be equipped with one to three (or more) pressure-sensitive and programmable buttons which assist in the selection of on-screen command paths.

Wireless Mouse
Connecting a mouse is usually done via USB or PS2 connectors. The wireless mouse, on the other hand, uses either radio frequencies or infrared signals to function. The radio frequency version will work up to 6 ft away, anywhere in a room, but the infrared version requires a line-of-sight channel between it and the infrared port on the side of the computer. The technology employed for the infrared mouse is identical to the technology used in television remote controls. The radio signal mouse functions much the same as a wired mouse and has a transmitter that connects to the USB port on the computer or transmits signals to a Bluetooth antenna in the computer. Wireless mice require batteries to function.

Optical Mouse

Laser or optical mice are devices that use light-emitting diodes (LEDs) and photosensors to detect movement as opposed to a mechanical ball. They are more accurate than mechanical mice, and they do not need internal cleaning. Some manufacturers assert that the optical mouse may reduce computer-related stress injuries.

Advances in ASICs (application-specific integrated circuits), imaging arrays, and embedded mathematics have reduced the costs of optical navigation, and this has meant that the optical mouse is now the most prevalent type of mouse used. One of the more interesting developments in mouse technology has also been the integration of a touchpad onto the body of the mouse, resulting in an optical mouse that can not only perform its traditional duties as a mouse but also support multitouch gestures.

3D Mouse

The 3D mouse (Fig. 11) is used in conjunction with a standard mouse. The key difference between the two is that the 3D mouse allows for movement in the third dimension. For example, with the addition of a 3D mouse, the user can pan, zoom, and orbit without having to use keyboard shortcuts or using the standard mouse for clicking, dragging, and so forth. The user can combine motions to perform more intricate changes to the views. Furthermore, increasing or decreasing pressure on the 3D mouse while performing an action will result in faster responses; to rotate an object faster, the user can simply apply more pressure on the 3D mouse while tilting it. It has been claimed that using a 3D mouse can increase productivity, due to the fact that the user can simultaneously navigate around and edit the model. This is further bolstered by the fact that many models feature programmable buttons for increased ease of use. Also, due to reduced mouse clicks, user comfort can increase.

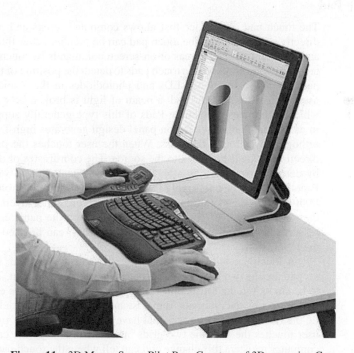

Figure 11   3D Mouse Space Pilot Pro. Courtesy of 3Dconnexion Corp.

3.3   Trackball

This device operates much like a mouse in reverse. The main components of the mouse are also present in the trackball. As in a mouse, the trackball uses a spherical roller which comes into contact with two orthogonally placed potentiometers, sending X- and Y-axis vector information to the CPU via a connecting wire. The difference between the mouse and the trackball lies in the placement of its spherical roller. In a trackball, the spherical roller rests on a base and is controlled directly by manual manipulation. As in a mouse, buttons may be present to facilitate the use of on-screen commands. A trackball is immobile, so when the user maneuvers the ball on the top, the movement is almost identical to the movement a mouse ball makes as the user moves the mouse. Since it is stationary, a trackball can also be useful on surfaces where it would be difficult to use a mouse, as well as being practical for users with minimal or limited desk space. Some desktop computers and notebook computers have trackballs built into their keyboards, both to save space and to allow the user to keep his or her hands on the keyboard at all times.

3.4   Pointing Stick

A pointing stick is a short joystick that is placed in the middle of the keyboard and controls the cursor. It is space efficient and requires minimal movement to use. In addition to the pointing stick, there are two buttons located on the bottom of the computer, allowing the pointing stick system to perform the identical functions as does a mouse. The user uses his or her finger to apply pressure to the pointing stick in the direction he or she wants the cursor to move. This device acts as a substitute for a touchpad on some notebook computers.

3.5   Touch Pad

The touch pad is a device that allows command inputs and data manipulation to take place directly on the screen. The touch pad can be mounted over the screen of the display terminal, and the user can select areas or on-screen commands by touching a finger to the pad. Various techniques are employed in touch pads to detect the position of the user's finger. Low-resolution pads employ a series of LEDs and photodiodes in the X and Y axes of the pad. When the user's finger touches the pad, a beam of light is broken between an LED and a photodiode, which determines a position. Pads of this type generally supply 10–50 resolvable positions in each axis. A high-resolution panel design generates high-frequency shock waves traveling orthogonally through the glass. When the user touches the panel, part of the waves in both directions is deflected back to the source. The coordinates of this input can then be calculated by determining how long after the wave was generated it was reflected back to the source. Panels of this type can supply resolution of up to 500 positions in each direction. A different high-resolution panel design uses two transparent panel layers. One layer is conductive while the other is resistive. The pressure of a finger on the panel causes the voltage to drop in the resistive layer, and the measurement of the drop can be used to calculate the coordinates of the input.

A touch pad is made of two different layers of material: first is the top, protective layer; second are the two layers of electrodes in a grid arrangement. The protective layer is designed to be smooth to the touch, allowing the user's finger to move effortlessly across the pad while still protecting the internals from dust and other harmful particles. Beneath the protective layer lie the two grids of electrodes that have alternating currents running through them. When the user touches the pad, the two grids interact and change the capacitance at that location. The computer registers this change and can tell where the touch pad was touched.

The touch pad was first used in PowerBook notebook computers, and the touch pad is the primary cursor control alternative to the mouse in notebook computers. Desktop computers also have the ability to use the touch pad as an input device. Due to advancements in multitouch technology, touchpads are commonly used to supplement other input devices. For example, some CAD software includes support for multitouch gestures such as pinch-to-zoom and two-finger dragging to pan.

3.6 Touch Screen

Touch screens are a special category of computer screens that allow the user to interface with the system other than with a traditional input device such as a keyboard or mouse. However, a touch-sensitive interface must first be installed into the screen. This is done in three primary ways: resistive, surface wave, and capacitive. The resistive method places an ultrathin metallic layer over the screen. This layer has both conductive and resistive properties, and touching the layer causes a variation in that conductivity and resistivity. The computer interprets this variation in the electrical charge of the metallic layer and the touched area becomes known. The advantage of a resistive-based touch screen is that they are cheaper than the other two options and are not affected by dust or moisture. However, they also do not have the picture clarity that the others have, require extra pressure on the screen compared to other methods, and are prone to damage from sharp objects.

The surface wave method has either ultrasonic waves or infrared light continually passed just over the surface of the screen. When an object interrupts the waves passing over the surface, the controller processes this information, and the location is determined. So far, this is the most advanced and expensive method used.

The final method, the capacitive method, requires that the screen be covered with an electrical charge storing material. When a user touches the screen, the charge is disrupted and can be interpreted as the location of the user's finger. The difficulty with the capacitive method is that it requires either a human finger or a special pointing device for the system to identify the touched portion of the screen. Nonetheless, this method is the least affected by external elements and can also boast the best clarity.

There are some key advantages to touch screens. In a factory setting with reduced footprint for the computer, data entry can be performed without the need for space for a mouse and a keyboard. Minor image manipulation can be performed in case a CAD design needs to be inspected prior to production. A stylus may be used to augment the user's fingers while using the touch screen. Touch screens are a staple characteristic of smaller computer devices like smart phones and touch pads. They cannot be used to perform CAD designs per se because the human finger is inaccurate but can be used for demonstration purposes and in a production environment.

3.7 Digitizer

A digitizer (Fig. 12) is an input device that consists of a large, flat surface coupled with an electronic tracking device, or cursor. The cursor is tracked by the tablet underneath it and buttons on the cursor act as switches to allow the user to input position data and commands.

Digitizing tablets apply different technologies to sense and track cursor position. The three most common techniques use electromagnetic, electrostatic, and magnetostrictive methods to track the cursor. Electromagnetic tablets have a grid of wires underlying the tablet surface. Either the cursor or the tablet generates an alternating current that is detected by a magnetic receiver in the complementary device. The receiver generates and sends a digital signal to the CPU, giving the cursor's position. Despite their use of electromagnetism, these types of digitizers are not compromised by magnetic or conductive materials on their surface.

Electrostatic digitizers generate a variable electric field which is detected by the tracking device. The frequency of the field variations and the time at which the field is sensed provide the information necessary to give accurate coordinates. Electrostatic digitizers function accurately

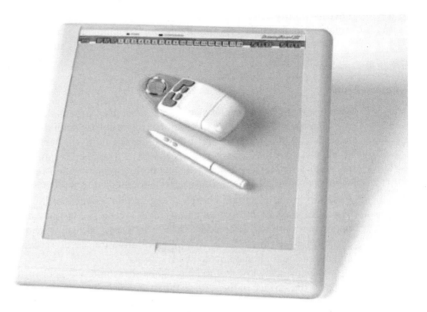

Figure 12   Digitizing tablet and cursor. Courtesy of GTCO CalComp Peripherals.

in contact with paper, plastic, or any other material with a small dielectric constant. They do lose accuracy, however, when even partially conductive materials are in close proximity to the tablet.

Magnetostrictive tablets use an underlying wire grid, similar to that used in electromagnetic tablets. These tablets, however, use magnetostrictive wires (i.e., wires which change dimension depending on a magnetic field) in the grid. A magnetic pulse initiated at one end of a wire propagates through the wire as a wave. The cursor senses the wave using a loop of wire and relays a signal to the CPU which then couples the time of the cursor signal with the time elapsed since the wave originated to give the position. These tablets require periodic remagnetization and recalibration to maintain their functional ability.

Digitizing tablets can usually employ various modes of operation. One mode allows the input of individual points. Other modes allow a continuous stream of points to be tracked into the CPU, either with or without one of the cursor buttons depressed, depending on the needs of the user. A digitizing-rate function, which enables the user to specify the number of points to be tracked in a given period of time, is also often present in CAD systems with digitizers. The rate can be adjusted as necessary to facilitate the accurate input of curves.

Whatever the type, digitizers are highly accurate graphical input devices and strongly suited to drafting original designs and to tracing existing designs from a hard-copy drawing. Resolution can be up to 2540 lines per linear inch (100 lines/mm). Tablet sizes typically range from 10 × 11 to 44 × 60 in. Often, plastic sheets with areas for command functions, such as switching between modes, are laid over the tablet to allow the designer to access software commands directly through the tablet using one of the buttons on the tracking device. Many of these commands deal with the generation of graphical elements like lines, circles, and other geometries.

Digitizing tablets are commonly used in imaging and illustration applications. Several different pointing devices are used, depending for what function the digitizing tablet is desired. Freehand artists often prefer a pen or stylus, due to the increased handling ease, as compared to a mouse. Instead of ink, a stylus generally either possesses sensors or is sensed when it makes contact with the digitizing tablet.

3.8 Scanners

A scanner converts information from its physical form to digital signals through the use of light-sensing equipment. When an object is scanned, it is first examined in small sections, and then the data are converted to a computer-friendly digital format. Systems that are able to transfer the digital signal include graphics software or OCR (optical character recognition) software; with the advent of 3D scanners for use in reverse engineering and CAD, PLM and 3D design software is increasingly capable of transferring digital signals into usable images. Text scans use a scanner and OCR software, which converts text to a word processor file. Graphical scanners translate the visual patterns from photographs, magazines, and other pictorial sources to bit-mapped files, where with the use of a computer one can then edit or display them. Most scanners are accompanied by imaging software, allowing the user to edit and resize the images, which is useful when one wishes to scan designs or keep permanent records of graphics.

2D Scanners

Different types of scanners are available, including flat-bed, sheet-fed, hand-held, and film scanners. All of these scanners are able to scan in either black and white or color. High-resolution printed images are scanned using a high-resolution scanner [4800 dots per inch (dpi)], while graphics for a computer display are scanned using a lower resolution (600 dpi).

Flat-bed scanners are capable of scanning a variety of objects, which is probably why they are the most popular type of scanner. Flat-bed scanners can scan books, photographs, papers, and 3D objects, although depth of field is not captured for the 3D objects. The scanner looks like a copy machine and has a glass sheet on which the scanned objects are placed facedown. A scanner head moves across the object being scanned to copy an image of the object.

Sheet-fed scanners scan individual pieces of paper at a high rate but, unlike flat-bed scanners, cannot scan books, bound papers, or 3D objects.

Film scanners were created to scan slides, transparencies, and negatives at a much higher resolution than flat-bed scanners. Film scanners cannot scan paper or 3D objects, and if a scanner is needed for both paper and occasional slides, then a transparency adapter is probably the most cost-effective way to scan both mediums.

Hand-held scanners are the most inexpensive form of scanner, but since they are much smaller than the other types, they can only scan areas that are a few inches wide. If a user only needs to scan small amounts of text or images, then the hand-held scanner is ideal for these purposes. Scanners usually connect to a computer via a USB port.

3D Scanning

Three-dimensional scanners use a variety of surface imaging methods to capture and reproduce objects in three dimensions. If the design process is taken as a continuum, 3D scanners provide the means to "hop the border" from the physical realm into the digital realm (Fig. 13). In essence, a 3D scanner analyzes physical objects and reconstructs them as 3D virtual models. 3D scanners in the CAD realm are used for "reverse engineering," which is the taking apart of existing physical objects and analyzing them further to determine their existing potential or their potential for improvement. When coupled with design software, 3D scanners make for an efficient reverse-engineering process.

3D scanners fall under two umbrellas: contact and noncontact:

Contact scanners probe the object by making physical contact with it, with the object either resting on a flat surface or being held firmly in place. An example of such a system is the coordinate measuring machine (CMM). The greatest advantage of systems like CMMs is their accuracy. However, a sizable drawback of such a system is that the act of scanning could damage or modify the object itself.

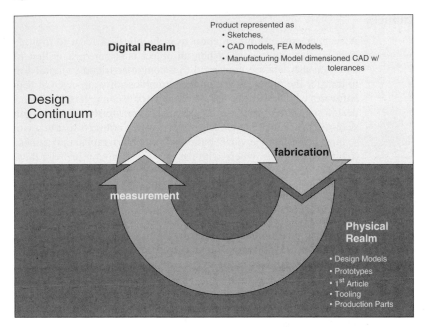

Figure 13   Design continuum. With permission from Martin Chader.

Noncontact scanners can be further divided into two categories, *active* or *passive*. *Active non-contact scanners* use the emission light beams, x rays, or ultrasound waves to determine the geometry of the object. For example, time-of-flight scanners simply time how long it takes for a beam of light to reflect off points on the object itself to determine the overall surface geometry. Another form is *structured light 3D scanning*, in which a pattern of light is projected onto the object, and with the use of a camera, the surface geometry of the object is determined. This is an especially quick method, as multiple point data can be collected simultaneously. *Passive noncontact scanners* also use a variety of technologies, but a particularly common example is a *stereoscopic* scanner which employs two video cameras set slightly apart from each other focusing on the same scene; upon analyzing the slight differences in the two images, the surface geometry is determined. This is not unlike human stereoscopic vision.

3D scanners use the above technologies to create scan data, which are generally in the form of a *point cloud*. A point cloud can be defined simply as a set of 3D vertices. Each point measured by the scanner is assigned its own coordinates in a preset system, and these coordinates are used to re-create a digital copy of the object.

It is important to note, however, that a point cloud is ultimately not enough to fully re-create and perform analyses on the object using 3D design software. In order to facilitate this, the point cloud can be used to create *NURBS curves*. NURBS curves are defined by Bezier curves through coplanar points in the point cloud. These NURBS curves define cross sections through the object, and multiple cross sections will point out with accuracy the extent of the object being scanned.

The points of the point cloud can also be joined by straight lines, forming flat faces of a *polygon mesh*. For reverse-engineering purposes, polygon meshes result in very memory-intensive files and computations, and polygon meshes are therefore more useful in FEM software. In order to make the model "lighter," the point cloud, instead of using polygon meshes, can use *NURBS surface patches*. These entail the connection of the points on the point cloud not with straight lines but with curves along its longitudinal and latitudinal directions, while laying a curved patch between these curves. This results in curved patches whose shape is driven by the point cloud as with a polygon mesh but which are much lighter in nature.

The latest, best-equipped CAD/CAM software takes full advantage of 3D scanning technologies to create the most detailed model possible. This means that, by using the scan data, the 3D modelers can provide editable discrete entities as well as the relationships between them. They can tolerate imperfect scan data, without reflecting those imperfections in the exported digital models. Furthermore, the exported models are not simply shapes but the CAD commands required to produce the shapes as well as the shapes themselves.

The use of 3D scanners with state-of-the-art 3D modelers yields several advantages. The first is time savings. With a 3D scanner, there is no need to re-create models from scratch; digital 3D models of existing models can be manipulated to suit one's design needs. With the latest 3D modeling technologies, it is also possible to attain higher accuracy in 3D models as well as yielding excellent models from marginal or imperfect physical object.

Just as a CAD system requires input devices to transfer information from the user to the CPU, output devices are also necessary to transfer data in visual form back to the user. Electronic displays provide real-time feedback to the user, enabling visualization and modification of information without hard-copy production. Often, however, a hard copy is required for presentation or evaluation, and the devices which use the data in the CPU or stored in memory to create the desired copy are a second category of output devices.

3.9 Electronic Displays

Many types of display technology have been used in display monitors. Up until the twenty-first century, the CRT was the most common. Historically, computer graphics displays used a CRT to generate an image on the display screen. The CRT heats a cathode to project a beam of electrons onto a phosphor-coated glass screen. The electron beam energizes the phosphor coating at the point of contact, causing the phosphor to glow (Fig. 14).

CRTs employ two different techniques, stroke writing and raster scan, to direct the beam onto the screen. A CRT using the stroke-writing technique directs the beam only along the

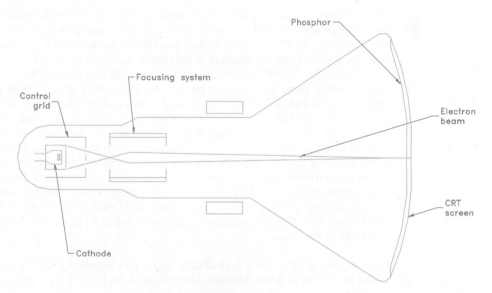

Figure 14 Diagrammatic representation of a CRT.

vectors given by the graphics data in the CPU. In a raster-scan CRT, the electron beam sweeps systematically from left to right and top to bottom at a continuous rate, employing what is known as "rasterization." The image is created by turning the electron beam on or off at various points along the sweep, depending on whether a light or a dark dot is required at those points to create a recognizable image.

Other forms of technology, such as the LCD (liquid crystal display) and LED, have essentially replaced the CRT. The LCD is the technology commonly seen in flat-screen televisions and computers and is the predominant display in use today. Still an emerging technology as far as CAD is concerned, OLED (organic LED) screens have been in use for smaller digital displays such as those of mobile phones and mp3 players. However, the television and computer monitor industries are making significant strides in developing larger OLED screens.

3.10   Printers and Plotters

The most commonly used plotters are:

1. Electrostatic plotter
2. Inkjet plotter
3. Laser plotter

Despite the ease with which design files can be managed using computer technology, a hard copy of the work is often required for record keeping, presentation, and production. Output devices have been developed to interface with the computer and produce a hard copy of the design or file. The various types of hard-copy devices are shown in the following outline:

Many plotters require rasterization, breaking up the image into a series of dots, which will then be reconstituted to re-create the image in much the same way that a raster-scan electronic display screen uses these dots to produce a recognizable image on the display terminal. Common raster plotters include the ink-jet plotter and the laser plotter. The time to produce a rasterized plot is independent of complexity; therefore raster plotters have the advantage of being significantly faster.

Inkjet plotters and printers fire tiny ink droplets at paper or a similar medium from thin nozzles in the printing head. Heat generated by a separate heating element almost instantaneously vaporizes the ink. The resulting bubble generates a pressure wave that ejects an ink droplet from the nozzle. Once the pressure pulse passes, ink vapor condenses and the negative pressure produced as the bubble contracts draws fresh ink into the nozzle. These plotters do not require special paper and can also be used for preliminary drafts. Inkjet plotters are available both as desktop units for 8.5 × 11-in. graphics and in wide format for engineering CAD drawings. Typical full-color resolution is 360 dpi, with black-and-white resolution rising to 700×720 dpi. These devices handle both roll-feed and cut-sheet media in widths ranging from 8.5 to as much as 15 ft. Also, ink capacity in recently developed plotters has increased, allowing these devices to handle large rolls of paper without depleting any one ink color. Inkjet plotters are very user friendly, often including sensors for the ink supply and ink flow, which warn users of an empty cartridge or of ink stoppage, allowing replacement without losing a print. Other sensors eliminate printing voids and unwanted marks caused by bubbles in the ink lines. Special print modes typically handle high-resolution printing by repeatedly going over image areas to smooth image lines.

Laser plotters produce fairly high quality hard copies in a short period of time. A laser housed within the plotter projects rasterized image data in the form of light onto a photostatic drum.

As the drum rotates further about its axis, it is dusted with an electrically charged powder known as toner. The toner adheres to the drum wherever the drum has been charged by the laser light. The paper is brought into contact with the drum and the toner is released onto the paper, where it is fixed by a heat source close to the exit point. Laser plotters can quickly produce images in black and white or in color, and resolution is high.

3.11   Mobile Devices

A mobile device is any computing device that is hand held, typically with a small display, capable of touch input and/or with a miniature keyboard, weighing less than 2 lb. Generally, devices that are termed "mobile devices" include PDAs (personal digital assistants), tablets, portable media players, mobile phones (smartphones), and hand-held gaming consoles. Modern mobile devices, especially tablets and smartphones, have hardware-specific operating systems; some of these include Android, Apple's iOS, and Microsoft's Windows Phone OS. These mobile devices have made their way onto corporate and research fields, as they can be used to take notes, send and receive invoices, banking, and entertainment via the Internet. Mobile devices have also found their way into CAD, with several advancements that provide engineers with the ability to design and alter designs "on the go." Some popular mobile platforms have applications which turn the mobile devices into supplemental input devices that can be used to perform otherwise cumbersome tasks, such as easily changing the views using gestures, and orbiting 3D models.

4   CAD SOFTWARE

Graphics software makes use of the CPU and its peripheral I/O devices to generate a design and represent it on screen and on paper or using other output devices like rapid prototyping. Analysis software uses the stored design data for dimensional modeling and various analytical methods such as interference checking using the computational speed of the CPU.

Traditional CAD software was limited to graphics software (2D and 3D design) and to analysis software such as FEA, property analysis, kinetic analysis, and rapid prototyping. CAD software has expanded to include applications such as PLM, collaborative product design (CPD), and PDM. Contemporary CAD software is often sold in "packages" which feature all of the programs needed for CAD applications. Modules and add-on packages can be acquired by the CAD software companies as the need arises.

4.1   2D Graphics Software

Traditional drafting has consisted of the creation of 2D technical drawings that operated in the synthesis stage of the general design process. However, contemporary computer graphics software, including that used in CAD systems, enables designs to be represented pictorially on the screen such that the human mind may create perspective, thus giving the illusion of three dimensions on a 2D screen. Regardless of the design representation, drafting only involves taking the conceptual solution for the previously recognized and defined problem and representing it pictorially. It has been asserted above that this "electronic drawing-board" feature is one of the advantages of computer-aided design.

The drawing board available through CAD systems is a result of the supporting graphics software. That software facilitates graphical representation of a design on screen by converting graphical input into Cartesian coordinates along x, y, and sometimes z axes. Design elements such as geometric shapes are often programmed directly into the software for simplified geometric representation. The coordinates of the lines and shapes created by the user can then be organized into a matrix and manipulated through matrix multiplication for operations such as scaling, rotation, and translation. The resulting points, lines, and shapes are relayed both to the graphics software and, finally, the display screen for simplified editing of designs. Because the whole process can take as little as a few nanoseconds, the user sees the results almost instantaneously.

While matrix mathematics provides the basis for the movement and manipulation of a drawing, much of CAD software is dedicated to simplifying the process of drafting itself because creating the drawing line by line and shape by shape is a lengthy and tedious process in itself. CAD systems offer users various techniques that can shorten the initial drafting time.

Geometric Definition.   All CAD systems offer defined geometric elements that can be called into the drawing by the execution of a software command. The user must usually indicate the variables specific to the desired element. For example, the CAD software might have, stored in the program, the mathematical definition of a circle. In the *x–y* coordinate plane, that definition is the following equation:

$$(x - m)^2 + (y - n)^2 = r^2$$

Here, the radius of the circle with its center at (m, n) is r. If the user specifies m, n, and r, a circle of the specified size will be represented on screen at the given coordinates. A similar process can be applied to many other graphical elements. Once defined and stored as an equation, the variables of size and location can be applied to create the shape on screen quickly and easily. This is not to imply that a user must input the necessary data in numerical form. Often, a graphical input device such as a mouse, trackball, digitizer, or light pen can be used to specify a point from which a line (sometimes referred to as a "rubber-band line" due to the variable length of the line as the cursor is moved toward or away from the given point) can be extended until the desired length is reached. A second input specifies that the desired endpoint has been reached, and variables can be calculated from the line itself. For a rectangle or square, the line might represent a diagonal from which the lengths of the sides could be extrapolated parallel to the *x* and *y* axes.

In the example of the circle above, the user would specify that a circle was to be drawn using a screen command or other input method. The first point could be established on screen as the center. Then, the line extending away from the center would define the radius. Often, the software will show the shape changing size as the line lengthens or shortens. When the radial line corresponds to the circle of desired size, the second point is defined. The coordinates of the two defined points give the variables needed for the program to draw the circle. The center is given by the coordinates of the first point, and the radius is easily calculated by determining the length of the line between points 1 and 2. Most engineering designs are much more complex than simple, whole shapes, and CAD systems are capable of combining shapes in various ways to create the desired design.

The combination of defined geometric elements enables the designer to create unique geometries quickly and easily on a CAD system. 2D combination such as addition and subtraction can be performed on the geometric elements. Once the desired geometric elements have been defined (identified as "cells"), they can be added as well as subtracted in any number of ways to create a desired design. For example, a rectangle might be defined as cell "A" and a circle might be defined as cell "B." When these designations have been made, the designer can add the two geometries or subtract one from the other using Boolean logic commands such as union, intersection, and subtraction. The concept for two dimensions is illustrated in Fig. 15. The new shape can also be defined as a cell "C" and combined in a similar manner to other

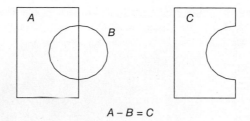

$$A - B = C$$

Figure 15   Two-dimensional example of Boolean cell difference.

primitives or conglomerate shapes. Cell definition, therefore, is recognized as a very powerful tool in CAD.

4.2   3D Graphics Software

Wire Frame Modeling

A wire frame model is a representation of an object in 3D computer graphics in which all surfaces are visibly outlined, including internal components and parts of the object normally hidden from view. A wire frame model allows for easy visualization of the underlying design of the object. It is, however, the least realistic method of representation of the 3D object. Wire frame models are mainly used in fields such as surveying, hydrology, geology, and mining.

Surface Modeling

Surface modeling (also called free-form surface modeling) is used in CAD to describe the outer surface, or skin, of a 3D object which does not necessarily have regular radial dimensions. Surface modeling generally uses NURBS mathematics to define surfaces. Free-form surfaces are defined in terms of their poles, degree, and patches (segments with spline curves) that determine their mathematical properties, shape, and smoothness of transition. This method is more complex than wire frame modeling but simpler than solid modeling. Applications are in areas such as designing turbine blades or automobile bodies.

Solid Modeling

Three-dimensional geometric or solid-modeling capabilities follow the same basic concept illustrated above, but with some other important considerations. First, there are various approaches to creating the design in three dimensions (Fig. 16). Second, different operators in solid-modeling software may be at work in constructing the 3D geometry. In CAD solid-modeling software there are various approaches that define the way in which the user creates the model. Since the introduction of solid-modeling capabilities into the CAD mainstream, various functional approaches to solid modeling have been developed. Many CAD software packages today support dimension-driven solid-modeling capabilities, which include variational design, parametric design, and feature-based modeling.

Dimension-Driven Design

Dimension-driven design denotes a system whereby the model is defined as sets of equations that are solved sequentially. These equations allow the designer to specify constraints, such as one plane must always be parallel to another plane. If the orientation of the first plane is

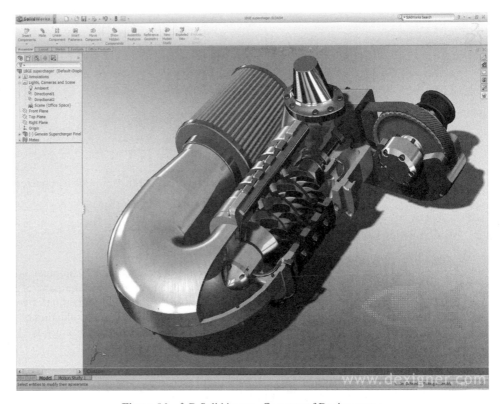

Figure 16   3-D Solid image.   Courtesy of Dexigner.com.

changed, the angle of the second plane will likewise be changed to maintain the parallel relationship. This approach gets its name from the fact that the equations used to define plane locations often define the distances between data points.

Variational Modeling

The variational modeling method describes the design in terms of a sketch that can later be readily converted to a 3D mathematical model with set dimensions. If the designer changes the design, the model must then be completely recalculated. This approach is quite flexible because it takes the dimension-driven approach of handling equations sequentially and makes it nonsequential. Dimensions can then be modified in any order, making it well suited for use early in the design process when the design geometry might change dramatically. Variational modeling also saves computational time (thus increasing the run speed of the program) by eliminating the need to solve any irrelevant equations. Variational sketching (Fig. 17) involves creating 2D profiles of the design that can represent end views and cross sections. Using this approach, the designer typically focuses on creating the desired shape with little regard toward dimensional parameters. Once the design shape has been created, a separate dimensioning capability can scale the design to the desired dimensions.

Parametric Modeling

Parametric modeling solves engineering equations between sets of parameters such as size parameters and geometric parameters. Size parameters are dimensions such as the diameter

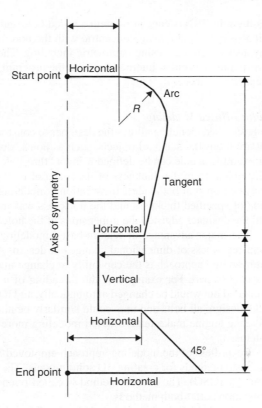

Figure 17   Variational sketching.

and depth of a hole. Geometric parameters are the constraints such as tangential, perpendicular, or concentric relationships. Parametric modeling approaches keep a record of operations performed on the design such that relationships between design elements can be inferred and incorporated into later changes in the design, thus making the change with a certain degree of acquired knowledge about the relationships between parts and design elements. For example, using the parametric approach, if a recessed area in the surface of a design should always have a blind hole in the exact center of the area and the recessed portion of the surface is moved, the parametric modeling software will also move the blind hole to the new center. In Fig. 18, if

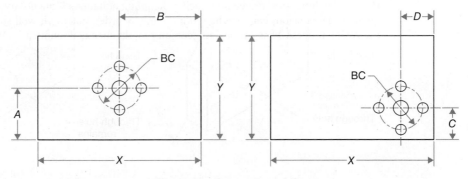

Figure 18   Parametric modeling.

a bolt circle (BC) is concentric with a bored hole and the bored hole is moved, the bolt circle will also move and remain concentric with the bored hole. The dimensions of the parameters may also be modified using parametric modeling. The design is modified through a change in these parameters either internally, within the program, or from an external data source, such as a separate database.

Feature-Based Modeling

Feature-based modeling allows the designer to construct solid models from geometric features that are industrial standard objects, such as holes, slots, shells, grooves, windows, and doors. For example, a hole can be defined using a "thru-hole" feature. Whenever this feature is used, independent from the thickness of the material through which the hole passes, the hole will always be open at both sides. In variational modeling, by contrast, if a hole was created in a plane of specified thickness and the thickness was increased, the hole would be a blind hole until the designer adjusted the dimensions of the hole to provide an opening at both ends.

The major advantage of feature-based modeling (Fig. 19) is the maintenance of design intent regardless of dimensional changes in design. Another significant advantage in using a feature-based approach is the capability to change many design elements relating to a change in a certain part. For example, if the threading of a bolt was changed, the threading of the associated nut would be changed automatically, and if that bolt design was used more than once in the design, all bolts and nuts could similarly be altered in one step. A knowledge base and inference engine make feature-based modeling more intelligent in some feature-based CAD systems.

Regardless of the modeling approach employed by a software package, there are usually two basic methods for creating 3D solid models. The first method is called constructive solid geometry (CSG). The second method is called boundary representation (B-rep). Most CAD applications offer both methods.

With the CSG method, using defined solid geometries such as those for a cube, sphere, cylinder, prism, etc., the user can combine them by subsequently employing a Boolean logic operator such as union, subtraction, difference, and intersection to generate a more complex part. In three dimensions, the Boolean difference between a cylinder and a cube might appear as shown in Fig. 20.

The boundary representation method is a modeling feature in 3-D representation. Using this technique, the designer first creates a 2D profile of the part. Then, using a linear, rotational, or compound sweep, the designer extends the profile along a line, about an axis, or along an

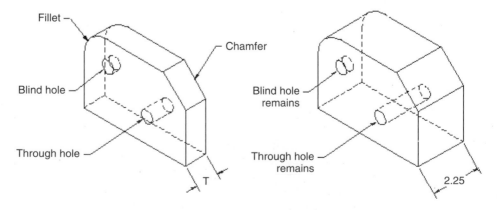

Figure 19 Feature-based modeling.

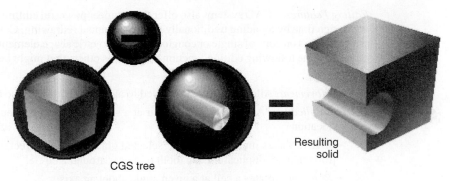

CGS tree

Resulting solid

Figure 20   Example of constructive solid geometry. Courtesy of SGI Corp.

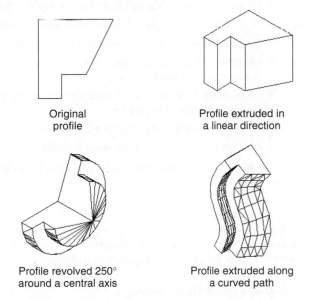

Original profile

Profile extruded in a linear direction

Profile revolved 250° around a central axis

Profile extruded along a curved path

Figure 21   Various common sweep methods in CAD software.

arbitrary curved path, respectively, to define a 3D image with volume. Figure 21 illustrates the linear, rotational, and compound sweep methods. Software manufacturers approach solid modeling differently. Nevertheless, every comprehensive solid modeler should have five basic functional capabilities. It should have interactive drawing, a solid modeler, a dimensional constraint engine, a feature manager, and assembly managers.

Solid modeling has advanced in that there are now parts libraries using CAD data files. As opposed to creating geometries from within the CAD program, CAD systems will accept geometries from other database systems and software. The Initial Graphics Exchange System (IGES) is an ANSI standard that defines a neutral form for the exchange of information between dissimilar CAD and CAM systems. Significant time can be saved when using models from differing sources. Often, corporations will supply drawing files online or through CD or DVD with catalogued listings of various parts and products. The engineer may then focus on the major design considerations without constantly redesigning small, common parts such as bearings, bolts, cogs, and sprockets. Libraries of parts are also available on the Internet.

Editing Features. CAD systems also offer the engineer powerful editing features which reduce the design time by avoiding traditionally required manual redrawing. Common editing features are performed on cells of single or conglomerate geometric shape elements. Most CAD systems offer all of the following editing functions, as well as others that might be specific to a program being used:

- *Movement.* Allows a cell to be moved to another location on the display screen.
- *Duplication.* Allows a cell to appear at a second location without deleting the original location
- *Array.* Makes multiple copies of selected objects, creating rectangular or polar arrays, for example, duplicating one tooth of a gear around a circle.
- *Rotation.* Rotates a cell at a given angle about an axis.
- *Mirroring.* Displays a mirror image of the cell about a plane.
- *Deletion.* Removes the cell from the display and the design data file.
- *Trim.* Removes any part of the cell extending beyond a defined point, line, or plane.
- *Scaling.* Enlarges or reduces the cell by a specified factor along x, y, and z axes.
- *Offsetting.* Creates a new object that is the same or a scaled version of the selected object at a specified distance.
- *Chamfering.* Connects two nonparallel objects by extending or trimming them to intersect or join with a beveled line.
- *Filleting.* Connects two objects with a smoothly fitted arc of a specified radius.
- *Hatching.* User can edit both hatch boundaries and hatch patterns.

Most of the editing features offered in CAD are transformations performed using algebraic matrix manipulations.

Transformations

A transformation refers to the movement or other manipulation of graphical data. Two-dimensional transformations are considered first in order to illustrate the basic concepts. Later, these concepts are applied to geometries with three dimensions.

Two-Dimensional Transformations

To locate a point in a two-axis Cartesian coordinate system, x and y values are specified. This 2D point can be modeled as a 1×2 matrix: (x, y). For example, the matrix $p = (3, 2)$ would be interpreted to be a point which is three units from the origin in the x direction and two units from the origin in the y direction.

This method of representation can be conveniently extended to define a line segment as a 2×2 matrix by giving the x and y coordinates of the two endpoints of the line. The notation would be

$$l = \begin{bmatrix} x_1 & y_1 \\ x_2 & y_2 \end{bmatrix} \tag{1}$$

Using the rules of matrix algebra, a point or line (or other geometric element represented in matrix notation) can be operated on by a transformation matrix to yield a new element.

There are several common transformations: translation, scaling, and rotation.

Translation involves moving the element from one location to another. In the case of a line segment, the operation would be

$$\begin{cases} x_1' = x_1 + \Delta x & y_1' = y_1 + \Delta y \\ x_2' = x_2 + \Delta x & y_2' = y_2 + \Delta y \end{cases} \tag{2}$$

where x', y' are the coordinates of the translated line segment, x, y are the coordinates of the original line segment, and Δx and Δy are the movements in the x and y directions, respectively.

In matrix notation this can be represented as

$$l' = l + T \tag{3}$$

where

$$T = \begin{bmatrix} \Delta x & \Delta y \\ \Delta x & \Delta y \end{bmatrix} \tag{4}$$

is the translation matrix.

Any other geometric element can be translated in space by adding Δx to the current x value and Δy to the current y value of each point that defines the element.

The scaling transformation enlarges or reduces the size of elements. Scaling of an element is used to enlarge it or reduce its size. The scaling need not necessarily be done equally in the x and y directions. For example, a circle could be transformed into an ellipse by using unequal x and y scaling factors.

A line segment can be scaled by the scaling matrix as follows:

$$l' = l \times S \tag{5}$$

where

$$S = \begin{bmatrix} \alpha & 0 \\ 0 & \beta \end{bmatrix} \tag{6}$$

is the scaling matrix. Note that the x scaling factor α and y scaling factor β are not necessarily the same. This would produce an alteration in the size of the element by the factor α in the x direction and by the factor β in the y direction. It also has the effect of repositioning the element with respect to the Cartesian system origin. If the scaling factors are less than 1, the size of the element is reduced and it is moved closer to the origin. If the scaling factors are larger than 1, the element is enlarged and removed farther from the origin. Scaling can also occur without moving the relative position of the element with respect to the origin. In this case, the element could be translated to the origin, scaled, and translated back to the original location.

In this transformation, the geometric element is rotated about the origin by an angle θ. For a positive angle, the rotation is in the counterclockwise direction. This accomplishes rotation of the element by the same angle, but it also moves the element. In matrix notation, the procedure would be as follows:

$$l' = l \times R \tag{7}$$

where

$$R = \begin{bmatrix} \cos\theta & \sin\theta \\ -\sin\theta & \cos\theta \end{bmatrix} \tag{8}$$

is the rotation matrix.

Besides rotating about the origin point (0, 0), it might be important in some instances to rotate the given geometry about an arbitrary point in space. This is achieved by first moving the center of the geometry to the desired point and then rotating the object. Once the rotation is performed, the transformed geometry is translated back to its original position.

The previous single transformations can be combined as a sequence of transformations called concatenation, and the combined transformations are called concatenated transformations. During the editing process, when a graphic model is being developed, the use of concatenated transformations is quite common. It would be unusual that only a single transformation would be needed to accomplish a desired manipulation of a cell. One example in which combinations of transformations would be required would be to uniformly scale a line l and then rotate the scaled geometry by an angle θ about the origin; the resulting new line is then

$$l' = l \times R \times S \tag{9}$$

where R is the rotation matrix and S is the scaling matrix. A concatenation matrix can then be defined as

$$C = RS \tag{10}$$

The above concatenation matrix cannot be used in the example of rotating geometry about an arbitrary point. In this case the sequence would be translation to the origin, rotation about the origin, then translation back to the original location. Note that the translation has to be done separately.

Three-Dimensional Transformations

Transformations by matrix methods can be extended to 3D space. The same three general categories defined in the preceding section are considered: translation, scaling, and rotation.

The translation matrix for a point defined in three dimensions would be

$$T = \left(\Delta x, \quad \Delta y, \quad \Delta z \right) \tag{11}$$

A cell would be translated by adding the increments Δx, Δy, and Δz to the respective coordinates of each of the points defining the 3D geometry element.

The scaling transformation is given by

$$S = \begin{bmatrix} \alpha & 0 & 0 \\ 0 & \beta & 0 \\ 0 & 0 & \gamma \end{bmatrix} \tag{12}$$

Equal values of α, β, and γ produce a uniform scaling in all three directions.

Rotation in 3D can be defined for each of the axes. Rotation about the z axis by an angle θ_z is accomplished by the matrix

$$R_z = \begin{bmatrix} \cos\theta_z & -\sin\theta_z & 0 \\ \sin\theta_z & \cos\theta_z & 0 \\ 0 & 0 & 1 \end{bmatrix} \tag{13}$$

Rotation about the y axis by the angle θ_y is accomplished similarly:

$$R_y = \begin{bmatrix} \cos\theta_y & 0 & \sin\theta_y \\ 0 & 1 & 0 \\ -\sin\theta_y & 0 & \cos\theta_y \end{bmatrix} \tag{14}$$

Rotation about the x axis by the angle θ_x is performed with an analogous transformation matrix:

$$R_x = \begin{bmatrix} 1 & 0 & 0 \\ 0 & \cos\theta_x & -\sin\theta_x \\ 0 & \sin\theta_x & \cos\theta_x \end{bmatrix} \tag{15}$$

All the three rotations about x, y, and z axes can be concatenated to form a rotation about an arbitrary axis.

Graphical Representation of Image Data

Wire frame representations, whether in 2D or 3D, can be ambiguous and difficult to understand. Because mechanical and other engineering designs often involve 3D parts and systems, CAD systems that offer 3D representation capabilities have quickly become the most popular in engineering design.

In order to generate a 2D view from a 3D model, the CAD software must be given information describing the viewpoint of the user. With this information, the computer can calculate angles of view and determine which surfaces of the design would be visible from the given viewpoint. The software typically uses surfaces that are closest to the viewer to block out surfaces that would be hidden from view. Then, applying this technique and working in a direction away from the viewer, the software determines which surfaces are visible. The next step determines the virtual distance between viewer and model, allowing those areas outside the boundaries of the screen to be excluded from consideration. Then, the colors displayed on each surface must be determined by combining considerations of the user's preferences for light source and surface color. The simulated light source can play a very important role in realistically displaying the image by influencing the values of colors chosen for the design and by determining reflection and shadow placements. Current high-end CAD software can simulate a variety of light sources including spot-lighting and sunlight, either direct or through some opening such as a door or window.

Some systems will even allow surface textures to be chosen and displayed. Once these determinations have been made, the software calculates the color and value for each pixel of the display. Since these calculations are computationally intensive, the choice of hardware is often just as important as the software used when employing solid-modeling programs with advanced surface representation features. The software manufacturers have minimum and optimum hardware system requirements listed for their CAD software.

4.3   Analysis Software

An important part of the design process is the simulation of the performance of a designed device. For example, a fastener may be designed to work under certain static or dynamic loads or the temperature distribution in a CPU chip may need to be calculated to determine the heat transfer behavior and possible thermal stress. Or the turbulent flow over a turbine blade controls cooling but may induce vibration and will have to be dealt with. Whatever the device being designed, there are many possible influences on the device's performance and CAD analysis software may help identify them.

The load types listed above can be calculated using FEA. The analysis divides a given domain into smaller, discrete fundamental parts called elements. An analysis of each element is then conducted using the required mathematics. Finally, the solution to the problem as a whole is determined through an aggregation of the individual solutions of the elements. Complex problems can thus be solved by dividing the problem into smaller and simpler problems upon which approximate solutions can be applied. General-purpose FEA software programs have been generalized such that users do not need to have detailed knowledge of FEA.

A FEM can be thought of as a system of solid blocks (elements) assembled together. Several types of elements that are available in the finite-element library are given below. Well-known general-purpose FEA packages, such as NASTRAN by NEi Software (formerly Noran Engineering, Inc.) and ANSYS by ANSYS Inc., provide an element library.

To demonstrate the concept of FEA, a 2D bracket is shown (Fig. 22) divided into quadrilateral elements each having four nodes. Although triangular nodes are sometimes used, recent work has shown that quadrilateral elements are generally more accurate. Elements are joined to each other at nodal points. When a load is applied to the structure, all elements deform until all forces balance. For each element in the model, equations can be written which relate displacement and forces at the nodes. Each node has a potential of displacement in x and y directions under F_x and F_y (x and y components of the nodal force) so that one element needs eight equations to express its displacement. The displacements and forces are identified by a coordinate numbering system for recognition by the computer program. For example, d_{xi}^I is the displacement in the x direction for element I at node i, while d_{yi}^I is the displacement in the y

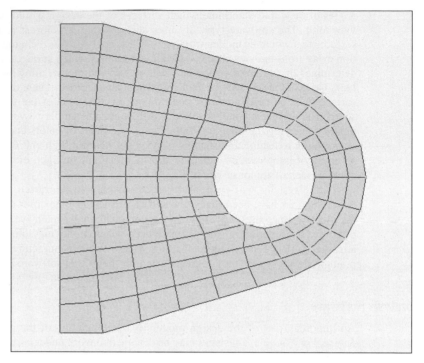

Figure 22   2D bracket divided into quadrilateral elements.

direction for the same node in element I. Forces are identified in a similar manner, so that F_{xi}^I is the force in the x direction for element I at the node i.

A set of equations relating displacements and forces for the element I should take the form of the basic spring equation, $F = kd$:

$$
\begin{aligned}
k_{11}d_{xi}^I + k_{12}d_{yi}^I + k_{13}d_{xj}^I + k_{14}d_{yj}^I + k_{15}d_{xk}^I + k_{16}d_{yk}^I + k_{17}d_{xl}^I + k_{18}d_{yl}^I &= F_{xi}^I \\
k_{21}d_{xi}^I + k_{22}d_{yi}^I + k_{23}d_{xj}^I + k_{24}d_{yj}^I + k_{25}d_{xk}^I + k_{26}d_{yk}^I + k_{27}d_{xl}^I + k_{28}d_{yl}^I &= F_{yi}^I \\
k_{31}d_{xi}^I + k_{32}d_{yi}^I + k_{33}d_{xj}^I + k_{34}d_{yj}^I + k_{35}d_{xk}^I + k_{36}d_{yk}^I + k_{37}d_{xl}^I + k_{38}d_{yl}^I &= F_{xj}^I \\
k_{41}d_{xi}^I + k_{42}d_{yi}^I + k_{43}d_{xj}^I + k_{44}d_{yj}^I + k_{45}d_{xk}^I + k_{46}d_{yk}^I + k_{47}d_{xl}^I + k_{48}d_{yl}^I &= F_{yj}^I \\
k_{51}d_{xi}^I + k_{52}d_{yi}^I + k_{53}d_{xj}^I + k_{54}d_{yj}^I + k_{55}d_{xk}^I + k_{56}d_{yk}^I + k_{57}d_{xl}^I + k_{58}d_{yl}^I &= F_{xk}^I \\
k_{61}d_{xi}^I + k_{62}d_{yi}^I + k_{63}d_{xj}^I + k_{64}d_{yj}^I + k_{65}d_{xk}^I + k_{66}d_{yk}^I + k_{67}d_{xl}^I + k_{68}d_{yl}^I &= F_{yk}^I \\
k_{71}d_{xi}^I + k_{72}d_{yi}^I + k_{73}d_{xj}^I + k_{74}d_{yj}^I + k_{75}d_{xk}^I + k_{76}d_{yk}^I + k_{77}d_{xl}^I + k_{78}d_{yl}^I &= F_{xl}^I \\
k_{81}d_{xi}^I + k_{82}d_{yi}^I + k_{83}d_{xj}^I + k_{84}d_{yj}^I + k_{85}d_{xk}^I + k_{86}d_{yk}^I + k_{87}d_{xl}^I + k_{88}d_{yl}^I &= F_{yl}^I
\end{aligned}
\tag{16}
$$

The k parameters are stiffness coefficients that relate the nodal deflections and forces. They are determined by the governing equations of the problem using given material properties such as Young's modulus and Poisson's ratio and from element geometry.

The set of equations can be written in matrix form for ease of operation as follows:

$$
\begin{Bmatrix}
k_{11} & k_{12} & k_{13} & k_{14} & k_{15} & k_{16} & k_{17} & k_{18} \\
k_{21} & k_{22} & k_{23} & k_{24} & k_{25} & k_{26} & k_{27} & k_{28} \\
k_{31} & k_{32} & k_{33} & k_{34} & k_{35} & k_{36} & k_{37} & k_{38} \\
k_{41} & k_{42} & k_{43} & k_{44} & k_{45} & k_{46} & k_{47} & k_{48} \\
k_{51} & k_{52} & k_{53} & k_{54} & k_{55} & k_{56} & k_{57} & k_{58} \\
k_{61} & k_{62} & k_{63} & k_{64} & k_{65} & k_{66} & k_{67} & k_{68} \\
k_{71} & k_{72} & k_{73} & k_{74} & k_{75} & k_{76} & k_{77} & k_{78} \\
k_{81} & k_{82} & k_{83} & k_{84} & k_{85} & k_{86} & k_{87} & k_{88}
\end{Bmatrix}
\times
\begin{Bmatrix}
d_{xi}^{l} \\ d_{yi}^{l} \\ d_{xj}^{l} \\ d_{yj}^{l} \\ d_{xk}^{l} \\ d_{yk}^{l} \\ d_{xl}^{l} \\ d_{yl}^{l}
\end{Bmatrix}
=
\begin{Bmatrix}
F_{xi}^{l} \\ F_{yi}^{l} \\ F_{xj}^{l} \\ F_{yj}^{l} \\ F_{xk}^{l} \\ F_{yk}^{l} \\ F_{xl}^{l} \\ F_{yl}^{l}
\end{Bmatrix}
\tag{17}
$$

When a structure is modeled, individual sets of matrix equations are automatically generated for each element. The elements in the model share common nodes so that individual sets of matrix equations can be combined into a global set of matrix equations. This global set relates all of the nodal deflections to the nodal forces. Nodal deflections are solved simultaneously from the global matrix. When displacements for all nodes are known, the state of deformation of each element is known and stress can be determined through stress–strain relations.

For a 2D structure problem each node displacement has three degrees of freedom: one translational in each of x and y directions and a rotational in the (x, y) plane. In a 3D structure problem the displacement vector can have up to six degrees of freedom for each nodal point. Each degree of freedom at a nodal point may be unconstrained (unknown) or constrained. The nodal constraint can be given as a fixed value or a defined relation with its adjacent nodes. One or more constraints must be given prior to solving a structure problem. These constraints are referred to as boundary conditions.

Finite-element analysis obtains stresses, temperatures, velocity potentials, and other desired unknown variables in the analyzed model by minimizing an energy function. The law of conservation of energy is a well-known principle of physics. It states that unless atomic energy is involved, the total energy of a system must be zero. Thus, the finite-element energy functional must equal zero. The FEM obtains the correct solution for any analyzed model by minimizing the energy functional. Thus, the obtained solution satisfies the law of conservation of energy.

The minimum of the functional is found by setting to zero the derivative of the functional with respect to the unknown nodal point potential. It is known from calculus that the minimum of any function has a slope or derivative equal to zero. Thus, the basic equation for FEA is

$$
\frac{dF}{dp} = 0
\tag{18}
$$

where F is the functional and p is the unknown nodal point potential to be calculated. The FEM can be applied to many different problem types. In each case, F and p vary with the type of problem.

Problem Types

Linear Statics. Linear static analysis represents the most basic type of analysis. The term "linear" means that the stress is proportional to strain (i.e., the materials follow Hooke's law). The term "static" implies that forces do not vary with time or that time variation is insignificant and can therefore be safely ignored. Assuming the stress is within the linear stress–strain range, a beam under constant load can be analyzed as a linear static problem. Another example of linear statics is a steady-state temperature distribution within a constant material property structure. The temperature differences cause thermal expansion, which in turn induces thermal stress.

Buckling.   In linear static analysis, a structure is assumed to be in a state of stable equilibrium. As the applied load is removed, the structure is assumed to return to its original, undeformed position. Under certain combinations of loadings, however, the structure continues to deform without an increase in the magnitude of loading. In this case the structure has buckled or has become unstable. For elastic, or linear, buckling analysis, it is assumed that there is no yielding of the structure and that the direction of applied forces does not change.

Elastic buckling incorporates the effect of differential stiffness, which includes higher order strain displacement relationships that are functions of the geometry, element type, and applied loads. From a physical standpoint, the differential stiffness represents a linear approximation of softening (reducing) the stiffness matrix for a compressive axial load and stiffening (increasing) the stiffness matrix for a tensile axial load.

In buckling analysis, eigenvalues are solved. These are scaling factors used to multiply the applied load in order to produce the critical buckling load. In general, only the lowest buckling load is of interest, since the structure will fail before reaching any of the higher order buckling loads. Therefore, usually only the lowest eigenvalue needs to be computed.

Normal Modes.   Normal-mode analysis computes the natural frequencies and mode shapes of a structure. Natural frequencies are the frequencies at which a structure will tend to vibrate if subjected to a disturbance. For example, the strings of a piano are each tuned to vibrate at a specific frequency. The deformed shape at a specific natural frequency is called the mode shape. Normal-mode analysis is also called real eigenvalue analysis.

Normal-mode analysis forms the foundation for a thorough understanding of the dynamic characteristics of the structure. In static analysis the displacements are the true physical displacements due to the applied loads. In normal-mode analysis, because there is no applied load, the mode shape components can all be scaled by an arbitrary factor for each mode.

Nonlinear Statics.   Nonlinear structural analysis must be considered if large displacements occur with linear materials (geometric nonlinearity), structural materials behave in a nonlinear stress–strain relationship (material nonlinearity), or a combination of large displacements and nonlinear stress–strain effects occurs. An example of geometric nonlinear statics is shown when a structure is loaded above its yield point. The structure will then tend to be less stiff and permanent deformation will occur, and Hooke's law will not be applicable anymore. In material nonlinear analysis, the material stiffness matrix will change during the computation. Another example of nonlinear analysis includes the contacting problem, where a gap may appear and/or sliding occurs between mating components during load application or load removal.

Dynamic Response.   Dynamic response in general consists of frequency response and transient response. Frequency response analysis computes structural response to steady-state oscillatory excitation. Examples of oscillatory excitation include rotating machinery, unbalanced tires, and helicopter blades. In frequency response, excitation is explicitly defined in the frequency domain. All of the applied forces are known at each forcing frequency. Forces can be in the form of applied forces and/or enforced motions. The most common engineering problem is to apply steady-state sinusoidally varying loads at several points on a structure and determine its response over a frequency range of interest. Transient response analysis is the most general method for computing forced dynamic response. The purpose of a transient response analysis is to compute the behavior of a structure subjected to time-varying excitation. All of the forces applied to the structure are known at each instant in time. The important results obtained from a transient analysis are typically displacements, velocities, and accelerations of the structure as well as the corresponding stresses.

5   CAD STANDARDS AND TRANSLATORS

In order for CAD applications to run across systems from various vendors, four main formats facilitate this data exchange:

- IGES (Initial Graphics Exchange Specification)
- STEP (Standard for the Exchange of Products)
- DXF (Drawing Exchange Format)
- ACIS (American Committee for Interoperable Systems)

IGES (Initial Graphics Exchange Specification)

IGES is an ANSI standard for the digital representation and exchange of information between CAD/CAM systems. 2D geometry and 3D CSG can be translated into the IGES format. Versions of IGES support B-rep solid-modeling capabilities. Common translators (IGES-in and IGES-out) functions available in the IGES library include IGES file parsing and formatting, general entity manipulation routines, common math utilities for matrix, vector, and other applications, and a robust set of geometry conversion routines and linear approximation facilities.

STEP (Standard for the Exchange of Products: ISO 10303)

STEP is an international collection of standards. It provides one natural format that can apply to CAD data throughout the life cycle of a product. STEP offers features and benefits that are absent from IGES. The user can pull out an IGES specification and get all the data required in one document. STEP can also transfer B-rep solids between CAD systems. STEP differs from IGES in how it defines data. In IGES, the user pulls out the specification, reads it, and implements what it says. In STEP, the implementor takes the definition and runs it through a special compiler, which then delivers the code. This process assures that there is no ambiguous understanding of data among implementors.

DXF (Drawing Exchange Format)

DXF, developed by Autodesk, Inc. for AutoCAD software, is the de facto standard for exchanging CAD/CAM data on a PC-based system. Only 2D drawing information can be converted into DXF either in ASCII or in binary format. DWG (the name of Autodesk's proprietary file format) is the evolution of the DXF format, one that allows both 2D and 3D data and associated parameters to be stored. DWG files are used natively by AutoCAD and other CAD software.

ACIS (American Committee for Interoperable Systems)

The ACIS modeling kernel is a set of software algorithms used for creating solid-modeling packages. Software developers license ACIS routines from the developer, Spatial Corp. (formerly Spatial Technology Corp.), to simplify the task of writing new solid modelers. The key benefit of this approach is that models created using software based on ACIS should run unchanged with other brands of ACIS-based modelers. This eliminates the need to use IGES translators for transferring model data back and forth among applications. ACIS-based packages have become commercially available for CAD/CAM and FEA software packages. Output files from ACIS have the suffix *.SAT or *.SAB.

6 APPLICATIONS OF CAD

6.1 Optimization Applications

As designs become more complex, engineers need fast, reliable tools. Finite-element analysis has become the major tool used to identify and solve design problems. Increased design efficiency provided by CAD has been augmented by the application of FEMs to analysis, but engineers still often use a trial-and-error method for correcting the problems identified through FEA. This method inevitably increases the time and effort associated with design because it increases the time needed for interaction with the computer, and solution possibilities are often limited by the designer's personal experiences.

Design optimization seeks to eliminate much of this extra time by applying a logical mathematical method to facilitate modification of complex designs. Optimization strives to minimize or maximize a characteristic, such as weight or physical size, which is subjected to constraints on one or more parameters. Either the size, shape, or both determines the approach used to optimize a design. Optimizing the size is usually easier than optimizing the shape of a design. The geometry of a plate does not significantly change by optimizing its thickness. On the other hand, optimizing a design parameter such as the radius of a hole does change the geometry during shape optimization.

Optimization approaches were difficult to implement in the engineering environment because the process was somewhat academic in nature and not viewed as easily applicable to design practices. However, if viewed as a part of the process itself, optimization techniques can be readily understood and implemented in the design process. Iterations of the design procedure occur as they normally do in design up to a point. At that point, the designer implements the optimization program. The objectives and constraints upon the optimization must first be defined. The optimization program then evaluates the design with respect to the objectives and constraints and makes automated adjustments in the design. Because the process is automatic, engineers should have the ability to monitor the progress of the design during optimization, stop the program if necessary, and begin again.

The power of optimization programs is largely a function of the capabilities of the design software used in earlier stages. Two- and three-dimensional applications require automatic and parametric meshing capabilities. Linear static, natural frequencies, mode shapes, linearized buckling, and steady-state analyses are required for other applications. Because the design geometry and mesh can change during optimization iterations, error estimate and adaptive control must be included in the optimization program. Also, when separate parts are to be assembled and analyzed as a whole, it is often helpful to the program to connect different meshes and element types without regard to nodal or elemental interface matches.

Preliminary design data are used to meet the desired design goals through evaluation, remeshing, and revision. Acceptable tolerances must then be entered along with imposed constraints on the optimization. The engineer should be able to choose from a large selection of design objectives and behavior constraints and use these with ease. Also, constraints from a variety of analytical procedures should be supported so that optimization routines can use the data from previously performed analyses.

Although designers usually find optimization of shape more difficult to perform than of size, the use of parametric modeling capabilities in some CAD software minimizes this difficulty. Shape optimization is an important tool in many industries, including shipbuilding, aerospace, and automotive manufacturing. The shape of a model can be designed using any number of parameters, but as few as possible should be used for the sake of simplicity. If the designer cannot define the parameters, neither design nor optimization can take place. Often, the designer will hold a mental note on the significance of each parameter. Therefore, designer input is crucial during an optimization run.

6.2   Virtual Prototyping

The creation of physical models for evaluation can often be time consuming and provide limited productivity. By employing kinematic and dynamic analyses on a design within the computer environment, time is saved, and often the result of the analysis is more useful than experimental results from physical prototypes. Physical prototyping often requires a great deal of manual work, not only to create the parts of the model, but also to assemble them and apply the instrumentation needed as well. Virtual prototyping uses kinematic and dynamic analytical methods to perform many of the same tests on a design model. The inherent advantage of virtual prototyping is that it allows the engineer to fine tune the design before a physical prototype is created. When the prototype is eventually fabricated, the designer is likely to have better information with which to create and test the model.

Physical models can provide the engineer with valuable design data, but the time required to create a physical prototype is long and must be repeated often through iterations of the process. A second disadvantage is that through repeated iterations the design is usually changed, so that time is lost in the process when parts are reconstructed as a working model. In some cases, the time invested in prototype construction and testing reveals less useful data than expected.

Virtual prototyping of a design is one possible solution to the problems of physical prototyping. Virtual prototyping employs computer-based testing so that progressive design changes can be incorporated quickly and efficiently into the prototype model. Also, with virtual prototyping, tests can be performed on the system or its parts in such a way that might not be possible in a laboratory setting. For example, the instrumentation required to test the performance of a small part in a system might disrupt the system itself, thus denying the engineer the accurate information needed to optimize the design. Virtual prototyping can also apply forces to the design that would be impossible to apply in the laboratory. For example, if a satellite is to be constructed, the design should be exposed to zero gravity in order to properly simulate its performance.

Engineers increasingly perform kinematic and dynamic analyses on a virtual prototype because a well-designed simulation leads to information that can be used to modify design parameters and characteristics that might not have otherwise been considered. Kinematic and dynamic analysis methods apply the laws of physics to a computerized model in order to analyze the motion of parts within the system and evaluate the overall interaction and performance of the system as a whole. At one time a mainframe computer was required to perform the necessary calculations to provide a realistic motion simulation. Today, microcomputers have the computational speed and memory capabilities necessary to perform such simulations on the desktop.

One advantage of kinematic/dynamic analysis software is that it allows the engineer to deliberately overload forces on the model. Because the model can be reconstructed in an instant, the engineer can take advantage of the destructive testing data. Physical prototypes would have to be fabricated and reconstructed every time the test was repeated. There are many situations in which physical prototypes must be constructed, but those situations can often be made more efficient and informative by the application of virtual prototyping analyses.

6.3   Rapid Prototyping and 3D Printing

One of the most recent applications of CAD technology has been in the area of rapid prototyping. Physical models traditionally have the characteristic of being one of the best evaluative tools for influencing the design process. Unfortunately, they have also represented the most time-consuming and costly stage of the design process. Rapid prototyping addresses this problem combining CAD data with sintering, layering, or deposition techniques to create a solid

Figure 23   Rapid prototype (right) and part cast from prototype (left). Courtesy of ProtoCam, Inc.

physical model of the design or part. The rapid prototyping industry is currently developing technology to enable the small-scale production of real parts as well as molds and dyes that can then be used in subsequent traditional manufacturing methods (Fig. 23). These two goals are causing the industry to become specialized into two major sectors. The first sector aims to create small rapid prototyping machines that one day might become as common in the design office as printers and plotters are today. The second branch of the rapid prototyping industry is specializing in the production of highly accurate, structurally sound parts to be used in the manufacturing process. The development of 3D printers has made prototyping quicker and more affordable than any other rapid prototyping method.

6.4   Additive Manufacturing

Computer-aided design is used to create the exact geometry required for the process of producing parts using an additive manufacturing (AM) approach. Essentially, AM is the process of producing parts by the successive addition of layers of material. In contrast, with conventional machining, material is removed to produce the final shape. AM allows for building parts with complex geometries without specialized tools or fixtures and without waste material.

Combining CAD with AM technology provides benefits for the entire production value chain. The geometric freedom enabled by CAD/AM allows for a significant reduction of traditional manufacturing constraints, such as weight limitations, complex contours, and intricate internal cavities. This can be utilized for lightweight designs, reduced part counts, or improved bone ingrowth of implants. For example, the production path from CAD to finished product is characterized by high material utilization and does not require expensive castings, forgings, or molds to be kept in stock, which is useful in mission-critical situations frequently occurring in the defense industry. In addition to its potential for cost efficiency, AM's high material utilization is energy efficient and provides for an environmentally friendly manufacturing route, reinforcing a sustainability competitive advantage.

CAD is also a key element of 3D printing. This is the process of making solid objects derived from CAD 3D model files. It is both technically and conceptually related to additive manufacturing. There is equipment available being marketed as "desktop 3D printers" with costs ranging from around $2000 and up. These printers reproduce part designs that can be created with 3D CAD software such as those of Solidworks and Autodesk Inventor.

There are numerous methods in which the 3D model is created from the original CAD data. Their differences include the processes used, the raw materials, and the cost of the finished product/prototype. These methods include stererolithography, laminated object manufacturing, selective laser sintering, fused deposition modeling, solid ground curing, and 3D inkjet printing.

A stereolithography machine divides a 3D CAD model into thin (0.0025-in.) slices. The information of each slice is sent to an ultraviolet laser beam which is used to trace the surface of the slice onto a container of photocurable liquid polymer. When the polymer is exposed to the laser, it solidifies into resin. This first resin layer is then lowered by the height of the next slice. The process of exposing the polymer to the laser is then repeated, and with each repetition, a new slice is solidified to the top of the existing slice and the solid is lowered to the height of the next slice, until the prototype has been completed. The workspace on one large stereolithography machine has a workspace volume of about 8000 in.^{3} ($20 \times 20 \times 20$). Larger parts may be made that are over 6 ft long given the right equipment.

Other materials for use with rapid prototyping are being tested and implemented by various companies involved in the growing rapid prototyping industry. The technology presents the opportunity for further automatization in the design and implementation environments.

Laminated object manufacturing was developed by Helisys of Torrance, California (currently Cubic Technologies). The process involves the bonding of layers of adhesive-coated material to form a model. Examples of construction material are paper, thermoplastics, or ferrous, nonferrous, and ceramics composites. The material is laminated with heat-activated glue. A feeder mechanism moves the laminated sheet over the building platform, and pressure is applied to bond the sheet to the base. A laser then cuts the outline of the first layer into the sheet and cross-hatches any excess area, which makes it easier to remove after completion of the build. The platform then is lowered by the thickness of the laminate, and the next layer is pressed to the first. The layer's outline is laser cut and the process of bonding and laser cutting is repeated until the model is built.

Selective laser sintering was developed by Carl Deckard and was patented in 1989. A laser is used to fuse powdered material into a solid. The models are built on a platform that sits in a heat-fusable powder. The laser sinters together the surface area of the first slice. Then the platform is lowered and powder is spread to the thickness of the next slice. This process repeats until the part is complete, and excess powder in each layer helps to support the part during the build. That excess powder can be brushed off or blown after completing the model.

Fused deposition modeling involves the extrusion of thin filaments of heated plastic from a tip that moves in the x–y plane. The deposition of these filaments onto the surface area of the first slice forms the first layer. The build platform is kept at a low temperature, so that the plastic hardens quickly. As the platform is lowered, a second layer is deposited, and so on, until the part is fully formed. Model supports may be similarly built during the process as needed.

Solid ground curing uses ultraviolet light to harden polymers one layer at a time. A photosensitive resin is first sprayed onto a build platform. Then, the machine develops a stencil/photomask of the desired layer. The mask is printed on a glass plate above the platform using a process similar to the electrostatic processes found in photocopiers and laser printers. The glass plate is then exposed to UV light to harden the shape of the entire desired layer in one step. The machine removes the excess resin and sprays wax in its place to support the model. The top surface is then milled flat, and the process repeats to finish the build. The finished part must be bathed in a solvent to remove the supporting wax.

3D inkjet printing was developed at MIT and it uses a spray of fluid to bind powdered material together. The unused powder supports the part as needed. The platform is then lowered, powder is added and leveled, and the process is repeated. When finished, the part is removed from the leftover powder, and excess powder is blown off. The finished parts can be infiltrated with wax, CA (cyanoacrylate) glue, or other sealant to improve durability and appearance.

6.5 Collaborative Product Design

Collaborative product design is an interactive design and product development methodology that involves designers from different geographical locations, with potentially different versions of CAD software, working together to concurrently design a product (Fig. 24). The development of specialized CAD software for collaborative product design has now made this process much easier through the use of open file format standards, such as IGES, STEP, VRML (Virtual Reality Modeling Language), and XML (Extensible Markup Language). CPD integrates several engineering steps such as conceptual design, detailed design, engineering analysis, assembly design, process design, and performance evaluation. CPD consists of acquiring requirements, defining the overall goal and task, decomposing the overall task into hierarchical subtasks, distributing subtasks among engineering designers, solving the subtasks, synthesizing the subsolutions, and finally providing the overall design solution. The goal of collaborating is to get more work accomplished efficiently in a shorter period of time by allowing a group to take advantage of skill sets possessed by all the people in that group.

CPD helps in optimizing the product development process, reducing errors between manufacturers and designers, increasing product quality, reducing production costs, as well as providing a platform for collaborators across the globe to work simultaneously on a project. CPD can help correct errors made during product design at an early stage, reducing product development costs, design iterations, lead times, and manufacturing costs.

To achieve collaborative product design, commercial systems such as engineering data management (EDM), PDM, product information management (PIM), technical document management (TDM), and technical information management (TIM) offer a structured way of efficiently storing, integrating, managing, and controlling data and engineering processes from design through manufacturing to distribution.

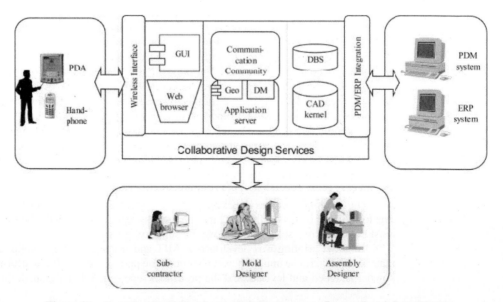

Figure 24 Scenario of collaborative product development. By permission from Elsevier.

CPD software may be used as either a cloud-based operation or a SaaS (software-as-a-service). In the cloud-based option, the CAD software package is installed at each individual workstation and each individual designer shares his or her work with others over the cloud (the cloud is a global network of computers communicating with each other over the Internet). The advantage of this approach is that speed of communication is optimized because engineers share only essential data over the cloud. CAD software needs to be installed at each workstation.

In the SaaS alternative, the software is not installed at each individual workstation but is instead subscribed to as a service (usually by paying a subscription fee) from a company that maintains the software on the cloud. Using the SaaS model may pose some issues such as clearly establishing ownership of any data or designs, managing continuity of software updates, and contingency planning for changes in vendors. The greatest advantage of SaaS is that only one copy of the software is necessary, greatly reducing maintenance costs. Some CAD software may include remote desktop applications to further reduce this cost. Remote desktop is an application that allows programs or applications to be run remotely while being accessed locally.

6.6 Product Life-Cycle Management

CAD is an integral part of PLM (Fig. 25). PLM is a comprehensive approach for innovation, new product development, introduction, and PDM from idea creation to end of life. Used together with CAM, entire product life cycles can be explored. Product functionality, cost, and sustainability are typically analyzed. PLM can be thought of as a repository for all information concerning the product (Fig. 26) and as a communication process among product stakeholders, such as marketing, engineering, manufacturing, and field service. The typical PLM system allows product information from marketing and design to be brought together and then be converted into information for manufacturing. The PLM approach was first introduced in mission-critical applications such as aerospace, medicine, military, and nuclear industries. PLM evolved from configuration management and electronic data management systems. PLM has now been adopted by various industries such as manufacturers of industrial machinery,

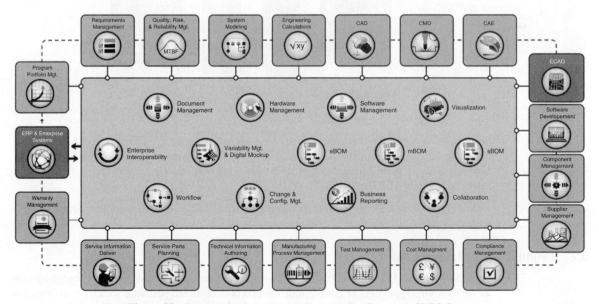

Figure 25 Product development system example. Courtesy of PTC Corp.

Figure 26   Typical elements associated with the product definition. Courtesy of PTC Corp.

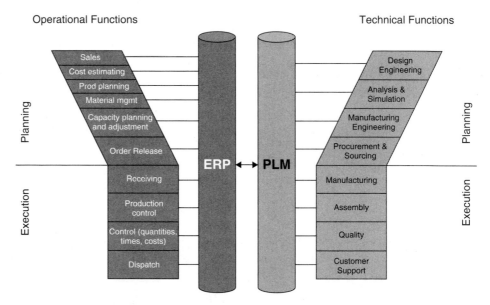

Figure 27   Technical aspects of PLM. Courtesy of PTC Corp.

biomedical equipment, consumer electronics, consumer goods, and other engineered products. The focus of PLM is on the technical aspects of the business such as engineering, design and manufacturing, and integration with the planning and order information in the enterprise resource planning (ERP) system (Fig. 27).

Typical providers of PLM systems focus on all business sizes, and they include Dassault Systèmes, Oracle/Agile, PTC, SAP, Siemens, and SofTech, among others.

6.7   Product Data Management

The traditional CAD system software structure has continued to evolve toward integration with a PDM system. Typically, a PDM system consists of centralized software which controls access to and is used to manage product data and process-related information. This information is

Figure 28   PDM example. Courtesy of PTC Corp.

derived from CAD data, models, parts information, manufacturing instructions, requirements, notes, and documents (Fig. 28). The ideal PDM system is accessible by multiple applications and multiple teams across an organization and supports business-specific needs.

As global enterprises continue to evolve, the CAD techniques and technologies presented in this chapter will continue to be critical in the evolution of these interconnected systems.

BIBLIOGRAPHY

H. Ali, "Optimization for Finite Element Applications," *Mech. Eng.*, December 1994, pp. 68–70.

F. Amirouche, *Principles of Computer Aided Design and Manufacturing*, Prentice-Hall, Englewood Cliffs, NJ, 2004.

S. Ashley, "Prototyping with Advanced Tools," *Mech. Eng.*, June 1994, pp. 48–55.

H. K. Ault, "3-D Geometric Modeling for the 21st Century," *Eng. Design Graphics J.*, **63**(2), 33–42, 2009.

"Basics of Design Engineering," *Machine Design*, April 6, 1995a, pp. 83–126.

"Basics of Design Engineering," *Machine Design*, April 4, 1996, pp. 47–83.

G. R. Bertoline, *Introduction to Graphics Communications for Engineers (B.E.S.T. Series)*, 2nd ed., Purdue University-West Lafayette, McGraw-Hill Higher Education, New York, 2002.

G. R. Bertoline and E. N. Wiebe, *Fundamentals of Graphics Communication*, 4th ed., McGraw-Hill Higher Education, New York, 2005.

J. D. Bethune, *Engineering Design Graphics with Autodesk Inventor 2011*, Prentice Hall, Boston, 2012.

R. Budynas and K. Nisbett, *Shigley's Mechanical Engineering Design*, 9th ed., McGraw-Hill Science/ Engineering/Math, New York, 2010.

"CAD/CAM Industry Report," *Machine Design*, May 23, 1994, pp. 38–98.

H. Cary and S. Helzer, *Modern Welding Technology*, 6th ed., Prentice-Hall, Englewood Cliffs, NJ, 2004.

M. Chader, "White Paper: Do it Once! The Value of 3rd Generation, Parametric Modeling from 3D Scan Data," SME Conference Rapid 2008, Society of Manufacturing Engineers, Orlando, Florida, April 18, 2008.

T.-C. Chang, H.-P. Wang, and R. A. Wysk, *Computer Integrated Manufacturing*, 3rd ed., Prentice-Hall, Englewood Cliffs, NJ, 2005.

Computing Encyclopedia, Smart Computing Reference Series, Vols 1–5, Sandhills Publishing Co., Lincoln, NE, 2002.

S. S. Condoor, *Mechanical Design Modeling Using Pro/Engineer*, 1st ed., Parks College of St. Louis University, McGraw-Hill Higher Education, New York, 2002.

T. Cook and R. H. Prater, *ABCs of Mechanical Drafting with an Introduction to AutoCAD 2000*, Prentice-Hall, Englewood Cliffs, NJ, 2002.

K. Crow, Product Data Management/Product Information Management, 2002, http://NPD-Solutions.com.

D. Deitz, "PowerPC: The New Chip on the Block," *Mech. Eng.*, January 1996, pp. 58–62.

A. D. Dimarogonas, *Machine Design, A CAD Approach*, Wiley, New York, 2000.

M. Dix and P. Riley, *Discovering AutoCAD 2012*, Prentice-Hall, Upper Saddle River, NJ, 2011.

B. A. Duffy and J. A. Leach, *AutoCAD 2002 Assistant*. McGraw-Hill Higher Education, New York, 2002.

P. Dvorak, "Windows NT Makes CAD Hum," *Machine Design*, January 10, 1994, pp. 46–52.

P. Dvorak, "Engineering on the Other Personal Computer," *Machine Design*, October 26, 1995, pp. 42–52.

R. J. Eggert, *Engineering Design*, 2nd ed., High Peak Press, Meridian, Idaho, 2010.

"Engineering Drives Document Management," Special Editorial Supplement, *Machine Design*, June 15, 1995b, pp. 77–84.

J. M. Escobar, J. M. Cascón, E. Rodriguez, and R. Montenegro, "A New Approach to Solid Modeling with Trivariate T-Splines Based on Mesh Optimization," *Comput. Methods Appl. Mech. Eng.*, **200**(45–46), 3210–3222, 2011.

S. Ethier and C. A. Ethier, *AutoCAD 2006 in 3 Dimensions Using AutoCAD 2006*, Prentice-Hall, Englewood Cliffs, NJ, 2007.

J. D. Foley, A. van Dam, S. K. Feiner, and J. Hughes, *Computer Graphics: Principles and Practice*, 2nd ed., Addison-Wesley, New York, 1995.

C. Freudenrich, "How OLEDs Work," March 24, 2005, available: HowStuffWorks.com, http://electronics.howstuffworks.com/oled.htm.

J. Y. H. Fuh and W. D. Li, "Advances in Collaborative CAD: The-State-of-the Art," *Computer-Aided Design*, **37**(5), 571–581, April 15, 2005.

T. A. Furness III, R. Earnshaw, R. Guedj, A. van Dam, and J. Vince (Eds.), "Toward Tightly Coupled Human Interfaces. Frontiers in Human-Centred Computing," *Online Communities and Virtual Environments*, 2001, pp. 80–98.

F. E. Giesecke, I. L. Hill, H. C. Spencer, A. E. Mitchell, and J. T. Dygdon, *Technical Drawing with Engineering Graphics*, 14th ed., Prentice-Hall, Upper Saddle River, NJ, 2011.

M. P. Groover and E. W. Zimmers, Jr., *CAD/CAM: Computer-Aided Design and Manufacturing*, Prentice-Hall, Englewood Cliffs, NJ, 1984.

P. J. Hanratty, "Making Solid Modeling Easier to Use," *Mech. Eng.*, March 1994, pp. 112–114.

D. Harrington, *Inside AutoCAD 2005*, Prentice-Hall, Englewood Cliffs, NJ, 2004.

J. D. Heisel, D. R. Short, and C. H. Jensen, *Engineering Draw Fundamental Version w/CDrom 2002*, 5th ed., McGraw-Hill, New York, 2002.

M. F. Hordeski, *Computer Integrated Manufacturing: Techniques and Applications*, TAB Professional and Reference Books, Blue Ridge Summit, PA, 1988.

W. E. Howard and J. C. Musto, *Introduction to Solid Modeling Using SolidWorks 2011*, McGraw-Hill Science/Engineering/Math, New York, 2012.

C. H. Jensen, J. D. Heisel, and D. R. Short, *Engineering Drawing and Design 2007*, 7th ed., McGraw-Hill, New York, 2011.

D. S. Kelley, *ProENGINEER 2001 Instructor with CD ROM*, Purdue University-West Lafayette, McGraw-Hill Higher Education, New York, 2002.

J. K. Krouse, *What Every Engineer Should Know About Computer-Aided Design and Computer-Aided Manufacturing*, Marcel Dekker, New York, 1982.

J. A. Leach and J. H. Dyer, *Autocad 2008 Companion: Essentials of AutoCAD Plus Solid Modeling*, McGraw-Hill, New York, 2008.

G. Lee, "Virtual Prototyping on Personal Computers," *Mech. Eng. J. Am. Soc. Mech. Eng.*, **117**(7), 70–73, 1995.

K. Lee, *Principles of CAD/CAM/CAE Systems*, Addison Wesley Longman, Reading, MA, 1999.

H. Lipson, "Design in the Age of 3-D Printing," *Mech. Eng.*, September 2012, pp. 30–35.

H. Livingston, "CAD at Your Fingertips," CADalyst, Jume 7, 2012, available: http://www.cadalyst.com/cad/cad-your-fingertips-14562.

D. A. Madsen, T. M. Shumaker, and D. P. Madsen, *Civil Drafting Technology*, Prentice-Hall, Upper Saddle River, NJ, 2010.

E. Marks, "Does PLM belong in the Cloud?" CADalyst, January 26, 2012, available: http://www.cadalyst.com/collaboration/product-lifecycle-management/does-plm-belong-cloud-14265.

R. Masson, "Parallel and Almost Personal," *Machine Design, April* **20**, 1995, pp. 70–76.

C. McMahon and J. Browne, *CADCAM: Principles, Practice, and Manufacturing Management*, Addison-Wesley, Harlow, England, 1998.

M. Middlebrook and D. Byrnes, *AutoCAD 'X' for Dummies*, Wiley, Hoboken, NJ, 2006.

X. G. Ming, J. Q. Yan, X. H Wang, S. N. Li, W. F. Lu, Q. J. Peng, and Y. S. Ma, "Collaborative Process Planning and Manufacturing in Product Lifecycle Management," *Comput. Industry*, **59**(2–3), 154–166, 2008.

M. Mitton, *Interior Design Visual Presentation: A Guide to Graphics, Models, and Presentation Techniques*, 4th ed., Wiley, Hoboken, NJ, 2012.

R. L. Norton, Jr. "Push Information, Not Paper," *Machine Design*, December 12, 1994, pp. 105–109.

G. E. Okudan, "A Multi-Criteria Decision-Making Methodology for Optimum Selection of a Solid Modeller for Design Teaching and Practice," *J. Eng. Design*, **17**.2, 159–175, 2006.

W. Palm, "Rapid Prototyping Primer," Penn State Learning Factory, July 30, 2002, available: http://www.me.psu.edu/lamancusa/rapidpro/primer/chapter2.htm

M. Puttre, "Taking Control of the Desktop," *Mech. Eng.*, **166**(9), 62–66, September 1994.

R. A. Reis, *Electronic Project Design and Fabrication*, Pearson, Upper Saddle River, NJ, 2005.

J. R. Rossignac and A. A. G. Requicha, "Solid Modeling," in J. Webster (Ed.), *Encyclopedia of Electrical and Electronics Engineering*, Wiley, New York, 1999.

R. A. Rutenbar, G. Gielen, and B. Antao (Eds.), *Computer-Aided Design of Analog Integrated Circuits and Systems*, Wiley, Hoboken, NJ, 2002.

T. Saka, *AutoCAD for Architecture*, Prentice-Hall, Englewood Cliffs, NJ, 2004.

O. W. Salomons, J. M. Kuipers, F. van Slooten, F. J. A. M. van Houten, and H. J. J. Kals, "Collaborative Product Development in CAD and CAPP, An Approach Based on Communication Due to Constraint Conflicts and User Initiative," in M. Mantyla (Ed.), *Proceedings IFIP WG5.2 workshop on Knowledge-Intensive CAD-1 (KIC-1)*, Chapman & Hall, London, 1995, pp. 81–104.

M. Sarfraz (Ed.), *Advances in Geometric Modeling*, Wiley, New York, 2004.

O. Shilovitsky, "Blog Post: CAD Data and PLM," October 5, 2010, available: http://beyondplm.com/2010/10/05/cad-data-and-plm/.

S. S.-F. Smith, "An Evaluation of Internet-Based CAD Collaboration Tools," *J. Technol. Stud.*, **30**(2), 79–85, Spring 2004.

R. R. Spencer and M. S. Ghausi, *Introduction to Electronic Circuit Design*, Prentice-Hall, Englewood, NJ, 2003.

V. Srinivasan, "An Integration Framework for Product Lifecycle Management," *Computer-Aided Design*, **43**.5, 464–478, 2011.

W. Stallings, *Computer Organization and Architecture, Designing for Performance*, 8th ed., Prentice-Hall, Upper Saddle River, NJ, 2010.

J. Tam, "Peer Software Facilitates CAD Collaboration," *CADalyst.*, 7, September 10, 2009, available: http://www.cadalyst.com/data-management/peer-software-facilitates-cad-collaboration-12903.

L. Teschler (Ed.), "Why PDM Projects Go Astray," *Machine Design*, **68**(4), 78–82, February 22, 1996,

J. Thilmany, "Focus on Design, Design Futures, Will You Sketch with a 3-D Pen? Or Crowd Source as an Independent from Home? Research Today Aims to Make It Easy Tomorrow," *Mech. Eng. J. Am. Soc. Mech. Eng.*, **130**(9), 26–29, 2008.

M. J. Wachowiak and B. V. Karas, "3D Scanning and Replication for Museum and Cultural Heritage Applications," *J. Am. Inst. Conserv.*, **48**(2), 141–158, 2009.

S. Wallach and J. Swanson, "Higher Productivity with Scalable Parallel Processing," *Mech. Eng.*, **116**(12), 72–74, December 1994.

K. B. Zandin and H. B. Maynard (Eds.), *Maynard's Industrial Engineering Handbook*, 5th ed., McGraw-Hill, New York, 2001.

CHAPTER 2

PRODUCT DESIGN FOR MANUFACTURING AND ASSEMBLY

Gordon Lewis
Digital Equipment Corporation
Maynard, Massachusetts

1 INTRODUCTION

Major changes in product design practices are occurring in all phases of the new product development process. These changes will have a significant impact on how all products are designed and the development of the related manufacturing processes over the next decade. The high rate of technology changes has created a dynamic situation that has been difficult to control for most organizations. There are some experts who openly say that if we have no new technology for the next five years, corporate America might just start to catch up. The key to achieving benchmark time to market, cost, and quality is in up-front technology, engineering, and design practices that encourage and support a wide latitude of new product development processes. These processes must capture modern manufacturing technologies and piece parts that are designed for ease of assembly and parts that can be fabricated using low-cost manufacturing processes. Optimal new product design occurs when the designs of machines and of the manufacturing processes that produce those machines are congruent.

The obvious goal of any new product development process is to turn a profit by converting raw material into finished products. This sounds simple, but it has to be done efficiently and economically. Many companies do not know how much it costs to manufacture a new product until well after the production introduction. *Rule 1*: The product development team must be given a cost target at the start of the project. We will call this cost the *unit manufacturing cost* (UMC) target. *Rule 3*: The product development team must be held accountable for this target cost. What happened to rule 2? We'll discuss that shortly. In the meantime, we should understand what UMC is.

$$UMC = BL + MC + TA$$

where BL = burdened assembly labor rate per hour; this is the direct labor cost of labor, benefits, and all appropriate overhead cost

  MC = material cost; this is the cost of all materials used in the product

  TA = tooling amortization; this is the cost of fabrication tools, molds and assembly tooling, divided by the forecast volume build of the product

UMC is the direct burdened assembly labor (direct wages, benefits, and overhead) plus the material cost. Material cost must include the cost of the transformed material plus piece part packaging plus duty, insurance, and freight (DIF). Tooling amortization should be included in the UMC target cost calculation based on the forecast product life volume.

Example UMC Calculation BL + MC + TA
Burdened assembly labor cost calculation (BL):

$$\underset{\text{Wages + Benefits overhead}}{BL = (\$18.75 + \overset{\text{Labor}}{138\%}) = \$44.06/\,hr}$$

Burdened assembly labor is made up of the direct wages and benefits paid to the hourly workers plus a percentage added for direct overhead and indirect overhead. The overhead added percentage will change from month to month based on plant expenses.

Material cost calculation (MC):

$$MC = \overset{\text{(Part cost+Packaging)+DIF+Mat. Acq. cost}}{(\$2.45 + \$.16) + 12\% + 6\%} =$$

$$MC = \$2.61 + \$.31 + \$.15 \underset{\text{Material FOB assm. plant}}{\quad = \$3.07}$$

Material cost should include the cost of the parts and all necessary packaging. This calculation should also include a percent adder for DIF and an adder for the acquisition of the materials (Mat. Acq.). DIF typically is between 4 and 12% and Mat. Acq. typically is in the range of 6–16%. It is important to understand the MC because material is the largest expense in the UMC target.

Tooling amortization cost calculations (TA):

$$TA = \overset{\text{(Tool cost)}}{TC} / \overset{\text{\# of parts}}{PL}$$

$$TA = \$56,000/10,000 = \$5.60 \text{ per assembly}$$

TC is the cost of tooling and PL is the estimated number of parts expected to be produced on this tooling. Tooling cost is the total cost of dies and mold used to fabricate the component parts of the new product. This also should include the cost of plant assembly fixtures and test and quality inspection fixtures.

The question is, "How can the product development team quickly and accurately measure UMC during the many phases of the project?" What is needed is a tool that provides insight into the product structure and at the same time exposes high-cost areas of the design.

2 DESIGN FOR MANUFACTURING AND ASSEMBLY

Designing for manufacturing and assembly (DFM&A) is a technique for reducing the cost of a product by breaking the product down into its simplest components. All members of the design team can understand the product's assembly sequence and material flow early in the design process.

DFM&A tools lead the development team in reducing the number of individual parts that make up the product and ensure that any additional or remaining parts are easy to handle and insert during the assembly process. DFM&A encourages the integration of parts and processes, which helps reduce the amount of assembly labor and cost. DFM&A efforts include programs to minimize the time it takes for the total product development cycle, manufacturing cycle, and product life-cycle costs. Additionally, DFM&A design programs promote team cooperation and supplier strategy and business considerations at an early stage in the product development process.

DATE:	January 26, 1997
TO:	Manufacturing Program Manager, Auto Valve Project
FROM:	Engineering Program Manager, Auto Valve Project
RE:	Design for Manufacturing & Assembly support for Auto Valve Project
CC:	Director, Flush Valve Division

Figure 1   Memorandum: *Ajax Bowl Corporation.*

The DFM&A process is composed of two major components: *design for assembly* (DFA) and *design for manufacturing* (DFM). DFA is the labor side of the product cost. This is the labor needed to transform the new design into a customer-ready product. DFM is the material and tooling side of the new product. DFM breaks the parts fabrication process down into its simplest steps, such as the type of equipment used to produce the part and fabrication cycle time to produce the part, and calculates a cost for each functional step in the process. The program team should use the DFM tools to establish the material target cost before the new product design effort starts.

Manufacturing costs are born in the early design phase of the project. Many different studies have found that as much as 80% of a new product's cost is set in concrete at the first drawing release phase of the product. Many organizations find it difficult to implement changes to their new product development process. The old saying applies: "only wet babies want to change, and they do it screaming and crying." Figure 1 is a memo that was actually circulated in a company trying to implement a DFM&A process. Only the names have been changed.

It is clear from this memo that neither the engineering program manager nor the manufacturing program manager understood what DFM&A was or how it should be implemented in the new product development process. It seems that their definition of concurrent engineering is "Engineering creates the design and manufacturing is forced to concur with it with little or no input." This is not what DFM&A is.

2.1   What Is DFM&A?

DFM&A is not a magic pill. It is a tool that, when used properly, will have a profound effect on the design philosophy of any product. The main goal of DFM&A is to lower product cost by examining the product design and structure at the early concept stages of a new product. DFM&A also leads to improvements in serviceability, reliability, and quality of the end product. It minimizes the total product cost by targeting assembly time, part cost, and the assembly process in the early stages of the product development cycle.

The life of a product begins with defining a set of product needs, which are then translated into a set of product concepts. Design engineering takes these product concepts and refines them into a detailed product design. Considering that from this point the product will most likely be in production for a number of years, it makes sense to take time out during the design phase to ask, "How should this design be put together?" Doing so will make the rest of the product life, when the design is complete and handed off to production and service, much smoother. To be truly successful, the DFM&A process should start at the early concept development phase of the project. True, it will take time during the hectic design phase to apply DFM&A, but the benefits easily justify additional time.

DFM&A is used as a tool by the development team to drive specific assembly benefits and identify drawbacks of various design alternatives, as measured by characteristics such as total number of parts, handling and insertion difficulty, and assembly time. DFM&A converts time into money, which should be the common metric used to compare alternative designs or redesigns of an existing concept. The early DFM&A analysis provides the product development team with a baseline to which comparisons can be made. This early analysis will help the designer to understand the specific parts or concepts in the product that require further improvement by keeping an itemized tally of each part's effect on the whole assembly. Once a user becomes proficient with a DFM&A tool and the concepts become second nature, the tool is still an excellent means of solidifying what is by now second nature to DFA veterans and helps them present their ideas to the rest of the team in a common language: cost.

Due to the intricate design constraints placed on the auto valve project engineering feels they will not have the resources to apply the design for manufacturing and assembly process. Additionally, this program is strongly schedule driven. The budget for the project is already approved as are other aspects of the program that require it to be on time in order to achieve the financial goals of upper management.

In the meeting on Tuesday, engineering set down the guidelines for manufacturing involvement on the auto valve project. This was agreed to by several parties (not manufacturing) at this meeting. The manufacturing folks wish to be tied early into the auto valve design effort:

1. Allow manufacturing to be familiar with what is coming
2. Add any ideas or changes that would reduce overall cost or help schedule
3. Work vendor interface early, and manufacturing owns the vendor issues when the product comes to the plant

Engineering folks like the concept of new ideas, but fear:

1. Inputs that get pushed without understanding of all properly weighted constraints
2. Schedule that drags on due to too many people asking to change things
3. Spending time defending and arguing the design. PROPOSAL—Turns out this is the way we will do it.

On a few planned occasions engineering will address manufacturing inputs through one manufacturing person. Most correspondence will be written and meeting time will be minimal. It is understood that this program is strongly driven by schedule, and many cost reduction efforts are already built into the design so that the published budget can be met.

Plan for Engineering (*ENG*):

- When the drawings are ready, the engineering program manager (EPM) will submit them to the manufacturing program manager (MPM).
- The MPM gathers inputs from manufacturing and submits them back in writing to the EPM. The MPM works questions through the EPM to minimize any attention units that engineering would have to spend.
- The EPM submits suggestions to engineering for one quick hour of discussion/acceptance/veto.
- The EPM submits a written response back to the MPM and works any design under ENG direction.
- When a prototype parts arrives, the EPM will allow the MPM to use it in manufacturing discussions.
- The MPM will submit a written document back to the EPM to describe issues and recommendations.
- Engineering will incorporate any changes that they can handle within the schedule that they see fit.

DFM&A is an interactive learning process. It evolves from applying a specific method to a change in attitude. Analysis is tedious at first, but as the ideas become more familiar and eventually ingrained, the tool becomes easier to use and leads to questions: questions about the assembly process and about established methods that have been accepted or existing design solutions that have been adopted. In the team's quest for optimal design solutions, the DFM&A process will lead to uncharted ways of doing things. Naturally, then, the environment in which DFA is implemented must be ripe for challenging pat solutions and making suggestions for new approaches. This environment must evolve from the top down, from upper management to the engineer. Unfortunately, this is where the process too often fails.

Figure 2 illustrates the ideal process for applying DFM&A. The development of any new product must go through four major phases before it reaches the marketplace: concept, design, development, and production. In the concept phase, product specifications are created and the design team creates a design layout of the new product. At this point, the first design for assembly analysis should be completed. This analysis will provide the design team with a theoretical minimum parts count and pinpoint high-assembly areas in the design.

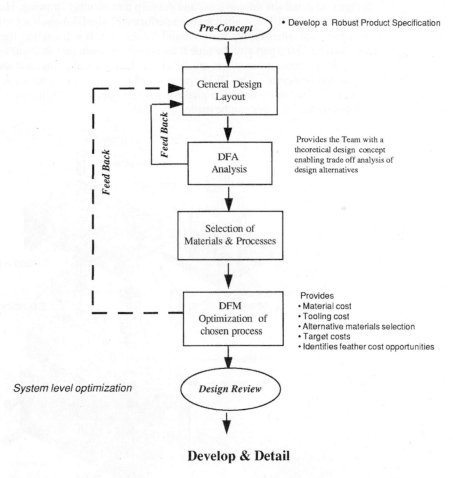

Figure 2   Key components of the DFM&A process.

At this point, the design team needs to review the DFA results and adjust the design layout to reflect the feedback of this preliminary analysis. The next step is to complete a design for manufacturing analysis on each unique part in the product. This will consist of developing a part cost and tooling cost for each part. It should also include doing a producibility study of each part. Based on the DFM analysis, the design team needs to make some additional adjustments in the design layout. At this point, the design team is now ready to start the design phase of the project. The DFM&A input at this point has developed a preliminary bill of material (BOM) and established a target cost for all the unique new parts in the design. It has also influenced the product architecture to improve the sequence of assembly as it flows through the manufacturing process.

The following case study illustrates the key elements in applying DFM&A. Figure 3 shows a product called the *motor drive assembly*. This design consists of 17 parts and assemblies. Outwardly it looks as if it can be assembled with little difficulty. The product is made up of two sheet metal parts and one aluminum machined part. It also has a motor assembly and a sensor, both bought from an outside supplier. In addition, the *motor drive assembly* has nine hardware items that provide other functions—or do they?

At this point, the design looks simple enough. It should take minimal engineering effort to design and detail the unique parts and develop an assembly drawing. Has a UMC been developed yet? Has a DFM&A analysis been performed? The DFA analysis will look at each process step, part, and subassembly used to build the product. It will analyze the time it takes to "get" and "handle" each part and the time it takes to insert each part in the assembly (see Table 1). It will point out areas where there are difficulties handling, aligning, and securing each and every part and subassembly. The DFM analysis will establish a cost for each part and estimate the cost of fabrication tooling. The analysis will also point out high-cost areas in the fabrication process so that changes can be made.

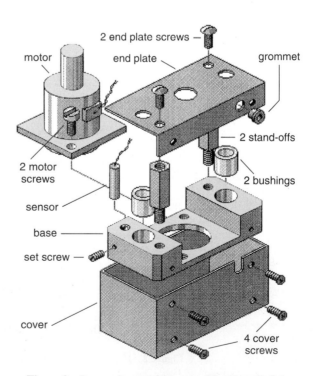

Figure 3 Proposed motor drive assembly. From Ref. 1.

Table 1   Motor Drive Assembly

Number of parts and assemblies	19
Number of reorientation or adjustment	1
Number of special operations	2
Total assembly time in seconds	213.4
Total cost of fabrication and assembly tooling	$3590
Tool amortization at 10K assemblies	$ 0.36
Total cost of labor at $74.50/h	$ 4.42
Total cost of materials	$ 42.44
Total cost of labor and materials	$ 46.86
Total UMC	$ 47.22

At this point, the DFA analysis suggested that this design could be built with fewer parts. A review of Table 2, column 5, shows that the design team feels it can eliminate the bushings, stand-offs, end-plate screws, grommet, cover, and cover screws. Also, by replacing the end plate with a new snap-on plastic cover, they can eliminate the need to turn the (reorientation) assembly over to install the end plate and two screws. Taking the time to eliminate parts and operations is the most powerful part of performing a DFA analysis. This is *rule 2*, which was left out above: DFM&A is a team sport. Bringing all members of the new product development team together and understanding the sequence of assembly, handling, and insertion time for each part will allow each team member to better understand the function of every part.

DFM Analysis

The DFM analysis provided the input for the fabricated part cost. As an example, the base is machined from a piece of solid aluminum bar stock. As designed, the base has 11 different holes drilled in it and 8 of them require taping. The DFM analysis (see Table 3) shows that it takes 17.84 min to machine this part from the solid bar stock. The finished machined base costs $10.89 in lots of 1000 parts. The ideal process for completing a DFM analysis might be as follows.

In the case of the base, the design engineer created the solid geometry in Matra Data's Euliked computer-aided design (CAD) system (see Fig. 4). The design engineer then sent the solid database as an STL file to the manufacturing engineer, who then brought the STL file into a viewing tool called *SolidView* (see Fig. 5). SolidView allowed the mechanical engineer to get all the dimensioning and geometry inputs needed to complete the Boothroyd Dewhurst design for manufacturing machining analysis of the base part. SolidView also allowed the mechanical engineer to take cut sections of the part and then step through it to ensure that no producibility rules had been violated.

Today all of the major CAD supplies provide the STL file output format. There are many new CAD viewing tools like SolidView available, costing about $500–$1000. These viewing tools will take STL or IGS files. The goal is to link all of the early product development data together so each member can have fast, accurate inputs to influence the design in its earliest stage.

In this example, it took the mechanical engineer a total of 20 min to pull the STL files into SolidView and perform the DFM analysis. Engineering in the past has complained that DFM&A takes too much time and slows the design team down. The mechanical engineer then analyzes the base as a die casting part, following the producibility rule. By designing the base as a die casting, it is possible to mold many of the part features into the part. This net shape die cast design will reduce much of the machining that was required in the original design. The die cast part will still require some machining. The DFM die casting analysis revealed that the base casting would cost $1.41 and the mold would cost $9050. Table 4 compares the two different fabrication methods.

Table 2 Motor Drive Assembly: Design for Assembly Analysis

1	2	3	4	5	6	7	8	9	10	11	12	13	14	15	16	17
Name	Sub No. Entry No.	Type	Repeat Count	Minimum Items	Tool Fetching Time, s	Handling Time, s	Insertion or Operation Time, s	Total Time, sec	Labor Cost, $	Assembly Tool or Fixture Cost, $	Item Cost, $	Total Item Cost, $	Manuf. Tool Cost, $	Target Cost, $	Part Number	Description
Base	1.1	Part	1	1	0	1.95	1.5	3.45	0.07	500	10.89	10.89	950	7.00	1P033-01	Add base to fixture
Bushing	1.2	Part	2	0	0	1.13	6.5	15.26	0.32	0	1.53	3.06	0	0.23	16P024-01	Add & press fit
Motor	1.3	Sub	1	1	0	7	6	13	0.27	0	18.56	18.56	0	12.00	121S021-02	Add & hold down
Motor screw	1.4	Part	2	2	2.9	1.5	9.6	25.1	0.52	0	0.08	0.16	0	0.08	112W0223-06	Add & thread
Sensor	1.5	Sub	1	1	0	5.6	6	11.6	0.24	0	2.79	2.79	0	2.79	124S223-01	Add & hold down
Set screw	1.6	Part	1	1	2.9	3	9.2	15.1	0.31	0	0.05	0.05	0	0.05	111W0256-02	Add & thread
Stand-off	1.7	Part	2	0	2.9	1.5	9.6	25.1	0.52	0	0.28	0.56	0	0.18	110W0334-07	Add & thread
End plate	1.8	Part	1	1	0	1.95	5.2	7.15	0.15	0	2.26	2.26	560	0.56	15P067-01	Add & hold down
End plate screw	1.9	Part	2	0	2.9	1.8	5.7	17.9	0.37	0	0.03	0.06	0	0.03	110W0777-04	Add & thread
Grommet	1.1	Part	1	0	0	1.95	11	12.95	0.27	0	0.12	0.03	0	0.12	116W022-08	Add & push fit
Dress wires—grommet	1.11	Oper	2	2	—	—	—	18.79	0.39	0	0.00	0.00	0	0.00		Library operation
Reorientation	1.12	Oper	1	0	—	—	4.5	4.5	0.09	350	0.00	0.00	0	0.00		Reorient & adjust
Cover	1.13	Part	1	0	0	2.3	8.3	10.6	0.22	0	3.73	3.73	1230	1.20	2P033-01	Add
Cover screw	1.14	Part	4	0	2.9	1.8	5.7	32.9	0.68	0	0.05	0.18	0	0.05	112W128-03	Add & thread
Totals =			22	9				213.4	4.42	850		42.33	2740	24.28		

UMC = 47.22

Notes: Production life volume = 10,000. Annual build volume = 3000. Assembly labor rate $/h = $74.50. The information presented in this table was developed from the Boothroyd Dewhurst DFA software program, version 8.0. See Ref. 2.

Table 3 Machining Analysis Summary Report

Setups	Time, min	Cost, $
Machine Tool Setups		
Setup	0.22	0.10
Nonproductive	10.63	4.87
Machining	6.77	3.10
Tool wear	—	0.31
Additional cost/part	—	0.00
Special tool or fixture	—	0.00
Library Operation Setups		
Setup	0.03	0.02
Process	0.20	0.13
Additional cost/part	—	0.03
Special tool or fixture	—	0.00
Material	—	2.34
Totals	17.84	10.89
Material		Gen aluminum alloy
Part number		5678
Initial hardness		55
Form of workpiece		Rectangular bar
Material cost, $/lb		2.75
Cut length, in.		4.000
Section height, in.		1.000
Section width, in.		2.200
Product life volume		10,000
Number of machine tool setups		3
Number of library operation setups		1
Workpiece weight, lb		0.85
Workpiece volume, in.^{3}		8.80
Material density, lb/in.^{3}		0.097

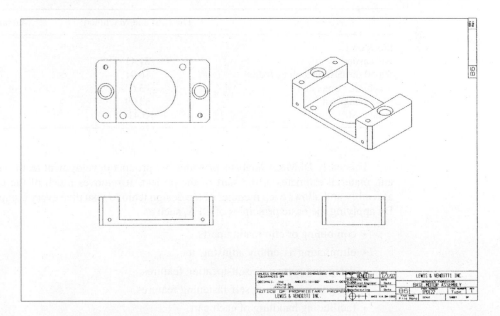

Figure 4

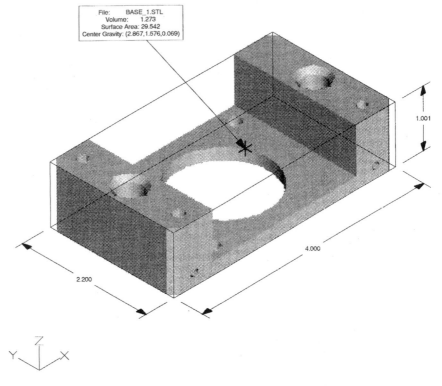

File:	BASE_1.STL
Volume:	1.273
Surface Area:	29.542
Center Gravity:	(2.867,1.576,0.069)

Figure 5

Table 4

	Die Cast and Machined	Machined from Bar Stock
Stock cost		$2.34
Die casting	$1.41	
$9050 die casting tooling/10,000	$0.91	
Machining time, min	3.6	17.84
Machining cost	$3.09	$8.55
Total cost	5.41	$10.89

This early DFM&A analysis provides the product development team with accurate labor and material estimates at the start of the project. It removes much of the complexity of the assembly and allows each member of the design team to visualize every component's function. By applying the basic principles of DFA, such as

- combining or eliminating parts,
- eliminating assembly adjustments,
- designing parts with self-locating features,
- designing parts with self-fastening features,
- facilitating handling of each part,

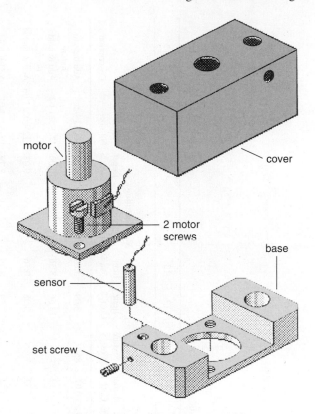

Figure 6   Redesign of motor assembly.

- eliminating reorientation of the parts during assembly, and
- specifying standard parts,

the design team is able to rationalize the motor drive assembly with fewer parts and assembly steps. Figure 6 shows a possible redesign of the original motor drive assembly. The DFM&A analysis (Table 5) provided the means for the design team to question the need and function of every part. As a result, the design team now has a new focus and an incentive to change the original design.

Table 6 shows the before-and-after DFM&A results.

If the motor drive product meets its expected production life volume of 10,000 units, the company will save $170,100. By applying principles of DFM&A to both the labor and material on the motor drive, the design team is able to achieve about a 35% cost avoidance on this program.

2.2   Getting the DFM&A Process Started

Management from All Major Disciplines Must Be on Your Side

In order for the DFM&A process to succeed, upper management must understand, accept, and encourage the DFM&A way of thinking. They must *want* it. It is difficult, if not impossible, for an individual or group of individuals to perform this task without management support, since the process requires the cooperation of so many groups working together. The biggest challenge

Table 5 Redesign of Motor Drive Assembly: Design for Assembly Analysis

1	2	3	4	5	6	7	8	9	10	11	12	13	14	15	16	17
	Sub No.						Insertion or			Assembly		Total	Manufacturing			
	Entry		Repeat	Minimum	Tool Fetching	Handling	Operation	Total	Labor	Tool or Fixture	Item	Item	Tool	Target	Part	
Name	No.	Type	Count	Items	Time, s	Time, s	Time, s	Time, s	Cost, $	Cost, $	Cost, $	Cost, $	Cost, $	Cost, $	Number	Description
Base, casting	1.1	Part	1	1	0	1.95	1.5	3.45	0.07	850	3.09	3.09	9,050	3.09	1P033-02	Add base to fixture
Motor	1.2	Sub	1	1	0	3	6	9	0.19		18.56	18.56		15.00	121S021-02	Add & hold down
Motor screw	1.3	Part	2	2	2.9	1.5	9.6	25.1	0.52		0.08	0.16		0.08	112W0223-06	Add & thread
Sensor	1.4	Sub	1	1	0	5.6	6	11.6	0.24		2.79	2.79		2.79	124S223-01	Add & hold down
Set screw	1.5	Part	1	1	2.9	2.55	9.2	14.65	0.30		0.05	0.05		0.05	111W0256-02	Add & thread
Push/pull wire—easy	1.6	Oper	2	2	—	—	—	18.79	0.39		0.00	0.00		0.00		Library operation
Cover	1.7	Part	1	1	0	1.95	1.8	3.75	0.08		1.98	1.98	7,988	1.70	2P033-02	Add & snap fit
Totals =			9	9				86.34	1.79	850		26.63	17,038	22.71		

UMC = 30.21

Notes: Production life volume =10,000. Annual build volume = 3000. Assembly labor rate =$74.50. The information presented in this table was developed from the Boothroyd Dewhurst DFA software program, version 8.0. See Ref. 2.

Table 6   Comparison of DFM&A Results

	Motor Drive Assembly	Redesign of Motor Drive Assembly
Number of parts and assemblies	19	7
Number of reorientation or adjustment	1	0
Number of special operations	2	2
Total assembly time in seconds	213.4	86.34
Total cost of labor at $74.50/h	$4.42	$1.79
Total cost of materials	$42.44	$26.63
Total cost of labor and material	$46.86	$28.42
Total cost of fabrication tooling	$3590	$17,888
Tool amortization at 10K assemblies	$0.36	$1.79
Total UMC	$47.22	$30.21
Savings =	$17.01	

Table 7   Human Factors Test (Check the Box Where Your Team Fits)

Yes	Team Environment	Yes	Work Group Environment
	Are the team members committed to the group's common goals?	✓	Are members loyal to outside groups with conflicting interests (functional managers)?
	Is there open communication with all members of the team?	✓	Is information unshared?
	Is there flexible, creative leadership?	✓	Is there a dominating leadership?
	Is the team rewarded as a group?	✓	Is there individual recognition?
	Is there a high degree of confidence and trust between members?	✓	Are you unsure of the group's authority?

of implementing DFM&A is the cooperation of so many individuals toward a common goal. This does not come naturally, especially if it is not perceived by the leaders as an integral part of the business's success. In many companies, management does not understand what DFM&A is. They believe it is a manufacturing process. It is not; it is a new product development process, which *must* include all disciplines (engineering, service, program managers, and manufacturing) to yield significant results. The simplest method to achieve cooperation between different organizations is to have the team members work in a common location (colocated team). The new product development team needs some nurturing and stimulation to become empowered. This is an area where most companies just do not understand the human dynamics of building a high-performance team. Table 7 should aid in determining whether you are working in a team environment or a work group environment.

Many managers will say that their people are working in a team environment, but they still want to have complete control over work assignments and time spent supporting the team. In their mind, the team's mission is secondary to the individual department manager's goals. This is not a team; it is a work group. The essential elements of a high-performance team are as follows:

- A clear understanding of the team's goals (a defined set of goals and tasks assigned to each individual team member)
- A feeling of openness, trust, and communication

- Shared decision making (consensus)
- A well-understood problem-solving process
- A leader who legitimizes the team-building process

Management must recognize that to implement DFM&A in their organization, they must be prepared to change the way they do things. Management's reluctance to accept the need for change is one reason DFM&A has been so slow to succeed in many companies. Training is one way of bringing DFM&A knowledge to an organization, but training alone cannot be expected to effect the change.

2.3   DFM&A Road Map

DFM&A Methodology (Product Development Philosophy)

- Form a multifunctional team.
- Establish the product goals through competitive benchmarking.
- Perform a design for assembly analysis.
- Segment the product into manageable subassemblies or levels of assembly.
- As a team, apply the design for assembly principles.
- Use creativity techniques to enhance the emerging design.
- As a team, evaluate and select the best ideas.
- Ensure economical production of every piece part.
- Establish a target cost for every part in the new design.
- Start the detailed design of the emerging product.
- Apply design for producibility guidelines.
- Reapply the process at the next logical point in the design.
- Provide the team with a time for reflection and sharing results.

This DFM&A methodology incorporates all of the critical steps needed to ensure a successful implementation.

Develop a multifunctional team of all key players before the new product architecture is defined. This team must foster a creative climate that will encourage ownership of the new product's design. The first member of this team should be the project leader, the person who has the authority for the project. This individual must control the resources of the organization, should hand-pick the people who will work on the team, and should have the authority to resolve problems within the team.

The team leader should encourage and develop a creative climate. It is of utmost importance to assemble a product development team that has the talent to make the right decisions, the ability to carry them out, and the persistence and dedication to bring the product to a successful finish. Although these qualities are invaluable, it is of equal importance that these individuals be allowed as much freedom as possible to germinate creative solutions to the design problem as early as possible in the product design cycle.

The product development team owns the product design and the development process. The DFM&A process is most successful when implemented by a multifunctional team, where each person brings to the product design process his or her specific area of expertise. The team should embrace group dynamics and the group decision-making process for DFM&A to be most effective.

Emphasis has traditionally been placed on the design team as the people who drive and own the product. Designers need to be receptive to team input and share the burden of the design process with other team members.

The team structure depends on the nature and complexity of the product. Disciplines that might be part of a product team include:

- Engineering
- Manufacturing
- Field service and support
- Quality
- Human factors or ergonomics
- Purchasing
- Industrial design and packaging
- Distribution
- Sales
- Marketing

Although it is not necessary for all of these disciplines to be present all of the time, they should have an idea of how things are progressing during the design process.

Clearly, there can be drawbacks to multidisciplinary teams, such as managing too many opinions, difficulty in making decisions, and factors in general that could lengthen the product development cycle. However, once a team has worked together and has an understanding of individual responsibilities, there is much to gain from adopting the team approach. Groups working together can pool their individual talents, skills, and insight so that more resources are brought to bear on a problem. Group discussion leads to a more thorough understanding of problems, ideas, and potential solutions from a variety of standpoints. Group decision making results in a greater commitment to decisions, since people are more motivated to support and carry out a decision that they helped make. Groups allow individuals to improve existing skills and learn new ones.

Having the team located together in one facility makes the process work even better. This colocation improves the team's morale and also makes communication easier. Remembering to call someone with a question or adding it to a meeting agenda is more difficult than mentioning it when passing in the hallway. Seeing someone reminds one of an issue that may have otherwise been forgotten. These benefits may seem trivial, but the difference that colocation makes is significant.

> *As a team, establish product goals through a competitive benchmarking process: concept development.* Competitive benchmarking is the continuous process of measuring your own products, services, and practices against the toughest competition, or the toughest competition in a particular area. The benchmarking process will help the team learn who the "best" are and what they do. It gives the team a means to understand how this new product measures up to other products in the marketplace. It identifies areas of opportunities that need changing in the current process. It allows the team to set targets and provides an incentive for change. Using a DFM&A analysis process for the competitive evaluation provides a means for relative comparison between those of your products and those of your competitors. You determine exactly where the competition is better.

Before performing a competitive teardown, decide on the characteristics that are most important to review, what the group wants to learn from the teardown, and the metrics that will be noted. Also keep the teardown group small. It is great to have many people walk through and view the results, but a small group can better manage the initial task of disassembly and analysis. Ideally, set aside a conference room for several days so the product can be left out unassembled, with a data sheet and metrics available.

Perform a design for assembly analysis of the proposed product that identifies possible candidate parts for elimination or redesign and pinpoints high-cost assembly operations. Early in this chapter, the motor drive assembly DFM&A analysis was developed. This example illustrates the importance of using a DFA tool to identify, size, and track the cost-savings opportunities. This leads to an important question: Do you need a formal DFA analysis software tool? Some DFM&A consultants will tell you that it is not necessary to use a formal DFA analysis tool. It is my supposition that these consultants want to sell you consulting services rather than teach the process. It just makes no sense *not* to use a formal DFA tool for evaluating and tracking the progress of the new product design through its evolution. The use of DFA software provides the team with a focus that is easily updated as design improvements are captured. The use of DFA software does not exclude the need for a good consultant to get the new team off to a good start. The selection of a DFA tool is a very important decision. The cost of buying a quality DFA software tool is easily justified by the savings from applying the DFA process on just one project.

At this point, the selection of the manufacturing site and type of assembly process should be completed. Every product must be designed with a thorough understanding of the capabilities of the manufacturing site. It is thus of paramount importance to choose the manufacturing site at the start of product design. This is a subtle point that is frequently overlooked at the start of a program, but to build a partnership with the manufacturing site, the site needs to have been chosen! Also, manufacturing facilities have vastly different processes, capabilities, strengths, and weaknesses that affect, if not dictate, design decisions. When selecting a manufacturing site, the process by which the product will be built is also being decided.

As a team, apply the design for assembly principles to every part and operation to generate a list of possible cost opportunities. The generic list of DFA principles includes the following:

- Designing parts with self-locating features
- Designing parts with self-fastening features
- Increasing the use of multifunctional parts
- Eliminating assembly adjustments
- Driving standardization of fasteners, components, materials, finishes, and processes

It is important for the team to develop its own set of DFA principles that relate to the specific product on which it is working. Ideally, the design team decides on the product characteristics it needs to meet based on input from product management and marketing. The product definition process involves gathering information from competitive benchmarking and teardowns, customer surveys, and market research. Competitive benchmarking illustrates which product characteristics are necessary.

Principles should be set forth early in the process as a contract that the team draws up together. It is up to the team to adopt many principles or only a few and how lenient to be in granting waivers.

Use brainstorming or other creativity techniques to enhance the emerging design and identify further design improvements. The team must avoid the temptation to start engineering the product before developing the DFM&A analysis and strategy. As a team, evaluate and select the best ideas from the brainstorming, thus narrowing and focusing the product goals.

With the aid of DFM software, cost models, and competitive benchmarking, establish a target cost for every part in the new design. Make material and manufacturing process selections. Start the early supplier involvement process to ensure economical production of every piece part. Start the detailed design of the emerging product. Model, test, and evaluate the new design for fit, form, and function. Apply design for producibility guidelines to the emerging parts design to ensure that cost and performance targets are met.

Provide the team with a time for reflection and sharing results. Each team member needs to understand that there will be a final review of the program, at which time members will be able to make constructive criticism. This time helps the team determine what worked and what needs to be changed in the process.

Use DFM&A Metrics

The development of some DFM&A metrics is important. The team needs a method to measure the before-and-after results of applying the DFM&A process, thus justifying the time spent on the project. Table 8 shows the typical DFM&A metrics that should be used to compare your old product design against a competitive product and a proposed new redesign.

The total number of parts in an assembly is an excellent and widely used metric. If the reader remembers only one thing from this chapter, let it be to strive to reduce the quantity of parts in every product designed. The reason limiting parts count is so rewarding is that when parts are reduced, considerable overhead costs and activities that burden that part also disappear. When parts are reduced, quality of the end product is increased, since each part that is added to an assembly is an opportunity to introduce a defect into the product. Total assembly time will almost always be lowered by reducing the quantity of parts.

A simple method to test for potentially unnecessary parts is to ask the following three questions for each part in the assembly:

1. During the products operation, does the part move relative to all other parts already assembled? (*answer yes or no*)

2. Does the part need to be made from a different material or be isolated from all other parts already assembled? (*answer yes or no*)

3. Must the part be separate from all other parts already assembled because of necessary assembly or disassembly of other parts? (*answer yes or no*)

You must answer the questions above for each part in the assembly. If your answer is "no" for all three questions, then that part is a candidate for elimination.

Table 8   DFM&A Metrics

	Old Design	Competitive	New Design
Number of parts and assemblies			
Number of separate assembly operations			
Total assembly time			
Total material cost			
Totals			

Table 9 DFM&A New Products Checklist

Design for Manufacturing and Assembly Consideration	Yes	No
Design for assembly analysis completed?	☐	☐
Has this design been analyzed for minimal part count?	☐	☐
Have all adjustments been eliminated?	☐	☐
Are more than 85% common parts and assemblies used in this design?	☐	☐
Has assembly sequence been provided?	☐	☐
Have assembly and part reorientations been minimized?	☐	☐
Have more than 96% preferred screws been used in this design?	☐	☐
Have all parts been analyzed for ease of insertion during assembly?	☐	☐
Have all assembly interferences been eliminated?	☐	☐
Have location features been provided?	☐	☐
Have all parts been analyzed for ease of handling?	☐	☐
Have part weight problems been identified?	☐	☐
Have special packaging requirements been addressed for problem parts?	☐	☐
Are special tools needed for any assembly steps?	☐	☐
Ergonomics Considerations	**Yes**	**No**
Does design capitalize on self-alignment features of mating parts?	☐	☐
Have limited physical and visual access conditions been avoided?	☐	☐
Does design allow for access of hands and tools to perform necessary assembly steps?	☐	☐
Has adequate access been provided for all threaded fasteners and drive tooling?	☐	☐
Have all operator hazards been eliminated (sharp edges)?	☐	☐
Wire Management	**Yes**	**No**
Has adequate panel pass-through been provided to allow for easy harness/cable routing?	☐	☐
Have harness/cable supports been provided?	☐	☐
Have keyed connectors been provided at all electrical interconnections?	☐	☐
Are all harnesses/cables long enough for ease of routing, tie-down, plug-in, and to eliminate strain relief on interconnects?	☐	☐
Does design allow for access of hands and tools to perform necessary wiring operations?	☐	☐
Does position of cable/harness impede air flow?	☐	☐
Design for Manufacturing and Considerations	**Yes**	**No**
Have all unique design parts been analyzed for producibility?	☐	☐
Have all unique design parts been analyzed for cost?	☐	☐
Have all unique design parts been analyzed for their impact of tooling/mold cost?	☐	☐
Assembly Process Consideration	**Yes**	**No**
Has assembly tryout been performed prior to scheduled prototype build?	☐	☐
Have assembly views and pictorial been provided to support assembly documentation?	☐	☐
Has opportunity defects analysis been performed on process build?	☐	☐
Has products cosmetics been considered (paint match, scratches)?	☐	☐

The total time it takes to assemble a product is an important DFM&A metric. Time is money, and the less time needed to assemble the product, the better. Since some of the most time-consuming assembly operations are fastening operations, discrete fasteners are always candidates for elimination from a product. By examining the assembly time of each and every part in the assembly, the designer can target specific areas for improvement. Total material cost is self-explanatory.

The new product DFM&A checklist (Table 9) is a good review of how well your team did with applying the DFM&A methodology. Use this check sheet during all phases of the product development process; it is a good reminder. At the end of the project you should have checked most of the *yes* boxes.

3   WHY IS DFM&A IMPORTANT?

DFM&A is a powerful tool in the design team's repertoire. If used effectively, it can yield tremendous results, the least of which is that the product will be easy to assemble! The most beneficial outcome of DFM&A is to reduce part count in the assembly, which in turn will simplify the assembly process, lower manufacturing overhead, reduce assembly time, and increase quality by lessening the opportunities for introducing a defect. Labor content is also reduced because with fewer parts, there are fewer and simpler assembly operations. Another benefit to reducing parts count is a shortened product development cycle because there are fewer parts to design. The philosophy encourages simplifying the design and using standard, off-the-shelf parts whenever possible. In using DFM&A, renewed emphasis is placed on designing each part so it can be economically produced by the selected manufacturing process.

REFERENCES

1. G. Boothroyd, P. Dewhurst, and W. Knight, *Product Design for Manufacturing and Assembly,* Marcel Dekker, New York, 1994.
2. Boothroyd Dewhurst Inc., *Design for Assembly Software,* Version 8.0, Wakefield, RI, 1996.

CHAPTER 3

DESIGN-FOR-ENVIRONMENT PROCESSES AND TOOLS

Daniel P. Fitzgerald and Thornton H. Gogoll
Stanley Black & Decker
Towson, Maryland

Linda C. Schmidt, Jeffrey W. Herrmann, and Peter A. Sandborn
University of Maryland
College Park, Maryland

1 INTRODUCTION

Throughout a product's life cycle, the activities associated with the product generate environmental impacts. Suppliers extract and process raw materials; the manufacturing firm creates components, assembles them, and packages and distributes the product; and the customer uses, maintains, and disposes of the product at the end of its life. Environmentally benign products are products that comply with environmental regulations and may have significant features that reduce their environmental impact. The ideal environmentally benign product is one that would not only be environmentally neutral to make and use but also actually mitigate whatever substandard conditions exist in its use environment. The ideal environmentally benign product would, instead of creating waste at the end its life cycle, become a useful input for another product.

In the past, manufacturing firms were concerned with meeting regulations that limited or prohibited the pollution and waste that is generated by manufacturing processes. More recent regulations, however, limit the material content of the products that are sold in an effort to control the substances that enter the waste stream.

The general trend is toward design for sustainability, an all-encompassing paradigm that reaches beyond a product's behavior and life cycle. Sustainability includes aspects of impact and influence throughout the entire supply chain and into the social fabric of everyday life. Sustainable development can be defined generally as economic growth done in a way that is compatible with the environment.

There are many ways to minimize a product's environmental impacts. Clearly, however, the greatest opportunity occurs during the product design phases, as discussed by many authors, including Refs. 1–4. Therefore, manufacturing firms that develop new products need to consider many factors related to the environmental impact of their products, including government regulations, consumer preferences, and corporate environmental objectives. Although this requires more effort than treating emissions and hazardous waste, it not only protects the environment but also reduces life-cycle costs by decreasing energy use, reducing raw material requirements, and avoiding costly pollution control.[5] Design-for-environment (DfE) tools, methods, and strategies have therefore become an important set of activities for manufacturing firms. This chapter reviews some relevant background information, discusses how a firm can systematically create a DfE process, and describes DfE tools and innovative products. This chapter is based on the authors' experiences and observations but does not describe the concerns or operations of Stanley Black & Decker[6,7] except where noted.

1.1 Design for Environment

Design for environment is "the systematic consideration of design performance with respect to environmental, health, and safety objectives over the full product and process life cycle" (Ref. 2). DfE is one of many possible design for X (DFX) activities that comprise concurrent engineering techniques that seek to address product life-cycle concerns early in the design phase. Thus, it is similar to design for manufacturing (DFM) and design for assembly (DFA).

DfE combines several design-related topics: disassembly, recovery, recycling, disposal, regulatory compliance, human health and safety impact, and hazardous material minimization.

Some designers view DfE as simply calculating environmental impact, similar to estimating cost. This perception is due to a trend in which manufacturing firms implement stand-alone DfE tools without explaining and motivating their use. As Lindahl (Ref. 8) states, "When the designers and actual users of the methods do not understand the reason why, or experience any benefits from using the DFE methods, there is a risk that they utilize the methods but do not use the results, or that they run through the method as quickly as possible." Therefore, designers produce an environmental output and consider their environmental management work finished. A more effective approach is to design and implement a sound DfE process, as this chapter explains.

1.2 Decision Making in New Product Development

Decision making is an important activity in new product development, and there exist a great variety of decisions that need to be made. Generally speaking, these fall into two types—design decisions and management decisions:

Design Decisions. What should the design be? Design decisions generate information about the product design by determining the product's shape, size, material, process, and components.

Management Decisions. What should be done? Management decisions control the progress of the design process. They affect the resources, time, and technologies available to perform development activities. They define which activities should happen, their sequence, and who should perform them, that is, what will be done, when will it be done, and who will do it. The clearest example is project management: planning, scheduling, task assignment, and purchasing.

In studying design projects, Krishnan and Ulrich[9] provide an excellent review of the decision making in new product development, organized around topics that follow the typical decomposition of product development. Herrmann and Schmidt[10,11] describe the decision-making view of new product development in more detail.

Traditionally, factors such as product performance and product cost have dominated design decisions, while time to market and development cost have influenced management decisions. Of course, many decisions involve combinations of these objectives.

Considering environmental issues during decision making in new product development, while certainly more important than ever before, has been less successful than considering and incorporating other objectives. Environmental objectives are not similar to the traditional objectives of product performance, unit cost, time to market, and development cost. These objectives directly affect profitability and are closely monitored. Unit cost, time to market, and development cost each use a single metric that is uncomplicated and well understood by all stakeholders. Although product performance may have multiple dimensions, these characteristics are quantifiable and clearly linked to the product design. Designers understand how changing the product design affects the product performance. As discussed in the next section, environmental objectives are very different.

1.3 Environmental Objectives

The increased prominence of global environment, social, and economic concerns was formalized by *Our Common Future*.[12] The World Commission on Environment and Economic Development (WCED) was created by the United Nations in 1983 to examine worldwide economic and environmental conditions for the purpose of determining strategies to manage global resources while accommodating an expanding population and a global environment displaying signs of deterioration. The WCED report was issued in 1987, and it is often called the Brundtland Report in honor of Gro Harlem Brundtland, the commission's chairman. The Brundtland Report summarized the aspirational goal for any business to engage in *sustainable development* to be environmentally responsible: "Sustainable development is development that meets the needs of the present without compromising the ability of future generations to meet their own needs" (Ref. 12).

The paradigm of sustainable development has expanded the set of environmental challenges confronting businesses by asserting that they are accountable for three impacts of their actions on the global population: environmental, social, and economic. If a manufacturing firm's actions could improve all three fronts at the same time, then sustainable development would be straightforward. However, it is routine that corporate actions are taken for the overall economic benefit of their long-term owners after taking into account primarily economic and performance trade-offs.[13]

Under pressure from various stakeholders to consider environmental issues when developing new products, manufacturing firms have declared their commitment to environmentally responsible product development and have identified the following relevant goals:

1. *Comply with legislation.* Products that do not comply with a nation's (or region's) environmental regulations cannot be sold in that nation (or region).

2. *Avoid liability.* Environmental damage caused by a product represents a financial liability.

3. *Satisfy customer demand.* Some consumers demand environmentally responsible products, and retailers, in turn, pass along these requirements to manufacturers.

4. *Participate in ecolabeling programs.* Products that meet requirements for ecolabeling are more marketable.

5. *Enhance profitability.* Certain environmentally friendly choices such as remanufacturing, recycling, and reducing material use make good business sense and have financial benefits.

6. *Behave ethically.* Being a good steward of the planet's resources by considering the environment during the product development process is the right thing to do.

Despite the high profile given to these objectives at the corporate level, product development teams often assign a "back-burner" status to environmental issues. Environmental objectives, for the most part, are driven by regulations and social responsibility, and reducing environmental impact does not clearly increase profit. Product managers are not willing to compromise profit, product quality, or time to market in order to create products that are more environmentally benign than required by regulations.

This unwillingness reflects a conflict between two engineering cultures: one that prioritizes constraining design activities so that the manufacturing firm adheres uniformly to regulations (including environmental regulations) and one that encourages the creativity and resourcefulness that can lead to more innovative, efficient, and environmentally benign designs. Some manufacturing firms can overcome this conflict if DfE is aligned with corporate strategy. Some manufacturing firms overtly court environmentally conscious consumers, for instance, while others adopt a remanufacturing strategy.

Environmental performance, however, is measured using multiple metrics, some of which are qualitative. Unfortunately, these metrics are often not closely correlated to a manufacturing firm's financial objectives. Measuring environmental performance, especially using life-cycle analysis (LCA), can require a great deal of effort. With environmental performance it is harder to make trade-offs.

At the product design level, it is not clear how to select between design alternatives because there is no aggregate measure to calculate. One designer presents an excellent example (Ref. 1):

> You have two ways of building a part. One option is based on metal. Metal is heavy (thus, it consumes more resources). It also creates waste during the actual manufacturing process (in form of sludge). However, it can be recycled when it reaches the end of its product life. In contrast, we make the product out of graphite. This part is lighter (which means it consumes less energy in use). In addition, it can be molded rather than machined (again resulting in less waste). However, when it reaches the end of its life, it must be disposed of in a landfill since it cannot be recycled. Which of these two options results in a greener product?

The triple bottom line (TBL) is a model of business impact that spans the elements of sustainability and describes the measures that a manufacturing firm has put into place to meet the demands of sustainable development. The term was first used by John Elkington in 1994.[14] The term has the familiar ring of an accounting measure, but it is not meant to be taken in a literal sense. Elkington's contention was that businesses should keep three separate balance sheets, one for each set of stakeholders impacted by its actions. Businesses could then detail separately their impact on the three areas necessary to assess sustainability performance: economic, environmental, and social improvement.

A balanced scorecard is one to begin implementing a TBL approach. The balanced scorecard assesses performance toward targets that cannot be reduced to the same quantitative scale, e.g., money. Although difficult to quantify, sustainability metrics are at the heart of documenting environmental progress of businesses.

Stanley Black & Decker (SBD) was placed in the North American Dow Jones Sustainability Index (DJSI North America) in the durable household products sector for the first time in 2011[6] and earned a place again in 2012. The process of determining membership in the DJSI requires a corporate benchmarking scorecard that compares the one manufacturing firm's TBL with the TBL of others in the same sector.[7] The corporate benchmarking scorecard criteria are divided into three categories with relative weightings: economic (46%), environmental (26%), and social (28%) dimensions. Among firms in the durable household products sector, SBD's performance rankings were in the 91st percentile for economic concerns, the 64th percentile for environmental impact, and the 82nd percentile for business dimensions.

2 CREATING A DFE PROGRAM

This section describes the process of creating a DfE program within a manufacturing firm.

2.1 Identifying and Understanding the Stakeholders

The first step in creating a DfE program is to identify and understand the environmental stakeholders. A stakeholder is defined as one who has a share or an interest, as in an enterprise.[15] Stakeholders ultimately define the objectives and resulting environmental metrics of the DfE program. Listed below are examples of typical stakeholders for manufacturing firms:

• *Board Members*. Internal stakeholders who directly define corporate policies and culture.

• *Socially Responsible Investors*. Stakeholders who invest in manufacturing firms that demonstrate socially responsible values such as environmental protection and safe working conditions.

• *Nongovernment Organizations*. Organizations such as the Global Reporting Initiative with specific environmental agendas.

• *Government Organizations*. Organizations [e.g., the U.S. Environmental Protection Agency (EPA)] that require meeting certain environmental regulations and provide incentives (e.g., Energy Star) for achieving an exceptional level of environmental compliance.

• *Customers*. A customer is anyone who purchases the manufacturing firm's product. This could be a retailer, another manufacturing firm, or an end user.

• *Competitors*. Competitors are the other manufacturing firms in the same market. It is important to benchmark competitors to understand the environmental issues and strategies within that market and to position the firm in the market effectively.

- *Community*. The community consists of people affected by the products throughout their life cycle. Depending on the scope of the assessment, this could be everyone in the world. More realistically, however, it is the community that surrounds the manufacturing facilities and directly interacts with the products at any point during their life cycle.

Each stakeholder has different environmental interests, which generates many environmental demands that must be met. Because manufacturing firms operate with limited resources, the stakeholders must be prioritized based on their influence. Influence generally correlates to the impact on profits due to not meeting a stakeholder's demand. Once the stakeholders are prioritized, the manufacturing firm will have a good idea of which environmental demands need to be met. A DfE program with objectives and metrics that support these demands is now possible.

2.2   Creating Environmental Objectives

After a thorough analysis of the stakeholders, it is possible to create environmental objectives for the DfE program. The environmental objectives will need to align with as many of the environmental demands of the stakeholders as possible. The objectives will also need to align with the values and culture of the manufacturing firm. Successfully implementing an innovation (in this case, a DfE program) depends upon "the extent to which targeted users perceive that use of the innovation will foster the fulfillment of their values" (Ref. 16). Because employees who want to be successful will adapt to the values of the firm, a DfE program that is aligned with its values will also align with employee values and should be successfully implemented.

When creating environmental objectives, it is important to use the correct level of specificity. The objectives should be broad enough to avoid frequent updates but specific enough to provide consistent direction for the DfE program. An environmental objective of "protect the earth" would be too broad while "eliminate the use of lead" would be too specific. Environmental objectives should have associated lower level targets that can be used to assess the manufacturing firm's progress toward the objectives. For example, "eliminate the use of lead" could be a lower level target for the environmental objective "reduce the use of hazardous materials."

The following paragraphs describe general environmental objectives that manufacturing firms may find useful. Within a large manufacturing firm, different business units (which focus on different types of products) may have different environmental objectives. These objectives listed here are not specifically the environmental objectives of SBD. For information about how SBD values sustainability as a guiding principle, see Refs. 6 and 7.

Practice Environmental Stewardship. A manufacturing firm can demonstrate environmental awareness through creating an environmental policy and publishing it on its website and including information about recycled content on packaging and its DfE program. A manufacturing firm can join environmental organizations such as the World Environmental Center, which contributes to sustainable development worldwide by strengthening industrial and urban environment, health, and safety policy and practices, and organizations that promote environmental stewardship, such as the Rechargeable Battery Recycling Corporation (RBRC) and RECHARGE, two organizations that promote rechargeable battery recycling.

Comply with Environmental Regulations. A manufacturing firm should comply with all applicable regulations of the countries in which its products were manufactured or sold. The European Union (EU) exerts significant influence on addressing environmental issues through regulations and directives.

There are many regulations that apply to American and European workers, and these are set by European, federal, and state agencies. The U.S. Occupational Safety and Health Administration (OSHA) limits the concentration of certain chemicals to which workers may be exposed. The EPA regulates the management of waste and emissions to the environment. The Resource Conservation and Recovery Act and the Hazardous Materials Transportation Act require that employees receive training on handling hazardous wastes. California's Proposition 65 (the Safe Drinking Water and Toxic Enforcement Act of 1986) requires a warning before potentially exposing a consumer to chemicals known to the State of California to cause cancer or reproductive toxicity. The legislation explicitly lists chemicals known to cause cancer and reproductive toxicity.

The EU Battery Directive (2006/66/EC[17]) places restrictions on the use of certain batteries. The EU Packaging Directive (2004/12/EC[18]) seeks to prevent packaging waste by requiring packaging reuse and recycling. EU directives on waste electrical and electronic equipment (WEEE, 2002/96/EC[19]) and on the restriction of the use of certain hazardous substances in electrical and electronic components (RoHS, 2002/95/EC[20]) address issues of product take-back and bans on hazardous materials, respectively. The EU Registration, Evaluation, Authorisation and Restriction of Chemicals regulation (REACH, EC No. 1907/2006[21]) is a complex regulation that limits the production and use of chemicals that are hazardous to human health and the environment.

Address Customer Concerns. Retail customers may express concern about the environmental impacts of the products they sell. These customer concerns included ensuring that timber comes from appropriate forests, increasing the recyclability and recycled content in packaging, and reducing the use of hazardous materials such as lead and cadmium. A manufacturing firm can use these concerns as guidelines when developing new products.

Some customers are also interested in developing ways to classify products on the basis of sustainability. The Sustainability Consortium (a partnership between business and academia) is one such organization working on this topic. Industry associations, such as the Association of Home Appliance Manufacturers (AHAM), are developing sustainability standards to address this concern as well. In June of 2012, the Sustainability Standard for Household Refrigeration Appliances (AHAM 7001-2012/CSA SPE-7001-12/UL 7001[22]) was published. Manufacturing firms can become involved with their appropriate customers and trade associations to understand how sustainability is affecting their business and participate in standards developing activities if applicable.

Mitigate Environmental Risks. An activity's environmental risk is the potential that the activity will adversely affect living organisms through its effluents, emissions, wastes, accidental chemical releases, energy use, and resource consumption.[23] A manufacturing firm can mitigate environmental risks by monitoring chemical emissions from manufacturing plants, reducing waste produced by its operations, ensuring safe use of chemicals in the workplace, and ensuring proper off-site waste management, for instance.

Reduce Financial Liability. There are different types of environmental liabilities[24]:

- Compliance obligations are the costs of coming into compliance with laws and regulations.

- Remediation obligations are the costs of cleaning up pollution posing a risk to human health and the environment.

- Fines and penalties are the costs of being noncompliant.

- Compensation obligations are the costs of compensating "damages" suffered by individuals, their property, and businesses due to use or release of toxic substances or other pollutants.
- Punitive damages are the costs of environmental negligence.
- Natural resource damages are the costs of compensating damages to federal, state, local, foreign, or tribal land.

Not all of these environmental liabilities apply to all manufacturing firms.

Report Environmental Performance. A manufacturing firm can report environmental performance to many different organizations with local, national, or global influence and authority. One such organization is the Investor Responsibility Research Center (IRRC).

Determine End-of-Life Strategy for Product Lines. All systems, devices, and products will age and wear to the point that they are no longer useful. This is the point in a product's life cycle where proper design decision making influences the ease with which an end-of-life transformation to a sustainable form can be made. Reflexively discarding products into garbage dumps does not meet environmental objectives. Ashby voices an accurate but outdated attitude toward products: "When stuff is useful, we show it respect and call it material. When the same stuff ceases to be useful, we lose respect for it and call it waste" (Ref. 25). There are a limited number of ways to handle products that are no longer wanted: reuse, remanufacture, recycle, combustion, or landfill.[26] A product developer can make design decisions that will increase the likelihood of product reuse, remanufacture, or recycling. These options are introduced in this section.

- *Reuse of Product.* Reuse means finding a new end user who sees value in the product as it exists when the first owner disposes of the product. Making a transfer of the product at this point in time is the challenge. For consumer products, for example, possible solutions include garage sales, rummage sales (usually organized by volunteer members of a nonprofit organization), and retail stores supported by donations and run by charitable or philanthropic organizations (e.g., Goodwill Industries International and Habitat for Humanity). The Internet has enabled the growth of marketing of used goods by enabling global marketplaces like eBay, which was founded in 1995 to create a global market to efficiently connect buyers and seller to promote sustainable commerce.

- *Reuse of Parts.* Parts of an existing product can be reused if recovering the desired part or subassembly is possible. Common cases of reuse are found in the construction industry where materials (especially those with historic or esthetic appeal) can be recovered from existing structures that are (or have been) demolished. Often a third party acquires the discarded product for harvesting reusable parts. The market for rebuilt automotive parts (e.g., alternators and carburetors) creates a market for used or otherwise disabled cars. Designing for easy recovery of reusable parts can aid in reuse in domains where reuse is acceptable and desirable.

- *Remanufacturing.* Another end-of-life strategy is remanufacturing, the restoring of a product to near-new condition. This process can be limited to the cleaning and replacement of worn parts (refurbishing) or the replacement of faulty subassemblies with new parts. Remanufacturing saves energy by reducing the need for the processing of raw materials into new products. A common example of remanufacturing is the refilling of inkjet cartridges for computer printers. The process is often called recycling because the user surrenders an empty cartridge to enter the supply chain again. The process actually reconditions the product, however, so that it can be restored to its original use. Envisioning how a product can be remanufactured and making it easy to do so are key strategies for improved DfE. For example, manufacturing firms such as Caterpillar that intend to remanufacture their products may make them more durable than necessary for one life so that the products will have multiple lives and generate additional revenue that offsets the increased initial unit cost.[27]

• *Recycling*. Recycling involves dipping into the waste stream to identify and recover discarded products that consist of valuable raw materials. Then the desired raw materials are separated from discarded product and put back into the original raw material stream to make the same or similar material. Recycling is the end-of-life strategy that may have the best chance of scoring environmental gains by adding to the supply of raw materials and reducing the amount of waste produced. In addition, energy is saved when producing raw material from recycled goods. Recycling requires energy with its accompanying gas emissions, but the recycle energy is small compared with the energy to make the material. For example, recycled aluminum requires only 5% of the energy required to produce it from the ore.[28] There are barriers to recovering material through recycling. It is an expensive process that requires (1) collection and transport, (2) separation, and (3) identification and sorting.[29] This process is also beyond the control of the product developer. The product developer can use design expertise to make a product more easily recyclable as part of their corporate strategy. This will be addressed in Section 3.

2.3 Metric Selection

After the environmental objectives are set, a manufacturing firm needs environmental metrics to measure its progress. Starting at the conceptual level, an overall metric was developed in the 1970s to conceptualize the impact that a technological society was having on the environment.[30] This metric details how the factors of population (P), affluence (A, meaning the supply of something usually a measure of value on a per-capita basis), and technology (T) influence the overall environment (I):

$$I = P \times A \times T \tag{1}$$

The term T is particularly important to engineers because it represents how technology, the result of engineering effort, can be used to change the effects of population growth and supply of goods and services on the environment.

The following equation is Allenby's interpretation of the $I–P–A–T$ equation for the environment[31]:

$$\text{Environmental impact} = \text{Population} \times \frac{\text{Resource use}}{\text{Person}} \times \frac{\text{Environmental impact}}{\text{Unit of resource use}} \tag{2}$$

A direct interpretation of this equation articulates the strategies for reducing the overall environmental impact. The strategies are (1) reduce population or slow its growth, (2) reduce the resource use per person, and (3) reduce the negative environmental impact per unit of resource. Naturally, designers and engineers will focus on redesigning existing technology to reduce any harm it may cause to the environment and developing new technologies that are environmentally benign or that will mitigate the harmful environmental impacts. DfE strategies, tools, and methods use metrics that quantify impact on the environment in meaningful ways for decision making.

There are many aspects that need to be considered when selecting metrics for a DfE program. First, each metric should directly relate to at least one of the environmental objectives. Metrics that relate to many objectives tend to be more desirable. Second, the manufacturing firm must have the capability of measuring the metric. A metric that can be easily measured ranks higher than a metric that requires costly changes and upgrades. Finally, the metrics should tailor to specific stakeholder reporting requests. An analysis of the most requested metrics can help prioritize the metrics. Because manufacturing firms operate with limited resources, the metrics will need to be prioritized based on these aspects. Although most metrics are quantitative, qualitative metrics (such as an innovation statement) do exist.

The following briefly describes eight product-level environmental metrics that product development teams can evaluate during a product development process. Note that a material is considered flagged if it is banned, restricted, or being watched with respect to regulations or customers.

Flagged Material Use in Product. This metric measures the mass of each flagged material contained in the product.

Total Product/Packaging Mass. This metric measures the mass of the product and packaging separately.

Flagged Material Used or Generated in Manufacturing Process. This is a list of each flagged material used or generated during the manufacturing process.

Recyclability and Separability Rating. This metric is the degree to which each component and subassembly in the product is recyclable. Recyclability and separability ratings can be calculated for each component, subassembly, and final assembly based on qualitative rankings (statements that describe the degree to which a component is recyclable or separable) and their associated values (which range from 1 to 6). Low values for recyclability and separability indicate a product design that facilitates disassembly and recycling.[32]

Disassembly Time. A measure of the time it will take to disassemble the product. Research has been conducted on how long it typically takes to perform certain actions. Charts with estimates for typical disassembly actions are provided to the design engineers who can then estimate how long it would take to disassemble a product.[32]

Energy Consumption. The total expected energy usage of a product during its lifetime. This metric can be calculated by multiplying the expected lifetime in hours by the energy that the product consumes during an hour of use. This metric is relevant only for products that consume a significant amount of energy.

Innovation Statement. This brief paragraph describes the ways that the product development team reduced the negative environmental impact of the product. This metric allows for a free-form method for explaining what the development team achieved that may not be apparent in quantitative metrics. It can also serve as a nice source of information to highlight achievements in a manufacturing firm's yearly sustainability report.

Application of DfE Approach. This binary measure (yes or no) is the answer to the following question: Did the product development team follow the DfE approach during the product development process? Following the DfE approach requires the team to review the DfE guidelines and evaluate the product-level environmental metrics. This metric is good to use when first implementing a DfE process to show the progress of adoption and may be dropped after the process is mature and widely used.

While these metrics do not measure every environmental impact, they provide designers with a simple way to compare how well different designs accomplish important environmental objectives.

2.4 Incorporating DfE into the Design Process

A DfE process that fits well into an existing product development process has significant potential to help a manufacturing firm achieve its environmental objectives. By researching the product development process and understanding the decision-making processes, information flow, and organizational and group values, it is possible to construct a DfE process that is customized and easy to implement. The product development process needs to be studied to ensure information availability for the desired metrics. Ideally, the DfE process should leverage existing processes in order to minimize time to market and should require little extra effort from the designers.

The safety review process is a common process that can be combined with the DfE process. Most manufacturing firms implement a formal safety review process to verify that the product is safe for consumer use (cf. Refs. 33 and 34). Typically, safety reviews are held at predetermined key points in the product development process. During these reviews, members from the design team and other safety specialists, such as liability and compliance representatives, meet to discuss the current product design. The meetings are run in a brainstorming format and can be guided by a checklist or a firm-specific agenda. Since product safety includes qualitative measures, it is necessary to assess the issue in a meeting format where ideas and concerns can be discussed with all interested parties.

There are major similarities between safety and environmental concerns. Both areas are important and involve subjective assessments on a variety of factors, many of which are qualitative and only indirectly linked to profitability. In the same way that safety testing and standards have reduced the subjective aspects of safety assessment, environmental metrics, standards, and design tools are reducing the subjective aspects of evaluating environmental impact. This suggests that most manufacturing firms should treat environmental objectives in the same way that they treat safety concerns. It is necessary to assess a product's environmental performance at key stages in the product development process. Furthermore, since assessing environmental performance requires information from multiple business units within the manufacturing firm, a meeting to discuss the issues with all interested parties is necessary. Thus, for manufacturing firms with similar corporate objectives for safety and for environmental issues, expanding the safety review process creates a practical DfE process that should be simple to implement.

On the other hand, manufacturing firms with elaborate safety evaluation (qualification) and verification (certification) procedures (used in areas such as aircraft manufacturing) may not require a similarly sophisticated DfE process (unless the product has many environmental concerns, as in automobile manufacturing). In manufacturing firms that do not explicitly consider safety during their new product development process, establishing a DfE process will be more work, but the need for a DfE process remains.

The safety and environmental objectives of manufacturing firms vary considerably, and each one uses different mechanisms for addressing these concerns. Certainly, practices that make sense in one domain may be impractical in another. This approach is based on the similarity of the safety objectives and environmental objectives. In manufacturing firms where the safety objectives and environmental objectives are quite different in scope, other types of DfE processes will be more effective.

2.5 Fitting the Pieces Together

A DfE program cannot be implemented in isolation from other programs within a manufacturing firm. The program needs to be integrated with other programs that fall under the corporate responsibility umbrella and carry the same weight. Typical corporate responsibility programs include giving back to the community, promoting diversity awareness, ensuring proper working conditions and benefits for employees, and environmental awareness. These programs have detailed plans and goals that are disseminated to all employees through a substantial medium such as a communications meeting. The employees then begin "living" these programs, which results in a corporate culture.

Most manufacturing firm's environmental awareness initiatives are based at the manufacturing level rather than the product level. A new DfE program will most likely be integrated with this preexisting portion of environmental awareness. Upon implementation, the program objectives and specific process need to be clearly presented to employees. The commitment from the upper management should be enough to get the program rolling. Ideally, product managers (who supervise the product development teams) will see that the DfE program is helping their engineers design products that are more successful on the most important metrics (like sales, profitability) and will insist on the engineers' executing it.

If there is resistance, the manufacturing firm may need to implement a system that rewards those who participate (and has consequences for those who do not). A manufacturing firm could, for instance, include environmental metrics in the list of criteria used to evaluate product managers. Only after seeing management's commitment and receiving appropriate direction will the product managers and engineers do their jobs and determine how to meet the goals.

The process itself can also be an obstacle. A product specification document is a statement of the customer requirements that a new product should meet. Typically, a manufacturing firm's marketing group will create this document at the beginning of the product development process, the managers will approve it, and the product development team's goal is to design a product that meets those requirements. Items not on the product specification document are considered optional, and product development teams have little incentive to meet environmental requirements that are not listed in their product specification documents, especially when increases in workload make meeting the important requirements promptly a challenge.

3   USING DFE TOOLS

This section will explore some general DfE tools and how they should be implemented within the product development process.

3.1   Guideline/Checklist Document

A guideline/checklist document is a simple DfE tool that forces designers to consider environmental issues when designing products. Integrating a guideline/checklist document within a new DfE process is a simple and effective way to highlight environmental concerns. However, it should be noted that the guideline/checklist document needs to be firm specific and integrated systematically into the product development process. Using an existing generic, stand-alone guideline/checklist document will most likely be ineffective. First, the point of a guideline/checklist document is to ensure that the designers are taking the proper steps toward achieving specific environmental objectives. A guideline/checklist document from one manufacturing firm may promote objectives that do not coincide with another's objectives. Second, obtaining a guideline/checklist document and simply handing it to designers will lead to confusion as to when and how to use the list. Specific procedures need to be implemented to ensure that the designers are exposed to the guideline/checklist document early in the product development process to promote environmental design decisions. The following are examples of DfE guidelines:

- Reduce the amount of flagged materials in the product by using materials not included on the manufacturing firm's *should-not-use* list.
- Reduce the raw material used in the product by eliminating or reducing components.

- Reduce the amount of flagged material released in manufacturing by choosing materials and processes that are less harmful.
- Increase the recyclability and separability of the product's components.
- Reduce the product's disassembly time.
- Reduce the amount of energy the product uses.

3.2 Product Design Matrix

The product design matrix[35] is a tool that was created with the Minnesota Office of Environmental Assistance and the Minnesota Technical Assistance Program (MnTAP). The matrix helps product designers determine where the most environmental impact of their product design occurs. Two different categories are explored within the matrix: environmental concerns and life stages. The environmental concerns (materials, energy use, solid residue, liquid residue, and gaseous residue) are listed across the top of the matrix and the life stages (premanufacture, product manufacture, distribution and packaging, product use and maintenance, and end of life) are listed on the left side of the matrix. Included in the matrix shown in Figure 1 is a series of questions for each block. Points are associated with each question, and the points are summed for each of the 25 blocks. The points in each row and column are summed to provide the designers with information regarding the largest environmental concern and the most environmentally detrimental stage of the product life cycle. It is possible that the product design matrix and the accompanying questions can be varied to suit a manufacturing firm's specific needs. This tool should be used during the design review stage of the product development process so that designers have an opportunity to make changes based on the results of the tool.

LIFE STAGE	Environmental Concern					Total
	1 Materials	2 Energy Use	3 Solid Residue	4 Liquid Residue	5 Gaseous Residue	
A Premanufacture	(A.1)	(A.2)	(A.3)	(A.4)	(A.5)	
B Product manufacture	(B.1)	(B.2)	(B.3)	(B.4)	(B.5)	
C Distribution, packaging	(C.1)	(C.2)	(C.3)	(C.4)	(C.5)	
D Product use, maintenance	(D.1)	(D.2)	(D.3)	(D.4)	(D.5)	
E End of life	(E.1)	(E.2)	(E.3)	(E.4)	(E.5)	
Total						

Figure 1 Product design matrix. *Source:* Adapted from Ref. 36.

3.3   GRANTA's CES Material Selector and Eco-Audit Tool

Material selection is an important issue in product design decision making because all physical systems derive their behavior from their material, their geometry, and the manufacturing processes used to convert the material into the specified geometry.[26] Material selection is even more critical to proper DfE methods and practices because the environmental impact of a product's material starts long before and ends long after the useful life of the product itself.

Material selection has more influence over a product's environmental cost because of the enormous energy it takes to transform raw materials into engineering stock. The production energy H_p is the amount of energy used to make 1 kg of a raw material. It is expressed in megajoules per kilogram. The production energy values of some common materials are as follows: 22.4–24.8 MJ/kg for low-carbon steel; 77.2–80.3 MJ/kg for stainless steel; 184–203 MJ/kg for aluminum alloys; and 63.5–70.2 MJ/kg for polyvinyl chloride (PVC).[37] It is important to note that the energy used to produce these materials uses fossil fuels that generate additional environmental costs in CO_2 emissions, for example. The CO_2 emissions created per kilogram of raw material produced are as follows: 1.9–2.1 for low-carbon steel; 4.8–5.4 for stainless steel; 11.6–12.8 for aluminum alloys; and 1.85–2.04 for PVC.[37]

There are many points in a product's life cycle that call for energy expenditure. There is the energy for manufacture, fabrication, and assembly, the energy required to use the product, and the energy required at the product's end of life. The energy required for manufacturing is normally an order of magnitude lower than the production energy.[26] It is clear that the first area to examine for energy savings is the selection of material.

Granta is an international company that provides information on materials for professionals and educational institutions. Mick Ashby (the developer of the material selection tool called the Ashby chart) and David Cebon collaborated on software to improve material selection, which led to the founding of Granta in 1994. Granta's CES Selector (in a variety of editions based on application area) software can provide materials information to aid in determining substitutions and cost analysis. The CES Selector suite of products includes the Eco-Audit Tool, which is designed to include material information to base selection decisions on environmental objectives. The Eco-Audit Tool allows a designer to estimate the environmental impact of a product over its entire life cycle. A designer can compare two different design alternatives with components, manufacturing processes, and useful life lengths and make more informed materials selection decisions.

3.4   Environmental Effect Analysis

The environmental effect analysis (EEA) was developed by multiple organizations, including the Swedish consulting agency HRM/Ritline, Volvo, and the University of Kalmar, Sweden. It is based on the quality assurance failure modes and effects analysis (FMEA), and the form looks much like a typical FMEA with environmental headings (Fig. 2). The tool was designed to be used early in the product development process by a product development team with the supervision of an environmental specialist to help with questions. First, the team identifies the key activities associated with each stage of the product's life cycle. Next, the team identifies the environmental aspects of the activities. Then, the team identifies the environmental impact associated with these environmental aspects. Some examples of environmental impacts are ozone depletion, resource depletion, and eutrophication. Next, the environmental impacts are evaluated to determine their significance. The evaluation technique is similar to that of the FMEA. An environmental priority number (EPN) is calculated using three variables: S—Controlling Documents, I—Public Image, and O—Environmental Consequences. The variables are given a ranking from 1 to 3 based on environmental compliance, where 1 is the best possible score and 3 is the worst possible score. The EPN is calculated by adding the

Environmental Effect Analysis - EEA   [The SIO-Method]

Part Name	Part Number	Drawing Number	Function		Date	Issue
Project	Supplier	Info			Follow-up Date	Page No.
EEA Leader	EEA Participants					

Inventory Life-cycle

No.	Life-cycle Phase	Activity	*Environmental Characteristics* Environmental Effect / Aspect	*Valuation* S	I	O	EPN / F	*Actions Proposals for Action* Recommended Actions	Environmental Effect / Aspect	*Valuation* S	I	O	EPN / F	*Realization* Remarks	Responsible

Figure 2   EEA form.[38]

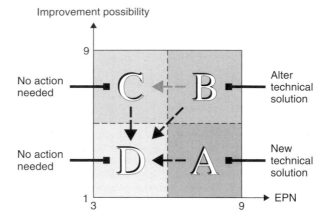

Figure 3   Evaluation matrix.[38]

three scores. A fourth variable, *F*—Improvement Possibly, is focused on the effort in time, cost, and technical possibility of improving the product. It is based on a 1–9 scale, where 1 represents no possibility for improvement and 9 represents a very large possibility for improvement. Detailed explanations of what each score means qualitatively for each variable can be found in Ref. 38. After the evaluation, designers can place the results into an evaluation matrix (Fig. 3) to determine what design changes should be made. Recommendations for design changes and actual design change decisions made are filled into the evaluation matrix, and the EPN and *F* are recalculated to ensure that improvement is achieved. The evaluation matrix provides a record of the aspects evaluated and design decisions made within the product development process.[8]

3.5   Life-Cycle Assessment

Life-cycle assessments are in-depth studies that research a product's environmental impacts and conduct tests to produce environmental impact quantities.[39] LCAs require extensive data, take a long time, are very expensive, and provide information only after the design is complete. LCAs do not help designers improve a current product's environmental impact, and they would be inappropriate during the product development process. Despite these limitations, LCAs are an effective DfE tool because they provide detailed information that can be used to determine the largest environmental impact of a current product, which can suggest opportunities to be more environmentally benign when the product is redesigned in the future.

3.6   ENVRIZ

ENVRIZ (a portmanteau word that combines "environment" and "TRIZ") is a methodology for resolving conflicts between product functionality and environmental impact.[40–42] ENVRIZ assists designers by providing them with knowledge about how product designs have overcome contradictions between product functionality and environmental impact. This knowledge includes the TRIZ principle demonstrated by each example. The results of an effectiveness study support that utilizing specific product examples from ENVRIZ provides better solutions compared to utilizing engineering principles from either ENVRIZ or TRIZ.

In the ENVRIZ approach, a product development team first identifies the functional and environmental parameters that they wish to improve and then identifies any conflicts between these parameters. For each conflict between functionality and environmental impact, the design team finds TRIZ principles and product examples that have overcome the conflict. The design team then uses design by analogy to exploit this knowledge to create conceptual designs that resolve the conflict. The design team continues until it resolves all of the conflicts.

4 EXAMPLES OF DFE INNOVATIONS

This section describes examples of products that have been designed to reduce adverse environmental impact. Most of these products introduce increased functionality in addition to being more environmentally friendly. It is important to recognize what has been accomplished in the field of environmental design and build on this existing knowledge. By combining ideas that have been implemented in the past with their own ingenuity, designers can create new products that have minimal adverse environmental impacts as well as adding to the environmental design knowledge base.

4.1 Forever Flashlight

The Forever Flashlight is a flashlight that does not require batteries or bulbs. Its power is generated by the user shaking the flashlight. When the user shakes the flashlight, a component wound with copper wire moves through a magnetic field and generates power that is stored in the flashlight. Fifteen to 30 s of shaking can provide up to 5 min of light. Also, instead of using a bulb, a blue light-emitting diode (LED) is used because it has greater longevity. This flashlight prevents environmental harm by reducing battery usage and provides more functionality than a typical flashlight because the user will never be left in the dark due to dead batteries.[43]

4.2 Battery-Free Remote Control

The Volvo Car Company and Delft University of Technology created a battery-free remote control for automobiles. The remote control utilizes the piezo effect (the charge created when crystals such as quartz are compressed). The remote has a button on top and a flexible bottom. When the user pushes the button, the top button and flexible bottom compress the crystal creating an electrical charge that powers a circuit to unlock the car. This prevents environmental harm by reducing battery usage.[44]

4.3 Toshiba's GR-NF415GX Refrigerator

The Toshiba GR-NF415GX refrigerator, which won the 2003 Grand Prize for Energy Conservation, is an excellent example of an environmentally benign product. Takehisa Okamoto, the engineer who designed the refrigerator, described the problem with previous refrigerators in this excerpt (Ref. 45):

> To review the mechanics of earlier refrigerators, previously both the refrigerator and freezer sections were cooled by a single cooling unit. Since the refrigerator section didn't require as much cooling as the freezer section, it tended to be over-cooled. To prevent this, a damper was attached to open and close vents automatically. This would close the vents when it got too cold and open them when it got too warm. However, in the area near the vents where the cold air came out, eggs would sometimes get too cold or tofu would sometimes freeze.

Takeshisa then described the solution to the problem and the advantages of the new refrigerator in these excerpts:

> Then the twin cooling unit refrigerator was developed. This involved two cooling units—one in the refrigerator section and one in the freezer section—using a single compressor. This system alternates

between cooling the refrigerator and cooling the freezer, which allows each section to be cooled to a more suitable temperature. While the freezer's being cooled, the frost that accumulates on the cooling unit in the refrigerator section, where coolant isn't flowing, is melted once again and returned to the refrigerator section using a fan for humidification. This prevented drying, so that cheese and ham wouldn't lose all their moisture.

The technology makes it possible to cool the refrigerator and freezer sections simultaneously and to maintain two temperatures, with a major difference in temperatures between the temperatures of the refrigerator cooling unit ($-3.5°C$) and the freezer cooling unit ($-24°C$). Now, the refrigerator section is cooled by $-2°C$ air to maintain a temperature of $1°C$. This technology uses an ultra-low-energy freeze cycle that makes it possible to cool using cold air at temperatures close to the ideal temperature for the food.

Since this is the first technology of this kind in the industry, some aspects were definitely difficult. At the same time, I think this innovation was really the key point of this development. The two-stage compressor distributes coolant compressed in two stages in two directions: to the refrigerator side and to the freezer side. For this reason, the flows of coolant to each cooling unit must be adjusted to ensure optimal flow. We achieved efficient simultaneous cooling using a pulse motor valve (PMV).

This dialogue highlights the improvement in the freshness of the food due to the accuracy of the air temperature being output into the refrigeration section. This innovation also conserves electricity because of the ultralow-energy freeze cycle. In addition, a typical engineering solution to this problem would require two compressors to achieve the final result, while this product only required one.

4.4   Matsushita Alkaline Ion Water Purifier

The Matsushita PJ-A40MRA alkaline ion water purifier[46] has more functionality and less environmental impact than the TK7505 alkaline ion water purifier. The new water purifier increases functionality by allowing the user to select seven kinds of water quality (as opposed to five), which allows the user to get the precise quality that is needed in a particular situation. By using two separate power sources for operation and standby, the new purifier reduces standby power from 6 to 0.7 W, which also decreases environmental impact.

4.5   Self-Cleaning Exhaust Hood

A self-cleaning exhaust hood (installed over a stovetop) actively deploys water vapors to intercept oil and dust before they reach the hood, whereas a traditional hood is equipped with a filter that needs to be regularly cleaned to prevent damage to the system.[43] Compared to standard exhaust hoods, this product not only removes the need for maintenance improvement but also increases the product's reliability and durability, which reduces its environmental impact.

4.6   Eco-Toaster

The Eco-Toaster uses 34% less energy than standard toasters due to its rather clever autoclose lid that keeps the heat around the toast and prevents it from escaping through the top.[47] It also toasts the bread more quickly and thus both improves performance and reduces environmental impact.

4.7   Eco-Kettle

The Eco-Kettle has a double chamber that directs a user to measure out exactly how much water they want to boil. By boiling only the water that is needed, the kettle heats the water more quickly and uses 31% less energy.[47] Like the Eco-Toaster, the Eco-Kettle does not resort to using more energy to accelerate a process; instead, it aims to use energy more efficiently.

5 CONCLUSIONS

This chapter reviewed DfE tools, methods, and strategies that have been developed and implemented to help manufacturing firms create environmentally benign products. From this review we draw the following general conclusions.

DfE tools vary widely with respect to the information they require and the analysis that they perform. Adopting a DfE tool does not automatically lead to environmentally benign products. It is important to have DfE tools that address relevant and important environmental metrics and that provide useful information to product development teams.

Manufacturing firms need DfE processes, not just DfE tools. However, a DfE process that adds a large amount of additional analysis, paperwork, and meetings (all of which add time and cost) is not desirable. Ideally, environmental objectives would be considered in every decision that occurs during new product development, like the objectives of product performance, unit cost, time to market, and development cost. However, environmental objectives are much different than these. Instead they more closely resemble safety objectives.

One possible approach to remedy this problem is for a manufacturing firm to create a DfE process by expanding the safety review process, which they may already have in place. In many manufacturing firms, the safety review process evaluates product safety at various points during the product development process. Therefore, combining the DfE process with the safety review process would require environmental performance to be assessed multiple times during the product development process.

This method of incorporating the DfE process into the product development process ensures that environmental performance will be evaluated at key points in the design process, not only when the design is complete.

The safety and environmental objectives of manufacturing firms vary considerably, and each one uses different mechanisms for addressing these concerns. Certainly, practices that make sense in one domain may be impractical in another. This chapter identified one way to create a DfE process, something that many manufacturing firms are now attempting to do, and discussed the use of this approach at a particular manufacturing firm. Our analysis of this approach is based on the similarity of the safety objectives and environmental objectives. In manufacturing firms where the safety objectives and environmental objectives are quite different in scope, other types of DfE processes will be more effective. Fundamentally, though, a manufacturing firm still needs a DfE process, not an isolated environmental assessment tool, to achieve its environmental objectives. More generally, a DfE process must be designed to fit within the existing patterns of information flow and decision making in the manufacturing firm, as discussed by Herrmann and Schmidt.[10,11]

The examples of DfE tools and DfE innovations illustrate the many approaches that creative engineers have developed to design environmentally benign products, and we hope that these will inspire future engineers as well.

REFERENCES

1. R. B. Handfield, S. A. Melnyk, R. J. Calantone, and S. Curkovic, "Integrating Environmental Concerns into the Design Process: The Gap between Theory And Practice," *IEEE Trans. Eng. Manag.*, **48**(2), 189–208, 2001.

2. J. Fiksel, *Design for Environment: Creating Eco-efficient Products and Processes*, McGraw-Hill, New York. 1996.

3. B. Bras, "Incorporating Environmental Issues in Product Design and Realization," *UNEP Industry and Environment*, **20**(1–2), 5–13, January–June, 1997.

4. S. Ashley, "Designing for the Environment," *Mech. Eng.*, **115**(3), 53–55. 1993.

5. D. Allen, D. Bauer, B. Bras, T. Gutowski, C. Murphy, T. Piwonka, P. Sheng, J. Sutherland, D. Thurston, and E. Wolff, "Environmentally Benign Manufacturing: Trends in Europe, Japan, and the USA,"

DETC2001/DFM-21204, in *Proceedings of DETC'01*, the ASME 2001 Design Engineering Technical Conference and Computers and Information in Engineering Conference, Pittsburgh, PA, September 9–12, 2001.

6. Stanley Black & Decker, "2011 Environmental Health & Safety Report," *Sustainability: Annual Reports*, 2012a, available: http://www.stanleyblackanddecker.com/company/sustainability/annual-reports, accessed October 29, 2012,

7. Stanley Black & Decker, "2012 SAM Company Benchmarking Scorecard," *Sustainability: Our Progress*, 2012b, available: http://www.stanleyblackanddecker.com/sites/www.stanleyblackanddecker.com/files/documents/2012_SAM_Company_Benchmarking_ScoreCard_StanleyBlackDeckerInc.PDF, accessed October 29, 2012.

8. M. Lindahl, "Designer's Utilization of DfE Methods," in *Proceedings of the 1st International Workshop on Sustainable Consumption*, Tokyo, Japan, The Society of Non-Traditional Technology (SNTT) and Research Center for Life Cycle Assessment (AIST), 2003.

9. V. Krishnan and K. T. Ulrich, "Product Development Decisions: A Review of the Literature," *Manag. Sci.*, **47**(1), 1–21, 2001.

10. J. W. Herrmann and L. C. Schmidt, "Viewing Product Development as a Decision Production System," DETC2002/DTM-34030, paper presented at the ASME 2002 Design Engineering Technical Conferences and Computers and Information in Engineering Conference, Montreal, Canada, September 29–October 2, 2002.

11. J. W. Herrmann and L. C. Schmidt, "Product Development and Decision Production Systems," in W. Chen, K. Lewis, and L. C. Schmidt (Eds.), *Decision Making in Engineering Design*, ASME Press, New York, 2006.

12. The World Commission on Environment and Development, *Our Common Future*, Oxford University Press, New York, 1987.

13. R. B. Pojasek, "Sustainability: The Three Responsibilities," *Environ. Quality Manag.*, Spring 2010, pp. 87–94.

14. J. Elkington, *Cannibals with Forks: The Triple Bottom Line of 21st Century Business*, Capstone, Gabriola Island, British Columbia, Canada, 1997.

15. The American Heritage Dictionary of the English Language, 4th ed., Houghton Mifflin, Boston, 2000.

16. K. J. Klein, and J. S. Sorra, "The Challenge of Innovation Implementation," *Acad. Manag. Rev.*, **21**(4), 1055–1080, 1996.

17. "Directive 2006/66/EC of the European Parliament and of the Council of the 6 September 2006 on batteries and accumulators and repealing Directive 91/157/EEC Article 1," http://eur-lex.europa.eu/LexUriServ/LexUriServ.do?uri=OJ:L:2006:266:0001:0014:EN:PDF, accessed April 16, 2014.

18. "Directive 2004/12/EC of the European Parliament and of the Council of 11 February 2004 Amending Directive 94/62/EC on Packaging and Packaging Waste," *Off. J. European Union*, available: http://www.europa.eu.int/eur-lex/pri/en/oj/dat/2004/l_047/l_04720040218en00260031.pdf.

19. "Directive 2002/96/EC of the European Parliament and of the Council of 27 January 2003 on waste electrical and electronic equipment (WEEE) - Joint declaration of the European Parliament, the Council and the Commission relating to Article 9,"http://eur-lex.europa.eu/legal-content/EN/TXT/?uri=CELEX:32002L0096, accessed April 16, 2014.

20. "Directive 2011/65/EU of the European Parliament and of the Council of 8 June 2011 on the restriction of the use of certain hazardous substances in electrical and electronic equipment," http://eur-lex.europa.eu/LexUriServ/LexUriServ.do?uri=OJ:L:2011:174:0088:0110:en:pdf, accessed April 16, 2014.

21. "Regulation (EC) No 1907/2006 of the European Parliament and of the Council of 18 December 2006 concerning the Registration, Evaluation, Authorisation and Restriction of Chemicals (REACH), establishing a European Chemicals Agency," http://eur-lex.europa.eu/legal-content/EN/TXT/?uri=uriserv:OJ.L_.2006.396.01.0001.01.ENG, accessed April 16, 2014.

22. Association of Home Appliance Manufacturers (AHAM), "Sustainability Standard for Household Refrigeration Appliances," AHAM 7001-2012, AHAM, Washington, DC, 2012.

23. U.S. Environmental Protection Agency (EPA), "Terms of Environment: Glossary, Abbreviations, and Acronyms," EPA175B97001, EPA, Washington, DC, December 1997.

24. U.S. Environmental Protection Agency (EPA), "Valuing Potential Environmental Liabilities for Managerial Decision-Making," EPA742-R-96-003, EPA, Washington, DC, December 1996.

25. M. F. Ashby, *Materials and the Environment: Eco-Informed Material Choice*, Elsevier, 2009.

26. G. Dieter and L. Schmidt, *Engineering Design*, 5th ed., McGraw-Hill, New York, 2012.

27. B. Hindo, "Everything Old Is New Again," *BusinessWeek*, September **25**, 2006, pp. 64–70.

28. "The Price of Virtue," *The Economist*, **383**(8532), 14, 2007.

29. "Design for Recycling and Life Cycle Analysis," *Metals Handbook*, Desk ed., 2nd ed., ASM International, Materials Park, OH, 1998, pp. 1196–1199.

30. M. R. Chertow, "The IPAT Equation and Its Variants," *J. Ind. Ecol.*, **4**(4), 13–29, 2000.

31. B. R. Allenby, *The Theory and Practice of Sustainable Engineering*, Prentice Hall, Upper Saddle River, NJ, 2011.

32. K. N. Otto, and K. L. Wood, *Product Design: Techniques in Reverse Engineering and New Product Development*, Prentice Hall, Upper Saddle River, NJ, 2001.

33. R. Goodden, *Product Liability Prevention—A Strategic Guide*, ASQ Quality Press, Milwaukee, WI, 2000.

34. W. F. Kitzes, "Safety Management and the Consumer Product Safety Commission," *Professional Safety*, **36**(4), 25–30. 1991.

35. J. Yarwood and P. D. Eagan, "Design for the Environment (DfE) Toolkit," Minnesota Office of Environmental Assistance, Minnesota Technical Assistance Program (MnTAP), 2001, available: http://www.pca.state.mn.us/index.php/view-document.html?gid=4683, accessed October 29, 2012.

36. T. E. Graedel and B. R. Allenby, *Industrial Ecology*, Prentice Hall, Englewood Cliffs, NJ, 1995.

37. M. F. Ashby, *Materials Selection in Mechanical Design*, 3th ed., Elsevier, Butterworth-Heinemann, Oxford, 2011.

38. M. Lindahl and J. Tingström, *A Small Textbook on Environmental Effect Analysis* (shorter version of the Swedish version), Dept. of Technology, University of Kalmar, Kalmar, Sweden, 2001.

39. M. R. Bohm, K. R. Haapala, K. Poppa, R. B. Stone, and I. Y. Turner, "Integrating Life Cycle Assessment into the Conceptual Phase of Design Using a Design Repository," *J. Mech. Design*, **132**(9), September 2010, Article 091005.

40. D. P. Fitzgerald, J. W. Herrmann, and L. C. Schmidt, "A Conceptual Design Tool for Resolving Conflicts between Product Functionality and Environmental Impact," *J. Mech. Design*, **132**(9), September 2010, Article 091006.

41. D. P. Fitzgerald, "ENVRIZ: A Methodology for Resolving Conflicts between Product Functionality and Environmental Impact," University of Maryland, December, 2011, available: http://hdl.handle.net/1903/12267, accessed February 24, 2012.

42. D. P. Fitzgerald, J. W. Herrmann, P. A. Sandborn, L. C. Schmidt, and T. Gogoll, "Beyond Tools: A Design for Environment Process," *Int. J. Performability Eng.*, **1**(2), 105–120, 2005.

43. H. -T. Chang and J. L. Chen, "Eco-Innovative Examples for 40 TRIZ Inventive Principles," *TRIZ J.*, pp. 1–16, August 2003.

44. Minnesota Office of Environmental Assistance, "Power for Products," 2006, available: http://www.moea.state.mn.us/p2/dfe/dfeguide/power.pdf.

45. A. Yoshino, "Interview with Takehisa Okamoto Concerning CFC-Free Refrigerator," available: http://kagakukan.toshiba.co.jp/en/08home/newtech131.html, accessed January 26, 2006.

46. Matsushita Group, "Alkaline Ion Water Purifier [TK7505] Factor X Calculation Data," available: http://panasonic.co.jp/eco/en/factor_x/m_pdf/fx_p21e.pdf, accessed January 26, 2006.

47. Nigel's Eco-Store, available: http://www.nigelsecostore.com/acatalog/about_us.html, accessed December 29, 2009.

CHAPTER 4

DESIGN OPTIMIZATION: AN OVERVIEW

A. Ravi Ravindran
The Pennsylvania State University
University Park, Pennsylvania

G. V. Reklaitis
Purdue University
West Lafayette, Indiana

1 INTRODUCTION

This chapter presents an overview of optimization theory and its application to problems arising in engineering. In the most general terms, optimization theory is a body of mathematical results and numerical methods for finding and identifying the best candidate from a collection of alternatives without having to enumerate and evaluate explicitly all possible alternatives. The process of optimization lies at the root of engineering since the classical function of the engineer is to design new, better, more efficient, and less expensive systems, as well as to devise plans and procedures for the improved operation of existing systems. The power of optimization methods to determine the best case without actually testing all possible cases comes through the use of a modest level of mathematics and at the cost of performing iterative numerical calculations using clearly defined logical procedures or algorithms implemented on computing machines. Because of the scope of most engineering applications and the tedium of the numerical calculations involved in optimization algorithms, the techniques of optimization are intended primarily for computer implementation.

2   REQUIREMENTS FOR THE APPLICATION OF OPTIMIZATION METHODS

In order to apply the mathematical results and numerical techniques of optimization theory to concrete engineering problems it is necessary to delineate clearly the boundaries of the engineering system to be optimized, to define the quantitative criterion on the basis of which candidates will be ranked to determine the "best," to select the system variables that will be used to characterize or identify candidates, and to define a model that will express the manner in which the variables are related. This composite activity constitutes the process of *formulating* the engineering optimization problem. Good problem formulation is the key to the success of an optimization study and is to a large degree an art. It is learned through practice and the study of successful applications and is based on the knowledge of strengths, weaknesses, and peculiarities of the techniques provided by optimization theory.

2.1   Defining the System Boundaries

Before undertaking any optimization study it is important to define clearly the boundaries of the system under investigation. In this context a system is the restricted portion of the universe under consideration. The system boundaries are simply the limits that separate the system from the remainder of the universe. They serve to isolate the system from its surroundings because, for purposes of analysis, all interactions between the system and its surroundings are assumed to be frozen at selected representative levels. Since interactions, nonetheless, always exist, the act of defining the system boundaries is the first step in the process of approximating the real system.

In many situations it may turn out that the initial choice of system boundary is too restrictive. In order to analyze a given engineering system fully it may be necessary to expand the system boundaries to include other subsystems that strongly affect the operation of the system under study. For instance, suppose a manufacturing operation has a paint shop in which finished parts are mounted on an assembly line and painted in different colors. In an initial study of the paint shop we may consider it in isolation from the rest of the plant. However, we may find that the optimal batch size and color sequence we deduce for this system are strongly influenced by the operation of the fabrication department that produces the finished parts. A decision thus has to be made whether to expand the system boundaries to include the fabrication department. An expansion of the system boundaries certainly increases the size and complexity of the composite system and thus may make the study much more difficult. Clearly, in order to make our work as engineers more manageable, we would prefer as much as possible to break down large complex systems into smaller subsystems that can be dealt with individually. However, we must recognize that this decomposition is in itself a potentially serious approximation of reality.

2.2   Performance Criterion

Given that we have selected the system of interest and have defined its boundaries, we next need to select a criterion on the basis of which the performance or design of the system can be evaluated so that the "best" design or set of operating conditions can be identified. In many engineering applications an economic criterion is selected. However, there is a considerable choice in the precise definition of such a criterion: total capital cost, annual cost, annual net profit, return on investment, cost to benefit ratio, or net present worth. In other applications a criterion may involve some technology factors, for instance, minimum production time, maximum production rate, minimum energy utilization, maximum torque, and minimum weight. Regardless of the criterion selected, in the context of optimization the best will always mean the candidate system with either the *minimum* or the *maximum* value of the performance index.

It is important to note that within the context of the optimization methods, only *one* criterion or performance measure is used to define the optimum. It is not possible to find a solution that, say, simultaneously minimizes cost and maximizes reliability and minimizes energy utilization. This again is an important simplification of reality because in many practical situations it would be desirable to achieve a solution that is best with respect to a number of different criteria. One way of treating multiple competing objectives is to select one criterion as primary and the remaining criteria as secondary. The primary criterion is then used as an optimization performance measure, while the secondary criteria are assigned acceptable minimum or maximum values and are treated as problem constraints. However, if careful consideration were not given while selecting the acceptable levels, a feasible design that satisfies all the constraints may not exist. This problem is overcome by a technique called *goal programming*, which is fast becoming a practical method for handling multiple criteria. In this method, all the objectives are assigned target levels for achievement and a relative priority on achieving these levels. Goal programming treats these targets as goals to aspire for and not as absolute constraints. It then attempts to find an optimal solution that comes as "close" as possible" to the targets in the order of specified priorities. Readers interested in multiple criteria optimization are directed to Chapter 5 of the *Operations Research Handbook* by Ravindran.[1]

2.3 Independent Variables

The third key element in formulating a problem for optimization is the selection of the independent variables that are adequate to characterize the possible candidate designs or operation conditions of the system. There are several factors that must be considered in selecting the independent variables. First, it is necessary to distinguish between variables whose values are amenable to change and variables whose values are fixed by external factors, lying outside the boundaries selected for the system in question. For instance, in the case of the paint shop, the types of parts and the colors to be used are clearly fixed by product specifications or customer orders. These are specified system parameters. On the other hand, the order in which the colors are sequenced is, within constraints imposed by the types of parts available and inventory requirements, an independent variable that can be varied in establishing a production plan.

Furthermore, it is important to differentiate between system parameters that can be treated as fixed and those that are subject to fluctuations that are influenced by external and uncontrollable factors. For instance, in the case of the paint shop equipment breakdown and worker absenteeism may be sufficiently high to influence the shop operations seriously. Clearly, variations in these key system parameters must be taken into account in the production planning problem formulation if the resulting optimal plan is to be realistic and operable.

Second, it is important to include in the formulation all of the important variables that influence the operation of the system or affect the design definition. For instance, if in the design of a gas storage system we include the height, diameter, and wall thickness of a cylindrical tank as independent variables, but exclude the possibility of using a compressor to raise the storage pressure, we may well obtain a very poor design. For the selected fixed pressure we would certainly find the least cost tank dimensions. However, by including the storage pressure as an independent variable and adding the compressor cost to our performance criterion, we could obtain a design that has a lower overall cost because of a reduction in the required tank volume. Thus the independent variables must be selected so that all important alternatives are included in the formulation. Exclusion of possible alternatives, in general, will lead to suboptimal solutions.

Finally, a third consideration in the selection of variables is the level of detail to which the system is considered. While it is important to treat all of the key independent variables, it is equally important not to obscure the problem by the inclusion of a large number of fine details of subordinate importance. For instance, in the preliminary design of a process involving a number of different pieces of equipment—pressure vessels, towers, pumps, compressors, and heat exchangers—one would normally not explicitly consider all of the fine details of the design of each individual unit. A heat exchanger may well be characterized by a heat transfer surface area as well as shell-side and tube-side pressure drops. Detail design variables such as number and size of tubes, number of tube and shell passes, baffle spacing, header type, and shell dimensions would normally be considered in a separate design study involving that unit by itself. In selecting the independent variables, a good rule to follow is to include only those variables that have a significant impact on the composite system performance criterion.

2.4   System Model

Once the performance criterion and the independent variables have been selected, then the next step in problem formulation is the assembly of the model that describes the manner in which the problem variables are related, and the performance criterion is influenced by the independent variables. In principle, optimization studies may be performed by experimenting directly with the system. Thus, the independent variables of the system or process may be set to selected values, the system operated under those conditions, and the system performance index evaluated using the observed performance. The optimization methodology would then be used to predict improved choices of the independent variable values and experiments continued in this fashion. In practice most optimization studies are carried out with the help of a model, a simplified mathematical representation of the real system. Models are used because it is too expensive or time consuming or risky to use the real system to carry out the study. Models are typically used in engineering design because they offer the cheapest and fastest way of studying the effects of changes in key design variables on system performance.

In general, the model will be composed of the basic material and energy balance equations, engineering design relations, and physical property equations that describe the physical phenomena taking place in the system. These equations will normally be supplemented by inequalities that define allowable operating ranges, specify minimum or maximum performance requirements, or set bounds on resource availabilities. In sum, the model consists of all of the elements that normally must be considered in calculating a design or in predicting the performance of an engineering system. Quite clearly the assembly of a model is a very time-consuming activity, and it is one that requires a thorough understanding of the system being considered. In simple terms, a model is a collection of equations and inequalities that define how the system variables are related and that constrain the variables to take on acceptable values.

From the preceding discussion, we observe that a problem suitable for the application of optimization methodology consists of a performance measure, a set of independent variables, and a model relating the variables. Given these rather general and abstract requirements, it is evident that the methods of optimization can be applied to a very wide variety of applications. We shall illustrate next a few engineering design applications and their model formulations.

3   APPLICATIONS OF OPTIMIZATION IN ENGINEERING

Optimization theory finds ready application in all branches of engineering in four primary areas:

1. Design of components of entire systems
2. Planning and analysis of existing operations
3. Engineering analysis and data reduction
4. Control of dynamic systems

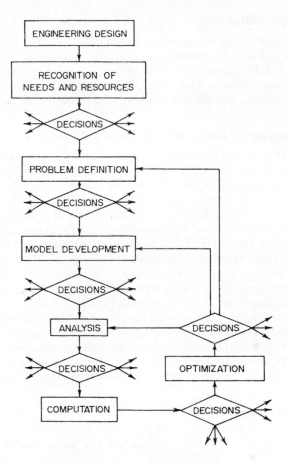

Figure 1 Optimal design process.

In this section we briefly consider representative applications from the first three areas.

In considering the application of optimization methods in design and operations, the reader should keep in mind that the optimization step is but one step in the overall process of arriving at an optimal design or an efficient operation. Generally, that overall process will, as shown in Fig. 1, consist of an iterative cycle involving synthesis or definition of the structure of the system, model formulation, model parameter optimization, and analysis of the resulting solution. The final optimal design or new operating plan will be obtained only after solving a series of optimization problems, the solution to each of which will have served to generate new ideas for further system structures. In the interest of brevity, the examples in this section show only one pass of this iterative cycle and focus mainly on preparations for the optimization step. This focus should not be interpreted as an indication of the dominant role of optimization methods in the engineering design and systems analysis process.

Optimization theory is but a very powerful tool that, to be effective, must be used skillfully and intelligently by an engineer who thoroughly understands the system under study. The primary objective of the following example is simply to illustrate the wide variety but common form of the optimization problems that arise in the design and analysis process.

3.1   Design Applications

Applications in engineering design range from the design of individual structural members to the design of separate pieces of equipment to the preliminary design of entire production facilities. For purposes of optimization the shape or structure of the system is assumed known and optimization problem reduces to the selection of values of the unit dimensions and operating variables that will yield the best value of the selected performance criterion.

Example 1 Design of an Oxygen Supply System

Description: The basic oxygen furnace (BOF) used in the production of steel is a large fed-batch chemical reactor that employs pure oxygen. The furnace is operated in a cyclic fashion: Ore and flux are charged to the unit, are treated for a specified time period and then are discharged. This cyclic operation gives rise to a cyclically varying demand rate for oxygen. As shown in Fig. 2, over each cycle there is a time interval of length t_1 of low demand rate, D_0, and a time interval $(t_2 - t_1)$ of high demand rate, D_1. The oxygen used in the BOF is produced in an oxygen plant. Oxygen plants are standard process plants in which oxygen is separated from air using a combination of refrigeration and distillation. These are highly automated plants, which are designed to deliver a fixed oxygen rate. In order to mesh the continuous oxygen plant with the cyclically operating BOF, a simple inventory system shown in Fig. 3 and consisting of a compressor and a storage tank must be designed. A number of design possibilities can be considered. In the simplest case, one could select the oxygen plant capacity to be equal to D_1, the high-demand rate. During the low-demand interval the excess oxygen could just be vented to the air. At the other extreme, one could also select the oxygen plant capacity to be just enough to produce the amount of oxygen required by the BOF over a cycle. During the low-demand interval, the excess oxygen production would then be compressed and stored for use during

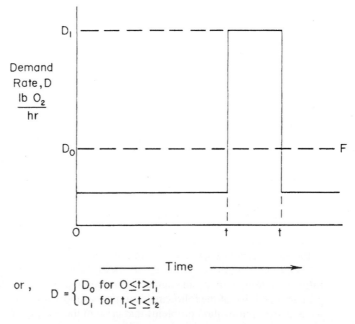

or, $D = \begin{cases} D_0 & \text{for } 0 \leq t \geq t_1 \\ D_1 & \text{for } t_1 \leq t \leq t_2 \end{cases}$

Figure 2   Oxygen demand cycle.

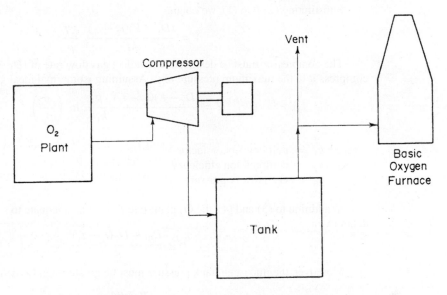

Figure 3 Design of oxygen production system.

the high-demand interval of the cycle. Intermediate designs could involve some combination of venting and storage of oxygen. The problem is to select the optimal design.

Formulation: The system of concern will consist of the O_2 plant, the compressor, and the storage tank. The BOF and its demand cycle are assumed fixed by external factors. A reasonable performance index for the design is the total annual cost, which consists of the oxygen production cost (fixed and variable), the compressor operating cost, and the fixed costs of the compressor and of the storage vessel. The key independent variables are the oxygen plant production rate F(lb O_2/h), the compressor and storage tank design capacities, H(hp) and V(ft^3), respectively, and the maximum tank pressure, p(psia). Presumably the oxygen plant design is standard, so that the production rate fully characterizes the plant. Similarly, we assume that the storage tank will be of a standard design approved for O_2 service.

The model will consist of the basic design equations that relate the key independent variables. If I_{max} is the maximum amount of oxygen that must be stored, then using the corrected gas law we have

$$V = \frac{I_{max}}{M}\frac{RT}{p}z \tag{1}$$

where R = gas constant
T = gas temperature (assume fixed)
Z = compressability factor
M = molecular weight of O_2

From Fig. 2, the maximum amount of oxygen that must be stored is equal to the area under the demand curve between t_1 and t_2 and D_1 and F. Thus

$$I_{max} = (D_1 - F)(t_2 - t_1) \tag{2}$$

Substituting (2) into (1), we obtain

$$V = \frac{(D_1 - F)(t_2 - t_1)}{M} \frac{RT}{p} z \qquad (3)$$

The compressor must be designed to handle a gas flow rate of $(D_1 - F)(t_2 - t_1)/t_1$ and to compress it to the maximum pressure of p. Assuming isothermal ideal gas compression,[2]

$$H = \frac{(D_1 - F)(t_2 - t_1)}{t_1} \frac{RT}{k_1 k_2} \ln\left(\frac{p}{p_0}\right) \qquad (4)$$

where k_1 = unit conversion factor
k_2 = compression efficiency
p_0 = O_2 delivery pressure

In addition to (3) and (4), the O_2 plant rate F must be adequate to supply the total oxygen demand, or

$$F \geq \frac{D_0 t + D_1 (t_2 - t_1)}{t_2} \qquad (5)$$

Moreover, the maximum tank pressure must be greater than O_2 delivery pressure

$$p \geq p_0 \qquad (6)$$

The performance criterion will consist of the oxygen plant annual cost,

$$C_1(\$/\text{yr}) = a_1 + a_2 F \qquad (7)$$

where a_1 and a_2 are empirical constants for plants of this general type and include fuel, water, and labor costs.

The capital cost of storage vessels is given by a power law correlation,

$$C_2(\$) = b_1 V^{b_2} \qquad (8)$$

where b_1 and b_2 are empirical constants appropriate for vessels of a specific construction.

The capital cost of compressors is similarly obtained from a correlation,

$$C_3(\$) = b_3 H^{b_4} \qquad (9)$$

The compressor power cost will, as an approximation, be given by $b_5 t_1 H$ where b_5 is the cost of power.

The total cost function will thus be of the form

$$\text{Annual cost} = a_1 + a_2 F + d(b_1 V^{b_2} + b_3 H^{b_4}) + N b_5 t_1 H \qquad (10)$$

where N = number of cycles per year
d = approximate annual cost factor

The complete design optimization problem thus consists of the problem of minimizing (10), by the appropriate choice of F, V, H, and p, subject to Eqs. (3) and (4) as well as inequalities (5) and (6).

The solution of this problem will clearly be affected by the choice of the cycle parameters $(N, D_0, D_1, t_1, \text{and } t_2)$, the cost parameters $(a_1, a_2, b_1 - b_5, \text{and } d)$, as well as the physical parameters $(T, p_0, k_2, z, \text{and } M)$.

In principle, we could solve this problem by eliminating V and H from (10) using (3) and (4), thus obtaining a two-variable problem. We could then plot the contours of the cost function (10) in the plane of the two variables F and p, impose the inequalities (5) and (6), and determine the minimum point from the plot. However, the methods discussed in subsequent sections allow us to obtain the solution with much less work. For further details and a study of solutions for various parameter values, the reader is invited to consult Ref. 3.

The preceding example presented a preliminary design problem formulation for a system consisting of several pieces of equipment. The next example illustrates a detailed design of a single structural element.

Example 2 Design of a Welded Beam

Description. A beam A is to be welded to a rigid support member B. The welded beam is to consist of 1010 steel and is to support a force F of 6000 lb. The dimensions of the beam are to be selected so that the system cost is minimized. A schematic of the system is shown in Fig. 4.

Formulation. The appropriate system boundaries are quite self-evident. The system consists of the beam A and the weld required to secure it to B. The independent or design variables in this case are the dimensions $h, l, t,$ and b as shown in Fig. 4. The length L is assumed to be specified at 14 in. For notational convenience we redefine these four variables in terms of the vector of unknowns $\mathbf{x}$,

$$\mathbf{x} = [x_1, x_2, x_3, x_4]^T = [h, l, t, b]^T$$

The performance index appropriate to this design is the cost of a weld assembly. The major cost components of such an assembly are (a) setup labor cost, (b) welding labor cost, and (c) material cost:

$$F(x) = c_0 + c_1 + c_2 \tag{11}$$

where $F(x)$ = cost function
   c_0 = setup cost
   c_1 = welding labor cost
   c_2 = material cost

Setup Cost: c_0. The company has chosen to make this component a weldment because of the existence of a welding assembly line. Furthermore, assume that fixtures for setup and holding of the bar during welding are readily available. The cost c_0 can, therefore, be ignored in this particular total cost model.

Welding Labor Cost: c_1. Assume that the welding will be done by machine at a total cost of $10 per hour (including operating and maintenance expense). Furthermore, suppose that the machine can lay down 1 in.^{3} of weld in 6 min. Therefore, the labor cost is

$$c_1 = \left(10 \; \frac{\$}{h}\right) \left(\frac{1\,h}{60\,\min}\right) \left(6 \; \frac{\min}{\text{in.}^3}\right) V_w = 1 \; \left(\frac{\$}{\text{in.}^3}\right) V_w$$

where V_w is the weld volume, in.^{3}

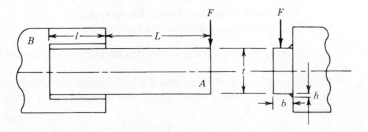

Figure 4   Welded beam.

Material Cost: c_2

$$c_2 = c_3 V_w + c_4 V_B$$

where c_3 = \$/volume of weld material = $(0.37)(0.283)$ (\$/in.^{3})
c_4 = \$/volume of bar stock = $(017)(0.283)$(\$/in.^{3})
V_B = volume of bar A(in.^{3})

From the geometry,

$$V_w = 2\left(\frac{1}{2}h^2 l\right) - h^2 l \quad \text{and} \quad V_B = tb(L + l)$$

so

$$c_2 = c_3 h^2 l + c_4 tb(L + l)$$

Therefore, the cost function becomes

$$F(x) = h^2 l + c_3 h^2 l + c_4 tb(L + l) \tag{12}$$

or, in terms of the x variables

$$F(x) = (l + c_3)x_1^2 x_2 + c_4 x_3 x_4 (L + x_2) \tag{13}$$

Note all combinations of x_1, x_2, x_3, and x_4 can be allowed if the structure is to support the load required. Several functional relationships between the design variables that delimit the region of feasibility must certainly be defined. These relationships, expressed in the form of inequalities, represent the design model. Let us first define the inequalities and then discuss their interpretation.

The inequities are

$$g_1(x) = \tau_d - \tau(x) \geq 0 \tag{14}$$

$$g_2(x) = \sigma_d - \sigma(x) \geq 0 \tag{15}$$

$$g_3(x) = x_4 - x_1 \geq 0 \tag{16}$$

$$g_4(x) = x_2 \geq 0 \tag{17}$$

$$g_5(x) = x_3 \geq 0 \tag{18}$$

$$g_6(x) = P_c(x) - F \geq 0 \tag{19}$$

$$g_7(x) = x_1 - 0.125 \geq 0 \tag{20}$$

$$g_8(x) = 0.25 - \text{DEL}(x) \geq 0 \tag{21}$$

where τ_d = design shear stress of weld
$\tau(x)$ = maximum shear stress in weld; a function of x
σ_d = design normal stress for beam material
$\sigma(x)$ = maximum normal stress in beam; a function of x
$P_c(x)$ = bar buckling load; a function of x
$\text{DEL}(x)$ = bar end deflection; a function of x

In order to complete the model it is necessary to define the important stress states.

Weld stress. $\tau(x)$. Following Shigley,[4] the weld shear stress has two components, τ' and τ'', where τ' is the primary stress acting over the weld throat area and τ'' is a secondary torsional stress:

$$\tau' = \frac{F}{\sqrt{2}x_1 x_2} \quad \text{and} \quad t'' = \frac{MR}{J}$$

with $M = F\left[L + \left(\frac{x_2}{2}\right)\right]$

$R = \left[\left(\frac{x_2^2}{4}\right) + \left(\frac{x_3 + x_1}{2}\right)^2\right]^{1/2}$

$J = 2\{0.707 x_1 x_2 [x_2^2/12 + (x_3 + x_1)/2^2]\}$

M = moment F about the center of gravity of the weld group

J = polar moment of inertia of the weld group

Therefore, the weld stress τ becomes

$$\tau(x) = [(\tau')^2 + 2\,\tau'\,\tau''\,\cos\,\theta + (\tau'')^2]^{1/2}$$

where $\cos\,\theta = x_2/2R$.

Bar Bending Stress: $\sigma(x)$. The maximum bending stress can be shown to be equal to

$$\sigma(x) = \frac{6FL}{x_4 x_3^2}$$

Bar Buckling Load: $P_c(x)$. If the ratio $t/b = x_3/x_4$ grows large, there is a tendency for the bar to buckle. Those combinations of x_3 and x_4 that will cause this buckling to occur must be disallowed. It has been shown[5] that for narrow rectangular bars, a good approximation to the buckling load is

$$P_c(x) = \frac{4.013\,\sqrt{EI\alpha}}{L^2}\left(1 - \frac{x_3}{2L}\sqrt{\frac{EI}{\alpha}}\right)$$

where E = Young's modulus = 30×10^6 psi

$I = \frac{1}{12} x_3 x_4^3$

$\alpha = \frac{1}{3} G x_3 x_4^3$

G = shearing modulus = 12×10^6 psi

Bar deflection: [DEL(x)]. To calculate the deflections assume the bar to be a cantilever of length L. Thus,

$$\text{DEL}_{(x)} = \frac{4FL^3}{E x_3^3 x_4}$$

The remaining inequalities are interpreted as follows.

Constraint g_3 states that it is not practical to have the weld thickness greater than the bar thickness; and g_4 and g_5 are nonnegativity restrictions on x_2 and x_3. Note that the nonnegativity of x_1 and x_4 are implied by g_3 and g_7. Constraint g_6 ensures that the buckling load is not exceeded. Inequality g_7 specifies that it is not physically possible to produce an extremely small weld.

Finally, the two parameters τ_d and σ_d in g_1 and g_2 depend on the material of construction. For 1010 steel $\tau_d = 13{,}600$ psi and $\sigma_d = 30{,}000$ psi are appropriate.

The complete design optimization problem thus consists of the cost function (13) and the complex system of inequalities that result when the stress formulas are substituted into (14) through (21). All of these functions are expressed in terms of four independent variables.

This problem is sufficiently complex that graphical solution is patently infeasible. However, the optimum design can readily be obtained numerically using the methods of subsequent sections. For a further discussion of this problem and its solution the reader is directed to Ref. 6.

3.2   Operations and Planning Applications

The second major area of engineering application of optimization is found in the tuning of existing operations. We shall discuss an application of goal programming model for machinability data optimization in metal cutting.[7]

Example 3 An Economic Machining Problem with Two Competing Objectives

Consider a single-point, single-pass turning operation in metal cutting wherein an optimum set of cutting speed and feed rate is to be chosen that balances the conflict between metal removal rate and tool life as well as being within the restrictions of horsepower, surface finish, and other cutting conditions. In developing the mathematical model of this problem, the following constraints will be considered for the machining parameters:

Constraint 1: *Maximum Permissible Feed*

$$f \leq f_{max} \tag{22}$$

where f is the feed in inches per revolution; and f_{max} is usually determined by a cutting force restriction or by surface finish requirements.[8]

Constraint 2: *Maximum Cutting Speed Possible*. If v is the cutting speed in surface feet per minute, then

$$v \leq v_{max} \tag{23}$$

where $v_{max} = \frac{\pi D N_{max}}{12}$.

$N_{max} =$ maximum spindle speed available on the machine

Constraint 3: *Maximum Horsepower Available*. If P_{max} is the maximum horsepower available at the spindle, then

$$v f^\alpha \leq \frac{P_{max}(33,000)}{c_t d_c^\beta}$$

where α, β, and c_t are constants,[8] and d_c is the depth of the cut in inches, which is fixed at a given value. For a given P_{max}, c_t, β, and d_c, the right-hand side of the above constraint will be a constant. Hence, the horsepower constraint can be written simply as

$$v f^\alpha \leq \text{constant} \tag{24}$$

Constraint 4: *Nonnegativity Restrictions on Feed Rate and Speed*

$$v, f \geq 0 \tag{25}$$

In optimizing metal cutting there are a number of optimality criteria that can be used. Suppose we consider the following objectives in our optimization: (i) maximize metal removal (MRR), and (ii) maximize tool life (TL). The expression for MRR is

$$\text{MRR} = 12vfd_c \text{ in.}^3/\text{min} \tag{26}$$

The TL for a given depth of cut is given by

$$\text{TL} = \frac{A}{v^{1/n}f^{1/n_1}} \tag{27}$$

where A, n, and n_1 are constants. We note that the MRR objective is directly proportional to feed and speed, while the TL objective is inversely proportional to feed and speed. In general, there is no single solution to a problem formulated in this way, since MRR and TL are competing objectives and their respective maxima must include some compromise between the maximum of MRR and the maximum of TL.

A Goal Programming Model. Goal programming is a technique specifically designed to solve problems involving complex, usually conflicting, multiple objectives. Goal programming requires the user to select a set of goals (which may or may not be realistic) that ought to be achieved (if possible) for the various objectives. It then uses preemptive weights or priority factors to rank the different goals and tries to obtain an optimal solution satisfying as many goals as possible. For this, it creates a single objective function that minimizes the deviations from the stated goals according to their relative importance.

Before we discuss the goal programming formulation of the machining problem, we should discuss the difference between the terms *real constraint* and *goal constraint* (or simply *goal*) as used in goal programming models. The real constraints are absolute restrictions placed on the behavior of the design variables, while the goal constraints are conditions one would like to achieve but are not mandatory. For instance, a real constraint given by

$$x_1 + x_2 = 3$$

requires all possible values of $x_1 + x_2$ to always equal 3. As opposed to this, if we simply had a goal requiring $x_1 + x_2 = 3$, then this is not mandatory, and we can choose values of x_1, x_2 such that $x_1 + x_2 \geq 3$ as well as $x_1 + x_2 \leq 3$. In a goal constraint positive and negative deviational variables are introduced as follows:

$$x_1 + x_2 + d_1^- - d_1^+ = 3d_1^- \text{ and } d_1^+ \geq 0$$

Note that if $d_1^- > 0$, then $x_1 + x_2 < 3$; and, if $d_1^+ > 0$, then $x_1 + x_2 > 3$. By assigning suitable preemptive weights on d_1^- and d_1^+, the model will try to achieve the sum $x_1 + x_2$ as close as possible to 3.

Returning to the machining problem with competing objectives, suppose that management considers that a given single-point, single-pass turning operation will be operating at an acceptable efficiency level if the following goals are met as closely as possible:

1. The MRR must be greater than or equal to a given rate $M_1(\text{in.}^3 / \text{min})$.

2. The tool life must equal $T_1(\text{min})$.

In addition, management requires that a higher priority be given to achieving the first goal than the second.

The goal programming approach may be illustrated by expressing each of the goals as goal constraints as shown below. Taking the MRR goal first,

$$12vfd_c + d_1^- - d_1^+ = M_1$$

where d_1^- represents the amount by which the MRR goal is underachieved, and d_1^+ represents any overachievement of the MRR goal. Similarly, the TL goal can be expressed as

$$\frac{A}{v^{1/n}f^{1/n1}} + d_2^- - d_2^+ = T_1$$

Since the objective is to have an MRR of at least M_1, the objective function must be set up so that a high penalty will be assigned to the underachievement variable d_1^-. No penalty will be assigned to d_1^+.

In order to achieve a tool of life of T_1, penalties must be associated with both d_2^- and d_2^+ so that both of these variables are minimized to their fullest extent. The relative magnitudes of these penalties must reflect the fact that the first goal is considered to be more important than the second. Accordingly, the goal programming objective function for this problem is

$$\text{Minimize } z = P_1 d_1^- + P_2(d_2^- + d_2^+)$$

where P_1 and P_2 are nonnumerical preemptive priority factors such that $P_1 >>> P_2$ (i.e., P_1 is infinitely larger than P_2). With this objective function every effort will be made to satisfy completely the first goal before any attempt is made to satisfy the second.

In order to express the problem as a linear goal programming problem, M_1 is replaced by M_2, where

$$M_2 = \frac{M_1}{12d_c}$$

The goal T_1 is replaced by T_2, where

$$T_2 = \frac{A}{T_1}$$

and logarithms are taken of the goals and constraints. The problem can then be stated as follows:

$$\text{Minimize} \quad z = P_1 d_1^- + P_2(d_2^- + d_2^+)$$

Subject to,

$$\text{(MRR goal)} \quad \log v + \log f + d_1^- - d_1^+ = \log M_2$$

$$\text{(TL goal)}$$

$$(1/n) \log v + (1/n_1) \log f + d_2^- - d_2^+ = \log T_2$$

$$(f_{max} \text{constraint}) \quad \log f \leq \log f_{max}$$

$$(V_{max} \text{constraint}) \quad \log v \leq \log v_{max}$$

$$\text{(Horsepower constraint)} \quad \log v + \alpha \log f \leq log\ constant$$

$$\log v, \log f, d_1^-, d_1^+, d_2^-, d_2^+ \geq 0$$

We would like to reemphasize here that the last three inequalities are real constraints on feed, speed, and horsepower that must be satisfied at all times, while the equations for MRR and TL are simply goal constraints. For a further discussion of this problem and its solution, see Ref. 7. An efficient algorithm and a computer code for solving linear goal programming problems is given in Ref. 9. Readers interested in other optimization models in metal cutting should see Ref. 10. The solution of nonlinear goal programming problems is discussed in Refs. 11 and 12. For a detailed discussion of other approaches to solving multiple objective optimization problems the reader is referred to Chapter 5 in Ref. 1.

3.3   Analysis and Data Reduction Applications

A further fertile area for the application of optimization techniques in engineering can be found in nonlinear regression problems as well as in many analysis problems arising in engineering science. A very common problem arising in engineering model development is the need to determine the parameters of some semitheoretical model given a set of experimental data. This data reduction or regression problem inherently transforms to an optimization problem because the model parameters must be selected so that the model fits the data as closely as possible.

Suppose some variable y is assumed to be dependent on an independent variable x and related to x through a postulated equation $y = f(x, \theta_1, \theta_2)$, which depends on two parameters θ_1 and θ_2. To establish the appropriate values of θ_1, θ_2, we run a series of experiments in which we adjust the independent variable x and measure the resulting y. As a result of a series of N experiments covering the range of x interest, a set of y and x values (y_i, x_i), $i = 1, \ldots, N$, is available. Using these data we now try to "fit" our function to the data by adjusting θ_1 and θ_2 until we get a "good fit." The most commonly used measure of a good fit is the *least squares criterion*,

$$L(\theta_1, \theta_2) = \sum_{i=1}^{N} [y_i - f(x_i, \theta_1, \theta_2)]^2 \tag{28}$$

The difference $y_i - f(x_i, \theta_1, \theta_2)$ between the experimental value y_i and the predicted value $f(x_i, \theta_1, \theta_2)$ measures how close our model prediction is to the data and is called the *residual*. The sum of the squares of the residuals at all the experimental points gives an indication of goodness of fit. Clearly, if $L(\theta_1, \theta_2)$ is equal to zero, then the choice of θ_1, θ_2 has led to a perfect fit; the data points fall exactly on the predicted curve. The data fitting problem can thus be viewed as an optimization problem in which $L(\theta_1, \theta_2)$ is minimized by appropriate choice of θ_1, θ_2.

Example 4 Nonlinear Curve Fitting

Description. The pressure–molar volume–temperature relationship of real gases is known to deviate from that predicted by the ideal gas relationship

$$Pv = RT$$

where P = pressure (atm)
v = molar volume (cm^3/g $\cdot$ mol)
T = temperature (K)
R = gas constant (82.06 atm $\bullet$ cm^3/g $\bullet$ mol K)

The semiempirical Redlich–Kwong equation is intended to direct for the departure from ideality but involves two empirical constants a and b whose values are best determined from experimental data.

$$P = \frac{RT}{v - b} - \frac{a}{T^{1/2}v(v + b)} \qquad (29)$$

A series of *PvT* measurements listed in Table 1 are made for CO_2, from which a and b are to be estimated using nonlinear regression.

Formulation. Parameters a and b will be determined by minimizing the least squares function (28). In the present case, the function will take the form

$$\sum_{i=1}^{8} \left[P_i - \frac{RT_i}{v_i - b} + \frac{a}{T^{1/2}v_i \left(v_i + b \right)} \right]^2 \qquad (30)$$

where P_i is the experimental value at experiment i, and the remaining two terms correspond to the value of P predicted from Eq. (29) for the conditions of experiment i for some selected

Table 1   Pressure–Molar Volume–Temperature Data for CO_2

Experiment Number	P(atm)	v(cm^3/ g.mol)	T(K)
1	33	500	273
2	43	500	323
3	45	600	373
4	26	700	273
5	37	600	323
6	39	700	373
7	38	400	273
8	63.6	400	373

value of the parameters a and b. For instance, the term corresponding to the first experimental point will be

$$\left(33 - \frac{82.06\,(273)}{500 - b} + \frac{a}{(273)^{1/2}(500)(500 + b)}\right)^2$$

Function (30) is thus a two-variable function whose value is to be minimized by appropriate choice of the independent variables a and b. If the Redlich–Kwong equation were to precisely match the data, then at the optimum the function (30) would be exactly equal to zero. In general, because of experimental error and because the equation is too simple to accurately model the CO_2 nonidealities, Eq. (30) will not be equal to zero at the optimum. For instance, the optimal values of $a = 6.377 \times 10^7$ and $b = 29.7$ still yield a squared residual of 9.7×10^{-2}.

4 STRUCTURE OF OPTIMIZATION PROBLEMS

Although the application problems discussed in the previous section originate from radically different sources and involve different systems, at root they have a remarkably similar form. All four can be expressed as problems requiring the minimization of a real-valued function $f(x)$ of an $N-$ component vector argument $x = (x_1, x_2,$ and $x_N)$ whose values are restriced to satisfy a number of real-valued equations $h_k(x) = 0$, a set of inequalities $g_j(x) \geq 0$, and the variable bounds $x_i^{(U)} \geq x_i \geq x_i^{(L)}$. In subsequent discussions we will refer to the function $f(x)$ as the *objective function*, to the equations $h_k(x) = 0$ as *equality constraints,* and to the inequalities $g_j(x) \geq 0$ as the *inequality constraints.* For our purposes, these problem functions will always be assumed to be real valued, and their number will always be finite.

The general problem,

$$\text{Minimize } f(x)$$

$$\text{Subject to } h_k(x) = 0 \qquad\qquad k = 1, \ldots, K$$

$$g_j(x) \geq 0 \qquad\qquad j = 1, \ldots, J$$

$$x_i^{(L)} \geq x_i \geq x_i^{(L)} \qquad\qquad i = 1, \ldots, N$$

is called *constrained* optimization problem. For instance, Examples 1, 2, and 3 are all constrained problems. The problem in which there are no constraints, that is,

$$J = K = 0$$

and

$$x_i^{(U)} = -x_i^{(L)} = \infty \qquad i = 1, \ldots, N$$

is called the *unconstrained* optimization problem. Example 4 is an unconstrained problem. Optimization problems can be classified further based on the structure of the functions $f, h_k,$ and g_j and on the dimensionality of x. Figure 5 illustrates on one such classification. The basic subdivision is between unconstrained and constrained problems. There are two important classes of methods for solving the unconstrained problems. The direct search methods require only that the objective function be evaluated at different points, at least through experimentation. Gradient-based methods require the analytical form of the objective function and its derivatives.

An important class of constrained optimization problems is *linear programming*, which requires both the objective function and the constraints to be linear functions. Out of all optimization models, linear programming models are the most widely used and accepted in practice. Professionally written software programs are available from all computer manufacturers for solving very large linear programming problems. Unlike the other optimization problems that require special solution methods based on the problem structure, linear programming has just one common algorithm, known as the *simplex method*, for solving all types of linear

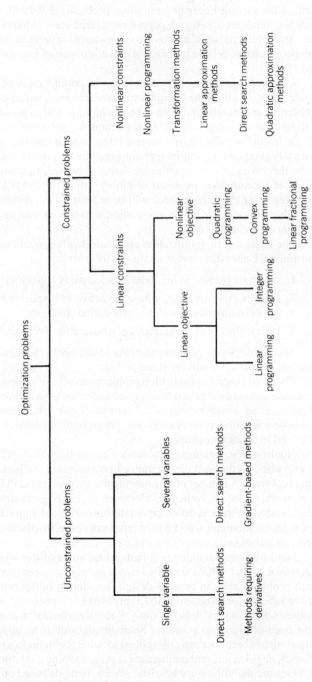

Figure 5 Classification of optimization problems.

113

programming problems. This essentially has contributed to the successful applications of linear programming model in practice. In 1984, Narendra Karmarkar,[13] an AT&T researcher, developed an interior point algorithm, which was claimed to be 50 times faster than the simplex method for solving linear programming problems. By 1990, Karmarkar's seminal work had spawned hundreds of research papers and a large class of interior point methods. It has become clear that while the initial claims are somewhat exaggerated, interior point methods do become competitive for very large problems. For a discussion of interior point methods, see Refs. 14, 15, and 16.

Integer programming (IP) is another important class of linearly constrained problems where some or all of the design variables are restricted to be integers. But solutions of IP problems are generally difficult, time-consuming, and expensive. Hence, a practical approach is to treat all the integer variables as continuous, solve the associated LP problem, and round off the fractional values to the nearest integers such that the constraints are not violated. This generally produces a good integer solution close to the optimal integer solution, particularly when the values of the variables are large. However, such an approach would fail when the values of the variables are small or binary valued (0 or 1). A good rule of thumb is to treat any integer variable whose value will be at least 20 as continuous and use special-purpose IP algorithms for the rest. For a complete discussion of integer programming applications and algorithms, see Refs. 16–19.

The next class of optimization problems involves nonlinear objective functions and linear constraints. Under this class we have the following:

1. Quatratic programming, whose objective is a quadratic function
2. Convex programming, whose objective is a special nonlinear function satisfying an important mathematical property called *convexity*.
3. Linear fractional programming, whose objective is the ratio of two linear functions

Special-purpose algorithms that take advantage of the particular form of the objective functions are available for solving these problems.

The most general optimization problems involve nonlinear objective functions and nonlinear constraints and are generally grouped under the term *nonlinear programming*. The majority of engineering design problems fall into this class. Unfortunately, there is no single method that is best for solving every nonlinear programming problem. Hence a host of algorithms are reviewed in the next section.

Nonlinear programming problems wherein the objective function and the constraints can be expressed as the sum of generalized polynomial functions are called *geometric programming* problems. A number of engineering design problems fall into the geometric programming framework. Since its earlier development in 1961, geometric programming has undergone considerable theoretical development, has experienced a proliferation of proposals for numerical solution techniques, and has enjoyed considerable practical engineering applications (see Refs. 20 and 21).

Nonlinear programming problems where some of the design variables are restricted to be discrete or integer valued are called *mixed integer nonlinear programming* (MINLP) problems. Such problems arise in process design, simulation optimization, industrial experimentation, and reliability optimization. MINLP problems are generally more difficult to solve since the problems have several local optima. Recently, *simulated annealing* and *genetic algorithms* have been emerging as powerful heuristic algorithms to solve MINLP problems. Simulated annealing has been successfully applied to solve problems in a variety of fields, including mathematics, engineering, and mathematical programming (see, for example, Refs. 19 and 22–24).

Genetic algorithms are heuristic search methods based on the two main principles of natural genetics, namely entities in a population reproduced to create offspring and the survival of the fittest (see, for example, Refs. 23, 25, and 26).

5   OVERVIEW OF OPTIMIZATION METHODS

Optimization methods can be viewed as nothing more than numerical hill-climbing procedures in which the objective function presenting the topology of the hill is searched to identify the highest point—or maximum—subject to constraining relations that might be equality constraints (stay on winding path) or inequality constraints (stay within fence boundaries). While the constraints do serve to reduce the area that must be searched, the numerical calculations required to ensure that the search stays on the path or within the fences generally do constitute a considerable burden. Accordingly, optimization methods for unconstrained problems and methods for linear constraints are less complex than those designed for nonlinear constraints. In this section, a selection of optimization techniques representative of the main families of methods will be discussed. For a more detailed presentation of individual methods the reader is invited to consult Ref. 27.

5.1   Unconstrained Optimization Methods

Methods for unconstrained problems are divided into those for single-variable functions and those appropriate for multivariable functions. The former class of methods are important because single-variable optimization problems arise commonly as subproblems in the solution of multivariable problems. For instance, the problem of minimizing a function $f(x)$ for a point x^0 in a direction of d (often called a *line search*) can be posed as a minimization problem in the scalar variable α:

$$\text{Minimize} f(x^0 + \alpha d)$$

Single Variable Methods.   The methods are roughly divided into region elimination methods and point estimation methods. The former use comparison of function values at selected trial points to reject intervals within which the optimum of the function does not lie. The latter typically use polynomial approximating functions to estimate directly the location of the optimum. The simplest polynomial approximating function is the quadratic

$$\bar{f}(x) = ax^2 + bx + c$$

whose coefficients a, b, c can be evaluated readily from those trial values of the actual function. The point at which the derivative of $\bar{f}$ is zero is used readily to predict the location of the optimum of the true function

$$\tilde{x} = -b/2a$$

The process is repeated using successively improved trial values until the differences between successive estimates $\tilde{x}$ become sufficiently small.

Multivariable Unconstrained Methods.   These algorithms can be divided into direct search methods and gradient-based methods. The former methods only use direct function values to guide the search, while the latter also require the computation of function gradient and, in some cases, second derivative values. Direct search methods in widespread use in engineering applications include the simplex search, the pattern search method of Hooke and Jeeves, random-sampling-based methods, and the conjugate directions method of Powell (see Chapter 3 of Ref. 27). All but the last of these methods make no assumptions about the smoothness of the function contours and hence can be applied to both discontinuous and discrete-valued objective functions.

Gradient-based methods can be grouped into the classical methods of steepest descent (Cauchy) and Newton's method and the modern quasi-Newton methods such as the conjugate gradient, Davidson–Fletcher–Powell, and Broyden–Fletche–Shamo algorithms. All gradient-based methods employ the first derivative or gradient of the function at the current best solution estimate $\bar{x}$ to compute a direction in which the objective function value is guaranteed to decrease (a descent direction). For instance, Cauchy's classical method used the direction.

$$d = -\nabla f(\bar{x})$$

followed by a line search from $\bar{x}$ in this direction. In Newton's method the gradient vector is premultiplied by the matrix of second derivatives to obtain an improved direction vector

$$d = -(\nabla^2 f(\bar{x})^{-1} \nabla f(\bar{x})$$

which in theory at least yields very good convergence behavior. However, the computation of $\nabla^2 f$ is often too burdensome for engineering applications. Instead, in recent years quasi-Newton methods have found increased application. In these methods, the direction vector is computed as

$$d = -H\nabla f(\bar{x})$$

where H is a matrix whose elements are updated as the iterations proceed using only values of gradient and function value difference from successive estimates. Quasi-Newton methods differ in the details of H updating, but all use the general form

$$H^{n+1} = H^n + C^n$$

where H^n is the previous value of H and C^n is a suitable correction matrix. The attractive feature of this family of methods is that the convergence rates approaching those of Newton's method are attained without the need for computing $\nabla^2 f$ or solving the linear equation set

$$\nabla^2 f(\bar{x})d = -f(\bar{x})$$

to obtain d. Recent developments in these methods have focused on strategies for eliminating the need for detailed line searching along the direction vectors and on enhancements for solving very large programs. For a detailed discussion of quasi-Newton methods the reader is directed to Refs. 27 and 28.

5.2   Constrained Optimization Methods

Constrained optimization methods can be classified into those applicable to totally linear or at least linearly constrained problems and those applicable to general nonlinear problems. The linear or linearly constrained problems can be well solved using methods of linear programming and extensions, as discussed earlier. The algorithms suitable for general nonlinear problems comprise four broad categories of methods:

1. Direct search methods that use only objective and constraint function values
2. Transformation methods that use constructions that aggregate constraints with the original objective function to form a single composite unconstrained function
3. Linearization methods that use linear approximations of the nonlinear problem functions to produce efficient search directions
4. Successive quadratic programming methods that use quasi-Newton constructions to solve the general problem via a series of subproblems with quadratic objective function and linear constraints

Direct Search.   The direct search methods essentially consist of extensions of unconstrained direct search procedures to accommodate constraints. These extensions are generally only possible with inequality constraints or linear equality constraints. Nonlinear equalities must be treated by implicit or explicit variable elimination. That is, each equality constraint is either explicitly solved for a selected variable and used to eliminate that variable from the search or the equality constraints are numerically solved for values of the dependent variables for each trial point in the space of the independent variables.

For example, the problem

$$\text{Minimize } f(x) = x_1\, x_2\, x_3$$

$$\text{Subject to } h_1(x) = x_1 + x_2 + x_3 - 1 = 0$$

$$= h_2(x) = x_1^2 x_3 + x_2 x_3^2 + x_2^{-1} - 2 = 0$$

$$= 0 \le (x_1,\, x_3) \le \frac{1}{2}$$

involves two equality constraints and hence can be viewed as consisting of two dependent variables and one independent variable. Clearly, h_1 can be solved for x_1 to yield

$$x_1 = 1 - x_2 - x_3$$

Thus, on substitution the problem reduces to

$$\text{Minimize } (1 - x_2 - x_3)x_2 x_3$$

$$\text{Subject to } (1 - x_2 - x_3)^2 x_3 + x_2 x_3^2 + x_2^{-1} - 2 = 0$$

$$0 \le 1 - x_2 - x_3 \le \frac{1}{2}$$

$$0 \le x_3 \le \frac{1}{2}$$

Solution of the remaining equality constraint for one variable, say x_3, in terms of the other is very difficult. Instead, for each value of the independent variable x_2, the corresponding value of x_3 would have to be calculated numerically using some root-finding method.

Some of the more widely used direct search methods include the adaptation of the simplex search due to Box (called the *complex method*), various direct random-sampling-type methods, and combined random sampling/heuristic procedures such as the combinatorial heuristic method[29] advanced for the solution of complex optimal mechanism design problems.

A typical direct sampling procedure is given by the formula

$$x_i p = \bar{x}_i \times z_i\, (2r - 1)^k \qquad \text{for each variable } x_i, = i = 1,\, \ldots, n$$

where $\bar{x}_i$ = the current best value of variable i

z_i = the allowable range of the variable i

r = a random variable uniformly distributed on the interval 0-1

k = an adaptive parameter whose value is adjusted based on past successes or failures in the search

For given $\bar{x}_i$, z, and k, r is sampled N times and the new point x^p evaluated. If x^p satisfies all constraints, it is retained; if it is infeasible, it is rejected and a new set of Nr values is generated. If x^p is feasible, $f(x^p)$ is compared to $f(\bar{x})$, and if improvement is found, x^p replaces $\bar{x}$. Otherwise x^p is rejected. The parameter k is an adaptive parameter whose value will regulate the contraction or expansion of the sampling region. A typical adjustment procedure for k might be to increase k by 2 whenever a specified number of improved points is found after a certain number of trials.

The general experience with direct search and especially random-sampling-based methods for constrained problems is that they can be quite effective for severely nonlinear problems that involve multiple local minima but are of low dimensionality.

Transformation Methods. This family consists of strategies for converting the general constrained problem to a parametrized unconstrained problem that is solved repeatedly for successive values of the parameters. The approaches can be grouped into the penalty/barrier function constructions, exact penalty methods, and augmented Lagrangian methods. The classical penalty function approach is to transform the general constrained problem to the form

$$P(x, R) = f(x) + \Omega(R, g(x), h(x))$$

where R = the penalty parameter
Ω = the penalty term

The ideal penalty function will have the property that

$$P(x, R) = \begin{cases} f(x) & \text{if } x \text{ is feasible} \\ \infty & \text{if } x \text{ is infeasible} \end{cases}$$

Given this idealized construction, $P(x, R)$ could be minimized using any unconstrained optimization method, and, hence, the underlying constrained problem would have been solved. In practice such radical discontinuities cannot be tolerated from a numerical point of view, and, hence, practical penalty functions use penalty terms of the form

$$\Omega(R, g, h) = R\left(\sum_k h_k(x)\right)^2 + R\left(\sum (\min(0, g_j(x)))^2\right)$$

A series of unconstrained minimizations of $P(x, R)$ with different values of R are carried out beginning with a low value of R(say $R = 1$) and progressing to very large values of R. For low values of R, the unconstrained minima of $P(x, R)$ obtained will involve considerable constraint violations. As R increases, the violations decrease until in the limit as $R \to \infty$, the violations will approach zero. A large number of the different forms of the Ω function have been proposed; however, all forms share the common feature that a sequence of problems must be solved and that, as the penalty parameter R becomes large, the penalty function becomes increasingly distorted and thus its minimization becomes increasingly more difficult. As a result the penalty function approach is best used for modestly sized problems (2–10 variables), few nonlinear equalities (2–5), and a modest number of inequalities. In engineering applications, the unconstrained subproblems are most commonly minimized using direct search methods, although successful use of quasi-Newton methods is also reported.

The exact penalty function and augmented Lagrangian approaches have been developed in an attempt to circumvent the need to force convergence by using increasing values of the penalty parameter. One typical representative of this type of method is the so-called method of multipliers.[30] In this method, once a sufficiently large value of R is reached, further increases are not required. However, the method does involve additional finite parameters that must be updated between subproblem solutions. Computational evidence reported to date suggests that, while augmented Lagrangian approaches are more reliable than penalty function methods, they, as a class, are not suitable for larger dimensionality problems.

Linearization Methods.   The common characteristic of this family of methods is the use of local linear approximations to the nonlinear problem functions to define suitable, preferably feasible, directions for search. Well-known members of this family include the method of feasible directions, the gradient projection method, and the generalized reduced gradient (GRG) method. Of these, the GRG method has seen the widest engineering application.

The key constructions of the GRG method are the following:

1. The calculation of the reduced objective function gradient $\widetilde{\nabla f}$.
2. The use of the reduced gradient to determine a direction vector in the space of the independent variables.

3. The adjustment of the dependent variable values using Newton's method so as to achieve constraint satisfaction.

Given a feasible point x^0, the gradients of the equality constraints are evaluated and used to form the constraint Jacobian matrix A. This matrix is partitioned into a square submatrix J and the residual rectangular matrix C where the variable associated with the columns of J are the dependent variables and those associated with C are the independent variables.

If J is selected to have nonzero determinant, then the reduced gradient is defined as

$$\nabla \tilde{f}(x^0) = \nabla \overline{f} - \nabla \hat{f} J^{-1} C$$

where $\nabla \overline{f}$ is the subvector of objective function partial derivatives corresponding to the (dependent) variables, and $\hat{f}$ is the corresponding subvector whose components correspond to the independent variables. The reduced gradient $\nabla \tilde{f}$ provides an estimate of the rate of change of $f(x)$ with respect to the independent variables when the dependent variables are adjusted to satisfy the linear approximations to the constraints.

Given $\nabla \tilde{f}$, in the simplest version of GRG algorithm, the direction subvector for the independent variables $\overline{d}$ is selected to be the reduced gradient descent direction

$$\overline{d} = -\nabla \tilde{f}$$

For a given step α in that direction, the constraints are solved iteratively to determine the value of the dependent variables $\hat{x}$ that will lead to a feasible point. Thus, the system

$$h_k(\overline{x}^0 + a\overline{d}, \hat{x}) = 0 \qquad k = 1, \ldots, K$$

is solved for the K unknown variables of $\hat{x}$. The new feasible point is checked to determine whether an improved objective value has been obtained and, if not, α is reduced and the solution for $\hat{x}$ repeated. The overall algorithm terminates when a point is reached at which the reduced gradient is sufficiently close to zero.

The GRG algorithm has been extended to accommodate inequality constraints as well as variable bound. Moreover, the use of efficient equation solving procedures, line search procedures for α, and quasi-Newton formulas to generate improved direction vectors $\overline{d}$ have been investigated. A commercial-quality GRG code will incorporate such developments and thus will constitute a reasonably complex software package. Computational testing using such codes indicates that GRG implementations are among the most robust and efficient general-purpose nonlinear optimization methods currently available.[31] One of the particular advantages of this algorithm, which can be critical in engineering applications, is that it generates *feasible* intermediate points; hence, it can be interrupted prior to final convergence to yield a feasible solution. Of course, this attractive feature and the general efficiency of the method are attained at the price of providing (analytically or numerically) the values of the partial derivatives of all of the model functions.

Successive Quadratic Programming (SQP) Methods.   This family of methods seeks to attain superior convergence rates by employing subproblems constructed using higher-order approximating functions than those employed by the linearization methods. The SQP methods are still the subject of active research; hence, developments and enhancements are proceeding apace. However, the basic form of the algorithm is well established and can be sketched out as follows.

At a given point x^0, a direction finding subproblem is constructed, which takes the form of a quadratic programming problem:

$$\text{Minimize} \quad \nabla^{\mathrm{T}} fd + \frac{1}{2} d^{\mathrm{T}} Hd$$

$$\text{Subject to} \quad h_k(x^0) + \nabla^{\mathrm{T}} h_k(x^0) d = 0$$

$$g_j(x^0) + \nabla^{\mathrm{T}} g_j(x^0) d \geq 0$$

The symmetric matrix H is a quasi-Newton approximation of the matrix of second derivatives of a composite function (the Lagrangian) containing terms corresponding to all of the functions f, h_k, and g_j.

Using only gradient differences H is updated as in the unconstrained case. The direction vector d is used to conduct a line search, which seeks to minimize a penalty function of the type discussed earlier. The penalty function is required because, in general, the intermediate points produced in this method will be infeasible. Use of the penalty function ensures that improvements are achieved in either the objective function values or the constraint violations or both. One major advantage of the method is that very efficient methods are available for solving large quadratic programming problems and, hence, that the method is suitable for large-scale applications. Recent computational testing indicates that the SQP approach is very efficient, outperforming even the best GRG codes.[32] However, it is restricted to models in which infeasibilities can be tolerated and will produce feasible solutions only when the algorithm has converged.

5.3   Optimization Software

With the exception of the direct search methods and the transformation-type methods, the development of computer programs implementing state-of-the-art optimization algorithms is a major effort requiring expertise in numerical methods in general and numerical linear algebra in particular. For that reason, it is generally recommended that engineers involved in design optimization studies take advantage of the number of good-quality implementations now available through various public sources.

Commercial computer codes for solving LP/IP/NLP problems are available from many computer manufacturers and private companies who specialize in marketing software for major computer systems. Depending on their capabilities, these codes vary in their complexity, ease of use, and cost (see, for example, Ref. 33). LP models with a few hundred constraints can now be solved on personal computers (PCs). There are now at least a hundred small companies marketing LP software for PCs. For a 2003 survey of LP software, see Ref. 34.

Nash[35] presents a 1998 survey of nonlinear programming software that will run on PC compatibles, Macintosh systems, and UNIX-based workstations. Detailed product descriptions, prices, and capabilities of NLP software are included in the survey. There are now LP/IP/NLP solvers that can be invoked directly from inside spreadsheet packages. For example, Microsoft Excel and Microsoft Office for Windows and Macintosh contain a general-purpose optimizer for solving small-scale linear, integer, and nonlinear programming problems. Developed by Frontline Systems for Excel users, the LP optimizer is based on the simplex algorithm, while the NLP optimizer is based on the GRG algorithm. Frontline Systems also offers more powerful Premium Solver Platforms for Excel that can solve larger problems. For a review, see Ref. 36. A number of other spreadsheet add-ins are also available now for solving optimization problems. For a 2002 survey, see Ref. 37.

There are now modeling languages that allow the user to express a model in a very compact algebraic form, with whole classes of constraints and variables defined over index sets. Models with thousands of constraints and variables can be defined in a couple of pages, in a syntax that is very close to standard algebraic notation. The algebraic form of the model is kept separate from the actual data for any particular instance of the model. The computer takes over the responsibility of transforming the abstract form of the model and the specific data into a specific constraint matrix. This has greatly simplified the building, and even more, the changing of the optimization models. There are several modeling languages available for PCs. The two high-end products are GAMS (General Algebraic Modeling System) and AMPL (A Mathematical Programming Language). For a discussion of GAMS, see Ref. 38. AMPL is marketed by AT&T. For a general introduction to modeling languages, see Refs. 33 and 39, and for an excellent discussion of AMPL, see Ref. 40.

For those who write their own programs, LINDO Systems offers a library of callable linear, quadratic, and MIP solvers[41]. Engineers, who are familiar with MATLAB[42] can use its Optimization Toolbox for solving LP, QP, IP, and NLP problems. For a MATLAB software review, see Ref. 43. MATLAB and the Optimization Toolbox provide users an interactive environment to define models, gather data, and analyze results.

For optimization on the Internet[44], users can get a complete list of optimization software available for LP, IP, and NLP problems at the NEOS website: www.neos-guide.org. The NEOS (Network Enabled Optimization System) guide on optimization software is based on the textbook by More and Wright,[45] an excellent resource for those interested in a broad review of the various optimization methods and their computer codes. The book is divided into two parts. Part I has an overview of algorithms for different optimization problems, categorized as unconstrained optimization, nonlinear least squares, nonlinear equations, linear programming, quadratic programming, bound-constrained optimization, network optimization, and integer programming. Part II includes product descriptions of 75 software packages that implement the algorithms described in Part I. Much of the software described in this book is in the public domain and can be obtained through the Internet. The NEOS guide also offers a number of engineering design optimization packages. Both MATLAB and the NEOS offer software for genetic algorithms also.

The NEOS server is a project of the University of Wisconsin at Madison. It allows users to submit optimization problems to be solved by state-of-the-art optimization software. The NEOS guide complements the NEOS server by providing information on the different optimization methods to solve both linear programming and nonlinear programming problems. The optimization tree in NEOS has descriptions of several methods for solving continuous and discrete optimization problems. It also has some case studies on optimization.

Another recent source of optimization software is the COIN-OR (Computational Infrastructure for OR), which can be accessed at www.coin-or.org. COIN-OR has a large collection of open-source optimization software that can be downloaded free. COIN-OR began as an IBM project in 2000 and was spun-off as a separate entity run by the COIN-OR nonprofit foundation. An excellent tutorial on COIN-OR is given by Martin[46]. The tutorial shows how COIN-OR software can be accessed and used for both research and applications. CoinEasy is recommended as the starting point for COIN-OR users.

6 SUMMARY

In this chapter an overview was given of the elements and methods comprising design optimization methodology. The key element in the overall process of design optimization was seen to be the engineering model of the system constructed for this purpose. The assumptions and formulation details of the model govern the quality and relevance of the optimal design obtained. Hence, it is clear that design optimization studies cannot be relegated to optimization software specialists but are the proper domain of the well-informed design engineer.

The chapter also gave a structural classification of optimization problems and a broad brush review of the main families of optimization methods. Clearly this review can only hope to serve as entry point to this broad field. For a more complete discussion of optimization techniques with emphasis on engineering applications, guidelines for model formulation, and practical solution strategies, the readers are referred to the text by Ravindran, Ragsdell, and Reklaitis.[27]

The Design Automation Committee of the Design Engineering Division of ASME has been sponsoring conferences devoted to engineering design optimization. Several of these presentations have subsequently appeared in the *Journal of Mechanical Design, ASME Transactions*. Ragsdell[47] presents a review of the papers published up to 1977 in the areas of machine design applications and numerical methods in design optimization. ASME published, in 1981, a special volume entitled *Progress in Engineering Optimization*, edited by Mayne and Ragsdell[48].

It contains several articles pertaining to advances in optimization methods and their engineering applications in the areas of mechanism design, structural design, optimization of hydraulic networks, design of helical springs, optimization of hydrostatic journal bearing and others. Finally, the persistent and mathematically oriented reader may wish to pursue the fine exposition given by Avriel[49], which explores the theoretical properties and issues of nonlinear programming methods.

6.1 Acknowledgment

The authors gratefully acknowledge the assistance provided by Subramanian Pazhani, an industrial engineering doctoral student at Penn State University.

REFERENCES

1. A. Ravi Ravindran, *Operations Research and Management Science Handbook,* CRC Press, Boca Raton, FL, 2008.

2. K. E. Bett, J. S. Rowlinson, and G. Saville, *Thermodynamics for Chemical Engineers*, MIT Press, Cambridge, MA, 1975.

3. F. C. Jen, C. C. Pegels, and T. M. Dupuis, "Optimal Capacities of Production Facilities," *Manage. Sci.* **14B**, 570–580, 1968.

4. J. E. Shigley, *Mechanical Engineering Design,* McGraw-Hill, New York, 1973, p. 271.

5. S. Timoshenko and J. Gere, *Theory of Elastic Stability*, McGraw-Hill, New York, 1961, p. 257.

6. K. M. Ragsdell and D. T. Phillips, "Optimal Design of a Class of Welded Structures Using Geometric Programming," *ASME J. Eng.Ind. Ser. B,* **98**(3), 1021–1025, 1975.

7. R. H. Philipson and A. Ravindran, "Application of Goal Programming to Machinability Data Optimization," *J. Mech. Design,Trans. of ASME*, **100**, 286–291, 1978.

8. E. J. A. Armarego and R. H. Brown, *The Machining of Metals,* Prentice-Hall, Englewood Cliffs, NJ, 1969.

9. J. L. Arthur and A. Ravindran, "PAGP-Partitioning Algorithm for (Linear) Goal Programming Problems," *ACM Trans.Math. Software,* **6**, 378–386, 1980.

10. R. H. Philipson and A. Ravindran, "Application of Mathematical Programming to Metal Cutting," *Math. Program. Study*, **11,** 116–134, 1979.

11. H. M. Saber and A. Ravindran, "A Partitioning Gradient Based Algorithm for Solving Nonlinear Goal Programming Problems," *Comput. Operat. Res.*, **23**, 141–152, 1996.

12. G. Kruiger and A. Ravindran, "Intelligent Search Methods for Nonlinear goal programs," *Informat. Syst. Operat. Res.*, **43**, 79–92, 2005.

13. N. K. Karmarkar, "A New Polynomial Time Algorithm for Linear Programming," *Combinatorica,* **4,** 373–395, 1984.

14. A. Arbel, *Exploring Interior Point Linear Programming: Algorithms and Software,* MIT Press, Cambridge, MA 1993.

15. S.-C. Fang and S. Puthenpura, *Linear Optimization and Extensions*, Prentice-Hall, Englewood Cliffs, NJ, 1993.

16. R. L. Rardin, *Optimization in Operations Research*, Prentice-Hall, Englewood Cliffs, NJ, 1998.

17. K. G. Murty, *Operations Research: Deterministic Optimization Models,* Prentice-Hall, Englewood Cliffs, NJ, 1995.

18. G. L. Nemhauser and L. A. Wolsey, *Integer and Combinatorial Optimization,* Wiley, New York, 1988.

19. A. D. Belegundu and T. R. Chandrupatla, *Optimization Concepts and Applications in Engineering*, Prentice-Hall, Englewood Cliffs, NJ, 1999.

20. C. S. Beightler and D. T. Phillips, *Applied Geometric Programming,* Wiley, New York, 1976.

21. M. J. Rijckaert, "Engineering Applications of Geometric Programming," in M. Avriel, M. J. Rijckaert, and D. J. Wilde (Eds.), *Optimization and Design*, Prentice-Hall, Englewood Cliffs, NJ, 1974.

22. I. O. Bohachevsky, M. E. Johnson, and M. L. Stein, "Generalized Simulated Annealing for Function Optimization," *Technometrics,* **28**, 209–217, 1986.

23. L. Davis (Ed.), *Genetic Algorithms and Simulated Annealing,* Pitman, London 1987.

24. S. Kirkpatrick, C. D. Gelatt, and M. P. Vecchi, "Optimization by Simulated Annealing," *Science*, **220**, 670–680, 1983.

25. D. E. Goldberg, *Genetic Algorithm in Search, Optimization, and Machine Learning,* Addison-Wesley, Reading, MA, 1989.

26. A. Maria, Genetic Algorithms for Multimodal Continuous Optimization Problems, Ph.D. Diss., University of Oklahoma, Norman, OK, 1995.

27. A. Ravindran, K. M. Ragsdell and G. V. Reklaitis, *Engineering Optimization: Methods and Applications*, 2 ed., Wiley, Hoboken, NJ, 2006.

28. R. Fletcher, *Practical Methods of Optimization*, 2nd ed., Wiley New York, 1987.

29. T. W. Lee and F. Fruedenstein, "Heuristic Combinatorial Optimization in the Kinematic Design of Mechanisms: Part 1: Theory," *J. Eng. Ind. Trans. ASME*, 1277–1280, 1976.

30. S. B. Schuldt, G. A. Gabriele, R. R. Root, E. Sandgren, and K. M. Ragsdell, "Application of a New Penalty Function Method to Design Optimization," *J. Eng. Ind Trans ASME*, 31–36, 1977.

31. E. Sandgren and K. M. Ragsdell, "The Utility of Nonlinear Programming Algorithms: A Comparative Study—Parts 1 and 2," *J. Mech. Design, Trans. ASME,* **102,** 540–541, 1980.

32. K. Schittkowski, *Nonlinear Programming Codes: Information, Tests, Performance, Lecture Notes in Economics and Mathematical Systems*, Vol. **183**, Springer, New York, 1980.

33. R. Sharda, *Linear and Discrete Optimization and Modeling Software: A Resource Handbook,* Lionheart, Atlanta, GA, 1993.

34. R. Fourer, "Linear Programming Software Survey," *OR/MS Today* **30,** December 2003.

35. S. G. Nash, "Nonlinear Programming Software Survey," *OR/MS Today* **25**, June, 1998.

36. C. Albright, "Premium Solver Platform for Excel," *OR/MS Today,* **28,** 58–63, June, 2001.

37. T. A. Grossman, "Spreadsheet Add-Ins for OR/MS," *OR/MS Today,* **29,** August, 2002.

38. B. D. Kendrick and A. Meeraus, *GAMS: A User's Guide,* Scientific Press, Redwood City, CA, 1988.

39. R. Sharda and G. Rampal, "Algebraic Modeling Languages on PC's," *OR/MS Today* **22**, 58–63, 1995.

40. R. Fourer, D. M. Gay and B. W. Kernighan, "A Modeling Language for Mathematical Programming," *Manage. Sci.* **36**, 519–554, 1990.

41. A. Felt, "LINDO API: Software Review," *OR/MS Today,* **29,** December, 2002.

42. R. Biran and M. Breiner, *MATLAB for Engineers,* Addison-Wesley, Reading, MA,1995.

43. J. J. Hutchinson, "MATLAB Software Review," *OR/MS Today,* **23,** October, 1996.

44. J. Czyzyk, J. H. Owen, and S. J. Wright, "Optimization on the Internet," *OR/MS Today,* **24,** October, 1997.

45. J. J. More and S. J. Wright, *Optimization Software Guide*, SIAM Philadelphia, 1993.

46. K. Martin, "Tutorial: COIN-OR: Software for the OR Community," *Interfaces*, **40**, 465–476, 2010.

47. K. M. Ragsdell, "Design and Automation," *J. Mech. Design, Trans. of ASME*, **102**, 424–429, 1980.

48. R. W. Mayne and K. M. Ragsdell (Eds.), *Progress in Engineering Optimization*, ASME, New York, 1981.

49. M. Avriel, *Nonlinear Programming: Analysis and Methods*, Prentice-Hall, Englewood Cliffs, NJ, 1976.

CHAPTER 5

TOTAL QUALITY MANAGEMENT IN MECHANICAL SYSTEM DESIGN

B. S. Dhillon
University of Ottawa
Ottawa, Ontario Canada

1 INTRODUCTION

In today's competitive environment, the age-old belief of many companies that "the customer is always right" has a new twist. In order to survive, companies are focusing their entire organization on customer satisfaction. The approach followed for ensuring customer satisfaction is known as total quality management (TQM). The challenge is to "manage" so that the "total" and the "quality" are experienced in an effective manner.[1]

The history of the quality-related efforts may be traced back to ancient times. For example, Egyptian wall paintings of around 1450 B.C. show some evidence of measurement and inspection activity.[2] In modern times, the total quality movement appears to have started in the early 1900s in the time and motion study works of Frederick W. Taylor, the father of scientific management.[3-5]

During the 1940s, the efforts of people such as W. E. Deming, J. Juran, and A. V. Feigenbaum greatly helped to strengthen the TQM movement.[4] In 1951, the Japanese Union of Scientists and Engineers established a prize named after W. E. Deming to be awarded to the organization that has implemented the most successful quality policies.[6] In 1987, on similar lines, the U.S. government established the Malcolm Baldrige award and Nancy Warren, an American behavioural scientist, coined the term total quality management.[7]

Quality cannot be inspected out of a product; it must be built in. The consideration of quality in design begins during the specification-writing phase. Many factors contribute to the success of the quality consideration in engineering or mechanical design. TQM is a useful tool for application during the design phase. It should be noted that the material presented in this chapter deals not specifically with mechanical system design but with system design in general. The same material is equally applicable to the design of mechanical items/systems.

This chapter presents some of the important aspects of TQM considered useful to mechanical system design.

2   TERMS AND DEFINITIONS

There are many terms and definitions used in TQM. This section presents some of them considered useful for mechanical design.[8-14]

- *Quality*. This is the totality of characteristics and features of an item or service that bear on its ability to satisfy stated requirements.
- *Total Quality Management*. This is a philosophy, a set of methods or approaches, and a process whose output leads to satisfied customers and continuous improvement.
- *Design*. This is a process of originating a conceptual solution to a need and expressing it in a form from which an item may be manufactured/produced or a service delivered.
- *Quality Plan*. This is the documented set of procedures that covers the in-process and final inspection of the product/item.
- *Quality Measure*. This is a quantitative measure of the features and characteristics of a product or a service.
- *Design Review*. This is a formal documented and systematic critical study of a given design by individuals other that the designer.
- *Relative Quality*. This is the degree of excellence of a product/service.
- *Design Rationale*. This is the justification to explain why specific design implementation approaches were or were not selected.
- *Quality Conformance*. This is the extent to which the item/service conforms with the stated requirements.
- *Quality Control*. This is the operational methods and activities used to satisfy requirements for quality.
- *Sampling Plan*. This is a plan that states the sample size to be inspected and provides rejection and acceptance numbers.

- *Quality Program.* This is the documented set of plans for implementing the quality system.
- *Process Inspection.* This is intermittent examination and measurement with emphasis on the checking of process-related variables.
- *Quality Problem.* This is the difference between the specified quality and the achieved quality.
- *Quality Requirements.* These are those requirements that pertain to the characteristics and features of an item/service and are required to be satisfied in order to meet a specified need.
- *Quality Performance Reporting System.* This is the system used to collect and report performance statistics of the product and service.
- *Software Quality.* This is the fitness for use of the software product/item.

3 TQM IN GENERAL

This section presents some basic and general aspects of TQM under four distinct sections.

3.1 Comparison between TQM and Traditional Quality Assurance Program

Table 1 presents a comparison between TQM and the traditional quality assurance program with respect to seven factors: objective, quality defined, customer, quality responsibility, decision making, cost, and definitions.[6,14-16]

3.2 TQM Elements

Seven important elements of TQM are as follows[17]:

Team Approach. Its main goal is to get everyone involved with TQM, including customers, subcontractors, and vendors. The quality team size may vary from 3 to 15 members and the team membership is voluntary. However, team members possess sufficient knowledge in areas such as cost–benefit analysis, planning and controlling projects, brainstorming, public relations, flow charting, and statistic presentation techniques. Furthermore, the leader of

Table 1 A Comparison between TQM and Traditional Quality Assurance

No.	Factor	Traditional quality assurance program	TQM
1	Objective	Discovers errors	Prevents errors
2	Quality defined	Creates manufactured goods that satisfy specifications	Produces goods suitable for consumer/customer use
3	Customer	Ambiguous comprehension of customer needs	Well-defined mechanisms to understand and meet customer needs
4	Quality responsibility	Quality control/inspection department	Involves all people in the organization
5	Decision making	Follows top-down management approach	Follows a team approach with worker groups
6	Cost	Better quality results in higher cost	Better quality lowers cost and enhances productivity
7	Definitions	Product driven	Customer driven

the team usually has a management background and possesses qualities such as group leadership skills, communication skills, skill in statistical methods and techniques, skill in group dynamics, and presentation skills.

Management Commitment and Leadership. They are crucial to the success of TQM and are normally achieved only after an effective understanding of the TQM concept by the senior management personnel. Consequently, the management establishes organization goals and directions and plays an instrumental role in achieving them.

Cost of Quality. This is basically a quality measurement tool. It is used to monitor the TQM process effectiveness, choose quality improvement projects, and justify the cost to doubters. Nonetheless, the cost of quality is expressed by[17]

$$C_q = QMC + DC$$
$$= PC + AC + DC \tag{1}$$

where

$$
\begin{aligned}
C_q &= \text{cost of quality} \\
QMC &= \text{quality management cost} \\
AC &= \text{appraisal cost} \\
PC &= \text{prevention cost} \\
DC &= \text{deviation cos.}
\end{aligned}
$$

Supplier Participation. This is important because an organization's ability to provide quality goods or services largely depends upon the types of relationships that exist among the parties involved in the process (e.g., the customer, the supplier, and the processor). Nowadays, companies increasingly require their suppliers to have well-established TQM programs as a precondition for potential business.[18]

Customer Service. The application of the TQM concept to this area in the form of joint teams usually leads to customer satisfaction. These teams are useful to determine customer satisfaction through interactions with customers. Nonetheless, these teams develop joint plans, goals, and controls.

Statistical Methods. These are used for purposes such as to identify and separate causes of quality problem, to verify, reproduce, and repeat measurements based on data, and to make decisions on facts based on actual data instead of opinions of various people.[17,20] Some examples of the statistical methods are Pareto diagrams, control charts, scatter diagrams, and cause-and-effect diagrams.[18,20,21]

Training. As per a Japanese axiom, quality begins with training and ends with training.[18] Nonetheless, under TQM, quality is the responsibility of all company employees; it means the training effort must be targeted for all hierarchy levels of the organization. Furthermore, it should cover areas such as fundamentals of TQM, team problem solving, and interpersonal communication.

3.3   TQM Principles and Barriers to TQM Success

The concept of TQM is based upon many principles.[22] The important ones in simple language are as follows:

- Quality can and must be managed effectively.
- Everyone in the company/organization is responsible for quality.

- Quality is a long-term investment.
- Avoid treating symptoms, carefully look for the cure.
- Quality must be clearly measurable.
- Quality-related improvements must be continuous.
- Processes, not humans, are the problem.

Over the years, professionals and researchers working in the TQM area have identified many barriers to success with TQM. The most frequently occurring eight of these barriers are as follows[23,24]:

- Inadequate management commitment
- Incorrect planning
- Poor continuous training and education
- Poor or improper usage of empowerment and teamwork
- Inability to make changes to organizational culture
- Giving poor attention to external and internal customers
- Poor measurement methods and inadequate access to data and results
- Incompatible organizational setup and isolated individuals, groups, and departments

4   THREE GENERAL APPROACHES TO TQM

Three individuals (i.e., W. E. Deming, J. M. Juran, and P. B. Crosby) have played an instrumental role in the development of TQM. Their approaches to TQM are presented below.[6,23,25–29]

4.1   Deming's Approach to TQM

Deming's approach is composed of the following 14 points:

- Establish consistency of purpose for improving services.
- Adopt the new philosophy for making the accepted levels of defects, delays, or mistakes unwanted.
- Stop reliance on mass inspection as it neither improves nor guarantees quality. Remember that teamwork between the firm and its suppliers is the way for the process of improvement.
- Stop awarding business with respect to the price.
- Discover problems. Management must work continually to improve the system.
- Take advantage of modern methods used for training. In developing a training program, take into consideration such items as:
  - Identification of company objectives
  - Identification of the training goals
  - Understanding of goals by everyone involved
  - Orientation of new employees
  - Training of supervisors in statistical thinking
  - Team building
  - Analysis of the teaching need
  - Institution of modern supervision approaches

- Eradication of fear so that everyone involved may work to his or her full capacity
  - Tearing down department barriers so that everyone can work as a team member
- Eliminate items such as goals, posters, and slogans that call for new productivity levels without the improvement of methods.
  - Make your organization free of work standards prescribing numeric quotas.
  - Eliminate factors that inhibit employee workmanship pride.
  - Establish an effective education and training program.
- Develop a program that will push the above 13 points every day for never-ending improvement.

4.2 Crosby's Approach to TQM

Crosby's approach to TQM is made up of the following 14 points:

- Ensure management commitment to quality.
- Establish quality improvement teams/groups having representatives from all departments.
- Identify locations of current and potential quality-related problems.
- Establish the quality cost and describe its application as a management tool.
- Promote the quality awareness and personal concern of all employees in the organization.
- Take appropriate measures to rectify problems highlighted through previous steps.
- Form a group or committee for the zero-defects program.
- Train all supervisors to actively perform their part of the quality improvement program.
- Hold a "zero-defects day" to let all organization personnel realize the change.
- Encourage all employees to develop improvement goals for themselves and their respective groups.
- Encourage all individuals to inform management about the obstacles faced by them to attain their set improvement goals.
- Recognize appropriately those who participate.
- Form quality councils for communicating on a regular basis.
- Repeat it to emphasize that the quality improvement program is a never-ending process.

4.3 Juran's Approach to TQM

Juran's approach is composed of the following 10 steps:

- Promote awareness of the requirement and opportunity for improvement.
- Establish appropriate goals for improvement.
- Organize for attaining the set goals (i.e., form a quality council, highlight problems, choose projects, form teams, and designate facilitators).
- Provide appropriate training.
- Carry out projects for finding solutions to problems.
- Report progress.

- Give appropriate recognition.
- Communicate results to all concerned individuals.
- Keep a record of scores.
- Maintain momentum by making yearly improvements part of the regular company systems and processes.

5   QUALITY IN DESIGN PHASE

Although TQM will help generally to improve design quality, specific quality-related steps are also necessary during the design phase. These additional steps will further enhance the product design.

An informal review during specification writing may be regarded as the beginning of quality assurance in the design phase. As soon as the first draft of the specification is complete, the detailed analysis begins.

This section presents topics considered directly or indirectly useful to improving quality in the design phase.[30–32]

5.1   Quality Design Characteristics and Their Examples

There are seven quality design characteristics.[30,31] These are performance, durability, conformance, features, serviceability, reliability, and aesthetics. For each of these factors, examples associated with a manufactured product (i.e., stereo amplifier) are signal-to-noise ratio (power), useful life, workmanship, remote control, ease to repair, mean time to failure, and oak cabinet, respectively. Similarly, for a service product (i.e., checking account) the corresponding examples are time to process customer requests, keeping pace with industry trends, accuracy, automatic bill paying, resolution of errors, variability of time to process, and appearance of bank lobby.

5.2   Steps for Controlling the Design

An effective control of the design activity can help to improve product quality. Any design activity can be controlled by following the steps listed below during the design process[8]:

- Establish customer requirements.
- Translate the customer requirements into a definitive specification document of the requirements.
- Perform a feasibility study to find if the requirements can be accomplished.
- Plan for accomplishing the requirements.
- Organize materials and resources for accomplishing the requirements.
- Perform a project definition study to find the most suitable solution out of many possible solutions.
- Develop a specification document that describes all the product's/service's features and characteristics.
- Develop a prototype/model of the proposed design.
- Conduct trials to determine the degree to which the product/service under development meets the design requirements and customer needs.
- Feed data back into the design and repeat the process until the product or service satisfies the specified requirements in an effective manner.

5.3   Product Design Review

Various types of design reviews are conducted during the product design phase. One reason for performing these reviews is to improve quality. Design reviews conducted during the design phase include preliminary design review, detailed design reviews (the number of which may vary from one project to another), critical design review (the purpose of this review is to approve the final design), preproduction design review (this review is performed after the prototype tests), postproduction design review, and operations and support design review.

The consideration of quality begins at the preliminary design review and becomes stronger as the design develops. The role of quality assurance in preliminary design review is to ensure that the new design is free of quality problems of similar existing designs. This requires a good knowledge of the strengths and weaknesses of the competing products.

5.4   Quality Function Deployment, Quality Loss Function, and Benchmarking

These three approaches are quite useful in ensuring quality during the design phase. Quality function deployment (QFD) is a value analysis tool used during product and process development. It is an extremely useful concept for developing test strategies and translating needs to specification.

QFD was developed in Japan. In the case of new product development, it is simply a matrix of consumer/customer requirements versus design requirements. Some of the sources for the input are market surveys, interviews, and brainstorming. To use the example of an automobile, customer needs include price, expectations at delivery (safety, perceived quality, service ability, performance, workmanship, etc.), and expectations over time (including customer support, durability, reliability, performance, repair part availability, low preventive maintenance and maintenance cost, mean time between failures within prediction, etc.).

Finally, QFD helps to turn needs into design engineering requirements.

The basis for the quality loss function is that if all parts are produced close to their specified values, then it is fair to expect best product performance and lower cost to society. According to Taguchi, quality cost goes up not only when the finished product is outside given specifications but also when it deviates from the set target value within the specifications.[4]

One important point to note, using Taguchi's philosophy, is that a product's final quality and cost are determined to a large extent by its design and manufacturing processes. It may be said that the loss function concept is simply the application of a life-cycle cost model to quality assurance. Taguchi expresses the loss function as follows:

$$L(x) = c\,(x - T_v)^2 \tag{2}$$

where

x	= a variable
$L(x)$	= loss at x
T_v	= targeted value of variable at which product is expected to show its best performance
C	= proportionality constant
$x - T_v$	= deviation from target value

In the formulation of the loss function, assumptions are made, such as zero loss at the target value and that the dissatisfaction of customers is proportional only to the deviation from the target value. The value of the proportionality constant, c, can be determined by estimating the loss value for an unacceptable deviation, such as the tolerance limit. Thus, the following relationship can be used to estimate the value of c:

$$c = \frac{L_a}{\Delta^2} \tag{3}$$

where

L_a = amount of loss expressed in dollars
Δ = deviation amount from target value T_v

Example

Assume that the estimated loss for the Rockwell hardness number beyond number 56 is $150 and the targeted value of the hardness number is 52. Estimate the value of the proportionality constant. Substituting the given data into Eq. 3, we get

$$c = \frac{150}{(56 - 52)^2}$$
$$= 9.375$$

Thus, the value of the proportionality constant is 9.375.

Benchmarking is a process of comparing in-house products and those that are most effective in the field and setting objectives for gaining a competitive advantage. The following steps are associated with benchmarking[33]:

- Identify items and their associated key features to benchmark during product planning.
- Select companies, industries, or technologies to benchmark against. Determine existing strengths of the items to benchmark against.
- Determine the best-in-class target from each selected benchmark item.
- Evaluate, as appropriate, in-house processes and technologies with respect to benchmarks.
- Set improvement targets remembering that the best-in-class target is always a moving target.

5.5 Process Design Review

Soon after the approval of a preliminary design, a process flowchart is prepared. In order to assume the proper consideration being given to quality, the quality engineer works along with process and reliability engineers.

For the correct functioning of the process, the quality engineer's expertise in variation control provides important input.

Lack of integration between quality assurance and manufacturing is one of the main reasons for the failure of the team effort. The performance of process failure mode and effect analysis (FMEA) helps this integration to take place early. The consideration of the total manufacturing process performance by the FMEA concept, rather than that of the mere equipment, is also a useful step in this regard. For FMEA to produce promising results, the quality and manufacturing engineers have to work as a team. Nevertheless, FMEA is a useful tool for performing analysis of a new process, including analysis of receiving, handling, and storing materials and tools. Also, the participation of suppliers in FMEA studies enhances FMEA's value. The following steps are associated with the process of FMEA[34]:

- Develop process flowchart that includes all process inputs: materials, storage and handling, transportation, etc.
- List all components/elements of the process.
- Write down each component/element description and identify all possible failure modes.
- Assign failure rate/probability to each component/element failure mode.

- Describe each failure mode cause and effect.
- Enter remarks for each failure mode.
- Review each critical failure mode and take corrective measures.

5.6 Plans for Acquisition and Process Control

The development of quality assurance plans for procurement and process control during the design phase is useful for improving product quality. One immediate advantage is the smooth transition from design to production. The equipment procurement plan should include such items as equipment performance verification, statistical tolerance analysis, testing for part interchangeability, and pilot runs. Similarly, the component procurement quality plans should address concerns and cooperation on areas including component qualification, closed-loop failure management, process control implementation throughout the production lines, and standard and special screening tests.

Prior to embarking on product manufacturing, there is a definite need for the identification of the critical points where the probability of occurrence of a serious defect is quite high. Thus, the process control plans should be developed by applying QFD and FMEA. These plans should include items such as:

- Acceptance of standard definitions
- Procedures to monitor defects
- Approaches for controlling critical process points

5.7 Taguchi's Quality Philosophy Summary and Kume's Approach for Process Improvement

Taguchi's approach was discussed earlier, but because of its importance, this section summarizes his quality philosophy again in seven basic steps[35,36]:

- A critical element of a manufactured item's quality is the total loss generated by that item to society as whole.
- In today's market, continuous cost reduction and quality improvement are critical for companies to stay in business.
- Design and its associated manufacturing processes determine, to a large extent, the ultimate quality and cost of a manufactured product.
- Unceasing reduction in a product performance characteristic's variation from its target values is part of a continuous quality improvement effort.
- The loss of customers due to variation in an item's performance is frequently almost proportional to the square of the performance characteristic's deviation from its target value.
- The identification of the product and process parameter settings that reduce performance variation can be accomplished through statistically designed experiments.
- Reduction in the performance variation of a product or process can be achieved by exploiting the product or process parameter nonlinear effects on performance characteristics.

To improve the process, Kume outlined a seven-step approach[37]:

- Select project.
- Observe the process under consideration.

- Perform process analysis.
- Take corrective measures.
- Evaluate effectiveness of corrective measures.
- Standardize the change.
- Review and make appropriate modifications, if applicable, in future plans.

5.8 Design for Six Sigma

The design for Six Sigma (DFSS) approach is quite useful to improve product or service quality and is composed of the following steps[38]:

- Identify the new product or service to be designed and establish a project plan and a project team.
- Plan and conduct research for understanding customer and other related requirements.
- Develop alternative design approaches, choose an approach for high-level design, and assess the design capability to satisfy the specified requirements.
- Develop the design, determine its capability, and plan a pilot test study.
- Perform the pilot test study, analyze the associated results, and make changes to the design as required.

This approach is described in detail in Ref. 40.

5.9 Design Guidelines for Quality Improvement

Some of these guidelines along with their advantages in parentheses are as follows[31,40]:

- Minimize number of part/component numbers (advantage: lesser variations of like parts or components).
- Select parts/components that can withstand process operations (advantages: less degradation of parts/components and less damage to parts/components).
- Eradicate adjustments altogether (advantages: elimination of assembly adjustment errors and adjustment parts with high failure rates).
- Utilize repeatable, well-understood processes (advantages: easy to control part quality and assembly quality).
- Eradicate all engineering changes on released items (advantages: lesser errors due to change-over and multiple revisions/versions).
- Design for robustness (advantage: low sensitivity to component/part (variability).
- Design for effective (i.e., efficient and adequate) testing (advantage: less mistaking "good" for "bad" item/product and vice-versa).
- Minimize number of parts/components (advantage: lesser parts/components to fail, fewer part and assembly drawings, fewer parts/components to hold to required quality characteristics, and less complicated assemblies).
- Make assembly easy, straightforward, and foolproof (advantages: self-securing parts, impossible to assemble parts incorrectly, no "force fitting" of parts, easy to spot missing parts, and assembly tooling designed into part).
- Lay out parts/components for reliable process completion (advantage: less damage to parts/components during handling and assembly).

5.10 Manager's Guide for Total Quality Software Design

As per past experiences, in order to have good quality software products, it is very important to take appropriate quality measures during the software development life cycle.[41],[42] A software development life cycle may be divided into five stages as follows[41]:

Stage I: Requirement Analysis. As per Ref. 44, about 60–80% of system development failures are due to poor understanding of user requirements. In this regard, normally major software vendors during the software development process use the QFD technique. Software quality function development (SQFD) is a front-end requirements collection method that quantifiably solicits and defines critical requirements of customers.

The main advantages of SQFD include fostering better attention to customers' requirements, establishing better communications among departments with customers, reaching features consensus faster, and quantifying qualitative customers' requirements.[41]

Stage II: Systems Design. This is considered the most critical stage of quality software development because a defect in design is very many times more costly to correct than a defect during the production stage. Concurrent engineering is a frequently used approach for changing systems design and also it is a very useful approach of implementing TQM.[41]

Stage III: Systems Development. Software TQM clearly calls to integrate quality into the total software development process. Thus, after the establishment of a quality process into the first two stages of the software development cycle, the task of coding becomes quite simple and straightforward.[41] The method of design and code inspections can be used for document inspections and control charts can be used to track the metrics of the effectiveness of code inspections.[43]

Stage IV: Testing. A TQM-based software testing process must have a very clear set of testing-related objectives. A metric-driven method can fit with such testing objectives. The six steps of the method are as follows[41]:

- Establish structured test objectives.
- Choose appropriate functional methods to drive test case suites.
- Run functional tests and assess the degree of structured coverage achieved.
- Extend the test suites until the achievement of the desired coverage.
- Calculate the test scores.
- Validate testing by recording errors not found during testing.

Stage V: Implementation and Maintenance. As per past experiences, most of the software maintenance-related activities are reactive. More clearly, involved programmers often zero in on the immediate problem, fix it, and wait until the next problem.[41],[44] In this case, statistical process control (SPC) can be employed to monitor the quality of software system maintenance. However, a TQM-based system must adapt to the SPC process to assure maintenance quality.

6 TQM METHODS

Over the years many methods and techniques that can be used to improve product or service quality have been developed by researchers and others. Effective application of these approaches can help to improve mechanical design/product quality directly or indirectly.

Reference 41 presents 100 TQM methods classified into four areas: idea generation, analytical, management, and data collection analysis and display.

The idea generation area has 17 methods, including brainstorming, buzz groups, idea writing, lateral thinking, list reduction, morphological forced connections, opportunity analysis, and snowballing.

The analytical area has 19 methods. Some of these methods are cause-and-effect analysis, failure modes and effect analysis, Taguchi methods, force field analysis, tolerance design, solution effect analysis, and process cost of quality.

The management area contains a total of 31 methods, including Deming wheel, benchmarking, Kaizen, Pareto analysis, quality circles, quality function deployment, error proofing (poka-yoke), and zero defects.

Finally, the data collection analysis and display area has 33 methods. Some of these methods are check sheets, histograms, Hoshin Kanri, Scatter diagrams, statistical process control, pie charts, spider web diagrams, and matrix diagram.

Seventeen methods belonging to the above four areas are presented below.[29,45]

6.1 Affinity Diagram

This diagram is used to organize large volumes of data into groups according to some kind of natural affinity.[45,46] The affinity diagram is used under circumstances when a group of people are ascertaining customer needs for the purpose of translating them into design requirements.

Each member of the group writes ideas about customer need on separate file cards. Then, these cards are laid on a table without conversing with each other on the matter. The same team members are asked to arrange the requirements on cards into natural groups they identify. Ultimately, the ideas having an affinity for each other are grouped together.

The main advantage of the affinity diagram approach is that, by organizing data in the form of natural affinity, associations are illustrated rather than the strictly logical connections between customer needs. This approach is described in detail in Ref. 47.

6.2 Deming Wheel

This is a management concept proposed by W. E. Deming for meeting the customer quality requirements by using the cycle made up of four elements: plan, do, check, and action.[25,45] More specifically, the wheel can be used when developing a new product based on the customer requirements.

The approach calls for teamwork among various groups of the company: product development, manufacturing, sales, and market research. The cycle or wheel is considered constantly rotating and its four elements are described below.

- *Plan*. This is concerned with determining the cause of the problem after its detection in the product development and planning corrective actions based on facts.
- *Do*. This calls for a quality improvement team to take responsibility for taking necessary steps to correct the problems after its detection.
- *Check*. This is concerned with determining, if the improvement process was successful.
- *Action*. This is concerned with accepting the new and better quality level if the improvement process is successful. If it is not, it (i.e., action) calls for repeating the Deming cycle.

This approach is described in detail in Ref. 26.

6.3 Fishbone Diagram

This approach, also known as the cause-and-effect or Ishikawa diagram, was originally developed by K. Ishikawa in Japan. The diagram serves as a useful tool in quality-related studies to perform cause-and-effect analysis for generating ideas and finding the root cause of a problem for investigation. The diagram somewhat resembles a fishbone; thus the name.

Major steps for developing a fishbone diagram are as follows[29]:

- Establish problem statement.
- Brainstorm to highlight possible causes.
- Categorize major causes into natural grouping and stratify by steps of the
- process.
- Insert the problem or effect in the "fish head" box on the right-hand side.

Develop the diagram by unifying the causes through following the necessary process steps. Refine categories by asking questions such as:

- What causes this?
- Why does this condition exist?

This method is described in detail in Ref. 30.

6.4   Pareto Diagram

An Italian economist, Vilfredo Pareto (1848–1923) developed a formula in 1897 to show that the distribution of income is uneven.[47,48] In 1907, a similar theory was put forward in a diagram by M. C. Lorenz, a U.S. economist. In later years, J. M. Juran applied Lorenz's diagram to quality problems and called it Pareto analysis.[49]

In quality control work, Pareto analysis simply means, for example, that there are always a few kinds of defects in the hardware manufacture that loom large in occurrence frequency and severity. Economically, these defects are costly and thus of great significance. Alternatively, it may simply be stated that on the average about 80% of the costs occur due to 20% of the defects.

The Pareto diagram, derived from the above reasoning, is helpful in identifying the spot for concerted effort. The Pareto diagram is a type of frequency chart with bars arranged in descending order from left to right, visually highlighting the major problem areas. The Pareto principle can be quite instrumental in the TQM effort, particularly in improving quality of product designs.

6.5   Kaizen Method

Kaizen means "improvement" in Japanese, and the Kaizen philosophy maintains that the current way of life deserves to be improved on a continuous basis. This philosophy is broader than TQM because it calls for ongoing improvement as workers, leaders, managers, and so on. Thus, Kaizen includes TQM, quality circles, zero defects, new product design, continuous quality improvement, customer service agreements, and so on.

Kaizen is often referred to as "the improvement movement" because it encompasses constant improvement in the social life, working life, and home life of everyone.

This method is described in detail in Ref. 30.

6.6   Force Field Analysis

This method was developed by Kurk Lewin to identify forces existing in a situation. It calls first for clear understanding of the driving and restraining forces and then for developing plans to implement change.[29]

The change is considered as a dynamic process and as the result of a struggle between driving forces (i.e., those forces seeking to upset the status quo) and restraining forces (i.e., those forces attempting to maintain the status quo). A change occurs only when the driving forces are stronger than the restraining forces.

The group brainstorming method serves as a useful tool to identify the driving forces and restraining forces in a given situation.

With respect to improving engineering design quality, the force field analysis facilitates changes in the following way:

- It forces the concerned personnel to identify and to think through the certain facets of an acquired change.
- It highlights the priority order of the involved driving and restraining forces.
- It leads to the establishment of a priority action plan.

An example of using the force field analysis technique is given in Ref. 30.

6.7 Customer Needs Mapping Method

This approach is used to identify consumer requirements and then to identify in-house processes' ability to meet those requirements satisfactorily. A process's two major customers are the external customer (the purchaser of the product or service) and the internal customer (the next step in the process of receiving the output). Past experience has shown that the internal customer is often overlooked by groups such as inventory control, accounting, facilities, and computer.

Both the external and internal customer's wants or requirements can be identified through brainstorming, customer interviews, and so on.

Some of the advantages of customer needs mapping are as follows:

- It enhances the understanding of the customer background.
- It highlights customer wants.
- It translates customer needs into design features.
- It focuses attention on process steps important to customers.
- It highlights overlooked customer needs.

6.8 Control Charts

Control charts were developed by Walter A. Shewhart of Bell Telephone Laboratories in 1924 for analyzing discrete or continuous data collected over a period of time.[50] A control chart is a graphical tool used for assessing the states of control of a given process. The variations or changes are inherent in all processes, but their magnitude may be large or very small. Minimizing or eliminating such variations will help to improve the quality of a product or service.

Physically, a control chart is composed of a center line (CL), upper control limit (UCL), and lower control limit (LCL). In most cases, a control chart uses control limits of plus or minus three standard deviations from the mean of the quality or other characteristic under study.

Following are some of the reasons for using control charts for improving product design quality[50]:

- To provide a visual display of a process
- To determine if a process is in statistical control
- To stop unnecessary process-related adjustments

- To provide information on trends over time
- To take appropriate corrective measures

Prior to developing a control chart, the following factors must be considered:

- Sample size, frequency, and the approach to be followed for selecting them
- Objective of the control chart under consideration
- Characteristics to be examined
- Required gauges or testing devices

6.9 Poka-Yoke Method

This is a mistake-proofing approach. It calls for designing a process or product so that the occurrence of any anticipated defect is eliminated. This is accomplished through the use of automatic test equipment that "inspects" the operations performed in manufacturing a product and then allows the product to proceed only when everything is correct. Poka-yoke makes it possible to achieve the goal of zero defects in the production process. The method is described in detail in Ref. 30.

6.10 Hoshin Planning Method

This method, also known as the "seven management tools," helps to tie quality improvement activities to the long-term plans of the organization.[51] Hoshin planning focuses on policy development issues: planning objective identification, management and employee action identification, and so on. The following are the three basic Hoshin planning processes:

- General planning begins with the study of consumers for the purpose of focusing the organization's attention on satisfying their needs.
- Intermediate planning begins after the general planning is over. It breaks down the general planning premises into various segments for the purpose of addressing them individually.
- Detailed planning begins after the completion of the intermediate planning and is assisted by the arrow diagram and by the process decision program chart.

The seven management tools related to or used in each of the above three areas are as follows:

General
- Interrelationship diagram
- Affinity chart

Intermediate
- Matrix data analysis
- Tree diagram
- Matrix diagram

Detailed
- Arrow diagram
- Process decision program chart

Each of the above management tools is discussed below.

The interrelationship diagram is used to identify cause-and-effect links among ideas produced. It is particularly useful in situations where there is a requirement to identify root causes. One important limitation of the interrelationship diagram is the overwhelming attempts to identify linkages between all generated ideas.

The affinity chart is used to sort related ideas into groups and then label each similar group. The affinity chart is extremely useful in handling large volumes of ideas, including the requirement to identify broad issues.

The matrix data analysis is used to show linkages between two variables, particularly when there is a requirement to show visually the strength of their relationships. The main drawback of this approach is that only two relationships can be compared at a time.

The tree diagram is used to map out required tasks into detailed groupings. This method is extremely useful when there is a need to divide broad tasks or general objectives into subtasks.

The matrix diagram is used to show relationships between activities, such as tasks and people. It is an extremely useful tool for showing relationships clearly.

The arrow diagram is used as a detailed planning and scheduling tool and helps to identify time requirements and relationships among activities. The arrow diagram is an extremely powerful tool in situations requiring detailed planning and control of complex tasks with many interrelationships.

The process decision program chart is used to map out contingencies along with countermeasures. The process decision program chart is an advantage in implementing a new plan with potential problems so that the countermeasures can be thought through.

6.11 Gap Analysis Method

This method is used to understand services offered from different perspectives. The method considers five major gaps that are evaluated so that when differences are highlighted between perceptions, corrective measures can be initiated to narrow the gap or difference.

- Consumer expectation and management perception gap
- Management perception of consumer expectation and service quality specification gap
- Service quality specifications and service delivery gap
- External communication and service delivery gap
- Consumer expectation concerning the service and the actual service received gap

This method is described in detail in Ref. 30.

6.12 Stratification

This method is used in defining a problem by identifying where it does and does not occur.[46] It is basically a technique of splitting data according to meeting or not meeting a set of criteria. More specifically, the stratification approach highlights patterns in the data. It is used before and after collecting data for designing the way to collect data and the way of focusing the analysis, respectively. The following steps are associated with the stratification approach[45]:

- Brainstorm to identify characteristics that could cause systematic differences in the data.
- Design data collection forms for incorporating these characteristics.
- Collect data and then examine them to identify any trends or patterns.

The main advantage of the stratification approach is that it ensures the collection of all necessary data the first time; thus it avoids wasting time and effort.

6.13   Opportunity Analysis

This is quite an efficient method and is used to evaluate a rather long list of options against desired goals and available resources. It can be used by an individual or a group. The method is composed of the following basic steps[45]:

- List all your desired goals in the situation under consideration.
- Rank each goal's importance with respect to satisfying customers and rate your capability to accomplish them.
- Determine if there are sufficient required resources available to accomplish these goals.
- Select the most promising options.

6.14   Gantt Charts

The purpose of the Gantt charts is to plan the steps necessary to implement quality improvement. A Gantt chart may simply be described as a horizontal bar chart used as a production control tool.

Gantt charts can be used as follows[45]:

- Break down the implementation plan under consideration into tasks and activities considered achievable.
- Estimate how much time each task or activity will take to complete and then set a realistically possible completion date.
- Decompose the steps into a logical sequence. Bars/lines signify when a task/activity is due to commence and end. The relationship over time between each and every task or activity is immediately visible.
- Assess each and every step on an individual basis, identifying:
  - Any issue that forbids the completion of a specified task (note this as a key issue)
  - Any dependent task/activity that must be accomplished prior to starting another task/activity

The main advantages of Gantt charts are that the visual representation of tasks or activities helps to highlight the key issues and brings into the open the steps necessary for successful completion as well as it (visual representation) makes easy to see when deadlines slip and changes to the plan have to be carried out.

Additional information on Gantt charts is available in Refs. 46, 53, and 54.

6.15   Concentration Diagrams

Concentration diagrams are a useful method for collecting data when the location of a defect/problem is important. It can be used either during problem definition when one is collecting data to determine what is happening, or after the implementation of a solution when one is collecting data to monitor the new situation.

A concentration diagram can be developed by following the five steps presented below[45].

- *Step 1*: Agree the data need to be collected.
- *Step 2*: Design the concentration diagram considered appropriate.

- *Step 3*: Test the diagram using an individual who was not involved in the design process. Modify the diagram, if it is considered necessary.
- *Step 4*: Design an appropriate master concentration diagram.
- *Step 5*: Collect the data considered appropriate.

Some of the advantages of concentration diagrams are as follows[45]:

- By establishing the real facts regarding the location of failure, a team can plan to highlight the failure causes and look for appropriate ways to remove them.
- Actions are executed with respect to evidence, not feeling.
- Concentration diagrams are an excellent tool to involve individuals in the area of administration in quality improvement effort.

6.16 Scatter Diagrams

A scatter diagram is considered a quite useful method for studying relationships between two variables. One variable is plotted on the horizontal axis and the other variable on the vertical axis. The relationship between the two variables is indicated by the slope of the diagram.[23] It is to be noted that although it is impossible for scatter diagrams to prove that one variable causes the other variable, they do indicate the existence of a relationship as well as its strength.

The following four steps are associated with the construction of a scatter diagram[48]:

- *Step 1*: Collect two pieces of data (i.e., a pair number) on a product/process and then develop the data summary table.
- *Step 2*: Construct a diagram labeling the horizontal and vertical axes.
- *Step 3*: Plot all the data pairs on the diagram by placing, say, a thick dot at the intersections of the coordinates of the variables X and Y for each pair of data under consideration.
- *Step 4*: Make effective interpretation of the final scatter diagram under consideration for direction and strength.

6.17 Potential Problem Analysis

The potential problem analysis (PPA) method is concerned with examining plans to highlight what can go wrong with them, so that appropriate preventive measures can be undertaken. PPA is a simple and straightforward method for a group of individuals to examine plans. The method is composed of the following eight steps[45]:

- *Step 1*: Draw up an appropriate plan in time order.
- *Step 2*: Flowchart the plan, identifying all the critical steps where certain outputs are necessary.
- *Step 3*: At each and every critical step brainstorm the problems that can happen.
- *Step 4*: Rate all the anticipated problems using the following scheme:
  - *Severity*: 10 (catastrophic) to 1 (mild)
  - *Likelihood*: 10 (very likely) to 1 (very unlikely)

In order to obtain the potential problem risk (PPR) number, multiply the likelihood by the severity.

- *Step 5*: Identify the likely causes for the occurrence of each potential problem. It is to be noted that all potential problems with a PPR over 50 or a likelihood or severity over 7 must be prevented.

- *Step 6*: For each and every cause, brainstorm the courses of measures that could be undertaken for preventing it from happening.
- *Step 7*: In the case of problems that could be prevented, take appropriate measures for removing the potential cause.
- *Step 8*: In the case of problems that are impossible to prevent, develop appropriate contingency plans to eradicate the problems when they occur.

REFERENCES

1. C. R. Farquhar and C. G. Johnston, "Total Quality Management: A Competitive Imperative," Report No. 60-90-E, Conference Board of Canada, Ottawa, Ontario, Canada.
2. D. C. Dague, *Quality: Historical Perspective*, in J. Evans and W. Lindsay (Eds.), *Quality Control in Manufacturing*, Society of Automotive Engineers, Warrendale, PA, 1981.
3. A. Rao et al., *Total Quality Management: A Cross Functional Perspective*, Wiley, New York, 1996.
4. C. D. Gevirtz, *Developing New Products with TQM*, McGraw Hill, New York, 1994.
5. D. L. Goetsch and S. Davis, *Implementing Total Quality*, Prentice Hall, Englewood Cliffs, NJ, 1995.
6. W. H. Schmidt and J. P. Finnigan, *The Race without a Finish Line: America's Quest for Total Quality*, Jossey-Bass, San Francisco, CA, 1992.
7. M. Walton, *Deming Management at Work*, Putnam, New York, 1990.
8. D. Hoyle, *ISO 9000 Quality Systems Handbook*, Butterworth-Heinemann, Boston, 1998.
9. T. P. Omdahl (Ed.), *Reliability, Availability, and Maintainability (RAM) Dictionary*, ASQC Quality Press, Milwaukee, WI, 1988.
10. ANSI/ASQC A3-1978, *Quality Systems Terminology*, American Society for Quality Control (ASQC), Milwaukee, WI, 1978.
11. *Glossary and Tables for Statistical Quality Control*, American Society for Quality Control, Milwaukee, WI, 1973.
12. J. L. Hradesky, *Total Quality Management Handbook*, McGraw Hill, New York, 1995.
13. B. S. Dhillon, *Quality Control, Reliability, and Engineering Design*, Marcel Dekker, New York, 1985.
14. B. S. Dhillon, *Engineering and Technology Management Tools and Applications*, Artech House, Norwood, MA, 2002.
15. C. N. Madu and K. Chu-Hua, "Strategic Total Quality Management (STQM)," in C. N. Madu (Ed.), *Management of New Technologies for Global Competitiveness*, Quorum Books, Westport, CT, 1993, pp. 3–25.
16. B. S. Dhillon, *Advanced Design Concepts for Engineers*, Technomic Publishing, Lancaster, PA, 1998.
17. J. L. Burati, M. F. Matthews, and S. N. Kalidindi, "Quality Management Organizations and Techniques," *J. Construct. Eng. Manag.*, **118**, 112–128, March 1992.
18. W. B. Ledbetter, *Measuring the Cost of Quality in Design and Construction*, Publication No. 10-2, The Construction Industry Institute, Austin, TX, 1989.
19. M. Imai, *Kaizen: The Key to Japan's Competitive Success*, Random House, New York, 1986.
20. J. Perisco, "Team Up for Quality Improvement," *Quality Prog.*, **22**(1), 33–37, 1989.
21. H. Kume, *Statistical Methods for Quality Improvement*, Association for Overseas Technology Scholarship, Tokyo, 1985.
22. K. Ishikawa, *Guide to Quality Control*, Asian Productivity Organization, Tokyo, 1982.
23. L. Martin, "Total Quality Management in the Public Sector," *Natl. Product. Rev.*, **10**, 195–213, 1993.

24. G. M. Smith, *Statistical Process Control and Quality Improvement*, Prentice Hall, Upper Saddle River, NJ, 2001.
25. R. J. Masters, "Overcoming the Barriers to TQM Success," *Quality Prog.*, May 1996, pp. 50–52.
26. W. E. Deming, *Out of the Crisis*, Center for Advanced Engineering Study, Massachusetts Institute of Technology, Cambridge, MA, 1986.
27. P. B. Crosby, *Quality without Tears*, McGraw Hill, New York, 1984.
28. J. S. Oakland, *Total Quality Management*, Butterworth-Heinemann, Boston, 2003.
29. J. Heizer and B. Render, *Production and Operations Management*, Prentice Hall, Upper Saddle River, NJ, 1995.
30. P. Mears, *Quality Improvement Tools and Techniques*, McGraw Hill, New York, 1995.
31. P. E. Pisek, "Defining Quality at the Marketing/Development Interface," *Quality Prog.*, June 1987, pp. 28–36.
32. J. R. Evans and W. M. Lindsay, *The Management and Control of Quality*, West Publishing, New York, 1989.
33. D. G. Raheja, *Assurance Technologies*, McGraw Hill, New York, 1991.
34. *Total Quality Management: A Guide for Implementation*, Document No. DOD 5000.51.6 (draft), U.S. Department of Defense, Washington, DC, March 23, 1989.
35. B. S. Dhillon and C., Singh, *Engineering Reliability: New Techniques and Applications*, Wiley, New York, 1981.
36. R. H. Lochner and J. E. Matar, *Designing for Quality*, ASQC Quality Press, Milwaukee, WI, 1990.
37. R. N. Kackar, "Taguchi's Quality Philosophy: Analysis Commentary," *Quality Prog.*, **19**, 21–29, 1986.
38. H. Kume, *Statistical Methods for Quality Improvement*, Japanese Quality Press, Tokyo, 1987.
39. G. J. Hahn, N. Doganaksoy, and R. Hoerl, "The Evolution of Six Sigma," *Quality Eng.*, **12**(3), 317–326, 2000.
40. F. M. Gryna, *Quality Planning and Analysis*, McGraw Hill, New York, 2001.
41. D. Daetz, "The Effect of Product Design on Product quality and Product Cost," *Quality Prog.*, June 1987, pp. 63–67.
42. B. Gong, D. C. Yen, and D. C. Chou, "A Manager's Guide to Total Quality Software Design," *Indust. Manag. Data Syst.*, **98**(3), 100–107.
43. B. S. Dhillon, *Applied Reliability and Quality: Fundamentals, Methods, and Procedures*, Springer, London, 2007.
44. M. E. Fagan, "Advances in Software Inspections," *IEEE Trans. Software Eng.*, **12**(7), 744–751, 1986.
45. W. M. Osborne, "All About Software Maintenance: 50 Questions and Answers," *J. Inform. Syst. Manag.*, **5**(3), 36–43, 1988.
46. G. K. Kanji and M. Asher, *100 Methods for Total Quality Management*, Sage Publications, London, 1996.
47. B. Bergman and B. Klefsjo, *Quality: From Customer Needs to Customer Satisfaction*, McGraw Hill, New York, 1994.
48. J. A. Burgess, *Design Assurance for Engineers and Managers*, Marcel Dekker, New York, 1984.
49. B. S. Dhillon, *Reliability, Quality, and Safety for Engineers*, CRC Press, Boca Raton, FL, 2005.
50. J. M. Juran, F. M. Gryna, and R. S. Bingham (Eds.), *Quality Control Handbook*, McGraw Hill, New York, 1979.
51. *Statistical Quality Control Handbook*, AT & T Technologies, Indianapolis, IN, 1956.
52. B. King, *Hoshin Planning: The Developmental Approach*, Methuen, Boston, MA, 1969.
53. W. Clark and H. Gantt, *The Gantt Chart, a Working Tool of Management*, Ronald Press, New York, 1922.

BIBLIOGRAPHY

TQM

W. Baker, *TQM: A Philosophy and Style of Managing*, Faculty of Administration, University of Ottawa, Ottawa, Ontario, 1992.

D. H. Besterfield, *Total Quality Management*, Prentice Hall, Upper Saddle River, NJ, 2003.

A. Chaiarini, *From Total Quality Control to Lean Six Sigma: Evolution of the Most Important Management Systems for the Excellence*, Springer, New York, 2012.

J. R. Evans, *Total Quality: Management, Organization, and Strategy*, Thomson/South-Western Publishing, Mason, OH, 2003.

C. R. Farquhar and C. G. Johnston, "Total Quality Management: A Competitive Imperative," Report No. 60-90-E, Conference Board of Canada, Ottawa, Ontario, 1990.

A. V. Feigenbaum, *Total Quality Control*, McGraw Hill, New York, 1983.

C. D. Gevirtz, *Developing New Products with TQM*, McGraw Hill, New York, 1994.

D. L. Goetsch, *Quality Management: Introduction to Total Quality Management for Production, Processing, and Services*, Pearson Prentice Hall, Upper Saddle River, NJ, 2006.

M. J. Gordon, *Total Quality Process Control for Injection Molding*, Wiley, Hoboken, NJ, 2010.

P. Mears, "TQM Contributors," in P. Mears (Ed.), *Quality Improvement Tools and Techniques*, McGraw Hill, New York, 1995, pp. 229–246.

J. S. Oakland (Ed.), *Total Quality Management: Proceedings of the Second International Conference*, IFS Publications, Kempston, Bedford, UK, 1989.

J. S. Oakland, *Total Quality Management: Text with Cases*, Butterworth-Heinemann, Boston, MA, 2003.

K. H. Pries and J. M. Quigley, *Total Quality Management for Project Management*, CRC Press, Boca Raton, FL, 2012.

H. K. Rampersad, *Total Quality Management: An Executive Guide to Continuous Improvement*, Springer, New York, 2000.

J. E. Ross, *Total Quality Management: Text, Cases, and Readings*, St. Lucie Press, Boca Raton, FL, 1999.

A. R. Shores, *Survival of the Fittest: Total Quality Control and Management*, ASQC Quality Press, Milwaukee, WI, 1988.

R. E. Stein, *The Next Phase of Total Quality Management*, Marcel Dekker, New York, 1994.

R. R. Tennerand and I. J. Detoro, *Total Quality Management: Three Steps to Continuous Improvement*, Addison-Wesley, Reading, MA, 1992.

Design Quality

J. A. Burgess, "Assuring the Quality of Design," *Machine Design*, February 1982, pp. 65–69.

A. P. Chaparian, "Teammates: Design and Quality Engineers," *Quality Prog.*, **10**(4), 16–17, April 1977.

Colloquium on Management of Design quality Assurance, IEE Colloquium Digest No. 1988/6, Institution of Electrical Engineers, London, 1988.

D. Daetz, "The Effect of Product Design on Product Quality and Product Cost," *Quality Prog.*, **20**, 63–67, June 1987.

J. R. Evans and W. M. Lindsay, "Quality and Product Design," in *Management and Control of Quality*, West, New York, 1982, pp. 188–221.

F. M. Gryna, Designing for Quality, in *Quality Planning and Analysis*, Edited by F.M. Gryna, McGraw Hill Book Company, New York, 2001, pp. 332–372.

J. J. Juran, *Juran on Quality by Design*, Free Press, New York, 1992.

R. H. Lockner and J. E. Matar, *Designing for Quality*, ASQC Quality Press, Milwaukee, WI, 1990.

J. M. Michalek and R. K. Holmes, "Quality Engineering Techniques in Product Design/Process", in *Quality Control in Manufacturing*, Society of Automotive Engineers, SP-483, pp. 17–22.

M. S. Phadke, *Quality Engineering Using Robust Design*, Prentice Hall, Englewood Cliffs, NJ, 1986.

J.J. Pignatiello, J.S. Ramberg, "Discussion on Off-line Quality Control, Parameter Design, and the Taguchi Method," *J. Quality Technol.*, Vol.**17**, 1985, pp., 198–206.

"Quality Assurance in the Design of Nuclear Power Plants: A Safety Guide," Report No. 50-SG-QA6, International Atomic Energy Agency, Vienna, 1981.

"Quality Assurance in the Procurement, Design, and Manufacture of Nuclear Fuel Assemblies: A Safety Guide," Report No. 50-SG-QA 11, International Atomic Agency, Vienna, 1983.

J. B. Revelle, J. W. Moran, and C. A. Cox, *The QFD Handbook*, John Wiley and Sons, New York, 1998.

P. J. Ross, *Taguchi Techniques for Quality Engineering*, McGraw Hill, New York, 1988.

M. Speegle, *Quality Concepts for the Process Industry*, Delmar Cengage Learning, Clifton Park, New York, 2010.

J. Turmel and L. Gartz, "Designing in Quality Improvement: A Systematic Approach to Designing for Six Sigma," in *Proceedings of the Annual Quality Congress*, 1997, pp. 391–398.

M. Vonderembse, T. Van Fossen, and K. Raghunathan, "Is Quality Function Deployment Good for Product Development? Forty Companies Say Yes," *Quality Manag. J.*, Vol. **4**, No. 3, 1997, pp. 65–79.

CHAPTER 6

RELIABILITY IN THE MECHANICAL DESIGN PROCESS

B.S. Dhillon
University of Ottawa
Ottawa, Ontario, Canada

1 INTRODUCTION

The history of the reliability field may be traced back to the early 1930s when probability concepts were applied to electric power system-related problems.[1-7] During World War II, German researchers applied the basic reliability concepts to improve reliability of their

V1 and V2 rockets. During the period from 1945 to 1950, the United States Department of Defense conducted various studies that revealed a definite need to improve equipment reliability. Consequently, the Department of Defense formed an ad hoc committee on reliability in 1950. In 1952, this committee was transformed to a permanent body: Advisory Group on the Reliability of Electronic Equipment (AGREE).[8]

The group released its report in 1957. In 1951, W. Weibull proposed a function to represent time to failure of various engineering items.[9] Subsequently, this function became known as the Weibull distribution, and it is regarded as the starting point of mechanical reliability along with the works of A.M. Freudenthal.[10,11]

In the early 1960s, National Aeronautics and Space Administration (NASA) played an important role in the development of the mechanical reliability field, basically due to the following three factors[12]:

- The loss of Syncom I in space in 1963 due to a busting high-pressure gas tank
- The loss of Mariner III in 1964 due to a mechanical failure
- The frequent failure of components such as valves, regulators, and pyrotechnics in the Gemini spacecraft systems

Consequently, NASA initiated and completed many projects concerned with mechanical reliability. A detailed history of mechanical reliability is given in Refs. 13–15 along with a comprehensive list of publications on the subject up to 1992.

2 STATISTICAL DISTRIBUTIONS AND HAZARD RATE MODELS

Various types of statistical distributions and hazard rate models are used in mechanical reliability to represent failure times of mechanical items. This section presents some of these distributions and models considered useful to perform various types of mechanical reliability analysis.

2.1 Statistical Distributions

This section presents three statistical or probability distributions: exponential, Weibull, and normal.

Exponential Distribution

This is probably the most widely used distribution in reliability work to represent failure behavior of various engineering items.[16] Moreover, it is relatively easy to handle in performing reliability analysis in the industrial sector. Its probability density function is expressed by[14,16]

$$f(t) = \lambda e^{-\lambda t} \quad \text{for } \lambda \rangle 0, t \geq 0 \tag{1}$$

where $f(t)$ = probability density function
   λ = distribution parameter (in reliability work, it is known as the constant failure rate)
   t = time

The cumulative distribution function is given by[14,16]

$$F(t) = \int_0^t f(t)\, dt = \int_0^t \lambda e^{-\lambda t} dt$$

$$= 1 - e^{-\lambda t} \tag{2}$$

where $F(t)$ is the cumulative distribution function.

Weibull Distribution

This distribution was developed by W. Weibull in the early 1950s, and it can be used to represent many different physical phenomena.[9] The distribution probability density function is expressed by[14,16]

$$f(t) = \frac{\alpha\, t^{\alpha-1}}{\theta^\alpha}\, e^{-(t/\theta)^\alpha} \qquad \text{for } t \geq 0, \alpha \rangle 0, \theta \rangle 0 \tag{3}$$

where σ and α are the distribution scale and shape parameters, respectively.

The cumulative distribution function is given by[14,16]

$$F(t) = \int_0^t f(t)\, dt = \int_0^t \frac{\alpha\, t^{\alpha-1}}{\theta^\alpha}\, e^{-(t/\theta)^\alpha}\, dt$$

$$= 1 - e^{-(t/\theta)^\alpha} \tag{4}$$

For $\alpha = 1$ and $\alpha = 2$, the Weibull distribution becomes exponential and Rayleigh distributions, respectively.

Normal Distribution

This is one of the most widely known distributions. In mechanical reliability, it is often used to represent an item's stress and strength. The probability density function of the distribution is expressed by

$$f(t) = \frac{1}{\sigma\sqrt{2\pi}}\, \exp\left(-\frac{t-\mu}{2\sigma^2}\right) - \infty \langle t \langle +\infty \tag{5}$$

where μ and σ are the distribution parameters (i.e., menu and standard deviation, respectively).

The cumulative distribution function is given by[14,16]

$$F(t) = \int_{-\infty}^t f(t)\, dt = \frac{1}{\sigma\sqrt{2\pi}} \int_{-\infty}^t \exp\left[-\frac{(x-\mu)^2}{2\sigma^2}\right]\, dx \tag{6}$$

2.2   Hazard Rate Models

In reliability studies, the term *hazard rate* is often used. It simply means constant or nonconstant failure rate of an item. Thus, the hazard rate of an item is expressed by

$$h(t) = \frac{f(t)}{1 - F(t)} \tag{7}$$

where $h(t)$ is the item hazard rate.

This section presents four hazard rate models considered useful to perform various types of mechanical reliability studies.

Exponential Distribution Hazard Rate Model

By substituting Eqs. (1) and (2) into Eq. (7), we get the following equation for the exponential distribution hazard rate function:

$$h(t) = \frac{\lambda\, e^{-\lambda t}}{1 - \{1 - e^{-\lambda t}\}}$$

$$= \lambda \tag{8}$$

As the right-hand side of Eq. (8) is independent of time, λ is called failure rate.

Weibull Distribution Hazard Rate Model

By substituting Eqs. (3) and (4) into Eq. (7), we get the following equation for the Weibull distribution hazard rate function:

$$h(t) = \frac{\dfrac{\alpha\, t^{\alpha-1}}{\theta^{\alpha}}\, e^{-(t/\theta)^{\alpha}}}{1 - [1 - e^{-(t/\theta)^{\alpha}}]} \qquad (9)$$

$$= \frac{\alpha\, t^{\alpha-1}}{\theta^{\alpha}}$$

For $\alpha = 1$ and $\alpha = 2$, Eq. (9) becomes the hazard rate function for exponential and Rayleigh distributions, respectively.

Normal Distribution Hazard Rate Model

By substituting Eqs. (5) and (6) into Eq. (7), we get the following equation for the normal distribution hazard rate function:

$$h(t) = \frac{\dfrac{1}{\sigma\sqrt{2\pi}} \exp\left(\dfrac{t-\mu}{2\sigma^2}\right)}{1 - \dfrac{1}{\sigma\sqrt{2\pi}} \displaystyle\int_{-\infty}^{t} \exp\left[-\dfrac{(x-\mu)^2}{2\sigma^2}\right] dx} \qquad (10)$$

General Distribution Hazard Rate Model

The distribution hazard rate function is defined by[17]

$$h(t) = c\lambda\, \alpha\, t^{\alpha-1} + (1-c)\, m\, t^{m-1}\, \theta\, e^{\theta\, t^{m}} \qquad \text{for } 0 \le c \le 1 \text{ and } \alpha, \theta, m, \lambda \rangle 0 \qquad (11)$$

where θ and λ = scale parameters
   α and m = shape parameters.
   t = time

The following distribution hazard rate functions are the special cases of Eq. (11):

- Bathtub; for $m = 1, \alpha = 0.5$.
- Makeham; for $m = 1, \alpha = 1$.
- Extreme value; for $c = 0, m = 1$.
- Weibull; for $c = 1$.
- Rayleigh; for $c = 1, \alpha = 2$.
- Exponential; for $c = 1, \alpha = 1$.

3   COMMON RELIABILITY NETWORKS

Components of a mechanical system can form configurations such as series, parallel, series–parallel, parallel–series, k-out-of-m, standby system, and bridge. Often, these configurations are referred to as the standard configurations. Sometime during the design process, it might be desirable to determine the reliability or the values of other related parameters of systems forming such configurations. All these configurations or networks are described below.[14,16]

3.1 Series Network

The block diagram of an m unit series network or configuration is shown in Fig. 1. Each block represents a system unit or component. If any one of the components fails, the system fails. It means, all of the series units must work normally for the system success.

For independent units, the reliability of the system shown in Fig. 1 is

$$R_S = R_1 R_2 R_3, \dots, R_m \tag{12}$$

where R_S = series system reliability
  m = number of units
  R_i = reliability of unit i; for $i = 1, 2, 3, \dots, m$

For constant unit failure rates of the units, Eq. (12) becomes[14]

$$R_S(t) = e^{-\lambda_1 t} e^{-\lambda_2 t} e^{-\lambda_3 t}, \dots, e^{-\lambda_m t}$$

$$= \exp\left(-\sum_{i=1}^{m} \lambda_i t\right) \tag{13}$$

where $R_S(t)$ = series system reliability at time t
  λ_i = constant failure rate of unit i; for $i = 1, 2, 3, \dots, m$

The system hazard rate is given by[14]

$$\lambda_S(t) = \frac{f_S(t)}{1 - F_S(t)} = -\frac{1}{R_S(t)} \cdot \frac{dR_S(t)}{dt}$$

$$= \sum_{i=1}^{m} \lambda_i \tag{14}$$

where $\lambda_S(t)$ = series system hazard rate or the total failure rate
  $f_S(t)$ = series system probability density function
  $F_S(t)$ = series system cumulative distribution function

It is to be noted that the system total failure rate given by Eq. (14) is the sum of the all units' failure rates. It simply means that whenever the failure rates of units are added, it is automatically assumed that the units are acting in series (i.e., if any one unit fails, the system fails). This is the worst-case assumption often practiced in the design of engineering systems.

The system mean time to failure (MTTF) is given by[14]

$$\mathrm{MTTF}_S = \int_0^\infty R_S(t)\,dt = \int_0^\infty \exp\left(-\sum_{i=1}^{m} \lambda_i t\right) dt$$

$$= \frac{1}{\sum_{i=1}^{m} \dfrac{1}{\lambda_i}} \tag{15}$$

where MTTF_S is the series system mean time to failure.

Figure 1 Series system block diagram.

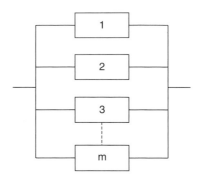

Figure 2 Parallel system block diagram.

3.2 Parallel Network

This type of configuration can be used to improve a mechanical system's reliability during the design phase. The block diagram of an *m* unit parallel network is shown in Fig. 2. Each block in the diagram represents a unit. This configuration assumes that all of its units are active and at least one unit must work normally for the system success.

For independently failing units, the reliability of the parallel network shown in Fig. 2 is expressed by[14]

$$R_p = 1 - \prod_{i=1}^{m}(1 - R_i) \tag{16}$$

where R_p = reliability of the parallel network
R_i = reliability of unit *i*; for $i = 1, 2, 3, \ldots, m$

For constant failure rates of the units, Eq. (16) becomes[14]

$$R_p(t) = 1 - \prod_{i=1}^{m}\left(1 - e^{-\lambda_i t}\right) \tag{17}$$

where $R_p(t)$ = parallel network reliability at time *t*
λ_I = constant failure rate of unit *i* ; for $i = 1, 2, 3, \ldots, m$

For identical units, the network mean time to failure is given by[14]

$$
\begin{aligned}
\mathrm{MTTF}_p &= \int_0^{\infty} R_p(t)\, dt \\
&= \frac{1}{\lambda} \sum_{i=1}^{m} \frac{1}{i}
\end{aligned}
\tag{18}
$$

where MTTF_p = parallel network mean time to failure
λ = unit constant failure rate

3.3 Series–Parallel Network

The network block diagram is shown in Fig. 3. Each block in the diagram represents a unit. This network represents a system having m number of subsystems in series. In turn, each subsystem

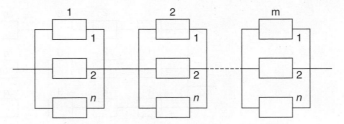

Figure 3   Series–parallel network block diagram.

contains n number of active units in parallel. All subsystems must operate normally for the system to succeed.

For independent units, the reliability of the series–parallel network shown in Fig. 3 is given by[18,19]

$$R_{\mathrm{Sp}} = \prod_{i=1}^{m} \left[1 - \prod_{j=1}^{n} F_{ij} \right] \tag{19}$$

where R_{Sp} = series–parallel network reliability
m = number of subsystems
n = number of units
F_{ij} = ith subsystem's jth unit's failure probability

For constant unit failure rates of the units, Eq. (19) becomes[18,19]

$$R_{\mathrm{Sp}}(t) = \prod_{i=1}^{m} \left[1 - \prod_{j=1}^{n} \left(1 - e^{-\lambda_{ij} t} \right) \right] \tag{20}$$

where $R_{\mathrm{Sp}}(t)$ = series–parallel network reliability at time t
λ_{ij} = constant failure rate of unit ij

For identical units, the network mean time to failure is given by[19]

$$\begin{aligned}
\mathrm{MTTF}_{\mathrm{Sp}} &= \int_{0}^{\infty} R_{\mathrm{Sp}}(t)\, dt \\[6pt]
&= \frac{1}{\lambda} \sum_{i=1}^{m} \left[(-1)^{i+1} \binom{m}{i} \sum_{j=1}^{in} \frac{1}{j} \right]
\end{aligned} \tag{21}$$

where $\mathrm{MTTF}_{\mathrm{Sp}}$ = series–parallel network mean time to failure
λ = unit failure rate

3.4   Parallel–Series Network

This network represents a system having m number of subsystems in parallel. In turn, each subsystem contains n number of units in series. At least one subsystem must function normally for the system to succeed. The network block diagram is shown in Fig. 4. Each block in the diagram represents a unit.

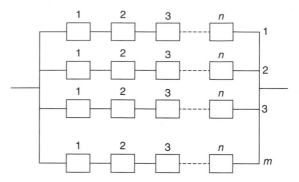

Figure 4   Parallel–series network block diagram.

For independent units, the reliability of the parallel–series network shown in Fig. 4 is expressed by[18,19]

$$R_{\text{pS}} = 1 - \prod_{i=1}^{m} \left(1 - \prod_{j=1}^{n} R_{ij} \right) \tag{22}$$

where R_{pS} = parallel–series network reliability
   m = number of subsystems
   n = number of units
   R_{ij} = ith subsystem's jth unit's reliability

For constant failure rates of the units, Eq. (23) becomes[18,19]

$$R_{\text{pS}}(t) = 1 - \prod_{i=1}^{m} \left(1 - \prod_{j=1}^{n} e^{-\lambda_{ij} t} \right) \tag{23}$$

where $R_{\text{pS}}(t)$ = parallel–series network reliability at time t
   λ_{ij} = constant failure rate of unit ij

For identical units, the network mean time to failure is given by[19]

$$\text{MTTF}_{\text{pS}} = \int_{0}^{\infty} R_{\text{pS}}(t)\, dt$$

$$= \frac{1}{n\lambda} \sum_{i=1}^{m} \frac{1}{i} \tag{24}$$

where MTTF_{pS} is the parallel–series network mean time to failure.

3.5   *K*-out-of-*m* Unit Network

This network is sometimes referred to as partially redundant network. It is basically a parallel network with a condition that at least K units out of the total of m units must operate normally for the system to succeed.

For independent and identical units, the network reliability is given by[14,16]

$$R_{k/m} = \sum_{i=K}^{m} \binom{m}{i} R^{i} (1-R)^{m-i} \tag{25}$$

where $\dbinom{m}{i} = \dfrac{m!}{i!\,(m-i)!}$

$R_{K/m}$ = K-out-of-m unit network reliability

R = unit reliability

For $K = 1$ and $K = m$, Eq. (25) becomes the reliability expression for parallel and series networks, respectively. More specifically, parallel and series networks are the special cases of the K-out-of-m unit network.

For constant failure rates of the units, Eq. (25) becomes[14,16]

$$R_{k/m} = \sum_{i=K}^{m} \binom{m}{i} e^{-i\lambda t} (1 - e^{-\lambda t})^{m-i} \tag{26}$$

where $R_{K/m}$ = K-out-of-m unit network reliability at time t

λ = unit constant failure rate

The network mean time to failure is given by[14,16]

$$\mathrm{MTTF}_{k/m} = \int_{0}^{\infty} R_{k/m}(t)\, dt \tag{27}$$

$$= \frac{1}{\lambda} \sum_{i=k}^{m} \frac{1}{i}$$

where $\mathrm{MTTF}_{K/m}$ is the K-out-of-m unit network mean time to failure.

3.6   Standby System

This is another configuration used to improve reliability. The block diagram of an $(m+1)$ unit standby system is shown in Fig. 5. Each block in the diagram represents a unit. In this configuration, one unit operates and m units are kept on standby. As soon as the operating unit fails, it is replaced by one of the standbys. The system fails when all of its units fail (i.e., operating plus all standbys)

For perfect switching, independent and identical units, and as good as new standby units, the standby system reliability is given by[14,16]

$$R_{\mathrm{SS}}(t) = \left\{ \sum_{i=0}^{m} \left[\int_{0}^{t} \lambda(t)\, dt \right]^{i} \exp\left[-\int_{0}^{t} \lambda(t)\, dt \right] \right\} \Big/ i! \tag{28}$$

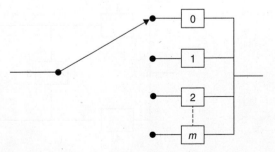

Figure 5   Block diagram of an $(m+1)$ unit standby system.

where $R_{SS}(t)$ = standby system reliability at time t
$\lambda(t)$ = unit hazard rate or time-dependent failure rate
m = number of standbys

For constant failure rates of the units [i.e., $\lambda(t) = \lambda$] Eq. (28) becomes

$$R_{SS}(t) = \left[\sum_{i=0}^{m} (\lambda t)^i \, e^{-\lambda t} \right] \Big/ i! \tag{29}$$

where λ is the unit constant failure rate.
The system mean time to failure is given by[14]

$$
\begin{aligned}
\text{MTTF}_{SS} &= \int_0^\infty R_{SS}(t)\,dt \\
&= \int_0^\infty \left\{ \left[\sum_{i=0}^{m} (\lambda t)^i \, e^{-\lambda t} \right] \Big/ i! \right\} dt \\
&= \frac{m+1}{\lambda}
\end{aligned} \tag{30}
$$

where MTTF_{SS} is the standby system mean time to failure.

3.7 Bridge Network

The block diagram of a bridge network is shown in Fig. 6. Each block in the diagram represents a unit. Mechanical components sometime can form this type of configuration.
For independent units, the bridge network shown in Fig. 6 reliability is[20]

$$
\begin{aligned}
R_{bn} &= 2 \prod_{i=1}^{5} R_i + \prod_{i=1}^{4} R_i + R_1 R_3 R_5 + R_1 R_4 + R_2 R_5 \\
&\quad - \prod_{i=2}^{5} R_i - \prod_{i=1}^{4} R_i - R_5 \prod_{i=1}^{3} R_i - R_1 \prod_{i=3}^{5} R_i \\
&\quad - R_1 R_2 R_4 R_5
\end{aligned} \tag{31}
$$

where R_{bn} = bridge network reliability
R_i = unit i reliability; for $i = 1, 2, 3, 4, 5$

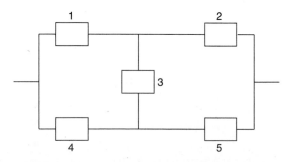

Figure 6   Block diagram of a five-unit bridge network.

For identical units and constant failure rates of the units, Eq. (31) becomes[14]

$$R_{\text{bn}}(t) = 2\,e^{-5\lambda t} - 5\,e^{-4\lambda t} + 2\,e^{-3\lambda t} + 2\,e^{-2\lambda t} \tag{32}$$

where λ is the unit constant failure rate.

The network mean time to failure is given by

$$\begin{aligned}
\text{MTTF}_{\text{bn}} &= \int_0^\infty R_{\text{bn}}(t)\,dt \\[2mm]
&= \frac{49}{60\,\lambda}
\end{aligned} \tag{33}$$

where MTTF_{bn} is the bridge network mean time to failure.

4 MECHANICAL FAILURE MODES AND CAUSES OF GENERAL AND GEAR FAILURES

A mechanical failure may be defined as any change in the shape, size, or material properties of a structure, piece of equipment, or equipment part that renders it unfit to carry out its specified mission adequately.[13] Thus, there are many different types of failure modes associated with mechanical items. Good design practices can reduce or eliminate altogether the occurrence of these failure modes. Nonetheless, some of these failure modes are as follows[14,21–23]:

- **Shear loading failure.** The ultimate and yield failure occurs when the shear stress surpasses the material strength when applying high torsion or shear loads. Usually this type of failure takes place along a 45° axis with respect to the principal axis.

- **Instability failure.** This type of failure takes place in structural members such as beams and columns, particularly the ones produced using thin material where the loading is generally in compression. However, instability failure may also occur due to torsion or by combined loading, i.e., including compression and bending.

- **Bending failure.** This type of failure may also be called a combined failure because it takes place when one outer surface is in compression and the other outer surface is in tension. The tensile rupture of the outer material is a good representative of the bending failure.

- **Compressive failure.** This type of failure causes permanent rupturing/cracking/deformation and is similar to the tensile failures but with one exception, i.e., under compressive loads.

- **Tensile yield strength failure.** This type of failure takes place under pure tension when the applied stress exceeds the material's yield strength.

- **Fatigue failure.** This failure occurs because of repeated unloading (or partial unloading) or loading of an item/unit.

- **Ultimate tensile strength failure.** This type of failure takes place when the applied stress exceeds the ultimate tensile strength and leads to a complete failure of the structure at a cross-sectional point.

- **Creep/rupture failure.** Generally, long-term loads make elastic materials stretch even when they are smaller than the material's normal yield strength. Nonetheless, material stretches (creeps) when the load is maintained on a continuous basis and generally it ultimately terminates in a rupture. Furthermore, at elevated temperatures, creep accelerates.

- **Stress concentration failure.** This failure takes place in circumstances of uneven stress "flow" through a mechanical design. Generally, the concentration of stress takes place at sudden variations in loading along a structure, at right-angle joints, at sudden transitions from thick gauges to thin gauges, or at various attachment conditions.

- **Material flaw failure.** This failure normally takes place because of factors such as weld defects, fatigue cracks, small cracks and flaws, and poor quality assurance.
- **Metallurgical failure.** This type of failure occurs because of extreme oxidation or operation in corrosive environment. Some of the environmental conditions that accelerate the occurrence of metallurgical failures are heat, nuclear radiation, corrosive media, and crosion.
- **Bearing failure.** This failure is quite similar in nature to the compressive failure, and generally it takes place because of a round cylindrical surface bearing on either a flat or a concave surface like roller bearings in a race. It is very important to consider the possibility of the occurrence of fatigue failures when repeated loading occurs.

There are many causes of product failures. Some of these are as follows[24]:

- Defective design
- Wear-out
- Defective manufacturing
- Wrong application
- Incorrect installation
- Failure of other parts

A study performed over a period of 35 years reported a total of 931 gear failures.[24] They were classified under four categories: breakage (61.2%), surface fatigue (20.3%), wear (3.2%), and plastic flow (5.3%). Furthermore, the causes of these failures were grouped under five categories as shown in Fig. 7.

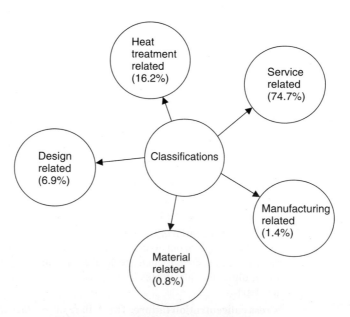

Figure 7   Classifications of the gear failure causes.

Each of these five categories were further broken into various elements. The elements of the service-related classification were continual overloading (25%), improper assembly (21.2%), impact loading (13.9%), incorrect lubrication (11%), foreign material (1.4%), abusive handling (1.2%), bearing failure (0.7%), and operator errors (0.3%).

The elements of the heat-treatment-related classification were incorrect hardening (5.9%), inadequate case depth (4.8%), inadequate core hardness (2%), excessive case depth (1.8%), improper tempering (1%), excessive core hardness (0.5%), and distortion (0.2%).

The elements of the design-related classification were wrong design (2.8%), specification of suitable heat treatment (2.5%), and incorrect material selection (1.6%). The elements of the manufacturing-related classification were grinding burns (0.7%) and tool marks or notches (0.7%). Finally, the three elements of the material-related classification were steel defects (0.5%), mixed steel or incorrect composition (0.2%), and forging defects (0.1%).

5   RELIABILITY-BASED DESIGN AND DESIGN-BY-RELIABILITY METHODOLOGY

It would be unwise to expect a system to perform to a desired level of reliability unless it is specifically designed for that reliability. The specification of desired system/equipment/part reliability in the design specifications due to factors such as well-publicized failures (e.g., the space shuttle *Challenger* disaster and the Chernobyl nuclear accident) has increased the importance of reliability-based design. The starting point for the reliability-based design is during the writing of the design specification. In this phase, all reliability needs and specifications are entrenched into the design specification. Examples of these requirements might include item mean time to failure (MTTF), mean time to repair (MTTR), test or demonstration procedures to be used, and applicable document.

The U.S. Department of Defense, over the years, has developed various reliability documents for use during the design and development of an engineering item. Many times, such documents are entrenched into the item design specification document. Table 1 presents some of these documents. Many professional bodies and other organizations have also developed documents on various aspects of reliability.[15,22,25] References 22 and 26 provide descriptions of documents developed by the U.S. Department of Defense.

Table 1   Selected Design-Related Documents Developed by U.S. Department of Defense

Document No.	Document Title
MIL-STD-721	Definitions of Terms for Reliability and Maintainability
MIL-STD-217	Reliability Prediction of Electronic Equipment
MIL-STD-781	Reliability Design, Qualification, and Production Acceptance Tests: Exponential Distribution
MIL-STD-756	Reliability Modeling and Prediction
MIL-STD-785	Reliability Program for Systems and Equipment
MIL-HDBK-251	Reliability/Design Thermal Applications
MIL-STD-1629	Procedures for Performing a Failure Mode, Effects, and Criticality Analysis
RADC-TR-75-22	Non-electronic Reliability Notebook
MIL-STD-965	Parts Control Program
MIL-STD-2074	Failure Classification for Reliability Testing

Reliability is an important consideration during the design phase. According to Ref. 27, as many as 60% of failures can be eliminated through design changes. There are many strategies the designer could follow to improve design:

- Eliminate failure modes.
- Focus design for fault tolerance.
- Focus design for fail-safe.
- Focus design to include mechanism for early warnings of failure through fault diagnosis.

During the design phase of a product, various types of reliability and maintainability analyses can be performed, including reliability evaluation and modeling, reliability allocation, maintainability evaluation, human factors/reliability evaluation, reliability testing, reliability growth modeling, and life-cycle costing. In addition, some of the design improvement strategies are zero-failure design, fault-tolerant design, built-in testing, derating, design for damage detection, modular design, design for fault isolation, and maintenance-free design. During design reviews, reliability and maintainability-related actions recommended/taken are to be thoroughly reviewed from desirable aspects.

A systematic series of steps are taken to design a reliable mechanical item. The design by methodology is composed of such steps[14,28,29]:

- Define the design problem under consideration.
- Identify and list all the associated design variables and parameters.
- Perform failure modes, effect and criticality analysis according to MIL-STD-1629.[30]
- Verify critical design parameter selection.
- Establish appropriate relationships between the failure-governing criteria and the critical parameters.
- Determine the failure-governing stress and strength functions and then the most appropriate failure-governing stress and strength distributions.
- Estimate the reliability utilizing the failure-governing stress and strength distributions for all critical failure modes.
- Iterate the design until the achievement of the set reliability goals.
- Optimize design with respect to factors such as safety, reliability, cost, performance, maintainability, weight, and volume.
- Repeat the design optimization process for all critical components.
- Estimate item reliability.
- Iterate the design until the item set reliability goals are fully satisfied.

6 DESIGN RELIABILITY ALLOCATION AND EVALUATION METHODS

Over the years, many reliability allocation and evaluation methods have been developed for use during the design phase.[14,29] This section presents some of these methods and techniques considered useful, particularly in designing mechanical items.

6.1 Failure Rate Allocation Method

This method is used to allocate failure rates to system components when the overall system required failure rate is given. The method is based on the following three assumptions[14]:

- All system components fail independently.
- Component failure rates are constant.
- System components form a series network.

Thus, the system failure rate using Eq. (14) is

$$\lambda_S = \sum_{i=1}^{m} \lambda_i \tag{34}$$

where λ_S = system failure rate
m = total number of system components
λ_I = failure rate of component i; for $i = 1, 2, 3, \ldots, m$

If the specified failure rate of the system is λ_{Sp}, then allocate component failure rate such that

$$\sum_{i=1}^{m} \lambda_i^* \leq \lambda_{\text{Sp}} \tag{35}$$

where λ_i^* is the failure rate allocated to component i; for $i = 1, 2, 3, \ldots, m$
The following three steps are associated with this approach:

- Estimate failure rates of the system components (i.e., λ_i for $i = 1, 2, 3, \ldots, m$) using the field data.
- Calculate the relative weight, α_i, of component i using the preceding step failure rate data and the following expression:

$$\alpha_i = \frac{\lambda_i}{\sum_{i=1}^{m} \lambda_i} \qquad \text{for } i = 1, 2, 3, \ldots, m \tag{36}$$

It is to be noted that α_i represents the relative failure vulnerability of component i and

$$\sum_{i=1}^{m} \alpha_i = 1 \tag{37}$$

- Allocate failure rate to part of component i by using the following equation:

$$\lambda_i^* = \alpha_i \lambda_{\text{Sp}}, \qquad \text{for } i = 1, 2, 3, \ldots, m \tag{38}$$

A solved example in Ref. 14 demonstrates the application of this method.

6.2 Hybrid Reliability Allocation Method

This method is the result of combining two reliability allocation methods known as the similar familiar systems and factors of influence. The method is more attractive because it incorporates benefits of both similar familiar systems and factors of influence methods[14,29].

Nonetheless, the basis for the similar familiar systems approach is the familiarity of the designer with similar systems as well as the utilization of failure data collected on similar systems from various sources, during the allocation process. The principal disadvantage of the similar familiar systems method is the assumption that reliability and life-cycle cost of similar systems were satisfactory or adequate.

The factors of influence method is based upon four factors that are considered to affect the item reliability: failure criticality, environment, complexity/time, and state of the art.

The failure criticality factor is concerned with the criticality of the item failure on the system (e.g., the failure of some auxiliary equipment in an aircraft may not be as critical as the failure of an engine). The environment factor takes into consideration the susceptibility of items to conditions such as vibration, humidity, and temperature.

The complexity/time factor relates to the number of item/subsystem components/parts and operating time of the item under consideration during the total system operating period. Finally, the state-of-the-art factor takes into consideration the advancement in the state of the art for an item under consideration.

During the reliability allocation process, each item is rated with respect to each of these four influence factors by assigning a number from 1 to 10. The assignment of 1 means the item is least affected by the influence factor under consideration and 10 means the item is most affected by the same factor. Subsequently, the reliability is allocated by using the weight of these assigned numbers for all the four influence factors.

6.3   Safety Factor and Safety Margin

Safety factor and safety margin, basically, are arbitrary multipliers, and they are used to ensure reliability of mechanical items during the design phase. These indexes can provide satisfactory design if they are established using considerable past experiences and data.

Safety Factor
A safety factor can be defined in many different ways.[13,31–35] Two commonly used definitions are as follows:

Definition I.   The safety factor is defined by[36]

$$SF = \frac{m_{Sh}}{m_{SS}} \geq 1 \qquad (39)$$

where   SF = safety factor
m_{Sh} = mean failure-governing strength
m_{SS} = mean failure-governing stress

This index is a good measure of safety when both stress and strength are normally distributed. However, when the spread of both strength and/or stress is large, the index becomes meaningless because of positive failure rate.[14]

Definition II.   The safety factor is defined by[32–37]

$$SF = \frac{US}{WS} \qquad (40)$$

where   SF = safety factor
WS = working stress expressed in pounds per square inch (psi)
US = ultimate strength expressed in psi

Safety Margin
This is defined by[14,31]

$$SM = SF - 1 \qquad (41)$$

where SM is the safety margin.

The negative value of this measure means that the item under consideration will fail. Thus, its value must always be greater than zero.

Safety margin for normally distributed stress and strength is expressed by[14,31]

$$SM = \frac{\mu_{as} - \mu_{ms}}{\sigma_{Sh}} \qquad (42)$$

where μ_{as} = average strength
μ_{ms} = maximum stress
σ_{Sh} = strength standard deviation

In turn the maximum stress, μ_{ms}, is expressed by

$$\mu_{ms} = \mu_{SS} + C\,\sigma_{SS} \qquad (43)$$

where μ_{SS} = mean stress
$\quad\quad\sigma_{SS}$ = stress standard deviation
$\quad\quad C$ = factor that takes values between 3 and 6

Example 1

Assume that the following data values are specified for a mechanical item/product under design:

- μ_{as} = 20,000 psi
- σ_{sh} = 1100 psi
- μ_{ss} = 10,000 psi
- σ_{ss} = 200 psi

For $C = 3$ and 6, estimate the values of the safety margin.
By substituting the given data values into Eq. (43), we get:
For $C = 3$

$$\mu_{ms} = 10,000 + 3(200)$$
$$= 10,600 \text{ psi}$$

For $C = 6$

$$\mu_{ms} = 10,000 + 6(200)$$
$$= 11,200 \text{ psi}$$

By substituting the above calculated values and the specified data values into Eq. (42), we obtain:
For $C = 3$

$$SM = \frac{20,000 - 10,600}{1100}$$
$$= 8.55$$

For $C = 6$

$$SM = \frac{20,000 - 11,200}{1100}$$
$$= 8$$

Thus, the values of the safety margin for $C = 3$ and $C = 6$ are 8.55 and 8, respectively.

6.4   Stress–Strength Interference Theory Method

This method is used to determine reliability of a mechanical item when its associated stress and strength probability density functions are known. The item reliability is defined by[13,14,29]

$$R = P(y < x) = P(x > y) \qquad (44)$$

where R = item reliability
$\quad\quad P$ = probability
$\quad\quad X$ = strength random variable
$\quad\quad Y$ = stress random variable

Equation (44) is rewritten to the following form[13,14,29]:

$$R = \int_{-\infty}^{\infty} f(y) \left[\int_{y}^{\infty} f(x)\,dx \right] dy \tag{45}$$

where $f(x)$ = strength probability density function
$f(y)$ = stress probability density function

Special Case Model I: Exponentially Distributed Stress and Strength
In this case, an item's stress and strength are defined by

$$f(y) = \theta\,e^{-\theta y} \qquad 0 \le y \langle \infty \tag{46}$$

and

$$f(x) - \lambda\,e^{-\lambda x} \qquad 0 \le x \langle \infty \tag{47}$$

where σ and λ are the reciprocals of the means values of stress and strength, respectively.
Using Eqs. (46) and (47) in Eq. (45) yields

$$R = \int_{0}^{\infty} \theta\,e^{-\theta y} \left(\int_{y}^{\infty} \lambda\,e^{-\lambda x}\,dx \right) dy$$
$$= \frac{\theta}{\theta + \lambda} \tag{48}$$

For $\theta = 1/\bar{y}$ and $\lambda = 1/\bar{x}$, Eq. (48) becomes

$$R = \frac{\bar{x}}{\bar{x} + \bar{y}} \tag{49}$$

where $\bar{x}$ = mean strength
$\bar{y}$ = mean stress

Special Case Model II: Exponentially Distributed Strength and Normally Distributed Stress
In this case, an item's strength and stress are defined by

$$f(x) = \lambda e^{-\lambda x} \qquad x \ge 0 \tag{50}$$

and

$$f(y) = \frac{1}{\sigma\sqrt{2\pi}} \exp\left[-\frac{1}{2} \left(\frac{y - \mu}{\sigma} \right)^2 \right] \qquad -\infty \langle y \langle \infty \tag{51}$$

where σ = stress standard deviation
μ = mean stress
λ = reciprocal of the mean value of strength

Using Eqs. (50) and (51) in Eq. (45) yields

$$R = \int_{-\infty}^{\infty} \left\{ \frac{1}{\sigma\sqrt{2\pi}} \exp\left[-\frac{1}{2}\left(\frac{y-\mu}{\sigma}\right)^2 \right] \right\} \left(\int_{y}^{\infty} \lambda e^{-\lambda x} dx \right) dy$$
$$= \int_{-\infty}^{\infty} \frac{1}{\sigma\sqrt{2\pi}} \exp\left\{ -\left[\frac{1}{2}\left(\frac{y-\mu}{\sigma}\right)^2 + \lambda y \right] \right\} dy \tag{52}$$

Since

$$\frac{y - \mu}{2\sigma^2} + \lambda y = \frac{2\mu\lambda\sigma^2 - \lambda^2\sigma^4 + (y - \mu + \lambda\sigma^2)^2}{2\sigma^2} \tag{53}$$

Equation (52) yields

$$R = \exp[-\tfrac{1}{2}(2\mu\lambda - \lambda^2\sigma^2)] \tag{54}$$

Similarly, special case models for other stress and strength probability distributions can be developed. A number of such models are presented in Ref. 13.

6.5   Failure Modes and Effect Analysis

Failure modes and effect analysis (FMEA) is a vital tool for evaluating system design from the point of view of reliability. It was developed in the early 1950s to evaluate the design of various flight control systems[38].

The difference between the FMEA and failure modes, effects, and criticality analysis (FMECA) is that FMEA is a qualitative technique used to evaluate a design, whereas FMECA is composed of FMEA and criticality analysis (CA). Criticality analysis is a quantitative method used to rank critical failure mode effects by taking into consideration their occurrence probabilities.

As FMEA is a widely used method in industry, there are many standards/documents written on it. In fact, Ref. 30 collected and evaluated 45 such publications prepared by organizations such as the U.S. Department of Defense (DOD), National Aeronautics and Space Administration (NASA), Institute of Electrical and Electronic Engineers (IEEE), and so on. These documents include[39]:

- DOD: MIL-STD-785A (1969), MIL-STD-1629 (draft) (1980), MIL-STD-2070 (AS) (1977), MIL-STD-1543 (1974), AMCP-706-196 (1976)
- NASA: NHB 5300.4 (1A) (1970), ARAC Proj. 79-7 (1976)
- IEEE: ANSI N 41.4 (1976)

Details of the above documents as well as a list of publications on FMEA are given in Ref. 24. The main steps involved in performing FMEA are as follows[29]:

- Define carefully all system boundaries and detailed requirements.
- List all parts/subsystems in the system under consideration.
- Identify and describe each part and list all its associated failure modes.
- Assign failure rate/probability to each failure mode.
- List effects of each failure mode on subsystem/system/plant.
- Enter remarks for each failure mode.
- Review each critical failure mode and take appropriate measures.

This method is described in detail in Ref. 14.

6.6   Fault Tree Analysis

Fault tree analysis (FTA), so called because it arranges fault events in a tree-shaped diagram, is one of the most widely used techniques for performing system reliability analysis. In particular, it is probably the most widely used method in the nuclear power industry. The technique is well suited for determining the combined effects of multiple failures.

The fault tree technique is more costly to use than the FMEA approach. It was developed in the early 1960s at Bell Telephone Laboratories to evaluate the reliability of the Minuteman

Launch Control System. Since that time, hundreds of publications on the method have appeared[15].

The fault tree analysis begins by identifying an undesirable event, called the "top event," associated with a system. The fault events that could cause the occurrence of the top event are generated and connected by logic gates known as AND, OR, and so on. The construction of a fault tree proceeds by generation of fault events (by asking the question "How could this event occur?") in a successive manner until the fault events need not be developed further. These events are known as primary or elementary events. In simple terms, the fault tree may be described as the logic structure relating the top event to the primary events. This method is described in detail in Ref. 14.

7 HUMAN ERROR AND RELIABILITY CONSIDERATION IN MECHANICAL DESIGN

Just like in the reliability of any other systems, human reliability and error play an important role in the reliability of mechanical systems. Over the years many times mechanical systems/equipment have failed due to human errors rather than hardware failures. A careful consideration to human error and reliability during the design of mechanical systems can help to eliminate or reduce the occurrence of non-hardware-related failures during the operation of such systems. Human errors may be classified under the following seven distinct categories[14,40–42]:

- **Operator errors.** This type of error occurs due to operator mistakes, and the conditions that lead to operator errors include poor environment, complex tasks, operator carelessness, lack of proper procedures, and poor personnel selection and training.

- **Inspection errors.** This type of error occurs because of involved operators' less than 100% accuracy. An example of an inspector error is rejecting and accepting in-tolerance and out-of-tolerance components/parts, respectively. Some studies indicate that an average inspection effectiveness is around 85%[14,42].

- **Design errors.** This type of error occurs because of poor design. Some of the main causes of the design errors are failure to implement human needs in the design, failure to ensure the effectiveness of the human–machine interaction, and assigning inappropriate functions to humans. One example of a design error is the placement of displays and controls so far apart that an operator is unable to use them effectively.

- **Maintenance errors.** This type of error takes place in the field environment due to oversights by the maintenance personnel. Some examples of maintenance errors are repairing the failed equipment incorrectly, calibrating equipment incorrectly, and applying the incorrect grease at appropriate points of the equipment.

- **Assembly errors.** This type of error occurs due to humans during the product assembly process. The main causes of assembly errors include poor illumination, poor blueprints and other related material, poor communication of related information, poorly designed work layout, excessive noise level, and excessive temperature in the work area.

- **Handling errors.** This type of error basically occurs because of poor transportation or storage facilities. More clearly, such facilities are not according to the ones specified by the equipment manufacturer.

- **Installation errors.** This type of error occurs due to various reasons including using the incorrect installation instructions or blueprints, or simply failing to install equipment as per the specification of the equipment manufacturer.

Additional information on the above categories of human errors is available in Ref. 42.

There are numerous causes for the occurrence of human errors including poor equipment design, complex tasks, poor work layout, poorly written operating and maintenance procedures, poor job environment (i.e., poor lighting, high/low temperature, crowded work space, high noise level, etc.), inadequate work tools, poor skill of involved personnel, and poor motivation of involved personnel.[40-42]

Human reliability of time-continuous tasks such as aircraft maneuvering, scope monitoring, and missile countdown can be calculated by using the following equation[42,43]:

$$R_h(t) = \exp\left[-\int_0^t \lambda(t)\,dt\right] \tag{55}$$

where $R_h(t)$ = human reliability at time t
  $\lambda(t)$ = time-dependent human error rate

For constant human error rate [i.e., $\lambda(t) = \lambda$], Eq. (55) becomes

$$R_h(t) = \exp\left(-\int_0^t \lambda\,dt\right) \tag{56}$$
$$= e^{-\lambda t}$$

where λ is the constant human error rate.

The subject of human reliability and error is discussed in detail in Ref. 42.

Example 2

A person is performing a certain time-continuous task. His/her error rate is 0.004 per hour. Calculate the person's probability of performing the task correctly during a 5-h period.

By using the specified data values in Eq. (56), we get

$$R_h(5) = e^{-(0.004)5}$$
$$= 0.98$$

It means there is approximately 98% chance that the person will perform the task correctly during the specified period.

8   FAILURE RATE ESTIMATION MODELS FOR VARIOUS MECHANICAL ITEMS

There are many mathematical models available in the published literature that can be used to estimate failure rates of items such as bearings, pumps, brakes, filters, compressors, and seals.[44-46] This section presents some of these models.

8.1   Brake System Failure Rate Estimation Model

The brake system failure rate is expressed by[44]

$$\lambda_{\text{brs}} = \sum_{i=1}^{6} \lambda_i \tag{57}$$

where λ_{brs} = brake system failure rate, expressed in failures/10^6h
  λ_1 = brake housing failure rate
  λ_2 = total failure rate of actuators
  λ_3 = total failure rate of seals

λ_4 = total failure rte of bearings
λ_5 = total failure rate of springs
λ_6 = total failure rate of brake friction materials

The values of λ_i for $i = 1, 2, \ldots, b$ are obtained through various means.[44,47]

8.2   Compressor System Failure Rate Estimation Model

The compressor system failure rate is expressed by[45]

$$\lambda_{comp} = \sum_{i=1}^{6} \lambda_i \tag{58}$$

where λ_{comp} = compressor system failure rate, expressed in failures/10^6 h
λ_1 = failure rate of all compressor bearings
λ_2 = compressor casing failure rate
λ_3 = failure rate due to design configuration
λ_4 = failure rate of valve assay (if any)
λ_5 = failure rate of all compressor seals
λ_6 = failure rate of all compressor shafts

Procedures for calculating $\lambda_1, \lambda_2, \lambda_3, \lambda_4, \lambda_5,$ and λ_6 are presented in Ref. 45.

8.3   Filter Failure Rate Estimation Model

The filter failure rate is expressed by[46]

$$\lambda_{ft} = \lambda_b \prod_{i=1}^{6} \theta_i \tag{59}$$

where λ_{ft} = filter failure rate, expressed in failures/10^6h
λ_b = filter base failure rate
θ_i = ith modifying factor; for $i = 1$ is for temperature effects, $i = 2$ water contamination effects, $i = 3$ cyclic flow effects, $i = 4$ differential pressure effects, $i = 5$ cold start effects, and $i = 6$ vibration effects.

Procedures for estimating $\lambda_b, \theta_1, \theta_2, \theta_3, \theta_4, \theta_5,$ and θ_6 are given in Ref. 46.

8.4   Pump Failure Rate Estimation Model

The pump failure rate is expressed by[46]

$$\lambda_{pm} = \sum_{i=1}^{5} \lambda_i \tag{60}$$

where λ_{pm} = pump failure rate, expressed in failures/10^6 cycles
λ_1 = pump fluid driver failure rate
λ_2 = pump casing failure rate
λ_3 = pump shaft failure rate
λ_4 = failure rate of all pump seals
λ_5 = failure rate of all pump bearings

Procedures for calculating $\lambda_1, \lambda_2, \lambda_3, \lambda_4,$ and λ_5 are presented in Ref. 46.

8.5 Bearing Failure Rate Estimation Model

The bearing failure rate is expressed by[22]

$$\lambda_{\text{bfr}} = \lambda_{\text{pfr}} \left(\frac{I_a}{L_s} \right)^y \left(\frac{E_a}{0.006} \right)^{2.36} \left(\frac{V_{s1}}{V_{01}} \right)^{0.54} L_{\text{ac}}^{2/3} \tag{61}$$

where λ_{bfr} = bearing failure rate (using actual conditions), expressed in failures/10^6 h of operation

V_{s1} = specification lubricant viscosity, lb $\cdot$ min /in^2

V_{01} = operating lubricant viscosity, I lb $\cdot$ min /in^2.

y = factor; $y = 4.0$ for ball bearings and $y = 3.3$ roller bearings

L_s = specification load, psi

L_a = actual load, psi

E_a = alignment error, radians

L_{ac} = actual contamination level μg/m^3/60 μg/m^3

λ_{pfr} = bearing pseudofailure rate

The bearing pseudofailure rate is expressed by

$$\lambda_{\text{pfr}} = \frac{10^6 N^{0.9}}{(5.45) L_r} \tag{62}$$

where L_r = rated life in revolutions

N = total number of bearings in system

The rated life in revolutions is expressed by

$$L_r = 16,700/\text{rpm} \, (L_{\text{br}}/L_{\text{er}})^\theta \tag{63}$$

where L_{br} = basic load rating, lb

L_{er} = equivalent radial, load, lb

rpm = shaft rotating velocity, expressed in rev/min

θ = constant; $\theta = 10/3$ for roller bearings and $\theta = 3$ for ball bearings

8.6 Clutch System Failure Rate Estimation Model

The clutch system failure rate is expressed by the following equation[44]:

$$\lambda_c = \lambda_{\text{cf}} + \lambda_{\text{sp}} + \lambda_{\text{bf}} + \lambda_{\text{ss}} + \lambda_{\text{ac}} \tag{64}$$

where λ_c = clutch system failure rate, expressed in failures/10^6 h

λ_{cf} = clutch friction material failure rate

λ_{sp} = failure rate of springs

λ_{bf} = failure rate of bearings

λ_{ss} = failure rate of seals

λ_{ac} = failure rate of actuators

The clutch friction material failure rate is given by

$$\lambda_{\text{cf}} = \lambda_{\text{bcf}} F_1 F_2 \tag{65}$$

where F_1 = factor that considers the effects on the base failure rate of ambient temperature

F_2 = factor that considers the effects on the base failure rate of multiple plates

λ_{bcf} = clutch friction material base failure rate

The clutch friction material base failure rate is expressed by[44, 48]

$$\lambda_{bcf} = \frac{n\text{AE}}{2M_{aw}A_{cfm}L_{th}}$$ (66)

where   AE = average energy dissipated per engagement, expressed in ft-lbf
  A_{fm} = total area of the clutch friction material on each disk, expressed in inches^2
  n = number of applications per hour
  M_{aw} = average wear of the material
  L_{th} = lining thickness, expressed in inches

9   FAILURE DATA AND FAILURE DATA COLLECTION SOURCES

Failure data provide invaluable information to reliability engineers, design engineers, management, and so on concerning the product performance. These data are the final proof of the success or failure of the effort expended during the design and manufacture of a product used under designed conditions. During the design phase of a product, past information concerning its failures plays a critical role in reliability analysis of that product. Some of the uses of the failure data are estimating item failure rate, performing effective design reviews, predicting reliability and maintainability of redundant systems, conducting tradeoff and life-cycle cost studies, and performing preventive maintenance and replacement studies. Table 2 presents failure rates for selected mechanical items.[13,49,50]

There are many different ways and means for collecting failure data. For example, during the equipment life cycle, there are eight identifiable data sources: repair facility reports, development testing of the item, previous experience with similar or identical items, customer's failure-reporting systems, inspection records generated by quality control and manufacturing

Table 2   Failure Rates for Some Mechanical Items[a]

Item Description	Failure Rate $10^{-6}h$
Roller bearing	8.323
Bellows (general)	13.317
Filter (liquid)	6.00
Compressor (general)	33.624
Pipe	0.2
Hair spring	1.0
Pump (vacuum)	10.610
Gear (spur)	3.152
Seal (o-ring)	0.2
Nut or bolt	0.02
Brake (electromechanical)	16.00
Knob (general)	2.081
Washer (lock)	0.586
Washer (flat)	0.614
Duct (general)	2.902
Guide pin	13
Hose, pneumatic	29.3
Heavy-duty ball bearing	14.4
Pressure regulator	2.4

[a]Use environment: ground fixed or general.

Table 3   Selected Failure Data Sources for Mechanical Items

Author(s)	Source title/ Document Ref. No./Year	Developed by
M. J. Rossi	Non-electronic Parts Reliability Data, (Rept. No. NPRD-3, 1985)	Reliability Analysis Center, Rome Air Development Center, Griffis Air Force Base, Rome, New York.
—	Component Reliability Data for Use in Probabilistic Safety Assessment (1998)	International Atomic Energy Agency, Vienna, Austria.
R. G. Arno	Non-electronic Parts Reliability Data (Rept. No. NPRD-2, 1981)	Reliability Analysis Center, Rome Air Development Center, Griffis Air Force Base, Rome, New York.
—	Government Industry Data Exchange Program (GIDEP)	GIDEP Operations Center, U.S. Dept. of Navy, Seal Beach, Corona, California.
R. E. Schafer, J. E. Angus, J.M. Finkelstein, M. Yerasi, and D. W. Fulton	RADC Non-electronic Reliability Notebook (Rept. No. RADC-TR-85-194, 1985)	Reliability Analysis Center, Rome Air Development Center, Griffis Air Force Base, Rome, New York.

groups, tests conducted during field demonstration, environmental qualification approval, and field installation, acceptance testing, and warranty claims[51]. Table 3 presents some sources for collecting mechanical items' failure data for use during the design phase.[13]

REFERENCES

1. W. J. Layman, "Fundamental Considerations in Preparing a Master System Plan," *Elect. World*, **101**, 778–792, 1933.
2. S. A. Smith, "Spare Capacity Fixed by Probabilities of Outage," *Elect. World*, **103**, 222–225, 1934.
3. S. A. Smith, "Probability Theory and Spare Equipment," *Edison Electric Inst. Bull.*, March, 310–314, 1934.
4. S. A. Smith, "Service Reliability Measured by Probabilities of Outage," *Elect. World*, **103**, 371–374, 1934.
5. P. E. Benner, "The Use of the Theory of Probability to Determine Spare Capacity," *Gen. Elect. Rev.*, **37**, 345–348, 1934.
6. S. M. Dean, "Considerations Involved in Making System Investments for Improved Service Reliability," *Edison Electric Inst. Bull.*, **6**, 491–496, 1938.
7. B. S. Dhillon, *Power System Reliability, Safety, and Management*, Ann Arbor Science, Ann Arbor, MI, 1983.
8. A. Coppola, "Reliability Engineering of Electronic Equipment: A Historical Perspective," *IEEE Trans.Rel.*, **33**, 29–35, 1984.
9. W. Weibull, "A Statistical Distribution Function of Wide Applicability," *J. Appl. Mech.*, **18**, 293–297, 1951.
10. A. M. Freudenthal, and E. J. Gumbel, "Failure and Survival in Fatigue," *J. Appl. Phys.*, **25**, 110–120, 1954.
11. A. M. Freudenthal, "Safety and Probability of Structural Failure," *Trans. Am. Soc. Civil Eng.*, **121**, 1337–1397, 1956.
12. W. M. Redler, "Mechanical Reliability Research in the National Aeronautics and Space Administration," *Proceedings of the Reliability and Maintainability Conference*, 763–768, 1966.
13. B. S. Dhillon, *Mechanical Reliability: Theory, Models, and Applications*, American Institute of Aeronautics and Astronautics, Washington, D.C., 1988.

14. B. S. Dhillon, *Design Reliability: Fundamentals and Applications*, CRC Press, Boca Raton, FL, 1999.

15. B. S. Dhillon, *Reliability and Quality Control: Bibliography on General and Specialized Areas*, Beta Publishers, Gloucester, Ontario, Canada, 1992.

16. P. Kales, *Reliability: for Technology, Engineering, and Management*, Prentice Hall, Upper Saddle River, NJ, 1998.

17. B. S. Dhillon, "A Hazard Rate Model," *IEEE Trans. Rel.*, **29**, 150, 1979.

18. B. S. Dhillon, *Systems Reliability, Maintainability, and Management*, Petrocelli Books, New York, 1983.

19. B. S. Dhillon, *Reliability, Quality, and Safety for Engineers*, CRC Press, Boca Raton, FL, 2004.

20. J. P. Lipp, "Topology of Switching Elements Versus Reliability," *Trans. IRE Rel. Qual. Control*, **7**, 21–34, 1957.

21. J. A. Collins, *Failure of Materials in Mechanical Design*, Wiley, New York, 1981.

22. W. Grant Ireson, C.F. Coombs, and R.Y. Moss, (Eds.), *Handbook of Reliability Engineering and Management*, McGraw-Hill, New York, 1996.

23. R. L. Doyle, *Mechanical System Reliability*, Tutorial Notes, Annual Reliability and Maintainability Symposium, 1992.

24. C. Lipson, Analysis and Prevention of Mechanical Failures, Course Notes No. 8007, University of Michigan, Ann Arbor, June, 1980.

25. S. S. Rao, *Reliability-Based Design*, McGraw-Hill, New York, 1992.

26. J. W. Wilbur, and N. B. Fuqua, A Primer for DOD Reliability, *Maintainability, and Safety*, Standards Document No. PRIM 1, Rome Air Development Center, Griffiss Air Force Base, Rome, New York, 1988.

27. D. G. Raheja, *Assurances Technologies*, McGraw-Hill, New York, 1991.

28. D. Kececioglu, "Reliability Analysis of Mechanical Components and Systems," *Nuclear Eng. Design*, **19**, 259–290, 1972.

29. B. S. Dhillon and C. Singh, *Engineering Reliability: New Techniques and Applications,* Wiley, New York, 1981.

30. MIL-STD-1629, Procedures for Performing Failure Mode, Effects, and Criticality Analysis, Department of Defense, Washington, DC.

31. D. Kececioglu and E. B. Haugen, "A Unified Look at Design Safety Factors, Safety Margin, and Measures of Reliability," *Proceedings of the Annual Reliability and Maintainability Conference*, pp. 522–530, 1968.

32. G. M. Howell, "Factors of Safety," *Machine Design*, July 12, 76–81, 1956.

33. R. B. McCalley, "Nomogram for Selection of Safety Factors", *Design News*, September 138–141, 1957.

34. R. Schoof, "How Much Safety Factor?" *Allis-Chalmers Elec. Rev.*, 21–24, 1960.

35. J. E. Shigley and L. D. Mitchell, *Mechanical Engineering Design*, McGraw-Hill, New York, 1983, pp. 610–611.

36. J. H. Bompass-Smith, *Mechanical Survival: The Use of Reliability Data*, McGraw-Hill London, 1973.

37. V. M. Faires, *Design of Machine Elements*, Macmillan, New York, 1955.

38. J. S. Countinho, "Failure Effect Analysis," *Trans. NY Acad. Sci.*, **26**, 564–584, 1964.

39. B. S. Dhillon, "Failure Modes and Effects Analysis: Bibliography," *Microelectr. Rel.*, **32**, 719–732, 1992.

40. D. Meister, "The Problem of Human-Initiated Failures," *Proceedings of the 8th National Symposium on Reliability and Quality Control*, 1962, pp. 234–239.

41. J. I. Cooper, "Human-Initiated Failures and Man-Function Reporting," *IRE Trans. Human Factors*, **10**, 104–109, 1961.

42. B. S. Dhillon, *Human Reliability: With Human Factors*, Pergamon Press, New York, 1986.

43. T. L. Regulinski and W. B. Askern, "Mathematical Modeling of Human Performance Reliability," *Proceedings of the Annual Symposium on Reliability*, 1969, pp. 5–11.

44. S. Rhodes, J. J. Nelson, J. D. Raze, and M. Bradley, "Reliability Models for Mechanical Equipment," *Proceedings of the Annual Reliability and Maintainability Symposium,* 1988, pp. 127–131.

45. J. D. Raze, J. J. Nelson, D. J. Simard, and M. Bradley, "Reliability Models for Mechanical Equipment", *Proceedings of the Annual Reliability and Maintainability Symposium*, 1987, pp. 130–134.

46. J. J. Nelson, J. D. Raze, J. Bowman, G. Perkins, and A. Wannamaker, "Reliability Models for Mechanical Equipment," Proceedings of the Annual Reliability and Maintainability Symposium, 1989, pp. 146–153.

47. T. D. Boone, "Reliability Prediction Analysis for Mechanical Brake Systems," NAVAIR/SYSCOM Report, Department of Navy, Department of Defense, Washington, D.C, August 1981.

48. R. B. Spokas, "Clutch Friction Material Evaluation Procedures," Society of Automotive Engineers (SAE), Paper No. 841066, SAE, Warrendale, PI, 1984.

49. M. J. Rossi, "Non-Electronic Parts Reliability Data," Report No. NRPD-3, Reliability Analysis Center, Rome Air Development Center, Griffiss Air Force Base, Rome, NY, 1985.

50. R. E. Schafer, J. E. Angus, J. M. Finkelstein, M. Yerasi, and D. W. Fulton, "RADC Non-electronic Reliability Notebook," Report No. RADC-TR-85-194, Reliability Analysis Center, Rome Air Development Center, Griffiss Air Force Base, Rome, NY, 1985.

51. B. S. Dhillon and H. C. Viswanath, "Bibliography of Literature on Failure Data," *Microelectro. Rel.*, **30**, 723–750, 1990.

Bibliography[*]

Y. Benhaim, *Robust Reliability in the Mechanical Sciences*, Springer, New York, 1996.

B. Bertsche, A. Schauz, and K. Pickard, *Reliability in Automotive and Mechanical Engineering: Determination of Component and System Reliability*, Springer, Berlin, 2008.

J. H. Bompas-Smith, *Mechanical Survival*, McGraw-Hill, London, 1973.

A. D. S. Carter, *Mechanical Reliability*, Macmillan Education, London, 1986.

A. D. S. Carter, *Mechanical Reliability and Design*, Wiley, New York, 1997.

T. A. Cruse, *Reliability-Based Mechanical Design*, Marcel Dekker, New York, 1996.

D. C. Cranmer and D. W. Richerson (Eds.), *Mechanical Testing Methodology for Ceramic Design and Reliability*, Marcel Dekker, New York, 1998.

J.W. Dally, P. Lall, and J. M. Dally, *Mechanical Design of Electronic Systems*, College House Enterprises, Knoxvile, TN, 2008.

B. S. Dhillon, *Robot Reliability and Safety*, Springer, New York, 1991.

E. G. Frankel, *Systems Reliability and Risk Analysis*, Martinus Nijhoff, The Hague, 1984.

E. B. Haugen, *Probabilistic Mechanical Design*, Wiley, New York, 1980.

K. C. Kapur and L. R. Lamberson, *Reliability in Engineering Design*, Wiley, New York, 1977.

G. Kivenson, *Durability and Reliability in Engineering Design*, Hayden, New York, 1971.

A. Little, *Reliability of Shell Buckling Predictions*, MIT Press, Cambridge, MA, 1964.

R. E. Little, *Mechanical Reliability Improvement: Probability and Statistics for Experimental Testing*, Marcel Dekker, New York, 2003.

Mechanical Reliability Concepts, ASME, New York, 1965.

W. H. Middendorf, *Design of Devices and Systems*, Marcel Dekker, New York, 1990.

W. D. Milestone (Ed.), *Reliability, Stress Analysis and Failure Prevention Methods in Mechanical Design*, ASME, New York, 1980.

N. W. Sachs, *Practical Plant Failure Analysis: A Guide to Understanding Machinery Deterioration and Improving Equipment Reliability*, CRC Press, Boca Raton, FL, 2007.

M. L. Shooman, *Probabilistic Reliability: An Engineering Approach*, R.E. Krieger, Melbourne, FL, 1990.

J. N. Siddell, *Probabilistic Engineering Design*, Marcel Dekker, New York, 1983.

O. G. Vinogradov, *Introduction to Mechanical Reliability: A Designer's Approach*, Hemisphere, New York, 1991.

G. S. Wasserman, *Reliability Verification, Testing, and Analysis in Engineering Design*, Marcel Dekker, New York, 2003.

---

[*]Additional publications on Mechanical Design Reliability may be found in Refs. 13 and 15.

CHAPTER 7

PRODUCT DESIGN AND MANUFACTURING PROCESSES FOR SUSTAINABILITY

I. S. Jawahir, P. C. Wanigarathne, and X. Wang
University of Kentucky
Lexington, Kentucky

1 INTRODUCTION

1.1 General Background on Sustainable Products and Processes

Sustainability studies in general have so far been focused on environmental, social, and economical aspects, including public health, welfare, and environment over their full commercial cycle, defined as the period from the extraction of raw materials to final disposition.[1] Sustainability requirements are based on the utilization of available, and the generation of new, resources for the needs of future generations. Sustainable material flow on our planet has been known to exist for over 3.85 billion years and, using the nature's simple framework in terms of *cyclic, solar,* and *safe means,* has been shown to offer the most efficient products for sustainability.[2,3] It is also generally known that sustainable products are fully compatible with nature throughout their entire life cycle. Designing and manufacturing sustainable products are major, high-profile challenges to the industry as they involve highly complex, interdisciplinary approaches and solutions. Most research and applications so far, however, have heavily focused on environmental sustainability. Sustainable products are shown to increase corporate profits while enhancing society as a whole, because they are cheaper to make, have fewer regulatory constraints and less liability, can be introduced to the market quicker, and are preferred by the public.[4] By designing a product with environmental parameters in mind, companies can increase profits by reducing material input costs, by extending product life cycles by giving them second and third life spans, or by appealing to a specific consumer base.[5] Recent effort on designing for environment includes the development of a customized software tool for determining the economic and environmental effects of "end-of-life" product disassembly process.[6]

In recent years, several sustainability product standards have emerged. Figure 1 shows a partial list of such standards.[7–18] While most standards are based on environmental benefits, some standards such as the Forest Stewardship Council Certified Wood Standards, the Sustainable Textile Standards, or the Global Sullivan Principles deal with social and economic criteria as well. The Institute for Market Transformation to Sustainability (MTS) has also recently produced a manual for standard practice for sustainable products economic benefits.[19] This profusion of competing standards may well become an obstacle to the management of product sustainability in the marketplace, leading to confusion among consumers and manufacturers alike. What is called for is the development of a sustainability management system that creates clear accountability methods across industries and market segments, and that determines not only *substantive* elements (e.g., "how much CO_2 was emitted in making the product?"), but also process elements (including the manufacturing systems and operations involved).

The idea of recycling, reuse, and remanufacturing has in recent times emerged with sound, innovative, and viable engineered materials, manufacturing processes, and systems to provide multiple life-cycle products. This is now becoming a reality in selected application areas of product manufacture. The old concept of "from cradle to grave" is now transforming into "from cradle to cradle,"[20] and this is a very powerful and growing concept in the manufacturing world, which takes its natural course to mature. Added to this is the awareness and the need for ecoefficiency and the environmental concerns often associated with minimum toxic emissions into the air, soil, and water; production of minimum amounts of useless waste; and minimum energy consumption at all levels. Finally, a future sustainability management system needs to identify how the public can be educated about sustainability, so that market incentives are created to persuade producers to follow more rigorous, evolving sustainability standards. Only at that point can a sustainability program be counted as successful.

Since the 1990s, environmental and energy factors have become an increasingly important consideration in design and manufacturing processes due to more stringent regulations promulgated by local, state, and federal governments as well as professional organizations in the United States and other industrial countries. The pressure on industry from the government as well as consumer sector has demanded new initiatives in environmentally benign design and manufacturing.[21] In the government sector, the U.S. Environmental Protection Agency (EPA) and the U.S. Department of Energy (DoE) have been the two leaders in these initiatives. The EPA has initiated several promotional programs, such as Design for the Environment Program, Product Stewardship Program, and Sustainable Industries Partnership Program, working with individual industry sectors to compare and improve the performance, human health, environmental risks, and costs of existing and alternative products, processes, and practices.[22] The EPA has also worked with selected industry sectors such as metal casting, metal finishing, shipbuilding and ship repair, and specialty-batch chemical industries to develop voluntary, multimedia performance improvement partnerships. Similarly, the DoE has launched a Sustainable

	Logo	Program	Website	1	2	3
PRODUCT SPECIFIC STANDARDS	FSC	Forest Stewardship Council Certified Wood	http://www.fscoax.org http://www.certifiedwood.org	X	X	X
	CLEAN CAR CAMPAIGN	Clean Vehicles	http://www.cleancarcampaign.org/standard.html http://www.environmentaldefense.org/greencar	X		
		Certified Organic Products Labeling	http://www.ota.com	X		
		Certified Green e Power	http://www.green-e.org	X		
	LEED	U. S. Green Building Council LEED Rating System	http://www.usgbc.org	X		
		Salmon Friendly Products	http://www.sustainableproducts.com/susproddef2.html#Salmonm	X		
	Certified Cleaner and Greener	Cleaner and Greener^{sm} Certification	http://www.cleanerandgreener.org	X		
OVERALL STANDARDS	the NATURAL STEP	Natural Step System Conditions	http://www.NaturalStep.org	X		
		Nordic Swan Ecolabel	www.ecolabel.no/ecolabel/english/about.html	X	X	X
	GREEN SEAL	Green Seal Product Standards	http://www.greenseal.org	X		
		Global Reporting Initiative (GRI) Sustainability Reporting Guidelines (2000) Social Equity Performance Indicators	http://www.sustainableproducts.com/susproddef2.html#Performance_Indicators		X	
		Life Cycle Assessment (LCA)	http://www.sustainableproducts.com/susproddef.html	X		
	MTS	Sustainable Textile Standard		X	X	X

Figure 1 Partial list of currently available sustainable products standards.

Design Program, which focuses on the systematic consideration, during the design process, of an activity, project, product, or facility's life-cycle impacts on the sustainable use of environmental and energy resources.[23,24] Recently, the DoE has also sponsored a series of new vision workshops and conferences, producing the Remanufacturing Vision Statement—2020 and Roadmaps, encouraging industry groups to work together in strategic relationships to produce more efficient production methods utilizing life-cycle considerations.[25,26]

The big three automotive companies, DaimlerChrysler, Ford, and General Motors, have been fierce competitors in the marketplace, but they have worked together on shared technological and environmental concerns under the umbrella of the United States Council for Automotive Research (USCAR), formed in 1992 by the three companies. USCAR has sought specific technologies in recycling, reuse, and recovery of auto parts, batteries, lightweight materials, engines, and other power sources as well as safety and emission reduction, sharing the results of joint projects with member companies.[27]

1.2 Projected Visionary Manufacturing Challenges

The National Research Council (NRC) report "Visionary Manufacturing Challenges for 2020" identifies six grand challenges for the future: concurrent manufacturing, integration of human and technical resources, conversion of information to knowledge, environmental compatibility, reconfigurable enterprises, and innovative processes.[28] Five of the 10 most important strategic technology areas identified by the NRC report for meeting the above six grand challenges involve sustainability applications for products and processes:

- *Waste-Free Processes.* Manufacturing processes that minimize waste and energy consumption.
- *New Materials Processes.* Innovative processes for designing and manufacturing new materials and components.
- *Enterprise Modeling and Simulation.* System synthesis, modeling, and simulation for all manufacturing operations.
- *Improved Design Methodologies.* Products and process design methods that address a broad range of product requirements.
- *Education and Training.* New educational and training methods that enable the rapid assimilation of knowledge.

1.3 Significance of Sustainable Product Design and Manufacture

Figure 2 shows the exponential increase in shareholder value when the *innovation-based sustainability* concepts are implemented against the traditional cost-cutting, substitution-based growth.[29] The business benefits of sustainability are built on the basis of 3Rs: Reduce, Reuse, and Recycle. A market-driven "logic of sustainability" is now emerging based on the growing expectations of stakeholders on performance. A compelling case for market transformation from short-term profit focus to innovation-based stakeholder management methods has been proposed in a well-documented book by Chris Laszlo.[30] This covers five major logics of sustainability:

1. Scientific (e.g., human-induced global climate change)
2. Regulatory (e.g., Title I of the Clean Air Act, amended in the United States in 1990)
3. Political (e.g., agenda of the green parties in Europe)
4. Moral, based on values and principles
5. Market, focusing on the shareholder value implications of stakeholder value.

The global challenge of sustainability may be restated as follows: Address the needs of a growing, developing global population without depleting our natural resources and without ruining the environment with our wastes. Fundamental knowledge must be developed and new innovative technologies established to meet this need. Engineers must move beyond their traditional considerations of functionality, cost, performance, and time to market to consider also sustainability. Engineers must begin thinking in terms of minimizing energy consumption, waste-free manufacturing processes, reduced material utilization, and resource recovery following the end of product use—all under the umbrella of a total life-cycle view. Of course, all this must be done with involvement of stakeholders and the development of innovative technologies, tools, and methods.

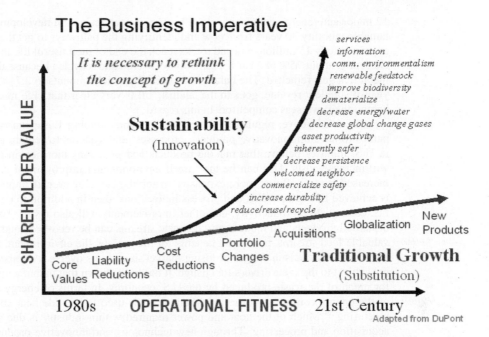

Figure 2 Exponential shareholding growth of innovation-based sustainability. *Source:* Adapted from Ref. 29.

Simply designing a green (environmentally friendly) product does not guarantee sustainable development for the following reasons: (a) a product cannot be green if the public does not buy it—business economics and marketing are critical for product acceptance—and (b) a green product is often just focused on the "use" stage of the product life cycle, with environmental burdens shifted to other life-cycle stages—sustainability requires a comprehensive, multi-life-cycle view. Certainly, industrialized countries have made some improvement in terms of being green with their use of materials, but waste generation continues to increase. As much as 75% of material resources used to manufacture goods are returned to the environment as wastes within a year.[31] This wasting of potential resources is disconcerting now, but over the next 50 years, as the demand for resources increases 10-fold and total waste increases by a comparable amount, this resource wasting could be viewed as tragic.

Countries around the world, especially in Western Europe and Japan, recognize that a concerted effort is needed to meet the global challenge of sustainability. The governments and manufacturers in these regions are well ahead of the United States in addressing the sustainability challenge through the development of energy/material-efficient technologies/products, low-impact manufacturing (value creation) processes, and postuse (value recovery) operations. The European Union (EU) has established the Waste Electrical and Electronic Equipment (WEEE) Directive to manage the recovery and postuse handling of these products.[32] While mandated recovery rates can be met economically by material recycling at present, remanufacturing and reuse are developing into very competitive alternatives. Another EU directive calls for the value recovery of end-of-life vehicles (ELVs) and their components, with 85% of the vehicle (by weight) to be reused or recycled by 2015.[33] If a company exports its products to the EU, it must conform to these directives. Japan is enacting regulations that closely follow those of the EU. Manufacturers in both the EU and Japan have begun to redesign their products to accommodate recycling.[34,35] The Sustainable Mobility Project, a sector project of the World Business Council for Sustainable Development (WBCSD), includes participation from

12 major auto/energy companies globally. The project deals with developing a vision for sustainable mobility 30 years from now and identifying the pathways to get there.[36,37] Each year, approximately 15 million cars and trucks reach the end of their useful life in the United States. Currently, about 75% of a car is profitably recovered and recycled because the majority of it is metal that gets remelted. The balance of materials, which amounts to 2.7–4.5 million tons per year of shredder residue, goes to the landfill.[38] It is very clear that U.S. manufacturers lag far behind their overseas competitors in this regard.

As noted above, regulatory drivers are currently forcing European and Japanese companies to develop innovative products, processes, and systems to remain competitive. Many in U.S. industry believe that making products and processes more environmentally friendly will increase costs. This can be the case if environmental improvement is achieved through increased control efforts, more expensive materials, etc. However, if improved sustainability is achieved through product and process innovations, then in addition to environmental benefits, cost, quality, productivity, and other improvements will also result. Through innovation, discarded products and manufacturing waste streams can be recovered and reengineered into valuable feed streams, producing benefits for the society, the environment, and U.S. industry. The United States is in danger of losing market share to its overseas competitors because it is not subject to the same drivers for change. It has been shown that manufacturing is responsible for much of the waste produced by the U.S. economy. In terms of energy usage, about 70% of the energy consumed in the industrial sector is used to provide heat and power for manufacturing.[39] Much of the heat and power required within industry is due simply to material acquisition and processing. Through new technology and innovative products and processes, utilizing previously processed materials for example, these energy requirements can be drastically reduced.

A significant effort has been undertaken by various groups from a range of disciplines to characterize, define, and formulate different forms and means of sustainable development. Continued progress in sustainable development heavily depends on sustained growth, primarily focusing on three major contributing areas of sustainability: environment, economy, and society (see Fig. 2). A relatively less-known and significantly impacting element of sustainability is sustainable manufacture, which includes sustainable products, processes, and systems in its core. The understanding of the integral role of these three functional elements of sustainability in product manufacture is important to develop quantitative predictive models for sustainable product design and manufacture. This integral role of sustainable manufacture, with its three major functional elements (innovative product development; value design, and manufacturing processes; and value creation and value recovery), all contributing to the sustained growth through the economic sustainability component, has been discussed[40] (Fig. 3).

2 NEED FOR SUSTAINABILITY SCIENCE AND ITS APPLICATIONS IN PRODUCT DESIGN AND MANUFACTURE

Sustainable development is now understood to encompass the full range of economic, environmental, and societal issues (often referred to as the "triple bottom line") that define the overall quality of life. These issues are inherently interconnected, and healthy survival requires engineered systems that support an enhanced quality of life and the recognition of this interconnectivity. Recent work, with details of integration requirements and sustainability indicators, shows that sustainability science and engineering are emerging as a metadiscipline.[41,42] We are already beginning to see the consequences of engineered systems that are inconsistent with the general philosophy of sustainability. Because of our indiscriminate release of global warming gases, the recent EPA report on global climate change forecasts some alarming changes in the

Sustainable Development

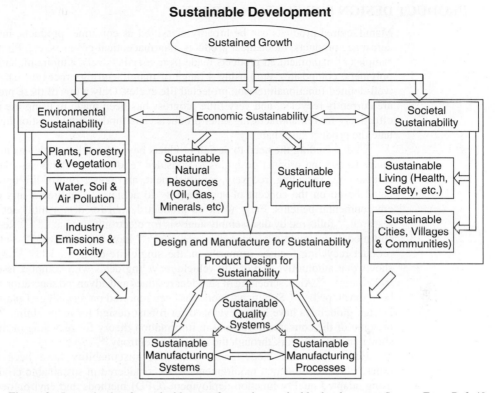

Figure 3 Integral role of sustainable manufacture in sustainable development. *Source:* From Ref. 40. Reprinted with permission of ASME International.

earth's temperature, with concomitant increases in the sea level of 1 m by 2100.[43] Obviously, fundamental changes are needed in engineered systems to reverse this trend.

The application of basic sustainability principles in product design and manufacture will serve as a catalyst for sustainable products to emerge in the marketplace. While the sustainable products make a direct contribution to economic sustainability, it also significantly contributes to environmental and societal sustainability. Building sustainability in manufactured products is a great challenge to the manufacturing world. The basic premise here is that, using the product sustainability principles comprehensively, all manufactured products can be designed, manufactured, assembled, used, and serviced/maintained/upgraded, and at the end of its life cycle, these products can also be effectively disassembled, recycled, reused/ remanufactured, and allowed to go through another cycle, and more. This multi-life-cycle approach and the associated need for product sustainability principles bring out an enormous technological challenge for the future. A cursory look at what would be required shows a long list of things to be performed; for example:

1. Known theories will be utilized while new theories emerge for sustainable product design.

2. Effective manufacturing processes with improved/enhanced sustainability applications will be developed and implemented.

3. Sustainable manufacturing systems will be developed to provide the overall infrastructure for sustainable product manufacture.

3   PRODUCT DESIGN FOR SUSTAINABILITY

Manufactured products can be broadly classified as consumer products, industrial products, aerospace products, biomedical products, pharmaceutical products, etc. Figure 4 shows some samples of manufactured products made from metals (steels, aluminum, hard alloys, plastics, polymers, composites, etc.) using a range of manufacturing processes. These products have well-defined functionalities and projected life cycles. Only a few of these products can be and are currently recycled, and very little progress has been made in using the recycled material effectively for remanufacturing other products. The fundamental question here is how to evaluate the product sustainability.

Understanding the need to design products beyond one life cycle has in recent times virtually forced the product designers to consider "end-of-life" status associated with product disassembly, recycling, recovery, refurbishment, and reuse. End-of-life options can be evaluated based on the concept of sustainability to achieve an optimum mix of economic and environmental benefits. Early work on product design for disassembly set the direction for research,[44] followed by disassembly analysis for electronic products.[45,46] Recovery methods[47] and models for materials separation methods[48,49] have been shown. End-of-life analysis for product recycling focuses primarily on the single-life-cycle model.[50–54] More recent work shows that automotive end-of-life vehicle recycling deals with complex issues of post-shred technologies.[55] Also, screening of shredder residues and advanced separation mechanisms have been developed.[56,57] Significant work has been reported on recycling of plastics and metals.[58] Design guidelines have been developed for robust design for recyclability.[59] The application of some of the concepts developed in information theory for recycling of materials has been shown in a recent work through the measure of entropy.[60]

Ecoefficient and biodesign products for sustainability have been urged in recent times.[61,62] Environmental requirements were considered in sustainable product development using adapted quality function deployment (QFD) methods and environmental performance

Figure 4   Samples of manufactured products. *Source:* From Ref. 40. Reprinted with permission of ASME International.

indicators (EPIs) for several industrial cases,[63] followed by a material grouping method for product life-cycle assessment.[64] A further extension of this work includes estimation of life-cycle cost and health impact in injection-molded parts.[65] Software development for environmentally conscious product design includes BEES (Building for Environmental and Economic Sustainability) by the National Institute for Standards and Technology (NIST)[66] and Design for Environment Software.[6] Recently, by extending the previously developed sustainability target method (STM) to the product's end-of-life stage, analytical expressions were derived for the effectiveness of material recycling and reuse and were also corrected with the product's performance.[67]

3.1 Measurement of Product Sustainability

Quantification of product sustainability becomes essential for comprehensive understanding of the "sustainability content" in a manufactured product. The societal appreciation, need, and even demand for such sustainability rating would become apparent with increasing awareness and the user value of all manufactured products more like the food labeling, energy efficiency requirements in appliances, and fuel efficiency rating in automotive vehicles.

Almost all previous research deals with a product's environmental performance and its associated economic and societal effects largely intuitively, and much of the qualitative descriptions offered are all, with the possible exception of a few recent efforts, difficult to measure and quantify. Thus, these analyses mostly remain nonanalytical and less scientific in terms of the need for quantitative modeling of product sustainability. The complex nature of the *systems property* of the term "product sustainability" seems to have limited the development of a science base for sustainability. Moreover, the partial treatment and acceptance of the apparent overall effects of several sustainability-contributing measures in relatively simplistic environmental, economic and societal impact categories have virtually masked the influence of other contributing factors such as a product's functionality, manufacturability, and reusability with multiple life cycles. Consideration of a total and comprehensive evaluation of product sustainability will always provide much cheaper consumer costs, over the entire life cycle of the product, while the initial product cost could be slightly higher. This benefit is compounded when a multiple life-cycle approach, as seen in the next section, is adopted. The overall economic benefits and the technological advances involving greater functionality and quality enhancement are far too much to outscore with the current practice. The technological and societal impact is great for undertaking such an innovative approach to define the scientific principles of the overall product sustainability.

Sustainability science is more than a reality and is inevitable. Sound theories and models involving the application of basic scientific principles for each contributing factor are yet to emerge, but the momentum for this is growing worldwide. The available wide range of manufactured products from the consumer electronics, automotive, aerospace, biomedical, pharmaceutical sectors, etc., need to be evaluated for product sustainability and the associated economic and societal benefits.

3.2 Impact of Multi–Life Cycles and Perpetual-Life Products

Figure 5 shows the various life-cycle stages for multi-life-cycle products leading toward the eventual "perpetual-life products." Even with innovative technology development, sustainability cannot be achieved in the absence of an engaged society. In the near future companies are envisioned to assume responsibility for the total product life cycle, but the consumer will remain responsible for preserving value during use and ensuring that postuse value enters the recovery stream. Education and knowledge transfer will play an important role in communicating systemic changes, such as ecolabeling and the establishment of product reclamation

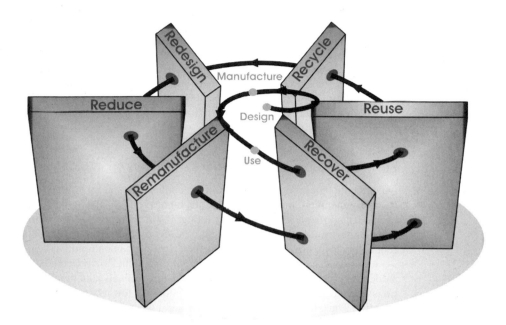

Figure 5 Generic multi-life-cycle products leading toward perpetual-life products.

centers. Engineers within industry must be educated to use innovative tools and methods for design/decision making and to apply novel processes for manufacturing, value recovery, recycling, reuse, and remanufacturing. A diverse cadre of future engineers must receive a broad education prior to entering the workforce and have an awareness of a wide variety of issues, such as public opinion, environmental indicators, and life-cycle design.

The impact of products on sustainability does not start and end with manufacturing. The material flow in the product life cycle includes all activities associated with the product, including design, material acquisition and manufacturing processes (value creation), and use (value retention) and postuse (value recovery) processes (e.g., reuse, remanufacturing, and recycling) as illustrated in Fig. 6.[68] Distribution and take-back logistics are other important elements of the

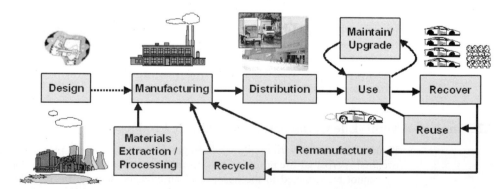

Figure 6 Material flow in the product life cycle for sustainability.[68] Republished with permission of Inderscience Publisher from International Journal of Vehicle Design. Permission conveyed through Copyright Clearance Center, Inc.

product life cycle. Obviously, some waste associated with the product life cycle is inevitable, and this waste is lost value and is associated with inefficiencies in the cycle. Increased use of innovative value recovery processes during the life cycle represents an underutilized potential business opportunity and means to be more sustainable.

3.3   Product Sustainability Assessment

While the concept of product sustainability continues to grow and become more compelling, the assessment of product sustainability has become difficult and challenging. There are no universally acceptable measurement methods for product sustainability as yet. This is largely due to the difficulty in quantifying and assessing some of the integral elements of product sustainability such as the societal and ethical aspects of sustainability. Also, the effects of the (social) use of products, in intended and unintended ways, are different from their material and production aspects, further complicating such an assessment. In 1987, the Brundtland Commission defined sustainability as "meeting the needs of present without compromising the ability of future generations to meet their own needs" (Ref. 69). This rather ambiguous definition has very much limited the establishment of meaningful goals and measurable metrics for sustainability. Consideration of the key aspects of business performance subsequently extended this definition to include the effects of economic, environmental, and societal factors, each providing several categories of sustainable product indicators.[70] The initial product rating system developed by VDI[71] was subsequently modified to include variable weighting factors for products.[72]

The ongoing work at the University of Kentucky (Lexington) within the Collaborative Research Institute for Sustainable Products (CRISP) involves a multidisciplinary approach for formulating the product sustainability level. A group of design, manufacturing/industrial, and materials engineers along with social scientists, economists, and marketing specialists are actively participating in a large program to establish the basic scientific principles for developing a product sustainability rating system. This includes the development of a science-based product sustainability index (PSI).[40]

3.4   Product Sustainability Index

The PSI will represent the "level of sustainability" built in a product by taking into account the following six major contributing factors:

1. **Product's environmental impact**
   - Life-cycle factor (including product's useful life span)
   - Environmental effect (toxicity, emissions, etc.)
   - Ecological balance and efficiency
   - Regional and global impact (CO_2 emission, ozone depletion, etc.)
2. **Product's societal impact**
   - Operational safety
   - Health and wellness effects
   - Ethical responsibility
   - Social impact (quality of life, peace of mind, etc.)
3. **Product's functionality**
   - Service life/durability
   - Modularity
   - Ease of use

- Maintainability/serviceability (including unitized manufacture and assembly effects)
- Upgradability
- Ergonomics
- Reliability
- Functional effectiveness

4. Product's resource utilization and economy

- Energy efficiency/power consumption
- Use of renewable source of energy
- Material utilization
- Purchase/market value
- Installation and training cost
- Operational cost (labor cost, capital cost, etc.)

5. Product's manufacturability

- Manufacturing methods
- Assembly
- Packaging
- Transportation
- Storage

6. Product's recyclability/remanufacturability

- Disassembly
- Recyclability
- Disposability
- Remanufacturing/reusability

Quantifiable and measurable means can be developed for each factor within each group and then be combined to produce a single rating for each group. This rating can be on a percentage basis on a 0–10 scale, with 10 being the best. Each product will be required to comply with appropriate ratings for all groups. Standards will be developed to establish an "acceptable" level of rating for each group. While the rating of each group contributes to the product's sustainability, the composite rating will represent the overall "sustainability index" of a product, the PSI. This overall product sustainability can be expressed in terms of a percentage level, on a 0–10 scale, or on a letter grade, such as A, B, C. Variations of these implementation methods of PSI are shown in Fig. 7.

4   PROCESSES FOR SUSTAINABILITY

The primary focus in identifying and defining the various contributing elements and subelements of manufacturing process sustainability is to establish a unified, standard scientific methodology to evaluate the degree of sustainability of a given manufacturing process. This evaluation can be performed irrespective of the product life-cycle issues, recycling, remanufacturability, etc., of the manufactured product. The overall goal of the new international standards [International Organization for Standardization (ISO) 14001] is to support environmental protection and prevention of pollution in balance with socioeconomic needs.[73] Requirements of sustainable manufacturing covering recycling and decision-making aspects such as supply chain, quality initiatives, environmental costing, and life-cycle assessment

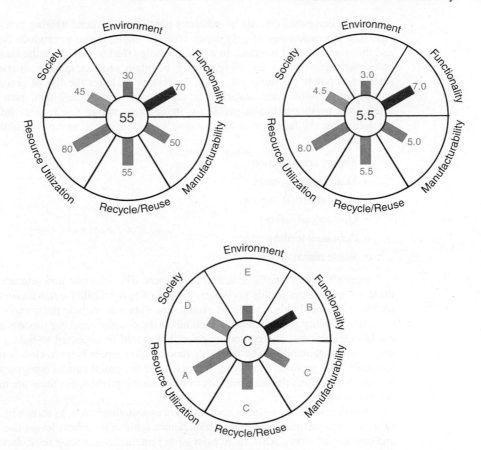

Figure 7 Variations of the proposed product label for a sustainable product. *Source:* From Ref. 40. Reprinted with permission of ASME International.

are well covered in a recent handbook on environmentally conscious manufacturing.[74] An early attempt to develop a sustainable process index was based on the assumption that in a truly sustainable society the basis of economy is the sustainable flow of solar energy.[75] More recent work predicts manufacturing wastes and energy for sustainable processes through a customized software system.[76] A modeling effort for the impact of fuel economy regulations on design decisions in automotive manufacturing was presented in a recent paper.[77]

4.1 Selection of Sustainability Measures for Manufacturing Operations

Manufacturing processes are numerous and, depending on the product being manufactured, the method of manufacture, and their key characteristics, these processes differ very widely. This makes the identification of the factors/elements involved in process sustainability and the demarcation of their boundaries complex. For example, the production process of a simple bolt involves a few clearly defined production stages: bolt design, tool/work material selection, metal removal/forming, finishing, packaging, transporting, storage, dispatching, etc. It is difficult to consider all these stages in evaluating manufacturing process sustainability, even though these processes either directly or indirectly contribute to manufacturing process sustainability. Also, the processing cost largely depends on the method used to produce the part/component and the work material selected. In a never-ending effort to minimize the manufacturing costs, industrial organizations are struggling to maintain product quality, operator and machine safety, and power consumption. If the processing includes the use of coolants or lubricants or the emission of toxic materials or harmful chemicals, this then poses environmental, safety, and personnel health problems. In general, among the various influencing factors, the following six factors can be regarded as significant to make a manufacturing process sustainable:

- Energy consumption
- Manufacturing costs
- Environmental impact
- Operational safety
- Personnel health
- Waste management

Figure 8 shows these interacting parameters. The selection and primary consideration of these six parameters at this preliminary stage of sustainability evaluation do not restrict the inclusion of other likely significant parameters. This may include parameters such as the product's functionality requirements, which affect the decision-making process and are related to machining costs and energy consumption, but would be expected to hold a secondary effect on process sustainability. The product's functionality aspect is more closely related to the sustainability of a product. Marketing strategies and the initial capital equipment investment can also indirectly affect the sustainability of a machining process, but these are not included in the present analysis.

All six selected parameters have different expectation levels, as shown in Table 1. But there is an obvious fact that all these factors cannot achieve their best levels due to technological and cost implications. Also, there exist strong interactions among these factors, often requiring a trade-off. Thus, only an optimized solution would be practical, and this would involve combinations of minimum and maximum levels attainable within the constraints imposed. The attainable level is again very relative and case specific. Measurement and quantification of the effect of contributing factors shown in Table 1 pose a significant technical challenge for use in an optimization system.

Figure 8 Major factors affecting sustainability of machining processes.

Table 1   Measurable Sustainability Factors in Machining
Processes and Their Desired Levels

Measurement Factor	Desired level
Energy consumption	Minimum
Environmental friendliness	Maximum
Machining costs	Minimum
Operational safety	Maximum
Personnel health	Maximum
Waste reduction	Maximum

Source: Ref. 84.

Energy Consumption

During manufacturing operations the power consumption level can be observed and evaluated against the theoretical values to calculate the efficiency of the power usage during the operation. Significant work has been done in this area to monitor the power consumption rate and to evaluate energy efficiency. Energy savings in manufacturing processes is a most needed sustainability factor, which needs to be considered for the entire operational duration of the machine, with significant overall savings in the long run. For any manufacturing operation, the energy consumed can be measured in real time. If the same task/operation is performed at two different machine shops or on two different machines, the power consumption may vary, due to the differences in the machines and the conditions used in manufacturing processes. Notably, the application of proper coolants and lubricants, the selection of cutting tool inserts, cutting conditions, and cutting tool-work material combinations, and facilitating improved tribological conditions can all reduce the power consumption, typically in a machining process. Also, the functional features built in a machine tool design contribute toward energy savings in a machining operation. For instance, the horizontal movement of the turret of a lathe may use less energy than a vertically slanted turret in machining centers, as more power is used to keep the location locked. This may amount to several kilowatt-hours of energy, in real value.

Hence, it is clear that setting a standard for power consumption is relative and complex in the industry. Use of minimum energy is, however, most desirable from the perspective of the global energy standards/requirements. In the case of machining, there is an attainable minimum energy level for every machining operation. The power used in the real operation can be compared against this in assessing the amount of excess energy utilized. Depending on the proximity of the two values, one can determine the relative efficiency of the power/energy consumption and then take measures to improve the process by reducing the gap. After these modifications and improvements, the specific process can be rated for sustainable use of energy. In sustainability assessment of energy/power consumption, it is generally anticipated that the preferred source of energy is environmentally friendly—solar or from a renewable source. If the renewable sources are available in abundance and are used in industry widely, the source-of-energy factor can be included in the process sustainability rating system.

Manufacturing Costs

Manufacturing cost involves a range of expenditures starting from the process planning activity until the part is dispatched to the next workstation, including the idling time and queuing time. Within the context of manufacturing process sustainability assessment, our interest is only on the manufacturing costs involved in and during the manufacturing operation time, including the cost of tooling. For example, in a machining operation, the material removal rate depends on the selected cutting conditions and the capabilities of the machine tools and cutting tools used. The criterion for selecting appropriate machine tools and cutting tools would generally facilitate a cost-effective machining operation. Numerous software tools are available for optimizing the machining cost through the use of proper cutting conditions. Recently, a new technique for multipass dry turning and milling operations has been developed. This technique employs analytical, experimental, and hybrid methods for performance-based machining optimization and cutting tool selection based on a tool life criterion involving minimum cost.[78,79] In addition, there are several other direct and indirect cost factors coming from the environmental effects and operator's health and safety aspects. The cost components for recycling and reusability of consumables such as the coolants also need to be considered in the overall machining cost along with the tooling cost contributions to the machining process.

Environmental Impact

Basic factors contributing to environmental pollution, such as emissions from metalworking fluids, metallic dust, and use of toxic, combustible, or explosive materials, contribute to this factor. Compliance with the EPA,[80] Occupational Safety and Health Administration (OSHA),[81] and National Institute for Occupational Safety and Health (NIOSH)[82] regulations is required. Measurable parameters have been defined and are continually updated. The ISO 14000 series of standards[83] has been designed to help enterprises meet and improve their environmental management system needs. The management system includes setting of goals and priorities, assignment of responsibility for accomplishing these goals, measuring and reporting of results, and external verification of claims. Since the standards have been designed as voluntary, the decision to implement will be a business decision. The motivation may come from the need to better manage compliance with environmental regulations, from the search for process efficiencies, from customer requirements, from community or environmental campaign group pressures, or simply from the desire to be good corporate citizens. The machining process sustainability rating will eventually force the industries to impose and show progress at every stage of the production process.

Operational Safety

The amount of unsafe human interaction during a manufacturing operation and the ergonomic design of the human interface are in focus for this category. Also, compliance with the regulatory safety requirements is made mandatory. Statistical data on safety violations and the associated corrective measures that are quantifiable are usually being reviewed and updated.

In general, safety aspects in relation to a manufacturing process can be divided into two broad categories: personnel safety and work safety. Safety of the operator and the occupants of the manufacturing station are considered paramount to the work safety. The amount of human interaction during a manufacturing operation and the safety precautions provided against the foreseeable accidents will be the primary focus in evaluating the operational safety as a sustainability parameter. The ergonomic design of the human interface with the work environment will be important in safety evaluations. The compliance to and the proper implementation of regulatory safety requirements will also be considered in assessing the personnel safety factor.

Personnel Health

Assessment of the personnel health element contributing to machining process sustainability is based on compliance with the regulatory requirements, imposed on industries by governmental and regulatory enforcement units such as the EPA,[80] OSHA,[81] and NIOSH[82] on emissions and waste from machining operations and their impact on directly exposed labor. One of the most prominent ways machine operators are affected is exposure to the mist and vapors of metalworking fluids, as most metalworking fluids used as coolants and lubricants in machining operations contain large amounts of chemicals added to "enhance" the machining performance. Over time, the fluid containers become an ideal environment for the growth of harmful bacteria. There are a few ways to avoid this problem, but it appears that none of these methods is in practice due to inadequacy of knowledge and implementation issues.

Waste Management

Recycling and the disposal of all types of manufacturing wastes, during and after the manufacturing process is complete, are accounted for in this category. Scientific principles are still emerging with powerful techniques such as lean principles being applied in quantifiable terms. Zero waste generation with no emissions into the environment is the ideal condition to be expected for products and processes, although it is technologically not feasible as yet. However, efforts to find means to reduce or eliminate wastes are continuing. For example, some cutting fluids can be forced to degrade biologically before being disposed of. The same technology can be used to control the growth of harmful bacteria in the cutting fluid containers as well as on the waste chip dumpers to make them biologically safe to handle and dispose of.

5 CASE STUDY

5.1 Assessment of Process Sustainability for Product Manufacture in Machining Operations

This case study provides a description of how sustainability measures can be selected along with an approximate method for modeling these measures for optimizing machining processes for maximum sustainability.[84] The machining processes used herein are to be considered a generic case for manufacturing processes.

Energy Consumption

Figure 9 shows a comparison of cutting force at varying coolant flow rates, ranging from no-coolant applied dry machining to flood cooling involving a large amount of coolant in the machining of automotive aluminum alloy A390 work material. This range includes three typical coolant flow rates of minimum quantity lubrication (MQL) conditions generally known as "near-dry" machining. As seen, the measurable cutting force component can serve as a direct

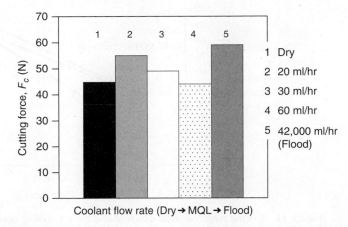

Figure 9 Cutting force variation when using different coolant flow rates. *Source:* From Ref. 84. Reprinted with permission of ASME International.

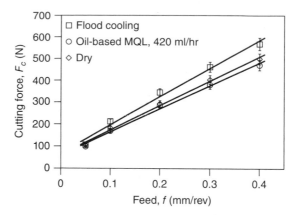

Figure 10 Cutting force variation with feed at different cooling conditions (AISI 4140 steel). *Source:* From Ref. 84. Reprinted with permission of ASME International.

indicator of the power/energy consumption rate. The optimum power consumed seems to lie beyond the largest coolant flow rate tested for near-dry machining (i.e., 60 mL/h). Obviously, the flood cooling method, despite the demonstrated major benefits of increased tool life, seems to show the highest power consumption for the test conditions used in the experimental work.

Figure 10 shows a comparison of cutting force generated in the machining of 4140 steel for three different conditions (flood cooling, oil-based MQL, and dry machining).[84] A fairly consistent trend is seen for this operation. In contrast, in machining aluminum A390 at a lower cutting speed of 300 m/min, a reverse trend is observed, with dry machining consuming the largest energy, as seen in Fig. 11. Comparison of Figs. 9 and 11 shows that cutting speed plays a major role in the power consumption rate with respect to cooling rates, all other conditions remaining constant. Complex relationships are also expected for varying tool geometry, work–tool material combinations, etc., thus making it essential to model the machining process for optimal cooling conditions to provide minimum power consumption. This technical challenge calls for a need to have full and comprehensive knowledge on all major influencing variables for the entire range of machining. Analytical modeling efforts must continue despite

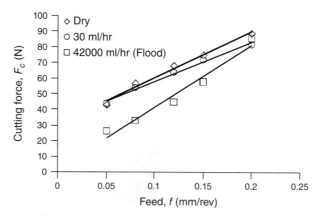

Figure 11 Cutting force variation with feed at different cooling conditions (aluminum A390). *Source:* From Ref. 84. Reprinted with permission of ASME International.

significant experimental difficulties involved. The sustainability contribution made by power consumption rate is too great to ignore, particularly when the environmental concerns are also addressed by reduced coolant applications and dry machining methods. Experimental work has shown that, in continued machining over time, oil-based MQL-assisted machining operations produced a slight decrease in cutting force, while the flood cooling method generated a steadily increasing trend for the cutting force.[84,85]

Machining Cost

Optimal use of machines and tooling, including jigs and fixtures, can provide reduced manufacturing costs. Limited analytical and empirical models are available for this evaluation, and accurate calculations are highly complex and customized applications would be necessary. Developing comprehensive analytical models to account for the overall machining cost is feasible as significant generic and rudimentary calculation methods are already in existence.

Environmental Impact

The environmental impact due to machining is an important contributing factor for machining sustainability evaluation. Basic factors affecting environmental pollution, such as emissions from metalworking fluids, metallic dust, machining of dangerous material (combustible or explosive), and amount of disposed untreated wastes, are among those considered under this category. The use of metalworking fluids in machining operations creates enormous health and environmental problems. Keeping the metalworking fluid tanks clean with no seepage to the environment and without harmful bacterial growth, e.g., is an important aspect to consider in the evaluation of environmental friendliness. All measures designed to minimize environmental pollution, such as the use of fume hoods and the treatment of metalworking fluids, must be given credit in the final assessment. Adherence and compliance to prevent emissions as vapor or mist, as regulated by the EPA[80] and OSHA,[81] are essential, and in consistent implementation practices, as per regulations, the machining environment needs to be inspected and certified.

Operational Safety

Some examples of this category are auto power doors, safety fences and guards, safety display boards, safety training, facilities to safe interactions with machines, methods of lifting and handling of work, mandatory requirement to wear safety glasses, hats, and coats, availability of fire safety equipment, and first aid facilities in-house. In addition to this routine inspection for operational safety and regular safety, specific training programs promoting safety precautions will be given due credit. Also, the routine maintenance of machines and availability of onsite fume detectors are considered desirable in assessing the operational safety measures for sustainability in machining operations.

Personnel Health

Compliance with the regulatory requirements according to the EPA, NIOSH, and OSHA on emissions from manufacturing operations and their impact on directly exposed labor can be the basis for this category of assessment. For example, the use of dry machining techniques or near-dry machining techniques can largely avoid and/or reduce the problems of mist generation and metalworking fluid handling typically encountered in flood cooling. Such measures will be assessed superior over traditional flood cooling methods in the personnel health factor assessment. In addition to those factors, such as compliance with regulations regarding space per machine and man-count in the factory, safety precautions in handling explosives or radioactive materials in machining will be considered significant, too.

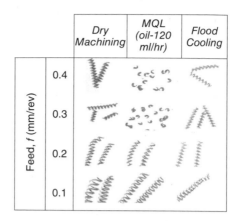

Figure 12 Chip form variation with different cooling methods. *Source:* From Ref. 84. Reprinted with permission of ASME International.

Waste Reduction and Management

Different types of wastes resulting from the machining process have to be treated, disposed of, and managed properly. Cost-effective and energy-efficient recycling of metal chips, debris, contaminated coolants, etc., would contribute to the sustainable machining process. Breaking chips into small, manageable sizes and shapes becomes a basic requirement for automated machining and for disposing of chips for recycling and/or reuse. Significant work has been done on chip breaking and control in metal machining,[86-89] and a new method developed for quantifying the chip form in terms of its size and shape[90] has been in wide use in practical applications. A computer-based predictive planning system developed for chip form predictions in turning operations includes the effects of chip groove configurations, tool geometry, and work material properties on chip forms/chip breakability.[91] More recent efforts to identify the variations of chip forms in machining with varying cooling conditions have shown a promising opportunity for MQL-assisted machining processes, as shown in Fig. 12, where at relatively high feeds the size and shapes of the chips produced are more suitable for chip disposal for recycling and reuse.[85]

5.2 Performance Measures Contributing to Product Sustainability in Machining

In designing a product for machining and in the subsequent process planning operations, an important consideration is assuring the product's functionality and specification requirements in terms of surface finish, part accuracy, etc. The anticipated surface roughness value and the related surface integrity issues contribute to the service life of a machined product. Maintaining low tool wear rates leading to increased tool life is essential in process modeling for surface roughness and surface integrity in machining.

Tool Life Evaluation in Dry and Near-Dry Machining and Tool Insert Recycling and Reuse Issues for Sustainability

Significant efforts have been made to improve tool life predictions in coated grooved tools, which provide the required increased tool life in dry machining with coated grooved tools.[92,93] More recent work on predictive modeling of tool life extended this work to include the effects of mist applications in near-dry machining, thus offering predictability of tool life in sustainable manufacturing.[94,95]

The life of a cutting tool insert can be as short as a few minutes and no more than an hour in most cases. Tool inserts are made from wear-resistant, hard materials and are coated with specialized hard materials. Once a tool insert is used up, with all its effective cutting edges worn, it is a normal practice to replace it. In large companies involved in a range of machining operations, the weight of tool inserts discarded per year may easily surpass a few hundred tons. Hence, new technologies need to be developed for tool insert recycling and/or reuse, given the large amount of waste in the form of worn tool inserts. Recent progress on tool recoating efforts, showing the technological feasibility with performance improvements and economic benefits, is encouraging.[96,97]

Surface Roughness and Surface Integrity Analysis for Sustainability

Comparison of surface roughness produced in machining under a range of cooling conditions (from dry to flood cooling) shows that flood cooling produces the least desirable surface roughness in turning operations of automotive alloys such as American Iron and Steel Institute (AISI) 4140 steel, despite the popular belief that coolants would provide better surface roughness (Fig. 13). However, in machining aluminum alloy A390, the trend is reversed, as seen in Fig. 14, where dry machining produces a rougher surface.

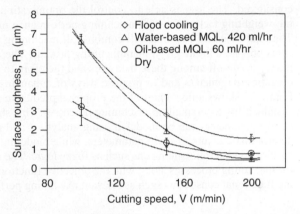

Figure 13   Variation of surface roughness with cutting speed in finish turning. *Source:* From Ref. 84. Reprinted with permission of ASME International.

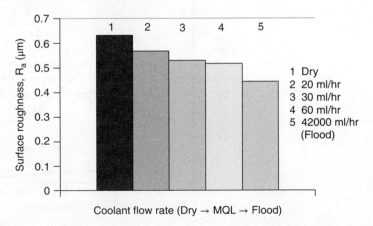

Figure 14   Surface roughness variation when using different cooling methods. *Source:* From Ref. 84. Reprinted with permission of ASME International.

In addition, the surface integrity is affected greatly by the residual stress formation in machining processes. Material behavior at a cutting tool with finite edge roundness has been modeled using a thermo–elasto–viscoplastic finite-element method to study the influence of sequential cuts, cutting conditions, etc., on the residual stress induced by cutting.[98,99] This work led to the conclusion that material fracture (or material damage) in machining is an important phenomenon to understand the actual material behavior on a finished surface and the surface integrity, both directly influencing product sustainability.

5.3   Optimized Operating Parameters for Sustainable Machining Processes

Figure 15 illustrates a method for selecting optimal cutting conditions in the rough pass of multipass dry turning operations providing sustainability benefits. Thick lines represent constraints of surface roughness, tool life, material removal rate, and chip form/chip breakability. Thin lines are the contours of the objective function in the optimization method. Points A and B represent different optimized results of cutting conditions subject to different initial requirement of the total depth of cut. A comprehensive criterion including major machining performance measures is used in the optimization process.

In the optimization method described here, an additional sustainability criterion will be considered. The user is able to control the optimization process by configuring and assigning weighting factors for both machining performance measures and sustainability measures. The total objective function will combine all machining performance and sustainability measures prevalent in the given manufacturing process. The aim of the optimization process is to make a trade-off among these measures and therefore to provide the optimal combinations of operating parameters and to propose ways of enhancing and improving sustainability level. Figure 16 shows a flow chart of the proposed predictive models and optimization method for sustainability assessment in machining processes. This shows that three of the six key sustainability parameters can be modeled using analytical and numerical techniques because of the deterministic nature of these parameters, while modeling of the other three parameters would require nondeterministic means such as fuzzy logic. The resulting hybrid sustainability model for machining processes along with the objective function and the relevant constraints, including functional constraints such as relevant machining performance measures and sustainability

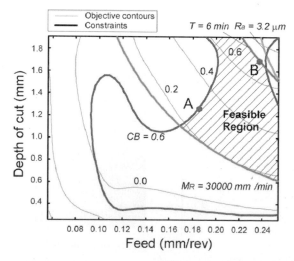

Figure 15   Optimized cutting conditions for rough turning. *Source:* From Ref. 84. Reprinted with permission of ASME International.

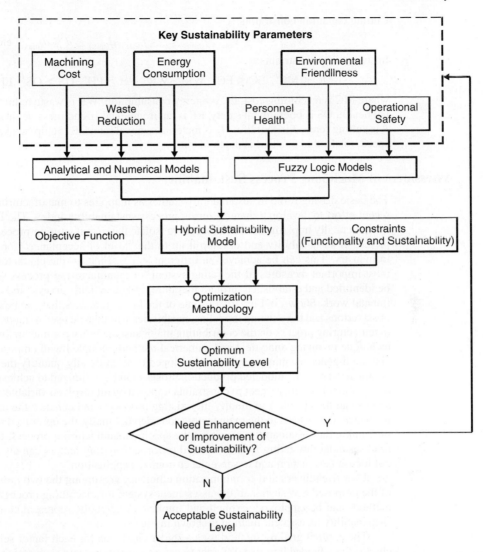

Figure 16 Flowchart of proposed predictive models and optimization method for process sustainability assessment of machining processes. *Source:* From Ref. 84. Reprinted with permission of ASME International.

constraints, can be used in the optimization module to provide the optimum sustainability level of the given machining process representing the actual, overall sustainability level in the form of a sustainability index. With a subsequent decision-making process, the process sustainability can either be improved using a feedback loop or be accepted as is. The optimization problem is formulated as follows:

$$\text{Maximize } U = U(\text{MC, EC, WR, PH, OS, EF})$$

With respect to cutting conditions and shop floor data subject to functional constraints:

$$R_a \leq R'_a, \quad F \leq F', \quad T \geq T', \quad M_R \geq M'_R, \quad \text{CB} \geq \text{CB}'$$

Cutting condition constraints:

$$V_{\min} \leq V \leq V_{\max}, f_{(\min)} \leq f \leq f_{(\max)}, d_{(\min)} \leq d \leq d_{(\max)}, \text{ etc.}$$

Sustainability constraints:

$$\text{MC} \leq \text{MC}', \ \text{EC} \leq \text{EC}', \ \text{WR} \geq \text{WR}', \ \text{PH} \geq \text{PH}', \ \text{OS} \geq \text{OS}', \ \text{EF} \geq \text{EF}'$$

where MC is machining cost, EC is energy consumption, WR is waste reduction, PH is personnel health, OS is operational safety, EF is environmental friendliness, R_a is surface roughness, F is cutting force, T is tool life, M_R is material removal rate, CB is chip breakability, V is cutting speed, f is feed rate, and d is depth of cut.

5.4 Assessment of Machining Process Sustainability

The basic driving force in sustainability studies as it applies to manufacturing processes is the recent effort to develop a manufacturing process sustainability index. The idea in developing this practically implementable concept is to isolate the manufacturing process from the global picture of sustainability and develop it up to the "level of acceptance" for common practice in industry. This can be achieved in different stages. First, in the characterization stage, the most important measures of the rating system for manufacturing process sustainability must be identified and established through literature, in-house/field surveys, and appropriate experimental work. Shown in Fig. 16 are some of the key parameters that can be considered. These observations and the existing modeling capabilities will then be used to model the impact of the manufacturing process on the contributing major sustainability parameters. A hybrid modeling technique involving analytical and numerical methods, coupled with empirical data and artificial intelligence techniques, must be developed to scientifically quantify the influence of each parameter. Then, the modeled production process can be optimized to achieve the desired level of sustainability with respect to constraints imposed by all involved variables. These optimized results can then be used to modify the existing processes and enhance the manufacturing performance with respect to the main factors considered. Finally, the optimized results can be used in defining the sustainability rating for the specific manufacturing process. In establishing the final sustainability rating for the selected process, weighing factors can also be used to bring out focused evaluation and to serve the customized application.

User friendliness and communication efficiency are among the two most-needed features of the proposed new sustainability assessment system for machining processes. Two proposed methods can be employed: explicit and implicit. A symbolic representation of the proposed sustainability assessment method is shown in Fig. 17.

The explicit grading method uses a spider chart axis for each factor selected. The spider chart is then divided into five different transforming regions, and each region is represented by a color. On this spider chart axis the relevant rating can be marked very clearly and the rating value can be indicated next to the point, on demand. The points closest to the outer periphery are the highest ranking categories, while the points closer to the origin are considered to be worst with respect to the expected sustainability rating and these areas need to be improved for enhanced sustainability.

To implicitly show the level of sustainability, a color-coding system—five shades of green—is used in the background. The darker the color, the further the factor is away from the expected level, and when the color turns bright green, it reaches the maximum. Colors are assigned with the darkest closer to the origin and the brightest closer to the outer periphery of the spider chart.

The proposed process sustainability assessment method will heavily involve science-based sustainability principles for product design and manufacture. The overall sustainability level of the machining process will be established through a new sustainability index to be developed comprehensively using the modeling and optimization method shown in Fig. 16.

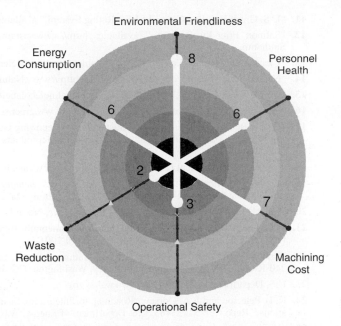

Figure 17 Example of proposed symbolic representation of sustainability rating system for six contributing factors. *Source:* From Ref. 84. Reprinted with permission of ASME International.

6 FUTURE DIRECTIONS

Continued trends in sustainability applications for products and processes indicate the need for identifying relevant sustainability metrics and for developing science-based methodologies for quantification of these factors.[40,84,100,101] Achieving global sustainability is a major challenge and requires international cooperative research and applications.[102]

REFERENCES

1. The Institute for Market Transformation to Sustainability (MTS), Sustainable Products Corporation, Washington, DC, 2004, http://MTS.sustainableproducts.com.

2. E. Datschefki, *Sustainable Products: Using Nature's Cyclic/Solar/Safe Protocol for Design, Manufacturing and Procurement,* E-Monograph, BioThinking International, Surrey, England, 2002.

3. E. Datschefki, "Cyclic, Solar, Safe—BioDesign's Solution Requirements for Sustainability," *J. Sustainable Product Design,* 42–51, 1999.

4. The Institute for Market Transformation to Sustainability (MTS), "Sustainable and Green Profits Are Starting to Drive the Economy," Sustainable Products Corporation, Washington, DC, available; http://MTS.sustainableproducts.com

5. Green Product Design at 1-2, BSR Ed. Fund, www.sustainableproducts.com.

6. W. Knight and M. Curtis, "Design for Environment Software Development," *J. Sustainable Product Design,* 36–44, 1999.

7. "Forest Stewardship Council Certified Wood," available: http://www.fscoax.org, http://www.certified wood.org.

8. "Clean Vehicles," available: http://www.cleancarcampaign.org/standard.html, http://www.environ mentaldefense.org/greencar.

9. "Certified Organic Products Labeling," available: http://www.ota.com.

10. "Certified Green e-Power," available: http://www.green-e.org.

11. "U.S. Green Building Council LEED Rating System," available: http://www.usgbc.org

12. "Salmon Friendly Products," available: http://www.sustainableproducts.com/susproddef2.html#Salmonm.

13. "Cleaner and Greener Certification," available: http://www.cleanerandgreener.org.

14. "Natural Step System Conditions," available: http://www.NaturalStep.org.

15. "Nordic Swan Ecolabel," available: www.ecolabel.no/ecolabel/english/about.html.

16. "Green Seal Product Standards," available: http://www.greenseal.org.

17. "Global Reporting Initiative (GRI) Sustainability Reporting Guidelines, Social Equity Performance Indicators (2002)," available: http://www.sustainableproducts.com/susproddef2.html#Performance Indicators.

18. "Life Cycle Assessment (LCA)," available: http://www.sustainableproducts.com/susproddef.html.

19. *Standard Practice for Sustainable Products Economic Benefits—Including Buildings and Vehicles,* The Institute for Market Transformation to Sustainability (MTS), Washington, DC, January 2003.

20. W. McDonough and M. Braungart, *Cradle to Cradle,* North Point Press, New York, 2002.

21. T. G. Gutowski et al., "WTEC Panel on Environmentally Benign Manufacturing—Final Report," ITRI-WTEC, Loyola College in Maryland, April 2001.

22. "Living the Vision—Accomplishments of National Metal Finishing Strategic Goal Program," EPA 240-R-00-007, U.S Environmental Agency, Washington, DC, January 2001.

23. U.S. Department of Energy, http://www.doe.gov/.

24. K. L. Peterson and J. A. Dorsey, "Roadmap for Integrating Sustainable Design into Site-Level Operations," Report PNNL 13183, U.S. Department of Energy, Washington, DC, March 2000.

25. "Visioning Workshop for Remanufacturing," Office of Industrial Technologies, U.S. Department of Energy, September 11, 1998, available: http://www.reman.rit.edu/reference/pubs/2020vision/DOC VisionStatement.htm.

26. D. Bauer and S. Siddhaye, "Environmentally Benign Manufacturing Technologies: Draft Summary of the Roadmaps for U.S. Industries," International Technology Research Institute, Loyola College in Maryland, 1999, available: http://www.itri.loyola.edu/ebm/usws/welcome.htm.

27. U.S. Council for Automotive Research, http://www.uscar.org/.

28. *Visionary Manufacturing Challenges for 2020,* National Academy of Sciences, National Academy Press, Washington, DC, 1998.

29. World Resource Institute, http://www.wri.org.

30. C. Laszlo, *The Sustainable Company: How to Create Lasting Value through Social and Environmental Performance,* Island Press, Washington, DC, 2003.

31. J. C. Crittenden, "Activities Report: October 1998–September 1999," National Center for Clean Industrial and Treatment Technologies (CenCITT), Houghton, MI, April 2000.

32. "Directive of the European Parliament and of the Council on Waste Electrical and Electronic Equipment" (WEEE), 2002/ /EC, European Union, Brussels.

33. http://www.plastics-in-elv.org/Regulatory_Comprehensive.htm.

34. http:// www.jama.org/library/position092303.htm.

35. http://europa.eu.int/comm/environment/waste/elv_index.htm.

36. http://www.wbcsd.org/templates/TemplateWBCSD5/layout.asp?MenuID=1.

37. http://wbcsdmobility.org.

38. E. J. Daniels, J. A. Carpenter, Jr., C. Duranceau, M. Fisher, C. Wheeler, and G. Wislow, "Sustainable End-of-Life Vehicle Recycling: R & D Collaboration Between Industry and the U.S. DOE," *JOM, A Publication of the Minerals, Metals & Materials Society,* August 2004, pp. 28–32.

39. U.S. Department of Energy, http://www.doe.gov/.

40. I. S. Jawahir and P. C. Wanigarathne, "New Challenges in Developing Science-Based Sustainability Principles for Next Generation Product Design and Manufacturing," Keynote Paper, in S. Ekinovic et al. (Eds.), *Proc TMT 2004,* DOM STAMPE zenica Neum, Bosnia and Herzegovina, September 2004, pp. 1–10.

41. J. W. Sutherland, V. Kumar, J. C. Crittenden, M. H. Durfee, J. K. Gershenson, H. Gorman, D. R. Hokanson, N. J. Hutzler, D. J. Michalek, J. R. Mihelcic, D. R. Shonnard, B. D. Solomon, and

S. Sorby, "An Education Program in Support of a Sustainable Future," *Proc. ASME / IMECE, MED,* **14**, 611–618, 2003.

42. J. R. Mihelcic, J. C. Crittenden, M. J. Small, D. R. Shonnard, D. R. Hokanson, Q. Zhang, H. Chen, S. A. Sorby, V. U. James, J. W. Sutherland, and J. L. Schnoor, "Sustainability Science and Engineering: The Emergence of a New Meta-discipline," *Environ. Sci. Technol.,* **37**(23), 5314–5324, 2003.

43. U.S. Environmental Protection Agency, http://www.epa.gov/(2004).

44. F. Jovane, L. Alting, A. Armillotta, W. Eversheim, K. Feldmann, G. Seliger, and N. Roth, "A Key Issue in Product Life Cycle: Disassembly," *Ann. CIRP,* **42**(2), 651–658, 1993.

45. H. C. Zhang, T. C. Kuo, H. Lu, and S. H. Huang, "Environmentally Conscious Design and Manufacturing: A State-of-the-Art Survey," *J. Manufact. Syst.,* **16**(5), 352–371, 1997.

46. S. Yu, K. Jin, H. C. Zhang, F. Ling, and D. Barnes, "A Decision Making Model for Materials Management of End-of-Life Electronic Products," *J. Manufact. Syst.,* **19**(2), 94–107, 2000.

47. G. Seliger and H. Perlewitz, "Disassembly Factories for the Recovery of Resources in Product and Material Cycles," in *Proceedings of Japanese Society of Precision Engineering,* Sapporo, Japan, 1998.

48. M. S. Sodhi, J. Young, and W. A. Knight, "Modeling Material Separation Processes in Bulk Recycling," *Int. J. Production Res.,* **37**(10), 2239–2252, 1997.

49. M. Sodhi and W. Knight, "Product Design for Disassembly and Bulk Recycling," *Ann. CIRP,* **47**(1), 115–118, 1998.

50. A. Gesing, "Assuring the Continued Recycling of Light Metals in End-of-Life Vehicles: A Global Perspective," *JOM,* August 2004, pp. 18–27.

51. A. van Schaik, M. A. Reuter, and K. Heiskanen, "The Influence of Particle Size Reduction and Liberation on the Recycling Rate of End-of-Life Vehicles," *Minerals Eng.,* **17**(2), 331–347, 2004.

52. E. J. Daniels, "Automotive Materials Recycling: A Status Report of U.S. DOE and Industry Collaboration," in H. Mostaghaci (Ed.), *Ecomaterials and Ecoprocesses,* CIM, Montreal, Canada, 2003, pp. 389–402.

53. M. Sodhi and B. Reimer, "Models for Recycling Electronics End-of-Life Products," *ORSPEKTRUM,* **23**, 97–115, 2001.

54. C. M. Rose, K. Ishii, and K. Masui, "How Product Characteristics Determine End-of-Life Strategies," in *Proceedings of the 1998 IEEE International Symposium on Electronics and the Environment,* Oak Brook, IL, 1998, pp. 322–327.

55. A. van Schaik and M. A. Reuter, "The Optimization of End-of-life Vehicles Recycling in the European Union", *JOM, A Publication of the Minerals, Metals & Materials Society,* August 2004, pp. 39–42.

56. V. Sendijarevec et al., "Screening Study to Evaluate Shredder Residue Materials," Annual SAE Congress, Paper No. 2004-01-0468, SAE, Detroit, MI, 2004.

57. G. R. Winslow et al., "Advanced Separation of Plastics from Shredder Residue," Annual SAE Congress, Paper No. 2004-01-0469, SAE, Detroit, MI, 2004

58. H. Antrekowitsch and F. Prior, "Recycling of Metals and Plastics from Electronic Scrap," in *Proceedings of Global Conference on Sustainable Product Development and Life Cycle Engineering,* Berlin, Germany, 2004, pp. 121–126.

59. B. H. Lee, S. Rhee, and K. Ishii, "Robust Design for Recyclability Using Demanufacturing Complexity Metrics," Paper No. 97-DET/DFM-4345, in *Proceedings of the 1997 ASME Design Engineering Technical Conference and Computers in Engineering Conference,* Sacramento, CA, 1997.

60. T. G. Gutowski and J. B. Dahmus, "Material Entropy and Product Recycling," in *Proceedings of the Global Conference on Sustainable Product Development and Life Cycle Engineering,* Berlin, 2004, pp.135–138.

61. M. Frei, "Eco-Effective Product Design: The Contribution of Environmental Management in Designing Sustainable Products," *J. Sustainable Product Design,* October 1998, pp. 16–25.

62. E. Datschefski, "Cyclic, Solar, Safe—BioDesign's Solution Requirements for Sustainability," *J. Sustainable Product Design,* January 1999, pp. 42–51.

63. H. Kaebernick, S. Kara, and M. Sun, "Sustainable Product Development and Manufacturing by Considering Environmental Requirements," *Robotics and Computer Integrated Manufacturing,* 461–468, 2003.

64. M. Sun, C. J. Rydh, and H. Kaebernick, "Material Grouping for Simplified Life Cycle Assessment," *J. Sustainable Product Design,* 45–58, 2004.

65. S. Kara, J. Hanafi, S. Manmek, and H. Kaebernick, "Life Cycle Cost and Health Impact Estimation of Injection Moulded Parts," in *Proceedings of Global Conference on Sustainable Product Development and Life Cycle Engineering,* Berlin, 2004, pp. 73–77.

66. http://www.bfrl.nist.gov/oae/software/bees.html.

67. D. A. Dickinson and R. J. Caudill, "Sustainable Product and Material End-of-Life Management: An Approach for Evaluating Alternatives," *IEEE Int. Symp. Electronics Environ.,* 153–158, 2003.

68. J. W. Sutherland, K. Gunter, D. Allen, B. Bras, T. Gutowski, C. Murphy, T. Piwonka, P. Sheng, D. Thurston, and E. Wolff, "A Global Perspective on the Environmental Challenges Facing the Automotive Industry: State-of-the-Art and Directions for the Future," *Int. J. Vehicle Design,* **35**(1/2), 86–102, 2004.

69. *Our Common Future: From One Earth to One World,* World Commission on Environment and Development, Bruntdland Commission Report, Oxford University Press, Oxford, 1987, pp. 22–23 IV.

70. J. Fiksel, J. McDaniel, and D. Spitzley, "Measuring Product Sustainability," *J. Sustainable Product Design,* July 1998, pp. 7–16.

71. "Design Engineering Methodics, Engineering Design and Optimum Cost, Valuation of Costs" (in German), VDI 2225 Sheet 3, Berlin, Beuth, 1998.

72. M. Voβ and H. Birkhofer, "How Much Ecology Does Your Company Want?—A Technique for Assessing Product Concepts Based on Variable Weighting Factors," in *Proceedings of the Global Conference on Sustainable Product Development and Life Cycle Engineering,* Berlin, 2004, pp. 301–304.

73. International Organization for Standardization (ISO), *Environmental Management,* ISO 14001, ISO, Geneva, 1996.

74. N. Madu (Ed.), *Handbook of Environmentally Conscious Manufacturing,* Kluwer Academic, Norwell, MA, 2001.

75. C. Krotscheck and M. Naradoslowsky, "The Sustainable Process Index: A New Dimension in Echological Evaluation," *Echol. Eng.,* **6**, 241–258, 1996.

76. K. R. Haapala, K. N. Khadke, and J. W. Sutherland, "Predicting Manufacturing Waste and Energy for Sustainable Product Development via WE-Fab Software," in *Proceedings of the Global Conference on Sustainable Product Development and Life Cycle Engineering,* Berlin, 2004, pp. 243–250.

77. J. J. Michalek, P. Y. Papalambros, and S. J. Skerlos, "Analytical Framework for the Evaluation of Government Policy and Sustainable Design: Automotive Fuel Economy Example," in *Proceedings of the Global Conference on Sustainable Product Development and Life Cycle Engineering,* Berlin, 2004, pp. 273–280.

78. X. Wang, Z. J. Da, A. K. Balaji, and I. S. Jawahir, "Performance-Based Optimal Selection of Cutting Conditions and Cutting Tools in Multi-Pass Turning Operations Using Genetic Algorithms," *Int. J. Production Res.,* **40**(9), 2053–2065, 2002.

79. X. Wang and I. S. Jawahir, "Web-Based Optimization of Milling Operations for the Selection of Cutting Conditions using Genetic Algorithms," *J. Eng. Manufact.—Proc. Inst. Mech. Eng. UK,* **218**, 647–655, 2004.

80. "A Case Study of the Kansas City Science & Technology," May 2003, available: http://www.epa.gov/oaintrnt/content/kc brochure.pdf.

81. "Final Report of the OSHA Metalworking Fluids Standards Advisory Committee," http://www.osha.gov/dhs/reports/metalworking/MWFSAC-FinalReportSummary.html, 1999.

82. Centers for Disease Control and Prevention, http://www.cdc.gov/niosh/homepage.html.

83. International Organization for Standardization (ISO), ISO 14000, ISO, Geneva.

84. P. C. Wanigarathne, J. Liew, X. Wang, O. W. Dillon, Jr., and I. S. Jawahir, "Assessment of Process Sustainability for Product Manufacture in Machining Operations," in *Proceedings of the Global Conference on Sustainable Product Development and Life Cycle Engineering,* Berlin, 2004, pp. 305–312.

85. P. C. Wanigarathne, K. C. Ee, and I. S. Jawahir, "Near-Dry Machining for Environmentally Benign Manufacturing: A Comparison of Machining Performance with Flood Cooling and Dry Machining," in *Proceedings of the Second International Conference on Design and Manufacture for Sustainable Development,* Cambridge, United Kingdom, September 2003, pp. 39–48.

86. I. S. Jawahir and C. A. van Luttervelt, "Recent Developments in Chip Control Research and Applications," *Ann. CIRP,* **42**(2), 659–693, 1993.

87. X. D. Fang, J. Fei, and I. S. Jawahir, "A Hybrid Algorithm for Predicting Chip-Form/Chip Breakability in Machining," *Int. J. Machine Tools & Manufacture,* **36**(10), 1093–1107, 1996.

88. R. Ghosh, O. W. Dillon, and I. S. Jawahir, "An Investigation of 3-D Curled Chip in Machining, Part 1: A Mechanics-Based Analytical Model," *J. Machining Sci. Technol.,* **2**(1), 91–116, 1998.

89. R. Ghosh, O. W. Dillon, and I. S. Jawahir, "An Investigation of 3-D Curled Chip in Machining, Part 2: Simulation and Validation Using FE Techniques," *J. Machining Sci. Technol.,* **2**(1), 117–145, 1998.

90. X. D. Fang and I. S. Jawahir, "Predicting Total Machining Performance in Finish Turning Using Integrated Fuzzy-Set Models of the Machinability Parameters," *Int. J. Production Res.,* **32**(4), 833–849, 1994.

91. Z. J. Da, M. Lin, R. Ghosh, and I. S. Jawahir, " Development of an Intelligent Technique for Predictive Assessment of Chip Breaking in Computer-Aided Process Planning of Machining Operations," in *Proceedings on Manufacturing and Materials Processing (IMMP),* Gold Coast, Australia, July 1997, pp. 1311–1320.

92. I. S. Jawahir, P. X. Li, R. Ghosh, and F. L. Exner, "A New Parametric Approach for the Assessment of Comprehensive Tool-Wear in Coated Grooved Tools," *Ann. CIRP,* **44**(1), pp. 49–54, 1995.

93. K. C. Ee, A. K. Balaji, P. X. Li, and I. S. Jawahir, "Force Decomposition Model for Tool-Wear in Turning with Grooved Cutting Tools," *Wear,* **249**, 985–994, 2002.

94. P. Marksberry, "An Assessment of Tool-Life Performance in Near-Dry Machining of Automotive Steel Components for Sustainable Manufacturing," Thesis, University of Kentucky, Lexington, 2004.

95. P. Marksberry and I. S. Jawahir, "A Comprehensive Tool-life Performance Model in Near-Dry Machining for Sustainable Manufacturing," *Int. J. Mach. Tools Mfg.,* **48** (7–8), 878–886, 2008.

96. M. Bromark, R. Cahlin, P. Hadenqvist, S. Hogmark, G. Hakansson, and G. Hansson, "Influence of Recoating on the Mechanical and Tribological Performance of TiN-Coated HSS," *Surface Coating Technol.,* 76–77, 481–486, 1995.

97. K. D. Bouzakis, G. Skordaris, S. Hadjiyiannis, A. Asimakopoulos, J. Mirisidis, N. Michailidis, G. Erkens, R. Kremer, F. Klocke, and M. Kleinjans "A Nanoindendation-Based Determination of Internal Stress Alterations in PVD Films and Their Cemented Carbides Substrates Induced by Re-Coating Procedures and Their Effect on the Cutting Performance," *Thin Solid Films,* 447–448, 264– 271, 2004.

98. K. C. Ee, O. W. Dillon, Jr., and I. S. Jawahir, "Finite Element Modeling of Residual Stresses in Machining Induced by Cutting Tool with a Finite Edge Radius," *Proceedings of the Sixth CIRP International Workshop on Modeling of Machining Operations,* McMaster University, Hamilton, ON, Canada, May 2003, pp. 101–112.

99. K. C. Ee, O. W. Dillon, Jr., and I. S. Jawahir, "An Analysis of the Effects of Chip-Groove Geometry on Residual Stress Formation in Machining Using Finite Element Methods," *Proceedings of the Seventh CIRP International Workshop on Modeling of Machining Operations,* ENSAM, Cluny, France, May 2004, pp. 267– 274.

100. S. K. Sikdar, "Sustainable Development and Sustainability Metrics," *AIChE J.,* **49**(8), 1928–1932, 2003.

101. D. Tanzil, G. Ma, and B. R. Beloff, "Sustainability Metrix," in *Proceedings of the Eleventh International Conference on Greening of Industry Network—Innovating for Sustainability,* October 2003.

102. G. Seliger, "Global Sustainability—A Future Scenario," in *Proceedings of the Global Conference on Sustainable Product Development and Life Cycle Engineering,* Berlin, 2004, pp. 29–35.

CHAPTER 8

LIFE-CYCLE DESIGN

Abigail Clarke and John K. Gershenson
Michigan Technological University
Houghton Michigan

Design for the life cycle (DfLC) is a dynamic and proactive means of improving the environment through product design. Historical events have inspired and influenced the way life-cycle design has been developed and adopted by industry and society. Market drivers and legislation have also motivated design for the life-cycle efforts. This chapter outlines the principles that characterize the goals of DfLC as well as the methods that provide the means to achieve these goals and the tools that support life-cycle-compatible design decisions. A few illustrative implementations of DfLC are shown to highlight the achievement of these goals. However, it is clear that the future holds new and important challenges to the current state of DfLC.

Designers actually have more potential to slow environmental degradation than economists, politicians, businesses and even environmentalists.[1]

1 HISTORICAL INFLUENCES ON DEVELOPMENT OF DFLC

"Green design has a long pedigree, and before the Industrial Revolution it was the norm for many cultures."[1] However, the Industrial Revolution forced a complete rethinking of the field. The rapid pace of industrialization took a great toll on the environment.[1,2] As the Industrial Revolution reached full flame in Britain, the British Arts and Crafts movement led one of the first outcries against the new industrial society and its environmental damage. Many critics of industrialization and champions of nature have famously stated their concerns for the environment and society, including William Wordsworth, William Blake, George Perkins Marsh, Henry David Thoreau, John Muir, Aldo Leopold, Buckminster Fuller, Rachel Carson, Victor Papanek, and Fritz Schumacher.[1,3] Organizations have also sprung up to protect nature from industry, including the Sierra Club, Environmental Defense, World Wildlife Federation, Natural Resources Defense Council, and BUND. In addition, environmental science and environmental engineering are direct outgrowths of the Industrial Revolution for the study and remediation of the effects of emissions to the environment.

More recently, the 1970 oil crisis created a public outcry leading to environmental legislation that created design for the environment (DfE) and energy-efficiency appliance labeling.[4,5] Meanwhile, life-cycle engineering and concurrent engineering entered the American business consciousness.[6] The discovery of the ozone hole over Antarctica in 1985 spurred phase-outs and bans of particularly hazardous compounds.[7] Then, the 1986 nuclear reactor meltdown in Chernobyl spread environmental concerns across Europe.[8] At about the same time, companies began to put business and the environment together; 3M started its Pollution Prevention Pays program, where business cost savings were derived from environmental improvements. In 1987, the United Nations sponsored the World Commission on Environment and Development to publish *Our Common Future*, which called for sustainable development.[3] Nils Peter Flint initiated the Global O2 Network in 1988, which spreads environmentally responsible design ideas to support the growing need.[8,9]

In the 1990s, landfill closings and public resistance to new landfill sites increased attention to solid-waste disposal and increased consumer demand for producer responsibility.[1,5] The 1992 United Nations Rio de Janeiro Conference tackled environmental problems and further inspired people to improve industrial impact.[3] On the design side, DfLC concepts were further developed by the American Electronics Association, yielding a DfE manual for electronics.[10] The field of industrial ecology surged in popularity in the 1990s,[11] and academia began to embrace the concept of design for X (DfX), including DfE. Development of DfE tools surged, in part through collaborations between diverse stakeholders such as Philips Electronics, the Dutch government, and the Technical University of Delft, which simplified life-cycle assessment through IdeMat software, leading to EcoScan, Eco-It, and SimaPro.[1] Events such as the 2002 World Summit on Sustainable Development in Johannesburg have kept the spotlight on environmental concerns,[1] and public concern about global climate change continues to increase.[12]

2 DFLC DEFINITIONS

The Industrial Revolution was devastating in its environmental impact. As greater environmental concerns arose, designers, writers, scientists, governments, and eventually industrialists initiated efforts to address these problems. These efforts led to the creation, adoption, and practice of life-cycle engineering and DfE programs.

DfLC considers the entire system within which the product is created, used, and discarded when making design decisions: "Life-cycle design is a proactive approach for integrating pollution prevention and resource conservation strategies into the development of more ecologically and economically sustainable product systems."[13] Typically, the product life cycle is considered from inception to retirement, or *cradle to grave*,[13] including product and packaging design, materials extraction and processing, product manufacture and assembly, product distribution, product use and service, and product end of life (see Fig. 1).[6] This product design approach aims to reduce environmental impact and risk through considering the effects of each and every stage of a product's life cycle[6] and to reduce environmental impact over the entirety of a product's life cycle, effectively striking a balance between improving individual life-cycle stages.[14] In the last decade, ecological damage, resource flows, liability concerns, and subsequent costs have become increasingly important considerations in each product life-cycle stage.[15] With the significant increase in all environmental concerns, the term *DfLC* has expanded to *DfE* across the product life cycle. *Life-cycle design, DfE, ecodesign, environmentally conscious design*, and *green design* all consider the life cycle of a product and its environmental effects holistically. Hence, these terms are for the most part interchangeable.[6]

2.1 Design for Environment

Instead of considering only the product and its environmental effects, DfE looks at the entire system within which the product is made, operates, and is discarded.[16,17] A product is evaluated over its entire life cycle and at each individual life-cycle phase to decrease environmental harm.[18] DfE is one of several life-cycle *DfX* initiatives that include design for assembly, manufacturability, reliability, serviceability, and cost.[18] DfE stems from concurrent engineering, which seeks to improve design characteristics over and above functionality, including environmental performance, through multiattribute design decision making.[2,4,13,19] DfE entails optimizing a design for its environmental improvement (minimizing environmental impact),

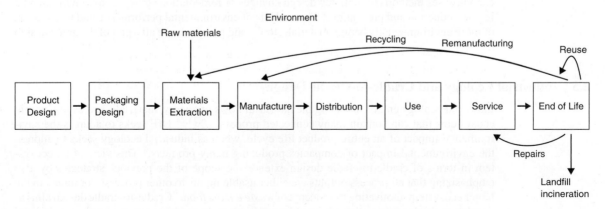

Figure 1 Product life cycle. *Source:* Adapted from Ref. 6.

instead of just fulfilling regulatory product environmental performance requirements.[4] The common aim of DfE, ecodesign, environmentally conscious design, and green design and engineering is to forecast and remedy environmental effects throughout the systematic, informed choices within a design process.

2.2 Ecodesign

Ecodesign is also part of the lexicon of DfLC. It is more commonly used throughout Europe than DfE, environmentally conscious design, or green design. Ecodesign is the synthesis of ecological concerns with traditional product concerns in the design of a product. The point of ecodesign is to satisfy both economic and ecological principles by considering the entirety of the product life cycle when designing products. Ecodesign aims to create products with less strain on material and energy resources or physical land mass and to reduce emissions and solid-waste production during their life cycle.[20]

2.3 Environmentally Conscious Design

Environmentally conscious design is an approach that is typically integrated with manufacturing: "Environmentally conscious design and manufacturing is a view of manufacturing which includes the social and technological aspects of the design, synthesis, processing, and use of products."[21] Perhaps more than other terms, environmentally conscious design places more emphasis on manufacturing than the design efforts undertaken.

2.4 Green Engineering and Green Design

Whereas DfE, ecodesign, and environmentally conscious design tend to consider mainly consumer products and industrial processes, green engineering seeks to improve environmental performance of products, processes, and systems encompassing issues ranging in scale from chemistry to land use planning.[22] Green engineering incorporates a holistic systems view that utilizes environmental performance as design targets, so that throughout the life cycle environmental objectives are given equal footing to other DfX requirements.[19] Green engineering addresses environmental problems by changing the underlying composition of a product, process, or system or by changing the context in which the system operates. Green product design encompasses methods of realizing design changes or remediation strategies for environmental impact reduction and prevention.[4] The analysis of environmental performance and the creation of more environmentally benign materials, tools, and processes are all a part of the green design methods.[2]

2.5 Industrial Ecology and Cradle-to-Cradle Design

Industrial ecology entails modeling the industrial flows of energy, materials, and capital within ecosystems that can contain many connected processes.[6,23,24] DfE seeks to address the environmental impact of an entire product life cycle, whereas industrial ecology seeks to address the environmental impact of companies producing many products.[6] This view of an ecosystem in terms of cradle-to-cradle design extends the scope of the previous strategies by also emphasizing that all process outputs are either usable inputs to other processes or inputs to the larger ecosystem, motivating the concept of *waste equals food*.[3] Cradle-to-cradle design aims to develop multiple regenerative product life cycles where byproducts can be cleanly metabolized by natural systems or fed back into the original processes.[3] Materials unsafe for ecosystems must be eliminated or stored until clean decomposition techniques are created.[3,25,26]

Practitioners use DfLC, DfE, green design, and ecodesign to denote tools, methods, and strategies that minimize the ecological impacts of the product life cycle.[27] Hence, throughout this chapter, DfLC, life-cycle design, and DfE are used interchangeably.

3 MOTIVATIONS FOR DFLC

The chief motivation for DfLC is to foresee environmental effects and mitigate them through product design. Many different players affect the environmental impact of products, motivated by a wide range of more specific concerns about emissions and wastes deposited in the air, on the land, and in the water. Resource depletion, rising human populations, climate change, and toxins are several of many important environmental problems that need addressing. Global, regional, and national governments have stated goals and enacted legislation that have had a profound impact on the implementation of DfLC. Corporations have also felt the push from competition, leadership, employees, cost pressures, liability, innovation needs, product performance, and production improvements to start implementing DfLC practices.[4] External pressures applied by consumers, the public, industry and trade organizations, governments, and nongovernmental organizations have resulted in increased regulation and competition, more complex and controlled supply chain dynamics, and a significant need for improving corporate images with respect to environmental impact. Many standards, certifications, and ecolabels have been put in place to encourage product design improvements and affect consumer buying habits. All of these motivators necessitate the adoption of DfLC approaches.

3.1 Who Is Responsible for Environmental Impact?

Humans are the major cause of environmental damage problems. Among the many issues, because of the ways in which industries produce the goods that society wants, substances are released that alter air, land, and water in ways that can harm people. The overuse of resources and the subsequent releases to ecosystems are what cause environmental damage.[6] People's actions at each stage of a product's life cycle, from the designers to the trash collectors, add to the total environmental impact of a product. Population increases have, in part, warranted the use of more materials and greater production levels.[23] As the human population continues to grow, many people desire an improved material quality of life. That desire presents many challenges to environmental health.[28] Likewise, poverty is a significant problem globally that leads to ecological damage.[29] However, it is those that enjoy the most benefits from industrial and consumer society who have the greatest ability to remedy the environmental destruction in the past and prevent harm in the present and future.

3.2 Effects to Environment

Graedel stated the following four widely held goals as preliminary aims that justify the creation of environmentally responsible practices: (1) prevent the extinction of people, (2) preserve the ability to increase quality of life for people, (3) sustain biodiversity, and (4) protect nature's beauty.[30] Graedel also identified several practices that promote these goals: employing renewable resources in ways that allow enough regeneration time, utilizing resources with finite availability in ways that allow enough time to discover renewable alternatives, causing no consequential impacts to global biodiversity, and emitting harmful substances at the rates that global ecosystems can metabolize the compounds.

Table 1 illustrates how different parts of the product life cycle can lead to different environmental impacts. Certain product life-cycle activities lead to environmental stresses that threaten the lives of organisms. The goal for environmental improvement is to allow

Table 1 Ecological Issues in Each Life-Cycle Phase

	Materials Choice	Energy Use	Solid Releases	Liquid Releases	Releases
Materials Extraction	—	Fossil fuel use Global climate change	—	—	—
Manufacture	—	Fossil fuel use Global climate change	—	Reductions in biodiversity Human health Water availability/ quality	Human health Ozone depletion
Distribution	—	Fossil fuel use Global climate change	—	—	—
Use	—	Fossil fuel use Global climate change	—	Reductions in biodiversity Human health	Ozone depletion
End of life	Fossil fuel use Global climate change	Fossil fuel use Global climate change	Fossil fuel use Global climate change	—	—

Note: Gray boxes are stresses that are always present; white boxes denote issues that are sometimes present.
Source: Adapted from Ref. 30.

industrial activity to benefit society while similarly allowing the environment to prosper. Since the environment provides the context within which industry and human society flourish, activities that balance environmental and human concerns are the ultimate aim of DfLC improvements.[2]

Environmental impacts can be characterized as primarily acting on a local, regional, or global scale. Several representative environmental impacts to land, air, water, and organism health are collated in Table 2. All effects have ramifications for humans. However, the seriousness differs by time scale and location as well as ability of particular individuals to handle burdening.[1,14,23,31–35]

Table 2 Representative Environmental Problems

		Scale of Effects		
		Local	Regional	Global
Effect Location	Land	Sludge siltation	Landfills	Resource depletion Soil dilapidation
	Air (climate)	Noise Light Smell Heat Photochemical smog	Visibility concerns	Ozone layer destruction Climate change
	Water	Eutrophication	Acidification Groundwater contaminants	Clean water scarcity
	Organisms	Indoor air quality	Toxins	Endocrine disrupters Habitat loss and biodiversity

3.3   Regulatory Motivators

Regulation has long been a motivation for behavioral changes with respect to the environment. Rather recently, legislative efforts have moved from polluter pays, also called *end-of-pipe efforts,* to producer responsibility legislation. *End-of-pipe legislation* pertains to environmental effects after processes are in place; hence, only small environmental improvements can be achieved.[14] *Producer responsibility legislation* can effectively motivate companies to work toward environmental improvement from the start of their design process.[36] Producer responsibility legislation necessitates setting goals for products during the beginning of the product development process, where environmental problems can be designed out.[14] Regulated product take-back or producer responsibility legislation is one such compelling force to adopt DfLC approaches.[31,36] Legislative efforts at the state, national, regional, and global levels all influence the adoption of DfLC. Table 3 lists several legislative initiatives that have had important ramifications on product design and the product development process.[5-7,37-42,84] For example, the German Packaging Ordinance of 1991 addressed lack of landfill space by putting extended producer responsibility into law through obligating producers to consider product end-of-life options involving reuse during the product development process.[5,6] Importantly, the efforts of one country or region influence other nations to adopt environmental regulations. The Waste Electrical and Electronic Equipment (WEEE) Directive makes electronic product manufacturers and retailers (including foreign producers and Internet retailers) financially responsible for electronic waste, motivating companies outside of the European Union to make product design changes and other countries to shun polluting practices by improving environmental legislation.[38,39] The goal of legislation is not to have companies meet the minimum criteria, but to spur industry to incorporate DfLC initiatives that involve the entire supply chain.[24] For environmental impacts to truly be reduced, designers, governments, producers, product users, and remanufacturers must work in concert.[14]

3.4   What Motivates Business?

Corporations make possible the creation and distribution of products and work hard to compel people to buy these products. These aims directly lead to environmental destruction through resource use and emissions.[30] Corporate leadership that acknowledges and accepts its role in creating ecological damage can steward DfLC aims.[31] Corporations stand to gain tremendously by improvements in business and product performance that come from implementing DfLC.[43] As shown in Table 4, a variety of external and internal motivations propel companies to adopt DfLC.[2,4,6,14,15,21,31,36,43-46] DfLC product standards and certifications are one example set of these motivators (Fig. 2).[23]

4   PRINCIPLES OF DFLC

The purpose of DfLC is to create products that positively affect the environment,[47] thus decreasing ecological damage. This is a broader aim than the elimination, reduction, and prevention of waste.[6,17,18] Each product life-cycle stage has its own guidelines or principles that come together to achieve this goal. Trade-offs among these stages, as well as between DfLC principles and other design objectives, must be balanced so that products are optimized for environmental performance over the entire life cycle. DfLC principles can act in concert to better environmental performance; addressing one aspect of product efficiency can lead to efficiency improvements in other aspects.[9]

Corporations and designers are responsible for the effects of products while in end users' hands and afterward.[6] According to Fuad-Luke: "Designers actually have more potential to slow environmental degradation than economists, politicians, businesses and even

Table 3 Representative Legislative Initiatives with Ramifications for Product Design

	Initiative Level				
	U.S. State	National	Regional	Global	Impacts of Legislation
Examples	1972 Oregon Beverage Container Act	—	—	—	Makes consumers pay deposits on bottles used for food packaging
	1987 Oregon Waste Tire Law	—	—	—	Makes consumers pay taxes on tires to pay for recycling and landfilling
		—	—	1987 Montreal Protocol (ratified 1989)	Phases out ozone-depleting compounds
		1990 U.S. Clean Air Act	—	—	Reduces emissions and creates phase-outs for particular chemicals
		1990 U.S. Corporate Average Fuel Economy	—	—	Reduces fuel use and emissions for vehicles
		1991 German Packaging Ordinance	—	—	Makes distributors take back packaging or else pay another organization to do so
			1994 European Commission Directive 94/62 on Packaging and Packaging Waste	—	Sets packaging take-back standards (recycling percentages including energy recovery) for every member country to meet
			—	1997 Kyoto Protocol (ratified 2005)	Reduces climate-changing gaseous emissions
			2003 European Commission Restriction of Hazardous Substances (RoHS) Directive	—	Limits and phases out the use of particular hazardous compounds
			2003 European Commission Waste Electrical and Electronic Equipment (WEEE) Directive	—	Makes electronic product manufacturers and retailers financially responsible for the collection, recycling, and disposal of electronic waste
			2005 European Commission Directive 2005/32/EC on the Ecodesign of Energy-using Products (EuP)	—	Mandates the environmentally responsible design of products that consume electricity and providing information to users about environmentally responsible use
	2006 Maine enacted electronic product take-back legislation	—	—	—	Makes producers financially responsible for the recycling and disposal of computer monitors and televisions

Note: End-of-pipe legislation is in gray, producer responsibility legislation is in white.

Table 4 Corporate Motivators

	Motivators	Impacts	Results
External	Reduce risk and liability	Eliminate dangers to workers and the environment to avoid lawsuits and regulations	Avoid lawsuits and regulations
	Better public relations	Improve public perception of the company	Maintain market share
	Meet consumer demands	Respond to consumer demand for environmentally preferable products	Retain and attract customers
	Beat competition	Create environmentally preferable products and gain more market share	Maintain market share
	Match supply chain demands	Implement environmental improvements dependent on all members of the supply chain, improving environmental performance	Maintain market share
	Attain standards	Set targets for products to reach to attain the credibility of a particular environmental designation	Attract customers
Internal	Enhance product performance	Improve products to increase customer satisfaction	Retain and attract customers
	Reduce costs	Reduce environmental violations and use resources more efficiently	Improve business performance
	Reinvigorate employee commitment	Form a rallying cause that expresses employee values	Improve business performance

Figure 2 Ecolabels from around the world. *Source:* From Ref. 23, Gradel, Thomas E.; Allenby, Braden R., Industrial Ecology, 1st Edition, © 1995. Reprinted by permission of Pearson Education, Inc., Upper Saddle River, NJ.

environmentalists."[1] Clearly, designers are responsible for applying DfLC principles, but changes in design alone have limited effects on reducing environmental impact of products and services.[27] Corporations need "to take responsibility both for the environmental consequences of [their] production and for the ultimate disposal of [their] products."[49] The stakeholders who have the greatest ability to enact changes (i.e., manufacturers and designers) have the greatest responsibility to design out environmental damage and accept liability for the consequences.[5]

4.1 Product Design Principles

The product-design process cements many details of a product and hence determines many of the possibilities for how other life-cycle design principles can be applied (see Table 5).[1,2,4,9,17,23,24,26,31,36,48–50] During the product design stage, designers must measure environmental performance iteratively and make design decisions accordingly. The costs of a product throughout its life cycle must be predicted, including environment-related expenditures. Lastly, by increasing the useful life of a product with appropriate technical and aesthetic life spans in mind, designers can attain many environmental benefits.[50]

4.2 Packaging Design Principles

The design of the product includes the design of packaging. Following environmentally responsible design principles such as those in Table 6[23,31] is an important step toward achieving

Table 5 Product Design Principles

	Explanation	Who Applies Principle	Example
Measure environmental performance	Assess resource use and risks	Design team	Use life-cycle assessment to identify and benchmark environmental impacts
Consider all costs	Determine all product life-cycle costs	Design team	Employ life-cycle costing to capture all costs incurred by a product
Minimize and eliminate	Choose designs that facilitate recycling	Design team	Upgrade the technology in a product, improve product durability, or employ aesthetics that people will enjoy long term

Table 6 Packaging Design Principles

	Explanation	Who Applies Principle	Example
Minimize and eliminate	Decrease packaging, decrease impact	Design team	Reduce the size and amount of material needed for packaging
Biodegrade	Create packaging that decomposes safely	Design team	Make packaging edible
Reuse	Design packaging for multiple uses	Design team	Create durable enough packaging for refilling
Recycle	Choose designs that facilitate recycling	Design team	Choose packaging materials that have established recycling markets
Use industry standards	Commonalize the packaging to make reuse or recycling more economically feasible	Design team	Select the industry preferred packaging

environmentally benign packaging. Besides the product covering and marketing materials, all transportation packaging must be considered as well. A good way to reduce packaging needs is for design and transportation engineers to communicate about product concerns and design packaging to fit both points of view.[23] Setting up a deposit or refund for packaging (e.g., bottle returns) or some type of return system between supplier, retailer, and user (e.g., pallet returns) encourages packaging reuse.[31]

4.3 Material Design Considerations

Guidelines for choosing the most environmentally responsible materials, such as those in Table 7, depend on product and packaging structure and requirements.[1,2,4,9,14,17,23,24,26,31,36,49–51] DfLC material considerations involve the types of materials chosen and how those materials should best be employed. Material choice can diminish or improve product performance and environmental impacts, both of which must be considered by designers.[9]

Table 7 Material Design Considerations

		Explanation	Who Applies Principle	Example
Material Selection	Choose abundant renewable resources	Avoid dependence on diminishing finite material and energy capital; instead use feedstocks that regenerate	Design team	Avoid dependence on diminishing finite material and energy resources like fossil fuels
	Choose sustainably harvested materials	Ensure that renewable resources remain available and viable	Design team	Pick materials that meet sustainable certification requirements, like wood with the Forest Stewardship Council label
	Choose recyclable materials	Extend the life of materials through several cycles	Design team	Avoid using composites; instead choose materials with economically viable recycling markets
	Choose recycled materials	Ensure that recyclable materials have a market	Design team	Keep recycled material quality high for multiple uses; use recycled materials in their original colors and textures
	Avoid hazardous substances	Ensure that products are safe for human and environmental health	Design team	Choose materials that cause no health or legal concerns
	Reduce material process energy	Account for material production effects in the environmental impact of a product	Material producers and design team	Consider the energy and impact differences for producing materials at a facility instead of outsourcing the finished substances before making manufacturing changes
Material Employment	Eliminate material waste	Decrease the amount of material that becomes waste during production	Design team	Design products to make manufacturing offcuts as small as possible
	Dematerialize	Remove material from a product	Design team	Reduce the weight and volume of materials in a product
	Simplify products	Eliminate the material waste of overdesign	Design team	Eliminate features that are not essential or necessary for a product to function or combine features

4.4   Product Manufacturing Design Principles

After setting goals for environmental improvement through material, product, and packaging design, designers must examine the production processes. Designers are expected to optimize products to eliminate inefficiencies and wastes during production as well as reduce process energy inputs for manufacturing.[50] This design goal can result in environmentally responsible process selection and energy-efficient production methods.[18] Designers must work with process engineers using the general design principles outlined in Table 8 to create methods that make ecologically benign objects.[4,9,18,31,44,50] Cleaner production can be achieved through optimization of product design, materials processing, and manufacturing using techniques from design for assembly and lean manufacturing.[31]

4.5   Product Distribution Design Principles

After production, the product is distributed. Again, there are design principles—principles for the design of the distribution system and principles for product design—that impact the distribution system (see Table 9)[1,23,31] and reduce a product's environmental impact. Management must consider the following factors to determine the best mode of transportation for each product: number of products, expense, time until a product is needed, length of travel and dependability, and ecological damage incurred. The National Research Council of Canada recommends having designers, shipper/receivers, and sales personnel compare the various modes

Table 8   Product Manufacturing Design Principles

		Explanation	Who Applies Principle	Example
	Choose cleaner production processes	Select the production processes with least environmental impact	Manufacturers	Employ lean manufacturing techniques to remove inefficiencies and waste from production and choose suppliers that use the most benign methods
	Improve quality	Ensure that production techniques and methods produce quality products	Manufacturers	Improve production to have fewer rejects and therefore less waste
	Choose clean power sources	Utilize power sources that create the least pollution	Manufacturers	Use renewable energy like wind power to generate needed electricity

Table 9   Product Distribution Design Principles

		Explanation	Who Applies Principle	Example
	Choose cleaner transportation methods	Select modes of transportation that create the least environmental impact	Design team and management	Distribute goods throughout a city by bicycle instead of truck, like Peace Coffee in Minneapolis, Minnesota
	Reduce transportation of products	Decrease transportation of goods to reduce environmental impact	Design team and management	Optimize distribution routes to deliver larger quantities of goods together, such as several product types, or use local suppliers

of transportation with these factors to select the most appropriate method for transporting products.[50] Transportation routes for products can also provide opportunities for packaging and product take-back.[23]

4.6 Product Use Design Principles

The use phase of a product can also contribute significantly to its impact. Designers are responsible for improving the energy efficiency of products.[4,26,50] However, ensuring that products are safe for users and their environment is also important for meeting DfLC principles. The principles in Table 10[1,2,9,23,24,26,31,50] highlight some of the general dos and don'ts that a product designer can control with respect to the environmental impact of product use. Including the hardware for users to reuse consumables with a product, such as rechargeable batteries and battery chargers, can reduce what a product consumes.[31]

4.7 Product Service Design Principles

Serviceable products likely have longer lifetimes than nonserviceable products. Increasing the product lifetime reduces the burden on material and energy resources and, hence, the environment. For this reason, companies should follow the principles outlined in Table 11 to provide facilities for servicing products and to design products with service in mind.[23,31]

Table 10   Product Use Design Principles

	Explanation	Who Applies Principle	Example
Reduce product energy use	Improve product energy efficiency to reduce waste and emissions	Design team	Fix leaks or energy losses and inform consumers how to best use products
Reduce what a product consumes	Eliminate or decrease the inputs a product needs over its lifetime to decrease waste and material and energy use	Design team	Use environmentally benign or reusable consumables and inform users how to best utilize consumables
Keep products clean	Create products that do not emit pollutants	Design team	Substitute materials used in adhesives to stop product off-gassing

Table 11   Product Service Design Principles

	Explanation	Who Applies Principle	Example
Make service easy	Design for ease of servicing to improve chances that maintenance will be performed and product life will be extended	Design team	Provide easy access to parts and clearly label parts that need different maintenance
Use benign servicing consumables	Ensure that all inputs used in servicing are safe for people and the environment	Product servicers	Use environmentally benign or reusable consumables and inform consumers how to best utilize consumables

4.8   Product End-of-Life Design Principles

At the end of a product's useful life, end-of-life systems for the collection of broken or unwanted products must be initiated or in place. Designing for product take-back and establishing a unique product take-back system increases the chances of a product being reused, remanufactured, or recycled.[26,31] Sometimes users discard the whole product when only one component fails, so designing all product components to fail at the same time can create less waste. Many end-of-life options exist for products; each has its own advantages and disadvantages. However, following the design principles in Table 12 will lead to reduced life-cycle environmental impact.[1,2,4,9,14,17,23,24,26,31,50]

4.9   Beyond the Principles

There are additional principles that lead to radically new ways in which DfLC can be realized. These principles fit in several broad categories, as shown in Table 13.[1-3,31,44,46,49-56] Using nature as inspiration for product design can lead to reduced environmental impact. Users can also enjoy the function of a product without possessing an object, reducing the necessary production volume while increasing utilization. Designing in multiple life cycles or industrial ecosystems is another worthy goal. Designing for sustainability incorporates DfLC principles as well as social and economic concerns, a challenge that corporations and designers recognize

Table 12   Product End-of-Life Design Principles

		Explanation	Who Applies Principle	Example
	Design for disassembly	Create a product to easily come apart into different materials and components to facilitate recycling, reuse, and remanufacturing	Design team and management	Design components in detachable modules that have similar characteristics, such as time to expected failure
	Design for recycling	Design products to facilitate the recycling of undesirables during the entire product life cycle	Design team and management	Label different components and materials with different colors for easy separation
	Design for reuse	Enable the reuse of products, components, and packaging to reduce waste and resource consumption	Design team and management	Create products for easy cleaning, fixing, or adapting to new improvements, uses, or aesthetics
	Design for remanufacturing	Recover, test, and use unwanted components in the same or different products to reduce resource use and waste	Design team and management	Consider packaging, transportation, and component design that facilitates shipping, tooling requirements, and processing for remanufacturing
	Design for biodegradation	Compost as a viable option for disposing of waste in an environmentally benign manner	Design team and management	Select benign, biodegradable materials for products

Table 13 Beyond Product Design for the Life-Cycle Principles

		Explanation	Who Applies Principle	Example
Nature-Inspired Design Principles	Design inspiration	Incorporating environmental concerns can lead to product innovation	Design team	Reconsider the underlying suppositions for material and energy use in products
	Sun fuels all	A set amount of matter cycles in our world, but net gains in energy come from the sun	Design team	Use the benefits of sun power through photosynthesis and plant metabolization or wind turbines and photovoltaic cells
	Use and render wisely	Industry impacts the environment when compounds are altered in form and dispersed more quickly than regeneration occurs	Material producers, design team and manufacturers	Replace scarce materials with resources that abound and use renewable resources within ecosystem capabilities for rehabilitation
	Consider the consequences	The creation of products has many ramifications for ecosystems	Marketers, design team, and management	Trace the beginnings and future of all resource uses and actions that go into a product
Functionality without Possession	Location Is Important	Acknowledge the particular character of each location that is affected by product life-cycle stages	Design team	Design to reflect the uniqueness of place while recognizing and preventing negative effects of design choices
	Products into services	Customers want the function provided, not an object	Marketers, design team, management, servicers, and product reclaimers	Create services, product systems, and life cycles of products that meet user expectations instead of designing products
	Immaterialize	Replacing the utilization of products with actions that do not involve resource use	Marketers, design team, and management	Create information and activities to occur electronically, eliminating infrastructure such as working or shopping over the Internet
	Design for sharing	Sharing reduces consumption and environmental damage	Marketers, design team, and management	Design organizational systems for sharing technical support and encouraging groups that facilitate sharing like libraries
Sustainable Design Principles	Use Local Resources	Using local resources supports the economies that lose when local ecosystems suffer	Design team and management	Substitute locally available materials or work with nearby suppliers
	Utilize natural advantages	Nature provides particular output energies that can be utilized easily	Design team, manufacturing, and distribution	Employ gravity-fed delivery or temporal temperature and climatic moisture differences for cooling
	Promote wellness of all people	Create products in such a way that all people benefit	Marketing, design team, and management	Set up fair trade systems for products, replacing toxins with benign materials

as *designing for the triple bottom line*.[3,6,57,58] All of these product design principles stretch the generally held body of thinking behind DfLC.

5 LIFE-CYCLE DESIGN METHODS

Life-cycle design—that is, product life-cycle design—is at the very heart of the development of the product life cycle. Many design methods are employed to incorporate DfLC concerns into product development. Including life-cycle design objectives in the design process increases the resources needed for product design and the number of stakeholders.[6] There are three facets to adding life-cycle considerations to the design requirements:

1. Managing and measuring material and energy streams within production processes and throughout the product life cycle
2. Integrating costs to the environment into the financial analysis
3. Considering the entire context in which a product design operates[4]

It is generally accepted that around 80% of a product's life-cycle costs and environmental impacts are decided during product design.[9,19,20] When DfLC is implemented early in product development, as part of a concurrent engineering process, more substantial and viable design impacts and therefore increased environmental benefits are possible.[4]

Various design methods are best applied (or only applicable) during different stages of the product design process. Typically, the product design process is described by four stages, each with unique opportunities for environmental improvements. These stages are described in Table 14 as product definition, conceptual design, embodiment, and detail design. Many design methods work to improve the environmental impact of a product during a single or multiple life-cycle phases, over the entire product life cycle, or with respect to concerns beyond a product's life cycle.[6] The design methods shown in Table 15 are general and hence can be used on a wide variety of products. Each method has different environmental impacts that are realized during life-cycle stages, and each method can be applied at different stages of the product life cycle. Source reduction can be achieved by analyzing a product design and removing excess material unnecessary for strength requirements.

5.1 Methods Applied during Product Definition and Conceptual Design

Methods that can be applied early in the design process have a greater effect on a product's overall environmental impact.[4] Customization involves designing products to be personal so that people develop the care and interest in maintaining and prolonging the life of a product.[59] Designing for modularity yields a product with clusters of components or modules by similar physical characteristics, such as the same recycling or disposal treatment.[26,60–63] Using expert systems, a company solicits information from product development experts about environmentally responsible products to set goals or requirements and develop a methodology toward producing successful products.[64] Sustainable product and service development examines the functional expectations, life-cycle stages, and supply chain of a product to convert the product into a service.[65]

5.2 Methods Applied during Embodiment or Detail Design

Methods applied during embodiment and detail design can still abate a product's environmental impact. Material substitution entails the exchange of ecologically damaging materials with alternatives that reduce environmental impact. Source reduction involves reducing the amount

Table 14 Design and Environmental Considerations during Product Design Process

	Product Design Application Stage	Product Life-Cycle Stages Affected	Environmental Benefits
Customization	Product definition	End of life	Extends product life, thereby reducing waste and conserving resources
Expert systems	Product definition	Materials extraction, manufacturing, distribution, use, service, and end of life	Sets particular environmental goals to apply throughout the product life cycle
Sustainable product and service development (SPSD) method	Product definition	Materials extraction, manufacturing, distribution, use, service, and end of life	Conserves energy and materials by extending product life, eliminates waste, reduces toxicity
Modular design	Product definition or conceptual design	Manufacturing, service, and end of life	Decreases manufacturing and servicing energy use, conserves resources by updating components of products, allows for safe disposal of toxins
Material substitution	Embodiment or detail design	Materials extraction, manufacturing, and distribution	Reduces toxicity and energy intensity and encourages reuse or recycling
Source reduction	Embodiment or detail design	Manufacturing, distribution, and end of life	Eliminates waste
Environmentally conscious product design: a collaborative Internet-based modeling approach	Embodiment or detail design	Materials extraction, manufacturing, distribution, use, service, and end of life	Variety of environmental benefits

Source: From Refs. 13, 50, and 105.

Table 15 Comparison of Representative Design for the Life-cycle Methods

		Product Definition	Conceptual Design	Embodiment Design	Detail Design
	Design Process Considerations	Define customer requirements and product functioning	Assess competing product concepts for customer satisfaction Select single product for further development	Design basic product architecture	Fully specify all aspects of the product and components
	Environmental Considerations	Set environmental performance objectives Challenge basic assumptions regarding product functioning and end-user needs	Determine end-of-life options	Engineer how to reduce impact during use Create efficient transport Enact cleaner production goals and strategies	Consider how to reduce the impact from materials

of waste caused by a product before it leaves the factory.[26] The environmentally conscious product design collaborative Internet-based modeling approach involves the exchange of design and environmental assessment information between designers and environmental knowledge experts, providing contemporaneous feedback to both parties.[66] All of these methods present different approaches to improving product environmental performance.

6 DESIGN FOR LIFE-CYCLE TOOLS

Design methods provide an overall approach to implementing DfLC in a design or throughout an organization. Design tools perform the specific tasks needed to implement a DfLC method. Tools can provide new information, organize existing information, or present information in a new light. Tools generally help provide logical, accurate, and repeatable predictions that can influence design decisions within multiple subject areas covered.[6,67] Tools perform trade-off analyses between design objectives and assess life-cycle impacts.[6] The tool must discern those design aspects in need of improvement and provide guidance for how to transcend the problems.[67] Tools should allow designers to gain a sense of what leads to better outcomes and how to redesign products in the future to meet more criteria and make better products.[6] The ability to describe and handle complex processes with many linked processes is essential for a good tool.[68] Tools work in different ways to aid designers in making better environmental design choices. Tools can aid designers' decision making by performing the following tasks: environmental assessment, priority setting, supporting idea generation, and integrating additional criteria besides environmental concerns. DfLC tools and indicators and environmental accounting tools all differ in the complexity and time required for use. Some tools help with a single task while others aid several tasks (Fig. 3).[69]

Design tools enable designers to make the engineering decisions essential to implementing DfLC. Indicators are a specific group of tools that assess and represent predicted environmental damage of product life-cycle actions. They are typically part of more comprehensive life-cycle design tools. Environmental accounting tools assess the total financial obligations of the product life cycle, including environment-related costs.

6.1 Design Tools

Design tools aid design engineers in many aspects of product development decision making. Tools are useful in different stages of the product development process, apply to different types of products, and consider distinct stages of the product life cycle. Tools that are applied in the early stages of product design have greater ability to improve the environmental performance of a product (see Fig. 4). Not all of the tools shown here are specific to DfLC. They are, however, tools that can greatly improve the efficiency and efficacy of the DfLC process.

6.2 Product Definition

At the product definition stage of product development, tools are needed to help gather, organize, and apply information leading to product design requirements for improved environmental impact. The tools in Table 16 aid in uncovering customer requirements, outlining the design process, and organizing innovation concepts.

Environmental customer requirements can be uncovered with *conjoint analysis*, where users rank products with different attributes, and *contingent valuation*, where willingness to pay for environmental attributes is assessed.[71] The *Kano technique* is a way to interpret the environmental "voice of the customer" into customer requirements through surveys of users. Similar to conjoint analysis and contingent valuation, the designer using the Kano technique must select

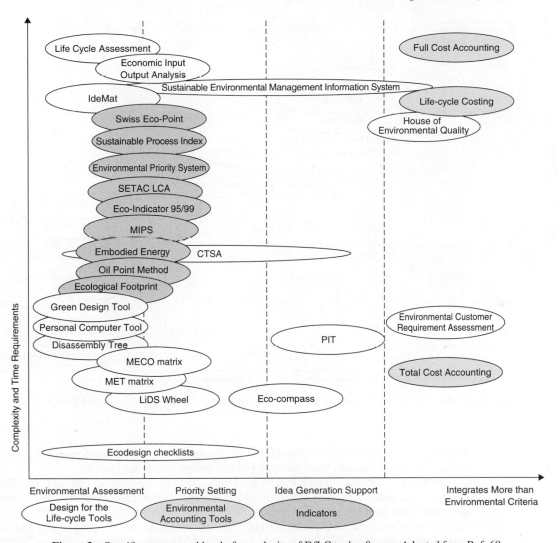

Figure 3 Specific purpose and level of complexity of DfLC tools. *Source:* Adapted from Ref. 69.

weightings of each design feature.[70,71] The *product ideas tree* (PIT) diagram (inspired by Mind Maps, the life-cycle design strategy (or LiDs) wheel, and the Eco-compass) provide a record and organization strategy for ideas to incorporate into a product concept generated during design brainstorming sessions where ideas are recorded by the most relevant design process stage and environmental impact category affected.[72]

Conceptual Design

Tools that can be applied during conceptual design often require more detailed information about material and energy flows with respect to a product than tools applied during product definition. In turn, these tools also result in greater specificity, as they help to narrow the design solution space. The tools in Table 17 help compare product concepts for environmental impact,

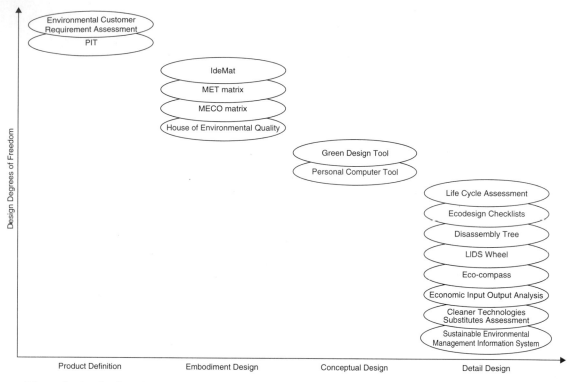

Figure 4   Application of different tools during the product design process stages. *Source:* Adapted from Ref. 14.

display potential resource flows throughout the product life cycle, and note potential conflicts between design parameters because more product design details are known.

IdeMat is a database (which can be used with life-cycle assessment software SimaPro's database) that aids selection of materials, processes, and components based on environmental indicators such as ecoindicator 95 and the Environmental Priority System.[73,74] The *material cycle, energy use, and toxic emissions (MET) matrix* can track and display inputs and outputs of materials, energy, and toxics that flow through the product life-cycle stages.[75,76] The *materials–energy–chemicals–other (MECO) matrix* creates an inventory of its namesake flows and ranks the impacts and feasibility of change of the flows throughout the product life cycle.[9] The *house of environmental quality* (derived from the house of quality)[77] illuminates potential conflicts between design and environmental criteria throughout the product life cycle using weightings of the importance of particular life-cycle criteria.[78]

Embodiment Design

Tools applied during embodiment design (see Table 18) can yield a more accurate assessment because the basic layout, components, and materials have already been chosen. At this point, designers can address deeper concerns about manufacturing, distribution, use, and end of life. Again, the fact that more product details are already set bars sweeping product design changes.

The *Green Design Tool* allows designers to see the effects of product design attribute choices on process waste creation by comparing environmental impact scores of designs based on end-of-life options, ease of disassembly, labeling, materials, and toxicity.[79] The *Personal*

Table 16 Product Definition: Design for the Life-Cycle Tools

Tools for Assessing Environmental Customer Requirements	Picture	Product Design Application Stage	Product Life-cycle Stage Affected	Impacts	Advantages or Disadvantages
		Product definition	Product design, packing design, materials extraction, manufacturing, distribution, use, service, end of life	Customer environmental concerns can be incorporated into product design	Difficult to interpret customer responses into requirements
Product Ideas Tree (PIT)		Product definition	Product design, packing design, materials extraction, manufacturing, distribution, use, service, end of life	Innovated ideas can be recorded and placed by what environmental benefit is espoused and when in the design process the idea can be implemented	Complicated

227

Table 17   Conceptual Design: Design for the Life-cycle Tools

	Picture			Product Design Application Stage	Product Life-cycle Stage Affected	Impacts	Advantages or Disadvantages
IdeMat				Conceptual, embodiment, or detail design	Product design, packing design, materials extraction, manufacturing, distribution, use, service, end of life	Promotes the selection of more environmentally benign materials, processes and components	Only provide a database
MET-matrix		Materials Cycle Input/Output	Energy Use Input/Output	Toxic Emissions Output	Conceptual, embodiment, or detail design	Product design, packing design, materials extraction, manufacturing, distribution, use, service, end of life	Displays and organizes resource flows throughout the life cycle	Simple
MECO-matrix					Conceptual, embodiment, or detail design	Product design, packing design, materials extraction, manufacturing, distribution, use, service, end of life	Records and designates relative impact of resource flows throughout the life cycle	Simple, but designers create relative impact weighting
House of Environmental Quality					Conceptual, embodiment, or detail design	Product design, packing design, materials extraction, manufacturing, distribution, use, service, end of life	Trade-offs between environmental customer requirements and design parameter choices are uncovered	Time consuming

Computer DfE tool rates the ability of a design to be recycled by scoring products for environmental impact of materials, disassembly ease, ease of recycling, and hazardous material content.[67]

Detail Design

Detail design is where all of the specifications of a part are decided. In this design process stage, the most extensive and accurate environmental impact predictions are made and some design changes can still be enacted. Table 19 showcases design tools that can characterize product environmental impact, offer suggestions for environmental improvement, or display product disassembly because product specifications are known.

Life-cycle assessment (LCA) sets a goal, scope, and product function (functional unit) based on an inventory of all inputs and outputs, impact assessment, and an interpretation of

Table 18   Embodiment Design: Design for Life-Cycle Tools

	Picture	Product Design Application Stage	Product Life-cycle Stage Affected	Impacts	Advantages or Disadvantages
Green Design Tool		Embodiment or detail design	Material extraction, manufacturing, end-of-life	A score is given for both the environmental soundness of product and process attributes, which designers can use to choose between design alternatives	Some life-cycle stages are ignored
Personal Computer DfE Tool		Embodiment or detail design	End of life	Recycling is improved	Focus is on single product and life-cycle stage

results. LCA is a relative measure for comparing environmental performance of two or more products useful for benchmarking environmental performance, setting environmental goals, aiding material/component design decisions, and uncovering environmental impacts that are not obvious.[9]

   Ecodesign checklists, of which there is an abundance, ask a series of questions related to DfLC strategies for each product life-cycle stage (e.g., for product end of life, is a product take-back strategy in place?) to compare different designs or highlight areas needing improvement.[31,80] *Disassembly trees* are diagrams that show the chronology of part removal to facilitate product disassembly for a variety of product end-of-life options.[64,68] Brezet and van Hemel[31] created the *LiDS wheel* for the Dutch Promise Manual and United Nations Environment Programme (UNEP) Ecodesign Manual to quickly characterize environmental concerns and to note the level of effort extended toward employing the environmental strategies at each life-cycle stage, allowing design engineers to compare which life-cycle stages need more design effort.[6,72,81] The *Eco-Compass* was developed by Dow Chemical Company in Europe for design decision making by denoting a numerical environmental impact score for a product with respect to mass and energy intensity, human and ecosystem health risk, reuse of wastes (revalorization), resource conservation, and improved product functioning (service extension).[72,81] *Economic Input Output Analysis Life-Cycle Assessment (EIO-LCA)*, a software tool that follows financial movements, resource intensity, and releases to the environment due to specific commodities in a national economy, was developed by the Carnegie Mellon Green Design Institute.[82] The *Cleaner Technologies Substitutes Assessment (CTSA)* was created and refined by the U.S. Environmental Protection Agency (EPA) in conjunction with industry, nonprofits, and academia to present companies with alternatives for meeting their product or process needs with different environmentally responsible technologies.[83] The Ricoh Group created the *Sustainable Environmental Management Information System* software for aiding the management of materials, purchasing, the supply chain, resource flows, environmental impact, and accounting product information.[84]

6.3   Indicators

Indicators assess environmental impact, which is then incorporated into comprehensive life-cycle design tools and approaches. Many of the life-cycle design indicators are used to

Table 19   Detail Design: Design for Life-Cycle Tools

	Picture	Product Design Application Stage	Product Life-cycle Stage Affected	Impacts	Advantages or Disadvantages
Life-cycle Assessment		Detail design	Product design, packaging design, materials extraction, manufacturing, distribution, use, service, end of life	Uncovers the more environmentally responsible choice of two or more options.	Detail and time intensive; it and has relative results.
Ecodesign Checklists	**Reuse/Recycling (closing technical material and energy cycles)** • recycling strategy in place? • guarantee for take back in place? • re-use of the complete product (e.g. second-hand, recycling cascade) • recycling of components (e.g. upgrading, reuse of components) • recycling of materials • dismantling of products • separability of different materials • low diversity of materials • low and materials energy input for reuse/recycling **Final Disposal** • compostable, fermentable products (closing biological cycles) • combustion characteristics • environmental aspects at deposition	Detail design	Product design, packaging design, materials extraction, manufacturing, distribution, use, service, end of life	Shows which design for environment strategies have not been employed.	Some questions may not be relevant to product and neglects trade-offs between employing different strategies.
Disassembly Trees		Detail design	End of life	Shows product disassembly process allows for optimization.	Time-consuming.
LiDS Wheel		Detail design	Product design, packaging design, materials extraction, manufacturing, distribution, use, service, end of life	Maps effort extended at each life-cycle stage and allows comparison to see which stages could use more effort.	Simple, but relationship between effort extended and environmental impact reduction is not clear.
Eco-Compass		Detail design	Product design, packaging design, materials extraction, manufacturing, distribution, use, service, end of life	Displays improvement in different categories of environmental impact relative to each other for a product or between design choices.	Simple, but all scores are relative and based on design team knowledge.
Economic Input Output Analysis		Detail design	Product design, packaging design, materials extraction, manufacturing, distribution, use, service, end of life	Show the environmental and financial significance of aggregate production within the U.S. economy of a particular commodity.	Provides big-picture view of industrial environmental impact, but data carry some uncertainty and represent U.S. only.
Cleaner Technologies Substitutes Assessment		Detail design	Product design, packaging design, materials extraction, manufacturing	Presents alternatives for meeting product or process need with environmentally responsible technologies.	Detail and time intensive.
Sustainable Environmental Management Information System		Detail design	Product design, packaging design, materials extraction, manufacturing, distribution, use, service, end of life	This software aids material and component selection, calculates environmental impact, suggests product end-of-life options and records environmentally related costs.	Helps organize many different types of environmental information for easy access.

establish the impact and weighting for LCA or similar tools. Indicators quantify particular categories of impact, resulting in a predicted environmental damage.[66] In general, indicators provide a single impact score by evaluating a single parameter or assessing many impact parameters and combining their values. The science used to measure environmental impacts is incomplete and complex. Hence, indicators can be inaccurate impact assessors.[9] Each indicator is a balance between the effort extended to achieve the indicator score and the knowledge provided from that result. Different indicators take more time and thought to complete while presenting more or less aid to the designer for choosing between product designs and attributes. When choosing to use an indicator, designers must balance the level of indicator complexity with the helpfulness of results gained (see Fig. 5).

Indicators can be characterized by focusing on a single or multiple categories of impact. Single-parameter indicators may be simpler to execute, but their results may miss crucial areas of environmental impact. Multiple-parameter indicators involve complex assessment and may present scores that are difficult to interpret. Both types of parameters can portray environmental impacts in a way that furthers some understanding of product design choices.

Single-Parameter Indicators

Single-parameter indicators have the potential to provide results that are more meaningful, since all impacts are rated by the effect to one unit of measure such as mass flow in kilograms for material input per service unit (MIPS). However, these indicators also have the disadvantage of overlooking environmental impacts not well characterized by that unit of measure. For MIPS, the differences in materials such as toxicity are ignored.[9] Table 20 shows how single-parameter indicators differ by units, environmental impact categories considered, and data requirements.[9,44,85,86]

The *ecological footprint*[87] assesses environmental impact by calculating the total area of land *bioproductivity*, or productive capacity used to support an activity. Factors relate different types of resource extraction (such as fossil fuel use) into areas of land productive capacity required for that activity. All effects can be represented by a single number with the unit hectares

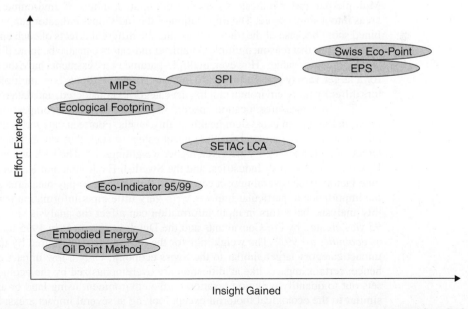

Figure 5   Comparison of indicators for insight gained from effort exerted. *Source:* Adapted from Ref. 9.

Table 20   Single-Parameter Indicators

	Unit of Comparison	Impact Categories Characterized	Data Requirements	Advantages or Disadvantages
Ecological footprint	Hectares of land	Land use	Product life-cycle inventory	Lacks data needed for analysis and only focuses on effects to land
Embodied energy	Energy	Energy use	Product life-cycle inventory	Simple measure allows easy comparison between products
Material input per service unit (MIPS)	Mass flow in kilograms	Resource use	Extensive data with high accuracy	Neglects differences between materials
Oil point method	Energy content of 1 kg of crude oil called an oil point (OP)	Resource use	Product life-cycle inventory	Gives qualitative results that needs careful holistic interpretation

of land required.[9,85] *Embodied energy* is a concept that comes from input/output analysis, which assesses the total amount of energy required for the product life cycle.[9] MIPS[88] accounts for specific material and energy flows throughout a product life cycle to reduce material through-put.[9,86] The *oil point method* (OPM) quantifies environmental impact by the energy content of 1 kg of crude oil, called an *oil point*. LCA methodology is used to uncover the energy used in each part of the product life cycle, and conversion information from energy into oil point indicators is provided.[44]

Multiple-Parameter Indicators

Multiple-parameter indicators combine designated values of environment impact in several areas into a single score. The implication of the individual indicator can be buried in the combined score because of the trade-offs among the different effects of each environmental impact category. For that reason, multiple-parameter indicators can also be more difficult to assess than single-indicator values. However, multiple-parameter assessments have the potential to account for a larger variety of environmental impacts. Table 21 shows how multiple-parameter indicators differ by units, environmental impact categories considered, and data requirements.[4,9,86,87]

The SEP measures location-specific impacts by a relative comparison measure, the *eco-point*, derived from ecosystem health quality levels. Fourteen categories are evaluated for relative distance from a target for the impact category such that values that lie further from the target value for a category are given higher weightings.[9,86] The EPS was created by Volvo, the Federation of Swedish Industries, and the Swedish Environmental Research Institute to combine factors from several impact categories. Willingness-to-pay measures are used to quantify the importance of particular impacts.[4,9,86] Very little input information is required to perform this analysis, but errors in input information can affect the analysis strongly.[86] *Ecoindicator 95* was created by Pré Consultants and the Dutch government in 1995 and improved in 1999 as *ecoindicator 99*.[89] The weightings for the indicator are measured by the distance from an impact category target similar to the Swiss ecopoint. Only a few impact categories are used; hence, certain impacts like acidification are overemphasized by the ecoindicator.[9,87] The SPI sets out to quantify pollution taxation on the environment using land or area as the measure, similar to the ecological footprint except looking at several impact areas. Data must be fairly accurate to get a pertinent outcome. The SETAC LCA characterizes impacts by assigning each

Table 21 Multiple-Parameter Indicators

	Unit of Comparison	Impact Categories Characterized	Data Requirements	Advantages or Disadvantages
Swiss ecopoint (SEP)	Ecopoint	Resource conservation, toxicity, global climate change, ozone generation, resource depletion, etc. (14 categories total)	Location-specific data with a fair amount of accuracy	Emissions outside of Switzerland are ignored
Environmental Priority System (EPS)	Environmental load unit per kilogram	Human health, biological diversity, manufacturing, resource conservation, and aesthetics	Little data needed, but high accuracy required	Errors in input information strongly affect analysis
Sustainable process index (SPI)	Meter squared	Resource conservation, toxicity, global climate change, ozone depletion, etc.	Location-specific data with a fair amount of accuracy	Robust to some errors in input information
Society of Environmental Toxicology and Chemistry's life-cycle impact assessment (SETAC LCA)	Relative scale	Global climate change, ozone creation and depletion, human and environmental toxicity, acidification, eutrophication, and effects to land, living creatures, and natural resources	Little data input needed	Gives consistent results
Ecoindicators 95 and 99	Numeric value	Human health, ecosystem health, and resources	Product life-cycle inventory	Incorporates fate of emissions and degree of effects on receiving ecosystems, but it overemphasizes acidification while deemphasizing land use and biodiversity concerns

a relative score in a particular impact category. Not much information is needed to perform this analysis.[86]

6.4 Environmental Cost Accounting Tools

Often, environmental costs are neglected by companies during accounting. Part of this omission comes from not including costs incurred by products over the entire product life cycle.[89] Environmental cost accounting tools differ from indicators because of the focus on the financial obligations of each part of the product life cycle. Table 22 shows where several environmental cost accounting tools are applied during the life cycle and what life-cycle stages these tools affect as well as the impacts using such tools can have.

Total cost accounting (TCA) incorporates financial obligations associated with liability and was created alongside the idea of cleaner production in the late 1980s. Some dynamic decision making is incorporated into the tool by trying to reflect how costs might differ when relationships with users and suppliers change. Most costs considered in TCA come from manufacturing.[89] *Life-cycle costing* (LCC) takes into account the dynamic effects of cost at all product life-cycle stages. Additionally, influences to costs are considered, such as the price of

Table 22   Environmental Cost Accounting Tools

	Product Design Application Stage	Product Life-Cycle Stages Affected	Impacts	Advantages or Disadvantages
Total cost accounting	Detail design	Product design, packaging design, materials extraction, and manufacturing	Incorporates financial obligations from liability and stakeholder dynamics into conventional accounting	Most costs included come from manufacturing life-cycle stage
Life-cycle costing	Detail design	Product design, packaging design, materials extraction, manufacturing, distribution, use, service, and end of life	Uncovers how costs affect different life-cycle stages	Cost advantages in the market due to life-cycle design measures can be revealed
Full cost accounting	Detail design	Product design, packaging design, materials extraction, manufacturing, distribution, use, service, and end of life	Shows how particular parties are affected by product life-cycle costs	Societal costs are difficult to assess, so willingness-to-pay measures are often used

capital, labor, materials, energy, and disposal. The overall aim of LCC is to reveal how influences on costs may create advantages for different parties because of how costs affect different life-cycle stages.[89] *Full cost accounting* (FCA) expands LCC by considering how particular parties are affected by the costs incurred during the product life cycle. Hence, FCA considers what costs are paid by society because of environment degradation. The cost of this detriment is hard to assess, so contingent valuation or willingness-to-pay measures are often used for assessment.[89]

7   IMPLEMENTATION OF DFLC

Implementing DfLC is complicated. There is no single path to incorporating DfLC concerns in a product, but approaches exist to help companies incorporate life-cycle design into their structure. According to Eyring: "The idea of green design seems simple, but there is no rigid formula or decision hierarchy for implementing it."[16] There are many different approaches to DfLC because there are many ways to view life-cycle design, and there are many different situations in which to apply it.

7.1   How to Implement DfLC within a Company

Most businesses start with a finished design and try to improve its environmental performance.[65] This strategy does not beget the most benefit from life-cycle design approaches for a product. Several approaches for implementing DfLC in a company are described in this section.

The International Organization for Standardization (ISO) *14000 standards* provide a standard for environmental management systems and allow companies to achieve and receive recognition.[90] This system helps companies identify and put in place their own environmental policy, including planning, actions, and reviews. All parts of the management system work together toward continuous environmental performance improvement.[64]

Integrated environmental management systems (IEMSs)[91] are management approaches that address the following actions:

- Evaluate changes with full cost accounting.
- Institute environmentally preferable processes.
- Improve handling and risk assessment of toxic compounds.
- Assess process and material inputs and outputs to improve operation performance.
- Decrease multimedia environmental harm created.
- Implement extended producer responsibility.
- Combine environment and employee health and safety requisites.

The EPA *Integrated Environmental Management Systems Guide* divides the task of creating these systems into 10 steps (Fig. 6).[91] The first step is to consider the fundamentals behind environmental management, including DfLC principles, quality management, and characterizing environmental impact at different life-cycle stages. Using this information, the organization then creates a corporate environmental policy. It sets goals and actions that take into account the risk of environmental performance currently. Substitute technologies or compounds are next

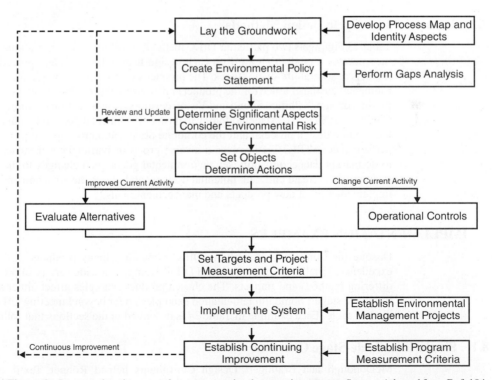

Figure 6   Integrated environmental management implementation process. *Source:* Adapted from Ref. 104.

Table 23   Industrial Life-cycle Design Methods

	Product Design Application Stage	Product Life-Cycle Stages Affected	Quantified Impact
Hewlett-Packard product steward	Detail design	Materials extraction, manufacturing, distribution, use, service, and end of life	Recycled material incorporation, resource conservation, waste and emissions reduction, energy efficiency, disassembly ease, and ability to recycle
STRETCH	Product definition, conceptual design, embodiment design, and detail design	Materials extraction, manufacturing, distribution, use, service, and end of life	Setting of environmental goals, assessment of goals are set by a diverse management team

considered for implementation. From the substitutions assessment, the organization creates specific aims and indicators of achievement. Operating procedures for instituting these aims are then decided upon. The organization chooses timelines for implementing the procedures, and responsibility for specific actions is designated throughout an organization.[91]

Each company must adapt methodologies for implementing DfLC to its own corporate and employee specifications. Some companies have created their own DfLC management systems. Electrolux and Philips include ecodesign in their product-oriented environmental management system.[65]

7.2   Industrial Life-Cycle Design Methods

Table 23 highlights two particular DfLC methods developed by corporations. Hewlett-Packard incorporates a product steward on each design team who considers potential environmental impacts at several life-cycle stages. The product steward is responsible for providing the environmental customer voice for the product and ensuring that environmental goals are included in product design.[92] Philips Sound and Vision developed the Selection of Strategic Environmental Challenges (STRETCH) methodology to push environmental goals into marketing approaches, the product design process, and the corporate organization to improve product environmental performance. STRETCH involves a diverse group of participants from within the company to do market data acquisition, set environmental goals, plan changes to the product development process, create and rank potential designs, and assess the environmental impact for final implementation of new products and management plans.[72]

8   IMPLEMENTATION EXAMPLES

Despite the difficulties in applying DfLC principles, many products exhibit environmental excellence. Finally, we present several DfLC examples, a wide variety of product designs with differing methods and impacts. The chosen product examples affect either single or multiple life-cycle stages. Some implementation examples go far beyond meeting DfLC principles over the life cycle. Each example in Table 24 is described in the sections that follow.

8.1   Single-Life-Cycle-Stage Concerns

McDonough and Braungart Design Consultants helped Röhner Textil create Climatex® Life-cycle™ fabric so that any waste fabric could be composted (instead of designated

Table 24   Design for Life-Cycle Implementation

	Key Players	Product Life-cycle Stages Affected	Impacts
Climatex® LifeguardFR™ Fabric	Rohner Textil, Design Tex (Steelcase), CD3A Specialty Chemicals, Clariant, Lenzing, Ply Designs, William McDonough, Michael Braungart	Manufacturing	Benefits include reducing toxicity of product and processing effluents and wastes and secondary use of processing waste
Re-Define Furniture	Wharrington International, MID Commercial Furniture, National Centre for Design at RMIT University, EcoRecycle Victoria	End of life	Benefits include incorporating recycled materials into product, creating completely recyclable products, and developing product take-back programs
Preserve® Toothbrush	Recycline, Stonyfield Farm®	Materials extraction, product design, and end of life	Benefits include using recycled materials, extending material life for a secondary product, taking products back at end of life
Patagonia Clothing	Patagonia, Teijin	Materials extraction, manufacturing, service, and end of life	Benefits include incorporating recycled and benignly grown materials, utilizing renewable energy during manufacturing, servicing products, performing product take-back to recycle products at end of life
Timberland Shoes	Timberland	Manufacturing	Customers are informed about social and environmental product attributes
Lifeline Aid/relief Radio	Free Play Energy	Use	Impoverished people receive radios, products can operate with human power or renewable energy

hazardous waste), and effluent water quality during manufacturing was improved by using natural fibers and replacing harmful dyes with help from Ciba Specialty Chemicals.[3,45] To improve the fabric, Braungart worked with the Clariant Chemical Corporation to create a flame retardant that is safe for people and the environment. Austrian fiber producer Lenzing helped to develop a way to apply the flame retardant to avoid off-gassing by incorporating the compound directly into a beechwood fiber. The new flame retardant and fiber together created a new fabric, Climatex® LifeguardFR™. Fabric offcuts are still compostable but now have a second life as the collapsible plyFOLD container (see Fig. 7).[93,94]

The Re-Define furniture project by the National Centre for Design at RMIT University developed a three-seat sofa (see Fig. 8) and lounge chair by following environmental guidelines. MID Commercial Furniture designed the furniture and Wharington International made the sofa mold from Recopol™ instead of wood. Recopol™ is recycled ABS (acrylonitrile butadiene styrene) from large appliances, automobiles, and electronics that are typically landfilled.[95,96] Wharington takes back the sofa shells for reuse, and the steel and recycled polyethylene terephthalate (PET) fabric are also recyclable.[9,96] No toxic materials are used in the furniture. Production costs are on par with similar furniture, though the development costs were greater.[96]

8.2 Multiple-Life-Cycle-Stage Concern

Recycline has several products that are made with 100% recycled handles, including the Preserve® toothbrushes, tongue cleaners, and razors. Preserve® toothbrushes are produced

Figure 7 PlyFOLD container. *Source:* From Ref. 93.

Figure 8 Re-Define sofa. *Source:* From Ref. 105. With kind permission from Springer Science+Business Media.

with polypropylene from 65% or more recycled Stonyfield Farm® yogurt cups, and the toothbrush bristles are made of virgin nylon. The price of the Recycline products includes the envelope and postage for returning to the company. Once the products are returned to Recycline, the handles become deck furniture.[45] Hence, the products have a guaranteed second use and two disposable product life cycles (yogurt cups and personal hygiene products) are successfully extended.

Patagonia, an outdoor clothing manufacturer, creates fleece with 90% postconsumer recycled polyester from soda bottles at a factory using 100% renewable wind energy: "Since 1993, Patagonia has diverted over 100 million plastic soda bottles from landfills."[45] The company also uses only organic cotton.[97] In addition, customers can now return worn-out Capilene® garments to Patagonia for recycling into new clothing by a Japanese textile production firm, Teijin, through the Common Threads Recycling Program.[98]

Concerns beyond Product Life Cycle
Timberland is going above and beyond life-cycle design principles by communicating its environmental efforts to consumers through a nutritional label look-alike placed on all shoeboxes.

The label consists of a table of environmental and social impacts and states where the shoe was manufactured. Environmental information provided considers production power requirements, including the renewable energy acquired. Social information on the label includes the time employees volunteer and the percentage of facilities that are evaluated for meeting ethical and child labor standards. Timberland added the label to display corporate values and change consumer opinions.

Free Play Energy produces radios, lights, cell phone chargers, and car battery chargers that are powered by human power and ac/dc suited for impoverished regions. The company encourages prospective buyers to consider using solar or wind energy if using electricity instead of muscle power.[100] In addition, the company distributes radios to impoverished areas throughout the world to facilitate education, health, emergency relief, peacemaking, and agriculture.[104]

9   FUTURE OF DFLC

There are some trends in product design that we believe will significantly impact life-cycle product design in the coming years. Deciding the end-of-life scenario of a product at the beginning of product design development will become standard.[14] More economically and technically feasible solutions to environmental problems will come about through *ecoinnovation*.[9] To achieve ecoinnovation, current products must be reconsidered from their very core, instead of sprucing up products already in existence.[24] A true concurrent, life-cycle engineering environment is necessary for such considerations during product development. Even with a properly structured design process, the information gap—the lack of appropriate information for making informed engineering decisions at each stage of the design—must be overcome in terms of databases and engineering information systems to efficiently access these databases. In addition to these general needs, we anticipate the rise of new design parameters and an increased prominence of new life-cycle design influences.

9.1   Design Paradigm Changes

The design process in the future will be influenced more greatly by life-cycle design. Change can occur through expanding the scope of DfLC to look beyond just the product life cycle, incorporating robust design and sharing into product design. Industrial ecology perspectives are likely to be further incorporated in many aspects of the design process. In addition, tools that can handle the uncertainty of environmental information and can incorporate DfLC information within computer-aided design tools will greatly change life-cycle design.

Expanded Robust Design

Robust products continue to perform by evolving when the systems and environments the products interact with change in two ways—over the life of the product as an artifact and through the product itself as a design. Fiksel uses the term *resilience,* "the ability to resist disorder," to mean stability in the face of perturbations.[44] According to Makower: "Although industrial systems are nonlinear and dynamic, most design and management methods are based on a linear, static worldview. As a result, our systems are brittle—vulnerable to small, unforeseen perturbations—and isolated from their environments."[99] The uncertainties in potential future changes to the environment and the need to minimize environmental impact in all environments are significant drivers for robust design. Thermodynamic efficiency, safety, and a supply chain's ability to cope with unpredicted changes must be addressed.[99] Designing for resilience seeks to address this deficiency by promoting diversity (many available options), efficiency (high output for low input), adaptability (ability to adjust to stimuli), and cohesion (unity between distinct elements).[44]

Life-cycle design tools must also become more robust. Early on in the design process, materials, product architecture, and other attributes are still undecided, and environmental impacts are hard to characterize. For the conceptual design stage, efficient and accurate product life-cycle characterization must be possible with little input data and over a variety of potential designs. In embodiment design, fewer designs are considered and more details are known, but still uncertainty in design exists. Life-cycle design tools must be capable of assessment with a low volume of information and many different design options and must increase in accuracy as product attributes become finalized and fewer design options are being considered.[102]

Uncertainty in DfLC

The lack of reliable data with regards to environmental impact and emissions for components, materials, and substances makes design decision making difficult. Gathering these data is expensive and can be subjective. Uncertainties in data must be cleared for more robust designs and better design decisions early in the design process. The vagueness of final product details during the very early stages of product development also makes design analysis for environmental concerns difficult. More robust life-cycle design tools would help address these uncertainties. An increase in intelligent computer-based life-cycle design tools and their gradual commercialization will increase design engineering capabilities. These tools will most likely be categorized by their application (e.g., automobile tools, computer hardware tools, and household electronics tools).[64]

Implementing Industrial Ecology

Designing product and industry ecologies is still a new challenge for design engineers.[23] Creating industrial systems patterned after ecosystems can yield unique goals for design, including "buildings that, like trees, produce more energy than they consume and purify their own wastewater[,] factories that produce effluents that are drinking water" and "products that, when their useful life is over, can be tossed onto the ground to decompose and become food for plants and animals and nutrients for soil."[3]

Moreover, it is essential for organizations to design industrial processes to be a part of the solution by not creating environmental problems in the first place and to participate in environmental clean-up. Creating systems that encourage and make possible the trading and use of byproducts between industries is one way of many that needs more work for the implementation of industrial ecology to become real.[24]

Designing Products for Multiple Lives

Designing products for multiple lives through reuse is part of the industrial ecology future: "Designing a product so that it has subsequent lives is a very difficult thing to do. It cannot be achieved generically."[102] Retrofitting old products and materials to meet human needs will become increasingly important. When designers are compelled to use available resources instead of creating new ones, they create results that incorporate component reclamation and reuse. Achieving products that are new in the marketplace that can reuse existing manufacturing technology also supports environmental and industrial ecology goals.[101] Currently, very few life-cycle design tools can evaluate environmental effects that span greater than one design, one producer, or one life cycle. Tools that have a broader perspective could support the kind of energy and materials sharing prescribed by industrial ecology or support sustainable development goals.[6]

Sharing

Many DfLC and consumer culture critics extol the value of sharing products. Examples of people sharing products already take place in society, such as libraries, which can be set up to lend everything from books to tools. Setting up community-based product lending/sharing or reuse centers is one way to encourage product sharing.[59] Sharing products engenders certain types of social attitudes and requires social changes. The cohousing movement in Denmark and Sweden sets up living situations where people live close to each other and in ways that facilitate the sharing of goods.[59] There is also a significant impact on product design, including the need for more robust products, longer life cycles, and increased end-user customization.

Designing for End-User Repair of Products

Products that require frequent maintenance and user interaction would lead to "owner-builder[s] [who] would develop an understanding of how the thing works, making trouble-shooting, repair, and replacement of parts easier."[59] Involving the end user in repairs will prolong the life of products and increase the products' value to the owner. However, designing products for end-user repair runs counter to most current product trends and requires significant design changes for ease of diagnostics, disassembly, repair, and assembly.

9.2   Future DfLC Influences

Changes to companies, the supply chain, and industry all affect life-cycle design implementation in the future. The drive to incorporate life-cycle design will increase, for reasons beyond the clear need to reduce environmental impact for the good of society. Companies and designers will be increasingly liable for product environmental effects. Industrializing nations and small manufacturers will press for help in adopting DfLC principles. In addition, the market risk of uncertainty in design will play a larger role in organizational decisions.

Increased Liability Concerns for Designers and Corporations

The future brings liability concerns for both corporations and designers, who will have to take responsibility for the entire product life cycle and its consequences because of changes in legislation and public expectations. Greater reconciliation between environmental and economic interests must be put into practice. Organizations must designate a responsible party to preserve designed intentions with respect to environmental impact, product take-back, and so on, and must protect designers' intentions, thereby reducing liability concerns and environmental damage.[103]

Product Information Sharing along Supply Chain

Future legal compliance will become more than merely meeting set environmental targets. Legislation will emphasize improvement through the sharing of information throughout the product life cycle among supply chain members. Information about how and where a product is manufactured and used can help companies optimize environmental impact over the life cycle. This will require new tools to facilitate the collection and sharing of product creation, use, and service information.[24]

Product Service Systems

The nature of business and transactions involving products will significantly change. A move exists to create and convert products as product service systems. Likely, fewer end users will own products. More user needs will be met by the service economy.[9] Changing products into services "can have significant environmental benefits in terms of reducing the volume of products manufactured while maintaining or increasing profits for the company through service provision."[65] In addition to rethinking their marketing, companies will have to change their product design, manufacturing, service, and end-of-life strategy to convert products into services. The conversion will involve new approaches, different life-cycle design tools, and new technologies.

DfLC Incorporation by Industrializing Nations

"The globalisation of markets is also extending the industrialised model of development (lifestyles, behavior and consumption patterns) to developing countries" while population increases globally.[9] A 90–95% reduction in consumption and resource use must occur in the next 50 years to maintain resource availability for the future. This reduction in resource use is possible if industrializing countries design a higher quality of life with fewer resources than industrialized countries have done so far. Industrialized countries must cut resource use to 5% of current levels. Clearly social, cultural, and technical changes must be made to achieve these goals.[9] These will have a significant impact on life-cycle product design.

Life-Cycle Design Help Needed for Small- and Medium-Sized Producers

Large companies with many resources are the most able to put DfLC changes into place. Small- and medium-sized companies lack the financial flexibility and resources to make sweeping changes in organizational structure and design processes. Small- and medium-sized producers need help in incorporating life-cycle design initiatives.[6] Governments and trade organizations may have to help the smaller companies adapt to meet expectations of improved product environmental performance. In addition, commercial tools that are generic in their application but accurate in their assessment must be made available to a wider community.

DfLC Becomes Sustainable

DfLC will begin to consider more than just a particular product's environmental performance.[24] DfLC must become a sustainable design with a systems view.[9,24] Industry cannot solve product environmental impact problems alone. According to Sun et al.: "It is important for the practice of DfLC to go beyond the confines of industry into the large society, since a fully successful implementation of DfLC is a cooperative effort of both industry and society."[64] The key to creating sustainable design is to involve perspectives from many different areas of study with a focus on natural science.[59]

Localization of DfLC Information

Characterizing how industrial systems affect specific ecological systems needs to be improved upon to really reduce industrial impacts on the environment. Considering localized toxicity instead of general large-scale environmental effects must occur. The localization of LCA and the connection of ecosystem models and LCA will support these needs. Similarly, it is important to recognize the connection between the product and production and their impacts on specific physical spaces and ecosystems.[24]

Localization of Production and Markets

Papanek calls for localization of manufacturing and product markets in the United States for societal and environmental gain, another need that is somewhat counter to current trends.[59] The localization of manufacturing and product markets could provide a variety of environmental benefits from reduced transportation and smaller geographical areas over which to characterize environmental impacts. The flexibility of a dispersed and diversified manufacturing infrastructure will also have significant economic benefits through the potential to increase product variety and regional modification of products. Localization will also require life-cycle design tools that can make design and manufacturing decisions based on locally available resources and environmental capital.

REFERENCES

1. A. Fuad-Luke, *The Eco-Design Handbook: A Complete Sourcebook for the Home and Office*, Thames and Hudson, London, 2002.
2. D. Navinchandra, "Design for Environmentability," *Design Theory Methodol. (DTM '91)*, **31**, 119–125, September 1991.
3. W. McDonough and M. Braungart, *Cradle to Cradle: Remaking the Way We Make Things*, North Point Press, New York, 2002.
4. S. Ashley, "Designing for the Environment," *Mech. Eng.*, **115**(3), 52–55, 1993.
5. G. A. Davis, C. A. Wilt, P. S. Dillon, and B. K. Fishbein, "Extended Product Responsibility: A New Principle for Product-oriented Pollution Prevention—Introduction and Chapters 1 through 4," 1997, University of Tennessee, Center for Clean Products and Clean Technologies, available: http://eerc.ra.utk.edu/clean/pdfs/eprn1-4.pdf, retrieved March 18, 2006.
6. B. Bras, "Incorporating Environmental Issues in Product Design and Realization," *Ind. Environ.*, **20**(1–2), 7–13, 1997.
7. Ozone Secretariat, "Montreal Protocol," 2004, United Nations Environment Programme, availale: http://ozone.unep.org/Treaties and Ratification/2B montreal protocol.asp, retrieved April, 20, 2006.
8. B. Kiser, "A Blast of Fresh Air: The History of O_2," 2000, O2 Global Network, available: http://www.o2.org/media/document/Kiser.pdf, retrieved April 26, 2006.
9. H. Lewis and J. Gertsakis, *Design + Environment: A Global Guide to Designing Greener Goods*, Greenleaf, Sheffield, 2001.
10. American Electronics Association, *The Hows and Whys of Design for the Environment—A Primer for Members of the American Electronics Association*, American Electronics Association, Washington, DC, 1993.
11. S. Erkman, "The Recent History of Industrial Ecology," in R. U. Ayres and L.W. Ayres (Eds.), *A Handbook of Industrial Ecology*, Edward Elgar, Cheltenham, UK, 2002, pp. 27–35.
12. J. Kluger, "Global Warming," *Time*, **167**(14), 28–42, 2006.
13. G. A. Keoleian and D. Menery, *Life Cycle Design Guidance Manual*, EPA Publication No. EPA 600/R-92/226, U.S. Government Publishing Office, Washington, DC, 1993.
14. C. M. Rose, "Design for Environment: A Method for Formulating Product End Of Life Strategies," Ph.D. Thesis, Stanford University, Stanford, CA, 2000.
15. C. L. Henn, "Design for Environment in Perspective," in J. Fiksel (Ed.), *Design for Environment: Creating Eco-Efficient Products and Processes*, McGraw-Hill, New York, 1996, pp. 473–490.
16. G. Eyring, "Policy Implications of Green Product Design," in *Proceedings of the 1993 IEEE International Symposium on Electronics and the Environment*, Arlington, VA, May 10–12, 1993, pp. 160–163.
17. J. Fiksel, "Design for Environment: An Integrated Systems Approach," in *Proceedings of the 1993 IEEE International Symposium on Electronics and the Environment*, Arlington, VA, May 10–12, 1993, pp. 126–131.
18. T. E. Graedel and B. R. Allenby, *Design for Environment*, Prentice Hall, Upper Saddle River, NJ, 1996.
19. S. B. Billatos and N. A. Basaly, *Green Technology and Design for the Environment*, Taylor and Francis, Washington, DC, 1997.
20. U. Tischner, "Introduction: Ecodesign in Practice," in U. Tischner, E. Schmincke, F. Rubik, M. Prösler, B. Dietz, S. Maßelter, and B. Hirschl (Eds.), *How to Do EcoDesign?* Verlag form GmbH, Frankfurt, 2000, pp. 9–14.
21. H. C. Zhang and T. C. Kuo, "Environmentally Conscious Design and Manufacturing: Concepts, Applications, and Perspectives," in *Proceedings of the 1997 ASME International Mechanical Engineering Congress and Exposition*, Dallas, TX, November 16–21, 1997, pp. 179–190.

22. P. T. Anastas and J. B. Zimmerman, "Design through the 12 Principles of Green engineering," *Environ. Sci. Technol.*, **37**(5), 94A–101A, 2003.

23. T. E. Graedel, and B. R. Allenby, *Industrial Ecology*, Pearson Education Inc., Upper Saddle River, NJ, 1995.

24. T. E. Graedel, and B. R. Allenby, *Industrial Ecology*, Pearson Education Inc., Upper Saddle River, NJ, 2003.

25. K. N. Blue, N. E. Davidson, and E. Kobayashi, "The 'Intelligent Product' System," *Business Econ. Rev.*, **45** (2), 15–20, 1999.

26. O. Mont, "Product-Service Systems," AFR-REPORT 288, Swedish Environmental Protection Agency, Stockholm, 2000.

27. W. J. Glantschnig, "Green Design: A Review of Issues and Challenges," in *Proceedings of the 1993 IEEE International Symposium on Electronics and the Environment*, Arlington, VA, May 10–12, 1993, pp. 74–78.

28. L. Alting and J. B. Legarth, "Life Cycle Engineering and Design," *Ann.CIRP-Manufact. Technol.*, **44**(2), 569–580, 1995.

29. World Commission on Environment and Development (WCED), *Our Common Future*, Oxford University Press, Oxford, UK, 1987.

30. T. E. Graedel, "The Grand Objectives: A Framework for Prioritized Grouping of Environmental Concerns in Life Cycle Assessment," *J. Ind. Ecol.,* **1**(2), 51–64, 1997.

31. H. Brezet and C. van Hemel, *Ecodesign: A Promising Approach to Sustainable Production and Consumption*, United Nations Environment Programme, Paris, 1997.

32. N. Bruce, R. Perez-Padilla, and R. Albalak, "Indoor Air Pollution in Developing Countries: A Major Environmental and Public Health Challenge," *Bull. World Health Org.*, **78**(9), 1078–1092, 2000.

33. T. Colborn, D. Dumanoski, and J. P. Myers, *Our Stolen Future*, Penguin Books, New York, 1997.

34. Committee on Health Risks of Exposure to Radon (BEIR VI), Board on Radiation Effect Research, Commission on Life Sciences, National Research Council, *Health Effects of Exposure to Radon*, National Academy Press, Washington, DC, 1999.

35. B. Watson et al., "Climate Change 2001: Synthesis Report," Intergovernmental Panel on Climate Change, 2001, available: http://www.ipcc.ch/ pub/un/syreng/spm.pdf, retrieved March 22, 2006.

36. D. Mackenzie, *Design for the Environment*, Rizzolli, New York, 1991.

37. Container Recycling Institute, "Beverage Container Deposit Systems in the United States: Key Features," in *Bottle Bill Resource Guide*, 2005, available: http://www.bottlebill.org/legislation/usa deposit.htm, retrieved May, 19, 2006.

38. European Parliament, "Directive 2002/95/EC of the European Parliament and of the Council of 27 January 2003 on the Restriction of the Use of Certain Hazardous Substances in Electrical and Electronic Equipment," *Off. J. Eur. Commun.*, L 037 (13/02/2003), 0019–0023, 2003, available: http:// europa.eu.int/smartapi/cgi/sga doc?smartapi!celexapi!prod!CELEXnumdocandlg=ENandnumdoc= 32002L0096andmodel=guichett, retrieved April 20, 2006.

39. European Parliament, "Directive 2002/96/EC of the European Parliament and of the Council of 27 January 2003 on Waste Electrical and Electronic Equipment (WEEE)," *Off. J. Eur. Commun.*, L 037 (13/02/2003), 0024–0039, 2003b, available: http://europa.eu.int/smartapi/cgi/sga_doc?smartap i!celexapi!prod!CELEXnumdocandlg=ENandnumdoc=32002L0095andmodel=guichett, retrieved April 20, 2006

40. European Parliament, "Directive 2005/32/EC of the European Parliament and of the Council of 6 July 2005 Establishing a Framework for the Setting of Ecodesign Requirements for Energy-Using Products and Amending Council Directive 92/42/EEC and Directives 96/57/EC and 2000/55/EC of the European Parliament and of the Council," *Off. J. Eur. Commun.*, L 191 (22/07/2005), 29–58, 2005, available: http://europa.eu.int/comm/enterprise/eco design/directive 2005 32.pdf, retrieved April 20, 2006.

41. United Nations Framework Convention on Climate Change (UNFCCC), "Kyoto Protocol," 2006, available: http://unfccc.int/kyoto protocol/items/2830.php, retrieved December 7, 2006.

42. E. Royte, "E-Waste@Large," *New York Times*, January 27, 2006, p. 23.

43. J. Fiksel, "Designing Resilient, Sustainable Systems," *Environ. Sci. Technol.,* **37**(23), 5330–5339, 2003.

44. C. Berner, B. Bauer, and J. Dahl, *New Tools for the Design of Green Products*, Danish Environmental Protection Agency, 2005, available: http:// www.mst.dk/publica/projects/2003/87-7972-585-6.htm, retrieved March 26, 2006,

45. E. Datschefski, *The Total Beauty of Sustainable Products*, Rotovision, Crans-PrèsCéligny, Switzerland, 2001.

46. C. Madu, *Handbook of Environmentally Conscious Manufacturing*, Kluwer Academic, Boston, 2000.

47. C. Beard and R. Hartmann, "Naturally Enterprising—Eco-Design, Creative Thinking, and the Greening of Business Products," *Eur. Bus. Rev.*, **97**(5), 237–243, 1997.

48. D. J. Richards, "Environmentally Conscious Manufacturing," *World Class Design to Manufacture,* **1**(3), 15–22, 1994.

49. D. Wann, *Biologic, Environmental Protection by Design*, Johnson Books, Boulder, CO, 1990.

50. National Research Council of Canada (NRC), "Design for Environment Guide," 2003, available: http://dfesce.nrc-cnrc.gc.ca/, retrieved May 17, 2006,

51. J. Todd, *From Eco Cities to Living Machines: Ecology as the Basis of Design*, North Atlantic Press, Berkeley, CA, 1994.

52. R. Bedrossian, "Green Design,"*Communication Arts*, May/June 2005, available: http://www.comm arts.com/CA/feadesign/green/, retrieved March 19, 2006,

53. J. Benyus, *Biomimicry: Innovation Inspired by Nature*. Morrow, New York, 1997.

54. Change Design, "What is D|MAT Design?" 2004c, available: http://www.changedesign.org/ DMat/DMatWhatMain.htm, retrieved March 20, 2006,

55. P. Hawken, A. Lovins, and L. H. Lovins, "The Next Industrial Revolution, in *Natural Capitalism,* Little, Brown and Co., Boston, 1999, pp. 1–21.

56. World Congress of Architects Chicago, "Chicago Declaration of Interdependence for a Sustainable Future," in *The Sustainable Design Resources Guide*, American Institute of Architects, 1993, available: http://www.aiasdrg.org/ sdrg.aspx?Page=5, rtrieved March 27, 2006,

57. J. R. Mihelcic, J. C. Crittenden, M. J. Small, D. R. Shonnard, D. R. Hokanson, Q. Zhang, H. Chen, S. A. Sorby, V. U. James, J. W. Sutherland, and J. L. Schnoor, "Sustainability Science and Engineering: The Emergence of a New Metadiscipline," *Environ. Sci. Technol.*, **37**, 5314–5324, 2003.

58. M. Z. Hauschild, J. Jeswiet, and L. Alting, "Design for Environment—Do We Get the Focus Right?" *Ann. CIRP,* **53**(1), 1–4, 2004.

59. V. J. Papanek, *The Green Imperative*, Thames and Hudson, New York, 1995.

60. F. Guo, and J. K. Gershenson, "Comparison of Modular Measurement Methods Based on Consistency Analysis and Sensitivity Analysis," in *Proceedings of the 2003 ASME Design Engineering Technical Conferences*, Chicago, IL, September 2–6, 2003, pp. 393–401.

61. F. Guo and J. K. Gershenson, "A Comparison of Modular Product Design Methods Based on Improvement and Iteration," in *Proceedings of the 2004 ASME Design Engineering Technical Conferences*, Salt Lake City, UT, September 28–Ocotber 2, 2004.

62. P. J. Newcomb, B. Bras, and D. W. Rosen, "Implications of Modularity on Product Design for the Life cycle," *J. Mech. Design,* **120**(3), 483–490, 1998.

63. X. Qian, Y. Yu, and H. Zhang, "A Semi-Quantitative Methodology of Environmentally Conscious Design for Electromechanical Products," in *Proceedings of the 2001 IEEE International Symposium on Electronics and the Environment*, Denver, CO, May 7–9, 2001, pp. 156–160.

64. J. Sun, B. Han, S. Ekwaro-Osire, and H. Zhang, "Design for Environment: Methodologies, Tools, and Implementation, *J. Integrated Design Process Sci.,* **7**(1), 59–75, 2003.

65. D. Maxwell and R. van der Vorst, "Developing Sustainable Products and Services," *J. Cleaner Production,* **11**(8), 883–895, 2003.

66. N. Borland and D. Wallace, "Environmentally Conscious Product Design: A Collaborative Internet-Based Modeling Approach," *J. Ind. Ecol.*, **3**(2, 3), 33–46, 1999.

67. H. C. Zhang and S. Y. Yu, "Environmentally Conscious Evaluation/Design Support Tool for Personal Computers," in *Proceedings of the* 1997 *IEEE International Symposium on Electronics and the Environment*, San Francisco, CA, May 5–7, 1997, pp. 131–136.

68. C. Mizuki, P. A. Sandborn, and G. Pitts, "Design for Environment—A Survey of Current Practices and Tools," in *Proceedings of IEEE International Symposium on Electronics and the Environment*, Dallas, TX, May 6–8, 1996, pp. 1–6.

69. U. Tischner and B. Dietz, "Checklists," in U. Tischner, E. Schmincke, F. Rubik, M. Prösler, B. Dietz, S. Maßelter, and B. Hirschl (Eds.), *How to Do EcoDesign?* Verlag GmbH, Frankfurt, 2000d, pp. 102–118.

70. D. L. Thurston and W. F. Hoffman III, "Integrating Customer Preferences into Green Design and Manufacturing," in *Proceedings of the 1999 IEEE International Symposium on Electronics and the Environment*, Danvers, MA, May 11–13, 1999, pp. 209–214.

71. M. Finster, P. Eagan, and D. Hussey, "Linking Industrial Ecology with Business Strategy: Creating Value for Green Product Design," *J. Ind. Ecol.*, **3**(1), 107–125, 2001.

72. E. Jones, D. Harrison, and J. McLaren, "Managing Creative Eco-innovation—Structuring Outputs from Eco-Innovation Projects," *J. Sustainable Product Design*, **1**(1), 27–39, 2001.

73. IdeMat Online, Design for Sustainability Program, Delft University of Technology, "Product Info," 2005, available: http://www.io.tudelft.nl/research/dfs/ idemat/Product/pi frame.htm, retrieved April 3, 2006.

74. S. Maßelter and U. Tischner, "Software Tools for ecodesign," in U. Tischner, E. Schmincke, F. Rubik, M. Prösler, B. Dietz, S. Maßelter, and B. Hirschl (Eds.), *How to Do EcoDesign?* Verlag GmbH, Frankfurt, 2000, pp. 147–149.

75. U. Tischner and B. Dietz, "Spider or Polar Diagrams," in U. Tischner, E. Schmincke, F. Rubik, M. Prösler, B. Dietz, S. Maßelter, and B. Hirschl (eds.), *How to Do EcoDesign?* Verlag GmbH, Frankfurt, 2000, pp. 91 97.

76. Environment Canada, "The Netherlands' Promise," *EcoCycle*, 5, 2003, available: http://www.ec .gc.ca/ecocyclc/issue5/en/p15.cfm, retrieved March 26, 2006,

77. Y. Akao, *Quality Function Deployment: Integrating Customer Requirements into Product Design*, G. Mazur, Trans., Productivity Press, Cambridge, MA 1990.

78. U. Tischner and B. Dietz, "Tools for Cost Estimation/Environmental Cost Accounting," in U. Tischner, E. Schmincke, F. Rubik, M. Prösler, B. Dietz, S. Maßelter, and B. Hirschl (Eds.), *How to Do EcoDesign?* Verlag GmbH, Frankfurt, 2000, pp. 142–146.

79. B. Kassahun, M. Saminathan, and J. C. Sekutowski, "Green Design Tool," in *Proceedings of the 1993 IEEE International Symposium on Electronics and the Environment*, Orlando, FL, May 10–12, 1995, pp. 118–125.

80. U. Tischner, and B. Dietz, "MET Matrix and Ecodesign Checklist," in U. Tischner, E. Schmincke, F. Rubik, M. Prösler, B. Dietz, S. Maßelter, and B. Hirschl (Eds.), *How to Do EcoDesign?* Verlag GmbH, Frankfurt, 2000, pp. 86–90.

81. U. Tischner and B. Dietz, "The Toolbox: Useful Tools for Ecodesign," in U. Tischner, E. Schmincke, F. Rubik, M. Prösler, B. Dietz, S. Maßelter, and B. Hirschl (Eds.), *How to Do EcoDesign?* Verlag GmbH, Frankfurt, 2000, pp. 65–70.

82. H. S. Matthews, J. Garrett, A. Horvath, C. Hendrickson, M. Legowski, M. Sin, J. Mayes, H. H. Ng, K. McCloskey, R. Ready, J. Knupp, V. Hodge, and S. Griffin, "eiolca.net," Carnegie Mellon University, Green Design Institute, 2005, available: http://www.eiolca.net/index.html, retrieved March 26, 2006.

83. U.S. Environmental Protection Agency (EPA) Design for the Environment (DfE) Program and the University of Tennessee Center for Clean Products and Clean Technologies, "Cleaner Technologies Substitutes Assessment—Executive Summary," 2006, available: http://www.epa.gov/opptintr/dfe/pubs/tools/ctsa/exsum/exsum.htm, retrieved March 26, 2006; "A Quick Reference Guide to State Scrap Tire Programs: 1999 Update," 1999, available: http://www.epa.gov/epaoswer/non-hw/muncpl/tires/scrapti.pdf, retrieved May, 19, 2006,

84. Ricoh, "Ricoh Group Sustainability Report (Environment)," 2005, available: http://www.ricoh.com/environment/report/pdf2005/49-50.pdf, retrieved April 4, 2006,

85. N. Chambers, C. Simmons, and M. Wackernagel, *Sharing in Nature's Interest: Ecological Footprints as an Indicator of Sustainability*, Routledge London, 2000.

86. E. G. Hertwich, W. S. Pease, and C. P. Koshland, "Evaluating the Environmental Impact of Products and Production Processes: A Comparison of Six Methods," *Sci. Total Environ.*, **196** (1), 13–29, 1997.

87. B. Jansen and A. Vercalsteren, "Eco-KIT: Web-Based Ecodesign Toolbox for SMEs," in *Proceedings of the 2nd International Symposium on Environmentally Conscious Design and Inverse Manufacturing (EcoDesign'01)*, Tokyo, Japan, December 11–15, 2001, pp. 234–239.

88. M. Wackernagel, "Ecological Footprint and Appropriated Carrying Capacity: A Tool for Planning Toward Sustainability," Ph.D. Thesis, University of British Columbia, 1994.

89. Pré Consultants, "Eco-Indicator 99," 2006, available: http://www.pre.nl/eco-indicator99/eco-indicator99introduction.htm, retrieved May 18, 2006.

90. F. Schmidt-Bleek and R. Klüting, *viel Umwelt braucht der Mensch?: MIPS, das Mass für ökologisches Wirtschaften*, Birkhäuser Verlag, Berlin, 1994.

91. U. Tischner and B. Dietz, "Tools for Setting Priorities, Making Decisions and Selecting," in U. Tischner, E. Schmincke, F. Rubik, M. Prösler, B. Dietz, S. Maßelter, and B. Hirschl (Eds.), *How to Do EcoDesign?* Verlag GmbH, Frankfurt, 2000, pp. 130–141.

92. International Organization for Standardization, "Environmental Management: The ISO 14000 Family of International Standards," 2002, available: http:// www.iso.org/iso/en/prods-services/otherpubs/iso14000/index.html, rtrieved May 18, 2006.

93. U.S. Environmental Protection Agency (EPA) Office of Pollution Prevention and Toxics (OPPT), Design for Environment (DfE), "Integrated Environmental Management Systems Partnership," 2006, available: http://www.epa.gov/opptintr/dfe/pubs/projects/iems/index.htm, retrieved March 26, 2006.

94. T. Korpalski, "Role of the 'Product Steward' in Advancing Design for Environment in Hewlett-Packard's Computer Products Organization," in *Proceedings of IEEE International Symposium on Electronics and the Environment*, Dallas, TX, May 6–8, 1996, pp. 37–41.

95. Röhner Textil AG, "*Product Climatex® Life Cycle*," available: http://www.climatex.com/en/products/felt_climatex_life-cycle.html, retrieved April 21, 2006.

96. P. Storey, "Exploring New Horizons in Product Design," McDonough Braungart Design Chemistry, 2002, available: http://www.mbdc. com/features/feature_june2002.htm, retrieved April 21, 2006.

97. Recycline, "Preserve," available: http:// www.recycline.com, retrieved March 29, 2006.

98. Patagonia, "Organic Cotton," 2006, available: http://www.patagonia.com/enviro/organic_cotton.shtml, retrieved April 21, 2006.

99. J. Makower, "Timberland Reveals Its 'Nutritional' Footprint,'" 2006, available: http://makower.typepad.com/joel_makower/2006/01/timber-land_reve.html, retrieved March 29, 2006.

100. Freeplay Foundation, "Lifeline Self-Powered Radios Provide Sustainable Access to Our Five Areas of Focus, " available: http://www.freeplayfoundation.org/," retrieved March 29, 2006.

101. Ohio State University, Center for Resilience, "Concepts," available: http://www.resilience.osu.edu/concepts.html, retrieved March 19, 2006.

102. P. Fitch and J. Cooper, "Life Cycle Modeling for Adaptive and Variant Design. Part 1: Methodology," *Res. Eng. Design*, **15**(4), 216–228, 2005.

103. Freeplay Energy, "Lifeline: Aid/Relief Radio," available: http://www.freeplayenergy.com/index.php?section=productsandsubsection=lifeline, retrieved March 29, 2006.

104. Change Design, Sustainments, "How We Came to Realize That There Was a Need for This Notion," 2004, available: http://www.changedesign/Sustainments/What_are/WhatAreMain.htm, retrieved March 29, 2006.

105. G. Pahl and W. Beitz, *Engineering Design: A Systematic Approach*, Springer, London, 1996.

CHAPTER 9

DESIGN FOR MAINTAINABILITY

O. Geoffrey Okogbaa and Wilkistar Otieno
University of South Florida
Tampa, Florida

1 INTRODUCTION

Environmental concerns have created the need for sustainable development and have brought a global understanding that there is economic benefit in promoting products that are environmentally responsible and profitable, with reduced health risks for the consumers. The idea, according to 1987 UN report, is to meet current needs without jeopardizing the needs of future generations. This has led to a renewed emphasis on *design for environment* (DfE).

Life-cycle analysis (LCA) is a development platform upon which to anchor the formalisms for evaluating environmental effects and economic impact of the different stages of a product life cycle under the overarching umbrella of DfE. Included in these different stages are product and process designs, material selection, product manufacture and demanufacture, assembly and disassembly, reliability, maintenance, recycling, material recovery, and disposal. Since the sequential analyses of these different product stages do not provide a comprehensive picture of the total effect in the context of product design efficiency and efficacy, a *concurrent* dynamism is employed in which the different stages are explored using the concept of *concurrent engineering*.

Thus, there are now different tools for addressing the different stages of LCA in support of design for environment, and these are typically referred to as design for X (DfX), such as design for manufacture (DfM) and design for assembly (DfA), design for product assurance (DfPA), and design for remanufacture (DfR), among others.

The need for self-diagnoses and fault tolerance and the need to conserve dwindling natural resources have led to the increased emphasis on DfPA, especially with respect to reusability, survivability, and maintainability. Thus, new frontiers in engineering design, beyond the issues of concurrent design and design for manufacturability, may well depend on a paradigm shift that encompasses maintenance intervention practices to sustain fault tolerance and reusability. The cornerstone of such a paradigm shift hinges on the ability to develop analytical techniques that realistically characterize the failure and renewal process distributions for a complex multiunit repairable system, with particular emphasis on the system's transient phenomenon, which characterizes its aging processes. This problem is especially amplified in autonomous systems where human interaction and intervention are minimal. For such systems, the issue of maintenance intervention has taken on renewed importance because these systems should, in general, experience little to no failures.

Maintenance has been long recognized as necessary to increase a system's reliability, availability, and safety. Maintenance is necessary not only for systems where high reliability is required or where the failure may result in a catastrophe such as in the case of airplane or space shuttle, but also for any system where availability is of concern. In fact, relatively few systems are designed to operate without maintenance of any kind.[1]

For small systems with very few components, maintenance can be implemented by using the operator's experience. However, for large, complex systems that are common today, personal experiences are no longer enough in maintenance planning. As a matter of fact, maintenance planning based solely on personal experience may lead to significant loss to the systems. As an example, components may be neglected until they fail, thereby causing significant losses. At the opposite extreme, brand-new components may be replaced for no reason. This is especially true for systems that consist of large numbers of complex and expensive components that experience random or gradual failures. For such systems, the issue of maintenance planning has taken on renewed importance because these systems generally represent large capital expenditures and should experience few or no failures.

In the past few decades, extensive research has been done with regard to maintenance planning. However, only a few of these have addressed the problem of multiunit systems with economic dependency. In a few of the existing research studies the authors conducted *steady-state analyses* based on two major assumptions[2–4]:

1. The planning horizon is infinite.
2. The long-run expected replacement rate is constant.

For systems with long lives, the steady-state results are appropriate approximation of system behavior. However, the steady-state analysis does not consider the inherent transient characteristics of the system failure behavior. After all, no system really has an infinite planning horizon, and most systems have only short life spans. Consequently, steady-state analysis, while useful, is biased and limited in its applications.

Currently, no unified approach exists in the literature that considers the realistic failure characteristics of a complex multiunit system under transient response (system with a finite horizon). In addition, there are no maintenance strategies that incorporate the system's renewal and potential aging process. These problems limit efforts to provide a framework necessary to integrate the intervention functions into the overall system availability as well as assurance estimates for complex systems. Very little research has been done on how the system transient behavior impacts its remaining life. Such analyses are important if effective preventive maintenance for complex systems is to be implemented. In the long run this would help in eventual realization of the completely autonomous systems where human intervention is reduced to the minimum.

1.1 Maintainability

Maintainability and reliability considerations are playing increasingly vital roles in virtually all engineering disciplines. As the demand for systems that perform better and cost less increase, there is a corresponding demand or requirement to minimize the probability of failures and the need to quickly bring the system back to normal operations when unavoidable failures occur.

Terms and Definitions

Maintainability is the probability that when operating under stated environmental conditions, the system (facility or device or component) having failed or in order to prevent expected failure will be returned to an operating condition (repaired) within a given interval of downtime. Maintainability is a characteristic of equipment design that is expressed in terms of ease and cost of maintenance, availability of equipment, safety, and accuracy in the performance of maintenance actions. Other related terms include the following:

- *Maintenance*. All actions necessary for retaining a system in, or restoring it to, a serviceable condition. Systems could be maintained by repair or replacement.
- *Corrective maintenance*. This is performed to restore an item to satisfactory condition after a failure.
- *Mean time to failure (MTTF)*. The mean time interval between two consecutive system failures is the MTTF.
- *Mean time between maintenance (MTBM)*. This is the mean time between two consecutive maintenance procedures (replacement or repair).
- *Maintenance downtime*. This is the portion of downtime attributed to both corrective and preventive maintenance.

Classes of Maintenance Policies

Preventive Maintenance. Preventive maintenance (PM) is performed to optimize the reliability and availability of a system prior to failure. Such maintenance involves overhaul, inspection, and condition verifications. It is widely considered as an effective strategy for reducing the number of system failures, thus lowering the overall life-cycle development cost.[5] PM is becoming increasingly beneficial, due to the need to increase the life cycle and performance of assets. Significant cost savings are achieved by predicting an impending failure in machinery. This might be in the form of savings achieved on spare parts, labor costs, downtime avoided, or prevention of damage due to accidents. Several parameters of the machinery are monitored to ensure their normal functioning. Such parameters might include vibration analysis, contaminant analysis, and monitoring of energy consumption, temperature, and noise levels.

A major effort in the development of a PM program is to determine the optimal replacement intervals. Operationally, maintenance management involves a transient and often uncertain environment with little data and proven tools that can assist in decision making. In particular, it is very difficult to develop an optimal maintenance schedule for a complex (multiunit) system with components that experience increasing failure rates in a dynamic environment and strong stochastic or economic dependencies.[2,6] Several approaches to this problem have fallen short because of the intractability of the problem, unrealistic assumptions about complex system behavior, and unnecessary restrictions on the maintenance model formulation and policy specification.

PM can be classified as condition-based and time-based policies. Time-based PM can be justified if and only if the following conditions hold:

1. The component under consideration has a significant aging behavior, which is characterized by increasing failure rate (IFR).

2. Preventive maintenance cost is much higher than the corrective maintenance cost.

3. Before failure, the component does not exhibit any abnormal behavior. In other words, it is not possible to detect any abnormal signal if such exists before the component fails.

Condition-based PM heavily depends on the available technology for early detection of imminent failures. Condition-monitoring techniques and equipment that detect failure effects are used. These are widely known as predictive techniques and can be categorized as dynamic, particle, chemical, physical, or electrical in nature.[7] Some of them include lubricant analysis, vibration analysis, thermography, penetrating liquids, radiography, ultrasound, and corrosion controls.

Corrective Maintenance (CM). This is performed to restore an equipment to satisfactory condition after a failure. It is assumed that in most cases the cost associated with CM is much higher than that associated with PM.

Opportunistic Maintenance (OM). The opportunistic maintenance concept originates from the fact that there can be economic and stochastic dependency between various components of a multiunit system. *Opportunism* here refers to the idea of jointly replacing several components at some point in time based on three conditions:

1. A component will be replaced when it fails.

2. A component will be actively replaced when it reaches age T_a (active replacement age).

3. A component will be replaced if it reaches age T_p (passive replacement age) and there is another component in the system that is being replaced, either due to failure or due to active preventive maintenance. In this case, $T_p < T_a$.

The Unit, System, or Component

In actual practical situations or considerations, maintainability may be viewed or defined differently for a system, components, and so on. However, the system or unit of interest normally determines what is being studied, and there is usually no ambiguity. From the system modeling point of view, a system can be considered either as a single-unit system or as a multiunit system. In a single-unit model, the entire system is viewed as one component, and its failure distribution, failure process, maintenance activities, and effects are well defined.

Most real-world systems are complex in nature and may consist of hundreds of different components. Thus, the assumption that such a system would follow a single failure distribution is too limiting and unrealistic. Hence, it would be of little use in maintenance program development. Rather, the first step should be to decompose the system into subsystems or components for which the failure distributions are more traceable, and the maintenance activities and associated costs and effects are well defined. For this reason, we define a complex system as a multiunit system.

Whether a group of components should be considered as several individual single-unit systems or as an integrated multiunit system depends on whether there exists economic or stochastic dependency between the components. *Stochastic dependency* implies that each component's transition probability depends on the status of other components in the systems and the notion that the failure of one component may increase the failure probability of the other components. *Economic dependency* implies an opportunity for group replacement of several components during a replacement event. This is justified by the fact that the joint replacement of several components will cost less than separate replacements of the individual components.[6]

On the one hand, weak economic and stochastic dependency means that decisions can be independently made for each component in the system. On the other hand, if stochastic and economic dependency between components is very strong, then an optimal decision on the repair or replacement of one component is not necessarily optimal for the whole system. Thus, for complex systems, maintenance intervention plans should be for the whole system rather than for each individual component.

Definition of Failure

A system or unit is commonly referred to as having *failed* when it ceases to perform its intended function. When there is total cessation of function, engines stop running, structures collapse, and so on—the system has clearly failed. However, a system can also be considered to be in a failed state when its deterioration function is within certain critical limits or boundaries. Such subtle forms of failure make it necessary to define or determine quantitatively what is meant by failure.

Time Element. The way in which time is specified can also vary with the nature of the system under consideration:

- In an intermittent system, one must specify whether calendar time or number of hours of operation is to be used in measuring time (car, shoes, etc.).
- If the system operation is cyclic (switch, etc.), then time is likely to be specified in terms of number of operations.
- If the maintainability is to be specified in calendar time, it may also be necessary to indicate the number or frequency of stops and gos.

Operating Condition

- Operating parameters include loads, weight, and electrical load.
- Environmental conditions include temperature extremes, dust, salt, vibrations, and similar factors.

1.2 Steady State versus Transient State

Steady State

If the system planning horizon is much longer than the life of the components, then it would be appropriate to develop a predictive maintenance program based on the long-run stable condition, or the steady state. In this regard, the expected replacement rate (ERR) would provide important information for maintenance management on spare parts inventory, size of maintenance workforce, and maintenance equipment. However, the reason for maintenance intervention is the inherent failure characteristics of a system that is transient by nature.[8] Thus, while the steady-state results are useful, the transient response provides the information that reflects the system's useful life profiles for planning, maintenance, and supportability requirements.

Transient State

A system is said to be in a transient state before it stabilizes into equilibrium of steady state, often due to exogenous environmental conditions or internal control factors. Transient state analysis is most important in systems whose homogenous stochastic behaviors converge very slowly to a steady state. In such cases the steady state would not be indicative of a system's actual behavior. Thus the transient state represents the true system behavior during the useful life. Analytical models that are based on the transient behavior of the system and that take the form of differential and integral equations can be used to model maintenance policies. Numerical methods can then be used to solve the resulting equations in an attempt to obtain a clear picture of the system's transient behavior. Several numerical methods have been proposed and have been used successfully in attacking this type of problem. However, the complex nature of manufacturing systems and the characteristics of the proposed transient models preclude the use of any of the existing methods. The Runge–Kutta and the Runge–Kutta–Gill methods are best to use because of their robustness relative to linear/nonlinear models. Afterward, simulation should be used to validate the developed transient state models.

1.3   Basic Preventive Maintenance Strategy

Considering an existing bottling system environment, the objective of preventive maintenance is to identify the right time and right subsystems with optimized schedule to perform the maintenance before the occurrence of failures. Failures of large systems or complex systems usually follow certain types of probability distribution, which are mostly determined by the system design reliability/availability and bill of materials (BOM). Practically before a failure happens, there often are some symptoms such as abnormal changes of system mechanic vibration, temperature, and so on.

The basic strategy of preventive maintenance is to decide on and execute the optimal maintenance plan, considering the designed system reliability features and real-time field-collected system data, as well as the constraint of maintenance cost.

1.4   Implementation of Preventive Maintenance

Considering the bottling process, the following four aspects should facilitate implementation of the optimal preventive maintenance:

1. System operation monitoring
2. Design maintainability/availability performance
3. Empirical performance history
4. Comprehensive evaluation and analysis that lead to decision making

Each element of the preventive maintenance procedure is covered in detail in this section.

System Operation Monitoring

This involves keeping track of the key system parameters during the bottling operation. It includes vibration analysis, infrared technology, and fluid analysis and tribology. These aspects of system operation monitoring can be explained as follows.

Vibration Measurement and Analysis. A vibration analyzer and some operator experience can be used to identify the causes of abnormal machinery conditions such as rotor imbalance and misalignment. Other problems, including bearing wear and the severity of the wear, can also be identified and quantified. Vibration monitoring provides earlier warning of machine deterioration than temperature monitoring. Small and gradual machine deterioration show up as significant vibration increases. Early detection usually permits continued operation until a scheduled shutdown. The vibration measurement and analysis can be implemented by deploying a handheld vibration meter PT908, solid-state vibration switch VS101, or TM101.

The PT908 vibration meter is one of the most useful tools in checking the overall condition on all types of rotating machinery. PredicTech's PT908 is portable and is economical, reliable, and rugged. The vibration meter can measure acceleration, velocity, and displacement in peak or root mean square (rms). Field operators don't need any vibration analysis knowledge to use this device. The displayed digital vibration level is an indicator of machine running condition.

The VS101 solid-state vibration switch is a direct upgrade of the mechanical vibration switch. This solid vibration switch measures excessive machine vibration and machine failure, with additional dry-contact relay for machine alarm. VS101 responds to destructive vibration by shutting down the machine when the vibration trip level is exceeded to preventing catastrophic damage and extensive repairs and downtime. VS101 can be used in rotation machines like electric motors, blowers, pumps, agitators, gear boxes, small compressors, and fans.

The TM101 transmitter monitor is used to measure any rotation machinery vibration such as case vibration, bearing housing vibration, or structural vibration. Output is in acceleration, velocity, or displacement. This system can be used in hazardous areas.

Infrared Thermography. Abnormal operation of mechanical systems usually results in excessive heat, which can be directly caused by friction, cooling degrading, material loss, or blockages. Infrared thermograph allows monitoring of the temperatures and thermal patterns while the equipment is online and running with full load. Unlike many other test methods, infrared can be used on a wide variety of equipment, including pumps, motors, bearings, generators, blower systems, pulleys, fans, drives, and conveyors. Infrared thermography should be deployed in key system components, such as motors and generators, blower systems, and bearings.

Identifying thermal pattern doesn't necessarily locate a problem. In mechanical applications, a thermograph is more useful in locating a problem area other than determining the root cause of the overheating, which usually is produced inside a component and is not visible to the camera. Other approaches such as vibration analysis, oil analysis, and ultrasound can be employed to determine the problem.

Fluid Analysis and Tribology. Tribology studies the interdisciplinary field of friction, lubrication, wear, and surface durability of materials and mechanical systems. Typical tribological applications include bearings, gears, bushings, brakes, clutches, chains, human body implants, floor tiles, seals, piston engine parts, and sports equipment.

Machine fluid (usually oil) analysis is used to evaluate two common contamination problems—namely, water and dust—which dramatically increase wear rates and inadequate lubrication. By monitoring, reporting, and recommending the correction of contamination problems, oil analysis is of a very proactive condition monitoring technology. It can also effectively predict impending catastrophic failure of mechanical and electrical systems through which the oil flows.

There are four typical categories of abnormal wear for mechanical systems: abrasive wear, adhesive wear, fatigue wear, and corrosive wear:

1. *Abrasive wear* particles are normally indicative of excessive dirt or other hard particles in the oil that are cutting away at the load-bearing surfaces.

2. *Adhesive wear* particles reveal problems with lubricant starvation as a result of low viscosity, high load, high temperature, slow speed, or inadequate lubricant delivery.

3. *Fatigue wear* particles are often associated with mechanical problems, such as improper assembly, improper fit, misalignment, imbalance, or other conditions.

4. *Corrosive wear* particles are the result of corrosive fluid such as water or process materials contacting metal surfaces.

Typical tribometers used for this purpose include the pin-on-disk tribometer, linear reciprocating tribometer, and high-temperature tribometer.

Design Maintainability/Availability Performance

These system parameters are used to calculate the expected the mean time to failure (MTTF) and the mean time to repair (MTTR), which are fundamental theoretic reference metrics of system maintenance plans.

To add maintainability into a design, designers need an intensive understanding of the system. This includes the system's configuration, topology, component interdependency, and failure distribution. Design for maintainability is well hinged in the ability to develop techniques that realistically characterize the failure and renewal process distributions for complex, multiunit repairable systems, with particular emphasis on a system's transient phenomenon as well as its aging process.

Empirical Performance History

System maintainability/availability performance varies when deployed in different physical operation environment such as at various humidity and voltage stability levels, with different load, under different human operators. Therefore, the empirical performance data also serve as a very important reference to predict early failures and schedule preventive maintenance for specific individual systems.

Comprehensive Evaluation and Analysis

To determine a preventive maintenance schedule, all the foregoing design and empirical reliability data and parameters must be comprehensively evaluated with more detailed scientific analysis. There are numerous statistical process control (SPC) methodologies and tools available to perform the comprehensive evaluation and analysis. This section covers several of the proven methods.

Design of Experiment (DOE)/Analysis of Variance (ANOVA). This is used as a mechanism to determine the significant factors and their interactions affecting the system reliability performance.

Control Charts. Control charts are useful as a tool to determine the boundary limits of process performance and related nonconformance levels.

Automation of Preventive Maintenance. The automation of scheduling the optimal preventive maintenance can be achieved by implementing the computer manufacturing integrated systems (CMIS), which is efficient yet costly. In the specific case of a bottling system, all the monitoring devices/terminals must be connected to a central station equipped with high-performance servers. The servers are programmed to trigger at different system thresholds the performance levels or status of the system and thus alert the decision maker to take action, based on comparisons of real-time data with preestablished thresholds of operation/maintainability performance. Typically, the software algorithms and protocols created from the system design reliability and historical failure data—as well as other customer-specified requirements—generate these limits or thresholds.

The rest of the chapter is sequenced as follows. Section 2 introduces the probability theory and probability distributions. The third section gives an introduction into system reliability functions, its measurement parameters, and how system maintenance is related to system reliability. The fourth section delves into maintenance, definitions and maintenance classifications, and system maintenance modeling. Finally, system availability is discussed as a way to measure the integrity of any maintenance policy of a reparable system.

2 REVIEW OF PROBABILITY AND RANDOM VARIABLES

Each realization of a manufacturing activity represents a random experimental trial—in other words, such an activity may be looked at as a process or operation that generates raw data, the nature or outcome of which cannot be predicted with certainty. Associated with each experiment or its realization is the sample space, which is the set of all possible outcomes of the experiment. An *outcome* of an experiment is defined as one of the set of possible observations that results from the experiment. One and only one outcome results from one realization of the experiment. Most quantities occurring in a manufacturing environment (a realization of a manufacturing experiment) are subject to random fluctuations. Because of the fluctuation and the randomness of the experiments, the outcomes are *random variables*. In order to characterize these random variables so that they can become useful tools in describing a manufacturing system, it is important to understand what random variables are.

2.1 Random Variables

An intuitive definition of a *random variable* is that it is a quantity that takes on real values randomly. It follows from this definition that a random vector is an *n*-tuple of real-valued random variables. An operating definition of a random variable is that it is a function that assigns a real value (a number on the real line) to each sample point in the sample space S. This assignment or mapping is one to one and is not bidirectional. In other words, each event in the sample space can only take on one value at a time. As an example, let the height of males in a region or country be a variable of interest in a study. Due to its nature, such a variable can be described as random. Thus, the random variable can be expressed as the function $f(x) =$ a numerical value equal to the height of a unique individual male named x. That is,

$$f(\text{John1}) = 6\ \text{ft} \quad f(\text{Paul20}) = 7\ \text{ft} \quad f(\text{Don10}) = 6\ \text{ft}$$

Note, however, that two elements from the sample space can have the same real values assigned to them. In our example, John1 and Don10 have the same height. However, John1 cannot have heights of 6 and 7 ft at the same time; hence the mapping is unique.

The domain of the random variable is the sample space (S) and the range of the random variable is the real line R. It is the range of the random variable that determines the types of values that are assigned to the random variable under consideration.

Random variables occurring in manufacturing situations or in industry in general are subject to random fluctuations that exhibit certain regularities, and they sometimes have well-defined forms. In some cases, they are also of a given form or belong to some class or family. Thus, depending on the type of random experiment that generated the domain of the random variable, the mapping can be generalized into closed-form expressions, formula, equations, rules, or graphs that describe how the values are assigned. These rules or equations are known as probability density functions (for continuous random variables) and probability mass functions (for discrete random variables). In other words, probability density or mass functions are simply closed-form expressions or rules that indicate how assignments to values on the real line are made from the domain of the random variable. The nature of the experiment ultimately determines the type of equations or rules that apply.

One of the problems of using the density function to characterize a manufacturing process is that in some cases such closed-form expressions or equations are difficult to come by. Hence, we are often left to examine the moments resulting from the data at hand using the moment-generating functions. Usually the first and second moments (the mean and variance) could provide useful insight as to the behavior of the random variable. However, in some cases these parameters are not enough to completely characterize the underlying distribution, and so we have to look to other approaches that would provide the confidence needed to ensure that indeed the distribution assumed is the appropriate one.

The following are some density or mass functions that are often used to describe random variables that typically occur in a manufacturing setting. These distributions can be categorized into discrete and continuous. The discrete distributions include binomial, negative binomial, geometric, hypergeometric, and Poisson distributions, while the continuous distributions that will be examined include normal, exponential, and Weibull distributions.

2.2   Discrete Distributions

Binomial

Consider a random experiment with the following conditions:

1. Each trial has only two possible outcomes; namely, the occurrence or nonoccurrence of an event (e.g., conforming/nonconforming, defective/nondefective, success/failure).
2. The probability (p) of occurrence of an event is constant and is the same for the each trial.
3. There are n trials (n is integer).
4. The trials are statistically independent.

Such an experiment is called a binomial experiment, which follows the Bernoulli sequence. The random variable of interest is the number of occurrences of a given outcome or event.

Example

Let x be the number of occurrences of an event (where n is known). The probability of having exactly x_0 occurrences in n trials, where p is the probability of an occurrence is given by

$$P(x = x_0) = {}^{n}C_{x_0}p^{x_0}(1 - p)^{n-x_0} \quad x_0 = 0, 1, \ldots n \tag{1}$$

where

$$^{n}C_{x_0} = \frac{n!}{(n - x_0)!x_o!}$$

The mean and variance of the binomial are $\mu = np$, and $\sigma^2 = np(1 - p)$, respectively, where $0 \leq p \leq 1$.

The probability that a certain wide column will fail under study is 0.05. If there are 16 such columns, what is the probability that the following will be true?

a. At most two will fail.
b. Between two and four will fail.
c. At least four will fail.

Solution

a. $P(X \leq 2 | n = 16 \quad p = 0.05) = \sum_{0}^{2} \binom{16}{x}(0.05)^x (0.95)^{16-x}$

$x = 0 \quad \binom{16}{0}(0.05)^0 (0.95)^{16-0} = 0.440$

$x = 1 \quad \binom{16}{1}(0.05)^1 (0.95)^{16-1} = 0.371$

$x = 2 \quad \binom{16}{2}(0.05)^2 (0.95)^{16-2} = 0.146$

$P(X \leq 2) = P(X = 0) + P(x = 1) + P(X = 2)$

therefore

$$P(X \leq 2) = 0.957$$

b. $P(X \leq 30) = \Phi(-1.81) = 0.035$

$$= \sum_{0}^{3} \binom{16}{x}(0.05)^x (0.95)^{16-x} - 0.957$$

$$= 0.957 + \binom{16}{3}(0.05)^3 (0.95)^{16-3} - 0.975$$

$$= 0.036$$

c. $P(X \geq 4 | n = 16, p = 0.05) = 1 - P(X \leq 3)$

$$= 1 - \sum_{0}^{2} \binom{16}{x}(0.05)^x (0.95)^{16-x}$$

$$= 1 - (0.036 + 0.957)$$

$$= 0.007$$

Negative Binomial

In this case, the random variable is the number of trials. For a sequence of independent trials with probability of the occurrence of an event equal to p, the number of trials x before exactly the rth occurrence is known as the negative binomial, or the Pascal distribution. The probability of exactly x_0 trials before the rth occurrences is given by

$$P(X = x_0) = \binom{x-1}{r-1} p^r (1-p)^{x_0-r} \quad x_0 = r, r+1, \ldots n \tag{2}$$

where the mean and variance is given, respectively, by

$$\mu = \frac{r}{p} \qquad \sigma^2 = \frac{r(1-p)}{p^2}$$

Example

The probability that on production, a critical defect is found is 0.3. Find the probability that the tenth item inspected on the line is the fifth critical defect.

Solution

x = sample size = 10, r = the outcome = 5:

$$P(x = 5) = \binom{10 - 1}{5 - 1} (0.3)^5 (0.7)^{10-5}$$

$$= \binom{9}{4} (0.3)^5 (0.7)^5 = 0.05145$$

Geometric Distribution

The random variable is the number of trials. In a Bernoulli sequence, the number of trials until a specified event occurs for the first time is governed by the geometric distribution. Thus, for a sequence of independent trials with probability of occurrence p, the number of trials x before the first success is the geometric distribution, which is also a member of the family of Pascal distributions. If the occurrence of the event is realized on the xth trial, then there must have been no occurrence of such an event in any of the prior $(x - 1)$ trials. Hence, it is same as the negative binomial distribution with $r = 1$:

$$P(X = x) = p(1 - p)^{x-1}$$

The mean and variance, respectively, are

$$E(X) = \mu = \frac{1}{p} \quad \sigma^2 = \frac{1 - p}{p^2}$$

In a time or space problem that can be modeled as a Bernoulli sequence, the number of time (or) space intervals until the first occurrence is called the *first occurrence time*. If the individual trials or intervals in the sequence are statistically independent, then the first occurrence time is also the time between any two consecutive occurrences of the same events—that is, the recurrence time is equal to the first occurrence time. The mean recurrence time, which is commonly known in engineering as the *average return period* or the *average run length* (ARL), is equal to the reciprocal of p, the probability of occurrence of the event within one time unit.

Example

A system maintainability analysis procedure requires an experiment to determine the number of defective parts before the system is shut down for corrective maintenance. The probability of a defect is 0.1. Let X be a random variable denoting the number of parts to be tested until the first defective part is found. Assuming that the trials are independent, what is the probability that the fifth test will result in a defective part?

Solution

$$P(X = 5) = 0.1(1 - 0.1)^{5-1}$$

$$= 0.066$$

Hypergeometric Distribution

The *hypergeometric distribution* is used to model events in a finite population of size N when a sample of size n is taken at random from the population without replacement and where the elements of the population can be dichotomized as belonging to one of two disjoint categories. Thus, in a finite population N with different categories of items (e.g., conforming/nonconforming, success/failure, defective/nondefective), if a sample is drawn in such a way that each successive drawing is not independent (i.e., the items are not replaced), then the underlying distribution of such an experiment is the hypergeometric. The random variable of interest is the number of occurrences X of a particular outcome for a classification or category a, with the sample size of n. The probability of exactly x_0 occurrences is given by

$$h(x = x_0 \mid n, a, N) = \frac{\binom{a}{x_0}\binom{N-a}{n-x_0}}{\binom{N}{n}} \qquad x_0 = 0, 1, \ldots n \tag{3}$$

The hypergeometric satisfies all the conditions of the binomial except for independence in trials and constant p.

where

$X =$ random variable representing the number of occurrences of a given outcome

$a =$ category or classification of N

$N =$ population size

$n =$ sample size

$\binom{N}{n} =$ number of samples of size n

$\binom{N}{n}\binom{N-a}{n-x_0} =$ number of samples having x_0 outcomes out of a

The mean $\mu = n(a/N)$, and the variance $\sigma^2 = n(a/N)(1 - a/N)$. Let $p = a/N$; then $\mu = n\,p$, and $\sigma^2 = n\,p(1-p)$.

Example
To test the reliability of an existing machine, a company is interested in evaluating a batch of 50 identical products. The procedure calls for taking a sample of 5 items from the lot of 50 and passing the batch if no more than 2 are found to be defective, therefore deeming the machine fit. Assuming that the batch is 20% defective, what is the probability of accepting the batch?

Solution
Given: $N = 50, n = 5, a = 20\%$ of $50 = 10$

$$= P(x \le 2) = \sum_{x=0}^{2}\left(\frac{\binom{10}{5}\binom{50-10}{5-x}}{\binom{50}{5}}\right)$$

For $x = 0, P(x = 0) = 0.3106$

For $x = 1, P(x = 1) = 0.4313$

For $x = 2, P(x = 2) = 0.2093$

$P(x \le 2) = P(x = 0) + P(x = 1) + P(x = 2) = 0.9517$

Poisson Distribution
Many physical problems of interest to engineers involve the occurrences of events in a continuum of time or space. A Poisson process involves observing discrete events in a continuum of time, length, or space, with μ as the average number of occurrence of the event. For example, in the manufacture of an aircraft frame, cracks could occur anywhere in the joint or on the surface of the frame. Also in the construction of a pipeline, cracks could occur along continuous

welds. In the manufacture of carpets, defects can occur anywhere in a given area of carpet. The lightbulb in a machine tool could burn out at any time. Examples abound of the types of situations where the occurrence rather than the nonoccurrence of events in a continuum is of interest. Such time–space problems can be modeled with the Bernoulli sequence by dividing the time or space into small time intervals, assuming that the event will either occur or not occur (only two possibilities). In the case of the Poisson, it is usually assumed that the event will occur at any time interval or any point in space and also that the event may occur no more than once at a given time or space interval. Four of the assumptions of the Poisson follow:

1. An event can occur at random and at any time or point in space.
2. The occurrence of an event in a given time or space interval is independent of that in any other nonoverlapping intervals.
3. The probability of occurrence of an event in a small interval Δt is proportional to Δt and is given by $\mu \Delta t$, where μ is the mean rate of occurrence of the event (μ is assumed a constant).
4. The probability of two or more occurrences in the interval Δt is negligible and numerically equal to zero (higher orders of Δt are negligible).

A random variable X is said to have a Poisson distribution with parameter μ if its density is given by

$$f(X = x) = \frac{\mu^x e^\mu}{x!} \quad \mu > 0, x = 0, 1, 2, \ldots \tag{4}$$

with mean = μ, and variance = $\sigma^2 = \mu$.

Example 1
A compressor is known to fail on average 0.2 times per hour.

a. What is the probability of a failure in 3 h?
b. What is the probability of at least two failures in 5 h?

Solution
Note: For discrete distributions,

$$P(X = x_0) = P[X \le x_0] - P[X \le (x_0 - 1)]$$

where

$$P(X \le x_0 | \mu) = F(x_0, \mu)$$

Thus:

$$(X = x_0) = F(x_0, \mu) - F(x_0 - 1, \mu)$$

a. $\mu = (0.2)(3) = 0.6$
$P(x = 1) = F(1, 0.6) - F(0, 0.6) = 0.878 - 0.549 = 0.324$
b. $\mu = (0.2)(5) = 1.0$
$P(x \ge 2) = 1 - P(x \le 1) = 1 - F(1, 1) = 1 - 0.736$

The F values can be found in the cumulative Poisson table available in most texts on statistics.

Example 2
An aircraft's landing gear has a probability of 10^{-5} per landing of being damaged from excessive impact. What is the probability that the landing gear will survive a 10,000 landing design life without damage?

Solution

Considering the problem as being Poisson in nature, $\lambda = np = 10^{-5} \times (10{,}000) = 0.1$:

$$P(x = 0) = \frac{0.1^0 e^{-0.1}}{0!} - 0.904837$$

2.3 Continuous Distributions

Normal Distribution

The normal probability is perhaps the most frequently used of all probability densities, and some of the most important statistical techniques are based on it. The normal random variable appears frequently in practical problems, and it provides a good approximation to a large number of other probability laws. The density function for a normal random variable is symmetric and bell-shaped. Due to the bell-shaped nature, a normally distributed random variable has a very high probability of taking on a value close to μ and correspondingly a lower probability of taking on values that are further away on either side of μ. Thus, a random variable X is normally distributed if its density function is of the form

$$f_X(x) = \frac{1}{\sigma\sqrt{2\pi}} e^{(x-\mu)^2/2\sigma^2} \tag{5}$$

where x is real, μ is any real number, and σ is positive.

Computing the probability associated with the occurrence of the normal probability event requires integrating the density function. Unfortunately, this integration cannot be carried out in closed form. However, numerical techniques can be used to evaluate the integral for specific values of μ and σ^2 after properly transforming the normal random variable with mean μ and variance σ^2 to a standard normal variable with mean $\mu = 0$, and $\sigma^2 = 1$. Tables of the standard normal are available in most statistics and probability texts. For the standard normal, the probability of the normal random variable X taking on a value less than x_0 is the cumulative distribution function defined as $P(X < x_0) = z$, where $z = (x_0 - \mu)/\sigma$ is the number of standard deviates between the mean and x_0.

The actual area represented by the value of z is given by $\Phi(z)$, where $\Phi(z)$ is the integral of the density function with boundaries from $-\infty$ to z. Most standardized normal tables have a sketch that shows the boundaries and the areas that result in values indicated on the tables. Because the normal is symmetric, the tail values of the areas are identical. The same is true of the corresponding standard deviation, except for a change in sign. Deviations above the mean are denoted as positive, while those below the mean are denoted as negative. This makes it possible to evaluate different values on both sides of the mean with just the table value for lower half of the normal function ($-\infty$ to z) or the upper half of the function (z to $+\infty$).

Example

The time to wear out of a cutting-tool edge is distributed normally with $\mu = 2.8$ h and $\sigma = 0.6$ h.

a. What is the probability that the tool will wear out in less than 1.5 h?

b. How often should the cutting edge be replaced to keep the failure rate less than 10% of the tools?

Solution

a. $P\{t < 1.5\} = F_t(1.5) = \Phi(z)$

where

$$z = \frac{t - \mu}{\sigma} = \frac{1.5 - 2.8}{0.6} = -2.1667$$

$$\Phi(-2.1667) = 0.0151$$

b. $P\{T < t\} = 0.10$; thus $\Phi(z) = 0.1$.

Interpolating from tables shows that $z \approx -1.28$.

Thus $t - u = -1.28\sigma$ and $t = 2.03$ h

Lognormal Distribution

Situations frequently arise where a random variable Y is a product of other independent random variables y_i —that is, $y = y_1 y_2 y_3, \ldots, y_N$. For instance, the failure of a shaft may be proportional to the product of forces of different magnitude. Taking the natural logarithm of the above equation,

$$\ln y = \ln y_1 + \ln y_2 + \cdots + \ln y_N$$

Let $\ln y$ be distributed normally, then y will be lognormally distributed. The lognormal density function for y is given by

$$f(y) = \frac{1}{y\sigma\sqrt{2\pi}} \exp\left[-\frac{1}{2\sigma^2}(\ln y - \mu)\right] \tag{6a}$$

Let $\omega = \sigma$, and let $\mu = \ln y_0$.

Then we can rewrite lognormal density function in a more common form as

$$f_y(y) = \frac{1}{\sqrt{2\pi}\omega y} \exp\left\{-\frac{1}{2\omega^2}\left[\ln\left(\frac{y}{y_0}\right)\right]^2\right\} \tag{6b}$$

and

$$F_y(y) = \Phi\left[\frac{1}{\omega}\ln\left(\frac{y}{y_0}\right)\right]$$

The parameters ω and y_0 are known for any scenario.

Example

Fatigue life data for a shaft are fit to a lognormal distribution with the following parameters: $y_0 = 2 \times 107$ cycles, and $\omega = 2.3$. To what value should the design life be set if the probability of failure is not to exceed 1.0%?

Solution

From the normal distribution tables, z such that $\Phi(z) = 0.01$ is -2.32.

$$-2.32 = \left[\frac{1}{2.3}\ln\left(\frac{y}{2 \times 10^7}\right)\right]$$

Thus

$$y = 9.63 \times 10^4 \text{ cycles}$$

Exponential Distribution

If events occur according to a Poisson process, then the time T between consecutive occurrences has an exponential distribution. Thus, in a Poisson process, if the rate of occurrence of the events in a continuum is λ, then time between occurrence is exponentially distributed with mean time to occur equal to θ and the mean occurrence rate $\lambda = 1/\theta$. The exponential distribution has been used to study life distributions in the physical and biological sciences as well as in engineering. If T is defined as the exponential random variable, then the density and cumulative distribution functions are given as follows:

$$f(t, \lambda) = \lambda \exp(-\lambda\tau) = \frac{1}{\theta}\exp\left(-\frac{t}{\theta}\right) \quad \text{where} \quad \lambda = \frac{1}{\theta} \tag{7}$$

$$F(T < t) = \int_0^t \frac{1}{\theta}\exp\left(-\frac{1}{\theta}\tau\right) d\tau = 1 - \exp\left(-\frac{1}{\theta}t\right) \tag{8}$$

The average value $\mu = \theta = s$, the standard deviation.

Example

The life in years of a certain kind of electrical switch has an exponential distribution with $\lambda = \frac{1}{2}(\theta = 2)$. If 100 switches are installed in a system, find the probability that at most 30 will fail during the first year.

Solution

First try to find the probability that one switch will fail in the first year ($t = 1$). $F(t) = (1 - e^{-\lambda t}) = 0.3935 =$ probability that one will fail during the first year. Now try to find $P(x \leq 30)$, with $n = 100, p = 0.3935$, using the binomial:

$$P(x \leq 30) = \sum_{0}^{30} \binom{100}{x} (0.3935)^x (0.6065)^{100-x}$$

Since $np > 5$ and $p > 0.1$, we can use normal approximation, where

$$\mu = np = (100)(0.3905) = 39.35$$

$$\sigma^2 = np(1 - p) = 23.86, \sigma = 4.485$$

Using the continuity for correction 30, where 30 lies between 29.5 and 30.5 such that:

For "at most 30" (30 or less), the appropriate value to use is 30.5.

For "at least 30" (30 or more), the appropriate value to use is 29.5.

$$Z = \left(\frac{30.5 - 39.35}{4.485} \right) = -1.81$$

$$P(X \leq 30) = \Phi(-1.81) = 0.035$$

Hence, the probability that at most 30 will fail during the first year is about 0.04.

Weibull Distribution

Advances in technology have made possible the design and manufacture of complex systems whose operation depends on the reliability and availability of the subsystems and components that make up such systems. The time to failure of the life of component measured from a specified point or time interval is a random variable. In 1951, W. Weibull introduced a distribution that has been found to be very useful in the study of reliability and maintenance of physical systems.

The most general form of the Weibull is the three-parameter form presented in Eqs. (9)–(10):

$$f(x) = \left(\frac{\beta}{\theta} \right) \left(\frac{\zeta - \gamma}{\theta - \gamma} \right)^{\beta} e \left[\frac{-(x - \gamma)}{(\theta - \gamma)} \right]^{\beta} \tag{9}$$

$$f(x) = (\alpha\beta)(x - \gamma)^{\beta-1} e^{-(\alpha x)\beta} \quad \text{where } \alpha = \frac{1}{\theta} \tag{10}$$

$$f(x) = 1 - \exp[-(\alpha x)^{\beta}] \tag{11}$$

where $\theta > 0$ and $\alpha > 0$:

$$x > \gamma$$
$$\beta > 0$$

where $\theta =$ characteristic life of the distribution. It is also referred to as the scale
parameter and is used to locate the distribution on the x axis
$\gamma =$ location parameter or the minimum life (useful for warranty specification)
$\beta =$ shape parameter or the slope

For the Weibull distribution, substituting $x = \theta$ in the cumulative distribution gives

$$f(x = \theta) = 1 - e^{-1} = 0.0632$$

So for any Weibull distribution, the probability of failure prior to θ is 0.632. Thus, θ will always divide the area under the probability density function (pdf) into 0.632 and 0.368 for all values of the slope β.

For the two-parameter Weibull, the minimum life γ is zero. A major benefit of modeling life distributions with the Weibull is that the distribution is robust enough that for different values of β it is possible to accommodate a host of other distribution types. For example, when $\beta = 1$, the Weibull becomes the exponential. Also when $\beta = 4$, the Weibull starts to resemble the normal distribution. It is also the distribution of choice to model the different regions of a component's or system's life such as the early, constant, and wear-out regions. The Weibull is particularly useful when the failure rate of the system is not constant but increases (IFR = increasing failure rate) or decreases (DFR = decreasing failure rate). The mean and variance are given as follows:

$$\mu = \theta \Gamma \left(1 + \frac{1}{\beta} \right) \tag{12}$$

$$\sigma^2 = \theta^2 \left[\Gamma \left(1 + \frac{2}{\beta} \right) \Gamma^2 \left(1 + \frac{1}{\beta} \right) \right] \tag{13}$$

where $\Gamma(n) = (n - 1)!$ is defined as the gamma function on an integer n.

Example

The life of a magnetic resonance imaging (MRI) machine is modeled by a Weibull distribution with parameters $\beta = 2$ and $\theta = 500$ h. Determine the mean life and the variance of the MRI. Also, what is the probability that the MRI will fail before 250 h?

Solution

   a. $\mu = 500\Gamma \left(1 + \frac{1}{2} \right) = 443.11$ h

$$\sigma^2 = 500^2 \left[\Gamma \left(1 + \frac{2}{2} \right) \Gamma^2 \left(1 + \frac{1}{2} \right) \right] = 53650.5$$

   b. $F(x) = 1 - \exp[-(250/500)^2] = 0.2212$

The gamma values can be found in the gamma tables available in most statistics texts.

2.4   Pareto Analysis for Data Segregation

Any systematic maintenance strategy is aimed at preventing the dominant causes of failure of critical equipment and, in turn, toward achieving acceptable system availability. In the natural scheme of things, all the major components of a system rarely have the same effect, magnitude-wise, on system performance or output. In social system theory, it is usually assumed that 80% of society's wealth is held by about only 20% of the people. The Pareto chart is a bar chart, or histogram, that demonstrates that the intensity of a given phenomenon (such as failure or nonconformance) due to one of the system's components is relative to failure caused by the other components in the system. Hence, the chart shows the contribution of each member to the total system failure. Pareto charts are powerful in identifying the critical failure contributors. Thus, in terms of evaluation, the available resources can be directed to the component that causes failure the most so as to decrease system failure.

Pareto chart analysis requires a statistical approach of measurement or aggregation for identifying the most important problems through different measuring scales or combination

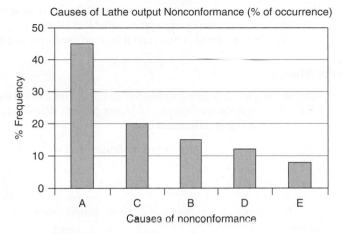

Figure 1   Pareto chart of various causes of lathe output nonconformance (% of occurrence).

of scales. Typical scales or units include frequency of occurrence, cost, labor use, and exposure factor. The idea is that while the sum total of a certain type of occurrence may be due to a number of different sources (or components), the vast majority of those occurrences are due to a small percentage of the sources.

The Pareto rule, also called the 80/20 rule (i.e., 80% of the occurrences are due to 20% of the sources of the occurrences) has been used very successfully in nonconformance analysis for quality assurance. Through the analyses of scrap information, the current number of significant nonconforming items and their causes are identified.

As an example, consider the problem of poor lathe performance for a certain operation. The following causes based on performance history and process experience have been identified:

Description	Code
Operator effect	A
Power surges	B
Material mixtures	C
Tool (wear, age)	D
Miscellaneous	E

Suppose a sampling system was set up whereby over the next week, parts coming out the lathe were examined and the defects were classified according to the categories identified earlier. The frequency of occurrence may be displayed as shown in Fig. 1. The plot shows the defect type on the horizontal axis and the frequency on the vertical axis. From this Pareto chart, it is obvious that the major contributors to nonconformance were operator and material problems. As possible solution options, operator training or retraining and vendor evaluation/qualification should be instituted as a means of reducing the problem of poor performance.

3   REVIEW OF SYSTEM RELIABILITY AND AVAILABILITY

System reliability is defined as the probability that a system will perform properly for a specified period of time under a given set of operating conditions. This definition implies that both the loading under which the system operates and the environmental conditions must be taken into consideration when modeling system reliability. However, perhaps the most important factor that must be considered is time, and it is often used to define the rate of failure.

3.1   Reliability Models

Define t = time to failure (random variable), and let T equal the age of the system. Then $F(t) = P(t \leq T)$ = distribution function of failure process $R(t)$ = reliability function = $1 - F(t)$,
Assume $F(t)$ is differentiable.

$$F'(t) = \text{failure density } f(t) = \frac{d}{dt} F(t) \tag{14}$$

$$R(t) = 1 - F(t) = 1 - \int_0^t f(s)\, ds = \int_t^\infty f(s)\, ds \tag{15}$$

If t is a negative exponential random variable, then

$$f(t) = \frac{1}{\theta} \exp\left(\frac{-t}{\theta}\right)$$

$$F(t) = 1 - \exp\left(-\frac{t}{\theta}\right)$$

$$R(t) = 1 - F(t) = \exp\left(-\frac{t}{\theta}\right)$$

Figure 2 gives a relative frequency of failure from the viewpoint of initial operation at time $t = 0$. The failure distribution function $F(t)$ is the special case when $t_1 = 0$ and t_2 is the argument of $F(t)$—that is, $F(t_2)$.

Failure Probability in the Interval (t_1, t_2)

$$\int_{t_1}^{t_2} f(t)\, dt = \int_0^{t_2} f(t)\, dt - \int_0^{t_1} f(t)\, dt$$

$$= F(t_2) - F(t_1)$$

$$= [1 - R(t_2)] - [1 - R(t_1)]$$

$$= R(t_1) - R(t_2) \tag{16}$$

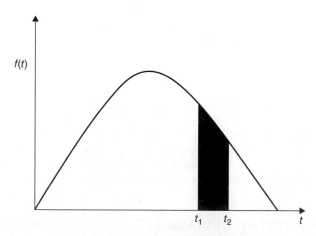

Figure 2   Failure density function.

Reliability of Component of Age t

The reliability (or survival probability) of a fresh unit corresponding to a mission of duration x is by definition $R(x) = \overline{F}(x) = 1 - F(x)$, where $F(x)$ is the life distribution of the unit. The corresponding conditional reliability of the unit of age t for an additional time duration x is

$$\overline{F}(x|t) = \frac{\overline{F}(t \cap x)}{\overline{F}(t)} \quad \overline{F}(x) > 0 \tag{17}$$

But $\overline{F}(t \cap x) = \overline{F}(t + x) = $ total life of the unit up to time$(t + x)$; therefore,

$$\overline{F}(x|t) = \frac{\overline{F}(t + x)}{\overline{F}(t)}$$

Similarly, the conditional probability of failure during the interval of duration x is $F(x|t)$ where $F(x|t) = 1 - \overline{F}(x|t)$ by definition.

But

$$\overline{F}(x|t) = \frac{\overline{F}(t + x)}{F(t)}$$

Thus,

$$F(x|t) = 1 - \frac{\overline{F}(t + x)}{\overline{F}(t)} = \frac{\overline{F}(t) - \overline{F}(t + x)}{\overline{F}(t)}$$

Conditional Failure Rate [Hazard or Intensity Function v(t)]

The conditional failure probability is given by $F(x|t)$. Hence, the conditional failure rate is given by $F(x|t)/x$ where

$$\frac{F(x|t)}{x} = \frac{1}{x}\left[\frac{\overline{F}(t) - \overline{F}(t + x)}{\overline{F}(t)}\right] = \left(\frac{1}{x}\right)\frac{R(t) - R(t + x)}{R(t)} \tag{18}$$

The hazard or intensity function is the limit of the failure rate as the interval (x, in this case) approaches zero. The hazard function is also referred to as instantaneous failure rate because the interval in question is very small. That is,

$$v(t) = \lim_{x \to 0} \frac{R(t) - R(t + x)}{x}\left[\frac{1}{R(t)}\right]$$

$$= -\lim_{x \to 0} \frac{R(t + x) - R(t)}{x}\left[\frac{1}{R(t)}\right]$$

$$= -\frac{1}{R(t)}\frac{dR(t)}{dt} \tag{19}$$

Note:

$$R(t) = 1 - \int_0^t f(s)\,ds = 1 - F(t)$$

Hence

$$\frac{d}{dt}R(t) = -\frac{d}{dt}F(t) = -f(t)$$

Therefore,

$$v(t) = \frac{1}{R(t)}[f(t)] = \frac{f(t)}{R(t)}$$

The quantity $v(t)$ represents the probability that a device of age t will fail in the small interval t to $t + x$.

Estimation of the Intensity Function Using Binomial Distribution Approach

Consider a population with N identical items and failure distribution function $F(t)$. If $N_s(t)$ is a random variable denoting the number of items functioning successfully at time t, then $N_s(t)$ is represented by

$$P[N_s(t) = n] = \frac{N!}{n!(N-n)!}(R(t)^n)(1 - R(t))^{N-n}$$

where $R(t)$ is the probability of success.

The expected value of $N_s(t)$ is $E(N_s(t))$ if x is binomially distributed. Then, $np = NR(t)$, where $E(N_s(t)) = NR(t) = \overline{N}(t)$.

Hence

$$R(t) = \frac{E(N_s(t))}{N} = \frac{\overline{N}(t)}{N}$$

$$F(t) = 1 - R(t) = 1 - \frac{\overline{N}(t)}{N} = \frac{N - \overline{N}(t)}{N}$$

But

$$f(t) = \frac{d}{dt}F(t) = -\frac{d}{N\,dt}\overline{N}(t)$$

$$= \lim_{\Delta t \to 0} \frac{N(t) - \overline{N}(t + \Delta t)}{\Delta t}\left(\frac{1}{N}\right)$$

Replacing N with $\overline{N}(t)$ where $\overline{N}(t)$ is the average number of successfully functioning at time t leads to

$$v(t) = \lim_{\Delta t \to 0} \frac{\overline{N}(t) - \overline{N}(t + \Delta t)}{\overline{N}(t)\,\Delta t} = -\frac{1}{\overline{N}(t)}\frac{d}{dt}\overline{N}(t) = \frac{N}{\overline{N}(t)}f(t) = \frac{f(t)}{R(t)} \tag{20}$$

Relationship between the Failure Density, the Intensity, and the Reliability Functions

$$v(t) = \frac{f(t)}{R(t)} = -\frac{1}{\overline{N}(t)}\frac{d}{dt}\overline{N}(t) = -\frac{d}{dt|}(\ln \overline{N}(t)) = -\frac{d}{dt}(\ln R(t))$$

$$R(t) = \exp\left[-\int_0^t v(\tau)\,d\tau\right]$$

$$f(t) = v(t)\exp\left[-\int_0^t v(\tau)\,d\tau\right]$$

Mean Time to Failure (MTTF and MTBF)

The MTTF is the expected value of the time to failure. By definition, the expected value of a density function $y = f(x)\{(f(x)\text{is continuous}\}$ is the following:

$$E(x) = \int_{-\infty}^{\infty} xf(x)\,dx \quad -\infty < x < \infty \tag{21}$$

For the mean time to failure $T, (T > 0)$,

$$E(t) = \int_0^{\infty} R(s)\,ds \quad 0 \le T \le \infty \tag{22}$$

By proper transformation and integration (integration by parts), the MTTF is equal to $\int_0^{\infty} sf(s)\,ds$. How?

$$E(T) = \int_0^{\infty} R(s)\,ds \quad 0 \le T \le \infty$$

Note:

$$\int u\, dv = uv - \int v\, du$$

Let $u = R(s)$, and let $dv = ds$. Then $v = s$, and

$$du = d(R(s)) = -f(s)\, ds$$

Thus:

$$\int_0^\infty R(s)\, ds \quad 0 \le T \le \infty = sR(s)\Big|_0^\infty + \int sf(s)\, ds$$

At $s = 0$, $sR(s) = 0$; at $s = \infty$, $sR(s) = 0$. Therefore,

$$\int_0^\infty R(s)\, ds = \int sf(s)\, ds = \text{expected value}$$

Example
Let us use the exponential density function to determine the MTTE:

$$f(t) = \frac{1}{\theta} \exp\left(\frac{-t}{\theta}\right)$$

$$E(t) = \int_0^\infty s\left(\frac{1}{\theta}\right) \exp\left(\frac{-s}{\theta}\right) ds$$

Let $s = u, \exp(-t/\theta)\, ds = dv$:

$$v = -\frac{\theta}{\theta} \exp\left(\frac{-t}{\theta}\right)$$

Using integration by parts, we get the following:

$$E(t) = \theta \quad \text{MTTF} = \theta$$

Therefore,

$$R(\theta) = \exp(-\theta/\theta) = \exp(-1)$$

$$R(\text{MTTF}) = \exp(-1) = 0.368$$

For normal density function:

$$Z = \frac{t - \mu}{\sigma} = \frac{t - \text{MTTF}}{\sigma}$$

when $t = \text{MTTF}$

$$Z_z = (0) \quad \Phi(0) = 0.5$$

Thus, even if MTTF is the same, reliability could change, depending on the underlying probability structure that governs the failure.

3.2 Intensity Functions for Some Commonly Used Distributions

The intensity functions of some commonly used distributions can be computed as follows:

Exponential Distribution

$$f(t) = \frac{1}{\theta e^{-t/\theta}} R(t) = e^{-t/\theta} \tag{23}$$

$$v(t) = \frac{f(t)}{R(t)} = \frac{1}{\theta} \tag{24}$$

Standardized Normal

$$f(t) = \frac{1}{\sqrt{2\pi}} \exp\left(-\frac{z^2}{2}\right) = \frac{\phi(z)}{\sigma} \tag{25}$$

$$R(t) = 1 - \int_{-\infty}^{Z} \frac{1}{\sqrt{2\pi}} \exp\left(-\frac{\tau^2}{2}\right) d\tau = 1 - F(t) = 1 - \phi(z) \tag{26}$$

$$v(t) = \frac{f(t)}{R(t)} = \frac{\left(z = \dfrac{t-\mu}{\sigma}\right)\Big/\sigma}{R(t)}$$
$$= \frac{\phi(z)}{\sigma(1 - \phi(z))} \tag{27}$$

where $\phi(z)$ = pdf for standard normal random variable
$\varphi(z)$ = cdf (cumulative density function) for standard normal random variable

Log Normal

$$f(t) = \frac{1}{\sigma t \sqrt{2\pi}} \exp\left[-\frac{1}{2}\left(\frac{\ln \tau - \mu}{\sigma}\right)^2\right] \tag{28}$$

$$R(t) = 1 - \int_0^t \frac{1}{\sigma t \sqrt{2\pi}} \exp\left[-\frac{1}{2}\left(\frac{\ln \tau - \mu}{\sigma}\right)^2\right] d\tau \tag{29}$$
$$= 1 - F(t) = 1 - P(T \le t) = 1 - P\left(Z \le \frac{\ln t - \mu}{\sigma}\right)$$

$$R(t) = 1 - P\left(z \le \frac{\ln t - \mu}{\sigma}\right) \tag{30}$$

$$v(t) = \frac{f(t)}{R(t)} = \frac{\phi\left(\dfrac{\ln t - \mu}{\sigma}\right)\Big/\sigma}{1 - F(t)} = \frac{\phi\left(\dfrac{\ln t - \mu}{\sigma}\right)}{t\sigma[1 - \phi(z)]}$$

Gamma Distribution

$$v(t) = \frac{\dfrac{\lambda^n}{\Gamma(n)} t^{n-1} e^{-\lambda t}}{\dfrac{\sum_{k=0}^{n-1} (\lambda t)^k \ \exp(-\lambda t)}{k!}} \tag{31}$$

Weibull Distribution

$$f(t) = \frac{\beta(t-\delta)^{\beta-1}}{(\theta-\delta)^\beta} \exp\left[\left(\frac{t-\delta}{\theta-\delta}\right)^\beta\right] \quad t \ge \delta \ge 0 \tag{32}$$

$$R(t) = 1 - F(t) = 1 - \int_0^t f(\tau)\delta\tau = \exp\left[\left(\frac{t-\delta}{\theta-\delta}\right)^\beta\right] \tag{33}$$

$$v(t) = \frac{\beta(t-\delta)^{\beta-1}}{(\theta-\delta)^\beta} \tag{34}$$

3.3 Three-Parameter Weibull Probability Distribution

The Weibull probability density function is an extremely important distribution to characterize the probabilistic behavior of a large number of real-world phenomena. It is especially useful as a failure model in analyzing the reliability of different types of systems. The complete three-parameter Weibull probability distribution, $W(a, b, c)$ is given by

$$f(x; \ a, b, c) = \frac{c(x-a)^{c-1}}{b^c} \exp\left\{-\left(\frac{x-a}{b}\right)^c\right\}$$

where $x \geq a, \ b > 0, c > 0$ and $a, \ b,$ and c are the location, scale, and shape parameters, respectively.

Successful application of the Weibull distribution depends on having acceptable statistical estimates of the three parameters. There is no existing method in obtaining closed-form estimates of the Weibull distribution parameters. Because of this difficulty, the three-parameter Weibull distribution is seldom used. Even the popular two-parameter model does not offer closed-form estimates of the parameters and relies on numerical procedures.

The Likelihood Function
The log likelihood function of the three-parameter model is given by

$$\log L\,(a, b, c) = n \log c - nc \log b + (c-1) \sum_{i=1}^{n} \log(x_i - a) - \sum_{i=1}^{n} \left(\frac{x_i - a}{b}\right)^c \quad (35)$$

The objective is to obtain estimates of a, b, and c so as to maximize $\log L(a, b, c)$ with respect to the constraints:

$$a \leq \min x_i \quad b > 0 \quad c > 0$$
$$1 \leq i \leq n$$

The following systems of likelihood equations are obtained:

$$\frac{\partial \log L}{\partial a} = -(c-1) \sum_{i=1}^{n} \log(x_i - a) + \frac{c}{b} \sum_{i=1}^{n} (x_i - a)^{c-1} = 0 \quad (36)$$

$$\frac{\partial \log L}{\partial b} = -\frac{nc}{b} + \frac{c}{b^{c+1}} \sum_{i=1}^{n} (x_i - a)^c = 0 \quad (37)$$

$$\frac{\partial \log L}{\partial c} = \frac{n}{c} - n \log b + \sum_{i=1}^{n} \log(x_i - a)$$

$$-\sum_{i=1}^{n} \left(\frac{x_i - a}{b}\right)^c \{\log(x_i - a) - \log b\} = 0 \quad (38)$$

Equations (36)–(38) constitute a system of nonlinear equations. There are no existing methods to solve them analytically. Moreover, to our knowledge, there are no effective numerical methods available. Based on this concern and the importance of the three-parameter Weibull distribution in practice, Tsokos and Qiao extended their converging two-parameter numerical solution to the three-parameter model.[9,10] They showed that the three-parameter solution also converges rapidly, and it does not depend on any other conditions. They briefly summarized their findings in the following theorem:

Theorem 1

The simple iterative procedure (SIP) always converges for any positive starting point, that is, c_k generated by SIP converges to the unique fixed point c^* of $q(c)$ as $k \to \infty$, for any starting point $c_0 > 0$. And the convergence is at least a geometric rate of $\frac{1}{2}$, where $c = q(c)$, $c_{k+1} = c_k + q(c_k)/2$, and $c = c_0$ for $k = 0, 1, 2, \ldots$.

3.4 Estimating $R(t)$, $v(t)$, $f(t)$ Using Empirical Data

Small Sample Size (n ≤ 10)

Consider the following ordered failure times:

$$_oT_1, \; _oT_2, \; _oT_3, \; _oT_4, \cdots, \; _oT_n$$

where $_oT_1 \le \, _oT_2 < \, _oT_3 \le \cdots \le \, _oT_n$, and the subscript o denotes ordered:

$$_nP_j = \widehat{F}(_oT_J)$$

and $_nP_j$ is the fraction of the population failing prior to the jth observation in a sample of size n.

The best estimate for $_nP_j$ is the median value. That is,

$$_nP_j = \widehat{F}(_oT_J) = \frac{j - 0.3}{n + 0.4} \tag{39}$$

Hence, the cumulative distribution at the jth ordered failure time t_j is estimated as

$$\widehat{F}(_oT_J) = \frac{j - 0.3}{n + 0.4} \tag{40}$$

$$\begin{aligned}
\widehat{R}(_oT_J) &= 1 - \widehat{F}(_oT_J) \\
&= 1 - \frac{j - 0.3}{n + 0.4} = \frac{n + 0.4 - j + 0.3}{n + 0.4} = \frac{n - j + 0.7}{n + 0.4}
\end{aligned} \tag{41}$$

$$\widehat{v}(_oT_j) = \frac{\widehat{R}(_oT_j) - \widehat{R}(_oT_{j+1})}{(_oT_{j+1} - _oT_j)\widehat{R}(_oT_j)} \tag{42}$$

$$\widehat{v}(_oT_j) = \frac{1}{(_oT_{j+1} - _oT_j)(n - j + 0.7)} \tag{43}$$

$$\widehat{f}(_oT_j) = \frac{\widehat{R}(_oT_j) - \widehat{R}(_oT_{j+1})}{(_oT_{j+1} - _oT_j)\widehat{R}(_oT_j)} = \frac{1}{(n + 0.4)(_oT_{j+1} - _oT_j)} \tag{44}$$

Large Sample Size

$$R(t) = \frac{\overline{N}(t)}{N} \tag{45}$$

$$v(t) = \frac{\overline{N}(t) - \overline{N}(t + x)}{\overline{N}(t)x} \tag{46}$$

$$f(t) = \frac{\overline{N}(t) - \overline{N}(t + x)}{N \cdot x} \tag{47}$$

where $x = \Delta t$.

3.5 Failure Process Modeling

The growing importance of maintenance has generated considerable interest in the development of maintenance policy models. The mathematical sophistication of these models has increased with the growth in the complexity of modern systems. If maintenance is performed too frequently or less frequently, the cost or system availability would become prohibitive. Thus, a major effort in the development of maintenance programs is to determine the optimal intervention intervals.

In order to develop any maintenance decision model, it is very important to understand and characterize the system's failure–repair process. The most important maintenance decision is to determine when a component needs to be replaced. To model a reparable system, the first question to be answered is, "What would the system look like after a maintenance activity?" In other words, what is the characteristic of the system failure process?

In making mathematical models for real-world phenomena, it is always necessary to make assumptions so as to make the calculations more tractable. However, too many simplifications and assumptions may render the models inapplicable. One simplifying assumption often made is that component failures are independent and identically distributed. From the stochastic process modeling point of view, the system failure process is generally classified as one of the three typical stochastic counting processes—homogenous Poisson process (HPP), nonhomogeneous Poisson process (NHPP), and renewal process (RP). A process $\{N(t), t \geq 0\}$ is said to a counting process if $N(t)$ represents the total number of events that occur by time t.[11] In addition, a counting process must fulfill the following conditions:

1. $N(t) \geq 0$.
2. $N(t)$ is an integer.
3. If $s < t$, then $N(t) \leq N(s)$.
4. For $s < t$, then $N(t) - N(s)$ equals the number of events that occur in the interval (s, t).

Specifically, a counting process is said to be Poisson in nature if, in addition to the above conditions, the following are also met:

1. $N(0) = 0$.
2. The process has independent increments.
3. The number of events in any interval of length t is Poisson distributed with a mean of λt:

$$P\{N(t + s) - N(s) = n\} = e^{\lambda t} \frac{(\lambda t)^n}{n!} \quad n = 0, 1, \ldots \quad \text{and} \quad E[N(t)] = \lambda t \qquad (48)$$

Independent increments mean that the numbers of events that occur in disjoint time intervals are independent.

A homogenous Poisson process, also known as stationary Poisson process, is a counting process in which the interarrival rate of two consecutive events is constant. This implies that the failure distribution of such a system is the exponential and that the failure rate or the hazard function is constant over time. Such an assumption on a failure process would mean that the system life is memoryless, and that a system that has already been in use for a time t is as good as new regarding the time it has until it fails. For such a process, preventive maintenance (PM) is ineffective since the failure process is random.

NHPP, also referred to as a nonstationary process, is a generalized Poisson process. The difference is that in NHPP the interarrival rate is a function of time. The following conditions must be fulfilled by an NHPP in addition to those of a Poisson process:

1. $P\{N(t + h) - N(t) \geq 2\} = o(h)$. That is, the probability that the number of events that occur in the interval $[t, t + h]$ is a function of the time elapsed.
2. $P\{N(t + h) - N(t) = 1\} = \lambda(t)h + o(h)$.

The major drawback of applying NHPP in maintenance decisions is that a single failure distribution would be used to describe the failure behavior of a complex system. As a result, it would be difficult to evaluate maintenance effects on different components.

Nonhomogeneous Poisson Process

Consider a complex system in the development and testing process. The system is tested until it fails, then it is repaired or redesigned if necessary, and then it is tested again until it fails. This process continues until we reach a desirable reliability, which would reflect the quality of the final design. This process of testing a system has been referred to as *reliability growth*. Likewise, the reliability of a repairable system will improve with time as component defects and flaws are detected, repaired, or removed. Consider a nonhomogeneous Poisson process, NHPP, with a failure intensity function given by the following:

$$v(t) = \frac{\beta}{\theta}\left(\frac{t}{\theta}\right)^{\beta-1} \quad t > 0 \tag{49}$$

This failure intensity function corresponds to the hazard rate function of the Weibull distribution in Eq. (34). NHPP is an effective approach to analyzing the reliability growth and predicting the failure behavior of a given system. In addition to tracking the reliability growth of a system, we can utilize such a modeling scheme for predictions since it is quite important to be able to determine the next failure time after the system has experienced some failures during the developmental process. Being able to have a good estimate as to when the system will fail again is important in strategically structuring maintenance policies. The probability of achieving n failures of a given system in the time interval $(0, t)$ can be written as

$$P(x = n;\ t) = \frac{\exp\left\{-\int_0^t v(x)\,dx\right\}\left\{-\int_0^t v(x)\,dx\right\}^n}{n!} \quad \text{for } t > 0 \tag{50}$$

When the failure intensity function $V(t) = \lambda$, Eq. (50) reduces into a homogeneous Poisson process,

$$P(x = n;\ t) = \frac{e^{-\lambda t}(\lambda t)^n}{n!} \quad 0 < t. \tag{51}$$

For tracking reliability growth of the system with the Weibull failure intensity,

$$V(t) = \frac{\beta}{\theta}\left(\frac{t}{\theta}\right)^{\beta-1} \quad 0 < t, 0 < \beta, \theta \tag{52}$$

Where β, θ are scale parameters, respectively, the Poisson density function reduces to

$$P(x = n;\ t) = \frac{1}{n!}\exp\left\{-\frac{t^\beta}{\theta^\beta}\right\}\left(\frac{t}{\theta}\right)^{n\beta} \tag{53}$$

Equation (53) is the NHPP or Weibull process. In reliability growth analysis it is important to be able to determine the next failure time after some failures have already occurred. With respect to this aim, the time difference between the expected failure time and the current failure time or the mean time between failures (MTBF) is of significant interest. The maximum likelihood estimate (MLE) for the shape and scale parameters β and θ are

$$\hat{\beta} = \frac{n}{\sum_{i=1}^n \log(t_n/t_i)} \quad \text{and} \quad \hat{\theta} = \frac{t_n}{n^{1/\beta}}$$

The MLE of the intensity function and its reciprocal can be approximated using the above estimates of β and θ. In reliability growth modeling, we would expect the failure intensity $v(t)$ to be decreasing with time; thus, $\beta < 1$. However, values of $\beta > 1$ indicate that the system is wearing out rapidly and would require intervention. Thus, since our goal is to improve reliability, we would need to establish the relationship between MTBF and $v(t)$:

$$\text{MTBF} = \frac{1}{v(t)} \quad \text{for} \quad \beta < 1$$

Nonparametric Kernel Density Estimate

Let $f(x)$ be the unknown probability density function (failure model). We shall nonparametrically estimate $f(x)$ with the following:

$$\widehat{f}_n(x) = \frac{1}{nh} \sum_{i=1}^{n} K\left(\frac{x - X_i}{h}\right) \tag{54}$$

The effectiveness of $\widehat{f}_n(x)$ depends on the selection of the kernel function K and bandwidth h. The kernel density estimate is probably the most commonly used estimate and is certainly the most studied mathematically. It does, however, suffer from a slight drawback when applied to data from long-tailed distributions. The kernel density estimate was first proposed by Rosenblatt in 1956. Since then, intensive work has been done to study its various properties, along with how to choose the bandwidth.[12-17] The kernel function is usually required to be a symmetric probability density function. This means that K satisfies the following conditions:

$$\int_{-\infty}^{\infty} K(u)\, du = 1 \qquad \int_{-\infty}^{\infty} uK(u)\, du = 0 \qquad \int_{-\infty}^{\infty} u^2 K(u)\, du = k_2 > 0$$

Commonly used kernel functions are listed in Table 1.

These kernel functions can be unified and put in a general framework, namely, symmetric beta family, which is defined by the following:

$$K(u) = \frac{1}{\text{Beta}\left(\frac{1}{2}, \gamma + 1\right)} (1 - u^2)_+^{\gamma} \qquad \gamma = 0, 1 \ldots \tag{55}$$

where the subscript + denotes the positive part, which is assumed to be taken before the exponentiation. The choices $\gamma = 0, 1, 2$, and 3 lead to, respectively, the uniform, the Epanechnikov, the biweight, and triweight kernel functions. As noted in Marron and Nolan,[17] this family includes the Gaussian kernel in the limit as $\gamma \to +\infty$.

Properties of the kernel function K determine the properties of the resulting kernel estimates, such as continuity and differentiability. From the definition of kernel estimate, we can see that if K is a density function, that is, K is positive and its integral over the entire line is 1, then $\widehat{f}_n(x)$ is also a probability density function. If K is n times differentiable, then also $\widehat{f}_n(x)$ is n times differentiable. From Table 1, we observe that only the Gaussian kernel will result in everywhere differentiable kernel estimate. This is one reason why the Gaussian kernel is most popular.

If $h \to 0$ with $nh \to \infty$, and the underlying density is sufficiently smooth (f'' absolutely continuous and f''' being squarely integrable), then we have the following:

$$\text{bias}[\widehat{f}_n(x)] = \frac{1}{2} h^2 k_2 f''(x) + o(h^2) \tag{56}$$

Table 1 Intensity Functions for Some Commonly Used Distributions

Kernel	1. Form	2. Inefficiency
Uniform	$\frac{1}{2}I(\lvert u \rvert \leq 1)$	1.0758
Epanechnikov	$\frac{3}{4}(1 - u^2)I(\lvert u \rvert \leq 1)$	1.0000
Biweight	$\frac{15}{16}(1 - u^2)^2 I(\lvert u \rvert \leq 1)$	1.0061
Triweight	$\frac{35}{32}(1 - u^2)^3 I(\lvert u \rvert \leq 1)$	1.0135
Gaussian	$\sqrt{\frac{1}{2\pi}} \exp\left(-\frac{1}{2}u^2\right)$	1.0513

and

$$\text{Var}[\widehat{f}_n(x)] = \frac{f(x)R(K)}{nh} + o((nh)^{-1}) \tag{57}$$

where $R(K) = \displaystyle\int_{-\infty}^{\infty} K^2(u)\,du$ is used. The derivation of the conclusions can be found in Silverman.[12] Combining the two expressions and integrating over the entire line, we have the asymptotic mean integrated squared error (AMISE):

$$\text{AMISE}[\widehat{f}_n(x)] = \frac{1}{4}h^4 k_2^2 R(f'') + \frac{R(K)}{nh} \tag{58}$$

It can be seen from Eq. (58) that it depends on four basic quantities: namely, the underlying probability density function $f(x)$, the sample size n, the bandwidth h, and the kernel function K. The first two quantities are usually out of our control. But we can select the other quantities, the bandwidth and the kernel function, to make the $\text{AMISE}[\widehat{f}_n(x)]$ minimal.

First, we fix the kernel function and find the best bandwidth. That is, for any fixed K, setting $[\partial \text{AMISE}(\widehat{f}_n(x))]/\partial h$ equal to 0 yields the optimal bandwidth:

$$h_o = \left[\frac{R(K)}{k_2^2 R(f'')} \right]^{1/5} n^{-1/5}$$

The corresponding minimal AMISE is given by

$$\text{AMISE}_o = \frac{5}{4}[\sqrt{k_2}R(K)]^{4/5} R(f'')^{1/5} n^{-4/5}$$

To determine the optimal kernel function, we proceed by minimizing AMISE_o with respect to K, which is equivalent to minimizing $\sqrt{k_2}R(K)$. The optimal kernel function obtained was the Epanechnikov kernel given by $\frac{3}{4}(1 - u^2)I(|u| \leq 1)$. The value of $\sqrt{k_2}R(K)$ for the Epanechnikov is $3/(5\sqrt{5})$. Thus, the ratio $\sqrt{k_2}R(K)/3/(5\sqrt{5})$ provides a measure of relative inefficiency of using other kernel functions. For example, Table 1 gives such values of this ratio for the common kernel functions.

Thus, we can conclude that we can select any of these kernels and obtain almost equal effective results. Thus, K should be chosen based on other issues, such as ease of computation and properties of $\widehat{f}_n$. The Gaussian kernel possess these properties and thus is commonly used.

4 REPAIRABLE SYSTEMS AND AVAILABILITY ANALYSIS

In many classes of systems where corrective maintenance plays a central role, reliability is no longer the central focus. In the case of repairable systems (as a result of corrective maintenance), we are interested in

- The probability of failure
- The number of failures
- The time required to make repairs

Under such considerations, two measures (parameters) of system performance in the context of system effectiveness become the focus—namely, *availability* and *maintainability*. Other related effectiveness measures of significance include the following:

- Serviceability
- Reparability
- Operational readiness
- Intrinsic availability

Definitions

O_t = operating time

I_t = idle time (scheduled system free time)

d_t = downtime (includes administrative and logistics time needed to marshal resources, active repair time, and additional administrative time needed to complete the repair process)

a_t = active repair time

m_d = mean maintenance downtime resulting from both corrective and preventive maintenance times

MTTF = mean time to failure

MTBF = mean time between failures

MTBM = mean time between repair or maintenance

CM = corrective maintenance

PM = preventive maintenance

In general, most practitioners consider MTTF and MTBF to be identical. In a strict reliability sense, MTTF is used in reference nonreparable systems such as satellites. By contrast, MTBF is used in reference to reparable systems that can entertain multiple failures and repairs in the system's life cycle. For practical purposes, both measures of system performance are identical, and the use of the MTTF and MTBF is indistinguishable from that viewpoint.

4.1 Definition of Systems Effectiveness Measures

Serviceability
This is the ease with which a system can be repaired. It is a characteristic of the system design and must be planned at the design phase. It is difficult to measure on a numeric scale.

Reparability
This is the probability that a system will be restored to a satisfactory condition in a specified interval of active repair time. This measure is very valuable to management since it helps quantify workload for the repair crew.

Operational Readiness (OR)
The probability that a system is operating or can operate satisfactorily when the system is used under stated conditions. This includes free (idle) time. It is defined as follows:

$$OR = \frac{O_t + I_t}{O_t + I_t + d_t}$$

Maintainability
This is the probability that a system can be repaired in a given interval of downtime.

Availability
Availability $A(t)$ is defined as the probability that a system is available when needed or the probability that a system is available for use at a given time. It is simply the proportion of time that the system is in an operating state and it considers only operating time and downtime.

$$A(t) = \frac{O_t}{O_t + d_t} \tag{59}$$

Intrinsic Availability

Intrinsic availability is defined as the probability that a system is operating in a satisfactory manner at any point in time. In this context, time is limited to operating and active repair time. Intrinsic availability, A_I, is more restrictive than availability and hence is always less than availability. It excludes free or idle time. It is defined as follows:

$$A_I = \frac{O_t}{O_t + a_t} \tag{60}$$

Inherent Availability

Inherent availability, A_{IN}, is defined as the probability that the system is operating properly given corrective maintenance activities. It excludes preventive maintenance times, administratively mandated free or idle time, logistics support times, and administrative time needed to inspect and ready the system after repair. Inherent availability is defined as follows:

$$A_{IN} = \frac{\text{MTBF}}{\text{MTBF} - \text{MTTR}} \tag{61}$$

Achieved Availability

According to Elsayed,[18] achieved availability is defined as the measure of availability that considers both corrective maintenance and preventive maintenance times.

It is a function of the frequency of maintenance (CM or PM) as well as the actual repair or maintenance times. Functionally, it is defined as

$$A_a = \frac{\text{MTBM}}{\text{MTBM} + m_d} \tag{62}$$

As indicated earlier, availability $A(t)$ = probability that a system is performing satisfactorily at a given time. It considers only operating time and downtime. This definition refers to point availability and is often not a true measure of the system performance. By definition, point availability is

$$A(T) = \frac{1}{T} \int_0^T A(t)\, dt \tag{63}$$

This is the value of the point availability averaged over some interval of time T. This time interval may represent the design life of the system or the time to accomplish some mission. It is often found that after some initial transient effects, the point availability assumes some time-independent value. This steady state or asymptotic availability is given by

$$A^*(\infty) = \lim_{T \to \infty} \frac{1}{T} \int_0^T A(t)\, dt \tag{64}$$

If the system or its components cannot be repaired, then the point availability at time t is simply the probability that it has not failed between time 0 and t. In this case,

$$A(t) = R(t)$$

Substituting for $A(t) = R(t)$ in Eq. (64), we get the following:

$$A^*(\infty) = \lim_{T \to \infty} \frac{1}{T} \int_0^T R(t)\, dt$$

As $T \to \infty$,

$$\int_0^T R(t)\, dt = \text{MTTF}$$

Hence:

$$A^*(\infty) = \frac{\text{MTTF}}{\infty} = 0$$

Therefore, $A^*(\infty) = 0$

This result is quite intuitive given our assumption. Since all systems eventually fail, and if there is no repair, then the availability averaged over an infinitely long time is zero. This is the same reasoning for reliability at time infinity being zero.

Example

A constant failure rate system is being examined to determine its availability, given a desired maximum design life. As in most constant failure rate systems, the design life is being measured against its average life, or as a function of the system's MTTF. The design question is to determine the system availability in the case where the design life is specified in terms of the MTTF for a nonreparable system whose MTTF is characterized by a system intensity function that is constant: $v(t) = c$.

Assuming that the design life $T = 1.2$ (MTTF), what system availability will sustain this design life?

Solution

For a constant failure rate system, with no repairs,

$$v(t) = c = \lambda$$

hence $R(t) = e^{-\lambda t}$

$$A^*(T) = \frac{1}{T} \int_0^T e^{-\lambda t} \, dt = \frac{1}{\lambda T}(1 - e^{-\lambda T})$$

Expanding by Taylor series,

$$A^*(T) = \frac{1}{\lambda T}\left(1 - 1 + \lambda T - \frac{1}{2}(\lambda T)^2 + \cdots\right) \quad \text{for } \lambda T <<< 1$$

Hence

$$A^*(T) \approx 1 - \tfrac{1}{2}(\lambda T) = 1 - p$$

$$p = \tfrac{1}{2}(\lambda T) \Rightarrow \quad \text{hence} \quad 2p = \lambda T$$

But for $v(t) =$ constant, $\lambda = 1/\text{MTTF}$:

$$T = 1.2 \text{ MTTF}$$

Thus

$$2p = \frac{1.2\text{MTTF}}{\text{MTTF}} \Rightarrow p = 0.6$$

Hence $A^*(T) = 0.4$.

This says that the maximum availability that can be expected in the system for a design life that is 20% more than the expected average life of the system is about 40%.

4.2   Repair Rate and Failure Rate

In order to estimate availability, one must take into account the repair rate, which is typically larger than the failure rate for system stability. For example, a repair time of 5 h is equal to a rate of $\left(\frac{1}{5}\right) = 0.2$, whereas an MTTF of 400 h is equal to a rate of $\left(\frac{1}{400}\right) = 0.0025$.

If the underlying probability structure of the repair process can be characterized as having a constant intensity function, then the pdf of the repair process is the exponential, by definition. Thus, if

$$v(t) = \mu$$

$$m(t) = f(t) = \mu e^{\mu t}$$

where $m(t)$ is the maintenance density function, then the MTTR $= 1/\mu$.

4.3   Modeling Maintainability

Using the definition of the maintenance density function $m(t)$ as our base, we can define the maintainability function $M(t)$ as follows[10]:

$$M(t) = \int_0^t m(\tau) \, d\tau \tag{65}$$

The corresponding mean repair time, or mean time to repair (MTTR), is given by

$$\text{MTTR} = \int_0^\infty tm(t) \, dt \tag{66}$$

Earlier in the analysis and development of the intensity function [see Eq. (17)] we observed that the corresponding conditional probability that a unit of age t will survive for an additional time duration Δt is given by

$$\overline{F}(\Delta t \,|\, t) = \frac{\overline{F}(t \cap \Delta t)}{\overline{F}(t)} \quad \overline{F}(t) > 0$$

where $\overline{F}(t) = 1 - F(t)$, given that $F(t)$ is the probability of failure in time t. But by the definition of conditional probability:

$$\overline{F}(t \cap \Delta t) = \overline{F}(t + \Delta t) = \text{Total life of a unit up to and including } \Delta t, \text{ that is } (t + \Delta t)$$

Therefore,

$$\overline{F}\left(\frac{\Delta t}{t}\right) = \frac{\overline{F}(t + \Delta t)}{\overline{F}t} \tag{67}$$

Similarly, the conditional probability of failure during the interval of duration Δt is $F(\Delta t | t)$ where $F(\Delta t \,|\, t) = 1 - \overline{F}(\Delta t \,|\, t)$.

But

$$\overline{F}(\Delta t \,|\, t) = \frac{\overline{F}(t + \Delta t)}{\overline{F}(t)} \Rightarrow F(t) = 1 - \frac{\overline{F}(t + \Delta t)}{\overline{F}(t)} = \frac{\overline{F}(t) - \overline{F}(t + \Delta t)}{\overline{F}(t)} \tag{68}$$

$$v(t) = -\lim_{\Delta t \to 0} \left[\frac{\overline{F}(t + \Delta t) - \overline{F}(t)}{\Delta t} \right] \left(\frac{1}{\overline{F}(t)} \right)$$

$$= \left(\frac{-1}{\overline{F}(t)} \right) \frac{d}{dt} \overline{F}(t) = -\frac{1}{R(t)} \frac{d}{dt} R(t) \tag{69}$$

Analogously, let the time to repair a system, from the point of failure, be given by t'. Then for a small interval Δt, the probability that repair will be completed in the time interval $t + \Delta t$ is given by

$$m(t)\Delta t = P(t \le t' \le t + \Delta t) \Rightarrow m(t) = \frac{P(t \le t' \le t + \Delta t)}{\Delta t} \tag{70}$$

The conditional probability of repair in the interval $t + \Delta t$ is given by

$$P(t \cap \Delta t) = \frac{P(t \le t' \le t + \Delta t)}{P(t' \ge t)} \tag{71}$$

The conditional repair rate is given by

$$\frac{P(t \cap \Delta t)}{\Delta t} = \frac{P(t \le t' \le t + \Delta t)}{P(t' \ge t)\Delta t} \tag{72}$$

Hence

$$w(t) = \frac{P(t \le t' \le t + \Delta t)}{P(t' \ge t)\,\Delta t} = \frac{m(t)}{1 - M(t)}$$

$$= \frac{m(t)}{1 - M(t)} \Rightarrow m(t) = w(t)[1 - M(t)] \tag{73}$$

but

$$m(t) = \frac{d}{dt} M(t)$$

$$w(t) = \lambda(t) = \frac{(d/dt)[M(t)]}{1 - M(t)}$$

Integrating:

$$\int_0^t w(\tau)\, d\tau = \int_0^{M(t)} \frac{dM}{1 - M} \tag{74}$$

Note that

$$\int \frac{1}{1 - x}\, dx = -\ln(1 - x)$$

Thus,

$$\int_0^{M(t)} \frac{dM}{1 - M} = -\ln[1 - M(t)]$$

$$\ln[1 - M(t)] = \int_0^t w(\tau)\, d\tau \Rightarrow 1 - M(t) = \exp\left[-\left(\int_0^t w(\tau)\, d\tau\right)\right]$$

Hence,

$$M(t) = 1 - \exp\left[-\int_0^t w(\tau)\, dt\right] \tag{75}$$

Also, since $m(t) = w(t)[1 - M(t)]$, then

$$m(t) = w(t)\left(1 - \left\{1 - \exp\left[-\int_0^t w(\tau)\, dt\right]\right\}\right) = w(t) \exp\left[-\int_0^t w(\tau)\, dt\right]$$

Hence,

$$m(t) = w(t)\, \exp\left[-\int_0^t w(\tau)\, dt\right] \tag{76}$$

Example

Most human performance and human factors experts believe the nature of the maintenance activity, as well as all other activities performed by humans, makes them amenable to be modeled by the lognormal probability structure. Consider an aluminum hot-roll line whose maintenance density function from Eq. (6a) is given by

$$m(t) = \frac{1}{t\sigma\sqrt{2\pi}} \exp\left[-\frac{1}{2\sigma^2}(\ln t - \mu)^2\right]$$

where t is the downtime delay in days, $\mu = 2$ months, and $\sigma = 1$ month. Also, the time to failure (in months) is exponentially distributed as follows:

$$f(t) = \frac{1}{\theta} e^{-(t/\theta)}$$

where $\theta = $ MBTF $= 30$ days

    a. Find the maintainability function $M(t)$.

    b. Find the steady-state availability.

Solution

    a. Given:

$$m(t) = \frac{1}{t\sigma\sqrt{2\pi}} \exp\left[-\frac{1}{2\sigma^2}(\ln t - \mu)^2\right]$$

        Let

$$y = \ln(t)$$

Then

$$dy = \frac{1}{t}dt$$

Hence,

$$f(y) = m(y) = \frac{1}{\sigma\sqrt{2\pi}} \exp\left[-\frac{1}{2\sigma^2}(y-\mu)^2\right]$$

$$M(t) = \int_0^t m(\tau)\,d\tau = \int_0^t \frac{1}{\sigma\sqrt{2\pi}} \exp\left[-\frac{1}{2\sigma^2}(y-\mu)^2\right]dt = \Phi\left(\frac{y-\mu}{\sigma}\right)$$

$$= \Phi\left(\frac{\ln t - \mu}{\sigma}\right)$$

$$M(t) = \Phi\left(\frac{\ln t - 2}{1}\right) = \Phi(\ln\,t - 2)$$

$$\text{MTTR} = E(t) = E(e^y) = \int_{-\infty}^{\infty} e^y f(y)$$

$$= \int_{-\infty}^{\infty} \frac{1}{\sigma\sqrt{2\pi}} e^y \exp\left[-\left(\frac{1}{2}\right)\left(\frac{y-\mu}{\sigma}\right)^2\right] dy$$

$$E(t) = \exp\left(\mu + \frac{\sigma^2}{2}\right) \int_{-\infty}^{\infty} \frac{1}{\sigma\sqrt{2\pi}} \exp\left(-\left\{\frac{[y-(\mu+\sigma^2)]^2}{2\sigma^2}\right\}\right) dy$$

But:

$$\int_{-\infty}^{\infty} \frac{1}{\sigma\sqrt{2\pi}} \exp\left(-\left\{\frac{[y-(\mu+\sigma^2)]^2}{2\sigma^2}\right\}\right) dy = 1 \Rightarrow \text{Std normal}$$

Hence

$$E(t) = \exp\left(\mu + \frac{\sigma^2}{2}\right) \Rightarrow \text{MTTR} = \exp\left(\mu + \frac{\sigma^2}{2}\right) = \exp\left(\mu + \frac{\sigma^2}{2}\right)$$

with $\mu = 2$ days, $\sigma = 1$ day:

$$\text{MTTR} = \exp\left(2 + \frac{1}{2}\right) = 12.812 \text{ days}$$

b. For ease of calculation, inherent availability is asymptotically equal to the steady-state availability. Using Eq. (61), we can find the steady-state availability as follows:

$$A^*(t) = \frac{\text{MTBF}}{\text{MTBF} + \text{MTTR}} = \frac{30}{12.812 + 30} = 0.707 \approx 71\%$$

Example

Let the maintenance density function of a system be given by

$$m(t) = te^{-(t^2/2)}$$

Also, let the underlying failure probability structure for the system be given by the Weibull distribution with characteristic life $\theta = 72$ months, and the slope $\beta = 2$.

a. Find the maintainability function, $M(t)$.

b. Find the MTTR.

c. What is the system steady-state availability?

Solution

a. From Eq. (65), we can find the maintainability function.

$$M(t) = \int_0^t m(t)\, dt$$

$$= \int_0^t t e^{-t^2/2}\, dt$$

Let

$$u = \frac{t^2}{2}$$

then

$$du = -t\, dt$$

$$M(t) = \int_0^t -e^{-u}\, du = 1 - e^{-t^2/2}$$

b. Equation (66) gives the following:

$$\text{MTTR} = \int_0^\infty t m(t)\, dt = \int_0^t t^2 e^{-t^2/2}\, dt$$

Let

$$u = t e^{-t^2/2}$$

Applying integration by part gives

$$\text{MTTR} = \left[-t e^{-t^2/2} + \int e^{-t^2/2}\, dt \right]_0^\infty = \int_0^\infty e^{-t^2/2} dt$$

Since

$$\int e^{a^2 t^2}\, dt = \frac{\sqrt{\pi}}{2a} \quad \text{for} \quad a > 0$$

Let

$$a = \frac{1}{\sqrt{2}} \quad \text{then } a^2 = \frac{1}{2}$$

$$\int_0^\infty e^{-t^2/2}\, dt = \frac{\sqrt{\pi}}{2\frac{1}{\sqrt{2}}} = \sqrt{\frac{\pi}{2}} = 1.254 \text{ h}$$

c. Equation (61) gives us the steady-state availability:

$$A^*(s) = \frac{\text{MTTF}}{\text{MTTF} + \text{MTTR}}$$

Given $\theta = 72$ months and $\beta = 2$,

$$\text{MTTF} = \theta \Gamma \left(1 + \frac{1}{\beta} \right) = 72 \Gamma \left(1 + \frac{1}{2} \right)$$

From the gamma tables:

$$\Gamma \left(1 + \frac{1}{2} \right) = 0.8862$$

Thus MTTF $= 72 \times 0.8862 = 63.82$ months

$$A^*(s) = \frac{63.82}{63.82 + 1.254} = 0.98$$

4.4   Preventive versus Corrective Maintenance

Unscheduled downtime due to system failure can render a line inoperable for some time, leading to reduced production output. Both preventive and corrective maintenance (PM and CM, respectively) strategies can be employed to reduce system downtime. Preventive maintenance (with added predictive capability) is a proactive strategy that attempts to keep the system in operation with minimal or limited number of breakdowns. Its implementation, when combined with predictive maintenance, which depends on a good understanding of the probability structure of the underlying failure process, is employed in order to reduce the episodic occurrences of the failure event. By contrast, corrective maintenance is employed as a tool to return the system to an operating state once it has experience a catastrophic event leading to system failure.

Corrective maintenance (CM) programs are designed to reduce equipment repair cycle time by ensuring that a trained technician is available to address equipment maintenance requirements, with emphasis on lowest mean time to repair. Ordinarily, PM is scheduled periodically to coincide with specific and equal time intervals. This is consistent with the idea of constant failure rate for the intensity function. Under such circumstance, PM is of little value in enhancing system performance. Thus, it is common (as we shall show later) to argue that *preventive* maintenance alone does not lead to improvement in system availability or reliability. By contrast, *predictive* maintenance provides the capability to extend preventive maintenance to the realm of intervention under the condition of nonconstant intensity function. In this domain, the intervention intervals are no longer constant but are determined through a predictive platform that models the underlying probability structure of the failure process to provide the intervention epochs. The better the predictive capability, the better is the efficacy of preventive interventions, and the lesser the episodic occurrences that require corrective maintenance actions.

4.5   Reliability and Idealized Maintenance

Typically, reliability is an appropriate measure of system performance when a component, system, or unit does not undergo repair or maintenance. In the case where there is repair or maintenance, the system effectiveness measure used is availability. However, in the case of idealized PM, where maintenance returns the system to an as-good-as-new condition, then the reliability of the maintained system can be considered the same as typical system reliability, for obvious reasons. This, in essence, is the same as the reliability measure of a system where the failure rate or the intensity function is a constant. Thus, assume that $v(t) = c = \lambda, t > 0$. *Then, by definition*, we can find the reliability function:

$$R(t) = \exp\left[-\int_0^t v(\tau)\, d\tau\right] = e^{-\lambda t} \tag{77}$$

Now assume that for preventive maintenance, the intervention interval is denoted by T and that t is the running time of the system less any downtime. Now if we perform preventive maintenance to the system at time T as called for by the PM schedule, then the question becomes, "What is the effect of this idealized maintenance on the intensity function and hence the system reliability?" In other words, does it lead to improved reliability or does the reliability decrease? The answer to that question depends on our assumption about the nature of the maintenance activity.

The reason we want to frame this question in the context of the intensity function is that most of the data collected for the purpose of intervention are typically in terms of the intensity function or the failure rate. For idealized maintenance, which brings the system to as-good-as-new condition, the system does not have any memory of any accumulated wear effect because the condition of the system is as good as new. Thus, it is assumed that intervention does not fundamentally change the underlying structure of the failure process.

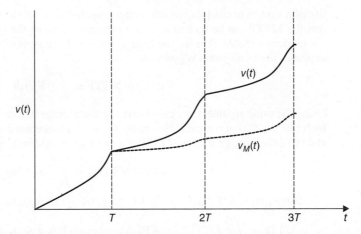

Figure 3   The effect of preventive maintenance on the failure rate/intensity function.

The graphs of the intensity function of an unmaintained system $v(t)$ and that of a maintained system $v_m(t)$ are shown on Figure 3. That is, for the maintained system at $t > T$,

$$v_M(t) = \lambda(t)$$

$$R_M(T) = \exp\left[-\int_0^T \lambda(t)\,dt\right] = R(t) \quad 0 < T < t$$

Define each maintenance epoch as an integer N, where $N = 0, 1, \dots, \infty$.

We will use the cumulative damage effect model to develop the resulting intensity function and hence the reliability function for the maintenance intervals.[1,20] The next maintenance epoch, beyond the first one at interval T, is denoted by the interval $T \le t \le 2T$. The intensity function $v_M(t)$ is given by

$$v_M(t) = v(T) + v(t - T), T \le t \le 2T \tag{78}$$

This is graphed in Fig. 3. Hence for all subsequent maintenance epochs, the intensity function will change to reflect the new system status. Thus, for system life cycle t, and assuming K equals the number of maintenance epochs during the life cycle, then the cumulative life-cycle intensity function is given by

$$v_M(t) = \sum_{K=0}^{\infty} [v(t - KT) + v(KT)] \quad KT \le t \le (K + 1)T \tag{79}$$

Correspondingly, we can determine the reliability of each maintenance epoch and hence the reliability of the maintained system using the cumulative intensity functions as follows:

$$R_M(t) = \exp\left(-\int_0^t v_M(t)\,dt\right) = \exp\left\{-\int_{KT}^{(K+1)T} \sum_{k=0}^{\infty}[v(t - KT) + v(KT)]\right\} \tag{80}$$

According to Lewis,[1] the reliability of a maintained system with equal intervention intervals T is given by the following:

$$R_M(t) = R(T)^K R(t - KT) \tag{81}$$

where $KT \le t \le (K + 1)T$, and $N = 0, 1, \dots, \infty$.

As shown previously, the MTTF for a system is defined as the average life of that system. The system reliability is really a measure of the life performance of the system. The average life of a system or random variable is the *expected value* of the life of that system. This means that the MTTF can be looked at as the expected value of the system. Thus, if we denote t as the random variable defining the time to failure of the system, then the expected value or the expected life of a system is given as

$$E(t) = \text{MTTF} = \int_0^\infty R(t)\,dt \tag{82}$$

Using the same argument, we can determine the average life, or MTTF, of a maintained system by replacing $R(t)$ of a regular system by $R_M(t)$ of a maintained system.[1] Hence from Eqs. (81) and (82), the average life, or MTTF, of a maintained system is given by the following:

$$E(t) = \text{MTTF}_M = \int_0^\infty R_M(t)\,dt$$

But $R_M(t) = R(T)^K R(t - KT)$, $KT \le t \le (K+1)T$ from Eq. (81). Thus

$$\text{MTTF} = \int_0^\infty R(T)^K R(t - KT)\,dt \quad \text{where } KT \le t \le (K+1)T \quad K = 0, 1, 2 \ldots, \infty$$

Let

$$t' = t - KT$$

when

$$t = KT \quad t' = 0$$

when

$$t = (K+1)T \quad t' = KT + T - KT = T.$$

Therefore,

$$\text{MTTF} = \sum_{K=0}^\infty R(T)^K \int_0^T R(t')\,dt'$$

But

$$\sum_0^\infty R(T)^K = \frac{1}{1 - R(t)} = \text{sum of an infinite series}$$

How? Let

$$\sum_0^\infty R(T)^K = Q$$

Then

$$Q = 1 + R(t) + R(t)^2 + R(t)^3 + \cdots + R(t)^K \tag{83}$$

$$R(t)Q = R(t) + R(t)^2 + R(t)^3 + \cdots + R(t)^{K+1} \tag{84}$$

Subtracting Eq. (84) from (83), we get the following:

$$Q - R(t)Q = 1 - R(t)^{K+1}$$

$$Q(1 - R(t)) = 1 - R(t)^{K+1}$$

Thus

$$Q = \frac{1 - R(t)^{K+1}}{1 - R(t)}$$

As

$$K \to \infty \Rightarrow Q = \frac{1 - 0}{1 - R(t)} \Rightarrow \sum_0^\infty R(T)^K = \frac{1}{1 - R(t)}$$

Thus

$$\text{MTTF} = \frac{\int_0^T R(t)\,dt}{1 - R(t)} \tag{85}$$

The MTTF for two important lifetime distributions, namely, the exponential and the Weibull, will be examined next to provide some insight as to how to manage the system under maintenance.

Exponential Life

For the exponential, the average life in the presence of maintenance intervention is given by

$$\text{MTTF}_{\text{Exponential}} = \frac{\int_0^T R(t)\,dt}{1 - R(T)} = \frac{\int_0^T e^{-(t/\theta)}\,dt}{[1 - e^{-(T/\theta)}]}$$

$$\int_0^T e^{-(t/\theta)}dt = -\theta[e^{-(t/\theta)}]\big|_0^T = -\theta[e^{-(T/\theta)} - 1] = \theta[1 - e^{-(T/\theta)}] \tag{86}$$

$$\text{MTTF}_{\text{PM}} \text{ (Exponential)} = \theta \left[\frac{1 - e^{-(T/\theta)}}{1 - e^{-(T/\theta)}}\right] = \theta = \text{MTTF}$$

Thus, in the case of the exponential, the MTTF is the same regardless of whether there is preventive maintenance. This is in agreement with our earlier notion that in the case of constant intensity function or failure rate, PM does not lead to any improvement in system performance.

Weibull Life

Considering a hot-roller with underlying failure distribution as the Weibull and life cycle of 15 years (i.e., 180 months, characteristic life $\theta = 72$ months, and slope $\beta = 2$), we want to compute the MTTF_{PM}(Weibull) given by the following equation:

$$\text{MTTF}_{\text{PM}} \text{ (Weibull)} = \frac{\int_0^T R(T)}{1 - R(T)} = \frac{\int_0^T e^{-(t/\theta)}\,dt}{1 - R(T)} = \frac{R_{\text{MTTF}}}{1 - R(T)} \tag{87}$$

Since the numerator (or R_{MTTF}) cannot easily be computed in closed form, numerical integration based on the Simpson's rule is used. The maintenance epochs (T) were considered to be fractions of the hot-roller's life cycle. The MTTF was then calculated by estimating the numerator using the numerical approximation for different planned maintenance epochs, as shown in Table 2.

The result presented in Fig. 4 is intuitive, that is, as the maintenance epochs' planning horizon increases, the MTTF exponentially decreases.

4.6   Imperfect Maintenance

Imperfect maintenance occurs mainly due to maintenance flaws induced by the human operator or, in the case of autonomous maintenance, by defective material or system architecture. The impact of imperfect maintenance is that the system could fail right after it has been repaired or it can fail during the process of repair. In recognition of the fact that this possibility exists, we will model the system reliability under imperfect maintenance by introducing the probability of system breakdown caused by a minute probability of failure.

Again, define the system reliability in the presence of maintenance as given in Eq. (81) as

$$R_M(t) = R(T)^K R(t - KT), KT < t < (K + 1)T$$

Table 2 MTTF for Different Maintenance Epochs

LC	f (fraction)	$T = f(\text{LC})$	$\int_0^T e^{(t/\theta)^\beta}\, dt$	MTTF
180	0.033	6	4.9266	726.3031
180	0.067	12	10.7731	401.1182
180	0.100	18	16.4628	277.0683
180	0.133	24	21.925	212.4933
180	0.167	30	27.98	173.1876
180	0.200	36	31.9312	146.9352
180	0.233	42	36.3857	128.3027
180	0.267	48	40.4357	114.5087
180	0.300	54	44.0683	103.9829
180	0.333	60	47.2825	95.7718
180	0.367	66	50.0881	89.2625
180	0.400	72	52.5039	84.043
180	0.433	78	54.556	79.8251
180	0.467	84	56.2756	76.4007
180	0.500	90	57.6972	73.6151
180	0.533	96	58.8564	71.3498
180	0.567	102	59.7891	69.5121
180	0.600	108	60.5292	68.0276
180	0.633	114	61.1087	66.8356
180	0.667	120	61.5562	65.8856
180	0.700	126	61.8972	65.135
180	0.733	132	62.1535	64.548
180	0.767	138	62.3435	64.0939
180	0.800	144	62.4825	63.7468
180	0.833	150	62.5828	63.4848
180	0.867	156	62.6541	63.2897
180	0.900	162	62.7043	63.1263
180	0.933	168	62.739	63.0425
180	0.967	174	62.7627	62.9684
180	1.000	180	62.7787	62.9163

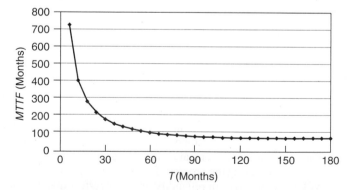

Figure 4 Graph of MTTF vs. T (maintenance epochs).

Now with the added possibility of failure, we define the probability of such minute failure occurrence as p, then the reliability of nonfailure is $(1 - p)$. Thus, for the system under imperfect maintenance, the reliability is given by the probability that the system will survive multiplied by the probability that the repair will not lead to failure:

$$R_M(t) = R(T)^K R(t - KT)(1 - p)^K \quad KT < t < (K + 1)T, \; p << 1 \tag{88}$$

Thus, the MTTF is given by the following:

$$\text{MTTF} = \int_0^\infty R_M(t) = \sum_{K=0}^\infty \int_{KT}^{(K+1)T} R(T)^K R(t - KT)(1 - p)^K$$

$$KT < t < (K + 1)T, P << 1 \tag{89}$$

But

$$(1 - p)^K \approx e^{-KP} \text{ for } p <<< 1$$

Thus

$$\text{MTTF} = \sum_0^\infty e^{-Kp} R(T)^K \int_0^T R(\tau) \, d\tau$$

where $\tau = t - KT$.
The series

$$\sum_0^\infty e^{-Kp} R(T)^K \approx \frac{e^{-Kp}}{1 - R(T)}$$

Thus,

$$\text{MTTF} = \frac{e^{-Kp} \int_0^T R(t) \, dt}{1 - R(T)} \tag{90}$$

For the Exponential

$$\text{MTTF} = e^{-Kp} \theta$$

Since the constant e^{-Kp} is always less than 1, the MTTF in the case of imperfect maintenance is less than for perfect maintenance. Although this is not a great revelation, it is important to note that every effort should be made to mitigate the errors of omission or commission that typically drive up the value of p. Table 3 demonstrates the dominant effect of p, given both p and K on the value of MTTF for imperfect maintenance.

4.7 Modeling Availability

Consider a two-state system classified as working (operational/available) or not working (non-operational/failed). In terms of availability, the probabilities of the two states can be defined as $A(t)$, and $\widetilde{A}(t)$, respectively, at any given time t, and where λ is defined as the failure rate and μ is the repair rate. The system initial conditions are given thus:

$$A(t) = 1, \widetilde{A}(t) = 0 \quad \text{and} \quad A(t) + \widetilde{A}(t) = 1$$

By using the differential equation approach, it is fairly straightforward to develop the equation for availability.

Assume that we are interested in looking at the system operation during the small time interval Δt given that the system has been in operation during time t. Now consider the change

Table 3   Effect of K and p on the MTTF for Imperfect Maintenance

K	p	Kp	e^{-Kp}
1	0.001	0.001	0.9990005
2	0.001	0.002	0.998002
3	0.001	0.003	0.9970045
10	0.001	0.01	0.99004983
40	0.001	0.04	0.96078944
1	0.005	0.005	0.99501248
2	0.005	0.01	0.99004983
3	0.005	0.015	0.98511194
10	0.005	0.05	0.95122942
40	0.005	0.2	0.81873075
1	0.1	0.1	0.90483742
2	0.1	0.2	0.81873075
3	0.1	0.3	0.74081822
10	0.1	1	0.36787944
40	0.1	4	0.01831564
1	0.5	0.5	0.60653066
2	0.5	1	0.36787944
3	0.5	1.5	0.22313016
10	0.5	5	0.00673795
40	0.5	20	2.0612E-09

in system availability $A(t)$ between the time t and $t + \Delta t$. There are two possibilities that affect the system state:

1. $\lambda \, \Delta t$ is the conditional probability of failure during t, given that the system was available at time t.

2. $\mu \, \Delta t$ is the conditional probability that the system has been repaired in the time interval of Δt given system failure.

Thus

$$P[\text{system failure during } \Delta t] = \lambda \, \Delta t$$

$$P[\text{repair during } \Delta t | \text{system failure}] = \mu \, \Delta t$$

Thus probability that the system is available at time $t + \Delta t$ is given by

$$A(t + \Delta t) = \{A(t)[1 - \lambda \, \Delta t]\} + \{[1 - A(t)]\mu \, \Delta t\} \tag{91}$$

This, in essence, means that the availability of the system in the interval $t + \Delta t$ is made of two components:

- The system was available in the time t and did not fail in the interval $t + \Delta t$.

  or

- It failed during Δt with probability $[1 - A(t)]$ and was thus repaired with probability $\mu \, \Delta t$.

Hence

$$A(t + \Delta t) = A(t)[1 - \lambda \Delta t] + [1 - A(t)]\mu \, \Delta t$$
$$= A(t) - \lambda A(t)\Delta t + \mu \, \Delta t - \mu A(t) \, \Delta t$$

Thus

$$\frac{A(t + \Delta t) - A(t)}{\Delta t} = -(\lambda + \mu)A(t) + \mu$$

As $\Delta t \to 0$,

$$\frac{A(t + \Delta t) - A(t)}{\Delta t} = \frac{d}{dt}[A(t)]$$

$$\frac{d}{dt}A(t) = -(\lambda + \mu)A(t) + \mu$$

$$\frac{d}{dt}A(t) + (\lambda + \mu)A(t) = \mu$$

Using the method of integration factor (IF), where

$$\text{IF} = \exp\left[\int (\mu + \lambda)\, dt\right] = e^{(\mu + \lambda)t}$$

$$\frac{d}{dt}[A(t)e^{(\mu + \lambda)t}] = \mu e^{(\mu + \lambda)t} \Rightarrow A(t)e^{(\mu + \lambda)t} = \mu \int e^{(\mu + \lambda)t}$$

$$A(t)e^{(\mu + \lambda)t} = \frac{\mu}{(\mu + \lambda)} e^{(\mu + \lambda)t} + C$$

where C is the integration constant. At $t = 0, A(0) = 1$.

Hence

$$C = 1 - \frac{\mu}{(\mu + \lambda)} = \frac{\lambda}{(\mu + \lambda)}$$

Therefore

$$A(t) = \frac{\mu}{(\mu + \lambda)} + \frac{\lambda}{(\mu + \lambda)} e^{(\mu + \lambda)t}$$

Thus the solution is given by

$$A(t) = \frac{\mu}{\mu + \lambda} + \frac{\lambda}{\mu + \lambda} e^{-(\mu + \lambda)t} \tag{92}$$

$$A^*(t \to \infty) = \lim_{t \to \infty} A(t) = \frac{\mu}{\lambda + \mu} \tag{93}$$

Define the following:

$$\text{Mean repair rate} = \text{MRR} = \mu$$

$$\text{Mean failure rate} = \text{MFR} = \lambda$$

$$\text{Also MTTF} = 1/\lambda, \text{MTTR} = 1/\mu$$

Then

$$A^*(t \to \infty) = \frac{\text{MRR}}{\text{MRR} + \text{MFR}} = A(t) \tag{94}$$

where t is the time interval.

Rewriting,

$$\frac{\mu}{\lambda + \mu} \quad \text{as} \quad \frac{1/\lambda}{(1/\mu + 1/\lambda)} = \frac{1}{\lambda}\left(\frac{1}{1/\mu + 1/\lambda}\right)$$

$$\frac{1}{\lambda}\left(\frac{1}{1/\mu + 1/\lambda}\right) = \frac{1}{\lambda}\left[\frac{1}{(\lambda + \mu)/\lambda\mu}\right] = \frac{\mu\lambda}{\lambda}\left(\frac{1}{\lambda + \mu}\right) = \frac{\mu}{\lambda + \mu}$$

Thus:

$$A^*(t \to \infty) = \frac{\mu}{\lambda + \mu} = \frac{\frac{1}{\lambda}}{(1/\mu + 1/\lambda)}$$

$$= \frac{\text{MTTF}}{\text{MTTR} + \text{MTTF}} = \text{interval availability} \tag{95}$$

Availability Computation Using Empirical Data

Example

An aluminum hot-roll line is used to reduce aluminum ingot through several processes to a final configuration for making aluminum cans. The line is prone to failure, and the plant engineer is interested in determining the availability of the line versus the design recommendations, as per the product end of manufacturing (EOM). Typically, the EOM stipulations aimed at value of $A(t)$ should be about 70% of the specified value, based on the complexity of the line and the fragile nature of the final product. Table 4 represents the times (in hours) over a 30-day period during which aluminum hot-roll line was down at time (td_i) and up at time (tu_i)—that is, repaired and put back into operation. The repair process is sensitive to the line content and so must be done rather quickly, or the product will have to be discarded. Determine the following:

a. MTTF

b. MTTR

Table 4   Availability Estimates Using Empirical Estimates of MTTF, MTTR, and Downtimes

i	t_{d_i}	CM downtime $d_i = (t_{u_i} - t_{d_i})$	t_{ui}	Uptime before failure $u_i = t_{d_i} - t_{u_i}$
0	0	0	0	0
1	80.80	5.24	86.04	80.80
2	99.00	2.07	101.06	12.95
3	107.02	5.34	112.36	5.96
4	124.89	6.07	130.96	12.53
5	136.82	5.89	142.71	5.86
6	155.11	7.93	163.05	12.40
7	176.06	4.53	180.59	13.01
8	203.99	5.75	209.74	23.40
9	225.51	1.70	227.20	15.77
10	239.53	1.99	241.52	12.33
11	249.82	4.21	254.03	8.30
12	299.66	4.45	304.11	45.63
13	322.67	3.77	326.44	18.56
14	390.34	3.09	393.43	63.90
15	424.93	3.46	428.39	31.50
16	430.75	3.69	434.45	2.36
17	456.36	5.29	461.65	21.92
18	474.25	6.21	480.46	12.60
19	509.19	4.69	513.87	28.73
20	516.86	3.55	520.41	2.98
21	534.10	2.85	536.95	13.69
22	544.78	4.05	548.82	7.82
23	557.20	2.33	559.53	8.37
24	565.27	5.56	570.83	5.74
25	596.56	5.90	602.46	25.73
26	610.47	3.87	614.34	8.00
27	631.04	4.33	635.37	16.70
28	653.52	3.95	657.47	18.15
29	679.21	4.64	683.86	21.74
30	715.50	4.50	720.00	31.64
Total		**130.92**		**589.08**

c. The 30-day (720-h) availability

d. Interval availability, $A(t)$

Solution

The main assumption is that the hot-roller runs for 24 h a day, 7 days a week.

Total number of hours of operation for the line considering a 30-day horizon $= 24 \times 30 = 720$ h. During this time, there were 30 occurrences (n) of failures and repairs. The times (td_i) and (tu_i) will be used to estimate MTTR and MTTF.

Uptime $u_i = (t_{d_i} - t_{u_i})$, where $tu_0 = 0$

Downtime $d_i = (t_{u_i} - t_{d_i})$

a. $\text{MTTF}_{\text{CM}} = \dfrac{1}{n} \sum_{i=1}^{n} (td_i - tu_{i-1}) = \dfrac{1}{n} \sum_{i=1}^{n} u_i = \dfrac{589.08}{30} = 19.65 \text{ h}$

b. $\text{MTTR}_{\text{CM}} = \dfrac{1}{n} \sum_{i=1}^{n} (tu_i - td_i) = \dfrac{1}{n} \sum_{i=1}^{n} d_i = \dfrac{130.92}{30} = 4.36 \text{ h}$

Mission Length $= T = 720$ h

c. $A(t) = \dfrac{1}{T} \sum_{i=1}^{n} (td_i - tu_{i-1}) = \dfrac{1}{T} \sum_{i=1}^{n} u_i = \left(\dfrac{589.08}{720} \right) \times 100 = 81.82\%$

On the other hand,

$$\widetilde{A}(t) = \text{unavailability} = 1 - A(t) = 100 - 81.82\% = 18.18\%$$

d. $A(T) = $ Interval availability

$$A(T) = \frac{\text{MTTF}}{\text{MTTF} + \text{MTTR}} = \frac{\lambda}{\lambda + \mu} = \frac{1}{1 + \mu/\lambda} = \frac{1}{1 + \text{MTTR}/\text{MTTF}}$$

$$= \frac{1}{1 + 4.36/19.65} = 81.81\%$$

4.8   Coherent Functions and System Availability

A coherent system structure ensures that only those system structures function that are sensible or reasonable, so that there are no irrelevant components. We can use the idea of coherent system structures to analyze system reliability or availability. For example, the reliability of a pure parallel configuration can be defined as

$$R = P(x_1 + x_2 + \cdots + x_n)$$

where x_i represents the event that path i is successful or operational. Given that

$$P(A_1 + A_2) = P(A_1) + P(A_2) - P(A_1 A_2)$$

for no mutually exclusive events, then

$$R = [P(x_1) + P(x_2) + \cdots + P(x_n)]$$
$$- [P(x_1 x_2) + P(x_1 x_3) + \cdots + P_{i \neq j}(x_i x_j)] + \cdots + (-1)^{n-1} P(x_1 x_2 \cdots x_n) \qquad (96)$$

This can also be expressed first in terms of system unreliability, where

$$R(t) = 1 - \prod_{i=1}^{n} (1 - R_i(t))$$

Thus, one can define the reliability of the system in terms of a coherent function using the system structure function developed in Eq. (96) earlier. For example, for an n component system, the system reliability using the coherent function structure approach can be defined as

$$R(t) = \phi \{ R_1(t) + R_2(t) \cdots + R_n(t) \} \qquad (97)$$

where ϕ is the coherent operator of the system. In the case of a three-component system, this is equal to

$$\phi(x) = 1 - [(1 - x_1)(1 - x_2)(1 - x_3)]$$
$$= x_1 + x_2 + x_3 - x_1 x_2 - x_1 x_3 - x_2 x_3 + x_1 x_2 x_3 \qquad (98)$$

Similarly, the system availability for the system can be defined as a coherent function defined as

$$A(t) = \phi\{A_1(t) + A_2(t) + \cdots + A_n(t)\} \qquad (99)$$

Example

Suppose the hot-roll line has three identical conveyor belts that operate in parallel such that they each feed into a washer. Each conveyor fails independently. The underlying probability structure for the intensity function for the time to failure of the conveyor belts is constant with an average life of 60 days. The system is run such that the probability of all three belts failing at the same time is very negligible. In addition, when a conveyor belt fails, the other belts will continue to feed the washers while the failed one(s) is (are) repaired. Assuming that the maintenance density function is given by

$$m(t) = \mu e^{-\mu t}$$

where μ = mean repair rate
           MTTR = 2 days

a. What is the steady-state availability of this conveyor system?

b. What is the probability that the conveyor system will fail during the production period of 30 days?

Solution

a. $A^*(t) = \dfrac{\mu}{\lambda + \mu}$

Since the conveyor belts are identical, then

$$A_i = \frac{\mu_i}{\lambda_i + \mu_i} \Rightarrow A_1 = A_2 = \frac{\mu}{\lambda + \mu}$$

But

$$A_{\text{System}}(t) = x_1 + x_2 + x_3 - x_1 x_2 - x_1 x_3 - x_2 x_3 + x_1 x_2 x_3$$

For steady state,

$$A_{\text{System}} = \left(\frac{3\mu}{\lambda + \mu}\right) - 3\left(\frac{\mu}{\lambda + \mu}\right)^2 + \left(\frac{\mu}{\lambda + \mu}\right)^3$$

$$= \frac{3\mu}{\lambda + \mu} - \frac{3\mu^2}{(\lambda + \mu)^2} + \frac{\mu^3}{(\lambda + \mu)^3}$$

$$A_{\text{System}} = \frac{\mu}{(\lambda + \mu)^3}[3(\lambda + \mu)^2 - 3\mu(\lambda + \mu) + \mu^2]$$

$$\mu = \frac{1}{2} \quad \lambda = \frac{1}{60} \Rightarrow u = 0.5 \quad \lambda = 0.01667$$

$$A_{\text{System}} = \frac{0.5}{(0.5 + 0.01667)^3}[3(0.516667)^2 - 3(0.5)(0.51667) + (0.5)^2]$$

$$A_{\text{System}}(t) = 0.999$$

b.
$$A_i(t) = \frac{\mu}{\lambda + \mu} + \frac{\lambda}{\lambda + \mu} e^{-(\lambda + \mu)t}$$
$$= A_1(t) = A_2(t) = A_3(t)$$
$$A_i(30) = \frac{0.5}{0.01667 + 0.5} + \frac{0.01667}{0.01667 + 0.5} e^{-(0.51667)(30)}$$
$$A_i(30) = \frac{0.5}{0.516667} + \frac{0.01667}{0.516667} e^{-(15.5)}$$
$$= 0.9677$$
$$A_{\text{System}}(30) = 1 - [1 - 0.9677]^3$$
$$= 0.9999$$

REFERENCES

1. E. E. Lewis, *Introduction to Reliability Engineering*, 2nd ed., Wiley, New York, 1994.
2. J. Huang, Preventive Maintenance Program Development for Multi-unit System with Economic Dependency—Stochastic Modeling and Simulation Study, Ph.D. Dissertation, University of South Florida, Tampa, Florida, 1993.
3. X. Zheng and N. Fard, "A Maintenance Policy for Repairable Systems based on Opportunistic Failure-rate Tolerance," *IEEE Trans. Rel.*, **40**, 237–244 (June 1991).
4. X. Zheng, and N. Fard, "Hazard-Rate Tolerance Method for an Opportunistic Replacement Policy," *IEEE Trans. Rel.*, **41**(1), 13–20, March, 1992.
5. O. G. Okogbaa and X. Peng, "A Methodology for Preventive Maintenance Analysis under Transient Response," Proceedings of 1996 Annual Reliability and Maintenance Symposium, Las Vegas, Nevada, pp. 335–340, 1996.
6. D. P. S. Sethi, "Opportunistic Replacement Policies," in C. P. Tsokos and I. N. Shimi (Eds.), *The Theory and Application of Reliability*, Academic, New York, 1977, pp. 433–447.
7. J. Moubray, *Reliability-Centered Maintenance*, 2nd ed., Industrial Press, NJ, South Norwalk, CT, 1997.
8. C. W. I. Watenhost and W. P. Groenendijk, "Transient Failure Behavior of Systems," *IMA J. Math. Appl. Business Ind.*, **3**(4), 1992.
9. C. P. Tsokos and H. Qiao, "Parameter Estimation of the Weibull Probability Distribution," *J. Math. Comput. Simul.*, **37**, 47–55, 1994.
10. C. P. Tsokos and H. Qiao, "Best Efficient Estimates of the Intensity Function of the Weibull Process," *J. Appl. Stat.*, **25**, 110–120, 1998.
11. M. S. Ross, *Introduction to Probability Models*, 8th ed., Academic, New York, 2003.
12. B. W. Silverman, *Density Function Estimation for Statistics and Data Analysis*, Chapman & Hall, London, 1986.
13. W. Hädle, *Smoothing Techniques with Implementation in S*, Springer, New York, 1991.
14. D. W. Scott, *Multivariate Density Estimation: Theory, Practice, and Visualization*, Wiley, New York, 1992.
15. M. P. Wand and M. C. Jones, *Kernel Smoothing*, Chapman and Hall, London, 1995.
16. J. S. Simonoff, *Smoothing Methods in Statistics*, Springer, New York, 1986.
17. J. S. Marron and D. Nolan, "Automatic Smoothing Parameter Selection: A Survey," *Empirical Econ.*, **13**, 187–208; "Canonical Kernels for Density Function," *Stat. Prob. Lett.*, **7**, 195–199, 1988.
18. A. E. Elsayed, *Reliability Engineering*, Addison Wesley Longman, New York, 1996.

BIBLIOGRAPHY

T. Aven, "Optimal Replacement under a Minimal Repair Strategy—A General Failure Model," *Adv. Appl. Prob.*, **15**, 198–211, 1983.

R. Barlow and L. C. Hunter, "Optimum Preventive Maintenance Policies," *Oper. Res.*, **8**, 90–100, 1960.

F. Beichelt, "A Generalized Block-Replacement Policy," *IEEE Trans. Rel.*, **30**, 171–172, 1981.

M. Berg, "Optimal Replacement Policies for Two-Unit Machines with Increasing Running Costs—I," *Stoch. Process. Appl.*, **4**, 89–106, 1976.

M. Berg, "Optimal Replacement Policies for Two-Unit Machines with Increasing Running Costs—II," *Stoch. Process. Appl.*, **5**, 315–322, 1977.

M. Berg and B. Epstein, "A Modified Block Replacement Policy," *Naval Res. Logist. Quart.*, **23**, 15–24, 1976.

M. Berg and B. Epstein, "Comparison of Age, Block, and Failure Replacement Policies," *IEEE Trans. Rel.*, **27**, 25–29, 1978.

M. Berg and B. Epstein, "A Note on a Modified Block Replacement Policy for Units with Increasing Marginal Running Costs," *Naval Res. Logi. Quart.*, **26**, 157–160, 1979.

P. J. Boland, "Periodical Replacement with Minimal Repair Costs Vary with Time," *Naval Res. Logi. Quart.*, **29**, 541–546, 1982.

P. J. Boland and F. Proschan, "Periodical Replacement with Increasing Minimal Repair Costs at Failure," *Oper. Res.*, **30**, 1183–1189, 1982.

D. I. Cho and M. Parlar, "A Survey of Maintenance Models for Multi-unit Systems," *Eur. J. Oper.l Res.*, **51**, 1–23, 1991.

R. Cleroux, S. Dubuc, and C. Tilquin, "The Age Replacement Problem with Minimal Repair Costs," *Oper. Res.*, **27**, 1158–1167, 1979.

S. Epstein and Y. Wilamowsky, "An Optimal Replacement Policy for Life Limited Parts," *Oper. Res.*, **23**, 152–163, 1986.

N. Fard and X. Zheng, "Approximate Method for Non-Repairable Systems based on Opportunistic Replacement Policy," *Rel. Eng. Syst. Saf.*, **33**(2), 277–288, 1991.

J. Huang and O. G. Okogbaa, "A Heuristic Replacement Scheduling Approach for Multi-Unit Systems with Economic Dependency," *Int. J. Rel. Qual. Saf. Eng.*, **3**(1), 1–10, 1996.

N. Jack, "Repair Replacement Modeling over Finite Time Horizons," *J. Oper. Res. Soc.*, **42**(9), 759–766, 1991.

N. Jack, "Costing a Finite Minimal Repair Replacement Policy," *J. Oper. Res. Soc.*, **43**(3), 271–275, 1992.

V. Jayabalan and D. Chaudhuri, "Cost Optimization of Maintenance Scheduling for a System with Assured Reliability," *IEEE Trans. Rel.*, **41**(1), 21–25, 1992.

D. A. Kadi and R. Cleroux, "Replacement Strategies with Mixed Corrective Actions at Failure," *Comput. Oper. Res.*, **18**(2), 141–149, 1991.

K. C. Kapur and L. R. Lamberson, *Reliability in Engineering Design*, Wiley, New York, 1977.

P. L'Ecuyer and A. Haurie, "Preventive Replacement for Multi-component Systems: An Opportunistic Discrete-time Dynamic Programming Model," *IEEE Trans. Rel.*, **32**, 117–118, 1983.

L. M. Leemis, *Reliability Probabilistic Models and Statistical Methods*, Prentice-Hall, Englewood Cliffs, NJ, 1995.

V. Makis and A. K. S. Jardine, "Optimal Replacement of a System with Imperfect Repair," *Microelectr. Rel.*, **31**(2–3), 381–388, 1991.

V. Makis and A. K. S. Jardine, "Note on Optimal Replacement Policy under General Repair," *Eur. J. Oper. Res.*, **69**(1), 75–82, 1993.

T. Nakagawa, "Optimal Maintenance Policies for a Computer System with Restart," *IEEE Trans. Rel.*, **33**, 272–276, 1994.

P. D. T. O'Conner, *Practical Reliability Engineering*, 2nd ed., Wiley, New York, 1985.

O. G. Okogbaa, and X. Peng, "Loss Function of Age Replacement Policy for IFR Unit under Transient Response," The NSF Design and Manufacturing Grantee Conference, Seattle, WA, January 1977, 7–10.

O. G. Okogbaa and X. Peng, "Time Series Intervention Analysis for Preventive/Predictive Maintenance Management of Multiunit Systems," *IEEE Int. Conf. Syst., Man, Cybernet.*, **5**, 4659–4664, 1998.

K. Okumoto and E. A. Elsayed, "An Optimal Group Maintenance Policy," *Naval Res. Logist. Quart.*, **30**, 667–674, 1983.

S. Ozekici, "Optimal Periodic Replacement of Multi-component Reliability Systems," *Oper. Res.*, **36**, 542–552, 1988.

K. S. Park, "Optimal Number of Minimal Repairs before Replacement," *IEEE Trans. Rel.*, **28**, 137–140, 1979.

W. P. Pierskalla and J. A. Voelker, "A Survey of Maintenance Models: The Control and Surveillance of Deteriorating Systems," *Naval Res. Logist. Quart.*, **23**, 353–388, 1976.

K. Pullen and M. Thomas, "Evaluation of an Opportunistic Replacement Policy for A Two-Unit System," *IEEE Trans. Rel.*, **35**, 320–324, August, 1986.

S. H. Sheu, "General Age Replacement Model with Minimal Repair and General Random Cost," *Microelectr. Rel.*, **31**(5), 1009–1017, 1991.

S. H. Sheu, "Generalized Model for Determining Optimal Number of Minimal Repairs before Replacement," *Eur. J. Oper. Res.*, **69**(1), 38–49, 1993.

B. D. Sivazlian and J. F. Mahoney, "Group Replacement of a Multi-component System Which Is Subject to Deterioration Only," *Adv. Appl. Prob.*, **10**, 867–885, 1978.

C. Tilquin and R. Cleroux, "Block Replacement Policies with General Cost Structures," *Technometrics*, **17**, 291–298, 1975.

C. Valdez-Flore, and R. M. Feldman, "A Survey of Preventive Maintenance Models for Stochastically Deteriorating Single-Unit Systems," *Naval Res. Logi.*, **36**, 419–446, 1989.

H. C. Young and S. L. Chang, "Optimal Replacement Policy for a Warranted System with Imperfect Preventive Maintenance Operations," *Microelectr. Rel.*, **32**(6), 839–843, 1992.

CHAPTER 10

DESIGN FOR REMANUFACTURING PROCESSES

Bert Bras
Georgia Institute of Technology
Atlanta, Georgia

This chapter discusses qualitative design for remanufacturing guidelines. It provides an overview of the industry, including typical facility-level processes, so we can better understand the rationale for the given design guidelines. A distinction will be made between overarching guidelines and specific component hardware-oriented design guidelines.

1   INTRODUCTION TO REMANUFACTURING

Many are unfamiliar with the basics of remanufacturing. In this section, we will focus on what these are and why remanufacturing is done.

1.1   Definitions

Remanufacturing is both a new and old phenomenon. It is receiving a lot of attention nowadays from an environmental point of view (as evident from this book) but has been practiced for a very long time. In remanufacturing, a nonfunctional or retired product is made like new through a series of industrial operations.[1] A minimum definition is "bringing a product back to sound working order." Remanufacturing basically is the process of disassembly of products, during which time parts are cleaned, repaired, or replaced and then reassembled to sound working condition. In that context, the Remanufacturing Institute (www.reman.org) considers a product remanufactured if the following conditions are met:

- Its primary components come from a used product.
- The used product is dismantled to the extent necessary to determine the condition of its components.
- The used product's components are thoroughly cleaned and made free from rust and corrosion.
- All missing, defective, broken, or substantially worn parts are either restored to sound, functionally good condition or replaced with new, remanufactured, or sound, functionally good used parts.
- To put the product in sound working condition, such machining, rewinding, refinishing, or other operations are performed as necessary.
- The product is reassembled, and a determination is made that it will operate like a similar new product.

Remanufacturing is viewed differently from recycling in that the geometry of the product is maintained, whereas in recycling the product's materials are separated, ground, shredded, and molten for use in new product manufacture. Remanufacturing is viewed by many as a special form of recycling. The U.S. Code of Federal Regulations, for example, allows remanufactured products to be claimed as recyclable (see 16 CFR 260.7), provided that certain conditions for such claims are met. The German Engineering Standard VDI 2243 uses the term "product recycling" to denote product remanufacture in contrast to *material recycling*.[2] And the European End of Life Vehicle (ELV) Directive allows reuse to count as a form of recycling.[3]

More complicated are the differences between *reconditioned*, *repaired*, *refurbished*, and *remanufactured*. These terms are often used synonymously, but they convey different meanings, dependent on the audience. The U.S. Code of Federal Regulations, Title 16—Commercial Practices, Part 20, "Guides for the Rebuilt, Reconditioned and Other Used Automobile Parts Industry," Paragraph 3—states:

> It is unfair or deceptive to use the words "Rebuilt," "Remanufactured," or words of similar import, to describe an industry product which, since it was last subjected to any use, has not been dismantled and reconstructed as necessary, all of its internal and external parts cleaned and made rust and corrosion free, all impaired, defective or substantially worn parts restored to a sound condition or replaced with new, rebuilt (in accord with the provisions of this paragraph) or unimpaired used parts, all missing parts replaced with new, rebuilt or unimpaired used parts, and such rewinding or machining and other operations performed as are necessary to put the industry product in sound working condition.

Similarly, the word *repair* is often considered as bringing a product back to a basic functional condition by removal a single fault condition, whereas remanufacturing goes much deeper and brings the product back to almost or better than new condition.

1.2 Potential Benefits

The primary benefits of remanufacturing arise from a reuse of resources. In contrast to recycling, remanufacturing tries to retain the geometric shape of parts. Hence, the need for a material forming process, and associated energy expenditures, is reduced. This is why remanufacturing is often seen as a key strategy for long-term sustainable development, and proponents tout huge energy savings compared to traditional manufacturing processes.[4] Benefits to society include:

- Cheaper goods and hence a higher standard of living
- The creation of new jobs, since remanufacture is heavily labor intensive[5]

A more technical potential benefit is that remanufacturing has the capability of bringing used products back to equal or *better* than new condition. Many argue that it is impossible to substantiate this claim in general, due to the fact that wear will degrade products. Nevertheless, remanufactured products often contain (new) components that are better than the original components the product came with from the original assembly line. Bearings, bushings, and motors are just some examples of parts that have increased in performance, quality, or life expectancy due to better materials and manufacturing processes. Engines may receive better emission control or fuel injectors for increased efficiency. Even large U.S. Navy ships that have undergone refits are arguably better than new when we consider their weapon systems and engine upgrades (consider the battleship *U.S.S. Missouri*, which served in both World War II and the first Gulf War). In such cases, the phrase *upward remanufacturing* is used to denote the better-than-new state of the product after remanufacturing.

1.3 Size of Industry

In the late 1990s, approximately 73,000 U.S. firms sold an estimated $53 billion worth of remanufactured products.[6] This number does not include remanufacturing operations within the U.S. Department of Defense, arguably the largest remanufacturer, which constantly remanufactures military equipment (from combat equipment to radar systems) to extend the service life of these products.[7] Remanufactured automobile parts such as engines, transmissions, alternators, and starter engines accounted already for almost 10% of the total production in Germany.[2] In the 1980s, replacement parts for automobiles were the largest application of remanufacturing in the United States.[5] These reports did not include the now-thriving business of toner cartridge remanufacturing. Even household appliances and industrial hand tools are currently remanufactured by third parties and (in some cases) original equipment manufacturers (OEMs). Wherever a product can be cost effectively remanufactured and can be priced significantly lower than a new product, there is potential for remanufacturing. Even the electronics industry is considering reuse and remanufacture more than in the past.

The market and business potential for remanufacturing is even larger if leasing is considered.[8] Of the $668 billion spent by business on productive assets by U.S. businesses in 2003, $208 billion was acquired through leasing.[9] For General Electric alone, leasing is a $10 billion business. Several manufacturers have realized the benefits of remanufacturing their own products after a (first) lease. The Xerox Corporation had already saved around $200 million in 1991 by remanufacturing copiers returned at the expiration of their lease contracts. Fuji-Xerox integrates remanufactured components into "new" copiers wherever appropriate.[10,11] In addition to selling its products, Caterpillar is also leasing them ("selling miles"), which allows penetration into markets that before could not afford a Caterpillar product.[10,11] Caterpillar now has a remanufacturing division that also provides its services to other companies. Some aircraft engine manufacturers can sell engines below manufacturing cost because they have integrated maintenance and financing contracts as part of the engine sales. All these examples are products that are relatively expensive and are of high capital value. Kodak, however, is a classic example of a company that has created a fully integrated reuse and remanufacture strategy around its (inexpensive) reusable Funsaver camera line. A key aspect is that users want to return the camera because they need the film to be developed, allowing for recovery of the product. Toner cartridges are another example of low-cost products/components that are being remanufactured.

Remanufacturing is most viable for products that have a high replacement cost or have valuable components that can cost-effectively be reused or reconditioned. The practice of remanufacturing is most predominant in the U.S. Department of Defense and most visible in the automotive industry. Remanufacturing is (still) mostly performed by third-party remanufacturers because OEMs do not consider it a core business. More and more OEMs, however, recognize the value of remanufacturing, especially if business strategies move from selling products to leasing products and/or selling services where the OEM retains ownership of the product.

1.4 Consumer Demand

Although prices vary for remanufactured parts and products, they are typically significantly lower than equivalent new products. The lower price is the primary reason for consumer demand for remanufactured parts. Early studies showed that customers preferred remanufactured parts over new if remanufactured parts were offered at about 57% of the new-part price.[12] To offer remanufactured parts at such a discount price compared to new parts, there must be a significant price margin between the sales price of the remanufactured product and the price that the remanufacturer has to pay for the used product/core. Some remanufacturers prefer to remanufacture foreign parts, because they usually sell for more in the local market than domestic components.

Despite the price advantage, some customers are still wary of remanufactured parts because they feel that these "used" parts are inferior compared to new parts. Many remanufacturers, therefore, back their work with a warranty, although it often covers only the cost of replacement parts and not labor.

Many times, however, the consumer does not have a choice because new parts may simply not be available anymore, and there is no other option than remanufactured parts. Original equipment manufacturers also rely on remanufacturing for parts supply, allowing them to stock fewer new parts and allowing their suppliers to free up tooling in production facilities for production of parts for newer products. Especially in the automotive industry, the OEM parts supply relies heavily on OEM-sanctioned remanufacturing efforts.

2 BASIC REMANUFACTURING BUSINESS PRACTICE

Remanufacturing is widespread and covers many industries, but some general observations can be made about the industry that will help us understand how and where changes in product design can be beneficial.

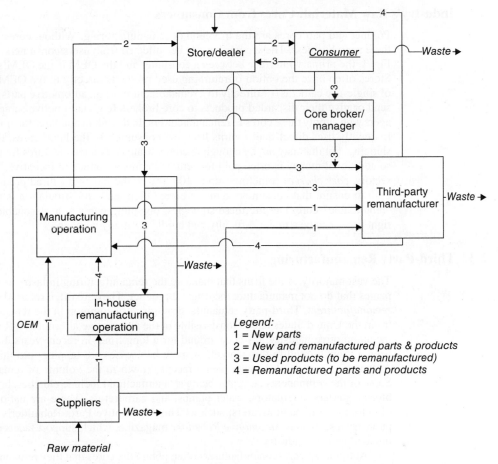

Figure 1 Simple schematic of possible part and product flows in remanufacturing industry.

2.1 Basic Business Scenarios

Different business practices exist with different combinations of actors. In Fig. 1, a schematic of possible product flows is shown between different actors in the remanufacturing business practice. Two basic scenarios exist:

1. OEM manufactures, sells (or leases), recovers, remanufactures, and resells products and parts.
2. OEM manufactures and sells products, but third-party actors independently capture, remanufacture, and resell the used products and parts.

The second scenario is the predominant business scenario, but the first scenario is gaining momentum in certain industries. A hybrid scenario—that is, direct collaboration between OEMs and third-party remanufacturers—is also possible and frequently seen in automotive parts remanufacturing.

2.2 Industry's Raw Material: Cores from Consumers

Product and part cores are the lifeblood of remanufacturing. Without cores to remanufacture, there is no business. These cores have to be collected from users/consumers. As can be seen in Fig. 1, the primary interface between a consumer and the OEM is the OEM's dealer or a store. Stores often have the option of returning used products directly to the OEM (e.g., in the case of single-use cameras) or directly to a remanufacturer (e.g., automotive parts). Consumers also send or give their discarded products to core brokers (e.g., automotive scrap-yards) and managers who supply these cores to remanufacturers. In the inkjet and laser toner industry, the OEM frequently uses direct-mail returns from the consumer. In the latter cases, the OEMs pay for shipping, but that may not be enough to compensate consumers or stores for the inconvenience so additional money is often paid for returned cores as an added incentive. In the automotive sectors, *core charges* sometimes up to 10–20% of the remanufactured product sales price are used to ensure that consumers return a core. Often, an extra stipulation is that it has to be a "rebuildable" core. Despite these strategies, obtaining cores at a reasonable cost (and at the right time) can be a major difficulty and hurdle for many companies.

2.3 Third-Party Remanufacturing

The vast majority of the firms that make up the remanufacturing industry are third-party companies that do not manufacture the original product. They are often referred to as *independent remanufacturers*. Third-party remanufacturing is very common in the remanufacturing industry in the United States and is most visible in the automotive aftermarket. Remanufacturing in the automotive industry has been around for a long time. In recent years, however, there has been a growth trend in this market, as more businesses are tapping into this profitable field. Automotive part retailers have seen a recent growth in the volume of remanufactured parts. Some of the components currently being remanufactured include clutches, brake shoes, engine blocks, starters, alternators, water pumps, and carburetors. There are national organizations of automotive remanufacturers, such as The Automotive Parts Rebuilder's Association, and publications, such as *Automotive Rebuilder* magazine, which support facets of the automotive remanufacturing industry.

Automotive part remanufacturers often obtain the cores that they remanufacture from core suppliers/brokers, from salvage yards, or by one-for-one core exchanges from stores and dealers. Parts are often purchased from (specialized) suppliers or even directly from the OEM (see Fig. 1).

Larger organizations frequently disassemble and rebuild the parts in large batches, in an assembly line manner, in order to obtain economies of scale. Smaller businesses often remanufacture parts on an individual basis in order to maintain a lean inventory of diverse models. For example, several years ago at a local Atlanta alternator and starter remanufacturer, alternators and starters were being remanufactured in batches—several of the same alternators on a given day—in an assembly line style. However, the business practice was changed such that each employee would remanufacture one product at a time, from start to finish. This resulted in a much leaner inventory, as well as employees who could perform all facets of the remanufacturing process.

Most remanufacturers will disassemble a unit, clean all functional parts, add grease, sealants, or paint to protect them, replace all worn parts, refurbish the exterior, reassemble it, and test the reassembled unit. However, there is a wide range of quality levels when it comes to remanufactured components. Remanufacturers who are conscientious about quality will frequently use higher quality replacement parts than the original (and called for by specifications), repaint assemblies (prevents corrosion and grease build-up), and monitor the quality of replacement parts in order to prevent warranty returns as well as to build their reputation as a quality remanufacturer.

Third-party remanufacturers typically sell their products directly to replacement parts stores. In some cases, they are contracted by OEMs to remanufacture replacement parts and, in that case, they tend to act as suppliers for OEMs, which sell the remanufactured parts through their existing dealer networks.

2.4   OEM Remanufacturing

Despite a growing interest in remanufacturing, most products are being remanufactured more out of serendipity than by design. Entrepreneurs tend to take advantage of the intrinsic value in (used) products by collecting and remanufacturing them profitably. Hence, unless an OEM is doing the remanufacture, there is little incentive to design products for remanufacture because that would only make it easier for entrepreneurs to start remanufacturing these products and potentially take away market share from new(er) products or OEM replacement parts.

A crucial issue for any organization considering remanufacture is the economic incentive. OEMs entering the remanufacturing market can gain a competitive edge by using their design capacity to make their products easy to remanufacture. OEMs can also offer a remanufactured warranty equivalent to the original warranty. And they can use existing store/dealer networks and distribution networks for marketing and sales. The combination of these factors would give OEMs an edge. Most importantly, in order for OEMs to enter into remanufacturing ventures, a paradigm shift must occur whereby remanufacturing is an important design and business driver.[13] This shift has already occurred in certain industries. For example, Pratt & Whitney is involved in remanufacture of the aircraft engines it manufactures.[5] General Electric remanufactures transformers, control systems, and other components. Caterpillar recently offered its internal remanufacturing services to external parties and formed a new remanufacturing division for this purpose. One of the most cited examples is the Xerox Corporation, which has been involved in the practice of remanufacturing for many years. Xerox (and its global partners) developed a program whereby it reclaims its copiers at the end of their service life (or lease). Xerox then remanufactures its copiers by reusing parts and replacing worn-out parts and returning them to service.[14] It assembles remanufactured and new copying machines on identical assembly lines with identical quality standards. Xerox has worked extensively in the design of its products to make its copiers easy to disassemble, modify, and reassemble. Xerox can do this by using modular attachment methods and by standardizing the parts used in the copiers. This practice allows Xerox to develop a broad variety of products using a large base of modular parts. One hurdle that OEMs face when promoting remanufactured products is that a number of states in the United States have policies blocking or limiting the use of remanufactured parts in products that are sold or leased as new.

In general terms, to successfully integrate remanufacturing into their business, OEMs must focus on providing value to consumers at different levels[15]: (1) initial sale/lease (compete based on features, performance, and price); (2) performance-sensitive (early) reuse (technology is still relatively current, higher price, testing and refurbishment required); (3) price-sensitive (later) reuse (older technology, lower price, not necessarily refurbished); (4) service and support (replacement parts); (5) second market reuse (other industries find another use for goods); and (6) recycle materials (lowest economic value but landfill is avoided). Benefits to the producer include:

- If price elasticity is significant, their market share grows.
- The trade-in value encourages customer loyalty and repeat business.
- Data on product failures can be used to create product improvements.
- The existing product distribution system used by the OEM can be converted into a two-way street in which products' used cores are returned.[5]

Good discussions on the opportunities, barriers, and steps needed for increasing the roles of OEMs in remanufacture are given in Refs. 5, 13, and 16–18.

2.5   Customer Returns—A Driver for OEM Remanufacturing

Customer returns can be an important initial driver for OEMs to consider remanufacturing as a business strategy option. When a customer returns a defective product to a retail or online store, it is invariably sent back to the original supplier, who can then simply replace it with a new one and discard the defective product or repair or remanufacture the defective product. In many cases, however, customers return products for other reasons than defects. Mostly, they can be restocked immediately in the store, but they may also be sent back to the original supplier/manufacturer if, for example, the packaging is damaged or a suspicion of damage exists. Such returns are inspected, repackaged, and often resold as "factory refurbished" by the original manufacturer or its dealers. Consumer laws prevent such items from being labeled *new*, even though they may never have been used at all. It is estimated that about 4% of total logistics costs are spent on returns,[19] which equated to over $35 billion per year in the late 1990s in the United States alone. This has led to an emergence of specialized firms that focus on *reverse* logistics. Large consumer product manufacturers (with correspondingly large return volumes), such as Dell and Black & Decker, have dedicated refurbish, recondition, and/or remanufacturing centers handling their returns. Because of the reliance on manual labor, many of these operations have been moved to low-cost countries such as Mexico and China.

3   REMANUFACTURING FACILITY PROCESSES

The preceding sections offered a high-level overview of the remanufacturing industry and its actors. In this section, we descend to the facility level to illustrate a typical remanufacturing process. The discussion is primarily focused on mechanical products (e.g., automotive and manufacturing equipment) as the main example due to their widespread remanufacture.

3.1   Typical Facility-Level Remanufacturing Process

In order to understand how to *design* for remanufacturing, one needs to know the basic processes. Remanufacturing spans many industry sectors. As in manufacturing, no single uniform process exists. The following 12 processes, however, can be found in any remanufacturing facility:

1. Warehousing of incoming cores, parts, and outgoing products
2. Sorting of incoming cores
3. Cleaning of cores
4. Disassembly of cores and subassemblies
5. Inspection of cores, subassemblies, and parts
6. Cleaning of specific parts and subassemblies
7. Parts repair or renewal
8. Testing of parts and subassemblies
9. Reassembly of parts, subassemblies, and products
10. Testing of subassemblies and finished products
11. Packaging
12. Shipping

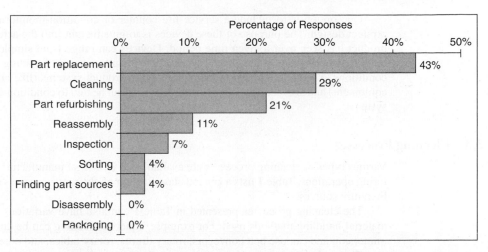

Figure 2 Most costly remanufacturing operations. *Source:* From Ref. 20.

Design for remanufacturing processes should be focused on reducing the cost, effort, and over-all resource expenditure of these processes by proper product design. In general, deeper insight is needed, however, and a detailed study of actual processes may need to be performed similar to design for manufacturing efforts. To give a flavor of what are critical-process issues, Hammond et al.[20] conducted a survey in which a number of automotive remanufacturers provided insight into their most costly remanufacturing operations. As shown in Fig. 2, part replacement tops the list, followed by cleaning and refurbishing (see Ref. 20 for more survey results). Although not exhaustive, such survey results are indicative of inherent product design problems. These processes, and associated issues, are discussed in more detail in the following sections.

3.2 Inspection and Testing Processes

Based on experience, technicians typically know what kind of failures are common to particular products and can estimate what it will take to remanufacture them. Most unforeseen variation costs are related to such items as hidden failures that have damaged major parts to the point that replacement of these parts would add too much cost to the component. Parts that have a damaged or broken core frequently cannot be remanufactured at all. In the case of starter and alternator remanufacture, there are usually one or two relatively inexpensive parts that cause the starter or alternator to fail. For example, certain alternators are installed very close to the exhaust manifold such that the bearing routinely fails because of the heat. Another type of alternator has a rectifier that consistently fails. Many remanufacturers have learned to recognize these defects and replace these parts every time, and by doing so, they can assure a quality product.

Material failures, if not at the surface, are often more difficult to inspect and require nondestructive inspection and testing methods. The American Society for Nondestructive Testing (ASNT) provides a good classification, overview, and primer of available nonde-structive testing (NDT) methods (www.asnt.org). Basically, NDT methods fall into six major categories: visual, penetrating radiation, magnetic–electrical, mechanical vibration, thermal, and chemical–electrochemical. Each has its specific purpose, advantages, and disadvantages. Depending on the product being remanufactured and its value, NDT may be required (e.g., aircraft parts). Sometimes, especially for low-value components, simple replacement (e.g., of bearings) may be preferred over the expense of testing.

One option is to include a service life counter or condition-monitoring device into the product design. The purpose of these devices is to give insight into the actual service that the product has seen over a given time period. Devices can range from simple page counters on copiers and printers to sophisticated systems that include sensing of loading and environmental conditions. This latter is used in many high-value product systems (like aircraft and military equipment) and is especially valuable if the devices can be tied to condition-based maintenance systems.

3.3   Cleaning Processes

Various types of cleaning processes are available for industrial manufacturing and remanufacturing operations. Table 1 lists a general classification of cleaning processes reported in various literature sources.

The cleaning processes presented in Table 1 can also have variations in application and material handling methods used. For example, aqueous cleaning can be done in spray and/or immersion and can use belt conveyors, rotary drums, rotary tables, and other methods.[21]

Thermal cleaning uses heat to loosen oil, grease, dirt, paint, adhesives, rust, and other contaminants from metal surfaces. Organic contaminants, such as oils and greases, oxidize and burn away. This is perhaps the most important benefit of thermal cleaning, because it eliminates the costs of disposal of these organic contaminants in liquid form that is typical of water-based washing operations. Blasting must be subsequently used, because the heat does not actually bake the foreign material off but dries it out, leaving ashes and surface oxides that require blasting to remove. Ovens to perform thermal cleaning have long operation cycle times, so batching is necessary. Thermal cleaning (using ovens), over the last decade of stricter environmental regulations, has become the cleaning method of choice for many remanufacturers of all sizes and product lines.[22] Plasma cleaning is an alternative that cleans surfaces by chemical reactions of activated gas species with surface contaminants, creating volatile products that will be pumped away.[23] The gases used for plasma (e.g., oxygen) are fairly inexpensive, accessible, and easy to handle. In oxygen plasma cleaning, contaminants are fully oxidized into CO, CO_2, and H_2O.[24] It leaves parts dry after treatment, is suitable for cleaning parts with complex shapes, and does not require any special safety measures. It also produces little or no waste.[25] However, it cannot clean inorganic materials or clean thick organic films effectively

Table 1   Classification of Cleaning Processes

Mechanical		Chemical	
Brushing	Wire	Aqueous	Acidic
	Fiber		Neutral
Steam-jet cleaning			Alkaline
Abrasive blasting	Sand	Solvent	Spray
	Metal shot		Immersion
	Silicon carbide		Vapor degreasing
	Plastic	Ultrasonic	
	Corncobs	Biological	
	Nut shells	CO_2	Liquid
	CO_2 pellets		Supercritical
	Ice pellets	Pickling	
Tumbling		Salt-bath	
Vacuum de-oiling		Thermal	Burn off
Scrubbing/wiping			Plasma

Figure 3 Part cleaning using ovens at Chinese remanufacturers.

or efficiently. For the latter reason, plasma cleaning is usually accompanied by wet cleaning to remove heavy soils from parts.[21] Clearly, no thermal cleaning process is possible when cleaning plastic components, lighter metal components, or heat-treated components. Figure 3 shows samples of (small) thermal cleaning ovens with cores that have been thermally cleaned as well.

Solvent-based chemical cleaning has become probably the least desirable technology for cleaning, especially if the solvents used are cataloged by the U.S. Environmental Protection Agency (EPA) as air, water, or land hazardous contaminants. For example, chemical degreasing operation using solvent-based solutions like perchloroethylene release volatile organic compounds (VOCs) into the atmosphere, which are major contributors to air pollution. The most benign chemical cleaning substances are aqueous-based solutions, and the trend in industry is to move away from solvent-based cleaning to aqueous cleaning that, if free of hazardous substances, does not pose an environmental problem for (re)manufacturers. Typical aqueous cleaning consists of a washing, rinsing, and drying step.[26] The cleaning capability generally comes from the temperature and/or chemistry of the water, whether it is acidic, alkaline, neutral, and/or emulsion,[27] and the mechanical form of application (e.g., immersion, spray, or both mechanisms). Additional differences include the following[21,28,29]:

- Emulsions can be chemically or mechanically induced.
- Immersion can be accompanied by part, liquid, and/or ultrasonic wave agitation.
- Spraying can vary by water pressure, spray configuration, and/or pattern.
- Drying can be carried out with steam, air, centrifuges, and/or vacuum.

Aqueous cleaning arose as a "greener" alternative to solvent cleaning because it does not use flammable, VOC-producing, ozone-depleting, and/or hazardous substances.[30] However, it can consume more water and energy and produce more wastewater and energy-related air emissions. The most efficient way to use water-based solutions is to have a closed-loop system that recycles the cleaning solution and reduces the need for make-up water and detergent. If the components to be cleaned have hazardous substances that will end up in the solution, such as heavy metals, a closed-loop system is the only option.

Biological cleaning combines aqueous emulsion cleaning with bioremediation.[27] Basically, oils and greases are removed by the emulsions, which are then consumed by bacteria in the bath. Since the cleaned oils and greases are consumed and oxidized into CO_2 and H_2O by the bacteria, the only waste is a sludge consisting of dead organisms.[31] Additional benefits include the ability to recycle surfactants (leading to a longer life for the cleaning solution), minimal downtime, and relatively small water, energy, and cleaning agent consumption.[27] A study conducted by the EPA on a biological cleaning system implemented by a manufacturing company found approximately 50% savings in water consumption, 80% savings in wastewater generation, and annual cost savings of $86,000 by the company due to the system.[32] Chemical CO_2

cleaning of metal parts is also possible and can be carried out in two ways: liquid or supercritical. In both cases, once the CO_2 dissolves the contaminants from the parts, the CO_2 can be easily separated and reused. The only waste produced with CO_2 cleaning is the contaminant itself, which can be easily treated, recycled, or disposed. However, it has high capital costs and raises safety concerns regarding the high-pressure components and potential increase in CO_2 levels in the worker area.[27]

Whether thermal cleaning or chemical cleaning is used, components also need to be abrasively cleaned to remove rust and scale as well as to improve surface finish and appearance. Abrasive cleaning is used by most remanufacturers to obtain like-new appearances, which is very important in this business. Most shops that remanufacture mechanical automotive parts (such as clutches, drive shafts, and engines) use airless centrifugal steel-shot abrasion technologies, whereas remanufacturers of electrical parts, such as starters and alternators, use air-blasting units with glass beads, aluminum oxide, and zinc oxide.[33] Airless centrifugal steel-shot abrasion technologies are self-destructive, which creates high maintenance and repair costs during the life of the equipment and induces higher work-in-process inventories due to downtime. In addition, they create noise and dust pollution. An alternative to using separate chemical and dry abrasive cleaning processes is to use wet blasting, which is a surface treatment process that performs simultaneous degreasing, deburring, surface cleaning, and descaling in one operation without the use of chemicals. Machines from, for example, Vapor-matt, use a recirculated water/abrasive media slurry with a high-volume vortex pump that feeds a nozzle or gun. Compressed air can be added at the nozzle to increase the effectiveness or aggressiveness of treatment to create a specific surface condition. A problem with abrasive cleaning technologies (e.g., using airless centrifugal steel shot) is that the technology can be too aggressive, and extra work is created when the core components are damaged.

One of the biggest problems with abrasive (and other) cleaning processes is that spent media will many times contain concentrations of heavy metals that exceed local and/or federal toxicity limits. Depending on the components being blasted, the threshold limit may be exceeded and spent media will need to be disposed of as hazardous waste. The administrative costs make hazardous waste disposal very costly.

Finally, abrasive cleaning shot retention plays a big role in the life of remanufactured products. For example, in the case of remanufactured engines,[22] 67% of engine wear occurs when it is first started after being remanufactured. The media left inside the engine can generate considerable wear. Shot-media retention is one of the biggest concerns for engine remanufacturers. Abraded parts that are not completely dry from a previous chemical cleaning process are especially prone to retaining shot media. Even compressed air blasting is susceptible to having wet media and subsequent shot retention, because of the moisture in the air lines. In high-humidity areas, moisture will cause a constant problem. For electrical components, when air blast media are propelled through the air, static electricity is generated, which can ruin electrical components. To prevent this problem, remanufacturers make sure their systems are completely grounded. One way remanufacturers deal with shot retention is by using tumblers and vibratory units that blow, shake, or wash out shot after the abrasion operation. One alternative to avoiding shot media retention is dry ice (CO_2) blasting. The dry ice dissipates into carbon dioxide gas after impacting the surface being abraded; thus only the heavy metals or contaminants are left over. Due to its expense, it is not used widely. Xerox has used it for cleaning copier parts.

3.4   Refurbishing Processes

In addition to simply replacing a damaged part (which is quite effective for cheap and commonly available parts), two basic strategies exist for refurbishing damaged or worn high-value and less-available parts: (1) cut out the damaged or worn area and/or (2) add (new) material to the damaged, worn, or cutout area. Often a combination is used in conjunction with smart selection of cheap replacement parts.

Cutting out a damaged or worn area is one of the most frequently used approaches, especially in refurbishing metal parts that have surface damage only. For example, damage to cylinder head surfaces (e.g., caused by cracked cylinder head gaskets) can often be removed by milling down the surface. The increase in compression ratio is often insignificant and/or can be offset by a slightly thicker head gasket. Damage to races in bearing housings can be removed by milling, turning, or other metal-appropriate cutting processes. The increased diameter can be offset by installing a new (replacement) bearing with a larger outer diameter. Here, the cost of the new bearing is much lower than the housing. Similarly, damage to crankshaft journals is often removed by turning processes and diameter differences are (again) offset through changes in bearing sizes. In this case, however, surface treatments typically also need to be performed because the original (hardened) surface has been removed. In some cases, surfaces may need to be refurbished using electroplating. This may be done on-site or outsourced, due to environmental permitting issues.

Parts that have internal material failures, such as fatigue cracks, require much more extensive refurbishing processes. If the part is valuable enough, remanufacturers go through great lengths to refurbish it. For example, Caterpillar can repair cylinder head cracks by cutting out the material containing a crack completely and filling the cavity using welding processes, almost analogous to dental cavity repair. Field testing and in-house experiments have verified this to be a viable refurbishment process.[10]

Adding weld material is one process option for *adding* material to worn or damaged surfaces. Another process option is to use *thermal spraying* technology where molten material is sprayed onto a surface. Once the desired thickness has been achieved, the surface is machined to correct surface tolerances. Thermal spraying is a process of particulate deposition in which molten or semimolten particles are deposited onto substrates, and the microstructure of the coating results from the solidification and sintering of the particles.[34] Various coatings can be achieved by using different combinations of equipment and consumables. Basic thermal spray systems typically consist of a spray gun, a power supply or gas controller, and a wire or powder feeder. There are several thermal spraying methods: flame spraying or combustion flame spraying, atmospheric plasma spraying, arc spraying, detonation-gun spraying, high-velocity oxy-fuel spraying, vacuum plasma spraying, and controlled atmosphere plasma spraying. Arc spraying is one of the fastest, simplest, and most energy efficient and inexpensive thermal spraying method.[35] Arc spraying machines consist of a wire feeder that pushes two wires through the arc spray gun. The heat zone created by the electric arc melts the wires, which are consumable electrodes. Compressed air blows the molten particles onto the substrate. Thermal spray equipment is used to apply coatings for building up worn areas, salvage improperly machined parts, improve characteristics of finished parts, and apply wear-resistant coatings. Depending on the material selected for resurfacing applications, the coating can result in parts that will outwear the originals by factors of 2 or more.[35]

3.5 Reliance on Employee Skills versus Automation

Remanufacturing relies heavily on human labor and skill due to the variation in product cores, condition, and subsequent uncertainties in processing. Employee skill can be a predominant issue in inspection, refurbishing, and reassembly.[20] Much of this is due to the diversity of unique products that the employee must be familiar with and the different assembly and disassembly techniques required for each. The process of inspection can be significantly affected by the availability of the operator to identify which quality standards the specific part must measure up to. This skill is extremely important when specifications are not available. Many aftermarket remanufacturers must often define their own part specifications, as these specifications are mostly not available from the manufacturer.

Consequently, the use of factory automation beyond basic material handling is typically low for third-party remanufacturers. Although automated (e.g., robotic) disassembly has been tried and experimented with, a skilled employee equipped with air tools and expertise to simultaneously inspect and sort parts while disassembling is hard to beat. OEMs can have higher levels of factory automation, especially when remanufactured parts are mixed with new parts in the (new) product assembly process. Automation, however, is pursued for processes where danger to human health may exist (e.g., in metal spraying and cleaning operations). Fuji-Xerox, for example, uses a small robotic cleaning cell for washing its copier chassis. The robot's chassis-specific cleaning program is initiated through reading a specific barcode that is embodied in a specific location on each chassis.[10] Caterpillar has used robotic metal spraying booths that shield workers from metal dust.

4 OVERARCHING DESIGN PRINCIPLES AND STRATEGIES ENHANCING REUSE

Products can be designed to facilitate remanufacturing. Prior to worrying about designing for facility-level remanufacturing processes, however, manufacturers should ensure that the actual product is even a candidate for reuse. Hence, the *first* step in effectively designing a remanufacturing process is to enhance the overall reusability of the original product (or specific components). In this context, market requirements are just as important as technical requirements. This section highlights a number of overarching issues that may enhance or hinder reuse and remanufacture.

4.1 Product or Component Remanufacture?

Remanufacturing should be part of a larger business strategy. As such, products should be designed not "just" for remanufacturing but also for functionality, initial manufacturability, and so on. Conflicts with other design guidelines can occur, and detailed design analyses may need to be performed.

When designing a product, keep in mind that remanufacturing the entire product may not be the best strategy and is more often an exception than the rule. Rather, remanufacture of certain product *subassemblies* is often more appropriate. A rather trivial example of this is an automobile. Powertrain components are commonly remanufactured, but interiors and bodies are not.

Similarly, remanufacturing entire products can be bad for the environment. Consider the fact if appliances and automobiles from the 1950s were kept in service as-is through remanufacturing. We would have much higher energy consumption due to their older and less efficient technology. Clearly, remanufacturing has its limitations. Leading OEMs that have internalized remanufacturing as part of their business, therefore, will spend significant time designing a product architecture that allows for technology upgrades. Fuji-Xerox, for example, looks five years ahead to see what technology may need to be incorporated in its copier systems and identifies which components should be designed as replaceable by upgrades versus which components should be designed for reuse. Also in manufacturing equipment, we see that such "upward" remanufacturing is done by adding new control systems. Hence, 100% reuse of all components is typically not feasible or even desirable. Finding what to reuse and what to replace by upgrades and how to design the architecture around that are the first major challenges for OEM designers.

4.2   Product Architecture Design Guidelines

Products become obsolete and are replaced because of five primary factors:

1. Degraded performance, including structural fatigue, caused by normal wear over repeated uses
2. Environmental or chemical degradation of (internal) components
3. Damage caused by accident or inappropriate use
4. Newer technology, prompting product replacement
5. Fashion changes

In general, the first three categories tend to be driving product returns and remanufacturing of mechanical engineering products. Replacement of information technology products (e.g., computers) is mostly caused by rapid technology changes. Consumer electronic products (e.g., cell phones) are examples where products are simply being replaced due to newer technology and/or changes in fashion. The replaced products are often fully functional and well within their operating specifications. In such cases, the remanufacturing process may collapse to a simple "collect, test, and resell or discard" operation.

To achieve a high degree of product and/or component reuse, designers and manufacturers must find a way to counter these factors. Components and subassemblies that are good candidates for reuse, therefore, have the following characteristics:

- Stable technology—not much change is expected in the product's lifetime.
- The product is resistant to damage.
- Aesthetics and fashion are (largely) irrelevant.

Given that we often do not know future technology or fashion demands, a critical issue is therefore the "openness" of the product design to future modifications and upgrades. Upgradeable products allow for a larger percentage to be salvaged. In creating product designs that address these characteristics, designers should set the following goals:

- *Strive for open systems and platform designs that have modular product structures to avoid technical obsolescence.* Platform design attempts to reduce component count by standardizing components and subassemblies while at the same time maximizing product diversity. Designing the product in *modules* allows for upgrading of function and performance (e.g., computers) and replacement of technically or aesthetically outdated modules (e.g., furniture covers). As mentioned before, Fuji-Xerox develops multiyear upgrade plans and associated product modules for its copier design. More information on modular design can be found[36] where a method is described to design products with consistent modularity with respect to life-cycle viewpoints such as servicing and recycling. The authors define *modularity* with respect to life-cycle concerns, beyond just a correspondence between form and function.
- *Strive for a "classic" design to avoid fashion obsolescence.* Aesthetically appealing and timeless designs usually are more desirable (higher priced), are better maintained, and have greater potential for long life spans and multiple reuse cycles. This is more in the realm of industrial design than mechanical design, but designing a product that does not become uninteresting or unpleasing quicker than its technical life will reduce the product's obsolescence and increase its desirability and potential for reuse.
- *Strive for damage-resistant designs.* Although this sounds like basic good engineering, lighter duty materials and smaller, more optimized, part sizes and geometries are

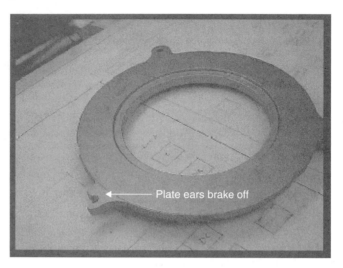

Figure 4 Damage to clutch pressure plate ear.

engineering design aspects that potentially reduce the number of service cycles and can become problems in various facets of remanufacturing. Both are directly related to design, as current designs are being optimized primarily to reduce weight, space, and cost. A good example is the reduction in wall thicknesses between cylinders in engine blocks. This reduces mass, but it also affects remanufacturability because damage due to, for example, scoring in the cylinder walls cannot be removed using machining. Instead a sleeve may have to be inserted, but this may not be possible due to the thin walls. Clearly, this practice benefits the manufacturer but can cause difficulty for remanufacturers. Figure 4 shows a clutch pressure plate that has a broken ear. Rough handling (e.g., by dropping it) in shipping or removal may have caused this failure. The plate (cast iron with machined surfaces) can only be salvaged using (expensive) welding and testing processes. This type of accidental failure will only increase if parts are designed closer to strength and endurance limits.

4.3   Product Maintenance and Repair Guidelines

The service life of products can be extended in two basic ways: (1) make the product stronger and more durable and (2) allow for good maintenance. Overall reliability and durability are enhanced by following solid engineering principles in developing a sound design and avoiding weak links. Methods such as failure mode and effect analysis are effective approaches to check the design.

Although maintenance is needed for many products, *incorrect* maintenance can have disastrous results. For example, car owners may add the wrong oil type to their automotive engines and transmissions. The design team can choose whether to allow for (user) maintenance and run the risk of unintended failures due to poor maintenance or to design the product so that it is either maintenance free or can *only* be maintained through specialized (OEM) personnel. In general, the latter is preferable when it is known that nonqualified personnel (such as users) will attempt prescribed maintenance operations. Maintenance by OEM personnel also adds a new business dimension for the OEM's overall business strategy.

Given that qualified personnel are available for maintenance, *designs should allow for easy maintenance and repair where needed*. Product design should follow available design for serviceability guidelines. Again, the best strategy is typically to design the product such that it needs little or no maintenance or only maintenance by expert personnel. If maintenance has to be done by users, it should be designed absolutely fool-proof. Some strategies for achieving easy maintenance follow:

- Indicate on the product how it should be opened for cleaning or repair.
- Indicate on the product itself which parts must be cleaned or maintained (e.g., by color-coding lubricating points).
- Indicate on the product which parts or subassemblies are to be inspected often due to rapid wear.
- Make the location of wear detectable so that repair or replacement can take place on time.
- Locate the parts that wear relatively quickly close to one another and within easy reach.
- Make the most vulnerable components easy to dismantle.

In addition, provide clear maintenance and repair manuals and communication. Consider including vital information on the actual product itself, too. Good examples are stickers or labels with tire pressure ratings and oil-type requirements placed in cars, which also aids service personnel.

4.4 Design for Reverse Logistics

If the remanufacturing process is part of an OEM's integrated strategy, core collection and reverse logistics also become crucial processes that can be aided by design. Core collection can be done by independent core managers or core brokers, through third-party subsidiaries or suppliers/customers (e.g., single-use cameras through photofinishers and automotive parts through parts stores), or through direct channels (e.g., direct mail-in of toner cartridges to OEMs). Although often overlooked, the design of easy-to-use and protective single or bulk packaging can greatly increase core returns. Good examples are toner cartridges that come in returnable boxes with prearranged return addresses and shipping labels.

4.5 Parts Proliferation versus Standardization

Product diversity (or part proliferation) is a significant problem, especially in automotive parts remanufacturing. In automotive remanufacturing, the term *part proliferation* refers to the practice of making many variations of the same product, differing in only one or two minor areas. However, these differences (such as electrical connectors) are distinct enough to prevent interchanging these similar products. For example, for a given model year, a car line may have one or more different alternators for each variation of the vehicle—the alternator for the two-door model would not be able to be used to replace the alternator for the four-door model. Not only can they not be used within the car line, but no other car line made by the manufacturer can use the part either.

Problems arising from this practice range from having to keep a large inventory of replacement parts to having to keep track of several, nonstandardized assembly and disassembly processes. An increase in the variety of assembly and disassembly processes also results in an increase of the number of process set-ups that have to be made, causing a reduction in throughput. Employee training also becomes a significant issue as a result, as they must be familiarized with all of the various, unique parts and the processes for each new product.

It is interesting to note that the trend of parts proliferation in the automotive sector started in the early 1980s. Consider the following numbers from an Atlanta-based large automotive remanufacturer. In 1983, there were approximately 3400 different part numbers for brake products. By 1995, there were approximately 16,500 different part numbers! This coincides with the move of major U.S. automakers to a platform organization and a move toward lean production. Between 1982 and 1990, Japanese automakers nearly doubled the number of models on the road, from 47 to 84 models. Reacting to this condition, U.S. automakers also increased their models on the road from 36 to 53 in the same period of time.[37] Furthermore, the independence of individual platforms within an automaker's organization seems to have led to a reduction of shared components among automotive models, resulting in decreased standardization and increased parts proliferation.

A good design practice to counter part proliferation is to use standard parts. Standardization always supports remanufacture as well as manufacturing operations and should be pursued wherever possible. Among other things, standardization reduces the number of different tools needed to assemble and disassemble, increases economies of scale in replacement part purchasing, and eases warehousing. Different product aspects can be standardized:

- *Components*. Use as much as possible standard, commonly and easily available components. Use of specialty components may render remanufacture of assemblies impossible if these specialty components cannot be obtained anymore.
- *Fasteners*. By standardizing the fasteners to be used in parts, the number of different fasteners can be reduced, thus reducing the complexity of assembly and disassembly as well as the material handling processes.
- *Interfaces*. By standardizing the interfaces of components, fewer parts are needed to produce a large variety of similar products. This helps to build economies of scale, which also improves remanufacturability. The PCI interface standard in computers is a good example of a standard interface.
- *Tools*. Ensure that the part can be remanufactured using commonly available tools. The use of specialty tools can also degrade serviceability.

4.6 Hazardous Materials and Substances of Concern

A critical issue is to avoid hazardous substances and materials of concern. Products that contain hazardous materials require specialized processing equipment (higher capital costs) and will be in lower demand, resulting in low(er) profit margins. Plastics that contain halogenated flame retardants are a good example of this in the material recycling domain. Although a large volume of these exist suitable for recycling, recyclers cannot find markets for these plastics. Sometimes, hazardous materials can be removed and retrofitted using nonhazardous materials during remanufacturing. Air-conditioning and refrigeration systems that used freon are examples where a new refrigerant can be substituted. Performance, however, may degrade slightly because the product design was not necessarily optimized for the new refrigerant. Regardless of the ability to retrofit, one should always strive to reduce the number of parts that contain environmentally hazardous materials. Also, machining or otherwise processing of parts with (heavy) metals like chromium, zinc, or lead may trigger EPA Toxic Release Inventory (TRI) reporting and require special air-handling equipment, as per federal and local regulations, adding to remanufacturing costs.

4.7 Intentional Use of Proprietary Technology

Use of technology that is proprietary or difficult to reverse engineer will block/limit the number of independent entrepreneurs remanufacturing OEM parts and products. This practice has started to emerge as certain OEMs have realized the value of remanufactured products and how third-party remanufacturers can take away the market share of OEM product and component sales. In the inkjet printing industry, some OEMs include chips that can only be reset by an authorized remanufacturer. Similarly, Kodak's single-use cameras have become more difficult to disassemble with common available tools in order to counter third-party film reloading and reuse. This strategy is counter to what many academics say should be done regarding product design for remanufacture, but this practice clearly makes sense from a higher level business strategy where an OEM wants to retain market share and sales.

4.8 Inherent Uncertainties

Last, but not least, in remanufacture, the number and range of uncertainties are higher than for original manufacture and logistics because many of the concerns are out of the control of the OEM and the designers. Some sample product uncertainties encountered follow:

- How long is a typical use and/or life span?
- What is its state after its each use?
- What changes have been made during use and throughout its life?

This affects organizational uncertainties:

- How many will be available for take-back, and when?
- How long will it take to reprocess the product?
- What is the demand?

Some remanufacturing operations have throughput yields as low as 40–60% (unheard of in manufacturing), due to a combination of poor quality cores and poor processing. Designers and product realization teams should be aware of these uncertainties and should ideally try to manage or even eliminate the uncertainties by smart product and process design. For example, changes can be avoided if the product design eliminates the possibility of user tampering.

5 HARDWARE DESIGN GUIDELINES

In the preceding, some specific design guidelines were given that enhance the overall suitability of remanufacturing a given product. In this section, some specific component and machine-design type guidelines will be given that primarily facilitate the facility level remanufacturing processes (as discussed in Section 3). Clearly, this discussion is not exhaustive; you are encouraged to use your own engineering insights as well to identify design guidelines for remanufacturing operations and product designs.

5.1 Basic Sources and Overviews

There are relatively few publications and sources with general design for remanufacturing guidelines in existence. The emergence of the waste electrical and electronic equipment (WEEE) and ELV take-back directives from the European Union,[3,38] however, has resulted in a number of design-for-recycling guidelines—some of which are applicable to remanufacturing. General design-for-recycling guidelines were formalized in the German engineering standard.[2] This guideline also contains directional criteria for the design of remanufacturable products. According to VDI 2243[2] and other sources,[4,5,7,14,39] remanufacturable assemblies should be designed with special emphasis on the following:

- *Ease of Disassembly.* Where disassembly cannot be bypassed, by making it easier, less time can be spent during this non-value-added phase. Permanent fastening such as welding or crimping should not be used if the product is intended for remanufacture. Also, it is important that no part be damaged by the removal of another.

- *Ease of Cleaning.* Parts that have seen use inevitably need to be cleaned. In order to design parts such that they may easily be cleaned, the designer must know what cleaning methods may be used and design the parts such that the surfaces to be cleaned are accessible and will not collect residue from cleaning (detergents, abrasives, ash, etc.).

- *Ease of Inspection.* As with disassembly, inspection is an important, yet non-value-added phase. The time that must be spent on this phase should be minimized.

- *Ease of Part Replacement.* It is important that parts that wear are capable of being replaced easily, not just to minimize the time required to reassemble the product but also to prevent damage during part insertion.

- *Ease of Reassembly.* As with the previous criteria, time spent on reassembly should be minimized using design-for-assembly guidelines.[40] Where remanufactured product is assembled more than once, this is very important. Tolerances also relate to reassembly issues.

- *Reusable Components.* As more parts in a product can be reused, it becomes more cost effective to remanufacture the product (especially if these parts are costly to replace).

In the following section, we will focus on a number of guidelines in more detail. Clearly, the inherent and underlying assumption is that the products are being designed for remanufacture by an OEM or friendly third party. Otherwise, there is no incentive to follow any of these design guidelines.

5.2   Sorting Guidelines

Sorting is the first step in any remanufacturing process. Mostly it is coupled with an initial inspection as well. Figure 5 is illustrative of how cores are received by many third-party remanufacturers. The container in Fig. 5 contains boxed and unboxed starters, alternators, and brake shoes of varying types, shapes, sizes, and conditions. In such cases, worker knowledge and expertise are key in the sorting process. Product and part design can facilitate the sorting process by following some guidelines:

- *Reduce product and part variety.* The less different parts need to be sorted, the less time it costs. This can also be achieved by remanufacturers themselves through specialization

Figure 5   Cores arrive at automotive remanufacturer.

on specific products and cores. This also implies for internal components. Standardization of fasteners, bearings, pulleys, and so on, will greatly speed up the initial core as well as subsequent part sorting.

- *Provide clear distinctive features that allow for easy recognition.* If different parts have to be used, make sure they are easily recognizable. For example, having two housings being exactly the same except for one different-sized hole may not be the best strategy because the sorter/inspector has to distinguish based on small size differences. A binary yes/no type distinction is much easier to do and can be achieved by, for example, changing the hole pattern.

- *Provide (machine) readable labels, text, and barcodes that do not wear off during product's service life.* Most products and parts have labels. Those that are exposed to the environment, however, tend to wear off during life unless they have been stamped, casted, or molded in. Even riveted serial plates and numbers can shear and wear off. Internal parts fare better provided they have part numbers. Some companies are experimenting with radio-frequency identification (RFID) tags to facilitate sorting, but that is the exception rather than the rule.

5.3 Disassembly Guidelines

A phrase often heard is, "If a remanufacturer can take a product apart, it can be remanufactured." At first, this statement would seem to indicate that the design should focus on disassembly to ensure that the product can be remanufactured. However, there is a hidden assumption in this statement. A more correct statement is, "If a remanufacturer can take a product apart *without damaging important parts*, it can be remanufactured." The two key ideas that designers should extract from this statement are nondestructive disassembly and preventing key parts from being damaged.

In remanufacture, the objective is to reuse cores and components. That means that (in contrast to material recycling) destructive disassembly techniques like shredding are not an option. Manual disassembly, supported by pneumatic or other hand-held mechanized means, is the general norm of the industry—for better or for worse (see, e.g., Fig. 6). Proper design can

Figure 6 Rivets being drilled out a clutch pressure plate—a cause for damage and high reject rates.

make disassembly easier so that less time can be spent during this non-value-added phase, but the goal of remanufacture is to salvage cores and components of value, and any damage must be repaired. Speedy disassembly is desired, but not at the expense of damaging cores. Avoiding and preventing damage, therefore, are often the more important objectives than increasing speed. Given this, we can define a number of simple overarching *guidelines for fasteners:*

Avoid and prevent damage:

- Avoid permanent fasteners that require destructive removal (such as rivets, welds, crimp joints, etc.).
- If fasteners require destructive removal, ensure that their removal will not result in damage to core and other reusable parts by incorporating breakpoints or appropriate strong lever points.
- Reduce number of fasteners prone to damage and breakage during removal (e.g., snap fits). For example, Phillips/Blade/Torx fasteners are more easily prone to head damage and removal difficulties than hex and Allen bolts. Molded-plastic snap fits often break due to the aging of the plastic, either causing a need for repair or resulting in the whole part being scrapped.
- Increase corrosion resistance of fasteners, where appropriate. This reduces damage and facilitates removal.

Increase speed:

- Reduce total number of fasteners in unit.
- Reduce the number of press-fits, which do not have "push-out" capability.
- Reduce number of fasteners without direct line of sight.
- Standardize fasteners by reducing the number of different types of fasteners (hex/ phillips/allen/torx, metric/SAE, etc.). Reducing the number of different *size* fasteners (i.e., length, diameter) will speed up reassembly and allow for larger economies of scale in purchasing fasteners.

5.4 Design for Reassembly

Reassembly, the last process in a typical remanufacturing process, is basically identical to assembly in manufacturing. To design for reassembly, follow common design-for-assembly (DfA) guidelines. Table 2 contains common DfA guidelines that can be found in the general literature.

Manufacturers tend to use design-for-assembly and manufacturing processes that make it difficult for parts to be reused or remanufactured. For example, solenoids for starter motors are crimped into their housings. Not only is it difficult to remove the crimps in order to remanufacture the solenoid, but crimped fasteners cannot be recrimped without degrading the strength of the crimp.

5.5 Cleaning

Parts that have seen use inevitably need to be cleaned. In order to design parts such that they may easily be cleaned, the designer must know what cleaning methods may be used and must design the parts such that the surfaces to be cleaned are accessible and will not collect residue from cleaning (detergents, abrasives, ash, etc.). The following guidelines capture the basic aspects:

- *Protect parts and surfaces against corrosion and dirt.* The best strategy is to minimize cleaning wherever and whenever. Proper corrosion coating and dirt protection

Table 2   Common Design for Assembly Guidelines

1. Overall component count should be minimized.
2. Minimize use of fasteners.
3. Design the product with a base for locating other components.
4. Do not require the base to be repositioned during assembly.
5. Design components to mate through straight-line assembly, all from the same direction.
6. Maximize component accessibility.
7. Make the assembly sequence efficient.
   - Assemble with the fewest steps.
   - Avoid risks of damaging components.
   - Avoid awkward and unstable component, equipment, and personnel positions.
   - Avoid creating many disconnected subassemblies to be joined later.
8. Avoid component characteristics that complicate retrieval (tangling, nesting, and flexibility).
9. Design components for a specific type of retrieval, handling, and insertion.
10. Design components for end-to-end symmetry when possible.
11. Design components for symmetry about their axes of insertion.
12. Design components that are not symmetric about their axes of insertion to be clearly asymmetric.
13. Make use of chamfers, leads, and compliance to facilitate insertion.

will support this. However, also consider that any coating (e.g., paint) may have to be removed if damaged. Hence, a balance may have to be found between protection and ease of removal.

- *Avoid product and/or part features that can be damaged during cleaning processes, or make them removable.* For example, when thermal cleaning is used, make sure all materials can withstand the heat without adverse effects. Abrasive cleaning methods can gouge surfaces.
- *Minimize geometric features that trap contaminants over the service life.* A sharp concave corner is an example of a geometric feature that traps contaminants. If a rib or plate is expected to trap dirt or grease, consider making it removable.
- *Reduce number of cavities/orifices that are capable of collecting residue (abrasives, chemicals, etc.) during cleaning operations.* Any orifice that can collect dirt or cleaning debris will have to be plugged or cleaned afterward.
- *Avoid contamination caused by wear.* Internal components can become dirty due to wear of other components. For example, oil seals may wear and the resulting leakage will cause contamination of other parts. Proper shielding or designing out such sources of wear can reduce the cleaning effort required.

5.6   Replacement, Reconditioning, and Repair

In general, remanufacturing tends to avoid the replacement of parts, but there are trade-offs as to whether to spend money to buy a new part or spend money to repair the part. For commonly available parts like bearings and fasteners, the choice is easy, but the higher the part price, the more incentive for refurbishment instead of replacement. The cost of replacement can be reduced by the following guidelines:

- Reduce number of parts subject to wear.
- Avoid materials that degrade through corrosion.

- Reduce numbers of parts to be removed to gain access to damaged parts to be replaced (or refurbished).
- Reduce number of independently functioning parts that are inseparably coupled.
- Reduce number of special parts (including aesthetic features).

As discussed in Section 3, there are a number of basic strategies for repairing damage and refurbishing surfaces. Proper material selection can aid remanufacturing, as can surface protection. An interesting problem with surface protection such as heavy-duty paint or powder coating is that it protects a part but can cause significant cleaning problems in remanufacturing when the coating needs to be removed for renewal. For such surface reconditioning, consider the following guidelines:

- Reduce number of parts whose surface finish cannot be refinished through commonly available and conventional means.
- Minimize number of orifices that must be masked prior to painting.
- Reduce the number of (exterior) parts that must be removed prior to painting.
- Minimize the number of parts that can retain dents/deformations.

5.7   Inspection and Testing

Inspection and testing can be facilitated by reducing the number of different testing and inspection equipment pieces needed as well as reducing the level of sophistication required. Although not in the realm of product design per se, good testing documentation and specifications should be provided to ensure that the correct specifications are achieved and tested for. This assumes (again) OEM involvement in the remanufacturing process.

6   DESIGN FOR REMANUFACTURING CONFLICTS

It should be noted that in some cases design for remanufacturing can conflict with design for manufacturing and even be not in the best interest for the environment. For example, increasing longevity by adding material can increase part weight, causing more upfront material expenditures (and cost) and potentially more fuel consumption and emissions in transportation systems. Some differences also exist between design for disassembly (DfD) and DfA. For example, complete nesting can slow disassembly by not providing a location for the disassembler to reach, grasp, or otherwise handle.[41] Table 3 broadly presents the influence of DfD on assembly compiled from the literature.[7,39,41,42] The guidelines are divided into three categories according to their hypothesized influence on assembly: (1) positive effect on assembly, (2) negative effect on assembly, and (3) relatively little effect on assembly. Most DfD guidelines affecting the product structure are placed in the "positive effect" category concerning assembly. The four negative effects on assembly for the most part deal with making easily separable joints. This would negatively affect assembly in the sense that the purpose of the assembly step could be easily negated during product use. A compromise solution would be to design joints that are very hard to disassemble during product use but are easy to dismantle after the customer use or for the purpose of servicing a product. The DfD guidelines with relatively little effect on assembly for the most part deal with material selection and identification.

Different DfD strategies will have different effects on the overall processing time, especially if coupled with reassembly processes. A study on single-use cameras suggested that a modular design was slightly more effective in improving disassembly efficiency than parts consolidation and much more effective than reducing orientation changes during disassembly.[44,45] *Clearly, as indicated in Section 4, good design-for-remanufacturing processes should take a life-cycle perspective—from both economical and environmental points of view.*

Table 3   DfD Guidelines and Effects on Assembly

Positive Effect on Assembly

1. Reduce the number of components.
2. Reduce the number of separate fasteners.
3. Provide open access and visibility for separation points.
4. Avoid orientation changes during disassembly.
5. Avoid nonrigid parts.
6. Ensure that disassembly can be done with common tools and equipment.
7. Design for ease of handling and cleaning of all components.

Negative Effect on Assembly

1. Design two-way snap fits or break points on snap fits.
2. Use joining elements that are detachable or easy to destroy.
3. Design for ease of separation of components.
4. Use water-soluble adhesives.

DfD Guidelines Having Relatively Little Effect on Assembly

1. Design products for reuse.
2. Eliminate need to separate parts.
3. Reduce number of different materials (for recycling).
4. Enable simultaneous separation and disassembly.
5. Place components in logical groups according to recycling group and disassembly sequence for modular design.
6. Identify separation points and materials.
7. Facilitate the sorting of noncompatible materials (for recycling).
8. Use molded-in material name in multiple locations to provide cut points (for recycling).
9. Provide techniques to safely dispose of hazardous waste.
10. Select an efficient disassembly sequence.

Source: From Refs. 43 and 44.

7   DESIGN DECISION SUPPORT TOOLS

Although some OEMs have gained significant experience in remanufacturing and have developed associated specialized computer tools, relatively few design-for-remanufacture decision support tools exist in the general domain. See Ref. 46 for a simple worksheet that can be implemented in a spreadsheet. A more sophisticated spreadsheet-based tool is described by Hammond and Bras,[47] who use a nondimensional metric to quantify and ultimately compare product designs for remanufacturability. A sample of the results worksheet is given in Fig. 7.

A model for assessing how product design characteristics, product development strategies, and different business conditions impact remanufacturing viability in terms of net present value for an OEM interested in integrated manufacture–remanufacture is discussed by McIntosh and Bras.[17] The model is not focused on the analysis of a single period and/or single product but focuses on the interplay between multiple products over multiple time periods. It illustrates the benefits of designing and remanufacturing a *family* of products with shared components.

It is encouraging to see that more and more research, including work in business schools, is focusing on providing OEMs the ability to assess (1) whether their business conditions are capable of implementing remanufacture and (2) how their decisions impact the success of remanufacture.

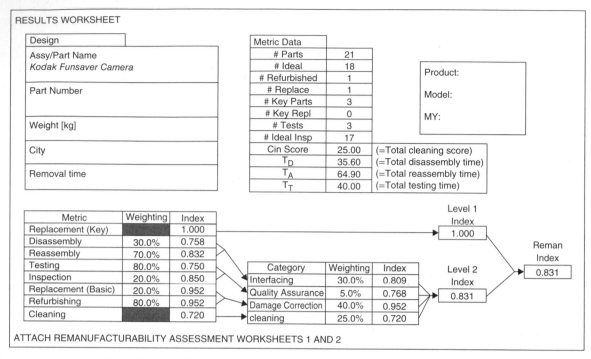

Figure 7 Spreadsheet-based remanufacturability assessment. *Source:* From Ref. 47.

8 SUMMARY

In this chapter, design-for-remanufacturing guidelines were presented and discussed. An overview of the industry, including typical facility-level processes, was given in order to better understand the rationale for the given design guidelines. It is encouraging to see that more and more research, including work in business schools, is focusing on providing OEMs the ability to assess whether their business conditions are capable of implementing remanufacture and how their decisions impact the success of remanufacture.

REFERENCES

1. N. Nasr, "*Remanufacturing—From Technology to Application*," paper presented at the Global Conference on Sustainable Product Development and Life Cycle Engineering, Berlin, Germany, Uni-Edition, September 29–October 1, 2004.

2. VDI, "Konstruieren Recyclinggerechter Technischer Produkte" ("Designing Technical Products for ease of Recycling"), VDI-Gesellschaft Entwicklung Konstruktion Vertrieb, Germany, 1993.

3. European Union, "Directive 2000/53/EC of the European Parliament and of the Council of September 18 2000 on End-Of Life Vehicles," *Off. J. Eur. Commun.*, L 269/34–42, 2000.

4. R. T. Lund, "Remanufacturing," *Technol. Rev.*, **87**, 18–23.

5. H. C. Haynsworth and R. T. Lyons, "Remanufacturing by Design, The Missing Link," *Production Inventory Manag.*, second quarter, 25–28, 1987.

6. U.S. Environmental Protection Agency (EPA), "Remanufactured Products: Good as New," *WasteWi$e Update*, EPA, Solid Waste and Emergency Response (5306 W), Washington, DC, 1997.

7. U.S. Congress, *Green Products by Design: Choices for a Cleaner Environment*, Office of Technology Assessment, Washington, DC, 1992.

8. B. K. Fishbein, L. S. McGarry, and P. S. Dillon, *Leasing: A Step toward Producer Responsibility*, Inform, New York, 2000.

9. Equipment Leasing Association (ELA), *Industry Research: Overview of the Equipment Leasing & Finance Industry*, 2004, available: http://www.elaonline.com/industryData/overview.cfm.

10. T. G. Gutowski, C. F. Murphy, D. T. Allen, D. J. Bauer, B. Bras, T. S. Piwonka, P. S. Sheng, J. W. Sutherland, D. L. Thurston, and E. E. Wolff, "Environmentally Benign Manufacturing," International Technology Research Institute, World Technology (WTEC) Division, Baltimore, MD, 2001.

11. D. T. Allen, D. J. Bauer, B. Bras, T. G. Gutowski, C. F. Murphy, T. S. Piwonka, P. S. Sheng, J. W. Sutherland, D. L. Thurston, and E. E. Wolff, "Environmentally Benign Manufacturing: Trends in Europe, Japan and the USA," *ASME J. Manufact. Sci.*, **124**(4), 908–920, 2002.

12. R. T. Lund and F. D. Skeels, "Start-up Guidelines for the Independent Remanufacturer," Center for Policy Alternatives, Massachusetts Institute of Technology, Cambridge, MA, 1983.

13. R. T. Lund and F. D. Skeels, "Guidelines for an Original Equipment Manufacturer Starting a Remanufacturing Operation," Center for Policy Alternatives, Massachusetts Institute of Technology, Cambridge, MA, 1983.

14. V. J. Berko-Boateng, J. Azar, E. DeJong, and G. A. Yander, "Asset Recycle Management—A Total Approach to Product Design for the Environment," in *International Symposium on Electronics and the Environment*, Arlington, VA, IEEE, New York, 1993.

15. B. Paton, "Market Considerations in the Reuse of Electronic Products," in *IEEE International Symposium on Electronics and the Environment*, San Francisco, CA, IEEE, New York, 1994.

16. M. W. McIntosh and B. A. Bras, "Determining the Value of Remanufacture in an Integrated Manufacturing-Remanufacturing Organization," paper presented at the 1998 ASME Design for Manufacture Conference, ASME Design Technical Conferences and Computers in Engineering Conference, Atlanta, GA, September 14–16, 1998.

17. M. McIntosh and B. A. Bras, "Product, Process, and Organizational Design for Remanufacture—An Overview of Research," *Robotics and Computer Integrated Manufacturing—Special Issue on Remanufacturing*, **15**, 167–178, 1999.

18. W. R. Stahel, "The Utilization-Focused Service Economy: Resource Efficiency and Product-Life Extension," in B. R. Allenby and D. J. Richards (Eds.), *The Greening of Industrial Ecosystems*, National Academy of Engineering, Washington, DC, 1994, pp. 178–190.

19. D. S. Rogers and R. S. Tibben-Lembke, *Going Backwards: Reverse Logistics Trends and Practices*, Reverse Logistics Executive Council, Pittsburgh, PA, 1999.

20. R. Hammond, T. Amezquita, and B. Bras, "Issues in Automotive Parts Remanufacturing Industry: Discussion of Results from Surveys Performed Among Remanufacturers," *J. Eng. Design Automation, Special Issue on Environmentally Conscious Design and Manufacturing*, **4**(1), 27–46, 1998.

21. A. J. Niksa, "Cleaning Equipment and Methods—Overview Presentation," paper presented at the 16th ASM Heat Treating Society Conference & Exposition, 1996.

22. D. Brooks, "Ovens," *Automotive Rebuilder*, **31**, 38–43, 1994.

23. A. Belkind, S. Zarrabian, and F. Engle, "Plasma Cleaning of Metals: Lubricant Oil Removal," *Metal Finishing*, **94**(7), 19–22, 1996.

24. W. Petasch, B. Kegel, H. Schmid, K. Lendenmann, and H. U. Keller, "Low-Pressure Plasma Cleaning: A Process for Precision Cleaning Applications," *Surface Coatings Technol.*, **97**, 176–181, 1997.

25. P. P. Ward, "Plasma Cleaning Techniques and Future Applications in Environmentally Conscious Manufacturing," *SAMPE J.*, **32**(1), 51–54, 1996.

26. J. Brown, *Advanced Machining Technology Handbook*, McGraw-Hill, New York, 1998.

27. C. Heaton, C. Northeim, and A. Helminger, "Pollution Prevention: Opting for Solvent-Free Cleaning Processes," *Chem. Eng.*, **111**(1), 13, 2004.

28. J. B. Durkee II, "Environmentally Conscious Cleaning for the Millennium," *SAE Int.*, 1–8, 2000.

29. C. Nelson, *Parts Cleaning Systems: Using Water and Steam as Cleaning Agents*, FabTech International, 1993.

30. D. B. LeBart and R. Sivakumar, "Aqueous Cleaning Technologies: Present and Future," paper presented at the 16th ASM Heat Treating Society Conference & Exposition, 1996.

31. J. B. Durkee II, "What's in Your Cleaning Tank?" *Metal Finishing*, **103**(2), 46–48, 2005.

32. G. Eskamani, D. Brown, and A. Daniels, "Evaluation of BioClean USA, LLC Biological Degreasing System for the Recycling of Alkaline Cleaners," in *Environmental Technology Verification*, U.S. Environmental Protection Agency, Washington, DC, 2000.

33. B. Bissler, "Blast & Tumble," *Automotive Rebuilder*, **31**, 44–47, 1994.

34. L. Pawlowski, *The Science and Engineering of Thermal Spray Coatings*, Wiley, West Sussex.

35. C. Howes, "Thermal Spraying: Processes, Preparation, Coatings and Applications," *Welding J.*, April 1994, pp. 47–51.

36. P. J. Newcomb, B. A. Bras, and D. W. Rosen, "Implications of Modularity on Product Design for the Life Cycle," *J. Mech. Design*, **120**(3), 483–490, 1998.

37. J. Womack, D. Jones, and D. Roos, *The Machine That Changed the World: The Story of Lean Production*, Harper Perennial, New York, 1991.

38. European Union, "Directive 2002/96/EC of the European Parliament and of the Council of 27 January 2003 on Waste Electrical and Electronic Equipment," *Off. J. Eur. Commun.*, L 37/24–38, 2003.

39. W. Beitz, "Designing for Ease of Recycling—General Approach and Industrial Applications," in *Ninth International Conference on Engineering Design*, Heurista, The Hague, Zurich, Switzerland, 1991,

40. G. Boothroyd, and P. Dewhurst, *Product Design for Assembly*, Boothroyd and Dewhurst, Wakefield, MA, 1991.

41. R. M. Noller, *Design for Disassembly Tactics*, Assembly, January 1992, pp. 24–26.

42. G. Boothroyd, and L. Alting, "Design for Assembly and Disassembly," *Ann. CIRP*, **41**(2), 625–636, 1992.

43. J. F. Scheuring, B. A. Bras, and K.-M. Lee, "Effects of Design for Disassembly on Integrated Disassembly and Assembly Processes," paper presented at the Fourth International Conference on Computer Integrated Manufacturing and Automation Technology, Rensselaer Polytechnic Institute, Troy, NY, October 10–12, IEEE, 1994.

44. J. F. Scheuring, B. A. Bras, and K.-M. Lee, "Significance of Design for Disassembly in Integrated Disassembly and Assembly Processes," *Int. J. Environ. Conscious Design Manufact.*, **3**(2), 21–33, 1994.

45. J. F. Scheuring, "Product Design for Disassembly," M.S. Thesis, George W. Woodruff School of Mechanical Engineering, Georgia Institute of Technology, Atlanta, GA, 1994.

46. T. Amezquita, R. Hammond, and B. Bras, "Design for Remanufacturing," paper presented at the Tenth International Conference on Engineering Design (ICED 95), Praha, Czech Republic, Heurista, Zurich, Switzerland, August 22–24, 1995.

47. R. C. Hammond and B. A. Bras, "Design for Remanufacturing Metrics," paper presented at the First International Workshop on Reuse, Eindhoven, The Netherlands, November 10–13, 1996.

CHAPTER 11

DESIGN FOR MANUFACTURE AND ASSEMBLY WITH PLASTICS

James A. Harvey
Under the Bridge Consulting, Inc.
Corvallis, Oregon

1 INTRODUCTION

This chapter is divided into three sections: plastic materials selection, plastic-joining techniques, and plastic part design. Our major focus will be on plastic materials selection. The information presented is based on both the lectures given and the information received in short courses taught to practicing engineers and scientists involved in all aspects of commercial plastics part designs and in graduate school courses to budding new materials scientists and engineers.

In the open literature for material selection you will find articles with titles similar to "The Science of Material Selection" or "The Art of Material Selection." Hopefully this chapter will eliminate some of the mystery or confusion regarding material selection.

2 PLASTIC MATERIALS SELECTION

2.1 Polymers

In the selection of plastic materials for a commercial part design the first step is as in all technology development: to learn the basic definitions, concepts, and principles of that technology.

Reprinted from *Handbook of Materials Selection,* Wiley, New York, 2002, by permission of the publisher.

The following terms will be defined as they are important in the selection of materials for plastics part design:

Polymer

Thermoplastics

Thermosets

Elastomers

Polymerization reactions

Molecular weight and distribution

Molecular structure of polymers

Five viscoelastic regions of polymers

Carothers equation

Additives

A *polymer* is a compound consisting of repeating structural units. A simple example of a repeat unit is the $-CH_2-$ chemical moiety. Two repeated units are equivalent to the organic compound ethane. Ethane is a gas at room temperature with a total molecular weight of 30 atomic mass units (amu). A polymer family with hundreds of thousands of these $-CH_2-$ repeat units represents the polyethylenes with molecular weights in the millions.

A *thermoplastic polymer* is a polymer that consists of linear polymer chains. Whenever you use a thermoplastic it is usually in its final molecular weight form. The major thermal event is to process it into the final part form. There are three types of thermoplastics: amorphous, semicrystalline, and liquid crystal.

A *thermosetting resin* is one that contains a highly crosslinked polymer network when processed. One has to "cook" or cure the resin before it can be formed into its final shape.

An *elastomer* is a very lightly crosslinked polymer with the ability to be extended to a high elongation and snap back to its original dimensions when the forces have been removed.

Polymerization reactions play a usual role in the process of material selection. From the name of the polymer and its polymerization reaction, one can make a reasonable first attempt to select a plastic material. But the reader must be cautioned that the preceding statement is a general rule. For example, polyethylene is named from the monomer from which it is made, ethylene. This monomer is polymerized through an addition reaction. Typically, addition polymers are water hating, or hydrophobic. For a first approximation, this type of material would be a good material to use in applications where exposure to water is required.

Now let us look at another polymer, polyethylene terephthalate. It is formed from the reaction of ethylene glycol with terephthalic acid or terephthalic acid ester. During the reaction, in order for the polymer, polyethylene terephthalate, to build up molecular weight, it loses either water or alcohol as a by-product. Polymers formed by adding two or more coreactants under conditions of time, temperature, and other reaction conditions with the formation of a byproduct such as water and alcohol are said to be formed by condensation and are named by the new chemical functional group formed. In general, these polymers are water loving, or hydrophilic.

As mentioned, polyesters are formed from the reaction of organic acids with organic alcohols with water as a byproduct. Nylons are formed from the condensation reaction of organic acids with organic amines with water as a byproduct. Polyimides are formed from the condensation reaction between acid anhydrides and organic amines with the release of water.

Organic chemistry plays a very important role in the selection of a polymer for a plastic part. Here we are only posing general rules, and to a first approximation one can make very reasonable selection in the early stages of plastic part design using these general observations.

For example, if you were assigned to design a plastic part that had to exist in a water environment, your first choice could be an addition polymer such as polyethylene rather than a condensation polymer such as polyethylene terephthalate. The polyethylene is water hating, or hydrophobic, and so should not be affected by water.

This writer can already imagine the "but what about this incident" remarks. Yes, water bottles are made from polyesters. The bottles are dated for lifetime and the companies that fill these bottles with their mountain spring fresh water want you to see how clear their water is. However, over time the polyester bottles will absorb water. Water (sport) bottles that are used over and over and filled by the consumer are made from the addition polymers. They are also opaque. Transparency in these sport plastic water containers is not important. This polymer character of being transparent or opaque will be discussed later.

Molecular weight and *molecular weight distributions* are other important parameters for a polymer. The polymerization reaction is complicated. The polymerization reaction does not result in a simple single molecule. The reaction yields many different sizes of polymer chains. The molecular mass of each chain refers to its molecular weight. And as mentioned, since many different sizes of polymer chains are formed thus, there will be a distribution of the molecular weights.

Knowledge of the molecular weight and its distribution aids in the selection of polymers for a plastics part and in the lifetime of plastic parts. The general techniques that can be used to determine molecular weights of polymers are achieved through viscosity measurements either in solution or using solid samples. Solution viscosities consist of timing the flow of polymer solutions of known concentrations through a fixed volume. The melt index or melt flow index (MFI) is derived from a standardized test in which a solid is used instead of a solution. A given amount of polymer is heated to a certain temperature, a known force is applied to the molten polymer, and its flow is timed. If all things are equal, the lower molecular weight polymer will flow through the given volume faster than a higher molecular weight polymer. Hence the higher the molecular weight of a polymer, the lower the MFI value. For the members of a given polymer family this is a reasonable way to distinguish between low- and high-molecular-weight versions. The final technique is gel permeation (size exclusion) chromatography. The polymer is dissolved in a solvent. The solution is then passed through a series of tubes (columns) packed with different porous particles. As the solution passes through, the polymer chains with the highest molecular weights pass through the fastest. A detector measures the polymer chains as they exit the instrument. Thus one ends up with a chromatograph, which shows the distribution of the different molecular weights of the polymer chain in the sample. Solution viscosities are usually by the polymer manufacturer. Melt flow index is used as an initial tool for material selection and as a tool to help determine the molding process. Gel permeation (size exclusion) chromatography was in the past treated as a research tool, but lately it has gained a great deal of popularity as a quality control technique.

Another important parameter for the different polymers refers to their thermal behavior. A typical thermoplastic is a solid at ambient temperatures. As the material is heated, it starts to soften; then it flows and in some cases it melts. When it is cool, it solidifies. Depending on the container (mold) used, the thermoplastic will retain the shape of that mold. This process should be repeatable. Thus thermoplastics are recyclable. Another thermal property of thermoplastics is creep. This property refers to the ability of the material to flow under a load as a function of temperature.

Thermosetting resin systems are quite different. When one processes a thermoplastic into a particular plastic part, its molecular weight has already been established by the manufacturer. With thermosetting resin systems one starts with low-molecular-weight reactants, and to process these ingredients, the reactants are cured, or "cooked," into the desired final shape. If the reactants have been fully reacted, the result is one giant molecule. To process a thermosetting resin system into a part, the thermal events consist of heating the ingredients so they start to soften followed by some of the ingredients melting. As the temperature is raised, the system is totally liquid. As the temperature continues to rise, the onset of curing (crosslinking) occurs. As the reaction proceeds, the viscosity increases and the part hardens. At the end of the curing reaction the part is solid; then it is cooled to ambient temperatures. Once formed the part cannot be reheated to change its shape. If the thermosetting resin system has been properly cured, it should not be affected by temperature or solvents. This behavior may be used to characterize the starting thermosetting material. Usually, as the material is curing, its solubility changes from soluble to insoluble. Good examples are the liquid epoxy resins. These materials are usually eutectic mixtures of three or more reactive species (oligomers). Such analyses as the number and amount of reactive species or epoxy equivalent weights affect the quality of the cure or the reproducibility of the final cure structure.

The behavior of elastomers is somewhat different than that of thermoplastics and thermosets. As a first approximation, it behaves more like a thermoplastic. We all know that car tires soften in the hot months of summer. Most elastomers will swell when placed in a solvent.

Thermal analyses are a set of techniques used to characterize the thermal behavior of the different types of polymers. In addition to providing the thermal characteristics of polymers, they can assist in determining a processing cycle. Differential scanning calorimetry (DSC) yields the thermal events of a sample, i.e., melting points, onsets, maximums, and offsets of curing, decomposition temperatures, crystallization temperatures, and glass transition temperatures (this will be discussed later). Thermogravimetric analyses (TGAs) give the changes in mass of a sample as a function of temperature and environment. Thermal mechanical analyses (TMAs) reveal the changes in volume of a sample (warpage and shrinkage) and glass transition temperatures. Dynamic mechanical analyses (DMAs) provide the modulus and changes in modulus and glass transition temperature as a function of temperature, time, and oscillation (dynamic load).

The *molecular structure of the polymer* will determine if it is transparent or opaque. This internal structure is called polymer morphology. Thermoplastic polymers can be divided into amorphous, semicrystalline, crystalline, and liquid crystal polymers. This classification is only reserved for thermoplastics. Morphology refers to how the polymer chains are arranged, in an ordered or disordered manner. Amorphous refers to total disorder. Crystalline refers to total order. Semicrystalline is a combination of disorder with domains of order within the structure. The liquid crystal polymers are a special class of thermoplastics that retain their order in the melt. Based upon chemical principles, as a material goes from the solid state to the liquid state, it goes from a state of order to one of disorder. The liquid crystal polymers lack this transition, and this unique characteristic has an enhanced effect on the processing of these materials.

We can examine the internal structure of amorphous, semicrystalline, and crystalline thermoplastics another way by viewing the polymer chains as spaghetti. We have cooked spaghetti (disordered) as one extreme and uncooked (ordered) spaghetti as the other extreme. Except for the liquid crystalline polymers, most thermoplastic polymers are either amorphous or semicrystalline (ordered polymer chains with crystalline domains). Due to the presence of crystalline domains, the semicrystalline polymers have a melting point and light will be scattered as it hits these domains, thus giving the material an opaque appearance. Thus amorphous polymers do not have a melting point and are transparent.

The next important definition involves the *five viscoelastic regions of polymers*. If we plot the modulus of a thermoplastic material as a function of temperature, we obtain a graph such as the one shown in Fig. 1.

Region 1 represents the behavior of the material at low temperatures. It is in its glassy state. The mobility of the polymer chains has decreased. The material is hard. As it is heated, it reaches region 3. This region is known as the rubbery region and the material loses strength in the range of three orders of magnitude. As the sample is heated to an even higher temperature,

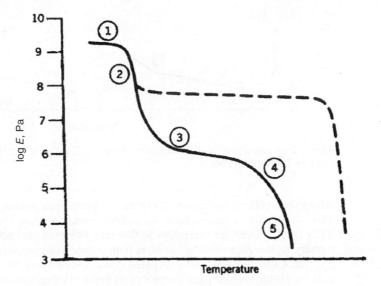

Figure 1   Five viscoelastic regions of a linear amorphous polymer. The dashed line represents the behavior of a semicrystalline polymer. *Source*: From Ref. 1.

the polymer (region 4) starts to decompose and finally, at region 5, decomposition occurs with loss of strength.

A semicrystalline thermoplastic has the appearance of the dashed line in Fig. 1. The drop in modulus from the glassy region to the rubbery region is not as drastic with semicrystalline polymers (region 2) as it is with amorphous polymers. As the semicrystalline thermoplastic reaches its melting point, its strength decreases sharply and, as expected, goes from a solid to a liquid.

The transition between the glassy region of a polymer to its rubbery region is known as its glass transition and the temperature at which it occurs at is its glass transition temperature.

Glass transition is defined as "the reversible change in an amorphous material or in amorphous regions of a partially crystalline material, from (or to) a viscous or rubbery condition to (or from) a hard and relatively brittle one."[2]

Some individuals use the term *glass transition temperature* while discussing cured thermosetting resin systems. To this writer, if the thermosetting resin is completely cured, it should not have a glass transition temperature. If it is completely cured as the material is heated over a temperature range, it should be unaffected by temperature until it reaches its decomposition temperature. In an analysis, the presence of a glass transition temperature may be due to either the thermoset not being completely cured or the thermoplastic nature of the crosslinked network. If you perform a thermal technique to determine glass transition temperature, cool the sample to ambient temperature and repeat the analysis on the same sample. If the material is not fully cured, the repeat run should indicate a higher apparent glass transition temperature and a lower drop in modulus.

Figure 2 shows the volume–temperature differences between the glass transition temperature and the crystalline melting point of a thermoplastic polymer. As the temperature is increased and a semicrystalline polymer (Fig. 2b) passes through its crystalline melting point (I_m), the change in volume is discontinuous—i.e., this is a step increase. As an amorphous polymer passes through its glass transition temperature (T_g), it goes from a rigid glass to an

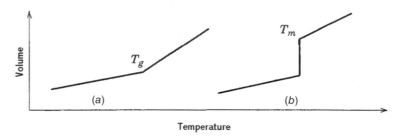

Figure 2 Comparison of glass transition temperature: (*a*) to a melting point; (*b*) of a thermoplastic polymer. *Source*: From Ref. 1.

amorphous, softer material, which exhibits an increase in volume with increasing temperature. Thus, in selecting a thermoplastic for a plastic part, it is always best to use the material below its T_g or T_m. There are exceptions to this rule. For example, polyethylenes are used in their rubbery region due to their subambient temperature glass transition temperatures.

The *Carothers equation* is included in our discussion of the selection of polymeric materials for plastic-molded parts because of its impact in thermoplastic polymerization:

$$\overline{X}_n = \frac{2}{2 - pr}$$

where $\overline{X}_n$ refers to the number-averaged degree of polymerization, p is the extent of the reaction, and r indicates either the ratio or the purity of the reactants. Basically what this equation teaches us is that one needs a high conversion of pure reactants and the correct stoichiometry to obtain the proper molecular weight. Small changes in purity, incomplete reaction, and incorrect ratio of reactants can have a drastic effect on the moldability of a part or its performance behavior. Thus the consistency and repeatability of a molded part shipped to a customer are highly dependent on the consistency and repeatability of the material from the polymer manufacturer, compounder, and molder.

2.2 Plastics

Plastics are polymers with *additives*. These additives perform many different functions. Some refer to these materials as "foo foo" dust. The addition of additives to the polymer enhances the manufacture of the part, the product performance, the lifetime of the part, and the appearance of the part. A partial list of these additives includes antioxidants, light stabilizers, acid scavengers, lubricants, polymer-processing aids, antiblocking additives, slip additives, antifogging additives, antistatic additives, antimicrobials, flame retardants, chemical blowing agents, polyolefins, colorants, fluorescent whitening agents, fillers, nucleating agents, and plasticizers.

From the name of an additive one can determine its function. There are exceptions to this, and in most cases we do not know what additive package compounders have put into a polymer. This information is treated as confidential. Thus one must be careful when one switches from one supplier to another. Even though the starting polymer may be the same, different additive packages will affect the performance behavior and lifetime of the designed plastic part. Thus once a material has been selected, this writer highly recommends to completely characterize that material in case the supplier changes the material or its consistency or you are involved in determining product failures. As a consultant, this writer has been involved in several failure analyses projects, and many problems are difficult and expensive to solve without baseline data.

Two types of additives will be defined in this section due to their importance: fillers and plasticizers. Fillers are added to polymers to affect the color and smoothness of the final molded part, to assist in the molding of the part by changing the flow behavior of the plastics, and to reduce the cost of the molded part.

Plasticizers are unique materials. They are added to a polymer to reduce the hardness of the polymer or make it more flexible. A good example is the "smell" we experience when we purchase a new car—a plasticizer causes this new-car smell as well as seats that are nice and soft. Over time the smell is gone, the seats become hard and brittle, and we have to clean the inside of the windshield. The reverse can occur with hydrophilic polymers in the presence of water. The polymer absorbs water, its glass transition temperature drops, and the material becomes softer. In the case of hydrophobic thermoplastic materials oils will have a similar effect.

Thus, it should be noted that these plastic systems are dynamic and are constantly changing. If you design a plastic part for a lifetime of five years, it would be "nice" to test the part for five years under the operating conditions of the plastic part assembly. However, based upon "time to market," this is not an option. One can retain samples of production.

2.3 Reinforced Plastics

Reinforced plastics are plastics containing reinforcing elements within a plastic matrix that are long or short discontinued (chapped) fibers and the parts are manufactured by injection molding. There are many different types of reinforced plastics. Reinforced plastics are in essence forms of composite materials. A composite is a heterogeneous mixture of matrix resin, reinforcement, and other components that act in concert with each other. The matrix resin protects the reinforcement from wear and "glues" the reinforcement in place. The reinforcement provides strength and stiffness to the plastic part and enhances the properties of the matrix resin. Also the reinforcement helps dissipate the energy throughout the structure when impacted.

Properties of reinforced plastics should be obtained from the supplier. There are many variations of reinforcement forms and sizes.

3 PLASTIC MATERIALS SELECTION TECHNIQUES

Selecting plastic materials for a particular design may be a difficult process. It is wishful thinking to hope that anyone can design a plastic part. It would be helpful to have an individual in the organization who has had experience in plastic material selection or a materials engineer that knows the material science of plastics.

This writer's first experience in selecting materials was relatively simple. The criteria were set by the equipment available for the project. Then the temperature, chemical requirements, and number of parts needed were used to pick the material. The eight main criteria were functionality, chemical resistance, external processing (your supplier), internal processing, lifetime, design margins, cost, and greenness of the part. As the demands on the materials increased, subcriteria were developed. The criteria and subcriteria are presented in Table 1.

The above scheme works reasonably well, but it does not include the most important factors in material selection—timing, schedule, or time to market. One factor that this treatment does give proper justification to is cost. However, sometimes cost decisions are not made by material scientists and engineers.

4 PLASTIC-JOINING TECHNIQUES

Table 2 contains a list of most of the techniques used to join two plastic parts together. Each technique has its own unique advantages and limitations. This writer has had more experience

Table 1 Criteria for Selection of Materials (Thermoplastics, Thermosets, Elastomers, and Adhesives)

Main Criteria	Subcriteria
Functionality	Purpose of part
	Type and magnitude of normal service stresses
	Loading pattern and time under load
	Fatigue resistance
	Overloads and abuse
	Impact resistance
	Normal range of operating temperatures
	Maximum and minimum service temperatures
	Electrical resistivity
	Dielectric loss
	Antistatic properties
	Tracking resistance
	Flammability
	Surface finish
	Color matching and color retention
	Tolerances and dimensional stability
	Weight factors
	Space limitations
	Allowable deflections
Materials acceptance	Compatibility with chemicals
	Solvent and vapor attack
	Reactions with acids, bases, water, etc.
	Water absorption effects
	Ultraviolet light exposure and weathering
	Oxidation
	Chemical erosion and/or corrosion (electrochemical effects)
	Attack by fungi, bacteria, or insects
	Leaching of additives from the part material into its environment
	Absorption of components into the part from its environment
	Permeability of vapors and gases
	Normal range of operating temperatures
	Maximum and minimum service temperatures
Environmental concerns	Scrap rates
	Recyclability
	Chloro- and fluoropolymers
Lifetime	Product lifetime
	Reliability
	Product specifications
	Acceptance codes and specifications
Margin	Safety design factors
Internal process	Normal range of processing temperatures
	Maximum and minimum processing temperatures
	Choice of processes
	Method of assembly
	Secondary processes
	Finishing and decorating
	Quality control and inspection
	Contamination

(continued)

Table 1   (*Continued*)

External process (supplier of parts)	Normal range of processing temperatures
	Maximum and minimum processing temperatures
	Choice of processes
	Method of assembly
	Secondary processes
	Finishing and decorating
	Quality control and inspection
	Contamination
	Timing for part design changes
	Timing for prototype molds
	Timing for production molds
	Technical support from supplier
	Contamination
Cost	Materials costs
	Materials availability
	Alternative material choices
	Supplier's availability
	Part costs
	Cost of capital plant: molds and processing machines
	Operation costs of component, including manufacturing and fuel consumption
	Capacity

Table 2   Plastic-Joining Techniques

Adhesive bonding	Infrared welding
Electrofusion bonding	Laser welding
Friction welding	Mechanical fastening
Linear	Beading
Rotational	Hot stakes
Heated tool welding	Interface fits
Hot plate	Molded-in and ultrasonic inserts
Hot shoe	Molded-in threads: riveting, self-threading screws, snap fit
High-frequency welding	Solvent joining
Hot-gas welding	Thermal impulse welding
Induction welding	Ultrasonic welding

with adhesive bonding. In the selection of adhesives one can follow most of the criteria and subcriteria listed in Table 1.

In the adhesive bonding of plastic parts several issues should be considered. First is the failure mechanism that one should stride for in the design. There are three failure mechanisms for an adhesive joint: adhesive, cohesive, and substrate failures. An adhesive failure is failure within the adhesive joint. Cohesive failure is failure between the adhesive and substrate. Substrate failure is failure within the plastic parts.

The golden rule of design using adhesives is to design for cohesive or substrate failure.

A chemical principle that one should consider in the selection of adhesives is "like dissolves like": The more similar the adhesive is to the plastic one is trying to bond together, the stronger the bond.

Another consideration before using adhesives involves surface treatment. Whether one employs mechanical or chemical etching, preparing the surface is a critical procedure in the manufacturing step. Specific surface treatments of the various polymers can be found by searching the Internet.

5 PLASTIC PART DESIGN

This writer recommends the use of suppliers' design guides and the two design books listed in the Suggested Readings. However, there are tricks one can employ when performing failure analyses on molded plastic parts. One technique is to ash a plastic molded part. One can obtain information as to the flow of the plastic and fiber orientation if the part is reinforced.

One can also section the molded reinforced plastic into a smaller specimen that can be analyzed by TGA to determine percent resin and reinforcement contents, which in turn will show the consistency of the molded part.

Sometimes if the plastic has a long distance to flow in a mold, the polymer chains can separate. The smaller polymer chains travel faster than the larger chain. To verify the consistency of the polymer molecular weight throughout the molded plastic part, one can section the part and subject the specimen to gel permeation chromatography.

6 PLASTIC PART MATERIAL SELECTION STRATEGY

After reading the first part of this chapter one may either be totally confused or have a headache. This writer understands perfectly. Materials selection is not an easy task. The following represents a series of suggestions and hints that hopefully make the task easier.

First, if you are part of a large organization, develop a team that may help you. This writer is in favor of at least a four-person team. The team should consist of a design engineer, a materials engineer (analytical chemists may be good substitutes if there are no materials engineers), an internal experience engineer, and a representative from the procurement department or an internal company buyer.

Some team selections are obvious. Each can handle part of the criteria and subcriteria listed in Table 1 or any other requirements that you develop.

The first step is probably the most difficult: selecting the polymer families to evaluate. Some consult *Modern Plastic Encyclopedia*.[3] This author is partial to Domininghaus's book on plastics for engineers for the selection of a thermoplastic.[4] *Modern Plastics Encyclopedia* is an excellent and well-respected source book. However, it only contains one data point within the total history of a particular thermoplastic. Domininghaus provides pressure–volume–temperature (P–V–T) graphs on the various thermoplastic families. These data are critical in the processing of materials.

The next action is to obtain samples and technical information from suppliers. As previously mentioned, most suppliers have design guides for their polymers. These are an excellent source of information that can be helpful in your efforts to design the plastic part. Also, obtain from the polymer supplier any analytical procedures as to how they characterize their materials. This will assist your internal analytical people to develop a material knowledge database.

Next you and your team should review all the available data on the polymer under consideration. A good literature search through a technical library may save you time, effort, and money. In addition, you may want to perform your own tests to fill in missing information. Polymer suppliers can be helpful in this area. They can provide molded, ASTM (American Society for Testing and Materials) test coupons to use for your own testing. If you are designing a plastic part to be in a certain chemical environment, you may wish to test your selection in a chemical-soak-type test. This can be achieved by soaking a test coupon in the chemical of concern. You may also want to soak the test coupon at different temperatures within the operating range of the designed plastic part or within a linear range of behavior of the polymer. If you can perform such a chemical soak at least three different temperatures, you can predict the lifetime of the polymer if it can be related to a chemical failure. However, data of this sort must be obtained using the principles of chemical kinetics.

An example of aging a part can be found within the different outcomes (failure mechanisms) of an egg. Take an egg and set it on a shelf and leave it alone. After several months the egg becomes rotten. Take a similar egg and place it under a hen and after a while we have a cute little chick. Take another egg and place it in boiling water and after about 10 min you have a hard-boiled egg. Never pick temperatures throughout the viscoelastic region of a thermoplastic; you will obtain three different responses (at its glassy region, at its glass transition temperature, and at its rubbery region) of the material.[5]

Now we can proceed with the plastic(s) of choice. This is the polymer with the magical foo-foo dust that the compounder puts in it for various reasons. Chemical soak tests are extremely important in the cases where the plastic part is used in a chemical environment. We do not want anything from the plastic to be extracted into the "chemical environment," thus either affecting the properties of the plastic or contaminating the chemical environment. The reverse is also true: We do not want the plastic to absorb chemicals from the chemical environment. This could cause the properties of the plastics to be lower due to a plasticization effect.

The same chemical soak tests should be conducted on the final molded part with the chemical soaking to exposed portions of the design. In addition, to develop a material knowledge base for the particular part you are designing, you will need these data in the event you or others have to perform failure analysis on the molded part.

At this stage you should be dealing with the molded plastic part assembly. The next step is to design tests that reflect the functionality of the plastic molded part assembly after it leaves your facility and is in the field.

7 CONCLUSION

The goal of this chapter was to provide hints, suggestions, and tricks to assist in the selection of materials for plastic parts design. These hints, suggestions, and tricks have helped this writer in various industrial positions and in consulting projects that have been completed. We have all heard the phrase "if you don't have time to do it right the first time, you won't have time to redo it." It is always more productive to do a project right the first time!

This author has participated in projects in which millions were spent to develop a plastic assembly but the project failed because nothing was spent on an analytical test. Doing such a test is critical to the success of any project.

We often hear "mechanical engineers can pick materials." This may not be true. But, again, this author has participated in projects where simply consulting a polymer handbook would have saved thousands of dollars and many months of time. For example, polyethylene terephthalate was selected as the material of choice for a plastic part that had to withstand an internal processing step of being adhesively bonded to another part for 2 min at 150°C. The grade of polyethylene terephthalate used was a recycled grade with a glass transition temperature in the vicinity of 60°C. Placing the final assembly in an oven at 60°C to simulate an aging test failed all parts due to changes in the dimensions of the part. Subjecting a polyethylene terephthalate coupon to 60°C testing or checking the literature would have been helpful.

Another example is the use of polystyrene as throw-away coffee cups. Several years ago a fast-food chain was sued for injuries a customer suffered for drinking coffee from one of these cups. Part of the injuries occurred because the cup was made from polystyrene. Polystyrene has a glass transition temperature in the vicinity of 100°C. This temperature is the same as the boiling point of water. Coffee that is extremely hot can reach temperatures close to its glass transition temperature or in the vicinity of when the material starts to transit from the glassy region to the rubbery region of the polymer. In one case it did, thus losing its structural integrity, causing the coffee to spill out of the cup, burning the customer, and resulting in legal action.

ACKNOWLEDGMENTS

This author has been fortunate to have the benefit of good books, good teachers, a good network of suppliers, and good co-workers who have been part of my team as well as engineers who were willing to learn and management that had faith in my methods. I am grateful to all of these.

REFERENCES

1. L. H. Sperling, *Polymeric Multicomponent Materials, an Introduction*, Wiley-Interscience, New York, 1997.
2. R. J. Seyler, "Opening Discussions," in R. J. Seyler (Ed.), *Assignment of the Glass Transition, STP 1249*, American Society for Testing and Materials, Philadelphia, 1994.
3. *Modern Plastics Encyclopedia*, McGraw-Hill/Modern Plastics, New york, various editions.
4. H. Domininghaus, *Plastics for Engineers—Materials, Properties, Applications*, Hanser/Gardner Publications, Cincinnati, OH, 1993.
5. K. T. Gillen, M. Celina, R. L. Clough, and J. Wise, "Extrapolation of Accelerated Aging Data—Arrhenius or Erroneous?," *Trends Polym. Sci.*, **5**(8), 250–257, August 1997.

SUGGESTED READINGS

American Society of Materials, handobooks on polymers, adhesives and adhesion, and composites, ASM International, Materials Park, OH.

W. Brostow (Ed.), *Performance of Plastics*, Hanser/Gardner Publications, Cincinnati, OH, 1995.

W. Brostow and R. D. Corneliussen (Eds.), *Failure of Plastics*, Hanser/Gardner Publications, Cincinnati, OH, 1986.

J. B. Dym, *Product Design with Plastics, a Practical Manual*, Industrial Press, New York, 1983.

M. Ezrin, *Plastics Failure Guide Cause and Prevention*, Hanser/Gardner Publications, Cincinnati, OH, 1996.

C. P. MacDermott and A. V. Shenoy, *Selecting Thermoplastics for Engineering Application*, 2nd expanded ed., Marcel Dekker, New York, 1997.

R. A. Malloy, *Plastic Part Design for Injection Molding*, Hanser/Gardner Publications, Cincinnati, OH, 1994.

D. H. Morton-Jones, *Polymer Processing*, Chapman & Hall, New York, 1989.

T. A. Osswald, *Polymer Processing Fundamentials*, Hanser/Gardner Publications, Cincinnati, OH, 1998.

T. A. Osswald and G. Menges, *Materials Science of Polymers for Engineers*, Hanser/Gardner Publications, Cincinnati, OH, 1995.

Plastics Design Library Staff, *Handbook of Plastics Joining, a Practical Guide*, Plastics Design Library, Norwich, NY, 1997.

A. Rubin, *The Elements of Polymer Science and Engineering, an Introductory Text for Engineers and Chemists*, Academic, New York, 1982.

R. J. Young and P. A. Lovell, *Introduction to Polymers*, 2nd ed., Chapman & Hall, New York, 1994.

CHAPTER 12

DESIGN FOR SIX SIGMA: A MANDATE FOR COMPETITIVENESS

James E. McMunigal
MCM Associates
Long Beach, California

H. Barry Bebb
ASI
San Diego, California

1 BACKGROUND

In contrast to the relatively standardized Six Sigma define, measure, analyze, improve, and control (DMAIC) process, an array of very different implementations of Design for Six Sigma (DFSS) has emerged. The different versions are identified by acronyms derived from the first character of the phase names, such as IDDOV, DIDOV, DMADV, DMEDI, ICOV, I^2DOV, and CDOV. While the various renditions of DFSS reveal significant differences, common themes can be found. In the broadest context, Six Sigma is characterized as an improvement process while DFSS is characterized as a creation process. At a somewhat lower level, most renditions of DFSS contain treatments of customer requirements, concept development, design and optimization, and verify and launch structured into the process phases like those mentioned above. For example, the five phases of IDDOV are identify project, define requirements, develop concept, optimize design, and verify and launch.

Conversely, the differences between implementations of DFSS can be large. The greatest difference resides in the choice of the foundation discipline, namely, statistical methods versus engineering methods. Renditions of DFSS that are tightly linked to Six Sigma process improvement methodologies are appropriately founded on statistical methods for consistency. Renditions of DFSS that are tightly linked to product development or improvement processes are best founded on engineering methods in order to capture the maximum possible benefits.

The IDDOV rendition of DFSS is described in the first book published on DFSS[1] and is chosen as the basic rendition for this chapter. While Chowdhury's DFSS book[1] was written for managers, this chapter is written for engineers and engineering managers.

2   INTRODUCTION

Few executives would dispute the observation that global competitiveness of a manufacturing corporation is largely determined by its engineering capability. Curiously, few of the same executives would dispute the observation that they do not spend a major portion of their time focusing on how to improve the engineering capability within their corporation. The authors believe that *the best products come from the best engineering organizations* and that *implementation of a properly formulated DFSS process is the fastest way to create "the best engineering organization."* These beliefs spawned the title of this chapter.

The Six Sigma DMAIC process has enjoyed unprecedented success, as illustrated by a recent article in *ASQ Six Sigma Forum Magazine*.[2] The Six Sigma process first designed by Mikel Harry and Richard Schroeder, the founders of the Six Sigma Academy, brings all of the key success factors together into a coherent process—top management commitment, a dedicated management structure of champions, and an effective DMAIC process that delivers rapid financial gains to the bottom line, as summarized in their book.[3] Chowdhury further emphasizes the combination of the power of people and the power of process in *The Power of Six Sigma*[4] and *Design for Six Sigma: The Revolutionary Process for Achieving Extraordinary Profits*.[1]

Design for Six Sigma would seem to be a natural sequel to Six Sigma. As Harry and Schroeder[3] point out, "Organizations that have adopted Six Sigma have learned that once they have achieved quality levels of roughly five sigma, the only way to get beyond the five sigma wall is to redesign the products and services from scratch using Design for Six Sigma (DFSS)." This and other broadly recognized and cited needs for DFSS have not yet led to the kind of pervasive application that Six Sigma continues to enjoy, perhaps because of a mixture of myths and truths.

Myths include:

DFSS is only used when needed in the improve phase of Six Sigma.

Six Sigma should be implemented prior to undertaking a DFSS initiative.

DFSS is a collection of contemporary engineering methodologies.

DFSS is not needed because all of the methods and tools are used in the product development process (PDP). (They may be available, but are they actually used?)

Truths include:

There is no single process that one can say is DFSS. Multiple renditions of DFSS have been published, none of which has been broadly accepted as the basic model of DFSS.

Payback from DFSS projects may be larger than from Six Sigma projects, but it takes longer.

Credible renditions of DFSS encompass Taguchi robust optimization methods that appear to be in opposition to statistical methods such as classical design of experiments. The unfortunate controversy that results creates unnecessary confusion for executives.

An objective of this chapter is to dispel the myths and provide a fundamentally sound model of DFSS which the authors believe should serve as the basis for current applications and future renditions of DFSS in engineering environments. Tailoring DFSS to meet the particular needs of different industries and corporations is expected, encouraged, and accommodated.

The authors' experience has been that the IDDOV methodology described in Chowdhury's book on Design for Six Sigma[1] is the best methodology. IDDOV is designed to complement, not replace, the engineering processes indigenous to a corporation's technology and product development processes.

The design of this rendition of IDDOV is not ad hoc. The design is based on benchmarking an array of DFSS processes, research-and-development (R&D) processes, and engineering processes that support PDPs and the authors' experience in dozens of corporations in a broad range of industries, including medical instrumentation, heavy equipment, office equipment, fixed and rotary wing aircraft, automotive OEMs (original equipment manufacturers) and suppliers, chemical, and others in North America, Europe, and Asia.

Major advancements in engineering processes introduced to the Western world within the past 25 years include quality function deployment (QFD), Pugh concept generation and selection methods, TRIZ, Taguchi methods , axiomatic design, Six Sigma, and the various renditions of DFSS. Many of these powerful methods and tools were not in the engineering curriculum when the more experienced engineers within corporations and other enterprises attended universities. Many, if not most, of the methodologies and tools are not yet taught in sufficient depth in most universities. It is left to hiring organizations to provide education and training in these contemporary methodologies. Six Sigma and DFSS provide the "Trojan horse" for effectively bringing these methodologies into an organization.

The IDDOV process selected is consistent with and embodies Taguchi methods, known as quality engineering in Japan. The portion of quality engineering, known as robust engineering in the West, encompasses three phases—concept/system design, parameter design, and tolerance design—which correspond to the develop concept and optimize design phases of IDDOV. Indeed, robust engineering is the heart of IDDOV.

All of these factors strongly influenced Chowdhury's DFSS book[1] and the rendition of IDDOV presented in this chapter.

DMAIC and IDDOV Application Domains. DMAIC is an improvement process that digs down into a particular portion of an existing process or product. IDDOV is a creation process that parallels a PDP. From a logical perspective, DMAIC works vertically between a high-level problem area and a low-level root-cause. IDDOV works horizontally along a time line through the various phases. This vertical-versus-horizontal logic leads to an unconventional representation of DMAIC, as illustrated in Fig. 1.

While IDDOV proceeds horizontally with a PDP, the Six Sigma DMAIC process works vertically down through the various system levels to find and correct root causes and then works back up through the system levels to complete the improve-and-control activities. DMAIC is a powerful process for improving existing processes. The intensity of DMAIC *find-and-fix* activities increases as products proceed toward production and launch.

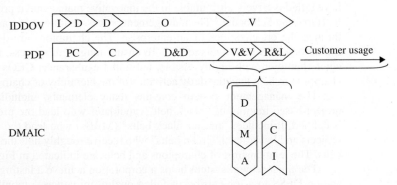

Figure 1 Unconventional representation of intersection of DMAIC and PDP shows DMAIC with a vertical orientation.

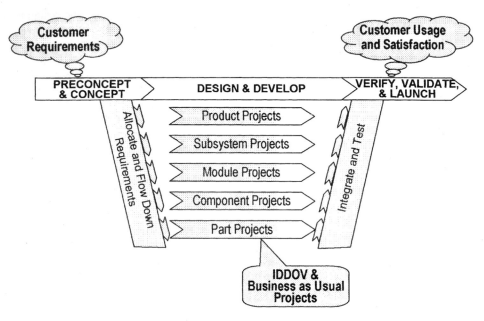

Figure 2 System perspective of a PDP for a complex system.

The Big Picture. A system perspective of a PDP for a complex system is provided in Fig. 2 that shows how IDDOV fits with product development. The sequence of chevrons positioned over a winged-U system diagram represents a typical PDP. The left leg of the winged U depicts the process of allocating and flowing down requirements to guide the development of the system architecture and the various lower level system elements within the architecture. Multiple IDDOV system element projects are depicted at the various system levels. As system element designs are completed, the process of integrating lower level system elements into higher and higher level system elements commences as depicted by the right leg of the winged U.

3 DFSS MANAGEMENT

The magic that has made Six Sigma and DFSS initiatives succeed where other quality initiatives have failed is largely attributable to the innovative management process created and published by Harry and Schroeder.[3] The management process features a hierarchy of "champions" spread through the organizations of an enterprise. The champions are dedicated to making the Six Sigma or DFSS initiative succeed. To ensure long-term success, the CEO must serve as the chief champion with highly visible, unbridled exuberance. CEOs often appoint an executive champion to help lead the daily activities of the hierarchy of champions.

The management system contains many elements, including intense multiple-week, project-based training of "black belt" candidates who lead the projects. A hierarchy of belts is defined. It includes "master black belts" (MBBs) who lead the training and manage larger projects and lower level "green belts" who receive roughly half the training provided for black belts. The relationships of champions and belts are indicated in Fig. 3.

The management system helps a corporation achieve a lasting transformation rather than adding DFSS to a long string of failed quality initiatives—monthly fads. The management system contains two components: (1) a high-level enterprise system for planning, launching,

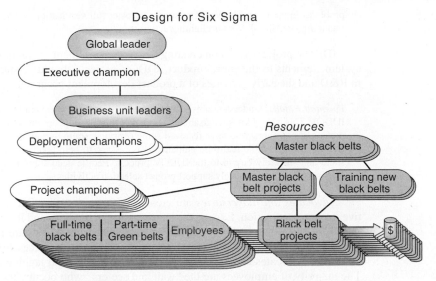

Design for Six Sigma

Figure 3   Relationships between line management, champions, and belts.

deploying, growing, sustaining, and periodically revitalizing the initiative as outlined above and (2) a project management system to ensure excellence of execution of the DFSS process that leads to superior outcomes.

Project selection is a central management task. Selecting the right projects is a primary determinant of success. Project selection is an ongoing process that remains the same from the first wave of training black belt candidates and subsequent waves through the life of the current version of DFSS when many projects are selected for trained and experienced black belts. Project selection is typically led by MBBs working under the guidance of champions, executives, and managers at levels within a big, medium, or small enterprise. During the initial portion of deployment, the MBBs are outside, experienced experts. Over time, an enterprise develops internal MBBs who assume the leadership role.

Criteria are established by top management against which the projects are evaluated. The criteria should be derived from an enterprise's strategic and tactical needs combined with the technical knowledge of the project selection team. In manufacturing corporations, tactical criteria include warranty cost problems; product cost problems; safety problems; regulatory problems; published customer satisfaction ratings from organizations such as J. D. Powers, *Consumer Reports,* and trade publications; and complaints from customers, dealers, service technicians, retail outlets, and others. Strategic criteria include resolution of known problems but focus more on creating new, innovative products and services that cause current customers to repurchase from the same corporation and attract swarms of new customers away from competitive offerings.

Projects Are Sorted into IDDOV, OV, and Other Buckets. Projects selected against strategic criteria tend to be full IDDOV projects. Projects selected against tactical criteria normally require only the OV phases of IDDOV or some other quality process. The OV projects focus on robust optimization of an existing, troublesome subsystem to eliminate the problems and prevent their reoccurrence in the next development program or in the field. The verify-and-launch portion of OV warrants some emphasis. Typically, find-and-fix activities late in a program cycle tend to "patch" problems with changes that do not go through the normal verification and validation processes due to time constraints. It should not be surprising that such patches often turn into warranty, safety, or regulatory problems in the field. The V of OV strives to foster full verification and validation of all changes, including late "fixes" from firefighting activities.

IDDOV projects focus on creating new system elements ranging in scope from low-level system elements to the entire product. IDDOV projects should be started as early as possible in R&D and the early activities of a product development program.

Potential Pitfalls. Understandably, people within the enterprise generally do not know much about IDDOV. This lack of knowledge about the IDDOV process impedes deployment efforts. Even champions and managers who have received some overview training may not yet be comfortable with IDDOV. Considerable real-time guidance is necessary to help people within an enterprise understand and sincerely buy in to the IDDOV process. People need some level of understanding and buy-in in order to effectively support project selection activities.

Advocates and adversaries are created with the introduction of any major, new initiative in any organization. Experience suggests that in medium and large organizations the ratio of adversaries to advocates is about 10 to 1. Visible adversaries are not necessarily obstructionist. However, invisible adversaries are a real threat to successful deployment of a new initiative. Advocates and adversaries occupy the leading and lagging tails of the distribution. The majority of employees are the "wait and see-ers" who occupy the middle 70–80% of the population. This human element is real and must be dealt with by the MBBs upfront. Pronouncements of firm commitment to the new initiative from on high can help, but taken alone, pronouncements—and even directions—from the CEO do not solve the daily human problems that arise during project selection activities.

Pet projects are often motivated more by emotion than logic, which can blur objectivity. Owners of pet projects may be advocates pushing for their project to be selected or adversaries who want no part of IDDOV and hide their projects from the MBBs. Some are good IDDOV candidate projects and some are duds. It can be very difficult for outside MBBs to tell the difference. The only protection against selecting the wrong projects is for the MBBs to dig relatively deeply into the details with the project team members. It is often hard to get people to take time away from what they regard as their real job to work with MBBs to select projects for inclusion in a process that they know nothing about.

Secrecy is another barrier to ferreting out the right projects. Employees are naturally and properly dubious about revealing corporate secrets, even with knowledge that nondisclosure agreements have been signed. Adversaries will use secrecy to hide their projects. Even conscientious advocates may tend to reveal the less sensitive elements of a project when full disclosure is needed to understand if a project is a suitable candidate for IDDOV.

Inadequate time may be allocated for project selection. The project selection process can take a few weeks or a few months depending on the size of the enterprise and the complexity of its products—but it must be done well, however long it takes.

The above pitfalls are couched in the context of initial deployment. However, deployment is a continuing activity over a period of time measured in years. Initial deployment does not involve the entire organization. Deployment is rolled out sequentially to different portions of an enterprise over time. An aged initiative is still new to the people downstream from earlier roll-outs. Even as internal MBBs begin to displace external consultants, the cited challenges continue to exist until the initiative is pervasive throughout the enterprise. Then, the challenge shifts toward maintaining momentum and excitement.

Project selection must be completed prior to the initiation of the project-based training of black belt candidates. The output of project selection is a set of project charters delivered to the team leaders (black belt candidates) selected to participate in training. The charter contains the information needed to define the scope of the project and to help the team develop project plans.

4   DFSS PROCESS

4.1   Introduction

Storyboards have become a common means of conveying a process in Six Sigma and DFSS. The storyboard of the IDDOV process is displayed in Fig. 4.

Abridged descriptions of the five phases that follow are derived from Ref. 5.

Identify project	Select project, refine its scope, develop project plan, and form high-powered team.
Define requirements	Understand customer, corporate, and regulatory requirements and translate them into technical requirements using QFD.
Develop concept	Generate concept alternatives using Pugh concept generation, TRIZ, and other innovative methods. Select the best concept design and technology set using Pugh concept selection methods and conduct failure mode and effects analysis (FMEA) to strengthen the selected concept/technology set. The concept may be developed at several levels starting with the system architecture for the entire product. Then, concepts are developed for the various system elements as needed.
Optimize design	Optimize technology set and concept design using robust optimization. Because Taguchi's robust engineering methodologies are central to the IDDOV formulation of DFSS, more detail is presented in this "O" phase.
Verify and launch	Finalize manufacturing process design, conduct prototype cycle, pilot run, ramp up to full production, and launch product.

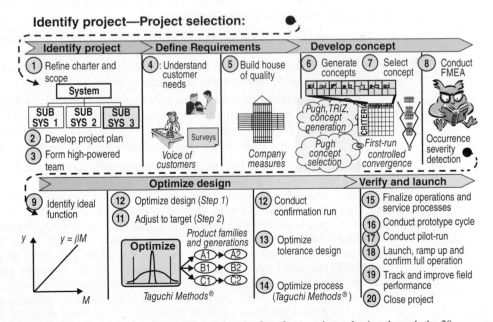

Figure 4   Visual storyboard of IDDOV showing the flow from project selection through the 20 process steps.

The first two phases, identify project and define requirements, focus on getting the right product. The last two phases, optimize design and verify and launch, focus on getting the product right. The middle phase, develop concept, is the bridge between getting the right product and getting the product right.

As the bridge, conceptual designs should simultaneously respond to upstream requirements and downstream engineering, manufacturing, and service requirements. Optimization is one of many downstream requirements. If a concept design cannot be optimized to provide the required performance, it may be necessary to loop back and seek a new concept. Concept designs that cannot be optimized should be discarded. Marginal concepts are a major source of chronic problems that emerge in program after program. Concepts that do not optimize well are very sensitive to sources of variation. Hence, any deviation from ideal conditions causes a marginal concept to fail.

4.2   Process Details

The IDDOV process is described step by step through the five phases of IDDOV as outlined in the storyboard.

Identify Project. the "I" of IDDOV is the first phase of activities executed by the black belt candidate and project leader together with the team on the selected project. Intense project training commences. A MBB instructor may explain in the first day that "Projects are the means to get things done. The famous and overused list of action items seldom caused anybody to do anything that was important. Until action items are turned into projects with objectives, resources, and plans, very little can be accomplished." The instructor may emphasize that his or her first project is expected to deliver meaningful results to the enterprise with financial benefits measured in millions of dollars.

The identify project phase contains three storyboard steps: (1) Refine the charter and scope of the project. (2) Develop the project plan. (3) Form a high-powered team.

Step 1: Refine Charter and Scope. Two scope tools are introduced—the *in–out of scope tool* and the *multigeneration plan (MGP) tool.*

> The in–out of scope tool generally elicits a range of humor on first encounter since it is nothing but a circle drawn on a sheet of paper. However, it is magic when put to use. This simple visualization tool greatly facilitates team participation in understanding, refining, and reaching consensus on the scope of the project.

> The MGP tool is a more involved tool to help the team determine what elements of scope should be included in the current project and what elements should be put off to future projects. The MGP typically contains three categories, such as product generation, platform generations, and technology generations, spread over three generations.

The primary purpose of both tools is to help define the scope of the current project, not to engage in long-range planning as the MGP tool might suggest.

Step 2: Develop Project Plan. Planning tools are more extensive. The foundation of all planning is the Shewhart–Deming plan–do–check–act (PDCA) circle. A special feature of the PDCA circle is the check–act step, which involves checking the progress against the plan and acting on the gaps on some predetermined cadence. The first use of the check–act pair should be to evaluate the completeness and doability of the completed project plan. Critical path/PERT (Program Evaluation Review Technique) is another important planning tool which combines the best elements of the two distinct methodologies into a single tool. A number of other planning tools like Gantt charts are introduced for use in the planning process.

Step 3: Form High-Powered Team. While project members should be identified prior to initiation of a project, some effort is usually required to bring them together as a team. Basic team-building methods for developing a high-powered team are introduced during training.

While this first critical phase, identify project, establishes the likelihood of the success of the project, the discussion is abbreviated since it is competently covered in broadly available Six Sigma and DFSS books and literature.

Define Requirements. The first "D" of IDDOV contains two critical-to-success actions, storyboard step 4 (understand customer requirements), and step 5 (build house of quality).

Step 4: Understand Customer Requirements. Customer, corporate, and regulatory requirements must be thoroughly understood if a team expects to deliver winning products to customers. Pare[6] suggests, "Don't overlook the most obvious way to get to know your customer—*talk to them.* Companies are exploring new ways of doing this." Peters[7] cites a study of 158 products in the electronics industry in which half were failures and half were successes:

- The unsuccessful products were technological marvels.
- The successes came from intense involvement with customers.

The successful firms had better and faster communications between customers and development teams. *Customer inputs were taken seriously!*

It is crucial to get the people responsible for delivering the products out and about with customers, with the support of marketing people. Throwing requirements over the wall from marketing to engineering is just as bad as engineering throwing drawings over the wall to manufacturing. Second-hand information cannot be as good as personal experience. Engineering and marketing people can be reluctant to have engineers talking with customers in prospect but are excitedly transformed in retrospect.

Don't Forget the Internal Customers. The primary internal customer is the recipient of the output of a project. Juran uses the phrase *Fit for use by next in line.* For the output of the project to be fit for use, the requirements of those next in line must be understood upfront. Stakeholders and sponsors are also customers. Treat internal customers as if they were external customers. In a formal sense, the boss is never the customer. Value-added activities flow horizontally. Vertical activities are overhead; they may be necessary, but they are not value adding.

The *Kano model*[8] was developed to classify requirements into basic needs, performance needs, and excitement needs. *Basic needs* are expected by customers and are often not spoken. Basic needs such as quality and reliability are established by industry standards. Customers also develop expectations about performance and features such as heating and cooling in automobiles. (Early automobiles did not have heaters or windshield defrosters. After all, buggies didn't!) *Performance needs* are items customers may desire that exceed the capability of their previous purchases. These needs are usually spoken in terms of more is better—larger flat-panel TV screens, easier to use alarm clocks in hotel rooms, more horsepower and less wind noise in vehicles. *Excitement needs* are seldom spoken because they are unknown. These needs are created by innovators within a corporation during concept development activities to gain competitive advantage.

Step 5: Build House of Quality. The house of quality (HoQ) is the centerpiece of QFD. The HoQ provides an effective methodology for translating customer requirements into technical requirements. A completed HoQ contains an enormous amount of easy-to-access information that a team needs to guide the creation of concept designs that will create customer excitement. Technical requirements are often called company measures to indicate that they are the internal responses to external customer requirements.

The HoQ gets its name from the shape of the gabled matrix made up of different rooms for the different types of information contained in the HoQ as summarized below (Fig. 5). The HoQ is structured so that all customer information is entered horizontally and all company information is entered vertically. It is important to provide all customer information prior

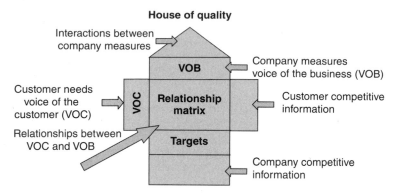

Figure 5   The HoQ with central rooms indicated.

to undertaking the development of company information. The information is developed and entered into the HoQ in the order presented below.

Customer Information (Horizontal Entries)

1. Customer needs (*Gather the voice of the customer to identify customer requirements.*)
2. Critical customer requirements (CCRs) consolidated from "raw customer needs" (*To keep the HoQ from becoming too unwieldy, it is important to limit the number of customer requirements to those that are truly critical.*)
3. Company and regulatory requirements
4. Customer importance ratings for each CCR
5. Customer competitive comparisons of the company's product with competitive products on a CCR-by-CCR basis
6. Customer complaint history

Company Information (Vertical Entries)

7. Critical company measures (CCMs) consolidated from company measures
8. Technical competitive comparisons for each CCM (*Tear down and compare competitive products.*)
9. Strengths of relationships between critical customer requirements and company measures (*A relationship matrix links company measures and customer measures. Strengths of the interactions between CCRs and CCMs are indicated by three weighting factors, typically 9, 3, 1 for strong, medium, and weak. The strength assigned each company measure indicates how much a change in the value of a company measure will change the degree of meeting customer needs.*)
10. Service history (*usually gathered at service centers*)
11. Target values for each company measure
12. Strengths of interactions between company measures (*The correlation matrix at the top of the HoQ indicates positive and negative interactions between the company measures. Negative interactions indicate a need for making trade-offs between company measures. A positive interaction indicates that the interacting company measures support each other in a way that can allow a reduction from the ideal of one of the*

company measures without a significant impact on customer needs. Positive interactions sometimes provide opportunities for cost reductions.)

13. Degree of organizational difficulty for each company measure (*Organizational difficulty relates to competency and capacity necessary to achieve the intent of the company measure.*)

14. Importance rating for each company measure (*This is a composite rating of the customer importance rating, the strength of relationships, and the degree of organizational difficulty to help guide the choice of company measures to be carried forward throughout the project. Selecting only the most critical company measures to carry forward is another opportunity to simplify the process.*)

Quality function deployment has fallen from favor in many corporations. The two main reasons for its fall from grace are as follows:

1. Teams try to develop so much information in the HoQ that it becomes unwieldy. It is difficult for people, especially engineers, to discard information even when it is pointed out that information not entered into the HoQ can be carried along with the HoQ.

2. The HoQ is best developed by a team of marketing and engineering people working together. In compartmentalized corporations, marketers and engineers seldom have either the opportunity or the desire to work together. Neither marketing nor engineering can gather customer information and build the HoQ without help from each other.

QFD remains the strongest method of gathering and interpreting customer information. The toughest competitors continue to pursue excellence in using QFD for competitive advantage. It is so important that most renditions of Six Sigma and DFSS include QFD as a means of reintroducing it into Western world corporations.

Develop Concept. The second "D" of IDDOV—the develop concept phase—provides the maximum opportunity for innovation. Many pundits have asserted that the only way to achieve and sustain competitive advantage is to innovate faster than the toughest competitors. *In long races, whether leading or lagging, those who go fastest win!* Going fastest carries internal risk. Going slower than competitors carries external risk. During Jack Welch's long tenure as CEO of GE, he often emphasized, "Change or die."[9]

Innovations range from clever problem solving to breakthrough innovations. Innovation is the accelerator pedal within any enterprise (Fig. 6). Small and large innovations help to increase the rate of improvement within an enterprise. Innovations are created through the conceptual design process. Concept development utilizes a creativity toolkit that contains a broad array of tools and methods for generating, synthesizing, and selecting ideas for many purposes,

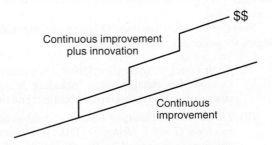

Figure 6 Innovation is accelerator pedal.

including products, manufacturing processes, service processes, business processes, business strategies and plans, and innovative solutions for technical and nontechnical problems.

Winning products can only come from winning concepts. The creation of product concepts is simultaneously critically important, very difficult, highly exciting, and often neglected. Pugh,[10] the creator of Pugh concept generation and selection methods, states, "The wrong choice of concept in a given design situation can rarely, if ever, be recouped by brilliant detailed design." The concept design is really the first step of detailed product and process design. The quality of the concept design is a major determinant of the quality of the final design.

Some of the rationale behind Pugh's often-quoted observation is that concept development:

- Provides the greatest opportunity for innovation
- Places the greatest demands on the project team
- Is where *the most important decisions* are made
- Is where 80–90% of the cost and performance are locked in

A good concept design provides the foundation for all downstream endeavors, including optimization, detailed design, manufacturing, service, and customer usage. How well a concept design optimizes using Taguchi methods is a key measure of its quality. A concept design that does not optimize well is a poor concept. Poor concepts cannot be fixed by end-of-process firefighting. Indeed, poor concepts create chronic problems that seem to reappear in every new program and create disappointments in product performance, cost, quality, reliability, and useful life.

The huge amount of information that is generated during conceptual design places enormous demands on the team. Concept development is, indeed, a most difficult step in the PDP. In traditional development processes, the opportunity to achieve competitive superiority has been either won or lost by the end of concept development.

Conceptual design can be exciting since most engineers find the opportunity to create something new far more fun than the tedium of detailed design. Whether seeking better mousetraps, car door weather strips, or flying machines, stretching one's imagination to create new widgets is fun, challenging, and rewarding. Where else in the PDP is a person likely to hear, "Eureka, we did it. I didn't think it was possible, but we did it!"

Conceptual design is often neglected because neither managers nor engineers appreciate its importance. Ever-increasing pressures to get product out the door on shorter and shorter schedules tend to disproportionately squeeze the time allocated for the unmanageable, "fluffy sandbox" of concept development. Managers seem to operate on the premise that good concepts can be developed in whatever abbreviated time is allocated to the task—the less time, the better.

Adequate time, resources, and support are needed to foster the pursuit of excellence of execution and the realization of meaningful results. The extra time needed to do things right up front is more than offset by time and resources saved later in the process. The hierarchy of champions can provide enormous support to ensure excellence of execution during the develop concept phase.

Step 6: Generate Concepts. This step encompasses two very different innovation methodologies for generating concept alternatives:

(i) *Pugh Concept Generation Using a Creativity Toolkit.* The toolkit contains familiar methods for creating ideas including brainstorming, brain-writing 6–3–5, painstorming, analogy, assumption busting, and other related methods.

(ii) *TRIZ or TIPS (Theory of Inventive Problem Solving) Created by the Brilliant Russian Engineer Genrich Altshuller.* TRIZ is a logical methodology to help anyone become creative, as suggested by the title of one of Altshuller's early books: *Suddenly, an Inventor Was Born.*

The output of the phrase define requirements—the HoQ—is the primary input into the develop concept phase. The HoQ information can be arranged and augmented as necessary to develop criteria for a good concept. Developing the criteria for a good concept should be completed prior to initiation of concept generation activities. The process flow for this phase is

HoQ >> criteria for a good concept >> concept generation (*Pugh and TRIZ*)

>> concept selection >> controlled convergence >> system FMEA

Develop criteria for a good concept. A good concept satisfies upstream customer requirements (*get the right product*) and downstream engineering requirements (*get the product right*). Typically, criteria are not developed until concept selection. Developing criteria prior to concept generation has several benefits. First, the process of developing the criteria causes the team to collectively focus on the details in the HoQ and develop a common understanding of the upfront customer requirements. Second, development of the criteria requires consideration of the downstream technical requirements. Third, the criteria provide guidance during concept generation.

General, high-level criteria establish the framework for developing more detailed requirements specific to the project that a product concept must satisfy:

Upstream customer requirements: Satisfy internal and external customer requirements (HoQ).

Downstream engineering requirements: (1) Do not depend on unproven technologies, (2) can be optimized to provide high quality/reliability/durability at low cost, and (3) can be manufactured, serviced, and recycled at low cost.

Company measures with target values from the portion of HoQ for an automobile door system are displayed in Fig. 7. The critical company measures—those selected to be carried forward against the criteria of new, important, or difficult—are boldface. Customer importance for door-closing effort from outside and inside was rated 5 and 4, respectively. However, easy door-closing effort and good weather strip sealing are in conflict, making the simultaneous achievement of easy closing and good sealing difficult. Hence, door-closing effort was selected as the critical company measure to be carried forward because of both its importance and difficulty.

Consideration of the four, general, high-level requirements provides the basis for developing criteria to guide the generation and selection of alternative concepts. A team starts with the deployed upstream requirements and adds downstream technical criteria:

Deployed Requirements		
Door-closing effort from outside	7.5 ft·lb	Closing effort
Pull force from inside	12 ft·lb	

Basic Door System

1	**Door closing effort outside**	**7.5 ft·lb**
2	Door opening effort outside	15 ft·lb
3	Door opening effort inside	8 ft·lb
4	Reach dist.-opening mech	26 in.
5	**Pull force inside**	**12 ft·lb**
6	Dynamic hold open force	15 ft·lb
7	Static hold open force	10 lb

Figure 7 Company measures for basic door system from HoQ.

Technical Criteria Added By Team	
No water leak	4-h soak/wind zero leak
Noise transmission	Test required
Wind noise	Test required
Design for assembly	Number of installation operations
Design for serviceability	Seal extraction/insertion forces
Robustness	Optimization experiment

Pugh's methodology involves three distinct activities:

(i) Concept generation is the process of creating a number (typically four to eight) of concept alternatives. TRIZ is an important complementary methodology for concept generation.

(ii) Concept selection is a process of synthesizing the best attributes of the alternatives into a smaller number of stronger alternatives and selecting a small number (one to three) of the stronger concepts.

(iii) Controlled convergence is the process of iterating between the first two activities with the objective of further synthesizing and winnowing down to the strongest one or two alternatives.

Concept generation, concept selection, and controlled convergence provide the framework for the develop concept phase of IDDOV.

For concept generation using Pugh methods, concept generation should be planned as an exciting activity that draws out the highest possible levels of creativity from team members. People who choose product or manufacturing engineering as a profession tend to be more analytical than creative. Many, probably most, technical people within an organization do not regularly practice serious creativity and have little interest in starting now. Engineering environments seldom foster flamboyant creativity.

It is best to employ a trained facilitator who knows how to foster a free-flowing, creative environment and helps team members effectively select and use the tools within the creativity toolkit. Basic tools include brainstorming, brain writing 6–3–5, pain storming, analogy, and assumption busting. With the exception of pain storming, these tools are broadly known and treated in numerous creativity and quality books.

Pain storming involves brainstorming the opposite of what you want to achieve. It is a way to help the team look at the problem from a different perspective. If the team gets stuck or reaches a dead point, the facilitator might suggest pain storming to change the perspective by focusing on the "antisolution." Suppose your topic is how to speed up invoice preparation. Change the topic to the antisolution, how to slow down invoice preparation. Then, use brainstorming to generate ideas for the "anti" topic. This simple process of examining the topic from the opposite perspective usually stimulates a flood of new ideas. Pain storming is sometimes called improvement's "evil twin."

Pugh's famous work hints at using three cycles for concept generation:

1. Group activity—gathering information and developing shared purpose
2. Individual activity—creating ideas
3. Group activity—combining, enhancing, improving, and refining ideas

Organizing concept generation efforts around these three separate steps dramatically enhances the process.

Display drawings with descriptors

1. A basic compression strip of sponge rubber (*the company's current concept, which is used as the datum*)

2. A deflection strip of sponge rubber

3. Combination of compression/deflection using thinner walled section sponge rubber

4. Double-sealing strips with body side compression and sponge rubber. Body strip attached with plastic nails.

5. Double-sealing strips with body side deflection and sponge rubber. Body strip attached with plastic nails.

6. Compression/deflection using foam rubber pressed into metal carrier

7. Thin-wall low-compression strip which inflates with air after door is closed. This concept reduces door-closing effort caused by "trapped" internal air.

Figure 8   Concept alternatives for automobile door weather strip.

The criteria for easy door closing and good sealing of an automobile weather strip was developed above. The team used Pugh concept generation methods to create several alternatives, shown in Fig. 8.

Concept 7 was the most innovative. It consisted of an inflatable seal which was deflated to provide easy door closing and inflated after the door was closed to provide good sealing. This study was conducted in the mid-980s. The inflatable seal was used by Mercedes about 10 years later.

For *concept generation using TRIZ methods,* the inflatable seal could have been derived using the TRIZ principles of separation. The separation principles include (1) separation in space, (2) separation in time, and (3) separation in scale. The inflatable seal is an example of separation in time—deflated at one time and inflated at another time (*before and after door closing*). (For more information on the TRIZ methodology, refer to Chapter 18.)

Step 7: Concept Selection. This is a process of synthesizing (*combining and separating*), enhancing the strengths, and attacking the weaknesses of the concept alternatives as an integral part of the process of selecting the best concept. Pugh's concept selection methodology is more than a process for simply selecting the best of the concept alternatives generated. It is as much a concept improvement methodology as it is a selection methodology. The steps in concept selection are as follows:

1. Prepare for first run.

   a. Prepare characterizations of concepts—drawings, models, word descriptions, videos, working prototypes, etc.

   b. Identify evaluation criteria for synthesis and selection of concepts.

   c. Prepare evaluation matrix with drawings of concepts across the top row and criteria down the first column.

   d. Select datum, usually the current concept. (*A better choice is the best competitive concept.*)

2. Conduct first run.

3. Conduct confirmation run.

4. Conduct controlled convergence runs (*additional runs to exhaustion of new concepts*).

   (a) *Pugh Step 1: Prepare for First Run.* The characterization of the concept alternatives is illustrated in Fig. 8. The second item, identify evaluation criteria, was done prior to concept generation. The next two items, prepare evaluation matrix and select

Criteria \ Concept	1	2	3	4	5	6	7
Closing effort	DATUM	+	+	+	+	s	+
Compression		+	+	+	+	−	+
Set		+	+	+	+	+	+
Meet freeze test		s	s	−	s	+	s
Durability		s	s	s	s	−	−
Section change at radius		−	s	+	+	+	+
Squeak		s	s	s	s	s	s
Water leak		s	s	+	+	s	+
Wind noise		−	s	+	+	s	+
Pleasing to customer		s	s	s	+	s	s
Accommodate mfg. var.		s	s	+	s	s	−
Process capability		s	−	−	−	+	−
Cost		s	s	−	−	s	−
No. installation operations		s	s	−	−	s	s
R & R for repair		s	s	−	−	+	s
Robustness		−	s	+	+	s	+
Total + / −		3 / 3	3 / 1	8 / 5	(8) / (4)	5 / 2	7 / 4

Figure 9   Evaluation matrix with +'s, −'s, s's entered into matrix.

datum, are combined with Pugh step 2 (conduct first run) in Fig. 8, which shows the information from the first run in the evaluation matrix.

(b) *Pugh Step 2: Conduct First Run.* This step generates the matrix depicted in Fig. 9. The entries in the matrix, pluses, minuses, and sames (+'s, −'s, s's) are determined by comparing each concept with the datum concept for each criterion and determining whether the alternative concept is better, worse, or the same as the datum concept. The sums are used to evaluate the alternatives. There is no value in summing the s's. Pugh makes a number of suggestions about attacking the negatives and enhancing the positives in the synthesis process of striving to use the best attributes of strong concepts to turn negatives into positives in weaker alternatives and enhancing the positives of the stronger alternatives. Alternative 5, with the circled eight positives and four negatives, was selected as the strongest alternative.

(c) *Pugh Step 3: Conduct Confirmation Run.* This step is carried out by using the selected alternative as the datum and running the matrix a second time. The output of steps 1–3 is typically one to three relatively strong concepts.

(d) *Pugh Step 4: Conduct Controlled Convergence Runs.* This step involves multiple iterations of steps 1–3 to further improve the concept. Often the most innovative ideas arise during controlled convergence runs when the team members have become very familiar with all of the concept alternatives and have generated new ideas during the synthesis process. It is a powerful methodology of intertwining divergent and convergent thinking to improve, often dramatically, product concepts. However, it is too often passed over as excessively time consuming under the pressures to get on with the real work of detailed design. Pugh countered arguments for shortcutting any of the concept selection process with "One thing is certain. It is extremely easy to select the wrong concept and very difficult to select the best one."[10]

Some teams insist on using weighted matrices in the concept selection process. Weighting unnecessarily complicates the process. Numerical weighting of criteria is good practice when making a single-pass decision about known alternatives such as selecting automobiles,

Controlled convergence

intertwines generation, synthesis, and selection in ways that strengthen both creativity and analysis.

It involves alternate convergent *(analysis)* and divergent *(synthesis)* thinking.

It helps the team to attack weaknesses and enhance strengths.

As team members learn more and gain new insights, they will consistently derive and create new, stronger concepts.

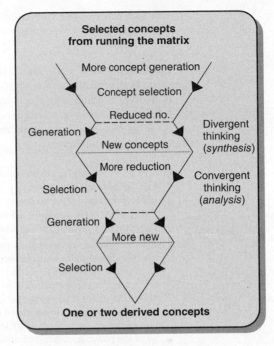

Figure 10 Visual depiction of controlled convergence process.

computers, cameras, or other items based on quantitative information about performance, features, price, etc. However, weighting criteria are not needed or advised when synthesizing attributes between alternatives to create improved or entirely new concepts, and weighting becomes unmanageable when conducting multiple runs in controlled convergence.

It is useful to rank order the criteria and look for concept alternatives with lots of +'s in the upper portion of the evaluation matrix.

Finally, note that robustness is the last criteria listed in the evaluation matrix. (*The team did not rank order the criteria.*) The final test of the strength of a concept selected is how well it optimizes. If the team selected more than one concept for further evaluation, conducting robust optimization experiments can be used to make a final selection.

The controlled convergence process is depicted in Fig. 10.

Step 8: Conduct FMEA. FMEA is a powerful methodology for strengthening the selected concept. At this stage the system FMEA is conducted. Refer to numerous books on FMEA for details.

Optimize Design. The "O" of IDDOV focuses on the Taguchi methods as the "heart" of IDDOV. Taguchi's robust optimization requires engineers and others to think differently in fundamental ways. Some of the premises in the Taguchi methods that require breaking traditional engineering thought patterns include the following:

A. *Work on the intended function, not the problems (unintended functions).* Problems such as heat, noise, vibration, degradation, soft failures (restart computer, clear paper jam in copier), high cost, etc., are only symptoms of poor performance of the intended function. Working on a problem such as audible noise does not necessarily improve the intended function. More often, another problem pops up. This is pejoratively called "whack-a-mole engineering," after the children's game. Whack one mole and another pops up somewhere else—whack one problem and another pops up somewhere else. The notion of optimizing the intended function to whack all moles with one whack rather than working directly on the problem (say, customer complaints about audible noise) is totally upside-down thinking for engineers who have spent much of their working life whacking problems.

Controlled convergence intertwines generation, synthesis, and selection in ways that strengthen both creativity and analysis.

It involves alternate convergent (*analysis*) and divergent (*synthesis*) thinking.

It helps the team to attack weaknesses and enhance strengths.

As team members learn more and gain new insights, they will consistently derive and create new, stronger concepts.

B. *Strive for robustness rather than meeting requirements and specifications.* This is another notion that demands thinking differently. Traditional engineering measures of performance include meeting requirements and specifications, reliability data, warranty information, customer complaints, Cp/Cpk (Cp is the capability without respect to target, and Cpk is the capability with respect to target process capability), sigma level, scrap, rework, percent defective, etc.

C *Use the signal-to-noise ratio as the measure of performance* (*robustness*). Increasing the signal-to-noise ratio simultaneously improves the performance against all traditional measures—whacks all moles with a single whack. The signal-to-noise ratio, borrowed from the communications industry, is simply a representation of the ratio of the power in the signal (the music) to the power in the noise (the static). The greater the signal-to-noise ratio, the greater the portion of total power that goes into making music compared to the power available to make static—by the law of conservation of energy. In an ideal AM/FM receiver or, more generally, the ideal function of any system, all of the power goes into making music, the intended function; no power remains to make static the unintended function.

D *Use multifactor-at-a-time (MFAT) experiments rather than the scientific method of changing one-factor-at-a-time (OFAT).* Engineering involves many factors that may interact with each other. OFAT experiments cannot reveal interactions. MFAT experiments are needed to characterize and optimize a system involving multiple interacting factors, $y = f\,(x_1, x_2, \ldots, x_n)$.

E *Strive for the ideal function, not just meeting requirements and specifications.* The ideal function is a basic concept in robust engineering. Striving to get a system to perform as closely as possible to the ideal rather than striving to meet requirements and specifications requires thinking differently. Striving for the ideal function may seem like the impossible dream of striving for a perpetual-motion machine. Nevertheless, the measure of how far the value of the actual function differs from the value of the ideal function is a useful measure of robustness. The objective of robust optimization is to maximize the relative volumes of music and static by getting the actual function as close as possible to the ideal function.

The ideal system response, y, is a linear relation to the input M, $y = \beta M$. An actual response to some input M is some function of M together with various system parameters, $x_1, x_2, \ldots, x_n, y = f\,(M, x_1, x_2, \ldots, x_n)$. The actual function may be rearranged into two parts, the ideal function (useful part) and the deviation from the ideal function (harmful part), by adding and subtracting the ideal function, βM:

$$y = f(M, x_1, x_2, \ldots, x_n) = \underset{\substack{\text{Ideal} \\ \text{function}}}{\beta M} + \underset{\substack{\text{Deviation from} \\ \text{ideal function}}}{[f(M, x_1, x_2, \ldots, x_n) - \beta M]}$$

The ideal function represents all the radio signal energy going into the music with no energy available to cause static. The deviation from ideal is the portion of energy in the actual function that goes into causing static. Robust optimization is the process of minimizing the deviation from the ideal function by finding the values of controllable parameters that move the actual function, $f\,(M, x_1, x_2, \ldots, x_n)$, as close as possible to the ideal function, βM.

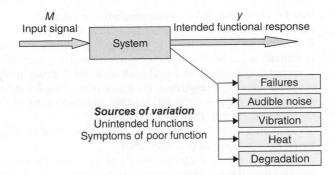

Figure 11 Energy transformations of a system into intended and unintended responses—*music and static.*

Some parameters are not controllable, such as environmental and usage conditions, variations in materials and part dimensions, and deterioration factors such as wear and aging. Uncontrollable factors are called noise factors. When the value of the actual function remains close to the values of the ideal function for all anticipated values of the noise factors, the system is robust in the presence of noise. Such a system is insensitive to sources of variation, the noise factors.

Taguchi defines good robustness as "the state where the technology, product, or process performance is minimally sensitive to factors causing variability (either in manufacturing or user's environment) and aging at the lowest unit manufacturing cost."[11]

Figure 11 provides a graphical representation of the above discussion.

Working on the problems, the symptoms of poor function, does not necessarily improve the intended function. In energy terms, all functions are energy transformations. Reducing the energy in the unintended functions does not necessarily increase the energy transformed into the intended function. It may just pop up as another unintended function.

A simple example is provided by a case study[12] of an automotive timing belt. The problem was excessive audible noise. The team worked for more than a year to successfully reduce the audible noise. However, the solution also reduced the already short life of the belt by a factor of 2. Subsequent use of robust optimization reduced the audible noise by a factor of 20 and simultaneously doubled the life of the belt. Robust optimization focused on the intended function of the belt, namely, to transfer torque energy from one pulley to another pulley. Similar results have been achieved with more complex systems such as internal combustion engines by optimizing energy efficiency to reduce audible noise—once again, work on the intended functions, not the problems.

Step 9. Identify Ideal Function. This is the first, and often the most difficult, step in robust optimization. The representation of the ideal function as displayed in Fig. 12 is deceptively

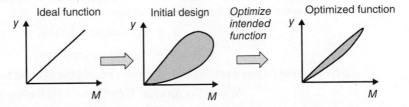

Figure 12 Comparisons between the ideal function and the variation from ideal in the actual function before and after optimization.

simple. The challenge is to determine what to measure. How should M and y be chosen for the optimization process? Said differently, what is the physics or chemistry of the system? What is the function to be optimized? Optimization is best performed on a single function. However, optimization can be conducted for systems involving multiple inputs and multiple outputs. The first step is to conduct functional analysis to break down the system to lower level, manageable elements which can range from the part level to complex subsystems. It is important to keep in mind that the focus is on variation of the function of the part or subsystem, not on variation in the part or parts that make up a subsystem.

Figure 12 illustrates the optimization process of striving to move as closely to the ideal function as possible as described above.

Step 10: Optimize Design (Step 1 of Two-Step Optimization). This step focuses on maximizing the signal-to-noise ratio (S/N) to minimize variation of the function from the ideal function:

$$S/N = 10 \log \left(\frac{\beta^2}{\sigma^2} \right)$$

where β^2 is a factor related to the energy in the signal and σ^2 is a factor related to the energy in the noise (sources of variation, not audible noise). Methods for calculating S/N from experimental data or math models are provided in numerous books.[13–15] Optimizing design involves a number of actions. A case study is used to illustrate the process.

Case Study. A published case study on wiper system chatter reduction[16] conducted by Ford Motor Company is used as an example. Windshield wiper chatter is the familiar, noisy skipping of the wiper blade that deteriorates the intended function of cleaning. The ideal function is based on the assumption that for an ideal system the actual time to reach a point D_n during a wiping cycle should be the same as the designed target time:

$$Y_n = \beta M_n$$

where Y_n = actual time that blade reaches fixed point on windshield at nth cycle
M_n = designed target time that blade reaches fixed point on windshield at nth cycle
$\beta = n$/RPM of motor

Three RPMs used in the experiments were M_n values of 40, 55, and 70 RPM.

NOISE STRATEGY. Noise is generally categorized into (1) part-to-part variation, (2) deterioration and aging, (3) customer usage conditions (duty cycle), and (4) environmental conditions. The team used these general categories to define the specific noise factors entered into the lower portion of the P-diagram (*the customer usage space*) displayed in Fig. 13. The P-diagram, or parameter diagram, provides a convenient and orderly way to organize and display control factors, noise factors, input signal, and output response.

To simplify the experiment, the Ford team compounded the noises at two levels. The team deemed that a wet or dry surface would be a dominant source of variation, so they kept it as a separate noise factor with the following definitions:

$$T_1 = \text{wet-surface condition}$$

$$T_2 = \text{dry-surface condition}$$

Several noise factors were compounded into two additional noise factors:

$$S_1 = -3° \text{at park}/20°\text{C}/50\% \text{ humidity/before aging}$$

$$S_2 = 1° \text{at park}/2°\text{C}/90\% \text{ humidity/after aging}$$

At park refers to the parked position of the wiper blade. *Aging* is typically accomplished by using old parts, presumably old rubber blade inserts. *Before aging* means the use of new parts.

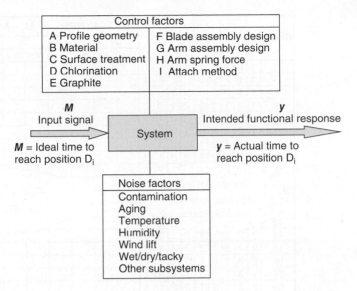

Figure 13 The P-diagram displays function inputs and outputs and control and noise factors in an orderly manner.

Table 1 Representative Portion of Control Factor Table

Control Factor	Level 1	Level 2	Level 3
A: Arm lateral rigidity	Design 1	Design 2	Design 3
B: Superstructure rigidity	Low	Medium	High
C: Vertebra shape	Straight	Concave	Convex
D: Spring force	Low	High	(None)
E: Profile geometry	Geo 1	Geo 2	Geo 3
F: Rubber material	Material 1	Material 2	Material 3

In robust optimization, it is not necessary to test at the extremes. All that is necessary is to ensure that the noise factors are significantly different to test the robustness of the system. This is a major advantage of robust optimization in shortening the length of time to conduct experiments and reducing the need to test to failure.

CONTROL FACTORS. The engineering design parameters are the control factors. The team brainstormed to select the design parameters that they believed had the largest impact on controlling wiping performance as the control parameters. The control parameters are entered into the upper portion of the P-diagram (the engineer's design space).

CONTROL FACTOR LEVELS. Robust optimization is a methodology for determining the values of the design parameters that optimize system performance. The team used their engineering knowledge and brainstorming to select values of the design parameters (control factors) that they believed covered the design space of optimum performance. A portion of the control factor levels is displayed in Table 1. The control factors not shown—G, graphite; H, chlorination; and I, attach method—have similar levels.

DESIGN OF THE EXPERIMENT. The team selected the popular L_{18} orthogonal array shown in Fig. 14. The popularity of this array stems from the fact that interactions between the control factors are reasonably balanced within the array. The standard L_{18} array contains up to eight control factors. The array in Fig. 14 was modified to accommodate nine control factors.

Run #	D	A	B	C	E	F	G	H	I	M₁ = 40 RPM				M₂ = 55 RPM				M₃ = 70 RPM			
										S_1		S_2		S_1		S_2		S_1		S_2	
										Wet	Dry	Wet	Dry	Wet	Dry	Wet	Dry	Wet	Dry	Wet	Dry
1	1	1	1	1	1	1	1	1	1												
2	1	1	2	2	2	2	2	2	1												
3	1	1	3	3	3	3	3	3	1												
4	1	2	1	1	2	2	1	3	2												
5	1	2	2	2	3	3	2	1	2												
6	1	2	3	3	1	1	2	2	2												
7	1	3	1	2	1	1	2	3	3												
8	1	3	2	3	2	2	3	1	3												
9								2	3												
									2												
									2												
13								2	3												
14	2	2	2	3	1	2	1	3	3												
15	2	2	3	1	2	3	2	1	3												
16	2	3	1	3	2	3	1	2	1												
17	2	3	2	1	3	1	2	3	1												
18	2	3	3	2	1	2	3	1	1												

Inner Array — Indicates levels of control factors

Record data in outer array

Figure 14 Modified L_{18} orthogonal array showing the inner array with levels 1, 2, or 3 indicated for each row representing one experimental run. The outer array is for collecting the data for each of the 18 runs under the inputs M_1, M_2, M_3, and combinations of the noise conditions S_1, S_2 and wet, dry, as shown at the top of the columns of the outer array.

Each of the 18 rows is one experimental run with the control factors set at the indicated levels—1, 2, or 3. When conducting experiments with a hardware/software system, 18 different sets with the indicated mixtures of levels are needed to conduct the 18 runs. For the wiper system, the only hardware changes were to park positions compounded with new and aged wiper blades. Two additional sets are needed to (1) establish the performance of the initial design and (2) experimentally confirm predictions made from the 18 runs.

Data are collected in the outer array for all combinations of wet/dry, S_1/S_2, and $M_1/M_2/M_3$ for a total of 12 data values for each of the 18 runs. If a particular combination in one of the runs is reasonably robust, the differences in the values of the data between different combinations of noise factors will be small. Less robust combinations of control factor levels will have larger differences in the values of data points. In any case, the trend in data values will track the RPM input values of 40, 55, and 70.

In this case study, a special test fixture was built with three sensors attached to the windshield to record a signal as the wiper blade passes by in order to determine the actual time to point D_n.

Typical experimental results from one run are shown in Fig. 15. The graphical representation of the values of β and σ are indicated on the chart. Their actual values are of course calculated from least-squares fits for the slope, β, and the square root of the variance for σ.

The values S/N and sensitivity (or efficiency) are calculated for each run using the following formulas:

$$\text{S/N} = 10\log\left(\frac{\beta^2}{\sigma^2}\right) \qquad \text{Sensitivity} = 10\log\beta^2$$

The S/N and β are calculated for each run using formulas that are not reproduced here. Columns for S/N and β are conveniently added to the right side of the outer array.

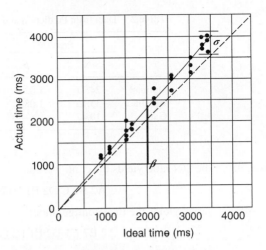

Figure 15   Typical data taken from one run. Solid line is the least square fit. The dashed line at 45° is the ideal function.

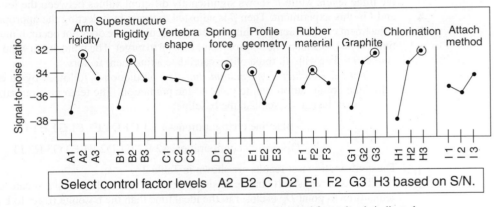

Figure 16   Response graph with selected control factor levels indicated.

To determine the S/N and β for a particular control factor at a particular level, a small bit of work is needed. For example, consider D_1. To determine S/N and β for D_1, find all of the D's in the L_{18} array at level 1. All D_1's appear in the first nine rows of the D column. Then average the nine S/N values of D_1. Repeat the process for β. The S/N for D_1 shown in Fig. 16 is about 36 dB. Continue the process until S/N and β are calculated for all levels of all control factors. The process leads to a response table. The S/N response graph shown is created from the response table data (not shown).

With a modest amount of arithmetic, a team can make a prediction about the new S/N ratio for the system. The Ford wiper team predicted an S/N ratio of about 25 dB.

Step 11: Adjust to Target (Step 2 of Two-Step Optimization). Two-step optimization is summarized as follows:

1. Maximize the S/N ratio.
2. Adjust β to target.

Table 2   Data from Prediction and Confirmation Run

	Predicted		Confirmed	
	S/N	β	S/N	β
Optimal	−24.71	1.033	−25.41	1.011
Baseline	−35.13	1.082	−35.81	1.01
Gain	10.42	—	11.4	—

Maximizing the S/N ratio yields the configuration

A2 B2 C D2 E1 F G3 H3 I

Adjusting to target yields the final configuration:

A2 B2 C1 D2 E1 F2 G3 H3 I3

Control factors that do not significantly impact the S/N are typically shown without their levels. These factors are often candidates for adjustment factors. Engineered systems usually provide a convenient control factor for adjusting β—one where the S/N remains relatively flat across the three levels while β shows significantly different values between the levels, such as *C, F,* and *I* in this experiment. Then β is adjusted to target by selecting the appropriate level for the adjustment control factor. If this convenient circumstance does not occur, it may be necessary to tinker with more than one control factor to set β to target. The windshield wiper team apparently chose to tinker with all three of the candidate adjustment factors.

Step 12: Conduct Confirmation Run. The prediction is followed by an experimental run, called the confirmation run, to validate the prediction. The team also ran an experiment with the original design to establish the baseline:

Baseline configuration A1 B1 C1 D1 E1 F1 G1 H1 I1

Optimized configuration A2 B2 C1 D2 E1 F2 G3 H3 I3

The confirmation test results are shown in Table 2.

The apparent small change in slope β from 1.082 to 1.01 is significant. It means that the actual time to point D_n is closer to the ideal time than the baseline (refer to Fig. 12).

Range of Variation. Figure 17 depicts the gain in S/N from a single robust optimization experiment assuming the average gain of 6 dB. The corresponding increase in σ level is indicated on the left side of the figure assuming that the baseline performance was at a σ level of 3.5. The formula, $\sigma_{OP}/\sigma_{BL} = (1/2)^{(\text{gain}/6)}$, can be written as $3.5\sigma_{OP} = 3.5\sigma_{BL}(1/2)^{(\text{gain}/6)}$ or, for a gain of 6 dB, as $7\sigma_{OP} = 3.5\sigma_{BL}$. The correlation of σ level and S/N is relative, not absolute. While the increase in σ level is a factor of 2 whatever the initial σ level, the baseline σ level is arbitrarily selected as 3.5 as an instructive example.

Returning to the case study of reducing windshield wiper system chatter, the range of variation of the baseline and optimized systems are related by $\sigma_{OP}/\sigma_{BL} = (1/2)^{(\text{gain}/6)}$. Hence, $\sigma_{OP} = \sigma_{BL}(1/2)^{(11.4/6)} = \sigma_{BL}/3.7$ or $\sigma_{BL} = 3.7\sigma_{OP}$.

Then, $3\sigma_{BL} = 11\sigma_{OP}$. If the baseline was at a 3σ level, the optimized system is at an 11σ level, an enormous improvement. Whatever the variation of the original wiper system, the range of variation was reduced by a factor of 3.7.

Cost Reduction. The S/N is relatively insensitive to the choice of the rubber material (factor *F*). Hence, the team is free to select the lowest cost material from the three tested. While this may not be very exciting for rubber wiper blades, it can be a significant cost opportunity in many situations.

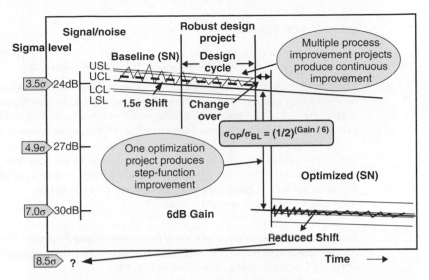

Figure 17 Depiction of improvement due to robust optimization experiment achieving 6 dB gain (30 dB − 24 dB = 6 dB), the average gain over thousands of experiments.

Conclusions from Case Study. The study indicated that chlorination, graphite, and arm and super-structure rigidity have significant impacts on the wiper system while vertebra shape (load distribution) has minimal impact. Hence, a low-friction, high-rigidity wiper system provides a robust windshield wiper system that will remain chatter free for an extended period of time. The increased durability of the system is achieved by including aged and worn wiper blades in the noise strategy.

Step 13: Optimize Tolerance Design. Tolerance design is not tolerancing. It is a methodology for balancing internal cost and external cost to either the customer or the company. It involves examination of the impact of the different control factors. A simple example that did not require tolerance design was the windshield wiper case study concerning the opportunity to use the lowest cost rubber material included in the study without significant impact on performance under customer usage conditions.

The methodology uses analysis of variance (ANOVA) to determine the percentage contribution of the various control factors to the performance of the system and Taguchi's quality loss function to determine internal and external costs. The combination facilitates the determination of whether to upgrade or degrade portions of the system. As another example of thinking differently, Taguchi goes even further to recommend starting with the lowest cost components and materials and upgrading only as necessary. This is opposite of the common practice of overspecifying things to be safe. Tolerance design can only be effectively conducted after optimization.

Step 14: Optimize Process. The same methodologies discussed above are effectively used to optimize manufacturing and service processes in the spirit of concurrent engineering.

Benefits of Robust Optimization
Improved Quality, Reliability, and Durability. A major strength of Taguchi's robust optimization is the ability to make the function of the system insensitive to aged and worn parts as well as new parts. This capability distinguishes robust optimization from methods such as classical design of experiments that focus on reducing variability. Failure rate over time of usage is used to illustrate the impact of robust optimization on reliability. Taguchi shows that failure rate is

reduced by the factor of $\left(\frac{1}{2}\right)^{\text{gain}/3}$:

$$\text{Failure rate}_{\text{optimized}} = \text{failure rate}_{\text{initial}}\ \left(\frac{1}{2}\right)^{\text{gain}/3}$$

The average gain over thousands of case studies is about 6 dB, which, on average, yields a factor of 4 reduction in failure rate. A more conservative prediction of failure rate after optimization is given by replacing gain/3 with gain/6, which yields a factor of 2 reduction in failure rate for a gain of 6 dB. The bathtub curves shown in Fig. 18 assume the more conservative improvement.

As emphasized in the develop concept phase, robust optimization is an important downstream measure of the quality of a conceptual design. Good concepts yield good gains such as the 11 dB gain in the wiper system case study or the average gain of about 6 dB. A concept that does not yield a good gain is a poor concept that will plague customers with problems and corporations with high warranty costs.

Reduce Development Time and Cost of Development and Product. The ability to optimize concept designs and technology sets combined with two-step optimization to increase the efficiency of product development is illustrated in Fig. 19.

After the function has been optimized, a subsystem can be set to different targets for different product requirements within a family of products. In addition, a good concept and final

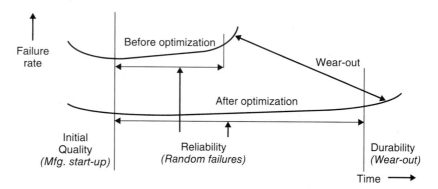

Figure 18 Good concepts can be optimized to provide superior quality, reliability, and durability. Poor concepts are not significantly improved through optimization.

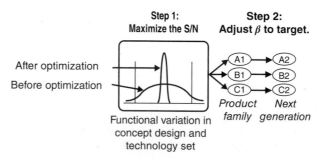

Figure 19 A good concept and technology set can be optimized and then set to different targets for different applications.

design can be carried over from generation to generation of product families. Multiple-use and carryover parts can dramatically reduce development cost and schedule.

Robust optimization enables some manufacturers to design only about 20% new parts and reuse about 80% of carryover parts in next-generation products for substantial competitive advantage. For other manufacturers, the percentages are reversed.

Verify and Launch. The "V" of IDDOV. The steps in the verify-and-launch phase are traditional:

> *Step 15: Finalize Operations and Service Processes.*
>
> *Step 16: Conduct Prototype Cycle.*
>
> *Step 17: Conduct Pilot Run.*
>
> *Step 18: Launch, Ramp, Up and Confirm Full Operation.*
>
> *Step 19: Track and Improve Field Performance.*
>
> *Step 20: Close Project.*

These steps are typical of any development process and are not discussed in detail. However, steps 15 and 16 warrant brief elaboration, starting with step 16. This is a manufacturing intent prototype cycle, in contrast with the many test rigs that have been built and tested throughout the IDDOV process.

The key recommendation in step 16 is to plan and execute a single manufacturing intent prototype cycle. Traditionally, teams use manufacturing intent prototype cycles to improve reliability to target by tinkering to correct problems as they emerge. The first cycle usually generates so many changes that configuration control is lost. Hence, the team redesigns and builds another set of manufacturing intent prototypes and repeats the test procedures. Often additional build–test–fix cycles are required to reach an acceptable level of performance and reliability at great expense in time and resources.

An upfront commitment that only a single prototype cycle will be executed fosters excellence of execution of all of the steps in the design process to complete the design prior to building the first set of prototypes. A strong concept design and robust optimization give credibility to the notion that performance and reliability requirements can be met with only a single manufacturing intent prototype cycle.

Step 15 is critical to the notion of a single prototype cycle. The manufacturing intent prototype should be assembled using full production processes and fully tested against all anticipated usage conditions, including completed operator and service documentation. This can only be done if all operations and service procedures are defined and documented.

An opportunity exists to improve manufacturing quality using Taguchi's on-line quality engineering. Unfortunately, on-line quality engineering has not received much attention in the Western world. A key attribute of on-line quality engineering is the focus on managing to target rather than to control limits. Managing the manufacturing process to target tends to yield a bell-shaped distribution where only a small number of parts and assemblies are near control limits. Managing to control limits tends to yield a flat distribution where a large portion of the parts and assemblies are near the control limits. Items near control limits are sometimes identified as latent defects just waiting to morph into active defects under customer usage. The portion of latent defects that morph into active defects depends on the ratio of specification limits (tolerance) to control limits, Cpk, or in contemporary terms, σ level. If the specification limits are significantly removed from control limits, say 6σ performance, the portion of latent defects that become active defects remains small. Obviously, at the other extreme of specifications set at the 3σ control limits, the portion of morphed defects becomes large. Managing to target leads to better products at lower overall cost.

5   SUMMARY AND CONCLUSIONS

The IDDOV process is all about increasing signal and decreasing noise, or in more traditional language, opening up design tolerances and closing down manufacturing variations in order to send superior products to market that droves of potential customers will purchase in preference to competitive offerings. The bottle model was created by the second author[17,18] in the early 1980s and published in 1988 to provide a visual representation of the role of robust engineering in the development process. The effort was motivated by numerous failed attempts to explain robust optimization to senior management. The bottle model is depicted in Fig. 20. It is presented here as a summary of the entire IDDOV process.

The interpretation of the bottle model curves differs between statistical language of capability indices, Cp and Cpk, and robust engineering language of S/N.

- The measure of statistical variation is the capability index, Cp, defined as the ratio of the design width (USL − LSL) and the process width (UCL − LCL):

$$Cp = \frac{USL - LSL}{UCL - LCL}$$

- The measure of robustness of function is the S/N, defined as the ratio of the energy in the intended function to the energy in the unintended functions:

$$S/N = 10 \log \left(\frac{\beta^2}{\sigma^2} \right)$$

Of course, the measurement scales and units are quite different between the two interpretations:

- Cp is a measure of variation of directly measured topics such as time, dimension, etc., in appropriate units such as seconds and centimeters.
- S/N is a measure of functional performance in terms of the energy ratios in units of decibels.

The caption in Fig. 20 includes the phrase "impact of noise." Sources of variation of noise are not necessarily reduced. Only its impact on functional variation is reduced. Recall that the definition of robustness is when technology, product, or process performance is minimally sensitive to factors which cause variability (either in manufacturing or a user's environment) and aging at the lowest unit manufacturing cost. Noise factors such as environment and customer

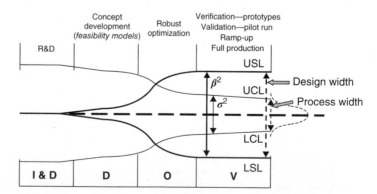

Figure 20   The bottle model provides a visual representation of the IDDOV process and engineering progress in opening up design tolerances (dark line) and closing down manufacturing variation (light line) or alternatively increasing the signal and increasing the impact of S/N.

usage conditions cannot be reduced. However, the system can be made minimally sensitive to such sources of variability. Part-to-part variation may even be allowed to increase to reduce manufacturing cost. Tolerance design is used to determine allowable variation that does not perceptibly impact the customer.

The distinction between statistical variation and robustness is critical. Statistics has to do with measuring and reducing statistical variation of parts and processes. Robustness has to do with measuring and minimizing functional variation in the presence of the statistical variation (i.e., sources of noise).

The robust optimization portion of the bottle model shows a rapid increase in signal (or design latitude) and rapid decrease in noise (or statistical variation) that results from conducting a series of robust optimization experiments on each of the system elements involved, as illustrated in Fig. 17 for a single system element.

The version of IDDOV described in Ref. 1 and in more technical detail herein has been deployed in a number of large, medium, and small corporations that have reaped "*extraordinary profits.*"

IDDOV is an engineering process to support product development processes. It is not a PDP as some renditions of DFSS apparently strive for. The entire IDDOV process is used within technology development, within the advanced product concept development phase, within the design and development phase of typical PDPs, and in OEM manufacturing operations and suppliers to OEMs.

IDDOV is also used to parallel the entire PDP. Within any PDP, many IDDOV projects will be in progress through all phases.

IDDOV is much more than DFSS for breaking through the Five Sigma wall that the Six Sigma DMAIC process encounters. It is a customer-to-customer engineering methodology that carries the voice of the customer throughout the engineering process and delivers superior, low-cost manufactured products that droves of people around the world purchase in preference to competitive offerings.

Last, deployment of IDDOV is not a cost. It is an investment with high returns (ROIs). In tomorrow's competitive world, it will be difficult to compete without pervasive implementation of a competent form of DFSS such as IDDOV.

ACKNOWLEDGMENTS

Special thanks for technical review of the chapter go to Ruth McMunigal, B. S., Green Belt, Black Belt candidate; Robert J. McMunigal, M.A.; and Kelly R. McMunigal, B.A., Yellow Belt, Green Belt candidate.

REFERENCES

1. S. Chowdhury, *Design for Six Sigma: The Revolutionary Process for Achieving Extraordinary Profits*, ASI, Dearborn, MI, 2002.
2. A. Fornari and G. Maszle, "Lean Six Sigma Leads Xerox," *ASQ Six Sigma Forum Mag.*, **3**(4), 57, August 2004.
3. M. Harry and R. Schroeder, *Six Sigma: The Breakthrough Management Strategy Revolutionizing the World's Top Corporations*, Doubleday, New York, 2000.
4. S. Chowdhury, *The Power of Six Sigma: An Inspiring Tale of How Six Sigma Is Transforming the Way We Work*, ASI, Dearborn, MI, 2001.
5. B. Bebb, "Role of Taguchi Methods in Design for Six Sigma," in G. Taguchi, S. Chowdhury, and Y. Wu (Eds.), *Taguchi's Quality Engineering Handbook*, Wiley, Hoboken, NJ, 2005, pp. 1492–1521.
6. T. P. Pare, "How to Find Out What They Want," *Fortune*, **128**(13), 39, Autumn/Winter 1993.

7. T. Peters, *Thriving on Chaos*, 1988, p. 185, available: www.fortune.com

8. J. Terninko, *Step-by-Step QFD: Customer-Driven Product Design*, St. Lucie Press, Boca Raton, FL, 1997, pp. 67–74.

9. N. Tichy and S. Sherman, *Control Your Destiny or Someone Else Will*, Harper Business, New York, 1994, p. 8.

10. S. Pugh, *Total Design: Integrated Methods for Successful Product Engineering*, Addison-Wesley, Workingham, England, 1990.

11. G. Taguchi, S. Chowdhury, and S. Taguchi, *Robust Engineering: Learn How to Boost Quality While Reducing Cost and Time to Market*, McGraw-Hill, New York, 2000, p. 4.

12. B. Bebb, ASI, personal communication, 2005.

13. G. Taguchi, S. Chowdhury, and S. Taguchi, *Robust Engineering: Learn How to Boost Quality While Reducing Cost and Time to Market*, McGraw-Hill, New York, 2000.

14. Y. Wu and A. Wu, *Taguchi Methods for Robust Design*, ASME, New York, 2000.

15. G. Taguchi, *Taguchi on Robust Technology Development: Bringing Quality Engineering Upstream*, American Society of Mechanical Engineers, New York, 1993.

16. G. Taguchi, S. Chowdhury, and S. Taguchi, *Robust Engineering: Learn How to Boost Quality While Reducing Cost and Time to Market*, McGraw-Hill, New York, 2000, pp. 82–92.

17. H. B. Bebb, "Responding to the Competitive Challenge, the Technological Dimension," *The Bridge*, **18**(1), 4–6, Spring 1988, published by the National Academy of Engineering.

18. H. B. Bebb, "Quality Design Engineering: The Missing Link in U.S. Competitiveness," in V. A. Tipnis (Ed.), *Tolerance and Deviation Information*, New York, 1992, pp. 113–128.

BIBLIOGRAPHY

C. M. Creveling, J. L. Slutsky, and D. Aritis, *Design for Six Sigma*, Prentice-Hall, Englewood Cliffs, NJ, 2003.

International Organization for Standardization/International Electrotechnical Commission (ISO/IEC), "Systems Thinking," 15288, available: www.iso.org

J. E. McMunigal, "American Agents of Change: Total Quality Management and Taguchi Method," paper presented at the Second Annual European Conference on Taguchi Method, London, UK, September 1989.

J. E. McMunigal, "TQM and Japanese Optimization Techniques," paper presented at the Third Annual European Conference on Japanese Optimization Technology, Wolfson College, Cambridge University, Cambridge, UK, October 1990.

J. E. McMunigal, "TQM and Japanese Optimization Techniques," in *Quality Technology Asia*, A.P. Publications Singapore, October 1991.

J. E. McMunigal, "The ABCs of Cost as Metric," in J. McMuniga (consulting Ed.), *Quality Technology Asia*, A.P. Publications, Singapore, March 1992.

J. E. McMunigal, Commissioned speaker and workshop director on "Malcolm Baldrige National Quality Award" for the Institute for International Research Conference on Benchmarking, Singapore, February 1992.

Minitab® standard software for most Six Sigma models, available: www.minitab.com

TRIZ Journal, available: http://www.triz-journal.com

K. Yang and El-Haik, *Design for Six Sigma*, McGraw-Hill, New York, 2003.

CHAPTER 13

ENGINEERING APPLICATIONS OF VIRTUAL REALITY

Wenjuan Zhu, Xiaobo Peng, and Ming C. Leu
Misssouri Univeristy of Science and Technology
Rolla, Missouri

1 INTRODUCTION

For over 50 years now, since the first virtual reality simulator, the "Sensorama Simulator," was invented in 1956,[1] the concept of virtual reality has fascinated engineering researchers. The continued development of virtual reality systems and the increased availability of more advanced computer hardware have made these systems increasingly more practically useful in recent years in terms of their ability to fulfill the requirements of end users. Virtual reality has grown into an exciting field that already has seen the development of numerous, diverse engineering applications.[2]

1.1 What Is Virtual Reality?

The term virtual reality (VR) has been used by many researchers and described in various ways. A broad definition of VR comes from the book "The Silicon Mirage"[3]: "Virtual Reality is a way for humans to visualize, manipulate and interact with computers and extremely complex data." Burdea and Coiffet[4] described VR as "a high-end user interface that involves real-time simulation and interactions through multiple sensorial channels. These sensorial modalities are visual, auditory, tactile, smell, taste, etc." Jayaram et al.[5] stated that VR often is regarded as an extension of three-dimensional computer graphics with advanced input and output devices. Generally speaking, VR acts as a medium of communication, requires physical immersion, provides synthetic sensory stimulation, can mentally immerse the user, and is interactive.[1]

VR is closely associated with an environment commonly known as a virtual environment (VE). A VE is an interactive graphic display enhanced by special processing and by nonvisual display modalities, such as auditory and haptic feedback, to convince users that they are immersed in a real physical space.[6] According to Kalawsky,[7] VR is the same as VE but more familiar to the public. VR also can be seen as a pinnacle of what is ultimately sought when implementing a VE system.

1.2 Types of VR Systems

Conventional Types of VR Systems

VR systems can be classified into three categories: fully immersive, semi-immersive, and nonimmersive, based on the physical configuration of the user interface and to what extent the user can perceive the real world during the simulation.[8]

Fully Immersive VR. The most common image of VR is that of an immersive VE, typically in which a user wears a head-mounted display (HMD) that attempts to isolate the user from the real world in order to increase the believability of the simulation. People more frequently suffer cybersickness[9] when wearing an HMD in the fully immersive VR system; this ailment cannot be prevented thoroughly and effectively.[8]

Semi-Immersive VR. The semi-immersive system, such as the CAVE shown in Fig. 1,[10] provides a VE that surrounds the user wholly or partially, but the user still can see him/herself and the other users. Images are displayed based on the main user's head position and orientation provided by a head tracker. Generally, the semi-immersive system allows several users to share the VE, making collaborative applications possible.

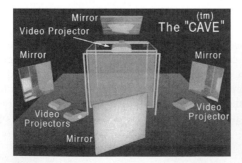
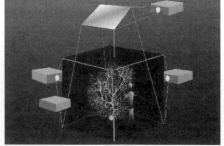

Figure 1 CAVE[TM] system.

Nonimmersive VR.   Nonimmersive VR systems normally run on standard desktop computers, bringing about the term "desktop VR." Desktop VR systems use the same three-dimensional (3D) computer graphics as immersive VR systems and have gained popularity because of their lower cost and ease of both use and installation. The best representative example is a video game, which involves users and holds their interest with appealing graphics and sound simulations. The Xbox 360 with Kinect (http://www.xbox.com/en-US/xbox360?xr=shellnav) does not even require a keyboard or controller for real-time interaction.

Extended Types of VR Systems

Many research areas are closely related to VR. They can be divided into the following categories.

Telepresence.   Telepresence is a computer-generated environment consisting of interactive simulations and computer graphics in which a human experiences presence in a remote location. An example is a pilot in a sophisticated simulator that controls a real airplane 500 miles away and provides to the pilot visual and other sensory feedback, as if the pilot were actually in the cockpit looking out through the windscreen and feeling the turbulence. Other applications of telepresence involve the use of remotely operated vehicles (e.g., robots) to handle dangerous conditions (e.g., nuclear accident sites) or for deep sea and space exploration.

Augmented Reality.   Augmented reality (AR) is the use of a transparent HMD to overlay computer-generated images onto the physical environment. Rapid head tracking is required to sustain the illusion. In the most popular AR systems, the user views the local real-world environment, but the system augments that view with virtual objects. For example, the Touring Machine system[11] acts as a campus information system, assisting users in finding places and allowing them to pose queries about items of interest, such as buildings and statues.

Distributed VR System.   In a distributed VR system, the simulation runs not only on one computer system but also on several systems connected over the Internet. Several users from different locations can share the same VE and interact with the environment and each other in real time. The main issue involved in building this distributed VR system is a certain amount of latency during the delivery of update information through the Internet. Moreover, the system has to be designed for portability due to the fact that people are working on different computer systems with different hardware and software.[12]

1.3 Aim of This Chapter

The aim of this chapter is to introduce the techniques involved in the design and development of VR systems. The chapter will describe the main techniques and applications of the exiting VR technology, including its hardware, software, and engineering applications.

2 VIRTUAL REALITY COMPONENTS

VR immerses the user into a sensory-rich, interactive experience. A typical VR configuration is shown in Fig. 2. Using input and output sensors, a virtual world is generated and then controlled by both the reality engine and the participant.[13] Software integrates various hardware elements into a coherent system that enables the user to interact with the VE. This chapter reviews the hardware and software tools commonly used for the creation of VEs and related issues. We will break down a VR system into its basic components and explain each one. This review is not intended to be exhaustive because VR hardware and software change constantly, with new developments taking place all the time.

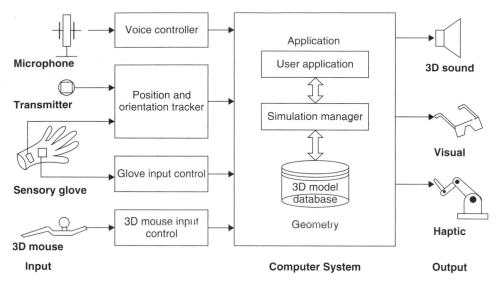

Figure 2   VR configuration.

2.1   VR Hardware

As shown in Fig. 2, VR hardware plays an extremely important role in a VR system. It provides the physical devices that comprise an immersive environment such as CAVE™, HMD, etc. Input and output devices transmit the information between the participant and the VE. They provide interactive ways to communicate information. This section will describe the hardware used in various VR systems, including tracking devices, input devices, visual stereo displays, haptic devices, and auditory displays.

Input Devices

An input device is the bridge that connects the real world and the virtual world. It gathers the input information from the user and transmits it to the virtual world. Using different input devices provides an intuitive and interactive environment, immersing the user in the virtual world.

Tracking Devices.   There are two types of tracking devices commonly used in VR systems: position trackers and body trackers. A position tracker is a sensor that reports an object's position (and possibly orientation) and maps it to the object's relative position in the VE. A body tracking device monitors the position and actions of the participant. Commonly used body tracking techniques include tracking the head, hand and fingers, eyes, torso, and feet. The current tracking devices used in VR are based primarily on electromagnetic, mechanical, optical, ultrasonic, inertial and gyroscopic, and neural and muscular sensing technologies. Some commercial devices derived from these different technologies are listed in Table 1 with comparisons between them.

Among the technologies listed in Table 1, markerless-based optical tracking devices excel in that no sensor needs to be attached to the tracking object, which makes it more flexible and easier to use. Figure 3 shows gamers using their body postures as input to play a game with the help of a markerless motion tracking device called Kinect.

Table 1 Various Types of Tracking Devices

Type	Product Model		Advantage	Disadvantage
	Position Tracker	Body Tracker		
Mechanical tracker	BOOM, Phantom		Using encoders and kinematic mechanisms provides very fast and accurate tracking.	Linkages restrict the user to a fixed location in the physical space.
Electromagnetic tracking	Flock of Birds, IS-600, StarTrak8	Ascension's MotionStar	No line-of-sight restriction. Wireless systems will reduce encumbrances on the participant.	Metal in the environment can distort data. The range of the generated magnetic field is short.
Optical tracking				
Marker based	HiBall Tracking System	Vicon, Motion Analysis, ART, Optitrack	No cable required for activating the marker; high accuracy.	Markers may interfere with human movement and vice versa; occlusion may occur.
Marker-less based		Kinect, Asus Xtion Pro	No marker needed to track objects.	Low accuracy and less reliable.
Ultrasonic tracking	Logitech Ultrasonic Tracker		Inexpensive because of use of commonly available speakers and microphones.	Limited range; easy to be interrupted in a noisy environment.
Inertial and gyroscopic	Xsens's MTi	Xsens's Moven	Easily portable with little latency.	A substantial measurement drift can accumulate over time.

Other Input Devices. Other than the tracking devices described above, 3D mice, joy sticks, sensory gloves, voice synthesizers, and force balls can be used as input devices for interaction with the user in the virtual world. Point input devices include the six-degree-of-freedom (6DOF) mouse and the force ball. In addition to functioning as a normal mouse, the 6DOF mouse can report height information as it is lifted in three dimensions. The force ball senses the forces and torques applied by the user in three directions. Figure 4 shows the SpaceMouse Pro (a 6DOF mouse) and a SpaceBall (a force ball device) from 3Dconnexion Inc.

The CyberGlove® is a fully instrumented glove that uses resistive bend-sensing elements to accurately transform hand and finger motions into real-time digital joint-angle data. Another device, the CyberForce, not only provides grounded force feedback to the hand and arm but also provides 6DOF positional tracking that accurately measures the translation and rotation of the hand in three dimensions. Figure 5 shows these two glove-type devices from CyberGlove

Figure 3 Kinect used as makerless tracking device.

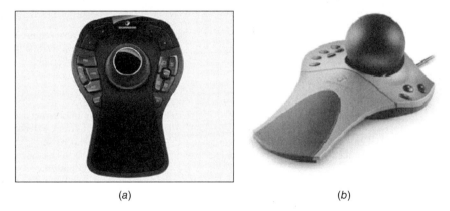

(a)   (b)

Figure 4 (*a*) SpaceMouse Pro; (*b*) SpaceBall from 3Dconnexion Inc. SpaceBall® and SpaceMouse® are registered trademarks of 3Dconnexion in the US and/or other countries.

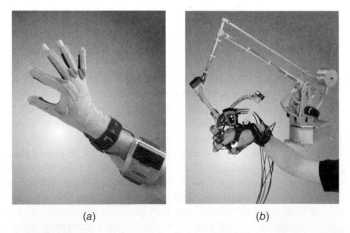

(a)   (b)

Figure 5 (*a*) CyberGlove II; (*b*) CyberForce. Images courtesy of CyberGlove Systems LLC.

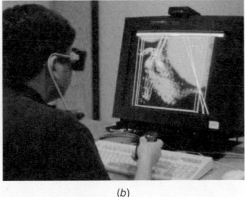

(a)   (b)

Figure 6   (a) Bird's-eye view from the flight chair in the Hemispherium™; (b) Fishtank VR display.

Systems LLC (http://www.cyberglovesystems.com/all-products). Voice input devices and bio-controllers also provide a convenient way for the user to interact with the VE.

Output Devices

How the user perceives the VR experience depends on what feedback the VE can provide. Output devices are applied in the VE to present the user with feedback about his or her actions. Three of the human perceptual senses, i.e., visual, aural, and haptic, are commonly presented to the VR user with synthetic stimuli through output devices.

Visual Display.   There are three general categories of VR visual displays: stationary, head based, and hand based.[14] *Stationary displays* can be divided into *large-screen stationary displays,* such as the CAVE; *wall diplays*, shaped like a cylinder, dome, or torus; and *desktop VR displays*, which integrate a standard computer monitor with the necessary tracking and other input devices and VR software to achieve the ability to display stereographically. Figure 6a shows an example of a large-screen display from Hemispherium™,[15] and Fig. 6b shows an example of a desktop VR display from.[16]

A *head-based* visual display is the most common VR display. The HMD, as shown in Fig. 7a, can provide a fairly immersive experience for the user. It is a helmet worn by the user with two small monitors in front of each eye to display images as a 3D view of the virtual world. Often, the HMD is combined with a head tracker so as to allow the participant a full 360° view. Without any weight being placed on the user's body, the binocular omni-orientation monitor (BOOM), as shown in Fig. 7b, adopts a mechanical arm-mounted display, which users can pull up in front of their faces.

Hand-based visual displays usually use a personal digital assistant (PDA) or mobile phone as a visual viewer and interface with the virtual or physical environment. Figure 8a shows a "mobile animator" that uses a handheld device to control virtual characters in a multi-user, semi-immersive VE.[8] Figure 8b shows an AR game in which a hand-held display superimposes the virtual model on a stationary scene from the real world (http://www.ardefender.com/).

Table 2 summarizes the three different types of visual display devices for VR systems by comparing their advantages and disadvantages.

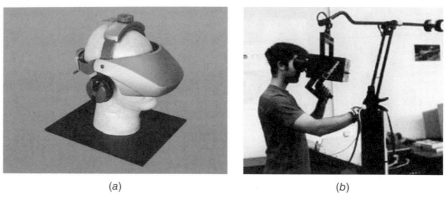

(a)   (b)

Figure 7   (*a*) HMD; (*b*) BOOM.

(a)   (b)

Figure 8   (*a*) Mobile animator; (*b*) AR game.

Haptic Display.   Haptics broadly refers to touch sensations that occur for the purpose of perception or physical contact with objects in the virtual world.[17] Haptics has played an important role in the application of VR. Haptic devices used in VR applications mainly render force and tactile sensations. Force sensations help the user estimate the hardness, weight, and inertia of a virtual object. Tactile sensations help the user feel the surface contact geometry, temperature, and smoothness of a virtual object. Haptic devices can be divided into two categories: *world grounded*, which has a base linked to some object in the real world and an arm with multiple linkages to exert an active or resistive force in a particular direction, and *self-grounded,* which is worn by the user, such as the CyberGrasp shown in Fig. 9*a*. The CyberGrasp is a lightweight, force-reflecting exoskeleton that fits over a CyberGlove® and adds resistive force feedback to each finger. The CyberForce, as shown in Fig. 5*b*, and PHANToM™, as shown in Fig. 9*b*, are world-grounded devices that allow the user to feel force when he or she manipulates a virtual object with a physical stylus that can be positioned and oriented in 6DOF.

The CyberTouch glove shown in Fig. 9*c*, which has tactors in the form of vibrators on each finger and the thumb, can allow tactile feedback. With the CyberGrasp force feedback system,

Table 2 Visual Display Devices

Category	Devices	Advantages	Disadvantages
Stationary			
Large-screen display	CAVE, wall displays	Big field-of-view, sharing VE among several users, reduced amount of hardware worn by users.	Incomplete view of the virtual world, expensive, difficult to mask the real world.
Desktop VR display	Fishtank VR	Can be used at user's desk, low cost.	Very limited field of view and field of regard.
Head based	HMD, BOOM	100% field of regard, less expensive than large-screen display, requires less space, portable.	Noticeable latency that can cause nausea or headache, limited field of view, lower resolution, allows only a single user at a time.
Hand based	Mobile phone, PDA	Can be combined with physical reality or other visual display like CAVE, not encumbering, portable.	Very limited field of view, less immersive.

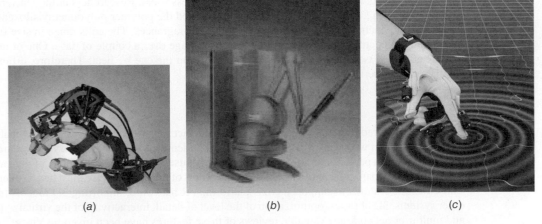

(a)   (b)   (c)

Figure 9 Haptic devices: (*a*) CyberGrasp from CyberGlove Systems; (*b*) PHANToM desktop device from SensAble; (*c*) Cyber-Touch from CyberGlove Systems.

the user can feel the size and shape of a computer-generated 3D object in a simulated virtual world.

Auditory Display.   In addition to visual and haptic information, sound can be incorporated in the VE to enhance the participant's sense of presence. Loudspeakers and headphones generally are used as the auditory display hardware in the VE. These two types of displays match well with stationary screen visual displays and head-based visual displays separately. One example of a potential application is to generate sound to confirm when an object is selected. Another example is to provide different pitches of sound to simulate machining with different material properties. There are two ways to generate sounds; one way is to play a recorded sound, and the other is to synthesize the sound from different parameters. The challenge of auditory display is

Table 3   Publicly Available VR Software

Type	Software	Website
SDK	CAVELib	http://www.vrco.com/products/cavelib/cavelib.html
	DI-Guy	http://www.bdi.com/content/sec.php?section=diguy
	DIVISION	http://www.ptc.com/products/index.htm
	WorldToolKit	http://www.sense8.com
	VR Juggler	http://www.vrjuggler.org/
Authoring tools	EON Studio	http://www.eonreality.com/
	WorldUp	http://www.sense8.com
	MultiGen Creator	http://www.multigen-paradigm.com/
	MotionBuilder	http://www.kaydara.com/products/

spatialization, which includes two steps. First, identify where the sound will emanate; in other words, attach the sound to some object. Second, render the sound to function of dual-speaker or multispeaker displays.[8]

Other Display.   Smell-o-vision can enhance the visual and auditory experiences in VR systems by releasing odors to match visual cues. However, to date, the limited range of smells, which can be deployed only once, largely has failed to impress users. Researchers at the University of California, San Diego, and at Samsung in Korea used the polymer polydimethylsiloxane to create a square matrix of cells that can hold different fragrances. The cells range in size from nanometers to micrometers and show no signs of leakage over a couple of days. One or more odors can be released from one or more cells based on the cue by heat. Therefore, different smells can be generated to enhance the VR experience.[18]

2.2   VR Software

Software is required to integrate various hardware devices into a coherent system that enables the user to interact with the VE. Bierbaum and Just[19] identified the three primary requirements of a VR system as performance, flexibility, and ease of use. When deciding which VR software is best suited to a particular application, various features of the VR software must be considered. These include features to support a cross-platform, VR hardware, importing 3D models from other systems, 3D libraries, optimization of the level of detail, interactivity of the virtual world, and multiuser networking. Valuable reviews of these features have been given by Vince[20] and Bierbaum and Just.[19]

VR implementation software can be classified into two major categories: software development kits (SDKs) and authoring tools.[21] SDKs are programming libraries (generally in C or C + +) that have a set of functions with which the developer creates VR applications. Authoring tools provide graphical user interfaces (GUIs) for the user to develop the virtual world without requiring tedious programming. Table 3 contains a list of publicly available VR software packages for the creation of VEs.

3   VR IN CONCEPT DESIGN

VR can be applied to an extensive range of fields. This and the following sections present VR applications in various areas of engineering. The list of applications reviewed here is not intended to be exhaustive but rather to provide some examples of VR applications in engineering.

In a conceptual design, the exact dimensions of a part are not determined initially; the designer is more interested in creating the part's shapes and features. Commercial computer-aided design (CAD) systems, such as Unigraphics, Catia, Creo Element/Pro, and SolidWorks, are powerful geometric modeling tools, but they require precise data for designing objects and thus do not allow users to implement their ideas regarding the design of shapes and features in an intuitive manner. Their user interface generally consists of windows, menus, icons, etc., which tend to prevent users from focusing on the design intent. Another limitation of the conventional CAD system concerns the use of input devices. Designers use a 2D input device, usually the mouse, to construct 3D objects. Concept design using VR techniques enables the user to create, modify, and manipulate 3D CAD models intuitively while simultaneously visualizing CAD models in a VE immersively. Moreover, VR techniques can be used to evaluate concept designs for improvement. Thus, the development time and cost can be reduced. Some systems that provide such capabilities are described in the following sections.

3.1 3-Draw

3-Draw, a system for interactive 3D shape design, was introduced at MIT.[22] Its user interface is based on a pair of Polhemus 6DOF tracking devices. One tracker is held as a palette-like sensor in the designer's left hand to specify a moving reference frame, and the other tracking device is used by the right hand as a stylus-like sensor to draw and edit 3D curves in space. A graphic display is used to visualize the scene from a virtual camera position. After drawing and editing the 3D curves, which form wireframe models, the next steps are to fit surfaces to groups of linked curves and to deform the surfaces until the required shape is obtained.

3.2 3DM

Built by researchers at the University of North Carolina,[23] this system uses an HMD to place the designer in a virtual modeling environment. An input device, such as a Polhemus 3-space Isotrak held in one hand, is used for all interactions, including selecting commands from a floating menu, selecting objects, scaling and rotating objects, and grabbing vertices to distort the surface of an object.

3.3 JDCAD

Researchers at the University of Alberta, Canada, developed a 3D modeling system called JDCAD,[24] which uses two 6DOF tracking devices, one to dynamically track the user's head and provide the kinetic 3D effect (e.g., correlation to the position and orientation of the head) and the other to be used as a hand-held "bat" to track hand movements.[25] A bat is a tracker that reports 3D position and orientation data. It has three buttons mounted on it for signaling events. Figure 10 shows the bat that is used with JDCAD. By switching modes, the bat can be used to rotate and translate the model under construction in order to select objects for subsequent operations as well as to orient and align individual pieces of the model.

3.4 Holosketch

Deering[26] at Sun Microsystems created the HoloSketch system. It uses a head-tracked stereo display with a desktop cathode ray tube (CRT), unlike many VR systems, which use HMDs. The user wears a pair of head-tracked stereo shutter glasses and manipulates the virtual world through a hand-held 3D mouse/wand. The HoloSketch system has a 3D multilevel fade-up circular menu, which is used to select the required drawing primitives or to perform

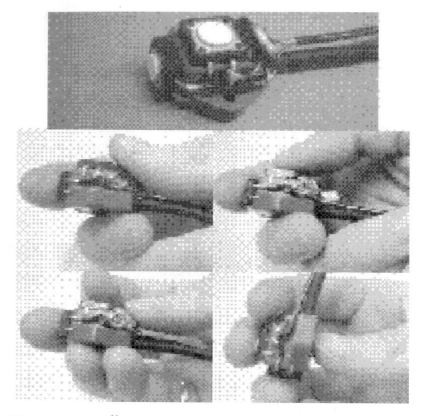

Figure 10 Bat input device.[25] © 1988 IEEE. Reprinted, with permission, from "Using the Bat: a Six Dimensional Mouse for Object Placement," IEEE Computer Graphics and Applications Magazine, by Ware, C., and Jessome, D.R.

one-shot actions, such as cut or paste. The HoloSketch system supports several types of 3D drawing primitives, including rectangular solids, spheres, cylinders, cones, freeform tubes, and many more.

3.5   COVIRDS

Dani and Gadh[27] presented an approach for creating shape designs in a VR environment called COVIRDS (Conceptual VIRtual Design System), as shown in Fig. 11. This system uses VR technology to provide a 3D virtual environment in which the designer can create and modify 3D shapes with an interface based on bimodal voice and hand tracking. A large-screen, projection-based system called the Virtual Design Studio (VDS) acts as an immersive VR-CAD environment. The designer creates 3D shapes using voice commands, hand motions, and finger motions and can grasp the shape to edit its features with his or her hands. Tests have been conducted to compare the efficiency of the COVIRDS with traditional CAD systems. Chu et al.[28] claimed that the COVIRDS system can achieve 10–30 times more productivity than conventional CAD systems.

3.6   Virtual Sculpting with Haptic Interface

Peng and Leu[29,30] developed a virtual sculpting system, as shown in Fig. 12. A VR approach was taken to employ more intuitive and interactive modeling tools. The virtual sculpting method

Figure 11   COVIRDS system.

Figure 12   Chair generated using the virtual sculpting system developed at the Missouri University of Science and Technology.

is based on the metaphor of carving a solid block into a 3D freeform object workpiece like a real sculptor would do with a piece of clay, wax, or wood. The VR interface includes stereo viewing and force feedback. The geometric modeling is based on the sweep differential equation method[31] to compute the boundary of the tool swept volume and on the ray-casting method to perform Boolean operations between the tool swept volume and the virtual stock in dexel data to simulate the sculpting process.[30] Incorporating a haptic interface into the virtual sculpting system gives the user a more realistic experience. Force feedback enables the user to feel the model creation process as would be the case when sculpting with physical materials. The PHANToM™ manipulator provides the position and orientation data of the sculpting tool while simultaneously providing haptic sensations to the user's hand during the virtual sculpting process. Multithreading is used in the geometry and force computations to address the different update rates required in the graphic and haptic displays.

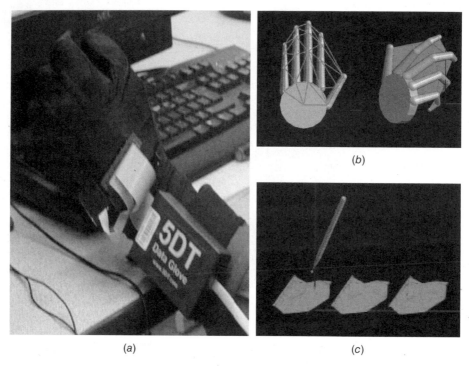

Figure 13 (*a*) 5DT data glove; (*b*) virtual hand covered with polygon mesh; (*c*) virtual pen used for selecting control point.

3.7 Freeform Sketching System for Product Concept Design

Yu[32] developed a freeform sketching system for product design using VR technology, which allows designers to perform conceptual design directly on a computer. A 5DT data glove is used to track the hand's motion, and the virtual hand is covered with polygon mesh, as shown in Figs. 13*a,b*. A virtual pen, as shown in Fig. 10*c*, is used to pick up the control point.

One example taken from Ref. 32 progresses through the following steps to design a car conceptually: (1) create a rough shape of the designed product using the designer's hands; (2) select a sufficient number of points using the virtual pen; (3) generate NURBS (nonuniform rational B-spline) surfaces interpolating these selected points; and (4) modify parametric surfaces by moving interpolating points using the virtual pen and editing the data of these interpolating points directly in the saved file. All four steps are shown in Fig. 14.

With VR technology, this system could significantly shorten the product development process; additionally, it takes advantage of the strength of both polygon mesh and parametric surfaces.

3.8 Concept Design Evaluation with VR Techniques

To satisfy customers, the design process must identify optimal concepts. With the help of VR techniques, concept design can be evaluated more efficiently and intensively without building a physical mock-up at a high cost. As shown in Fig. 15, at Ford, the tester is immersed in a virtual world assisted by a physical, configurable vehicle sled, which can match the physical dimensions of nearly any vehicle.[33] The tester, wearing HMD and data gloves, sits in the sled to

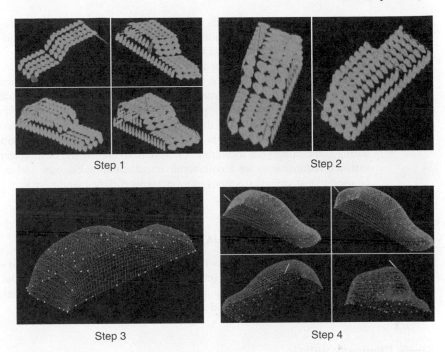

Figure 14   Freeform sketching procedure.

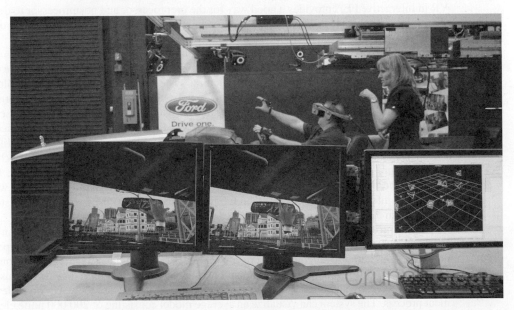

Figure 15   Concept design evaluation with VR technique at Ford.

implement a variety of motions in order to evaluate the conceptual design, such as the seating position, steering wheel position, blind spots, instrument occlusion and accessibility, effect of occupant height variations, and human ergonomic factors.

Gironimo et al.[34] employed a VR environment to evaluate the concept design of a coffee maker and the subassembly of a new minicar in order to improve their quality by involving experts and customers in a dynamic simulation and stereoscopic visualization. This research achieved a more realistic and impressive interaction with virtual prototypes than in a CAD environment; moreover, it reduced the cost by overcoming the need for several physical prototypes. Ingrassia et al.[35] used Kansei methodology to identify the users' emotional needs before evaluating some prototypes through a VR system to choose the best concept. This method better satisfied requirements for a collaborative and emotional design, and more particularly, it can be used to test various solutions in a very powerful, fast, and easy way.

4 VIRTUAL REALITY IN DATA VISUALIZATION

VR is beginning to have a major impact on engineering design because it streamlines the process of converting analysis results into design solutions.[36] By immersing the user in the visualized solution, VR reveals the spatially complex structures in computational science in a way that makes them easy to understand and study.[37]

4.1 Finite-Element Analysis

The two important finite-element analysis (FEA) procedures include preprocessing and postprocessing, the former for setting up the model and the latter for displaying the results. Applying VR in interactive FEA provides a good way for the preprocessor and postprocessor to describe the model and results in a 3D immersive environment.[38] VR offers a more intuitive environment and a real-time, full-scale interaction technique in the design.[39]

Yeh and Vance[40] incorporated finite-element data into a VE such that the designer can interactively modify the transverse force using an instrumented glove and then view the resulting deformation and stress contour of a structural system. Methods are applied to allow the designer to modify design variables and immediately view the effects without performing a separate reanalysis. A head-mounted display and a BOOM are used with a 3D mouse as an interface device to interact with the variables in the VE. Later, Ryken and Vance[41] applied their research inside C2, a second-generation CAVE-like device. Surrounded by this CAVE-like VE, designers can better review the analyzed part, understand how the stress changed with changes to the design, and determine how the part interferes with the surrounding geometry. In this research, the deformation and stress differences are represented by different colors without directional information. Moreover, when running the application, the designer has to stop the VR program to run the FE analysis, which cannot be done in real time.

The University of Erlangen partnered with BMW[42,43] to develop a VE for car crash simulation, as shown in Fig. 16. BMW used the system to analyze a wide variety of time-dependent numerical simulations, ranging from crash worthiness to sheet-metal forming, vibrations, and acoustics. The fully immersive environment Fakespace BOOM was used in which the user can intuitively navigate and interact with different FEA methods relevant to the car body development process. The system can visualize an average model consisting of 200,000 finite elements for each time step up to 60 times steps. The system can load simulation files into the VE directly, achieving immediate polygon reduction without any special preprocessing. This system also can generate VRML files for low-cost workstation and Internet applications. However, it cannot provide FEA simulation and visualization in real time.

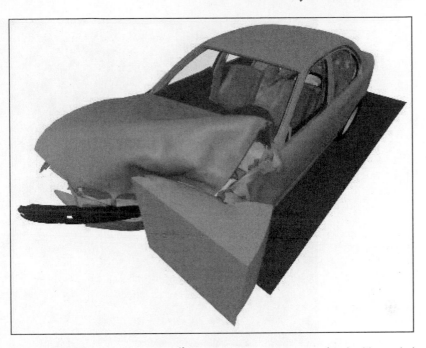

Figure 16   Full car crash simulation in a VE.[43] Copyright © 1998, IEEE. Reprinted, with permission, from Schulz, M., Ertl, T., and Reuding, T., 1998b, "Crashing in Cyberspace—Evaluating Structural Behaviour of Car Bodies in a Virtual Environment," *Virtual Reality Annual International Symposium Proceedings*, March 14–18, Atlanta, Georgia, pp. 160–166.

Connell and Tullberg[44,45] developed a framework for performing interactive finite-element simulation within a VE as well as a concrete method for its implementation. The developed software provides the ability to interchange large-scale components within the system, including the FEA code, visualization code, and VR hardware support; that is to say, this system integrates the simulation and visualization into a single user environment, as opposed to the system developed by Taylor et al.,[46] which only couples the customized software components closely using Standard VR-tools (dVISE) and FEA tools (ABAQUS) to build the VR and FEA modules, respectively. The real-time FEA results are visualized in a CAVE. Unfortunately, this real-time FEA simulation capability comes at the cost of accuracy. Figure 17 shows a stress contour over the surface of the bridge deck under different loading conditions, with color information.

Real-time FEA allows engineers to observe a simulation as it is calculated, so they can stop the simulation at any point before its success is hindered. This capability saves both money and time. Simplified simulation models usually are adopted to achieve real-time FEA[38,45]; however, the simulation accuracy is relatively low. Lian et al.[47] proposed a real-time FEA toolkit to reduce the time required to implement linear-elastic FEA simulation in a desktop-based VR environment. Tactile gloves and virtual hands were used to increase the human–computer interaction. In addition to the default FEA solver from ANSYS, a stiffness matrix inversion and key-frame method were implemented in the system to improve the performance of real-time analysis. Figure 18 shows the real-time FEA simulation from this system, which only solved linear structural problems.

In previous applications, all of the FEA simulation results (force, stress, etc.) were represented by traditional color coding of model nodes, which only displays parameter comparisons but does not include a visualization of the parameter direction. Scherer and Wabner[48]

Figure 17   Concrete case screenshot.

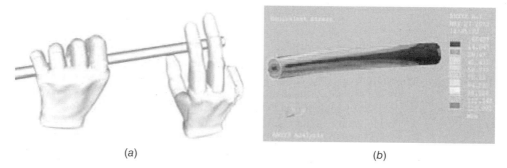

(a)   (b)

Figure 18   Real-time FEA simulation.

used 3D glyphs to map the stress direction for visualizing the FEA result in a VR environment. Figure 19 shows a VR–based FEA of an assembly (stress direction displayed using cuboid-shaped glyphs).

4.2   Computational Fluid Dynamics

Computational fluid dynamics (CFD) software is a powerful tool that can be used to provide fluid velocity, temperature, and other relevant variables; it commonly is used in industry to predict flow behavior. VR provides a fully interactive, 3D interface in which users can interact with computer-generated geometric models for CFD analysis.

The NASA Virtual Windtunnel (VWT)[49] is a pioneering application of VR technology to visualize the results of modern CFD simulations. Figure 20 shows the user interacting with the

Figure 19   Visualization of FEA result with cuboid-shaped glyphs.

simulation via a BOOM display and a data glove. The data glove effectively acts as a source of "smoke," allowing the user to observe local flow lines. The computation of unsteady flow for visualization occurs in real time, and the user can view the dynamic scene from many points of view and scales. The advantage of the VR approach is that, unlike a real user in a flow field, the presence of the user does not disturb the flow.

CAVEvis,[50] designed by the National Center for Supercomputing Applications, is a scientific visualization tool developed to study particle flow in vector and scalar fields using the CAVE[TM]. The scientific simulations include CFD simulations of tornados, hurricanes, and smog as well as the flow of gas and fluid around airplanes, cars, and other vehicles. To visualize very large and complex numerical datasets in real time, the system distributes visualizations using asynchronous computation and rendering modules over multiple machines. The CAVEvis system addresses important issues related to the management of time-dependent data, module synchronization, and interactivity bottlenecks. Figure 21 is a sample screen showing a number of different visualizations when a tornado touches down.

VR-CFD ,[51] developed at Iowa State University and shown in Fig. 22, is a VR interface for the visualization of CFD data. A CAVE-type VR facility is utilized to provide a collaborative environment in which multiple users can investigate the entire flow field together. The system provides the features needed for the creation and manipulation of flow visualization entities, including streamlines, rakes, cutting planes, isosurfaces, and vector fields, in real time.

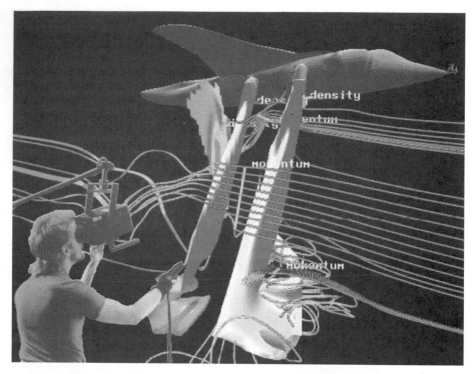

Figure 20 NASA Virtual Windtunnel.

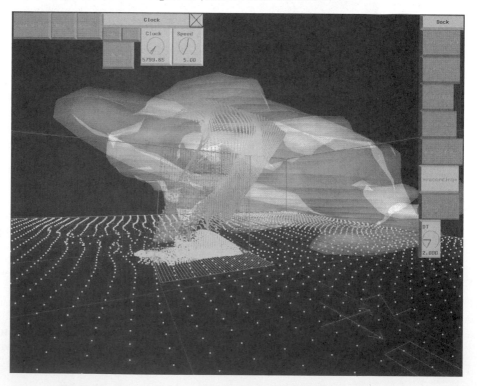

Figure 21 Sample screen of CAVEvis showing the touch-down of a tornado.

Figure 22 Visualization of and immersion in a vector field.

ViSTA FlowLib[52] is a framework that provides an interactive exploration of unsteady fluid flows in a VE. It combines efficient rendering techniques and parallel computations with intuitive multimodal user interfaces into a single, powerful, cross-platform CFD data visualization library. Special care was taken to achieve high scalability with respect to computing power, projection technology, and input–output device availability. Figure 23 shows the ViSTA environment. Van Reimersdahl et al.[53] introduced the haptic rendering subsystem of ViSTA FlowLib. The haptic rendering facilitates visual rendering and the exploration of CFD simulation data.

Other examples of data visualization and data representation in VR environments include the interactive FEA of shell structures, called VRFEA,[54] stress analysis of a tractor lift arm,[41] a VE called COVISE for large-scale scientific data analysis,[55] and a virtual scientific visualization tool called MSVT.[56]

5 VIRTUAL REALITY IN DRIVING SIMULATION

A driving simulator is a VR tool that gives the user an impression that he or she is driving an actual vehicle. It can be based on a physical mock-up or a fully virtual vehicle. The physical mock-up can be equipped with a steering wheel, gearshift, and pedals with corresponding visual, motion, and audio feedback cues. The virtual vehicle only allows the user to interact with the vehicle through some VR hardware systems.

The advantages of using a driving simulator vs. driving a real vehicle include a wide range of possible configurations, repeatable conditions, easy-to-change tasks and parameters, and good experimental efficiency. A driving simulator can be used to study the design and evaluation of vehicles, driver training, driver behaviors, highway systems, in-vehicle information and

Figure 23   Interactive exploration of airflow into the cylinder of a spark ignition engine at the HoloDesk.

warning systems, traffic management systems, and much more. Some physically based driving simulators are reviewed in the following paragraphs.

The National Advanced Driving Simulator (http://www.nads-sc.uiowa.edu/overview.php) was developed at the University of Iowa. Currently, it has the following three different versions, as shown in Fig. 24: (a) NADS-1, the world's highest-fidelity simulator; (b) NADS-2, a fixed-based simulator; and (c) the NADS MiniSim, a low-cost PC-based portable simulator. The applications of these driving simulators have included the study of driver crash avoidance, evaluation of advanced in-vehicle systems and control technologies, and highway design and engineering research related to traffic safety. NADS-1 uses a simulation dome, which is 24 ft in diameter, with the interchangeable car cabs sitting inside of the dome. Within the dome are eight liquid crystal display (LCD) projectors that provide a 360° photorealistic VE. At the same time, the motion subsystem, on which the dome is mounted, provides 64 ft of horizontal and longitudinal travel and 330° of rotation, with a total of 13° of freedom. The VR effect is that the driver can feel vibration, acceleration, and braking and steering cues and can hear wind, tire, engine, and other vehicle noises as if he or she were driving a real car, SUV, truck, or bus. The system supports the generation and control of traffic within the VE. The NADS-2 is a fixed-based version of NADS-1 with a limited forward field of view, which is suitable for simulations without motion and wrap-around visuals. The NADS-MiniSim is a portable, high-performance driving simulator.

Ford Motor Company developed a motion-based driving simulator, VIRTTEX (VIRtual Test Track EXperience), to test the reactions of sleepy drivers.[57] This simulator can generate

(a)

(b)   (c)

Figure 24   National Advanced Driving Simulator (NADS): (*a*) NADS-1; (*b*) NADS-2; (*c*) NADS-MiniSim.

forces that would be experienced by a person while driving a car. The simulator dome houses five projectors, three for the forward view and two for the rear view, that rotate with the dome and provide a 300° computer-generated view of the road. Different car cabs in the simulator are attached to a hydraulic motion platform, called the hexapod Stewart platform, that can simulate the motion associated with more than 90% of the typical miles in the United States including spinouts. A continuation of many years of driver drowsiness research is conducted by Volvo using VIRTTEX. The new safety technology developed through this research will be integrated into the new car design. In Fig. 25, the left image shows the driving simulator dome, inside which the actual car body is placed; the right image shows the hexapod system.

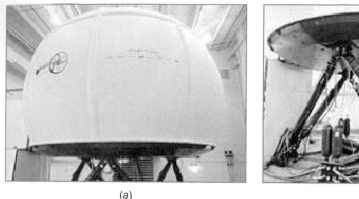
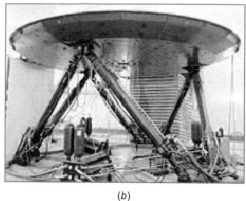

(a)   *(b)*

Figure 25   VIRTTEX driving simulator.

These driving simulators, the most advanced available, make use of more than one linear or hexapod motion system. Other driving simulators are considered semiadvanced if they make use of one linear or hexapod motion system or use the vibrating platform for motion cues.

Researchers at the Kookmin University in Seoul, Korea, have developed a driving simulator[58] that uses a 6DOF Stewart platform as the base. This platform helps the user to experience realistic force feedback. It has 32° of horizontal and vertical field of view for visualization. The applications of this simulator include vehicle system development, safety improvement, and human factor study. An advanced application is the integration of an unmanned vehicle system with the driving simulator.

Researchers at the State University of New York at Buffalo have conducted a project to study and enhance the ability of a driver to cope with inclement conditions using a VE.[59] The goal was to develop strategies for supplementing a driver's natural cues with synthetic cues derived from sensors and control strategies. A simulator was developed to evaluate driver assistive technologies for inclement conditions.

The Drunk Driving Simulator developed at the University of Missouri-Rolla[60] is the initial development of a VR-based driving simulator that can be used as a tool to educate people, especially college and high school students, about the consequences of drunk driving. The VE initially is built on a desktop monitor with an HMD. The developed system consists of real-time vehicle dynamics, a hardware control interface, force feedback, test environments, and driving evaluation. Drunkenness is simulated by implementing the control time lag effect and visual effects. Testing and evaluation techniques have been developed and carried out by students and the results are presented in the paper. The simulator has been extended to the CAVE™ system to provide sufficient immersion and realism.[61]

Other examples of driving simulators include Iowa State University's Driving Simulator[62] and the UMTRI Driving Simulator.[63]

For virtual prototyping, a physical mock-up cannot be built for every design change due to the substantial time and cost. Instead, a virtual vehicle simulator can be used, especially to evaluate design elements such as lay-out efficiency, visibility of instruments, reachability and accessibility, clearance and collision, human performance, and appeal, by immersing a person in the virtual car's interior to study the design and interact with the virtual car.[8] Haptic feedback plays an important role in this simulation. Therefore, the Haptic Workstation™ from CyberGlove Systems Company was employed in developing a virtual driving simulator.[64,65] Figure 26 shows a virtual cockpit used in the virtual driving simulator.

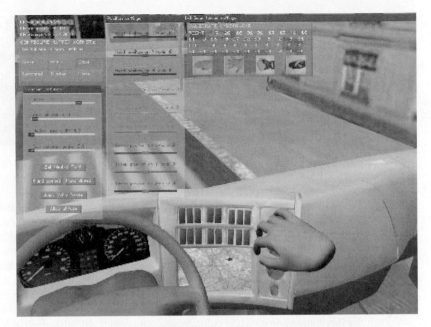

Figure 26   Virtual cockpit.

6   VIRTUAL REALITY IN MANUFACTURING

Generally speaking, there are four areas of VR in manufacturing applications: process simulation, factory layout, assembly method prototyping, and part flow simulation.

6.1   Process Simulation

A suitable process by which to manufacture a product's final shape can be achieved by analyzing manufacturability, final shape, residual stress levels, material, etc., through simulation in an immersive and interactive environment. The optimized process can eliminate material waste and faulty designs. Process simulation also can provide a better understanding of a process without costly "build and break" prototype testing. Moreover, it can train new operators on a virtual machine without having to waste precious machine time. In short, process simulation can reduce manufacturing costs and the time-to-market while dramatically improving productivity.[66]

In a remote VE integrated with some machine and process simulations, Kumar and Annamalai[66] performed a series of case studies, including optimizing the roll-forming process, revealing manufacturing defects through deep drawing simulation, redesigning a rubber boot through stress analysis to reduce the repair cost, optimizing the connecting rod forging process, and conducting an impact analysis of car doors to reduce injuries. For example, in the case of optimizing the roll-forming process, the dimensional tolerances, angular tolerance, longitudinal bow, twist sheet edge waviness, and profile end deformation were analyzed in the VE. The results were used to optimize the manufacturing process; the manufacturer was able to manufacture the right tools and run a test, which avoided the high costs arising from improperly designed tools that end up requiring adjustment and retesting. Figure 27 illustrates this example.

Figure 27 Optimizing the roll-forming process through process simulation.

Process simulation also was used to optimize a hybrid layered manufacturing technology to reduce the risk and cost of making this type of physical machine.[67] Such a machine can manufacture a component featuring an optimized combination of different materials in its different portions for a specific application. A virtual prototype of the manufacturing facility was developed and is shown in Fig. 28; it includes a grinding/milling workstation (to mill the superfluous material from the layer to obtain precise layer boundaries, or a required thickness), the spraying station (to spray required materials), and the laser beam engraving station (to engrave voids for the required microstructure). The virtual prototype was used to simulate the process by fabricating some virtual components, which were visualized in the VE and analyzed in terms of geometric accuracy and material volume fraction accuracy. The results of the analysis were

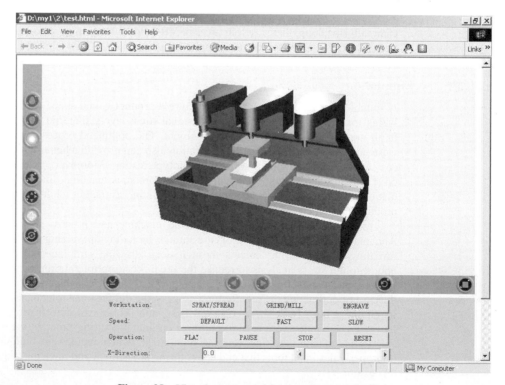

Figure 28 Virtual prototype of the manufacturing facility.

further used to optimize technological parameters, therefore providing a more reliable base for physical facility development.

ESI Group (www.esi-group.com) has noted in its special report that end-to-end virtual prototyping will be a future trend. This means that several manufacturing processes will be integrated in the virtual simulation so that interactions among them can be evaluated and then further used to optimize the design. For example, simulating the hemming process after stamping can reveal valuable information about the whole chain.

6.2   Factory and Process Models

The biggest advantage of using VR for factory design is that it supports the user in planning space or logistical issues by allowing the interactive moving and relocation of the machines after the simulation has been conducted. By integrating all levels of product development and the production cycle, the virtual factory provides a unique environment for achieving the above goals.[68]

VR-Fact! developed at the University of Buffalo provides an interactive virtual factory environment that explores the applications of VEs in the area of manufacturing automation.[68] The VR-Fact! simulation, shown in Fig. 29, can be visualized using stereo HMDs and CrystalEye stereo glasses with a Silicon Graphics ONYX 2 computer. The system also supports viewing on a Fakespace Boom3C. VR-Fact! can be used to create a digital mock-up of a real factory shop floor for a given product mix and set of machines. By intuitively dragging and placing modular machines in the factory, designers can study issues such as plant layout, cluster formation, and part flow analysis. This software's VR walk-through environment provides a unique tool for studying physical aspects of machine placements. It uses mathematical algorithms to generate independent manufacturing cells. This approach simultaneously identifies part families and machine groups and is particularly useful for large and sparse matrices.

Chawla and Banerjee[69,70] at Texas A&M University presented a 3D virtual environment to simulate basic manufacturing operations (unload, load, process, move, and store). It enables a 3D facility to be reconstructed using space-filling curves (SFCs) from a 2D layout. The facility's layout is represented in 3D as a scenegraph structure, which is a directed acyclic graph. SGI's IRIS Performer is used to construct and manipulate the scenegraph structure. Immersadesk[TM] of Fakespace Systems is used to visualize the VE developed with CAVELIB software. The scenegraph structure automatically encapsulates the static and dynamic behavior of the manufacturing system. These properties of the 3D objects include position and orientation, material properties, texture information, spatial sound, elementary sensory data, precedence relationships, event control lists, etc. There are two distinct types of manufacturing objects in terms of their ability to move within the facility—static and dynamic. The simulation of basic manufacturing process operations is represented by suitable object manipulation within the scenegraph.

Kibira and McLean[71] presented their work on the design of a production line for a mechanically assembled product in a VR environment. The simulation was developed as part of the Manufacturing Simulation and Visualization Program at the National Institute of Standards and Technology. Their research focused primarily on the partitioning and analysis of the assembly operation of the prototype product into different tasks and the allocation of those tasks to different assembly workstations. The manufacturing process design was constructed using three software applications. The geometry of components was created using AutoCAD[TM]. IGRIP[TM] was used to provide the graphical ergonomic modeling of workstation operations. QUEST[TM] provided discrete-event modeling of the overall production line. In addition, the difficulties of using simulation modeling for simultaneous graphical simulation of assembly operations and discrete-event analysis of a production process were addressed.

Kelsick et al.[72] presented an immersive virtual factory environment called VRFactory. This system was developed with an interface to the commercial discrete-event simulation

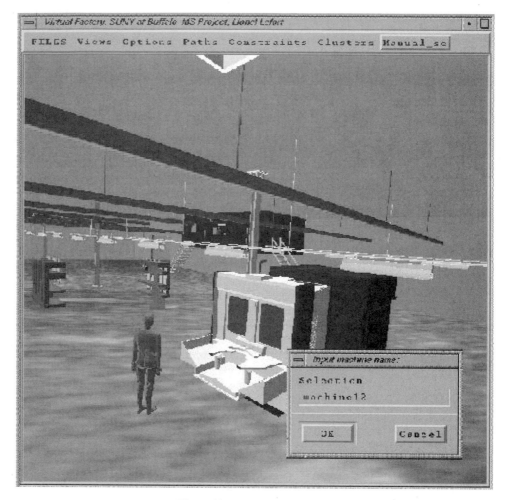

Figure 29   VR-Fact! simulation.

program SLAM II to investigate how potential changes to a manufacturing cell affect part production. The results of the discrete-event simulation of a manufacturing cell were integrated with a virtual model of the cell. The VE was built in a CAVE[TM]-like projection screen based facility called the C2. The interaction device was a Fakespace PinchGlove equipped with a Flock of Birds tracker. This factory allowed the user to explore the effect of various product mixes, inspection schedules, and worker experience on productivity through immersion in a visual, three-dimensional space. The geometric models of all the machines were created using Pro/Engineer CAD software and WorldUP[TM] modeling software and then loaded in a virtual world. Figure 30 illustrates the models in the VRFactory. The output of the simulation software AweSim was used to control the flow of parts through the factory.

In addition to the research described above, Anthony et al.,[73] in cooperation with Ford Motor Company and Lanner Group, automatically generated a VR factory using the WITNESS VR simulation package. Mueller-Wittig et al.[74] presented an interactive visualization environment for the electronics manufacturing industry. The behavior of different manufacturing

Figure 30   VRFactory models.

stages, including fabrication, assembly, testing, and packing, can be simulated in the VR environment.

6.3   VR Assembly

VR provides a means of immersing production engineers in a computer-synthesized work environment. Rather than using physical mock-ups or hand drawings, VR-based assembly can reduce the manufacturing lead time by simulating assembly sequences with virtual prototypes. It also provides an immersive, interactive environment for planning assembly paths and part layout for assembly.[75] However, virtual assembly simulation is one of the most challenging VR applications, mostly because of the very high degree of interactivity needed in the virtual assembly environment. In addition to the high degree of functionality needed, many of the interactions must be as natural as possible.

The Virtual Assembly Design Environment (VADE) was designed at Washington State University in collaboration with NIST.[76] It is a VR-based assembly design environment that allows engineers to plan, evaluate, and verify the assembly of mechanical systems. The VE was built with an SGI Onyx2 (with six processors) and two Infinite Reality pipes, a Flock of Birds motion tracker, a CyberGlove, and an HMD. The mechanical system can be designed using any parametric CAD software (e.g., Pro/Engineer) and exported to the VADE system automatically through a user-selected option. The user specifies the initial location and orientation of the parts and assembly tools based on the actual setup being simulated. The system provides an interactive, dynamic simulation of parts using physically based modeling techniques while the user performs the assembly. The assembly path can be edited interactively in the VE directly by the user. Subsequently, the assembly design information can be generated and made available to the designer in the CAD system. The VADE system was used to conduct industry case studies, demonstrating the value of virtual assembly simulation in various applications, such as ergonomics, process planning, assembly installation, and serviceability.[75,77–79]

The Fraunhofer-Institute for Industrial Engineering in Germany has developed a planning system to carry out assembly planning operations.[80] The system applies a virtual model of a person (VirtualANTHROPOS) to execute assembly operations. It provides interfaces to connect with a CAD system and to present CAD objects in a virtual production system. After the user decides the assembly operations, a precedence graph can be generated with the assembly time and cost determined.

Zachmann and Rettig[81] at the University of Bonn, Germany, investigated robust and natural interaction techniques to perform assembly tasks in the VE. They found that interaction metaphors for virtual assembly must be balanced between naturalness, robustness, precision, and efficiency. Multimodal input techniques were utilized to achieve robust and efficient interaction with the system, including speech input, gesture recognition, tracking, and menus. The precise positioning of parts was addressed by constraining interactive object motions and abstract positioning by employing command interfaces. A natural grasping algorithm was presented to provide intuitive interaction. In addition, a robust, physically based simulation was proposed to create collision-free assembly paths. All of the algorithms were integrated into the VR assembly system. Figure 31 shows the interface of the virtual assembly simulation.

Yuan and Yang[82] presented a biologically inspired intelligence approach for virtual assembly. A biologically inspired neural network was incorporated into the development of a virtual assembly system. The neural network proved useful in the planning of real-time optimal robot motions in dynamic situations without the need for any learning procedures. This work could lead to improvements in flexible product manufacturing, i.e., automatically producing alternative assembly sequences with robot-level instructions for evaluation and optimization.

Haptic force feedback improves system effectiveness and user performance in virtual assembly environments, as proved by providing grasping force feedback through the CyberGrasp^{TM} device and collision force feedback through the CyberForce^{TM} device. Seth[83] developed a dual-handed haptic interface for assembly/disassembly which allowed users to simultaneously manipulate and orient CAD models to simulate dual-handed assembly

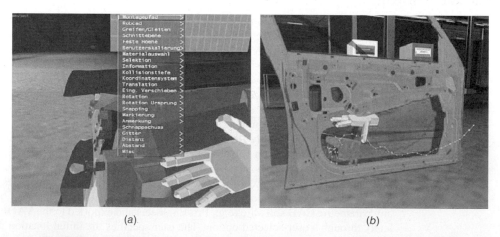

(a)   (b)

Figure 31   Virtual assembly simulation: (*a*) multimodal user interface; (*b*) assembly path.

Figure 32   Dual-handed haptic interface.

operations. The interface is shown in Fig. 32. Ritchie et al.[84] developed the Haptic Assembly, Manufacturing and Machining System to explore the use of immersive technology and haptics in assembly planning.

In addition to its use in assembly sequence planning, VR shows great potential to improve and enrich training applications.[85] Users can visualize the assembly process and manipulate virtual models intuitively in a VE, which can promote their understanding of the assembly task.[86] An AR assembly interface can provide assembly training instructions and guidance to assembly operators in real time and in the actual workplace where the operations are performed without the operators having to alternate their attention between the assembly work itself and the instructions available on computers or in paper manuals. Thus, AR can improve assembly efficiency and lower the cost of products.

Zhang et al.[87] developed an AR-based assembly training system where the 3D-to-2D point matching method is implemented with the object tracking information to render 3D assembly information, assisting assembly operator. RFID and infrared (IR)–enhanced motion tracking techniques have been integrated to identify the assembly activities and to retrieve or generate 3D and 2D point sets. Wearing a see-through HMD, the operator gets the sense of immersion with free hands for assembly operations. A wireless AAD (assembly activity detector) is used to detect the assembly movements that are of interest by attaching to the operator. The computing unit estimates the camera pose and identifies the relevant component(s), retrieves and analyzes data from the AAD module, and then dynamically renders the 2D assembly information on a screen or the 3D assembly information on the relevant component(s). This information assists the operator in running through the entire assembly sequence. Figure 33 shows the scenes captured during this case study in order of the assembly steps.

Manufacturing and assembly processes usually involve a number of manual operations performed by human operators working on the shop floor. For example, fastening is one of the major operations performed at aircraft assembly plants. During the installation of the belly skin of an aircraft fuselage, approximately 2000 fasteners are installed. The mechanic performing the fastening operation at awkward postures may risk ergonomic injuries. Nearly one-third of workplace injuries are related to ergonomics and require compensation.[88] To design safe workplaces, the probable causes of injuries must be identified by simulating the work conditions and quantifying the risk factors. Researches at Missouri University of Science and Technology have developed a methodology using a motion capture system to track the assembly operation using both an immersive VE and a physical mock-up as well as captured motion data for ergonomic analysis. They have demonstrated using this system for the fastening operation in the aircraft industry.[89] Figure 34a shows the data captured inside the CAVE and its simulation in Jack. Figure 34b shows the data captured on the mock-up with the help of the Optitrack motion capture system and simulation in Jack. Figure 34c shows a RULA (rapid upper limb assessment) analysis for fastening operations with different postures with the help of Jack software.

7 VR IN CIVIL ENGINEERING AND CONSTRUCTION

VR techniques have been explored widely and successfully in the area of civil engineering and construction. VR offers considerable benefits for many stages of the construction process, such as (1) enabling the designer to evaluate and modify the design by immersing himself in a building, (2) virtually disassembling and reassembling the components to design the construction process, (3) enabling construction management through 4D planning techniques, which promote the interaction between the geometric model and the planning of the construction activity by integrating the 3D model with the time factor, and (4) providing AR assistance to help students learn and field workers work on-site. This section will present some examples of recent successes of VR applications in construction.

VIRCON (The VIrtual CONstruction Site)[90] was a collaborative project between the University College London, Teeside University, the University of Wolverhampton, and 11 construction companies in the United Kingdom. The aim of this project was to provide a decision support system that would allow planners to trade off the temporal sequencing of tasks with their spatial distribution using VR technologies. The research involved the development of a space scheduling tool called "critical space analysis," its combination with critical path analysis in a space-time broker, and the development of advanced visualization tools for both of these analyses. This system applied visual 4D planning techniques that combine solid CAD models with the construction schedule (time), as shown in Fig. 35. The 4D tools developed visualized not only the construction products but also the movement of plants/temporary objects, highlighting spatial overload. This allows for a better evaluation and communication of activity dependency in spatial and temporal aspects.

DIVERCITY (Distributed Virtual Workspace for Enhancing Communication within the Construction Industry) was a project funded by the European Union Information Society Technologies.[91] This project developed virtual workplaces that enabled three key phases, client briefing, design reviews, and site operations and constructability, to be visualized and manipulated, therefore improving productivity and design, reducing building costs and waste, and improving safety both in the final building and during the construction process, all with more efficient communication and collaboration between all stakeholders.

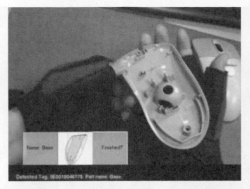

(a) Assembly starts: recognition of the first component

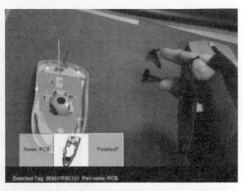

(b) Rendering of the PCB circuit board on the real base component

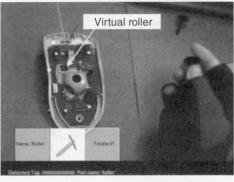

(c) Rendering or the first roller (recognition emulated by pressing a key)

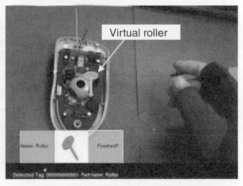

(d) Rendering or the second roller (recognition emulated by pressing a key)

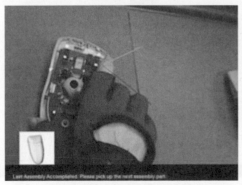

(e) Wrong camera pose estimation when some of the points are occluded

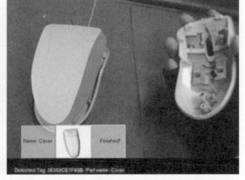

(f) Rendering of the last component, the mouse cover

Figure 33   Case study of a computer mouse assembly.

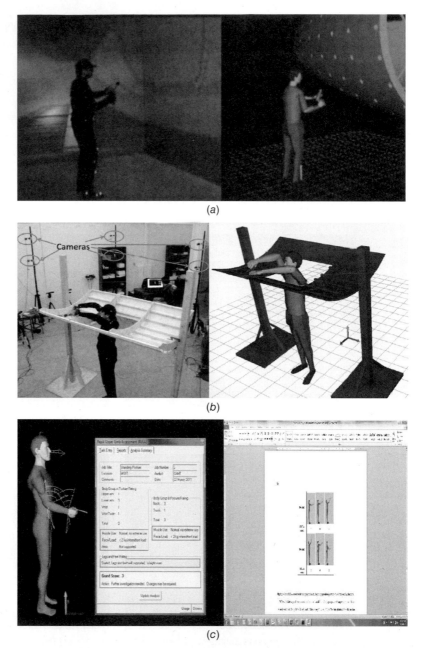

Figure 34 Virtual assembly-based ergonomic analysis.

Figure 36 shows the DIVERCITY visual workspace with the following six virtual applications: client briefing, acoustic simulation, thermal simulation, lighting simulation, site planning, and visual product chronology. It allows the user to produce designs and simulate them in a VE. It also allows teams based in different geographic locations to collaboratively design, test, and validate shared virtual projects.

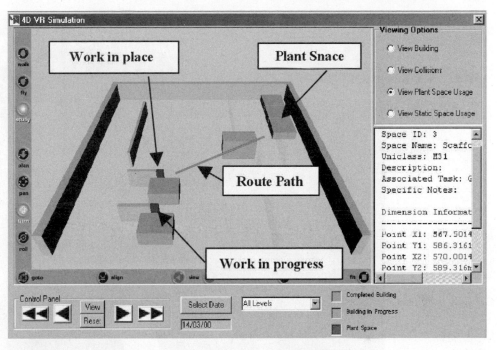

Figure 35   Visualization of construction in VIRCON.

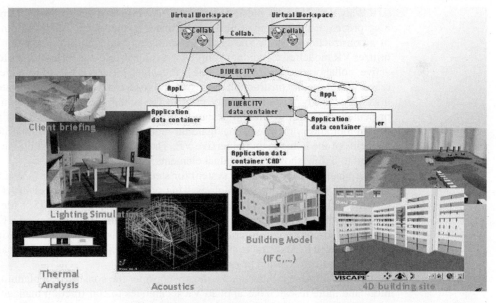

Figure 36   The DIVERCITY visual workplace.

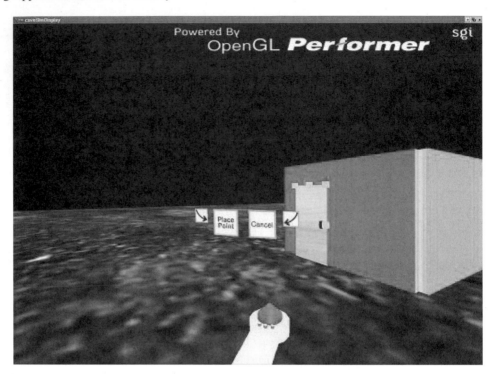

Figure 37 Interactions within a virtual home design tool.

Waly and Thabet[92] presented a virtual construction environment framework. The planning process using this environment contributes to the successful development and execution of a construction project. The virtual planning tool integrates a new planning approach that utilizes VR modeling techniques coupled with object-oriented technologies to develop an interactive, collaborative environment. The VE enables the project team to undertake inexpensive rehearsals of major construction processes and to test various execution strategies in surroundings very closely mirroring reality prior to the actual start of construction. Cowden et al.[93] presented an immersive home design tool using CAVE™, as shown in Fig. 37. The home design tool allows the user to navigate through the model of a residential home and redesign elements of the home in an immersive VE. The interactions provided by the tool currently are limited to hiding/showing individual elements.

Lipman[94] developed a VR system for steel structures using the CAVE™ at the National Institute of Standards and Technology. This system can load the steel structures in the format of CIMsteel Integration Standards (CIS/2) and then convert the CIS/2 file to a VRML file, which can then be visualized in the CAVE™. Yerrapathruni et al.[95] used a CAD system and a CAVE™ to improve construction planning. A clean-room piping design using VR techniques was introduced by Dunston et al.[96] Kamat and Martinez[97] described a VR-based discrete-event simulation of construction processes.

The examples provided above mostly consider the construction planning stage. Recently, more and more research has applied VR techniques to construction management and maintenance during real activities. Sampaio and Santos[98] built 4D models which connected 3D interactive virtual models to construction planning schedules in order to monitor the development

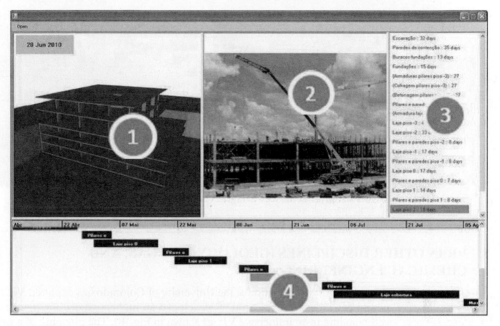

Figure 38   Application interface.

of the construction activity. EON Studio[99] was used to program actions associated with different objects so that the final user can experience the interaction and VR in the presentation. Figure 38 shows the application interface. This project clearly demonstrates the constrictive process that limits inaccuracies and building errors and improves the communication between partners in the construction process. Moreover, Sampaio et al.[100] created VR models to help in the maintenance of exterior closures and interior wall finishes and in the construction of buildings.

Different from VR, AR is a powerful visualization technique that allows real and virtual objects to coexist; it has been applied in diverse scientific and engineering fields. Schall and Schmalstieg[101] developed a hand-held AR device for utility field workers based on a mobile PC which can provide sufficient computing power for 3D graphics and several hours of battery-powered operation in the field. A global positioning system (GPS) was attached to the device for position tracking. When a field worker started the system, the mobile AR client queried a geospatial database with the hand-held GPS position to retrieve data corresponding to the current location. Thus, the worker could visualize both hidden underground objects, such as cables, pipes, and joints, and abstract information, such as legal boundaries, safety buffers, or corresponding semantic object data on-site, as shown in Fig. 39. This system provided a more intuitive way to convey information than the conventional 2D map, consequently saving time.

A collaborative, augmented, reality-based modeling environment was developed by Behzadan et al.[102] to help students develop a comprehensive understanding of construction equipment, processes, and operational safety. Wang and Dunston[103] studied the potential of AR as an assistant viewer for computer-aided drawing in piping systems. However, these simulations were static. Anumba et al.[104] developed an AR animation authoring language that can smoothly and accurately generate animations of arbitrary length and complexity at an operations level of detail.

(a)   (b)

Figure 39 Hand-held AR device used by utility field worker.

8 VR IN OTHER DISCIPLINES (GEOLOGY, OIL, GAS, AND CHEMICAL ENGINEERING)

The Center for Visualization at the University of Colorado has explored VR applications in the oil and gas industry. Dorn et al.[105] developed an immersive drilling planner for platform and well planning in an immersive VE, as shown in Fig. 40. The planning of a well requires a wide variety of data, geometric accuracy of the data display, interaction with the data, and collaborative efforts of an interdisciplinary team. The system can import and visualize critical surface and subsurface data in the CAVE™ system. The data include seismic data, horizons, faults, rock properties, existing well paths and surveys, log data, bathymetry, pipeline maps, and drilling hazards. With these data, the user can design the platform and other well properties interactively. The immersive drilling planner has been applied to several projects, reducing the development planning cycle times from several months to one week or less. Furthermore, the immersive VE has enhanced the quality of results and has encouraged more effective communication among multidisciplinary team members.

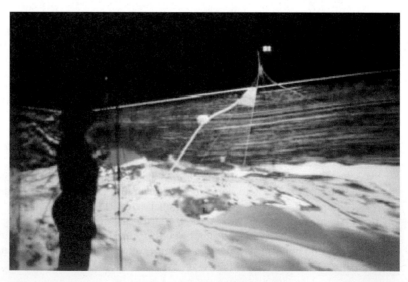

Figure 40 Immersive well path editing.

Figure 41 Project team discussing mine plan in Virtual Reality Laboratory.

The Virtual Reality Laboratory at Laurentian University in Canada[106] offers a collaborative, immersive VR environment for mine planning and design. As shown in Fig. 41, the VE is built on a large, spherical, stereoscopic projection system with advanced earth-modeling software. The system has a 9-ft × 22-ft curved screen and can hold up to 20 people. Because of the constantly changing work environment and enormous overflow of data, mine design and planning are difficult, time-consuming, and expensive exercises. Large volumes of mine geometry, geology, geomechanics, and mining data can be interpreted and evaluated efficiently by utilizing a large-scale, immersive visualization environment. The collaborative environment also speeds up the planning process. The key benefits of using VR in mine design are data fusion, knowledge transfer, technical conflict resolution, and collaboration.

Cowgill et al.[107] applied interactive VR visualization to analyze massive amounts (67 GB) of multiresolution, large-footprint, airborne LiDAR terrain data during the rapid scientific response to the 2010 Haiti earthquake. A four-sided CAVE and a software tool, LiDAR Viewer, were used to analyze LiDAR point cloud data, and Crusta was used to map a 2.5-dimensional surficial geology on a bare-earth digital elevation model. With the help of VR-based data visualization, geologists could remotely observe tectonically produced and modified landforms imaged by LiDAR within hours of the data being released.

Bell and Fogler[108] presented a VR-based educational tool in the field of chemical engineering. A series of interactive virtual chemical plant simulations have been developed to demonstrate the chemical kinetics and reactor design. Figure 42 shows one of the simulations focusing on nonisothermal effects in chemical kinetics. The simulations allow the user to explore domains that otherwise are inaccessible using traditional educational tools, such as the interiors of operating reactors and microscopic reaction mechanics. Different virtual laboratory accidents have been simulated to allow users to experience the consequences of not following proper laboratory safety procedures. However, this application does not provide a realistic interaction between the user and the environment.

Given financial constraints and the increasing popularity of online learning programs, more and more online VR learning simulators are being developed with realistic online interaction capabilities. ViRLE (Virtual Reality Interactive Learning Environment)[109] is one such piece of software that simulates the configuration and operation of a polymerization plant; educators and students can play with and explore new possibilities and alternative configurations through interactive virtual experiments. Figure 43 shows an interaction with a virtual processing plant.

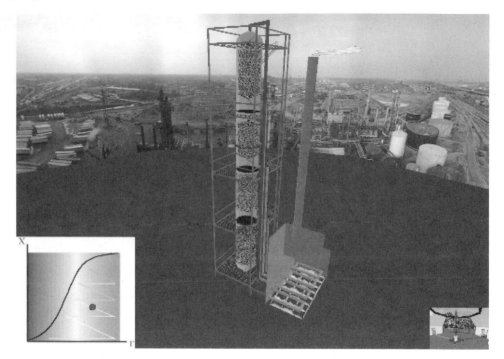

Figure 42 Vicher 2 snapshot.

Figure 43 Interaction with a virtual processing plant.

Calongne[110] and Jarmon et al.[111] have researched and investigated the pedagogical benefits of using this interactive environment as an online educational tool.

9 CONCLUSION

Because of the availability of more advanced computers and other hardware, as well as increasing attention from users and researchers, VR technology and applications have made considerable progress in the past decade. This chapter first reviewed VR software and hardware and their integration for a meaningful engineering application. Special devices, such as head-mounted displays, the haptic device, and the data glove, allow the user to interact with the VE while feeling like it is the real world.

VR has been applied to many aspects of modern society, such as engineering, architecture, entertainment, art, surgical training, and homeland security. This chapter reviewed various VR applications in engineering, including concept design, data visualization, virtual manufacturing, driving simulation, and construction engineering. VR techniques provide intuitive interactions between the VE and the engineer. They enable the user to create, modify, and manipulate 3D models intuitively while visualizing and analyzing data in an immersive VE. By immersing the user in the VE, VR reveals spatially complex structures in a way that makes them easy to understand and study. Visualizing large-scale models is imperative for some engineering applications, such as driving simulation and assembly layout design, but is difficult to achieve using traditional simulation methods. VR techniques conquer this hurdle by providing large-scale visualization devices, such as the CAVE™. This has been demonstrated in driving simulation, virtual manufacturing, construction engineering, and other applications.

REFERENCES

1. W. R. Sherman and A. B. Craig, *Understanding Virtual Reality: Interface, Application, and Design,* Morgan Kaufmann Publishers, Burlington, MA, 2003, Chapter 1.
2. K. Warwick, J. Gray, and D. Roberts, *Virtual Reality in Engineering*, The Institution of Electrical Engineers, London, 1993.
3. S. Aukstakalnis and D. Blatner, *Silicon Mirage: The Art and Science of Virtual Reality*, Peachpit Press, Berkeley, CA. 1992.
4. G. Burdea and P. Coiffet, *Virtual Reality Technology*, Wiley, New York, 1994, Chapter 1.
5. S. Jayaram, J. Vance, R. Gadh, U. Jayaram, and H. Srinivasan, "Assessment of VR Technology and Its Applications to Engineering Problems," *J. Comput. Inf. Sci. Eng.*, **1**(1), 72–83. 2001.
6. P. Banerjee and D. Zetu, *Virtual Manufacturing*, Wiley, Hoboken, NJ, 2001, Chapter 1.
7. R. Kalawsky, *The Science of Virtual Reality and Virtual Environments*, Addison-Wesley, 1991.
8. M. Guiterrez, F. Vexo, and D. Thallmann, *Stepping into Virtual Reality*, Springer. 2008.
9. J. J. LaViola, Jr., "A Discussion of Cybersickness in Virtual Environments," *SIGCHI Bull.*, **32**(1), 47–56, 2000.
10. VTCAVE Information Center, http://www.cave.vt.edu/.
11. S. Güven and S. Feiner, "Authoring 3D Hypermedia for Wearable Augmented and Virtual Reality," in *Proceedings of the ISWC '03 (Seventh International Symposium on Wearable Computers)*, White Plains, NY, October 21–23, 2003, pp. 118–226.
12. B. Roehl, *Distributed Virtual Reality—An Overview*, VRML, 1995.
13. K. Pimentel and K. Teixeira, *Virtual Reality: Through the New Looking Glass*, Intel/Windcrest/McGraw-Hill, 1993.
14. A. Craig, W. Sherman, and J. Will, *Developing Virtual Reality Applications*, Morgan Kaufmann. 2009.
15. J. Webster, "The Hemispherium Experience: Fasten Your Seat Belts," in *Proceedings of Trends, Technology, Theming and Design in Leisure Environments Conference*, London, 1999.

16. Prabhat, A. Forsberg, M. Katzourin, K. Wharton, and M. Slater, "A Comparative Study of Desktop, Fishtank, and CAVE Systems for the Exploration of Volume Rendered Confocal Data Sets," *IEEE Trans. Vis. Computer Graphics*, **14**(3), 551–563, 2008,

17. K. Salisbury, F. Conti, and F.,Barbagli, "Haptic Rendering: Introductory Concepts," *IEEE Computer Graphics Appl.*, **24**(2), 24–32, 2004,

18. Chemistry World, "Making Smell-o-vision a Reality Using a Polymer Matrix," 2011, available: http://www.rsc.org/chemistryworld/News/2011/June/21061101.asp.

19. A. Bierbaum and C. Just, "Software Tools for Virtual Reality Application Development," SIGGRAPH 98 Course 14: Applied Virtual Reality, SIGGRAPH, Orlando, FL, July 19–24, 1998, pp. 3-2 to 3-22.

20. J. Vince, *Essential Virtual Reality Fast: How to Understand the Techniques and Potential of Virtual Reality?* Springer-Verlag, 1998.

21. J. Isdale, "What is Virtual Reality?" 1998, available: http://vr.isdale.com/WhatIsVR.html.

22. E. Sachs, A. Roberts, and D. Stoops, "3-Draw: A Tool for Designing 3D Shapes," *Proc. IEEE Computer Graphics Appl.*, 18–24. 1991.

23. J. Butterworth, A. Davidson, S. Hench, and M. Olano, "3DM: A Three Dimensional Modeler Using a Head-Mounted Display," in *Proceedings of the 1992 Symposium on Interactive 3D Graphics,* 1992, pp. 135–138.

24. J. Liang and M. Green, "JDCAD: A Highly Interactive 3D Modeling System," *Comput. Graphics*, **18**(4), 499–506. 1994,

25. C. Ware and D. R. Jessome, "Using the Bat: A Six Dimensional Mouse for Object Placement," *Proc. Graphics Interf.*, 119–124, 1988.

26. M. F. Deering, "The Holosketch VR Sketching System," *Commun. ACM*, **39**(5), 54–61, 1996.

27. T. H. Dani and R. Gadh, "Creation of Concept Shape Designs via a Virtual Reality Interface," *Comput.-Aided Des.*, **29**(8), 555–563, 1997.

28. C. Chu, T. Dani, and R. Gadh, "Evaluation of Virtual Reality Interface for Product Shape Design," *IIE Trans. Design Manufact.*, **30**, 629–643, 1998.

29. X. Peng and M. C. Leu, "Interactive Virtual Sculpting with Force Feedback," in *Proceedings of 2004 Japan-USA Symposium on Flexible Automation*, Denver, CO, July 19–21, 2004.

30. X. Peng and M. C., Leu, "Interactive Solid Modeling in a Virtual Environment with Haptic Interface," in *Virtual and Augmented Reality Applications in Manufacturing*, Springer-Verlag, London, 2004.

31. D. Blackmore and M. C. Leu, "Analysis of Swept Volume via Lie Groups and Differential Equations," *Int. J. Robot. Res.*, 516–537, 1992.

32. H. Yu, "Free Form Sketching System for Product Design Using Virtual Reality Technology," Thesis, Concordia University, Montreal, Quebec, Canada, 2005.

33. M. Burns, "An Insider's Look at Ford's Virtual Reality Design Tools," 2010, available: http://techcrunch.com/2010/05/22/an-insiders-look-at-fords-virtual-reality-design-tools/.

34. G. D. Gironimo, A, Lanzotti, and A. Vanacore, "Concept Design for Quality in Virtual Environment," *Comput. Graphics*, **30**, 1011–1019, 2006.

35. T. Ingrassia, E. Lombardo, V. Nigrelli, and G. Sabatino, " Kansei Engineering and Virtual Reality in Conceptual Design," 2008, available: http://www.google.com/url?sa=t&rct=j&q=&esrc=s&frm=1&source=web&cd=1&cad=rja&ved=0CDUQFjAA&url=http%3A%2F%2Fwww.ep.liu.se%2Fecp%2F033%2F043%2Fecp0803343.pdf&ei=4AqbUJ-hO-Sb2QWCuIHABg&usg=AFQjCNGsr7mRkvHezSEMaub68nLMHxBy6A&sig2=t0KX9XFd7JdorhLhKeTJ8w.

36. K. M. Bryden, "Virtual Reality Helps Convert Fluid Analysis Results to Solutions," ME Magazine Online, July 22, 2003, available: http://www.memagazine.org/contents/current/webonly/webex722.html.

37. S. Bryson, "Virtual Reality in Scientific Visualization," *Commun. ACM*, **39**(5), 62–71, 1996.

38. L. Lu, M. Connell, and O. Tullberg, "The Use of Virtual Reality in Interactive Finite Element Analysis: State of the Art Report," paper presented at The AVR II and CONVR Conference, Gothenburg, Sweden, October 4–5, 2001, pp. 60–68.

39. L. Rosenblum and M. Macedonia, "Integrating VR and CAD," in *1999 IEEE Projects in VR*, September/October 1999, pp. 14–19.

40. T. P. Yeh and J. M. Vance, "Combining MSC/NASTRAN, Sensitivity Methods, and Virtual Reality to Facilitate Interactive Design," *Finite Elem. Anal. Design*, **26**, 161–169, 1997.

41. M. Ryken and J. Vance, "Applying Virtual Reality Techniques to the Interactive Stress Analysis of a Tractor Lift Arm," *Finite Elem. Anal. Design*, **35**(2), 141–155, 2000.

42. M. Schulz, T. Reuding, and T. Ertl, "Analyzing Engineering Simulation in a Virtual Environment," *IEEE Comput. Graphics Appl.*, **18**, (6), 46–52, 1998.

43. M. Schulz, T. Ertl, and T. Reuding, "Crashing in Cyberspace—Evaluating Structural Behaviour of Car Bodies in a Virtual Environment," in *Virtual Reality Annual International Symposium Proceedings*, Atlanta, GA, March 14-18, 1998, pp. 160–166.

44. M. Connell and O. Tullberg, "A Framework for the Interactive Investigation of Finite Element Simulations within Virtual Environments," in *Proceedings of the 2nd International Conference on Engineering Computational Technology*, Leuben, Belgium, September 6–8, 2000.

45. M. Connell and O. Tullberg, "A Framework for Immersive FEM Visualization Using Transparent Object Communication in a Distributed Network Environment," *Adv. Eng. Software*, **33**, 453–459, 2002.

46. V. E. Taylor, J. Cehn, M. Huang, T. Canfield, and R. Stevens, "Identifying and Reducing Critical Lag in Finite Element Simulation," *IEEE Comput. Graphics Appl.*, **16**(4), 67–71, 1996.

47. D. Lian, L. A. Oraifige, and F. R. Hall, *Real-Time Finite Element Analysis with Virtual Hands—An Introduction*, WACS, 2004,

48. S. Scherer and M. Wabner, "Advanced Visualization for Finite Elements Analysis in Virtual Reality Environment," *Int. J. Interact. Des. Manuf.*, **2**, 169-173, 2008.

49. S. Bryson and C. Levit, "The Virtual Wind Tunnel," *IEEE Comput. Graphics Appl.*, **12**(4), 25–34, 1992.

50. V. Jaswa, "CAVEvis: Distributed Real-Time Visualization of Time-Varying Scalar and Vector Fields Using the CAVE Virtual Reality Theater," *IEEE Vis. Proc.*, 301–308, 1997.

51. V. Shahnawaz, J. Vance, and S. Kutti, "Visualization of Post-Process CFD Data in a Virtual Environment," Paper No. DETC99/CIE-9042, in *Proceedings of the ASME Computers in Engineering Conference*, 1999.

52. M. Schirski, A. Gerndt, T. van Reimersdahl, T. Kuhlen, P. Adomeit, O. Lang, S. Pischinger, and C. Bischof, "ViSTA FlowLib—A Framework for Interactive Visualization and Exploration of Unsteady Flows in Virtual Environments," in *Proceedings of the 7th International Immersive Projection Technologies Workshop* and *The 9th Eurographics Workshop on Virtual Environments*, ACM Siggraph, Zurich, Switzerland, 2003, pp. 77–85.

53. T. Van Reimersdahl, F. Bley, T. Kuhlen, and C. Bischof, "Haptic Rendering Techniques for the Interactive Exploration of CFD Datasets in Virtual Environments," in *Proceedings of the 7th International Immersive Projection Technologies Workshop* and *The 9th Eurographics Workshop on Virtual Environments*, ACM Siggraph, Zürich, Switzerland, 2003, pp. 241–246.

54. A. Liverani, F. Kuester, and B. Hamann, "Towards Interactive Finite Element Analysis of Shell Structures in VR," *Proc. IEEE Inf. Vis.*, 340–346, 1999.

55. D. Rantzau and U. Lang, "A Scalable Virtual Environment for Large Scale Scientific Data Analysis," *Future Gen. Comput. Syst.*, **14**(3–4), 215–222. 1998,

56. J. J. Laviola, "MVST: A Virtual Reality-Based Multimodal Scientific Visualization Tool," in *Proceedings of the Second IASTED International Conference on Computer Graphics and Imaging*, 1999, pp. 221–225.

57. P. Grant, B. Artz, J. Greenberg, and L. Cathey, "Motion Characteristics of the VIRTTEX Motion System," in *Proceedings of the 1st Human-Centered Transportation Simulation Conference*, IA, 2001.

58. D. Yun, J. Shim, M. Kim, Y. Park, and J. Kim, "The System Development of Unmanned Vehicle for the Tele-Operated System Interfaced with Driving Simulator," in *Proceedings of IEEE International Conference on Robotics & Automation*, Seoul, Korea, 2001.

59. T. Singh, T. Kesavadas, R. M. Mayne, J. Kim, and A. Roy, *"Design of Hardware/Algorithms for Enhancement of Driver-Vehicle Performance in Inclement Conditions Using a Virtual Environment,"* SAE Trans. J. Passenger Car Mech. Syst., 2001.

60. M.,Sirdeshmukh, "Development of Personal Computer Based Driving Simulator for Education to Prevent Drunk Driving," M.S. Thesis, University of Missouri-Rolla, Rolla, MO, 2004.

61. K. R. Nandanoor, "Development and Testing of Driving Simulator with a Multi-Wall Virtual Environment," M.S. Thesis, University of Missouri-Rolla, Rolla, MO, 2004.

62. X. Fang, H. A. Pham, and S. Tan, "Driving in the Virtual World," in *Proceedings of the 1st Human-Centered Transportation Simulation Conference*, 2001.

63. P. Green, C. Nowakowski, K. Mayer, and O. Tsimhoni, "Audio-Visual System Design Recommendations from Experience with the UMTRI Driving Simulator," in *Proceedings of Driving Simulator Conference North America*, Dearborn, MI, 2003.

64. R. Ott, M. Gutierrez, D. Thalmann, and F. Vexo, "VR Haptic Interfaces For Teleoperation: An Evaluation Study," in *IV'2005*, 2005, pp. 788–93.

65. P. Salamin, D. Thalmann, and F. Vexo, "Visualization Learning for Visually Impaired People," paper present at The 2nd International Conference of E-Learning and Games: Edutainment, 2007, pp. 171–181.

66. E. R. Kumar and K. Annamalai, "An Overview of Virtual Manufacturing with Case Studies," *Int. J. Eng. Sci. Technol.*, **3**(4), 2720–2727, 2011.

67. F. Wang, K. Chen, and X. Feng, "Virtual Manufacturing to Design a Manufacturing Technology for Components Made of a Multiphase Perfect Material," *Comput.-Aided Design Appl.*, **5**(1–4), 110–120, 2008.

68. T. Kesavadas and M. Ernzer, "Design of an Interactive Virtual Factory Using Cell Formation Methodologies," in *Proceedings of the ASME Symposium on Virtual Environment for Manufacturing*, MH-Vol. 5, Nashville, TN, November 14–19, 1999, pp. 201–208.

69. R. Chawla and A. Banerjee, "A Virtual Environment for Simulating Manufacturing Operations in 3D," in *Proceedings of the* 2001 Winter *Simulation Conference*, Vol. 2, 2001, pp. 991–997.

70. R. Chawla and A. Banerjee, "An Automated 3D Facilities Planning and Operations Model Generator for Synthesizing Generic Manufacturing Operations in Virtual Reality," *J. Adv. Manufact. Syst.*, **1**(1), 5–17, 2002.

71. D. Kibira and C. McLean, "Virtual Reality Simulation of a Mechanical Assembly Production Line," in *Proceedings of the Winter Simulation Conference*, San Diego, CA, 2002, pp. 1130–1137.

72. J. Kelsick, J. M. Vance, L. Buhr, and C. Moller, "Discrete Event Simulation Implemented in a Virtual Environment," *J. Mechan. Des.*, **125**(3), 428-433, 2003.

73. A. P. Waller and J. Ladbrook, "Experiencing Virtual Factories of the Future," paper presented at the Winter Simulation Conference, San Diego, CA, December 8–11, 2002.

74. W. Mueller-Wittig, R. Jegathese, M. Song, J. Quick, H. Wang, and Y. Zhong, "Virtual Factory—Highly Interactive Visualization for Manufacturing," in *Proceedings of the Winter Simulation Conference*, San Diego, CA, 2002, pp. 1061–1064.

75. F. Taylor, S. Jayaram, U. Jayaram, and T. Mitsui, "Functionality to Facilitate Assembly of Heavy Machines in a Virtual Environment," in *Proceedings of ASME Design Engineering Technical Conference*, 2000.

76. S. Jayaram, U. Jayaram, Y. Wang, K. Lyons, and P. Hart, "VADE: A Virtual Assembly Design Environment," *IEEE Comput. Graphics Appl.*, **19**(6), 44–50, 1999.

77. I. Shaikh, U. Jayaram, and C. Palmer, "Participatory Ergonomics Using VR Integrated with Analysis Tools," paper presented at the 2004 Winter Simulation Conference, Washington, DC, 2004.

78. S. Jayaram, U. Jayaram, Y. Kim, C. DeChenne, K. Lyons, C. Palmer, and T. Mitsui, "Industry Case Studies in the Use of Immersive Virtual Assembly," *Virtual Reality*, **11**(4), 217–228, 2007.

79. U. Jayaram, S. Jayaram, I. Shaikh, Y. Kim, and C. Palmer, "Introducing Quantitative Analysis Methods into Virtual Environments for Real-Time and Continuous Ergonomic Evaluations," *Comput. Ind.*, **57**(3), 283–296, 2006.

80. H. J. Bullinger, M. Richter, and K. A. Seidel, "Virtual Assembly Planning," *Human Factors Ergon. Manufact.*, **10**(3), 331–341, 2000.

81. G. Zachmann and A. Rettig, "Natural and Robust Interaction in Virtual Assembly Simulation," in *Proceedings of the 8th ISPE International Conference on Concurrent Engineering: Research and Applications*, Anaheim, CA, 2001.

82. X. Yuan and S. Yang, "Virtual Assembly with Biologically Inspired Intelligence," *IEEE Trans. Syst. Man Cybernet., Part C*, **33**(2), 159–167, 2003.

83. A. Seth, "SHARP: A System for Haptic Assembly and Realistic Prototyping," in *Proceedings of the DETC'06*, 2006.

84. J. M. Ritchie, T. Lim, R. S. Sung, J. R. Corney, and H. Rea, "The Analysis of Design and Manufacturing Tasks Using Haptic and Immersive VR: Some Case Studies," in *Product Design*, Springer, 2008, pp. 507–522.

85. A. Hirsch, V. Wittstock, and F. Pürzel, "Use of Virtual Reality Technology for Engineering Education at Universities: Chances and Challenges," paper presented at The International Technology, Education and Development Conference, 2011, pp. 3438–3444.

86. Y. D. Lang, Y. Yao, and P. Xia, "A Survey of Virtual Assembly Technology," *Appl. Mechan. Mater.*, **10–12**, 711–716, 2008.

87. J. Zhang, S. K. Ong, and A. Y. C. Nee, "RFID-Assisted Assembly Guidance System in an Augmented Reality Environment," *Int. J. Product. Res.*, **49**(13), 3919–3938, 2011.

88. Bureau of Labor Statistics, 2007, available: http://www.bls.gov/opub/ted/2008/dec/wk1/art02.htm.

89. S. C. Puthenveetil, C. P. Daphalalpurkar, W. Zhu, M. C. Leu, X. F. Liu, A. M. Chang, S. D. Snodgrass, J. K. Gilpin-Mcminn, and P. H. Wu, "Computer Automated Ergonomic Analysis Based on Motion Capture and Assembly Simulation," *Virtual Reality*, 2012.

90. N. Dawood, E. Sriprasert, Z. Mallasi, and B. Hobb, "Implementation of Space Planning and Visualisation in a Real-Life Construction Case Study: VIRCON Approach," paper presented at the Conference on Construction Applications of Virtual Reality, Virginia Tech, Blacksburg, VA, September 24–26, 2003.

91. M. Sarshar and P. Christiansson, "Towards Virtual Prototyping in the Construction Industry: The Case Study of the DIVERCITY Project," in *Proceedings of Incite2004—International Conference on Construction Information Technology*, Langkawi, Malaysia, February 18–21, 2004.

92. A. F. Waly and W. Y. Thabet, "A Virtual Construction Environment for Preconstruction Planning," *Automat. Construct.*, **12**(2), 139–154, 2003.

93. J. Cowden, D. Bowman, and W. Thabet, "Home Design in an Immersive Virtual Environment," in *Proceedings of CONVR Conference on Construction Applications of Virtual Reality*, Blacksburg, VA, 2003, pp. 1–7.

94. R. R. Lipman, "Immersive Virtual Reality for Steel Structures," in *Proceedings of CONVR Conference on Construction Applications of Virtual Reality*, Blacksburg, VA, 2003.

95. S. Yerrapathruni, J. I. Messner, A. J. Baratta, and M. J. Horman, "Using 4D CAD and Immersive Virtual Environments to Improve Construction Planning," in *Proceedings of CONVR Conference on Construction Applications of Virtual Reality*, Blacksburg, VA, 2003.

96. P. S. Dunston, X. Wang, T. Y. Lee, and L. M. Chang, "Benefits of CAD and Desktop Virtual Reality Integration for Cleanroom Piping Design," in *Proceedings of CONVR Conference on Construction Applications of Virtual Reality*, Blacksburg, VA, 2003.

97. V. R. Kamat and J. C. Martinez, "Interactive Discrete-Event Simulation of Construction Processes in Dynamic Immersive 3D Virtual Worlds," in *Proceedings of CONVR Conference on Construction Applications of Virtual Reality*, Blacksburg, VA, 2003.

98. A. Z. Sampaio and J. P. Santos, "Construction Planning Supported in 4D Interactive Virtual Models," *J. Civil Eng. Construct. Technol.*, **2**(6), 125–137, 2011.

99. EON, 2012, available: http://www.eonreality.com/, accessed October 2012.

100. A. Z. Sampaio, A. R. Gomes, and J. P. Santos, "Management of Building Supported on Virtual Interactive Models: Construction Planning and Preventive Maintenance," *Electronic J. Info. Technol. Construction,***17**, 121–133, 2012.

101. G. Schall and D. Schmalstieg, "Handheld Augmented Reality in Civil Engineering," in *Proceedings of Rosus09 Conference*, 2009, pp. 19–25.

102. A. H. Behzadan, A. Iqbal, and V. R. Kamat, "A Collaborative Augmented Reality Based Modeling Environment for Construction Engineering and Management Education," in *Proceedings of the 2011 Winter Simulation Conference*, 2011, pp. 3573–3581.

103. X. Wang and P. S. Dunston, "Potential of Augmented Reality As an Assistant Viewer for Computer-aided Drawing," *Comput. Civil Eng.*, **20**(6), 437–441, 2006.

104. C. J. Aunumba, X. Wang, A. H. Behzadan, S. Dong, and V. R. Kamat, *Mobile and Pervasive Construction Visualization Using Outdoor Augmented Reality*, Wiley. Hobken, NJ, 2012.

105. G. A. Dorn, K. Touysinhthiphonexay, J. Bradley, and A. Jamieson, "Immersive 3-D Visualization Applied to Drilling Planning," *The Leading Edge*, **20**(12), 1389–1392, 2001.

106. P. K. Kaiser, J. Henning, L. Cotesta, and A. Dasys, "Innovations in Mine Planning and Design Utilizing Collaborative Immersive Virtual Reality (CIVR)," in *Proceedings of the 104th CIM Annual General Meeting*, Vancouver, Canada, April 28 –May 1, 2002.

107. E. Cowgill et al., "Interactive Terrain Visualization Enables Virtual Field Work during Rapid Scientific Response to the 2010 Haiti earthquake," Geological Society of America, 2012, available: eosphere.gsapubs.org.

108. J. T. Bell and H. S. Fogler, "The Application of Virtual Reality to Chemical Engineering Education," in *Proceedings of 2004* IEEE Virtual Reality, March 27–31, 2004, pp. 217–218.

109. D. Schofield, "Mass Effect: A Chemical Engineering Education Application of Virtual Reality Simulator Technology," *MERLOT J. Online Learning Teaching*, **8**(1), 63–78, 2012.

110. C. M. Calongne, "Educational Frontiers: Learning in a Virtual World," *EDUCAUSE review*, **43**(5), 2008.

111. L. Jarmon, T. Traphagan, M. Mayrath, and A. Trivedi, "Virtual World Teaching, Experiential Learning, and Assessment: An Interdisciplinary Communication Course in Second Life," *Comput. Ed.*, **53**(1), 169–182, 2009.

112. T. H. Dani and R. Gadh, "Virtual Reality—A New Technology for the Mechanical Engineer," in M. Kutz, *Mechanical Engineer's Handbook*, 2nd ed., Wiley, New York, 1998, pp. 319–327.

CHAPTER 14

PHYSICAL ERGONOMICS

Maury A. Nussbaum
Virginia Tech
Blacksburg, Virginia

Jaap H. van Dieën
Free University
Amsterdam, The Netherlands

1 WHAT IS PHYSICAL ERGONOMICS?

1.1 Definitions and a Brief History

Ergonomics is derived from the Greek *ergon* (work) and *nomos* (principle or law). Original attribution is debated but usually given to the Polish professor Wojciech Jastrzębowski in a treatise published in 1857 and intended to represent the "Science of Work." The same term seems to have been reinvented in 1949 by the British professor K.F.H. Murrell and with the same general intended meaning. In the past decades ergonomics as a scientific and/or engineering discipline has seen a dramatic increase in research and application, as well as attention from the general public. As such, the original term has, in cases, lost some of its original meaning. Returning to these origins, and in order to provide a focus for this chapter, the following definition does well to encompass the current state of ergonomics as having a theoretical and multidisciplinary basis, being concerned with humans in systems, and ultimately driven by real-world application:

> Ergonomics produces and integrates knowledge from the human sciences to match jobs, products, and environments to the physical and mental abilities and limitations of people. In doing so it seeks to safeguard safety, health and well being whilst optimising efficiency and performance.[1]

417

As a recognized professional domain, ergonomics is quite young, becoming formalized only after World War II. Related efforts were certainly conducted much earlier, such as Borelli's mechanical analysis of physical efforts in the late seventeenth century, and the identification of work-related musculoskeletal illnesses by Ramazzini in the early eighteenth century. It was only in the 1950s, however, that researchers with expertise in engineering, psychology, and physiology began to come together to realize their common goals and approaches. Ergonomics continues to be an inherently multidisciplinary and interdisciplinary field, which will be evident in the remaining material presented in this chapter.

1.2 Focus of This Chapter

As with any technical domain, the evolution and development of ergonomics has led to increasing specialization and branching. At a high level, three domains presently exist.[2] *Cognitive* ergonomics addresses human mental processes, including information presentation, sensation and perception, and memory. Related applications include mental workload, skill and training, reliability assessment, and human–computer interaction. *Organizational* ergonomics (also called organizational design or macroergonomics) focuses on humans in social-technical systems, addressing the design and evaluation of organization structures and group processes.

The present chapter focuses solely on the third domain of *physical* ergonomics (PE), primarily because of its relevance to mechanical engineering and the design of mechanical processes and systems. PE addresses diverse physical characteristics of humans, including anatomical, dimensional, mechanical, and physiological, as they affect physical activity (or the consequences thereof). Among a variety of applications, PE is commonly used to address, for example, forceful and/or repetitive motions, workplace layout, tool and equipment design, and environmental stresses.

In the remainder of this section, the fundamental basis of PE is provided, along with its technical foundation and role in the design process, and a summary of key needs motivating PE research and application. In the subsequent section, several common PE analytical methods are described. The last section demonstrates two applications of the material addressing practical occupational design concerns.

This chapter is intended as both an introduction to and overview of PE, and three goals have motivated our presentation. Throughout, we wish to demonstrate the utility, if not the necessity, of PE in the mechanical design process, with a main focus on occupational applications and design for this context. We also intend to provide the reader with a level of familiarity with common contemporary methods, and with sufficient information to allow for at least a rough understanding of application to actual problems. Finally, we have limited our formal citations to the literature in order to improve readability but have provided numerous sources, in the references, of additional information on the topics presented as well as related topics.

1.3 Basis of Physical Ergonomics

At its most basic level, PE is concerned with two aspects of physical effort: (1) the physical *demands* placed on the human and (2) the physical *capabilities* of the human in the situation where the demands are present. Simply, the goal is to ensure that demands do not exceed capacity, as is typical for the design of any mechanical system.

A major challenge in PE is the *measurement* of both demands and capacity in the wide variety of circumstances where human physical exertions are performed (or required, as in the occupational context). Physical requirements vary widely, and include dimensions such as force, torque, repetition, duration, posture, etc. Similarly, human attributes vary widely across related dimensions (e.g., strength, endurance, mobility).

Along with the measurement problem is the related problem of *matching*. Given the variability in both demands and capacity, potential mismatches are likely (and prevalent in many cases). Occupationally, poor matching can lead to low productivity and quality, worker dissatisfaction and turnover, and in many cases to musculoskeletal illness and injury.

Ongoing research is being widely conducted to address both the measurement and matching problems. In the material presented below, we have given examples of contemporary approaches, but the reader is advised that more advanced methods are currently available and that the technology is rapidly advancing.

1.4 Disciplines Contributing to Physical Ergonomics

As noted earlier, ergonomics is inherently a multidisciplinary field. Indeed, it can be argued that ergonomics is essentially the intersection of several more primary fields, both receiving basic information from them and providing applications, tools, and procedures to them. Among the primary disciplines contributing to ergonomics, and in which the ergonomist must have a fair degree of knowledge, are anatomy, physiology, and mechanics. As will be seen in the subsequent presentation of methods, each are needed at some level to adequately address the measurement and matching problems earlier noted.

1.5 Ergonomics in the Design Process

Where are ergonomists and how do they contribute to the design process? As with any profession, the answer is necessarily diverse. In smaller enterprises, if any individual has responsibility for ergonomics it is often along with safety and other related topics. In larger enterprises, there may be one or more persons with specific ergonomics training and expertise. In the largest, such as automotive manufacturers, whole departments may exist. Ergonomics consultants are also widely employed, as are ergonomists in academic centers.

Methods by which ergonomics is part of the design process are similarly quite varied. Broadly, ergonomics in design is either proactive or reactive. In the former, procedures such as those described below are applied early in a product or process life cycle, such as in the conceptual or prototype phases. In reactive situations, ergonomics methods are employed only after a design problem has been identified (e.g., a control cannot be reached, required torques cannot be generated, or work-related injuries have occurred).

A natural question arises as to the need for ergonomics in design. Justification comes from a range of documented case studies, formal experiments, and economic evaluations. Though somewhat oversimplified, ergonomic design attempts to optimize design by minimizing adverse physical consequences and maximizing productivity or efficiency. It should be noted that these goals are at times in conflict, but ergonomic methods provide the tools whereby an optimal balance can be achieved.

An important justification for ergonomics is provided by the high costs of mismatches between demands and capabilities. Occupational musculoskeletal illnesses and injuries (e.g., sprains, strains, low back pain, carpal tunnel syndrome) results in tens of billions of dollars (U.S.) in worker compensation claims, additional costs related to worker absenteeism and turnover, and unnecessary human pain and suffering. Increasingly, there are also legal requirements and/or expectations for ergonomics in design (e.g., national standards or collective bargaining contracts). With respect to economics, it has been well documented[3-5] that occupational ergonomic programs have led to reductions in overall injuries and illnesses and work days missed, with concurrent improvements in morale, productivity, and work quality. Further, ergonomic controls usually require small to moderate levels of investment and resources[6] and do not drastically change jobs, tasks, or operations.

2 PHYSICAL ERGONOMICS ANALYSES

2.1 Overview

In this section we present several of the more common ergonomic methods, tools, and procedures. A focus is maintained on material that is directly relevant to design. In each section, a review of the underlying theory is given, followed by exemplary approaches (e.g., empirical equations or models). In some sections, sample applications to occupational tasks or scenarios are given. Sources of additional information are provided at the end of this chapter.

2.2 Anthropometry

Fundamentals of Anthropometry and Measurement

Anthropometry is the science that addresses the measurement and/or characterization of the human body, either individually or for populations. Engineering anthropometry is more application oriented, specifically incorporating human measures in design. Examples include the following: placement of a control so that most individuals can reach it, grip sizing for a handheld tool, and height of a conveyor. Within ergonomics, anthropometric measures can be classified as either *static* or *functional*. The former are fundamental and generally fixed measures, such as the length of an arm or a body segment moment of inertia. Such static data is widely available from public and commercial sources. Functional measures are obtained during performance of some task or activity and may thus depend on several individual factors (e.g., training, experience, motivation). These latter measures are specific to the measurement situation and are hence relatively limited in availability. Yet, it is the functional measures that are more directly relevant in design. The remainder of this section provides an overview of applied anthropometric methods. Results from anthropometry will also be critical in subsequent sections that address mechanical loading during task performance.

Static anthropometric measures are of four types: linear dimensions (e.g., body segment lengths), masses or weights, mass center locations, and moments of inertia. Linear dimensions can be obtained quite simply using tape measures or calipers, with more advanced recent approaches using three-dimensional (3D) laser scanning. A key issue with respect to linear dimensions is the differentiation between surface landmarks and underlying joint centers of rotation. The former are easily located (e.g., the lateral and medial boney "knobs" above the ankle joint), and methods have been developed to translate these to estimates of underlying joint centers that are required for biomechanical modeling (Section 2.4).

Mass (and/or volume) measures were historically obtained using liquid immersion, though as noted above recent scanning methods are also being employed. Locations of segment (or whole-body) center of mass can also be obtained using liquid immersion and also a number of segmental balance methods. Segment moments of inertia are usually obtained using dynamical tests, where oscillatory frequencies are obtained during natural swinging or following a quick release. Representative geometric solids (e.g., a truncated cone) can also be used to model body parts and obtain analytical estimates.

Anthropometric Data and Use

Several large-scale anthropometric studies were conducted in the 1960s and 1970s, mostly in industrialized countries. Contemporary studies are typically of smaller scope, with the exception of the ongoing CAESAR (Civilian American and European Surface Anthropometry Resource) project (http://store.sae.org/caesar/). Anthropometric data is generally presented in tabular form, with some combination of means, standard deviations, and population percentiles. A normal statistical distribution is usually assumed, a simplification that is reasonable in most cases, though which also leads to larger magnitudes of errors at extremes of populations (e.g., the largest and smallest individuals).

Standard statistical methods can be employed directly for a number of applications. If, for example, we wish to design the height of a doorway to allow 99% of males to pass through unimpeded, we can estimate this height from the mean (μ) and standard deviation (σ) as follows (again assuming a normal distribution). Male stature has, roughly, $\mu = 175.58$ and $\sigma = 6.68$ cm. The standard normal variate, z, is then used along with a table of cumulative normal probabilities to obtain the desired value:

$$z_A = \sigma / Y - \mu \quad \text{or} \quad Y = \mu + z_A \sigma$$

where z_A is the z value corresponding to a cumulative area A, and Y is the value to be estimated. Here, $z_{0.99} = 2.326$, and thus $Y = 191.1$ cm, or the height of a 99th percentile male. Clearly, however, further consideration is needed to address a number of practical issues. These include the relevance of the tabular values, whether this static value is applicable to a functional situation, if/how allowances should be made for clothing, gait, etc.

Percentile calculations, as given above, are straightforward only for single measures. With multiple dimensions, such as several contiguous body segments, the associated procedures become more involved. To combine anthropometric measures, it is necessary to create a new distribution for the combination. In general, means add, but variances (or standard deviations) do not. Equations are given below for two measures, X and Y (a statistics source should be consulted for methods appropriate for $n > 2$ values):

$$\mu_{X \pm Y} = \mu_X \pm \mu_Y$$
$$\sigma_{X \pm Y} = [\sigma_X^2 + \sigma_Y^2 \pm 2 \, \text{cov}(X, Y)]^{1/2}$$

or

$$\sigma_{X \pm Y} = [\sigma_X^2 + \sigma_Y^2 \pm 2 \, (r_{XY})(\sigma_X)(\sigma_Y)]^{1/2}$$

where $\pm$ indicates addition if measures are to be added and subtraction otherwise, cov is the covariance, and r is the correlation coefficient. As can be seen from these equations, the variance (σ^2) of the combined measure reduces to the sum of the individual variances when the two measures are independent, or when $\text{cov}(X, Y) = r_{XY} = 0$. Human measures are generally moderately correlated, however, with r on the order of $0.2 - 0.8$ depending on the specific measures.

Anthropometry in Design
Anthropometric data are often estimated using predictive formulas or standardized manikins. Most often, these approaches are intended as indicators of the "average" human. As such, their utility can be limited in that there is no individual who is truly average across multiple dimensions, and relationships between measures may not be linear nor the same between people. For example, a person with a 50th percentile arm length likely does not have a 50th percentile leg length (it may be close or quite different). Further, many anthropometric tables only present mean values (e.g., for center-of-mass location), making estimates of individual differences impossible.

A second limitation in the application of anthropometry arises from potential biases. As noted above, most of the larger data sets were derived several decades ago, thus not accounting for general and nontrivial secular trends toward larger body sizes across all populations. Many of these studies were also performed on military populations, and questions arise as to whether the values are representative in general. Additional biases can arise due to ethnic origins, age, and gender. Overall, application of anthropometric data requires careful attention to minimize such sources of bias.

Three traditional approaches have been employed when using anthropometry in design. Each may have value, depending on the circumstances, and differ in their emphasis on a specific portion of a population. The first, and most straightforward, is *design for extremes*. In this approach, one "tail" of the distribution of a measure is the focus. In the example above for door height, the tall males were of interest since if those individuals are accommodated then all shorter males and nearly all females will as well. Alternatively, the smaller individual may be of interest, as when specifying locations where reaching is required: If the smallest individual can reach it, so will larger ones.

The second approach, *design for average*, focuses on the middle of the distribution. This has also been termed the "min-max" strategy, as it addresses the minimal dimension needed for small individuals and the maximal dimensions for large individuals. A typical nonadjustable seat or workstation is an example of designing for the average. In this case, both the smallest and largest users may not be accommodated (e.g., unable to find a comfortable posture).

Design for adjustability is the third approach, which seeks to accommodate the largest possible proportion of individuals. For example, an office chair may be adjustable in height and/or several other dimensions. While this approach is generally considered the best among the three, increasing levels or dimensions of adjustability come at increasing costs. In practice, designers must balance these costs with those resulting from failure to accommodate some users.

In all cases, the design strategy usually involves a goal or criterion for accommodation. Where the large individual is of concern (e.g., for clearance) it is common practice to design for the 95th percentile males. Similarly, the 5th percentile female is often used when the smaller individual is of concern (e.g., for reaching). When the costs of failure to accommodate individuals are high, the "tails" are typically extended. From the earlier example, it might be desirable to ensure that 99.99% (or more) of the population can fit through a doorway.

Application of anthropometry in the design process usually involves a number of steps. Key anthropometric attributes need to be identified first, and then appropriate sources of population data found (or collected if unavailable). Targets for accommodation are usually defined early (e.g., 99%) but may change as costs dictate. Mockups and/or prototypes are often built, which allow for estimating whether allowances are needed (e.g., for shoe height or gait in the doorway example). Testing may then be conducted, specifically with extremes of the population, to determine whether accommodations meet the targets.

2.3 Range-of-Motion and Strength

A number of measures are required to describe the capacity of an individual (or population) to achieve task performance (e.g., reach, lift, pull). Joint range of motion (also called mobility or flexibility), and joint (or muscle) strength begin to describe capacity and are especially relevant for tasks performed briefly or infrequently. Additional information will be required for highly demanding, prolonged, or frequent tasks, as well as additional types of measures (e.g., fatigue and environmental stress as described below).

Range of Motion

Joint range of motion (ROM) refers to the limits of joint motion and is represented as rotations about a given joint or of body segments (e.g., torso flexion). Two different forms of ROM are commonly measured. The first, passive (or assisted), involves external sources of force or moment to achieve joint motion. Examples include the use of gravity during a squat, to assess knee flexion, or forces/moments applied by an experimenter or device. The second, active ROM, requires muscle contraction to achieve joint motion, and is associated with narrower motion limits than passive. In practice, the relevant type of ROM is determined by task requirements.

Measuring ROM from individuals is possible using a variety of equipment, from low-cost goniometers (for measuring included angles) to high-cost and sophisticated marker tracking systems. More often, population ROM data is obtained from a number of accessible sources (often in conjunction with anthropometric data). A number of factors can be expected to have an influence on ROM. Although ROM decreases with age, the changes are usually minimal in healthy individuals until the end of typical working life (i.e., 65). Women generally have higher ROM ranges, although gender differences are typically < 10%. Little association has been found between anthropometry and ROM, although ROM does decrease with obesity. In simple cases, such as those involving one joint, application of ROM data is straightforward and follows similar methods described in anthropometry (e.g., using percentiles). When multiple joints are involved, it is common to use human modeling software to assess the potential limitations due to ROM.

Strength

In common terminology, strength is an obvious concept. In practice, strength is not a single measure but instead is specific to the kinematic requirements of a task. Depending on these requirements, strength might be measured as a force (e.g., capacity to pull) or a moment (e.g., shoulder elevation), for an isolated body segment, multiple segments, or the body as a whole. The capacity of an individual muscle to generate force depends on its length and velocity, with associated length–tension and velocity–tension relationships. Temporal dependencies (e.g., duration of effort) are discussed in Section 2.5.

Strength is typically measured using maximum voluntary efforts. An individual is placed in a specified posture, or performs a specified motion, while a load cell is used to record the relevant kinetics. Extensive data is available on population strengths, in many cases broken down by gender and/or age. Similar procedures as those described for anthropometry can again be used for design purposes (i.e., determining population percentages with sufficient strength). These procedures usually involve comparing strength (as a measure of capacity) with task demands (determined using biomechanics as described in the next section).

The vast majority of compiled population strength data are for static exertions, wherein postures are fixed (hence no length or velocity effects). Compared to dynamic exertions, static data collection and interpretation is simpler, faster, and less expensive. Unfortunately, real-world efforts normally involve motion. Thus, there is a fundamental mismatch between available measures of capacity and task demands. Further, static tests usually have low associations with dynamic performance capabilities. When tasks do not involve substantial dynamics, use of static data should involve relatively minimal errors. When dynamics are substantial, assessing both capacity and demands becomes quite complex (and is currently a topic of extensive research).

Several factors can have important influences on strength. Women, on average, are weaker than males by approximately 35%, though the differences are highly joint and task specific. In addition, there is extensive overlap between the two gender distributions. Strength normally peaks in the late 20s to early 30s, and decreases with increasing rapidity afterwards. By age 40, 50, and 70, typically strength reductions are on the order of 5, 20, and 40%, respectively. Strength has only a minor to moderate relationship with body size, and most attempts at predicting strength from anthropometry have been relatively inaccurate.

As noted above, there are length and velocity effects on muscle capacity, and these, in turn, are seen as posture and motion effects on strength. For most joints, strength is maximal near the middle of the ROM (e.g., an elbow angle of 90°), though several exceptions do exist. When muscles shorten during an exertion (a concentric effort), the capacity for force decreases with increasing velocity, and there is a small increase in force capacity when muscles lengthen (an eccentric effort). When combined with the inertial effects of body segments, strength declines with increasing motion speed.

2.4 Biomechanics

Biomechanics in Ergonomics

Biomechanics is the application of classical mechanics to biological systems such as the human body. Although this comprises many branches of mechanics, in the context of ergonomics, we will deal only with dynamics and to a limited extent the mechanics of materials. In physical ergonomics, the human body can be seen as a mechanical system, or rather a part of a mechanical system comprising also the tools, objects, and environment with which the human operator interacts. Inverse dynamics can be applied to the analysis of this mechanical system to estimate forces and moments being produced by or acting on the human body, whereas mechanics of materials contributes toward understanding the effects of these mechanical loads on the body. Direct measurement of forces and moments in the human is not feasible for practical and ethical reasons. Consequently, the vast majority of available methods and data rely to some extent on inverse dynamical models.

Estimating Joint Moments

The principles of inverse dynamics will probably be known to any mechanical engineer and can be found in dynamics textbooks or specialty books on biomechanics (see references). To summarize, application of the equations of motion based on Newton's second law ($\Sigma F = ma$) and Euler's extension of this law to angular motions ($\Sigma M = I\alpha$), to a system of linked segments, is used to yield forces and moments acting at each of the segments in each of the links. In biomechanical analyses, moments about the joints of the human body are usually of interest since these reflect the combined effect of all muscles spanning the joint. Note that physically these moments are the effect of muscle forces. In actual analysis, the moments of force of the muscles appear as a lumped moment in the moment equilibrium equation, and they do not appear in the force equilibrium equation. To perform this type of analysis, masses and moments of inertias of the segments and the accelerations of the segments need to be known. Finally, if two or more external forces act on the system, all but one of these need to be known. If all external forces are known, redundant information is available that can be used to validate the model with respect to anthropometric assumptions (see below) or to decrease the estimation errors that would result from assuming accelerations to be zero (see below).

A simplified example of a linked segment model, which was used to estimate the moment on the knee while climbing ladders with different rung separations, is given in Fig. 1. Video data were used to approximate the positions of centers of mass of the foot and lower leg, as well as the joint rotation centers of the knee and ankle. Forces on a rung were measured with a force transducer. Segment masses were estimated on the basis of anthropometric data. Initially, a free-body diagram of the foot was created, and the reaction force at the ankle was calculated by equating the sum of the forces on this segment to its mass times acceleration. Next, the moment at the ankle was calculated from equating the sum of the moments to inertia times angular acceleration. Subsequently, the opposites of this force and moment were used as input for a free-body diagram of the lower leg and the reaction force and moment at the knee were calculated.

Segment masses and moments of inertias cannot be directly measured when dealing with the human body. Therefore, estimations need to be made on the basis of anthropometric models (see Section 2.2.1). Obviously, these estimations may introduce errors, and the magnitude of such errors can be gauged by making use of redundant information when all external forces have been measured. For example, the moment about the low back in lifting can be calculated on the basis of a model of the lower body (legs and pelvis) using measured ground reaction forces on each foot as input. This same moment can also be estimated using a model of the upper body (arms and trunk) and the object lifted. It has been shown that with a careful choice of anthropometric assumptions, errors in moment estimates will generally be below 10 Nm.

Accelerations can be measured or calculated from position data by double differentiation. This involves labor-intensive measurements and is only feasible when at least a mockup of the

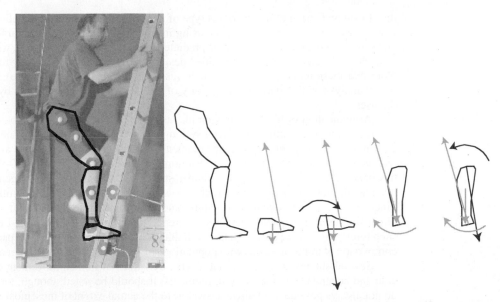

Figure 1 Example of a linked segment model used to estimate the moment on the knee while climbing ladders with different rung separations.

situation to be analyzed is available. Consequently, in many ergonomic applications accelerations are assumed to be zero, in which case only the static configuration of the human body needs to be known or predicted. This simplification will lead to underestimation of mechanical loads on the human body in dynamic tasks, which in some cases can be substantial (e.g., in manual lifting, the moments around the low back may be underestimated by a factor of 2). Such errors may lead to questionable conclusions, even in a comparative analysis. For instance, earlier studies comparing stoop and squat lifting techniques appear to have been biased toward favoring the squat technique due to the application of static models. Consequently, early studies have often reported a lower low-back load in squat lifting as compared to stoop lifting, whereas more recent studies using dynamic models have reported the opposite.[7] If an analysis of a dynamic task is performed assuming acceleration to be zero but inputting the measured external forces on the body into the analysis—the so-called quasi-dynamic approach—reasonable estimates of joint moments result.

A second simplification often used is to assume that all movement and force exertion takes place in a single plane, which allows application of a 2D model. In analyses of asymmetric lifting tasks, this can cause significant and substantial errors in estimated moments around the low back (roughly 20% when the load is placed 30° outside of the primary plane of movement). At 10° of asymmetry, differences between 2D and 3D analyses have been found to be insignificant.

Data collection required for the estimation of net moments is usually not prohibitive in a comparative analysis of working methods and techniques, since this can be done in a laboratory mockup setting. However, for monitoring and identification of the most stressful tasks or task elements, field measurements covering long periods are desirable. In this case, use of an inverse dynamics approach usually is prohibitive. Methods have been developed to estimate moments based on measurements of the electrical activity of muscles, the latter using electrodes applied on the skin overlying the muscle group of interest (electromyography or EMG). However, additional kinematic data and extensive calibrations are needed to obtain valid estimates. Currently, miniature kinematic sensors and efficient calibration procedures are being developed and tested to facilitate this type of measurements. Finally estimation of mechanical loads in the design stage can be done using inverse dynamics when external forces are known and postures (and movements) can be predicted. Several software programs, which can in some cases be integrated with computer-aided design (CAD) applications, allow for such analyses. Note that these models usually are static (assume accelerations to be zero), and the validity of the analysis will depend on the validity of the posture predictions made by the software or the user.

An indication of how load magnitude, as expressed by the moment about a joint, relates to the capacity of the musculoskeletal system, can be obtained by comparison of the moments during a task to maximum voluntary moments. Usually such comparisons are made with the results of isometric strength tests (see Section 2.3). For example, lifting a 20-kg load manually has been predicted to exceed the shoulder strength of about 30% of the general population, whereas the same task performed with a hoist allowed over 95% to have sufficient strength.[8] Since many tasks are dynamic in nature, and both joint angle and angular velocity strongly affect the moment capacity, dynamic reference data are needed. As noted earlier, however, such reference data are only partially available. Some commercial software packages provide a comparison of joints moments with population strength data, though the latter are usually static.

Several studies have shown that inverse dynamics of human movement can provide reliable and accurate estimates of joint moments. It should be noted, though, that joint moments do not always provide a definitive answer as to the actual extent of musculoskeletal loading, as will be discussed in the next paragraph. Further, since human motor behavior can be quite variable, the reliability of moment estimates derived from limited numbers of measurements, and more so when derived from model simulations, should be considered with care. When comparative analyses are done, substantial differences in net moments (e.g., > 10%) will usually allow conclusions to be drawn with respect to musculoskeletal loading. For normative interpretation of joint moments with data on muscle strength, it is recommended that a margin of safety be included in view of the variability and sources of error both in moment estimates and strength data.

Estimating Muscle Forces

As discussed above, the joint moment reflects the combined effect of all muscle forces (and of elastic forces in passive structures like ligaments). A fundamental problem in biomechanics is that muscle forces cannot generally be calculated from the net moment since the number of muscles spanning a joint creates an indeterminate problem. In order to relate mechanical loads imposed by a task to, for example, the mechanical tolerance of a joint, this problem needs to be solved.

Let's consider as an example the compression force acting on the contact surface of a joint. Since muscles act with small leverage about a joint relative to external forces and inertial forces on body segments, muscle forces greatly exceed the reaction forces calculated through inverse dynamics. Forces in ligaments (and other passive structures) are low, except at extreme joint positions. Consequently, the compression force on a joint is mainly determined by muscle forces. However, their magnitude will depend on which muscles (with what lever arms) produce the joint moment, and they can also be strongly affected by the level of antagonistic co-contraction (i.e., activity of muscles producing moments opposite to the total moment around the joint).

Several types of models have been developed to obtain estimates of muscle force, all of which use the joint moment as a starting point of the estimation procedure. In addition, all rely on an anatomical model of the joint and surrounding muscles. Models can be differentiated on the basis of the way in which muscle activation is estimated. One class of models estimates muscle activation from measures of the electrical activity of muscles (EMG). Another class of models uses static optimization to simulate assumed optimal control of muscle activation.

Models used in design software are mostly of the latter class. Usually these models rely on a cost function related to the efficiency of moment production. Consequently, antagonistic co-contraction is predicted to be absent. This has been shown not to produce large errors in predictions of forces on the spine for a wide range of different tasks, provided that a realistic cost function is used. However, effects of specific task conditions, which affect levels of co-contraction, will go unnoticed. At present, only EMG-based models account for modulations of the level of co-contraction, but other methods may be made to predict co-contraction, for instance, by imposing constraints on stability of the joint.

Finally, it should be noted that close relationships between joint moments and joint compression force estimates have been found for the shoulder as well as the low back. This suggests that joint moments can be sufficient for comparative analyses of joint and muscle loading in many cases, though this conclusion may not hold for other joints.

Estimating Tissue Tolerances and Their Use in Design

To allow a normative interpretation of estimated mechanical loads, such as joint compression forces, information on the strength of the loaded structures and tissues is needed. This will, in theory, also allow setting workload or design standards, but such an approach has had only limited application. Setting such standards is hampered by difficulties in obtaining estimates of tissue loading, as well as difficulties in acquiring valid data on mechanical tissue strength. Tissue strength estimates obviously can be obtained only from cadaver material or nonhuman tissues. The validity thereof is fundamentally unsure, and availability is limited. In addition, biological materials display viscoelasticity, making strength dependent on loading rate and loading history. Furthermore, it has been a matter of debate whether normative data should be based on ultimate strength or values below that (e.g., yield strength). It is conceivable that subfailure damage can cause musculoskeletal disorders, suggesting that yield strength might be the correct criterion. Conversely, subfailure damage may be reparable and even a prerequisite for maintenance and adaptation of biological tissues. Finally, the strength of biological structures is highly variable between specimens (and individuals), making it quite difficult to define thresholds.

For spinal loading, many ergonomic studies and some guidelines have proposed using spine compression to establish whether a certain load would be acceptable or not for occupational lifting. Compression of spinal motion segments (vertebrae and intervening disks) can cause a fracture of the endplate, the boundary layer between the vertebrae and the intervertebral disk, at forces between 2 and 10 kN. Since extensive data are available on compression strength and the underlying properties of the spine, a population distribution of spinal strength can be estimated. This population distribution can be used to estimate the percentage of the population for which a given task, with a known peak compression force, is theoretically above injury threshold. Based on this approach, a widely cited threshold (based on the 25th percentile of the general population) of 3.4 kN has been suggested by NIOSH.[9] This threshold has been criticized, though, because it does not account for the wide variability in compression strength. It may be too limiting, especially for young males but also may not provide enough protection for some groups, especially older females. Moreover, experiments using repetitive compression have shown that spinal motion segments fail at much lower forces versus one-cycle loading. Models based on creep failure and fatigue failure have been able to describe strength in repetitive loading, allowing a similar approach for cyclic tasks.

In view of the questionable validity of tolerance data, limiting the application to comparative analyses seems indicated. Specifically, tissue tolerance data can be used to assess the effects of a design toward reducing physical loading. For example, given the 2–10 kN range in compression strength of the human lumbar spine, a 50-N reduction in compression can be considered negligible. In contrast, spinal compression forces in manipulating a 20-kg load manually or with a hoist have been estimated at approximately 3 and 1.2 kN, respectively.[8]

A reduction in compression of such magnitude as this is likely to substantially reduce the probability of injury (of course, assuming that other task factors remain comparable).

2.5   Whole-body and Localized Fatigue

Earlier (Section 2.3), strength was considered as the ability to generate a force or moment. This capacity, however, has a temporal dependency: the larger the effort the smaller the duration over which it can be maintained. Fatigue, in turn, can be defined as the loss of capacity to generate a physical exertion as a result of prior exertions. Limits to physical exertion capacity can be broader discriminated as those involving the whole body (or large portions thereof) or more localized (e.g., a single joint).

Whole-Body Fatigue

For repetitive whole-body efforts, fatigue typically involves a loss of capacity in the body's metabolic system or an inability to generate the necessary energy. Over a short duration, less than ~5 min, aerobic capacity (AC) provides a relevant measure and is defined as the maximum energy production rate using aerobic means (using oxygen, versus anaerobic metabolism). AC can be measure directly using a stress test (on a treadmill or bicycle ergometer) or indirectly estimated using submaximal tests. Average AC values for males and females are roughly 15 and 10.5 kcal/min, respectively, and both show consistent declines with age past ~30 years.

Of more relevance in practice is the ability to generate energy for prolonged periods, or physical work capacity (PWC). PWC is directly related to AC, so that the noted effects of gender and age also apply. Further, both AC and PWC are task dependent since different muscle groups have different energetic capacities. As an example, AC is about 30% lower for tasks isolated to the upper versus lower extremities. PWC can be represented as a fraction of AC, with the fraction declining as the task duration is prolonged. Roughly 50% of AC can be maintained for up to one hour and 25–33% of AC for up to 8 hours. PWC can be estimated from individual or population AC as follows (units of minutes and kcal/minute[8]:

$$PWC = [(\log 4400 - \log \text{duration})(AC)]/3.0$$

With an estimate of individual or population PWC, design evaluation requires an assessment of task energetic demands. A range of approaches exists, differing in both complexity and accuracy. Tabular values and subjective evaluations can provide quick and easy, but rough, estimations. Alternatively, direct measurement (of oxygen uptake), estimation from heart rate, or use of predictive equations can be used for more precise estimates.

Localized Muscle Fatigue

Localized muscle fatigue (LMF), in contrast to the above, involves an overload of the specific (or regional) capacity of a muscle or muscle group. For relatively simple efforts (e.g., static holds), the relationship between effort and endurance time has been well described. If the effort level is represented as a percentage of strength (F = %MVE, maximum voluntary exertion), then endurance time (T, in minutes) can be estimated as follows[10]:

$$T = [1.2/(0.01F - 0.15)0.618] - 1.21$$

It should be noted, however, that the equation provided is only valid for efforts between roughly 20 and 95% MVE (being an empirical fit to a data set). Endurance times at the lower end have not been well characterized, and endurance times at the higher end are quite brief. The relationship can also vary substantially,[11,12] including between individuals, muscle groups, with training, etc. Furthermore, it is important to note that endurance represents the culmination of an ongoing fatiguing process, specifically the time at which an effort can no longer be maintained because of declining muscle capacity. Nonetheless, this relationship has important design implications. If muscular efforts are required for more than brief times, the tradeoff between effort level and duration on capacity can be estimated. If the duration is fixed, endurance times can be increased by reducing the relative effort (as a %MVE), either by reducing task kinetics or by increasing capacity (e.g., using large muscle groups or keeping joints near the middle of the ROM).

It has proven to be difficult to expand the above results to more complex (and more realistic) exertions. Some research has led to guidelines for intermittent static tasks (periods of fixed efforts interrupted by fixed rest periods), but these have yet to be verified for general application. Little, if any, reliable guidelines exist for more complex, time-varying exertions.

2.6 Environmental Stress

Human capacity can be both facilitated and compromised by environmental conditions. In this section, we focus primarily on the adverse consequences, with specific emphasis on exposure to heat and vibration and implications for design. A number of other important environmental exposures exist, such as light and sound, for which references are provided at the end of this chapter (see General References).

Heat Stress

The body has a complex control mechanism that maintains core temperature in a narrow range (near 37°C). This control is achieved except under conditions of extreme exposures, with adverse outcomes including heat exhaustion and fainting. Even with less extreme exposures, exposure to heat can be expected to compromise performance, on the order of 7 and 15% with prolonged exposure above 27 and 32°C, respectively.[13]

More quantitative estimates of the effects of heat are conducted by determining the rate of change in body heat content. Important sources to these *heat balance* calculations are metabolic, radiant, convective, and evaporative. The absolute and relative magnitudes of each heat source are dependent on the workload and specific environmental conditions (e.g., airflow, exposure surface areas, air vapor pressure, etc.). Individuals differ greatly in their tolerance to hot environments, with major effects due to physical fitness, age, gender, and level of obesity. In addition, individuals can become acclimated through repeated exposure, a process that takes roughly 1–2 weeks, with deacclimation occurring over a similar time period.

Assessment of environmental conditions requires measures that reflect the different sources of heat noted earlier. A common guideline[14] determines recommended exposure limits (in minutes per hour) based on metabolic heat and a compound measure of environmental heat, the wet-bulb–globe temperature (WBGT). WBGT, in turn, is determined as weighted average of dry-bulb (air temperature), wet-bulb (accounting for humidity and airflow), and globe (accounting for radiant heat) temperatures.

Additional quantitative assessments can be done by estimation of heat balance. Empirical equations have been developed[14] to predict rates of body heat gain or loss from radiation, convection, and evaporation. Together with estimates (or direct measurement) of metabolic heat, the rate of net body heat storage can be obtained. Allowable levels of heat storage can be estimated (approximately 63 kcal for an individual with average body mass). From these estimates, one can determine an allowable exposure time as the ratio of maximum heat storage to rate of heat gain. If heat exposure is determined to be excessive, control measures can be applied based on the sources of heat storage (e.g., improve airflow to maximize evaporative heat loss, shielding to reduce radiant heat transfer, etc.).

Vibration

While historically used in a therapeutic mode (due to analgesic effects), exposure to vibration has been consistently associated with several adverse health outcomes. These include discomfort, reflexive muscle contraction, and decreased fine motor control in the short term and low back pain and hand/finger circulation impairment in the long term.

Vibration exposure is quantified in both the temporal [peak, average, root mean square (rms)] and frequency domains (typically using octave or one-third octave bands). Spectral analysis is of particular utility, as the human response to vibration is strongly dependent on the frequencies of exposure. Different tissues, organs, and body segments also have distinct transmission and resonance responses as a function of frequency.

A number of vibration exposure guidelines have been developed by the International Standards Organization (ISO). The differing guidelines address the site of exposure (e.g., whole body or hand/arm), the direction of exposure (axis), posture (e.g., seated or standing), and the vibration measures (e.g., acceleration or velocity). Exposure limits are obtained based on a spectral analysis of exposures. Controls for vibration exposure can be categorized as occurring at the source, path, or receiver. Examples of source control include avoiding resonances and using balancing to reduce vibration magnitudes. Control in the vibration path can be achieved by limiting exposure time and using vibration isolation. At the receiver (the human), vibration exposure can be reduced using damping apparel (e.g., gloves) and reducing contact forces and areas.

3 EXAMPLE APPLICATIONS

3.1 Overview

In this section, we present two practical applications of physical ergonomics in design. Each demonstrates the use of several of the methods and procedures presented above. Only the general approach and results are given, with references provided for additional detail.

3.2 Manual Material Handling Systems

Associations between physically demanding manual handling of materials and work-related musculoskeletal injury have been well documented by ergonomists and epidemiologists. Mechanized assistive devices are widely used to control such problems. Material handling systems can be generally categorized as positioners (e.g., lift tables, conveyors), used to place or orient objects, and manipulators (articulated arms, hoists), used to move and/or support objects. The latter were the focus of a detailed ergonomic/biomechanical analysis, with a goal of understanding the impact of using manipulators on task performance and the user's physical demands.

Only minimal investigation has been conducted on manipulators, mainly focusing on biomechanical modeling of the static gravitational component. While use of manipulators reduces the static component (i.e., object weight), substantial kinetic loads may result from body segment dynamics and the inertial dynamics of the manipulator and object handled. In addition, use of a manipulator frequently requires complex body postures and motions.

In the first study,[15] two mechanical manipulators (an articulated arm and an overhead hoist; Fig. 2) were used to move objects with moderate masses (10–40 kg) for several short-distance transfers. Differences between the two devices and manual (unassisted) movements were obtained for motion times, applied hand forces, and torso kinematics. Use of either manipulator increased motions times by 36–63% for symmetric motions (in the y-z plane) and by 62–115% for asymmetric motions (in the x-z plane). Hand forces were substantially lower when using either manipulator (by 40–50%), while torso motions were generally similar regardless of how the transfers were conducted. For self-paced job tasks (see below for effects of pacing), it was concluded that moderate mass objects require significant increases in motion times, but with substantially reduced levels of upper extremity exertion.

A second study[8] examined physical loads in more detail, using the same manipulators and motions. A 3D inverse dynamics biomechanical model was developed to allow for estimation

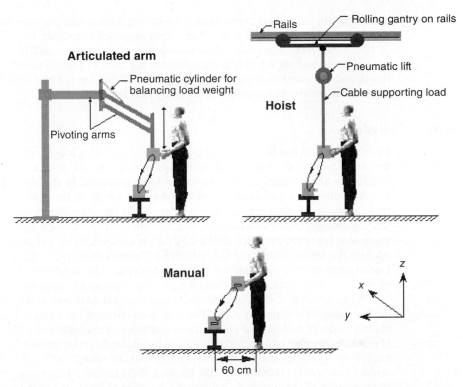

Figure 2 Illustration of object transfers performed manually or with a material handling manipulator. *Source:* Ref. 17. Reprinted by permission of Sage Publications.

of the strength demands (required moments versus joint strength) at the shoulders and low back, as well as compressive and shear forces in the lower spine (the latter as estimates of the relative risk of low back injury). Strength demands decreased substantially when using either manipulator in comparison to transfers done manually. Strength demands were also much less affected by increases in object mass. Spine compression and shear were reduced by roughly 40% when using either manipulator, primarily because of the decreased hand forces noted above, and the resulting spinal moments. Estimates of trunk muscle activity (using electromyography) suggested that the use of a hoist imposes higher demands on coordination and stability, particularly at extreme heights or with asymmetric motions. In contrast, the articulated arm imposed higher demands on strength and resulted in increased spinal compression, likely because of the higher system inertia. From both studies, it was concluded that use of either type of manipulator can reduce physical demands compared with manual methods, though performance (motion times) will be compromised. In addition, either a hoist or articulated arm may be preferable depending on the task requirements (e.g., height and asymmetry).

Two companion studies were conducted to address additional aspects of manipulators as used in practice. Effects of short-term practice (40 repetitions) were addressed in the first study,[16] when using the two manipulators to lift and lower a 40-kg object. Nonlinear decreases in several measures were found (e.g., low back moments and spine compression). These decreases, however, were fairly slow and comparable to rates of learning for manual lifting. Familiarization with material handling manipulators is thus recommended, and the learning process may be somewhat lengthy. Effects of pacing (speed of object transfer) were determined in the second experiment,[17] by having participants perform transfers 20% more rapidly than those presented above. This requirement led to roughly 10% higher hand forces, 5–10% higher torso moments, similar torso postures and motions, and 10% higher spine loads (compression and shear). When manipulators are used in paced operations (e.g., assembly lines), and the noted increases in required motion times are not accounted for, these results suggest that the risk of musculoskeletal injury may be increased.

3.3   Refuse Collection

Refuse collection can be considered a physically demanding job. Several studies have shown that many refuse collectors indeed suffer from musculoskeletal disorders. Especially low back complaints and shoulder complaints appear to be prevalent. In the Netherlands, this has led to introduction of a job-specific guideline that sets limits on the amount of refuse a collector is allowed to collect per day. This guideline was based on energy demands of the task of refuse collecting as discussed in Section 2.5.1. When this guideline was introduced, additional ergonomic measures to reduce workload were encouraged, and it was stated that insofar these additional measures would lead to a reduction of energy consumption during the task, this could lead to adjustment of the guideline to allow a larger amount of refuse to be collected. A series of studies were performed to evaluate the efficacy of proposed ergonomic measures.

A first series of studies concentrated on mechanical workload in relation to the design of the two-wheeled containers used by Dutch refuse collectors. As a first step, a mechanical simulation model of pushing and pulling these containers was constructed. This 2D model served to evaluate the effect of design parameters on the moments acting around the shoulders and low back. It was predicted that mechanical loading could be strongly affected by the location of the center of mass of the container and the location of the handle. Furthermore, the anthropometric characteristics of the user (body height), the direction of the force exerted on the container, and the angle over which it is tilted did strongly affect predictions of mechanical workload. Since the direction of the force exerted and the tilt angle cannot be predicted, it was considered necessary to do an experimental study in a mockup situation. Four experienced refuse collectors, with a wide range of body heights, symmetrically tilted, pushed, and pulled a container of which the center of mass and handle positions were systematically varied over 9 and 11 positions, respectively. The position of the center of mass had substantial effect on workload, whereas changes in handle location had limited effect because the study participants adjusted the tilt angle of the container to accommodate different handle positions. Furthermore, it was found that replacing the center of mass toward the axis could reduce workload relative to that with the standard container in use in the Netherlands.[18]

A second more comprehensive experiment was performed with nine refuse collectors, again with a wide range of body heights and including asymmetric tasks.[19] A standard container was compared to a redesigned container with an optimized handle position and larger wheels. In addition, for each of these containers three different positions of the center of mass were evaluated. Participants performed four tasks on a brick-paved road built in the laboratory. The tasks, selected on the basis of systematic observations of refuse collectors at work, were tilting the container and pulling it with one hand, tilting the container and pushing it with two hands, rotating the container, and pulling an empty container up the pavement. A 3D biomechanical model was used to estimate moments about the low back and shoulders (Fig. 3). In addition, models of the shoulder and the low back were used to estimate compression forces on these joints. Participants also gave a subjective rating of the ease of handling of each of the containers.

The use of the redesigned container resulted in a decrease of the force exerted on the container, a decrease of moments around the shoulder and low back and a decrease of shoulder compression. No significant decrease in lower spine compression was found. Only when

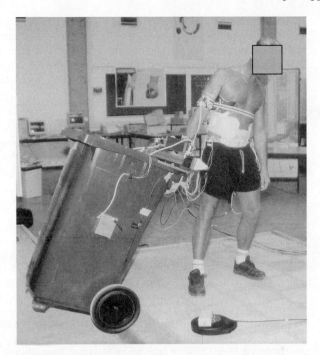

Figure 3 A refuse collector during data collection for estimation of mechanical workload in pulling a refuse container up the pavement.

pulling the empty container up the pavement was an increase in shoulder joint moment found. In general, the participants rated the redesigned container as more easy to use. It was concluded that the redesign would reduce mechanical workload especially when the task of pulling the container back up the pavement would be eliminated, by leaving it placed against the kerb as is already often done in practice.

Energetic workload in refuse collecting could be reduced by gathering more containers at one collection site in a street than is currently the practice. To study the effect of this measure, a mockup study was performed.[20] Gathering sites with 2, 16, and 32 containers were compared. Oxygen consumption, heart rate, and perceived exertion were measured in 18 men. In addition, the time it took to collect 32 containers was measured. The size of the gathering point had no effect on oxygen consumption, heart rate, or perceived exertion. However, the time it took to collect 32 containers was lower for the gathering points with 16 and 32 containers as compared to 2 containers. Since this would allow for more recovery during the shift, average oxygen consumption over the day would be decreased, and, in the context of the guideline described above, this would imply that more refuse can be collected.

Finally, job rotation between refuse collection and driving the refuse truck could be implemented as a measure to reduce workload in refuse collectors. A field study was performed to evaluate the efficacy of this measure.[21] Three teams of three workers participated in the study. Each team member worked at least 1 week as a truck driver only, 1 week as a refuse collector only, and 2 weeks rotating between these jobs. Of the latter 2 weeks, in 1 week the participants rotated between jobs on the same day; in the other week they rotated between days. Physical workload was quantified by measuring heart rate and subjective ratings. From the heart rate measurements the average oxygen consumption as a percentage of the individual maximum was estimated.

Job rotation yielded lower average oxygen consumption over the workday and subjective workload in comparison to refuse collecting only. In contrast, driving was associated with a lower oxygen consumption, heart rate, and perceived exertion as compared to job rotation. No differences were found between rotating on or between days. It was concluded that job rotation can contribute to a reduction of physical workload in refuse collectors.

REFERENCES

1. International Organization for Standardization, *ISO Working Draft, Ergonomics Principles in the Design of Work Systems* (ISO/CD6385): TC/159/SC1, 1999.

2. International Ergonomics Association, "The Discipline of Ergonomics," accessed July 27, 2004, available: http://www.iea.cc/01_what/What is Ergonomics.html.

3. General Accounting Office, *Worker Protection: Private Sector Ergonomics Programs Yield Positive Results*, GAO/HEHS Publication 97-163, Washington, DC, 1997.

4. H. W. Hendrick, *Good Ergonomics is Good Economics*, Human Factors and Ergonomics Society, Santa Monica, CA, 1996.

5. E. Tompa, R. Dolinschi, C. de Oliveira, B. C. Amick, 3rd, and E. Irvin, "A Systematic Review of Workplace Ergonomic Interventions with Economic Analyses," *J. Occupat. Rehab.*, **20**, 220–234, 2010.

6. S. Lahiri, P. Markkanen, and C. Levenstein, "The Cost Effectiveness of Occupational Health Interventions: Preventing Occupational Back Pain," *Am. J. Ind. Med.*, **48**, 515–529, 2005.

7. J. H. van Dieën, M. J. M. Hoozemans, and H. M. Toussaint, "Stoop or Squat: A Review of Biomechanical Studies on Lifting Technique," *Clin. Biomech.*, **14**, 685–696, 1999.

8. M. A. Nussbaum, D. B. Chaffin, and G. Baker, "Biomechanical Analysis of Materials Handling Manipulators in Short Distance Transfers of Moderate Mass Objects: Joint Strength, Spine Forces and Muscular Antagonism," *Ergonomics*, **42**, 1597–1618, 1999.

9. National Institute for Occupational Safety and Health, *Work Practices Guide for Manual Lifting*, DHHS (NIOSH) Publication 81-122, Washington, DC, 1981.

10. B. W. Niebel, and A. Freivalds, *Methods, Standards, and Work Design*, 10th ed., McGraw-Hill, Boston, 1999.

11. J. H. van Dieën, and H. H. E. Oude Vrielink, "The Use of the Relation between Force and Endurance Time," *Ergonomics*, **37**, 231–243, 1994.

12. L. A. Frey Law, and K. G. Avin, "Endurance Time Is Joint-Specific: A Modelling and Meta-analysis Investigation," *Ergonomics*, **53**, 109–129, 2010.

13. J. J. Pilcher, E. Nadler, and C. Busch, "Effects of Hot and Cold Temperature Exposure on Performance: A Meta-analytic Review," *Ergonomics*, **45**, 682–698, 2002.

14. National Institute for Occupational Safety and Health, *Occupational Exposure to Hot Environments*, DHHS (NIOSH) Publication 86-113, Washington, DC, 1986.

15. M. A. Nussbaum, D. B. Chaffin, B. S. Stump, G. Baker, and J. Foulke, "Motion Times, Hand Forces, and Trunk Kinematics When Using Material Handling Manipulators in Short-Distance Transfers of Moderate Mass Objects," *Appl. Ergon.*, **31**, 227–237, 2000.

16. D. B. Chaffin, B. S. Stump, M. A. Nussbaum, and G. Baker, "Low Back Stresses When Learning to Use a Materials Handling Device," *Ergonomics*, **42**, 94–110, 1999.

17. Maury A. Nussbaum, and Don B. Chaffin, "Effects of Pacing When Using Material Handling Manipulators," *Human Factors*, **41:2**, 12, 1999.

18. I. Kingma, P. P. F. M. Kuijer, M. J. M. Hoozemans, J. H. van Dieën, A. J. van der Beek, and M. H. W. Frings-Dresen, "Effect of Design of Two-Wheeled Containers on Mechanical Loading," *Int. J. Ind. Ergon.*, **31**, 73–86, 2003.

19. P. P. F. M. Kuijer, M. J. M. Hoozemans, I. Kingma, J. H. van Dieën, W. H. K. de Vries, H. E. J. Veeger, A. J. van der Beek, B. Visser, and M. H. W. Frings-Dresen, "Effect of a Redesigned Two-Wheeled Container for Refuse Collecting on Mechanical Loading of Low Back and Shoulders," *Ergonomics*, **46**, 543–560, 2003.

20. P. P. F. M. Kuijer, M. H. W. Frings-Dresen, A. J. van der Beek, J. H. van Dieën, and B. Visser, "Effect of the Number of Two-Wheeled Containers at a Gathering Point on the Energetic Workload and Work Efficiency in Refuse Collecting," *Appl. Ergon.*, **33**, 571–577, 2002.

21. P. P. F. M. Kuijer, M. H. W. Frings-Dresen, W. H. K. de Vries, A. J. van der Beek, J. H. van Dieën, and B. Visser, "Effect of Job Rotation on Work Demands, Physical and Mental Workload, and Recovery of Refuse Truck Drivers and Collectors," *Human Factors*, **46**, 437–448, 2004.

BIBLIOGRAPHY

General References

R. S. Bridger, *Introduction to Ergonomics*, 3rd ed., CRC Press, Boca Raton, FL, 2008.

D. B. Chaffin, G. B. J. Andersson, and B. J. Martin, *Occupational Biomechanics*, 4th ed., Wiley, Hoboken, NJ, 2006.

S. N. Chengular, S. H. Rodgers, and T. E. Bernard, *Kodak's Ergonomic Design for People at Work*, 2nd ed., Wiley, Hoboken, NJ, 2004.

N. J. Delleman, C. M. Haslegrave, and D. B. Chaffin, *Working Postures and Movements. Tools for Evaluation and Engineering*, CRC Press, Boca Raton, FL, 2004.

S. Konz, and S. Johnson, *Work Design: Occupational Ergonomics*, 7th ed., Holcomb Hathaway, Scottsdale, AZ, 2007.

K. Kroemer, H. Kroemer, and K. Kroemer-Elbert, *Ergonomics: How to Design for Ease and Efficiency*, 2nd ed., Prentice Hall, Upper Saddle River, NJ, 2001.

A. Mital, A. Kilbom, S. Kumar, *Ergonomics Guidelines and Problem Solving*, Elsevier Ergonomics Book Series, Vol. 1, Elsevier, Amsterdam, 2000.

Anthropometry, Range-of-Motion, and Strength

R. N. Haber, and L. Haber, "One Size Fits All?" *Ergon. Design.*, **5**, 10–17, 1997.

A. E. Majoros, and S. A. Taylor, "Work Envelopes in Equipment Design," *Ergon. Design.*, **5**, 18–24, 1997.

S. Pheasant, and C. M. Haslegrave, *Bodyspace: Anthropometry, Ergonomics and the Design of Work*, 3rd ed., CRC Press, Boca Raton, FL, 2005.

J. A. Roebuck, Jr., *Anthropometric Methods: Designing to Fit the Human Body*, Human Factors and Ergonomics Society, Santa Monica, CA, 1995.

Biomechanics

S. Kumar, (Ed.), *Biomechanics in Ergonomics*, 2nd ed., CRC Press, Boca Raton, FL, 2007.

D. G. E. Robertson, G. E. Caldwell, J. Hamill, G. Kamen, and S. N. Whittlesey, *Research Methods in Biomechanics*, Human Kinetics, Champaign, IL, 2004.

D. A. Winter, *Biomechanics and Motor Control of Human Movement*, 3rd ed., Wiley, Hoboken, NJ, 2005.

V. M. Zatsiorsky, *Kinetics of Human Motion*, Human Kinetics, Champaign, IL, 2002.

Manual Material Handling

M. M. Ayoub, and A. Mital, *Manual Materials Handling*, Taylor and Francis, London, 1989.

R. Burgess-Limerick, "Squat, Stoop, or Something in between?" *Int. J. Ind. Ergon.*, **31**, 143–148, 2003.

M. J. M. Hoozemans, A. J. van der Beek, M. H. W. Frings-Dresen, F. J. H. van Dijk, and L. H. V. van der Woude, "Pushing and Pulling in Relation to Musculoskeletal Disorders: A Review of Risk Factors," *Ergonomics*, **41**, 757–781, 1998.

L. Straker, "Evidence to Support Using Squat, Semi-squat and Stoop Technique to Lift Low-Lying Objects," *Int. J. Ind. Ergon.*, **31**, 149–160, 2003.

J. H. van Dieën, M. J. M. Hoozemans, and H. M. Toussaint, "Stoop or Squat: A Review of Biomechanical Studies on Lifting Technique," *Clin. Biomech.*, **14**, 685–696, 1999.

Fatigue

D. B. Chaffin, "Localized Muscle Fatigue: Definition and Measurement," *J. Occupat. Med.*, **15**, 346–354, 1973.

R. M. Enoka, and D. G. Stuart, "Neurobiology of Muscle Fatigue," *J. Appl. Physiol.*, **72**, 1631–1648, 1992.

G. Sjøgaard. "Muscle Fatigue," *Med. Sport Sci.*, **26**, 98–109, 1987.

Environmental Stress

International Organization for Standardization, *Mechanical Vibration and Shock—Evaluation of Human Exposure to Whole-Body Vibrations* (ISO 2631-1), 1997.

International Organization for Standardization, *Mechanical Vibration—Measurement and Evaluation of Human Exposure to Hand-Transmitted Vibration* (ISO 5349-1), 2001.

K. Parsons, *Human Thermal Environments*, 2nd ed., Taylor & Francis, London, 2003.

PART 2
INSTRUMENTATION, SYSTEMS, CONTROLS, AND MEMS

CHAPTER **15**

ELECTRIC CIRCUITS

Albert J. Rosa
University of Denver
Denver, Colorado

1 INTRODUCTION

1.1 Overview

The purpose of this chapter is to introduce the analysis and design of linear circuits. Circuits are important in electrical engineering because they process electrical signals that carry energy and information. For the present a *circuit* is defined as an interconnection of electrical devices and a *signal* as a time-varying electrical quantity. A modern technological society is intimately dependent on the generation, transfer, and conversion of electrical energy. Recording media such as CDs, DVDs, thumb drives, hard drives, and tape-based products; communication systems such as radar, cell phones, radio, television, and the Internet; information systems such as computers and the world wide web; instrumentation and control systems; and the national electrical power grid X all involve circuits that process and transfer signals carrying either energy or information or both.

This chapter will focus on *linear circuits*. An important feature of a linear circuit is that the amplitude of the output signal is proportional to the input signal amplitude. The proportionality property of linear circuits greatly simplifies the process of circuit analysis and design. Most circuits are only linear within a restricted range of signal levels. When driven outside this range, they become nonlinear and proportionality no longer applies. Hence only circuits operating within their linear range will be studied.

An important aspect of this study involves interface circuits. An *interface* is defined as a pair of accessible terminals at which signals may be observed or specified. The interface concept is especially important with integrated circuit (IC) technology. Integrated circuits involve many thousands of interconnections, but only a small number are accessible to the user. Creating systems using ICs involves interconnecting large circuits at a few accessible terminals in such a way that the circuits are compatible. Ensuring compatibility often involves relatively small circuits whose purpose is to change signal levels or formats. Such interface circuits are intentionally introduced to ensure that the appropriate signal conditions exist at the connections between two larger circuits.

In terms of signal processing, *analysis* involves determining the output signals of a given circuit with known input signals. Analysis has the compelling feature that a unique solution exists in linear circuits. Circuit analysis will occupy the bulk of the study of linear circuits since it provides the foundation for understanding the interaction of signals and circuits. *Design* involves devising circuits that perform a prescribed signal-processing function. In contrast to analysis, a design problem may have no solution or several solutions. The latter possibility leads to *evaluation*. Given several circuits that perform the same basic function, the alternative designs are rank ordered using factors such as cost, power consumption, and part counts. In reality the engineer's role involves analysis, design, and evaluation, and the boundaries between these functions are often blurred.

There are some worked examples to help the reader understand how to apply the concepts needed to master the concepts covered. These examples describe in detail the steps needed to obtain the final answer. They usually treat analysis problems, although design examples and application notes are included where appropriate.

Symbols and Units

This chapter uses the International System (SI) of units. The SI units include six fundamental units: meter (m), kilogram (kg), second (s), ampere (A), kelvin (K), and candela (cd). All the other units can be derived from these six. Table 1 contains the quantities important to this chapter.

Numerical values encountered in electrical engineering range over many orders of magnitude. Consequently, the system of standard decimal prefixes in Table 2 is used. These prefixes on the unit abbreviation of a quantity indicate the power of 10 that is applied to the numerical value of the quantity.

Circuit Variables

The underlying physical quantities in the study of electronic systems are two basic variables, *charge* and *energy*. The concept of electrical charge explains the very strong electrical forces that occur in nature. To explain both attraction and repulsion, we say there are two kinds of charge—positive and negative. Like charges repel while unlike charges attract. The symbol q is used to represent charge. If the amount of charge is varying with time, we emphasize the fact by writing $q(t)$. In SI charge is measured in *coulombs* (abbreviated C). The smallest quantity of charge in nature is an electron's charge ($q_E = 1.6 \times 10^{-19}$ C). There are 6.24×10^{18} electrons in 1 C.

Table 1 Some Important Quantities, Symbols, and Unit Abbreviations

Quantity	Symbol	Unit	Unit Abbreviation
Time	t	Second	s
Frequency	f	Hertz	Hz
Radian frequency	ω	Radians per second	rad/s
Phase angle	θ, φ	Degree or radian	° or rad
Energy	w	Joule	J
Power	p	Watt	W
Charge	q	Coulomb	C
Current	i	Ampere	A
Electric field	ε	Volt per meter	V/m
Voltage	v	Volt	V
Impedance	Z	Ohm	Ω
Admittance	Y	Siemen	S
Resistance	R	Ohm	Ω
Conductance	G	Siemens	S
Reactance	X	Ohm	Ω
Susceptance	B	Siemen	S
Inductance, self	L	Henry	H
Inductance, mutual	M	Henry	H
Capacitance	C	Farad	F
Magnetic flux	ϕ	Weber	Wb
Flux linkages	λ	Weber-turns	Wb-t
Power ratio	$\log_{10}(p_2/p_1)$	Bel	B

Table 2 Standard Decimal Prefixes

Multiplier	Prefix	Abbreviation
10^{18}	Exa	E
10^{15}	Peta	P
10^{12}	Tera	T
10^{9}	Giga	G
10^{6}	Mega	M
10^{3}	Kilo	k
10^{-1}	Deci	d
10^{-2}	Centi	c
10^{-3}	Milli	m
10^{-6}	Micro	μ
10^{-9}	Nano	n
10^{-12}	Pico	p
10^{-15}	Femto	f
10^{-18}	Atto	a

Electrical charge is a rather cumbersome variable to work with in practice. Moreover, in many situations the charges are moving, and so it is more convenient to measure the amount of charge passing a given point per unit time. To do this in differential form, a signal variable i called *current* is defined as follows:

$$i = \frac{dq}{dt} \tag{1}$$

Current is a measure of the flow of electrical charge. It is the time rate of change of charge passing a given point. The physical dimensions of current are coulombs per second. The unit of current is the *ampere* (abbreviated A). That is,

$$1 \text{ coulomb per second} = 1 \text{ ampere}$$

In electrical engineering it is customary to define the direction of current as the direction of the net flow of *positive* charges, that is, the opposite of electron flow.

A second signal variable called *voltage* is related to the change in energy that would be experienced by a charge as it passes through a circuit. The symbol w is commonly used to represent energy. Energy carries the units of *joules* (abbreviated J). If a small charge dq were to experience a change in energy dw in passing from point A to point B, then the voltage v between A and B is defined as the change in energy per unit charge. One can express this definition in differential form as

$$v = \frac{dw}{dq} \tag{2}$$

Voltage does not depend on the path followed by the charge dq in moving from point A to point B. Furthermore, there can be a voltage between two points even if there is no charge motion (i.e., no current), since voltage is a measure of how much energy dw would be involved if a charge dq were moved. The dimensions of voltage are joules per coulomb. The unit of voltage is the *volt* (abbreviated V). That is,

$$1 \text{ joule per coulomb} = 1 \text{ volt}$$

A third signal variable, *power*, is defined as the time rate of change of energy:

$$p = \frac{dw}{dt} \tag{3}$$

The dimensions of power are joules per second, which is called a *watt* (abbreviated W). In electrical situations, it is useful to have power expressed in terms of current and voltage. Using the chain rule, Eq. (3) and Eqs. (1) and (2) can be combined as

$$p = \left(\frac{dw}{dq} \right) \left(\frac{dq}{dt} \right) = v \cdot i \tag{4}$$

This shows that the electrical power associated with a situation is determined by the product of voltage and current.

Signal References

The three signal variables (current, voltage, and power) are defined in terms of two basic variables (charge and energy). Charge and energy, like mass, length, and time, are basic concepts of physics that provide the scientific foundation for electrical engineering. However, engineering problems rarely involve charge and energy directly but are usually stated in terms of the signal variables because current and voltage are much easier to measure.

A signal can be either a current or a voltage, but it is essential that the reader recognize that current and voltage, while interrelated, are quite different variables. Current is a measure of the time rate of charge passing a point. Since current indicates the direction of the flow of electrical charge, one thinks of current as a *through* variable. Voltage is best thought as an *across* variable because it inherently involves two points. Voltage is a measure of the net change in energy involved in moving a charge from one point to another. Voltage is measured not at a single point but rather between two points or across an element.

Figure 1 shows the notation used for assigning reference directions to current and voltage. The reference mark for current [the arrow below $i(t)$] does not indicate the actual direction of the current. The actual direction may be reversing a million times per second. However, when the actual direction coincides with the reference direction, the current is positive. When the

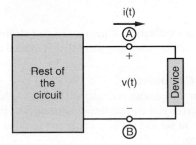

Figure 1 Voltage and current reference marks for two-terminal device.[1]

opposite occurs, the current is negative. If the net flow of positive charge in Fig. 1 is to the right, the current $i(t)$ is positive. Conversely, if the current $i(t)$ is positive, then the net flow of positive charge is to the right.

Similarly, the voltage reference marks (+ and – symbols) in Fig. 1 do not imply that the potential at the positive terminal is always higher than the potential at the B terminal. However, when this is true, the voltage across the device is positive. When the opposite is true, the voltage is negative.

The importance of relating the reference directions (the plus and minus voltage signs and the current arrows) to the actual direction of the current (in and out) and voltage (high and low) can be used to determine the power associated with a device. That is, if the actual direction of the current is the same as the reference arrow drawn on the device, the current goes "in" and comes "out" of the device in the same direction as the reference arrow. Also, the voltage is "high" at the positive reference and "low" at the negative reference. If the actual and reference directions agree and i and v have the same sign, the power associated with this device is positive since the product of the current and voltage is positive. A positive sign for the associated power indicates that the device absorbs or consumes power. If the actual and reference direction disagrees for either voltage or current so i and v have opposite signs, $p = i \cdot v$ is negative and the device provides power. This definition of reference marks is called the *passive-sign convention*. Certain devices such as heaters (e.g., a toaster) can only absorb power. On the other hand, the power associated with a battery is positive when it is charging (absorbing power) and negative when it is discharging (delivering power). The passive-sign convention is used throughout electrical engineering. It is also the convention used by computer circuit simulation programs.

Ground

Voltage as an across variable is defined and measured between two points. It is convenient to identify one of the points as a reference point commonly called *ground*. This is similar to measuring elevation with respect to mean sea level. The heights of mountains, cities, and so on are given relative to sea level. Similarly, the voltages at all other points in a circuit are defined with respect to ground. Circuit references are denoted using one of the "ground" symbols shown in Fig. 2. The voltage at the ground point is always taken to be 0 V.

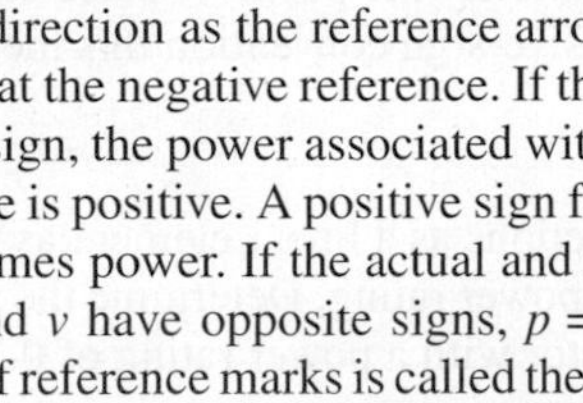

$$v_A(t) \qquad v_B(t) \qquad v_C(t)$$
$$+\circ \qquad\quad +\circ \qquad\quad +\circ$$
$$\text{A} \qquad\qquad \text{B} \qquad\qquad \text{C}$$

$$-\circ\, \text{G}$$

Figure 2 Ground symbol indicates a common voltage reference point.[1]

1.2 Fundamentals

A *circuit* is a collection of interconnected electrical devices that performs a useful function. An electrical *device* is a component that is treated as a distinct entity.

Element Constraints

A two-terminal device is described by its *i–v characteristic*, that is, the relationship between the voltage across and current through the device. In most cases the relationship is complicated and nonlinear so we use simpler linear models that adequately approximate the dominant features of a device.

Resistor. A resistor is a linear device described by a simple *i–v* characteristic as follows:

$$v = Ri \quad \text{or} \quad i = Gv \tag{5}$$

where R and G are positive constants related as $G = 1/R$. The power rating of the resistor determines the range over which the *i–v* characteristic can be represented by this linear relation. Equations (5) are collectively known as *Ohm's law*. The parameter R is called *resistance* and has the unit *ohms* (Ω). The parameter G is called *conductance* with the unit *siemens* (S).

The power associated with the resistor can be found from $p = v \cdot i$. Using Eqs. (5) to eliminate v or i from this relationship yields

$$p = i^2 R = v^2 G = \frac{v^2}{R} \tag{6}$$

Since the parameter R is positive, these equations state that the power is always nonnegative. Under the passive-sign convention this means the resistor always absorbs power.

Example 1

A resistor functions as a linear element as long as the voltage and current are within the limits defined by its power rating. Determine the maximum current and voltage that can be applied to a 47-k Ω resistor with a power rating of 0.25 W and remain within its linear operating range.

Solution

Using Eq. (6) to relate power and current, we obtain

$$I_{\text{MAX}} = \sqrt{\frac{P_{\text{MAX}}}{R}} = \sqrt{\frac{0.25}{47 \times 10^3}} = 2.31 \text{ mA}$$

Similarly, using Eq. (6) to relate power and voltage,

$$V_{\text{MAX}} = \sqrt{R P_{\text{MAX}}} = \sqrt{47 \times 10^3 \times 0.25} = 108 \text{ V}$$

A resistor with infinite resistance, that is, $R = \infty\ \Omega$, is called an *open circuit*. By Ohm's law no current can flow through such a device. Similarly, a resistor with no resistance, that is, $R = 0\ \Omega$, is called a *short circuit*. The voltage across a short circuit is always zero. In circuit analysis the devices in a circuit are assumed to be interconnected by zero-resistance wire, that is, by short circuits. Figure 3 shows the circuit symbols for a resistor and open and short circuits.

Ideal Sources

The signal and power sources required to operate electronic circuits are modeled using two elements: voltage sources and current sources. These sources can produce either constant or time-varying signals. The circuit symbols of an ideal voltage source and an ideal current source are shown in Fig. 4.

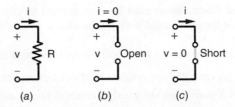

Figure 3 (a) Resistor symbol, (b) open circuit, and (c) short circuit.

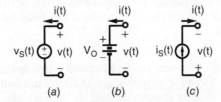

Figure 4 (a) Voltage source, (b) battery (traditional symbol), and (c) current source.

The i–v characteristic of an *ideal voltage source* in Fig. 4 is described by the element equations

$$v = v_S \quad \text{and} \quad i = \text{any value} \tag{7}$$

The element equations mean the ideal voltage source produces v_S volts across its terminals and will supply whatever current may be required by the circuit to which it is connected.

The i–v characteristic of an *ideal current source* in Fig. 4 is described by the element equations

$$i = i_S \quad \text{and} \quad v = \text{any value} \tag{8}$$

The ideal current source supplies i_S amperes in the direction of its arrow symbol and will furnish whatever voltage is required by the circuit to which it is connected.

In practice, circuit analysis involves selecting an appropriate model for the actual device. Figure 5 shows the practical models for the voltage and current sources. These models are called practical because they more accurately represent the properties of real-world sources than do the ideal models. The resistances R_S in the practical source models in Fig. 5 do not represent physical resistors but represent circuit elements used to account for resistive effects within the devices being modeled.

Connection Constraints

The previous section dealt with individual devices and models while this section considers the constraints introduced by interconnections of devices to form circuits. *Kirchhoff's laws* are

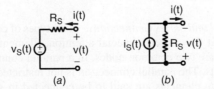

Figure 5 Circuit symbols for practical independent sources: (a) practical voltage source and (b) practical current source.[1]

derived from conservation laws as applied to circuits and are called *connection constraints* because they are based only on the circuit connections and not on the specific devices in the circuit.

The treatment of Kirchhoff's laws uses the following definitions.

A *circuit* is any collection of devices connected at their terminals.

A *node* is an electrical juncture of two or more devices.

A *loop* is a closed path formed by tracing through a sequence of devices without passing through any node more than once.

While it is customary to designate a juncture of two or more elements as a node, it is important to realize that a node is not confined to a point but includes all the wire from the point to each element.

Kirchhoff's Current Law. Kirchhoff's first law is based on the principle of conservation of charge. *Kirchhoff's current law* (KCL) states that the algebraic sum of the currents entering a node is zero at every instant. In forming the algebraic sum of currents, one must take into account the current reference directions associated with the devices. If the current reference direction is into the node, a positive sign is assigned to the algebraic sum of the corresponding current. If the reference direction is away from the node, a negative sign is assigned.

There are two signs associated with each current in the application of KCL. The first is the sign given to a current in writing a KCL connection equation. This sign is determined by the orientation of the current reference direction relative to a node. The second sign is determined by the actual direction of the current relative to the reference direction.

The following general principle applies to writing KCL equations: *In a circuit containing N nodes there are only $N - 1$ independent KCL connection equations.* In general, to write these equations, we select one node as the reference or ground node and then write KCL equations at the remaining $N - 1$ nonreference nodes.

Kirchhoff's Voltage Law. The second of Kirchhoff's circuit laws is based on the principle of conservation of energy. *Kirchhoff's voltage law* (KVL) states that the algebraic sum of all of the voltages around a loop is zero at every instant. There are two signs associated with each voltage. The first is the sign given the voltage when writing the KVL connection equation. The second is the sign determined by the actual polarity of a voltage relative to its assigned reference polarity.

The following general principle applies to writing KVL equations: *In a circuit containing E two-terminal elements and N nodes there are only $E - N + 1$ independent KVL connection equations.* Voltage equations written around $E - N + 1$ different loops contain all of the independent connection constraints that can be derived from KVL. A sufficient condition for loops to be different is that each contains at least one element that is not contained in any other loop.

Parallel and Series Connections. Two types of connections occur so frequently in circuit analysis that they deserve special attention. Elements are said to be connected in *parallel* when they share two common nodes. In a parallel connection KVL forces equal voltages across the elements. The parallel connection is not restricted to two elements.

Two elements are said to be connected in *series* when they have one common node to which no other current-drawing element is connected. A series connection results in equal current through each element. Any number of elements can be connected in series.

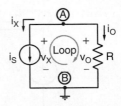

Figure 6   Circuit used to demonstrate combined constraints.[1]

Combined Constraints

The usual goal of circuit analysis is to determine the currents or voltages at various places in a circuit. This analysis is based on constraints of two distinctly different types. The element constraints are based on the models of the specific devices connected in the circuit. The connection constraints are based on Kirchhoff's laws and the circuit connections. The element equations are independent of the circuit in which the device is connected. Likewise, the connection equations are independent of the specific devices in the circuit. But taken together, the *combined constraints* from the element and connection equations provide the data needed to analyze a circuit.

The study of combined constraints begins by considering the simple but important example in Fig. 6. This circuit is driven by the current source i_S and the resulting responses are current/voltage pairs (i_X, v_X) and (i_O, v_O). The reference marks for the response pairs have been assigned using the passive-sign convention.

To solve for all four responses, four equations are required. The first two are the element equations:

$$\text{Current source:} \quad i_X = i_S$$
$$\text{Resistor:} \quad v_O = R \cdot i_O \tag{9}$$

The first element equation states that the response current i_X and the input driving force i_S are equal in magnitude and direction. The second element equation is Ohm's law relating v_O and i_O under the passive-sign convention.

The connection equations are obtained by applying Kirchhoff's laws. The circuit in Fig. 6 has two elements ($E = 2$) and two nodes ($N = 2$); hence for a total solution $E - N + 1 = 1$ KVL equation and $N - 1 = 1$ KCL equation are required. Selecting node B as the reference or ground node, a KCL at node A and a KVL around the loop yield

$$\text{KCL:} \quad -i_X - i_O = 0$$
$$\text{KVL:} \quad -v_X + v_O = 0 \tag{10}$$

With four equations and four unknowns all four responses can be found. Combining the KCL connection equation and the first element equations yields $i_O = -i_X = -i_S$. Substituting this result into the second element equations (Ohm's law) produces $v_O = -Ri_S$. The minus sign in this equation does not mean v_O is always negative. Nor does it mean the resistance is negative since resistance is always positive. It means that when the input driving force i_S is positive, the response v_O is negative, and vice versa.

Example 2

Find all of the element currents and voltages in Fig. 7 for $V_O = 10$ V, $R_1 = 2$ kΩ, and $R_2 = 3$ kΩ.

Solution

Substituting the element constraints into the KVL connection constraint produces

$$-V_O + R_1 i_1 + R_2 i_2 = 0$$

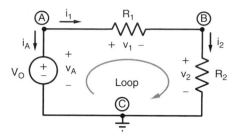

Figure 7   Element currents and voltages.[1]

This equation can be used to solve for i_1 since the second KCL connection equation requires that $i_2 = i_1$. Hence

$$i_1 = \frac{V_O}{R_1 + R_2} = \frac{10}{2000 + 3000} = 2 \text{ mA}$$

By finding the current i_1, all currents can be found from

$$-i_A = i_1 = i_2$$

since all three elements are connected in series. Substituting all of the known values into the element equations gives

$$v_A = 10\text{V} \quad v_1 = R_1 i_1 = 4\text{V} \quad v_2 = R_2 i_2 = 6\text{V}$$

Assigning Reference Marks

In all previous examples and exercises the reference marks for the element currents (arrows) and voltages (+ and −) were given. When reference marks are not shown on a circuit diagram, they must be assigned by the person solving the problem. Beginners sometimes wonder how to assign reference marks when the actual voltage polarities and current directions are as yet unknown. It is important to remember that reference marks do not indicate the actual polarities and directions. They are benchmarks assigned in an arbitrary way at the beginning of the analysis. If it turns out the actual direction and reference direction agree, then the numerical value of the response will be positive. If they disagree, the numerical value will be negative. In other words, the sign of the answer together with arbitrarily assigned reference marks tells us the actual voltage polarity or current direction. When assigning reference marks in this chapter the passive-sign convention will always be used. By always following the passive-sign convention any confusion about the direction of power flow in a device will be avoided. In addition, Ohm's law and other device i–v characteristics assume the voltage and current reference marks follow the passive-sign convention. Always using this convention follows the practice used in all SPICE-based computer circuit analysis programs.

Equivalent Circuits

The analysis of a circuit can often be simplified by replacing part of the circuit with one that is equivalent but simpler. The underlying basis for two circuits to be equivalent is contained in their i–v relationships: *Two circuits are said to be equivalent if they have identical i–v characteristics at a specified pair of terminals.*

Equivalent Resistance.   Resistances connected in series simply add, while conductances connected in parallel also simply add. Since conductance is not normally used to describe a resistor, two resistors R_1 and R_2 connected in parallel result in the expression

$$R_1 || R_2 = R_{EQ} = \frac{1}{G_{EQ}} = \frac{1}{G_1 + G_2} = \frac{1}{1/R_1 + 1/R_2}$$

$$= \frac{R_1 R_2}{R_1 + R_2} \tag{11}$$

where the symbol $||$ is shorthand for "in parallel." The expression on the far right in Eq. (11) is called the product over the sum rule for *two* resistors in parallel. The product-over-the-sum rule only applies to two resistors connected in parallel. When more than two resistors are in parallel, the following must be used to obtain the equivalent resistance:

$$R_{EQ} = \frac{1}{G_{EQ}} = \frac{1}{1/R_1 + 1/R_2 + 1/R_3 + \cdots} \tag{12}$$

Example 3
Given the circuit in Fig. 8:

    a. Find the equivalent resistance R_{EQ1} connected between terminals A and B.

    b. Find the equivalent resistance R_{EQ2} connected between terminals C and D.

Solution
First resistors R_2 and R_3 are connected in parallel. Applying Eq. (11) results in

$$R_2 || R_3 = \frac{R_2 R_3}{R_2 + R_3}$$

    a. The equivalent resistance between terminals A and B equals R_1 and the equivalent resistance $R_2 || R_3$ connected in series. The total equivalent resistance R_{EQ1} between terminals A and B thus is

$$R_{EQ1} = R_1 + (R_2 || R_3)$$

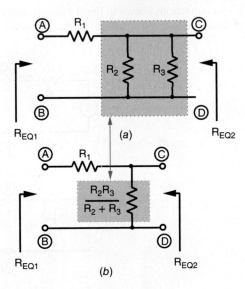

Figure 8 Circuit.[1]

$$R_{EQ1} = R_1 + \frac{R_2 R_3}{R_2 + R_3}$$

$$R_{EQ1} = \frac{R_1 R_2 + R_1 R_3 + R_2 R_3}{R_2 + R_3}$$

b. Looking into terminals C and D yields a different result. In this case R_1 is not involved since there is an open circuit (an infinite resistance) between terminals A and B. Therefore, only $R_2 || R_3$ affects the resistance between terminals C and D. As a result R_{EQ2} is simply

$$R_{EQ2} = R_2 || R_3 = \frac{R_2 R_3}{R_2 + R_3}$$

This example shows that equivalent resistance depends upon the pair of terminals involved.

Equivalent Sources

The practical source models shown in Fig. 9 consist of an ideal voltage source in series with a resistance and an ideal current source in parallel with a resistance.

If $R_1 = R_2 = R$ and $v_S = i_S R$, the two practical sources have the same $i–v$ relationship, making the two sources equivalent. When equivalency conditions are met, the rest of the circuit is unaffected regardless if driven by a practical voltage source or a practical current source.

The source transformation equivalency means that either model will deliver the same voltage and current to the rest of the circuit. It does not mean the two models are identical in every way. For example, when the rest of the circuit is an open circuit, there is no current in the resistance of the practical voltage source and hence no $i^2 R$ power loss. But the practical current source model has a power loss because the open-circuit voltage is produced by the source current in the parallel resistance.

Y–Δ Transformations

The Y–Δ connections shown in Fig. 10 occasionally occur in circuits and are especially prevalent in three-phase power circuits. One can transform from one configuration to the other by

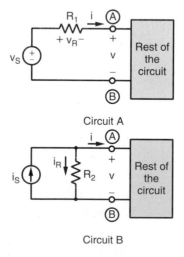

Circuit A

Circuit B

Figure 9   Equivalent practical source models.[1]

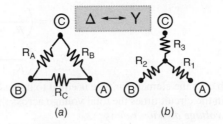

Figure 10   Y–Δ transformation.[2]

the following set of transformations:

$$R_A = \frac{R_1R_2 + R_2R_3 + R_1R_3}{R_1} \qquad R_1 = \frac{R_BR_C}{R_A + R_B + R_C}$$

$$R_B = \frac{R_1R_2 + R_2R_3 + R_1R_3}{R_2} \qquad R_2 = \frac{R_AR_C}{R_A + R_B + R_C} \qquad (13)$$

$$R_C = \frac{R_1R_2 + R_2R_3 + R_1R_3}{R_3} \qquad R_3 = \frac{R_BR_A}{R_A + R_B + R_C}$$

Solving Eqs. (13) for R_1, R_2, and R_3 yields the equations for a Δ-to-Y transformation while solving Eqs. (13) for R_A, R_B, and R_C yields the equations for a Y-to-Δ transformation. The Y and Δ subcircuits are said to be *balanced* when $R_1 = R_2 = R_3 = R_Y$ and $R_A = R_B = R_C = R_A$. Under balanced conditions the transformation equations reduce to $R_Y = R_\Delta/3$ and $R_\Delta = 3R_Y$.

Voltage and Current Division
These two analysis tools find wide application in circuit analysis and design.

Voltage Division.   Voltage division allows us to solve for the voltage across each element in a series circuit. Figure 11 shows a circuit that lends itself to solution by voltage division. Applying KVL around the loop in Fig. 11 yields $v_S = v_1 + v_2 + v_3$. Since all resistors are connected in series, the same current i exists in all three. Using Ohm's law yields $v_S = R_1i + R_2i + R_3i$. Solving for i yields $i = v_S/(R_1 + R_2 + R_3)$. Once the current in the series circuit is found, the voltage across each resistor is found using Ohm's law:

$$v_1 = R_1i = \left(\frac{R_1}{R_1 + R_2 + R_3}\right)v_S$$

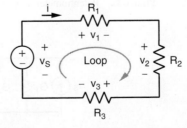

Figure 11   Voltage divider circuit.[1]

$$v_2 = R_2 i = \left(\frac{R_2}{R_1 + R_2 + R_3} \right) v_S$$

$$v_3 = R_3 i = \left(\frac{R_3}{R_1 + R_2 + R_3} \right) v_S \qquad (14)$$

In each case the element voltage is equal to its resistance divided by the equivalent series resistance in the circuit times the total voltage across the series circuit. Thus, the general expression of the *voltage division rule* is

$$v_k = \left(\frac{R_k}{R_{EQ}} \right) v_{total} \qquad (15)$$

The operation of a potentiometer is based on the voltage division rule. The device is a three-terminal element that uses voltage (potential) division to meter out a fraction of the applied voltage. Figure 12 shows the circuit symbol of a potentiometer. Simply stated, a potentiometer is an adjustable voltage divider.

The voltage v_O in Fig. 12 can be adjusted by turning the shaft on the potentiometer to move the wiper arm contact. Using the voltage division rule, v_O is found as

$$v_O = \left(\frac{R_{total} - R_1}{R_{total}} \right) v_S \qquad (16)$$

Moving the movable wiper arm all the way to the top makes R_1 zero, and voltage division yields v_S. In other words, 100% of the applied voltage is delivered to the rest of the circuit. Moving the wiper to the other extreme delivers zero voltage. By adjusting the wiper arm position we can obtain an output voltage anywhere between zero and the applied voltage v_S. Applications of a potentiometer include volume control, voltage balancing, and fine-tuning adjustment.

Current Division.   Current division is the dual of voltage division. By duality current division allows for the solution of the current through each element in a parallel circuit. Figure 13 shows a parallel circuit that lends itself to solution by current division. Applying KCL at node A yields $i_S = i_1 + i_2 + i_3$. The voltage v appears across all three conductances since they are connected in parallel. So using Ohm's law we can write $i_S = vG_1 + vG_2 + vG_3$ and solve for v as

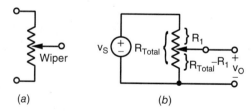

(a)   (b)

Figure 12   Potentiometer: (*a*) circuit symbol and (*b*) an application.

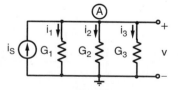

Figure 13   Current divider circuit.[1]

$v = i_S/(G_1 + G_2 + G_3)$. Given the voltage v, the current through any element is found using Ohm's law as

$$i_1 = vG_1 = \left(\frac{G_1}{G_1 + G_2 + G_3} \right) i_S$$

$$i_2 = vG_2 = \left(\frac{G_2}{G_1 + G_2 + G_3} \right) i_S$$

$$i_3 = vG_3 = \left(\frac{G_3}{G_1 + G_2 + G_3} \right) i_S$$

These results show that the source current divides among the parallel resistors in proportion to their conductances divided by the equivalent conductances in the parallel connection. Thus, the general expression for the *current division rule* is

$$i_k = \left(\frac{G_k}{G_{EQ}} \right) i_{total} = \frac{1/R_k}{1/R_{EQ}} i_{total} \qquad (17)$$

Circuit Reduction

The concepts of series/parallel equivalence, voltage/current division, and source transformations can be used to analyze *ladder circuits* of the type shown in Fig. 14. The basic analysis strategy is to reduce the circuit to a simpler equivalent in which the desired voltage or current is easily found using voltage and/or current division and/or source transformation and/or Ohm's law. There is no fixed pattern to the reduction process, and much depends on the insight of the analyst.

When using circuit reduction, it is important to remember that the unknown voltage exists between two nodes and the unknown current exists in a branch. The reduction process must not eliminate the required node pair or branch; otherwise the unknown voltage or current cannot be found. The next example will illustrate circuit reduction.

Example 4

Find the output voltage v_O and the input current i_S in the ladder circuit shown in Fig. 14a.

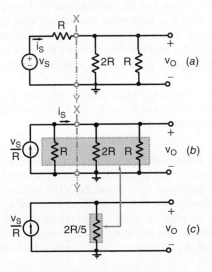

Figure 14 Ladder circuits.[1]

Solution

Breaking the circuit at points X and Y produces voltage source v_S in series with a resistor R: Using source transformation this combination can be replaced by an equivalent current source in parallel with the same resistor, as shown in Fig. 14b. Using current division the input current i_S is

$$i_S = \frac{R}{\left(\frac{2}{3}\right)R + R} \times \frac{v_S}{R} = \frac{v_S}{\left(\frac{5}{3}\right)R} = \frac{3}{5}\frac{v_S}{R}$$

The three parallel resistances in Fig. 14b can be combined into a single equivalent resistance without eliminating the node pair used to define the output voltage v_F:

$$R_{EQ} = \frac{1}{1/R + 1/(2R) + 1/R} = \frac{2R}{5}$$

which yields the equivalent circuit in Fig. 14c. The current source v_S/R determines the current through the equivalent resistance in Fig. 14c. The output voltage is found using Ohm's law:

$$v_O = \left(\frac{v_S}{R}\right) \times \left(\frac{2R}{5}\right) = \frac{2}{5}v_S$$

Several other analysis approaches are possible.

2 DIRECT-CURRENT (DC) CIRCUITS

This section reviews basic dc analysis using traditional circuits theorems with application to circuit analysis and design.

2.1 Node Voltage Analysis

Using node voltage instead of element voltages as circuit variables can greatly reduce the number of equations that must be treated simultaneously. To define a set of node voltages, a reference node or ground is first selected. The *node voltages* are then defined as the voltages between the remaining nodes and the reference node. Figure 15 shows a reference node indicated by the ground symbol as well as the notation defining the three nonreference node voltages. The node voltages are identified by a voltage symbol adjacent to the nonreference nodes. This notation means that the positive reference mark for the node voltage is located at the node in question while the negative mark is at the reference node. Any circuit with N nodes involves $N - 1$ node voltages.

The following is a fundamental property of node voltages: *If the Kth two-terminal element is connected between nodes X and Y, then the element voltage can be expressed in terms of the two node voltages as*

$$v_K = v_X - v_Y \tag{18}$$

where X is the node connected to the positive reference for element voltage vK.

Equation (18) is a KVL constraint at the element level. If node Y is the reference node, then by definition $v_Y = 0$ and Eq. (18) reduces to $v_K = v_X$. On the other hand, if node X is the reference node, then $v_X = 0$ and therefore $v_K = -v_Y$. The minus sign here comes from the fact that the positive reference for the element is connected to the reference node. In any case, the important fact is that the voltage across any two-terminal element can be expressed as the difference of two node voltages, one of which may be zero.

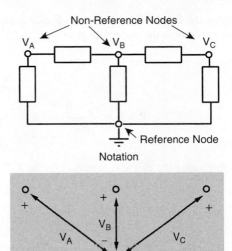

Figure 15 Node voltage definition and notation.[1]

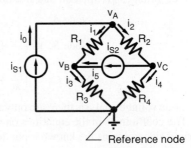

Figure 16 Bridge circuit for node voltage example.[1]

To formulate a circuit description using node voltages, device and connection analysis is used, except that the KVL connection equations are not explicitly written down. Instead the fundamental property of node analysis is used to express the element voltages in terms of the node voltages.

The circuit in Fig. 16 will demonstrate the formulation of node voltage equations. The ground symbol identifies the reference node, six element currents ($i_0, i_1, i_2, i_3, i_4,$ and i_5), and three node voltages ($v_A, v_B,$ and v_C).

The KCL constraints at the three nonreference nodes are

$$\text{Node } A: \quad -i_0 + i_1 + i_2 = 0$$

$$\text{Node } B: \quad -i_1 + i_3 - i_5 = 0$$

$$\text{Node } C: \quad -i_2 + i_5 + i_4 = 0$$

Using the fundamental property of node analysis, device equations are used to relate the element currents to the node voltages:

$$R_1: i_1 = \frac{v_A - v_B}{R_1} \qquad R_2: \quad i_2 = \frac{v_A - v_C}{R_2}$$

$$\text{Voltage source: } v_S = v_A$$

$$R_3: i_3 = \frac{v_B - 0}{R_3} \qquad R_4: \quad i_4 = \frac{v_C - 0}{R_4}$$

$$\text{Current source: } i_5 = i_{S2}$$

Substituting the element currents into the KCL equations yields

$$\text{Node } A: \quad -i_0 + \frac{v_A - v_B}{R_1} + \frac{v_A - v_C}{R_2} = 0$$

$$\text{Node } B: \qquad \frac{v_B - v_A}{R_1} + \frac{v_B}{R_3} - i_{S2} = 0$$

$$\text{Node } C: \qquad \frac{v_C - v_A}{R_2} + \frac{v_C}{R_4} + i_{S2} = 0$$

But since the reference is connected to the negative side of the voltage source, $v_A = v_S$. Thus at node A the voltage is already known and this reduces the number of equations that need to be solved. The equation written above can be used to solve for the current through the voltage source if that is desired. Writing these equations in standard form with all of the unknown node voltages grouped on one side and the independent sources on the other yields

$$\text{Node } A: \qquad v_A \qquad = v_S$$

$$\text{Node } B: v_B \left(\frac{1}{R_1} + \frac{1}{R_3} \right) = i_S + \frac{v_S}{R_1}$$

$$\text{Node } C: v_C \left(\frac{1}{R_2} + \frac{1}{R_4} \right) = -i_S + \frac{v_S}{R_2}$$

Using node voltage analysis there are only two equations and two unknowns (v_B and v_C) to be solved. The coefficients in the equations on the left side depend only on circuit parameters, while the right side contains the known input driving forces.

Supernodes

When neither node of a voltage source can be selected as the reference, a *supernode* must be used. The fact that KCL applies to the currents penetrating a boundary can be used to write a node equation at the supernode. Then node equations at the remaining nonreference nodes are written in the usual way. This process reduces the number of available node equations to $N - 3$ plus one supernode equation, leaving us one equation short of the $N - 1$ required. The voltage source inside the supernode constrains the difference between the node voltages to be the value of the voltage source. The voltage source constraint provides the additional relationship needed to write $N - 1$ independent equations in the $N - 1$ node voltages.

Example 5

For the circuit in Fig. 17:

    a. Formulate node voltage equations.

    b. Solve for the output voltage v_O using $R_1 = R_4 = 2$ kΩ and $R_2 = R_3 = 4$ kΩ.

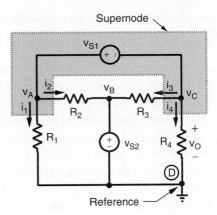

Figure 17 Circuit.[1]

Solution

a. The voltage sources in Fig. 17 do not have a common node so a reference node that includes both sources cannot be selected. Choosing node D as the reference forces the condition $v_B = v_{S_2}$ but leaves the other source v_{S1} ungrounded. The ungrounded source and all wires leading to it are encompassed by a supernode boundary as shown in the figure. Kirchhoff's current law applies to the four-element currents that penetrate the supernode boundary and we can write

$$i_1 + i_2 + i_3 + i_4 = 0$$

These currents can easily be expressed in terms of the node voltages:

$$\frac{v_A}{R_1} + \frac{v_A - v_B}{R_2} + \frac{v_C - v_B}{R_3} + \frac{v_C}{R_4} = 0$$

But since $v_B = v_{S2}$, the standard form of this equation is

$$v_A \left(\frac{1}{R_1} + \frac{1}{R_2} \right) + v_C \left(\frac{1}{R_3} + \frac{1}{R_4} \right) = v_{S2} \left(\frac{1}{R_2} + \frac{1}{R_3} \right)$$

We have one equation in the two unknown node voltages v_A and v_C. Applying the fundamental property of node voltages inside the supernode, we can write

$$v_A - v_C = v_{S1}$$

That is, the ungrounded voltage source constrains the difference between the two unknown node voltages inside the supernode and thereby supplies the relationship needed to obtain two equations in two unknowns.

b. Substituting in the given numerical values yields

$$(7.5 \times 10^{-4})v_A + (7.5 \times 10^{-4})v_C = (5 \times 10^{-4})v_{S2}$$

$$v_A - v_C = v_{S1}$$

To find the output v_O, we need to solve these equations for v_C. The second equation yields $v_A = v_C + v_{S1}$, which when substituted into the first equation yields the required output:

$$v_O = v_C = 1/3\, v_{S2} - 1/2\, v_{S1}$$

Node voltage equations are very useful in the analysis of a variety of electronic circuits. These equations can always be formulated using KCL, the element constraints, and the fundamental property of node voltages. The following guidelines summarize this approach:

1. Simplify the circuit by combining elements in series and parallel wherever possible.

2. If not specified, select a reference node so that as many dependent and independent voltage sources as possible are directly connected to the reference.

3. Label a node voltage adjacent to each nonreference node.

4. Create supernodes for dependent and independent voltage sources that are not directly connected to the reference node.

5. Node equations are required at supernodes and all other nonreference nodes except op amp outputs and nodes directly connected to the reference by a voltage source.

6. Write symmetrical node equations by treating dependent sources as independent sources and using the inspection method.

7. Write expressions relating the node and source voltages for voltage sources included in supernodes.

8. Substitute the expressions from step 7 into the node equations from step 6 and place the result in standard form.

2.2   Mesh Current Analysis

Mesh currents are an alternative set of analysis variables that are useful in circuits containing many elements connected in series. To review terminology, a loop is a sequence of circuit elements that forms a closed path that passes through each element just once. A mesh is a special type of loop that does not enclose any elements.

The following development of mesh analysis is restricted to planar circuits. A *planar circuit* can be drawn on a flat surface without crossovers in a "window pane" fashion. To define a set of variables, a *mesh current* $(i_A, i_B, i_C, \ldots)$ is associated with each window pane and a reference direction assigned customarily in a clockwise sense. There is no momentous reason for this except perhaps tradition.

Mesh currents are thought of as circulating through the elements in their respective meshes; however, this viewpoint is not based on the physics of circuit behavior. There are not different types of electrons flowing that somehow get assigned to mesh currents i_A or i_B. Mesh currents are variables used in circuit analysis. They are only somewhat abstractly related to the physical operation of a circuit and may be impossible to measure directly. Mesh currents have a unique feature that is the dual of the fundamental property of node voltages. In a planar circuit any given element is contained in at most two meshes. When an element is in two meshes, the two mesh currents circulate through the element in opposite directions. In such cases KCL declares that the net element current through the element is the difference of the two mesh currents.

These observations lead to the fundamental property of mesh currents: *If the Kth two-terminal element is contained in meshes X and Y, then the element current can be expressed in terms of the two mesh currents as*

$$i_K = i_X - i_Y \tag{19}$$

where X is the mesh whose reference direction agrees with the reference direction of i_K.

Equation (19) is a KCL constraint at the element level. If the element is contained in only one mesh, then $i_K = i_X$ or $i_K = -i_Y$ depending on whether the reference direction for the element current agrees or disagrees with the mesh current. The key fact is that the current through every two-terminal element in a planar circuit can be expressed as the difference of at most two mesh currents.

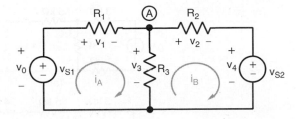

Figure 18   Circuit demonstrating mesh current analysis.[1]

Mesh currents allow circuit equations to be formulated using device and connection constraints, except that the KCL constraints are not explicitly written down. Instead, the fundamental property of mesh currents is used to express the device constraints in terms of the mesh currents, thereby avoiding using the element currents and working only with the element voltages and mesh currents.

For example, the planar circuit in Fig. 18 can be analyzed using the mesh current method. In the figure two mesh currents are shown as well the voltages across each of the five elements. The KVL constraints around each mesh using the element voltages yield

$$\text{Mesh } A: \quad -v_0 + v_1 + v_3 = 0$$

$$\text{Mesh } B: \quad -v_3 + v_2 + v_4 = 0$$

Using the fundamental property of mesh currents, the element voltages in terms of the mesh currents and input voltages are written as

$$v_1 = R_1 i_A v_0 = v_{S1}$$

$$v_2 = R_2 i_B v_4 = v_{S2}$$

$$v_3 = R_3 (i_A - i_B)$$

Substituting these element equations into the KVL connection equations and arranging the result in standard form yield

$$(R_1 + R_3) i_A - R_3 i_B = v_{S1}$$

$$-R_3 i_A + (R_2 + R_3) i_B = -v_{S2}$$

This results in two equations in two unknown mesh currents. The KCL equations $i_1 = i_A, i_2 = i_B$, and $i_3 = i_A - i_B$ are implicitly used to write mesh equations. In effect, the fundamental property of mesh currents ensures that the KCL constraints are satisfied. Any general method of circuit analysis must satisfy KCL, KVL, and the device i–v relationships. Mesh current analysis appears to focus on the latter two but implicitly satisfies KCL when the device constraints are expressed in terms of the mesh currents.

Solving for the mesh currents yields

$$i_A = \frac{(R_2 + R_3) v_{S1} - R_3 v_{S2}}{R_1 R_2 + R_1 R_3 + R_2 R_3} \quad \text{and} \quad i_B = \frac{R_3 v_{S1} - (R_1 + R_3) v_{S2}}{R_1 R_2 + R_1 R_3 + R_2 R_3}$$

The results for i_A and i_B can now be substituted into the device constraints to solve for every voltage in the circuit. For instance, the voltage across R_3 is

$$v_A = v_3 = R_3 (i_A - i_B) = \frac{R_2 R_3 v_{S1} + R_1 R_3 v_{S2}}{R_1 R_2 + R_1 R_3 + R_2 R_3}$$

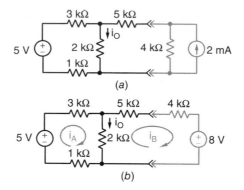

Figure 19   Circuit.[1]

Example 6

Use mesh current equations to find i_O in the circuit in Fig. 19a.

Solution

The current source in this circuit can be handled by a source transformation. The 2-mA source in parallel with the 4-kΩ resistor in Fig. 19a can be replaced by an equivalent 8-V source in series with the same resistor as shown in Fig. 19b. In this circuit the total resistance in mesh A is 6 kΩ, the total resistance in mesh B is 11 kΩ, and the resistance contained in both meshes is 2 kΩ. By inspection the mesh equations for this circuit are

$$(6000)i_A - (2000)i_B = 5$$
$$-(2000)i_A + (11000)i_B = -8$$

Solving for the two mesh currents yields

$$i_A = 0.6290 \text{ mA} \qquad \text{and} \qquad i_B = 0.6129 \text{ mA}$$

By KCL the required current is

$$i_0 = i_A - i_B = 1.2419 \text{ mA}$$

Supermesh

If a current source is contained in two meshes and is not connected in parallel with a resistance, then a *supermesh* is created by excluding the current source and any elements connected in series with it. One mesh equation is written around the supermesh using the currents i_A and i_B. Then mesh equations of the remaining meshes are written in the usual way. This leaves the solution one equation short because parts of meshes A and B are included in the supermesh. However, the fundamental property of mesh currents relates the currents i_S, i_A, and i_B as

$$i_S = i_A - i_B$$

This equation supplies the additional relationship needed to get the requisite number of equations in the unknown mesh currents. This approach is obviously the dual of the supernode method for modified node analysis. The following example demonstrates the use of a supermesh.

Example 7

Use mesh current equations to find the v_O in Fig. 20.

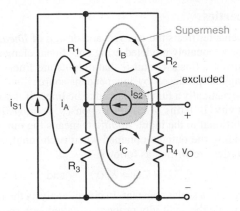

Figure 20 Example of supermesh.

Solution
The current source i_{S_2} is in both mesh B and mesh C, so we exclude this element and create the supermesh shown in the figure. The sum of voltages around the supermesh is

$$R_1(i_B - i_A) + R_2(i_B) + R_4(i_C) + R_3(i_C - i_A) = 0$$

The supermesh voltage constraint yields one equation in the three unknown mesh currents. Applying KCL to each of the current sources yields

$$i_A = i_{S1} \quad i_B - i_C = i_{S2}$$

Because of KCL, the two current sources force constraints that supply two more equations. Using these two KCL constraints to eliminate i_A and i_B from the supermesh KVL constraint yields

$$(R_1 + R_2 + R_3 + R_4)i_C = (R_1 + R_3)i_{S1} - (R_1 + R_2)i_{S2}$$

Hence, the required output voltage is

$$v_O = R_4 i_C = R_4 \times \left[\frac{(R_1 + R_3)\, i_{S1} - (R_1 + R_2)i_{S2}}{R_1 + R_2 + R_3 + R_4} \right]$$

Mesh current equations can always be formulated from KVL, the element constraints, and the fundamental property of mesh currents. The following guidelines summarize an approach to formulating mesh equations for resistance circuits:

1. Simplify the circuit by combining elements in series or parallel wherever possible.
2. Assign a clockwise mesh current to each mesh.
3. Create a supermesh for dependent and independent current sources that are contained in two meshes.
4. Write symmetrical mesh equations for all meshes by treating dependent sources as independent sources and using the inspection method.
5. Write expressions relating the mesh and source currents for current sources contained in only one mesh.
6. Write expressions relating the mesh and source currents for current sources included in supermeshes.
7. Substitute the expressions from steps 5 and 6 into the mesh equations from step 4 and place the result in standard form.

2.3   Linearity Properties

This chapter treats the analysis and design of *linear circuits*. A circuit is said to be linear if it can be adequately modeled using only linear elements and independent sources. The hallmark feature of a linear circuit is that outputs are linear functions of the inputs. Circuit *inputs* are the signals produced by independent sources and *outputs* are any other designated signals. Mathematically a function is said to be linear if it possesses two properties—homogeneity and additivity. In terms of circuit responses, *homogeneity* means the output of a linear circuit is proportional to the input. *Additivity* means the output due to two or more inputs can be found by adding the outputs obtained when each input is applied separately. Mathematically these properties are written as follows:

$$f(Kx) = Kf(x) \qquad \text{and} \qquad f(x_1 + x_2) = f(x_1) + f(x_2) \tag{20}$$

where K is a scalar constant. In circuit analysis the homogeneity property is called *proportionality* while the additivity property is called *superposition*.

Proportionality Property

The *proportionality property* applies to linear circuits with one input. For linear resistive circuits proportionality states that every input–output relationship can be written as

$$y = K \cdot x$$

where x is the input current or voltage, y is an output current or voltage, and K is a constant. The block diagram in Fig. 21 describes a relationship in which the input x is multiplied by the scalar constant K to produce the output y. Examples of proportionality abound. For instance, using voltage division in Fig. 22 produces

$$v_O = \left(\frac{R_2}{R_1 + R_2} \right) v_S$$

which means

$$x = v_S y = v_O$$

$$K = \frac{R_2}{R_1 + R_2}$$

In this example the proportionality constant K is dimensionless because the input and output have the same units. In other situations K could carry the units of ohms or siemens when the input or output does not have the same units.

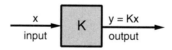

Figure 21   Block diagram representation of proportionality property.[1]

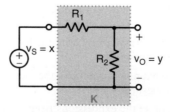

Figure 22   Example of circuit exhibiting proportionality.[1]

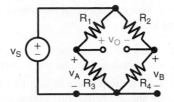

Figure 23   Bridge circuit.[1]

Example 8
Given the bridge circuit of Fig. 23:

   a. Find the proportionality constant K in the input–output relationship $v_O = Kv_S$.

   b. Find the sign of K when $R_2R_3 > R_1R_4$, $R_2R_3 = R_1R_4$, and $R_2R_3 < R_1R_4$.

Solution

   a. Note that the circuit consists of two voltage dividers. Applying the voltage division rule to each side of the bridge circuit yields

$$v_A = \frac{R_3}{R_1 + R_3}v_S \qquad v_B = \frac{R_4}{R_2 + R_4}v_S$$

The fundamental property of node voltages allows us to write $v_O = v_A - v_B$. Substituting the equations for v_A and v_B into this KVL equation yields

$$v_O = \begin{cases} \left(\dfrac{R_3}{R_1 + R_3} - \dfrac{R_4}{R_2 + R_4}\right)v_S \\[2ex] \left(\dfrac{R_2R_3 - R_1R_4}{(R_1 + R_3)(R_2 + R_4)}\right)v_S \\[2ex] (K)v_S \end{cases}$$

   b. The proportionality constant K can be positive, negative, or zero. Specifically:

$$\text{If } R_2R_3 > R_1R_2, \text{ then } K > 0.$$
$$\text{If } R_2R_3 = R_1R_2, \text{ then } K = 0.$$
$$\text{If } R_2R_3 < R_1R_2, \text{ then } K < 0.$$

When the product of the resistances in opposite legs of the bridge are equal, $K = 0$ and the bridge is said to be balanced.

Superposition Property
The *additivity property*, or superposition, states that any output current or voltage of a linear resistive circuit with multiple inputs can be expressed as a linear combination of several inputs:

$$y = K_1x_1 + K_2x_2 + K_3x_3 + \cdots$$

where $x_1, x_2, x_3, \ldots$ are current or voltage inputs and $K_1, K_2, K_3 \ldots$ are constants that depend on the circuit parameters.

   Since the output y above is a linear combination, the contribution of each input source is independent of all other inputs. This means that the output can be found by finding the contribution from each source acting alone and then adding the individual response to obtain the total response. This suggests that the output of a multiple-input linear circuit can be found by the following steps:

Step 1: "Turn off" all independent input signal sources except one and find the output of the circuit due to that source acting alone.

Step 2: Repeat the process in step 1 until each independent input source has been turned on and the output due to that source found.

Step 3: The total output with all sources turned on is then a linear combination (algebraic sum) of the contributions of the individual independent sources.

A voltage source is turned off by setting its voltage to zero ($v_S = 0$) and replacing it with a short circuit. Similarly, turning off a current source ($i_S = 0$) entails replacing it with an open circuit. Figure 24a shows that the circuit has two input sources. Figure 24b shows the circuit with the current source set to zero. The output of the circuit v_{O1} represents that part of the total output caused by the voltage source. Using voltage division yields

$$v_{O1} = \frac{R_2}{R_1 + R_2} v_S$$

Next the voltage source is turned off and the current source is turned on, as shown in Fig. 24c. Using Ohm's law, $v_{O2} = i_{O2}R_2$. Then using current division to express i_{O2} in terms of i_S yields

$$v_{O2} = i_{O2} \times R_2 = \left[\frac{R_1}{R_1 + R_2} i_S \right] R_2 = \frac{R_1 R_2}{R_1 + R_2} i_S$$

Applying the superposition theorem, the response with both sources "turned on" is found by adding the two responses v_{O1} and v_{O2}:

$$v_O = v_{O1} + v_{O2}$$

$$v_O = \left[\frac{R_2}{R_1 + R_2} \right] v_S + \left[\frac{R_1 R_2}{R_1 + R_2} \right] i_S$$

Superposition is an important property of linear circuits and is used primarily as a conceptual tool to develop other circuit analysis and design techniques. It is useful, for example, to determine the contribution to a circuit by a certain source.

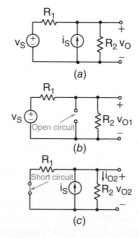

Figure 24 Circuit analysis using superposition: (a) current source off and (b) voltage source off.[1]

2.4 Thévenin and Norton Equivalent Circuits

An *interface* is a connection between circuits that perform different functions. Circuit interfaces occur frequently in electrical and electronic systems so special analysis methods are used to handle them. For the two-terminal interface shown in Fig. 25, one circuit can be considered as the source S and the other as the load L. Signals are produced by the source circuit and delivered to the load. The source–load interaction at an interface is one of the central problems of circuit analysis and design.

The Thévenin and Norton equivalent circuits shown in Fig. 25 are valuable tools for dealing with circuit interfaces. The conditions under which these equivalent circuits exist can be stated as a theorem: *If the source circuit in a two-terminal interface is linear, then the interface signals v and i do not change when the source circuit is replaced by its Thévenin or Norton equivalent circuit.*

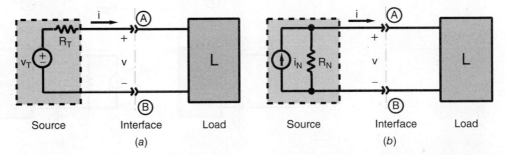

Figure 25   Equivalent circuits for source circuit: (*a*) Thévenin equivalent and (*b*) Norton equivalent.[1]

The equivalence requires the source circuit to be linear but places no restriction on the linearity of the load circuit. The Thévenin equivalent circuit consists of a voltage source (v_T) in series with a resistance (R_T). The Norton equivalent circuit is a current source (i_N) in parallel with a resistance (R_N). The Thévenin and Norton equivalent circuits are equivalent to each other since replacing one by the other leaves the interface signals unchanged. In essence the Thévenin and Norton equivalent circuits are related by the source transformation covered earlier under equivalent circuits.

The two parameters can often be obtained using open-circuit and short-circuit loads. If the actual load is disconnected from the source, an open-circuit voltage v_{OC} appears between terminals A and B. Connecting an open-circuit load to the Thévenin equivalent produces $v_{OC} = v_T$ since the open circuit causes the current to be zero, resulting in no voltage drop across R_T. Similarly, disconnecting the load and connecting a short circuit as shown produce a current i_{SC}. Connecting a short-circuit load to the Norton equivalent produces $i_{SC} = i_N$ since all of the source current i_N is diverted through the short-circuit load.

In summary, the parameters of the Thévenin and Norton equivalent circuits at a given interface can be found by determining the open-circuit voltage and the short-circuit current:

$$v_T = v_{OC}$$
$$i_N = i_{SC}$$
$$R_N = R_T = \frac{v_{OC}}{i_{SC}} \tag{21}$$

General Applications

Since even complex linear circuits can be replaced by their Thévenin or Norton equivalent, the chore of designing circuits that interface with these complex circuits is greatly simplified. Suppose a load resistance in Fig. 26*a* needs to be chosen so the source circuit to the left of the interface A–B delivers 4 V to the load.

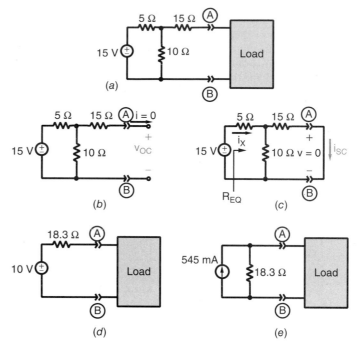

Figure 26 Example of finding Thévenin and Norton equivalent circuits: (*a*) given circuit, (*b*) open circuit yields Thévenin voltage, (*c*) short circuit yields Norton current, (*d*) Thévenin equivalent circuit, and (*e*) Norton equivalent circuit.[1]

The Thévenin and Norton equivalents v_{OC} and i_{SC} are first found. The open-circuit voltage v_{OC} is found by disconnecting the load at terminals $A–B$ as shown in Fig. 26*b*. The voltage across the 15-Ω resistor is zero because the current through it is zero due to the open circuit. The open-circuit voltage at the interface is the same as the voltage across the 10-Ω resistor. Using voltage division, this voltage is

$$v_T = v_{OC} = \frac{10}{10+5} \times 5 = 10\text{V}$$

Then the short-circuit current i_{SC} is calculated using the circuit in Fig. 26*c*. The total current i_X delivered by the 15-V source is $i_X = 15/R_{EQ}$, where R_{EQ} is the equivalent resistance seen by the voltage source with a short circuit at the interface:

$$R_{EQ} = 5 + \frac{10 \times 15}{10 + 15} = 11\ \Omega$$

The source current i_X can now be found: $i_X = 15/11 = 1.36$ A. Given i_X, current division is used to obtain the short-circuit current,

$$i_N = i_{SC} = \frac{10}{10 + 15} \times i_X = 0.545\ \text{A}$$

Finally, we compute the Thévenin and Norton resistances:

$$R_T = R_N = \frac{v_{OC}}{i_{SC}} = 18.3\ \Omega$$

The resulting Thévenin and Norton equivalent circuits are shown in Figs. 26*d,e*.

It now is an easy matter to select a load R_L so 4 V is supplied to the load. Using the Thévenin equivalent circuit, the problem reduces to a voltage divider,

$$4 \text{ V} = \frac{R_L}{R_L + R_T} \times V_T = \frac{R_L}{R_L + 18.3} \times 10$$

Solving for R_L yields $R_L = 12.2 \ \Omega$.

The Thévenin or Norton equivalent can always be found from the open-circuit voltage and short-circuit current at the interface. Often they can be measured using a multimeter to measure the open-circuit voltage and the short-circuit current.

Application to Nonlinear Loads

An important use of Thévenin and Norton equivalent circuits is finding the voltage across, current through, and power dissipated in a two-terminal nonlinear element (NLE). The method of analysis is a straightforward application of device i–v characteristics. An interface is defined at the terminals of the nonlinear element and the linear part of the circuit is reduced to the Thévenin equivalent in Fig. 27a. Treating the interface current i as the dependent variable, the i–v relationship of the Thévenin equivalent is written in the form

$$i = \left(-\frac{1}{R_T} \right) v + \left(\frac{v_T}{R_T} \right)$$

This is the equation of a straight line in the i–v plane shown in Fig. 27b. The line intersects the i axis ($v = 0$) at $i = i_{SC} = v_T / R_T$ and intersects the v axis ($i = 0$) at $v = v_{OC} = v_T$. This line is called the *load line*.

The nonlinear element has the i–v characteristic shown in Fig. 27c. Mathematically this nonlinear characteristic has the form $i = f(v)$. Both the nonlinear equation and the load line equation must be solved simultaneously. This can be done by numerical methods when $f(v)$ is known explicitly, but often a graphical solution is adequate. By superimposing the load line on the i–v characteristic curve of the nonlinear element in Fig. 27d, the point or points of intersection represent the values of i and v that satisfy the source constraints given in the form of the Thévenin equivalent above, and nonlinear element constraints. In the terminology of electronics the point of intersection is called the operating point, or Q *point*, or the quiescent point.

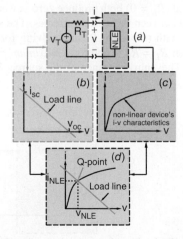

Figure 27 Graphical analysis of nonlinear circuit: (*a*) given circuit, (*b*) load line, (*c*) nonlinear device i–v characteristics, and (*d*) Q point.[1]

2.5 Maximum Signal Transfer

Circuit interfacing involves interconnecting circuits in such a way that they are compatible. In this regard an important consideration is the maximum signal levels that can be transferred across a given interface. This section defines the maximum voltage, current, and power available at an interface between a *fixed source* and an *adjustable load*.

The source can be represented by its Thévenin equivalent and the load by an equivalent resistance R_L, as shown in Fig. 28. For a fixed source the parameters v_T and R_T are given and the interface signal levels are functions of the load resistance R_L. By voltage division, the interface voltage is

$$v = \frac{R_L}{R_L + R_T} v_T$$

For a fixed source and a variable load, the voltage will be a maximum if R_L is made very large compared to R_T. Ideally R_L should be made infinite (an open circuit), in which case

$$v_{\text{MAX}} = v_T = v_{\text{OC}}$$

Therefore, the maximum voltage available at the interface is the source open-circuit voltage v_{OC}. The current delivered at the interface is

$$i = \frac{v_T}{R_L + R_T}$$

Again, for a fixed source and a variable load, the current will be a maximum if R_L is made very small compared to R_T. Ideally R_L should be zero (a short circuit), in which case

$$i_{\text{MAX}} = \frac{v_T}{R_T} = i_N = i_{\text{SC}}$$

Therefore, the maximum current available at the interface is the source short-circuit current i_{SC}.

The power delivered at the interface is equal to the product vi. Using interface voltage, and interface current results found above, the power is

$$p = v \times i = \frac{R_L v_T^2}{(R_T + R_L)^2}$$

For a given source, the parameters v_T and R_T are fixed and the delivered power is a function of a single variable R_L. The conditions for obtaining maximum voltage ($R_L \to \infty$) or maximum current ($R_L = 0$) both produce zero power. The value of R_L that maximizes the power lies somewhere between these two extremes. The value can be found by differentiating the power expression with respect to R_L and solving for the value of R_L that makes $dp/dR_L = 0$. This occurs when $R_L = R_T$. Therefore, *maximum power transfer* occurs when the load resistance equals the Thévenin resistance of the source. When $R_L = R_T$, the source and load are said to

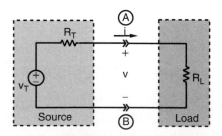

Figure 28 Two-terminal interface for deriving maximum signal transfer conditions.[1]

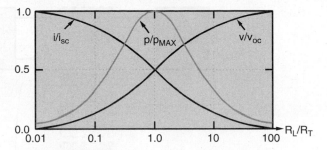

Figure 29   Normalized plots of current, voltage, and power versus R_L/R_T.[1]

be *matched*. Substituting the condition $R_L = R_T$ back into the power equation above shows the maximum power to be

$$p_{\text{MAX}} = \frac{v_T^2}{4R_T} = \frac{i_N^2 R_T}{4}$$

These results are consequences of what is known as the maximum power transfer theorem: *A fixed source with a Thévenin resistance RT delivers maximum power to an adjustable load RL when RL = RT*.

Figure 29 shows plots of the interface voltage, current, and power as functions of R_L/R_T. The plots of v/v_{OC}, i/i_{SC}, and p/p_{MAX} are normalized to the maximum available signal levels so the ordinates in Fig. 29 range from 0 to 1.

The plot of the normalized power p/p_{MAX} in the neighborhood of the maximum is not a particularly strong function of R_L/R_T. Changing the ratio R_L/R_T by a factor of 2 in either direction from the maximum reduces p/p_{MAX} by less than 20%. The normalized voltage v/v_{OC} is within 20% of its maximum when $R_L/R_T = 4$. Similarly, the normalized current is within 20% of its maximum when $R_L/R_T = 3$. In other words, for engineering purposes maximum signal levels can be approached with load resistances that only approximate the theoretical requirements.

2.6   Interface Circuit Design

The maximum signal levels discussed in the previous section place bounds on what is achievable at an interface. However, those bounds are based on a fixed source and an adjustable load. In practice there are circumstances in which the source or the load or both can be adjusted to produce prescribed interface signal levels. Sometimes it is necessary to insert an interface circuit between the source and load. Figure 30 shows the general situations and some examples of resistive interface circuits. By its very nature the inserted circuit has two terminal pairs, or interfaces, at which voltage and current can be observed or specified. These terminal pairs are also called *ports*, and the interface circuit is referred to as a *two-port network*. The port connected to the source is called the input and the port connected to the load the output. The purpose of this two-port network is to ensure that the source and load interact in a prescribed way.

Basic Circuit Design Concepts

This section introduces a limited form of circuit design, as contrasted with circuit analysis. Although circuit analysis tools are essential in design, there are important differences. A linear circuit analysis problem generally has a unique solution. A circuit design problem may have many solutions or even no solution. The maximum available signal levels found above provide

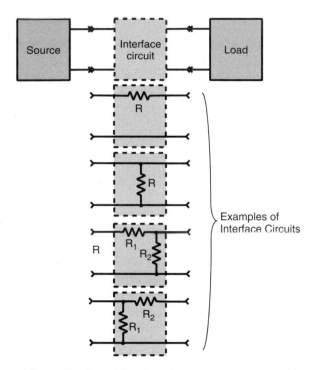

Figure 30 General interface circuit and some examples.[1]

bounds that help test for the existence of a solution. Generally there will be several ways to meet the interface constraints, and it then becomes necessary to evaluate the alternatives using other factors such as cost, power consumption, or reliability. Currently only the resistor will be used to demonstrate interface design. In subsequent sections other useful devices such as op amps and capacitors and inductors will be used to design suitable interfaces. In a design situation the engineer must choose the resistance values in a proposed circuit. This decision is influenced by a host of practical considerations such as standard values, standard tolerances, manufacturing methods, power limitations, and parasitic elements.

Example 9
Select the load resistance in Fig. 31 so the interface signals are in the range defined by $v \geq 4$ V and $i \geq 30$ mA.

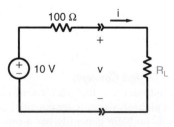

Figure 31 Circuit.[1]

Solution

In this design problem the source circuit is given and a suitable load needs to be selected. For a fixed source the maximum signal levels available at the interface are

$$v_{\text{MAX}} = v_T = 10 \text{ V} \qquad i_{\text{MAX}} = \frac{v_T}{R_T} = 100 \text{ mA}$$

The bounds given as design requirements are below the maximum available signal levels a suitable resistor can be found. Using voltage division, the interface voltage constraint requires

$$v = \frac{R_L}{100 + R_L} \times 10 \geq 4 \quad \text{or} \quad v = 10R_L \geq 4R_L + 400$$

This condition yields $R_L \geq 400/6 = 66.7 \ \Omega$. The interface current constraint can be written as

$$i = \frac{10}{100 + R_L} \geq 0.03 \quad \text{or} \quad i = 10 \geq 3 + 0.03R_L$$

which requires $R_L \leq 7/0.03 = 233 \ \Omega$. In principle any value of R_L between 66.7 and 233 Ω will work. However, to allow for circuit parameter variations, choose $R_L = 150 \ \Omega$ because it lies at the arithmetic midpoint of allowable range and is a standard value.

Example 10

Design the two-port interface circuit in Fig. 32 so that the 10-A source delivers 100 V to the 50-Ω load.

Solution

The problem requires that the current delivered to the load is $i = 100/50 = 2$ A, which is well below the maximum available from the source. In fact, if the 10-A source is connected directly to the load, the source current divides equally between two 50-Ω resistors producing 5 A through the load. Therefore, an interface circuit is needed to reduce the load current to the specified 2-A level. Two possible design solutions are shown in Fig. 32. Applying current division to the parallel-resistor case yields the following constraint:

$$i = \frac{1/50}{1/50 + 1/50 + 1/R_{\text{PAR}}} \times 10$$

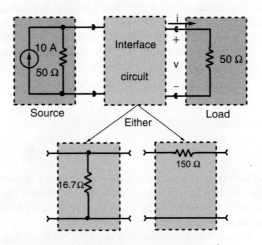

Figure 32 Two-port interface circuit.[1]

For the $i = 2$-A design requirement this equation becomes

$$2 = \frac{10}{2 + 50/R_{\text{PAR}}}$$

Solving for R_{PAR} yields

$$R_{\text{PAR}} = \frac{50}{3} = 16.7 \ \Omega$$

Applying the two-path current division rule to the series-resistor case yields the following constraint:

$$i = \frac{50}{50 + (50 + R_{\text{SER}})} \times 10 = 2 \text{ A}$$

Solving for R_{SER} yields

$$R_{\text{SER}} = 150 \ \Omega$$

Both these two designs meet the basic $i = 2$-A requirement. In practice, engineers evaluate alternative designs using additional criteria. One such consideration is the required power ratings of the resistors in each design. The voltage across the parallel resistor is $v = 100$ V, so the power loss is

$$p_{\text{PAR}} = \frac{100^2}{50/3} = 600 \text{ W}$$

The current through the series-resistor interface is $i = 2$ A so the power loss is

$$p_{\text{SER}} = 2^2 \times 150 = 600 \text{ W}$$

In either design the resistors must have a power rating of at least 600 W. The series resistor is a standard value whereas the parallel resistor is not. Other factors besides power rating and standard size could determine which design should be selected.

Example 11

Design the two-port interface circuit in Fig. 33 so the load is a match to 50 Ω between terminals C and D, while simultaneously the source matches to a load resistance of 300 Ω between A and B.

Solution

No single-resistor interface circuit could work. Hence try an interface circuit containing two resistors. Since the load must see a smaller resistance than the source, it should "look" into a parallel resistor. Since the source must see a larger resistance than the load, it should look into a series resistor. A configuration that meets these conditions is the L circuit shown in Figs. 33*b,c*.

The above discussion can be summarized mathematically. Using the L circuit, the design requirement at terminals C and D is

$$\frac{(R_1 + 300)R_2}{R_1 + 300 + R_2} = 50 \ \Omega$$

At terminals A and B the requirement is

$$R_{11} + \frac{50R_2}{R_2 + 50} = 300 \ \Omega$$

The design requirements yield two equations in two unknowns — what could be simpler? It turns out that solving these nonlinear equations by hand analysis is a bit of a chore. They can easily be solved using a math solver such as MATLAB or MathCad. But a more heuristic approach might serve best.

Given the L circuits in Fig. 33*b*, such an approach goes as follows. Let $R_2 = 50 \ \Omega$. Then the requirement at terminals C and D will be met, at least approximately. Similarly,

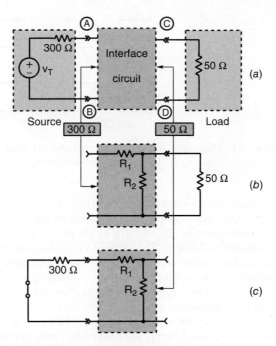

Figure 33 Two-port interface circuit.[1]

if $R_1 + R_2 = 300\ \Omega$, the requirements at terminals A and B will be approximately satisfied. In other words, try $R_1 = 250\ \Omega$ and $R_2 = 50\ \Omega$ as a first cut. These values yield equivalent resistances of $R_{CD} = 50||550 = 45.8\ \Omega$ and $R_{AB} = 250 + 50||50 = 275\ \Omega$. These equivalent resistances are not the exact values specified but are within $\pm 10\%$. Since the tolerance on electrical components may be at least this high, a design using these values could be adequate. The exact values found by a math solver yields $R_1 = 273.861\ \Omega$ and $R_2 = 54.772\ \Omega$.

3 LINEAR ACTIVE CIRCUITS

This section treats the analysis and design of circuits containing active devices such as transistors or operational amplifiers (op amps). An *active device* is a component that requires an external power supply to operate correctly. An *active circuit* is one that contains one or more active devices. An important property of active circuits is that they are capable of providing signal amplification, one of the most important signal-processing functions in electrical engineering. Linear active circuits are governed by the proportionality property so their input–output relationships are of the form $y = Kx$. The term *signal amplification* means the proportionality factor $K > 1$ when the input x and output y have the same dimensions. Thus, active circuits can deliver more signal voltage, current, and power at their output than they receive from the input signal. The passive resistance circuits studied thus far cannot produce voltage, current, or power gains greater than unity.

3.1 Dependent Sources

When active devices operate in a linear mode, they can be modeled using resistors and one or more of the four dependent source elements shown in Fig. 34. The dominant feature of a dependent source is that the strength or magnitude of the voltage source (VS) or current source (CS)

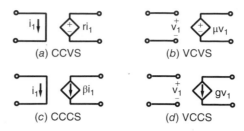

Figure 34   Dependent source circuit symbols: (*a*) current-controlled voltage source, (*b*). voltage-controlled voltage source, (*c*). current-controlled current source, and (*d*). voltage-controlled current source.[1]

is proportional to—that is, controlled by—a voltage (VC) or current (CC) appearing elsewhere in the circuit. For example, the dependent source model for a current-controlled current source (CCCS) is shown in Fig. 34*c*. The output current βi_1 is dependent on the input current i_1 and the dimensionless factor β. This dependency should be contrasted with the characteristics of the independent sources studied earlier. The voltage (current) delivered by an independent voltage (current) source does not depend on the circuit to which it is connected. Dependent sources are often but not always represented by the diamond symbol, in contrast to the circle symbol used for independent sources.

Each dependent source is characterized by a single parameter, μ, β, r, or g. These parameters are often called simply the *gain* of the controlled source. Strictly speaking, the parameters μ and β are dimensionless quantities called the *open-loop voltage gain* and *open-loop current gain*, respectively. The parameter r has the dimensions of ohms and is called the *transresistance*, a contraction of transfer resistance. The parameter g is then called *transconductance* and has the dimensions of siemens.

In every case the defining relationship for a dependent source has the form $y = Kx$, where x is the controlling variable, y is the controlled variable, and K is the element gain. It is this linear relationship between the controlling and controlled variables that make dependent sources linear elements.

Although dependent sources are elements used in circuit analysis, they are conceptually different from the other circuit elements. The linear resistor and ideal switch are models of actual devices called resistors and switches. But dependent sources are not listed in catalogs. For this reason dependent sources are more abstract, since they are not models of identifiable physical devices. Dependent sources are used in combination with other resistive elements to create models of active devices.

A voltage source acts like a short circuit when it is turned off. Likewise, a current source behaves like an open circuit when it is turned off. The same results apply to dependent sources, with one important difference. Dependent sources cannot be turned on and off individually because they depend on excitation supplied by independent sources. When applying the super-position principle or Thévenin's theorem to active circuits, the state of a dependent source depends on excitation supplied by independent sources. In particular, for active circuits the superposition principle states that the response due to all independent sources acting simultaneously is equal to the sum of the responses due to each independent source acting one at a time.

Analysis with Dependent Sources

With certain modifications the circuit analysis tools developed for passive circuits apply to active circuits as well. Circuit reduction applies to active circuits, but the control variable for a dependent source must not be eliminated. Applying a source transformation to a dependent

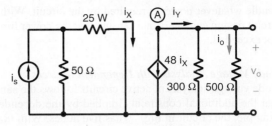

Figure 35 Circuit with dependent source.[1]

source is sometimes helpful. Methods like node and mesh analysis can be adapted to handle dependent sources as well. But the main difference is that the properties of active circuits can be significantly different from those of the passive circuits.

In the following example the objective is to determine the current, voltage, and power delivered to the 500-Ω output load in Fig. 35. The control current i_X is found using current division in the input circuit:

$$i_X = \left(\frac{50}{50 + 25}\right) i_S = \frac{2}{3} i_S$$

Similarly the output current i_O is found using current division in the output circuit:

$$i_O = \left(\frac{300}{300 + 500}\right) i_Y = \frac{3}{8} i_Y$$

But at node A KCL requires that $i_Y = -48 i_X$. Combining this result with the equations for i_X and i_O yields the output current:

$$i_O = (3/8)(-48)i_X = -18(2/3\ i_S) = -12 i_S \tag{22}$$

The output voltage v_O is found using Ohm's law:

$$v_O = i_O \times 500 = -6000 i_S \tag{23}$$

The input–output relationships in Eqs. (22) and (23) are of the form $y = Kx$ with $K < 0$. The proportionality constants are negative because the reference direction for i_O in Fig. 35 is the opposite of the orientation of the dependent source reference mark. Active circuits often produce negative values of K. As a result the input and output signals have opposite signs, a result called *signal inversion*. In the analysis and design of active circuits it is important to keep track of signal inversions.

The delivered output power is

$$p_O = v_O i_O = (-6000 i_S)(-12 i_S) = 72,000 i_S^2$$

The input independent source delivers its power to the parallel combination of 50 and 25 Ω. Hence, the power supplied by the independent source is

$$p_S = (50 || 25)i_S^2 = (50/3)i_S^2$$

Given the input power and output power, we find the power gain in the circuit:

$$\text{Power gain} = \frac{p_O}{p_S} = \frac{72,000 i_S^2}{(50/3)i_S^2} = 432$$

A power gain greater than unity means that the circuit delivers more power at its output than it receives from the input source. At first glance, this appears to be a violation of energy conservation, but dependent sources are models of active devices that require an external power supply to operate. In general, circuit designers do not show the external power supply in circuit diagrams. Control source models assume that the external supply and the active device can handle whatever power is required by the circuit. With real devices this is not the case, and in circuit design engineers must ensure that the power limits of the device and external supply are not exceeded.

Node Voltage Analysis with Dependent Sources

Node voltage analysis of active circuits follows the same process as for passive circuits except that the additional constraints implied by the dependent sources must be accounted for. For example, the circuit in Fig. 36 has five nodes. With node E as the reference both independent voltage sources are connected to ground and force the condition $v_A = v_{S_1}$ and $v_B = v_{S2}$.

Node analysis involves expressing element currents in terms of the node voltages and applying KCL at each unknown node. The sum of the currents *leaving* node C is

$$\frac{v_C - v_{S1}}{R_1} + \frac{v_C - v_{S2}}{R_2} + \frac{v_C}{R_B} + \frac{v_C - v_D}{R_p} = 0$$

Similarly, the sum of currents leaving node D is

$$\frac{v_D - v_C}{R_P} + \frac{v_D}{R_E} - \beta i_B = 0$$

These two node equations can be rearranged into the forms

$$\text{Node } C: \quad v_C \left(\frac{1}{R_1} + \frac{1}{R_2} + \frac{1}{R_B} + \frac{1}{R_P} \right) - \frac{1}{R_P} v_D = \frac{1}{R_1} v_{S1} + \frac{1}{R_2} v_{S2}$$

$$\text{Node } D: \quad -\frac{1}{R_P} v_C + \left(\frac{1}{R_P} + \frac{1}{R_E} \right) v_D = \beta i_B$$

Applying the fundamental property of node voltages and Ohm's law, the current i_B can be written in terms of the node voltages as

$$i_B = \frac{v_C - v_D}{R_P}$$

Substituting this expression for i_B into the above node equation and putting the results in standard form yield

$$\text{Node } C: \quad \left[\left(\frac{1}{R_1} + \frac{1}{R_2} + \frac{1}{R_B} + \frac{1}{R_P} \right) v_C - \frac{1}{R_P} v_D \atop R_P + \frac{1}{R_E} \right] = \frac{1}{R_1} v_{S1} + \frac{1}{R_2} v_{S2}$$

$$\text{Node } D: \quad -(\beta + 1)\frac{1}{R_P} v_C + \left[(\beta + 1)\frac{1}{R_P} + \frac{1}{R_E} \right] v_D = 0$$

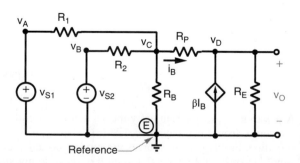

Figure 36 Circuit used for node voltage analysis with dependent sources.[1]

The final result involves two equations in the two unknown node voltages and includes the effect of the dependent source.

This example illustrates a general approach to writing node voltage equations for circuits with dependent sources. Dependent sources are initially treated as if they are independent sources and node equations written for the resulting passive circuit. This step produces a set of symmetrical node equations with the independent and dependent source terms on the right side. Next the dependent source terms are expressed in terms of the unknown node voltages and moved to the left side with the other terms involving the unknowns. The last step destroys the coefficient symmetry but leads to a set of equations that can be solved for the active circuit response.

Mesh Current Analysis with Dependent Sources

Mesh current analysis of active circuits follows the same pattern noted for node voltage analysis. Treat the dependent sources initially as independent sources and write the mesh equations of the resulting passive circuit. Then account for the dependent sources by expressing their constraints in terms of unknown mesh currents. The following example illustrates the method.

Example 12

a. Formulate mesh current equations for the circuit in Fig. 37.

b. Use the mesh equations to find v_O and R_{IN} when $R_1 = 50\ \Omega, R_2 = 1\ k\Omega, R_3 = 100\ \Omega, R_4 = 5\ k\Omega$, and $g = 100\ mS$.

Solution

a. Applying source transformation to the parallel combination of R_3 and gv_X in Fig. 37a produces the dependent voltage source $R_{3gv_X} = \mu v_X$ in Fig. 37b. In the modified circuit we have identified two mesh currents. Initially treating the dependent source $(gR_3)v_x$ as an independent source leads to two symmetrical mesh equations:

$$\text{Mesh } A: \quad (R_1 + R_2 + R_3)i_A - R_3 i_B = v_S - (gR_3)v_X$$

$$\text{Mesh } B: \quad -R_3 i_A + (R_3 + R_4)i_B = (gR_3)v_X$$

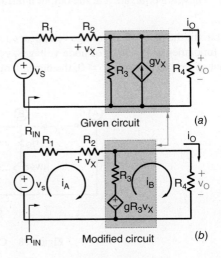

Figure 37   Circuit.[1]

The control voltage v_x can be written in terms of mesh currents as

$$v_X = R_2 i_A$$

Substituting this equation for v_x into the mesh equations and putting the equations in standard form yield

$$(R_1 + R_2 + R_3 + gR_2R_3)i_A - R_3i_B = v_S$$
$$-(R_3 + gR_2R_3)i_A + (R_3 + R_4)i_B = 0$$

The resulting mesh equations are not symmetrical because of the controlled source.

b. Substituting the numerical values into the mesh equations gives

$$(1.115 \times 10^4)i_A - 10^2 i_B = v_S$$
$$-(1.01 \times 10^4)i_A + (5.1 \times 10^3)i_B = 0$$

Using Cramer's rule the mesh currents are found to be

$$i_A = (0.9131 \times 10^{-4})v_S \quad \text{and} \quad i_B = (1.808 \times 10^{-4})v_S$$

The output voltage and input resistance are found using Ohm's law:

$$v_O = R_4 i_B = 0.904 v_S R_{IN} = \frac{v_S}{i_A} = 10.95 \text{ k}\Omega$$

Thévenin Equivalent Circuits with Dependent Sources

To find the Thévenin equivalent of an active circuit, the independent sources are left on or else one must supply excitation from an external test source. This means that the Thévenin resistance cannot be found by the "look-back" method, which requires that all independent sources be turned off. Turning off the independent sources deactivates the dependent sources as well and can result in a profound change in input and output characteristics of an active circuit. Thus, Thévenin equivalents of active circuits can be found using the open-circuit voltage and short-circuit current at the interface.

Example 13

Find the Thévenin equivalent at the output interface of the circuit in Fig. 38.

Solution

In this circuit the controlled voltage v_X appears across an open circuit between nodes A and B. By the fundamental property of node voltages, $v_X = v_S - v_O$. With the load disconnected and the input source turned off, $v_x = 0$, the dependent voltage source μv_X acts like a short circuit,

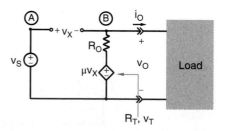

Figure 38   Circuit.[1]

and the Thévenin resistance looking back into the output port is R_O. With the load connected and the input source turned on, the sum of currents leaving node B is

$$\frac{v_O - \mu v_X}{R_O} + i_O = 0$$

Using the relationship $v_X = v_S - v_O$ to eliminate v_X and then solving for v_O produce the output i–v relationship of the circuit as

$$v_O = \frac{\mu v_S}{\mu + 1} - i_O \left[\frac{R_O}{\mu + 1} \right]$$

The i–v relationship of a Thévenin circuit is $v = v_T - iR_T$. By direct comparison, the Thévenin parameters of the active circuit are found to be

$$v_T = \frac{\mu v_S}{\mu + 1} \quad \text{and} \quad R_T = \frac{R_O}{\mu + 1}$$

The circuit in Fig. 38 is a model of an op amp circuit called a voltage follower. The resistance R_O for a general-purpose op amp is on the order of 100 Ω, while the gain μ is about 10^5. Thus, the active Thévenin resistance of the voltage follower is not 100 Ω, as the look-back method suggests, but around a milliohm!

3.2 Operational Amplifier

The operational amplifier (op amp) is the premier linear active device made available by IC technology. John R. Ragazzini apparently first used the term operational amplifier in a 1947 paper and his colleagues who reported on work carried out for the National Defenses Research Council during World War II. The paper described high-gain dc amplifier circuits that perform mathematical operations (addition, subtraction, multiplication, division, integration, etc.)—hence the name "operational" amplifier. For more than a decade the most important applications were general- and special-purpose analog computers using vacuum tube amplifiers. In the early 1960s general-purpose, discrete-transistor, op amp became readily available and by the mid-1960s the first commercial IC op amps entered the market. The transition from vacuum tubes to ICs resulted in a decrease in size, power consumption, and cost of op amps by over three orders of magnitude. By the early 1970s the IC version became the dominant active device in analog circuits. The device itself is a complex array of transistors, resistors, diodes, and capacitors all fabricated and interconnected on a single silicon chip. In spite of its complexity, the op amp can be modeled by rather simple i–v characteristics.

Op-Amp Notation
Certain matters of notation and nomenclature must be discussed before developing a circuit model for the op amp. The op amp is a five-terminal device, as shown in Fig. 39a. The "+" and "−" symbols identify the input terminals and are a shorthand notation for the noninverting and inverting input terminals, respectively. These + and − symbols identify the two input terminals and have nothing to do with the polarity of the voltages applied. The other terminals are the output and the positive and negative supply voltage, usually labeled $+V_{CC}$ and $-V_{CC}$. While some op amps have more than five terminals, these five are always present. Figure 39b shows how these terminals are arranged in a common eight-pin IC package.

While the two power supply terminals in Fig. 39 are not usually shown in circuit diagrams, they are always there because the external power supplies connected to these terminals make the op amp an active device. The power required for signal amplification comes through

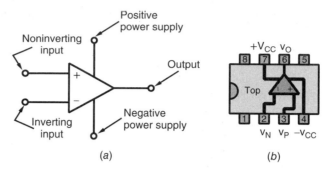

Figure 39 Op amp: (*a*) circuit symbol and (*b*) pin out diagram for eight-pin package.[1]

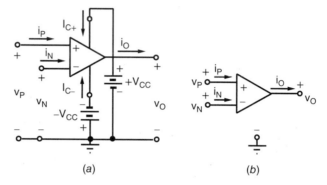

Figure 40 Op-amp voltage and current definitions: (*a*) complete set and (*b*) shorthand set.[1]

these terminals from an external power source. The $+V_{CC}$ and $-V_{CC}$ voltages applied to these terminals also determine the upper and lower limits on the op-amp output voltage.

Figure 40*a* shows a complete set of voltage and current variables for the op amp, while Fig. 40*b* shows the typical abbreviated set of signal variables. All voltages are defined with respect to a common reference node, usually ground. Voltage variables v_P, v_N, and v_O are defined by writing a voltage symbol beside the corresponding terminals. This notation means the + reference mark is at the terminal in question and the − reference mark is at the reference or ground terminal. The reference directions for the currents are directed in at input terminals and out at the output. A global KCL equation for the complete set of variable in Fig. 40*a* is $i_O = I_{C+} + I_{C-} + i_P + i_N$, NOT $i_O = i_N + i_P$, as might be inferred from Fig. 40*b*, since it does not include all of the currents. More importantly, it implies that the output current comes from the inputs. In fact, this is wrong. The input currents are very small, ideally zero. The output current comes from the supply voltages even though these terminals are not shown on the abbreviated circuit diagram.

Transfer Characteristics

The dominant feature of the op amp is the transfer characteristic shown in Fig. 41. This characteristic provides the relationships between the *noninverting input* v_P, the *inverting input* v_N, and the *output voltage* v_O. The transfer characteristic is divided into three regions or modes called +saturation, −saturation, and *linear*. In the linear region the op amp is a *differential amplifier* because the output is proportional to the difference between the two inputs. The slope of

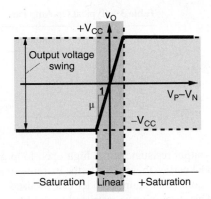

Figure 41   Op-amp transfer characteristics.[1]

the line in the linear range is called the *open-loop gain*, denoted as μ. In the linear region the input–output relation is $v_O = \mu(v_P - v_N)$. The open-loop gain of an op amp is very large, usually greater than 10^5. As long as the net input $v_P - v_N$ is very small, the output will be proportional to the input. However, when $\mu|v_P - v_N| > V_{CC}$, the op amp is saturated and the output voltage is limited by the supply voltages (less some small internal losses).

The op amp has three operating modes:

1. $+$ Saturation mode when $\mu(v_P - v_N) > +V_{CC}$ and $v_O = +V_{CC}$.
2. $-$ Saturation mode when $\mu(v_P - v_N) < -V_{CC}$ and $v_O = -V_{CC}$.
3. Linear mode when $\mu|v_P - v_N| < V_{CC}$ and $v_O = \mu(v_P - v_N)$.

Usually op-amp circuits are analyzed and designed using the linear mode model.

Ideal Op-Amp Model

A controlled source model of an op amp operating in its linear range is shown in Fig. 42. This model includes an input resistance (R_I), an output resistance (R_O), and a voltage-controlled voltage source whose gain is the open-loop gain μ. Some typical ranges for these op-amp parameters are given in Table 3, along with the values for the ideal op amp. The high input

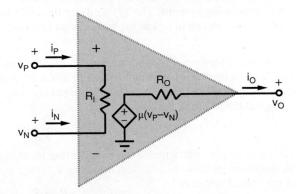

Figure 42   Dependent source model of op amp operating in linear mode.[1]

Table 3 Typical Op-Amp Parameters

Name	Parameter	Range	Ideal Values
Open-loop gain	μ	$10^5 - 10^8$	∞
Input resistance	R_I	$10^6 - 10^{13}\,\Omega$	$\infty\,\Omega$
Output resistance	R_O	$10 - 100\ \Omega$	$0\ \Omega$
Supply voltages	V_{CC}	± 5 to $\pm 40\,\mathrm{V}$	

and low output resistances and high open-loop gain are the key attributes of an op amp. The ideal model carries these traits to the extreme limiting values.

The controlled source model can be used to develop the i–v relationships of the ideal model. This discussion is restricted to the linear region of operation. This means the output voltage is bounded as

$$-V_{CC} \le v_O \le +V_{CC} - \frac{V_{CC}}{\mu} \le (v_P - v_N) \le +\frac{V_{CC}}{\mu}$$

The supply voltages V_{CC} are most commonly ± 15 V, although other supply voltages are available, while μ is a very large number, usually 10^5 or greater. Consequently, linear operation requires that $v_P \cdot v_N$. For the ideal op amp the open-loop gain is infinite ($\mu \to \infty$), in which case linear operation requires $v_P = v_N$. The input resistance R_I of the ideal op amp is assumed to be infinite, so the currents at both input terminals are zero. In summary, the i–v relationships of the *ideal model* of the op amp are

$$v_P = v_N i_P = i_N = 0 \tag{24}$$

At first glance the element constraints of the ideal op amp appear to be fairly useless. They actually look more like connection constraints and are totally silent about the output quantities (v_O and i_O), which are usually the signals of interest. In fact, they seem to say that the op-amp input terminals are simultaneously a short circuit ($v_P = v_N$) and an open circuit ($i_P = i_N = 0$). The ideal model of the op amp is useful because in linear applications feedback is always present. That is, in order for the op amp to operate in a linear mode, it is necessary that there be feedback paths from the output to one or both of the inputs. These feedback paths ensure that $v_P = v_N$ and allow for analysis of op-amp circuits using the ideal op-amp element constraints.

Op-Amp Circuit Analysis

This section introduces op-amp circuit analysis using circuits that are building blocks for analog signal-processing systems. The key to using the *building block* approach is to recognize the feedback pattern and to isolate the basic circuit as a building block.

Noninverting Op Amp. To illustrate the effects of feedback, consider the circuit in Fig. 43. This circuit has a feedback path from the output to the inverting input via a voltage divider. Since the ideal op amp draws no current at either input ($i_P = i_N = 0$), voltage division determines the voltage at the inverting input as

$$v_N = \frac{R_2}{R_1 + R_2} v_O$$

The input source connection at the noninverting input requires the condition $v_P = v_S$. But the ideal op-amp element constraints demand that $v_P = v_N$; therefore, the input–output relationship of the overall circuit is

$$v_O = \frac{R_1 + R_2}{R_2} v_S \tag{25}$$

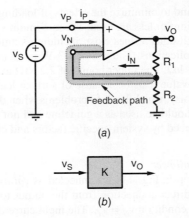

Feedback path

(a)

(b)

Figure 43 Noninverting amplifier circuit.[1]

The circuit in Fig. 43*a* is called a *noninverting amplifier*. The input–output relationship is of the form $v_O = Kv_S$, a linear relationship. Figure 43*b* shows the functional building block for this circuit, where the proportionality constant K is

$$K = \frac{R_1 + R_2}{R_2} \tag{26}$$

The constant K is called the *closed-loop* gain since it includes the effect of the feedback path. When discussing op-amp circuits, it is necessary to distinguish between two types of gains. The first is the open-loop gain μ provided by the op-amp device. The gain μ is a large number with a large uncertainty tolerance. The second type is the closed-loop gain K of the op-amp circuit with a feedback path. The gain K must be smaller than μ, typically no more than 1/100 of μ, and its value is determined by the resistance elements in the feedback path.

For example, the closed-loop gain in Eq. (26) is really the voltage division rule upside down. The uncertainty tolerance assigned to K is determined by the quality of the resistors in the feedback path and not the uncertainty in the actual value of the closed-loop gain. In effect, feedback converts a very large but imprecisely known open-loop gain into a much smaller but precisely controllable closed-loop gain.

Example 14
Design an amplifier with a closed-loop gain $K = 10$.

Solution
Using a noninverting op-amp circuit, the design problem is to select the values of the resistors in the feedback path. From Eq. (26) the design constraint is

$$10 = \frac{R_1 + R_2}{R_2}$$

This yields one constraint with two unknowns. Arbitrarily selecting $R_2 = 10\ \text{k}\Omega$ makes $R_1 = 90\ \text{k}\Omega$. These resistors would normally have high precision ($\pm 1\%$ or less) to produce a precisely controlled closed-loop gain.

Comment. The problem of choosing resistance values in op-amp circuit design problems deserves some discussion. Although resistances from a few ohms to several hundred megohms are commercially available, generally designers limit themselves to the range from about 1 kΩ to perhaps 2.2 MΩ. The lower limit of 1 kΩ exists in part because of power dissipation in the resistors and to minimize the effects of loading (discussed later). Typically resistors with 3-W power ratings or less are used. The maximum voltages in op-amp circuits are often around ±15 V although other values exist, including single-sided op amps, with a 0–5-V V_{CC} for use in digital applications. The smallest 3-W resistance we can use is $R_{MIN} > (15)^2/0.25 = 900\ \Omega$, or about 1 k$\Omega$. The upper bound of 2.2 MΩ exists because it is difficult to maintain precision in a high-value resistor because of surface leakage caused by humidity. High-value resistors are also noisy, which leads to problems when they are in the feedback path. The range 1 kΩ to 2.2 MΩ should be used as a guideline and not an inviolate design rule. Actual design choices are influenced by system-specific factors and changes in technology.

Voltage Follower

The op amp in Fig. 44a is connected as *voltage follower* or *buffer*. In this case the feedback path is a direct connection from the output to the inverting input. The feedback connection forces the condition $v_N = v_O$. The input current $i_P = 0$ so there is no voltage across the source resistance R_S. Applying KVL results in $v_P = v_S$. The ideal op-amp model requires $v_P = v_N$, so that $v_O = v_S$. By inspection the closed-loop gain is $K = 1$. The output exactly equals the input, that is, the output follows the input, and hence the name voltage follower.

The voltage follower is used in interface circuits because it isolates the source and load—hence its other name, *buffer*. Note that the input–output relationship $v_O = v_S$ does not depend on the source or load resistance. When the source is connected directly to the load as in Fig. 44b, the voltage delivered to the load depends on R_S and R_L. The source and load interaction limits the signals that can transfer across the interface. When the voltage follower is inserted between the source and load, the signal levels are limited by the capability of the op amp.

Inverting Amplifier

The circuit in Fig. 45 is called an *inverting amplifier*. The key feature of this circuit is that the input signal and the feedback are both applied at the inverting input. Note that the noninverting input is grounded, making $v_P = 0$. Using the fundamental property of node voltages and KCL,

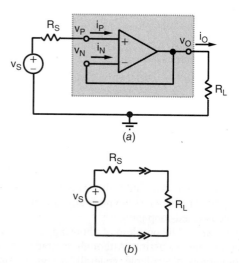

Figure 44 (a) Source–load interface with voltage follower and (b) interface without voltage follower.[1]

Feedback path

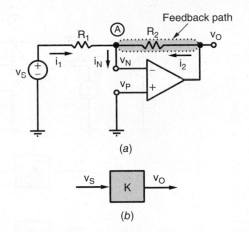

(a)

(b)

Figure 45   Inverting amplifier circuit.[1]

the sum of currents entering node A can be written as

$$\frac{v_S - v_N}{R_1} + \frac{v_O - v_N}{R_2} - i_N = 0$$

The element constraints for the op amp are $v_P = v_N$ and $i_P = i_N = 0$. Since $v_P = 0$, it follows that $v_N = 0$. Substituting the op amp constraints and solving for the input–output relationship yield

$$v_O = -\left(\frac{R_2}{R_1}\right) v_S \tag{27}$$

This result is of the form $v_O = Kv_S$, where K is the closed-loop gain. However, in this case the closed-loop gain $K = -R_2/R_1$ is negative, indicating a signal inversion—hence the name inverting amplifier. The block diagram symbol shown in Fig. 45b is used to indicate either the inverting or noninverting op-amp configuration since both circuits provide a gain of K.

The op amp constraints mean that the input current i_1 in Fig. 43a is

$$i_1 = \frac{v_S - v_N}{R_1} = \frac{v_S}{R_1}$$

This in turn means that the input resistance seen by the source v_S is

$$R_{\mathrm{IN}} = \frac{v_S}{i_1} = R_1 \tag{28}$$

In other words, the inverting amplifier has as finite input resistance determined by the external resistor R_1. This finite input resistance must be taken into account when analyzing circuits with op amps in the inverting amplifier configuration.

Summing Amplifier

The *summing amplifier* or *adder* circuit is shown in Fig. 46a. This circuit has two inputs connected at node A, which is called the *summing point*. Since the noninverting input is grounded, $v_P = 0$. This configuration is similar to the inverting amplifier; hence a similar analysis yields the circuit input–output relationship

$$v_O = \begin{cases} \left(-\frac{R_F}{R_1'}\right) v_1 + \left(-\frac{R_F}{R_2}\right) v_2 (1) \\ (K_1)v_1 + (K_2)v_2 \end{cases} \tag{29}$$

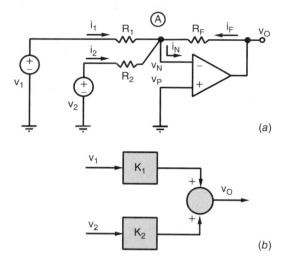

Figure 46   Inverting summer.[1]

The output is a weighted sum of the two inputs. The scale factors, or gains as they are called, are determined by the ratio of the feedback resistor R_F to the input resistor for each input: that is, $K_1 = -R_F/R_1$ and $K_2 = -R_F/R_2$. In the special case $R_1 = R_2 = R$, Eq. (29) reduces to

$$v_O = K(v_1 + v_2)$$

where $K = -R_F/R$. In this special case the output is proportional to the negative sum of the two inputs—hence the name inverting summing amplifier or simply adder. A block diagram representation of this circuit is shown in Fig. 46b.

Example 15

Design an inverting summer that implements the input–output relationship $v_O = -(5v_1 + 13v_2)$.

Solution

The design problem involves selecting the input and feedback resistors so that

$$\frac{R_F}{R_1} = 5 \quad \text{and} \quad \frac{R_F}{R_2} = 13$$

One solution is to arbitrarily select $R_F = 65$ kΩ, which yields $R_1 = 13$ kΩ and $R_2 = 5$ kΩ. The resulting circuit is shown in Fig. 47a. The design can be modified to use standard resistance values for resistors with ±5% tolerance. Selecting the standard value $R_F = 56$ kΩ requires $R_1 = 11.2$ kΩ and $R_2 = 4.31$ kΩ. The nearest standard values are 11 and 4.3 kΩ. The resulting circuit shown in Fig. 47b uses only standard value resistors and produces gains of $K_1 = 56/11 = 5.09$ and $K_2 = 56/4.3 = 13.02$. These nominal gains are within 2% of the values in the specified input–output relationship.

Differential Amplifier

The circuit in Fig. 48a is called a *differential amplifier* or *subtractor*. Like the summer, this circuit has two inputs, but unlike the summer, one is applied at the inverting input and one at the noninverting input of the op amp. The input–output relationship can be obtained using the superposition principle.

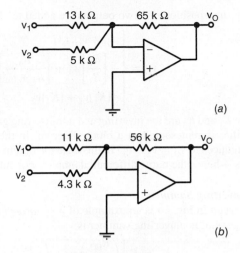

Figure 47 Circuit.[1]

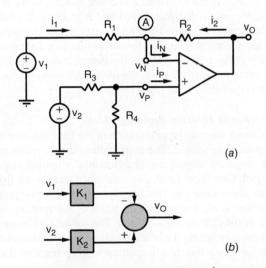

Figure 48 Differential amplifier.[1]

With source v_2 off there is no excitation at the noninverting input and $v_P = 0$. In effect, the noninverting input is grounded and the circuit acts like an inverting amplifier with the result that

$$v_{O1} = -\frac{R_2}{R_1}v_1$$

Now turning v_2 back on and turning v_1 off, the circuit looks like a noninverting amplifier with a voltage divider connected at its input. Thus

$$v_{O2} = \left[\frac{R_4}{R_3 + R_4}\right]\left[\frac{R_1 + R_2}{R_1}\right]v_2$$

Using superposition, the two outputs are added to obtain the output with both sources on:

$$v_O = \begin{cases} v_{O1} + v_{O2} \ (3) \\[2mm] -\left[\dfrac{R_2}{R_1}\right]v_1 + \left[\dfrac{R_4}{R_3 + R_4}\right]\left[\dfrac{R_1 + R_2}{R_1}\right]v_2 \ (4) \\[2mm] -[K_1]v_1 + [K_2]v_2 \end{cases} \qquad (30)$$

where K_1 and K_2 are the inverting and noninverting gains. Figure 48*b* shows how the differential amplifier is represented in a block diagram. In the special case of $R_1 = R_2 = R_3 = R_4$, Eq. (30) reduces to $v_O = v_2 - v_1$. In this case the output is equal to the difference between the two inputs—hence the name differential amplifier or subtractor.

Noninverting Summer

The circuit in Fig. 49 is an example of a *noninverting summer*. The input–output relationship for a general noninverting summer is

$$v_O = K\left[\left(\dfrac{R_{EQ}}{R_1}\right)v_1 + \left(\dfrac{R_{EQ}}{R_2}\right)v_2 + \cdots + \left(\dfrac{R_{EQ}}{R_n}\right)v_n\right] \qquad (31)$$

where R_{EQ} is the Thévenin resistance looking to the left at point P with all sources turned off (i.e., $R_{EQ} = R_1||R_2||R_3\ldots||R_n$) and K is the gain of the noninverting amplifier circuit to the right of point P. Comparing this equation with the general inverting summer result in Eq. (29), we see several similarities. In both cases the weight assigned to an input voltage is proportional to a resistance ratio in which the denominator is its input resistance. In the inverting summer the numerator of the ratio is the feedback resistor R_F and in the noninverting case the numerator is the equivalent of all input resistors R_{EQ}.

Design with Op-Amp Building Blocks

The block diagram representation of the basic op-amp circuit configurations were developed in the preceding section. The noninverting and inverting amplifiers are represented as gain blocks. The summing amplifier and differential amplifier require both gain blocks and the summing symbol. One should exercise care when translating from a block diagram to a circuit, or vice versa, since some gain blocks may involve negative gains. For example, the gain of the inverting amplifier is negative, as are the gains of the common inverting summing amplifier and the K_1 gain of the differential amplifier. The minus sign is sometimes moved to the summing symbol and the gain within the block changed to a positive number. Since there is no standard convention for doing this, it is important to keep track of the signs associated with gain blocks and summing point symbol.

Operational amplifier circuit design generally requires that a given equation or block diagram representation of a signal-processing function be created to implement that function. Circuit design can often be accomplished by interconnecting the op amp, summer, and subtractor

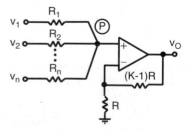

Figure 49   Noninverting summer.[2]

building blocks. The design process is greatly simplified by the near one-to-one correspondence between the op-amp circuits and the elements in a block diagram. However, the design process is not unique since often there are several ways to use basic op-amp circuits to meet the design objective. Some solutions are better than others. The following example illustrates the design process.

Example 16
Design an op-amp circuit that implements the block diagram in Fig. 50.

Solution
The input–output relationship represented by the block diagram is $v_O = 5v_1 + 10v_2 + 20v_3$. An op-amp adder can implement the summation required in this relationship. A three-input adder implements the relationship

$$v_O = -\left[\frac{R_F}{R_1}v_1 + \frac{R_F}{R_2}v_2 + \frac{R_F}{R_3}v_3\right]$$

The required scale factors are realized by first selecting $R_F = 100$ kΩ and then choosing $R_1 = 20$ kΩ, $R_2 = 10$ kΩ, and $R_3 = 5$ kΩ. However, the adder involves a signal inversion. To correctly implement the block diagram, we must add an inverting amplifier $(K = -R_2/R_1)$ with $R_1 = R_2 = 100$ kΩ. The final implementation is shown in Fig. 51a. An alternate solution avoiding the second inverting op amp by using a noninverting summer is shown in Fig. 51b.

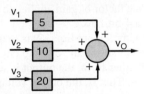

Figure 50   Block diagram.[2]

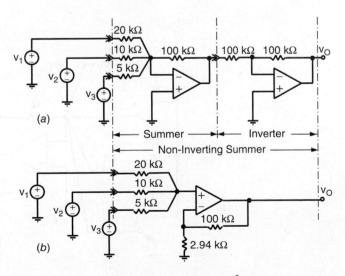

Figure 51   Block diagram.[2]

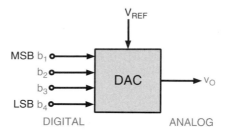

Figure 52 A DAC.[1]

Digital-to-Analog Converters

Operational amplifiers play an important role in the interface between digital systems and analog systems. The parallel 4-bit output in Fig. 52 is a digital representation of a signal. Each bit can only have two values: (a) a high or 1 (typically +5 V) and (b) a low or 0 (typically 0 V). The bits have binary weights so that v_1 is worth $2^3 = 8$ times as much v_4, v_2 is worth $2^2 = 4$ times as much as v_4, v_3 is worth $2^1 = 2$ times as much v_4, and v_4 is equal to $2^0 = 1$ times itself. In a 4-bit digital-to-analog converter (DAC) v_4 is the least significant bit (LSB) and v_1 the most significant bit (MSB). To convert the digital representation of the signal to analog form, each bit must be weighted so that the analog output v_O is

$$v_O = \pm K(8v_1 + 4v_2 + 2v_3 + 1v_4) \tag{32}$$

where K is a scale factor or gain applied to the analog signal. Equation (32) is the input–output relationship of a 4-bit DAC.

One way to implement Eq. (32) is to use an inverting summer with binary-weighted input resistors. Figure 53 shows the op-amp circuit and a block diagram of the circuit input–output relationship. In either form, the output is seen to be a binary-weighted sum of the digital input scaled by $-R_F/R$. That is, the output voltage is

$$v_O = \frac{-R_F}{R}(8v_1 + 4v_2 + 2v_3 + v_4)$$

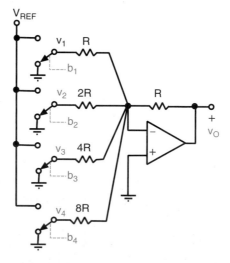

Figure 53 Binary-weighted summer DAC.[1]

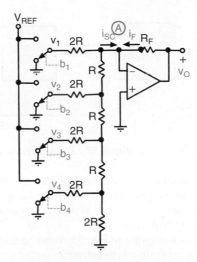

Figure 54 An R–$2R$ ladder DAC.[1]

The R–$2R$ ladder in Fig. 54 also implements a 4-bit DAC. The resistance seen looking back into the R–$2R$ ladder at point A with all sources turned off is seen to be $R_T = R$. A Thévenin equivalent voltage of the R–$2R$ network is

$$v_T = \tfrac{1}{2}v_1 + \tfrac{1}{4}v_2 + \tfrac{1}{8}v_3 + \tfrac{1}{16}v_4$$

The output voltage is found using the inverting amplifier gain relationship:

$$v_O = \frac{-R_F}{R}v_T = \frac{-R_F}{R}\left(\frac{v_1}{2} + \frac{v_2}{4} + \frac{v_3}{8} + \frac{v_4}{16}\right)$$

Using $R_F = 16R$ yields

$$v_O = -(8v_1 + 4v_2 + 2v_3 + v_4)$$

which shows the binary weights assigned to the digital inputs.

In theory the circuits in Figs. 53 and 54 perform the same signal-processing function— 4-bit digital-to-analog conversion. However, there are important practical differences between the two circuits. The inverting summer in Fig. 53 requires precision resistors with four different values spanning an 8 : 1 range. A more common 8-bit converter would require eight precision resistors spanning a 256 : 1 range. Moreover, the digital voltage sources in Fig. 53 see input resistances that span an 8 : 1 range; therefore, the source–load interface is not the same for each bit. On the other hand, the resistances in the R–$2R$ ladder converter in Fig. 54 span only a 2 : 1 range regardless of the number of digital bits. The R–$2R$ ladder also presents the same input resistance to each binary input. The R–$2R$ ladder converters are readily made on integrated or thin-film circuits and are the preferred DAC type.

Instrumentation Systems

One of the most interesting and useful applications of op-amp circuits is in instrumentation systems that collect and process data about physical phenomena. In such a system an *input transducer* (a device that converts some physical quantity, such as temperature, strain, light intensity, acceleration, wavelength, rotation, velocity, pressure, or whatever, into an electrical signal) generates an electrical signal that describes some ongoing physical process. In a simple system the transducer signal is processed by op amp circuits and displayed on an *output*

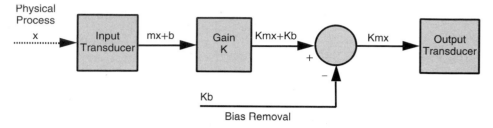

Figure 55   Block diagram of instrumentation system.[2]

transducer such as a meter or an oscilloscope or more commonly sent into a DAC for further processing or analysis by a microprocessor or digital computer. The output signal can also be used in a feedback control system to monitor and regulate the physical process itself or to control a robotic device.

The block diagram in Fig. 55 shows an instrumentation system in its simplest form. The objective of the system is to deliver an output signal that is directly proportional to the physical quantity measured by the input transducer. The input transducer converts a physical variable x into an electrical voltage v_{TR}. For many transducers this voltage is of the form $v_{TR} = mx + b$, where m is a calibration constant and b is a constant offset or bias. The transducer voltage is often quite small and must be amplified by the gain K, as indicated in Fig. 55. The amplified signal includes both a signal component $K(mx)$ and a bias component $K(b)$. The amplified bias $K(b)$ is then removed by subtracting a constant electrical signal. The resulting output voltage $K(mx)$ is directly proportional to the quantity measured and goes to an output transducer for display. The required gain K can be found from the relation

$$K = \frac{\text{desired output range}}{\text{available input range}} \tag{33}$$

Example 17
Design a light intensity detector to detect 5–20 lm of incident light using a photocell serving as the input transducer. The system output is to be displayed on a 0–10-V voltmeter. The photocell characteristics are shown in Fig. 56a. The design requirements are that 5 lm indicates 0 V and 20 lm indicates 10 V on the voltmeter.

Solution
From the transducer's characteristics the light intensity range $\Delta L = 20 - 5 = 15$ lm will produce an available range of $\Delta v = (0.6 \text{ m} - 0.2 \text{ m}) = 0.4$ mV at the system input. This 0.4-mV change must be translated into a 0–10-V range at the system output. To accomplish this, the transducer voltage must be amplified by a gain of

$$K = \frac{\text{desired output range}}{\text{available input range}}$$

$$= \frac{10 - 0}{0.6 \times 10^{-3} - 0.2 \times 10^{-3}} = 2.5 \times 10^4$$

When the transducer's output voltage range (0.2–0.6 mV) is multiplied by the gain K found above, we obtain a voltage range of 5–15 V. This range is shifted to the required 0–10-V range by subtracting the 5-V bias from the amplified signal. A block diagram of the required signal-processing functions is shown in Fig. 56b.

A cascade connection of op-amp circuits is used to realize the signal-processing functions in the block diagram. Figure 57 shows one possible design using an inverting amplifier and an

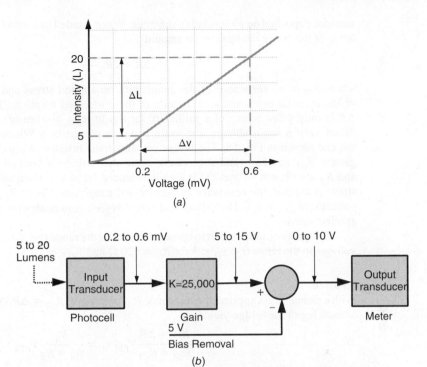

Figure 56   (*a*) Photocell characteristics and (*b*) signal-processing functions.[2]

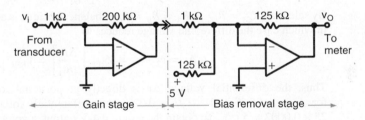

Figure 57   Possible design using an inverting amplifier and an inverting adder.[2]

inverting adder. This design includes two inverting circuits in cascade so the signal inversions cancel in the output signal. Part of the overall gain of $K = 2.5 \times 10^4$ is realized in the inverting amplifier ($K_1 = -200$) and the remainder by the inverting summer ($K_2 = -125$). Dividing the overall gain between the two stages avoids trying to produce too large of a gain in a single stage. A single-stage gain of $K = 25,000$ is not practical since the closed-loop gain is not small compared to the open-loop gain μ of most op amps. The high gain would also require a very low input resistance that could load the input and an uncommonly large feedback resistance, for example, 100 Ω and 2.5 MΩ.

Example 18

A strain gauge is a resistive device that measures the elongation (strain) of a solid material caused by applied forces (stress). A typical strain gauge consists of a thin film of conducting material deposited on an insulating substrate. When bonded to a member under stress, the resistance of the gauge changes by an amount

$$\Delta R = 2R_G \frac{\Delta L}{L}$$

where R_G is the resistance of the gauge with no applied stress and $\Delta L/L$ is the elongation of the material expressed as a fraction of the unstressed length L. The change in resistance ΔR is only a few tenths of a milliohm, far too little to be measured with an ohmmeter. To detect such a small change, the strain gauge is placed in a Wheatstone bridge circuit like the one shown in Fig. 58. The bridge contains fixed resistors R_A and R_B, two matched strain gauges R_{G1} and R_{G2}, and a precisely controlled reference voltage v_{REF}. The values of R_A and R_B are chosen so that the bridge is balanced ($v_1 = v_2$) when no stress is applied. When stress is applied, the resistance of the stressed gauge changes to $R_{G2} + \Delta R$ and the bridge is unbalanced ($v_1 \neq v_2$). The differential signal ($v_2 - v_1$) indicates the strain resulting from the applied stress.

Design an op amp circuit to translate strains on the range $0 < \Delta L/L < 0.02\%$ into an output voltage on the range $0 < v_O < 4$ for $R_G = 120\ \Omega$ and $v_{REF} = 25$ V.

Solution

With external stress applied, the resistance R_{G2} changes to $R_{G2} + \Delta R$. Applying voltage division to each leg of the bridge yields

$$v_2 = \frac{R_{G2} + \Delta R}{R_{G1} + R_{G2}} V_{REF} \qquad v_1 = \frac{R_B}{R_A + R_B} V_{REF}$$

The differential voltage ($\Delta v = v_2 - v_1$) can be written as

$$\Delta v = v_2 - v_1 = V_{REF} \left[\frac{R_{G1} + \Delta R}{R_{G1} + R_{G2}} - \frac{R_A}{R_A + R_B} \right]$$

By selecting $R_{G1} = R_{G2} = R_A = R_B = R_G$, a balanced bridge is achieved in the unstressed state, in which case the differential voltage reduces to

$$\Delta v = v_2 - v_1 = V_{REF} \left[\frac{\Delta R}{2R_G} \right] = V_{REF} \left[\frac{\Delta L}{L} \right]$$

Thus, the differential voltage Δv is directly proportional to the strain $\Delta L/L$. However, for $V_{REF} = 25$ V and $\Delta L/L = 0.02\%$ the differential voltage is only $(V_{REF})(\Delta L/L) = 25 \times 0.0002 = 5$ mV. To obtain the required 4-V output, a voltage gain of $K = 4/0.005 = 800$ is required.

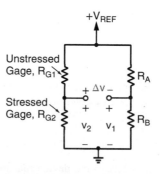

Figure 58   Wheatstone bridge.[2]

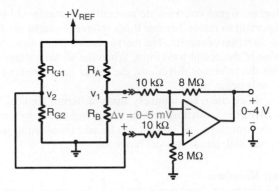

Figure 59 Circuit.[2]

The op-amp subtractor is specifically designed to amplify differential signals. Selecting $R_1 = R_3 = 10$ kΩ and $R_2 = R_4 = 8$ MΩ produces an input–output relationship for the subtractor circuit of

$$v_O = 800(v_2 - v_1)$$

Figure 59 shows the selected design.

The input resistance of the subtractor circuit must be large to avoid loading the bridge circuit. The Thévenin resistance look-back into the bridge circuit is

$$R_T = R_{G1}||R_{G2} + R_A||R_B$$

$$= R_G||R_G + R_G||R_G$$

$$= R_G = 120 \ \Omega$$

which is small compared to 10-kΩ input resistance of the subtractor's inverting input.

Comment. The transducer in this example is the resistor R_{G2}. In the unstressed state the voltage across this resistor is $v_2 = 12.5$ V. In the stressed state the voltage is $v_2 = 12.5$ V plus a 5-mV signal. In other words, the transducer's 5-mV signal component is accompanied by a very large bias. It is important to amplify the 12.5-V bias component by $K = 800$ before subtracting it out. The bias is eliminated at the input by using a bridge circuit in which $v_1 = 12.5$ V and then processing the differential signal $v_2 - v_1$. The situation illustrated in this example is actually quite common. Consequently, the first amplifier stage in most instrumentation systems is a differential amplifier.

4 ALTERNATING CURRENT (AC) CIRCUITS

4.1 Signals

Electrical engineers normally think of a signal as an electrical current $i(t)$, voltage $v(t)$, or power $p(t)$. In any case, the time variation of the signal is called a waveform. More formally, a *waveform* is an equation or graph that defines the signal as a function of time.

Waveforms that are constant for all time are called *dc signals*. The abbreviation dc stands for direct current, but it applies to either voltage or current. Mathematical expressions for a dc voltage $v(t)$ or current $i(t)$ take the form

$$v(t) = V_0 \quad i(t) = I_0 \quad \text{for} \ -\infty < t < \infty$$

Although no physical signal can remain constant forever, it is still a useful model, however, because it approximates the signals produced by physical devices such as batteries. In a circuit diagram signal variables are normally accompanied by reference marks $(+, -, \rightarrow$ or $\leftarrow$). It is important to remember that these reference marks *do not* indicate the polarity of a voltage or the direction of current. The marks provide a baseline for determining the sign of the numerical value of the actual waveform. When the actual voltage polarity or current direction coincides with the reference directions, the signal has a positive value. When the opposite occurs, the value is negative.

Since there are infinitely many different signals, it may seem that the study of signals involves the uninviting task of compiling a lengthy catalog of waveforms. Fortunately, most of the waveforms of interest can be addressed using just three basic signal models: the step, exponential, and sinusoidal functions.

Step Waveform

The first basic signal in our catalog is the step waveform. The general step function is based on the *unit step function* defined as

$$u(t) \equiv \begin{cases} 0 & \text{for } t < 0 \\ 1 & \text{for } t \geq 0 \end{cases}$$

Mathematically, the function $u(t)$ has a jump discontinuity at $t = 0$. While it is impossible to generate a true step function since signal variables like current and voltage cannot transition from one value to another in zero time, it is possible to generate very good approximations to the step function. What is required is that the transition time be short compared with other response times in the circuit. The step waveform is a versatile signal used to construct a wide range of useful waveforms. It often is necessary to turn things on at a time other than zero and with an amplitude different from unity. Replacing t by $t - T_S$ produces a waveform $V_A u(t - T_S)$ that takes on the values

$$V_A u(t - T_S) = \begin{cases} 0 & \text{for } t < T_S \text{ (8)} \\ V_A & \text{for } t \geq T_S \text{(9)} \end{cases} \tag{34}$$

The *amplitude* V_A scales the size of the step discontinuity, and the *time shift* parameter T_S advances or delays the time at which the step occurs.

Amplitude and time shift parameters are required to define the general step function. The amplitude V_A carries the units of volts. The amplitude of the step function in an electric current is I_A and carries the units of amperes. The constant T_S carries the units of time, usually seconds. The parameters V_A (or I_A) and T_S can be positive, negative, or zero, as shown in Fig. 60.

Example 19

Express the waveform in Fig. 61*a* in terms of step functions.

Solution

The amplitude of the pulse jumps to a value of 3 V at $t = 1$ s; therefore, $3u(t - 1)$ is part of the equation for the waveform. The pulse returns to zero at $t = 3$ s, so an equal and opposite

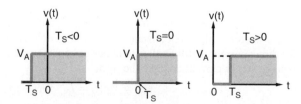

Figure 60   Effect time shifting on step function waveform.[1]

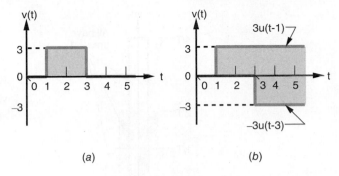

Figure 61   Waveform.[1]

step must occur at $t = 3$ s. Putting these observations together, we express the rectangular pulse as

$$v(t) = 3u(t - 1) - 3u(t - 3)$$

Figure 61b shows how the two step functions combine to produce the given rectangular pulse.

Impulse Function

The generalization of Example 19 is the waveform

$$v(t) = V_A[u(t - T_1) - u(t - T_2)] \tag{35}$$

This waveform is a rectangular pulse of amplitude V_A that turns on at $t = T_1$ and off at $t = T_2$. Pulses that turn on at some time T_1 and off at some later time T_2 are sometimes called *gating functions* because they are used in conjunction with electronic switches to enable or inhibit the passage of another signal.

A rectangular pulse centered on $t = 0$ is written in terms of step functions as

$$v_1(t) = \frac{1}{T}\left[u\left(t + \frac{T}{2}\right) - u\left(t - \frac{T}{2}\right)\right] \tag{36}$$

The pulse in Eq. (36) is zero everywhere except in the range $-T/2 \le t \le T/2$, where its amplitude is $1/T$. The area under the pulse is 1 because its amplitude is inversely proportional to its duration. As shown in Fig. 62a, the pulse becomes narrower and higher as T decreases but maintains its unit area. In the limit as $T \to 0$ the amplitude approaches infinity but the area remains unity. The function obtained in the limit is called a *unit impulse*, symbolized as $\delta(t)$. The graphical representation of $\delta(t)$ is shown in Fig. 62b. The impulse is an idealized model of a large-amplitude, short-duration pulse.

A formal definition of the unit impulse is

$$\delta(t) = 0 \quad \text{for } t \ne 0 \quad \text{and} \quad \int_{-\infty}^{t} \delta(x)\,dx = u(t) \tag{37}$$

The first condition says the impulse is zero everywhere except at $t = 0$. The second condition implies that the impulse is the derivative of a step function, although it cannot be justified using elementary mathematics since the function $u(t)$ has a discontinuity at $t = 0$ and its derivative at that point does not exist in the usual sense. However, the concept can be justified using limiting conditions on continuous functions as discussed in texts on signals and systems.

The strength of an impulse is defined by its area since amplitude is infinite. An impulse of strength K is denoted $K\delta(t)$, where K is the *area* under the impulse. In the graphical representation of the impulse the value of K is written in parentheses beside the arrow, as shown in Fig. 62b.

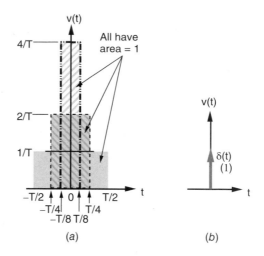

Figure 62 Rectangular pulse waveforms and impulse.[1]

Ramp Function

The *unit ramp* is defined as the integral of a step function:

$$r(t) = \int_{-\infty}^{t} u(x)\,dx = tu(t) \tag{38}$$

The unit ramp waveform $r(t)$ is zero for $T \leq 0$ and is equal to t for $t > 0$. The slope of $r(t)$ is unity. The general ramp waveform is written $Kr(t - T_S)$. The general ramp is zero for $t \leq T_S$ and equal to $K(t - T_S)$ for $t > 0$. The scale factor K defines the slope of the ramp for $t > 0$. By adding a series of ramps the triangular and sawtooth waveforms can be created.

Singularity Functions

The impulse, step, and ramp form a triad of related signals that are referred to as *singularity functions*. They are related by integration or by differentiation as

$$\delta(t) = \int_{-\infty}^{t} \delta'(x)\,dx \; \delta'(t) = \frac{d\delta(t)}{dt} u(t) = \int_{-\infty}^{t} \delta(x)\,dx \; \delta(t) = \frac{du(t)}{dt}$$

$$r(t) = \int_{-\infty}^{t} u(x)\,dx \; u(t) = \frac{dr(t)}{dt} \tag{39}$$

These signals are used to generate other waveforms and as test inputs to linear systems to characterize their responses. When applying the singularity functions in circuit analysis, it is important to remember that $u(t)$ is a dimensionless function. But $\delta(t)$ carries the units of reciprocal seconds and $r(t)$ carries units of seconds. Here, $\delta'(t)$ is called a doublet and is included for completeness. It is the derivative of an impulse function and carries the units of reciprocal seconds squared.

Exponential Waveform

The *exponential signal* is a step function whose amplitude gradually decays to zero. The equation for this waveform is

$$v(t) = [V_A e^{-t/T_C}]u(t) \tag{40}$$

A graph of $v(t)$ versus t/T_C is shown in Fig. 63. The exponential starts out like a step function. It is zero for $t < 0$ and jumps to a maximum amplitude of V_A at $t = 0$. Thereafter it monotonically

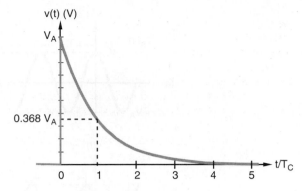

Figure 63   Exponential waveform.

decays toward zero versus time. The two parameters that define the waveform are the *amplitude* V_A (in volts) and the *time constant* T_C (in seconds). The amplitude of a current exponential would be written I_A and carry the units of amperes.

The time constant is of special interest since it determines the rate at which the waveform decays to zero. An exponential decays to about 37% of its initial amplitude $v(0) = V_A$ in one time constant because, at $t = T_C$, $v(T_C) = V_A e^{-1}$ or approximately $0.368V_A$. At $t = 5T_C$, the value of the waveform is $V_A e^{-5}$ or approximately $0.00674V_A$. An exponential signal decays to less than 1% of its initial amplitude in a time span of five time constants. In theory an exponential endures forever, but practically speaking after about $5T_C$ the waveform amplitude becomes negligibly small. For this reason we define the *duration* of an exponential waveform to be $5T_C$.

Decrement Property of Exponential Waveforms.   The *decrement property* describes the decay rate of an exponential signal. For $t > 0$ the exponential waveform is given by $v(t) = V_A e^{-t/T_C}$. At time $t + \Delta t$ the amplitude is

$$v(t + \Delta t) = V_A e^{-(t+\Delta t)/T_C} = V_A e^{-t/T_C} e^{-\Delta t/T_C}$$

The ratio of these two amplitudes is

$$\frac{v(t + \Delta t)}{v(t)} = \frac{V_A e^{-t/T_C} e^{-\Delta t/T_C}}{V_A e^{-t/T_C}} = e^{-\Delta t/T_C} \tag{41}$$

The decrement ratio is independent of amplitude and time. In any fixed time period Δt, the fractional decrease depends only on the time constant. The decrement property states that the same percentage decay occurs in equal time intervals.

Slope Property of Exponential Waveforms.   The slope of the exponential waveform (for $t > 0$) is found by differentiating Eq. (40) with respect to time:

$$\frac{dv(t)}{dt} = -\frac{V_A}{T_C} e^{-t/T_C} = -\frac{v(t)}{T_C} \tag{42}$$

The *slope property* states that the time rate of change of the exponential waveform is inversely proportional to the time constant. Small time constants lead to large slopes or rapid decays, while large time constants produce shallow slopes and long decay times.

Sinusoidal Waveform
The cosine and sine functions are important in all branches of science and engineering. The corresponding time-varying waveform in Fig. 64 plays an especially prominent role in electrical engineering.

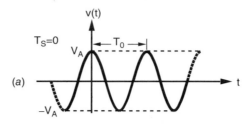

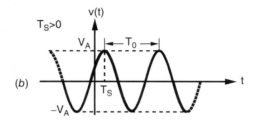

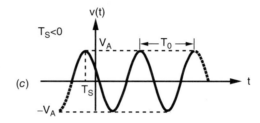

Figure 64   Effect of time shifting on sinusoidal waveform.[1]

In contrast with the step and exponential waveforms studied earlier, the sinusoid extends indefinitely in time in both the positive and negative directions. The sinusoid in Fig. 64 is an endless repetition of identical oscillations between positive and negative peaks. The *amplitude* V_A defines the maximum and minimum values of the oscillations. The *period* T_0 is the time required to complete one cycle of the oscillation. Using these two parameters, a voltage sinusoid can be expressed as

$$v(t) = V_A \cos\left(\frac{2\pi t}{T_0}\right) \text{ V} \qquad (43)$$

The waveform $v(t)$ carries the units of V_A (volts in this case) and the period T_0 carries the units of time t (usually seconds). Equation (43) produces the waveform in Fig. 64, which has a positive peak at $t = 0$ since $v(0) = V_A$.

As in the case of the step and exponential functions, the general sinusoid is obtained by replacing t by $t - T_S$. Inserting this change in Eq. (43) yields a general expression for the sinusoid as

$$v(t) = V_A \cos\left[\frac{2\pi\left(t - T_S\right)}{T_0}\right] \qquad (44)$$

where the constant T_S is the time shift parameter. The sinusoid shifts to the right when $T_S > 0$ and to the left when $T_S < 0$. In effect, time shifting causes the positive peak nearest the origin to occur at $t = T_S$.

The time-shifting parameter can also be represented by an angle:

$$v(t) = V_A \cos\left[\frac{2\pi t}{T_0} + \phi\right] \tag{45}$$

The parameter ϕ is called the *phase angle*. The term phase angle is based on the circular interpretation of the cosine function where the period is divided into 2π radians or $350°$. In this sense the phase angle is the angle between $t = 0$ and the nearest positive peak. The relation between T_S and ϕ is

$$\phi = -2\pi\frac{T_S}{T_0} \tag{46}$$

An alternative form of the general sinusoid is obtained by expanding Eq. (45) using the identity $\cos(x + y) = \cos(x)\cos(y) - \sin(x)\sin(y)$. This results in the general sinusoid being written as

$$v(t) = a\cos\left(\frac{2\pi t}{T_0}\right) + b\sin\left(\frac{2\pi t}{T_0}\right) \tag{47}$$

The two amplitude-like parameters a and b have the same units as the waveform (volts in this case) and are called Fourier coefficients. By definition the Fourier coefficients are related to the amplitude and phase parameters by the equations

$$
\begin{aligned}
a = V_A\cos\phi \qquad V_A = \sqrt{a^2 + b^2} \\
b = -V_A\sin\phi \qquad \phi = \tan^{-1}\frac{-b}{a}
\end{aligned}
\tag{48}
$$

It is customary to describe the time variation of the sinusoid in terms of a frequency parameter. *Cyclic frequency* f_0 is defined as the number of periods per unit time. By definition the period T_0 is the number of seconds per cycle; consequently the number of cycles per second is

$$f_0 = \frac{1}{T_0} \tag{49}$$

where f_0 is the cyclic frequency or simply the frequency. The unit of frequency (cycles per second) is the *hertz* (Hz). Because there are 2π radians per cycle, the *angular frequency* ω_0 in radians per second is related to cyclic frequency by the relationship

$$\omega_0 = 2\pi f_0 = \frac{2\pi}{T_0} \tag{50}$$

In summary, there are several equivalent ways to describe the general sinusoid:

$$
v(t) = \begin{cases}
V_A\cos\left[\dfrac{2\pi(t - T_S)}{T_0}\right] = V_A\cos\left(\dfrac{2\pi t}{T_0} + \phi\right) \\
\qquad\qquad = a\cos\left(\dfrac{2\pi t}{T_0}\right) + b\sin\left(\dfrac{2\pi t}{T_0}\right) \\
V_A\cos[2\pi f_0(t - T_S)] = V_A\cos(2\pi f_0 t + \phi) \\
\qquad\qquad = a\cos(2\pi f_0 t + \phi) + b\sin(2\pi f_0 t + \phi) \\
V_A\cos[\omega_0(t - T_S)] = V_A\cos(\omega_0 t + \phi) \\
\qquad\qquad = a\cos(\omega_0 t) + b\sin(\omega_0 t)
\end{cases}
\tag{51}
$$

Additive Property of Sinusoids. The *additive property* of sinusoids states that summing two or more sinusoids with the same frequency yields a sinusoid with different amplitude and phase parameters but the same frequency.

Derivative and Integral Property of Sinusoids.   The *derivative* and *integral* properties of the sinusoid state that a sinusoid maintains its wave shape when differentiated or integrated. These operations change the amplitude and phase angle but do not change the basic sinusoidal wave shape or frequency. The fact that the wave shape is unchanged by differentiation and integration is a key property of the sinusoid. No other periodic waveform has this shape-preserving property.

Waveform Partial Descriptors

An equation or graph defines a waveform for all time. The value of a waveform $v(t)$, $i(t)$, or $p(t)$ at time t is called the *instantaneous value* of the waveform. Engineers often use numerical values or terminology that characterizes a waveform but do not give a complete description. These waveform *partial descriptors* fall into two categories: (a) those that describe temporal features and (b) those that describe amplitude features.

Temporal Descriptors.   Temporal descriptors identify waveform attributes relative to the time axis. A signal $v(t)$ is *periodic* if $v(t + T_0) = v(t)$ for all t, where the period T_0 is the smallest value that meets this condition. Signals that are not periodic are called *aperiodic*. The fact that a waveform is periodic provides important information about the signal but does not specify all of its characteristics. The period and periodicity of a waveform are partial descriptors. A sine wave, square wave, and triangular wave are all periodic. Examples of aperiodic waveforms are the step function, exponential, and damped sine.

Waveforms that are identically zero prior to some specified time are said to be *causal*. A signal $v(t)$ is *casual* if $v(t)/0$ for $t < T$; otherwise it is *noncausal*. It is usually assumed that a causal signal is zero for $t < 0$, since time shifting can always place the starting point of a waveform at $t = 0$. Examples of causal waveforms are the step function, exponential, and damped sine. An infinitely repeating periodic waveform is noncausal.

Causal waveforms play a central role in circuit analysis. When the input driving force $x(t)$ is causal, the circuit response $y(t)$ must also be causal. That is, a physically realizable circuit cannot anticipate and respond to an input before it is applied. Causality is an important temporal feature but only a partial description of the waveform.

Amplitude Descriptors.   Amplitude descriptors are generally positive scalars that identify size features of the waveform. Generally a waveform's amplitude varies between two extreme values denoted as V_{MAX} and V_{MIN}. The *peak-to-peak* value (V_{pp}) describes the total excursion of $v(t)$ and is defined as

$$V_{pp} = V_{MAX} - V_{MIN} \qquad (52)$$

Under this definition V_{pp} is always positive even if V_{MAX} and V_{MIN} are both negative. The *peak value* (v_p) is the maximum of the absolute value of the waveform. That is,

$$V_P = \max\{|V_{MAX}|, |V_{MIN}|\} \qquad (53)$$

The peak value is a positive number that indicates the maximum absolute excursion of the waveform from zero. Figure 65 shows examples of these two amplitude descriptors.

The *average value* (v_{avg}) smoothes things out to reveal the underlying waveform baseline. Average value is the area under the waveform over some period of time T divided by that time period:

$$V_{avg} = \frac{1}{T} \int_{t_0}^{t_0+T} v(x)\,dx \qquad (54)$$

For periodic signals the averaging interval T equals the period T_0. The average value measures the waveform's baseline with respect to the $v = 0$ axis. In other words, it indicates whether the

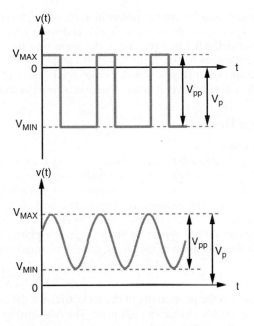

Figure 65   Peak value (V_p) and peak-to-peak value (V_{pp}).[1]

waveform contains a constant, non-time-varying component. The average value is also called the *dc component* of the waveform because dc signals are constant for all t.

Root-Mean-Square Value.   The *root-mean-square value* (v_{rms}) of a waveform is a measure of the average power carried by the signal. The instantaneous power delivered to a resistor R by a voltage $v(t)$ is

$$p(t) = \frac{1}{R}[v(t)]^2$$

The average power delivered to the resistor in time span T is defined as

$$P_{avg} = \frac{1}{T}\int_{t_0}^{t_0+T} p(t)\, dt$$

Combining the above equations yields

$$P_{avg} = \frac{1}{R}\left[\frac{1}{T}\int_{t_0}^{t_0+T}[v(t)]^2\, dt\right]$$

The quantity inside the large brackets is the average value of the square of the waveform. The units of the bracketed term are volts squared. The square root of this term defines the amplitude descriptor v_{rms}:

$$V_{rms} = \sqrt{\frac{1}{T}\int_{t_0}^{t_0+T}[v(t)]^2\, dt} \tag{55}$$

For periodic signals the averaging interval is one cycle since such a waveform repeats itself every T_0 seconds. The average power delivered to a resistor in terms of v_{rms} is

$$P_{avg} = \frac{1}{R}V_{rms}^2 \tag{56}$$

The equation for average power in terms of v_{rms} has the same form as the instantaneous power. For this reason the rms value is also called the *effective value*, since it determines the average power delivered to a resistor in the same way that a dc waveform $v(t) = v_{dc}$ determines the instantaneous power. If the waveform amplitude is doubled, its rms value is doubled, and the average power is quadrupled. Commercial electrical power systems use transmission voltages in the range of several hundred kilovolts (rms) to transfer large blocks of electrical power.

4.2 Energy Storage Devices

Capacitor

A capacitor is a dynamic element involving the time variation of an electric field produced by a voltage. Figure 66a shows the parallel-plate capacitor, which is the simplest physical form of a capacitive device, and two common circuit symbols for the capacitor are shown in Fig. 66b.

Electrostatics shows that a uniform electric field (t) exists between the metal plates when a voltage exists across the capacitor. The electric field produces charge separation with equal and opposite charges appearing on the capacitor plates. When the separation d is small compared with the dimension of the plates, the electric field between the plates is

$$\mathscr{E}(t) = \frac{q(t)}{\varepsilon A}$$

where ε is the permittivity of the dielectric, A is the area of the plates, and $q(t)$ is the magnitude of the electric charge on each plate. The relationship between the electric field and the voltage across the capacitor $v_C(t)$ is given by

$$\mathscr{E}(t) = \frac{v_C(t)}{d}$$

Setting both equations equal and solving for the charge $q(t)$ yields

$$q(t) = \left[\frac{\varepsilon A}{d}\right] v_C(t) = C v_C(t) \tag{57}$$

The proportionality constant inside the bracket in this equation is the *capacitance C*. The unit of capacitance is the farad (F), a term that honors the British physicist Michael Faraday. Values of capacitance range from picofarads (10^{-12}F) in semiconductor devices to tens of millifarads (10^{-3}F) in industrial capacitor banks. Differentiating Eq. (57) with respect to time t and realizing that $i_C(t)$ is the time derivative of $q(t)$ result in the capacitor $i-v$ relationship

$$i_C(t) = \frac{dq(t)}{dt} = \frac{d[C v_C(t)]}{dt} = C \frac{dv_C(t)}{dt} \tag{58}$$

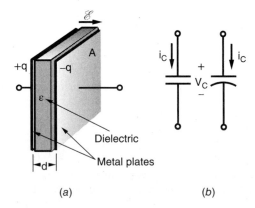

(a)   (b)

Figure 66   Capacitor: (a) parallel-plate device and (b) circuit symbol.[1]

The time derivative in Eq. (58) means the current is zero when the voltage across the capacitor is constant, and vice versa. In other words, the capacitor acts like an open circuit ($i_C = 0$) when dc excitations are applied. The capacitor is a dynamic element because the current is zero unless the voltage is changing. However, a discontinuous change in voltage requires an infinite current, which is physically impossible. Therefore, the capacitor voltage must be a continuous function of time.

Equation (59a) is the integral form of the capacitor i–v relationship where x is a dummy integration variable:

$$v_C(t) = v_C(0) + \frac{1}{C}\int_0^t i_C(x)\,dx \tag{59a}$$

With the passive-sign convention the power associated with the capacitor is

$$p_C(t) = i_C(t) \times v_C(t) \tag{59b}$$

This equation shows that the power can be either positive or negative because the capacitor voltage and its time rate of change can have opposite signs. The ability to deliver power implies that the capacitor can store energy. Assuming that zero energy is stored at $t = 0$, the capacitor energy is expressed as

$$w_C(t) = \tfrac{1}{2}Cv_C^2(t) \tag{60}$$

The stored energy is never negative since it is proportional to the square of the voltage. The capacitor absorbs power from the circuit when storing energy and returns previously stored energy when delivering power to the circuit.

Inductor

The inductor is a dynamic circuit element involving the time variation of the magnetic field produced by a current. Magnetostatics shows that a magnetic flux ϕ surrounds a wire carrying an electric current. When the wire is wound into a coil, the lines of flux concentrate along the axis of the coil as shown in Fig. 67a. In a linear magnetic medium the flux is proportional to both the current and the number of turns in the coil. Therefore, the total flux is

$$\phi(t) = k_1 N i_L(t)$$

where k_1 is a constant of proportionality involving the permeability of the physical surroundings and dimensions of the wire.

The magnetic flux intercepts or links the turns of the coil. The flux linkages in a coil is represented by the symbol λ, with units of webers (Wb), named after the German scientist Wilhelm Weber (1804–1891). The number of flux linkages is proportional to the number of turns in the coil and to the total magnetic flux, so λ is given as

$$\lambda(t) = N\phi(t)$$

Substituting for $\varphi(t)$ gives

$$\lambda(t) = [k_1 N^2]i_L(t) = Li_L(t) \tag{61}$$

The $k_1 N^2$ inside the brackets in this equation is called the *inductance L* of the coil. The unit of inductance is the henry (H) (plural henrys), a name that honors American scientist Joseph Henry. Figure 67b shows the circuit symbol for an inductor.

Equation (61) is the inductor element constraint in terms of current and flux linkages. Differentiating Eq. (61) with respect to time t and realizing that according to Faraday's law $v_L(t)$ is the time derivative of $\lambda(t)$ result in the inductor i–v relationship

$$v_L(t) = \frac{d[\lambda(t)]}{dt} = \frac{d[Li_L(t)]}{dt} = L\frac{di_l(t)}{dt} \tag{62}$$

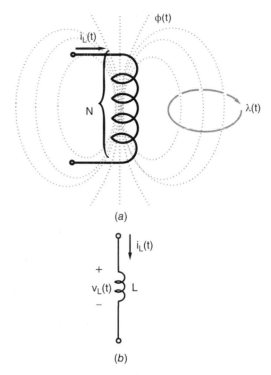

$\phi(t)$

$i_L(t)$

N

$\lambda(t)$

(a)

$i_L(t)$

$+$

$v_L(t)$

L

$-$

(b)

Figure 67 (*a*) Magnetic flux surrounding current-carrying coil. (*b*) Circuit symbol for inductor.[1]

The time derivative in Eq. (62) means that the voltage across the inductor is zero unless the current is time varying. Under dc excitation the current is constant and $v_L = 0$ so the inductor acts like a short circuit. The inductor is a dynamic element because only a changing current produces a nonzero voltage. However, a discontinuous change in current produces an infinite voltage, which is physically impossible. Therefore, the current $i_L(t)$ must be a continuous function of time t.

Equation (63) is the integral form of the inductor i–v relationship where x is a dummy integration variable:

$$i_L(t) = i_L(0) + \frac{1}{L} \int_0^t v_L(x)\, dx \tag{63}$$

With the passive-sign convention the inductor power is

$$p_L(t) = i_L(t) \times v_L(t) \tag{64}$$

This expression shows that power can be positive or negative because the inductor current and its time derivative can have opposite signs. The ability to deliver power indicates that the inductor can store energy. Assuming that zero energy is stored at $t = 0$, the inductor energy is expressed as

$$w_L(t) = \tfrac{1}{2} L i_L^2(t) \tag{65}$$

The energy stored in an inductor is never negative because it is proportional to the square of the current. The inductor stores energy when absorbing power and returns previously stored energy when delivering power.

Equivalent Capacitance and Inductance

Resistors connected in series or parallel can be replaced by equivalent resistances. The same principle applies to connections of capacitors and inductors. N capacitors connected in parallel can be replaced by a single capacitor equal to the sum of the capacitance of the parallel capacitors, that is,

$$C_{EQ} = C_1 + C_2 + \cdots + C_N \text{ (parallel connection)} \tag{66}$$

The initial voltage, if any, on the equivalent capacitance is $v(0)$, the common voltage across all of the original N capacitors at $t = 0$. Likewise, N capacitors connected in series can be replaced by a single capacitor equal to

$$C_{EQ} = \frac{1}{1/C_1 + 1/C_2 + \cdots + 1/C_N} \quad \text{(series connection)} \tag{67}$$

The equivalent capacitance of a parallel connection is the sum of the individual capacitances. The reciprocal of the equivalent capacitance of a series connection is the sum of the reciprocals of the individual capacitances. Since the capacitor and inductor are dual elements, the corresponding results for inductors are found by interchanging the series and parallel equivalence rules for the capacitor. That is, in a series connection the equivalent inductance is the sum of the individual inductances:

$$L_{EQ} = L_1 + L_2 + \cdots + L_N \quad \text{(series connection)} \tag{68}$$

For the parallel connection the reciprocals add to produce the reciprocal of the equivalent inductance:

$$L_{EQ} = \frac{1}{1/L_1 + 1/L_2 + \cdots + 1/L_N} \quad \text{(parallel connection)} \tag{69}$$

Example 20

Find the equivalent capacitance and inductance of the circuit in Fig. 68a.

Solution

The circuit contains both inductors and capacitors. The inductors and the capacitors are combined separately. The 5-pF capacitor in parallel with the 0.1-μF capacitor yields an equivalent capacitance of 0.100005 μF. For all practical purposes the 5-pF capacitor can be ignored, leaving two 0.1-μF capacitors in series with equivalent capacitance of 0.05 μF. Combining this equivalent capacitance in parallel with the remaining 0.05-μF capacitor yields an overall equivalent capacitance of 0.1 μF. The parallel 700- and 300-μH inductors yield an equivalent

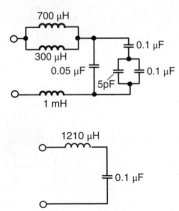

Figure 68   Circuit.[1]

inductance of $1/(1/700 + 1/300) = 210 \ \mu\text{H}$. This equivalent inductance is effectively in series with the 1-mH inductor at the bottom, yielding $1000 + 210 = 1210 \ \mu\text{H}$ as the overall equivalent inductance. Figure 68*b* shows the simplified equivalent circuit.

Mutual Inductance

The *i*–*v* characteristics of the inductor result from the magnetic field produced by current in a coil of wire. The magnetic flux spreads out around the coil forming closed loops that cut or link with the turns in the coil. If the current is changing, then Faraday's law states that voltage across the coil is equal to the time rate of change of the total flux linkages.

Now suppose that a second coil is brought close to the first coil. The flux from the first coil will link with the turns of the second coil. If the current in the first coil is changing, then these flux linkages will generate a voltage in the second coil. The coupling between a changing current in one coil and a voltage across a second coil results in *mutual inductance*.

If there is coupling between the two coils in Fig. 69, there are two distinct effects occurring in the coils. First there is the self-inductance due to the current flowing in each individual coil and the voltage induced by that current in that coil. Second, there are the voltages occurring in the second coil caused by current flowing through the first coil and vice versa. A double-subscript notation is used because it clearly identifies the various cause-and-effect relationships. The first subscript indicates the coil in which the effect takes place, and the second identifies the coil in which the cause occurs. For example, $v_{11}(t)$ is the voltage across coil 1 due to causes occurring in coil 1 itself, while $v_{12}(t)$ is the voltage across coil 1 due to causes occurring in coil 2. The self-inductance is

$$\text{Coil 1:} \quad v_{11}(t) = \frac{d\lambda_{11}(t)}{dt} = N_1 \frac{d\phi_1(t)}{dt}$$

$$= [k_1 N_1^2] \frac{di_1(t)}{dt}$$

$$\text{Coil 2:} \quad v_{22}(t) = \frac{d\lambda_{22}(t)}{dt} = N_2 \frac{d\phi_2(t)}{dt} \tag{70}$$

$$= [k_2 N_2^2] \frac{di_2(t)}{dt}$$

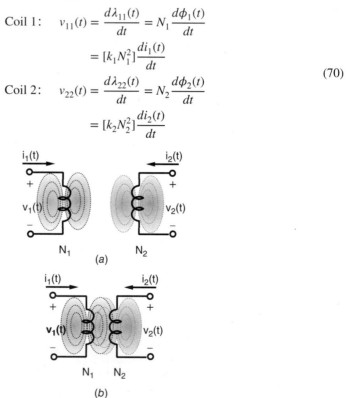

Figure 69   (*a*) Inductors separated, only self-inductance present. (*b*) Inductors coupled, both self- and mutual inductance present.[1]

Equations (70) provide the i–v relationships for the coils when there is no mutual coupling. The mutual inductance is

$$\text{Coil 1:} \quad v_{12}(t) = \frac{d\lambda_{12}(t)}{dt} = N_1 \frac{d\phi_{12}(t)}{dt}$$

$$= [k_{12}N_1N_2]\frac{di_2(t)}{dt}$$

$$\text{Coil 2:} \quad v_{21}(t) = \frac{d\lambda_{21}(t)}{dt} = N_2 \frac{d\phi_{21}(t)}{dt} \tag{71}$$

$$= [k_{21}N_1N_2]\frac{di_1(t)}{dt}$$

The quantity $\phi_{12}(t)$ is the flux intercepting coil 1 due to the current in coil 2 and $\phi_{21}(t)$ is the flux intercepting coil 2 due to the current in coil 1. The expressions in Eq. (71) are the i–v relationships describing the cross coupling between coils when there is mutual coupling.

When the magnetic medium supporting the fluxes is linear, the superposition principle applies, and the total voltage across the coils is the sum of the results in Eqs. (70) and (71):

$$\text{Coil 1:} \quad v_1(t) = v_{11}(t) + v_{12}(t)$$

$$\text{Coil 2:} \quad v_2(t) = v_{21}(t) + v_{22}(t)$$

There are four inductance parameters in these equations. Two *self-inductance* parameters $L_1 = k_1N_1^2$ and $L_2 = k_2N_2^2$ and two *mutual inductances* $M_{12} = k_{12}N_1N_2$ and $M_{21} = k_{21}N_2N_1$. In a linear magnetic medium $k_{12} = k_{21} = k_M$, there is a single mutual inductance parameter M defined as $M = M_{12} = M_{21} = k_MN_1N_2$. Putting these all together yields

$$\text{Coil 1:} \quad v_1(t) = L_1\frac{di_1(t)}{dt} \pm M\frac{di_2(t)}{dt}$$

$$\text{Coil 2:} \quad v_2(t) = \pm M\frac{di_1(t)}{dt} + L_2\frac{di_2(t)}{dt} \tag{72}$$

The coupling across coils can be additive or subtractive. This gives rise to the $\pm$ sign in front of the mutual inductance M. Additive (+) coupling means that a positive rate of change of current in coil 2 induces a positive voltage in coil 1, and vice versa for subtractive coupling (−).

When applying these element equations, it is necessary to know when to use a plus sign and when to use a minus sign. Since the additive or subtractive nature of a coupled-coil set is predetermined by the manufacturer of the windings, a dot convention is used. The dots shown near one terminal of each coil are special reference marks indicating the relative orientation of the coils. Figure 70 shows the dot convention.

The correct sign for the mutual inductance term hinges on how the reference marks for currents and voltages are assigned relative to the coil dots: *Mutual inductance is additive when both current reference directions point toward or both point away from dotted terminals; otherwise, it is subtractive.*

Ideal Transformer

A *transformer* is an electrical device that utilizes mutual inductance coupling between two coils. Transformers find application in virtually every type of electrical system, especially in power supplies and commercial power grids.

In Fig. 71 the transformer is shown as an interface device between a source and a load. The coil connected to the source is called the *primary winding* and the coil connected to the load the *secondary winding*. In most applications the transformer is a coupling device that transfers signals (especially power) from the source to the load. The basic purpose of the device is to change voltage and current levels so the signal conditions at the source and load are compatible.

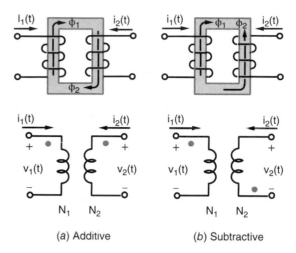

Figure 70 Winding orientations and corresponding reference dots: (*a*) additive and (*b*) subtractive.

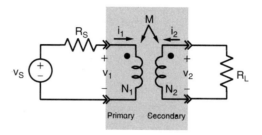

Figure 71 Transformer connected at source–load interface.[1]

Transformer design involves two primary goals: (a) to maximize the magnetic coupling between the two windings and (b) to minimize the power loss in the windings. The first goal produces near-perfect coupling ($k = 1$) so that almost all of the flux in one winding links the other. The second goal produces nearly zero power loss so that almost all of the power delivered to the primary winding transfers to the load. The *ideal transformer* is a circuit element in which coupled coils are assumed to have perfect coupling and zero power loss.

Perfect coupling assumes that all the coupling coefficients are equal to each other, that is, $k_{11} = k_{22} = k_{12} = k_{21} = k_M = 1$. Dividing the two equations in Eq. (72) and using the concept of perfect coupling result in the equation

$$\frac{v_2(t)}{v_1(t)} = \pm\frac{N_2}{N_1} = \pm n \tag{73}$$

where n is the *turns ratio*. With perfect coupling the secondary voltage is proportional to the primary voltage so they have the same wave shape. For example, when the primary voltage is $v_1(t) = V_A \sin \omega t$, the secondary voltage is $v_2(t) = \pm n V_A \sin \omega t$. When the turns ratio $n > 1$, the secondary voltage amplitude is larger than the primary and the device is called a *step-up transformer*. Conversely, when $n < 1$, the secondary voltage is smaller than the primary and the device is called a *step-down transformer*. The ability to increase or decrease ac voltage levels is a basic feature of transformers. Commercial power systems use transmission voltages of several hundred kilovolts. For residential applications the transmission voltage is reduced to safer levels (typically 220/110 V_{rms}) using step-down transformers.

The $\pm$ sign in Eq. (73) depends on the reference marks given the primary and secondary currents relative to the dots indicating the relative coil orientations. The rule for the ideal transformer is a corollary of the rule for selecting the sign of the mutual inductance term in coupled-coil element equations.

The ideal transformer model also assumes that there is no power loss in the transformer. With the passive-sign convention, the power in the primary winding and secondary windings is $v_1(t)i_1(t)$ and $v_2(t)i_2(t)$, respectively. Zero power loss requires

$$v_1(t)i_1(t) + v_2(t)i_2(t) = 0$$

which can be rearranged in the form

$$\frac{i_2(t)}{i_1(t)} = -\frac{v_1(t)}{v_2(t)}$$

But under the perfect coupling assumption $v_2(t)/v_1(t) = \pm n$. With zero power loss and perfect coupling the primary and secondary currents are related as

$$\frac{i_2(t)}{i_1(t)} = \mp\frac{1}{n} \tag{74}$$

The correct sign in this equation depends on the orientation of the current reference directions relative to the dots describing the transformer structure.

With both perfect coupling and zero power loss, the secondary current is inversely proportional to the turns ratio. A step-up transformer ($n > 1$) increases the voltage and decreases the current, which improves transmission line efficiency because the i^2R losses in the conductors are smaller.

Using the ideal transformer model requires some caution. The relationships in Eqs. (73) and (74) state that the secondary signals are proportional to the primary signals. These element equations appear to apply to dc signals. This is, of course, wrong. The element equations are an idealization of mutual inductance, and mutual inductance requires time-varying signals to provide the coupling between two coils.

Equivalent Input Resistance.   Because a transformer changes the voltage and current levels, it effectively changes the load resistance seen by a source in the primary circuit. Consider the circuit shown in Fig. 72. The device equations are

$$\text{Resistor:} \quad v_2(t) = R_L i_L(t)$$

$$\text{Transformer:} \quad v_2(t) = nv_1(t)$$

$$i_2(t) = -\frac{1}{n}i_1(t)$$

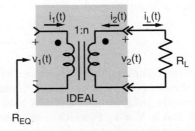

Figure 72   Equivalent resistance seen in primary winding.[1]

Dividing the first transformer equation by the second and inserting the load resistance constraint yield

$$\frac{v_2(t)}{i_2(t)} = \frac{i_L(t)R_L}{i_2(t)} = -n^2\frac{v_1(t)}{i_1(t)}$$

Applying KCL at the output interface tells us $i_L(t) = -i_2(t)$. Therefore, the equivalent resistance seen on the primary side is

$$R_{EQ} = \frac{v_1(t)}{i_1(t)} = \frac{1}{n^2}R_L \tag{75}$$

The equivalent load resistance seen on the primary side depends on the turns ratio and the load resistance. Adjusting the turns ratio can make R_{EQ} equal to the source resistance. Transformer coupling can produce the resistance match condition for maximum power transfer when the source and load resistances are not equal.

4.3   Phasor Analysis of Alternating Current Circuits

Those ac circuits that are excited by a single frequency, for example, power systems, can be easily and effectively analyzed using sinusoidal steady-state techniques. Such a technique was first proposed by Charles Steinmetz (1865–1923) using a vector representation of sinusoids called *phasors*.

Sinusoids and Phasors

The phasor concept is the foundation for the analysis of linear circuits in the sinusoidal steady state. Simply put, a *phasor* is a complex number representing the amplitude and phase angle of a sinusoidal voltage or current. The connection between sine waves and complex numbers is provided by Euler's relationship:

$$e^{j\theta} = \cos\theta + j\sin\theta$$

To develop the phasor concept, it is necessary to adopt the point of view that the cosine and sine functions can be written in the form

$$\cos\theta = \text{Re}\{e^{j\theta}\} \quad \text{and} \quad \sin\theta = \text{Im}\{e^{j\theta}\}$$

where Re stands for the "real part of" and Im for the "imaginary part of." Development of the phasor concept begins with reference of phasors to the cosine function as

$$v(t) = V_A\cos(\omega t + \phi) = V_A\text{Re}\{e^{j(\omega t+\phi)}\}$$
$$= V_A\text{Re}\{e^{j\omega t}e^{j\phi}\} = \text{Re}\{(V_Ae^{j\phi})e^{j\omega t}\} \tag{76}$$

Moving the amplitude V_A inside the real-part operation does not change the final result because it is real constant. By definition, the quantity $V_Ae^{j\phi}$ in Eq. (76) is the *phasor representation* of the sinusoid $v(t)$. The phasor **V**—a boldface **V**—or sometimes written with a tilde above the variable, $\tilde{V}$, can be represented in polar or rectangular form as

$$\mathbf{V} = \underbrace{V_Ae^{j\phi}}_{\text{polar form}} = \underbrace{V_A(\cos\phi + j\sin\phi)}_{\text{rectangular form}} \tag{77}$$

Note that **V** is a complex number determined by the amplitude and phase angle of the sinusoid. Figure 73 shows a graphical representation commonly called a phasor diagram. An alternative way to write the polar form of a phasor is to replace the exponential $e^{j\phi}$ by the shorthand notation ϕ, that is, $\mathbf{V} = V_A \phi$, which is equivalent to the polar form in Eq. (77). It is important to realize that a phasor is determined by its amplitude and phase angle and does not contain any information about the frequency of the sinusoid.

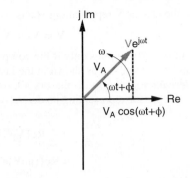

Figure 73   Complex exponential $Ve^{j\omega t}$.[1]

The first feature points out that signals can be described in different ways. Although the phasor **V** and waveform $v(t)$ are related concepts, they have quite different physical interpretations and one must clearly distinguish between them. The absence of frequency information in the phasors results from the fact that in the sinusoidal steady state all currents and voltages are sinusoids with the same frequency. Carrying frequency information in the phasor would be redundant since it is the same for all phasors in any given steady-state circuit problem.

In summary, given a sinusoidal waveform $v(t) = V_A \cos(\omega t + \phi)$, the corresponding phasor representation is $\mathbf{V} = V_A e^{j\phi}$. Conversely, given the phasor $\mathbf{V} = V_A e^{j\phi}$, the corresponding sinusoid waveform is found by multiplying the phasor by $e^{j\omega t}$ and reversing the steps in Eq. (76) as follows:

$$v(t) = \text{Re}\{\mathbf{V}e^{j\omega t}\} = \text{Re}\{(V_A e^{j\phi})e^{j\omega t}\}$$
$$= V_A \text{Re}\{e^{j(\omega t + \phi)}\} = V_A \cos(\omega t + \phi)$$

The frequency ω in the complex exponential $\mathbf{V}e^{j\omega t}$ in Eq. (76) must be expressed or implied in a problem statement since by definition it is not contained in the phasor. Figure 73 shows a geometric interpretation of the complex exponential $\mathbf{V}e^{j\omega t}$ as a vector in the complex plane of length v_A, which rotates counterclockwise with a constant angular velocity ω. The real-part operation projects the rotating vector onto the horizontal (real) axis and thereby generates $v(t) = V_A \cos(\omega t + \phi)$. The complex exponential is sometimes called a *rotating phasor*, and the phasor **V** is viewed as a snapshot of the situation at $t = 0$.

Properties of Phasors

Phasors have two properties. The *additive property* states that the phasor representing a sum of sinusoids of the same frequency is obtained by adding the phasor representations of the component sinusoids. To establish this property, we write the expression

$$v(t) = v_1(t) + v_2(t) + \cdots + v_N(t)$$
$$v(t) = \text{Re}\{\mathbf{V}_1 e^{j\omega t}\} + \text{Re}\{\mathbf{V}_2 e^{j\omega t}\} + \cdots + \text{Re}\{\mathbf{V}_N e^{j\omega t}\} \tag{78}$$

where $v_1(t), v_2(t), \ldots, v_N(t)$ are sinusoids of the same frequency whose phasor representations are $\mathbf{V}_1, \mathbf{V}_2, \ldots, \mathbf{V}_N$. The real-part operation is additive, so the sum of real parts equals the real part of the sum. Consequently, Eq. (78) can be written in the form

$$v(t) = \text{Re}\{\mathbf{V}_1 e^{j\omega t} + \mathbf{V}_2 e^{j\omega t} + \cdots + \mathbf{V}_N e^{j\omega t}\}$$
$$= \text{Re}\{(\mathbf{V}_1 + \mathbf{V}_2 + \cdots + \mathbf{V}_N)e^{j\omega t}\} \tag{79}$$

Hence the phasor $\mathbf{V}$ representing $v(t)$ is

$$\mathbf{V} = \mathbf{V}_1 + \mathbf{V}_2 + \cdots + \mathbf{V}_N \tag{80}$$

The result in Eq. (80) applies only if the component sinusoids all have the same frequency so that $e^{j\omega t}$ can be factored out as shown in the last line in Eq. (79).

The *derivative property* of phasors allows us to easily relate the phasor representing a sinusoid to the phasor representing its derivative. Differentiating Eq. (76) with respect to time t yields

$$\frac{dv(t)}{dt} = \frac{d}{dt}\mathrm{Re}\{\mathbf{V}e^{j\omega t}\} = \mathrm{Re}\left\{\mathbf{V}\frac{d}{dt}e^{j\omega t}\right\}$$

$$= \mathrm{Re}\{(j\omega\mathbf{V})e^{j\omega t}\} \tag{81}$$

From the definition of a phasor we see that the quantity $j\omega\mathbf{V}$ on the right side of this equation is the phasor representation of the time derivative of the sinusoidal waveform.

In summary, the *additive property* states that adding phasors is equivalent to adding sinusoidal waveforms of the same frequency. The *derivative property* states that multiplying a phasor by $j\omega$ is equivalent to differentiating the corresponding sinusoidal waveform.

Phasor Circuit Analysis

Phasor circuit analysis is a method of finding sinusoidal steady-state responses directly from the circuit without using differential equations.

Connection Constraints in Phasor Form.   Kirchhoff's laws in phasor form are as follows:

KVL: The algebraic sum of phasor voltages around a loop is zero.

KCL: The algebraic sum of phasor currents at a node is zero.

Device Constraints in Phasor Form.   The device constraints of the three passive elements are

$$
\begin{aligned}
\text{Resistor:} \quad & v_R(t) = Ri_R(t) \quad (11) \\[6pt]
\text{Inductor:} \quad & v_L(t) = L\frac{di_L(t)}{dt} \quad (12) \\[6pt]
\text{Capacitor:} \quad & i_C(t) = C\frac{dv_C(t)}{dt}
\end{aligned}
\tag{82}
$$

Now in the sinusoidal steady state all of these currents and voltages are sinusoids. In the sinusoidal steady state the voltage and current of the resistor can be written in terms of phasors as $v_R(t) = \mathrm{Re}\{\mathbf{V}_R e^{j\omega t}\}$ and $i_R(t) = \mathrm{Re}\{\mathbf{I}_R e^{j\omega t}\}$. Consequently, the resistor i–v relationship in Eq. (82) can be expressed in terms of phasors as follows:

$$\mathrm{Re}\{\mathbf{V}_R e^{j\omega t}\} = R \times \mathrm{Re}\{\mathbf{I}_R e^{j\omega t}\}$$

Moving R inside the real-part operation on the right side of this equation does not change things because R is a real constant:

$$\mathrm{Re}\{\mathbf{V}_R e^{j\omega t}\} = \mathrm{Re}\{R\mathbf{I}_R e^{j\omega t}\}$$

This relationship holds only if the phasor voltage and current for a resistor are related as

$$\mathbf{V}_R = R\mathbf{I}_R \tag{83a}$$

If the current through a resistor is $i_R(t) = I_A\cos(\omega t + \phi)$. Then the phasor current is $\mathbf{I}_R = I_A e^{j\phi}$ and, according to Eq. (83a), the phasor voltage across the resistor is

$$\mathbf{V}_R = RI_A e^{j\phi} \tag{83b}$$

This result shows that the voltage has the same phase angle (ϕ) as the current. Phasors with the same phase angle are said to be *in phase*; otherwise they are said to be *out of phase*.

In the sinusoidal steady state the voltage and phasor current for the inductor can be written in terms of phasors as $v_L(t) = \text{Re}\{\mathbf{V}_L e^{j\omega t}\}$ and $i_L(t) = \text{Re}\{\mathbf{I}_L e^{j\omega t}\}$. Using the derivative property of phasors, the inductor i–v relationship can be expressed as

$$\text{Re}\{\mathbf{V}_L e^{j\omega t}\} = L \times \text{Re}\{j\omega \mathbf{I}_L e^{j\omega t}\}$$
$$= \text{Re}\{j\omega L \mathbf{I}_L e^{j\omega t}\} \tag{84}$$

Moving the real constant L inside the real-part operation does not change things, leading to the conclusion that phasor voltage and current for an inductor are related as

$$\mathbf{V}_L = j\omega L \mathbf{I}_L \tag{85}$$

When the current is $i_L(t) = i_A \cos(\omega t + \phi)$, the corresponding phasor is $\mathbf{I}_L = I_A e^{j\phi}$ and the i–v constraint in Eq. (85) yields

$$\mathbf{V}_L = j\omega L \mathbf{I}_L = (\omega L e^{j90\text{deg}})(I_A e^{j\phi})$$
$$= \omega L I_A e^{j(\phi+90\text{deg})}$$

The resulting phasor diagram in Fig. 74 shows that the inductor voltage and current are 90° out of phase. The voltage phasor is advanced by 90° counterclockwise, which is in the direction of rotation of the complex exponential $e^{\omega t}$. When the voltage phasor is advanced counterclockwise, that is, ahead of the rotating current phasor, the voltage phasor *leads* the current phasor by 90° or equivalently the current *lags* the voltage by 90°.

Finally, the capacitor voltage and current in the sinusoidal steady state can be written in terms of phasors as $v_C(t) = \text{Re}\{\mathbf{V}_C e^{j\omega t}\}$ and $i_C(t) = \text{Re}\{\mathbf{I}_C e^{j\omega t}\}$. Using the derivative property of phasors, the i–v relationship of the capacitor becomes

$$\text{Re}\{\mathbf{I}_C e^{j\omega t}\} = C \times \text{Re}\{j\omega \mathbf{V}_C e^{j\omega t}\}$$
$$= \text{Re}\{j\omega C \mathbf{V}_C e^{j\omega t}\} \tag{86}$$

Moving the real constant C inside the real-part operation does not change the final results, so we conclude that the phasor voltage and current for a capacitor are related as

$$\mathbf{I}_C = j\omega C \mathbf{V}_C \quad \text{or} \quad \mathbf{V}_C = \frac{1}{j\omega C}\mathbf{I}_C \tag{87}$$

When $i_C(t) = I_A \cos(\omega t + \varphi)$, then Eq. (87), the phasor voltage across the capacitor, is

$$\mathbf{V}_C = \frac{1}{j\omega C}\mathbf{I}_C = \left(\frac{1}{\omega C}e^{-j90°}\right)(I_A e^{j\phi})$$
$$= \frac{I_A}{\omega C}e^{j(\phi-90°)}$$

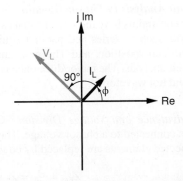

Figure 74   Phasor i–v characteristics of inductor.[1]

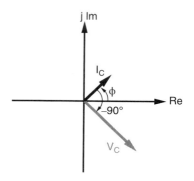

Figure 75   Phasor $i-v$ characteristics of capacitor.[1]

The resulting phasor diagram in Fig. 75 shows that voltage and current are 90° out of phase. In this case the voltage phasor is retarded by 90° clockwise, which is in a direction opposite to the rotation of the complex exponential $e^{j\omega t}$. When the voltage is retarded clockwise, that is, behind the rotating current phasor, we say the voltage phasor *lags* the current phasor by 90° or equivalently the current *leads* the voltage by 90°.

Impedance Concept.   The **I–V** constraints in Eqs. (83a), (85), and (87) are all of the form

$$\mathbf{V} = Z\mathbf{I} \tag{88}$$

where Z is called the impedance of the element. Equation (88) is analogous to Ohm's law in resistive circuits. *Impedance* is the proportionality constant relating phasor voltage and phasor current in linear, two-terminal elements. The impedances of the three passive elements are

$$
\begin{aligned}
\text{Resistor:} \quad & Z_R = R \tag{18}\\
\text{Inductor:} \quad & Z_L = j\omega L \tag{19}\\
\text{Capacitor:} \quad & Z_C = \frac{1}{j\omega C} = -\frac{j}{\omega C}
\end{aligned}
\tag{89}
$$

Since impedance relates phasor voltage to phasor current, it is a complex quantity whose units are ohms. Although impedance can be a complex number, it is not a phasor. Phasors represent sinusoidal signals while impedances characterize circuit elements in the sinusoidal steady state.

Basic Circuit Analysis in Phasor Domain
The phasor constraints have the same format as the constraints for resistance circuits; therefore, familiar tools such as series and parallel equivalence, voltage and current division, proportionality and superposition, and Thévenin and Norton equivalent circuits are applicable to phasor circuit analysis. The major difference is that the circuit responses are complex numbers (phasors) and not waveforms.

Series Equivalence and Voltage Division.   Consider a simple series circuit with several impedances connected to a phasor voltage. The same phasor responses **V** and **I** exist when the series-connected elements are replaced by equivalent impedance Z_{EQ}:

$$Z_{EQ} = \frac{\mathbf{V}}{\mathbf{I}} = Z_1 + Z_2 + \cdots + Z_N$$

In general, the equivalent impedance Z_{EQ} is a complex quantity of the form

$$Z_{EQ} = R + jX$$

where R is the real part and X is the imaginary part. The real part of Z is called *resistance* and the imaginary part $(X$, not $jX)$ is called *reactance*. Both resistance and reactance are expressed in ohms. For passive circuits resistance is always positive while reactance X can be either positive or negative. A positive X is called an *inductive* reactance because the reactance of an inductor is ωL, which is always positive. A negative X is called a *capacitive* reactance because the reactance of a capacitor is $-1/\omega C$, which is always negative.

The phasor voltage across the kth element in the series connection is

$$\mathbf{V}_k = Z_k \mathbf{I}_k = \frac{Z_k}{Z_{EQ}} \mathbf{V} \tag{90}$$

Equation (90) is the phasor version of the voltage division principle. The phasor voltage across any element in a series connection equals the ratio of its impedance to the equivalent impedance of the connection times the total phasor voltage across the connection.

Example 21
The circuit in Fig. 76a is operating in the sinusoidal steady state with $v_S(t) = 35 \cos 1000\,t$ volts.

    a. Transform the circuit into the phasor domain.

    b. Solve for the phasor current **I**.

    c. Solve for the phasor voltage across each element.

    d. Find the waveforms corresponding to the phasors found in (b) and (c).

Solution
    a. The phasor representing the input source voltage is $\mathbf{V}_S = 35 < 0°$. The impedances of the three passive elements are

$$Z_R = R = 50\ \Omega$$

$$Z_L = j\omega L = j1000 \times 25 \times 10^{-3} = j25\ \Omega$$

$$Z_C = \frac{1}{j\omega C} = \frac{1}{j1000 \times 10^{-5}} = -j100\ \Omega$$

Using these, results we obtain the phasor domain circuit in Fig. 76b.

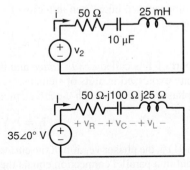

Figure 76   Circuit.[1]

b. The equivalent impedance of the series connection is

$$Z_{EQ} = 50 + j25 - j100 = 50 - j75$$
$$= 90.1\angle - 56.3 \deg \Omega$$

The current in the series circuit is

$$\mathbf{I} = \frac{\mathbf{V}_S}{Z_{EQ}} = \frac{35\angle 0°}{90.1\angle - 56.3°} = 0.388\angle 56.3° \text{ A}$$

c. The current **I** exists in all three series elements so the voltage across each passive element is

$$\mathbf{V}_R = Z_R\mathbf{I} = 50 \times 0.388\angle 56.3°$$
$$= 19.4\angle 56.3° \text{ V}$$
$$\mathbf{V}_L = Z_L\mathbf{I} = j25 \times 0.388\angle 56.3°$$
$$= 9.70\angle 146.3° \text{ V}$$
$$\mathbf{V}_C = Z_C\mathbf{I} = -j100 \times 0.388\angle 56.3°$$
$$= 38.8\angle - 33.7° \text{ V}$$

d. The sinusoidal steady-state waveforms corresponding to the phasors in (b) and (c) are

$$i(t) = \text{Re}\{0.388e^{j56.3°}e^{j1000t}\}$$
$$= 0.388\cos(1000t + 56.3°) \text{ A}$$
$$v_R(t) = \text{Re}\{19.4e^{j56.3°}e^{j1000t}\}$$
$$= 19.4\cos(1000t + 56.3°) \text{ V}$$
$$v_L(t) = \text{Re}\{9.70e^{j146.3°}e^{j1000t}\}$$
$$= 9.70\cos(1000t + 146.3°) \text{ V} v_C(t) = \text{Re}\{38.8e^{-j33.7°}e^{j1000t}\}$$
$$= 38.8\cos(1000t - 33.7°)\text{V}$$

Parallel Equivalence and Current Division

Consider a number of impedances connected in parallel so the same phasor voltage **V** appears across them. The same phasor responses **V** and **I** exist when the parallel-connected elements are replaced by equivalent impedance Z_{EQ}:

$$\frac{1}{Z_{EQ}} = \frac{\mathbf{I}}{\mathbf{V}} = \frac{1}{Z_1} + \frac{1}{Z_2} + \cdots + \frac{1}{Z_N}$$

These results can also be written in terms of admittance Y, which is defined as the reciprocal of impedance:

$$Y = \frac{1}{Z} = G + jB$$

The real part of Y is called *conductance* and the imaginary part B is called *susceptance*, both of which are expressed in units of siemens.

The phasor current through the kth element of the parallel connection is

$$\mathbf{I}_k = Y_k\mathbf{V}_k = \frac{Y_k}{Y_{EQ}}\mathbf{I} \tag{91}$$

Equation (91) is the phasor version of the current division principle. The phasor current through any element in a parallel connection equals the ratio of its admittance to the equivalent admittance of the connection times the total phasor current entering the connection.

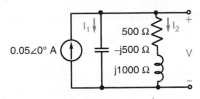

Figure 77 Circuit.[1]

Example 22
For the circuit in Fig. 77 solve for the phasor voltage **V** and for the phasor current through each branch.

Solution

    a. The admittances of the two parallel branches are

$$Y_1 = \frac{1}{-j500} = j2 \times 10^{-3} \text{ S}$$

$$Y_2 = \frac{1}{500 + j1000} = 4 \times 10^{-4} - j8 \times 10^{-4} \text{ S}$$

The equivalent admittance of the parallel connection is

$$Y_{EQ} = Y_1 + Y_2 = 4 \times 10^{-4} + j12 \times 10^{-4}$$

$$= 12.6 \times 10^{-4} \angle 71.6° \text{ S}$$

and the voltage across the parallel circuit is

$$\mathbf{V} = \frac{\mathbf{I}_S}{Y_{EQ}} = \frac{0.05\angle 0°}{12.6 \times 10^{-4} \angle 71.6°}$$

$$= 39.7 \angle -71.6° \text{ V}$$

    b. The current through each parallel branch is

$$\mathbf{I}_1 = Y_1 \mathbf{V} = j2 \times 10^{-3} \times 39.7 \angle -71.6°$$

$$= 79.4\angle 18.4° \text{ mA}$$

$$\mathbf{I}_2 = Y_2 \mathbf{V} = (4 \times 10^{-4} - j8 \times 10^{-4})$$

$$\times 39.7\angle -71.6° = 35.5\angle -135° \text{ mA}$$

Y–Δ *Transformations*
In Section 1.2 in the discussion of equivalent circuits the equivalence of Δ- and Y-connected resistors to simplify resistance circuits with no series- or parallel-connected branches was covered. The same basic concept applies to the Δ- and Y-connected impedances (see Fig. 78).

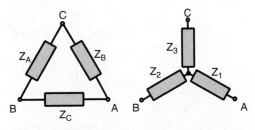

Figure 78 Y–Δ impedance transformation.[2]

The equations for the Δ–Y transformation are

$$Z_1 = \frac{Z_B Z_C}{Z_A + Z_B + Z_C} \qquad Z_2 = \frac{Z_C Z_A}{Z_A + Z_B + Z_C}$$

$$Z_3 = \frac{Z_A Z_B}{Z_A + Z_B + Z_C}$$

(92)

The equations for a Y–Δ transformation are

$$Z_A = \frac{Z_1 Z_2 + Z_2 Z_3 + Z_1 Z_3}{Z_1} \qquad Z_B = \frac{Z_1 Z_2 + Z_2 Z_3 + Z_1 Z_3}{Z_2}$$

$$Z_C = \frac{Z_1 Z_2 + Z_2 Z_3 + Z_1 Z_3}{Z_3}$$

(93)

The equations have the same form except that here they involve impedances rather than resistances.

Example 23
Find the phasor current $\mathbf{I}_{IN}$ in Fig. 79a.

Solution
One cannot use basic reduction tools on the circuit because no elements are connected in series or parallel. However, by replacing either the upper Δ (A, B, C) or lower Δ (A, B, D) by an equivalent Y subcircuit, series and parallel reduction methods can be applied. Choosing the upper Δ because it has two equal resistors simplifies the transformation equations. The sum of the impedance in the upper Δ is $100 + j200$ Ω. This sum is the denominator in the expression in Δ–Y transformation equations. The three Y impedances are found to be

$$Z_1 = \frac{(50)(j200)}{100 + j200} = 40 + j20 \ \Omega$$

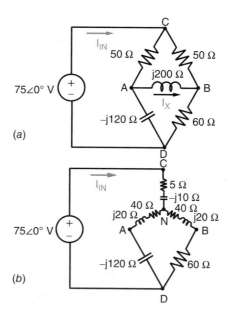

(a)

(b)

Figure 79 Phasor Current.[2]

$$Z_2 = \frac{(50)(j200)}{100 + j200} = 40 + j20 \ \Omega$$

$$Z_3 = \frac{(50)(50)}{100 + j200} = 5 - j10 \ \Omega$$

Figure 79*b* shows the revised circuit with the equivalent Y inserted in place of the upper Δ. Note that the transformation introduces a new node labeled N. The revised circuit can be reduced by series and parallel equivalence. The total impedance of the path *NAD* is $40 - j100 \ \Omega$. The total impedance of the path *NBD* is $100 + j20 \ \Omega$. These paths are connected in parallel so the equivalent impedance between nodes N and D is

$$Z_{ND} = \frac{1}{1/(40 - j100) + 1/(100 + j20)}$$
$$= 60.6 - j31.1 \ \Omega$$

The impedance Z_{ND} is connected in series with the remaining leg of the equivalent Y, so the equivalent impedance seen by the voltage source is

$$Z_{EQ} = 5 - j10 + Z_{ND} = 65.6 - j41.1 \ \Omega$$

The input current then is

$$\mathbf{I}_{IN} = \frac{\mathbf{V}_S}{Z_{EQ}} = \frac{75\angle 0°}{65.6 - j41.1} = 0.891 + j0.514$$
$$= 968\angle 32.0° \ \text{mA}$$

Circuit Theorems in Phasor Domain

Phasor analysis does not alter the linearity properties of circuits. Hence all of the theorems that are applied to resistive circuits can be applied to phasor analysis. These include proportionality, superposition, and Thévenin and Norton equivalence.

Proportionality.   The *proportionality* property states that phasor output responses are proportional to the input phasor. Mathematically proportionality means that $\mathbf{Y} = K\mathbf{X}$, where $\mathbf{X}$ is the input phasor, $\mathbf{Y}$ the output phasor, and K the proportionality constant. In phasor circuit analysis the proportionality constant is generally a complex number.

Superposition.   Care needs to be taken when applying superposition to phasor circuits. If the sources all have the same frequency, then one can transform the circuit into the phasor domain (impedances and phasors) and proceed as in dc circuits with the superposition theorem. If the sources have different frequencies, then superposition can still be used but its application is different. With different frequency sources each source must be treated in a separate steady-state analysis because the element impedances change with frequency. The phasor response for each source must be changed into waveforms and then superposition applied in the time domain. In other words, the superposition principle always applies in the time domain. It also applies in the phasor domain when all independent sources have the same frequency. The following example illustrates the latter case.

Example 24

Use superposition to find the steady-state current $i(t)$ in Fig. 80 for $R = 10 \ \text{k}\Omega, L = 200 \ \text{mH}, v_{S1} = 24 \cos 20,000t$ V, and $v_{S2} = 8 \cos (60,000t + 30°)$ V.

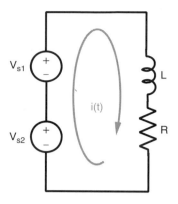

Figure 80 Steady-state current.[1]

Solution

In this example the two sources operate at different frequencies. With source 2 off, the input phasor is $\mathbf{V}_{S1} = 24\angle 0°$ V at a frequency $\omega = 20$ krad/s. At this frequency the equivalent impedance of the inductor and resistor is

$$Z_{EQ1} = R + j\omega L = (10 + j4)\ \text{k}\Omega$$

The phasor current due to source 1 is

$$\mathbf{I}_1 = \frac{\mathbf{V}_{S1}}{Z_{EQ1}} = \frac{24\angle 0°}{10,000 + j4000} = 2.23\angle -21.8°\ \text{mA}$$

With source 1 off and source 2 on, the input phasor $\mathbf{V}_{S2} = 8\angle 30°$ V at a frequency $\omega = 60$ krad/s. At this frequency the equivalent impedance of the inductor and resistor is

$$Z_{EQ2} = R + j\omega L = (10 + j12)\ \text{k}\Omega$$

The phasor current due to source 2 is

$$\mathbf{I}_2 = \frac{\mathbf{V}_{S2}}{Z_{EQ2}} = \frac{8\angle 30°}{10,000 + j12,000}$$

$$= 0.512\angle -20.2°\ \text{mA}$$

The two input sources operate at different frequencies so the phasor responses $\mathbf{I}_1$ and $\mathbf{I}_2$ cannot be added to obtain the overall response. In this case the overall response is obtained by adding the corresponding time domain waveforms:

$$i(t) = \text{Re}\{\mathbf{I}_1 e^{j20,000t}\} + \text{Re}\{\mathbf{I}_2 e^{j60,000t}\}$$

$$i(t) = 2.23\cos(20,000t - 21.8°)$$

$$+ 0.512\cos(60,000t - 20.2°)\ \text{mA}$$

Thévenin and Norton Equivalent Circuits

In the phasor domain a two-terminal circuit containing linear elements and sources can be replaced by Thévenin or Norton equivalent circuits. The general concept of Thévenin's and Norton's theorems and their restrictions are the same as in the resistive circuit studied earlier. The important difference here is that the signals $\mathbf{V}_T, \mathbf{I}_N, \mathbf{V}$, and $\mathbf{I}$ are phasors and $Z_T = 1/Y_N$ and Z_L are complex numbers representing the source and load impedances.

Thévenin equivalent circuits are useful to address the maximum power transfer problem. Consider the source–load interface as shown in Fig. 81. The source circuit is represented by a

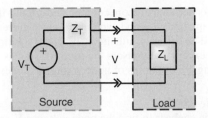

Figure 81 Source–load interface in the sinusoidal steady state.[1]

Thévenin equivalent circuit with source voltage $\mathbf{V}_T$ and source impedance $Z_T = R_T + jX_T$. The load circuit is represented by an equivalent impedance $Z_L = R_L + jX_L$. In the maximum power transfer problem the source parameters $\mathbf{V}_T, R_T$, and X_T are given, and the objective is to adjust the load impedance R_L and X_L so that average power to the load is a maximum.

The average power to the load is expressed in terms of the phasor current and load resistance:

$$P = \tfrac{1}{2} R_L |\mathbf{I}|^2$$

Then, using series equivalence, the magnitude of the interface current is

$$|\mathbf{I}| = \left| \frac{\mathbf{V}_T}{Z_T + Z_L} \right| = \frac{|\mathbf{V}_T|}{|(R_T + R_L) + j(X_T + X_L)|}$$

$$= \frac{|\mathbf{V}_T|}{\sqrt{(R_T + R_L)^2 + (X_T + X_L)^2}}$$

Combining the last two equations yields the average power delivered across the interface:

$$P = \frac{1}{2} \frac{R_L |\mathbf{V}_T|^2}{(R_T + R_L)^2 + (X_T + X_L)^2}$$

Since the quantities $|\mathbf{V}_T|, R_T$, and X_T are fixed, P will be maximized when $X_L = -X_T$. This choice of X_L always is possible because a reactance can be positive or negative. When the source Thévenin equivalent has an inductive reactance ($X_T > 0$), the load is selected to have a capacitive reactance of the same magnitude and vice versa. This step reduces the net reactance of the series connection to zero, creating a condition in which the net impedance seen by the Thévenin voltage source is purely resistive. In summary, to obtain maximum power transfer in the sinusoidal steady state, we select the load resistance and reactance so that $R_L = R_T$ and $X_L = -X_T$. The condition for maximum power transfer is called a *conjugate match* since the load impedance is the conjugate of the source impedance $Z_L = Z_T^*$. Under conjugate-match conditions the maximum average power available from the source circuit is

$$P_{\text{MAX}} = \frac{|\mathbf{V}_T|^2}{8R_T}$$

where $|\mathbf{V}_T|$ is the peak amplitude of the Thévenin equivalent voltage.

It is important to remember that conjugate matching applies when the source is fixed and the load is adjustable. These conditions arise frequently in power-limited communication systems. However, conjugate matching does not apply to electrical power systems because the power transfer constraints are different.

Node Voltage and Mesh Current Analysis in Phasor Domain
The previous sections discuss basic analysis methods based on equivalence, reduction, and circuit theorems. These methods are valuable because they work directly with element impedances and thereby allow insight into steady-state circuit behavior. However, node and mesh analysis

allows for solution of more complicated circuits than the basic methods can easily handle. There general methods use node voltage or mesh current variables to reduce the number of equations that must be solved simultaneously. These solution approaches are identical to those in resistive circuits except that phasors are used for signals and impedances in lieu of only resistors. The following are examples of node voltage and mesh current problems.

Example 25

Use node analysis to find the node voltages $\mathbf{V}_A$ and $\mathbf{V}_B$ in Fig. 82a.

Solution

The voltage source is connected in series with an impedance consisting of a resistor and inductor connected in parallel. The equivalent impedance of this parallel combination is

$$Z_{EQ} = \frac{1}{1/50 + 1/(j100)} = 40 + j20 \ \Omega$$

Applying a source transformation produces an equivalent current source of

$$\mathbf{I}_{EQ} = \frac{10\angle - 90°}{40 + j20} = -0.1 - j0.2 \text{ A}$$

Figure 82b shows the circuit produced by the source transformation. The node voltage equation at the remaining nonreference node in Fig. 82b is

$$\left(\frac{1}{-j50} + \frac{1}{j100} + \frac{1}{50} \right) \mathbf{V}_A = 0.1\angle 0° - (-0.1 - j0.2)$$

Solving for $\mathbf{V}_A$ yields

$$\mathbf{V}_A = \frac{0.2 + j0.2}{0.02 + j0.01} = 12 + j4 = 12.6\angle 18.4° \text{ V}$$

Referring to Fig. 82a, KVL requires $\mathbf{V}_B = \mathbf{V}_A + 10\angle - 90°$. Therefore, $\mathbf{V}_B$ is found to be

$$\mathbf{V}_B = (12 + j4) + 10\angle - 90° = 12 - j6$$

$$= 13.4\angle - 26.6° \text{ V}$$

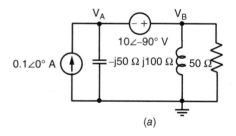

(a)

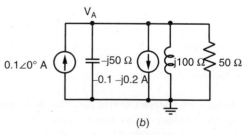

(b)

Figure 82 Node voltages.[1]

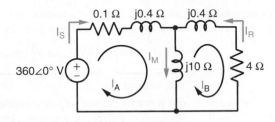

Figure 83 Circuit.[1]

Example 26

The circuit in Fig. 83 is an equivalent circuit of an ac induction motor. The current $\mathbf{I}_S$ is called the stator current, $\mathbf{I}_R$ the rotor current, and $\mathbf{I}_M$ the magnetizing current. Use the mesh current method to solve for the branch currents $\mathbf{I}_S$, $\mathbf{I}_R$, and $\mathbf{I}_M$.

Solution

Applying KVL to the sum of voltages around each mesh yields

$$\text{Mesh } A: \quad -360\angle 0° + [0.1 + j0.4]\mathbf{I}_A + j10[\mathbf{I}_A - \mathbf{I}_B] = 0$$

$$\text{Mesh } B: \quad j10[\mathbf{I}_B - \mathbf{I}_A] + [4 + j0.4]\mathbf{I}_B = 0$$

Solving these equations for $\mathbf{I}_A$ and $\mathbf{I}_B$ produces

$$\mathbf{I}_A = 79.0 - j48.2 \ \mathbf{A} \mathbf{I}_B = 81.7 - j14.9 \ \text{A}$$

The required stator, rotor, and magnetizing currents are related to these mesh currents as follows:

$$\mathbf{I}_S = \mathbf{I}_A = 92.5\angle 31.4° \ \text{A}$$

$$\mathbf{I}_R = -\mathbf{I}_B = -81.8 + j14.9 = 83.0\angle 170° \ \text{A}$$

$$\mathbf{I}_M = \mathbf{I}_A - \mathbf{I}_B = -2.68 - j33.3 = 33.4\angle - 94.6° \ \text{A}$$

4.4 Power in Sinusoidal Steady State

Average and Reactive Power

In power applications it is normal to think of one circuit as the source and the other as the load. It is important to describe the flow of power across the interface between source and load when the circuit is operating in the sinusoidal steady state. The interface voltage and current in the time domain are sinusoids of the form

$$v(t) = V_A\cos(\omega t + \theta) i(t) = I_A\cos \omega t$$

where v_A and i_A are real, positive numbers representing the peak amplitudes of the voltage and current, respectively. The forms of $v(t)$ and $i(t)$ above are completely general. The positive maximum of the current $i(t)$ occurs at $t = 0$ whereas $v(t)$ contains a phase angle θ to account for the fact that the voltage maximum may not occur at the same time as the current's. In the phasor domain the angle $\theta = \phi_V - \phi_I$ is the angle between the phasors $\mathbf{V} = V_A\angle\phi_V$ and $\mathbf{I} = i_A\angle\phi_I$. In effect, choosing $t = 0$ at the current maximum shifts the phase reference by an amount $-\varphi_I$ so that the voltage and current phasors become $\mathbf{V} = v_A\angle\theta$ and $\mathbf{I} = I_A\angle 0°$. The instantaneous power in the time domain is

$$p(t) = v(t) \times i(t) = V_A I_A\cos(\omega t + \theta)\cos \omega t \ \text{W}$$

This expression for instantaneous power contains both dc and ac components. Using the identities $\cos^2 x = 2(1 + \cos 2x)$ and $\cos x \sin x = 2 \sin 2x$, $p(t)$ can be written as

$$p(t) = \underbrace{\left[\tfrac{1}{2} V_A I_A \cos \theta \right]}_{\text{dc component}} + \underbrace{\left[\tfrac{1}{2} V_A I_A \cos \theta \right] \cos 2\omega t - \left[\tfrac{1}{2} V_A I_A \sin \theta \right] \sin 2\omega t}_{\text{ac component}} \qquad (94)$$

The instantaneous power is the sum of a dc component and a double-frequency ac component. That is, the instantaneous power is the sum of a constant plus a sinusoid whose frequency is 2ω, which is twice the angular frequency of the voltage and current. The instantaneous power in Eq. (94) is periodic and its average value is

$$P = \frac{1}{T} \int_0^T p(t)\, dt$$

where $T = 2\pi/2\omega$ is the period of $p(t)$. Since the average value of a sinusoid is zero, the *average value* of $p(t)$, denoted P, is equal to the constant or dc term in Eq. (94):

$$P = \tfrac{1}{2} V_A I_A \cos \theta \qquad (95)$$

The amplitude of the $\sin 2\omega t$ term in Eq. (94) has a form much like the average power in Eq. (95), except it involves $\sin \theta$ rather than $\cos \theta$. This amplitude factor is called the *reactive power* of $p(t)$, where reactive power Q is defined as

$$Q = \tfrac{1}{2} V_A I_A \sin \theta \qquad (96)$$

The instantaneous power in terms of the average power and reactive power is

$$p(t) = \underbrace{P(1 + \cos 2\omega t)}_{\text{unipolar}} - \underbrace{Q \sin 2\omega t}_{\text{bipolar}} \qquad (97)$$

The first term in Eq. (97) is said to be unipolar because the factor $1 + \cos 2\omega t$ never changes sign. As a result, the first term is either always positive or always negative depending on the sign of P. The second term is said to be bipolar because the factor $\sin 2\omega t$ alternates signs every half cycle.

The energy transferred across the interface during one cycle $T = 2\pi/2\omega$ of $p(t)$ is

$$W = \int_0^T p(t)\, dt$$

$$= \underbrace{P \int_0^T (1 + \cos 2\omega t)\, dt}_{\text{net energy}} - \underbrace{Q \int_0^T \sin 2\omega t\, dt}_{\text{no net energy}} \qquad (98)$$

$$= P \times T - 0$$

Only the unipolar term in Eq. (97) provides any net energy transfer and that energy is proportional to the average power P. With the passive-sign convention the energy flows from source to load when $W > 0$. Equation (98) shows that the net energy will be positive if the average power $P > 0$. Equation (95) points out that the average power P is positive when $\cos \theta > 0$, which in turn means $|\theta| < 90°$.

The bipolar term in Eq. (97) is a power oscillation that transfers no net energy across the interface. In the sinusoidal steady state the load borrows energy from the source circuit during part of a cycle and temporarily stores it in the load's reactance, namely its inductance or capacitance. In another part of the cycle the borrowed energy is returned to the source unscathed. The amplitude of the power oscillation is called reactive power because it involves periodic energy storage and retrieval from the reactive elements of the load. The reactive power can be either positive or negative depending on the sign of $\sin \theta$. However, the sign of Q says nothing about the net energy transfer, which is controlled by the sign of P.

Consumers are interested in average power since this component carries net energy from source to load. For most power system customers the basic cost of electrical service is proportional to the net energy delivered to the load. Large industrial users may also pay a service charge for their reactive power as well. This may seem unfair, since reactive power transfers no net energy. However, the electric energy borrowed and returned by the load is generated within a power system that has losses. From a power company's viewpoint the reactive power is not free because there are losses in the system connecting the generators in the power plant to the source–load interface at which the lossless interchange of energy occurs.

In ac power circuit analysis, it is necessary to keep track of both the average power and reactive power. These two components of power have the same dimensions, but because they represent quite different effects, they traditionally are given different units. The average power is expressed in watts while reactive power is expressed in volt-amperes reactive (VARs).

Complex Power

It is important to relate average and reactive power to phasor quantities because ac circuit analysis is conveniently carried out using phasors. The magnitude of a phasor represents the peak amplitude of a sinusoid. However, in power circuit analysis it is convenient to express phasor magnitudes in rms values. In this chapter phasor voltages and currents are expressed as

$$\mathbf{V} = V_{rms}e^{j\phi_V} \text{ and } \mathbf{I} = I_{rms}e^{j\phi_I}$$

Equations (95) and (96) express average and reactive power in terms of peak amplitudes v_A and i_A. The peak and rms values of a sinusoid are related by $V_{rms} = V_A/\sqrt{2}$. The expression for average power can be easily converted to rms amplitudes, Eq. (95), as

$$P = \frac{V_A I_A}{2}\cos \theta = \frac{V_A}{\sqrt{2}}\frac{I_A}{\sqrt{2}}\cos \theta$$

$$= V_{rms}I_{rms}\cos \theta \tag{99}$$

where $\theta = \phi_V - \phi_I$ is the angle between the voltage and current phasors. By similar reasoning, Eq. (96) becomes

$$Q = V_{rms}I_{rms}\sin \theta \tag{100}$$

Using rms phasors, we define the *complex power* (S) at a two-terminal interface as

$$S = \mathbf{VI}^* = V_{rms}e^{j\phi_V}I_{rms}e^{-j\phi_I} = [V_{rms}I_{rms}]e^{j(\phi_V-\phi_I)} \tag{101}$$

That is, the complex power at an interface is the product of the voltage phasor times the conjugate of the current phasor. Using Euler's relationship and the fact that the angle $\theta = \phi_V - \phi_I$, complex power can be written as

$$S = [V_{rms}I_{rms}]e^{j\theta} = [V_{rms}I_{rms}]\cos \theta + j[V_{rms}I_{rms}]\sin \theta = P + jQ \tag{102}$$

The real part of the complex power S is the average power while the imaginary part is the reactive power. Although S is a complex number, it is not a phasor. However, it is a convenient variable for keeping track of the two components of power when voltage and current are expressed as phasors.

The power triangles in Fig. 84 provide a convenient way to remember complex power relationships and terminology. Considering those cases in which net energy is transferred from source to load, $P > 0$ and the power triangles fall in the first or fourth quadrant.

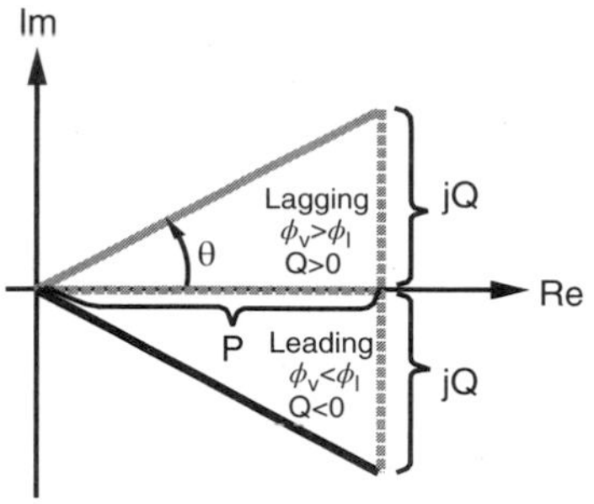

Figure 84 Power triangles.[1]

The magnitude $|S| = V_{rms}I_{rms}$ is called *apparent power* and is expressed using the unit volt-ampere (VA). The ratio of the average power to the apparent power is called the *power factor* (pf):

$$\text{pf} = \frac{P}{|S|} = \frac{V_{rms}I_{rms}\cos\theta}{V_{rms}I_{rms}} = \cos\theta$$

Since $\text{pf} = \cos\theta$, the angle θ is called the *power factor angle*.

When the power factor is unity, the phasors $\mathbf{V}$ and $\mathbf{I}$ are in phase ($\theta = 0°$) and the reactive power is zero since $\sin\theta = 0$. When the power factor is less than unity, the reactive power is not zero and its sign is indicated by the modifiers lagging or leading. The term *lagging power factor* means the current phasor lags the voltage phasor so that $\theta = \phi_V - \phi_I > 0$. For a lagging power factor S falls in the first quadrant in Fig. 84 and the reactive power is positive since $\sin\theta > 0$. The term *leading power factor* means the current phasor leads the voltage phasor so that $\theta = \phi_V - \phi_I < 0$. In this case S falls in the fourth quadrant in Fig. 84 and the reactive power is negative since $\sin\theta < 0$. Most industrial and residential loads have lagging power factors.

The apparent power rating of electrical power equipment is an important design parameter. The ratings of generators, transfomers, and transmission lines are normally stated in kilovolt-amperes. The rating of most loads is stated in kilowatts and power factor. The wiring must be large enough to carry the required current and insulated well enough to withstand the rated voltage. However, only the average power is potentially available as useful output since the reactive power represents a lossless interchange between the source and device. Because reactive power increases the apparent power rating without increasing the available output, it is desirable for electrical devices to operate as close as possible to unity power factor (zero reactive power).

In many cases power circuit loads are described in terms of their power ratings at a specified voltage or current level. In order to find voltages and current elsewhere in the circuit, it is necessary to know the load impedance. In general, the load produces the element constraint $\mathbf{V} = Z\mathbf{I}$. Using this constraint in Eq. (101), we write the complex power of the load as

$$S = \mathbf{V} \times \mathbf{I} = Z\mathbf{I} \times \mathbf{I}^* = Z|\mathbf{I}|^2$$

$$= (R + jX)I_{rms}^2$$

where R and X are the resistance and reactance of the load, respectively. Since $S = P + jQ$, we conclude that

$$R = \frac{P^2}{I_{rms}} \quad \text{and} \quad X = \frac{Q}{I_{rms}^2} \tag{103}$$

The load resistance and reactance are proportional to the average and reactive power of the load, respectively.

The first condition in Eq. (103) demonstrates that resistance cannot be negative since P cannot be negative for a passive circuit. The second condition points out that when the reactive power is positive the load is inductive since $X_L = \omega L$ is positive. Conversely, when the reactive power is negative the load is capacitive since $X_C = -1/\omega C$ is negative. The terms inductive load, lagging power factor, and positive reactive power are synonymous, as are the terms capacitive load, leading power factor, and negative reactive power.

Example 27

At 440 V (rms) a two-terminal load draws 3 kVA of apparent power at a lagging power factor of 0.9. Find i_{rms}, P, Q, and the load impedance.

Solution

$$I_{rms} = \frac{|S|}{V_{rms}} = \frac{3000}{440} = 6.82 \text{A (rms)}$$

$$P = V_{rms}I_{rms}\cos\theta = 3000 \times 0.9 = 2.7 \text{ kW}$$

For $\cos\theta = 0.9$ lagging, $\sin\theta = 0.436$ and $Q = v_{rms}I_{rms}\sin\theta = 1.31$ kVAR.

$$Z = \frac{P + jQ}{(I_{rms})^2} = \frac{2700 + j1310}{46.5} = 58.0 + j28.2 \ \Omega$$

Three-Phase Circuits

The three-phase system shown in Fig. 85 is the predominant method of generating and distributing ac electrical power. The system uses four lines (A, B, C, N) to transmit power from the source to the loads. The symbols stand for the three phases A, B, and C and a neutral line labeled N. The three-phase generator in Fig. 85 is modeled as three independent sources, although the physical hardware is a single unit with three separate windings. Similarly, the loads are modeled as three separate impedances, although the actual equipment may be housed within a single container.

The terminology Y connected and Δ connected refer to the two ways the source and loads can be electrically connected. In a Y connection the three elements are connected from line to neutral, while in the Δ connection they are connected from line to line. In most systems the source is Y connected while the loads can be either Y or Δ, although the latter is more common.

Three-phase sources usually are Y connected because the Δ connection involves a loop of voltage sources. Large currents may circulate in this loop if the three voltages do not exactly

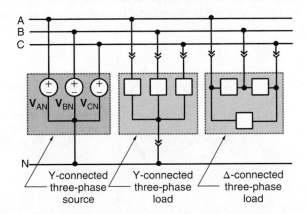

Figure 85 Three-phase source connected to three-phase Y connection and to three-phase Δ connection.[2]

sum to zero. In analysis situations, a Δ connection of ideal voltage sources is awkward because it is impossible to uniquely determine the current in each source.

A double-subscript notation is used to identify voltages in the system. The reason is that there are at least six voltages to deal with: three line-to-line voltages and three line-to-neutral voltages. The two subscripts are used to define the points across which a voltage is defined. For example, $\mathbf{V}_{AB}$ means the voltage between points A and B with an implied plus reference mark at the first subscript (A) and an implied minus at the second subscript (B).

The three line-to-neutral voltages are called the *phase voltages* and are written in double-subscript notation as $\mathbf{V}_{AN}$, $\mathbf{V}_{BN}$, and $\mathbf{V}_{CN}$. Similarly, the three line-to-line voltages, called simply the *line voltages*, are identified as $\mathbf{V}_{AB}$, $\mathbf{V}_{BC}$, and $\mathbf{V}_{CA}$. From the definition of the double-subscript notation it follows that $\mathbf{V}_{XY} = -\mathbf{V}_{YX}$. Using this result and KVL we derive the relationships between the line voltages and phase voltages:

$$\mathbf{V}_{AB} = \mathbf{V}_{AN} + \mathbf{V}_{NB} = \mathbf{V}_{AN} - \mathbf{V}_{BN}$$

$$\mathbf{V}_{BC} = \mathbf{V}_{BN} + \mathbf{V}_{NC} = \mathbf{V}_{BN} - \mathbf{V}_{CN} \qquad (104)$$

$$\mathbf{V}_{CA} = \mathbf{V}_{CN} + \mathbf{V}_{NA} = \mathbf{V}_{CN} - \mathbf{V}_{AN}$$

A balanced three-phase source produces phase voltages that obey the following two constraints:

$$|\mathbf{V}_{AN}| = |\mathbf{V}_{BN}| = |\mathbf{V}_{CN}| = V_P$$

$$\mathbf{V}_{AN} + \mathbf{V}_{BN} + \mathbf{V}_{CN} = 0 + j0$$

That is, the phase voltages have equal amplitudes (v_P) and sum to zero. There are two ways to satisfy these constraints:

Positive Phase Sequence		Negative Phase Sequence
$\mathbf{V}_{AN}$	$=$	$V_P \angle 0°$
$\mathbf{V}_{BN}$	$=$	$V_P \angle -120°$
$\mathbf{V}_{CN}$	$=$	$V_P \angle -240°$
$\mathbf{V}_{AN}$	$=$	$V_P \angle 0°$
$\mathbf{V}_{BN}$	$=$	$V_P \angle -240°$
$\mathbf{V}_{CN}$	$=$	$V_P \angle -120°$

(105)

Figure 86 shows the phasor diagrams for the positive and negative phase sequences. It is apparent that both sequences involve three equal-length phasors that are separated by an angle of 120°. As a result, the sum of any two phasors cancels the third. In the positive sequence

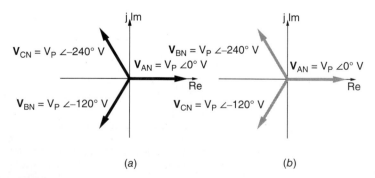

Figure 86   Two possible phase sequences: (*a*) positive and (*b*) negative.[1]

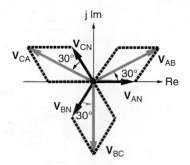

Figure 87 Phasor diagram showing phase and line voltages for positive phase sequence.[1]

the phase B voltage lags the phase A voltage by $120°$. In the negative sequence phase B lags by $240°$. It also is apparent that one phase sequence can be converted into the other by simply interchanging the labels on lines B and C. From a circuit analysis viewpoint there is no conceptual difference between the two sequences.

However, the reader is cautioned that "no conceptual difference" does not mean phase sequence is unimportant. It turns out that three-phase motors run in one direction when the positive sequence is applied and in the opposite direction for the negative sequence. In practice, it is essential that there be no confusion about which line A, B, and C is and whether the source phase sequence is positive or negative.

A simple relationship between the line and phase voltages is obtained by substituting the positive-phase-sequence voltages from Eq. (105) into the phasor sums in Eq. (104):

$$\mathbf{V}_{AB} = \mathbf{V}_{AN} - \mathbf{V}_{BN} = \sqrt{3}V_P\angle 30°$$

$$\mathbf{V}_{BC} = \sqrt{3}V_P\angle -90°$$

$$\mathbf{V}_{CA} = \sqrt{3}V_P\angle -210°$$

Figure 87 shows the phasor diagram of these results. The line voltage phasors have the same amplitude and are displaced from each other by $120°$. Hence, they obey equal-amplitude and zero-sum constraints like the phase voltages.

If the amplitude of the line voltages is v_L, then $V_L = \sqrt{3}V_P$. In a balanced three-phase system the line voltage amplitude is $\sqrt{3}$ times the phase voltage amplitude. This ratio appears in equipment descriptions such as 277/480 V three phase, where 277 is the phase voltage and 480 the line voltage. It is necessary to choose one of the phasors as the zero-phase reference when defining three-phase voltages and currents. Usually the reference is the line A phase voltage (i.e., $\mathbf{V}_{AN} = V_P\angle 0°$), as illustrated in Figs. 86 and 87.

5 TRANSIENT RESPONSE OF CIRCUITS

5.1 First-Order Circuits

First-order RC and RL circuits contain linear resistors and a single capacitor or a single inductor. Figure 88 shows RC and RL circuits divided into two parts: (a) the dynamic element and (b) the rest of the circuit containing only linear resistors and sources.

Dealing first with the RC circuit in Fig. 88a, a KVL equation is

$$R_T i(t) + v(t) = v_T(t) \tag{106}$$

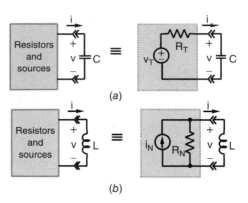

Figure 88   First-order circuits: (*a*) *RC* circuit and (*b*) *RL* circuit.[1]

The capacitor i–v constraint is

$$i(t) = C\frac{dv(t)}{dt}$$

Substituting the i–v constraint into the source constraint produces the equation governing the *RC* series circuit:

$$R_T C\frac{dv(t)}{dt} + v(t) = v_T(t)$$

The unknown in Eq. (106) is the capacitor voltage $v(t)$, which is called the *state variable* because it determines the amount or state of energy stored in the capacitive element.

Writing a KCL equation for the *RL* circuit in Fig. 88*b* yields

$$\frac{1}{R_N}v(t) + i(t) = i_N(t) \tag{107}$$

The element constraint for the inductor is

$$v(t) = L\frac{di(t)}{dt}$$

Combining the element and source constraints produces the differential equation for the *RL* circuit:

$$\frac{L}{R_N}\frac{di(t)}{dt} + i(t) = i_N(t)$$

The unknown in Eq. (107) is the inductor current, also called the state variable because it determines the amount or state of energy stored in the inductive element.

Note that Eqs. (106) and (107) have the same form. In fact, interchanging the following quantities converts one equation into the other:

$$G \leftrightarrow R$$
$$L \leftrightarrow C$$
$$i \leftrightarrow v$$
$$i_N \leftrightarrow v_T$$

This interchange is an example of the principle of duality. Because of duality there is no need to study the *RC* and *RL* circuits as independent problems. Everything learned solving the *RC* circuit can be applied to the *RL* circuit as well.

Step Response of RL and RC Circuits

For the *RC* circuit the response $v(t)$ must satisfy the differential equation (106) and the initial condition $v(0)$. The initial energy can cause the circuit to have a nonzero response even when the input $v_T(t) = 0$ for $t \geq 0$.

When the input to the RC circuit in Fig. 88 is a step function, the source can be written as $v_T(t) = v_A u(t)$. The circuit differential equation (106) then becomes

$$R_T C \frac{dv(t)}{dt} + v(t) = V_A u(t)$$

The step response of this circuit is a function $v(t)$ that satisfies this differential equation for $t \geq 0$ and meets the initial condition $v(0)$. Since $u(t) = 1$ for $t \geq 0$, then

$$R_T C \frac{dv(t)}{dt} + v(t) = V_A \quad \text{for } t \geq 0 \tag{108}$$

The solution $v(t)$ can be divided into two components:

$$v(t) = v_N(t) + v_F(t)$$

The first component $v_N(t)$ is the *natural response* and is the general solution equation (108) when the input is set to zero. The natural response has its origin in the physical characteristic of the circuit and does not depend on the form of the input. The component $v_F(t)$ is the *forced response* and is a particular solution of Eq. (108) when the input is the step function.

Finding the natural response requires the general solution of Eq. (108) with the input set to zero:

$$R_T C \frac{dv_N(t)}{dt} + v_N(t) = 0 \quad \text{for } t \geq 0$$

But this is the homogeneous equation that produces the zero-input response. Therefore, the form of the natural response is

$$v_N(t) = K e^{-t/(R_T C)} t \geq 0$$

This is a general solution of the homogeneous equation because it contains an arbitrary constant K. To evaluate K from the initial condition the total response is needed since the initial condition applies to the total response (natural plus forced).

Turning now to the forced response, a particular solution of the equation needs to be found:

$$R_T C \frac{dv_F(t)}{dt} + v_F(t) = V_A \quad \text{for } t \geq 0 \tag{109}$$

The equation requires that a linear combination of $v_F(t)$ and its derivative equal a constant v_A for $t \geq 0$. Setting $v_F(t) = K_F$ meets this condition since $dv_F/dt = dV_A/dt = 0$. Substituting $v_F = K_F$ into Eq. (109) results in $K_F = v_A$.

Now combining the forced and natural responses yields

$$v(t) = v_N(t) + v_F(t)$$
$$= K e^{-t/(R_T C)} + V_A t \geq 0$$

This equation is the general solution for the step response because it satisfies Eq. (106) and contains an arbitrary constant K. This constant can now be evaluated using the initial condition, $v(0) = V_0 = K e^0 + V_A = K + V_A$. The initial condition requires that $K = V_0 - V_A$. Substituting this conclusion into the general solution yields the step response of the RC circuit:

$$v(t) = (V_0 - V_A) e^{-t/(R_T C)} + V_A t \geq 0 \tag{110}$$

A typical plot of the waveform of $v(t)$ is shown in Fig. 89. The RL circuit in Fig. 88 is the dual of the RC circuit, so the development of its step response is similar. The result is

$$i(t) = (I_0 - I_A) e^{-R_N t/L} + I_A t \geq 0 \tag{111}$$

The RL circuit step response has the same form as the RC circuit step response in Eq. (110). At $t = 0$ the starting value of the response is $i(0) = I_0$ as required by the initial condition. The final value is the forced response $i(\infty) = i_F = i_A$ since the natural response decays to zero as time increases.

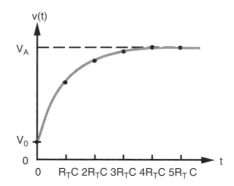

Figure 89   Step response of first-order *RC* circuit.[1]

Initial and Final Conditions.   The state variable responses can be written in the form

$$v_c(t), i_L(t) = [\text{IC} - \text{FC}]e^{-t/T_C} + \text{FC} \quad t \geq 0 \tag{112}$$

where IC stands for the initial condition ($t = 0$) and FC for the final condition ($t = 4$). To determine the step response of any first-order circuit, only three quantities, IC, FC, and T_C, are needed.

The final condition can be calculated directly from the circuit by observing that for $t > 5T_C$ the step responses approach a constant, or dc, value. Under the dc condition a capacitor acts like an open circuit and an inductor acts like a short circuit, so the final value of the state variable can be calculated using resistance circuit analysis methods.

Similarly the dc analysis method can be used to determine the initial condition in many practical situations. One common situation is a circuit containing a switch that remains in one state for a period of time that is long compared with the circuit time constant. If the switch is closed for a long period of time, then the state variable approaches a final value determined by the dc input. If the switch is now opened at $t = 0$, a transient occurs in which the state variable is driven to a new final condition.

The initial condition at $t = 0$ is the dc value of the state variable for the circuit configuration that existed before the switch was opened at $t = 0$. The switching action cannot cause an instantaneous change in the initial condition because capacitor voltage and inductor current are continuous functions of time. In other words, opening a switch at $t = 0$ marks the boundary between two eras. The dc condition of the state variable for the $t < 0$ era is the initial condition for the $t > 0$ era that follows. The parameters IC, FC, and T_C in switched dynamic circuits are found using the following steps:

Step 1: Find the initial condition IC by applying dc analysis to the circuit configuration for $t < 0$.

Step 2: Find the final condition FC by applying dc analysis to the circuit configuration for $t \geq 0$.

Step 3: Find the time constant *TC* of the circuit with the switch in the position for $t \geq 0$.

Step 4: Write the step response directly using Eqs. (112) without formulating and solving the circuit differential equation.

Example 28

For the circuit shown in Fig. 90a the switch has been closed for a long time. At $t = 0$ it opens. Find the capacitor voltage $v(t)$ and current $i(t)$ for $t \geq 0$.

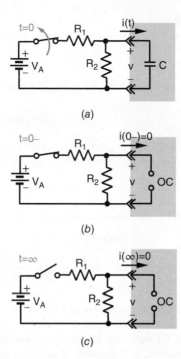

Figure 90   Solving switched dynamic circuit using initial and final conditions.[1]

Solution

Step 1: The initial condition is found by dc analysis of the circuit configuration in Fig. 90*b* where the switch is closed. Using voltage division, the initial capacitor voltage is found to be

$$v_C(0-) = \text{IC} = \frac{R_2 V_0}{R_1 + R_2}$$

Step 2: The final condition is found by dc analysis of the circuit configuration in Fig. 90*c* where the switch is open. Five time constants after the switch is opened the circuit has no practical dc excitation, so the final value of the capacitor voltage is zero.

Step 3: The circuit in Fig. 90*c* also is used to calculate the time constant. Since R_1 is connected in series with an open switch, the capacitor sees an equivalent resistance of only R_2. For $t \geq 0$ the time constant is $R_2 C$. Using Eq. (112) the capacitor voltage for $t \geq 0$ is

$$v_C(t) = (\text{IC} - \text{FC})e^{-t/T_C} + \text{FC}\, t \geq 0$$

$$= \frac{R_2 V_A}{R_1 + R_2} e^{-t/(R_2 C)} t \geq 0$$

This result is a zero-input response since there is no excitation for $t \geq 0$. To complete the analysis, the capacitor current is found by using its element constraint:

$$i_C(t) = C\frac{dv_C}{dt} = -\frac{V_0}{R_1 + R_2} e^{1/(R_2 C)} t \geq 0$$

For $t < 0$ the initial-condition circuit in Fig. 90*b* points out that $i_C(0-) = 0$ since the capacitor acts like an open circuit.

Example 29

The switch in Fig. 91a has been open for a "long time" and is closed at $t = 0$. Find the inductor current for $t > 0$.

Solution

The initial condition is found using the circuit in Fig. 91b. By series equivalence the initial current is

$$i(0-) = \text{IC} = \frac{V_0}{R_1 + R_2}$$

The final condition and the time constant are determined from the circuit in Fig. 91c. Closing the switch shorts out R_2 and the final condition and time constant for $t > 0$ are

$$i(\infty) = \text{FC} = \frac{V_0}{R_1} \qquad T_C = \frac{L}{R_N} = \frac{L}{R_1}$$

Using Eq. (112) the inductor current for $t \geq 0$ is

$$i(t) = (\text{IC} - \text{FC})e^{-t/T_C} + \text{FC} \qquad t \geq 0$$

$$= \left[\frac{V_0}{R_1 + R_2} - \frac{V_0}{R_1} \right] e^{-R_1 t/L} + \frac{V_0}{R_1} \qquad \text{A} \quad t \geq 0$$

First-Order Circuit Response to Other Than dc Signals

The response of linear circuits to a variety of signal inputs is an important concept in electrical engineering. Of particular importance is the response to a step reviewed in the previous section to the exponential and sinusoid. If the input to the *RC* circuit in Fig. 88 is an exponential or a

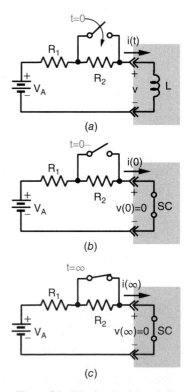

Figure 91 RL circuit with switch.

sinusoid, then the circuit differential equation is written as

$$\text{Exponential input:} \quad R_T C \frac{dv(t)}{dt} + v(t) = V_A e^{-\alpha t} u(t)$$

$$\text{Sinusoidal input:} \quad R_T C \frac{dv(t)}{dt} + v(t) = V_A \cos \omega t u(t)$$

(113)

The inputs on the right side of Eq. (113) are signals that start at $t = 0$ through some action such as closing a switch. A solution function $v(t)$ is needed that satisfies Eq. (113) for $t \geq 0$, and that meets the prescribed initial condition $v(0) = V_0$.

As with the step response, the solution is divided into two parts: natural response and forced response. The natural response is of the form

$$v_N(t) = K e^{-t/(R_T C)}$$

The natural response of a first-order circuit always has this form because it is a general solution of the homogeneous equation with input set to zero. The form of the natural response depends on physical characteristics of the circuit and is independent of the input.

The forced response depends on both the circuit and the nature of the forcing function. The forced response is a particular solution of the equation

$$\text{Exponential input:} \quad R_T C \frac{dv_F(t)}{dt} + v_F(t) = V_A e^{-\alpha t} t \geq 0$$

$$\text{Sinusoidal input:} \quad R_T C \frac{dv_F(t)}{dt} + v_F(t) = V_A \cos \omega t t \geq 0$$

(114)

This equation requires that $v_F(t)$ plus $R_T C$ times its first derivative add to produce either an exponential or a sinusoidal waveform for $t \geq 0$. The only way this can happen is for $v_F(t)$ and its derivative to be either an exponential of the same decay or sinusoids of the same frequency. This requirement brings to mind the derivative property of the exponential or the sinusoid. Hence one chooses a solution in the form of

$$\text{Exponential:} \quad v_F(t) = K_F e^{-\alpha t}$$

$$\text{Sinusoidal:} \quad v_F(t) = K_A \cos \omega t + K_B \sin \omega t$$

In this expression the constant K_F or the Fourier coefficients K_A and K_B are unknown. The approach we are using is called the method of undetermined coefficients. The unknown coefficients are found by inserting the forced solution $v_F(t)$ into the differential equation and equating the coefficients of the exponential in that case or of the sine and cosine terms. This yields the following:

$$\text{Exponential:} \quad K_F = \frac{V_A}{1 - \alpha R_T C}$$

$$\text{Sinusoidal:} \quad K_A = \frac{V_A}{1 + (\omega R_T C)^2}$$

$$K_B = \frac{\omega R_T C V_A}{1 + (\omega R_T C)^2}$$

The undetermined coefficients are now known since these equations express constants in terms of known circuit parameters $(R_T C)$ and known input signal parameters $(\omega$ and $v_A)$.

The forced and natural responses are combined and the initial condition used to find the remaining unknown constant K:

$$\text{Exponential:} \quad K = V_0 + \frac{V_A}{R_T C \alpha - 1}$$

$$\text{Sinusoidal:} \quad = V_0 - \frac{V_A}{1 + (\omega R_T C)^2}$$

Combining these together yields the function $v(t)$ that satisfies the differential equation and the initial conditions:

$$v(t) = \left\{ \underbrace{\left[V_0 + \frac{V_A}{R_T C \alpha - 1} \right] e^{-t/(R_T C)}}_{\text{Natural response}} - \underbrace{\frac{V_A}{R_T C \alpha - 1} e^{-\alpha t}}_{\text{Forced response}} \right\} u(t) \mathrm{V}$$

$$= \left\{ \overbrace{\left[V_0 - \frac{V_A}{1 + (\omega R_T C)^2} \right] e^{-t/(R_T C)}}^{\text{Natural response}} + \overbrace{\frac{V_A}{1 + (\omega R_T C)^2} (\cos \omega t + \omega R_T C \sin \omega t)}^{\text{Forced response}} \right\}$$

$$\times u(t) \mathrm{V} \tag{115}$$

Equations (115) are the complete responses of the RC circuit for an initial condition V_0 and either an exponential or a sinusoidal input.

5.2   Second-Order Circuits

Second-order circuits contain two energy storage elements that cannot be replaced by a single equivalent element. They are called *second-order circuits* because the circuit differential equation involves the second derivative of the dependent variable. The series *RLC* circuit will illustrate almost all of the basic concepts of second-order circuits.

The circuit in Fig. 92*a* has an inductor and a capacitor connected in series. The source–resistor circuit can be reduced to the Thévenin equivalent shown in Fig. 92*b*. Applying KVL around the loop and applying the two *i–v* characteristics of the inductor and capacitor yields

$$v_L(t) + v_R(t) + v_C(t) = v_T(t)$$

$$LC \frac{d^2 v_C(t)}{dt^2} + R_T C \frac{dv_C(t)}{dt} + v_C(t) = v_T(t) \tag{116}$$

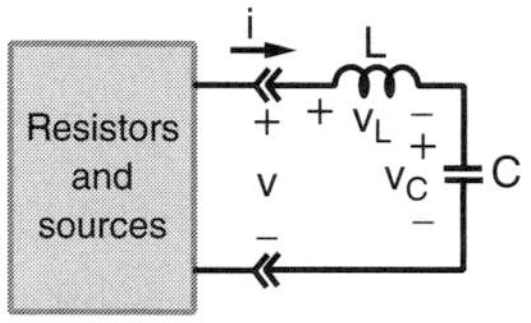

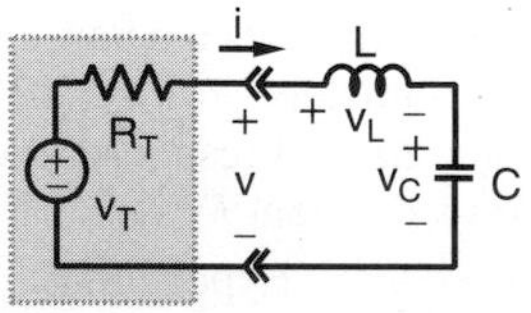

(a)

(b)

Figure 92   Series *RLC* circuit.[1]

In effect, this is a KVL equation around the loop in Fig. 92b, where the inductor and resistor voltages have been expressed in terms of the capacitor voltage.

The Thévenin voltage $v_T(t)$ is a known driving force. The initial conditions are determined by the values of the capacitor voltage and inductor current at $t = 0$, that is, V_0 and I_0:

$$v_C(0) = V_0 \quad \text{and} \quad \frac{dv_C}{dt}(0) = \frac{1}{C}i(0) = \frac{I_0}{C}.$$

The circuit dynamic response for $t \geq 0$ can be divided into two components: (1) the zero-input response caused by the initial conditions and (2) the zero-state response caused by driving forces applied after $t = 0$. With $v_T = 0$ (zero input) Eq. (116) becomes

$$LC\frac{d^2v_C(t)}{dt^2} + R_T C\frac{dv_C(t)}{dt} + v_C(t) = 0$$

This result is a second-order homogeneous differential equation in the capacitor voltage. Inserting a trial solution of $v_{CN}(t) = Ke^{st}$ into the above equation results in the following *characteristic equation* of the series *RLC* circuit:

$$LCs^2 + R_T Cs + 1 = 0$$

In general, the above quadratic characteristic equation has two roots:

$$s_1, s_2 = \frac{-R_T C \pm \sqrt{(R_T C)^2 - 4LC}}{2LC}$$

The roots can have three distinct possibilities:

Case A: If $(R_T C)^2 - 4LC > 0$, the discriminant is positive and there are two real, unequal roots ($s_1 = -\alpha_1 \neq s_2 = -\alpha_2$).

Case B: If $(R_T C)^2 - 4LC = 0$, the discriminant vanishes and there are two real, equal roots ($s_1 = s_2 = -\alpha$).

Case C: If $(R_T C)^2 - 4LC < 0$, the discriminant is negative and there are two complex conjugate roots ($s_1 = -\alpha - j\beta$ and $s_2 = -\alpha + j\beta$).

Second-Order Circuit Zero-Input Response

Since the characteristic equation has two roots, there are two solutions to the homogeneous differential equation:

$$v_{C1}(t) = K_1 e^{s_1 t} \quad \text{and} \quad v_{C2}(t) = K_2 e^{s_2 t}$$

Therefore, the general solution for the zero-input response is of the form

$$v_C(t) = K_1 e^{s_1 t} + K_2 e^{s_2 t} \tag{117}$$

The constants K_1 and K_2 can be found using the initial conditions:

$$v_c(t) = \frac{s_2 V_0 - I_0/C}{s_2 - s_1}e^{s_1 t} + \frac{-s_1 V_0 + I_0/C}{s_2 - s_1}e^{s_2 t} \qquad t \geq 0 \tag{118}$$

Equation (118) is the general zero-input response of the series *RLC* circuit. The response depends on two initial conditions, V_0 and I_0, and the circuit parameters R_T, L, and C since s_1 and s_2 are the roots of the characteristic equation $LCs^2 + R_T Cs + 1 = 0$. The response has different waveforms depending on whether the roots s_1 and s_2 fall under case A, B, or C.

For case A the two roots are real and distinct. Using the notation $s_1 = -\alpha_1$ and $s_2 = -\alpha_2$, the form of the zero-input response for $t \geq 0$ is

$$v_c(t) = \left[\frac{\alpha_2 V_0 + I_0/C}{\alpha_2 - \alpha_1}\right]e^{-\alpha_1 t} - \left[\frac{\alpha_1 V_0 + I_0/C}{\alpha_2 - \alpha_1}\right]e^{-\alpha_2 t} \qquad t \geq 0$$

This form is called the *overdamped response*. The waveform has two time constants $1/\alpha_1$ and $1/\alpha_2$.

With case B the roots are real and equal. Using the notation $s_1 = s_2 = -\alpha$, the general form becomes

$$v_C(t) = V_0 e^{-\alpha t} + \left(\alpha V_0 + \frac{I_0}{C}\right) te^{-\alpha t} \qquad t \geq 0$$

This special form is called the *critically damped response*. The critically damped response includes an exponential and a damped ramp waveform.

Case C produces complex-conjugate roots of the form $s_1 = -\alpha - j\beta$ and $s_2 = -\alpha + j\beta$. The form of case C is

$$v_C(t) = V_0 e^{-\alpha t}\cos \beta t + \left(\frac{\alpha V_0 + I_0/C}{\beta}\right) e^{-\alpha t}\sin \beta \qquad t \geq 0$$

This form is called the *underdamped response*. The underdamped response contains a damped sinusoid waveform where the real part of the roots (α) provides the damping term in the exponential, while the imaginary part (β) defines the frequency of the sinusoidal oscillation.

Second-Order Circuit Step Response
The general second-order linear differential equation with a step function input has the form

$$a_2 \frac{d^2 y(t)}{dt^2} + a_1 \frac{dy(t)}{dt} + a_0 y(t) = Au(t)$$

where $y(t)$ is a voltage or current response, $Au(t)$ is the step function input, and a_2, a_1, and a_0 are constant coefficients. The step response is the general solution of this differential equation for $t \geq 0$. The step response can be found by partitioning $y(t)$ into forced and natural components:

$$y(t) = y_N(t) + y_F(t)$$

The natural response $y_N(t)$ is the general solution of the homogeneous equation (input set to zero), while the forced response $y_F(t)$ is a particular solution of the equation

$$a_2 \frac{d^2 y_F(t)}{dt^2} + a_1 \frac{dy_F(t)}{dt} + a_0 y_F(t) = A \qquad t \geq 0$$

The particular solution is simply $y_F = A/a_0$.

In a second-order circuit the zero-state and natural responses take one of the three possible forms: overdamped, critically damped, or underdamped. To describe the three possible forms, two parameters are used: ω_0, the *undamped natural frequency*, and ζ, the *damping ratio*. Using these two parameters, the general homogeneous equation is written in the form

$$\frac{d^2 y_N(t)}{dt^2} + 2\zeta\omega_0 \frac{dy_N(t)}{dt} + \omega_0^2 y_N(t) = 0$$

The above equation is written in *standard form* of the second-order linear differential equation. When a second-order equation is arranged in this format, its damping ratio and undamped natural frequency can be readily found by equating its coefficients with those in the standard form. For example, in the standard form the homogeneous equation for the series *RLC* circuit is

$$\frac{d^2 v_C(t)}{dt^2} + \frac{R_T}{L}\frac{dv_c(t)}{dt} + \frac{1}{LC}v_C(t) = 0$$

Equating like terms yields

$$\omega_0^2 = \frac{1}{LC} \quad \text{and} \quad 2\zeta\omega_0 = \frac{R_T}{L}$$

for the series RLC circuit. Note that the circuit elements determine the values of the parameters ω_0 and ζ. The characteristic equation is

$$s^2 + 2\zeta\omega_0 s + \omega_0^2 = 0$$

and its roots are

$$s_1, s_2 = \omega_0(-\zeta \pm \sqrt{\zeta^2 - 1})$$

The expression under the radical defines the form of the roots and depends only on the damping ratio ζ:

Case A: For $\zeta > 1$ the discriminant is positive and there are two unequal, real roots

$$s_1, s_2 = -\alpha_1, -\alpha_2 = \omega_0(-\zeta \pm \sqrt{\zeta^2 - 1})$$

and the natural response is of the form

$$y_N(t) = K_1 e^{-\alpha_1 t} + K_2 e^{-\alpha_2 t} \quad t \geq 0 \tag{119}$$

Case B: For $\zeta = 1$ the discriminant vanishes and there are two real, equal roots,

$$s_1 = s_2 = -\alpha = -\zeta\omega_0$$

and the natural response is of the form

$$y_N(t) = K_1 e^{-\alpha t} + K_2 t e^{-\alpha t} \quad t \geq 0 \tag{120}$$

Case C: For $\zeta < 1$, the discriminant is negative leading to two complex, conjugate roots $s_1, s_2 = -\alpha \pm j\beta$, where $\alpha = \zeta\omega_0$ and $\beta = \omega_0\sqrt{1 - \zeta^2}$ and the natural response is of the form

$$y_N(t) = e^{-\alpha t}(K_1 \cos \beta t + K_2 \sin \beta t) \quad t \geq 0 \tag{121}$$

In other words, for $\zeta > 1$ the natural response is overdamped, for $\zeta = 1$ the natural response is critically damped, and for $\zeta < 1$ the response is underdamped.

Combining the forced and natural responses yields the step response of the general second-order differential equation in the form

$$y(t) = y_N(t) + \frac{A}{a_0} \quad t \geq 0$$

The factor A/a_0 is the forced response. The natural response $y_N(t)$ takes one of the forms in Eqs. (119)–(121) depending on the value of the damping ratio. The constants K_1 and K_2 in the natural response can be evaluated from the initial conditions.

Example 30
The series RLC circuit in Fig. 93 is driven by a step function and is in the zero state at $t = 0$. Find the capacitor voltage for $t \geq 0$.

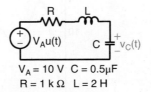

$V_A = 10 \text{ V} \quad C = 0.5\mu\text{F}$
$R = 1\,\text{k}\Omega \quad L = 2\,\text{H}$

Figure 93   Series RLC Circuit.[1]

Solution

This is a series *RLC* circuit so the differential equation for the capacitor voltage is

$$10^{-6}\frac{d^2 v_C(t)}{dt^2} + 0.5 \times 10^{-3}\frac{d v_C(t)}{dt} + v_C(t) = 10 \quad t \geq 0$$

By inspection the forced response is $v_{CF}(t) = 10$ V. In standard format the homogeneous equation is

$$\frac{d^2 v_{CN}(t)}{dt^2} + 500\frac{d v_{CN}(t)}{dt} + 10^6 v_{CN}(t) = 0 \quad t \geq 0$$

Comparing this format, the standard form yields

$$\omega_0^2 = 10^6 \quad \text{and} \quad 2\zeta\omega_0 = 500$$

so that $\omega_0 = 1000$ and $\zeta = 0.25$. Since $\zeta < 1$, the natural response is underdamped (case C) and has the form

$$\alpha = \zeta\omega_0 = 250 \text{ Np}$$

$$\beta = \omega_0\sqrt{1 - \zeta^2} - 968 \text{ rad/s}$$

$$v_{CN}(t) = K_1 e^{-250t}\cos 968t + K_2 e^{-250t}\sin 968t$$

The general solution of the circuit differential equation is the sum of the forced and natural responses:

$$v_C(t) = 10 + K_1 e^{-250t}\cos 968t$$
$$+ K_2 e^{-250t}\sin 968 \quad t \geq 0$$

The constants K_1 and K_2 are determined by the initial conditions.

The circuit is in the zero state at $t = 0$, so the initial conditions are $v_C(0) = 0$ and $i_L(0) = 0$. Applying the initial-condition constraints to the general solution yields two equations in the constants K_1 and K_2:

$$v_C(0) = 10 + K_1 = 0$$

$$\frac{dv_C}{dt}(0) = -250 K_1 + 968 K_2 = 0$$

These equations yield $K_1 = -10$ and $K_2 = -2.58$. The step response of the capacitor voltage step response is

$$v_C(t) = 10 - 10e^{-250t}\cos 968t$$
$$- 2.58 e^{-250t}\sin 968t \text{V} \quad t \geq 0$$

A plot of $v_C(t)$ versus time is shown in Fig. 94. The waveform and its first derivative at $t = 0$ satisfy the initial conditions. The natural response decays to zero so the forced response determines the final value of $v_C(\infty) = 10$ V. Beginning at $t = 0$ the response climbs rapidly but

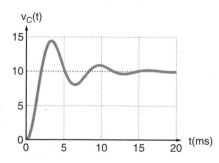

Figure 94 Plot of $v_C(t)$ versus time.[1]

overshoots the final value several times before eventually settling down. The damped sinusoidal behavior results from the fact that $\zeta < 1$, producing an underdamped natural response.

6 FREQUENCY RESPONSE

Linear circuits are often characterized by their behavior to sinusoids, in particular, how they process signals versus frequency. Audio, communication, instrumentation, and control systems all require signal processing that depends at least in part on their frequency response.

6.1 Transfer Functions and Input Impedance

The proportionality property of linear circuits states that the output is proportional to the input. In the *phasor* domain the proportionality factor is a rational function of $j\omega$ called a *transfer function*. More formally, in the phasor domain a transfer function is defined as the ratio of the output phasor to the input phasor with *all initial conditions set to zero*:

$$\text{Transfer function} = \frac{\text{Output phasor}}{\text{Input phasor}} = H(j\omega)$$

To study the role of transfer functions in determining circuit responses is to write the phasor domain input–output relationship as

$$Y(j\omega) = H(j\omega) \cdot X(j\omega) \tag{122}$$

where $H(j\omega)$ is the transfer function, $X(j\omega)$ is the input signal transform (a voltage or a current phasor), and $Y(j\omega)$ is the output signal transform (also a voltage or current phasor). Figure 95 shows a block diagram representation of the phasor domain input–output relationship.

In an *analysis* problem the circuit defined by $H(j\omega)$ and the input $X(j\omega)$ are known and the response $Y(j\omega)$ is sought. In a *design* problem the circuit is unknown. The input and the desired output or their ratio $H(j\omega) = Y(j\omega)/X(j\omega)$ are given, and the objective is to devise a circuit that realizes the specified input–output relationship. A linear circuit analysis problem has a unique solution, but a design problem may have one, many, or even no solution. Choosing the best of several solutions is referred to as an *evaluation* problem.

There are two major types of functions that help define a circuit: input impedance and transfer functions. *Input impedance* relates the voltage and current at a pair of terminals called a port. The input impedance $Z(j\omega)$ of the one-port circuit in Fig. 96 is defined as

$$Z(j\omega) = \frac{V(j\omega)}{I(j\omega)} \tag{123}$$

When the one port is driven by a current source, the response is $V(j\omega) = Z(j\omega)I(j\omega)$. On the other hand, when the one port is driven by a voltage source, the response is $I(j\omega) = [Z(j\omega)]^{-1} V(j\omega)$.

The term input impedance means that the circuit is driven at one port and the response is observed at the same port. The impedances of the three basic circuit elements $Z_R(j\omega)$, $Z_L(j\omega)$, and $Z_C(j\omega)$ are elementary examples of input impedances. The equivalent impedances found by combining elements in series and parallel are also effectively input impedances. The terms input impedance, *driving-point impedance*, and *equivalent impedance* are synonymous. Input

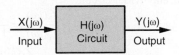

Figure 95 Block diagram for phasor domain input–output relationship.[1]

544 Electric Circuits

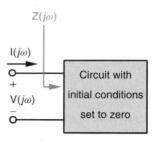

Figure 96 One-port circuit.[1]

impedance is useful in impedance-matching circuits at their interface and to help determine if *loading* will be an issue.

Transfer functions are usually of greater interest in signal-processing applications than input impedances because they describe how a signal is modified by passing through a circuit. A *transfer function* relates an input and response (or output) at different ports in the circuit. Since the input and output signals can be either a current or a voltage, four kinds of transfer functions can be defined:

$$H_V(j\omega) = \text{voltage transfer function} = \frac{V_2(j\omega)}{V_1(j\omega)}$$

$$H_Y(j\omega) = \text{transfer admittance} = \frac{I_2(j\omega)}{V_1(j\omega)}$$

$$H_I(j\omega) = \text{current transfer function} = \frac{I_2(j\omega)}{I_1(j\omega)} \qquad (124)$$

$$H_Z(j\omega) = \text{transfer impedance} = \frac{V_2(j\omega)}{I_1(j\omega)}$$

The functions $H_V(j\omega)$ and $H_I(j\omega)$ are dimensionless since the input and output signals have the same units. The function $H_Z(j\omega)$ has units of ohms and $H_Y(j\omega)$ has unit of siemens.

Transfer functions always involve an input applied at one port and a response observed at a different port in the circuit. It is important to realize that a transfer function is only valid for a given input port and the specified output port. They cannot be turned upside down like the input impedance. For example, the voltage transfer function $H_V(j\omega)$ relates the voltage $V_1(j\omega)$ applied at the input port to the voltage response $V_2(j\omega)$ observed at the output port in Fig. 95. The voltage transfer function for signal transmission in the opposite direction is usually *not* $1/H_V(j\omega)$.

Determining Transfer Functions

The divider circuits in Fig. 97 occur so frequently that it is worth taking time to develop their transfer functions in general terms. Using phasor domain analysis the voltage transfer function of a voltage divider circuit is

$$H_V(j\omega) = \frac{V_2(j\omega)}{V_1(j\omega)} = \frac{Z_2(j\omega)}{Z_1(j\omega) + Z_2(j\omega)}$$

Similarly, using phasor domain current division in Fig. 97b results in the current transfer function of a current divider circuit:

$$H_I(j\omega) = \frac{I_2(j\omega)}{I_1(j\omega)} = \frac{1/[Z_2(j\omega)]}{1/[Z_1(j\omega)] + 1/[Z_2(j\omega)]}$$

$$= \frac{Z_1(j\omega)}{Z_1(j\omega) + Z_2(j\omega)}$$

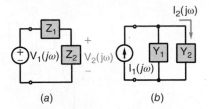

Figure 97 Basic divider circuits: (*a*) voltage divider and (*b*) current divider.[1]

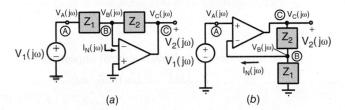

Figure 98 Basic op-amp circuits: (*a*) inverting amplifier and (*b*) noninverting amplifier.[1]

By series equivalence the driving-point impedance at the input of the voltage divider is $Z_{EQ}(j\omega) = Z_1(j\omega) + Z_2(j\omega)$. By parallel equivalence the driving-point impedance at the input of the current divider is $Z_{EQ}(j\omega) = 1/[1/Z_1(j\omega) + 1/Z_2(j\omega)]$.

Two other useful circuits are the inverting and noninverting op-amp configurations shown in Fig. 98.

The voltage transfer function of the inverting circuit in Fig. 98*a* is

$$H_V(j\omega) = \frac{V_2(j\omega)}{V_1(j\omega)} = -\frac{Z_2(j\omega)}{Z_1(j\omega)}$$

The input impedance of this circuit is simply $Z_1(j\omega)$ since $v_B(j\omega) = 0$. The effect of $Z_1(j\omega)$ should be studied when connecting it to another circuit or a nonideal source since it can cause undesired loading.

For the noninverting circuit in Fig. 98*b* the voltage transfer function is

$$H_V(j\omega) = \frac{V_2(j\omega)}{V_1(j\omega)} = \frac{Z_1(j\omega) + Z_2(j\omega)}{Z_1(j\omega)}$$

The ideal op-amp draws no current at its input terminals, so theoretically the input impedance of the noninverting circuit is infinite; in practice it is quite high, upward of $10^{10}\,\Omega$.

Example 31

For the circuit in Fig. 99, find (a) the input impedance seen by the voltage source and (b) the voltage transfer function $H_V(j\omega) = V_2(j\omega)/V_1(j\omega)$.

Solution

a. The circuit is a voltage divider. First find the equivalent impedances of the two legs of the divider. The two elements in parallel combine to produce the series leg impedance $Z_1(j\omega)$:

$$Z_1(j\omega) = \frac{1}{C_1 j\omega + 1/R_1} = \frac{R_1}{R_1 C_1 j\omega + 1}$$

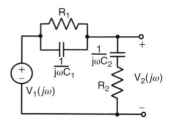

Figure 99   Circuit.[1]

The two elements in series combine to produce shunt (parallel) leg impedance $Z_2(j\omega)$:

$$Z_2(j\omega) = R_2 + \frac{1}{C_2 j\omega} = \frac{R_2 C_2 j\omega + 1}{C_2 j\omega}$$

Using series equivalence, the input impedance seen at the input is

$$Z_{EQ}(j\omega) = Z_1(j\omega) + Z_2(j\omega)$$

$$= \frac{R_1 C_1 R_2 C_2 (j\omega)^2 + (R_1 C_1 + R_2 C_2 + R_1 C_2)j\omega + 1}{C_2 j\omega (R_1 C_1 j\omega + 1)}$$

b. Using voltage division, the voltage transfer function is

$$H_V(j\omega) = \frac{Z_2(j\omega)}{Z_{EQ}(j\omega)}$$

$$= \frac{(R_1 C_1 j\omega + 1)(R_2 C_2 j\omega + 1)}{R_1 C_1 R_2 C_2 (j\omega)^2 + (R_1 C_1 + R_2 C_2 + R_1 C_2)j\omega + 1}$$

Example 32

a. Find the driving-point impedance seen by the voltage source in Fig. 100.

b. Find the voltage transfer function $H_V(j\omega) = V_2(j\omega)/V_1(j\omega)$ of the circuit.

c. If $R_1 = 1 \text{ k}\Omega$, $R_2 = 10 \text{ k}\Omega$, $C_1 = 10 \text{ nF}$, and $C_2 = 1 \text{ }\mu\text{F}$, evaluate the driving-point impedance and the transfer function.

Solution

The circuit is an inverting op-amp configuration. The input impedance and voltage transfer function of this configuration are

$$Z_{IN}(j\omega) = Z_1(j\omega) \quad \text{and} \quad H_V(j\omega) = -\frac{Z_2(j\omega)}{Z_1(j\omega)}$$

a. The input impedance is

$$Z_1(j\omega) = R_1 + \frac{1}{j\omega C_1} = \frac{R_1 C_1 j\omega + 1}{j\omega C_1}$$

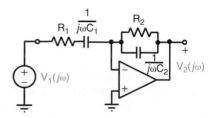

Figure 100   Voltage source.[1]

b. The feedback impedance is

$$Z_2(j\omega) = \frac{1}{j\omega C_2 + 1/R_2} = \frac{R_2}{R_2 C_2 j\omega + 1}$$

and the voltage transfer function is

$$H_V(j\omega) = -\frac{Z_2(j\omega)}{Z_1(j\omega)}$$

$$= -\frac{R_2 C_1 j\omega}{(R_1 C_1 j\omega + 1)(R_2 C_2 j\omega + 1)}$$

c. For the values of R's and C's given

$$Z_1(j\omega) = \frac{1000(j\omega + 10^5)}{j\omega}$$

and

$$H_V(j\omega) = -\frac{1000 j\omega}{(j\omega + 100)(j\omega + 10^5)}$$

6.2 Cascade Connection and Chain Rule

Signal-processing circuits often involve a *cascade connection* in which the output voltage of one circuit serves as the input to the next stage. In some cases, the overall voltage transfer function of the cascade can be related to the transfer functions of the individual stages by a *chain rule*:

$$H_V(j\omega) = H_{V1}(j\omega) \times H_{V2}(j\omega) \times \cdots \times H_{Vk}(j\omega) \qquad (125)$$

where $H_{V1}, H_{V2}, \ldots, H_{Vk}$ are the voltage transfer functions of the individual stages when operated separately. It is important to understand when the chain rule applies since it greatly simplifies the analysis and design of cascade circuits.

Figure 101 shows two *RC* circuits or stages connected in cascade at an interface. When disconnected and operated separately, the transfer functions of each stage are easily found using voltage division as follows:

$$H_{V1}(j\omega) = \frac{R}{R + 1/j\omega C} = \frac{Rj\omega C}{Rj\omega C + 1}$$

$$H_{V2}(j\omega) = \frac{1/j\omega C}{R + 1/j\omega C} = \frac{1}{Rj\omega C + 1}$$

When connected in cascade the output of the first stage serves as the input to the second stage. If the chain rule applies, the overall transfer function would be expected to be

$$H_V(j\omega) = \frac{V_3(j\omega)}{V_1(j\omega)} = \left(\frac{V_2(j\omega)}{V_1(j\omega)}\right)\left(\frac{V_3(j\omega)}{V_2(j\omega)}\right)$$

$$= H_{V1}(j\omega) \times H_{V2}(j\omega)$$

$$= \underbrace{\left(\frac{Rj\omega C}{Rj\omega C + 1}\right)}_{\text{1st stage}} \underbrace{\left(\frac{1}{Rj\omega C + 1}\right)}_{\text{2nd stage}}$$

$$= \underbrace{\frac{Rj\omega C}{(Rj\omega C)^2 + 2Rj\omega C + 1}}_{\text{combined}}$$

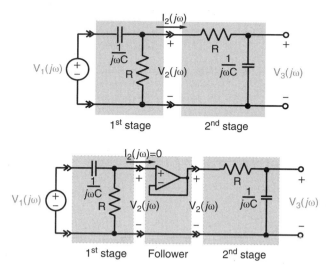

Figure 101   (*a*) Two-part circuits connected in cascade. (*b*) Cascade connection with voltage follower isolation.[1]

However, the overall transfer function of this circuit is actually found to be

$$H_V(j\omega) = \frac{Rj\omega C}{(Rj\omega C)^2 + 3Rj\omega C + 1}$$

which disagrees with the chain rule result.

The reason for the discrepancy is that when they are connected in cascade the second circuit "loads" the first circuit. That is, the voltage divider rule requires the current $I_2(j\omega)$ in Fig. 101*a* be zero. The no-load condition $I_2(j\omega) = 0$ is valid when the stages operate separately, but when connected together the current is no longer zero. The chain rule does not apply here because loading caused by the second stage alters the transfer function of both stages.

The loading problem goes away when an op-amp voltage follower is inserted between the *RC* circuit stages (Fig. 101*b*). With this modification the chain rule in Eq. (125) applies because the voltage follower isolates the two *RC* circuits. Recall that ideally a voltage follower has infinite input resistance and zero output resistance. Therefore, the follower does not draw any current from the first *RC* circuit [$I_2(j\omega) = 0$] and its transfer function of "1" allows $V_2(j\omega)$ to be applied directly across the input of the second *RC* circuit.

The chain rule in Eq. (125) applies if connecting a stage does not change or load the output of the preceding stage. Loading can be avoided by connecting an op amp voltage follower between stages. More importantly, loading does not occur if the output of the preceding stage is the output of an op amp or controlled source unless the load resistance is very low. These elements act very close to ideal voltage sources whose outputs are unchanged by connecting the subsequent stage.

For example, in the top representation the two circuits in Fig. 102 are connected in a cascade with circuit C1 appearing first in the cascade followed by circuit C2. The chain rule applies to this configuration because the output of circuit C1 is an op amp that can handle the load presented by circuit C2. On the other hand, if the stages are interchanged so that the op amp circuit C1 follows the *RC* circuit C2 in the cascade, then the chain rule would not apply because the input impedance of circuit C1 would then load the output of circuit C2.

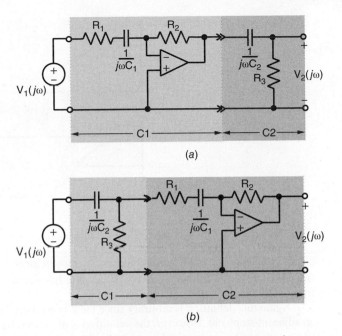

Figure 102 Effects of stage location on loading.[1]

6.3 Frequency Response Descriptors

The relationships between the input and output sinusoids are important to frequency-sensitive circuits and can be summarized in the following statements.

Realizing the circuit transfer function is usually a complex function of ω, its effect on the sinusoidal steady-state response can be found through its *gain* function $|H(j\omega)|$ and *phase* function $< H(j\omega)$ as follows:

$$\text{Magnitude of } H(j\omega) = |H(j\omega)| = \frac{\text{output amplitude}}{\text{input amplitude}}$$

$$\text{Angle of } H(j\omega) = \angle H(j\omega)$$

$$= \text{output phase} - \text{input phase}$$

Taken together the gain and phase functions show how the circuit modifies the input amplitude and phase angle to produce the output sinusoid. These two functions define the *frequency response* of the circuit since they are frequency-dependent functions that relate the sinusoidal steady-state input and output. The gain and phase functions can be expressed mathematically or presented graphically as in Fig. 103, which shows an example frequency response plot called a Bode diagram. These diagrams can be constructed by hand but are readily and more accurately produced by simulation and mathematical software products.

The terminology used to describe the frequency response of circuits and systems is based on the form of the gain plot. For example, at high frequencies the gain in Fig. 103 falls off so that output signals in this frequency range are reduced in amplitude. The range of frequencies over which the output is significantly attenuated is called the *stopband*. At low frequencies the gain is essentially constant and there is relatively little attenuation. The frequency range over which there is little attenuation is called a *passband*. The frequency associated with the boundary between a passband and an adjacent stopband is called the *cutoff frequency* ($\omega_C = 2\pi f_C$). In general, the transition from the passband to the stopband is gradual so the precise location of the

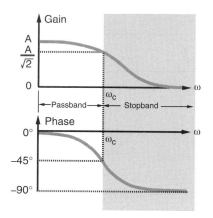

Figure 103 Frequency response plots.[1]

cutoff frequency is a matter of definition. The most widely used definition specifies the cutoff frequency to be the frequency at which the gain has decreased by a factor of $1/\sqrt{2} = 0.707$ from its maximum value in the passband.

Again this definition is arbitrary since there is no sharp boundary between a passband and an adjacent stopband. However, the definition is motivated by the fact that the power delivered to a resistance by a sinusoidal current or voltage waveform is proportional to the square of its amplitude. At a cutoff frequency the gain is reduced by a factor of $1/\sqrt{2}$ and the square of the output amplitude is reduced by a factor of $\frac{1}{2}$. For this reason the cutoff frequency is also called the *half-power frequency*. In filter design the region from where the output amplitude is reduced by 0.707 and a second frequency wherein the output must have decayed to some specified value is called the transition region. This region is where much of the filter design attention is focused. How rapidly a filter transitions from the cutoff frequency to some necessary attenuation is what occupies much of the efforts of filter designers.

Additional frequency response descriptors are based on the four prototype gain characteristics shown in Fig. 104. A *low-pass* gain characteristic has a single passband extending from zero frequency (dc) to the cutoff frequency. A *high-pass* gain characteristic has a single passband extending from the cutoff frequency to infinite frequency. A *bandpass* gain has a single passband with two cutoff frequencies neither of which is zero or infinite. Finally, the *bandstop* gain has a single stopband with two cutoff frequencies neither of which is zero or infinite.

The *bandwidth* of a gain characteristic is defined as the frequency range spanned by its passband. The bandwidth (BW) of a low-pass circuit is equal to its cutoff frequency (BW = ω_C). The bandwidth of a high-pass characteristic is infinite since passband extends to infinity. For the bandpass and bandstop cases in Fig. 104 the bandwidth is the difference in the two cutoff frequencies:

$$\text{BW} = \omega_{C2} - \omega_{C1} \tag{126}$$

For the bandstop case Eq. (126) defines the width of the stopband rather than the passband.

The gain responses in Fig. 104 have different characteristics at zero and infinite frequency:

Prototype	Gain at $\omega = 0$	Gain at $\omega = 4$
Low pass	Finite	0
High pass	0	Finite
Bandpass	0	0
Bandstop	Finite	Finite

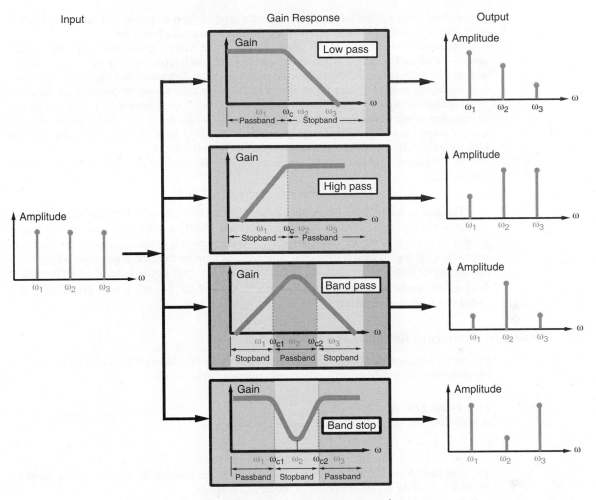

Input Gain Response Output

Figure 104 Four basic gain responses.[1]

Since these extreme values form a unique pattern, the type of gain response can be inferred from the values of $|H(0)|$ and $|H(\infty)|$. These endpoint values in turn are usually determined by the impedance of capacitors and inductors in the circuit. In the sinusoidal steady state the impedances of these elements are

$$Z_C(j\omega) = \frac{1}{j\omega C} \quad \text{and} \quad Z_L(j\omega) = j\omega L$$

These impedances vary with respect to frequency. An inductor's impedance increases linearly with increasing frequency, while that of a capacitor varies inversely with frequency. They form a unique pattern at zero and infinite frequency:

Element	Impedance (Ω) at $\omega = 0$ (dc)	Impedance (Ω) at $\omega = 4$
Capacitor ($1/j\omega C$)	Infinite (open circuit)	0 (short circuit)
Inductor ($j\omega L$)	0 (short circuit)	Infinite (open circuit)
Resistor (R)	R	R

These observations often allow one to infer the type of gain response and hence the type of filter directly from the circuit itself without finding the transfer functions.

Frequency response plots are almost always made using logarithmic scales for the frequency variable. The reason is that the frequency ranges of interest often span several orders of magnitude. A logarithmic frequency scale compresses the data range and highlights important features in the gain and phase responses. The use of a logarithmic frequency scale involves some special terminology. Any frequency range whose endpoints have a 2 : 1 ratio is called an *octave*. Any range whose endpoints have a 10 : 1 ratio is called a *decade*. For example, the frequency range from 10 to 20 Hz is one octave, as is the range from 20 to 40 MHz. The standard UHF (ultrahigh frequency) band spans one decade from 0.3 to 3 GHz.

In frequency response plots the gain $|H(j\omega)|$ is often expressed in *decibels* (dB), defined as

$$|H(j\omega)|_{dB} = 20\log_{10}|H(j\omega)|$$

The gain in decibels can be positive, negative, or zero. A gain of 0 dB means that $|H(j\omega)| = 1$; that is, the input and output amplitudes are equal. Positive decibel gains mean the output amplitude exceeds the input since $|H(j\omega)| > 1$ and the circuit is said to *amplify* the signal. A negative decibel gain means the output amplitude is smaller than the input since $|H(j\omega)| < 1$ and the circuit is said to *attenuate* the signal. A cutoff frequency occurs when the gain is reduced from its maximum passband value by a factor $1/\sqrt{2}$ or 3 dB. For this reason the cutoff is also called the *3-dB down frequency*.

6.4 First-Order Frequency Response and Filter Design

Frequency-selective circuits are fundamental to all types of systems. First-order filters are simple to design and can be effective for many common applications.

First-Order Low-Pass Response
A first-order low-pass transfer function can be written as

$$H(j\omega) = \frac{K}{j\omega + \alpha}$$

The constants K and α are real. The constant K can be positive or negative, but α must be positive so that the natural response of the circuit is stable.

The gain and phase functions are given as

$$|H(j\omega)| = \frac{|K|}{\sqrt{\omega^2 + \alpha^2}}$$

$$\angle H(j\omega) = \angle K - \tan^{-1}\left(\frac{\omega}{\alpha}\right) \tag{127}$$

The gain function is a positive number. Since K is real, the angle of K ($< K$) is either $0°$ when $K > 0$ or $\pm 180°$ when $K < 0$. An example of a negative K occurs in an inverting op amp configuration where $H(j\omega) = -Z_2(j\omega)/Z_1(j\omega)$.

Figure 105 shows the gain and phase functions versus normalized frequency ω_c/α. The maximum passband gain occurs at $\omega = 0$ where $|H(0)| = |K|/\alpha$. As frequency increases, the gain gradually decreases until at $\omega = \alpha$:

$$|H(j\alpha)| = \frac{|K|}{\sqrt{\alpha^2 + \alpha^2}} = \frac{|K| \to \alpha}{\sqrt{2}} = \frac{|H(0)|}{\sqrt{2}}$$

That is, the cutoff frequency of the first-order low-pass transfer function is $\omega_C = \alpha$. The graph of the gain function in Fig. 105a displays a low-pass characteristic with a finite dc gain and zero infinite frequency gain.

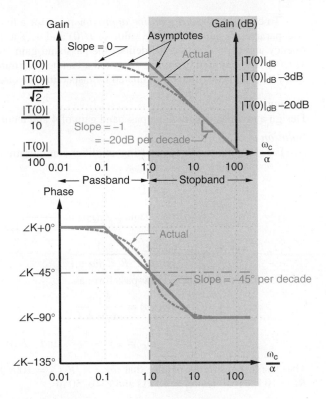

Figure 105 First-order low-pass Bode plots.[1]

The low- and high-frequency gain asymptotes shown in Fig. 105a are especially important. The low-frequency asymptote is the horizontal line and the high-frequency asymptote is the sloped line. At low frequencies ($\omega \ll \alpha$) the gain approaches $|H(j\omega)| \to |K|/\alpha$. At high frequencies ($\omega \gg \alpha$) the gain approaches $|H(j\omega)| \to |K|/\omega$. The intersection of the two asymptotes occurs when $|K|/\alpha = |K|/\omega$. The intersection forms a "corner" at $\omega = \alpha$, so the cutoff frequency is also called the *corner frequency*.

The high-frequency gain asymptote decreases by a factor of 10 (-20 dB) whenever the frequency increases by a factor of 10 (one decade). As a result, the high-frequency asymptote has a slope of -1 or -20 dB/decade, and the low-frequency asymptote has a slope of 0 or 0 dB/decade. These two asymptotes provide a straight-line approximation to the gain response that differs from the true response by a maximum of 3 dB at the corner frequency.

The semilog plot of the phase shift of the first-order low-pass transfer function is shown in Fig. 105b. At $\omega = \alpha$ the phase angle in Eq. (127) is $< K - 45°$. At low frequency ($\omega < \alpha$) the phase angle approaches $< K$ and at high frequencies ($\omega > \alpha$) the phase approaches $< K - 90°$. Almost all of the $-90°$ phase change occurs in the two-decade range from $\omega/\alpha = 0.1$ to $\omega/\alpha = 10$. The straight-line segments in Fig. 105b provide an approximation of the phase response. The phase approximation below $\omega/\alpha = 0.1$ is $\theta = < K$ and above $\omega/\alpha = 10$ is $H < \theta = < K - 90°$. Between these values the phase approximation is a straight line that begins at $H < \theta = < K$, passes through $H < \theta = < K - 45°$ at the cutoff frequency, and reaches $H < \theta = < K - 90°$ at $\omega/\alpha = 10$. The slope of this line segment is $-45°$/decade since the total phase change is $-90°$ over a two-decade range.

To construct the *straight-line approximations* for a first-order low-pass transfer function, two parameters are needed, the value of $H(0)$ and α. The parameter α defines the cutoff frequency and the value of $H(0)$ defines the passband gain $|H(0)|$ and the low-frequency phase $< H(0)$. The required quantities $H(0)$ and α can be determined directly from the transfer function $H(j\omega)$ and can often be estimated by inspecting the circuit itself.

Example 33

Design a low-pass filter with a passband gain of 4 and a cutoff frequency of 100 rad/s.

Solution

See Fig. 106. Start with an inverting amplifier configuration since a gain is required:

$$H(j\omega) = -\frac{Z_2(j\omega)}{Z_1(j\omega)}$$

$$Z_1(j\omega) = R_1 \quad \text{and} \quad Z_2(j\omega) = \frac{1}{j\omega C_2 + 1/R_2} = \frac{R_2}{R_2 C_2 j\omega + 1}$$

$$H(j\omega) = -\frac{R_2}{R_1} \times \frac{1}{R_2 C_2 j\omega + 1}$$

Rearrange the standard low-pass form as

$$H(j\omega) = \frac{K/\alpha}{j\omega/\alpha + 1}$$

$$\omega_C = \alpha = \frac{1}{R_2 C_2} \quad \text{and} \quad H(0) = -\frac{R_2}{R_1}$$

The design constraints require that $\omega_C = 1/R_2 C_2 = 100$ and $|H(0)| = R_2/R_1 = 4$. Selecting $R_1 = 10\,\text{k}\Omega$ requires $R_2 = 40\,\text{k}\Omega$ and $C = 250\,\text{nF}$.

First-Order High-Pass Response

A first-order high-pass transfer function is written as

$$H(j\omega) = \frac{Kj\omega}{j\omega + \alpha}$$

The high-pass function differs from the low-pass case by the introduction of a $j\omega$ in the numerator, resulting in the function becoming zero at $\omega = 0$. Solving for the gain and phase functions yields

$$|H(j\omega)| = \frac{|K|\omega}{\sqrt{\omega^2 + \alpha^2}}$$

$$\angle H(j\omega) = \angle K + 90° - \tan^{-1}\left(\frac{\omega}{\alpha}\right) \tag{128}$$

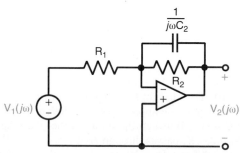

Figure 106   Inverting amplifier configuration.[1]

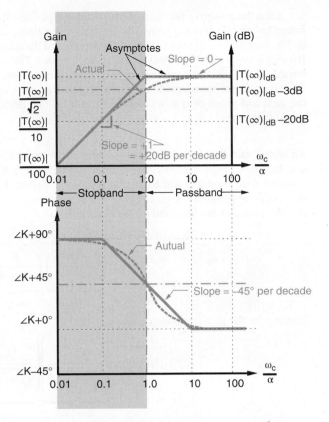

Figure 107 First-order high-pass Bode plots.[1]

Figure 107 shows the gain and phase functions versus normalized frequency ω/α. The maximum gain occurs at high frequency ($\omega > \alpha$) where $|H(j\omega)|$ 6 $|K|$. At low frequency ($\omega < \alpha$) the gain approaches $|K|\omega/\alpha$. At $\omega = \alpha$ the gain is

$$|H(j\alpha)| = \frac{|K|\alpha}{\sqrt{\alpha^2 + \alpha^2}} = \frac{|K|}{\sqrt{2}}$$

which means the cutoff frequency is $\omega_C = \alpha$. The gain response plot in Fig. 107a displays a high-pass characteristic with a passband extending from $\omega = \alpha$ to infinity and a stopband between zero frequency and $\omega = \alpha$.

The low- and high-frequency gain asymptotes approximate the gain response in Fig. 107a. The high-frequency asymptote ($\omega > \alpha$) is the horizontal line whose ordinate is $|K|$ (slope = 0 or 0 dB/decade). The low-frequency asymptote ($\omega < \alpha$) is a line of the form $|K|\omega/\alpha$ (slope = +1 or +20 dB/decade). The intersection of these two asymptotes occurs when $|K| = |K|\omega/\alpha$, which defines a corner frequency at $\omega = \alpha$.

The semilog plot of the phase shift of the first-order high-pass function is shown in Fig. 107b. The phase shift approaches < K at high frequency, passes through < $K + 45°$ at the cutoff frequency, and approaches < $K + 90°$ at low frequency. Most of the 90° phase change occurs over the two-decade range centered on the cutoff frequency. The phase shift can be approximated by the straight-line segments shown in the Fig. 107b. As in the low-pass case, < K is 0° when K is positive and $\pm 180°$ when K is negative.

Like the low-pass function, the first-order high-pass frequency response can be approximated by straight-line segments. To construct these lines, we need two parameters, $H(\infty)$, and α. The parameter α defines the cutoff frequency and the quantity $H(\infty)$ gives the passband gain $|H(\infty)|$ and the high-frequency phase angle $< H(\infty)$. The quantities $H(\infty)$ and α can be determined directly from the transfer function or estimated directly from the circuit in some cases. The straight line shows the first-order high-pass response can be characterized by calculating the gain and phase over a two-decade band from one decade below to one decade above the cutoff frequency.

Bandpass and Band-Stop Responses Using First-Order Circuits

The first-order high- and low-pass circuits can be used in a building block fashion to produce a circuit with bandpass and band-stop responses. Figure 108 shows a cascade connection of first-order high- and low-pass circuits. When the second stage does not load the first, the overall transfer function can be found by the chain rule:

$$H(j\omega) = H_1(j\omega) \times H_2(j\omega)$$

$$= \underbrace{\left(\frac{K_1 j\omega}{j\omega + \alpha_1} \right)}_{\text{high pass}} \underbrace{\left(\frac{K_2}{j\omega + \alpha_2} \right)}_{\text{low pass}}$$

Solving for the gain response yields

$$|H(j\omega)| = \underbrace{\left(\frac{|K_1|\,\omega}{\sqrt{\omega^2 + \alpha_1^2}} \right)}_{\text{high pass}} \underbrace{\left(\frac{|K_2|}{\sqrt{\omega^2 + \alpha_2^2}} \right)}_{\text{low pass}}$$

Note the gain of the cascade is zero at $\omega = 0$ and at infinite frequency.

When $\alpha_1 < \alpha_2$ the high-pass cutoff frequency is much lower than the low-pass cutoff frequency, and the overall transfer function has a bandpass characteristic. At low frequencies ($\omega < \alpha_1 < \alpha_2$) the gain approaches $|H(j\omega)| \rightarrow |K_1 K_2|\omega/\alpha_1\alpha_2$. At midfrequencies ($\alpha_1 < \omega < \alpha_2$) the gain approaches $|H(j\omega)| \rightarrow |K_1 K_2|/\alpha_2$. The low- and midfrequency asymptotes intersect when $|K_1 K_2|\omega/\alpha_1\alpha_2 = |K_1 K_2|/\alpha_2$ at $\omega = \alpha_1$, that is, at the cutoff frequency of the high-pass stage. At high frequencies ($\alpha_1 < \alpha_2 < \omega$) the gain approaches $|H(j\omega)| \rightarrow |K_1 K_2|/\omega$. The high- and midfrequency asymptotes intersect when $|K_1 K_2|/\omega = |K_1 K_2|/\alpha_2$ at $\omega = \alpha_2$, that is, at the cutoff frequency of the low-pass stage. The plot of these asymptotes in Fig. 109 shows that the asymptotic gain exhibits a passband between α_1 and α_2. Input sinusoids whose frequencies are outside of this range fall in one of the two stop bands.

In the bandpass cascade connection the input signal must pass both a low- and a high-pass stage to reach the output. In the parallel connection in Fig. 110 the input can reach the output

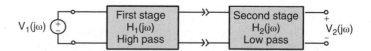

Figure 108   Cascade connection of high- and low-pass circuits.[1]

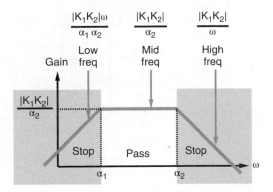

Figure 109   Bandpass gain characteristic.[1]

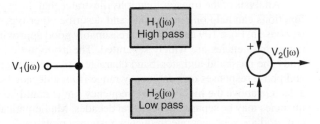

Figure 110   Parallel connection of high- and low-pass circuits.[1]

via either a low- or a high-pass path. The overall transfer function is the sum of the low- and high-pass transfer functions:

$$|H(j\omega)| = \underbrace{\left(\frac{|K_1|\,\omega}{\sqrt{\omega^2 + \alpha_1^2}}\right)}_{\text{high pass}} + \underbrace{\left(\frac{|K_2|}{\sqrt{\omega^2 + \alpha_2^2}}\right)}_{\text{low pass}}$$

Any sinusoid whose frequency falls in either passband will find its way to the output unscathed. An input sinusoid whose frequency falls in both stop bands will be attenuated.

When $\alpha_1 > \alpha_2$, the high-pass cutoff frequency is much higher than the low-pass cutoff frequency, and the overall transfer function has a band-stop gain response as shown in Fig. 111. At low frequencies ($\omega < \alpha_2 < \alpha_1$) the gain of the high-pass function is negligible and the overall gain approaches $|H(j\omega)| \to |K_2|/\alpha_2$, which is the passband gain of the low-pass function. At high frequencies ($\alpha_2 < \alpha_1 < \omega$) the low-pass function is negligible and the overall gain approaches $|H(j\omega)| \to |K_1|$, which is the passband gain of the high-pass function. With a band-stop function the two passbands normally have the same gain, hence $|K_1| = |K_2|/\alpha_2$. Between these two passbands there is a stop band. For $\omega > \alpha_2$ the low-pass asymptote is $|K_2|/\omega$, and for $\omega < \alpha_1$ the high-pass asymptote is $|K_1|\omega/\alpha_1$. The asymptotes intersect at $\omega^2 = \alpha_1|K_2|/|K_1|$. But equal gains in the two passband frequencies requires $|K_1| = |K_2|/\alpha_2$, so the intersection frequency is $\omega = \sqrt{\alpha_1\alpha_2}$. Below this frequency the stop-band attenuation is

558 Electric Circuits

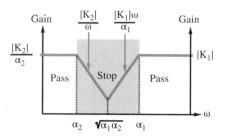

Figure 111 Band-stop gain characteristic.[1]

determined by the low-pass function and above this frequency the attenuation is governed by the high-pass function.

Analysis of the transfer functions illustrates that the asymptotic gain plots of the first-order functions can help one understand and describe other types of gain response. The asymptotic response in Figs. 109 and 111 are a reasonably good approximation as long as the two first-order cutoff frequencies are widely separated. The asymptotic analysis gives insight to see that to study the passband and stop-band characteristics in greater detail one needs to calculate gain and phase responses on a frequency range from a decade below the lowest cutoff frequency to a decade above the highest. This frequency range could be very wide since the two cutoff frequencies may be separated by several decades. Mathematical and simulation software packages can produce very accurate frequency response plots.

Example 34
Design a first-order bandpass circuit with a passband gain of 10 and cutoff frequencies at 20 Hz and 20 kHz.

Solution
A cascade connection of first-order low- and high-pass building blocks will satisfy the design. The required transfer function has the form

$$H(j\omega) = H_1(j\omega) \times H_2(j\omega)$$

$$= \underbrace{\left(\frac{K_1 j\omega}{j\omega + \alpha_1}\right)}_{\text{high pass}}\underbrace{\left(\frac{K_2}{j\omega + \alpha_2}\right)}_{\text{low pass}}$$

with the following constraints:

Lower cutoff frequency: $\alpha_1 = 2\pi(20)(11)$
$$= 40\pi \text{ rad/s}(12)$$

Upper cutoff frequency: $\alpha_2 = 2\pi(20 \times 10^3)(13)$
$$= 4\pi \times 10^4 \text{ rad/s}(14)$$

Midband gain: $-10pt\dfrac{|K_1 K_2|}{\alpha_2} = 10$

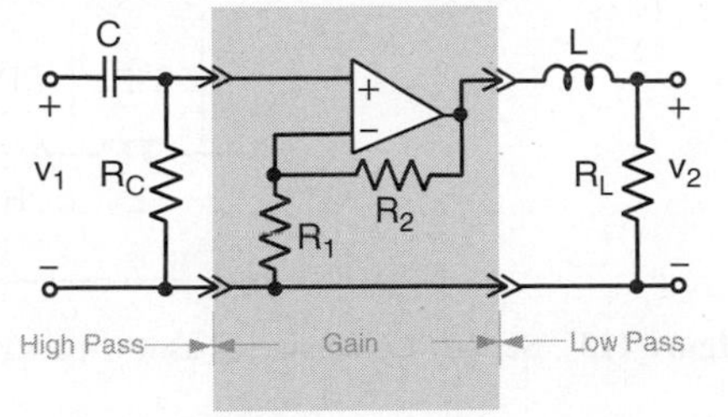

Figure 112 High-pass/low-pass cascade circuit.[1]

With numerical values inserted, the required transfer function is

$$H(j\omega) = \underbrace{\left[\frac{j\omega}{j\omega + 40\pi}\right]}_{\text{high pass}} \underbrace{[10]}_{\text{gain}} \underbrace{\left[\frac{40\pi \times 10^4}{j\omega + 40\pi \times 10^4}\right]}_{\text{low pass}}$$

This transfer function can be realized using the high-pass/low-pass cascade circuit in Fig. 112. The first stage is a passive RC high-pass circuit, and the third stage is a passive RL low-pass circuit. The noninverting op-amp second stage serves two purposes: (a) It isolates the first and third stages, so the chain rule applies, and (b) it supplies the midband gain. Using the chain rule, the transfer function of this circuit is

$$H(j\omega) = \underbrace{\left[\frac{j\omega}{j\omega + 1/RC}\right]}_{\text{high pass}} \underbrace{\left[\frac{R_1 + R_2}{R_1}\right]}_{\text{gain}} \underbrace{\left[\frac{R/L}{j\omega + R/L}\right]}_{\text{low pass}}$$

Comparing this to the required transfer function leads to the following design constraints:

High-pass stage: $R_C C = 1/40\pi$. Let $R_C = 100$ kΩ. Then $C = 79.6$ nF.

Gain stage: $(R_1 + R_2)/R_1 = 10$. Let $R_1 = 10$ kΩ. Then $R_2 = 90$ kΩ.

Low-pass stage: $R_L/L = 40000\pi$. Let $R_L = 200$ kΩ. Then $L = 0.628$ H.

6.5 Second-Order *RLC* Filters

Simple second-order low-pass, high-pass, or bandpass filters can be made using series or parallel *RLC* circuits. Series or parallel *RLC* circuits can be connected to produce the following transfer functions:

$$H(j\omega)_{\text{LP}} = \frac{K}{-\omega^2 + 2\zeta\omega_0 j\omega + \omega_0^2}$$

$$H(j\omega)_{\text{HP}} = \frac{-K\omega^2}{-\omega^2 + 2\zeta\omega_0 j\omega + \omega_0^2} \qquad (129)$$

$$H(j\omega)_{\text{BP}} = \frac{Kj\omega}{-\omega^2 + 2\zeta\omega_0 j\omega + \omega_0^2}$$

where for a series *RLC* circuit

$$\omega_0 = \sqrt{LC} \quad \text{and} \quad \zeta = \frac{R}{2}\sqrt{\frac{C}{L}}.$$

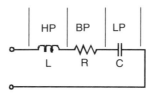

Figure 113   Series *RLC* connected as low-pass (LP), high-pass (HP), or bandpass (BP) filter.

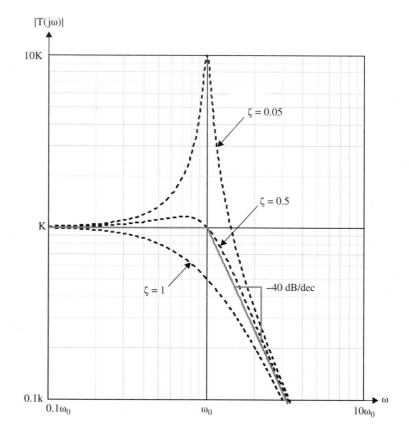

Figure 114   Second-order low-pass gain responses.[1]

The undamped natural frequency ω_0 is related to the cutoff frequency in the high- and low-pass cases and is the center frequency in the bandpass case. Zeta (ζ) is the damping ratio and determines the nature of the roots of the equation that translates to how quickly a transition is made from the passband to the stopband. In the bandpass case ζ helps define the bandwidth of the circuit, that is, $B = 2\zeta\omega_0$. Figure 113 shows how a series *RLC* circuit can be connected to achieve the transfer functions given in Eq. (129). The gain $|H(j\omega)|$ plots of these circuits are shown in Figs. 114–116.

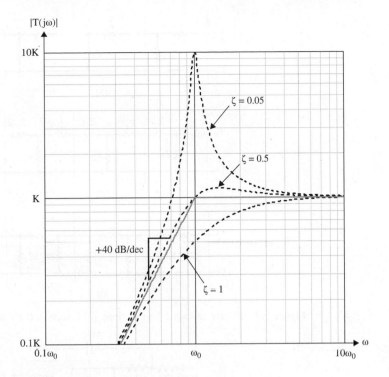

Figure 115 Second-order high-pass gain responses.[1]

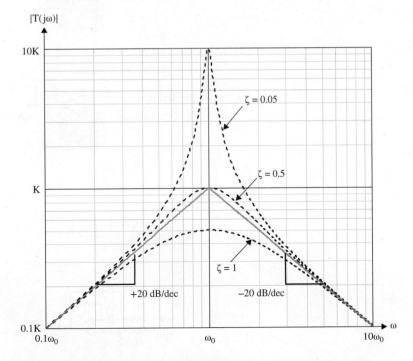

Figure 116 Second-order bandpass gain responses.[1]

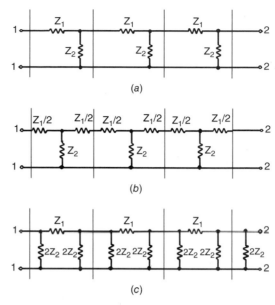

Figure 117   Passive cascaded filter sections: (*a*) L section, (*b*) T section, and (*c*) π section.

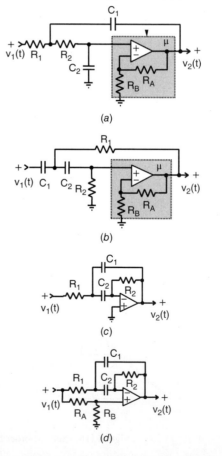

Figure 118   Second-order configurations: (*a*) LP, (*b*) HP, (*c*) tuned, and (*d*) notch.[1]

6.6 Compound Filters

Compound filters are higher order filters obtained by cascading lower order designs. *Ladder circuits* are an important class of compound filters. Two of the more common passive ladder circuits are the *constant-k* and the *m-derived* filters, either of which can be configured using a *T section*, *π section*, *L section* (Fig. 117), or combinations thereof, the bridge-T network and parallel-T network.

Active filters are generally designed using first-, second-, or third-order modules such as the *Sallen–Key* configurations shown in Fig. 118a and b, and the Delyannies–Friend configurations shown in Fig. 118c and d. The filters are then developed using an algorithmic approach following the *Butterworth*, *elliptical*, or *Tchebycheff* realizations.

REFERENCES

1. R. Thomas, and A. J. Rosa, *The Analysis and Design of Linear Circuits*, 5th ed., Wiley, Hoboken, NJ, 2005.
2. R. Thomas, and A. J. Rosa, *The Analysis and Design of Linear Circuits*, 2nd ed., Prentice-Hall, Englewood Cliffs, NJ, 1998.

CHAPTER 16

MEASUREMENTS

E. L. Hixson and E. A. Ripperger
University of Texas
Austin, Texas

1 STANDARDS AND ACCURACY

1.1 Standards

Measurement is the process by which a quantitative comparison is made between a standard and a measurand. The measurand is the particular quantity of interest—the thing that is to be quantified. The standard of comparison is of the same character as the measurand, and so far as mechanical engineering is concerned the standards are defined by law and maintained by the National Institute of Standards and Technology (NIST, formerly known as the National Bureau of Standards). The four independent standards that have been defined are length, time, mass, and temperature.[1] All other standards are derived from these four. Before 1960 the standard for length was the international prototype meter, kept at Sevres, France. In 1960 the meter was redefined as 1,650,763.73 wavelengths of krypton light. Then in 1983, at the Seventeenth General Conference on Weights and Measures, a new standard was adopted: A meter is the distance traveled in a vacuum by light in 1/299,792,458 seconds.[2] However, there is a copy of the international prototype meter, known as the national prototype meter, kept by NIST. Below that level there are several bars known as national reference standards and below that there are the working standards. Interlaboratory standards in factories and laboratories are sent to NIST for comparison with the working standards. These interlaboratory standards are the ones usually available to engineers.

Standards for the other three basic quantities have also been adopted by NIST, and accurate measuring devices for those quantities should be calibrated against those standards.

The standard mass is a cylinder of platinum–iridium, the international kilogram, also kept at Sevres, France. It is the only one of the basic standards that is still established by a prototype. In the United States the basic unit of mass is the basic prototype kilogram No. 20. Working copies of this standard are used to determine the accuracy of interlaboratory standards. Force is not one of the fundamental quantities, but in the United States the standard unit of force is the pound, defined as the gravitational attraction for a certain platinum mass at sea level and 45° latitude.

Absolute time, or the time when some event occurred in history, is not of much interest to engineers. Engineers are more likely to need to measure time intervals, that is, the time between two events. The basic unit for time measurements is the *second*. At one time the second was defined as 1/86,400 of the average period of rotation of the Earth on its axis, but that is not a practical standard. The period varies and the Earth is slowing up. Consequently, a new standard based on the oscillations associated with a certain transition within the cesium atom was defined and adopted. That standard, the cesium clock, has now been superceded by the cesium fountain atomic clock as the primary time and frequency standard of the United States.[3] Although this cesium "clock" is the basic frequency standard, it is not generally usable by mechanical engineers. Secondary standards such as tuning forks, crystals, electronic oscillators, and so on are used, but from time to time access to time standards of a higher order of accuracy may be required. To help meet these requirements, NIST broadcasts 24 h per day, 7 days per week time and frequency information from radio stations WWV, WWVB, and WWVL located in Fort Collins, Colorado, and WWVH located in Hawaii. Other nations also broadcast timing signals. For details on the time signal broadcasts, potential users should consult NIST.[4]

Temperature is one of four fundamental quantities in the international measuring system. Temperature is fundamentally different in nature from length, time, and mass. It is an intensive quantity, whereas the others are extensive. Join together two bodies that have the same temperature and you will have a larger body at that same temperature. If you join two bodies that have a certain mass, you will have one body of twice the mass of the original body. Two bodies are said to be at the same temperature if they are in thermal equilibrium. The international practical temperature scale, adopted in 1990 (ITS-90) by the International Committee on Weights and Measurement is the one now in effect and the one with which engineers are primarily concerned. In this system the kelvin (K) is the basic unit of temperature. It is 1/273.16 of the temperature at the triple point of water, the temperature at which the solid, liquid, and vapor phases of water exist in equilibrium.[5] Degrees Celsius (°C) is related to degrees kelvin by the equation

$$t = T - 273.15$$

where t = degrees Celsius
         T = degrees kelvin

1.2 Accuracy and Precision

In measurement practice four terms are frequently used to describe an instrument. They are accuracy, precision, sensitivity, and linearity. Accuracy, as applied to an instrument, is the closeness with which a reading approaches the true value. Since there is some error in every reading, the "true value" is never known. In the discussion of error analysis that follows, methods of estimating the "closeness" with which the determination of a measured value approaches the true value will be presented. Precision is the degree to which readings agree among themselves. If the same value is measured many times and all the measurements agree very closely, the instrument is said to have a high degree of precision. It may not, however, be a very accurate instrument. Accurate calibration is necessary for accurate measurement. Measuring instruments must, for accuracy, be from time to time compared to a standard. These will usually be laboratory or company standards that are in turn compared from time to time with a working standard at NIST. This chain can be thought of as the pedigree of the instrument, and the calibration of the instrument is said to be traceable to NIST.

1.3 Sensitivity and Resolution

These two terms, as applied to a measuring instrument, refer to the smallest change in the measured quantity to which the instrument responds. Obviously, the accuracy of an instrument will depend to some extent on the sensitivity. If, for example, the sensitivity of a pressure transducer is 1 kPa, any particular reading of the transducer has a potential error of at least 1 kPa. If the readings expected are in the range of 100 kPa and a possible error of 1% is acceptable, then the transducer with a sensitivity of 1 kPa may be acceptable, depending upon what other sources of error may be present in the measurement. A highly sensitive instrument is difficult to use. Therefore, a sensitivity significantly greater than that necessary to obtain the desired accuracy is no more desirable than one with insufficient sensitivity.

Many instruments today have digital readouts. For such instruments the concepts of sensitivity and resolution are defined somewhat differently than they are for analog-type instruments. For example, the resolution of a digital voltmeter depends on the "bit" specification and the voltage range. The relationship between the two is expressed by the equation

$$R = \frac{V}{2^n}$$

where R = resolution in volts
V = voltage range
n = number of bits

Thus an 8-bit instrument on a 1-V scale would have a resolution of 1/256, or 0.004, volt. On a 10-V scale that would increase to 0.04 V. As in analog instruments, the higher the resolution, the more difficult it is to use the instrument, so if the choice is available, one should use the instrument that just gives the desired resolution and no more.

1.4 Linearity

The calibration curve for an instrument does not have to be a straight line. However, conversion from a scale reading to the corresponding measured value is most convenient if it can be done by multiplying by a constant rather than by referring to a nonlinear calibration curve or by computing from an equation. Consequently, instrument manufacturers generally try to produce instruments with a linear readout, and the degree to which an instrument approaches this ideal is indicated by its *linearity*. Several definitions of linearity are used in instrument specification practice.[6] The so-called independent linearity is probably the most commonly used in specifications. For this definition the data for the instrument readout versus the input are plotted and then a "best straight line" fit is made using the method of least squares. Linearity is then a measure of the maximum deviation of any of the calibration points from this straight line. This deviation can be expressed as a percentage of the actual reading or a percentage of the full-scale reading. The latter is probably the most commonly used, but it may make an instrument appear to be much more linear than it actually is. A better specification is a combination of the two. Thus, linearity equals +A percent of reading or +B percent of full scale, whichever is greater. Sometimes the term independent linearity is used to describe linearity limits based on actual readings. Since both are given in terms of a fixed percentage, an instrument with A percent proportional linearity is much more accurate at low reading values than an instrument with A percent independent linearity.

It should be noted that although specifications may refer to an instrument as having A percent linearity, what is really meant is A percent nonlinearity. If the linearity is specified as independent linearity, the user of the instrument should try to minimize the error in readings by selecting a scale, if that option is available, such that the actual reading is close to full scale. A reading should never be taken near the low end of a scale if it can possibly be avoided.

For instruments that use digital processing, linearity is still an issue since the analog-to-digital converter used can be nonlinear. Thus linearity specifications are still essential.

2   IMPEDANCE CONCEPTS

Two basic questions that must be considered when any measurement is made are: How has the measured quantity been affected by the instrument used to measure it? Is the quantity the same as it would have been had the instrument not been there? If the answers to these questions are no, the effect of the instrument is called *loading*. To characterize the loading, the concepts of *stiffness* and *input impedance* are used.[7] At the input of each component in a measuring system there exists a variable q_{i1}, which is the one we are primarily concerned with in the transmission of information. At the same point, however, there is associated with q_{i1} another variable q_{i2} such that the product $q_{i1}q_{i2}$ has the dimensions of power and represents the rate at which energy is being withdrawn from the system. When these two quantities are identified, the generalized input impedance Z_{gi} can be defined by

$$Z_{gi} = \frac{q_{i1}}{q_{i2}} \tag{1}$$

if q_{i1} is an *effort variable*. The effort variable is also sometimes called the *across variable*. The quantity q_{i2} is called the *flow variable* or *through variable*. In the dynamic case these variables can be represented in the frequency domain by their Fourier transform. Then the quantity Z is a complex number. The application of these concepts is illustrated by the example in Fig. 1. The output of the linear network in the blackbox (Fig. 1a) is the open-circuit voltage E_o until the load Z_L is attached across the terminals A–B. If Thévenin's theorem is applied after the load Z_L is attached, the system in Fig. 1b is obtained. For that system the current is given by

$$i_m = \frac{E_o}{Z_{AB} + Z_L} \tag{2}$$

and the voltage E_L across Z_L is

$$E_L = i_m Z_L = \frac{E_o Z_L}{Z_{AB} + Z_L}$$

or

$$E_L = \frac{E_o}{1 + Z_{AB}/Z_L} \tag{3}$$

Equations (2)–(7) are frequency-domain equations.

In a measurement situation E_L would be the voltage indicated by the voltmeter, Z_L would be the input impedance of the voltmeter, and Z_{AB} would be the output impedance of the linear

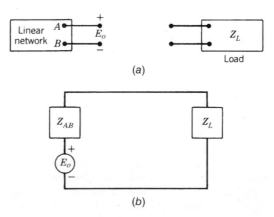

Figure 1   Application of Thévenin's theorem.

network. The true output voltage, E_o, has been reduced by the voltmeter, but it can be computed from the voltmeter reading if Z_{AB} and Z_L are known. From Eq. (3) it is seen that the effect of the voltmeter on the reading is minimized by making Z_L as large as possible.

If the generalized input and output impedances Z_{gi} and Z_{go} are defined for nonelectrical systems as well as electrical systems, Eq. (3) can be generalized to

$$q_{im} = \frac{q_{iu}}{1 + Z_{go}/Z_{gi}} \tag{4}$$

where q_{im} is the measured value of the effort variable and q_{iu} is the undisturbed value of the effort variable. The output impedance Z_{go} is not always defined or easy to determine; consequently Z_{gi} should be large. If it is large enough, knowing Z_{go} is unimportant.

If q_{i1} is a flow variable rather than an effort variable (current is a flow variable, voltage an effort variable), it is better to define an input admittance

$$Y_{gi} = \frac{q_{i1}}{q_{i2}} \tag{5}$$

rather than the generalized input impedance

$$Z_{gi} = \frac{\text{effort variable}}{\text{flow variable}}$$

The power drain of the instrument is

$$P = q_{i1}q_{i2} = \frac{q_{i2}^2}{Y_{gi}} \tag{6}$$

Hence, to minimize power drain, Y_{gi} must be large. For an electrical circuit

$$I_m = \frac{I_u}{1 + Y_o/Y_i} \tag{7}$$

where I_m = measured current
I_u = actual current
Y_o = output admittance of circuit
Y_i = input admittance of meter

When the power drain is zero and the deflection is zero, as in structures in equilibrium, for example, when deflection is to be measured, the concepts of impedance and admittance are replaced with the concepts of *static stiffness* and *static compliance*. Consider the idealized structure in Fig. 2.

To measure the force in member K_2, an elastic link with a spring constant K_m is inserted in series with K_2. This link would undergo a deformation proportional to the force in K_2. If the link is very soft in comparison with K_1, no force can be transmitted to K_2. On the other hand, if the link is very stiff, it does not affect the force in K_2 but it will not provide a very good measure of the force. The measured variable is an effort variable, and in general, when it is measured, it

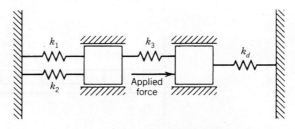

Figure 2   Idealized elastic structure.

is altered somewhat. To apply the impedance concept, a flow variable whose product with the effort variable gives power is selected. Thus,

$$\text{Flow variable} = \frac{\text{power}}{\text{effor variable}}$$

Mechanical impedance is then defined as force divided by velocity, or

$$Z = \frac{\text{force}}{\text{velocity}}$$

where force and velocity are dynamic quantities represented by their Fourier transform and Z is a complex number. This is the equivalent of electrical impedance. However, if the static mechanical impedance is calculated for the application of a constant force, the impossible result

$$Z = \frac{\text{force}}{0} = \infty$$

is obtained.

This difficulty is overcome if energy rather than power is used in defining the variable associated with the measured variable. In that case the static mechanical impedance becomes the *stiffness*:

$$\text{Stiffness} = S_g = \frac{\text{effort}}{\int \text{flow } dt}$$

In structures,

$$S_g = \frac{\text{effort variable}}{\text{displacement}}$$

When these changes are made, the same formulas used for calculating the error caused by the loading of an instrument in terms of impedances can be used for structures by inserting S for Z. Thus

$$q_{im} = \frac{q_{iu}}{1 + S_{go}/S_{gi}} \qquad (8)$$

where q_{im} = measured value of effort variable
q_{iu} = undisturbed value of effort variable
S_{go} = static output stiffness of measured system
S_{gi} = static stiffness of measuring system

For an elastic force-measuring device such as a load cell S_{gi} is the spring constant K_m. As an example, consider the problem of measuring the reactive force at the end of a propped cantilever beam, as in Fig. 3.

According to Eq. (8), the force indicated by the load cell will be

$$F_m = \frac{F_u}{1 + S_{go}/S_{gi}}$$

$$S_{gi} = K_m \quad \text{and} \quad S_{go} = \frac{3EI}{L^3}$$

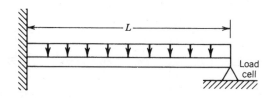

Figure 3   Measuring the reactive force at the tip.

The latter is obtained by noting that the deflection at the tip of a tip-loaded cantilever is given by

$$\delta = \frac{PL^3}{3EI}$$

The stiffness is the quantity by which the deflection must be multiplied to obtain the force producing the deflection.

For the cantilever beam

$$F_m = \frac{F_u}{1 + 3EI / K_m L^3} \tag{9}$$

or

$$F_u = F_m \left(1 + \frac{3EI}{K_m L^3} \right) \tag{10}$$

Clearly, if $K_m >> 3EI/L^3$, the effect of the load cell on the measurement will be negligible.

To measure displacement rather than force, the concept of compliance is introduced and defined as

$$C_g = \frac{\text{flow variable}}{\int \text{effort variable } dt}$$

Then

$$q_m = \frac{q_u}{1 + C_{go} / C_{gi}} \tag{11}$$

If displacements in an elastic structure are considered, the compliance becomes the reciprocal of stiffness, or the quantity by which the force must be multiplied to obtain the displacement caused by the force. The cantilever beam in Fig. 4 again provides a simple illustrative example.

If the deflection at the tip of this cantilever is to be measured using a dial gauge with a spring constant K_m,

$$C_{gi} = \frac{1}{K_m} \quad \text{and} \quad C_{go} = \frac{L^3}{3EI}$$

Thus

$$\delta_m = \delta_u \left(1 + \frac{K_m L^3}{3EI} \right) \tag{12}$$

Not all interactions between a system and a measuring device lend themselves to this type of analysis. A pitot tube, for example, inserted into a flow field distorts the flow field but does not extract energy from the field. Impedance concepts cannot be used to determine how the flow field will be affected.

There are also applications in which it is not desirable for a force-measuring system to have the highest possible stiffness. A subsoil pressure gauge is an example. Such a gauge, if it is much stiffer than the surrounding soil, will take a disproportionate share of the total load and will consequently indicate a higher pressure than would have existed in the soil if the gauge had not been there.

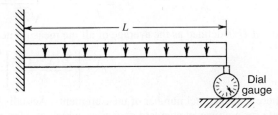

Figure 4 Measuring the tip deflection.

3 ERROR ANALYSIS

It may be accepted as axiomatic that there will always be errors in measured values. Thus if a quantity X is measured, the correct value q, and X will differ by some amount e. Hence

$$\pm(q - X) = e$$

or

$$q = X \pm e \tag{13}$$

It is essential, therefore, in all measurement work that a realistic estimate of e be made. Without such an estimate the measurement of X is of no value. There are two ways of estimating the error in a measurement. The first is the external estimate, or ϵ_E, where $\epsilon = e/q$. This estimate is based on knowledge of the experiment and measuring equipment and to some extent on the internal estimate ϵ_I.

The internal estimate is based on an analysis of the data using statistical concepts.

3.1 Internal Estimates

If a measurement is repeated many times, the repeat values will not, in general, be the same. Engineers, it may be noted, do not usually have the luxury of repeating measurements many times. Nevertheless, the standardized means for treating results of repeated measurements are useful, even in the error analysis for a single measurement.[8]

If some quantity is measured many times and it is assumed that the errors occur in a completely random manner, that small errors are more likely to occur than large errors, and that errors are just as likely to be positive as negative, the distribution of errors can be represented by the curve

$$F(X) = \frac{Y_o e^{-(X-U)}}{2\sigma^2} \tag{14}$$

where
$F(X)$ = number of measurements for a given value of $(X - U)$
Y_o = maximum height of curve or number of measurements for which $X = U$
U = value of X at point where maximum height of curve occurs σ determines lateral spread of the curve

This curve is the normal, or Gaussian, frequency distribution. The area under the curve between X and δX represents the number of data points that fall between these limits, and the total area under the curve denotes the total number of measurements made. If the normal distribution is defined so that the area between X and $X + \delta X$ is the probability that a data point will fall between those limits, the total area under the curve will be unity and

$$F(X) = \frac{\exp -(X-U)^2/2\sigma^2}{\sigma\sqrt{2\Pi}} \tag{15}$$

and

$$P_x = \int \frac{\exp -(X-U)^2/2\sigma^2}{\sigma\sqrt{2\Pi}}\, dx \tag{16}$$

Now if U is defined as the average of all the measurements and s as the standard deviation,

$$\sigma = \left[\frac{\sum (X-U)^2}{N}\right]^{1/2} \tag{17}$$

where N is the total number of measurements. Actually this definition is used as the best estimate for a universal standard deviation, which it is for a very large number of measurements.

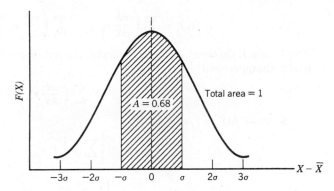

Figure 5 Probability curve.

For smaller subsets of measurements the best estimate of σ is given by

$$\sigma = \left[\frac{\sum (X - U)^2}{n - 1} \right]^{1/2} \tag{18}$$

where n is the number of measurements in the subset. Obviously, the difference between the two values of σ becomes negligible as n becomes very large (or as $n \to N$).

The probability curve based on these definitions is shown in Fig. 5.

The area under this curve between $-\sigma$ and $+\sigma$ is 0.68. Hence 68% of the measurements can be expected to have errors that fall in the range of $\pm\delta$. Thus the chances are 68/32, or better than 2 to 1, that the error in a measurement will fall in this range. For the range $\pm 2\sigma$ the area is 0.95. Hence 95% of all the measurement errors will fall in this range, and the odds are about 20:1 that a reading will be within this range. The odds are about 384:1 that any given error will be in the range of $\pm 3\sigma$.

Some other definitions related to the normal distribution curve are as follows:

1. *Probable Error.* The error likely to be exceeded in half of all the measurements and not reached in the other half of the measurements. This error in Fig. 5 is about $0.67\,\delta$.

2. *Mean Error.* The arithmetic mean of all the errors regardless of sign. This is about $0.8\,\sigma$.

3. *Limit of Error.* The error that is so large it is most unlikely ever to occur. It is usually taken as $4\,\delta$.

3.2 Use of Normal Distribution to Calculate Probable Error in X

The foregoing statements apply strictly only if the number of measurements is very large. Suppose that n measurements have been made. That is a sample of n data points out of an infinite number. From that sample U and σ are calculated as above. How good are these numbers? To determine that, we proceed as follows: Let

$$U = F(X_1, X_2, X_3, \ldots, X_n) = \frac{\sum X_i}{n} \tag{19}$$

$$e_u = \sum \frac{\partial F}{\partial X_i} e_{xi} \tag{20}$$

where e_u = error in U
$\quad\quad\quad e_{xi}$ = error in X_i

$$(e_u)^2 = \sum \left(\frac{\partial F}{\partial X_i} e_{xi} \right)^2 + \sum \left(\frac{\partial F}{\delta X_i} e_{xi} \right) \left(\frac{\delta F}{\delta X_j} e_{xj} \right) \tag{21}$$

where $I \neq j$. If the errors e_i to e_n are independent and symmetrical, the cross-product terms will tend to disappear and

$$(e_u)^2 = \sum \left(\frac{\partial F}{\partial X_i} e_{xi} \right)^2 \tag{22}$$

Since $\partial F / \partial X_i = 1/n$,

$$e_u = \left[\sum \left(\frac{1}{n} \right)^2 e_{xi}^2 \right]^{1/2} \tag{23}$$

or

$$e_u = \left[\left(\frac{1}{n} \right)^2 \sum (e_{xi})^2 \right]^{1/2} \tag{24}$$

from the definition of σ

$$\sum (e_{xi})^2 = n\sigma^2 \tag{25}$$

and

$$e_u = \frac{\sigma}{\sqrt{n}}$$

This equation must be corrected because the real errors in X are not known. If the number n were to approach infinity, the equation would be correct. Since n is a finite number, the corrected equation is written as

$$e_u = \frac{\sigma}{(n-1)^{1/2}} \tag{26}$$

and

$$q = U \pm \frac{\sigma}{(n-1)^{1/2}} \tag{27}$$

This says that if one reading is likely to differ from the true value by an amount σ, then the average of 10 readings will be in error by only $\sigma/3$ and the average of 100 readings will be in error by $\sigma/10$. To reduce the error by a factor of 2, the number of readings must be increased by a factor of 4.

3.3   External Estimates

In almost all experiments several steps are involved in making a measurement. It may be assumed that in each measurement there will be some error, and if the measuring devices are adequately calibrated, errors are as likely to be positive as negative. The worst condition insofar as accuracy of the experiment is concerned would be for all errors to have the same sign. In that case, assuming the errors are all much less than 1, the resultant error will be the sum of the individual errors, that is,

$$\epsilon_E = \epsilon_1 + \epsilon_2 + \epsilon_3 + \cdots \tag{28}$$

It would be very unusual for all errors to have the same sign. Likewise it would be very unusual for the errors to be distributed in such a way that

$$\epsilon_E = 0$$

A general method follows for treating problems that involve a combination of errors to determine what error is to be expected as a result of the combination.

Suppose that

$$V = F(a, b, c, d, e, \dots, x, y, z) \tag{29}$$

where a, b, c, ... , x, y, z represent quantities that must be individually measured to determine V. Then

$$\delta V = \sum \left(\frac{\partial F}{\partial n}\right)\delta n$$

and

$$\epsilon_E = \sum \left(\frac{\partial F}{\partial n}\right)e_n \tag{30}$$

The sum of the squares of the error contributions is given by

$$e_E^2 = \left[\sum \left(\frac{\partial F}{\partial n}\right)e_n\right]^2 \tag{31}$$

Now, as in the discussion of internal errors, assume that errors e_n are independent and symmetrical. This justifies taking the sum of the cross products as zero:

$$\sum \left(\frac{\partial F}{\partial n}\right)\left(\frac{\partial F}{\partial m}\right)e^n e^m = 0 \qquad n \neq m \tag{32}$$

Hence

$$(\epsilon_E)^2 = \sum \left(\frac{\partial F}{\partial n}\right)^2 e_n^2$$

or

$$e_E = \left[\sum \left(\frac{\partial F}{\partial n}\right)^2 e_n^2\right]^{1/2} \tag{33}$$

This is the *most probable value* of e_E. It is much less than the worst case:

$$\epsilon_e = [|\epsilon_a| + |\epsilon_b| + |\epsilon_c| \cdots + |\epsilon_z|] \tag{34}$$

As an application, the determination of g, the local acceleration of gravity, by use of a simple pendulum will be considered:

$$g = \frac{4\Pi^2 L}{T^2} \tag{35}$$

where L = length of pendulum
       T = period of pendulum

If an experiment is performed to determine g, the length L and the period T would be measured. To determine how the accuracy of g will be influenced by errors in measuring L and T write

$$\frac{\partial g}{\partial L} = \frac{4\Pi^2}{T^2} \qquad \text{and} \qquad \frac{\partial g}{\partial T} = \frac{-8\Pi^2 L}{T^3} \tag{36}$$

The error in g is the variation in g written as follows:

$$\delta g = \left(\frac{\partial g}{\partial L}\right)\Delta L + \left(\frac{\partial g}{\partial T}\right)\Delta T \tag{37}$$

or

$$\delta g = \left(\frac{4\Pi^2}{T^2}\right)\Delta L - \left(\frac{8\Pi^2 L}{T^3}\right)\Delta T \tag{38}$$

It is always better to write the errors in terms of percentages. Consequently, Eq. (38) is rewritten as

$$\delta g = \frac{(4\Pi^2 L/T^2)\,\Delta L}{L} - \frac{2(4\Pi^2 L/T^2)\,\Delta T}{T} \tag{39}$$

or

$$\frac{\delta g}{g} = \frac{\Delta L}{L} - \frac{2\,\Delta T}{T} \tag{40}$$

Then

$$e_g = [e_L^2 + (2e_T)^2]^{1/2} \tag{41}$$

where e_g is the *most probable error* in the measured value of g. That is,

$$g = \frac{4\Pi^2 L}{T^2} \pm e_g \qquad (42)$$

where L and T are the measured values. Note that even though a positive error in T causes a negative error in the calculated value of g, the contribution of the error in T to the most probable error is taken as positive. Note also that an error in T contributes four times as much to the most probable error as an error in L contributes. It is fundamental in measurements of this type that those quantities that appear in the functional relationship raised to some power greater than unity contribute more heavily to the most probable error than other quantities and must, therefore, be measured with greater care.

The determination of the most probable error is simple and straightforward. The question is how are the errors, such as $\Delta L/L$ and $\Delta T/T$, determined. If the measurements could be repeated often enough, the statistical methods discussed in the internal error evaluation could be used to arrive at a value. Even in that case it would be necessary to choose some representative error such as the standard deviation or the mean error. Unfortunately, as was noted previously, in engineering experiments it usually is not possible to repeat measurements enough times to make statistical treatments meaningful. Engineers engaged in making measurements will have to use what knowledge they have of the measuring instruments and the conditions under which the measurements are made to make a reasonable estimate of the accuracy of each measurement. When all of this has been done and a most probable error has been calculated, it should be remembered that the result is not the actual error in the quantity being determined but is, rather, the engineer's best estimate of the magnitude of the uncertainty in the final result.[9,10]

Consider again the problem of determining g. Suppose that the length L of the pendulum has been determined by means of a meter stick with 1-mm calibration marks and the error in the calibration is considered negligible in comparison with other errors. Suppose the value of L is determined to be 91.7 cm. Since the calibration marks are 1 mm apart, it can be assumed that ΔL is no greater than 0.5 mm. Hence the maximum

$$\frac{\Delta L}{L} = 5.5 \times 10^{-4}$$

Suppose T is determined with the pendulum swinging in a vacuum with an arc of $\pm 5°$ using a stop watch that has an inherent accuracy of one part in 10,000. (If the arc is greater than $\pm 5°$, a nonisochronous swing error enters the picture.) This means that the error in the watch reading will be no more than 10^{-4} s. However, errors are introduced in the period determination by human error in starting and stopping the watch as the pendulum passes a selected point in the arc. This error can be minimized by selecting the highest point in the arc because the pendulum has zero velocity at that point and timing a large number of swings so as to spread the error out over that number of swings. Human reaction time may vary from as low as 0.2 s to as high as 0.7 s. A value of 0.5 s will be assumed. Thus the estimated maximum error in starting and stopping the watch will be 1 s (± 0.5 s at the start and ± 0.5 s at the stop). A total of 100 swings will be timed. Thus the estimated maximum error in the period will be 1/100 s. If the period is determined to be 1.92 s, the estimated maximum error will be $0.01/1.92 = 0.005$. Compared to this, the error in the period due to the inherent inaccuracy of the watch is negligible. The nominal value of g calculated from the measured values of L and T is 982.03 cm/s^2. The most probable error [Eq. (29)] is

$$[4(0.005)^2 + (5.5 \times 10^{-4})^2]^{1/2} = 0.01 \qquad (43)$$

The uncertainty in the value of g is then ± 9.82 cm/s^2, or in other words the value of g will be somewhere between 972.21 and 991.85 cm/s^2.

Often it is necessary for the engineer to determine in advance how accurately the measurements must be made in order to achieve a given accuracy in the final calculated result. For example, in the pendulum problem it may be noted that the contribution of the error in T to the most probable error is more than 300 times the contribution of the error in the length measurement. This suggests, of course, that the uncertainty in the value of g could be greatly reduced if the error in T could be reduced. Two possibilities for doing this might be (1) find a way to do the timing that does not involve human reaction time or (2) if that is not possible, increase the number of cycles timed. If the latter alternative is selected and other factors remain the same, the error in T timed over 200 swings is 1/200 or 0.005, second. As a percentage the error is $0.005/1.92 = 0.0026$. The most probable error in g then becomes

$$e_g = [4 \times (2.6 \times 10^{-3})^2 + (5.5 \times 10^{-4})^2]^{1/2} = 0.005 \tag{44}$$

This is approximately half of the most probable error in the result obtained by timing just 100 swings. With this new value of e_g the uncertainty in the value of g becomes ± 4.91 cm/s^2 and g then can be said to be somewhere between 977.12 and 986.94 cm/s^2. The procedure for reducing this uncertainty still further is now self-evident.

Clearly the value of this type of error analysis depends upon the skill and objectivity of the engineer in estimating the errors in the individual measurements. Such skills are acquired only by practice and careful attention to all the details of the measurements.

REFERENCES

1. W. A. Wildhack, "NBS Source of American Standards," *ISA J.* **8**(2), February, 1961.
2. P. Giacomo "News from the IBPM," *Metrologia* **20**(1), April, 1984.
3. NIST-F1 Cesium Fountain Atomic Clock.
4. "NIST Time and Frequency Services," NIST Special Publication 432, 2002.
5. R. E. Bentley (Ed.), *Handbook of Temperature Measurement,* CSIRO Springer.
6. E. A. Doebelin, *Measurement Systems—Application and Design,* 5th ed., McGraw-Hill, New York, 2004, pp. 85–91.
7. C. M. Harris and A. G. Piersol, "Mechanical Impedance," in *Shock and Vibration Handbook,* 5th ed. McGraw-Hill, New York, 2002, Chapter 10, pp. 10.1–10.14.
8. N. H. Cook and E. Rabinowicz, *Physical Measurement and Analysis,* Addison Wesley, Readng, MA, 1963, pp. 29–68.
9. S. J. Kline and F. A. McClintock, "Describing Uncertainties in Single Sample Experiments," *Mech. Eng.,* **75**(3), January, 1953.
10. B. N. Taylor and C. E. Kuyatt, "Guidelines for Evaluating and Expressing the Uncertainty of NIST Measurement Results," NIST Technical Note 1297, 1994.

CHAPTER 17

SIGNAL PROCESSING

John Turnbull
Case Western Reserve University
Cleveland, Ohio

1 FREQUENCY-DOMAIN ANALYSIS OF LINEAR SYSTEMS

Signals are any carriers of information. Our objective in signal processing involves the encoding of information for the purpose of transmission of information or decoding the information at the receiving end of the transmission. Unfortunately, the signal is often corrupted by noise during our transmission, and hence it is our objective to extract the information from the noise. The standard method most commonly used for this involves filters that exploit some separation of the signal and noise in the frequency domain. To this end, it is useful to use frequency-domain tools such as the Fourier transform and the Laplace transform in designing and analyzing various filters. The Fourier transform of a function of a time is

$$\mathscr{F}\{f(t)\} = F(\omega) = \frac{1}{\sqrt{2\pi}} \int_{-\infty}^{\infty} f(t)e^{j\omega t}\, dt \qquad j^2 = -1 \tag{1}$$

For continuous systems, the transfer characteristics of a filter system is a function that gives information of the gain versus frequency. The Laplace transform for a given time-domain function is

$$\mathscr{L}\{f(t)\} = F(s) = \int_{0}^{\infty} f(t)e^{-st}\, dt \tag{2}$$

The steady-state Laplace transform (i.e., neglecting transients) for the derivative and integral of a given function is

$$\mathscr{L}\left\{\frac{df(t)}{dt}\right\} = sF(s) \qquad \mathscr{L}\left\{\int f(t)\,dt\right\} = \frac{F(s)}{s} \tag{3}$$

By convention, functions in the time domain use t as the independent variable and functions in the Laplace domain use s as the independent variable. For this reason, the Laplace domain is commonly called the *S domain*. The Fourier transform and the Laplace transform are similar but different in two respects: (1) The Fourier transform integrates the signal over all time while the Laplace transform integrates for positive times only, and (2) the exponent of the kernel in the Laplace transform is complex with both real and imaginary components while the exponent in the Fourier transform has imaginary component and no real component. By using Euler's identity, $e^{j\omega} = \cos(\omega) + j\sin(\omega)$, we see that

$$\mathscr{F}\{f(t)\} = F(\omega) = \int_{-\infty}^{\infty} f(t)e^{j\omega}\,dt$$

$$= \int_{-\infty}^{\infty} f(t)\cos(\omega t)\,dt + j\int_{-\infty}^{\infty} f(t)\sin(\omega t)\,dt$$

$$= \langle\cos(\omega t)\rangle + j\langle\sin(\omega t)\rangle \tag{4}$$

We approximate the Fourier transform from discrete samples $f(k) \leftrightarrow F(k)$, where $0 \leq k \leq N$ and $F(k) = \alpha_k + j\beta_k$:

$$\alpha_k = \sum_{i=0}^{N-1} f(i)\cos\left(2\pi\frac{i\cdot k}{N}\right) \qquad \beta_k = \sum_{i=0}^{N-1} f(i)\sin\left(2\pi\frac{i\cdot k}{N}\right) \tag{5}$$

However, if the number of points we transform is a composite number and not a prime number, we can restructure our calculations to eliminate some of the calculations. Furthermore, of the factors that are themselves composite, we can further factor and eliminate more calculations. For this reason, the most efficient vector sizes are those that are highly composite. As an example, consider the simple case of transforming four points. We can express Eq. (5) for this case in matrix form as

$$\begin{bmatrix} c_0 \\ c_1 \\ c_2 \\ c_3 \end{bmatrix} = \begin{bmatrix} 1 & 1 & 1 & 1 \\ 1 & \rho & \rho^2 & \rho^3 \\ 1 & \rho^2 & \rho^4 & \rho^6 \\ 1 & \rho^3 & \rho^6 & \rho^9 \end{bmatrix} \begin{bmatrix} f_0 \\ f_1 \\ f_2 \\ f_3 \end{bmatrix}$$

where ρ is the principal root of 1. The nth principal root of 1 is $\cos(2\pi/n) + j\sin(2\pi/n)$. We can then factor the matrix into two sparse matrices:

$$\begin{bmatrix} c_0 \\ c_1 \\ c_2 \\ c_3 \end{bmatrix} = \begin{bmatrix} 1 & 1 & 0 & 0 \\ 1 & \rho^2 & 0 & 0 \\ 0 & 0 & 1 & \rho \\ 0 & 0 & 1 & \rho^3 \end{bmatrix} \begin{bmatrix} 1 & 0 & 1 & 0 \\ 0 & 1 & 0 & 1 \\ 1 & 0 & \rho^2 & 0 \\ 0 & 1 & 0 & \rho^2 \end{bmatrix} \begin{bmatrix} f_0 \\ f_1 \\ f_2 \\ f_3 \end{bmatrix}$$

Although it is possible to implement this efficient algorithm—known as the fast Fourier transform (FFT)—for any vector size that is a composite number, it is most commonly implemented for vector sizes in which all factors are 2. The effort in evaluating Eq. (5) increases with the square of the number of points to transform, or $O(n^2)$. In contrast, the effort for the FFT of order 2^p increases proportionately to $O(n\ \log(n))$. Finally, we can use the FFT to estimate the power spectral density (PSD) of a given discrete signal by computing the square of the magnitude of the FFT. The following algorithm describes the method for computing the fast Fourier transform.[1]

FAST FOURIER TRANSFORM 1 *Given a vector* $\{x_1, ..., x_n\}$ *of complex numbers, where n is some integer and there exists some integer p such that* $n = 2^p$. *This algorithm outputs the Fourier transform overwriting the input vector* **x**.

```
k ← 1
For i = 1 To n
If k > i
swap(x_i,x_k)
End If
m = n/2
While m ≥ 1 And k > m
k ← k - m
m ← m/2
End While
k ← k + m
Next i
M_max ← 1
While n > M_max
i_c ← 2·M_max
θ ← π/M_max
wp ← -2 sin² (θ /2) - j·sin(θ)
w ← 1
For m = 1 To M_max
For i = m To n By i_c
k ← i + M_max
xtemp ← x_i - w·x_k
x_i ← x_i + w·x_k
x_k ← xtemp
Next i
w ← w·(wp + 1)
Next m
M_max = i_c
End While
```

2 BASIC ANALOG FILTERS

Linear filters apply frequency-specific gains to a signal. This is often done to enhance desired portions of the spectrum while attenuating or eliminating other portions. Four common filters are low pass, high pass, bandpass, and band reject. The objective of an ideal low-pass filter is to eliminate a range of undesired high frequencies from a signal and leave the remaining portion undistorted. To this end, an ideal low-pass filter will have a gain of 1 for all frequencies less than some desired cutoff frequency f_c and a gain of 0 for all frequencies greater than f_c, as seen in Fig. 1. There are various rational functions that approximate this ideal. But because of the discontinuity in the ideal low-pass response, all realizations of this ideal will be an approximation. The various approximation functions generally trade off between three characteristics: passband ripple, stop-band ripple, and the transition width, shown in Fig. 2. Four common rational function approximations for low-pass filters are the Butterworth, the Tchebyshev types I and II, and the elliptical filter. By convention, we use $H(s)$ as the transfer function from which we determine the frequency response, where

$$H(s) = \frac{V_{\text{output}}(s)}{V_{\text{input}}(s)} \tag{6}$$

is the output-to-input gain of a standard two-port system.

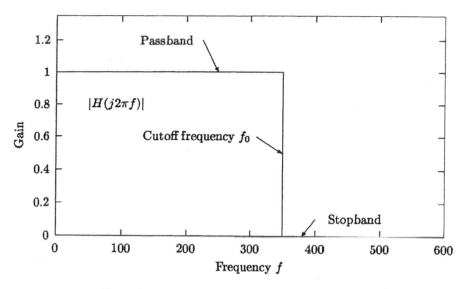

Figure 1   Frequency response for an ideal low-pass filter.

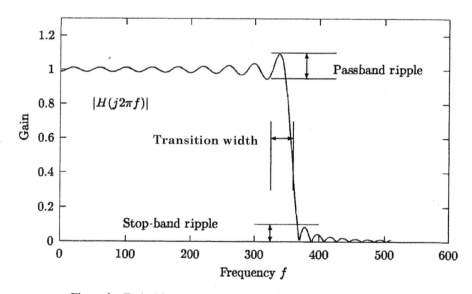

Figure 2   Typical frequency response for an approximate low-pass filter.

2.1   Butterworth

The Butterworth filter has a smooth passband region (frequencies less than f_c hertz) and a smooth stop band (frequencies greater than f_c) and a comparatively wide transition region as shown in Figs. 3 and 4. Let S_c be the cutoff frequency; then a low-pass rational function approximation is as follows:

$$|H(s)|^2 = \frac{1}{1 + (s/s_c)^{2N}} \qquad (7)$$

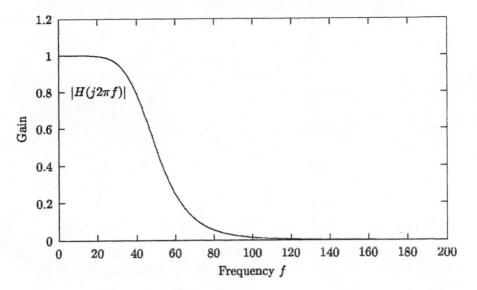

Figure 3 Frequency response to a third-order Butterworth filter.

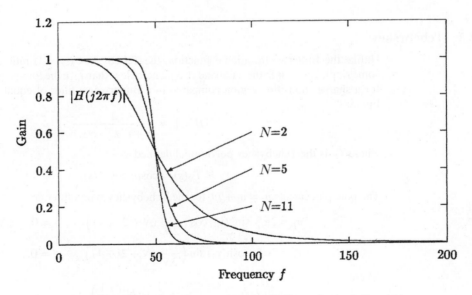

Figure 4 Frequency response to Butterworth filters of order 2, 5, and 11.

where $2\pi f_c = s_c$. In factored form,

$$H(s) = \frac{1}{(s - p_0)(s - p_1)\cdots(s - p_{N-1})} \tag{8}$$

where $p_i = \alpha_i + j\beta_i$ and

$$\alpha_i = 2\pi f_c \, \cos\left(\frac{\pi}{2N}\,(N + 2i + 1)\right) \qquad i = 0, \ldots, N - 1$$

$$\beta_i = 2\pi f_c \, \sin\left(\frac{\pi}{2N}\,(N + 2i + 1)\right) \qquad i = 0, \ldots, N - 1 \tag{9}$$

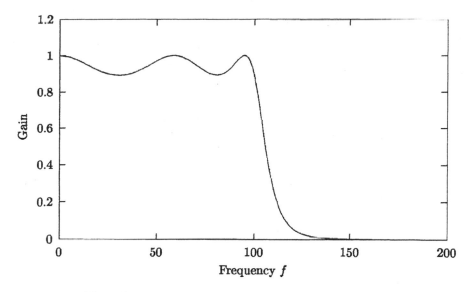

Figure 5   Frequency response to a fifth-order Tchebyshev filter.

2.2   Tchebyshev

Unlike the Butterworth rational function, the Tchebyshev (type I) rational function permits some ripple to occur in the passband (frequencies less than f_c in the low-pass filter), in exchange for a sharper transition region compared to a Butterworth filter of equal order N, as seen in Fig. 5:

$$|H(s)|^2 = \frac{1}{1 + \epsilon^2 T_N^2 (s/s_c)} \tag{10}$$

where T_N is the Tchebyshev polynomial defined as

$$T_N(s) = \cos[n \cos^{-1}(s)] \tag{11}$$

The pole placements $p_i = \alpha_i + j\beta_i$ for the Tchebyshev type I filter are

$$\alpha_i = 2\pi f_c \, \sinh(v_0) \cos\left(\frac{\pi}{2N} (N + 2i + 1)\right) \qquad i = 0, \ldots, N - 1$$

$$\beta_i = 2\pi f_c \, \cosh(v_0) \sin\left(\frac{\pi}{2N} (N + 2i + 1)\right) \qquad i = 0, \ldots, N - 1 \tag{12}$$

where

$$v_0 = \frac{1}{N} \sinh^{-1}\left(\frac{1}{\epsilon}\right) \tag{13}$$

$$\epsilon = \sqrt{\frac{1}{(1 - r)^2} - 1} \qquad 0 < r < 1 \tag{14}$$

where r is this amplitude of the ripple in proportion to the gain of the passband.

2.3   Inverse Tchebyshev

The inverse Tchebyshev (or Tchebyshev type II) filter has a smooth passband, ripple in the stop band (frequencies less than f_c for the low-pass filter), and a sharper transition region compared

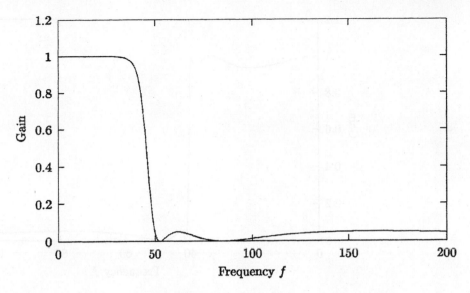

Figure 6 Frequency response to a fifth-order inverse Tchebyshev filter.

to the Butterworth function of equal order N, as seen in Fig. 6:

$$|H(s)|^2 = \frac{\epsilon^2 T_N^2 \left(s_c/s\right)}{1 + \epsilon^2 T_N^2 \left(s_c/s\right)} \tag{15}$$

The zero placements $\zeta_i = \alpha_i + \beta_i$ for the Tchebyshev type II filter are

$$\zeta_i = \frac{1}{\sin(i\pi/2N)} \qquad 0 \le i \le N - 1 \tag{16}$$

The pole placements $p_i = \alpha_i + j\beta_i$ for the Tchebyshev type II filter are simply the reciprocals of the pole placements computed for the Tchebyshev type I filter.

2.4 Elliptical

The elliptical filter has ripple in both the passband and the stop band but, in exchange, has the narrowest transition region for equal filter order N, as seen in Fig. 7. The derivation and implementation for the determination of the poles and zeros involve the Jacobian elliptic function:

$$f(t, k) = \int_0^t \frac{dx}{\sqrt{1 - k^2 \sin^2(x)}} \tag{17}$$

The method is beyond the scope of this chapter. The interested reader is referred to Refs. 2 and 3.

2.5 Arbitrary Frequency Response Curve Fitting by Method of Least Squares

It is possible to design a filter to approximate a desired frequency response $F(\omega)$ by the method of least squares. Consider a transfer function in factored form as

$$H(s) = G \frac{\prod_{i=i}^{M} s - \zeta_i}{\prod_{k=1}^{N} s - p_k} \tag{18}$$

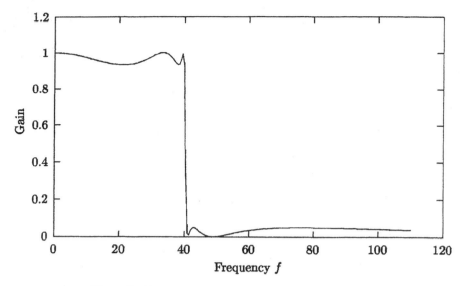

Figure 7 Frequency response to a fifth-order elliptical filter.

Minimize

$$\chi^2(\boldsymbol{G}; \boldsymbol{\zeta}; \mathbf{p}) = \int_D [|H(j\omega)| - F(\omega)]^2 \, d\omega \tag{19}$$

subject to

$$\mathscr{R}e(p_k) < 0 \text{ for all } k = 1, ..., N \tag{20}$$

Unfortunately, this results in a nonlinear system of equations. Furthermore, the topography of this objective function is generally complicated with many local minima, making standard gradient-descent methods unfeasible. It is usually best to use finite-impulse-response (FIR) filtering or filtering in the frequency domain for these types of problems. If an infinite-impulse-response (IIR) filter is desired, the reader is referred to Prony's method, which linearizes this system and finds an approximate optimal solution.[4]

2.6 Circuit Prototypes for Pole and Zero Placement for Realization of Filters Designed from Rational Functions

The voltage–current relation for a resistor (R), inductor (L), and capacitor (C) is

$$v_r = iR \qquad v_l = L\frac{di}{dt} \qquad v_c = \frac{1}{C}\int_{-\infty}^{t} i \, dr \tag{21}$$

These relationships are represented in the S domain as

$$V_r(s) = RI(s) \qquad V_l(s) = sLI(s) \qquad V_c = \frac{I(s)}{sC} \tag{22}$$

Thus, the general transfer function for any linear circuit involving standard passive, active, and reactive devices is a rational function, that is, a ratio of polynomials:

$$H(s) = \frac{\sum_{i=0}^{M-1} a_i s^i}{\sum_{k=0}^{N-1} b_k s^k}$$

$$= \frac{V_0}{V_i} = -\frac{Z_2}{Z_1} \tag{23}$$

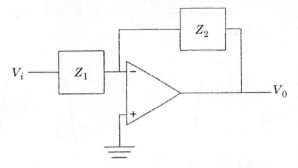

Figure 8   Prototype circuit element for construction of analog filter.

Table 1   Circuit Elements for the Construction of Basic
Transfer Function Prototypes

Single pole	$Z_1 \leftarrow$ resistor
	$Z_2 \leftarrow RC$ in parallel
Single zero	$Z_1 \leftarrow RC$ in parallel
	$Z_2 \leftarrow$ resistor
Complex-conjugate pole pair	$Z_1 \leftarrow LRC$ in series
	$Z_2 \leftarrow$ capacitor
Complex-conjugate zero pair	$Z_1 \leftarrow$ capacitor
	$Z_2 \leftarrow LRC$ series

Thus one can construct any arbitrary transfer function through a serial placement of this building block circuit prototype shown in Fig. 8. Table 1 gives circuit elements for Z_1 and Z_2 to construct the basic prototype circuits.

3   BASIC DIGITAL FILTER

Basic linear digital filters are of two types: those that have a finite response to an impulse, or FIR, and those that have an infinite response to an impulse (IIR). The general form of a linear digital filter is

$$y_k = b_1 y_{k-1} + b_2 y_{k-2} + \cdots + b_{k-m} y_{k-m} + a_0 x_k + a_1 x_{k-1} + \cdots + a_n x_{k-n} \qquad (24)$$

where $k - i$ represents the ith delay. That is, the kth output from a linear digital filter is some linear combination of previous inputs and outputs. The filter will have a finite response to an impulse if $b_1 = b_2 = \cdots = b_m = 0$; otherwise, the filter is of type IIR.

3.1 z Transforms

The z transform is used to analyze the frequency response and stability of a system of difference equations in much the same way that the Laplace transform is used to analyze the frequency response and stability of a system of differential equations. The z transform of Eq. (24) is

$$H(z) = \frac{a_0 + a_1 z^{-1} + \cdots + a_M z^{-M}}{1 - b_1 z^{-1} - \cdots b_N z^{-N}} = \frac{\sum_0^M a_j z^{-j}}{1 - \sum_{j=1}^N b_j z^{-j}} \qquad (25)$$

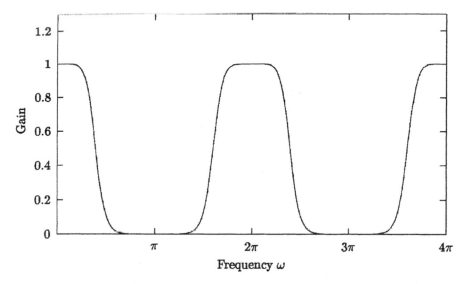

Figure 9 Demonstration of aliasing from digital filter response.

We determine the frequency response by $|H(e^{j\omega})|$, where $0 \le \omega \le 2\pi$ and corresponds to the scaled frequencies of our sampled system from 0 Hz up to the sampled frequency. The function $e^{j\omega}$ is a periodic signal. For $\pi < \omega \le 2\pi$, $e^{j\omega} = e^{j(\pi - \omega)}$, and for $2\pi < \omega \le 4\pi$, $e^{j\omega} = e^{j(\omega - 2\pi)}$. Thus frequencies greater than π, corresponding to the Nyquist frequency, or one-half of the sampling frequency assume an identical characteristic to an analogous frequency less than π. This phenomenon is called *aliasing* and is illustrated in Fig. 9.

3.2 Design of FIR Filters

It is possible to use the Fourier transform to determine the coefficients to an FIR filter. However, the Fourier coefficients are generally complex numbers, and when working with real signals, it is desirable to have real coefficients in our filter. To do this, we apply Euler's identity and observe from Eq. (4) that the coefficients will be real if the inner products with the sine function are all zero. We can artificially construct our desired frequency response so that this will be so. To do this, suppose $F(\omega)$ is the desired frequency response where $0 \le \omega \le \pi$:

Step 1. Augment the domain of the function over $0 \le \omega \le 2\pi$.

Step 2. Augment the function values from $\pi \le \omega \le 2\pi$ as $F(\omega) = F(2\pi - \omega)$.

Step 3. Compute the discrete cosine transform [α_i coefficients from Eq. (5)] over the range $0 \le \omega \le 2\pi$.

For a discrete system with a desired FIR filter of length N:

Step 1. Augment the domain of the function over $1 \le i \le 2N$.

Step 2. Augment the function values from $N + 1 \le i \le 2N$ as $F(i) = F(2N - i)$.

Step 3. Construct the FIR filter with $2N$ points from the discrete cosine transform of the desired frequency response.

Step 4. Keep the first N coefficients and truncate the remaining coefficients.

The discrete cosine transform of an ideal low-pass filter is a sinc function, defined as

$$\text{sinc}(x) = \begin{cases} \dfrac{\sin(\pi x)}{\pi x} & \text{if } x \neq 0 \\ 1 & \text{if } x = 0 \end{cases} \tag{26}$$

Therefore, the coefficients to a low-pass FIR filter with cutoff frequency f_c and length $2N-1$ are determined using Eq. (27), where ω_c is the desired cutoff frequency divided by the Nyquist frequency:

$$h(i) = \omega_c \ \text{sinc}\left[\omega_c\left(i - \left\lfloor\frac{N}{2}\right\rfloor\right)\right] \qquad 1 \leq i \leq N \tag{27}$$

Where $\lfloor x \rfloor$ is the greatest integer less than or equal to x and ω_c is the desired cutoff frequency divided by the Nyquist frequency.

Windowing

This FIR filter will have ripple in the passband and in the stop band. It is possible to suppress these ripples and smooth the frequency response, but the tradeoff will be an increased transition width. The method for suppressing these ripples is with the application of a windowing function. There is a large class of windowing functions that allow the designer to determine how he or she wishes to trade off the transition width and how much ripple is to be tolerated. The design of an FIR filter with windowing involves the use of Eq. (26) for the determination of the FIR coefficients followed by the component-by-component product of the coefficients with the windowing values, that is, $h'(i) = h(i)\ \text{win}(i)$. Below, is a list of common windows[5]:

Rectangular:

$$\text{win}(i) = 1 \qquad 0 \leq i \leq N-1 \tag{28}$$

Bartlett:

$$\text{win}(i) = \begin{cases} \dfrac{2i}{N-1} & 0 \leq i \leq \dfrac{N-1}{2} \\ 2 - \dfrac{2i}{N-1} & \dfrac{N-1}{2} \leq i \leq N-1 \end{cases} \tag{29}$$

Hanning:

$$\text{win}(i) = \frac{1}{2}\left[1 - \cos\left(\frac{2\pi i}{N-1}\right)\right] \qquad 0 \leq i \leq N-1 \tag{30}$$

Hamming:

$$\text{win}(i) = 0.54 - 0.46\ \cos\left(\frac{2\pi i}{N-1}\right) \qquad 0 \leq i \leq N-1 \tag{31}$$

Blackman:

$$\text{win}(i) = 0.42 - 0.5\ \cos\left(\frac{2\pi i}{N-1}\right) + 0.08\cos\left(\frac{4\pi i}{N-1}\right) \qquad 0 \leq i \leq N-1 \tag{32}$$

Table 2 gives a list of several common windowing functions together with their characteristics.[5]

Figure 10 demonstrates the effect of a Hanning window.

FIR High-Pass and Bandpass Design

The design of a high-pass filter is simply $1 - H(z)$. In the time domain, this is

$$h(i) = \omega_c\ \text{sinc}\left[\omega_c\left(i - \left\lfloor\frac{N}{2}\right\rfloor\right)\right] - \omega_c \delta\left(\left\lfloor\frac{N}{2}\right\rfloor\right) \qquad i \leq i \leq N \tag{33}$$

Table 2   Comparison of Characteristics for Commonly
Used Windowing Functions

Window Name	Minimum Stop-Band Attenuation (dB)
Rectangular	−21
Bartlett	−25
Hanning	−44
Hamming	−53
Blackman	−74

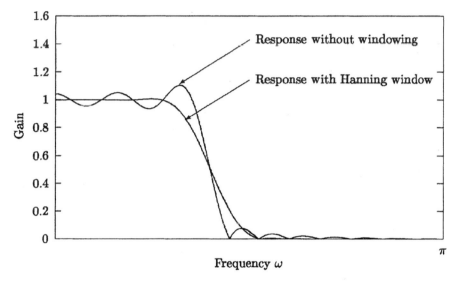

Figure 10   Comparison of FIR filter with Hanning window and with windowing.

where

$$\delta(i) = \begin{cases} 1 & i = 0 \\ 0 & i \neq 0 \end{cases} \qquad (34)$$

We construct a bandpass by filtering the data with a high-pass filter, then filtering the output with a low-pass filter. Or we can combine the two filters together into a single filter by convolving the coefficients.

3.3   Design of IIR Filters

The common strategy in designing IIR filters is as follows:

Step 1.   Design a rational function in the S domain in factored form that best approximates the desired frequency response characteristics (using Butterworth, Tchebyshev, elliptical, etc.).

Step 2.   Transform the poles and zeros into the z domain.

Step 3. Reconstruct the rational function.

Step 4. Realize the difference equation by inverse z transform of the rational polynomials.

There are different S to z transforms. The two most common are the impulse-invariant and the bilinear transformation. The impulse-invariant transformation is

$$z = e^{sT} \tag{35}$$

where T is the sampling period. This method is usually not used because it can cause aliasing. The bilinear transformation avoids this but distorts the mapping in other ways for which we need to compensate. The bilinear transformation is

$$z = \frac{2/T + s}{2/T - s} \tag{36}$$

The S to z transformation is not exact and always involves various tradeoffs. Because of this, the actual placement of the cutoff in a designed digital filter is misplaced and the error increases for cutoff frequencies closer to the Nyquist frequency. To compensate for this effect, we apply Eq. (37) in the design of our filter:

$$f_c' = \frac{1}{\pi T} \tan(\pi f_c T) \tag{37}$$

This process is called "prewarping."

Example IIR Design

For a digital system with a sampling rate of 100 Hz, a third-order low-pass Tchebyshev filter with a cutoff frequency at 15 Hz and with 20% ripple in the passband is designed as follows:

Step 1. Compute the Nyquist frequency $f_N = 100 \text{ Hz}/2 = 50 \text{ Hz}$.

Step 2. Compute f' using Eq. (37) to prewarp the cutoff frequency:

$$f' = 16.22 \text{ Hz}$$

Step 3. Compute v_0 using Eqs. (13) and (14):

$$\epsilon = 0.75 \quad v_0 = 0.3662$$

Step 4. Determine the poles and zeros of the analog system using Eq. (12):

$$p_1 = -19.0789 + j94.2364$$
$$p_2 = -38.1578$$
$$p_3 = -19.0789 - j94.2364$$

Step 5. Map the poles from the S domain into the z domain using the bilinear transformation equation (36) and $T = (1/\text{sampling rate}) = 0.01$ s:

$$pz_1 = 0.5407 + j0.6627$$
$$pz_2 = 0.6796$$
$$pz_3 = 0.5407 - j0.6627$$

Step 6. Map the zeros from the S domain into the z domain also using Eq. (36). Although we may be tempted to conclude that there are no zeros in the S domain, since our numerator is constant, we note that $H(s) \to 0$ as $s \to \infty$. Thus mapping the bilinear transformation for this case yields

$$\lim_{s \to \infty} \frac{2/T + s}{2/T - s} = -1$$

And since this is a third-order system,

$$z_1 = -1 \quad z_2 = -1 \quad z_3 = -1$$

Step 7. Expand the numerator and denominator polynomials:

$$\frac{(z - \zeta_1)(z - \zeta_2)(z - \zeta_3)}{(z - pz_1)(z - pz_2)(z - pz_3)} = \frac{1 + 3z^{-1} + 3z^{-2} + z^{-3}}{-1 + 1.7611z^{-1} - 1.466z^{-2} + 0.4972z^{-3}}$$

Step 8. Normalize the filter so that the gain in the passband will be 1. In this case, we know that the gain should be 1 at $\omega = 0$. Hence, we evaluate $|H(e^{j\omega})|$ for $\omega = 0$: $|H(e^0)| = 38.4$. The normalized transfer function is then

$$H(z) = \frac{0.026 + 0.078z^{-1} + 0.078z^{-2} + 0.026z^{-3}}{-1 + 1.7611z^{-1} - 1.466z^{-2} + 0.4972z^{-3}}$$

Step 9. Realize the difference equation from inverse z transformation of the derived transfer function:

$$y_n = 1.7611y_{n-1} - 1.466y_{n-2} + 0.4972y_{n-3} + 0.026x_n + 0.078x_{n-1}$$
$$+ 0.078x_{n-2} + 0.026x_{n-3}$$

3.4   Design of Various Filters from Low-Pass Prototypes

The procedure for designing a high-pass, bandpass, or band-reject IIR filter is as follows: First, design, by pole–zero placement, a low-pass filter (Butterworth, Tchebyshev, etc.) with an arbitrary cutoff frequency (though for practical considerations, it is best to choose a value midway between 0 Hz and Nyquist), transform the poles and zeros according to the following formulas, reconstitute a new transfer function from the transformed poles and zeros, then realize the digital filter by taking the inverse z transform of the new transfer function. The formulas with an example are given below, where ω_L and ω_H are the low- and high-frequency cutoffs, respectively, normalized between 0 and π, that is, $\omega_L = 2\pi f_L$/sample rate, where f_L is the cutoff frequency in hertz; ϕ_L is the normalized cutoff frequency of the low-pass prototype.[5]

High Pass

$$z' = \frac{1 + AZ}{Z + A} \tag{38}$$

where Z is the pole or zero to be transformed and z' is the transformed pole or zero and

$$A = -\frac{\cos[(\omega_H + \phi_L)/2]}{\cos[(\omega_H - \phi_L)/2]} \tag{39}$$

Bandpass

The bandpass filter has two transitions: a rising edge and a falling edge. For this reason, we need twice as many coefficients for the same approximate transition width as the prototype filter. Thus the order of these polynomials will be twice the order of polynomials in the prototype. Each pole from the prototype will transform into a pair of poles (z_1' and z_2'). Likewise, each zero will transform into a pair of zeros:

$$z_1' = \frac{(A + AZ) + \sqrt{(A + AZ)^2 - 4(Z + B)(BZ + 1)}}{2(Z + B)} \tag{40}$$

$$z_2' = \frac{(A + AZ) - \sqrt{(A + AZ)^2 - 4(Z + B)(BZ + 1)}}{2(Z + B)} \tag{41}$$

$$A = \frac{2CD}{D+1} \tag{42}$$

$$B = \frac{D-1}{D+1} \tag{43}$$

$$C = \frac{\cos[(\omega_H + \omega_L)/2]}{\cos[(\omega_H - \omega_L)/2]} \tag{44}$$

$$D = \cot\left(\frac{\omega_H - \omega_L}{2}\right) \tan\frac{\phi_L}{2} \tag{45}$$

Band Reject

$$z'_1 = \frac{(AZ - A) + \sqrt{(AZ - A)^2 - 4(Z - B)(BZ - 1)}}{2(Z - B)} \tag{46}$$

$$z'_2 = \frac{(AZ - A) - \sqrt{(AZ - A)^2 - 4(Z - B)(BZ - 1)}}{2(Z - B)} \tag{47}$$

$$C = \frac{\cos[(\omega_H + \omega_L)/2]}{\cos[(\omega_H - \omega_L)/2]} \tag{48}$$

$$D = \tan\left(\frac{\omega_H - \omega_L}{2}\right) \tan\frac{\phi_L}{2} \tag{49}$$

As an example, design a Tchebyshev (type I) bandpass filter for a system sampled at 100 Hz with a low cutoff frequency of 20 Hz and a high cutoff frequency of 35 Hz.

Step 1. We can use the poles and zeros designed in Section 3.3 as the low-pass prototype.

Step 2. Convert the low and high cutoff frequencies to values normalized between 0 and π, where π corresponds to the Nyquist frequency:

$$\omega_L = 2\pi \cdot \frac{20}{100} = 1.2566 \quad \omega_H = 2\pi \cdot \frac{35}{100} = 2.1991 \tag{50}$$

Step 3. Map the poles from the prototype using Eqs. (40) and (41):

$$p_1 \to \begin{cases} -0.6095 - j0.9105 \\ 0.2940 + j1.0722 \end{cases} \tag{51}$$

$$p_2 \to \begin{cases} -0.2302 - j1.2527 \\ -0.2302 + j1.2527 \end{cases} \tag{52}$$

$$p_3 \to \begin{cases} -0.6095 + j0.9105 \\ -0.2940 - j1.0722 \end{cases} \tag{53}$$

Then map the zeros from the prototype using Eqs. (40) and (41):

$$\zeta_1 \to \begin{cases} -1 \\ 1 \end{cases} \tag{54}$$

$$\zeta_2 \to \begin{cases} -1 \\ 1 \end{cases} \tag{55}$$

$$\zeta_3 \to \begin{cases} -1 \\ 1 \end{cases} \tag{56}$$

Step 4. Expand the numerator and denominator polynomials:

$$\frac{(z - \zeta_1) \cdots (z - \zeta_6)}{(z - pz_1) \cdots (z - pz_6)}$$

$$= \frac{1 - 3z^{-2} + 3z^{-4} - z^{-6}}{(1 + 1.091z^{-1} + 3.632z^{-2} + 2.616z^{-3} + 4.642z^{-4} + 1.982z^{-5} + 2.407z^{-6})} \quad (57)$$

Step 5. Normalize the transfer function so that it will have unity gain in the passband. For this, we estimate

$$M = \max_{0 \le \omega \le \pi} |H(e^{j\omega})| \quad (58)$$

Then compute

$$H_{\text{normalized}}(z) = \frac{1}{M} H(z) \quad (59)$$

For this example, $M = 14.45$.

Step 6. Realize the difference equation from inverse z transformation of the derived transfer function:

$$y_n = -1.091y_{n-1} - 3.632y_{n-2} - 2.616y_{n-3} - 4.642y_{n-4} - 1.982y_{n-5}$$
$$- 2.407y_{n-6} + 0.0692 * x_n - 0.2076x_{n-2} + 0.2076x_{n-4} - 0.0692x_{n-6} \quad (60)$$

3.5 Frequency-Domain Filtering

It is possible to filter the data in the frequency domain. The method involves the use of the Fourier transform. We Fourier transform the data, multiply by the desired frequency response, then inverse Fourier transform the data. This is similar to the FIR filters discussed earlier. Deriving the FIR coefficients by performing a discrete cosine transform (DCT) of the desired frequency response and then convolving the coefficients with the data is equivalent to filtering the data in the frequency domain. One difference, however, is that the frequency domain filtering is generally done on blocks of data and not on streaming data, as is done in the time domain, which can be of concern when processing highly nonstationary data with abrupt transients. The inverse Fourier transform is

$$\mathcal{F}^{-1}\{F(\omega)\} = \frac{1}{\sqrt{2\pi}} \int_{-\infty}^{\infty} f(\omega)e^{-j\omega t} \, d\omega \quad (61)$$

4 STABILITY AND PHASE ANALYSIS

4.1 Stability Analysis

Consider a transfer function 1

$$H(s) = \frac{1}{s - p} \quad (62)$$

where the pole p is a complex number $\alpha + j\beta$. The inverse Laplace transform of this function is $e^{\alpha t}[\cos(\beta t) + j \sin(\beta t)]$. This function is bounded as $t \to \infty$ if and only if $\alpha < 0$. From this, we can determine the stability of a function by inspecting the real components of all poles of a given transfer function. The procedure for a rational function (a ratio of polynomials) would be to factor the polynomials in the denominator and inspect to ensure that the real components to all of the poles are less than zero. Suppose

$$H(s) = \frac{a_0 + a_1 s + \cdots + a_m s^m}{b_0 + b_1 s + \cdots + b_n s^2}$$

$$= \frac{(s - \zeta_0)(s - \zeta_1) \cdots (s - \zeta_m)}{(s - p_0)(s - p_1) \cdots (s - p_n)} \quad (63)$$

In a similar way, by inspection of the S-to-z transformation $z = e^s$, we see that the entire left half of the plane in the S domain maps inside the unit circle in the z domain. For this reason, we analyze the stability of systems in the z domain by inspecting the poles of the transfer function. The system is stable if the norm of all poles is less than 1.

4.2 Phase Analysis

While processing the data in real time, our filters must act on the signal history. For this reason, there will always be some delay in the output of our process. Worse, certain filters will delay some frequency components by more or less than other frequency components. This results in a phase distortion of the filter. For a certain class of FIR filters, it is possible to design filters that shift each frequency component by a time delay in proportion to the frequency. In this way, all frequency components are shifted by an equal time delay. Though it is possible to design certain non-real-time, noncausal IIR filters that are phase shift distortionless, in general, IIR filters will produce some phase shift distortion. We can determine the actual phase shift for each frequency component by computing

$$\arg H(j\omega) = \angle H_{\text{real}} + j H_{\text{imag}} = \tan^{-1} \frac{H_{\text{imag}}}{H_{\text{real}}}$$

The arctan will produce the principal value of the phase shift, not necessarily the cumulative phase shift, since the arctan function produces principal value $-\pi \leq \tan^{-1}(\phi) \leq \pi$. It is possible to recover the accumulated phase shift by factoring the rational function into its binomial parts, then expressing this in exponential form as a summation. One can then determine the principal angle on each part and the accumulated phase shift by summing the parts.

4.3 Comparison of FIR and IIR Filters

There are various factors when deciding on a particular filter for a given application. Table 3 summarizes these.

5 EXTRACTING SIGNAL FROM NOISE

The PSD of white noise is uniformly distributed over all frequencies. Therefore, it is possible to detect the PSD signature of a signal corrupted by white noise by inspecting spectral components that rise above some baseline. From this we can design a matching filter to optimally extract the signal from the noise. Figures 11 and 12 illustrate this procedure.

Table 3 Comparison of FIR and IIR Characteristics

	FIR	IIR
Run-time efficiency	Less efficient; requires high-order filter	Higher efficiency; usually possible to achieve a desired design specification in fewer computations
Stability	Always stable	Stable if all poles are inside the unit circle
Phase shift distortion	Can be designed to be phase shift distortionless	Generally distorts phase
Ease of design	Simpler design process, usually involving Fourier transforms or solving linear systems	Design is more complex, involving special functions or solving nonlinear systems

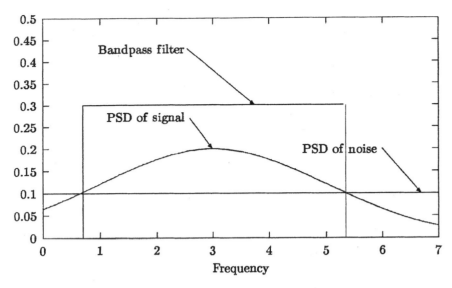

Figure 11 Use of bandpass filter for discriminating signal from noise.

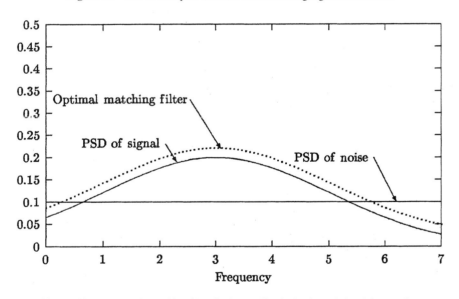

Figure 12 Improved matching filter for better discrimination of signal from noise.

REFERENCES

1. W. Press, B. Flannery, A. Teukolsky, and W. Vetterline, *Numerical Recipes in C*, Cambridge University Press, Cambridge, UK, 1988.
2. T. Parks and C. Burrus, *Digital Filter Design*, Wiley, New York, 1987.
3. M. Vlček and R. Unbehauen, "Analytical Solutions for Design of IIR Equiripple Filters," *IEEE Trans. Acoustics, Speech, Signal Process.*, **37**(10) (October 1989).
4. S. Marple, *Digital Spectral Analysis with Applications*, Prentice-Hall, Englewood Cliffs, NJ, 1987.
5. A. Oppenheim and R. Schafer, *Digital Signal Processing*, Prentice-Hall, Englewood Cliffs, NJ, 1975.

CHAPTER 18

DATA ACQUISITION AND DISPLAY SYSTEMS

Philip C. Milliman
Weyerhaeuser Company
Federal Way, Washington

1 INTRODUCTION

The industry has changed significantly since this chapter was first written in the months before 1990. The personal computer has become part and parcel of everyday life. Control systems have become increasingly based on standard systems and interfaces; sensors themselves are often based on just smaller versions of the same operating system as large manufacturing systems. This has tended to change the focus from the technology of data acquisition to the software and systems to support data acquisition.

The trend has been away from requiring the engineer to understand the science of how sensors work and the lowest levels of data acquisition and more toward the engineer understanding the collection, coordination, storage, access, and manipulation of data. With that in mind, this chapter has been updated to focus more on the latter and less on the former. Other chapters in this book cover details of the electronics, transducers, sampling, and calibration.

To control any process or understand what occurs during the life cycle of a process, the system (a human or machine) must have information about what is occurring. In the simplest of control loops, the measured variable must be converted to usable units, comparison in some form to a target occurs, and a response is determined. At the plant level, improvement of plant operation relies upon understanding the relationships between processes within the plant (not only current, but historical), which in turn requires collecting data throughout the plant, characterizing the relationship of the data with other data, storing the data in such a way as to be retrievable in a useful, timely way, and manipulating the data for presentation and hopefully providing an aid to understanding the relationships between processes. In today's competitive environment, focusing on local control and ignoring the interaction between processes, both internal to the plant and external, can be disastrous. If one is not focused on improvement, one can bet the competitor is. Larger corporations, especially, can bring analytical tools to bear to improve local processes, plantwide processes, and their relationships to external influences, such as the supply chain. On the other hand, today's computing tools bring very powerful data acquisition and analysis capability within the reach of the average technician with a little bit of motivation.

Data acquisition and display systems have changed dramatically. Twenty years ago, terms referring to specialized systems such as SCADA (supervisory control and data acquisition) and data loggers were common terms. Now, with the proliferation and broadening role of computer systems and their intrusion into every aspect of manufacturing, many of the features that used to be in specialized instruments and systems are part of the everyday tools available to anyone with a computer. This chapter attempts to cover aspects of data acquisition and manipulation that may help the engineer better understand issues and give a foundation for using and even constructing tools. The organization is as follows:

- The initial sections cover the nature of data and the acquisition and conversion of data to usable units and includes some discussion of useful display techniques. The discussion attempts to identify issues of which the engineer should be aware and give guidelines on how to manage data.

- The latter sections cover the coordination, storage, access, and manipulation of data. A discussion of pros and cons of different strategies should help the reader understand the trade-offs in system selection and construction. It is difficult to do this without describing specific technologies and brands, but the author has endeavored to level the discussion in such a way that changes in technology will not change the value of the discussion. Time will tell if the approach is effective.

2   DATA ACQUISITION

Data acquisition includes the following: (1) acquiring raw data from the process being measured and (2) converting data to usable units. Included in this section are also some topics of data display closely related to the nature of the data being acquired. Other aspects of data display will be covered in later sections.

In process industries, much of the data are analog in nature, such as pressure, temperature, and flow rate. The values acquired are sampled representations of process data that have a scale and a range, with various issues around effective range and whether values over a range are linear or more complex. When acquired in a data acquisition system, there are a number of issues that must be addressed related to how data are sampled and represented in the computer as digital values but still able to be manipulated as analog numbers and how continuously changing values can be stored without exceeding the capacity of storage or computation of the acquisition system. Discrete manufacturing still has a number of analog data sources, but a larger proportion involves discrete data, such as motor stops, starts, and pulses. These have their own issues of acquisition and storage and are often related to attributes of the process. The next section deals primarily with process data, additionally covering some discrete data and issues around data collection, representation, and storage.

As businesses begin to broaden the scope of optimization and understand their global processes, the context of the data in terms of product, plant conditions, market conditions, and other environmental aspects has increasingly added discrete data to the set of data to be obtained. The interaction of the process with factors such as which crew is managing the process, which customer's needs are highest priority, legal controls such as environmental limits impacting allowable process rates, operator decisions, which product is being manufactured, grade achieved, and a large number of other factors become important when a company is competing with other companies that have already achieved excellent local control of processes. These data involve less understanding of how to deal with continuous data and more with the coordination of data within and between processes. These can be termed manufacturing attributes to emphasize their importance in providing an environment around process data. Later sections deal with manufacturing attribute data and issues around their collection and coordination with process data.

3   PROCESS DATA ACQUISITION

Most modern data acquisition is via digital systems that may have a lower level analog collection mechanism but is now so removed from the engineer that the engineer is only concerned with the digital portion of the system. The ability to use digital microprocessors as building blocks for data collection, the prevalence of computer tools, and the creation of widely available operating systems that operate on small footprints have virtually eliminated the need for analog instrumentation. While the data may be analog in nature, the technology has been developed to such a degree that the engineer decreasingly needs to pay attention to the analog aspects of the data.

A digital-to-analog (D/A) and analog-to-digital (A/D) converter performs the actual processing required to bring analog information from or to the process. While the resulting signal may be digital, it is a representation of a continuous number that has characteristics that, if not understood, can result in erroneous conclusions from data, including missing data, misinterpreting trends, or improperly weighting certain values.

The engineer should be aware of several features of analog data to ensure that the data are used properly. An understanding of sampling interval, scaling, and linearization will facilitate the use of data once collected.

3.1   Sampling Interval

One of the important steps with any data collection process includes the proper choice of sampling interval. As an example of the impact of selection of sampling interval, or frequency, a sine wave with a period of 1 s (Fig. 1a) is measured with several sampling intervals. Intervals of 0.5 s (Fig. 1b) and 0.1 s (Fig. 1c) both provide different impressions of what is actually happening. The 1-s sampling rate being in phase with the sine wave yields the impression that we are measuring a nonvarying level. The 0.5-s sampling rate yields several different results depending on what phase shift is encountered. This is known as aliasing (see Ref. 1, pp. 122–125). If a 0.1-s period is used, we finally begin to obtain a realistic idea of what the waveform truly looks like.

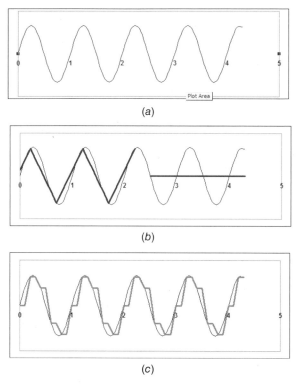

Figure 1   Sampling: (*a*) sine wave form; (*b*) aliasing of data; (*c*) sampled every tenth second.

The sampling interval has a different impact when collecting manufacturing attribute data. With manufacturing attribute data, every change in value has importance. They provide a context to the process that assists with the tieback to business interactions. The values are often coded. The sampling must be close enough to the time of occurrence to allow determination of state in relation to other events. The sampling interval must be fast enough to capture any change in state. Sampling at slower than the change rate of the data will mean lost events—potentially critical in a situation where the count of items processed is important. Sampling at a rate slightly faster than the maximum change rate of the manufacturing attribute data assures that no change will be missed. Another important consideration is to know at what time an event occurred. For instance, if a value changes only infrequently but other related manufacturing attribute values are changing at a faster rate, then the scan rate has to be fast enough to match the fastest change rate of all the related manufacturing attribute data. This is sometimes called the master scan rate, meaning that the frequency of scanning must be fast enough to capture faster events and determine the state of other variables relative to those events. Similarly, if there are related analog data, the scan rate may have to be fast enough to even characterize the curve of the analog data (remember Fig. 1).

The capacity of the target system must also be taken into account. Storing large volumes of data is becoming more feasible as systems increase in speed and power, but the retrieval and organization of those data may become a time-consuming, overly complex task with too much data or poorly organized data. Consequently, even though storage itself is less of an issue, other factors impact how much data are retained and how the data are organized for later retrieval. Later sections examine approaches for organizing and retrieving data.

3.2 Accuracy and Precision of Data

Accuracy and precision are dependent on the sampling interval as well as the resolution of the system (Ref. 2; Ref. 3, pp. 78–80; and Chapter 1 in this volume). When dealing with the A/D conversion process, the step size or number of bits used is critical when determining the system precision and accuracy (Ref. 1, pp. 78–81). Figure 2 illustrates the difference between accuracy and precision of data. Table 1 illustrates the effect the number of bits has on the precision.

This also interacts with range, which will be discussed later, since having an accurate number over a small percent of the desired range would not allow the ability to fully characterize the process. For example, highly accurate readings with 1% moisture resolution over a range from 10 to 20% moisture content would be inadequate if one were attempting to measure moisture over a 5–40% range. When selecting transducers, it is necessary that they have both the accuracy and range needed for the process being observed. When selecting converters, one should be aware of the settling time (governs how often readings can be obtained), resolution of the converter (affects range and detail of measurements), and accuracy of sensor.

Chapter 1 includes some characteristics of transducers, including calibration, and the sampling of data.

One should be aware that an event that has been stabilized in a data collection system may be offset in time, resulting in a potential discrepancy between events or values when values are compared from different sources or from multiplexed data. It should be verified that the transducer is collecting the data fast enough to allow one to have relevant times of collection in the data acquisition system. Also, one should ensure that the potential relationship between events from different sources and their intended use is understood when considering the speed and accuracy of transducers.

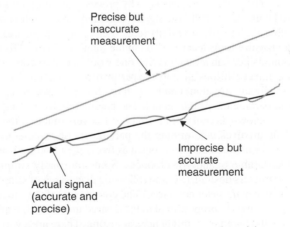

Figure 2 Difference between accuracy and precision.

Table 1 Relationship between Number of Bits and Precision

Number of Bits	Steps	Resolution on 5-V Measurement	Percent Resolution
8	256	0.01950	0.3900
10	1024	0.00488	0.0980
12	4096	0.00122	0.0240
16	65,536	0.00008	0.0015

3.3 Time-Based versus Event-Driven Collection

There are two major approaches when collecting data with a general-purpose data acquisition system. In one approach data are collected on a regular frequency based on time, such as once per second. This is easy to institute and it is relatively easy to analyze the data and their relationships after the fact. This approach tends to require more data storage and can make it difficult to identify events or the interactions with manufacturing attribute data. The other approach is event-based acquisition. An event is identified, such as when a package is dropped onto a platform, the time of that event is recorded, and the values of related variables are collected for that time. The sampling rate of the transducers to acquire the other variables may be important, as their values may become irrelevant if too long a time interval has passed after the event has occurred when the related variables are sampled. Batch processes, such as mixing a tankful of chemicals, often have some data collected only at the start and end of the process. Other data may be recorded at fixed time intervals during the batch process. Depending on the needs, the data during the actual reaction may be of great or of little use. The time between sampled events may be several minutes, hours, or even days in length, but the time of the event may be critical, as well as detailed data at the time of the event, resulting in a common tactic of using high-speed scanning to detect the occurrence of an infrequent event. Approaches for combining and analyzing data will be covered in a later section.

4 DATA CONDITIONING

Often the data obtained from a process are not in the form or units desired. This section describes several methods of transforming data to produce proper units, reduce storage quantity, and reduce noise.

There are many reasons why process measurements might need to be transformed in order to be useful. Usually the signals obtained will be values whose units (e.g., voltage, current) are other than the desired units (e.g., temperature, pressure). For example, the measurement from a pressure transducer may be in the range 4–20 mA. To use this as a pressure measurement in pounds per square inch (PSI), one would need to convert it using some equation. As environmental conditions change, the performance characteristics of many sensors change. A parametric model (equation) can be used to convert between types of units or to correct for changes in the parameters of the model. The parameters for this equation may be derived through a process known as calibration (Chapter 1 covers much of the process of calibration and sampling). This involves determining the parameters of some equation by placing the sensor in known environmental conditions (such as freezing or boiling water) and recording the voltage or other measurable quantity it produces. Some (normally simple) calculations will then produce the parameters desired. (See the following discussion of simple linear fit for the procedure for a simple two-parameter equation.) The complexity of the model increases when the measured value is not directly proportional to the desired units (nonlinear). Additionally, as sensors get dirty or age, the parameters might need adjusting. There are a variety of control techniques to assist with compensating for changes in the environment around the sensor, including adaptive control.

4.1 Simple Linear Fit

The simplest formula for converting a measured value to the desired units is a simple linear equation. The form of the equation is $y = ax + b$, where x is the measured value, y represents the value in the units desired, and a and b are parameters to adjust the slope and offset, respectively. The procedure for finding a and b is as follows:

1. Create a known state for the sensor in the low range. An example would be to put a temperature sensor in ice water.
2. Determine the value obtained from the sensor.

3. Create a known state for the sensor in the high range. An example would be to immerse the sensor in boiling water.

4. Determine the value obtained from the sensor.

5. Calculate the values of a and b from these values using the equations

$$a = \frac{\text{actual high value} - \text{actual low value}}{\text{measured high value} - \text{measured low value}} \qquad (1)$$

$$b = \text{actual low value} - (a \times \text{measured low value}) \qquad (2)$$

Figure 3 demonstrates example relationships between measured values and engineering units.

4.2 Nonlinear Relationships

Often, there is not a simple linear relationship between the engineering units and the measured units (Fig. 4). Instead, for a constantly rising pressure or temperature, the measured value would form some curve. If possible, we use a portion of the sensor's range where it is linear, and we can use Eqs. (1) and (2). When this is not possible, we have to characterize the sensor by a different equation, which could be a polynomial, a transcendental, or a combination of a series of functions.

One can imagine several sensors which are linear in different ranges to be used in conjunction to create a larger range of operational data. This variety of formulas should make one point clear: Without an understanding of the basic model of the sensor, one cannot know what type of conversion to use. Many sensors have known differences in output depending on the range of sensed data. Be aware of the effect environmental conditions have on the sensor readings. If the characteristics of a sensor are unknown, then the sensor must be measured under a variety of conditions to determine the basic relationship between the measured values and engineering

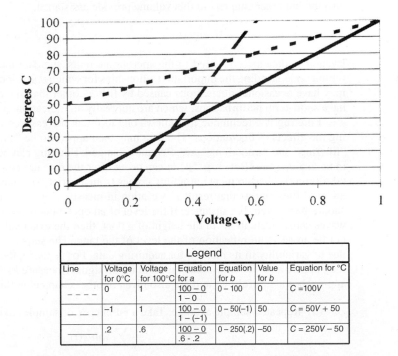

Legend						
Line	Voltage for 0°C	Voltage for 100°C	Equation for a	Equation for b	Value for b	Equation for °C
- - - -	0	1	$\frac{100-0}{1-0}$	$0-100$	0	$C = 100V$
- - -	-1	1	$\frac{100-0}{1-(-1)}$	$0-50(-1)$	50	$C = 50V + 50$
———	.2	.6	$\frac{100-0}{.6-.2}$	$0-250(.2)$	-50	$C = 250V - 50$

Figure 3 Relationship between measured values and engineering units.

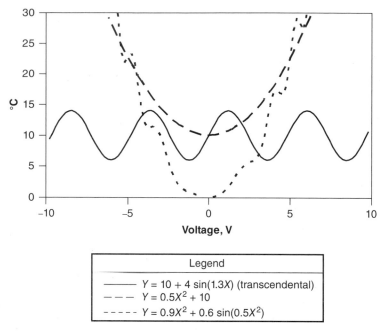

Figure 4   More complex data relationships.

units. Some knowledge of the theory of the sensor's mechanism will help to give an idea of which model to use. The development and evaluation of a model are beyond the scope of this chapter, but other chapters in this volume provide assistance.

4.3   Filtering

Even after data are converted to the appropriate units, the data may have characteristics that inhibit understanding the important relationships for which one is looking. For instance, the data may have occasional fluctuations caused by factors other than the process or the process may have short-term perturbations, which are not really an indication of the major process factors.

Filtering is a technique that allows one to retain the essence of the data while minimizing the effects of fluctuations. The data may then appear to be "smoothed." In fact, the terms "filtering" and "smoothing" are often interchanged. Filtering can occur when the data are still in an analog state (Ref. 1, p. 54) or can occur after the data are converted to digital data (digital filtering). Measurement variability comes from a variety of sources. The process itself may undergo fluctuations that result in variation in measurement but that are only temporary and should be ignored. For instance, if the level of an open tank of water were to be measured but waves cause fluctuations in the height of a float, then the exact value at any given time would not be an accurate reflection of the level of the tank. The sensor itself may have fluctuations due to variability in its method for acquiring data. For instance, the proximity of a 60-Hz line may induce a 60-Hz sinusoidal variation in the signal (measured value). Examples of filtering approaches are as follows and are also given in other chapters in this volume (Ref. 4, p. 538):

(a) Repeated sample average: Take a number N of samples at once and average them:

$$\frac{1}{N}\sum Value(i)$$

(b) Finite-length average (moving): Take the average of the last N measurements, averaging them to obtain a current calculated value.

(c) Digital filters:

$$y = (1 - \alpha)y_{i-1} + \alpha x_{i-1}$$

The simple average is useful when repeated samples are taken at approximately the same point in time. The more samples, the more random noise is removed. Chapter 1 addresses some of the issues with sampling and the concept of population distribution. The formula for an average is shown in (a) above.

However, if the noise appeared for all the samples (as when all the samples are taken at just the time that a wave ripples through a tank), then this average would still have the noise value. A moving average can be taken over time [see (b) above] with the same formula, but each value would be from the same sensor, only displaced in time. A disadvantage of this approach is that one has to keep a list of previous values at least as long as the time span one wishes to average.

A simpler approach is the first-order digital filter, where a portion of the current sample is combined with a portion of previous samples. This composite value, since it contains more than one measured value, will tend to discard transitory information and retain information that has existed over more than one scan. The formula for a first-order digital filter is described in (c) above, where α is a factor selected by the user. The more one wants the data filtered, the smaller the choice of α; the less one wants filtered, the larger the choice of α. Alpha must be between 0 and 1 inclusive.

The moving average or digital filter can tend to make the appearance of important events to be later than the event occurred in the real world. This can be mitigated somewhat for moving averages by including data centered on the point of time of interest in the moving-average calculation. These filters can be cascaded, that is, the output of a filter can be used as the input to another filter.

A danger with any filter is that valuable information might be lost. This is related to the concept of compression, which is covered in the next section. When data are not continuous, with peaks or exceptions being important elements to record, simple filters such as moving-average or digital filters are not adequate. Some laboratory instruments such as a gas chromatograph may have profiles that correspond to certain types of data (a peak may correspond to the existence of an element). The data acquisition system can be trained to look for these profiles through a preexisting set of instructions or the human operator could indicate which profiles correspond to an element and the data acquisition system would build a set of rules. An example is to record the average of a sample of a set of data but also record the minimum and maximum. In situations where moisture or other physical attributes are measured, this is a common practice. Voice recognition systems often operate on a similar set of procedures. The operator speaks some words on request into the computer and it builds an internal profile of how the operator speaks to use later on new words. One area where pattern matching is used is in error-correcting serial data transmission. When serial data are being transmitted, a common practice to reduce errors is to insert known patterns into the data stream before and after the data. What if there is noise on the line? One can then look for a start-of-message pattern and an end-of-message pattern. Any data coming over the line that are not bracketed by these characters would be ignored or flagged as extraneous transmission.

4.4 Compression Techniques

For high-speed or long-duration data collection sessions there may be massive amounts of data collected. It is a difficult decision to determine how much detail to retain. The trade-offs are not just in space but also in the time required to store and later retrieve the data. Sampling techniques, also covered in other chapters, provide a way of retaining much of the important features of the data while eliminating the less important noise or redundant data. As an example, 1000 points of data collected each second for one year in a database could easily approach 0.5 terabyte when index files and other overhead are taken into account. Even if one has a large disk farm, the time required to get to a specific data element can be prohibitive. Often, systems are indexed by time of collection of the data point. This speeds up retrieval when a specific time frame is desired but is notoriously slow when relationships between data are explored or when events are searched for based on value and not on time. Approaches to reducing data volume include the following:

- Reduce the volume of data stored by various compression tools, such as discovering repeating data, storing one copy, and then indicating how many times the data are repeated. There are many techniques for compressing data, covered elsewhere. Zip files are a common instance of compression to make the data consume less storage and to take less time in transmission from one computer to another. Compression tends to increase the storage and retrieval time slightly. Increasingly, file systems associated with common operating systems include compression as a standard option or feature of mass storage. These systems are quite good at compressing repeated data but are less effective when data vary but have a mathematical relationship, such as a straight line between two points.

- Normalize the data. When the developer knows relationships between data, redundancy can be avoided by normalizing the data—following some basic principles to organize the data in such a way that redundancy is avoided.[5] C. J. Date describes the levels of normalization of data in a relational database.[5] For instance, if a person has several addresses, then one could store the person's name once, store each address, and store the links from the person to the address. While very similar to compression, it relies on the developer identifying and taking advantage of the relationships between data to eliminate redundancy and reduce space. This creates significant effort in planning for acquisition and storage of data. It pays off in reduced storage and significantly improved retrieval and analysis times.

- Eliminate nonessential data. If one is not interested in the shape of a sinusoidal signal, for instance, but only interested in how many cycles occurred during a given time frame, then sampling techniques can be used to characterize the data without having to store significant data.

The engineer or researcher has to make assumptions about how the data will be used and factor those into the acquisition and storage system. A project attempting to discover the relationships between waveforms would require high-frequency sampling and probably time-based storage. A project attempting to record the number of times a boiler went over a certain temperature level might have a high-speed scanning capability but only store those values that were above the temperature limit. An inventory tracking system may have triggers that cause scanning only when some event occurs.

Often, a batch or pallet of product may contain a large number of items. The items can be sampled for some process attribute. The customer may want to know summary statistics about the pallet, but the storage of all the data may not be feasible. In this case, statistical results can normally be derived from summary data. Average, standard deviation, total, correlation, maximum, and minimum are easily calculated from summary, accumulated data[*]:

Averages. Keep a running sum of data and count of readings:

$$\text{Average} = \text{Sum of values}/\text{count of values}$$

---

[*] Standard deviation and correlation from Ref. 6, pp. 473, 477.

Totals. Keep a running sum of data.

Standard Deviation

$$\sqrt{\frac{n \sum_{i=1}^{n} x_i^2 - (\sum_{i=1}^{n} x_i)^2}{n(n-1)}}$$

Correlation

$$r = \frac{n \sum x_i y_i - (\sum x_i)(\sum y_i)}{\sqrt{[n \sum x_i^2 - (\sum x_i)^2][n \sum y_i^2 - (\sum y_i)^2]}}$$

Range. Save largest and smallest values.

Median. Find the middle value of a distribution, which requires keeping all values.

The median (the true center of the data) requires the raw data to be calculated. A compromise for depicting the distribution of data without having to store the full details is to store a distribution of the data. For instance, the range of possible important data can be broken into a series of totals, reflecting the count of items that fit into the particular total. A histogram representing the distribution of data can be created from the totals without requiring the full set of original data. In addition, the median can be approximated using this technique. The distribution can also be used to supply data for statistics based on distribution of data, such as the Taguchi loss function (Ref. 7, pp. 397–400).

4.5 More on Sampling and Compression

Rather than just sample the data, why not save all the changed values of the data, discarding values which are the same or within some limits of the previous reading? This really applies best to continuous processes. Quite significant space reduction can be maintained in processes that are slowly changing and have only occasional large upsets. Variations of this technique can provide additional improvements. For instance, rather than just checking to see if the current value is the same as or within some limits from the previous reading, see if it is on the same line or curve as the previous value. This can result in a great reduction of storage requirements at the loss of a slight amount of accuracy in reconstruction. The more flexible the compression technique, the more work must be done to reconstruct the data later for examination. For instance, if the user of the data acquisition system wants to retrieve a data point within data that have been reduced to a line segment, the user or the system must determine which line segment is wanted using the time stamp for the beginning and ending of the line segment interval and then recalculate the point from the equation. This is referred to as a boxcar algorithm. Values that are close to the line segment can be treated as on the line segment if one can afford to lose some accuracy.[8] The formula for a boxcar has to take into account the length of the interval (maximum), the height of the box (how much noise is allowed), and how peak or exception values are treated. For instance, Table 2 presents a set of data with several types of compression applied for data sampled at a constant interval.

In Table 2, the simple repeating-value compression will not lose data but will result in little or no compression if the data are changing value frequently, including having any noise. The boxcar compression technique results in much higher compression for slowly changing data with only the loss of fine detail data (depending on the height of the window). For data that are nonlinear or changing frequently, the boxcar compression method results in little compression. Process information systems often use the boxcar compression method. If data are slow moving with occasional bursts of activity, the boxcar and repeating-value methods can result in dramatic reductions in space required. If data changes tend to be linear, then the boxcar algorithm tends to be superior to the repeated-value approach. For an extreme example, see Table 3.

The raw data would have resulted in 631 data points being stored. The boxcar method would result in 5 data points being stored, less than 1% of the storage required. In the example

Table 2 Examples of Compression Techniques

Time Stamp and Raw Value

		Simple Repeating Value			Boxcar Compression		
Time	Value	Time	Value	Count	Start Time	Start Value	Slope
12:00	1	12:00	1.0	3	12:00	1	0
12:01	1	12:03	2.0	1	12:03	2	1
12:02	1	12:04	3.0	1	12:08	7	−2
12:03	2	12:05	4.0	1	12:11	1	4
12:04	3	12:06	5.0	1	12:12	5	0
12:05	4	12:07	6.0	1	12:16	5	−4
12:06	5	12:08	7.0	1			
12:07	6	12:09	5.0	1			
12:08	7	12:10	3.0	1			
12:09	5	12:11	1.0	1			
12:10	3	12.12	5.0	5			
12:11	1	12:17	1.0	1			
12:12	5						
12:13	5						
12:14	5						
12:15	5						
12:16	5						
12:17	1						

Table 3 Data Compression: Raw Data versus Boxcar

Raw Data[a]		Boxcar Method		
Start Time	Start Value	Start Time	Start Value	Slope
12:00	1	12:00	1	0
12:01	1	15:00	1	1
...		15:01	2	3
14:59	1	15:02	5	−4
15:00	1	15:03	1	0
15:01	2			
15:02	5			
15:03	1			
15:04	1			
...				
18:59	1			
19:00	1			

[a]Gaps represent no change in data.

above, the repeated-value method would have resulted in almost exactly the same compression as the boxcar. However, if there had been a 0.1% slope in data throughout the period, the boxcar would remain the same but the repeated-value method would have resulted in no compression. A deadband (the height of the boxcar) around the repeated value (meaning two values within some small deviation from each other would be counted as the same value) would result in very high compression in the above example. A long ramp-up of the value during that time frame would have further differentiated the two compression methods.

There are many techniques for compression of data. As mentioned above, many rely on assumptions about the underlying nature of the data, such as being continuous data. Where data are directly related to events, more traditional compression techniques which look for repeating patterns in the data may be used. These are typically performed by operating systems and database systems and can therefore be taken advantage of with little or no work on the part of the engineer.

5   DATA STORAGE

In whatever ways data are sampled, collected, filtered, smoothed, and/or compressed, at some point the data must be stored on some media if long-term data are to be analyzed (covered later in this chapter). There are several approaches to data storage that will be discussed in brief here.

5.1   In-Memory Storage

There are normally limitations on how much data can be stored, particularly when low-frequency events have high-frequency data surrounding them that are of interest. For example, if scientists are monitoring Mt. St. Helens for seismic data, it would be prohibitive to capture millisecond data for years while waiting for an eruption. It would be of interest to capture data at high density just before, during, and after each eruption but not in the quiet times in the intervening years. Collecting the millisecond data on many sensors would overflow the storage capability of most systems. There are techniques to store subsets of the data that allow high-density data from constrained time intervals to be stored.

An approach for collecting and later reporting high-density data around an event of interest is to collect the data continuously using the triggered snapshot method. High-speed data are temporarily retained for a fixed time interval or memory capacity, with the start and end time of the data moving forward with time. Older data are discarded as the time range moves past it. This moving window is useful for creating trends and summary data for that interval. The user can be shown dynamic displays that update over time and reflect characteristics of the moving window of the process. Periodically, a set of the data can be extracted to mass storage, especially triggered by some event of interest. An event is recognized by some means (automatic or user generated) that the engineer has preconfigured to cause the transfer of the current instance of the moving window to permanent storage.

The relationship between the trigger and the moving window can be configured several ways, as depicted in Fig. 5. The handling of the data involves moving values through a data array, adding more recent values at the end and pushing the rest toward the beginning—a queue.

The triggered snapshot is particularly useful when knowledge about the sequence of events just before the event of interest can help discover problems. As an example in manufacturing, in sawmills there are often very high speed sequences of events, such as where a board may come out of one conveyor and is transferred to another conveyor and some event such as the board leaving the conveyor occurs. High-speed video can be always in progress, and the detection of the board leaving the system can be used to trigger the transfer to permanent storage of the video. Events of concern can be safety issues and the triggered snapshot method can be used to help eliminate potential life-threatening situations. The triggered snapshot method is particularly useful for discovering the causes of unusual events.

This has the advantage of allowing monitoring and analysis of high-speed events and still capturing some data to enable determining some data relationships. This is most useful if the engineer has some idea of what events may yield valuable relationships. It is much less useful when events, triggers, or relationships are unknown or unexpected. Sampling data at slower intervals may serve to allow accidental capture and identification of useful relationships, but

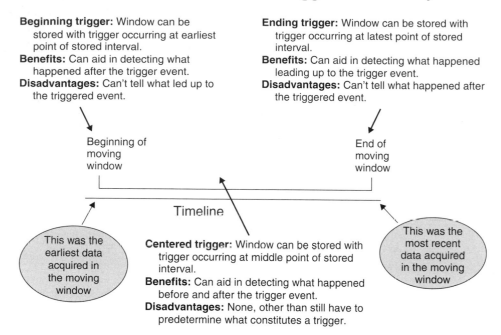

Figure 5 Relationship between trigger and moving window.

the work required to find those relationships is much higher and of questionable probability of success.

As an example, in the sawmill many variables are changing state at high speed. For diagnostic purposes, it is valuable to see a high-speed snapshot of states of photo eyes compared to saw drops, gate openings, and grade decisions. However, the volume of data is normally too great for storage and later analysis. There are some events that are of more importance than others, such as when a gate is opening early or a saw is failing to drop consistently. These can often be recognized and the data captured in the window of time before and after the event can be stored, allowing later analysis of what led to the event and what happened shortly afterward. Some characteristic data can be summarized for each time window, stored, and used later for analysis, such as the number of photo eye changes, number of saw drops, and number of gate openings. More complex relationships can be tallied to aid in diagnostics, such as number of gate openings for grade 2. The more complex the relationship, the more difficult the programming task to ensure capturing the incidence to storage. A typical pattern is to collect process variables that may be of interest, often from a programmable logic controller. As a given problem begins to be identified, additional logic can be added to examine relationships between process inputs and sequences of events, creating a new variable that reflects some attribute of that relationship. The data collection system can store the results of that variable, such as the sum of the number of times it occurred in some larger time interval, allowing it to be low enough in volume to be mass stored for later analysis.

5.2 File Storage

An easy way to store data is in a file, often a comma-delimited file. This is easy to program and can easily be imported into analysis tools such as spreadsheets. It is not well suited to the

Table 4   Example of Storing Multiple Variables in Files

Filename	ProcessData20040312142305
Data in file	Timestamp,TemperatureA,TemperatureB,TemperatureC,Rate1,Rate2,Rate3
	2004-03-20 12:34,15,14,15,12,13,13
	2004-03-20 12:35,15,14,14,12,13,13
	2004-03-20 12:36,16,14,13,12,13,13
	2004-03-20 12:37,15,14,12,12,13,13
	2004-03-20 12:38,15,14,11,12,13,13
	2004-03-20 12:39,14,14,11,12,13,13
	2004-03-20 12:40,15,14,10,12,13,13
	2004-03-20 12:41,15,14,10,12,13,13
	2004-03-20 12:42,16,14,10,12,13,13
	2004-03-20 12:43,15,14,09,12,13,13

compression techniques mentioned above because of the complexity of storing and interpreting data. However, for storing records of multiple variables collected at a time interval, this can be a very useful technique. An example is shown in Table 4.

In the above example, the filename was chosen so that it would be unique. The date and time were concatenated together, eliminating every invalid character with an underscore. This helps to prevent files from being overwritten accidentally and facilitates store-and-forward techniques described below. Files can be sorted by date or name. Often, they are stored in a directory structure so the number of files in any one directory does not get too great. This speeds up file search in a given directory but can make programs more complex that search for files across directories. An example directory structure is the following:

```
C:
 DataDirectory
  2003
   11-files created during November 2003 are stored here
   12
  2004
  01
  02
  03
```

Archiving files to backup media is easy in this file organization because one only needs to reference the particular directory for the time frame desired. Files can fill up mass media and so either a manual process to check for file limits, back up old files, and delete them must be instituted or a program to provide the same functions would need to be created.

A major deficiency with a file-based system is that when the time range of a search is larger than one file, the analysis can become very difficult. One may be searching for events, specific time frames that go across file boundaries, or relationships between variables that may not be effectively evaluated within the time frame of one file. The analysis task usually consists of importing a number of files into some analysis tool and then using the analysis tool to look for relationships. This means that the importation process has to include the organizing of data and identifying relationships between events, a difficult task at best. A common tactic is to import data via a script or macro for a given time range, so the user only has to specify a beginning and ending time.

5.3   Database Storage

Database technology has been improving for many years, resulting in database management systems being increasingly the storage tool of choice for data acquisition systems. Database

Table 5   Example Database Structure

Time Data Table		Batch Event Table	
Time stamp	datetime	Time stamp	datetime
Temperature A	float	Batch number	integer
Temperature B	float	Mix percent	float
Temperature C	float	Input material A quantity	integer
Rate 1	integer	Input material B quantity	integer
Rate 2	integer	Output product type	varchar(50)
Rate 3	integer	Output product quantity	integer
		Begin batch time	datetime
		End batch time	datetime

management systems provide organization tools, compression of data, access aids in the form of indexes, and easy access for analysis tools. A special benefit of database management systems is that they allow the combination of discrete data and time-based data collected on different time intervals. Relational databases are now the dominant database management system type. Data are organized in tables. Each table is composed of a set of rows, each row having a fixed set of columns. Indexes are provided to speed access to data. In data acquisition systems, the designer often adds a time stamp column to each row to facilitate retrieval and analysis of data. An example of a simple database is given in Table 5.

The time data table contains data that are sampled every second, while the batch event table contains a record for each batch that has occurred. The Structured Query Language (SQL) is a common language used to examine and extract data in the tables. An example of its power is that a query can be constructed to use the begin batch time and end batch time to extract data from the time data table and combine it with related batch events in the batch event table. A sample query to combine event- and time-based data is as follows:

```
Select BatchEvent.Timestamp, Batchevent.BatchNumber,
Batchevent.MixPercent,
Min(TimeData.TemperatureA),Max(TimeData.TemperatureA)
From BatchEvent, TimeData
Where BatchEvent.OutputProductType = 'BENCH' and TimeData.Timestamp
between BatchEvent.BeginBatchTime and BatchEvent.EndBatchTime
Group by BatchEvent.BatchNumber
Order by BatchEvent.BatchNumber
```

The above query searches for batches that created a certain output product type (BENCH) and reports the maximum and minimum temperatures from those batches. This can facilitate research, for example, on what conditions lead to the best yield of a particular product.

The ease of performing this operation is a particular advantage of relational databases. There are some disadvantages, including overhead due to the access methods, extra space requirements due to the creation of indexes, and costs and complexity associated with the database management system. Indexes may add as much or more than 100% to the size of a database. Old data must be managed and removed as with any other storage system. This typically is via an automated program, since the structure is not as simple as just looking for the file creation date of a file.

5.4   Using Third-Party Data Acquisition Systems

When data storage is fairly simple, it is not hard to store data in the above-mentioned methods, but when one is using sophisticated methods of data compression, it is recommended that systems that have robust implementations of those methods be used rather than attempting to reinvent the wheel. They can be quite complex to implement reliably. Transfer of data to those systems can occur through a variety of methods, including creation of files that are captured by the other systems, insertion of data into standard interfaces such as OPC or message buses, or insertion into database tables which are monitored by the other systems (often ODBC links). Third-party systems often have software development kit (SDK) interfaces that allow the engineer with some programming skills to store data directly into the system.

Process historians are optimized for storing time-based data. A technique used to provide some relational capability to the data is the following:

- Store related data at exactly the same time stamp (time stored in the database for when the data elements were collected).
- Treat data stored with the same time stamp as being part of the same record.
- Select a set of these "records" based on a time range.
- Search a variable for some attribute value (e.g., having some value or range of values).
- Provide to the display system the values of some related variable in the same record having the same time stamp as the desired attribute variables.

This is functionally the same as performing a relational database query on a set of records in a table, with criteria based on values in some columns.

6 DATA DISPLAY AND REPORTING

There are a variety of ways to reference and display data acquired from a sensor and stored in suitable media. The current value can be inspected, values can be stored for inspection later, values can be trended, alarm conditions can be detected and reported, or some output back to the process can be performed.

6.1 Current-Value Inspection

Often, one wants to see the data as they are being collected. This can be of critical importance in experiments which are hard or costly to repeat, allowing the researcher to react to situations as they occur. As it is collected, each data item will be called the current value for that sensor. Current data are usually stored in high-speed storage (the computer main memory). As new values are obtained, they replace the value from the last reading. The collection rate can vary widely (Table 6).

Table 6 Data Collection Rates: Examples

Type of Operation	Time per Event
Discrete manufacturing operations	
Assembly line manufacturing/assembly	0.01 to multiple seconds
Video image processing	0.001 s
Parts machining	0.002–0.02 s
Continuous processes	
Paper machine	1–60 s
Boiler	Several seconds to several minutes
Refinery	Seconds to minutes
Dissolving operations	1 s–20 min

For instance, detecting the profile at 10-mm intervals for a log moving at 100 m/min requires values to be obtained for each sensor 167 times per second. In continuous processes, data may only need to be acquired once per minute, as in monitoring the level of a large vat. It is useful to remember that human reaction time is in terms of tenths of a second, so displaying data at a faster rate would only be useful if it were easier to program the data. Do not waste time and energy attempting to record data at a high frequency if the only reason is for display to an operator, even if the operator must immediately react to an alert. Human–machine interfaces often show data changes at the time the new value arrives from a sensor. They have display elements that are tied to sensing points. Process historians (data acquisition systems architected for the long-term storage of process data) provide tools to extract data and present the data to the analyst. Their display systems normally provide update tools that automatically refresh the user's display at some display refresh rate (often in terms of seconds, such as 10 s). The data being collected by the historian may be changing faster, and the data may be stored at a faster rate, but the display normally is still refreshed at the standard refresh rate. It is useful to have the time displayed when the value was collected, as there is often a time delay between the collection and the display of data values. This is especially true where the data collection system may be disconnected temporarily from the display system. The data may come back in a rush when the link is reconnected and is useful for the user (and systems performing analysis) to provide a context for what time the data represent.

6.2   Display of Individual Data Points

Display of the data is normally in text or some simple bar graph representation. Other techniques include button or light indicators where color may represent the state of some value or range of values of the current value. Coded values may take the current value and translate it into some form that provides more value to the user such as zero being translated to the string "FALSE" on the display.

Often, one can better understand the data being obtained by using an analog representation (Ref. 9, pp. 243–254, and Ref. 10). This involves representing the measured quantity by some other continuously variable quantity such as position, intensity, or rotation. A common example is the traditional wristwatch. The hours and minutes are determined by the position of a line indicator on a circle. A common analog representation for data acquisition is the faceplate. This is a bar graph where the height of the bar corresponds to the value being measured. Often, lines or symbols may be overlaid on the bar to indicate high or low ranges. A frequent indicator is the meter. A needle rotates in a circle with the degrees of movement corresponding to the value obtained from a sensor. Many voltmeters use this technique (Fig. 6). Increasingly sophisticated calculations can be established to translate a flow rate, for example, into a cost number, providing the user with immediate feedback on the costs being incurred by the current process rate.

A common technique for representing trends of current value is to create a simple array and plot it as a trend line on the display. As new values are gathered, the array values are shifted through the array, with old values shifted out at one end of the array while the new values are shifted in at the other. This is a simple technique that provides some of the benefits of data storage without requiring the complexity of actually storing data in mass storage and managing it.

6.3   Display of Historical Data

There are two main issues with display of historical data:

- Selection of the data
- The representation the data will have

Digital   Coded   Analog

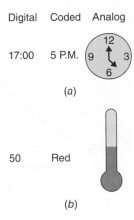

17:00   5 P.M.

(a)

50   Red

(b)

Figure 6   Comparison of digital, coded, and analog data representation: (*a*) time; (*b*) temperature, °F.

Selection of Historical Data

Selection of historical data involves several factors, including time frame and attributes of the data. Identifying a time frame is probably the most common activity in selecting historical data. How the data are updated can be an important consideration when comparing third-party historians.

The time frame is often referenced by a span (such as a number of hours) and a starting point, which can be absolute time (e.g., 2004-12-14 16:22:03) or relative time (e.g., −4H for starting 4 h in the past). Another option is to provide an absolute start time and an absolute end time. It is common to have the time frame updating (moving forward with time) if the start point is relative (but check your particular vendor's software for their practice) and to be fixed if the start point is absolute. For example, if the span is 2 h and the start point is −2 h at 15:00, the starting point on a trend line would be 13:00 and the ending point would be 15:00. Ten minutes later the starting point on the trend line would be 13:10 and the ending point would be 15:10.

For relational data, there is often a desired time frame as described above (to restrict the size of the data to be searched) and some selection criteria based on characteristics of the data itself or of related data, including events. For example, one may wish to find those manufactured units for the past month that were for product *X* and see how many had quality defects. As described above under third-party data acquisition systems, there are techniques for selecting data from process history databases that approach (but do not equal) relational capability.

Representation of Historical Data

The primary difference between current data and historical data is that there are multiple data points, normally with an order defined by the time they were acquired (for process history data) and/or by their relationship to other variables and events (for relational data). Once the time frame and relationships are selected, the data must have some method of representation on a display.

Historical data have a number of potential representation techniques. Multiple data values can be combined into a single number such as average, standard deviation, mode, maximum, minimum, range, variance, and so on. These can then be represented by techniques for individual data elements as described above. The equations in Section 4.4 are examples of summary statistics formulas.

The simplest form to represent historical data is the list. Create a column for each variable of interest. If they are all collected at the same time (the "records" described above), then each time data were collected can be used as the first value on the left in each row. The data for each

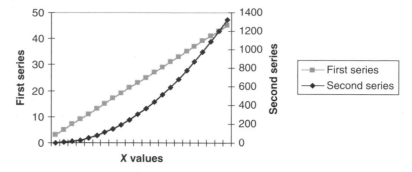

Figure 7   Multiple-axis chart.

variable of interest can be placed on the row corresponding to its time stamp, similar to that for files shown in Table 4.

Where data were collected with different time stamps, a new row can be created whenever a new value is obtained and values only placed in the column–row combinations where there is a corresponding time stamp between the row's time stamp and the time stamp of the variable in question. While useful, the problem with this is that there are now holes in the data. This list is very useful when viewing a trend line or graphical tool to validate numbers, verify time stamps, and diagnose problems with collection.

Often, one desires to view the relationships between data and time or other variables. Trend plots typically are used to show the relationship between variables and time. The X axis typically represents the time range and the Y axis represents the value of the desired variable. The range of data on the Y axis can vary depending on how one wishes to view the data. The maximum and minimum of the data can be used to set the top and bottom of the range of the Y axis, respectively. This can have two undesired effects. It may make small movements in data appear to be very large when the range is small. In cases where there are spikes in the data where a value is disproportionately high or low, representing the Y axis based on the maximum and minimum could make it difficult to view normal variation in the data. One has to understand the potential use of the data to choose the Y-axis scale appropriately.

When more than one variable is shown on a trend chart, the selection of scale of the Y axis becomes more complex. If all the variables are representative of the same domain, such as all temperatures, then perhaps the same Y axis can be used for all of them. Often, however, the viewer is attempting to compare relative variations in data, sort of a poor man's correlation analysis. In this case it may be useful to have multiple Y axes and select the range of each of them such that they represent the range of one or more of the variables being viewed (Fig. 7).

Another approach to compare variation between two variables is to use one variable for the X value and the other variable for the Y value (an X–Y chart). This is useful when two variables are related by sample time or some other selection technique that results in a paired relationship between the two variables. The correlation function in Section 4.4 represents the mathematical correlation between two variables and can be used to determine the strength of that relationship. Chapter 1 discusses correlation and the calculation of the line through a distribution of data.

7   DATA ANALYSIS

7.1   Distributed Systems

Distributed systems are a powerful approach to data acquisition systems because they combine some of the best of both stand-alone and host-based systems. The data acquisition portion is located on a small processor that has communication capability to a host computer system. The small system collects the data, possibly reducing some to a more compact form, and then sends the data to the host systems for analysis. The host system can analyze the data when it has the available time to do so. Only the data acquisition portion needs to be very responsive to the process. If the data acquisition task gets too big for the small system, the cost of expansion is limited to moving the data acquisition software to a new computer or splitting it up over several computers and changes to the host computer portion are not required. The major disadvantage of distributed systems is that they suffer from a more complex overall architecture even though the individual parts are simple. This leads to problems with understanding error sources and increases the potential errors because of more parts. Unless the communications are designed carefully, the messages sent between the small systems and the host system may be inflexible, causing increased effort when one wants to change the type of data being collected. Distributed systems may be expensive because of the number of individual components and the complexity required but often fit well with environments where one already has a host computer.

7.2   System Error Analysis

The errors that can occur at different stages in the data acquisition process must be analyzed, as they can add up to make the data meaningless. For instance, one may have very accurate sensors, but by the time the data reach the host computer they might have been converted to integer data or from real to integer and back to real again. This can lead to dangerous assumptions about the accuracy of the received numbers, because each conversion can cause rounding or other errors. It is the responsibility of the person setting up the acquisition system and the analyst to examine each source of potential error, discover its magnitude, and reduce it to the point where it will not have a significant impact on the conclusions to be derived from the data. Use of the filtering techniques described above under data collection can be of use to eliminate random error. It is not within the scope of this chapter to cover system error analysis, but Chapter 1 gives some foundation.

8   DATA COMMUNICATIONS

Data communications are involved in many aspects of data acquisition systems. The communications between the sensing and control elements and data acquisition devices, as well as the communications between the data acquisition system and other computer systems, can be carried out in many ways. This section will cover some aspects of communications, especially as they pertain to computer systems.

8.1   Serial Communications

A serial communication link means that data sent over a communications line are spread out over time on one physical data path. For instance, if a character is sent from a sensor to a computer, each bit making up the character (normally eight bits) will be sent one after the other (Table 7). This is often useful for low-cost, low-speed (usually less than 10,000 cps) rates of data transfer.

8.2   Parallel Communications

A serial communication link may not require very many wires, but the time spent to transfer data can add up. A way to improve the speed of communications is to use parallel communication links. This is done by having a number of wires to carry data. For instance, sending the same "A" over a nine-wire bus would only require one transfer (Table 8).

Table 7 Time Sequence of Bits Sent over Serial Communication Line

Character "A" is (in bit form)
Bit number 6 5 4 3 2 1 0
Bit value   1 0 0 0 0 0 1
The communications are using the RS232C communications standard
and sending the ASCII character A.

Bit Value	Time
Start bit	0 s after start
1 (bit 0 of A)	1/9600 s after start
0 (bit 1 of A)	2/9600 s after start
0 (bit 2 of A)	3/9600 s after start
0 (bit 3 of A)	4/9600 s after start
0 (bit 4 of A)	5/9600 s after start
0 (bit 5 of A)	6/9600 s after start
1 (bit 6 of A)	7/9600 s after start
Parity bit	8/9600 s after start
Stop bit	9/9600 s after start

Table 8 Time Sequence of Bits Sent over Parallel Communications
Interface

Character "A" is (in bit form)
Bit number 6 5 4 3 2 1 0
Bit value   1 0 0 0 0 0 1
The communications are using a hypothetical nine-line parallel
communications bus sending the ASCII character A.

Bit Value	Time
Start bit	0 s after start
1 (bit 0 of A)	0/9600 s after start
0 (bit 1 of A)	0/9600 s after start
0 (bit 2 of A)	0/9600 s after start
0 (bit 3 of A)	0/9600 s after start
0 (bit 4 of A)	0/9600 s after start
0 (bit 5 of A)	0/9600 s after start
1 (bit 6 of A)	0/9600 s after start
parity bit	0/9600 s after start

If the bus could handle the same rate of change of bits as the serial interface,
then the next character could be sent 1/9600 s after the first character (the A).

8.3 Networks

Ethernet with Transmission Control Protocol/Internet Protocol (TCP/IP) has become the dominant communications network protocol for data collection. There are proprietary process control and data acquisition networks that serve special purposes, but Ethernet has proven to be versatile for everything from office communications to data collection from smart sensors. Many computers can be connected to the same network segments. The use of switches and routers provides ways to isolate and limit communications to improve performance and security. Firewalls provide filters and protection for entire classes of messages and sources.

While Ethernet cannot guarantee delivery (being based on a collision detection and retransmit strategy), it has been shown to provide excellent response to moderate network activity. Communication speeds are regularly being improved to provide an even greater range of applicability.

Table 9 Open Systems Interconnect Model

Layer	Principle	Example
7. Application	Application	Millwide reporting
6. Presentation	Display, format, edit	Convert ASCII to EBCDIC
5. Session	Establish communications	Log onto remote computer
4. Transport	Virtual circuits	Make sure all message parts got there in order
3. Network	Route to other networks	Talk to Internet
2. Data link	Correct errors	Send character downline
	Synchronize communications	Send Ack-Nak
1. Physical	Electrical interface	Wire and voltages

8.4 OSI Standard

The International Standards Organization has developed a set of standards for discussing communications between cooperating systems called the Open Systems Interconnect (OSI) model (see Table 9). This defines communications protocols in terms of seven layers.[11] While not providing for specific interface protocols, the OSI model has had a significant impact on communications because it has provided a framework for compartmentalizing aspects of communications to allow the handoff of information from one device to another in a standard way. For instance, the transmission of data from one media type to another (such as copper wire to fiber to satellite to copper wire and then to wireless) is a result of standards enabling the seamless transfer of messages in a way that is transparent to the user.

8.5 OPC Standard

A recent standard of use in manufacturing is the OPC (OLE for process control) standard, which provides for a standard way of communicating with process equipment. It is sponsored by the OPC Foundation and originated as an extension for process control from the Microsoft OLE functionality (www.opcfoundation.org).

Functions provided by OPC include ability to browse the variable database of a device and monitor data on demand or when events occur. The capability of OPC has been expanded to work with Web communication methods such as XML and cover complex data such as record structures. The power of OPC is that data from an instrument can be available via a standard network interface so that any data acquisition program that uses the OPC interface can gain access to any OPC device. The need to know the protocol of each device or adhere to the wiring of specialized communications or use custom database access methods is eliminated through the use of a standard protocol. Multiple programs can be simultaneously monitoring the same piece of data, performing different functions at different time intervals, as events occur. The last point is particularly significant, because much of the work of data acquisition systems is spent in polling for changes in data or otherwise attempting to determine when an event has taken place. A program can subscribe to an OPC item and it will be notified when the item changes value, reducing the complexity of monitoring data dramatically.

As an example of the power of this approach, consider the following example (Fig. 8). A device can collect the identification from a unit of material such as unit number, color, and manufacturing date and make it available via OPC. As the unit is processed in a manufacturing center, another device collects defect counts and makes the unit number available via OPC. A human–machine interface program can monitor both sources with the same interface protocol and software and display it live for an operator to see. Simultaneously, another program can collect the defect counts and summarize them into totals. Yet another program can monitor the totals and wait for the unit number to change, triggering a transaction to a database or an email

Multiple server/clients with OPC

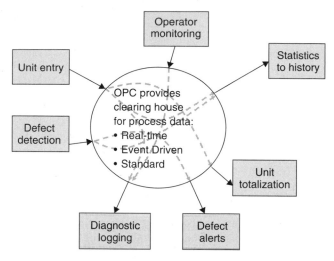

Figure 8 Example use of OPC communications.

if there was a problem. The power of the OPC interface is that it provides real-time access to the data from each of the sources and multiple programs can monitor the same OPC sources to perform work. Diagnostics can monitor the same data to evaluate system processing, downtime, or quality issues. Other programs can sample and store data to log files or diagnostic files for further analysis. All can be operating in parallel with no need to understand the internals of the other programs via a standard interface and standard protocols that support asynchronous delivery of data.

8.6 Benefits of Standard Communications

When implementing data acquisition and display systems, the ability to communicate in a standard fashion can play a large role in the cost of the system. This is realized in a variety of ways:

1. Different sensors can be connected to a system without having to buy a whole new system.
2. Data can be sent to other systems as needed for further processing.
3. As technology or need changes, portions of a system can be mixed and matched.
4. Increased competition from vendors tends to bring prices down.
5. A standard will have many people using products based on the standard, resulting in more vendors and greater availability of parts with a greater variety of options.

9 OTHER DATA ACQUISITION AND DISPLAY TOPICS

9.1 Data Chain

As materials and parts move through a manufacturing operation, the data collected are separated by time and type of data. Combining that data together presents a number of challenges. One can consider the first piece of data collected about some object to be the beginning of a chain of data and each successive data acquisition point in the manufacturing process to be a link in that chain until the end of the chain, where the last piece of data about a manufactured item is collected. Often, the steps of the manufacturing process may proceed from raw material to some intermediate work-in-process unit, to some other step that may be time based, then to some other step that may be finished unit based. Many varieties of the above exist. Each link can represent one of the following:

- Data collected from a start time to an end time
- A set of attributes about a particular manufacturing unit with an associated time of processing

Often referred to as the genealogy, the steps can be linked together through:

- Some assumptions based on time stamp relationship of one link back to the previous one
- Recording of units that entered or left the time-based portion of the process and the beginning and ending of the entry time
- Some assumptions about the mixing of elements of the manufactured item

Using the techniques described above for combination of time-based and event data, a set of data for the whole life cycle of a manufactured item can be created. Where mixing occurs, the data will be less accurate but may provide clues as to the factors that went into the final product, such as proportions of additives. As an example, Table 10 shows various queries that can be combined to provide one picture of all data sources for a manufactured unit.

Starting at the finished-good item, the batch ID of the previous step acts as a link into the range of time data in the previous step. If the list of items broken down is retained from the prior step, then those can be used to link back to the previous time frame.

Depending on the amount of mixing, the results will be more or less indicative of what actually happened. The smaller the batch sizes, the easier the tracking back to source data will be. Time lags between process steps can dramatically impact the ability to assume when the raw materials for a particular item were processed.

9.2 Web Programs and Interfaces

Web interfaces have improved to the point that user interfaces can be written in Web browser screens. This eliminates the effort and organization required to deploy code across a company. The application is written for a Web interface. When the user uses a Web browser to access the page, functionality is downloaded to the user's page or is executed in such a fashion as to obtain the results of a query and transmit a data page to the user. The developer does not have to get involved in the process of manually installing software on the user's machine. This reduces the demand on the desktop computer and allows the developer to make a change and have it proliferated to all users when they next reference the Web.

9.3 Configuration versus Implementation

As a general rule of thumb, third-party data acquisition and storage systems provide configurable tools for acquisition and display. For simple applications which do not require great flexibility in program functions, such as generating alarms, unusual graphics types, control of the process, or integration into larger systems, it is appropriate to use these, often simple, question–answer or menu-type systems. When the system must be very flexible or customized, it may be more appropriate to write a custom program. When considering this approach, be cautious, for the cost of implementing a program is often much higher than expected. For instance,

Table 10 Data Chain Sample Queries

State in Process	Important Data (Typical)	Example Data	Simplified Query to Combine Data with Previous Step
Raw material inventory	Identifiers: raw material batch ID Data: supplier, quality, time in inventory	Lot 1: 5% rejects; lot 2: 7% rejects	
Intermediate goods processing	Identifiers: raw material lots consumed; raw material batch ID: start and end time of entry to process; time delay through process Data: piece count, rejects, downgrades, process characteristics such as rate, temperature, modifications to materials	3:00–4:00 raw materials from lot 1: 1000 pieces; 10 rejects; 200 degrees 4:00–5:00 raw materials from lot 2: 1045 pieces; 14 rejects; 205 degrees; :10 residence time in process	Use raw material batch ID to match to raw materials characteristics; material run at 4:00–5:00 had 7% rejects when delivered to plant
Intermediate package creation	Identifiers: raw package ID, start/end time of creation Data: piece count, package dimensions	Package PR1 created: 3:30–4:10; 100 pieces; package PR2 created: 4:10–5:10; 150 pieces	Use time of manufacture, time lag through process to identify characteristics from intermediate process and raw materials lots; package PR1 was processed at 200 degrees, was created from lot 1 and was from a lot that had 5% rejects detected when delivered to the plant
Intermediate inventory	Identifier: intermediate package ID Data: location in inventory	Package PR1 location warehouse 1	Use ID of intermediate package to match to package at intermediate package creation; package PR1 had 100 pieces
Finished goods processing	Identifier: finishing batch ID, intermediate package consumed, start/end time of consumption; time delay through process Data: piece count, rejects, downgrades, process characteristics such as rate, temperature, modification to materials	Intermediate packages consumed at breakdown: 5:30–6:00 PR1; 95 pieces; 400 degrees 6:00–6:30 PR2; 147 pieces; 390 degrees 6:30–6:40 PR3; 25 pieces; 395 degrees 6:40–7:00 PR4; 40 pieces; 410 degrees Typical residence time :15	Use ID of intermediate package to match to intermediate inventory; in this example, package PR1 came from warehouse 1, lost 5 pieces in consumption
Finished goods package creation	Identifiers: finished package ID; start/end time of creation Data: piece count, grade, package dimensions, customer order	Finished package PF1 created from 6:00 to 6:30; 40 pieces; prime grade; order PO5670	Use time of manufacture, time lag through process to identify characteristics from breakdown, and which raw packages sourced this finished package. There may be significant mixing. In this example, PR1 and PR2 would be sources for PF1. PF1 was possibly created at 400 degrees, stored in warehouse 1, lost 5 pieces when loading into finished goods process, was probably processed at 200 degrees, and was probably from lot 1.

if one wanted to perform simple data acquisition and storage from a sensor, the cost to write a program would probably be higher than buying a small off-the-shelf system and entering the parameters for data collection. Writing a program involves analysis, design, development, debugging the program, and testing of results. The cost of documenting a program is often a large unplanned cost. If the results are intended to be used to make economic or process-related decisions, then the program must be tested carefully. Additionally, maintenance of the program can be quite expensive. Someone must be trained in the technologies used to build the program, the logic of the program, and the installation of the program. Another factor to consider is that costs of improvement of third-party software are borne by many customers and driven by many customers. The net result to the user of this software is that it is normally constantly improving, constantly tested, and maintained by a group of developers whose primary job is software development. One reason to build and maintain software internally is that a company can keep special knowledge within the company and thus maintain competitive advantage.

9.4 Store and Forward

When data acquisition and data storage are on two separate machines, it is important to provide methods to retain data in case the link between systems is broken. Message buses provide automated methods of maintaining a link between data acquisition systems. The developer inserts data into the message bus. If the link between the two systems is broken, the message bus queues up the data messages on the collection machine. When the data storage machine connection is reestablished, the message bus passes on the data to the data storage machine.

When a message bus is not available or feasible, a simplified mechanism can be created where a file representing each sample of data is created. If the data collection and data storage system are linked, then the data storage system monitors the directory of the collector for a new data file. If the data storage system detects one or more files on the data collection computer, then it will process them into storage. If the link is broken, then the files build up until the link is reestablished. A related technique is to store data in a database or similar mechanism on the data collection computer and scan it periodically for missing data from the storage computer. This is particularly useful when connection to the data collection computer is unreliable.

9.5 Additional Communications Topics

When considering transmission media, some points may provide value to the engineer. Fiber-optic cabling is less sensitive to noise than other transmission media. Wireless access points provide increased flexibility in positioning of sensors and greatly reduce wiring costs. Particularly, if one wishes to collect data from sites that may move, such as environmental sampling sites, the costs of wiring and rewiring can be quite significant. Using wireless transmission technology eliminates much of the wiring costs and facilitates moving the sensors from one location to another. Wireless transmission has a set of concerns that must be taken into account by the engineer, including security, since other units can monitor signals (still evolving) and ability to be jammed.

10 SUMMARY

The tremendous change in technology for data acquisition and display systems since this chapter was first written has driven us to take a different approach than with the first edition. The technologies for data acquisition and display have become more standardized. Engineers are increasingly reliant upon and versed in computing technologies. The combination of data from various sources into an integrated view of the process has facilitated process improvement and leads to competitive advantage.

This chapter has attempted to provide tools and techniques to aid in the acquisition, storage, and manipulation of process data, expanding from the previous edition into techniques to aid in the manipulation of data for integration and analysis.

REFERENCES

1. C. D. Johnson, *Microprocessor-Based Process Control,* Prentice-Hall, Englewood Cliffs, NJ, 1984.

2. P. W. Murrill, *Fundamentals of Process Control Theory,* Instrument Society of America, Research Triangle Park, NC, 1981.

3. B. G. Liptak, "System Accuracy," in B. G. Liptak (Ed. in Chief), *Instrument Engineer's Handbook*, 4th ed., Vol. **1**: *Process Measurement and Analysis,* CRC, Boca Raton, FL, 2003.

4. J. D. Wright and T. F. Edgar, "Digital Computer Control and Signal Processing Algorithms," in D. A. Mellichamp (Ed.), *Real-Time Computing,* Van Nostrand Reinhold, New York, 1983.

5. C. J. Date, *An Introduction to Database Systems,* 5th ed., Addison-Wesley, Boston, MA, 1990.

6. W. H. Beyer (Ed.), *CRC Standard Mathematical Tables,* 24th ed., CRC, Boca Raton, FL, 1976.

7. M. L. Crossley, *The Desk Reference of Statistical Quality Methods,* ASQ Quality Press, Milwaukee, WI, 2000.

8. http://www.aspentech.com/publication_files/White_Paper_for_IP_21.pdf.

9. R. W. Bailey, *Human Performance Engineering: A Guide for Systems Designers,* Prentice-Hall, Englewood Cliffs, NJ, 1982.

10. E. R. Tufte, *The Visual Display of Quantitative Information,* Graphics, Cheshire, England, 1983.

11. "Open Systems Interconnection—Basic Reference Model," Draft Proposal 7498, 97/16 N719, American National Standards Institute, New York, 1981.

MAGAZINES THAT CARRY RELEVANT INFORMATION

Control Engineering International: http://www.controleng.com/.

Design Engineering: http://www.designengineering.co.uk/.

IEEE Control Systems Magazine: http://www.ieee.org/organizations/pubs/magazines/cs.htm.

Industrial Technology: http://www.industrialtechnology.co.uk/.

Instrumentation and Automation News: http://www.ianmag.com/.

Pollution Engineering Online: http://www.pollutionengineering.com/.

Scientific Computing and Instrumentation: http://www.scamag.com/.

Sensors Magazine: www.sensorsmag.com.

CHAPTER **19**

SYSTEMS ENGINEERING: ANALYSIS, DESIGN, AND INFORMATION PROCESSING FOR ANALYSIS AND DESIGN

Andrew P. Sage
George Mason University
Fairfax, Virginia

1 INTRODUCTION

Systems engineering is a management technology. Technology involves the organization and delivery of science for the (presumed) betterment of humankind. Management involves the interaction of the organization, and the humans in the organization, with the environment. Here, we interpret environment in a very general sense to include the complete external milieu surrounding individuals and organizations. Hence, systems engineering as a management technology involves three ingredients: science, organizations, and their environments. Information and knowledge are ubiquitous throughout systems engineering and management efforts and are, in reality, a fourth ingredient. Systems engineering is thus seen to involve science, organizations and humans, environments, technologies, and information and knowledge.

The process of systems engineering involves working with clients in order to assist them in the organization of information and knowledge to aid in judgment and choice of activities that lead to the engineering of trustworthy systems. These activities result in the making of decisions and associated resource allocations through enhanced efficiency, effectiveness, equity, and explicability as a result of systems engineering efforts.

This set of action alternatives is selected from a larger set, in accordance with a value system, in order to influence future conditions. Development of a set of rational policy or action alternatives must be based on formation and identification of candidate alternative policies and objectives against which to evaluate the impacts of these proposed activities, such as to enable selection of efficient, effective, and equitable alternatives for implementation.

In this chapter, we are concerned with the engineering of large-scale systems, or *systems engineering*.[1] We are especially concerned with strategic-level systems engineering, or *systems management*.[2] We begin by first discussing the need for systems engineering and then providing some definitions of systems engineering. We next present a structure describing the systems engineering process. The result of this is a *life-cycle model* for systems engineering processes. This is used to motivate discussion of the functional levels, or considerations, involved in systems engineering efforts: *measurements, systems engineering methods and tools, systems methodology or processes,* and *systems management*. Considerably more details are presented in Refs. 1 and 2, which are the sources from which most of this chapter is derived.

Systems engineering is an appropriate combination of mathematical, behavioral, and management theories in a useful setting appropriate for the resolution of complex real-world issues of large scale and scope. As such, systems engineering consists of the use of management, behavioral, and mathematical constructs to identify, structure, analyze, evaluate, and interpret generally incomplete, uncertain, imprecise, and otherwise imperfect information. When associated with a value system, this information leads to knowledge to permit decisions that have been evolved with maximum possible understanding of their impacts. A central need, but by no means the only need, in systems engineering is to select an appropriate life cycle, or process, that is explicit, rational, and compatible with the implementation framework extant and the perspectives and knowledge bases of those responsible for decision activities. When this is accomplished, an appropriate choice of systems engineering methods and tools may be made to enable full implementation of the life-cycle process.

Information is a very important quantity that is assumed to be present in the management technology that is systems engineering. This strongly couples notions of systems engineering with those of technical direction or systems management of technological development, rather than exclusively with one or more of the methods of systems engineering, important as they may be for the ultimate success of a systems engineering effort. It suggests that *systems engineering is the management technology that controls a total system life-cycle process, which involves and which results in the definition, development, and deployment of a system that is of high quality, trustworthy, and cost-effective in meeting user needs.* This process-oriented notion of systems engineering and systems management will be emphasized here.

Among the appropriate conditions for use of systems engineering are the following:

- There are many considerations and interrelations.
- There are far-reaching and controversial value judgments.
- There are multidisciplinary and interdisciplinary considerations.
- The available information is uncertain, imprecise, incomplete, or otherwise flawed.
- Future events are uncertain and difficult to predict.
- Institutional and organizational considerations play an important role.
- There is a need for explicit and explicable consideration of the efficiency, effectiveness, and equity of alternative courses of action.

There are a number of results potentially attainable from use of systems engineering approaches. These include:

- Identification of perceived needs in terms of identified objectives and values of a client group
- Identification or definition of a set of user or client requirements for the product system or service system that will ultimately be fielded

- Enhanced identification of a wide range of proposed alternatives or policies that might satisfy these needs, achieve the objectives of the clients in a high-quality and trustworthy fashion, and fulfill the requirements definition
- Increased understanding of issues that led to the effort and the impacts of alternative actions upon these issues
- Ranking of these identified alternative courses of action in terms of the utility (benefits and costs) in achieving objectives, satisfying needs, and fulfilling requirements
- A set of alternatives that is selected for implementation, generally by a group of content specialists responsible for detailed design and implementation, and an appropriate plan for action to achieve this implementation

Ultimately these action plans result in a working product or service, each of which is maintained over time in subsequent phases of the postdeployment efforts that also involve systems engineering.

To develop professionals capable of coping satisfactorily with diverse factors involved in wide-scope problem solving is a primary goal of systems engineering and systems engineering education. This does not imply that a single individual or even a small group can, despite its strong motivation, solve all of the problems involved in a systems study. Such a requirement would demand total and absolute intellectual maturity on the part of the systems engineer and such is surely not realistic. It is also unrealistic to believe that issues can be resolved without very close association with a number of people who have stakes, and who thereby become stakeholders, in problem solution efforts. Consequently, systems engineers must be capable of facilitation and communication of knowledge between the diverse groups of professionals, and their publics, that are involved in wide-scope problem solving. This requires that systems engineers be knowledgeable and able to use not only the technical methods-based tools that are needed for issue and problem resolution but also the behavioral constructs and management abilities that are needed for resolution of complex, large-scale problems. Intelligence, imagination, and creativity are necessary but not sufficient for proper use of the procedures of systems engineering. Facility in human relations and effectiveness as a broker of information among parties of interest in a systems engineering program are very much needed as well.

It is this blending of the technical, managerial, and behavioral that is a normative goal of success for systems engineering education and for systems engineering professional practice. Thus, systems engineering involves:

- The sciences and the various methods, analysis, and measurement perspectives associated with the sciences
- Life-cycle process models for definition, development, and deployment of systems
- The systems management issues associated with choice of an appropriate process
- Organizations and humans and the understanding of organizational and human behavior
- Environments and understanding of the diverse interactions of organizations of people, technologies, and institutions with their environments
- Information and the way in which it can and should be processed to facilitate all aspects of systems engineering efforts

Successful systems engineering must be practiced at three levels: systems methods and measurements, systems processes and methodology, and systems management. Systems engineers must be aware of a wide variety of methods that assist in the formulation, analysis, and interpretation of contemporary issues. They must be familiar with systems engineering process life cycles (or methodology, as an open set of problem-solving procedures) in order to be able to select eclectic approaches that are best suited to the task at hand. Finally, a knowledge of

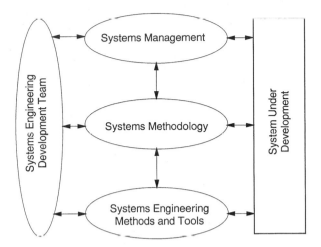

Figure 1   Conceptual illustration of the three levels for systems engineering.

systems management is necessary in order to be able to select life-cycle processes that are best matched to behavioral and organizational concerns and realities.

All three of these levels, suggested in Fig. 1, are important. To neglect any of them in the practice of systems engineering is to invite failure. It is generally not fully meaningful to talk only of a method or algorithm as a useful system-fielding or life-cycle process. It is ultimately meaningful to talk of a particular process as being useful. A process or product line that is truly useful for the fielding of a system will depend on the methods that are available, the operational environment, and leadership facets associated with use of the system and the system-fielding process. Thus systems management, systems engineering processes, and systems engineering methods and measurements do, separately and collectively, play a fundamental role in systems engineering.

2   SYSTEM LIFE-CYCLE AND FUNCTIONAL ELEMENTS OF SYSTEMS ENGINEERING

We have provided one definition of systems engineering thus far. It is primarily a structural and process-oriented definition. A related definition, in terms of purpose, is that systems engineering is management technology to assist and support policy-making, planning, decision-making, and associated resource allocation or action deployment for the purpose of acquiring a product desired by customers or clients. Systems engineers accomplish this by quantitative and qualitative formulation, analysis, and interpretation of the impacts of action alternatives upon the needs perspectives, the institutional perspectives, and the value perspectives of their clients or customers. Each of these three steps is generally needed in solving systems engineering problems. Issue *formulation* is an effort to identify the needs to be fulfilled and the requirements associated with these in terms of objectives to be satisfied, constraints and alterables that affect issue resolution, and generation of potential alternative courses of action. Issue *analysis* enables us to determine the impacts of the identified alternative courses of action, including possible refinement of these alternatives. Issue *interpretation* enables us to rank in order the alternatives in terms of need satisfaction and to select one for implementation or additional study. This particular listing of three systems engineering steps and their descriptions is rather formal. Often,

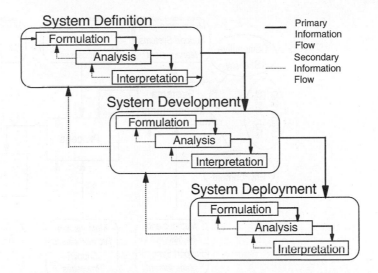

Figure 2   One representation of three systems engineering steps within each of three life-cycle phases.

issues are resolved this way. The steps of formulation, analysis, and interpretation may also be accomplished on an as-if basis by application of a variety of often useful heuristic approaches. These may well be quite appropriate in situations where the problem solver is experientially familiar with the task at hand and the environment into which the task is imbedded.[1]

The key words in this definition are "formulation," "analysis," and "interpretation." In fact, all of systems engineering can be thought of as consisting of formulation, analysis or assessment, and interpretation efforts, together with the systems management and technical direction efforts necessary to bring this about. We may exercise these in a formal sense throughout each of the several phases of a systems engineering life cycle or in an as-if or experientially based intuitive sense. These formulation, analysis, and interpretation efforts are the stepwise or microlevel components that comprise a part of the structural framework for systems methodology. They are needed for each phase in a systems engineering effort, although the specific formulation methods, analysis methods, and interpretation methods may differ considerably across the phases.

We can also think of a functional definition of systems engineering: Systems engineering is the art and science of producing a product, based on phased efforts, that satisfies user needs. The system is functional, reliable, of high quality, and trustworthy, and has been developed within cost and time constraints through use of an appropriate set of methods and tools.

Systems engineers are very concerned with the appropriate *definition, development,* and *deployment* of product systems and service systems. These comprise a set of phases for a systems engineering life cycle. There are many ways to describe the life-cycle phases of the systems engineering process, and we have described a number of them in Refs. 1 and 2. Each of these basic life-cycle models, and those that are outgrowths of them, is comprised of these three phases of definition, development, and deployment. For pragmatic reasons, a typical life cycle will almost always contain more than three phases. Often, it takes on the "waterfall" pattern illustrated in Fig. 2, although there are a number of modifications of the basic waterfall, or "grand-design," life cycles that allow for incremental and evolutionary development of systems life-cycle processes.[2]

A successful approach to systems engineering as an intellectual and action-based approach for increased innovation and productivity and other contemporary challenges must be capable

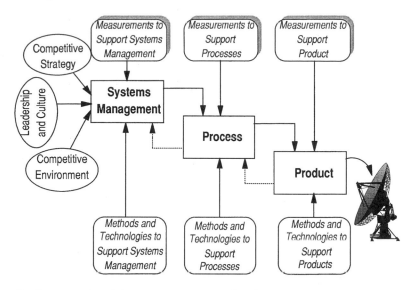

Figure 3   Representation of the structure systems engineering and management functional efforts.

of issue formulation, analysis, and interpretation at the level of institutions and values as well as at the level of symptoms. Systems engineering approaches must allow for the incorporation of need and value perspectives as well as technology perspectives into models and postulates used to evolve and evaluate policies or activities that may result in technological and other innovations.

In actual practice, the steps of the systems process (formulation, analysis, and interpretation) are applied iteratively, across each of the phases of a systems engineering effort, and there is much feedback from one step (and one phase) to the other. This occurs because of the learning that is accomplished in the process of problem solution. Underlying all of this is the need for a general understanding of the diversity of the many systems engineering methods and algorithms that are available and their role in a systems engineering process. The knowledge taxonomy for systems engineering, which consists of the major intellectual categories into which systems efforts may be categorized, is of considerable importance. The categories include systems methods and measurements, systems engineering processes or systems methodology, and systems management. These are used, as suggested in Fig. 3, to produce a *system,* which is a generic term that we use to describe a product or a service.

The methods and metrics associated with systems engineering involve the development and application of concepts that form the basis for problem formulation and solution in systems engineering. Numerous tools for mathematical systems theory have been developed, including operations research (linear programming, nonlinear programming, dynamic programming, graph theory, etc.), decision and control theory, statistical analysis, economic systems analysis, and modeling and simulation. Systems science is also concerned with psychology and human factors concepts, social interaction and human judgment research, nominal group processes, and other behavioral science efforts. Of very special significance for systems engineering is the interaction of the behavioral and the algorithmic components of systems science in the choice-making process. The combination of a set of systems science and operations research methods and a set of relations among these methods and activities constitutes what is known as a *methodology*. References 3 and 4 discuss a number of systems engineering methods and associated methodologies for systems engineering.

As we use it here, a methodology is an open set of procedures that provides the means for solving problems. The tools or the content of systems engineering consists of a variety of algorithms and concepts that use words, mathematics, and graphics. These are structured in ways that enable various problem-solving activities within systems engineering. Particular sets of relations among tools and activities, which constitute the framework for systems engineering, are of special importance here. Existence and use of an appropriate systems engineering process are of considerable utility in dealing with the many considerations, interrelations, and controversial value judgments associated with contemporary problems.

Systems engineering can be and has been described in many ways. Of particular importance is a morphological description, that is, in terms of form. This description leads to a specific methodology that results in a process* that is useful for fielding a system and/or issue resolution. We can discuss the knowledge dimension of systems engineering. This would include the various disciplines and professions that may be needed in a systems team to allow it to accomplish intended purposes of the team, such as provision of the knowledge base. Alternatively, we may speak of the phases or time dimension of a system's effort. These include system definition, development, and deployment. The deployment phase includes system operation, maintenance, and finally modification or reengineering or ultimate retirement and phase-out of the system. Of special interest are the steps of the logic structure or logic dimension of systems engineering:

- Formulation of issues, or identification of problems or issues in terms of needs and constraints, objectives, or values associated with issue resolution, and alternative policies, controls, hypotheses, or complete systems that might resolve or ameliorate issues
- Analysis of impacts of alternative policies, courses of action, or complete systems
- Interpretation or evaluation of the utility of alternatives and their impacts upon the affected stakeholder group and selection of a set of action alternatives for implementation

We could also associate feedback and learning steps to interconnect these steps one to another. The systems process is typically very iterative. We shall not explicitly show feedback and learning in our conceptual models of the systems process, although it is ideally always there.

Here we have described a three-dimensional morphology of systems engineering. There are a number of systems engineering morphologies or frameworks. In many of these, the logic dimension is divided into a larger number of steps that are iterative in nature. A particular seven-step framework, due to Hall,[5,6] involves:

1. *Problem definition,* in which a descriptive and normative scenario of needs, constraints, and alterables associated with an issue is developed. Problem definition clarifies the issues under consideration to allow other steps of a systems engineering effort to be carried out.

2. *Value system design,* in which objectives and objectives measures or attributes with which to determine success in achieving objectives are determined. Also, the interrelationship between objectives and objectives measures and the interaction between objectives and the elements in the problem definition step are determined. This establishes a measurement framework, which is needed to establish the extent to which the impacts of proposed policies or decisions will achieve objectives.

---

* As noted in Refs. 1 and 2, there are life cycles for systems engineering efforts in research, development, test, and evaluation (RDT&E); systems acquisition, production, or manufacturing; and systems planning and marketing. Here, we restrict ourselves to discussions of the life cycle associated with acquisition or production of a system.

3. *System synthesis,* in which candidate or alternative decisions, hypotheses, options, policies, or systems that might result in needs satisfaction and objective attainment are postulated.

4. *Systems analysis and modeling,* in which models are constructed to allow determination of the consequences of pursuing policies. Systems analysis and modeling determine the behavior or subsequent conditions resulting from alternative policies and systems. Forecasting and impact analysis are, therefore, the most important objectives of systems analysis and modeling.

5. *Optimization or refinement* of each alternative, in which the individual policies and/or systems are tuned, often by means of parameter adjustment methods, so that each individual policy or system is refined in some "best" fashion in accordance with the value system that has been identified earlier.

6. *Evaluation and decision making,* in which systems and/or policies and/or alternatives are evaluated in terms of the extent to which the impacts of the alternatives achieve objectives and satisfy needs. Needed to accomplish evaluation are the attributes of the impacts of proposed policies and associated objective and/or subjective measurement of attribute satisfaction for each proposed alternative. Often this results in a prioritization of alternatives, with one or more being selected for further planning and resource allocation.

7. *Planning for action,* in which implementation efforts, resource and management allocations, or plans for the next phase of a systems engineering effort are delineated.

More often than not, the information required to accomplish these seven steps is not perfect due to information uncertainty, imprecision, or incompleteness effects. This presents a major challenge to the design of processes and for systems engineering practice as well.

Figure 4 illustrates a not-untypical 49-element morphological box for systems engineering. This is obtained by expanding our initial three systems engineering steps of formulation, analysis, and interpretation to the seven just discussed. The three basic phases of definition,

			Problem Definition	Value System Design	System Synthesis	Systems Analysis	Alternative Refinement	Decision Making	Planning for Action
			Formulation			Analysis		Interpretation	
	Definition	Program Planning	1	2	3	4	5	6	7
		Project Planning	8						
	Develop-ment	System Development			16				
		Production							
	Deployment	Distribution							
		Operations			38				
		Reengineering or Retirement						48	49

Figure 4 Phases and steps in one 49-element two-dimensional systems engineering framework with activities shown sequentially for waterfall implementation of effort.

development, and deployment are expanded to a total of seven phases. These seven steps, and the seven phases that we associate with them, are essentially those identified by Hall in his pioneering efforts in systems engineering.[5,6] The specific methods we need to use in each of these seven steps are clearly dependent upon the phase of activity that is being completed, and there are a plethora of systems engineering methods available.[3,4] Using a seven-phase, seven-step framework raises the number of activity cells to 49 for a single life cycle. A very large number of systems engineering methods may be needed to fill in this matrix, especially since more than one method will almost invariably be associated with many of the entries.

The requirements and specification phase of the systems engineering life cycle has as its goal the identification of client or stakeholder needs, activities, and objectives for the functionally operational system. This phase should result in the identification and description of preliminary conceptual design considerations for the next phase. It is necessary to translate operational deployment needs into requirement specifications so that these needs may be addressed by the system design efforts. As a result of the requirement specification phase, there should exist a clear definition of development issues such that it becomes possible to make a decision concerning whether to undertake preliminary conceptual design. If the requirement specification effort indicates that client needs can be satisfied in a functionally satisfactory manner, then documentation is typically prepared concerning system-level specifications for the preliminary conceptual design phase. Initial specifications for the following three phases of effort are typically also prepared, and a concept design team is selected to implement the next phase of the life-cycle effort. This effort is sometimes called *system-level architecting*.[7,8] Many[9,10] have discussed technical-level architectures. It is only recently that the need for major attention to architectures at the systems level has also been identified.

Preliminary conceptual system design typically includes or results in an effort to specify the content and associated architecture and general algorithms for the system product in question. The desired product of this phase of activity is a set of detailed design and architectural specifications that should result in a useful system product. There should exist a high degree of user confidence that a useful product will result from detailed design or the entire design effort should be redone or possibly abandoned. Another product of this phase is a refined set of specifications for the evaluation and operational deployment phases of the life cycle. In the third phase, these are translated into detailed representations in logical form so that system development may occur. A product, process, or system is produced in the fourth phase of the life cycle. This is not the final system design, but rather the result of implementation of the design that resulted from the conceptual design effort.

Evaluation of the detailed design and the resulting product, process, or system is achieved in the sixth phase of the systems engineering life cycle. Depending upon the specific application being considered, an entire systems engineering life-cycle process could be called *design,* or *manufacturing,* or some other appropriate designator. *System acquisition* is an often-used term to describe the entire systems engineering process that results in an operational systems engineering product. Generally, an acquisition life cycle primarily involves knowledge practices or standard procedures to produce or manufacture a product based on established practices. An RDT&E life cycle is generally associated with an emerging technology and involves knowledge principles. A marketing life cycle is concerned with product planning and other efforts to determine market potential for a product or service and generally involves knowledge perspectives.

The intensity of effort needed for the steps of systems engineering varies greatly with the type of problem being considered. Problems of large scale and scope will generally involve a number of perspectives. These interact and the intensity of their interaction and involvement with the issue under consideration determines the scope and type of effort needed in the various steps of the systems process. Selection of appropriate algorithms or approaches to enable completion of these steps and satisfactory transition to the next step, and ultimately to completion of each phase of the systems engineering effort, are major systems engineering tasks.

Each of these phases of a systems engineering life cycle is very important for sound development of physical systems or products and such service systems as information systems. Relatively less attention appears to have been paid to the requirement specification phase than to the other phases of the systems engineering life-cycle process. In many ways, the requirement specification phase of a systems engineering design effort is the most important. It is this phase that has as its goal the detailed definition of the needs, activities, and objectives to be fulfilled or achieved by the process to be ultimately developed. Thus, this phase strongly influences all the phases that follow. It is this phase that describes preliminary design considerations that are needed to achieve successfully the fundamental goals underlying a systems engineering study. It is in this phase that the information requirements and the method of judgment and choice used for selection of alternatives are determined. Effective systems engineering, which inherently involves design efforts, must also include an operational evaluation component that will consider the extent to which the product or service is useful in fulfilling the requirements that it is intended to satisfy.

3 SYSTEMS ENGINEERING OBJECTIVES

Ten performance objectives appear to be of primary importance to those who desire to evolve quality plans, forecasts, decisions, or alternatives for action implementation:

1. Identify needs, constraints, and alterables associated with the problem, issue, or requirement to be resolved (problem definition).

2. Identify a planning horizon or time interval for alternative action implementation, information flow, and objective satisfaction (planning horizon, identification).

3. Identify all significant objectives to be fulfilled, values implied by the choice of objectives, and objectives measures or attributes associated with various outcome states, with which to measure objective attainment (value system design).

4. Identify decisions, events, and event outcomes and the relations among them such that a structure of the possible paths among options, alternatives, or decisions and the possible outcomes of these emerge (impact assessment).

5. Identify uncertainties and risks associated with the environmental influences affecting alternative decision outcomes (probability identification).

6. Identify measures associated with the costs and benefits or attributes of the various outcomes or impacts that result from judgment and choice (worth, value, or utility measurement).

7. Search for and evaluate new information, and the cost-effectiveness of obtaining this information, relevant to improved knowledge of the time-varying nature of event outcomes that follow decisions or choice of alternatives (information acquisition and evaluation).

8. Enable selection of a best course of action in accordance with a rational procedure (decision assessment and choice making).

9. Reexamine the expected effectiveness of all feasible alternative courses of action, including those initially regarded as unacceptable, prior to making a final alternative selection (sensitivity analysis).

10. Make detailed and explicit provisions for implementation of the selected action alternative, including contingency plans, as needed (planning for implementation of action).

These objectives are, of course, very closely related to the aforementioned steps of the framework for systems engineering. To accomplish them requires attention to and knowledge of the methods of systems engineering such that we are able to design product systems and service systems and also enable systems to support products and services. We also need to select an appropriate process, or product line, to use for management of the many activities associated with fielding a system. Also required is much effort at the level of systems management so that the resulting process is efficient, effective, equitable, and explicable. Thus, it is necessary to ensure that those involved in systems engineering efforts be concerned with technical knowledge of the issue under consideration, able to cope effectively with administrative concerns relative to the human elements of the issue, interested in and able to communicate across those actors involved in the issue, and capable of innovation and out-scoping of relevant elements of the issue under consideration. These attributes (technical knowledge, human understanding and administrative ability, communicability, and innovativeness) are, of course, primary attributes of effective management.

4   SYSTEMS ENGINEERING METHODOLOGY AND METHODS

A variety of methods are suitable to accomplish the various steps of systems engineering. We shall briefly describe some of them here.

4.1   Issue Formulation

As indicated above, issue formulation is the step in the systems engineering effort in which the problem or issue is defined (problem definition) in terms of the objectives of a client group (value system design) and where potential alternatives that might resolve needs are identified (system synthesis). Many studies have shown that the way in which an issue is resolved is critically dependent on the way in which the issue is formulated or framed. The issue formulation effort is concerned primarily with identification and description of the elements of the issue under consideration, with, perhaps, some initial effort at structuring these in order to enhance understanding of the relations among these elements. Structural concerns are also of importance in the analysis effort. The systems process is iterative and interactive, and the results of preliminary analysis are used to refine the issue formulation effort. Thus, the primary intent of issue formulation is to identify relevant elements that represent and are associated with issue definition, the objectives that should be achieved in order to satisfy needs, and potential action alternatives.

There are at least four ways to accomplish issue formulation, or to identify requirements for a system, or to accomplish the initial part of the definition phase of systems engineering:

1. Asking stakeholders in the issue under consideration for the requirements
2. Descriptive identification of the requirements from a study of presently existing systems
3. Normative synthesis of the requirements from a study of documents describing what "should be," such as planning documents
4. Experimental discovery of requirements, based on experimentation with an evolving system

These approaches are neither mutually exclusive nor exhaustive. Generally, the most appropriate efforts will use a combination of these approaches.

There are conflicting concerns with respect to which blend of these requirement identification approaches is most appropriate for a specific task. The asking approach seems very appropriate when there is little uncertainty and imprecision associated with the issue under consideration, so that the issue is relatively well understood and may be easily structured, and where members of the client group possess much relevant expertise concerning the issue and the environment in which the issue is embedded. When these characteristics of the issue—lack of imprecision and presence of expert experiential knowledge—are present, then a direct declarative approach based on direct "asking" of "experts" is a simple and efficient approach. When there is considerable imprecision or a lack of experiential familiarity with the issue under concern, the other approaches take on greater significance. The asking approach is also prone to a number of human information-processing biases, as will be discussed in Section 4.5. This is not as much of a problem in the other approaches.

Unfortunately, however, there are other difficulties with each of the other three approaches. Descriptive identification, from a study of existing systems of issue formulation elements, will very likely result in a new system that is based or anchored on an existing system and tuned, adjusted, or perturbed from this existing system to yield incremental improvements. Thus, it is likely to result in incremental improvements to existing systems but not to result in major innovations or totally new systems and concepts.

Normative synthesis from a study of planning documents will result in an issue formulation or requirement identification effort that is based on what have been identified as desirable objectives and needs of a client group. A plan at any given phase may well not exist or it may be flawed in any of several ways. Thus, the information base may well not be present or may be flawed. When these circumstances exist, it will not be a simple task to accomplish effective normative synthesis of issue formulation elements for the next phase of activity from a study of planning documents relative to the previous phase.

Often it is not easily possible to determine an appropriate set of issue formulation elements or requirements. Often it will not be possible to define an appropriate set of issue formulation efforts prior to actual implementation of a preliminary system design. There are many important issues where there is an insufficient experiential basis to judge the effectiveness and completeness of a set of issue formulation efforts or requirements. Often, for example, clients will have difficulty in coping with very abstract formulation requirements and in visualizing the system that may ultimately evolve. Thus, it may be useful to identify an initial set of issue formulation elements and accomplish subsequent analysis and interpretation based on these, without extraordinary concern for completeness of the issue formulation efforts. A system designed with ease of adaptation and change as a primary requirement is implemented on a trial basis. As users become familiar with this new system or process, additions and modifications to the initially identified issue formulation elements result. Such a system is generally known as a *prototype*. One very useful support for the identification of requirements is to build a prototype and allow the users of the system to be fielded to experiment with the prototype and, through this experimentation, to identify system requirements.[11] This heuristic approach allows users to identify the requirements for a system by experimenting with an easily changeable set of system design requirements and to improve their identification of these issue formulation elements as their experiential familiarity with the evolving prototype system grows.

The key parts of the problem definition step of issue formulation involve identification of needs, constraints, and alterables and determination of the interactions among these elements and the group that they impact. Need is a condition requiring supply or relief or is a lack of something required, desired, or useful. In order to define a problem satisfactorily, we must determine the alterables or those items pertaining to the needs that can be changed. Alterables can be separated into those over which control is or is not possible. The controllable alterables are of special concern in systems engineering since they can be changed or modified to assist in achieving particular outcomes. To define a problem adequately, we must also determine the limitations or constraints under which the needs can or must be satisfied and the range over which it is permissible to vary the controllable alterables. Finally, we must determine relevant groups of people who are affected by a given problem.

Value system design is concerned with defining objectives, determining their interactions, and ordering these into a hierarchical structure. Objectives and their attainment are, of course, related to the needs, alterables, and constraints associated with problem definition. Thus, the objectives can and should be related to these problem definition elements. Finally, a set of measures is needed whereby to measure objective attainment. Generally, these are called *attributes of objectives* or *objectives measures*. It is necessary to ensure that all needs are satisfied by attainment of at least one objective.

The first step in system synthesis is to identify activities and alternatives for attaining each of the objectives or the postulation of complete systems to this end. It is then desirable to determine interactions among the proposed activities and to illustrate relationships between the activities and the needs and objectives. Activities measures are needed to gauge the degree of accomplishment of proposed activities. Systemic methods useful for problem definition are generally useful for value system design and system synthesis as well. This is another reason that suggests the efficacy of aggregating these three steps under a single heading: *issue formulation*.

Complex issues will have a structure associated with them. In some problem areas, structure is well understood and well articulated. In other areas, it is not possible to articulate structure in such a clear fashion. There exists considerable motivation to develop techniques with which to enhance structure determination, as a system structure must always be dealt with by individuals or groups, regardless of whether the structure is articulated or not. Furthermore, an individual or a group can deal much more effectively with systems and make better decisions when the structure of the underlying system is well defined and exposed and communicated clearly. One of the fundamental objectives of systems engineering is to structure knowledge elements such that they are capable of being better understood and communicated.

We now discuss several formal methods appropriate for "asking" as a method of issue formulation. Most of these and other approaches are described in Refs. 1, 3, and 4. Then we shall very briefly contrast and compare some of these approaches. The methods associated with the other three generic approaches to issue formulation also involve approaches to analysis that will be discussed in the next section.

Several of the formal methods that are particularly helpful in the identification, through asking, of issue formulation elements are based on principles of collective inquiry, in which interested and motivated people are brought together to stimulate each other's creativity in generating issue formulation elements. We may distinguish two groups of collective-inquiry methods:

1. *Brainwriting, Brainstorming, Synectics, Nominal Group Technique,* and *Charette.* These approaches typically require a few hours of time, a group of knowledgeable people gathered in one place, and a group leader or facilitator. Brainwriting is typically better than brainstorming in reducing the influence of dominant individuals. Both methods can be very productive: 50–150 ideas or elements might be generated in less than an hour. Synectics, based on problem analogies, might be appropriate if there is a need for truly unconventional, innovative ideas. Considerable experience with the method is a requirement, however, particularly for the group leader. The nominal group technique is based on a sequence of idea generation, discussion, and prioritization. It can be very useful when an initial screening of a large number of ideas or elements is needed. Charette offers a conference or workshop-type format for generation and discussion of ideas and/or elements.

2. *Questionnaires, Surveys,* and *Delphi.* These three methods of collective-inquiry modeling do not require the group of participants to gather at one place and time, but they typically take more time to achieve results than the first group of methods. In questionnaires and surveys, a usually large number of participants are asked, on an individual

basis, for ideas or opinions, which are then processed to achieve an overall result. There is no interaction among participants. Delphi usually provides for written interaction among participants in several rounds. Results of previous rounds are fed back to participants, who are asked to comment, revise their views as desired, and so on. A Delphi exercise can be very instructive but usually takes several weeks or months to complete.

Use of most structuring methods, in addition to leading to greater clarity of the problem formulation elements, will also typically lead to identification of new elements and revision of element definitions. As we have indicated, most structuring methods contain an analytical component; they may, therefore, be more properly labeled analysis methods. The following element-structuring aids are among the many modeling aids available:

- *Interaction matrices* may be used to identify clusters of closely related elements in a large set, in which case we have a self-interaction matrix, or to structure and identify the couplings between elements of different sets, such as objectives and alternatives. In this case, we produce cross-interaction matrices, such as shown in Fig. 5. Interaction matrices are useful for initial, comprehensive exploration of sets of elements. Learning about problem interrelationships during the process of constructing an interaction matrix is a major result of use of these matrices.

- *Trees* are graphical aids particularly useful in portraying hierarchical or branching-type structures. They are excellent for communication, illustration, and clarification. Trees may be useful in all steps and phases of a systems effort. Figure 6 presents an attribute tree that represents those aspects that will be formally considered in the evaluation and prioritization of a set of proposals.

- *Causal loop diagrams,* or influence diagrams, represent graphical pictures of causal interactions between sets of variables. They are particularly helpful in making explicit one's perception of the causes of change in a system and can serve very well as

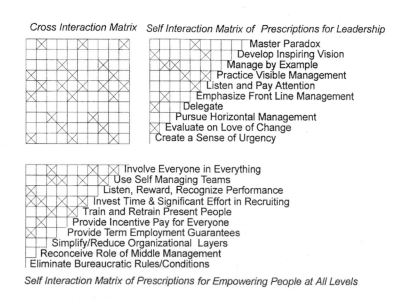

Figure 5   Hypothetical self- and cross-interaction matrices for prescriptions for leadership and for empowering people at all levels.

1. Understanding of Problem
   1.1 Navy Cost Credentialing Process
   1.2 NAVELEX Cost Analysis Methodology
   1.3 DoD Procurement Procedures
2. Technical Approach
   2.1 Establishment of a Standard Methodology
   2.2 Compatibility with Navy Acquisition Process
3. Staff Experience
   3.1 Directly Related Experience in Cost Credentialing, Cost Analysis, and Procurement Procedures
   3.2 Direct Experience with Navy R&D Programs
4. Corporate Qualification
   Relevant Experience in Cost Analysis for Navy R&D Programs
5. Management Approach
   5.1 Quality/Relevance
   5.2 Organization and Control Effectiveness
6. Cost
   6.1 Manner in which Elements of Cost Contribute Directly to Project Success
   6.2 Appropriate of Cost Mix to the Technical Effort

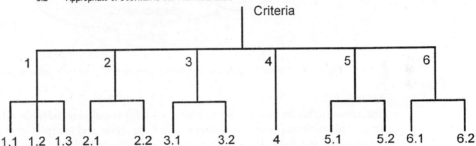

Figure 6 Possible attribute tree for evaluation of proposals concerning cost credentialing.

communication aids. A causal loop diagram is also useful as the initial part of a detailed simulation model. Figure 7 represents a causal loop diagram of a belief structure.

Two other descriptive methods are potentially useful for issue formulation:

- The *system definition matrix,* options profile, decision balance sheet, or checklist provides a framework for specification of the essential aspects, options, or characteristics of an issue, a plan, a policy, or a proposed or existing system. It can be helpful for the design and specification of alternative policies, designs, or other options or alternatives. The system definition matrix is just a table that shows important aspects of the options that are important for judgment relative to selection of approaches to issue formulation or requirement determination.

- *Scenario writing* is based on narrative and creative descriptions of existing or possible situations or developments. Scenario descriptions can be helpful for clarification and communication of ideas and obtaining feedback on those ideas. Scenarios may also be helpful in conjunction with various analysis and forecasting methods, where they may represent alternative or opposing views.

Clearly, successful formulation of issues through "asking" requires creativity. Creativity may be much enhanced through use of a structured systems engineering framework. For example, group meetings for issue formulation involve idea formulation, idea analysis, and idea interpretation. The structure of a group meeting may be conceptualized within a systems engineering

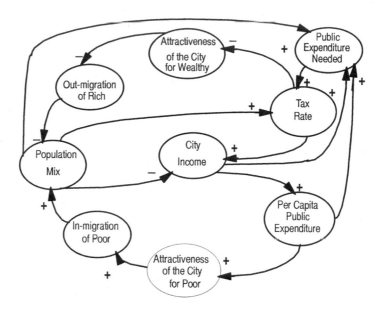

Figure 7 Causal loop diagram of belief structure in a simple model of urban dynamics.

framework. This framework is especially useful for visualizing the trade-offs that must be made among allocation of resources for formulation, analysis, and interpretation of ideas in the issue formulation step itself. If there is an emphasis on idea formulation, we shall likely generate too many ideas to cope with easily. This will lead to a lack of attention to detail. On the other hand, if idea formulation is deemphasized, we shall typically encourage defensive avoidance through undue efforts to support the present situation or a rapid unconflicted change to a new situation. An overemphasis on analysis of ideas is usually time consuming and results in a meeting that seems to drown in details. There is inherent merit in encouraging a group to reach consensus, but the effort may also be inappropriate, since it may encourage arguments over concerns that are ineffective in influencing judgments.

Deemphasizing the analysis of identified ideas will usually result in disorganized meetings in which hasty, poorly thought-out ideas are accepted. Postmeeting disagreements concerning the results of the meeting are another common disadvantage. An emphasis on interpretation of ideas will produce a meeting that is emotional and people centered. Misunderstandings will be frequent as issues become entrenched in an adversarial, personality-centered process. On the other hand, deemphasizing the interpretation of ideas results in meetings in which important information is not elicited. Consequently, the meeting is awkward and empty, and routine acceptance of ideas is a likely outcome.

4.2   Issue Analysis

In systems engineering, issue analysis involves forecasting and assessing of the impacts of proposed alternative courses of action. In turn, this suggests construction, testing, and validation of models. Impact assessment in systems engineering includes system analysis and modeling and optimization and ranking or refinement of alternatives. First, the options or alternatives defined in issue formulation are structured, often as part of the issue formulation effort, and then analyzed in order to assess the anticipated impacts that may result from their implementation.

Second, a refinement or optimization effort is often desirable. This is directed toward refinement or fine-tuning a viable alternative and parameters within an alternative, so as to obtain maximum needs satisfaction, within given constraints, from a proposed policy.

To determine the structure of systems in the most effective manner requires the use of quantitative analysis to direct the structuring effort along the most important and productive paths. This is especially needed when time available to construct structural models is limited. Formally, there are at least four types of self-interaction matrices: nondirected graphs, directed graphs (or digraphs), signed digraphs, and weighted digraphs. The theory of digraphs and structural modeling is authoritatively presented in Ref. 12 and a number of applications to what is called *interpretative structural modeling* are described in Refs. 3 and 13–15. Cognitive map structural models are considered in Ref. 16. A development of structural modeling concepts based on signed digraphs is discussed in Ref. 17. Geoffrion has been especially concerned with the development of a structured modeling methodology[18,19] and environment. He has noted[20] that a modeling environment needs five quality- and productivity-related properties. A modeling environment should:

1. Nurture the entire modeling life cycle, not just a part of it
2. Be hospitable to decision- and policy-makers as well as to modeling professionals
3. Facilitate the maintenance and ongoing evolution of those models and systems that are contained therein
4. Encourage and support those who use it to speak the same paradigm-neutral language in order to best support the development of modeling applications
5. Facilitate management of all the resources contained therein

Structured modeling is a general conceptual framework for modeling. Cognitive maps, interaction matrices, intent structures, Delta charts, objective and attribute trees, causal loop diagrams, decision outcome trees, signal flow graphs, and so on, are all structural models that are very useful graphic aids to communications. The following are requirements for the processes of structural modeling:

1. An object system, which is typically a poorly defined, initially unstructured set of elements to be described by a model
2. A representation system, which is a presumably well-defined set of relations
3. An embedding of perceptions of some relevant features of the object system into the representation system

Structural modeling, which has been of fundamental concern for some time, refers to the systemic iterative application of typically graph-theoretic notions such that an easily communicable directed-graph representation of complex patterns of a particular contextual relationship among a set of elements results. There are a number of computer software realizations of various structural modeling constructs, such as cognitive policy evaluation (COPE), interpretive structural modeling (ISM), and various multiattribute utility theory-based representations, as typically found in most decision-aiding software.

Transformation of a number of identified issue formulation elements, which typically represent unclear, poorly articulated mental models of a system, into visible, well-defined models useful for many purposes is the object of systems analysis and modeling. The principal objective of systems analysis and modeling is to create a process with which to produce information concerning consequences of proposed actions or policies. From the issue formulation steps of problem definition, value system design, and system synthesis, we have various descriptive and normative scenarios available for use. Ultimately, as a part of the issue interpretation step, we wish to evaluate and compare alternative courses of action with respect to the value system through use of a systems model. A model is always a substitute for reality but is, one hopes, descriptive enough of the system elements under consideration to be useful. By posing a variety of questions using the model, we can, from the results obtained, learn how to cope with that subset of the real world being modeled.

A model must depend on much more than the particular problem definition elements being modeled; it must also depend strongly on the value system and the purpose behind construction and utilization of the model. These influence, generally strongly, the structure of the situation and the elements that comprise this structure. Which elements a client believes important enough to include in a model depends on the client's value system.

We wish to be able to determine correctness of predictions and forecasts that are based on model usage. Given the definition of a problem, a value system, and a set of proposed policies, we wish to be able to design a simulation model consisting of relevant elements of these three steps and to determine the results or impacts of implementing proposed policies. Following this, we wish to be able to validate a simulation model to determine the extent to which it represents reality sufficiently to be useful. Validation must, if we are to have confidence in what we are doing with a model, precede actual use of model-based results.

There are three essential steps in constructing a simulation model:

1. Determination of those problem definitions, value systems, and system synthesis elements that are most relevant to a particular problem

2. Determination of the structural relationships among these elements

3. Determination of parametric coefficients within the structure

There are three uses to which models may normally be put. Model categories corresponding to these three uses are descriptive models, predictive or forecasting models, and policy or planning models. Representation and replication of important features of a given problem are the objects of a descriptive model. Good descriptive models are of considerable value in that they reveal much about the substance of complex issues and how, typically in a retrospective sense, change over time has occurred. One of the primary purposes behind constructing a descriptive model is to learn about the past. Often the past will be a good guide to the future.

In building a predictive or forecasting model, we must be especially concerned with determination of proper cause-and-effect relationships. If the future is to be predicted with integrity, we must have a method with which to determine exogenous variables, or input variables that result from external causes, accurately. Also, the model structure and parameters within the structure must be valid for the model to be valid. Often, it will not be possible to predict accurately all exogenous variables; in that case, conditional predictions can be made from particular assumed values of unknown exogenous variables.

The future is inherently uncertain. Consequently, predictive or forecasting models are often used to generate a variety of future scenarios, each a conditional prediction of the future based on some conditioning assumptions. In other words, we develop an "if–then" model.

Policy or planning models are much more than predictive or forecasting models, although any policy or planning model is also a predictive or forecasting model. The outcome from a policy or planning model must be evaluated in terms of a value system. Policy or planning efforts must not only predict outcomes from implementing alternative policies but also present these outcomes in terms of the value system that is in a form useful and suitable for alternative ranking, evaluation, and decision making. Thus, a policy model must contain some provision for impact interpretation.

Model usefulness cannot be determined by objective truth criteria alone. Well-defined and well-stated functions and purposes for the simulation model are needed to determine simulation model usefulness. Fully objective criteria for model validity do not typically exist. Development of a general-purpose, context-free simulation model appears unlikely; the task is simply far too complicated. We must build models for specific purposes, and thus the question of model validity is context dependent.

Model credibility depends on the interaction between the model and model user. One of the major potential difficulties is that of building a model that reflects the outlook of the modeler. This activity is proscribed in effective systems engineering practice, since the purpose of a model is to describe systematically the "view" of a situation held by the client, not that held by the analyst.

A great variety of approaches have been designed and used for the forecasting and assessment that are the primary goals of systems analysis. There are basically two classes of methods that we describe here: expert-opinion methods and modeling and/or simulation methods.

Expert-opinion methods are based on the assumption that knowledgeable people will be capable of saying sensible things about the impacts of alternative policies on the system, as a result of their experience with or insight into the issue or problem area. These methods are generally useful, particularly when there are no established theories or data concerning system operation, precluding the use of more precise analytical tools. Among the most prominent expert-opinion-based forecasting methods are surveys and Delphi. There are, of course, many other methods of asking experts for their opinion—for example, hearings, meetings, and conferences. A particular problem with such methods is that cognitive bias and value incoherence are widespread, often resulting in inconsistent and self-contradictory results. There exists a strong need in the forecasting and assessment community to recognize and ameliorate, by appropriate procedures, the effects of cognitive bias and value incoherence in expert-opinion-modeling efforts. Expert-opinion methods are often appropriate for the "asking" approach to issue formulation. They may be of considerably less value, especially when used as stand-alone approaches, for impact assessment and forecasting.

Simulation and modeling methods are based on the conceptualization and use of an abstraction or model of the real world intended to behave in a similar way to the real system. Impacts of policy alternatives are studied in the model, which will, it is hoped, lead to increased insight with respect to the actual situation.

Most simulation and modeling methods use the power of mathematical formulations and computers to keep track of many pieces of information at the same time. Two methods in which the power of the computer is combined with subjective expert judgments are cross-impact analysis and workshop dynamic models. Typically, experts provide subjective estimates of event probabilities and event interactions. These are processed by a computer to explore their consequences and fed back to the analysts and thereafter to the experts for further study. The computer derives the resulting behavior of various model elements over time, giving rise to renewed discussion and revision of assumptions.

Expert judgment is virtually always included in all modeling methods. Scenario writing can be an expert-opinion-modeling method, but typically this is done in a less direct and explicit way than in Delphi, survey, ISM, cross-impact, or workshop dynamic models. As a result, internal inconsistency problems are reduced with those methods based on mathematical modeling. The following list describes six additional forecasting methods based on mathematical modeling and simulation. In these methods, a structural model is generally formed on the basis of expert opinion and physical or social laws. Available data are then processed to determine parameters within the structure. Unfortunately, these methods are sometimes very data intensive and, therefore, expensive and time consuming to implement.

- Trend extrapolation/time-series forecasting is particularly useful when sufficient data about past and present developments are available, but there is little theory about underlying mechanisms causing change. The method is based on the identification of a mathematical description or structure that will be capable of reproducing the data into the future, typically over the short to medium term.

- Continuous-time dynamic simulation is based on postulation and qualification of a causal structure underlying change over time. A computer is used to explore long-range behavior as it follows from the postulated causal structure. The method can be very useful as a learning and qualitative forecasting device, but its application may be rather costly and time consuming.

- Discrete-event digital simulation models are based on applications of queuing theory to determine the conditions under which system outputs or states will switch from one condition to another.

- Input–output analysis has been specially designed for study of equilibrium situations and requirements in economic systems in which many industries are interdependent. Many economic data fit in directly to the method, which mathematically is relatively simple and can handle many details.

- Econometrics is a method mainly applied to economic description and forecasting problems. It is based on both theory and data, with, usually, the main emphasis on specification of structural relations based on macroeconomic theory and the derivation of unknown parameters in behavioral equations from available economic data.

- Microeconomic models represent an application of economic theories of firms and consumers who desire to maximize the profit and utility of their production and consumption alternatives.

Parameter estimation is a very important subject with respect to model construction and validation. Observation of basic data and estimation or identification of parameters within an assumed structure, often denoted as system identification, are essential steps in the construction and validation of system models. The simplest estimation procedure, in both concept and implementation, appears to be the least-squares error estimator. Many estimation algorithms to accomplish this are available and are in actual use. The subjects of parameter estimation and system identification are being actively explored in both economics and systems engineering. There are numerous contemporary results, including algorithms for system identification and parameter estimation in very-large-scale systems representative of actual physical processes and organizations.

Verification of a model is necessary to ensure that the model behaves in a fashion intended by the model builder. If we can determine that the structure of the model corresponds to the structure of the elements obtained in the problem definition, value system design, and system synthesis steps, then the model is verified with respect to behaving in a gross, or structural, fashion, as the model builder intends.

Even if a model is verified in a structural as well as parametric sense, there is still no assurance that the model is valid in the sense that predictions made from the model will occur. We can determine validity only with respect to the past. That is all that we can possibly have available at the present. Forecasts and predictions inherently involve the future. Since there may be structural and parametric changes as the future evolves, and since knowledge concerning results of policies not implemented may never be available, there is usually no way to validate a model completely. Nevertheless, there are several steps that can be used to validate a model. These include a reasonableness test in which we determine that the overall model, as well as model subsystems, responds to inputs in a reasonable way, as determined by "knowledgeable" people. The model should also be valid according to statistical time series used to determine parameters within the model. Finally, the model should be epistemologically valid, in that the policy interpretations of the various model parameters, structure, and recommendations are consistent with ethical, professional, and moral standards of the group affected by the model.

Once a model has been constructed, it is often desirable to determine, in some best fashion, various policy parameters or controls that are subject to negotiation. The optimization or refinement-of-alternatives step is concerned with choosing parameters or controls to maximize or minimize a given performance index or criterion. Invariably, there are constraints that must be respected in seeking this extremum. As previously noted, the analysis step of systems engineering consists of systems analysis and modeling and optimization or refinement of alternatives and related methods that are appropriate in aiding effective judgment and choice.

There exist a number of methods for fine tuning, refinement, or optimization of individual specific alternative policies or systems. These are useful in determining the best (in terms of needs satisfaction) control settings or rules of operation in a well-defined, quantitatively describable system. A single scalar indicator of performance or desirability is typically needed. There are, however, approaches to multiple objective optimization that are based on welfare-type optimization concepts. It is these individually optimized policies or systems that are an input to the evaluation and decision-making effort in the interpretation step of systems engineering.

Among the many methods for optimization and refinement of alternatives are:

- *Mathematical programming,* which is used extensively for operations research and analysis practice, resource allocation under constraints, resolution of planning or scheduling problems, and similar applications. It is particularly useful when the best equilibrium or one-time setting has to be determined for a given policy or system.

- *Optimum systems control,* which addresses the problem of determining the best controls or actions when the system, the controls or actions, the constraints, and the performance index may change over time. A mathematical description of system change is necessary to use this approach. Optimum systems control is particularly suitable for refining controls or parameters in systems in which changes over time play an important part.

Application of the various refinement or optimization methods, like those described here, typically requires significant training and experience on the part of the systems analyst. Some of the many characteristics of analysis that are of importance for systemic efforts include the following:

1. Analysis methods are invaluable for understanding the impacts of proposed policy.
2. Analysis methods lead to consistent results if cognitive bias issues associated with expert forecasting and assessment methods are resolved.
3. Analysis methods may not necessarily lead to correct results since "formulation" may be flawed, perhaps by cognitive bias and value incoherence.

Unfortunately, however, large models and large optimization efforts are often expensive and difficult to understand and interpret. There are a number of possibilities for "paralysis through analysis" in the unwise use of systems analysis. On the other hand, models and associated analysis can help provide a framework for debate. It is important to note that small "back-of-the-envelope" models can be very useful. They have advantages that large models often lack, such as cost, simplicity, and ease of understanding and, therefore, explicability.

It is important to distinguish between analysis and interpretation in systems engineering efforts. Analysis cannot substitute or will generally be a foolish substitute for judgment, evaluation, and interpretation as exercised by a well-informed decision-maker. In some cases, refinement of individual alternative policies is not needed in the analysis step. But evaluation of alternatives is always needed, since, if there is but a single policy alternative, there really is no alternative at all. The option to do nothing at all must always be considered as a policy alternative. It is especially important to avoid a large number of cognitive biases, poor judgment heuristics, and value incoherence in the activities of evaluation and decision making. The efforts involved in evaluation and choice making interact strongly with the efforts in the other steps of the systems process, and these are also influenced by cognitive bias, judgment heuristics, and value incoherence. One of the fundamental tenets of the systems process is that making the complete issue resolution process as explicit as possible makes it easier to detect and connect these deficiencies than it is in holistic intuitive processes.

4.3   Information Processing by Humans and Organizations

After completion of the analysis step, we begin the evaluation and decision-making effort of interpretation. Decisions must typically be made and policies formulated, evaluated, and applied in an atmosphere of uncertainty. The outcome of any proposed policy is seldom known with certainty. One of the purposes of analysis is to reduce, to the extent possible, uncertainties associated with the outcomes of proposed policies. Most planning, design, and resource allocation issues will involve a large number of decision-makers who act according to their varied preferences. Often, these decision-makers will have diverse and conflicting data available to them and the decision situation will be quite fragmented. Furthermore, outcomes resulting from actions can often only be adequately characterized by a large number of incommensurable attributes. Explicit informed comparison of alternatives across these attributes by many stakeholders in an evaluation and choice-making process is typically most difficult.

As a consequence of this, people will often search for and use some form of a dominance structure to enable rejection of alternatives that are perceived to be dominated by one or more other alternatives. An alternative is said to be "dominated" by another alternative when the other alternative has attribute scores at least as large as those associated with the dominated alternative and at least one attribute score that is larger. However, biases have been shown to be systematic and prevalent in most unaided cognitive activities. Decisions and judgments are influenced by differential weights of information and by a variety of human information-processing deficiencies, such as base rates, representativeness, availability, adjustment, and anchoring. Often it is very difficult to disaggregate values of policy outcomes from causal relations determining these outcomes. Often correlation is used to infer causality. Wishful thinking and other forms of selective perception encourage us not to obtain potentially disconfirming information. The resulting confounding of values with facts can lead to great difficulties in discourse and related decision making.

It is especially important to avoid the large number of potential cognitive biases and flaws in the process of formulation, analysis, and interpretation for judgment and choice. These may well occur due to flaws in human information processing associated with the identification of problem elements, structuring of decision situations, and the probabilistic and utility assessment portions of the judgmental tasks of evaluation and decision making.

Among the cognitive biases and information-processing flaws that have been identified are several that affect information formulation or acquisition, information analysis, and interpretation. These and related material are described in Ref. 21 and the references contained therein. Among these biases, which are not independent, are the following:

1. *Adjustment and Anchoring.* Often a person finds that difficulty in problem solving is due not to the lack of data and information but rather to an excess of data and information. In such situations, the person often resorts to heuristics, which may reduce the mental efforts required to arrive at a solution. In using the anchoring and adjustment heuristic when confronted with a large number of data, the person selects a particular datum, such as the mean, as an initial or starting point or anchor and then adjusts that value improperly in order to incorporate the rest of these data, resulting in flawed information analysis.

2. *Availability.* The decision-maker uses only easily available information and ignores sources of significant but not easily available information. An event is believed to occur frequently, that is, with high probability, if it is easy to recall similar events.

3. *Base Rate*. The likelihood of occurrence of two events is often compared by contrasting the number of times the two events occur and ignoring the rate of occurrence of each event. This bias often arises when the decision-maker has concrete experience with one event but only statistical or abstract information on the other. Generally, abstract information will be ignored at the expense of concrete information. A base rate determined primarily from concrete information may be called a *causal base rate*, whereas that determined from abstract information is an *incidental base rate*. When information updates occur, this individuating information is often given much more weight than it deserves. It is much easier for the impact of individuating information to override incidental base rates than causal base rates.

4. *Conservatism*. The failure to revise estimates as much as they should be revised, based on receipt of new significant information, is known as *conservatism*. This is related to data saturation and regression effects biases.

5. *Data Presentation Context*. The impact of summarized data, for example, may be much greater than that of the same data presented in detailed, nonsummarized form. Also, different scales may be used to change the impact of the same data considerably.

6. *Data Saturation*. People often reach premature conclusions on the basis of too small a sample of information while ignoring the rest of the data, which is received later, or stopping acquisition of data prematurely.

7. *Desire for Self-Fulfilling Prophecies*. The decision-maker values a certain outcome, interpretation, or conclusion and acquires and analyzes only information that supports this conclusion. This is another form of selective perception.

8. *Ease of Recall*. Data that can easily be recalled or assessed will affect perception of the likelihood of similar events reoccurring. People typically weigh easily recalled data more in decision making than those data that cannot easily be recalled.

9. *Expectations*. People often remember and attach higher validity to information that confirms their previously held beliefs and expectations than they do to disconfirming information. Thus, the presence of large amounts of information makes it easier for one to selectively ignore disconfirming information such as to reach any conclusion and thereby prove anything that one desires to prove.

10. *Fact–Value Confusion*. Strongly held values may often be regarded and presented as facts. That type of information is sought that confirms or lends credibility to one's views and values. Information that contradicts one's views or values is ignored. This is related to wishful thinking in that both are forms of selective perception.

11. *Fundamental Attribution Error* (success/failure error). The decision-maker associates success with personal inherent ability and associates failure with poor luck in chance events. This is related to availability and representativeness.

12. *Habit*. Familiarity with a particular rule for solving a problem may result in reuse of the same procedure and selection of the same alternative when confronted with a similar type of problem and similar information. We choose an alternative because it has previously been acceptable for a perceived similar purpose or because of superstition.

13. *Hindsight*. People are often unable to think objectively if they receive information that an outcome has occurred and they are told to ignore this information. With hindsight, outcomes that have occurred seem to have been inevitable. We see relationships much more easily in hindsight than in foresight and find it easy to change our predictions after the fact to correspond to what we know has occurred.

14. *Illusion of Control*. A good outcome in a chance situation may well have resulted from a poor decision. The decision-maker may assume an unreasonable feeling of control over events.

15. *Illusion of Correlation.* This is a mistaken belief that two events covary when they do not covary.

16. *Law of Small Numbers.* People are insufficiently sensitive to quality of evidence. They often express greater confidence in predictions based on small samples of data with nondisconfirming evidence than in much larger samples with minor disconfirming evidence. Sample size and reliability often have little influence on confidence.

17. *Order Effects.* The order in which information is presented affects information retention in memory. Typically, the first piece of information presented (primacy effect) and the last presented (recency effect) assume undue importance in the mind of the decision-maker.

18. *Outcome-Irrelevant Learning System.* Use of an inferior processing or decision rule can lead to poor results that the decision-maker can believe are good because of inability to evaluate the impacts of the choices not selected and the hypotheses not tested.

19. *Representativeness.* When making inference from data, too much weight is given to results of small samples. As sample size is increased, the results of small samples are taken to be representative of the larger population. The "laws" of representativeness differ considerably from the laws of probability and violations of the conjunction rule $P(A \cap B) < P(A)$ are often observed.

20. *Selective Perceptions.* People often seek only information that confirms their views and values and disregard or ignore disconfirming evidence. Issues are structured on the basis of personal experience and wishful thinking. There are many illustrations of selective perception. One is "reading between the lines"—for example, to deny antecedent statements and, as a consequence, accept "if you don't promote me, I won't perform well" as following inferentially from "I will perform well if you promote me."

Of particular interest are circumstances under which these biases occur and their effects on activities such as the identification of requirements for a system or for planning and design. Through this, it may be possible to develop approaches that might result in debiasing or amelioration of the effects of cognitive bias. A number of studies have compared unaided expert performance with simple quantitative models for judgment and decision making. While there is controversy, most studies have shown that simple quantitative models perform better in human judgment and decision-making tasks, including information processing, than holistic expert performance in similar tasks. There are a number of prescriptions that might be given to encourage avoidance of possible cognitive biases and to debias those that do occur:

1. Sample information from a broad database and be especially careful to include databases that might contain disconfirming information.

2. Include sample size, confidence intervals, and other measures of information validity in addition to mean values.

3. Encourage use of models and quantitative aids to improve upon information analysis through proper aggregation of acquired information.

4. Avoid the hindsight bias by providing access to information at critical past times.

5. Encourage people to distinguish good and bad decisions from good and bad outcomes.

6. Encourage effective learning from experience. Encourage understanding of the decision situation and methods and rules used in practice to process information and make decisions so as to avoid outcome-irrelevant learning systems.

A definitive discussion of debiasing methods for hindsight and overconfidence is presented by Fischhoff, a definitive chapter in an excellent edited work.[22] He suggests identifying faulty judges, faulty tasks, and mismatches between judges and tasks. Strategies for each of these situations are given.

Not everyone agrees with the conclusions just reached about cognitive human information processing and inferential behavior. Several arguments have been advanced for a decidedly less pessimistic view of human inference and decision. Jonathan Cohen,[23,24] for example, argues that all of this research is based upon a conventional model for probabilistic reasoning, which Cohen calls the "Pascalian" probability calculus. He expresses the view that human behavior does not appear "biased" at all when it is viewed in terms of other equally appropriate schemes for probabilistic reasoning, such as his own "inductive probability" system. Cohen states that human irrationality can never be demonstrated in laboratory experiments, especially experiments based upon the use of what he calls "probabilistic conundrums."

There are a number of other contrasting viewpoints as well. In their definitive study of behavioral and normative decision analysis, von Winterfeld and Edwards[25] refer to these information-processing biases as "cognitive illusions." They indicate that there are four fundamental elements to every cognitive illusion:

1. A *formal operational* rule that determines *the* correct solution to an intellectual question

2. An intellectual question that almost invariably includes all of the information required to obtain the correct answer through use of the formal rule

3. A human judgment, generally made without the use of these analytical tools, that is intended to answer the posed question

4. A systematic and generally large and unforgivable discrepancy between the correct answer and the human judgment

They also, as does Phillips,[26] describe some of the ways in which subjects might have been put at a disadvantage in this research on cognitive heuristics and information-processing biases. Much of this centers around the fact that the subjects have little experiential familiarity with the tasks that they are asked to perform. It is suggested that as inference tasks are decomposed and better structured, it is very likely that a large number of information-processing biases will disappear. Thus, concern should be expressed about the structuring of inference and decision problems and the learning that is reflected by revisions of problem structure in the light of new knowledge. In any case, there is strong evidence that humans are very strongly motivated to understand, to cope with, and to improve themselves and the environment in which they function. One of the purposes of systems engineering is to aid in this effort.

4.4   Interpretation

While there are a number of fundamental limitations to systems engineering efforts to assist in bettering the quality of human judgment, choice, decisions, and designs, there are also a number of desirable activities. These have resulted in several important holistic approaches that provide formal assistance in the evaluation and interpretation of the impacts of alternatives, including the following:

- *Decision analysis,* which is a very general approach to option evaluation and selection, involves identification of action alternatives and possible consequence identification of the probabilities of these consequences, identification of the valuation placed by the decision-maker on these consequences, computation of the expected utilities of the consequences, and aggregating or summarizing these values for all consequences of each action. In doing this, we obtain an expected utility evaluation of each alternative act. The one with the highest value is the most preferred action or option. Figure 7 presents some of the salient features involved in the decision analysis of a simplified problem.

- *Multiattribute utility theory* (MAUT) has been designed to facilitate comparison and ranking of alternatives with many attributes or characteristics. The relevant attributes are identified and structured and a weight or relative utility is assigned by the decision-maker to each basic attribute. The attribute measurements for each alternative are used to compute an overall worth or utility for each attribute. Multiattribute utility theory allows for explicit recognition and incorporation of the decision-maker's attitude toward risk in the utility computations. There are a number of variants of MAUT; many of them are simpler, more straightforward processes in which risk and uncertainty considerations are not taken into account. The method is very helpful to the decision-maker in making values and preferences explicit and in making decisions consistent with those values. The tree structure of Fig. 6 also indicates some salient features of the MAUT approach for the particular case where there are no risks or uncertainties involved in the decision situation. We simply need to associate importance weights with the attributes and then provide scores for each alternative on each of the lowest level attributes.

- *Policy capture* (or social judgment theory) has also been designed to assist decision-makers in making their values explicit and their decisions consistent with their values. In policy capture, the decision-maker is asked to rank order a set of alternatives in a gestalt or holistic fashion. Alternative attributes and associated attribute measures are then determined by elicitation from the decision-maker. A mathematical procedure involving regression analysis is used to determine that relative importance weight of each attribute that will lead to a ranking as specified by the decision-maker. The result is fed back to the decision-maker, who, typically, will express the view that his or her values are different. In an iterative learning process, preference weights and/or overall rankings are modified until the decision-maker is satisfied with both the weights and the overall alternative ranking.

There are many advantages to formal interpretation efforts in systems engineering, including the following:

1. Developing decision situation models to aid in making the choice-making effort explicit helps one both to identify and to overcome the inadequacies of implicit mental models.

2. The decision situation model elements, especially the attributes of the outcomes of alternative actions, remind us of information we need to obtain about alternatives and their outcomes.

3. We avoid such poor information-processing heuristics as evaluating one alternative on attribute A and another on attribute B and then comparing them without any basis for compensatory trade-offs across the different attributes.

4. We improve our ability to process information and, consequently, reduce the possibilities for cognitive bias.

5. We can aggregate facts and values in a prescribed systemic fashion rather than by adopting an agenda-dependent or intellect-limited approach.

6. We enhance brokerage, facilitation, and communication abilities among stakeholders to complex technological and social issues.

There is a plethora of literature describing the decision assessment or decision-making part of the interpretation step of systems engineering. In addition to the discussions in Refs. 1, 2, and 3, excellent discussions are to be found in Refs. 27–29.

4.5   Central Role of Information in Systems Engineering

Information is certainly a key ingredient supporting quality decisions; all systems engineering efforts are based on appropriate acquisition and use of information. There are three basic types of information, which are fundamentally related to the three-step framework of systems engineering:

1. Formulation information
    a. Information concerning the problem and associated needs, constraints, and alterables
    b. Information concerning the value system
    c. Information concerning possible option alternatives
    d. Information concerning possible future alternative outcomes, states, and scenarios
2. Analysis information
    a. Information concerning probabilities of future scenarios
    b. Information concerning impacts of alternative options
    c. Information concerning the importance of various criteria or attributes
3. Interpretation information
    a. Information concerning evaluation and aggregation of facts and values
    b. Information concerning implementation

We see that useful and appropriate formulation, analysis, and interpretation of information is one of the most important and vital tasks in systems engineering efforts, since it is the efficient processing of information by the decision-maker that produces effective decisions. A useful definition of information for our purposes is that it is data of value for decision making. The decision-making process is influenced by many contingency and environmental influences. A purpose of the management technology that is systems engineering is to provide systemic support processes to further enhance efficient decision-making activities.

After completion of evaluation and decision-making efforts, it is generally necessary to become involved in planning for action to implement the chosen alternative option or the next phase of a systems engineering effort. More often than not, it will be necessary to iterate the steps of systems engineering several times to obtain satisfactory closure upon one or more appropriate action alternatives. Planning for action also leads to questions concerning resource allocation, schedules, and management plans. There are, of course, a number of methods from systems science and operations research that support determination of schedules and implementation plans. Each of the steps is needed, with different focus and emphasis, at each phase of a systems effort. These phases depend on the particular effort under consideration but will typically include such phases as policy and program planning, project planning, and system development.

There are a number of complexities affecting "rational" planning, design, and decision making. We must cope with these in the design of effective systemic processes. The majority of these complexities involve systems management considerations. Many have indicated that the capacity of the human mind for formulating, analyzing, and interpreting complex large-scale issues is very small compared with the size and scope of the issues whose resolution is required for objective, substantive, and procedurally rational behavior. Among the limits to rationality are the fact that we can formulate, analyze, and interpret only a restricted amount of information; can devote only a limited amount of time to decision making; and can become involved in many more activities than we can effectively consider and cope with simultaneously. We must therefore necessarily focus attention only on a portion of the major competing concerns. The direct effect of these is the presence of cognitive bias in information acquisition and processing and the use of cognitive heuristics for evaluation of alternatives.

Although in many cases these cognitive heuristics will be flawed, this is not necessarily so. One of the hoped-for results of the use of systems engineering approaches is the development of effective and efficient heuristics for enhanced judgment and choice through effective decision support systems.[30]

There are many cognitive biases prevalent in most information acquisition activities. The use of cognitive heuristics and decision rules is also prevalent and necessary to enable us to cope with the many demands on our time. One such heuristic is satisficing or searching for a solution that is "good enough." This may be quite appropriate if the stakes are small. In general, the quality of cognitive heuristics will be task dependent, and often the use of heuristics for evaluation will be both reasonable and appropriate. Rational decision making requires time, skill, wisdom, and other resources. It must, therefore, be reserved for the more important decisions. A goal of systems engineering is to enhance information acquisition, processing, and evaluation so that efficient and effective use of information is made in a process that is appropriate to the cognitive styles and time constraints of management.

5   SYSTEM DESIGN

This section discusses several topics relevant to the design and evaluation of systems. In order to develop our design methodology, we first discuss the purpose and objectives of systems engineering and systems design. Development of performance objectives for quality systems is important, since evaluation of the logical soundness and performance of a system can be determined by measuring achievement of these objectives with and without the system. A discussion of general objectives for quality system design is followed by a presentation of a five-phase design methodology for system design. The section continues with leadership and training requirements for use of the resulting system and the impact of these requirements upon design considerations. While it is doubtless true that not every design process should, could, or would precisely follow each component in the detailed phases outlined here, we feel that this approach to systems design is sufficiently robust and generic that it can be used as a normative model of the design process and as a guide to the structuring and implementation of appropriate systems evaluation practices.

5.1   Purposes of Systems Design

Contemporary issues that may result in the need for systems design are invariably complex. They typically involve a number of competing concerns, contain much uncertainty, and require expertise from a number of disparate disciplines for resolution. Thus, it is not surprising that intuitive and affective judgments, often based on incomplete data, form the usual basis used for contemporary design and associated choice making. At the other extreme of the cognitive inquiry scale are the highly analytical, theoretical, and experimental approaches of the mathematical, physical, and engineering sciences. When intuitive judgment is appropriately skill based, it is generally effective and appropriate. One of the major challenges in system design engineering is to develop processes that are appropriate for a variety of process users, some of whom may approach the design issue from a skill-based perspective, some from a rule-based perspective, and some from a knowledge-based perspective.

A central purpose of systems engineering and management is to incorporate appropriate methods and metrics into a methodology for problem solving, or a systems engineering process or life cycle, such that, when it is associated with human judgment through systems management, it results in a high-quality systems design procedure. By high-quality design, we mean one that will, with high probability, produce a system that is effective and efficient and trustworthy.

A systems design procedure must be specifically related to the operational environment for which the final system is intended. Control group testing and evaluation may serve many useful purposes with respect to determination of many aspects of algorithmic and behavioral efficacy of a system. Ultimate effectiveness involves user acceptability of the resulting system, and evaluation of this process effectiveness will often involve testing and evaluation in the environment, or at least a closely simulated model of the environment, in which the system would be potentially deployed.

The potential benefits of systems engineering approaches to design can be interpreted as attributes or criteria for evaluation of the design approach itself. Achievement of many of these attributes may often not be experimentally measured except by inference, anecdotal, or testimonial and case study evidence taken in the operational environment for which the system is designed. Explicit evaluation of attribute achievement is a very important part of the overall systemic design process. This section describes the following:

1. A methodological framework for the design of systems, such as planning and decision support systems
2. An evaluation methodology that may be incorporated with or used independently of the design framework

A number of characteristics of effective systems efforts can be identified. These form the basis for determining the attributes of systems and systemic design procedures. Some of these attributes will be more important for a given environment than others. Effective design must typically include an operational evaluation component that will consider the strong interaction between the system and the situational issues that led to the systems design requirement. This operational evaluation is needed in order to determine whether a product system or a service consisting of humans and machines:

1. Is logically sound
2. Is matched to the operational and organizational situation and environment extant
3. Supports a variety of cognitive skills, styles, and knowledge of the humans who must use the system
4. Assists users of the system to develop and use their own cognitive skills, styles, and knowledge
5. Is sufficiently flexible to allow use and adaptation by users with differing cognitive skills, styles, and knowledge
6. Encourages more effective solution of unstructured and unfamiliar issues, allowing the application of job-specific experiences in a way compatible with various acceptability constraints
7. Promotes effective long-term management

It is certainly possible that the product, or system, that results from a systems engineering effort may be used as a process or life cycle in some other application. Thus, what we have to say here refers both to the design of products and to the design of processes.

5.2   Operational Environments and Decision Situation Models

In order to develop robust scenarios of planning and design situations in various operational environments and specific instruments for evaluation, we first identify a mathematical and situational taxonomy:

- Algorithmic constructs used in systemic design
- Performance objectives for quality design
- Operational environments for design

One of the initial goals in systems design engineering is to obtain the conceptual specifications for a product such that development of the system will be based on customer or client information, objectives, and existing situations and needs. An aid to the process of design should assist in or support the evaluation of alternatives relative to some criteria. It is generally necessary that design information be described in ways that lead to effective structuring of the design problem. Of equal importance is the need to be aware of the role of the affective in design tasks such as to support different cognitive styles and needs, which vary from formal knowledge-based to rule-based to skill-based behavior.[31] We desire to design efficient and effective physical systems, problem-solving service systems, and interfaces between the two. This section is concerned with each of these.

Not all of the performance objectives for quality systems engineering will be or need be fully attained in all design instances, but it is generally true that the quality of a system or of a systems design process necessarily improves as more and more of these objectives are attained. Measures of quality or effectiveness of the resulting system, and therefore systems design process quality or effectiveness, may be obtained by assessing the degree of achievement of these performance criteria by the resulting system, generally in an operational environment. In this way, an evaluation of the effectiveness of a design decision support system may be conducted.

A taxonomy based on operational environments is necessary to describe particular situation models through which design decision support may be achieved. We are able to describe a large number of situations using elements or features of the three-component taxonomy described earlier. With these, we are able to evolve test instruments to establish quantitative and qualitative evaluations of a design support system within an operational environment. The structural and functional properties of such a system, or of the design process itself, must be described in order that a purposeful evaluation can be accomplished. This purposeful evaluation of a systemic process is obtained by embedding the process into specific operational planning, design, or decision situations. Thus, an evaluation effort also allows iteration and feedback to ultimately improve the overall systems design process. The evaluation methodology to be described is useful, therefore, as a part or phase of the design process. Also, it is useful, in and of itself, to evaluate and prioritize a set of systemic aids for planning, design, and decision support. It is also useful for evaluation of resulting system designs and operational systems providing a methodological framework both for the design and evaluation of physical systems and for systems that assist in the planning and design of systems.

5.3   Development of Aids for Systems Design Process

This section describes five important phases in the development of systems and systemic aids for the design process. These phases serve as a guide not only for the sound design and development of systems and systemic aids for design decision support but also for their evaluation and ultimate operational deployment:

- Requirements specification
- Preliminary conceptual design and architecting
- Detailed design, integration, testing, and implementation
- Evaluation (and potential modification)
- Operational deployment

These five phases are applicable to design in general. Although the five phases will be described as if they are to be sequenced in a chronological fashion, sound design practice will generally necessitate iteration and feedback from a given phase to earlier phases.

Requirements Specification Phase

The requirements specification phase has as its goal the detailed definition of those needs, activities, and objectives to be fulfilled or achieved by the system or process that is to result from the system design effort. Furthermore, the effort in this phase should result in a description of preliminary conceptual design considerations appropriate for the next phase. This must be accomplished in order to translate operational deployment needs, activities, and objectives into requirements specifications if, for example, that is the phase of the systems engineering design effort under consideration.

Among the many objectives of the requirements specifications phase of systems engineering are the following:

1. To define the problem to be solved, or range of problems to be solved, or issue to be resolved or ameliorated, including identification of needs, constraints, alterables, and stakeholder groups associated with operational deployment of the system or the systemic process

2. To determine objectives for operational system or the operational aids for planning, design, and decision support

3. To obtain commitment for prototype design of a system or systemic process aid from user group and management

4. To search the literature and seek other expert opinions concerning the approach that is most appropriate for the particular situation extant

5. To determine the estimated frequency and extent of need for the system or the systemic process

6. To determine the possible need to modify the system or the systemic process to meet changed requirements

7. To determine the degree and type of accuracy expected from the system or systemic process

8. To estimate expected effectiveness improvement or benefits due to the use of the system or systemic process

9. To estimate the expected costs of using the system or systemic process, including design and development costs, operational costs, and maintenance costs

10. To determine typical planning horizons and periods to which the system or systemic process must be responsive

11. To determine the extent of tolerable operational environment alteration due to use of the system or systemic process

12. To determine what particular planning, design, or decision process appears best

13. To determine the most appropriate roles for the system or systemic process to perform within the context of the planning, design, or decision situation and operational environment under consideration

14. To estimate potential leadership requirements for use of the final system itself

15. To estimate user group training requirements

16. To estimate the qualifications required of the design team

17. To determine preliminary operational evaluation plans and criteria

18. To determine political acceptability and institutional constraints affecting use of an aided support process and those of the system itself

19. To document analytical and behavioral specifications to be satisfied by the support process and the system itself

20. To determine the extent to which the user group can require changes during and after system development

21. To determine potential requirements for contractor availability after completion of development and operational tests for additional needs determined by the user group, perhaps as a result of the evaluation effort

22. To develop requirements specifications for prototype design of a support process and the operational system itself

As a result of this phase, to which the four issue requirements identification approaches of Section 4.1 are fully applicable, there should exist a clear definition of typical planning, design, and decision issues, or problems requiring support, and other requirements specifications, so that it is possible to make a decision whether to undertake preliminary conceptual design. If the result of this phase indicates that the user group or client needs can potentially be satisfied in a cost-effective manner, by a systemic process aid, for example, then documentation should be prepared concerning detailed specifications for the next phase, preliminary conceptual design, and initial specifications for the last three phases of effort. A design team is then selected to implement the next phase of the system life cycle. This discussion emphasizes the inherently coupled nature of these phases of the system life cycle and illustrates why it is not reasonable to consider the phases as if they are uncoupled.

Preliminary Conceptual Design and Architecting Phase

The preliminary conceptual design and architecting phase includes specification of the mathematical and behavioral content and associated algorithms for the system or process that should ultimately result from the effort as well as the possible need for computer support to implement these. The primary goal of this phase is to develop conceptualization of a prototype system or process in response to the requirements specifications developed in the previous phase. Preliminary design according to the requirements specifications should be achieved. Objectives for preliminary conceptual design include the following:

1. To search the literature and seek other expert opinion concerning the particular approach to design and implementation that is likely to be most responsive to requirements specifications

2. To determine the specific analytic algorithms to be implemented by the system or process

3. To determine the specific behavioral situation and operational environment in which the system or process is to operate

4. To determine the specific leadership requirements for use of the system in the operational environment extant

5. To determine specific hardware and software implementation requirements, including type of computer programming language and input devices

6. To determine specific information input requirements for the system or process

7. To determine the specific type of output and interpretation of the output to be obtained from the system or process that will result from the design procedure

8. To reevaluate objectives obtained in the previous phase, to provide documentation of minor changes, and to conduct an extensive reexamination of the effort if major changes are detected that could result in major modification and iteration through requirements specification or even termination of effort

9. To develop a preliminary conceptual design of, or architecture for, prototype aid that is responsive to the requirements specifications

The expected product of this phase is a set of detailed design and testing specifications that, if followed, should result in a usable prototype system or process. User group confidence that an ultimately useful product should result from detailed design should be above some threshold or the entire design effort should be redone. Another product of this phase is a refined set of specifications for the evaluation and operational deployment phases.

If the result of this phase is successful, the detailed design, testing, and implementation phase is begun. This phase is based on the products of the preliminary conceptual design phase, which should result in a common understanding among all interested parties about the planning and decision support design effort concerning the following:

1. Who the user group or responsive stakeholder is
2. The structure of the operational environment in which plans, designs, and decisions are made
3. What constitutes a plan, a design, or a decision
4. How plans, designs, and decisions are made without the process or system and how they will be made with it
5. What implementation, political acceptability, and institutional constraints affect the use of the system or process
6. What specific analysis algorithms will be used in the system or process and how these algorithms will be interconnected to form the methodological construction of the system or process

Detailed Design, Integration, Testing, and Implementation Phase

In the third phase of design, a system or process that is presumably useful in the operational environment is produced. Among the objectives to be attained in this phase are the following:

1. To obtain and design appropriate physical facilities (physical hardware, computer hardware, output device, room, etc.)
2. To prepare computer software
3. To document computer software
4. To integrate these effectively
5. To prepare a user's guide to the system and the process in which the system is embedded
6. To prepare a leader's guide for the system and the associated process
7. To conduct control group or operational (simulated operational) tests of the system and make minor changes in the aid as a result of the tests
8. To complete detailed design and associated testing of a prototype system based on the results of the previous phase
9. To implement the prototype system in the operational environment as a process

The products of this phase are detailed guides to use of the system as well as, of course, the prototype system itself. It is very important that the user's guide and the leader's guide address, at levels appropriate for the parties interested in the effort, the way in which the performance objectives identified in Section 5.3 are satisfied. The description of system usage and leadership topics should be addressed in terms of the analytic and behavioral constructs of the system and the resulting process as well as in terms of operational environment situation concerns. These concerns include the following:

1. Frequency of occurrence of need for the system or process
2. Time available from recognition of need for a plan, design, or decision to identification of an appropriate plan, design, or decision

3. Time available from determination of an appropriate plan, design, or decision to implementation of the plan, design, or decision

4. Value of time

5. Possible interactions with the plans, designs, or decisions of others

6. Information base characteristics

7. Organizational structure

8. Top-management support for the resulting system or process

It is especially important that the portion of this phase that concerns implementation of the prototype system specifically address important questions concerning cognitive style and organizational differences among parties of interest and institutions associated with the design effort. Stakeholder understanding of environmental changes and side effects that will result from use of the system is critical for ultimate success. This need must be addressed. Evaluation specification and operational deployment specifications will be further refined as a result of this phase.

Evaluation Phase

Evaluation of the system in accordance with evaluation criteria, determined in the requirements specification phase and modified in the subsequent two design phases, is accomplished in the fourth phase of systems development. This evaluation should always be assisted to the extent possible by all parties of interest to the systems design effort and the resultant systemic process. The evaluation effort must be adapted to other phases of the design effort so that it becomes an integral functional part of the overall design process. As noted, evaluation may well be an effort distinct from design that is used to determine usefulness or appropriateness for specified purposes of one or more previously designed systems. Among the objectives of system or process evaluation are the following:

1. To identify a methodology for evaluation

2. To identify criteria on which the success of the system or process may be judged

3. To determine effectiveness of the system in terms of success criteria

4. To determine an appropriate balance between the operational environment evaluation and the control group evaluation

5. To determine performance objective achievement of the system

6. To determine behavioral or human factor effectiveness of the system

7. To determine the most useful strategy for employment of the existing system

8. To determine user group acceptance of the system

9. To suggest refinements in existing systems for greater effectiveness of the process in which the new system has been embedded

10. To evaluate the effectiveness of the system or process

These objectives are obtained from a critical evaluation issue specification or evaluation need specification, which is the first, or problem definition, step of the evaluation methodology. Generally, the critical issues for evaluation are minor adaptations of the elements that are present in the requirements specifications step of the design process outlined in the previous section. A set of specific evaluation test requirements and tests is evolved from these objectives and needs. These must be such that each objective measure and critical evaluation issue component can be determined from at least one evaluation test instrument.

If it is determined that the system and the resulting process support cannot meet user needs, the systems design process iterates to an earlier phase and development continues. An important by-product of evaluation is the determination of ultimate performance limitations and the establishment of a protocol and procedure for use of the system that results in maximum user group satisfaction. A report is written concerning results of the evaluation process, especially those factors relating to user group satisfaction with the designed system. The evaluation process should result in suggestions for improvement in design and in better methodologies for future evaluations.

Section 5.6 will present additional details of the methodologies framework for evaluation. These have applicability to cases where evaluation is a separate and independent effort as well as cases where it is one of the phases of the design process.

Operational Deployment Phase

The last phase of design concerns operational deployment and final implementation. This must be accomplished in such a way that all user groups obtain adequate instructions in use of the system and complete operating and maintenance documentation and instructions. Specific objectives for the operational deployment phase of the system design effort are as follows:

1. To enhance operational deployment
2. To accomplish final design of the system
3. To provide for continuous monitoring of postimplementation effectiveness of the system and the process into which the system is embedded
4. To provide for redesign of the system as indicated by effectiveness monitoring
5. To provide proper training and leadership for successful continued operational use of the system
6. To identify barriers to successful implementation of the final design product
7. To provide for "maintenance" of the system

5.4 Leadership Requirements for Design

The actual use, as contrasted with potential usefulness, of a system is directly dependent on the value that the user group of stakeholders associates with use of the system and the resulting process in an operational environment. This in turn is dependent, in part, on how well the system satisfies performance objectives and on how well it is able to cope with one or more of the pathologies or pitfalls of planning, design, and/or decision making under potentially stressful operational environment conditions.

Quality planning, design, and decision support are dependent on the ability to obtain relatively complete identification of pertinent factors that influence plans, designs, and decisions. The careful, comprehensive formulation of issues and associated requirements for issue resolution will lead to identification of pertinent critical factors for system design. These factors are ideally illuminated in a relatively easy-to-understand fashion that facilitates the interpretation necessary to evaluate and subsequently select plans, designs, and decisions for implementation. Success in this is, however, strongly dependent on adroitness in use of the system. It is generally not fully meaningful to talk only of an algorithm or even a complete system—which is, typically, a piece of hardware and software but which may well be a carefully written set of protocols and procedures—as useful by itself. It is meaningful to talk of a particular systemic process as being useful. This process involves the interaction of a methodology with systems management at the cognitive process or human judgment level. A systemic process depends on the system, the operational environment, and leadership associated with use of the system. A process involves design integration of a methodology with the behavioral concerns of human cognitive judgment in an operational environment.

Operational evaluation of a systemic process that involves human interaction, such as an integrated manufacturing complex, appears to be the only realistic way to extract truly meaningful information concerning process effectiveness of a given system design. This must necessarily include leadership and training requirements to use the system. There are necessary trade-offs associated with leadership and training for using a system and these are addressed in operational evaluation.

5.5   System Evaluation

Previous sections have described a framework for a general system design procedure. They have indicated the role of evaluation in this process. Successful evaluation, especially operational evaluation, is strongly dependent on explicit development of a plan for evaluation developed prior to, and perhaps modified and improved during the course of, an actual evaluation. This section will concern itself with development of a methodological framework for system evaluation, especially for operational evaluation of systemic processes for planning, design, and decision support.

Evaluation Methodology and Evaluation Criteria
Objectives for evaluation of a system concern the following:

1. Identification of a methodology for operational evaluation
2. Establishing criteria on which the success of the system may be judged
3. Determining the effectiveness of the support in terms of these criteria
4. Determining the most useful strategy for employment of an existing system and potential improvements such that effectiveness of the newly implemented system and the overall process might be improved

Figure 8 illustrates a partial intent structure or objectives tree, which contributes to system evaluation. The lowest level objectives contribute to satisfaction of the 10 performance objectives for systems engineering and systems design outlined in Section 3. These lowest level elements form pertinent criteria for the operational system evaluation. They concern the algorithmic effectiveness or performance objective achievement of the system, the behavioral or human factor effectiveness of the system in the operational environment, and the system efficacy. Each of these three elements become top-level criteria or attributes and each should be evaluated to determine evaluation of the system itself.

Subcriteria that support the three lowest level criteria of Fig. 8 may be identified. These are dependent on the requirements identified for the specific system that has been designed.

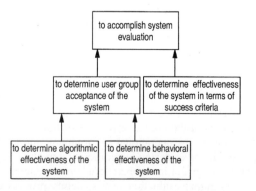

Figure 8   Objectives tree for evaluation of decision support system.

Attainment of each of these criteria by the system may be measured by observation of the system within the operational environment and by test instruments and surveys of user groups involved with the operational system and process.

Algorithmic Effectiveness of Performance Objectives Achievement Evaluation

A number of performance objectives can be cited that, if achieved, should lead to a quality system. Achievement of these objectives is measured by logical soundness of the operational system and process; improved system quality as a result of using the system; and improvements in the way an overall process functions, compared to the way it typically functions without the system or with an alternative system.

Behavioral or Human Factors Evaluation

A system may be well structured algorithmically in the sense of achieving a high degree of satisfaction of the performance objectives, yet the process incorporating the system may seriously violate behavioral and human factor sensibilities. This will typically result in misuse or underuse. There are many cases where technically innovative systems have failed to achieve broad-scope objectives because of human factor failures. Strongly influencing the acceptability of system implementation in operational settings are such factors as organizational slack; natural human resistance to change; and the present sophistication, attitude, and past experience of the user group and its management with similar systems and processes. Behavioral or human factor evaluation criteria used to evaluate performance include political acceptability, institutional constraint satisfaction, implementability evaluation, human workload evaluation, management procedural change evaluation, and side-effect evaluation.

Efficacy Evaluation

Two of the three first-level evaluation criteria concern algorithmic effectiveness or performance objective achievement and behavioral or human factors effectiveness. It is necessary for a system to be effective in each of these for it to be potentially capable of truly aiding in terms of improving process quality and being acceptable for implementation in the operational environment for which it was designed. There are a number of criteria or attributes related to usefulness, service support, or efficacy to which a system must be responsive. Thus, evaluation of the efficacy of a system and the associated process are important in determining the service support value of the process. There are seven attributes of efficacy:

1. *Time Requirements*. The time requirements to use a system form an important service support criterion. If a system is potentially capable of excellent results but the results can only be obtained after critical deadlines have passed, the overall process must be given a low rating with respect to a time responsiveness criterion.

2. *Leadership and Training*. Leadership and training requirements for use of a system are important design considerations. It is important that there be an evaluation component directed at assessing leadership and training needs and trade-offs associated with the use of a system.

3. *Communication Accomplishments*. Effective communication is important for two reasons. (1) Implementation action is often accomplished at a different hierarchical level, and therefore by a different set of actors, than the hierarchical level at which selection of alternative plans, designs, or decisions was made. Implementation action agents often behave poorly when an action alternative is selected that they regard as threatening or arbitrary, either personally or professionally, on an individual or a group basis. Widened perspectives of a situation are made possible by effective communication.

Enhanced understanding will often lead to commitment to successful action implementation as contrasted with unconscious or conscious efforts to subvert implementation action. (2) Recordkeeping and retrospective improvements to systems and processes are enhanced by the availability of well-documented constructions of planning and decision situations and communicable explanations of the rationale for the results of using the system.

4. *Educational Accomplishments.* There may exist values to a system other than those directly associated with improvement in process quality. The participating group may, for example, learn a considerable amount about the issues for which a system was constructed. The possibility of enhanced ability and learning with respect to the issues for which the system was constructed should be evaluated.

5. *Documentation.* The value of the service support provided by a system will be dependent on the quality of the user's guide and its usefulness to potential users of the system.

6. *Reliability and Maintainability.* To be operationally useful, a planning-and-decision-support system must be, and be perceived by potential users to be, reliable and maintainable.

7. *Convenience of Access.* A system should be readily available and convenient to access or usage will potentially suffer. While these last three service support measures are not of special significance with respect to justification of the need for a system, they may be important in determining operational usage and, therefore, operational effectiveness of a system and the associated process.

5.6   Evaluation Test Instruments

Several special evaluation test instruments to satisfy test requirements and measure achievement of the evaluation criteria will generally need to be developed. These include investigations of effectiveness in terms of performance objective attainment; selection of appropriate scenarios that affect use of the system, use of the system subject to these scenarios by a test group, and completion of evaluation questionnaires; and questionnaires and interviews with operational users of the system.

Every effort must be made to ensure, to the extent possible, that evaluation test results will be credible and valid. Intentional redundancy should be provided to allow correlation of results obtained from the test instruments to ensure maximum supportability and reliability of the facts and opinions to be obtained from test procedures.

The evaluation team should take advantage of every opportunity to observe use of the system within the operational environment. Evaluation of personnel reactions to the aid should be based on observations, designed to be responsive to critical evaluation issues, and the response of operational environment personnel to test questionnaires. When any of a number of constraints make it difficult to obtain real-time operational environment observation, experiential and anecdotal information becomes of increased value. Also, retrospective evaluation of use of a system is definitely possible and desirable if sufficiently documented records of past usage of an aided process are available.

Many other effectiveness questions will likely arise as an evaluation proceeds. Questions specific to a given evaluation are determined after study of the particular situation and the system being evaluated. It is, however, important to have an initial set of questions to guide the evaluation investigation and a purpose of this section to provide a framework for accomplishing this.

One of the important concerns in evaluation is that of those parts of the efficacy evaluation that deal with various "abilities" of a system. These include producibility, reliability,

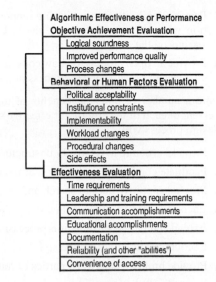

Figure 9 Attribute tree and criteria for evaluation of decision support system for design and other end uses.

maintainability, and marketability. Figure 9 presents a listing of attributes that may be used to "score" the performance of systems on relevant effectiveness criteria.

6 CONCLUSIONS

In this chapter, we have discussed salient aspects concerning the systems engineering of large and complex systems. We have been concerned especially with systems design engineering and associated information-processing and analysis efforts. To this end, we suggested a process for the design and evaluation of systems and how we might go about fielding a design decision support system. There are a number of effectiveness attributes or aspects of effective systems. Design of an effective large-scale system necessarily involves integration of operational environment concerns involving human behavior and judgment with mechanistic and physical science concerns. An effective systemic design process should:

1. Allow a thorough and carefully conducted requirements specification effort to determine and specify needs of stakeholders prior to conceptual design of a system process to accomplish the desired task

2. Be capable of dealing with both quantitative and qualitative criteria representing costs and effectiveness from their economic, social, environmental, and other perspectives

3. Be capable of minimizing opportunities for cognitive bias and provide debiasing procedures for those biases that occur

4. Allow separation of opinions and facts from values, and separation of ends from means, or values from alternative acts

5. Provide an objective communicable framework that allows identification, formulation, and display of the structure of the issue under consideration as well as the rationale of the choice process

6. Allow for considerations of trade-offs among conflicting and incommensurate criteria

7. Provide flexibility and monitoring support to allow design process evaluation rule selection with due consideration to the task structure and operational environment constraints on the decision-maker

8. Provide an open process to allow consideration of new criteria and alternatives as values change and broad-scope awareness of issues grows

There are a number of potential benefits of the systems approach that should follow high achievement of each of the criteria for effective systems design processes. An appropriate systems design process will:

1. Provide structure to relatively unstructured issues

2. Facilitate conceptual formulation of issues

3. Provide cognitive cues to search and discovery

4. Encourage parsimonious collection, organization, and utilization of relevant data

5. Extend and debias information-processing abilities

6. Encourage vigilant cognitive style

7. Provide brokerage between parties of interest

There are many imperfections and limits to processes designed using the methodologies from what we know as systems engineering and systems analysis, design, and integration.[32] Some of these have been documented in this chapter. Others are documented in the references provided here and in a recent handbook of systems engineering and management.[33] But what are the alternatives to appropriate systemic processes for the resolution of issues associated with the design of complex large-scale systems and are not the fundamental limitations to these alternatives even greater?

REFERENCES

1. A. P. Sage, *Systems Engineering*, Wiley, New York, 1992.

2. A. P. Sage, *Systems Management for Information Technology and Software Engineering*, Wiley, New York, 1995.

3. A. P. Sage, *Methodology for Large Scale Systems*, McGraw-Hill, New York, 1977.

4. J. E. Armstrong and A. P. Sage, *Introduction to Systems Engineering*, Wiley, New York, 2000.

5. A. D. Hall, *A Methodology for Systems Engineering*, Van Nostrand, New York, 1962.

6. A. D. Hall, "A Three Dimensional Morphology of Systems Engineering," *IEEE Trans. Syst. Sci. Cybernet.*, **5**(2), 156–160, April 1969.

7. E. Rechtin, *Systems Architecting: Creating and Building Complex Systems*, Prentice-Hall, Englewood Cliffs, NJ, 1991.

8. E. Rechtin, "Foundations of Systems Architecting," *Syst. Eng. J. Natl. Council Syst. Eng.*, **1**(1), 35–42, July/September 1994.

9. W. R. Beam, *Systems Engineering: Architecture and Design*, McGraw-Hill, New York, 1990.

10. D. N. Chorfas, *Systems Architecture and Systems Design*, McGraw-Hill, New York, 1989.

11. A. P. Sage and J. D. Palmer, *Software Systems Engineering*, Wiley, New York, 1990.

12. F. Harary, R. Z. Norman, and D. Cartwright, *Structural Models: An Introduction to the Theory of Directed Graphs*, Wiley, New York, 1965.

13. J. N. Warfield, *Societal Systems: Planning, Policy, and Complexity*, Wiley, New York, 1976.

14. D. V. Steward, *Systems Analysis and Management: Structure, Strategy, and Design*, Petrocelli, New York, 1981.

15. G. G. Lendaris, "Structural Modeling: A Tutorial Guide," *IEEE Trans. Syst. Man Cybernet.*, SMC 10, 807–840, 1980.

16. C. Eden, S. Jones, and D. Sims, *Messing about in Problems*, Pergamon, Oxford, 1983.

17. F. M. Roberts, *Discrete Mathematical Models*, Prentice-Hall, Englewood Cliffs, NJ, 1976.

18. A. M. Geoffrion, "An Introduction to Structured Modeling," *Manag. Sci.*, **33**, 547–588, 1987.

19. A. M. Geoffrion, "The Formal Aspects of Structured Modeling," *Oper. Res.*, **37**(1), 30–51, January 1989.

20. A. M. Geoffrion, "Computer Based Modeling Environments," *Eur. J. Oper. Res.*, **41**(1), 33–43, July 1989.

21. A. P. Sage (Ed.), *Concise Encyclopedia of Information Processing in Systems and Organizations*, Pergamon, Oxford, 1990.

22. D. Kahneman, P. Slovic, and A. Tversky (Eds.), *Judgment Under Uncertainty: Heuristics and Biases*, Cambridge University Press, New York, 1981.

23. L. J. Cohen, "On the Psychology of Prediction: Whose Is the Fallacy," *Cognition*, **7**, 385–407, 1979.

24. L. J. Cohen, "Can Human Irrationality Be Experimentally Demonstrated?" *Behav. Brain Sci.*, **4**, 317–370, 1981.

25. D. von Winterfeldt and W. Edwards, *Decision Analysis and Behavioral Research*, Cambridge University Press, Cambridge, 1986.

26. L. Phillips, "Theoretical Perspectives on Heuristics and Biases in Probabilistic Thinking," in P. C. Humphries, O. Svenson, and O. Vari (Eds.), *Analyzing and Aiding Decision Problems*, North Holland, Amsterdam, 1984.

27. R. T. Clemen, *Making Hard Decisions: An Introduction to Decision Analysis*, Duxbury, Belmont, CA, 1986.

28. R. L. Keeney, *Value Focused Thinking: A Path to Creative Decision Making*, Harvard University Press, Cambridge, MA, 1992.

29. C. W. Kirkwood, *Strategic Decision Making: Multiobjective Decision Analysis with Spreadsheets*, Duxbury, Belmont, CA, 1997.

30. A. P. Sage, *Decision Support Systems Engineering*, Wiley, New York, 1991.

31. J. Rasmussen, *Information Processing and Human–Machine Interaction*, North Holland, Amsterdam, 1986.

32. D. M. Buede, *The Engineering Design of Systems: Models and Methods*, Wiley, New York, 2000.

33. A. P. Sage and W. B. Rouse (Eds.), *Handbook of Systems Engineering and Management*, Wiley, New York, 1999.

CHAPTER 20

MATHEMATICAL MODELS OF DYNAMIC PHYSICAL SYSTEMS

K. Preston White Jr.
University of Virginia
Charlottesville, Virginia

1 RATIONALE

The design of modern control systems relies on the formulation and analysis of mathematical models of dynamic physical systems. This is simply because a model is more accessible to study than the physical system the model represents. Models typically are less costly and less time consuming to construct and test. Changes in the structure of a model are easier to implement, and changes in the behavior of a model are easier to isolate and understand. A model often can be used to achieve insight when the corresponding physical system cannot because experimentation with the actual system is too dangerous or too demanding. Indeed, a model can be used to answer "what if" questions about a system that has not yet been realized or actually cannot be realized with current technologies.

The type of model used by the control engineer depends upon the nature of the system the model represents, the objectives of the engineer in developing the model, and the tools the engineer has at his or her disposal for developing and analyzing the model. A mathematical model is a description of a system in terms of equations. Because the physical systems of primary interest to the control engineer are dynamic in nature, the mathematical models used to represent these systems most often incorporate difference or differential equations. Such equations, based on physical laws and observations, are statements of the fundamental relationships among the important variables that describe the system. Difference and differential equation models are expressions of the way in which the current values assumed by the variables combine to determine the future values of these variables.

Mathematical models are particularly useful because of the large body of mathematical and computational theory that exists for the study and solution of equations. Based on this theory, a wide range of techniques has been developed specifically for the study of control systems. In recent years, computer programs have been written that implement virtually all of these techniques. Computer software packages are now widely available for both simulation and computational assistance in the analysis and design of control systems.

It is important to understand that a variety of models can be realized for any given physical system. The choice of a particular model always represents a tradeoff between the fidelity of the model and the effort required in model formulation and analysis. This tradeoff is reflected in the nature and extent of simplifying assumptions used to derive the model. In general, the more faithful the model is as a description of the physical system modeled, the more difficult it is to obtain general solutions. In the final analysis, the best engineering model is not necessarily the most accurate or precise. It is, instead, the simplest model that yields the information needed to support a decision. A classification of various types of models commonly encountered by control engineers is given in Section 8.

A large and complicated model is justified if the underlying physical system is itself complex, if the individual relationships among the system variables are well understood, if it is important to understand the system with a great deal of accuracy and precision, and if time and budget exist to support an extensive study. In this case, the assumptions necessary to formulate the model can be minimized. Such complex models cannot be solved analytically, however. The model itself must be studied experimentally, using the techniques of computer simulation. This approach to model analysis is treated in Section 7.

Simpler models frequently can be justified, particularly during the initial stages of a control system study. In particular, systems that can be described by linear difference or differential equations permit the use of powerful analysis and design techniques. These include the transform methods of classical control theory and the state-variable methods of modern control theory. Descriptions of these standard forms for linear systems analysis are presented in Sections 4, 5, and 6.

During the past several decades, a unified approach for developing lumped-parameter models of physical systems has emerged. This approach is based on the idea of idealized system elements, which store, dissipate, or transform energy. Ideal elements apply equally well to the many kinds of physical systems encountered by control engineers. Indeed, because control engineers most frequently deal with systems that are part mechanical, part electrical, part fluid, and/or part thermal, a unified approach to these various physical systems is especially useful and economic. The modeling of physical systems using ideal elements is discussed further in Sections 2, 3, and 4.

Frequently, more than one model is used in the course of a control system study. Simple models that can be solved analytically are used to gain insight into the behavior of the system and to suggest candidate designs for controllers. These designs are then verified and refined in more complex models, using computer simulation. If physical components are developed

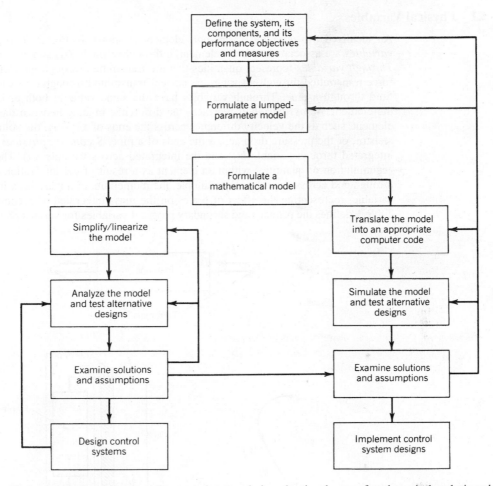

Figure 1 An iterative approach to control system design, showing the use of mathematical analysis and computer simulation.

during the course of a study, it is often practical to incorporate these components directly into the simulation, replacing the corresponding model components. An iterative, evolutionary approach to control systems analysis and design is depicted in Fig. 1.

2 IDEAL ELEMENTS

Differential equations describing the dynamic behavior of a physical system are derived by applying the appropriate physical laws. These laws reflect the ways in which energy can be stored and transferred within the system. Because of the common physical basis provided by the concept of energy, a general approach to deriving differential equation models is possible. This approach applies equally well to mechanical, electrical, fluid, and thermal systems and is particularly useful for systems that are combinations of these physical types.

2.1 Physical Variables

An idealized *two-terminal* or *one-port* element is shown in Fig. 2. Two *primary physical variables* are associated with the element: a through variable $f(t)$ and an across variable $v(t)$. *Through variables* represent quantities that are transmitted through the element, such as the force transmitted through a spring, the current transmitted through a resistor, or the flow of fluid through a pipe. Through variables have the same value at both ends or terminals of the element. *Across variables* represent the difference in state between the terminals of the element, such as the velocity difference across the ends of a spring, the voltage drop across a resistor, or the pressure drop across the ends of a pipe. *Secondary physical variables* are the integrated through variable $h(t)$ and the integrated across variable $x(t)$. These represent the accumulation of quantities within an element as a result of the integration of the associated through and across variables. For example, the momentum of a mass is an integrated through variable, representing the effect of forces on the mass integrated or accumulated over time. Table 1 defines the primary and secondary physical variables for various physical systems.

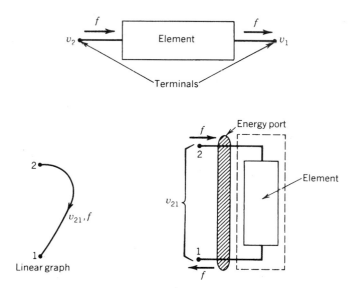

Figure 2 A two-terminal or one-port element, showing through and across variables.[1] Shearer, Introduction to System Dynamic, 1st Edition, © 1967. Reprinted by permission of Pearson Education, Inc., Upper Saddle River, NJ.

Table 1 Primary and Secondary Physical Variables for Various Systems[1]

System	Through Variable f	Integrated through Variable h	Across Variable v	Integrated across Variable x
Mechanical–translational	Force F	Translational momentum p	Velocity difference v_{21}	Displacement difference x_{21}
Mechanical–rotational	Torque T	Angular momentum h	Angular velocity difference Ω_{21}	Angular displacement difference Θ_{21}
Electrical	Current i	Charge q	Voltage difference v_{21}	Flux linkage λ_{21}
Fluid	Fluid flow Q	Volume V	Pressure difference P_{21}	Pressure–momentum Γ_{21}
Thermal	Heat flow q	Heat energy $\mathcal{H}$	Temperature difference θ_{21}	Not used in general

Shearer, Introduction to System Dynamic, 1st Edition, © 1967. Reprinted by permission of Pearson Education, Inc., Upper Saddle River, NJ.

2.2   Power and Energy

The flow of *power* $P(t)$ into an element through the terminals 1 and 2 is the product of the through variable $f(t)$ and the difference between the across variables $v_2(t)$ and $v_1(t)$. Suppressing the notation for time dependence, this may be written as

$$P = f(v_2 - v_1) = f v_{21}$$

A negative value of power indicates that power flows out of the element. The *energy* $E(t_a, t_b)$ transferred to the element during the time interval from t_a to t_b is the integral of power, that is,

$$E = \int_{t_a}^{t_b} P \, dt = \int_{t_a}^{t_b} f v_{21} \, dt$$

A negative value of energy indicates a net transfer of energy out of the element during the corresponding time interval.

Thermal systems are an exception to these generalized energy relationships. For a thermal system, power is identically the through variable $q(t)$, heat flow. Energy is the integrated through variable $\mathcal{H}(t_a, t_b)$, the amount of heat transferred.

By the *first law of thermodynamics*, the net energy stored within a system at any given instant must equal the difference between all energy supplied to the system and all energy dissipated by the system. The generalized classification of elements given in the following sections is based on whether the element stores or dissipates energy within the system, supplies energy to the system, or transforms energy between parts of the system.

2.3   One-Port Element Laws

Physical devices are represented by idealized system elements, or by combinations of these elements. A physical device that exchanges energy with its environment through one pair of across and through variables is called a *one-port* or *two-terminal* element. The behavior of a one-port element expresses the relationship between the physical variables for that element. This behavior is defined mathematically by a *constitutive relationship*. Constitutive relationships are derived empirically, by experimentation, rather than from any more fundamental principles. The *element law*, derived from the corresponding constitutive relationship, describes the behavior of an element in terms of across and through variables and is the form most commonly used to derive mathematical models.

Table 2 summarizes the element laws and constitutive relationships for the one-port elements. Passive elements are classified into three types. *T-type* or *inductive storage* elements are defined by a single-valued constitutive relationship between the through variable $f(t)$ and the integrated across-variable difference $x_{21}(t)$. Differentiating the constitutive relationship yields the element law. For a linear (or ideal) *T*-type element, the element law states that the across-variable difference is proportional to the rate of change of the through variable. Pure translational and rotational compliance (springs), pure electrical inductance, and pure fluid inertance are examples of *T*-type storage elements. There is no corresponding thermal element.

A-type or *capacitive storage elements* are defined by a single-valued constitutive relationship between the across-variable difference $v_{21}(t)$ and the integrated through variable $h(t)$. These elements store energy by virtue of the across variable. Differentiating the constitutive relationship yields the element law. For a linear *A*-type element, the element law states that the through variable is proportional to the derivative of the across-variable difference. Pure translational and rotational inertia (masses) and pure electrical, fluid, and thermal capacitance are examples.

It is important to note that when a nonelectrical capacitance is represented by an *A*-type element, one terminal of the element must have a constant (reference) across variable, usually assumed to be zero. In a mechanical system, for example, this requirement expresses the fact

Table 2   Element Laws and Constitutive Relationships for Various One-Port Elements[1]

Type of element	Physical element	Linear graph	Diagram	Constitutive relationship	Energy or power function	Ideal elemental equation	Ideal energy or power
T-type energy storage $\varepsilon \geq 0$ Pure $\;\; x_{21}=f(f)$ Ideal $\;\; x_{21}=Lf$ $\varepsilon=\int_0^f f\,dx_{21}$ $\varepsilon=\frac{1}{2}Lf^2$	Translational spring			$x_{21}=f(F)$	$\varepsilon=\int_0^F F\,dx_{21}$	$v_{21}=\frac{1}{k}\frac{dF}{dt}$	$\varepsilon=\frac{1}{2}\frac{F^2}{k}$
	Rotational spring			$\Theta_{21}=f(T)$	$\varepsilon=\int_0^T T\,d\Theta_{21}$	$\Omega_{21}=\frac{1}{K}\frac{dT}{dt}$	$\varepsilon=\frac{1}{2}\frac{T^2}{K}$
	Inductance			$\lambda_{21}=f(i)$	$\varepsilon=\int_0^i i\,d\lambda_{21}$	$v_{21}=L\frac{di}{dt}$	$\varepsilon=\frac{1}{2}Li^2$
	Fluid inertance			$\Gamma_{21}=f(Q)$	$\varepsilon=\int_0^Q Q\,d\Gamma_{21}$	$P_{21}=I\frac{dQ}{dt}$	$\varepsilon=\frac{1}{2}IQ^2$
A-type energy storage $\varepsilon \geq 0$ Pure $\;\; h=f(v_{21})$ Ideal $\;\; h=Cv_{21}$ $\varepsilon=\int_0^{v_{21}} v_{21}\,dh$ $\varepsilon=\frac{1}{2}Cv_{21}^2$	Translational mass		$v_1=\text{const}$	$p=f(v_2)$	$\varepsilon=\int_0^{v_2} v_2\,dp$	$\bar{F}=m\frac{dv_2}{dt}$	$\varepsilon=\frac{1}{2}mv_2^2$
	Inertia		$\Omega_1=\text{const}$	$h=f(\Omega_2)$	$\varepsilon=\int_0^{\Omega_2} \Omega_2\,dh$	$T=J\frac{d\Omega_2}{dt}$	$\varepsilon=\frac{1}{2}J\Omega_2^2$
	Electrical capacitance			$q=f(v_{21})$	$\varepsilon=\int_0^{v_{21}} v_{21}\,dq$	$i=C\frac{dv_{21}}{dt}$	$\varepsilon=\frac{1}{2}Cv_{21}^2$
	Fluid capacitance		$P_1=\text{const}$	$V=f(P_2)$	$\varepsilon=\int_0^{P_2} P_2\,dV$	$Q=C_f\frac{dP_2}{dt}$	$\varepsilon=\frac{1}{2}C_f P_2^2$
	Thermal capacitance		$\theta_1=\text{const}$	$\mathcal{H}=f(\theta_2)$	$\varepsilon=\int_0^{\theta_2} \mathbf{q}\,dt=\mathcal{H}$	$\mathbf{q}=C_t\frac{d\theta_2}{dt}$	$\varepsilon=C_t\theta_2$

D-type energy dissipators $\mathscr{P} \gg 0$	Graph	Pure — Elemental	Pure — Power	Ideal — Elemental	Ideal — Power
(definition)	$v_2 \,/\!\!\!\backslash\!\!\backslash\!\, v_1,\ f$	$f = f(v_{21})$	$\mathscr{P} = v_{21}f(v_{21})$	$f = \dfrac{1}{R}v_{21}$	$\mathscr{P} = \dfrac{1}{R}v_{21}^2 = Rf^2$
Translational damper	$2\ \dfrac{b}{v_{21}}\ 1;\ F;\ v_2,v_1$	$F = f(v_{21})$	$\mathscr{P} = Fv_{21}$	$F = bv_{21}$	$\mathscr{P} = bv_{21}^2$
Rotational damper	$2\ \dfrac{B}{\Omega_{21}}\ 1;\ T;\ \Omega_2,\Omega_1$	$T = f(\Omega_{21})$	$\mathscr{P} = T\Omega_{21}$	$T = B\Omega_{21}$	$\mathscr{P} = B\Omega_{21}^2$
Electrical resistance	$2\ \dfrac{R}{v_{21}}\ 1;\ i;\ v_2,v_1$	$i = f(v_{21})$	$\mathscr{P} = iv_{21}$	$i = \dfrac{1}{R}v_{21}$	$\mathscr{P} = \dfrac{1}{R}v_{21}^2$
Fluid resistance	$2\ \dfrac{R_f}{P_{21}}\ 1;\ Q;\ P_2,P_1$	$Q = f(P_{21})$	$\mathscr{P} = QP_{21}$	$Q = \dfrac{1}{R_f}P_{21}$	$\mathscr{P} = \dfrac{1}{R_f}P_{21}^2$
Thermal resistance	$2\ \dfrac{R_t}{\theta_{21}}\ 1;\ q;\ \theta_2,\theta_1$	$q = f(\theta_{21})$	$\mathscr{P} = q$	$q = \dfrac{1}{R_t}\theta_{21}$	$\mathscr{P} = \dfrac{1}{R_t}\theta_{21}$
Energy sources $\mathscr{P} \gtrless 0$, $\varepsilon \gtrless 0$					
A-type across-variable source	$2\ (v)\ 1;\ {+}\,v\,{-};\ v_2,v_1$	$v_{21} = f(t)$	$\mathscr{P} = fv_{21}$		
T-type through-variable source	$2\ (f)\ 1;\ f;\ v_2,v_1$	$f = f(t)$	$\mathscr{P} = fv_{21}$		

Shearer, Introduction to System Dynamic, 1st Edition, © 1967. Reprinted by permission of Pearson Education, Inc., Upper Saddle River, NJ.

Nomenclature

$\mathscr{E}$ = energy, $\mathscr{P}$ = power

f = generalized through variable, F = force, T = torque, i = current, Q = fluid flow rate, q = heat flow rate

h = generalized integrated through variable, p = translational momentum, h = angular momentum, q = charge, l' = fluid volume displaced, $\mathscr{H}$ = heat

v = generalized across variable, v = translational velocity, Ω = angular velocity, v = voltage, P = pressure, θ = temperature

x = generalized integrated across variable, x = translational displacement, Θ = angular displacement, λ = flux linkage, Γ = pressure–momentum

L = generalized ideal inductance, $1/k$ = reciprocal translational stiffness, $1/K$ = reciprocal rotational stiffness, L = inductance, I = fluid inertance

C = generalized ideal capacitance, m = mass, J = moment of inertia, C = capacitance, C_f = fluid capacitance, C_t = thermal capacitance

R = generalized ideal resistance, $1/b$ = reciprocal translational damping, $1/B$ = reciprocal rotational damping, R = electrical resistance, R_f = fluid resistance,

R_t = thermal resistance

that the velocity of a mass must be measured relative to a noninertial (nonaccelerating) reference frame. The constant-velocity terminal of a pure mass may be thought of as being attached in this sense to the reference frame.

D-type or *resistive elements* are defined by a single-valued constitutive relationship between the across and the through variables. These elements dissipate energy, generally by converting energy into heat. For this reason, power always flows into a *D*-type element. The element law for a *D*-type energy dissipator is the same as the constitutive relationship. For a linear dissipator, the through variable is proportional to the across-variable difference. Pure translational and rotational friction (dampers or dashpots) and pure electrical, fluid, and thermal resistance are examples.

Energy storage and energy-dissipating elements are called *passive* elements because such elements do not supply outside energy to the system. The fourth set of one-port elements are *source elements,* which are examples of *active* or power-supplying elements. Ideal sources describe interactions between the system and its environment. A pure *A-type source* imposes an across-variable difference between its terminals, which is a prescribed function of time, regardless of the values assumed by the through variable. Similarly, a pure *T-type source* imposes a through-variable flow through the source element, which is a prescribed function of time, regardless of the corresponding across variable.

Pure system elements are used to represent physical devices. Such models are called *lumped-element models.* The derivation of lumped-element models typically requires some degree of approximation since (1) there rarely is a one-to-one correspondence between a physical device and a set of pure elements and (2) there always is a desire to express an element law as simply as possible. For example, a coil spring has both mass and compliance. Depending on the context, the physical spring might be represented by a pure translational mass, or by a pure translational spring, or by some combination of pure springs and masses. In addition, the physical spring undoubtedly will have a nonlinear constitutive relationship over its full range of extension and compression. The compliance of the coil spring may well be represented by an ideal translational spring, however, if the physical spring is approximately linear over the range of extension and compression of concern.

2.4 Multiport Elements

A physical device that exchanges energy with its environment through two or more pairs of through and across variables is called a *multiport element.* The simplest of these, the idealized *four-terminal* or *two-port* element, is shown in Fig. 3. Two-port elements provide for transformations between the physical variables at different energy ports, while maintaining instantaneous continuity of power. In other words, net power flow into a two-port element is always identically zero:

$$P = f_a v_a + f_b v_b = 0$$

The particulars of the transformation between the variables define different categories of two-port elements.

A *pure transformer* is defined by a single-valued constitutive relationship between the integrated across variables or between the integrated through variables at each port:

$$x_b = f(x_a) \quad \text{or} \quad h_b = f(h_a)$$

For a linear (or ideal) transformer, the relationship is proportional, implying the following relationships between the primary variables:

$$v_b = n v_a \quad f_b = -\frac{1}{n} f_a$$

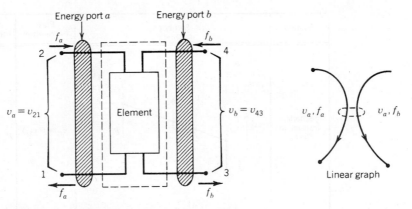

Figure 3 A four-terminal or two-port element, showing through and across variables.

where the constant of proportionality n is called the *transformation ratio*. Levers, mechanical linkages, pulleys, gear trains, electrical transformers, and differential-area fluid pistons are examples of physical devices that typically can be approximated by pure or ideal transformers. Figure 4 depicts some examples. *Pure transmitters,* which serve to transmit energy over a distance, frequently can be thought of as transformers with $n = 1$.

A *pure gyrator* is defined by a single-valued constitutive relationship between the across variable at one energy port and the through variable at the other energy port. For a linear gyrator, the following relations apply:

$$v_b = rf_a \qquad f_b = \frac{-1}{r} v_a$$

where the constant of proportionality is called the *gyration ratio* or *gyrational resistance*. Physical devices that perform pure gyration are not as common as those performing pure transformation. A mechanical gyroscope is one example of a system that might be modeled as a gyrator.

In the preceding discussion of two-port elements, it has been assumed that the type of energy is the same at both energy ports. A *pure transducer,* on the other hand, changes energy from one physical medium to another. This change may be accomplished as either a transformation or a gyration. Examples of *transforming transducers* are gears with racks (mechanical rotation to mechanical translation) and electric motors and electric generators (electrical to mechanical rotation and vice versa). Examples of *gyrating transducers* are the piston-and-cylinder (fluid to mechanical) and piezoelectric crystals (mechanical to electrical).

More complex systems may have a large number of energy ports. A common *six-terminal* or *three-port element* called a *modulator* is depicted in Fig. 5. The flow of energy between ports a and b is controlled by the energy input at the modulating port c. Such devices inherently dissipate energy, since

$$P_a + P_c \geq P_b$$

although most often the modulating power P_c is much smaller than the power input P_a or the power output P_b. When port a is connected to a pure source element, the combination of source and modulator is called a *pure dependent source*. When the modulating power P_c is considered the input and the modulated power P_b is considered the output, the modulator is called an *amplifier*. Physical devices that often can be modeled as modulators include clutches, fluid valves and couplings, switches, relays, transistors, and variable resistors.

System	Symbol	Pure transformer	Ideal transformer	Transformation ratio
Mechanical translation (lever)	(diagram with v_1, x_1; v_2, x_2; v_4, x_4; r_a; points 1, 2, 4; F_a, F_b, F_c)	$x_{41}=f(x_{21})$	$v_{41}=nv_{21}$ $F_b=-\dfrac{1}{n}F_a$	$n=-\dfrac{r_b}{r_a}$ Lever ratio
Mechanical rotational (gears)	(diagram with Ω_2,Θ_2; T_a; T_c; Ω_4,Θ_4; Ω_1,Θ_1; T_b; N_a; N_b)	$\Theta_{41}=f(\Theta_2)$	$\Omega_{41}=n\Omega_{21}$ $T_b=-\dfrac{1}{n}T_a$	$n=-\dfrac{N_a}{N_b}$ Gear ratio
Electrical (magnetic)	(diagram with i_a, i_b; v_2,λ_2 at 2; v_1,λ_1 at 1; v_4,λ_4 at 4; v_3,λ_3 at 3; N_a; N_b)	$\lambda_{43}=f(\lambda_{21})$	$v_{43}=nv_{21}$ $i_b=-\dfrac{1}{n}i_a$	$n=\dfrac{N_b}{N_a}$ Turns ratio
Fluid (differential piston)	(diagram with P_2, P_1, P_4; Q_a, V_a; Q_b, V_b; A_a; A_b; V_c, Q_c)	$V_b=f(V_a)$	$P_{41}=nP_{21}$ $Q_b=-\dfrac{1}{n}Q_a$	$n=\dfrac{A_a}{A_b}$ Area ratio

(a)

Figure 4(a) Examples of transforms and transducers: pure transformers.[1] Shearer, Introduction to System Dynamic, 1st Edition, © 1967. Reprinted by permission of Pearson Education, Inc., Upper Saddle River, NJ.

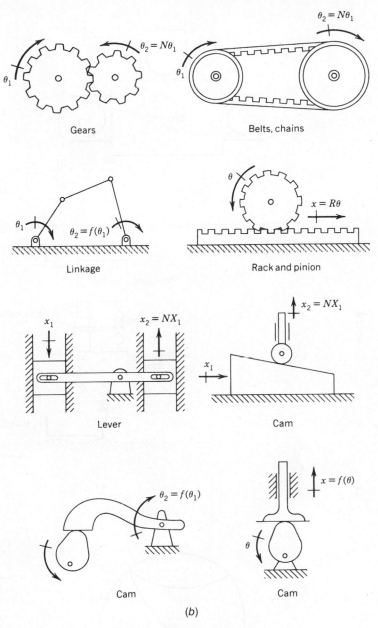

Figure 4(b)   Examples of transformers and transducers: pure mechanical transformers and transforming transducers.[2] Reprinted with permission of Taylor and Francis LLC.

3   SYSTEM STRUCTURE AND INTERCONNECTION LAWS

3.1   A Simple Example

Physical systems are represented by connecting the terminals of pure elements in patterns that approximate the relationships among the properties of component devices. As an example, consider the mechanical–translational system depicted in Fig. 6a, which might represent an

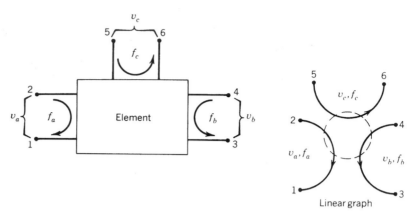

Figure 5   A six-terminal or three-port element, showing through and across variables.

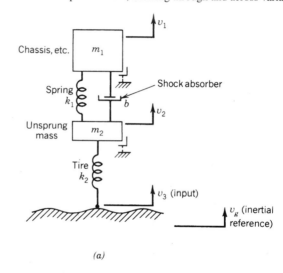

(a)

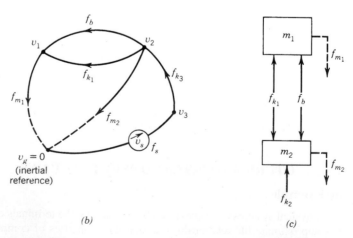

(b)   *(c)*

Figure 6   An idealized model of an automobile suspension system: (*a*) lumped-element model, (*b*) system graph, and (*c*) free-body diagram.

idealized automobile suspension system. The inertial properties associated with the masses of the chassis, passenger compartment, engine, and so on, all have been lumped together as the pure mass m_1. The inertial properties of the unsprung components (wheels, axles, etc.) have been lumped into the pure mass m_2. The compliance of the suspension is modeled as a pure spring with stiffness k_1 and the frictional effects (principally from the shock absorbers) as a pure damper with damping coefficient b. The road is represented as an input or source of vertical velocity, which is transmitted to the system through a spring of stiffness k_2, representing the compliance of the tires.

3.2 Structure and Graphs

The *pattern of interconnections* among elements is called the *structure* of the system. For a one-dimensional system, structure is conveniently represented by a *system graph*. The system graph for the idealized automobile suspension system of Fig. 6a is shown in Fig. 6b. Note that each distinct across variable (velocity) becomes a distinct *node* in the graph. Each distinct through variable (force) becomes a *branch* in the graph. Nodes coincide with the terminals of elements and branches coincide with the elements themselves. One node always represents *ground* (the constant velocity of the inertial reference frame v_g), and this is usually assumed to be zero for convenience. For nonelectrical systems, all the A-type elements (masses) have one terminal connection to the reference node. Because the masses are not physically connected to ground, however, the convention is to represent the corresponding branches in the graph by dashed lines.

System graphs are oriented by placing arrows on the branches. The orientation is arbitrary and serves to assign reference directions for both the through-variable and the across-variable difference. For example, the branch representing the damper in Fig. 6b is directed from node 2 (tail) to node 1 (head). This assigns $v_b = v_{21} = v_2 - v_1$ as the across-variable difference to be used in writing the damper elemental equation

$$f_b = bv_b = bv_{21}$$

The reference direction for the through variable is determined by the convention that power flow $P_b = f_b v_b$ into an element is positive. Referring to Fig. 6a, when v_{21} is positive, the damper is in compression. Therefore, f_b must be positive for compressive forces in order to obey the sign convention for power. By similar reasoning, tensile forces will be negative.

3.3 System Relations

The structure of a system gives rise to two sets of *interconnection laws* or *system relations*. Continuity relations apply to through variables and compatibility relations apply to across variables. The interpretation of system relations for various physical systems is given in Table 3.

Continuity is a general expression of dynamic equilibrium. In terms of the system graph, continuity states that the algebraic sum of all through variables entering a given node must

Table 3 System Relations for Various Systems

System	Continuity	Compatibility
Mechanical	Newton's first and third laws (conservation of momentum)	Geometrical constraints (distance is a scalar)
Electrical	Kirchhoff's current law (conservation of charge)	Kirchhoff's voltage law (potential is a scalar)
Fluid	Conservation of matter	Pressure is a scalar
Thermal	Conservation of energy	Temperature is a scalar

be zero. Continuity applies at each node in the graph. For a graph with n nodes, continuity gives rise to n continuity equations, $n-1$ of which are independent. For node i, the continuity equation is

$$\sum_j f_{ij} = 0$$

where the sum is taken over all branches (i,j) incident on i.

For the system graph depicted in Fig. 6b, the four continuity equations are

Node 1: $f_{k_1} + f_b - f_{m_1} = 0$

Node 2: $f_{k_2} - f_{k_1} - f_b - f_{m_2} = 0$

Node 3: $f_s - f_{k_2} = 0$

Node g: $f_{m_1} + f_{m_2} - f_s = 0$

Only three of these four equations are independent. Note, also, that the equations for nodes 1–3 could have been obtained from the conventional *free-body diagrams* shown in Fig. 6c, where f_{m_1} and f_{m_2} are the *D'Alembert forces* associated with the pure masses. Continuity relations are also known as *vertex, node, flow*, and *equilibrium relations*.

Compatibility expresses the fact that the magnitudes of all across variables are scalar quantities. In terms of the system graph, compatibility states that the algebraic sum of the across-variable differences around any closed path in the graph must be zero. Compatibility applies to any closed path in the system. For convenience and to ensure the independence of the resulting equations, continuity is usually applied to the *meshes* or "windows" of the graph. A one-part graph with n nodes and b branches will have $b-n+1$ meshes, each mesh yielding one independent compatibility equation. A planar graph with p separate parts (resulting from multiport elements) will have $b-n+p$ independent compatibility equations. For a closed path q, the compatibility equation is

$$\sum_q v_{ij} = 0$$

where the summation is taken over all branches (i,j) on the path.

For the system graph depicted in Fig. 6b, the three compatibility equations based on the meshes are

Path $1 \to 2 \to g \to 1$: $-v_b + v_{m_2} - v_{m_1} = 0$

Path $1 \to 2 \to 1$: $-v_{k_1} + v_b = 0$

Path $2 \to 3 \to g \to 2$: $-v_{k_2} - v_s - v_{m_2} = 0$

These equations are all mutually independent and express apparent geometric identities. The first equation, for example, states that the velocity difference between the ends of the damper is identically the difference between the velocities of the masses it connects. Compatibility relations are also known as *path, loop,* and *connectedness* relations.

3.4 Analogs and Duals

Taken together, the element laws and system relations are a complete mathematical model of a system. When expressed in terms of generalized through and across variables, the model applies not only to the physical system for which it was derived but also to any physical system with the same generalized system graph. Different physical systems with the same generalized model are called *analogs*. The mechanical rotational, electrical, and fluid analogs of the mechanical translational system of Fig. 6a are shown in Fig. 7. Note that because the original system contains an inductive storage element, there is no thermal analog.

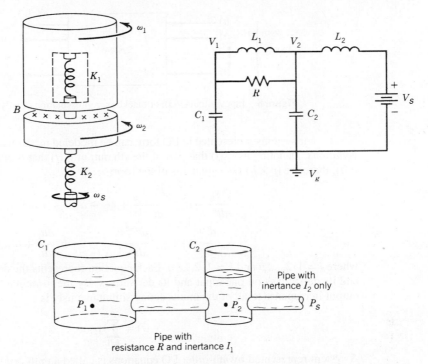

Figure 7   Analogs of the idealized automobile suspension system depicted in Fig. 6.

Systems of the same physical type but in which the roles of the through variables and the across variables have been interchanged are called *duals*. The analog of a dual—or, equivalently, the dual of an analog—is sometimes called a *dualog*. The concepts of analogy and duality can be exploited in many different ways.

4   STANDARD FORMS FOR LINEAR MODELS

The element laws and system relations together constitute a complete mathematical description of a physical system. For a system graph with n nodes, b branches, and s sources, there will be $b-s$ element laws, $n-1$ continuity equations, and $b-n+1$ compatibility equations. This is a total of $2b-s$ differential and algebraic equations. For systems composed entirely of linear elements, it is always possible to reduce these $2b-s$ equations to either of two standard forms. The *input/output,* or *I/O, form* is the basis for *transform* or so-called *classical linear systems analysis*. The *state-variable form* is the basis for *state-variable* or so-called *modern linear systems analysis*.

4.1   I/O Form

The classical representation of a system is the "blackbox," depicted in Fig. 8. The system has a set of p inputs (also called *excitations* or *forcing functions*), $u_j(t), j = 1, 2, ..., p$. The system also has a set of q outputs (also called *response variables*), $y_k(t), k = 1, 2, ..., q$. Inputs correspond to sources and are assumed to be known functions of time. Outputs correspond to physical variables that are to be measured or calculated.

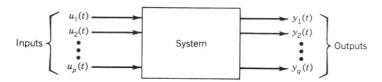

Figure 8 Input/output (I/O), or blackbox, representation of a dynamic system.

Linear systems represented in I/O form can be modeled mathematically by *I/O differential equations.* Denoting as $y_{kj}(t)$ that part of the kth output $y_k(t)$ that is attributable to the jth input $u_j(t)$, there are $(p \times q)$ I/O equations of the form

$$\frac{d^n y_{kj}}{dt^n} + a_{n-1} \frac{d^{n-1} y_{kj}}{dt^{n-1}} + \cdots + a_1 \frac{dy_{kj}}{dt} + a_0 y_{kj}(t)$$

$$= b_m \frac{d^m u_j}{dt^m} + b_{m-1} \frac{d^{m-1} u_j}{dt^{m-1}} + \cdots + b_1 \frac{du_j}{dt} + b_0 u_j(t)$$

where $j = 1, 2, ..., p$ and $k = 1, 2, ..., q$. Each equation represents the dependence of one output and its derivatives on one input and its derivatives. By the *principle of superposition,* the kth output in response to all of the inputs acting simultaneously is

$$y_k(t) = \sum_{j=1}^{p} y_{kj}(t)$$

A system represented by nth-order I/O equations is called an *nth-order system.* In general, the order of a system is determined by the number of *independent* energy storage elements within the system, that is, by the combined number of T-type and A-type elements for which the initial energy stored can be independently specified.

The coefficients $a_0, a_1, ..., a_{n-1}$ and $b_0, b_1, ..., b_m$ are parameter groups made up of algebraic combinations of the system physical parameters. For a system with constant parameters, therefore, these coefficients are also constant. Systems with constant parameters are called *time-invariant* systems and are the basis for classical analysis.

4.2 Deriving the I/O Form—An Example

I/O differential equations are obtained by combining element laws and continuity and compatibility equations in order to eliminate all variables except the input and the output. As an example, consider the mechanical system depicted in Fig. 9a, which might represent an idealized milling machine. A rotational motor is used to position the table of the machine tool through a rack and pinion. The motor is represented as a torque source T with inertia J and internal friction B. A flexible shaft, represented as a torsional spring K, is connected to a pinion gear of radius R. The pinion meshes with a rack, which is rigidly attached to the table of mass m. Damper b represents the friction opposing the motion of the table. The problem is to determine the I/O equation that expresses the relationship between the input torque T and the position of the table x.

The corresponding system graph is depicted in Fig. 9b. Applying continuity at nodes 1, 2, and 3 yields

$$\text{Node 1:} \qquad T - T_J - T_B - T_K = 0$$

$$\text{Node 2:} \qquad T_K - T_p = 0$$

$$\text{Node 3:} \qquad -f_r - f_m - f_b = 0$$

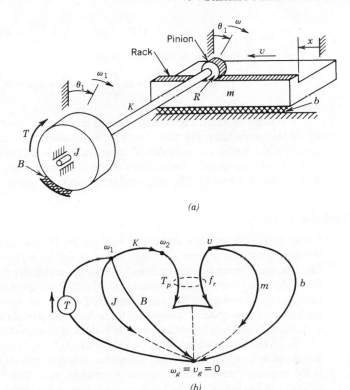

$$(a)$$

$$(b)$$

Figure 9 An idealized model of a milling machine: (*a*) lumped-element model[3] and (*b*) system graph.

Substituting the elemental equation for each of the one-port elements into the continuity equations and assuming zero ground velocities yield

$$\text{Node 1:} \quad T - J\dot{\omega}_1 - B\omega_1 - K \int (\omega_1 - \omega_2)\, dt = 0$$

$$\text{Node 2:} \quad K \int (\omega_1 - \omega_2)\, dt - T_p = 0$$

$$\text{Node 3:} \quad -f_r - m\dot{v} - bv = 0$$

Note that the definition of the across variables for each element in terms of the node variables, as above, guarantees that the compatibility equations are satisfied. With the addition of the constitutive relationships for the rack and pinion

$$\omega_2 = \frac{1}{R}v \quad \text{and} \quad T_p = -Rf_r$$

there are now five equations in the five unknowns $\omega_1, \omega_2, v, T_p$, and f_r. Combining these equations to eliminate all of the unknowns except v yields, after some manipulation,

$$a_3 \frac{d^3 v}{dt^3} + a_2 \frac{d^2 v}{dt^2} + a_1 \frac{dv}{dt} + a_0 v = b_1 T$$

where

$$a_3 = Jm \quad a_1 = \frac{JK}{R^2} + Bb + mK \quad b_1 = \frac{K}{R}$$

$$a_2 - Jb + mB \qquad a_0 - \frac{BK}{R^2} + Kb$$

Differentiating yields the desired I/O equation

$$a_3 \frac{d^3x}{dt^3} + a_2 \frac{d^2x}{dt^2} + a_1 \frac{dx}{dt} + a_0 x = b_1 \frac{dT}{dt}$$

where the coefficients are unchanged.

For many systems, combining element laws and system relations can best be achieved by ad hoc procedures. For more complicated systems, formal methods are available for the orderly combination and reduction of equations. These are the so-called *loop method* and *node method* and correspond to procedures of the same names originally developed in connection with electrical networks. The interested reader should consult Ref. 1.

4.3   State-Variable Form

For systems with multiple inputs and outputs, the I/O model form can become unwieldy. In addition, important aspects of system behavior can be suppressed in deriving I/O equations. The "modern" representation of dynamic systems, called the *state-variable form,* largely eliminates these problems. A state-variable model is the maximum reduction of the original element laws and system relations that can be achieved without the loss of any information concerning the behavior of a system. State-variable models also provide a convenient representation for systems with multiple inputs and outputs and for systems analysis using computer simulation.

State variables are a set of variables $x_1(t), x_2(t), ..., x_n(t)$ internal to the system from which any set of outputs can be derived, as depicted schematically in Fig. 10. A set of state variables is the minimum number of independent variables such that by knowing the values of these variables at any time t_0 and by knowing the values of the inputs for all time $t \geq t_0$, the values of the state variables for all future time $t \geq t_0$ can be calculated. For a given system, the number n of state variables is unique and is equal to the order of the system. The definition of the state variables is not unique, however, and various combinations of one set of state variables can be used to generate alternative sets of state variables. For a physical system, the state variables summarize the *energy state* of the system at any given time.

A complete state-variable model consists of two sets of equations, the *state* or *plant equations* and the *output equations*. For the most general case, the state equations have the form

$$\dot{x}_1(t) = f_1[x_1(t), x_2(t), \ ... \ , x_n(t), u_1(t), u_2(t), \ ... \ , u_p(t)]$$

$$\dot{x}_2(t) = f_2[x_1(t), x_2(t), \ ... \ , x_n(t), u_1(t), u_2(t), \ ... \ , u_p(t)]$$

$$\vdots$$

$$\dot{x}_n(t) = f_n[x_1(t), x_2(t), \ ... \ , x_n(t), u_1(t), u_2(t), \ ... \ , u_p(t)]$$

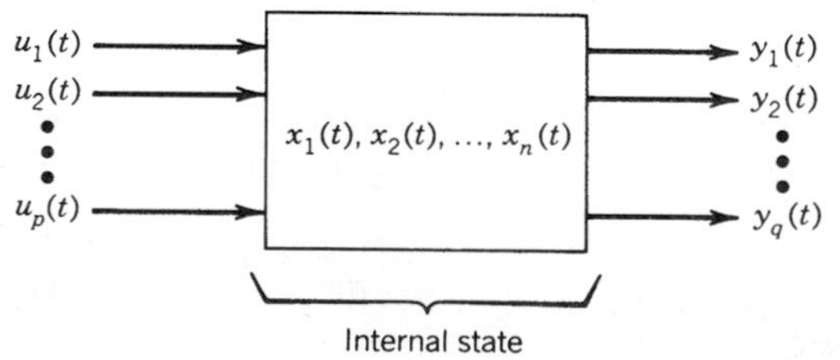

Figure 10   State-variable representation of a dynamic system.[4]

and the output equations have the form

$$y_1(t) = g_1[x_1(t), x_2(t), \ldots, x_n(t), u_1(t), u_2(t), \ldots, u_p(t)]$$

$$y_2(t) = g_2[x_1(t), x_2(t), \ldots, x_n(t), u_1(t), u_2(t), \ldots, u_p(t)]$$

$$\vdots$$

$$y_q(t) = g_q[x_1(t), x_2(t), \ldots, x_n(t), u_1(t), u_2(t), \ldots, u_p(t)]$$

These equations are expressed more compactly as the two vector equations

$$\dot{x}(t) = f[x(t), u(t)]$$

$$y(t) = g[x(t), u(t)]$$

where

$$\dot{x}(t) = n \times 1 \text{ state vector}$$
$$u(t) = p \times 1 \text{ input or control vector}$$
$$y(t) = q \times 1 \text{ output or response vector}$$

and f and g are vector-valued functions.

For linear systems, the state equations have the form

$$\dot{x}_1(t) = a_{11}(t)x_1(t) + \cdots + a_{1n}(t)x_n(t) + b_{11}(t)u_1(t) + \cdots + b_{1p}(t)u_p(t)$$

$$\dot{x}_2(t) = a_{21}(t)x_1(t) + \cdots + a_{2n}(t)x_n(t) + b_{21}(t)u_1(t) + \cdots + b_{2p}(t)u_p(t)$$

$$\vdots$$

$$\dot{x}_n(t) = a_{n1}(t)x_1(t) + \cdots + a_{nn}(t)x_n(t) + b_{n1}(t)u_1(t) + \cdots + b_{np}(t)u_p(t)$$

and the output equations have the form

$$y_1(t) = c_{11}(t)x_1(t) + \cdots + c_{1n}(t)x_n(t) + d_{11}(t)u_1(t) + \cdots + d_{1p}(t)u_p(t)$$

$$y_2(t) = c_{21}(t)x_1(t) + \cdots + c_{2n}(t)x_n(t) + d_{21}(t)u_1(t) + \cdots + d_{2p}(t)u_p(t)$$

$$\vdots$$

$$y_n(t) = c_{q1}(t)x_1(t) + \cdots + c_{qn}(t)x_n(t) + d_{q1}(t)u_1(t) + \cdots + d_{qp}(t)u_p(t)$$

where the coefficients are groups of parameters. The linear model is expressed more compactly as the two linear vector equations

$$\dot{x}(t) = A(t)x(t) + B(t)u(t)$$

$$y(t) = C(t)x(t) + D(t)u(t)$$

where the vectors x, u, and y are the same as the general case and the matrices are defined as

$$A = [a_{ij}] \text{ is the } n \times n \text{ system matrix}$$

$$B = [b_{jk}] \text{ is the } n \times p \text{ control, input, or distribution matrix}$$

$$C = [c_{lj}] \text{ is the } q \times n \text{ output matrix}$$

$$D = [d_{lk}] \text{ is the } q \times p \text{ output distribution matrix}$$

For a time-invariant linear system, all of these matrices are constant.

4.4   Deriving the "Natural" State Variables—A Procedure

Because the state variables for a system are not unique, there are an unlimited number of alternative (but equivalent) state-variable models for the system. Since energy is stored only in generalized system storage elements, however, a natural choice for the state variables is the set of through and across variables corresponding to the independent T-type and A-type elements, respectively. This definition is sometimes called the set of *natural state variables* for the system.

For linear systems, the following procedure can be used to reduce the set of element laws and system relations to the natural state-variable model.

Step 1. For each independent T-type storage, write the element law with the derivative of the through variable isolated on the left-hand side, that is, $\dot{f} = L^{-1}v$.

Step 2. For each independent A-type storage, write the element law with the derivative of the across variable isolated on the left-hand side, that is, $\dot{v} = C^{-1}f$.

Step 3. Solve the compatibility equations, together with the element laws for the appropriate D-type and multiport elements, to obtain each of the across variables of the independent T-type elements in terms of the natural state variables and specified sources.

Step 4. Solve the continuity equations, together with the element laws for the appropriate D-type and multiport elements, to obtain the through variables of the A-type elements in terms of the natural state variables and specified sources.

Step 5. Substitute the results of step 3 into the results of step 1; substitute the results of step 4 into the results of step 2.

Step 6. Collect terms on the right-hand side and write in vector form.

4.5   Deriving the "Natural" State Variables—An Example

The six-step process for deriving a natural state-variable representation, outlined in the preceding section, is demonstrated for the idealized automobile suspension depicted in Fig. 6:

Step 1
$$\dot{f}_{k_1} = k_1 v_{k_1} \qquad \dot{f}_{k_2} = k_2 v_{k_2}$$

Step 2
$$\dot{v}_{m_1} = m_1^{-1} f_{m_1} \qquad \dot{v}_{m_2} = m_2^{-1} f_{m_2}$$

Step 3
$$v_{k_1} = v_b = v_{m_2} - v_{m_1} \qquad v_{k_2} = -v_{m_2} - v_s$$

Step 4
$$f_{m_1} = f_{k_1} + f_b = f_{k_1} + b^{-1}(v_{m_2} - v_{m_1})$$
$$f_{m_2} = f_{k_2} - f_{k_1} - f_b = f_{k_2} - f_{k_1} - b^{-1}(v_{m_2} - v_{m_1})$$

Step 5
$$\dot{f}_{k_1} = k_1(v_{m_2} - v_{m_1}) \qquad \dot{v}_{m_1} = m_1^{-1}[f_{k_1} + b^{-1}(v_{m_2} - v_{m_1})]$$
$$\dot{f}_{k_2} = k_2(-v_{m_2} - v_s) \qquad \dot{v}_{m_2} = m_2^{-1}[f_{k_2} - f_{k_1} - b^{-1}(v_{m_2} - v_{m_1})]$$

Step 6
$$\frac{d}{dt}\begin{bmatrix} f_{k_1} \\ f_{k_2} \\ v_{m_1} \\ v_{m_2} \end{bmatrix} = \begin{bmatrix} 0 & 0 & -k_1 & k_1 \\ 0 & 0 & 0 & -k_2 \\ 1/m_1 & 0 & -1/m_1 b & 1/m_1 b \\ -1/m_2 & 1/m_2 & 1/m_2 b & -1/m_2 b \end{bmatrix}\begin{bmatrix} f_{k_1} \\ f_{k_2} \\ v_{m_1} \\ v_{m_2} \end{bmatrix} + \begin{bmatrix} 0 \\ -k_2 \\ 0 \\ 0 \end{bmatrix} v_s$$

4.6   Converting from I/O to "Phase-Variable" Form

Frequently, it is desired to determine a state-variable model for a dynamic system for which the I/O equation is already known. Although an unlimited number of such models is possible, the easiest to determine uses a special set of state variables called the *phase variables*. The phase variables are defined in terms of the output and its derivatives as follows:

$$x_1(t) = y(t)$$

$$x_2(t) = \dot{x}_1(t) = \frac{d}{dt}\,y(t)$$

$$x_3(t) = \dot{x}_2(t) = \frac{d^2}{dt^2}y(t)$$

$$\vdots$$

$$x_n(t) = \dot{x}_{n-1}(t) = \frac{d^{n-1}}{dt^{n-1}}y(t)$$

This definition of the phase variables, together with the I/O equation of Section 4.1, can be shown to result in a state equation of the form

$$\frac{d}{dt}\begin{bmatrix} x_1(t) \\ x_2(t) \\ \vdots \\ x_{n-1}(t) \\ x_n(t) \end{bmatrix} = \begin{bmatrix} 0 & 1 & 0 & \cdots & 0 \\ 0 & 0 & 1 & \cdots & 0 \\ \vdots & \vdots & \vdots & \cdots & \vdots \\ 0 & 0 & 0 & \cdots & 1 \\ -a_0 & -a_1 & -a_2 & \cdots & -a_{n-1} \end{bmatrix}\begin{bmatrix} x_1(t) \\ x_2(t) \\ \vdots \\ x_{n-1}(t) \\ x_n(t) \end{bmatrix} + \begin{bmatrix} 0 \\ 0 \\ \vdots \\ 0 \\ 1 \end{bmatrix}u(t)$$

and an output equation of the form

$$y(t) = \begin{bmatrix} b_0 & b_1 \cdots b_m \end{bmatrix}\begin{bmatrix} x_1(t) \\ x_2(t) \\ \vdots \\ x_n(t) \end{bmatrix}$$

This special form of the system matrix, with 1s along the upper off-diagonal and 0s elsewhere except for the bottom row, is called a *companion matrix*.

5   APPROACHES TO LINEAR SYSTEMS ANALYSIS

There are two fundamental approaches to the analysis of linear, time-invariant systems. *Transform methods* use rational functions obtained from the Laplace transformation of the system I/O equations. Transform methods provide a particularly convenient algebra for combining the component submodels of a system and form the basis of so-called *classical control theory*. *State-variable methods* use the vector state and output equations directly. State-variable methods permit the adaptation of important ideas from linear algebra and form the basis for so-called *modern control theory*. Despite the deceiving names of "classical" and "modern," the two approaches are complementary. Both approaches are widely used in current practice and the control engineer must be conversant with both.

5.1   Transform Methods

A *transformation* converts a given mathematical problem into an equivalent problem, according to some well-defined rule called a *transform*. Prudent selection of a transform frequently results in an equivalent problem that is easier to solve than the original. If the solution to the original problem can be recovered by an inverse transformation, the three-step process of (1) transformation, (2) solution in the *transform domain,* and (3) inverse transformation may prove more attractive than direct solution of the problem in the original problem domain. This is true for fixed linear dynamic systems under the *Laplace transform,* which converts differential equations into equivalent algebraic equations.

Laplace Transforms: Definition

The one-sided Laplace transform is defined as

$$F(s) = \mathcal{L}[f(t)] = \int_0^\infty f(t)e^{-st}\, dt$$

and the inverse transform as

$$f(t) = \mathcal{L}^{-1}[F(s)] = \frac{1}{2\pi j} \int_{\sigma - j\omega}^{\sigma + j\omega} F(s)e^{-st} ds$$

The Laplace transform converts the function $f(t)$ into the transformed function $F(s)$; the inverse transform recovers $f(t)$ from $F(s)$. The symbol $\mathcal{L}$ stands for the "Laplace transform of"; the symbol $\mathcal{L}^{-1}$ stands for "the inverse Laplace transform of."

The Laplace transform takes a problem given in the *time domain,* where all physical variables are functions of the *real variable t,* into the *complex-frequency domain,* where all physical variables are functions of the complex frequency $s = \sigma + j\omega$, where $j = \sqrt{-1}$ is the imaginary operator. Laplace transform pairs consist of the function $f(t)$ and its transform $F(s)$. Transform pairs can be calculated by substituting $f(t)$ into the defining equation and then evaluating the integral with s held constant. For a transform pair to exist, the corresponding integral must converge, that is,

$$\int_0^\infty |f(t)|e^{-\sigma^* t}\, dt < \infty$$

for some real $\sigma^* > 0$. Signals that are physically realizable always have a Laplace transform.

Tables of Transform Pairs and Transform Properties

Transform pairs for functions commonly encountered in the analysis of dynamic systems rarely need to be calculated. Instead, pairs are determined by reference to a *table of transforms* such as that given in Table 4. In addition, the Laplace transform has a number of properties that are useful in determining the transforms and inverse transforms of functions in terms of the tabulated pairs. The most important of these are given in a *table of transform properties* such as that given in Table 5.

Poles and Zeros

The response of a dynamic system most often assumes the following form in the complex-frequency domain:

$$F(s) = \frac{N(s)}{D(s)} = \frac{b_m s^m + b_{m-1} s^{m-1} + \cdots + b_1 s + b_0}{s^n + a_{n-1} s^{n-1} + \cdots + a_1 s + a_0} \tag{1}$$

Functions of this form are called *rational functions* because these are the ratio of two polynomials $N(s)$ and $D(s)$. If $n \geq m$, then $F(s)$ is a *proper rational function;* if $n > m$, then $F(s)$ is a *strictly proper rational function.*

In factored form, the rational function $F(s)$ can be written as

$$F(s) = \frac{N(s)}{D(s)} = \frac{b_m(s - z_1)(s - z_2) \cdots (s - z_m)}{(s - p_1)(s - p_2) \cdots (s - p_n)} \tag{2}$$

Table 4 Laplace Transform Pairs

$F(s)$	$f(t), t \geq 0$
1. 1	$\delta(t)Z$, the unit impulse at $t = 0$
2. $\dfrac{1}{s}$	1, the unit step
3. $\dfrac{n!}{s^{n+1}}$	t^n
4. $\dfrac{1}{s+a}$	e^{-at}
5. $\dfrac{1}{(s+a)^n}$	$\dfrac{1}{(n-1)!}t^{n-1}e^{-at}$
6. $\dfrac{a}{s(s+a)}$	$1 - e^{-at}$
7. $\dfrac{1}{(s+a)(s+b)}$	$\dfrac{1}{b-a}(e^{-at} - e^{-bt})$
8. $\dfrac{s+p}{(s+a)(s+b)}$	$\dfrac{1}{b-a}[(p-a)e^{-at} - (p-b)e^{-bt}]$
9. $\dfrac{1}{(s+a)(s+b)(s+c)}$	$\dfrac{e^{-at}}{(b-a)(c-a)} + \dfrac{e^{-bt}}{(c-b)(a-b)} + \dfrac{e^{-ct}}{(a-c)(b-c)}$
10. $\dfrac{s+p}{(s+a)(s+b)(s+c)}$	$\dfrac{(p-a)e^{-at}}{(b-a)(c-a)} + \dfrac{(p-b)e^{-bt}}{(c-b)(a-b)} + \dfrac{(p-c)e^{-ct}}{(a-c)(b-c)}$
11. $\dfrac{b}{s^2 + b^2}$	$\sin bt$
12. $\dfrac{s}{s^2 + b^2}$	$\cos bt$
13. $\dfrac{b}{(s+a)^2 + b^2}$	$e^{-at}\sin bt$
14. $\dfrac{s+a}{(s+a)^2 + b^2}$	$e^{-at}\cos bt$
15. $\dfrac{\omega_n^2}{s^2 + 2\zeta\omega_n s + \omega_n^2}$	$\dfrac{\omega_n}{\sqrt{1-\zeta^2}}e^{-\zeta\omega_n t}\sin \omega_n\sqrt{1-\zeta^2}\,t, \quad \zeta < 1$
16. $\dfrac{\omega_n^2}{s(s^2 + 2\zeta\omega_n s + \omega_n^2)}$	$1 + \dfrac{1}{\sqrt{1-\zeta^2}}e^{-\zeta\omega_n t}\sin(\omega_n\sqrt{1-\zeta^2}\,t + \phi)$ $\phi = \tan^{-1}\dfrac{\sqrt{1-\zeta^2}}{\zeta} + \pi \text{ (third quadrant)}$

The roots of the numerator polynomial $N(s)$ are denoted by $z_j, j = 1, 2, ..., m$. These numbers are called the *zeros* of $F(s)$, since $F(z_j) = 0$. The roots of the denominator polynomial are denoted by p_i, $1, 2, ..., n$. These numbers are called the *poles* of $F(s)$ since $\lim_{s \to pi} F(s) = \pm\infty$.

Inversion by Partial-Fraction Expansion
The *partial-fraction expansion theorem* states that a strictly proper rational function $F(s)$ with distinct (*nonrepeated*) poles p_i, $i = 1, 2, ..., n$, can be written as the sum

$$F(s) = \frac{A_1}{s - p_1} + \frac{A_2}{s - p_2} + \cdots + \frac{A_n}{s - p_n} = \sum_{i=1}^{n} A_i \left(\frac{1}{s - p_i}\right) \tag{3}$$

Table 5 Laplace Transform Properties

$f(t)$	$F(s) = \int_0^\infty f(t)e^{-st}\, dt$
1. $af_1(t) + bf_2(t)$	$aF_1(s) + bF_2(s)$
2. $\dfrac{df}{dt}$	$sF(s) - f(0)$
3. $\dfrac{d^2 f}{dt^2}$	$s^2 F(s) - sf(0) - \left.\dfrac{df}{dt}\right|_{t=0}$
4. $\dfrac{d^n f}{dt^n}$	$s^n F(s) - \displaystyle\sum_{k=1}^{n} s^{n-k} g_{k-1}$
	$g_{k-1} = \left.\dfrac{d^{k-1} f}{dt^{k-1}}\right|_{t=0}$
5. $\displaystyle\int_0^t f(t)\, dt$	$\dfrac{F(s)}{s} + \dfrac{h(0)}{s}$
	$h(0) = \left.\displaystyle\int f(t)\, dt\right|_{t=0}$
6. $\left\{ \begin{matrix} 0, & t < D \\ f(t-D), & t\ \dots\ D \end{matrix} \right\}$	$e^{-sD} F(s)$
7. $e^{-at} f(t)$	$F(s+a)$
8. $f\left(\dfrac{t}{a}\right)$	$aF(as)$
9. $f(t) = \displaystyle\int_0^t x(t-\tau)y(\tau)\, d\tau$ $= \displaystyle\int_0^t y(t-\tau)x(\tau)\, d\tau$	$F(s) = X(s)Y(s)$
10. $f(\infty) = \displaystyle\lim_{s\to 0} sF(s)$	
11. $f(0+) = \displaystyle\lim_{s\to\infty} sF(s)$	

where the A_i, $i = 1, 2, ..., n$, are constants called *residues*. The inverse transform of $F(s)$ has the simple form

$$f(t) = A_1 e^{p_1 t} + A_2 e^{p_2 t} + \cdots + A_n e^{p_n t} = \sum_{i=1}^{n} A_i e^{p_i t}$$

The *Heaviside expansion theorem* gives the following expression for calculating the residue at the pole p_i,

$$A_i = (s - p_i)F(s)|_{s=p_i} \qquad \text{for } i = 1, 2, ..., n$$

These values can be checked by substituting into Eq. (3), combining the terms on the right-hand side of Eq. (3), and showing that the result yields the values for all the coefficients $b_j, j = 1, 2, ..., m$, originally specified in the form of Eq. (3).

Repeated Poles

When two or more poles of a strictly proper rational function are identical, the poles are said to be *repeated* or *nondistinct*. If a pole is repeated q times, that is, if $p_i = p_{i+1} = \cdots = p_{i+q-1}$, then the pole is said to be of *multiplicity* q. A strictly proper rational function with a pole of multiplicity q will contain q terms of the form

$$\frac{A_{i1}}{(s - p_i)^q} + \frac{A_{i2}}{(s - p_i)^{q-1}} + \cdots + \frac{A_{iq}}{s - p_i}$$

in addition to the terms associated with the distinct poles. The corresponding terms in the inverse transform are

$$\left[\frac{1}{(q-1)!} A_{i1} t^{(q-1)} + \frac{1}{(q-2)!} A_{i2} t^{(q-2)} + \cdots A_{iq} \right] e^{p_i t}$$

The corresponding residues are

$$A_{i1} = (s - p_i)^q F(s)|_{s=p_i}$$

$$A_{i2} = \left(\frac{d}{ds} \left[(s - p_i)^q F(s) \right] \right) \bigg|_{s=p_i}$$

$$\vdots$$

$$A_{iq} = \frac{1}{(q-1)!} \left(\frac{d^{(q-1)}}{ds^{(q-1)}} [(s - p_i)^q F(s)] \right) \bigg|_{s=p_i}$$

Complex Poles

A strictly proper rational function with complex-conjugate poles can be inverted using partial-fraction expansion. Using a method called *completing the square,* however, is almost always easier. Consider the function

$$F(s) = \frac{B_1 s + B_2}{(s + \sigma - j\omega)(s + \sigma + j\omega)}$$

$$= \frac{B_1 s + B_2}{s^2 + 2\sigma s + \sigma^2 + \omega^2}$$

$$= \frac{B_1 s + B_2}{(s + \sigma)^2 + \omega_2}$$

From the transform tables the Laplace inverse is

$$f(t) = e^{-\sigma t}[B_1 \cos \omega t + B_3 \sin \omega t]$$

$$= K e^{-\sigma t} \cos(\omega t + \phi)$$

where

$B_3 = (1/\omega)(B_2 - aB_1)$
$K = \sqrt{B_1^2 + B_3^2}$
$\phi = -\tan^{-1}(B_3/B_1)$

Proper and Improper Rational Functions

If $F(s)$ is not a strictly proper rational function, then $N(s)$ must be divided by $D(s)$ using *synthetic division*. The result is

$$F(s) = \frac{N(s)}{D(s)} = P(s) + \frac{N^*(s)}{D(s)}$$

where $P(s)$ is a polynomial of degree $m - n$ and $N^*(s)$ is a polynomial of degree $n - 1$. Each term of $P(s)$ may be inverted directly using the transform tables. The strictly proper rational function $N^*(s)/D(s)$ may be inverted using partial-fraction expansion.

Initial-Value and Final-Value Theorems

The limits of $f(t)$ as time approaches zero or infinity frequently can be determined directly from the transform $F(s)$ without inverting. The *initial-value theorem* states that

$$f(0_+) = \lim_{s \to \infty} sF(s)$$

where the limit exists. If the limit does not exist (i.e., is infinite), the value of $f(0_+)$ is undefined. The *final-value theorem* states that

$$f(\infty) = \lim_{s \to 0} sF(s)$$

provided that (with the possible exception of a single pole at $s = 0$) $F(s)$ has no poles with nonnegative real parts.

Transfer Functions

The Laplace transform of the system I/O equation may be written in terms of the transform $Y(s)$ of the system response $y(t)$ as

$$Y(s) = \frac{G(s)N(s) + F(s)D(s)}{P(s)D(s)}$$

$$= \left(\frac{G(s)}{P(s)} \right) \left(\frac{N(s)}{D(s)} \right) + \frac{F(s)}{P(s)}$$

where

a. $P(s) = a_n s^n + a_{n-1} + \cdots + a_1 s + a_0$ is the *characteristic polynomial* of the system

b. $G(s) = b_m s^m + b_{m-1} s^{m-1} + \cdots + b_1 s + b_0$ represents the *numerator dynamics* of the system

c. $U(s) = N(s)/D(s)$ is the transform of the input to the system, $u(t)$, assumed to be a rational function

d. $F(s) = a_n y(0)s^{n-1} + \left(a_n \frac{dy}{dt}(0) + a_{n-1}y(0) \right) s^{n-2} + \cdots$
$\qquad + \left(a_n \frac{d^{n-1}y}{dt^{n-1}}(0) + a_{n-1}\frac{d^{n-2}y}{dt}(0) + \cdots + a_1 y(0) \right)$

reflects the initial system state [i.e., the initial conditions on $y(t)$ and its first $n-1$ derivatives]

The transformed response can be thought of as the sum of two components,

$$Y(s) = Y_{zs}(s) + Y_{zi}(s)$$

where

e. $Y_{zs}(s) = [G(s)/P(s)][N(s)/D(s)] = H(s)U(s)$ is the transform of the *zero-state response,* that is, the response of the system to the input alone

f. $Y_{zi}(s) = F(s)/P(s)$ is the transform of the *zero-input response,* that is, the response of the system to the initial state alone

The rational function

g. $H(s) = Y_{zs}(s)/U(s) = G(s)/P(s)$ is the *transfer function* of the system, defined as the Laplace transform of the ratio of the system response to the system input, assuming zero initial conditions

The transfer function plays a crucial role in the analysis of fixed linear systems using transforms and can be written directly from knowledge of the system I/O equation as

$$H(s) = \frac{b_m s^m + \cdots + b_0}{a_n s^n + a_{n-1}s^{n-1} + \cdots + a_1 s + a_0}$$

Impulse Response

Since $U(s) = 1$ for a unit impulse function, the transform of the zero-state response to a unit impulse input is given by the relation (g) as

$$Y_{zs}(s) = H(s)$$

that is, the system transfer function. In the time domain, therefore, the unit *impulse response* is

$$h(t) = \begin{cases} 0 & \text{for } t \leq 0 \\ \mathscr{L}^{-1}\left[H(s)\right] & \text{for } t > 0 \end{cases}$$

This simple relationship is profound for several reasons. First, this provides for a direct characterization of time-domain response $h(t)$ in terms of the properties (poles and zeros) of the rational function $H(s)$ in the complex-frequency domain. Second, applying the convolution transform pair (Table 5) to relation (e) above yields

$$Y_{zs}(t) = \int_0^t h(\tau)u(t - \tau)\, d\tau$$

In words, the zero-state output corresponding to an arbitrary input $u(t)$ can be determined by convolution with the impulse response $h(t)$. In other words, the impulse response completely characterizes the system. The impulse response is also called the system *weighing function*.

Block Diagrams

Block diagrams are an important conceptual tool for the analysis and design of dynamic systems because block diagrams provide a graphic means for depicting the relationships among system variables and components. A block diagram consists of unidirectional blocks representing specified system components or subsystems interconnected by arrows representing system variables. Causality follows in the direction of the arrows, as in Fig. 11, indicating that the output is caused by the input acting on the system defined in the block.

Combining transform variables, transfer functions, and block diagrams provides a powerful graphical means for determining the overall transfer function of a system when the transfer functions of its component subsystems are known. The basic blocks in such diagrams are given in Fig. 12. A block diagram comprising many blocks and summers can be reduced to a single transfer function block by using the diagram transformations given in Fig. 13.

5.2   Transient Analysis Using Transform Methods

Basic to the study of dynamic systems are the concepts and terminology used to characterize system behavior or performance. These ideas are aids in *defining* behavior in order to consider for a given context those features of behavior that are desirable and undesirable; in *describing* behavior in order to communicate concisely and unambiguously various behavioral attributes

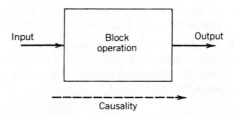

Figure 11   Basic block diagram, showing assumed direction of causality or loading.[4]

	Input–Output Relations		
Type	Time Domain	Transform Domain	Symbol
(a) Multiplier	$y(t) = Kv(t)$	$Y(s) = KV(s)$	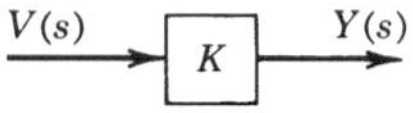 $V(s) \rightarrow \boxed{K} \rightarrow Y(s)$
(b) General transfer function	$y(t) = \mathscr{L}^{-1}[T(s)V(s)]$	$Y(s) = T(s)V(s)$	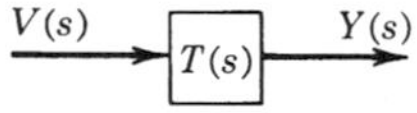 $V(s) \rightarrow \boxed{T(s)} \rightarrow Y(s)$
(c) Summer	$y(t) = v_1(t) + v_2(t)$	$Y(s) = V_1(s) + V_2(s)$	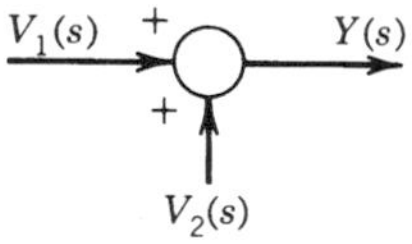 $V_1(s) \xrightarrow{+} \bigcirc \rightarrow Y(s)$, $\xleftarrow{+} V_2(s)$
(d) Comparator	$y(t) = v_1(t) - v_2(t)$	$Y(s) = V_1(s) - V_2(s)$	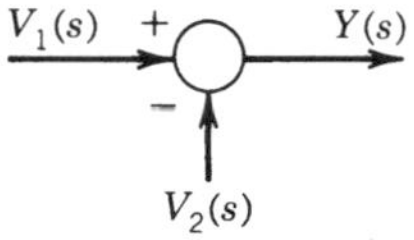 $V_1(s) \xrightarrow{+} \bigcirc \rightarrow Y(s)$, $\xleftarrow{-} V_2(s)$
(e) Takeoff point	$y(t) = v(t)$	$Y(s) = V(s)$	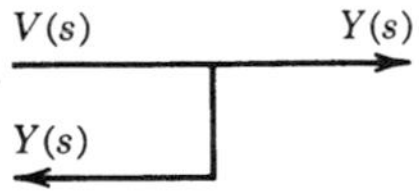 $V(s) \rightarrow Y(s)$, $\downarrow Y(s)$

Figure 12 Basic block diagram elements.[4]

of a given system; and in *specifying* behavior in order to formulate desired behavioral norms for system design. Characterization of dynamic behavior in terms of standard concepts also leads in many cases to analytical shortcuts since key features of the system response frequently can be determined without actually solving the system model.

Parts of the Complete Response

A variety of names are used to identify terms in the response of a fixed linear system. The complete response of a system may be thought of alternatively as the sum of the following:

1. The *free response* (or complementary or homogeneous solution) and the *forced response* (or particular solution). The free response represents the natural response of a system when inputs are removed and the system responds to some initial stored energy. The forced response of the system depends on the form of the input only.

2. The *transient response* and the *steady-state response*. The transient response is that part of the output that decays to zero as time progresses. The steady-state response is that part of the output that remains after all the transients disappear.

3. The *zero-state response* and the *zero-input response*. The zero-state response is the complete response (both free and forced responses) to the input when the initial state is zero. The zero-input response is the complete response of the system to the initial state when the input is zero.

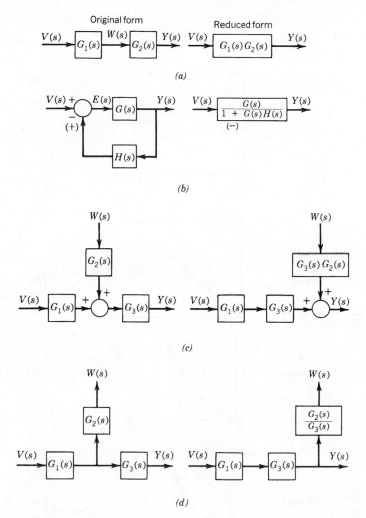

Figure 13   Representative block diagram transformations: (*a*) series or cascaded elements, (*b*) feedback loop, (*c*) relocated summer, and (*d*) relocated takeoff point.[4]

Test Inputs or Singularity Functions

For a stable system, the response to a specific input signal will provide several measures of system performance. Since the actual inputs to a system are not usually known a priori, characterization of the system behavior is generally given in terms of the response to one of a standard set of *test input signals*. This approach provides a common basis for the comparison of different systems. In addition, many inputs actually encountered can be approximated by some combination of standard inputs. The most commonly used test inputs are members of the family of *singularity functions,* depicted in Fig. 14.

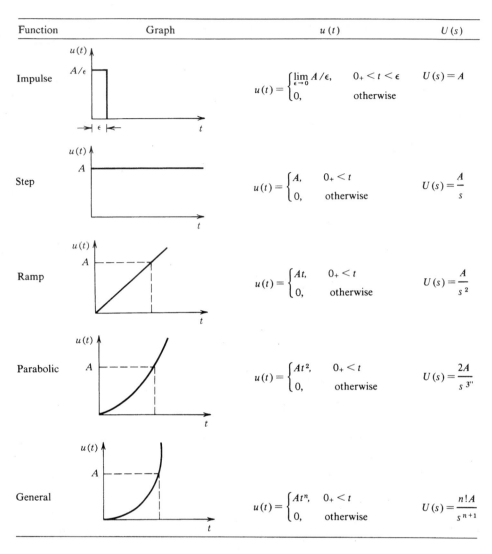

Figure 14   Family of singularity functions commonly used as test inputs.

First-Order Transient Response

The standard form of the I/O equation for a first-order system is

$$\frac{dy}{dt} + \frac{1}{\tau}y(t) = \frac{1}{\tau}u(t)$$

where the parameter τ is called the system *time constant*. The response of this standard first-order system to three test inputs is depicted in Fig. 15, assuming zero initial conditions on the output $y(t)$. For all inputs, it is clear that the response approaches its steady state monotonically (i.e., without oscillations) and that the *speed of response* is completely characterized by the time constant τ. The transfer function of the system is

$$H(s) = \frac{Y(s)}{U(s)} = \frac{1/\tau}{s + 1/\tau}$$

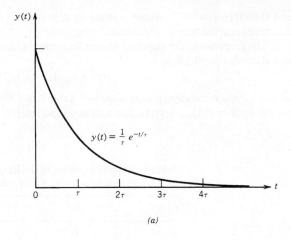

(a)

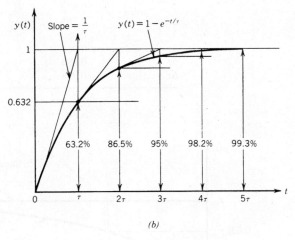

(b)

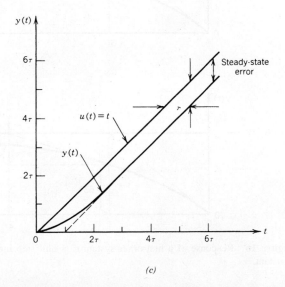

(c)

Figure 15   Response of a first-order system to (a) unit impulse, (b) unit step, and (c) unit ramp inputs.

and therefore $\tau = -p^{-1}$, where p is the system pole. As the absolute value of p increases, τ decreases and the response becomes faster.

The response of the standard first-order system to a step input of magnitude u for arbitrary initial condition $y(0) = y_0$ is

$$y(t) = y_{ss} - [y_{ss} - y_0]e^{-t/\tau}$$

where $y_{ss} = u$ is the steady-state response. Table 6 and Fig. 16 record the values of $y(t)$ and $\dot{y}(t)$ for $t = k\tau, k = 0, 1, ..., 6$. Note that over any time interval of duration τ, the response increases

Table 6 Tabulated Values of the Response of a First-Order System to a Unit Step Input

t	$y(t)$	$\dot{y}(t)$
0	0	τ^{-1}
τ	0.632	$0.368\tau^{-1}$
2τ	0.865	$0.135\tau^{-1}$
3τ	0.950	$0.050\tau^{-1}$
4τ	0.982	$0.018\tau^{-1}$
5τ	0.993	$0.007\tau^{-1}$
6τ	0.998	$0.002\tau^{-1}$

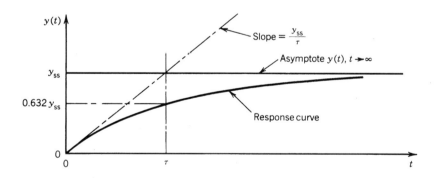

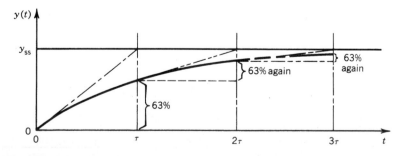

Figure 16 Response of a first-order system to a unit step input, showing the relationship to the time constant.

approximately 63% of the difference between the steady-state value and the value at the beginning of the time interval, that is,

$$y(t + \tau) - y(t) \approx 0.63212[y_{ss} - y(t)]$$

Note also that the slope of the response at the beginning of any time interval of duration τ intersects the steady-state value y_{ss} at the end of the interval, that is,

$$\frac{dy}{dt}(t) = \frac{y_{ss} - y(t)}{\tau}$$

Finally, note that after an interval of four time constants, the response is within 98% of the steady-state value, that is,

$$y(4\tau) \approx 0.98168(y_{ss} - y_0)$$

For this reason, $T_s = 4\tau$ is called the (2%) *setting time*.

Second-Order Transient Response

The standard form of the I/O equation for a second-order system is

$$\frac{d^2y}{dt^2} + 2\zeta\omega_n\frac{dy}{dt} + \omega_n^2 y(t) = \omega_n^2 u(t)$$

with transfer function

$$H(s) = \frac{Y(s)}{U(s)} = \frac{\omega_n^2}{s^2 + 2\zeta\omega_n s + \omega_n^2}$$

The system poles are obtained by applying the quadratic formula to the characteristic equation as

$$p_{1,2} = -\zeta\omega_n \pm j\omega_n\sqrt{1 - \zeta^2}$$

where the following parameters are defined: ζ is the *damping ratio*, ω_n is the *natural frequency*, and $\omega_d = \omega_n\sqrt{1 - \zeta^2}$ is the *damped natural frequency*.

The nature of the response of the standard second-order system to a step input depends on the value of the damping ratio, as depicted in Fig. 17. For a stable system, four classes of response are defined.

1. *Overdamped Response* ($\zeta > 1$). The system poles are real and distinct. The response of the second-order system can be decomposed into the response of two cascaded first-order systems, as shown in Fig. 18.

2. *Critically Damped Response* ($\zeta > 1$). The system poles are real and repeated. This is the limiting case of overdamped response, where the response is as fast as possible without overshoot.

3. *Underdamped Response* ($1 > \zeta > 0$). The system poles are complex conjugates. The response oscillates at the damped frequency ω_d. The magnitude of the oscillations and the speed with which the oscillations decay depend on the damping ratio ζ.

4. *Harmonic Oscillation* ($\zeta = 0$). The system poles are pure imaginary numbers. The response oscillates at the natural frequency ω_n and the oscillations are undamped (i.e., the oscillations are sustained and do not decay).

The Complex s Plane

The location of the system poles (roots of the characteristic equation) in the *complex s plane* reveals the nature of the system response to test inputs. Figure 19 shows the relationship between the location of the poles in the complex plane and the parameters of the standard second-order system. Figure 20 shows the unit impulse response of a second-order system corresponding to various pole locations in the complex plane.

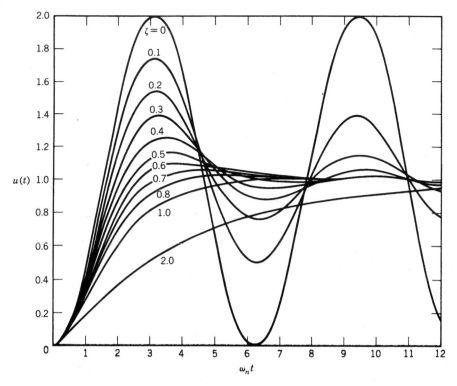

Figure 17   Response of a second-order system to a unit step input for selected values of the damping ratio.

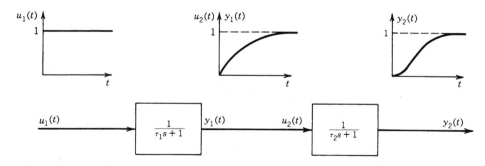

Figure 18   Overdamped response of a second-order system decomposed into the responses of two first-order systems.

Transient Response of Higher Order Systems

The response of third- and higher order systems to test inputs is simply the sum of terms representing component first- and second-order responses. This is because the system poles must either be real, resulting in first-order terms, or complex, resulting in second-order underdamped terms. Furthermore, because the transients associated with those system poles having the largest

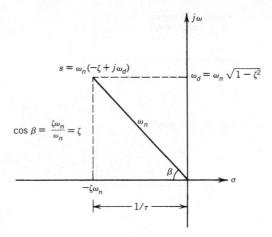

Figure 19 Location of the upper complex pole in the s plane in terms of the parameters of the standard second-order system.

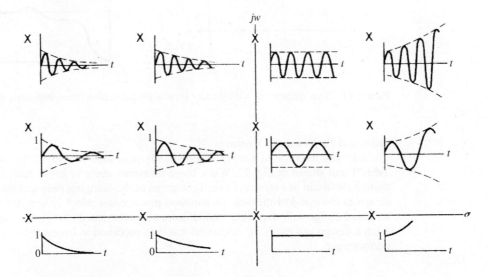

Figure 20 Unit impulse response for selected upper complex pole locations in the s plane.

real part decay the most slowly, these transients tend to dominate the output. The response of higher order systems therefore tends to have the same form as the response to the *dominant poles,* with the response to the *subdominant poles* superimposed over it. Note that the larger the relative difference between the real parts of the dominant and subdominant poles, the more the output tends to resemble the dominant mode of response.

For example, consider a fixed linear third-order system. The system has three poles. Either the poles may all be real or one may be real while the other pair is complex conjugate. This leads to the three forms of step response shown in Fig. 21, depending on the relative locations of the poles in the complex plane.

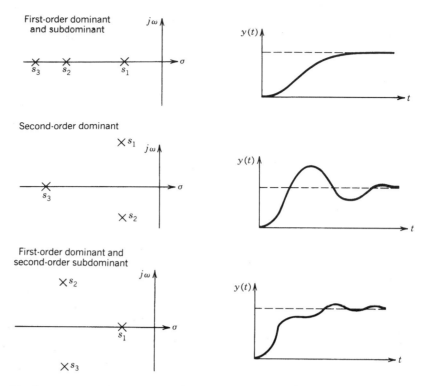

Figure 21   Step response of a third-order system for alternative upper complex pole locations in the s plane.

Transient Performance Measures

The transient response of a system is commonly described in terms of the measures defined in Table 7 and shown in Fig. 22. While these measures apply to any output, for a second-order system these can be calculated exactly in terms of the damping ratio and natural frequency, as shown in column 3 of the table. A common practice in control system design is to determine an initial design with dominant second-order poles that satisfy the performance specifications. Such a design can easily be calculated and then modified as necessary to achieve the desired performance.

The Effect of Zeros on the Transient Response

Zeros arise in a system transfer function through the inclusion of one or more derivatives of $u(t)$ among the inputs to the system. By sensing the rate(s) of change of $u(t)$, the system in effect *anticipates* the future values of $u(t)$. This tends to increase the speed of response of the system relative to the input $u(t)$.

The effect of a zero is greatest on the modes of response associated with neighboring poles. For example, consider the second-order system represented by the transfer function

$$H(s) = K \frac{s - z}{(s - p_1)(s - p_2)}$$

If $z = p_1$, then the system responds as a first-order system with $\tau = -p_2^{-1}$; whereas if $z = p_2$, then the system responds as a first-order system with $\tau = -p_1^{-1}$. Such *pole–zero cancellation* can only be achieved mathematically, but it can be approximated in physical systems. Note that

Table 7   Transient Performance Measures Based on Step Response

Performance Measure	Definition	Formula for a Second-Order System
Delay time, t_d	Time required for the response to reach half the final value for the first time	
10–90% rise time, t_r	Time required for the response to rise from 10 to 90% of the final response (used for overdamped responses)	
0–100% rise time, t_r	Time required for the response to rise from 0 to 100% of the final response (used for underdamped responses)	$t_r = \frac{\pi - \beta}{\omega_d}$ where $\beta = \cos^{-1}\zeta$
Peak time, t_p	Time required for the response to reach the first peak of the overshoot	$t_p = \frac{\pi}{\omega_d}$
Maximum overshoot, M_p	The difference in the response between the first peak of the overshoot and the final response	$M_p = e^{-\zeta\pi/\sqrt{1-\zeta^2}}$
Percent overshoot, PO	The ratio of maximum overshoot to the final response expressed as a percentage	$PO = 100e^{-\zeta\pi/\sqrt{1-\zeta^2}}$
Setting time, t_s	The time required for the response to reach and stay within a specified band centered on the final response (usually 2 or 5% of final response band)	$t_s = \frac{4}{\zeta\omega_n}$ (2%band) $t_s = \frac{3}{\zeta\omega_n}$ (5%band)

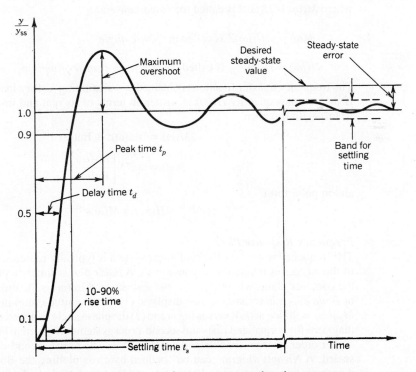

Figure 22   Transient performance measures based on step response.

by diminishing the residue associated with the response mode having the larger time constant, the system responds more quickly to changes in the input, confirming our earlier observation.

5.3   Response to Periodic Inputs Using Transform Methods

The response of a dynamic system to periodic inputs can be a critical concern to the control engineer. An input $u(t)$ is *periodic* if $u(t + T) = u(t)$ for all time t, where T is a constant called the period. Periodic inputs are important because these are ubiquitous: rotating unbalanced machinery, reciprocating pumps and engines, alternating-current (ac) electrical power, and a legion of noise and disturbance inputs can be approximated by periodic inputs. Sinusoids are the most important category of periodic inputs because these are frequently occurring and easily analyzed and form the basis for analysis of general periodic inputs.

Frequency Response

The *frequency response* of a system is the steady-state response of the system to a sinusoidal input. For a linear system, the frequency response has the unique property that the response is a sinusoid of the same frequency as the input sinusoid, differing only in amplitude and phase. In addition, it is easy to show that the amplitude and phase of the response are functions of the input frequency, which are readily obtained from the system transfer function.

Consider a system defined by the transfer function $H(s)$. For an input

$$u(t) = A \sin \omega t$$

the corresponding steady-state output is

$$y_{ss}(t) = AM(\omega) \sin[\omega t + \phi(\omega)]$$

where $M(\omega) = H(j\omega)|$ is called the *magnitude ratio*

$$\phi(\omega) = \angle H(j\omega)Z \text{ is called the } phase\ angle$$

$H(j\omega) = H(s)|_{s=j\omega}$ is called the *frequency transfer function*

The frequency transfer function is obtained by substituting $j\omega$ for s in the transfer function $H(s)$. If the complex quantity $H(j\omega)$ is written in terms of its real and imaginary parts as $H(j\omega) = \text{Re}(\omega) + j\text{Im}(\omega)$, then

$$M(\omega) = [\text{Re}(\omega)^2 + \text{Im}(\omega)^2]^{1/2}$$

$$\phi(\omega) = \tan^{-1}\left[\frac{\text{Im}(\omega)}{\text{Re}(\omega)}\right]$$

and in polar form

$$H(j\omega) = M(\omega)e^{j\phi(\omega)}$$

Frequency Response Plots

The frequency response of a fixed linear system is typically represented graphically using one of three types of frequency response plots. A *polar plot* is simply a plot of the vector $H(j\omega)$ in the complex plane, where $\text{Re}(\omega)$ is the abscissa and $\text{Im}(\omega)$ is the ordinate. A *logarithmic plot* or *Bode diagram* consists of two displays: (1) the magnitude ratio in decibels $M_{dB}(\omega)$ [where $M_{dB}(\omega) = 20 \log M(\omega)$] versus $\log \omega$ and (2) the phase angle in degrees $\phi(\omega)$ versus $\log \omega$. Bode diagrams for normalized first- and second-order systems are given in Fig. 23. Bode diagrams for higher order systems are obtained by adding these first- and second-order terms, appropriately scaled. A *Nichols diagram* can be obtained by cross plotting the Bode magnitude and phase diagrams, eliminating $\log \omega$. Polar plots and Bode and Nichols diagrams for common transfer functions are given in Table 8.

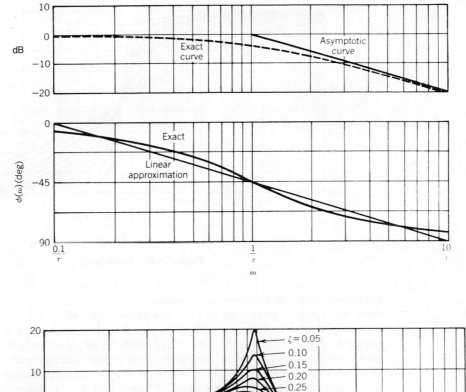

dB

$\phi(\omega)$ (deg)

$20 \log |G|$

$u = \omega/\omega_n$ = Frequency ratio

(a)

Figure 23(a) Bode diagrams for normalized (*a*) first-order and (*b*) second-order systems.

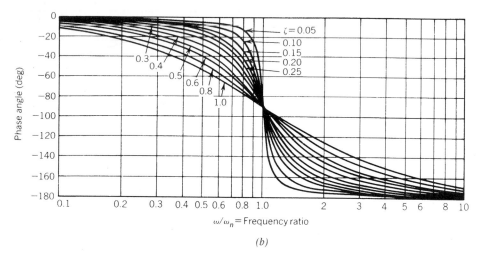

(b)

Figure 23(b)   *(continued)*

Frequency Response Performance Measures

Frequency response plots show that dynamic systems tend to behave like *filters,* "passing" or even amplifying certain ranges of input frequencies while blocking or attenuating other frequency ranges. The range of frequencies for which the amplitude ratio is no less than 3 dB of its maximum value is called the *bandwidth* of the system. The bandwidth is defined by upper and lower *cutoff frequencies* ω_c, or by $\omega = 0$ and an upper cutoff frequency if $M(0)$ is the maximum amplitude ratio. Although the choice of "down 3 dB" used to define the cutoff frequencies is somewhat arbitrary, the bandwidth is usually taken to be a measure of the range of frequencies for which a significant portion of the input is felt in the system output. The bandwidth is also taken to be a measure of the system speed of response, since attenuation of inputs in the higher frequency ranges generally results from the inability of the system to "follow" rapid changes in amplitude. Thus, a narrow bandwidth generally indicates a sluggish system response.

Response to General Periodic Inputs

The *Fourier series* provides a means for representing a general periodic input as the sum of a constant and terms containing sine and cosine. For this reason the *Fourier series,* together with the superposition principle for linear systems, extends the results of frequency response analysis to the general case of arbitrary periodic inputs. The Fourier series representation of a periodic function $f(t)$ with period $2T$ on the interval $t^* + 2T \geq t \geq t^*$ is

$$f(t) = \frac{a_0}{2} + \sum_{n=1}^{\infty} \left(a_n \cos \frac{n\pi t}{T} + b_n \sin \frac{n\pi t}{T} \right)$$

where

$$a_n = \frac{1}{T} \int_{t^*}^{t^*+2T} f(t) \cos \frac{n\pi t}{T} \, dt$$

$$b_n = \frac{1}{T} \int_{t^*}^{t^*+2T} f(t) \sin \frac{n\pi t}{T} \, dt$$

If $f(t)$ is defined outside the specified interval by a periodic extension of period $2T$ and if $f(t)$ and its first derivative are piecewise continuous, then the series converges to $f(t)$ if t is a point

Table 8 Transfer Function Plots for Representative Transfer Functions[5]

$G(s)$	Polar Plot	Bode Diagram
1. $\dfrac{K}{s\tau_1 + 1}$		
2. $\dfrac{K}{(s\tau_1 + 1)(s\tau_2 + 1)}$		
3. $\dfrac{K}{(s\tau_1 + 1)(s\tau_2 + 1)(s\tau_3 + 1)}$		
4. $\dfrac{K}{s}$		

(continued)

Table 8   (*continued*)

Nichols Diagram	Root Locus	Comments
		Stable; gain margin $= \infty$
		Elementary regulator; stable; gain margin $= \infty$
		Regulator with additional energy storage component; unstable, but can be made stable by reducing gain
		Ideal integrator; stable

Table 8 (*continued*)

$G(s)$	Polar Plot	Bode Diagram
5. $\dfrac{K}{s(s\tau_1 + 1)}$	$\omega \to 0$ $-\omega$ -1 $\omega = \infty$ $+\omega$ $\omega \to 0$	-6 db/oct $-90°$ $\phi\ M$ Phase margin $-180°$ 0 db $\frac{1}{\tau_1}$ log ω -12 db/oct
6. $\dfrac{K}{s(s\tau_1 + 1)(s\tau_2 + 1)}$	$\omega \to 0$ $-\omega$ -1 $\omega = \infty$ $+\omega$ $\omega \to 0$	$-90°$ $\phi\ M$ -6 Phase margin Gain margin $1/\tau_2$ $-180°$ 0 db $\frac{1}{\tau_1}$ -12 log ω $-270°$ -18 db/oct
7. $\dfrac{K(s\tau_a + 1)}{s(s\tau_1 + 1)(s\tau_2 + 1)}$	$\omega \to 0$ $-\omega$ -1 $\omega = \infty$ $+\omega$ $\omega \to 0$	-6 db oct $-90°$ $\phi\ M$ Phase margin -12 ϕ $1/\tau_2$ $-180°$ 0 db $\frac{1}{\tau_1}$ $\frac{1}{\tau_a}$ -6 log ω -12 db/oct
8. $\dfrac{K}{s^2}$	$\omega \to$ $-\omega$ -1 $\omega = \infty$ $+\omega$	Gain margin $= 0$ Phase margin $= 0$ $\phi\ M$ -12 db/oct $-180°$ 0 db $\dfrac{\phi}{\log \omega} \to$

(*continued*)

Table 8   (*continued*)

Nichols Diagram	Root Locus	Comments
		Elementary instrument servo; inherently stable; gain margin $= \infty$
		Instrument servo with field control motor or power servo with elementary Ward–Leonard drive; stable as shown but may become unstable with increased gain
		Elementary instrument servo with phase lead (derivative) compensator; stable
		Inherently unstable; must be compensated

Table 8 (*continued*)

$G(s)$	Polar Plot	Bode Diagram
9. $\dfrac{K}{s^2(s\tau_1 + 1)}$		
10. $\dfrac{K(s\tau_a + 1)}{s^2(s\tau_1 + 1)}$ $\tau_a > \tau_1$		
11. $\dfrac{K}{s^3}$		
12. $\dfrac{K(s\tau_a + 1)}{s^3}$		

(*continued*)

Table 8   (*continued*)

Nichols Diagram	Root Locus	Comments
		Inherently unstable; must be compensated
		Stable for all gains
		Inherently unstable
		Inherently unstable

Table 8   (*continued*)

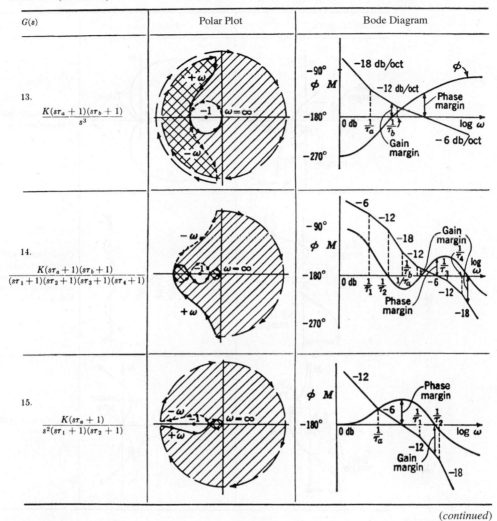

$G(s)$	Polar Plot	Bode Diagram
13. $\dfrac{K(s\tau_a + 1)(s\tau_b + 1)}{s^3}$		
14. $\dfrac{K(s\tau_a + 1)(s\tau_b + 1)}{(s\tau_1 + 1)(s\tau_2 + 1)(s\tau_3 + 1)(s\tau_4 + 1)}$		
15. $\dfrac{K(s\tau_a + 1)}{s^2(s\tau_1 + 1)(s\tau_2 + 1)}$		

(*continued*)

of continuity or to $\frac{1}{2}[f(t_+) + f(t_-)]$ if t is a point of discontinuity. Note that while the Fourier series in general is infinite, the notion of bandwidth can be used to reduce the number of terms required for a reasonable approximation.

6   STATE-VARIABLE METHODS

State-variable methods use the vector state and output equations introduced in Section 4 for analysis of dynamic systems directly in the time domain. These methods have several advantages over transform methods. First, state-variable methods are particularly advantageous for the study of multivariable (multiple-input/multiple-output) systems. Second, state-variable

Table 8 *(continued)*

Nichols Diagram	Root Locus	Comments
		Conditionally stable; becomes unstable if gain is too low
		Conditionally stable; stable at low gain, becomes unstable as gain is raised, again becomes stable as gain is further increased, and becomes unstable for very high gains
		Conditionally stable; becomes unstable at high gain

methods are more naturally extended for the study of linear time-varying and nonlinear systems. Finally, state-variable methods are readily adapted to computer simulation studies.

6.1 Solution of the State Equation

Consider the vector equation of state for a fixed linear system:

$$\dot{x}(t) = Ax(t) + Bu(t)$$

The solution to this system is

$$x(t) = \Phi(t)x(0) + \int_0^t \Phi(t - \tau)Bu(\tau)\,d\tau$$

where the matrix $\Phi(t)$ is called the *state transition matrix*. The state transition matrix represents the free response of the system and is defined by the matrix exponential series

$$\Phi(t) = e^{At} = I + At + \frac{1}{2!}A^2t^2 + \cdots = \sum_{k=0}^{\infty} \frac{1}{k!}A^kt^k$$

where I is the identity matrix. The state transition matrix has the following useful properties:

$$\Phi(0) = I$$

$$\Phi^{-1}(t) = \Phi(-t)$$

$$\Phi^k(t) = \Phi(kt)$$

$$\Phi(t_1 + t_2) = \Phi(t_1)\Phi(t_2)$$

$$\Phi(t_2 - t_1)\Phi(t_1 - t_0) = \Phi(t_2 - t_0)$$

$$\dot{\Phi}(t) = A\Phi(t)$$

The Laplace transform of the state equation is

$$sX(s) - x(0) = AX(s) + BU(s)$$

The solution to the fixed linear system therefore can be written as

$$x(t) = \mathcal{L}^{-1}[X(s)]$$

$$= \mathcal{L}^{-1}[\Phi(s)]x(0) + \mathcal{L}^{-1}[\Phi(s)BU(s)]$$

where $\Phi(s)$ is called the *resolvent matrix* and

$$\Phi(t) = \mathcal{L}^{-1}[\Phi(s)] = \mathcal{L}^{-1}[sI - A]^{-1}$$

6.2   Eigenstructure

The internal structure of a system (and therefore its free response) is defined entirely by the system matrix A. The concept of matrix *eigenstructure*, as defined by the eigenvalues and eigenvectors of the system matrix, can provide a great deal of insight into the fundamental behavior of a system. In particular, the system eigenvectors can be shown to define a special set of first-order subsystems embedded within the system. These subsystems behave independently of one another, a fact that greatly simplifies analysis.

System Eigenvalues and Eigenvectors

For a system with system matrix A, the system *eigenvectors* v_i and associated *eigenvalues* λ_i are defined by the equation

$$Av_i = \lambda_i v_i$$

Note that the eigenvectors represent a set of special directions in the state space. If the state vector is aligned in one of these directions, then the homogeneous state equation becomes $\dot{v}_i = Av_i = \lambda v_i$, implying that each of the state variables changes at the *same* rate determined by the eigenvalue λ_i. This further implies that, in the absence of inputs to the system, a state vector that becomes aligned with an eigenvector will remain aligned with that eigenvector.

The system eigenvalues are calculated by solving the nth-order polynomial equation

$$|\lambda I - A| = \lambda^n + a_{n-1}\lambda^{n-1} + \cdots + a_1\lambda + a_0 = 0$$

This equation is called the *characteristic equation*. Thus the system eigenvalues are the roots of the characteristic equation, that is, the system eigenvalues are identically the system poles defined in transform analysis.

Each system eigenvector is determined by substituting the corresponding eigenvalue into the defining equation and then solving the resulting set of simultaneous linear equations. Only $n-1$ of the n components of any eigenvector are independently defined, however. In other words, the magnitude of an eigenvector is arbitrary, and the eigenvector describes a direction in the state space.

Diagonalized Canonical Form

There will be one linearly independent eigenvector for each distinct (nonrepeated) eigenvalue. If all of the eigenvalues of an nth order system are distinct, then the n independent eigenvectors form a new basis for the state space. This basis represents new coordinate axes defining a set of state variables $z_i(t), i = 1, 2, ..., n$, called the *diagonalized canonical variables*. In terms of the diagonalized variables, the homogeneous state equation is

$$\dot{z}(t) = \Lambda z$$

where Λ is a diagonal system matrix of the eigenvectors, that is,

$$\Lambda = \begin{bmatrix} \lambda_1 & 0 & \cdots & 0 \\ 0 & \lambda_2 & \cdots & 0 \\ \vdots & \vdots & \cdots & \vdots \\ 0 & 0 & \vdots & \lambda_n \end{bmatrix}$$

The solution to the diagonalized homogeneous system is

$$zt = e^{\Lambda t}z(0)$$

where $e^{\Lambda t}$ is the diagonal state transition matrix

$$e^{\Lambda t} = \begin{bmatrix} e^{\lambda_1 t} & 0 & \cdots & 0 \\ 0 & e^{\lambda_2 t} & \cdots & 0 \\ \vdots & \vdots & \cdots & \vdots \\ 0 & 0 & \cdots & e^{\lambda_n t} \end{bmatrix}$$

Modal Matrix

Consider the state equation of the nth-order system

$$\dot{x}(t) = Ax(t) + Bu(t)$$

which has real, distinct eigenvalues. Since the system has a full set of eigenvectors, the state vector $x(t)$ can be expressed in terms of the canonical state variables as

$$x(t) = v_1 z_1(t) + v_2 z_2(t) + + v_n z_n(t) = Mz(t)$$

where M is the $n \times n$ matrix whose columns are the eigenvectors of A, called the *modal matrix*. Using the modal matrix, the state transition matrix for the original system can be written as

$$\Phi(t) = e^{\Lambda t} = Me^{\Lambda t}M^{-1}$$

where $e^{\Lambda t}$ is the diagonal state transition matrix. This frequently proves to be an attractive method for determining the state transition matrix of a system with real, distinct eigenvalues.

Jordan Canonical Form

For a system with one or more repeated eigenvalues, there is not in general a full set of eigenvectors. In this case, it is not possible to determine a diagonal representation for the system.

Instead, the simplest representation that can be achieved is block diagonal. Let $L_k(\lambda)$ be the $k \times k$ matrix

$$L_k(\lambda) = \begin{bmatrix} \lambda & 1 & 0 & \cdots & 0 \\ 0 & \lambda & 1 & \cdots & 0 \\ \vdots & \vdots & \lambda & \cdots & 0 \\ \vdots & \vdots & \vdots & \cdots & 1 \\ 0 & 0 & 0 & 0 & \lambda \end{bmatrix}$$

Then for any $n \times n$ system matrix A there is certain to exist a nonsingular matrix T such that

$$T^{-1}AT = \begin{bmatrix} L_{k1}(\lambda_1) & & \\ & L_{k2}(\lambda_2) & \\ & & L_{kr}(\lambda_r) \end{bmatrix}$$

where $k_1 + k_2 + \cdots + k_r = n$ and $\lambda_i, i = 1, 2, ..., r$, are the (not necessarily distinct) eigenvalues of A. The matrix $T^{-1}AT$ is called the *Jordan canonical form*.

7   SIMULATION

7.1   Simulation—Experimental Analysis of Model Behavior

Closed-form solutions for nonlinear or time-varying systems are rarely available. In addition, while explicit solutions for time-invariant linear systems can always be found, for high-order systems this is often impractical. In such cases it may be convenient to study the dynamic behavior of the system using *simulation*.

Simulation is the *experimental* analysis of model behavior. A *simulation run* is a controlled experiment in which a specific realization of the model is manipulated in order to determine the response associated with that realization. A *simulation study* comprises *multiple runs,* each run for a different combination of model parameter values and/or initial conditions. The generalized solution of the model must then be inferred from a finite number of simulated data points.

Simulation is almost always carried out with the assistance of computing equipment. *Digital simulation* involves the *numerical solution* of model equations using a digital computer. *Analog simulation* involves solving model equations by analogy with the behavior of a physical system using an analog computer. *Hybrid simulation* employs digital and analog simulation together using a hybrid (part digital and part analog) computer.

7.2   Digital Simulation

Digital continuous-system simulation involves the approximate solution of a state-variable model over successive time steps. Consider the general state-variable equation

$$\dot{x}(t) = f[x(t), u(t)]$$

to be simulated over the time interval $t_0 \le t \le t_K$. The solution to this problem is based on the repeated solution of the single-variable, single-step subproblem depicted in Fig. 24. The subproblem may be stated formally as follows:

Given:

1. $\Delta t(k) = t_k - t_{k-1}$, the length of the kth *time step.*
2. $x_i(t) = f_i[x(t), u(t)]$ for $t_{k-1} \le t \le t_k$, the ith equation of state defined for the state variable $x_i(t)$ over the kth time step.

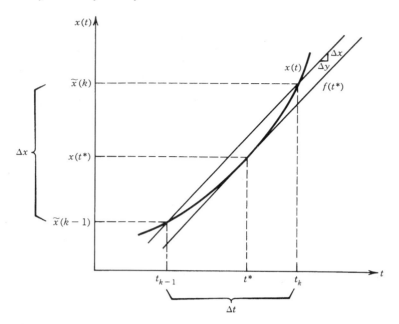

Figure 24 Numerical approximation of a single variable over a single time step.

3. $u(t)$ for $t_{k-1} \leq t \leq t_k$, the input vector defined for the kth time step.

4. $\widetilde{x}(k-1) \simeq x(t_{k-1})$, an initial approximation for the state vector at the beginning of the time step.

Find:

5. $\widetilde{x}_i(k) \simeq x_i(t_k)$, a final approximation for the state variable $x_i(t)$ at the end of the kth time step.

Solving this single-variable, single-step subproblem for each of the state variables $x_i(t)$, $i = 1, 2, ..., n$, yields a final approximation for the state vector $\widetilde{x}(k) \simeq x(t_k)$ at the end of the kth time step. Solving the complete single-step problem K times over K time steps, beginning with the initial condition $\widetilde{x}(0) = x(t_0)$ and using the final value of $\widetilde{x}(t_k)$ from the kth time step as the initial value of the state for the $(k + 1)$st time step, yields a discrete succession of approximations $\widetilde{x}(1) \simeq x(t_1), \widetilde{x}(2) \simeq x(t_2), \ldots, \widetilde{x}(K) \simeq x(t_k)$ spanning the solution time interval.

The basic procedure for completing the single-variable, single-step problem is the same regardless of the particular integration method chosen. It consists of two parts: (1) calculation of the average value of the ith derivative over the time step as

$$\dot{x}_i(t^*) = f_i[x(t^*), u(t^*)] = \frac{\Delta x_i(k)}{\Delta t(k)} \simeq \widetilde{f}_i(k)$$

and (2) calculation of the final value of the simulated variable at the end of the time step as

$$\widetilde{x}_i(k) = \widetilde{x}_i(k-1) + \Delta x_i(k)$$

$$\simeq \widetilde{x}_i(k-1) + \Delta t(k)\widetilde{f}_i(k)$$

If the function $f_i[x(t), u(t)]$ is continuous, then t^* is guaranteed to be on the time step, that is, $t_{k-1} \leq t^* \leq t_k$. Since the value of t^* is otherwise unknown, however, the value of $x(t^*)$ can only be approximated as $\widetilde{f}(k)$.

Different *numerical integration* methods are distinguished by the means used to calculate the approximation $\tilde{f}_i(k)$. A wide variety of such methods is available for digital simulation of dynamic systems. The choice of a particular method depends on the nature of the model being simulated, the accuracy required in the simulated data, and the computing effort available for the simulation study. Several popular classes of integration methods are outlined in the following sections.

Euler Method

The simplest procedure for numerical integration is the Euler method. The standard Euler method approximates the average value of the *i*th derivative over the *k*th time step using the derivative evaluated at the beginning of the time step, that is,

$$\tilde{f}_i(k) = f_i[\tilde{x}(k-1), u(t_{k-1})] \simeq f_i(t_{k-1})$$

$i = 1, 2, ..., n$ and $k = 1, 2, ..., K$. This is shown geometrically in Fig. 25 for the scalar single-step case. A modification of this method uses the newly calculated state variables in the derivative calculation as these new values become available. Assuming the state variables are computed in numerical order according to the subscripts, this implies

$$\tilde{f}_i(k) = f_i[\tilde{x}_1(k), \ldots, \tilde{x}_{i-1}(k), \tilde{x}_i(k-1), \ldots, \tilde{x}_n(k-1), u(t_{k-1})]$$

The modified Euler method is modestly more efficient than the standard procedure and, frequently, is more accurate. In addition, since the input vector $u(t)$ is usually known for the entire time step, using an average value of the input, such as

$$u(k) = \frac{1}{\Delta t\,(k)} \int_{t_{k-1}}^{t_k} u(\tau)\,d\tau$$

frequently leads to a superior approximation of $\tilde{f}_i(k)$.

The Euler method requires the least amount of computational effort per time step of any numerical integration scheme. Local truncation error is proportional to Δt^2, however, which means that the error within each time step is highly sensitive to step size. Because the accuracy of the method demands very small time steps, the number of time steps required to implement

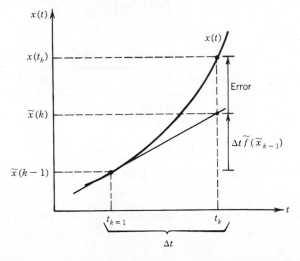

Figure 25   Geometric interpretation of the Euler method for numerical integration.

the method successfully can be large relative to other methods. This can imply a large computational overhead and can lead to inaccuracies through the accumulation of roundoff error at each step.

Runge–Kutta Methods

Runge–Kutta methods precompute two or more values of $f_i[x(t), u(t)]$ in the time step $t_{k-1} \le t \le t_k$ and use some weighted average of these values to calculate $\tilde{f}_i(k)$. The *order* of a Runge–Kutta method refers to the number of derivative terms (or *derivative calls*) used in the scalar single-step calculation. A Runge–Kutta routine of order N therefore uses the approximation

$$\tilde{f}_i(k) = \sum_{j=1}^{N} w_j f_{ij}(k)$$

where the N approximations to the derivative are

$$f_{i1}(k) = f_i[\tilde{x}(k-1), u(t_{k-1})]$$

(the Euler approximation) and

$$f_{ij} = f_i \left[\tilde{x}(k-1) + \Delta t \sum_{t=1}^{j-1} I b_{jt} f_{il}, u \left(t_{k-1} + \Delta t \sum_{i=1}^{j-1} b_{jl} \right) \right]$$

where I is the identity matrix. The weighting coefficients w_j and b_{jl} are not unique, but are selected such that the error in the approximation is zero when $x_i(t)$ is some specified Nth-degree polynomial in t. Coefficients commonly used for Runge–Kutta integration are given in Table 9.

Among the most popular of the Runge–Kutta methods is fourth-order Runge–Kutta. Using the defining equations for $N = 4$ and the weighting coefficients from Table 9 yields the derivative approximation

$$\tilde{f}_i(k) = \frac{1}{6}[f_{i1}(k) + 2f_{i2}(k) + 2f_{i3}(k) + f_{i4}(k)]$$

based on the four derivative calls

$$f_{i1}(k) = f_i[\tilde{x}(k-1), u(t_{k-1})]$$

$$f_{i2}(k) = f_i \left[\tilde{x}(k-1) + \frac{\Delta t}{2} I f_{i1}, u \left(t_{k-1} + \frac{\Delta t}{2} \right) \right]$$

Table 9   Coefficients Commonly Used for Runge–Kutta Numerical Integration[6]

Common Name	N	b_{jl}	w_j
Open or explicit Euler	1	All zero	$w_1 = 1$
Improved polygon	2	$b_{21} = \frac{1}{2}$	$w_1 = 0$
			$w_2 = 1$
Modified Euler or Heun's method	2	$b_{21} = 1$	$w_1 = \frac{1}{2}$
			$w_2 = \frac{1}{2}$
Third-order Runge–Kutta	3	$b_{21} = \frac{1}{2}$	$w_1 = \frac{1}{6}$
		$b_{31} = -1$	$w_2 = \frac{2}{3}$
		$b_{32} = 2$	$w_3 = \frac{1}{6}$
Fourth-order Runge–Kutta	4	$b_{21} = \frac{1}{2}$	$w_1 = \frac{1}{6}$
		$b_{31} = 0$	$w_2 = \frac{1}{3}$
		$b_{32} = \frac{1}{2}$	$w_3 = \frac{1}{3}$
		$b_{43} = 1$	$w_4 = \frac{1}{6}$

$$f_{i3}(k) = f_i \left[\widetilde{x}(k-1) + \frac{\Delta t}{2} I f_{i2}, u \left(t_{k-1} + \frac{\Delta t}{2} \right) \right]$$

$$f_{i4}(k) = f_i[\widetilde{x}(k-1) + \Delta t \, I f_{i3}, u(t_k)]$$

where I is the identity matrix.

Because Runge–Kutta formulas are designed to be exact for a polynomial of order N, local truncation error is of the order Δt^{N+1}. This considerable improvement over the Euler method means that comparable accuracy can be achieved for larger step sizes. The penalty is that N derivative calls are required for each scalar evaluation within each time step.

Euler and Runge–Kutta methods are examples of *single-step methods* for numerical integration, so-called because the state $x(k)$ is calculated from knowledge of the state $x(k-1)$, without requiring knowledge of the state at any time prior to the beginning of the current time step. These methods are also referred to as *self-starting methods,* since calculations may proceed from any known state.

Multistep Methods

Multistep methods differ from the single-step methods previously described in that multistep methods use the stored values of two or more previously computed states and/or derivatives in order to compute the derivative approximation $\widetilde{f}_i(k)$ for the current time step. The advantage of multistep methods over Runge–Kutta methods is that these require only one derivative call for each state variable at each time step for comparable accuracy. The disadvantage is that multistep methods are not self-starting, since calculations cannot proceed from the initial state alone. Multistep methods must be started, or restarted in the case of discontinuous derivatives, using a single-step method to calculate the first several steps.

The most popular of the multistep methods are the *Adams–Bashforth predictor methods* and the *Adams–Moulton corrector methods*. These methods use the derivative approximation

$$\widetilde{f}_i(k) = \sum_{j=0}^{N} b_j f_i[\widetilde{x}(k-j), u(k-j)]$$

where the b_j are weighting coefficients. These coefficients are selected such that the error in the approximation is zero when $x_i(t)$ is a specified polynomial. Table 10 gives the values of the weighting coefficients for several Adams–Bashforth–Moulton rules. Note that the predictor methods employ an *open* or *explicit rule*, since for these methods $b_0 = 0$ and a prior estimate of $x_i(k)$ is not required. The corrector methods use a *closed* or *implicit rule*, since for these methods $b_i \neq 0$ and a prior estimate of $x_i(k)$ is required. Note also that for all of these methods $\Sigma_{j=0}^{N} b_j = 1$, ensuring unity gain for the integration of a constant.

Table 10 Coefficients Commonly Used for Adams–Bashforth–Moulton Numerical Integration[6]

Common Name	Predictor or Corrector	Points	b_{-1}	b_0	b_1	b_2	b_3
Open or explicit Euler	Predictor	1	0	1	0	0	0
Open trapezoidal	Predictor	2	0	$3/2$	$-1/2$	0	0
Adams three-point predictor	Predictor	3	0	$23/12$	$-16/12$	$5/12$	0
Adams four-point predictor	Predictor	4	0	$55/24$	$-59/24$	$37/24$	$-9/24$
Closed or implicit Euler	Corrector	1	1	0	0	0	0
Closed trapezoidal	Corrector	2	$1/2$	$1/2$	0	0	0
Adams three-point corrector	Corrector	3	$5/12$	$8/12$	$-1/2$	0	0
Adams four-point corrector	Corrector	4	$9/24$	$19/24$	$-5/24$	$1/24$	0

Predictor–Corrector Methods

Predictor–corrector methods use one of the multistep predictor equations to provide an initial estimate (or "prediction") of $x(k)$. This initial estimate is then used with one of the multistep corrector equations to provide a second and improved (or "corrected") estimate of $x(k)$ before proceeding to the next step. A popular choice is the four-point Adams–Bashforth predictor together with the four-point Adams–Moulton corrector, resulting in a prediction of

$$\widetilde{x}_i(k) = \widetilde{x}_i(k-1) + \frac{\Delta t}{24}[55\,\widetilde{f}_i(k-1) - 59\widetilde{f}_i(k-2) + 37\widetilde{f}_i(k-3) - 9\widetilde{f}_i(k-4)]$$

for $i = 1, 2, ..., n$ and a correction of

$$\widetilde{x}_i(k) = \widetilde{x}_i(k-1) + \frac{\Delta t}{24}\{9f_i[\widetilde{x}(k), u(k)] + 19\widetilde{f}_i(k-1) - 5\widetilde{f}_i(k-2) + \widetilde{f}_i(k-3)\}$$

Predictor–corrector methods generally incorporate a strategy for increasing or decreasing the size of the time step depending on the difference between the predicted and corrected $x(k)$ values. Such *variable time step methods* are particularly useful if the simulated system possesses local time constants that differ by several orders of magnitude or if there is little a priori knowledge about the system response.

Numerical Integration Errors

An inherent characteristic of digital simulation is that the discrete data points generated by the simulation $x(k)$ are only approximations to the exact solution $x(t_k)$ at the corresponding point in time. This results from two types of errors that are unavoidable in the numerical solutions. *Round-off errors* occur because numbers stored in a digital computer have finite word length (i.e., a finite number of bits per word) and therefore limited precision. Because the results of calculations cannot be stored exactly, round-off error tends to increase with the number of calculations performed. For a given total solution interval $t_0 \le t \le t_K$, therefore, round-off error tends to increase (1) with increasing integration rule order (since more calculations must be performed at each time step) and (2) with decreasing step size Δt (since more time steps are required).

Truncation errors or numerical approximation errors occur because of the inherent limitations in the numerical integration methods themselves. Such errors would arise even if the digital computer had infinite precision. *Local* or *per-step truncation error* is defined as

$$e(k) = x(k) - x(t_k)$$

given that $x(k-1) = x(t_{k-1})$ and that the calculation at the kth time step is infinitely precise. For many integration methods, local truncation errors can be approximated at each step. *Global* or *total truncation error* is defined as

$$e(K) = x(K) - x(t_k)$$

given that $x(0) = x(t_0)$ and the calculations for all K time steps are infinitely precise. Global truncation error usually cannot be estimated, neither can efforts to reduce local truncation errors be guaranteed to yield acceptable global errors. In general, however, truncation errors can be decreased by using more sophisticated integration methods and by decreasing the step size Δt.

Time Constants and Time Steps

As a general rule, the step size Δt for simulation must be less than the smallest local time constant of the model simulated. This can be illustrated by considering the simple first-order system

$$\dot{x}(t) = \lambda x(t)$$

and the difference equation defining the corresponding Euler integration

$$x(k) = x(k-1) + \Delta t\,\lambda x(k-1)$$

The continuous system is stable for $\lambda < 0$, while the discrete approximation is stable for $|1 + \lambda \Delta t| < 1$. If the original system is stable, therefore, the simulated response will be stable for

$$\Delta t \leq 2 \left| \frac{1}{\lambda} \right|$$

where the equality defines the *critical step size*. For larger step sizes, the simulation will exhibit *numerical instability*. In general, while higher order integration methods will provide greater per-step accuracy, the critical step size itself will not be greatly reduced.

A major problem arises when the simulated model has one or more time constants $|1/\lambda_i|$ that are small when compared to the total solution time interval $t_0 \leq t \leq t_K$. Numerical stability will then require very small Δt, even though the transient response associated with the higher frequency (larger λ_i) subsystems may contribute little to the particular solution. Such problems can be addressed either by neglecting the higher frequency components where appropriate or by adopting special numerical integration methods for *stiff systems*.

Selecting an Integration Method

The best numerical integration method for a specific simulation is the method that yields an acceptable global approximation error with the minimum amount of roundoff error and computing effort. No single method is best for all applications. The selection of an integration method depends on the model simulated, the purpose of the simulation study, and the availability of computing hardware and software.

In general, for well-behaved problems with continuous derivatives and no stiffness, a lower order Adams predictor is often a good choice. Multistep methods also facilitate estimating local truncation error. Multistep methods should be avoided for systems with discontinuities, however, because of the need for frequent restarts. Runge–Kutta methods have the advantage that these are self-starting and provide fair stability. For stiff systems where high-frequency modes have little influence on the global response, special stiff-system methods enable the use of economically large step sizes. Variable-step rules are useful when little is known a priori about solutions. Variable-step rules often make a good choice as general-purpose integration methods.

Roundoff error usually is not a major concern in the selection of an integration method, since the goal of minimizing computing effort typically obviates such problems. Double-precision simulation can be used where roundoff is a potential concern. An upper bound on step size often exists because of discontinuities in derivative functions or because of the need for response output at closely spaced time intervals.

Continuous-System Simulation Languages

Digital simulation can be implemented for a specific model in any high-level language such as FORTRAN or C. The general process for implementing a simulation is shown in Fig. 26. In addition, many special-purpose continuous-system simulation languages are commonly available across a wide range of platforms. Such languages greatly simplify programming tasks and typically provide for good graphical output.

8 MODEL CLASSIFICATIONS

Mathematical models of dynamic systems are distinguished by several criteria that describe fundamental properties of model variables and equations. These criteria in turn prescribe the theory and mathematical techniques that can be used to study different models. Table 11 summarizes these distinguishing criteria. In the following sections, the approaches adopted for the analysis of important classes of systems are briefly outlined.

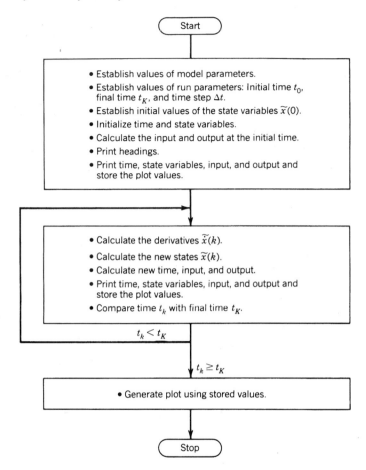

Figure 26 General process for implementing digital simulation (adapted from Close and Frederick[3]).

8.1 Stochastic Systems

Systems in which some of the dependent variables (input, state, output) contain random components are called *stochastic systems*. Randomness may result from environmental factors, such as wind gusts or electrical noise, or simply from a lack of precise knowledge of the system model, such as when a human operator is included within a control system. If the randomness in the system can be described by some rule, then it is often possible to derive a model in terms of probability distributions involving, for example, the means and variances of model variables or parameters.

State-Variable Formulation

A common formulation is the fixed, linear model with additive noise

$$\dot{x}(t) = Ax(t) + Bu(t) + w(t)$$
$$y(t) = Cx(t) + v(t)$$

Table 11 Classification of Mathematical Models of Dynamic Systems

Criterion	Classification	Description
Certainty	Deterministic	Model parameters and variables can be known with certainty. Common approximation when uncertainties are small.
	Stochastic	Uncertainty exists in the values of some parameters and/or variables. Model parameters and variables are expressed as random numbers or processes and are characterized by the parameters of probability distributions.
Spatial characteristics	Lumped	State of the system can be described by a finite set of state variables. Model is expressed as a discrete set of point functions described by ordinary differential or difference equations.
	Distributed	State depends on both time and spatial location. Model is usually described by variables that are continuous in time and space, resulting in partial differential equations. Frequently approximated by lumped elements. Typical in the study of structures and mass and heat transport.
Parameter variation	Fixed or time invariant	Model parameters are constant. Model described by differential or difference equations with constant coefficients. Model with same initial conditions and input delayed by t_d has the same response delayed by t_d.
	Time varying	Model parameters are time dependent.
Superposition property	Linear	Superposition applies. Model can be expressed as a system of linear difference or differential equations.
	Nonlinear	Superposition does not apply. Model is expressed as a system of nonlinear difference or differential equations. Frequently approximated by linear systems for analytical ease.
Continuity of independent variable (time)	Continuous	Dependent variables (input, output, state) are defined over a continuous range of the independent variable (time), even though the dependence is not necessarily described by a mathematically continuous function. Model is expressed as differential equations. Typical of physical systems.
	Discrete	Dependent variables are defined only at distinct instants of time. Model is expressed as difference equations. Typical of digital and nonphysical systems.
	Hybrid	System with continuous and discrete subsystems, most common in computer control and communication systems. Sampling and quantization typical in A/D (analog-to-digital) conversion; signal reconstruction for D/A conversion. Model frequently approximated as entirely continuous or entirely discrete.
Quantization of dependent variables	Nonquantized	Dependent variables are continuously variable over a range of values. Typical of physical systems at macroscopic resolution.
	Quantized	Dependent variables assume only a countable number of different values. Typical of computer control and communication systems (sample data systems).

where $w(t)$ is a zero-mean Gaussian disturbance and $v(t)$ is a zero-mean Gaussian measurement noise. This formulation is the basis for many *estimation problems,* including the problem of *optimal filtering.* Estimation essentially involves the development of a rule or algorithm for determining the best estimate of the past, current, or future values of measured variables in the presence of disturbances or noise.

Random Variables

In the following, important concepts for characterizing random signals are developed. A *random variable x* is a variable that assumes values that cannot be precisely predicted a priori. The likelihood that a random variable will assume a particular value is measured as the *probability* of that value. The probability *distribution function $F(x)$* of a continuous random variable x is defined as the probability that x assumes a value no greater than x, that is,

$$F(x) = \Pr(X \le x) = \int_{-\infty}^{x} f(x) \, dx$$

The probability *density function $f(x)$* is defined as the derivative of $F(x)$.

The *mean* or *expected value* of a probability distribution is defined as

$$E(X) = \int_{-\infty}^{\infty} xf(x) \, dx = \overline{X}$$

The mean is the first moment of the distribution. The *nth moment* of the distribution is defined as

$$E(X^n) = \int_{-\infty}^{\infty} x^n f(x) \, dx$$

The mean square of the difference between the random variable and its mean is the *variance* or *second central moment* of the distribution,

$$\sigma^2(X) = E(X - \overline{X})^2 = \int_{-\infty}^{\infty} (x - \overline{X})^2 f(x) \, dx = E(X^2) - [E(X)]^2$$

The square root of the variance is the *standard deviation* of the distribution:

$$\sigma(X) = \sqrt{E(X^2) - [E(X)]^2}$$

The mean of the distribution therefore is a measure of the average magnitude of the random variable, while the variance and standard deviation are measures of the variability or dispersion of this magnitude.

The concepts of probability can be extended to more than one random variable. The *joint distribution* function of two random variables x and y is defined as

$$F(x, y) = \Pr(X < x \text{ and } Y < y) = \int_{-\infty}^{x} \int_{-\infty}^{y} f(x, y) \, dy \, dx$$

where $f(x, y)$ is the joint distribution. The *ij*th moment of the joint distribution is

$$E(X^i Y^j) = \int_{-\infty}^{\infty} x^i \int_{-\infty}^{\infty} y^i (x, y) \, dy \, dx$$

The *covariance* of x and y is defined to be

$$E[(X - \overline{X})(Y - \overline{Y})]$$

and the normalized covariance or *correlation coefficient* as

$$\rho = \frac{E[(X - \overline{X})(Y - \overline{Y})]}{\sqrt{\sigma^2(X)\sigma^2(Y)}}$$

Although many distribution functions have proven useful in control engineering, far and away the most useful is the *Gaussian* or *normal distribution*

$$F(x) = \frac{1}{\sigma\sqrt{2\pi}} \exp\left[\frac{(-x-\mu)^2}{2\sigma^2}\right]$$

where μ is the mean of the distribution and σ is the standard deviation. The Gaussian distribution has a number of important properties. First, if the input to a linear system is Gaussian, the output also will be Gaussian. Second, if the input to a linear system is only approximately Gaussian, the output will tend to approximate a Gaussian distribution even more closely. Finally, a Gaussian distribution can be completely specified by two parameters, μ and σ, and therefore a zero-mean Gaussian variable is completely specified by its variance.

Random Processes

A *random process* is a set of random variables with time-dependent elements. If the statistical parameters of the process (such as σ for the zero-mean Gaussian process) do not vary with time, the process is *stationary*. The *autocorrelation function* of a stationary random variable $x(t)$ is defined by

$$\phi_{xx}(\tau) = \lim_{T\to\infty} \frac{1}{2T} \int_{-T}^{T} x(t)x(t+\tau)\, dt$$

a function of the fixed time interval τ. The autocorrelation function is a quantitative measure of the sequential dependence or time correlation of the random variable, that is, the relative effect of prior values of the variable on the present or future values of the variable. The autocorrelation function also gives information regarding how rapidly the variable is changing and about whether the signal is in part deterministic (specifically, periodic). The autocorrelation function of a zero-mean variable has the properties

$$\sigma^2 = \phi_{xx}(0) \geq \phi_{xx}(\tau) \qquad \phi_{xx}(\tau) = \phi_{xx}(-\tau)$$

In other words, the autocorrelation function for $\tau = 0$ is identically the variance and the variance is the maximum value of the autocorrelation function. From the definition of the function, it is clear that (1) for a purely random variable with zero mean, $\phi_{xx}(\tau) = 0$ for $\tau \neq 0$, and (2) for a deterministic variable, which is periodic with period T, $\phi_{xx}(k2\pi T) = \sigma^2$ for k integer. The concept of time correlation is readily extended to more than one random variable. The *cross-correlation function* between the random variables $x(t)$ and $y(t)$ is

$$\phi_{xy}(\tau) = \lim_{T\to\infty} \int_{-\infty}^{\infty} x(t)y(t+\tau)\, dt$$

For $\tau = 0$, the cross-correlation between two zero-mean variables is identically the covariance. A final characterization of a random variable is its *power spectrum,* defined as

$$G(\omega, x) = \lim_{T\to\infty} \frac{1}{2\pi T} \left| \int_{-T}^{T} x(t)\, e^{-j\omega t}\, dt \right|^2$$

For a stationary random process, the power spectrum function is identically the Fourier transform of the autocorrelation function

$$G(\omega, x) = \frac{1}{\pi} \int_{-\infty}^{\infty} \phi_{xx}(\tau)e^{-j\omega t}\, dt$$

with

$$\phi_{xx}(0) = \int_{-\infty}^{\infty} G(\omega, x)\, d\omega$$

8.2   Distributed-Parameter Models

There are many important applications in which the state of a system cannot be defined at a finite number of points in space. Instead, the system state is a continuously varying function of both time and location. When continuous spatial dependence is explicitly accounted for in a model, the independent variables must include spatial coordinates as well as time. The resulting *distributed-parameter model* is described in terms of *partial differential equations*, containing partial derivatives with respect to each of the independent variables.

Distributed-parameter models commonly arise in the study of mass and heat transport, the mechanics of structures and structural components, and electrical transmission. Consider as a simple example the unidirectional flow of heat through a wall, as depicted in Fig. 27. The temperature of the wall is not in general uniform but depends on both the time t and position within the wall x, that is, $\theta = \theta(x, t)$. A distributed-parameter model for this case might be the first-order partial differential equation

$$\frac{d}{dt}\theta(x,t) = \frac{1}{C_t}\frac{\partial}{\partial x}\left[\frac{1}{R_t}\frac{\partial}{\partial x}\theta(x,t)\right]$$

where C_t is the thermal capacitance and R_t is the thermal resistance of the wall (assumed uniform).

The complexity of distributed-parameter models is typically such that these models are avoided in the analysis and design of control systems. Instead, distributed-parameter systems are approximated by a finite number of spatial "lumps," each lump being characterized by some average value of the state. By eliminating the independent spatial variables, the result is a *lumped-parameter (or lumped-element) model* described by coupled ordinary differential equations. If a sufficiently fine-grained representation of the lumped microstructure can be achieved, a lumped model can be derived that will approximate the distributed model to any desired degree of accuracy. Consider, for example, the three temperature lumps shown in Fig. 28, used to approximate the wall of Fig. 27. The corresponding third-order lumped approximation is

$$\frac{d}{dt}\begin{bmatrix}\theta_1(t)\\\theta_2(t)\\\theta_3(t)\end{bmatrix} = \begin{bmatrix}-\dfrac{9}{C_tR_t} & \dfrac{3}{C_tR_t} & 0\\[2mm]\dfrac{3}{C_tR_t} & \dfrac{6}{C_tR_t} & \dfrac{3}{C_tR_t}\\[2mm]0 & \dfrac{3}{C_tR_t} & \dfrac{6}{C_tR_t}\end{bmatrix}\begin{bmatrix}\theta_1(t)\\\theta_2(t)\\\theta_3(t)\end{bmatrix} + \begin{bmatrix}\dfrac{6}{C_tR_t}\\[2mm]0\\[2mm]0\end{bmatrix}\theta_0(t)$$

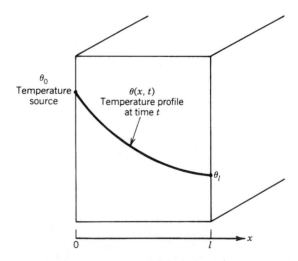

Figure 27   Uniform heat transfer through a wall.

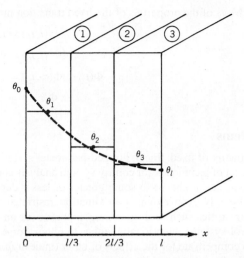

Figure 28 Lumped-parameter model for uniform heat transfer through a wall.

If a more detailed approximation is required, this can always be achieved at the expense of adding additional, smaller lumps.

8.3 Time-Varying Systems

Time-varying systems are those with characteristics that change as a function of time. Such variation may result from environmental factors, such as temperature or radiation, or from factors related to the operation of the system, such as fuel consumption. While in general a model with variable parameters can be either linear or nonlinear, the name time varying is most frequently associated with linear systems described by the following state equation:

$$\dot{x}(t) = A(t)x(t) + B(t)u(t)$$

For this linear time-varying model, the superposition principle still applies. Superposition is a great aid in model formulation but unfortunately does not prove to be much help in determining the model solution.

Paradoxically, the form of the solution to the linear time-varying equation is well known[7]:

$$x(t) = \Phi(t, t_0)x(t_0) + \int_{t_0}^{t} \Phi(t, \tau)B(\tau)u(\tau)\, dt$$

where $\Phi(t, t_0)$ is the time-varying state transition matrix. This knowledge is typically of little value, however, since it is not usually possible to determine the state transition matrix by any straightforward method. By analogy with the first-order case, the relationship

$$\Phi(t, t_0) = \exp\left(\int_{t_0}^{t} A(\tau)\, d\tau\right)$$

can be proven valid *if and only if*

$$A(t) \int_{t_0}^{t} A(\tau)\, d\tau = \int_{t_0}^{t} A(\tau)\, d\tau\, A(t)$$

that is, if and only if $A(t)$ and its integral commute. This is a very stringent condition for all but a first-order system and, as a rule, it is usually easiest to obtain the solution using simulation.

Most of the properties of the fixed transition matrix extend to the time-varying case:

$$\Phi(t, t_0) = I$$

$$\Phi^{-1}(t, t_0) = \Phi(t_0, t)$$

$$\Phi(t_2, t_1)\Phi(t_1, t_0) = \Phi(t_2, t_0)$$

$$\dot{\Phi}(t, t_0) = A(t)\Phi(t, t_0)$$

8.4 Nonlinear Systems

The theory of fixed, linear, lumped-parameter systems is highly developed and provides a powerful set of techniques for control system analysis and design. In practice, however, all physical systems are nonlinear to some greater or lesser degree. The linearity of a physical system is usually only a convenient approximation, restricted to a certain range of operation. In addition, nonlinearities such as dead zones, saturation, or on–off action are sometimes introduced into control systems intentionally, either to obtain some advantageous performance characteristic or to compensate for the effects of other (undesirable) nonlinearities.

Unfortunately, while nonlinear systems are important, ubiquitous, and potentially useful, the theory of nonlinear differential equations is comparatively meager. Except for specific cases, closed-form solutions to nonlinear systems are generally unavailable. The only universally applicable method for the study of nonlinear systems is *simulation*. As described in Section 7, however, simulation is an experimental approach, embodying all of the attending limitations of experimentation.

A number of special techniques are available for the analysis of nonlinear systems. All of these techniques are in some sense approximate, assuming, for example, either a restricted range of operation over which nonlinearities are mild or the relative isolation of lower order subsystems. When used in conjunction with more complex simulation models, however, these techniques often provide insights and design concepts that would be difficult to discover through the use of simulation alone.[8]

Linear versus Nonlinear Behaviors

There are several fundamental differences between the behavior of linear and nonlinear systems that are especially important. These differences not only account for the increased difficulty encountered in the analysis and design of nonlinear systems, but also imply entirely new types of behavior for nonlinear systems that are not possible for linear systems.

The fundamental property of linear systems is *superposition*. This property states that if $y_1(t)$ is the response of the system to $u_1(t)$ and $y_2(t)$ is the response of the system to $u_2(t)$, then the response of the system to the linear combination $a_1 u_1(t) + a_2 u_2(t)$ is the linear combination $a_1 y_1(t) + a_2 y_2(t)$. An immediate consequence of superposition is that the responses of a linear system to inputs differing only in amplitude is qualitatively the same. Since superposition does not apply to nonlinear systems, the responses of a nonlinear system to large and small changes may be fundamentally different.

This fundamental difference in linear and nonlinear behaviors has a second consequence. For a linear system, interchanging two elements connected in series does not affect the overall system behavior. Clearly, this cannot be true in general for nonlinear systems.

A third property peculiar to nonlinear systems is the potential existence of *limit cycles*. A linear oscillator oscillates at an amplitude that depends on its initial state. A limit cycle is an oscillation of fixed amplitude and period, independent of the initial state, that is unique to the nonlinear system.

A fourth property concerns the response of nonlinear systems to sinusoidal inputs. For a linear system, the response to sinusoidal input is a sinusoid of the same frequency, potentially differing only in magnitude and phase. For a nonlinear system, the output will in general contain other frequency components, including possibly harmonics, subharmonics, and aperiodic terms. Indeed, the response need not contain the input frequency at all.

Linearizing Approximations

Perhaps the most useful technique for analyzing nonlinear systems is to approximate these with linear systems. While many linearizing approximations are possible, linearization can frequently be achieved by considering small excursions of the system state about a reference trajectory. Consider the nonlinear state equation

$$\dot{x}(t) = f[x(t), u(t)]$$

together with a reference trajectory $x^0(t)$ and reference input $u^0(t)$ that together satisfy the state equation

$$x^0(t) = f[x^0(t), u^0(t)]$$

Note that the simplest case is to choose a static equilibrium or *operating point* $\bar{x}$ as the reference "trajectory" such that $0 = t(\bar{x}, 0)$. The actual trajectory is then related to the reference trajectory by the relationships

$$x(t) = x^0(t) + \delta x(t)$$

$$u(t) = u^0(t) + \delta u(t)$$

where $\delta x(t)$ is some small perturbation about the reference state and $\delta u(t)$ is some small perturbation about the reference input. If these perturbations are indeed small, then applying Taylor's series expansion about the reference trajectory yields the linearized approximation

$$\delta \dot{x}(t) = A(t)\, \delta x(t) + B(t)\, \delta u(t)$$

where the state and distribution matrices are the *Jacobian matrices*

$$A(t) = \begin{bmatrix} \frac{\partial f_i}{\partial x_1} & \frac{\partial f_1}{\partial x_2} & \cdots & \frac{\partial f_1}{\partial x_n} \\ \frac{\partial f_2}{\partial x_1} & \frac{\partial f_2}{\partial x_2} & \cdots & \frac{\partial f_2}{\partial x_n} \\ \vdots & \vdots & & \vdots \\ \frac{\partial f_n}{\partial x_1} & \frac{\partial f_n}{\partial x_2} & \cdots & \frac{\partial f_n}{\partial x_n} \end{bmatrix}_{x(t)=x^0(t);u(t)=u^0(t)}$$

$$B(t) = \begin{bmatrix} \frac{\partial f_1}{\partial u_1} & \frac{\partial f_1}{\partial u_2} & \cdots & \frac{\partial f_1}{\partial u_m} \\ \frac{\partial f_2}{\partial u_1} & \frac{\partial f_2}{\partial u_2} & \cdots & \frac{\partial f_2}{\partial u_m} \\ \vdots & \vdots & & \vdots \\ \frac{\partial f_n}{\partial u_1} & \frac{\partial f_n}{\partial u_2} & \cdots & \frac{\partial f_n}{\partial u_m} \end{bmatrix}_{x(t)=x^0(t);u(t)=u^0(t)}$$

If the reference trajectory is a fixed operating point $\bar{x}$, then the resulting linearized system x, is time invariant and can be solved analytically. If the reference trajectory is a function of time, however, then the resulting system is linear but time varying.

Describing Functions

The describing function method is an extension of the frequency transfer function approach of linear systems, most often used to determine the stability of limit cycles of systems

Figure 29   General nonlinear system for describing function analysis.

containing nonlinearities. The approach is approximate and its usefulness depends on two major assumptions:

1. All the nonlinearities within the system can be aggregated mathematically into a single block, denoted as $N(M)$ in Fig. 29, such that the equivalent gain and phase associated with this block depend only on the amplitude M_d of the sinusoidal input $m(\omega t) = M \sin(\omega t)$ and are independent of the input frequency ω.

2. All the harmonics, subharmonics, and any direct-current (dc) component of the output of the nonlinear block are filtered out by the linear portion of the system such that the effective output of the nonlinear block is well approximated by a periodic response having the same fundamental period as the input.

Although these assumptions appear to be rather limiting, the technique gives reasonable results for a large class of control systems. In particular, the second assumption is generally satisfied by higher order control systems with symmetric nonlinearities, since (a) symmetric nonlinearities do not generate dc terms, (b) the amplitudes of harmonics are generally small when compared with the fundamental term and subharmonics are uncommon, and (c) feedback within a control system typically provides low-pass filtering to further attenuate harmonics, especially for higher order systems. Because the method is relatively simple and can be used for systems of any order, describing functions have enjoyed wide practical application.

The describing function of a nonlinear block is defined as the ratio of the fundamental component of the output to the amplitude of a sinusoidal input. In general, the response of the nonlinearity to the input

$$m(\omega t) = M \sin \omega t$$

is the output

$$n(\omega t) = N_1 \sin(\omega t + \phi_1) + N_2 \sin(2\omega t + \phi_2) + N_3 \sin(3\omega t + \phi_3) + \cdots$$

and, hence, the describing function for the nonlinearity is defined as the complex quantity

$$N(M) = \frac{N_1}{M} e^{j\phi_1}$$

Derivation of the approximating function typically proceeds by representing the fundamental frequency by the Fourier series coefficients

$$A_1(M) = \frac{2}{T} \int_{-T/2}^{T/2} n(\omega t) \cos \omega t \, d(\omega t)$$

$$B_1(M) = \frac{2}{T} \int_{-T/2}^{T/2} n(\omega t) \sin \omega t \, d(\omega t)$$

The describing function is then written in terms of these coefficients as

$$N(M) = \frac{B_1(M)}{M} + j\frac{A_1(M)}{M} = \left[\left(\frac{B_1(M)}{M} \right)^2 + \left(\frac{A_1(M)}{M} \right)^2 \right]^{1/2} \exp\left[j\tan^{-1}\left(\frac{A_1(M)}{B_1(M)} \right) \right]$$

Note that if $n(\omega t) = -n(-\omega t)$, then the describing function is odd, $A_1(M) = 0$, and there is no phase shift between the input and output. If $n(\omega t) = n(-\omega t)$, then the function is even, $B_1(M) = 0$, and the phase shift is $\pi/2$.

The describing functions for a number of typical nonlinearities are given in Fig. 30. Reference 9 contains an extensive catalog. The following derivation for a dead-zone nonlinearity demonstrates the general procedure for deriving a describing function. For the saturation

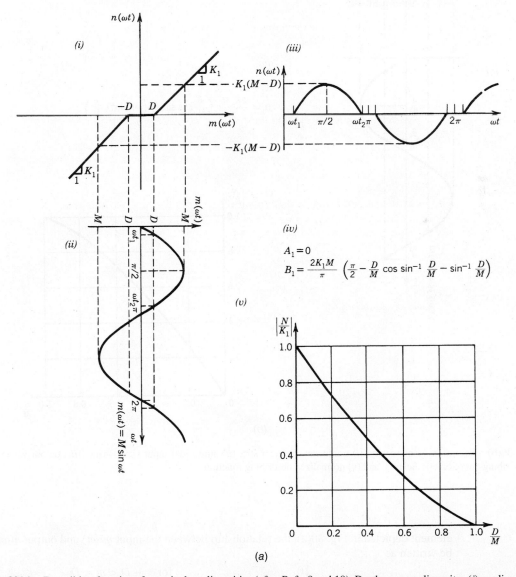

(a)

Figure 30(a)   Describing functions for typical nonlinearities (after Refs. 9 and 10). Dead-zone nonlinearity: (*i*) nonlinear characteristic, (*ii*) sinusoidal input wave shape, (*iii*) output wave shape, (*iv*) describing-function coefficients, and (*v*) normalized describing function. From *Nonlinear Automatic Control* by J. E. Gibson. © McGraw-Hill Education.

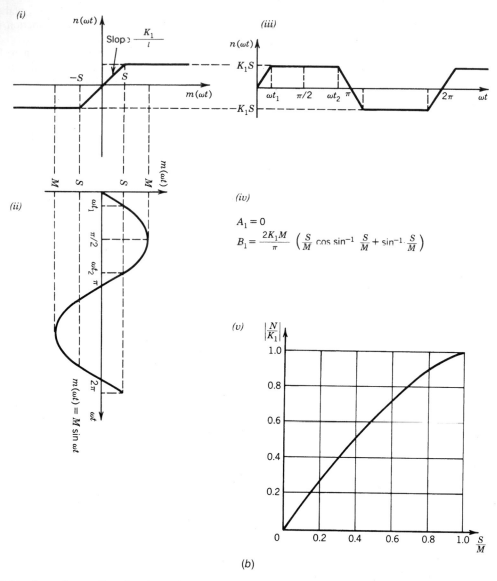

Figure 30(b) Saturation nonlinearity: (*i*) nonlinear characteristic, (*ii*) sinusoidal input wave shape, (*iii*) output wave shape, (*iv*) describing-function coefficients, and (*v*) normalized describing function.

element depicted in Fig. 30(a), the relationship between the input $m(\omega t)$ and output $n(\omega t)$ can be written as

$$
n(\omega t) = \begin{cases}
0 & \text{for} \quad -D < m < D \\
K_1 M \left(\sin \omega t - \sin \omega_1 t \right) & \text{for} \quad m > D \\
K_1 M (\sin \omega t + \sin \omega_1 t) & \text{for} \quad m < -D
\end{cases}
$$

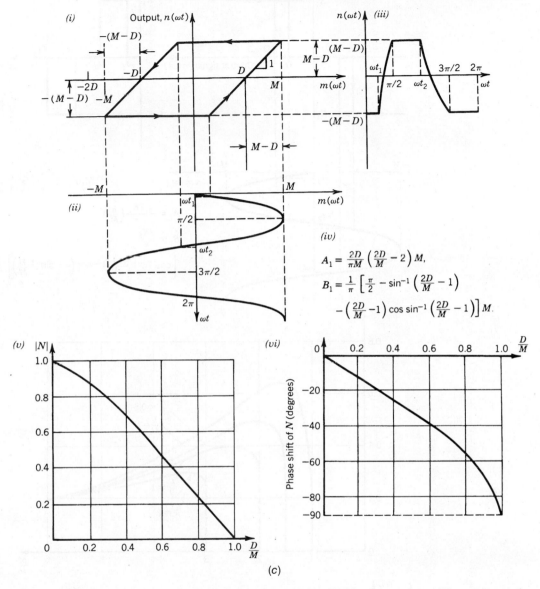

Figure 30(c)   Backlash nonlinearity: (*i*) nonlinear characteristic, (*ii*) sinusoidal input wave shape, (*iii*) output wave shape, (*iv*) describing-function coefficients, (*v*) normalized amplitude characteristics for the describing function, and (*vi*) normalized phase characteristics for the describing function.

Since the function is odd, $A_1 = 0$. By the symmetry over the four quarters of the response period,

$$B_1 = 4 \left[\frac{2}{\pi/2} \int_0^{\pi/2} n(\omega t) \sin \omega t \, d(\omega t) \right]$$

$$= \frac{4}{\pi} \left[\int_0^{\omega t_1} (0) \sin \omega t \, d(\omega t) + \int_{\omega t_1}^{\pi/2} K_1 M(\sin \omega t - \sin \omega_1 t) \sin \omega t \, d(\omega t) \right]$$

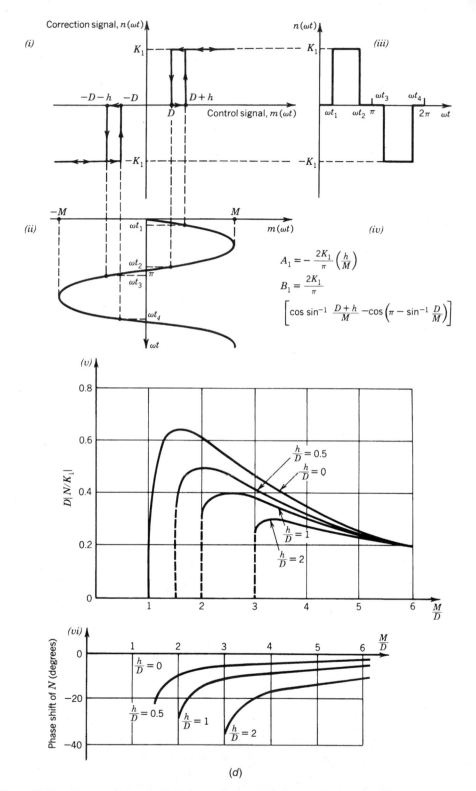

Figure 30(d) Three-position on–off device with hysteresis: (*i*) nonlinear characteristic, (*ii*) sinusoidal input wave shape, (*iii*) output wave shape, (*iv*) describing-function coefficients, (*v*) normalized amplitude characteristics for the describing function, and (*vi*) normalized phase characteristics for the describing function.

where $\omega t_1 = \sin^{-1}(D/M)$. Evaluating the integrals and dividing by M yields the describing function listed in Fig. 30.

Phase-Plane Method

The *phase-plane method* is a graphical application of the state-space approach used to characterize the free response of second-order nonlinear systems. While any convenient pair of state variables can be used, the *phase variables* originally were taken to be the displacement and velocity of the mass of a second-order mechanical system. Using the two state variables as the coordinate axis, the transient response of a system is captured on the *phase plane* as the plot of one variable against the other, with time implicit on the resulting curve. The curve for a specific initial condition is called a *trajectory* in the phase plane; a representative sample of trajectories is called the *phase portrait* of the system. The phase portrait is a compact and readily interpreted summary of the system response. Phase portraits for a sample of typical nonlinearities are shown in Fig. 31(a).

Four methods can be used to construct a phase portrait: (1) direct solution of the differential equation, (2) the graphical *method of isoclines,* (3) transformation of the second-order system (with time as the independent variable) into an equivalent first-order system (with one of the phase variables as the independent variable), and (4) numerical solution using simulation. The first and second methods are usually impractical; the third and fourth methods are frequently used in combination. For example, consider the second-order model

$$\frac{dx_1}{dt} = f_1(x_1, x_2) \qquad \frac{dx_2}{dt} = f_2(x_1, x_2)$$

Dividing the second equation by the first and eliminating the dt terms yield

$$\frac{dx_2}{dx_1} = \frac{f_2(x_1, x_2)}{f_1(x_1, x_2)}$$

This first-order equation describes the phase-plane trajectories. In many cases it can be solved analytically. If not, it always can be simulated.

The phase-plane method complements the describing-function approach. A describing function is an approximate representation of the sinusoidal response for systems of any order, while the phase plane is an exact representation of the (free) transient response for first- and second-order systems. Of course, the phase-plane method theoretically can be extended for higher order systems, but the difficulty of visualizing the nth-order state space typically makes such a direct extension impractical. An approximate extension of the method has been used with some considerable success,[8] however, in order to explore and validate the relationships among pairs of variables in complex simulation models. The approximation is based on the assumptions that the paired variables define a second-order subsystem that, for the purposes of analysis, is weakly coupled to the remainder of the system.

8.5 Discrete and Hybrid Systems

A *discrete-time system* is one for which the dependent variables are defined only at distinct instants of time. Discrete-time models occur in the representation of systems that are inherently discrete, in the analysis and design of digital measurement and control systems, and in the numerical solution of differential equations (see Section 7). Because most control systems are now implemented using digital computers (especially microprocessors), discrete-time models are extremely important in dynamic systems analysis. The discrete-time nature of a computer's sampling of continuous physical signals also leads to the occurrence of *hybrid systems,* that is, systems that are in part discrete and in part continuous. Discrete-time models of hybrid systems are called *sampled-data systems.*

Name	Roots	Sketch
Stable focus or spiral	Damped complex conjugate 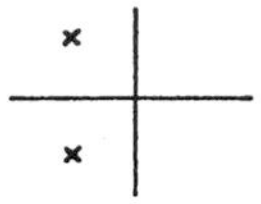	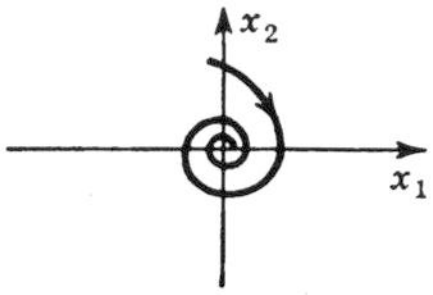 Trajectories spiral asymptotically to focus
Stable node	Stable real roots 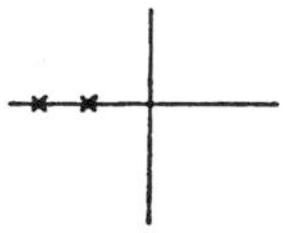	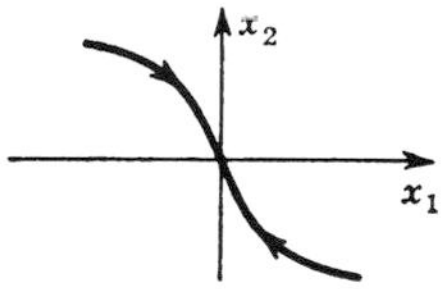 Trajectories approach node monotonically
Vortex or center (structurally unstable)	Imaginary roots 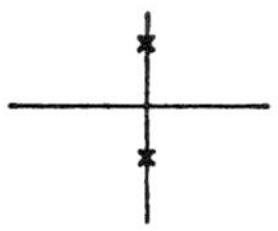	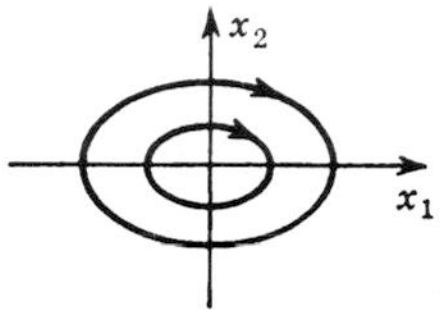 Conservative system or oscillator
Unstable focus	Complex conjugate with positive real part 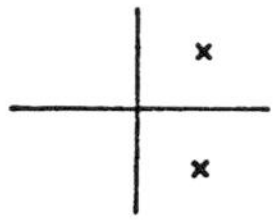	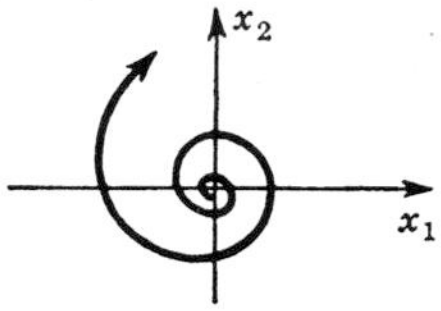
Saddle point	A point of unstable equilibrium 	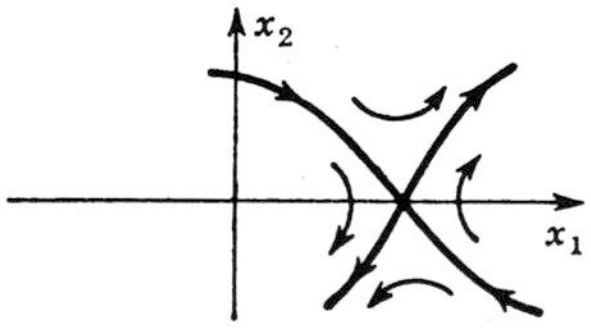
Unstable node	Unstable real roots 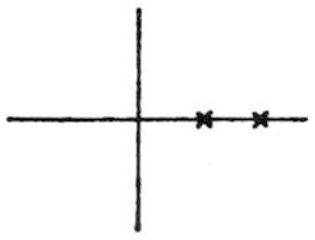	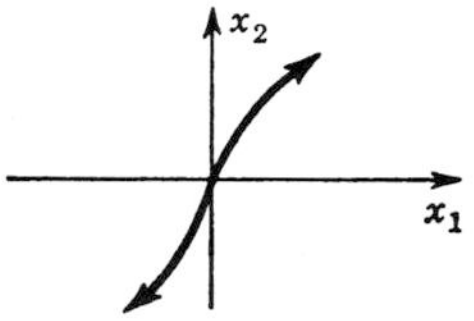 Trajectories diverge monotonically from node

Figure 31(a) Typical phase-plane plots for second-order systems.[9] From *Nonlinear Automatic Control* by J. E. Gibson. © McGraw-Hill Education.

Name	Roots	Sketch
Spiral with nonlinearity viscous damping and coulomb friction		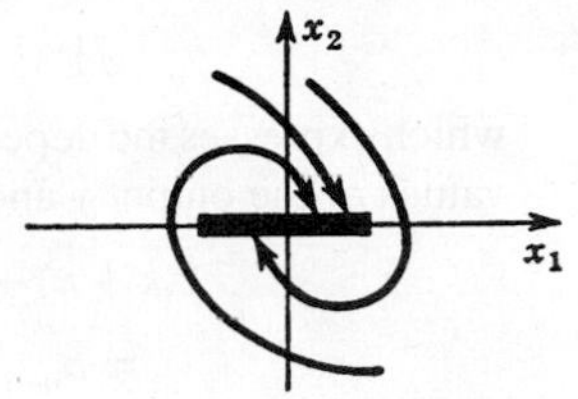 Spirals are warped by coulomb friction
Nonlinear system with coulomb damping and no viscous damping		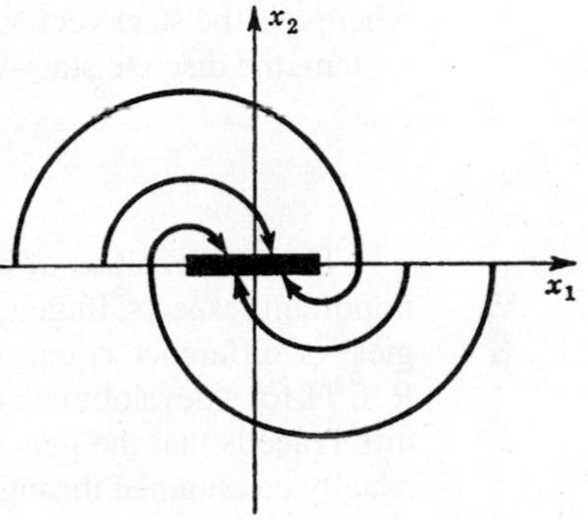 Semicircles center at ends of coulomb line
Spiral with backlash		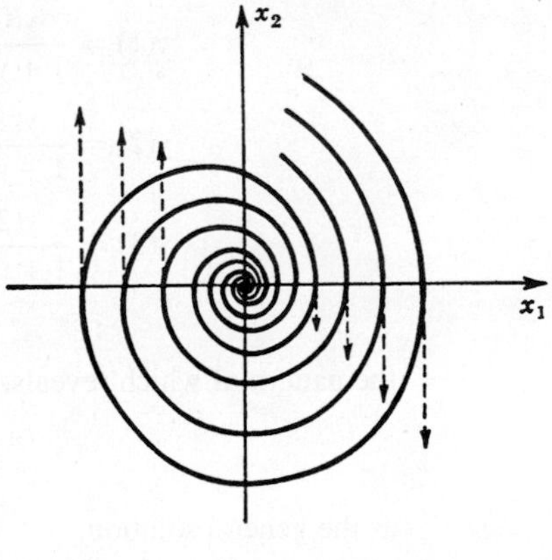

Figure 31(b) (*continued*)

Difference Equations

Dynamic models of discrete-time systems most naturally take the form of *difference equations.* The I/O form of an *n*th-order difference equation model is

$$f[y(k+n), y(k+n-1), ..., y(k), u(k+n-1), ..., u(k)] = 0$$

which expresses the dependence of the $(k+n)$th value of the output, $y(k+n)$, on the *n* preceding values of the output *y* and input *u*. For a linear system, the I/O form can be written as

$$y(k+n) + a_{n-1}(k)y(k+n-1) + \cdots + a_1(k)y(k+1) + a_0(k)y(k)$$
$$= b_{n-1}(k)u(k+n-1) + \cdots + b_0(k)u(k)$$

In state-variable form, the discrete-time model is the vector difference equation

$$x(k+1) = f[x(k), u(k)]$$

$$y(k) = g[x(k), u(k)]$$

where *x* is the state vector, *u* is the vector of inputs, and *y* is the vector of outputs. For a linear system, the discrete state-variable form can be written as

$$x(k+1) = A(k)x(k) + B(k)u(k)$$

$$y(k) = C(k)x(k) + D(k)u(k)$$

The mathematics of difference equations parallels that of differential equations in many important respects. In general, the concepts applied to differential equations have direct analogies for difference equations, although the mechanics of their implementation may vary (see Ref. 11 for a development of dynamic modeling based on difference equations). One important difference is that the general solution of nonlinear and time-varying difference equations can usually be obtained through *recursion.* For example, consider the discrete nonlinear model

$$y(k+1) = \frac{y(k)}{1 + y(k)}$$

Recursive evaluation of the equation beginning with the initial condition $y(0)$ yields

$$y(1) = \frac{y(0)}{1 + y(0)}$$

$$y(2) = \frac{y(1)}{1 + y(1)} = \left[\frac{y(0)}{1 + y(0)}\right] \bigg/ \left[1 + \frac{y(0)}{1 + y(0)}\right] = \frac{y(0)}{1 + 2y(0)}$$

$$y(3) = \frac{y(2)}{1 + y(2)} = \frac{y(0)}{1 + 3y(0)}$$

$$\vdots$$

the pattern of which reveals, by induction,

$$y(k) = \frac{y(0)}{1 + ky(0)}$$

as the general solution.

Uniform Sampling

Uniform sampling is the most common mathematical approach to *A/D conversion,* that is, to extracting the discrete-time approximation $y^*(k)$ of the form

$$y^*(k) = y(t = kT)$$

from the continuous-time signal $y(t)$, where T is a constant interval of time called the *sampling period*. If the sampling period is too large, however, it may not be possible to represent the continuous signal accurately. The *sampling theorem* guarantees that $y(t)$ can be reconstructed from the uniformly sampled values $y^*(k)$ if the sampling period satisfies the inequality

$$T \leq \frac{\pi}{\omega_u}$$

where ω_u is the highest frequency contained in the Fourier transform $Y(\omega)$ of $y(t)$, that is, if

$$Y(\omega) = 0 \text{ for all } \omega > \omega_u$$

The Fourier transform of a signal is defined to be

$$\mathcal{F}[y(t)] = Y(\omega) = \int_{-\infty}^{\infty} y(t)e^{-j\omega t}\, dt$$

Note that if $y(t) = 0$ for $t \geq 0$, and if the region of convergence for the Laplace transform includes the imaginary axis, then the Fourier transform can be obtained from the Laplace transform as

$$Y(\omega) = [Y(s)]_{s=j\omega}$$

For cases where it is impossible to determine the Fourier transform analytically, such as when the signal is described graphically or by a table, numerical solution based on the *fast Fourier transform (FFT) algorithm* is usually satisfactory.

In general, the condition $T \leq \pi/\omega_u$ cannot be satisfied exactly since most physical signals have no finite upper frequency ω_u. A useful approximation is to define the upper frequency as the frequency for which 99% of the signal "energy" lies in the frequency spectrum $0 \leq \omega \leq \omega_u$. This approximation is found from the relation

$$\int_0^{\omega_u} |Y(\omega)|^2\, d\omega = 0.99 \int_0^{\infty} |Y(\omega)|^2\, d\omega$$

where the square of the amplitude of the Fourier transform $|Y(\omega)|^2$ is said to be the *power spectrum* and its integral over the entire frequency spectrum is referred to as the "energy" of the signal. Using a sampling frequency 2–10 times this approximate upper frequency (depending on the required factor of safety) and inserting a low-pass filter (called a *guard filter*) before the sampler to eliminate frequencies above the *Nyquist frequency* π/T usually lead to satisfactory results.[4]

The z Transform

The *z transform* permits the development and application of transfer functions for discrete-time systems, in a manner analogous to continuous-time transfer functions based on the Laplace transform. A discrete signal may be represented as a series of impulses

$$y^*(t) = y(0)\delta(t) + y(1)\delta(t - T) + y(2)\delta(t - 2T) + \cdots$$

$$= \sum_{k=0}^{N} y(k)\delta(t - kT)$$

where $y(k) = y^*(t = kT)$ are the values of the discrete signal, $\delta(t)$ is the unit impulse function, and N is the number of samples of the discrete signal. The Laplace transform of the series is

$$Y^*(s) = \sum_{k=0}^{N} y(k)e^{-ksT}$$

where the shifting property of the Laplace transform has been applied to the pulses. Defining the *shift* or *advance operator* as $z = e^{sT}$, $Y^*(s)$ may now be written as a function of z:

$$Y^*(z) = \sum_{k=0}^{N} \frac{y(k)}{z^k} = Z\,[y(t)]$$

where the transformed variable $Y^*(z)$ is called the z transform of the function $y^*(t)$. The inverse of the shift operator $1/z$ is called the *delay operator* and corresponds to a time delay of T.

The z transforms for many sampled functions can be expressed in closed form. A listing of the transforms of several commonly encountered functions is given in Table 12. Properties of the z transform are listed in Table 13.

Pulse Transfer Functions

The transfer function concept developed for continuous systems has a direct analog for sampled-data systems. For a continuous system with sampled output $u(t)$ and sampled input $y(t)$, the *pulse* or *discrete transfer function* $G(z)$ is defined as the ratio of the z-transformed

Table 12 z-Transform Pairs

	$X(s)$	$x(t)$ or $x(k)$	$X(z)$
1	1	$\delta(t)$	1
2	e^{-kTs}	$\delta(t - kT)$	z^{-k}
3	$\dfrac{1}{s}$	$1(t)$	$\dfrac{z}{z - 1}$
4	$\dfrac{1}{s^2}$	t	$\dfrac{Tz}{(z - 1)^2}$
5	$\dfrac{1}{s + a}$	e^{-at}	$\dfrac{z}{z - e^{-aT}}$
6	$\dfrac{a}{s(s + a)}$	$1 - e^{-at}$	$\dfrac{(1 - e^{-aT})z}{(z - 1)(z - e^{-aT})}$
7	$\dfrac{\omega}{s^2 + \omega^2}$	$\sin \omega t$	$\dfrac{z \sin \omega T}{z^2 - 2z \cos \omega T + 1}$
8	$\dfrac{s}{s^2 + \omega^2}$	$\cos \omega t$	$\dfrac{z(z - \cos \omega T)}{z^2 - 2z \cos \omega T + 1}$
9	$\dfrac{1}{(s + a)^2}$	te^{-at}	$\dfrac{Tze^{-aT}}{(z - e^{-aT})^2}$
10	$\dfrac{\omega}{(s + a)^2 + \omega^2}$	$e^{-at} \sin \omega t$	$\dfrac{ze^{-aT} \sin \omega T}{z^2 - 2ze^{-aT} \cos \omega T + e^{-2aT}}$
11	$\dfrac{s + a}{(s + a)^2 + \omega^2}$	$e^{-at} \cos \omega t$	$\dfrac{z^2 - ze^{-aT} \cos \omega T}{z^2 - 2ze^{-aT} \cos \omega T + e^{-2aT}}$
12	$\dfrac{2}{s^3}$	t^2	$\dfrac{T^2 z(z + 1)}{(z - 1)^3}$
13		a	$\dfrac{z}{z - a}$
14		$a^k \cos k\pi$	$\dfrac{z}{z + a}$

Table 13   *z*-Transform Properties

	$x(t)$ or $x(k)$	$Z[x(t)]$ or $Z[x(k)]$
1	$ax(t)$	$aX(z)$
2	$x_1(t) + x_2(t)$	$X_1(z) + X_2(z)$
3	$x(t+T)$ or $x(k+1)$	$zX(z) - zx(0)$
4	$x(t+2T)$	$z^2X(z) - z^2x(0) - zx(T)$
5	$x(k+2)$	$z^2X(z) - z^2x(0) - zx(1)$
6	$x(t+kT)$	$z^kX(z) - z^kx(0) - z^{k-1}x(T) - \cdots - zx(kT-T)$
7	$x(k+m)$	$z^mX(z) - z^mx(0) - z^{m-1}x(1) - \cdots - zx(m-1)$
8	$tx(t)$	$-Tz\dfrac{d}{dz}[X(z)]$
9	$kx(k)$	$-z\dfrac{d}{dz}[X(z)]$
10	$e^{-at}x(t)$	$X(ze^{aT})$
11	$e^{-ak}x(k)$	$X(ze^a)$
12	$a^kx(k)$	$X\left(\dfrac{z}{a}\right)$
13	$ka^kx(k)$	$-z\dfrac{d}{dz}\left[X\left(\dfrac{z}{a}\right)\right]$
14	$x(0)$	$\lim_{z\to\infty} X(z)$ if the limit exists
15	$x(\infty)$	$\lim_{z\to1}[(z-1)X(z)]$ if $\dfrac{z-1}{z}X(z)$ is analytic on and outside the unit circle
16	$\sum_{k=0}^{\infty} x(k)$	$X(1)$
17	$\sum_{k=0}^{n} x(kT)y(nT-kT)$	$X(z)Y(z)$

output $Y(z)$ to the z-transformed input $U(z)$, assuming zero initial conditions. In general, the pulse transfer function has the form

$$G(z) = \frac{Y(z)}{U(z)} = \frac{b_0 + b_1z^{-1} + b_2z^{-2} + \cdots + b_mz^{-m}}{1 + a_1z^{-1} + a_2z^{-1} + \cdots + a_nz^{-n}}$$

Zero-Order Hold

The *zero-order data hold* is the most common mathematical approach to *D/A conversion,* that is, to creating a piecewise continuous approximation $u(t)$ of the form

$$u(t) = u^*(k) \text{ for } kT \le t < (k+1)T$$

from the discrete-time signal $u^*(k)$, where T is the period of the hold. The effect of the zero-order hold is to convert a sequence of discrete impulses into a staircase pattern, as shown in Fig. 32. The transfer function of the zero-order hold is

$$G(s) = \frac{1}{s}(1 - e^{-Ts}) = \frac{1 - z^{-1}}{s}$$

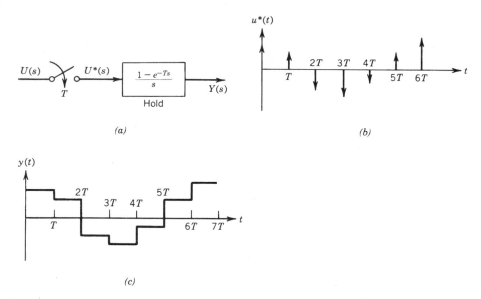

Figure 32   Zero-order hold: (*a*) block diagram of hold with a sampler, (*b*) sampled input sequence, and (*c*) analog output for the corresponding input sequence.[4]

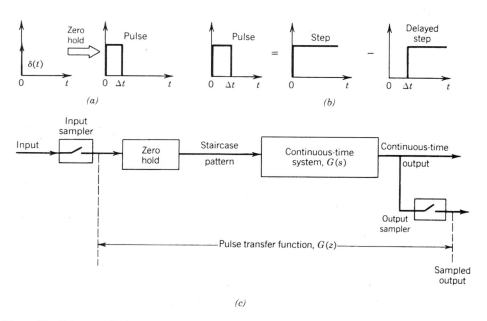

Figure 33   Pulse transfer function of a continuous system with sampler and zero hold.[12] From *Introducing Systems and Control* by D. M. Auslander, Y. Takahashi, and M. J. Rabins. © McGraw-Hill Education.

Using this relationship, the pulse transfer function of the sampled-data system shown in Fig. 33 can be derived as

$$G(z) = (1 - z^{-1}) \mathscr{Z} \left[+^{-1} \frac{G(s)}{s} \right]$$

The continuous system with transfer function $G(s)$ has a sampler and a zero-order hold at its input and a sampler at its output. This is a common configuration in many computer control applications.

References

1. J. L. Shearer, A. T. Murphy, and H. H. Richardson, *Introduction to System Dynamics*, 1st Edition, © 1967. Pearson Education, Inc., Upper Saddle River, NJ.

2. E. O. Doebelin, *System Dynamics: Modeling, Analysis, Simulation, and Design*, Merrill, Columbus, OH, 1998.

3. C. M. Close, D. K. Frederick, and J. C. Newell, *Modeling and Analysis of Dynamic Systems*, 3rd ed., Wiley, Boston, 2001.

4. W. J. Palm III, *Modeling, Analysis, and Control of Dynamic Systems*, 2nd ed., Wiley, New York, 2000.

5. G. J. Thaler and R. G. Brown, *Analysis and Design of Feedback Control Systems*, 2nd ed., McGraw-Hill, New York, 1960.

6. G. A. Korn and J. V. Wait, *Digital Continuous System Simulation*, Prentice-Hall, Englewood Cliffs, NJ, 1975.

7. B. C. Kuo and F. Golnaraghi, *Automatic Control Systems*, 8th ed., Prentice-Hall, Englewood Cliffs, NJ, 2002.

8. W. Thissen, "Investigation into the World3 Model: Lessons for Understanding Complicated Models," *IEEE Transactions on Systems, Man, and Cybernetics*, SMC-8(3) 183–193, 1978.

9. J. E. Gibson, *Nonlinear Automatic Control*, McGraw-Hill, New York, 1963.

10. S. M. Shinners, *Modern Control System Theory and Design*, 2nd ed., Wiley, New York, 1998.

11. D. G. Luenberger, *Introduction to Dynamic Systems: Theory, Models, and Applications*, Wiley, New York, 1979.

12. D. M. Auslander, Y. Takahashi, and M. J. Rabins, *Introducing Systems and Control*, McGraw-Hill, New York, 1974.

Bibliography

R. N., Bateson, *Introduction to Control System Technology*, Prentice-Hall, Englewood Cliffs, NJ, 2001.

R. H., Bishop, and R. C. Dorf, *Modern Control Systems*, 10th ed., Prentice-Hall, Upper Saddle River, 2004.

W. L., Brogan, *Modern Control Theory*, 3rd ed., Prentice-Hall, Englewood Cliffs, NJ, 1991.

R. H., Cannon, Jr., *Dynamics of Physical Systems*, McGraw-Hill, New York, 1967.

C. W., DeSilva, *Control Sensors and Actuators*, Prentice-Hall, Englewood Cliffs, NJ, 1989.

E. O., Doebelin, *Measurement Systems*, 5th ed., McGraw-Hill, New York, 2004.

G. F., Franklin, J. D. A., Powell, and A. Emami-Naeini, *Feedback Control of Dynamic Systems*, 4th ed., Addison-Wesley, Reading, MA, 2001.

Grace, A. J. Laub, J. N. Little, and C. Thompson, *Control System Toolbox Users Guide*, The Math-Works, Natick, MA, 1990.

T. T., Hartley, G. O. Beale, and S. P. Chicatelli, *Digital Simulation of Dynamic Systems: A Control Theory Approach*, Prentice-Hall, Englewood Cliffs, NJ, 1994.

N. A., Kheir, *Systems Modeling and Computer Simulation*, 2nd ed., Marcel Dekker, New York, 1996.

W. D., Kelton, R. P. Sadowski, and D. T. Sturrock, *Simulation with Arena*, 3rd ed., McGraw-Hill, New York, 2004.

D. C., Lay, *Linear Algebra and Its Applications*, 3rd ed., Addison-Wesley, Reading, MA, 2006.

L., Ljung, and T. Glad, *Modeling Simulation of Dynamic Systems*, Prentice-Hall, Englewood Cliffs, NJ, 1994.

The MathWorks on-line at ⟨www.mathworks.com⟩ NJ, 1995.

C. L., Phillips, and R. D. Harbor, *Feedback Control Systems*, 4th ed., Prentice-Hall, Englewood Cliffs, NJ, 1999.

C. L., Phillips, and H. T. Nagle, *Digital Control System Analysis and Design*, 3rd ed., Prentice-Hall, Englewood Cliffs, NJ, 1995.

C., Van Loan, *Computational Frameworks for the Fast Fourier Transform*, SIAM, Philadelphia, PA, 1992.

S., Wolfram, *Mathematica Book*, 5th ed., Wolfram Media, Inc., Champaign, IL, 2003.

CHAPTER 21

BASIC CONTROL SYSTEMS DESIGN

William J. Palm III
University of Rhode Island
Kingston, Rhode Island

1 INTRODUCTION

The purpose of a *control system* is to produce a desired *output*. This output is usually specified by the command *input* and is often a function of time. For simple applications in well-structured situations, *sequencing* devices like timers can be used as the control system. But most systems are not that easy to control, and the controller must have the capability of reacting to disturbances, changes in its environment, and new input commands. The key element that allows a control system to do this is *feedback,* which is the process by which a system's output is used to influence its behavior. Feedback in the form of the room temperature measurement is used to control the furnace in a thermostatically controlled heating system. Figure 1 shows the *feedback loop* in the system's *block diagram,* which is a graphical representation of the system's control structure and logic. Another commonly found control system is the pressure regulator shown in Fig. 2.

Feedback has several useful properties. A system whose individual elements are nonlinear can often be modeled as a linear one over a wider range of its variables with the proper use of feedback. This is because feedback tends to keep the system near its reference operation condition. Systems that can maintain the output near its desired value despite changes in the environment are said to have good *disturbance rejection*. Often we do not have accurate values for some system parameter or these values might change with age. Feedback can be used to minimize the effects of parameter changes and uncertainties. A system that has both good disturbance rejection and low sensitivity to parameter variation is *robust*. The application that resulted in the general understanding of the properties of feedback is shown in Fig. 3. The electronic amplifier gain A is large, but we are uncertain of its exact value. We use the resistors R_1 and R_2 to create a feedback loop around the amplifier and pick R_1 and R_2 to create a feedback loop around the amplifier and R_1 and R_2 so that $AR_2/R_1 \gg 1$. Then the input–output relation becomes $e_o \approx R_1 e_i/R_2$, which is independent of A as long as A remains large. If R_1 and R_2 are known accurately, then the system gain is now reliable.

Figure 4 shows the block diagram of a *closed-loop* system, which is a system with feedback. An *open-loop* system, such as a timer, has no feedback. Figure 4 serves as a focus for

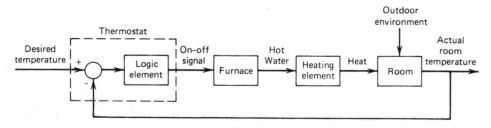

Figure 1   Block diagram of the thermostat system for temperature control.

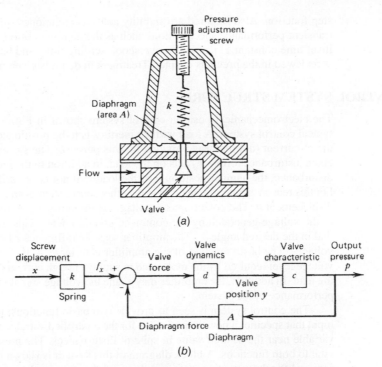

Figure 2 Pressure regulator: (*a*) cutaway view and (*b*) block diagram.

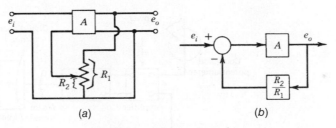

Figure 3 A closed-loop system.

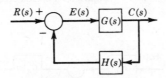

Figure 4 Feedback compensation of an amplifier.

outlining the prerequisites for this chapter. The reader should be familiar with the *transfer function* concept based on the Laplace transform, the *pulse transfer* function based on the z transform, for digital control, and the differential equation modeling techniques needed to obtain them. It is also necessary to understand block diagram algebra, characteristic roots, the final-value theorem, and their use in evaluating system response for common inputs like the step function. Also required are stability analysis techniques such as the Routh criterion and transient performance specifications such as the damping ratio ζ, natural frequency ω_n, dominant time constant τ, maximum overshoot, settling time, and bandwidth. The above material is reviewed in the previous chapter. Treatment in depth is given in Refs. 1–4.

2   CONTROL SYSTEM STRUCTURE

The electromechanical position control system shown in Fig. 5 illustrates the structure of a typical control system. A load with an inertia I is to be positioned at some desired angle θ_r. A direct-current (dc) motor is provided for this purpose. The system contains viscous damping, and a disturbance torque T_d acts on the load, in addition to the motor torque T. Because of the disturbance, the angular position θ of the load will not necessarily equal the desired value θ_r. For this reason, a potentiometer, or some other sensor such as an encoder, is used to measure the displacement θ. The potentiometer voltage representing the controlled position θ is compared to the voltage generated by the command potentiometer. This device enables the operator to dial in the desired angle θ_r. The amplifier sees the difference e between the two potentiometer voltages. The basic function of the amplifier is to increase the small error voltage e up to the voltage level required by the motor and to supply enough current required by the motor to drive the load. In addition, the amplifier may shape the voltage signal in certain ways to improve the performance of the system.

The control system is seen to provide two basic functions: (1) to respond to a command input that specifies a new desired value for the controlled variable and (2) to keep the controlled variable near the desired value in spite of disturbances. The presence of the feedback loop is vital to both functions. A block diagram of this system is shown in Fig. 6. The power supplies required for the potentiometers and the amplifier are not shown in block diagrams of control system logic because they do not contribute to the control logic.

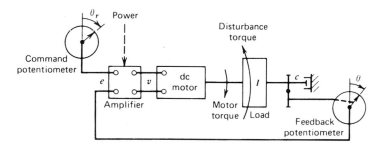

Figure 5   Position control system using a dc motor.[1]

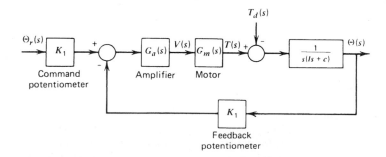

Figure 6   Block diagram of the position control system shown in Fig. 5.[1]

2.1   A Standard Diagram

The electromechanical positioning system fits the general structure of a control system (Fig. 7). This figure also gives some standard terminology. Not all systems can be forced into this format, but it serves as a reference for discussion.

The controller is generally thought of as a logic element that compares the command with the measurement of the output and decides what should be done. The input and feedback elements are transducers for converting one type of signal into another type. This allows the error detector directly to compare two signals of the same type (e.g., two voltages). Not all functions show up as separate physical elements. The error detector in Fig. 5 is simply the input terminals of the amplifier.

The control logic elements produce the control signal, which is sent to the final control elements. These are the devices that develop enough torque, pressure, heat, and so on to influence the elements under control. Thus, the final control elements are the "muscle" of the system, while the control logic elements are the "brain." Here we are primarily concerned with the design of the logic to be used by this brain.

The object to be controlled is the plant. The manipulated variable is generated by the final control elements for this purpose. The disturbance input also acts on the plant. This is an input over which the designer has no influence and perhaps for which little information is available as to the magnitude, functional form, or time of occurrence. The disturbance can be a random input, such as wind gust on a radar antenna, or deterministic, such as Coulomb friction effects. In the latter case, we can include the friction force in the system model by using a nominal value for the coefficient of friction. The disturbance input would then be the deviation of the friction force from this estimated value and would represent the uncertainty in our estimate.

Several control system classifications can be made with reference to Fig. 7. A *regulator* is a control system in which the controlled variable is to be kept constant in spite of disturbances.

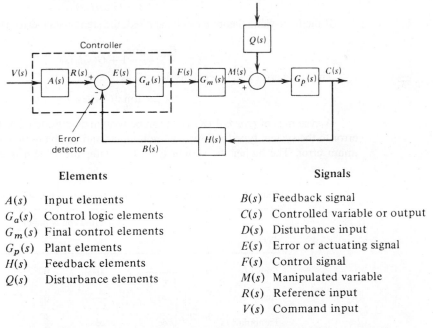

Elements		Signals	
$A(s)$	Input elements	$B(s)$	Feedback signal
$G_a(s)$	Control logic elements	$C(s)$	Controlled variable or output
$G_m(s)$	Final control elements	$D(s)$	Disturbance input
$G_p(s)$	Plant elements	$E(s)$	Error or actuating signal
$H(s)$	Feedback elements	$F(s)$	Control signal
$Q(s)$	Disturbance elements	$M(s)$	Manipulated variable
		$R(s)$	Reference input
		$V(s)$	Command input

Figure 7   Terminology and basic structure of a feedback control system.[1]

The command input for a regulator is its *set point*. A *follow-up system* is supposed to keep the control variable near a command value that is changing with time. An example of a follow up system is a machine tool in which a cutting head must trace a specific path in order to shape the product properly. This is also an example of a *servomechanism,* which is a control system whose controlled variable is a mechanical position, velocity, or acceleration. A thermostat system is not a servomechanism, but a *process control system,* where the controlled variable describes a thermodynamic process. Typically, such variables are temperature, pressure, flow rate, liquid level, chemical concentration, and so on.

2.2 Transfer Functions

A transfer function is defined for each input–output pair of the system. A specific transfer function is found by setting all other inputs to zero and reducing the block diagram. The primary or command transfer function for Fig. 7 is

$$\frac{C(s)}{V(s)} = \frac{A(s)G_a(s)G_m(s)G_p(s)}{1 + G_a(s)G_m(s)G_p(s)H(s)} \tag{1}$$

The disturbance transfer function is

$$\frac{C(s)}{D(s)} = \frac{-Q(s)G_p(s)}{1 + G_a(s)G_m(s)G_p(s)H(s)} \tag{2}$$

The transfer functions of a given system all have the same denominator.

2.3 System-Type Number and Error Coefficients

The error signal in Fig. 4 is related to the input as

$$E(s) = \frac{1}{1 + G(s)H(s)} R(s) \tag{3}$$

If the final-value theorem can be applied, the steady-state error is

$$E_{ss} = \lim_{s \to 0} \frac{sR(s)}{1 + G(s)H(s)} \tag{4}$$

The static error coefficient c_i is defined as

$$C_i = \lim_{s \to 0} s^i G(s)H(s) \tag{5}$$

A system is of *type n* if $G(s)H(s)$ can be written as $s^n F(s)$. Table 1 relates the steady-state error to the system type for three common inputs and can be used to design systems for minimum error. The higher the system type, the better the system is able to follow a rapidly

Table 1 Steady-State Error e_{ss} for Different System-Type Numbers

	System-Type Number n			
$R(s)$	0	1	2	3
Step $1/s$	$\dfrac{1}{1 + C_0}$	0	0	0
Ramp $1/s^2$	∞	$\dfrac{1}{C_1}$	0	0
Parabola $1/s^3$	∞	∞	$\dfrac{1}{C_2}$	0

changing input. But higher type systems are more difficult to stabilize, so a compromise must be made in the design. The coefficients c_0, c_1, and c_2 are called the *position, velocity,* and *acceleration error coefficients.*

3   TRANSDUCERS AND ERROR DETECTORS

The control system structure shown in Fig. 7 indicates a need for physical devices to perform several types of functions. Here we present a brief overview of some available transducers and error detectors. Actuators and devices used to implement the control logic are discussed in Sections 4 and 5.

3.1   Displacement and Velocity Transducers

A *transducer* is a device that converts one type of signal into another type. An example is the potentiometer, which converts displacement into voltage, as in Fig. 8. In addition to this conversion, the transducer can be used to make measurements. In such applications, the term *sensor* is more appropriate. Displacement can also be measured electrically with a *linear variable differential transformer* (LVDT) or a *synchro*. An LVDT measures the linear displacement of a movable magnetic core through a primary winding and two secondary windings (Fig. 9). An alternating-current (ac) voltage is applied to the primary. The secondaries are connected together and also to a detector that measures the voltage and phase difference. A phase difference of $0°$ corresponds to a positive core displacement, while $180°$ indicates a negative displacement. The amount of displacement is indicated by the amplitude of the ac voltage in the secondary. The detector converts this information into a dc voltage e_o, such that $e_o = Kx$. The LVDT is sensitive to small displacements. Two of them can be wired together to form an error detector.

A synchro is a rotary differential transformer, with angular displacement as either the input or output. They are often used in pairs (a *transmitter* and a *receiver*) where a remote indication

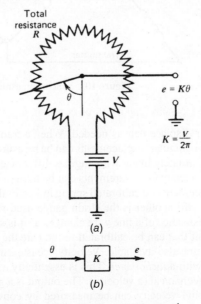

Figure 8   Rotary potentiometer.[1]

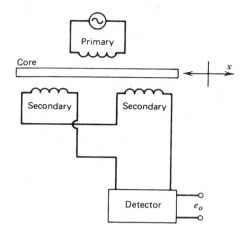

Figure 9 Linear variable differential transformer.[1]

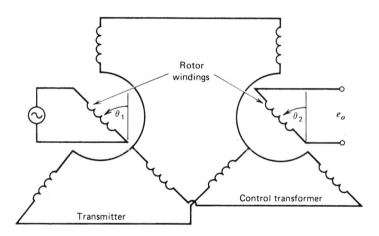

Figure 10 Synchro transmitter control transformer.[1]

of angular displacement is needed. When a transmitter is used with a synchro *control transformer,* two angular displacements can be measured and compared (Fig. 10). The output voltage *e* is approximately linear with angular difference within $\pm 70°$, so that $e_o = K(\theta_1 - \theta_2)$

Displacement measurements can be used to obtain forces and accelerations. For example, the displacement of a calibrated spring indicates the applied force. The accelerometer is another example. Still another is the strain gauge used for force measurement. It is based on the fact that the resistance of a fine wire changes as it is stretched. The change in resistance is detected by a circuit that can be calibrated to indicate the applied force. Sensors utilizing piezoelectric elements are also available. Velocity measurements in control systems are most commonly obtained with a *tachometer.* This is essentially a dc generator (the reverse of a dc motor). The input is mechanical (a velocity). The output is a generated voltage proportional to the velocity. Translational velocity can be measured by converting it to angular velocity with gears, for example. Tachometers using ac signals are also available.

Other velocity transducers include a magnetic pickup that generates a pulse every time a gear tooth passes. If the number of gear teeth is known, a pulse counter and timer can be used to compute the angular velocity. This principle is also employed in turbine flowmeters. A similar principle is employed by *optical encoders,* which are especially suitable for digital control purposes. These devices use a rotating disk with alternating transparent and opaque elements whose passage is sensed by light beams and a photosensor array, which generates a binary (on–off) train of pulses. There are two basic types: the absolute encoder and the incremental encoder. By counting the number of pulses in a given time interval, the incremental encoder can measure the rotational speed of the disk. By using multiple tracks of elements, the absolute encoder can produce a binary digit that indicates the amount of rotation. Hence, it can be used as a position sensor.

Most encoders generate a train of transistor–transistor logic (TTL) voltage level pulses for each channel. The incremental encoder output contains two channels that each produce N pulses every revolution. The encoder is mechanically constructed so that pulses from one channel are shifted relative to the other channel by a quarter of a pulse width. Thus, each pulse pair can be divided into four segments called *quadratures.* The encoder output consists of $4N$ *quadrature counts per revolution.* The pulse shift also allows the direction of rotation to be determined by detecting which channel leads the other. The encoder might contain a third channel, known as the zero, index, or marker channel, that produces a pulse once per revolution. This is used for initialization.

The gain of such an incremental encoder is $4N/2\pi$. Thus, an encoder with 1000 pulses per channel per revolution has a gain of 636 counts per radian. If an absolute encoder produces a binary signal with n bits, the maximum number of positions it can represent is $2n$, and its gain is $2^n/2\pi$. Thus, a 16-bit absolute encoder has a gain of $2^{16}/2\pi = 10,435$ counts per radian.

3.2   Temperature Transducers

When two wires of dissimilar metals are joined together, a voltage is generated if the junctions are at different temperatures. If the reference junction is kept at a fixed, known temperature, the thermocouple can be calibrated to indicate the temperature at the other junction in terms of the voltage v. Electrical resistance changes with temperature. Platinum gives a linear relation between resistance and temperature, while nickel is less expensive and gives a large resistance change for a given temperature change. Semiconductors designed with this property are called *thermistors.* Different metals expand at different rates when the temperature is increased. This fact is used in the bimetallic strip transducer found in most home thermostats. Two dissimilar metals are bonded together to form the strip. As the temperature rises, the strip curls, breaking contact and shutting off the furnace. The temperature gap can be adjusted by changing the distance between the contacts. The motion also moves a pointer on the temperature scale of the thermostat. Finally, the pressure of a fluid inside a bulb will change as its temperature changes. If the bulb fluid is air, the device is suitable for use in pneumatic temperature controllers.

3.3   Flow Transducers

A flow rate q can be measured by introducing a flow restriction, such as an orifice plate, and measuring the pressure drop Δp across the restriction. The relation is $\Delta p = Rq^2$, where R can be found from calibration of the device. The pressure drop can be sensed by converting it into the motion of a diaphragm. Figure 11 illustrates a related technique. The Venturi-type flowmeter measures the static pressures in the constricted and unconstricted flow regions. Bernoulli's principle relates the pressure difference to the flow rate. This pressure difference produces the diaphragm displacement. Other types of flowmeters are available, such as turbine meters.

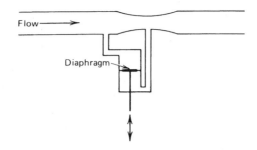

Figure 11 Venturi-type flowmeter. The diaphragm displacement indicates the flow rate.[1]

3.4 Error Detectors

The error detector is simply a device for finding the difference between two signals. This function is sometimes an integral feature of sensors, such as with the synchro transmitter–transformer combination. This concept is used with the diaphragm element shown in Fig. 11. A detector for voltage difference can be obtained, as with the position control system shown in Fig. 5. An amplifier intended for this purpose is a *differential amplifier*. Its output is proportional to the difference between the two inputs. In order to detect differences in other types of signals, such as temperature, they are usually converted to a displacement or pressure. One of the detectors mentioned previously can then be used.

3.5 Dynamic Response of Sensors

The usual transducer and detector models are static models and as such imply that the components respond instantaneously to the variable being sensed. Of course, any real component has a dynamic response of some sort, and this response time must be considered in relation to the controlled process when a sensor is selected. If the controlled process has a time constant at least 10 times greater than that of the sensor, we often would be justified in using a static sensor model.

4 ACTUATORS

An *actuator* is the final control element that operates on the low-level control signal to produce a signal containing enough power to drive the plant for the intended purpose. The armature-controlled dc motor, the hydraulic servomotor, and the pneumatic diaphragm and piston are common examples of actuators.

4.1 Electromechanical Actuators

Figure 12 shows an electromechanical system consisting of an armature-controlled dc motor driving a load inertia. The rotating armature consists of a wire conductor wrapped around an iron core. This winding has an inductance L. The resistance R represents the lumped value of the armature resistance and any external resistance deliberately introduced to change the motor's behavior. The armature is surrounded by a magnetic field. The reaction of this field with the armature current produces a torque that causes the armature to rotate. If the armature voltage v is used to control the motor, the motor is said to be *armature controlled*. In this case, the field is produced by an electromagnet supplied with a constant voltage or by a permanent magnet. This motor type produces a torque T that is proportional to the armature current i_a:

$$T = K_T i_a \tag{6}$$

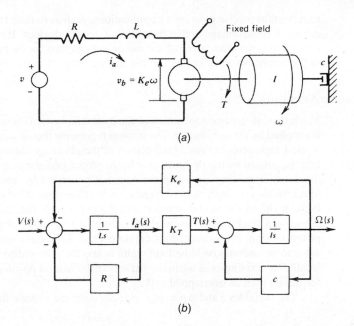

Figure 12   Armature-controlled dc motor with a load and the system's block diagram.[1]

The torque constant K_T depends on the strength of the field and other details of the motor's construction. The motion of a current-carrying conductor in a field produces a voltage in the conductor that opposes the current. This voltage is called the *back emf* (electromotive force). Its magnitude is proportional to the speed and is given by

$$e_b = K_e\omega \tag{7}$$

The transfer function for the armature-controlled dc motor is

$$\frac{\Omega(s)}{V(s)} = \frac{K_T}{LIs^2 + (RI + cL)s + cR + K_eK_T} \tag{8}$$

Another motor configuration is the *field-controlled* dc motor. In this case, the armature current is kept constant and the field voltage v is used to control the motor. The transfer function is

$$\frac{\Omega(s)}{V(s)} = \frac{K_T}{(Ls + R)(Is + c)} \tag{9}$$

where R and L are the resistance and inductance of the field circuit and K is the torque constant. No back emf exists in this motor to act as a self-braking mechanism.

Two-phase ac motors can be used to provide a low-power, variable-speed actuator. This motor type can accept the ac signals directly from LVDTs and synchros without demodulation. However, it is difficult to design ac amplifier circuitry to do other than proportional action. For this reason, the ac motor is not found in control systems as often as dc motors. The transfer function for this type is of the form of Eq. (9).

An actuator especially suitable for digital systems is the *stepper motor*, a special dc motor that takes a train of electrical input pulses and converts each pulse into an angular displacement of a fixed amount. Motors are available with resolutions ranging from about 4 steps per revolution to more than 800 steps per revolution. For 36 steps per revolution, the motor will rotate by 10° for each pulse received. When not being pulsed, the motors lock in place. Thus, they are excellent for precise positioning applications, such as required with printers and computer tape drives. A disadvantage is that they are low torque devices. If the input pulse frequency is not near the resonant frequency of the motor, we can take the output rotation to be directly related to the number of input pulses and use that description as the motor model.

4.2   Hydraulic Actuators

Machine tools are one application of the hydraulic system shown in Fig. 13. The applied force f is supplied by the servomotor. The mass m represents that of a cutting tool and the power piston, while k represents the combined effects of the elasticity naturally present in the structure and that introduced by the designer to achieve proper performance. A similar statement applies to the damping c. The valve displacement z is generated by another control system in order to move the tool through its prescribed motion. The spool valve shown in Fig. 13 had two *lands*. If the width of the land is greater than the port width, the valve is said to be *overlapped*. In this case, a dead zone exists in which a slight change in the displacement z produces no power piston motion. Such dead zones create control difficulties and are avoided by designing the valve to be *underlapped* (the land width is less the port width). For such valves there will be a small flow opening even when the valve is in the neutral position at $z = 0$. This gives it a higher sensitivity than an overlapped valve.

The variables z and $\Delta p = p_2 - p_1$ determine the volume flow rate, as

$$q = f(z, \Delta p)$$

For the reference equilibrium condition ($z = 0, \Delta p = 0, q = 0$) a linearization gives

$$q = c_1 z - c_2 \Delta p \tag{10}$$

The linearization constants are available from theoretical and experimental results.[5] The transfer function for the system is[1,2]

$$T(s) = \frac{X(s)}{Z(s)} = \frac{c_1}{C_2 m/As^2 + (cC_2/A + A)C_2 k/A} \tag{11}$$

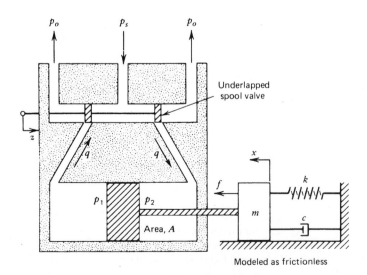

Figure 13   Hydraulic servomotor with a load.[1]

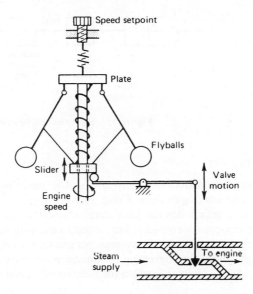

Figure 14 James Watt's flyball governor for speed control of a steam engine.[1]

The development of the steam engine led to the requirement for a speed control device to maintain constant speed in the presence of changes in load torque or steam pressure. In 1788, James Watt of Glasgow developed his now-famous flyball governor for this purpose (Fig. 14). Watt took the principle of sensing speed with the centrifugal pendulum of Thomas Mead and used it in a feedback loop on a steam engine. As the motor speed increases, the flyballs move outward and pull the slider upward. The upward motion of the slider closes the steam valve, thus causing the engine to slow down. If the engine speed is too slow, the spring force overcomes that due to the flyballs, and the slider moves down to open the steam valve. The desired speed can be set by moving the plate to change the compression in the spring. The principle of the flyball governor is still used for speed control applications. Typically, the pilot valve of a hydraulic servomotor is connected to the slider to provide the high forces required to move large supply valves.

Many hydraulic servomotors use multistage valves to obtain finer control and higher forces. A *two-stage valve* has a *slave valve,* similar to the pilot valve but situated between the pilot valve and the power piston.

Rotational motion can be obtained with a *hydraulic motor,* which is, in principle, a pump acting in reverse (fluid input and mechanical rotation output). Such motors can achieve higher torque levels than electric motors. A hydraulic pump driving a hydraulic motor constitutes a *hydraulic transmission.*

A popular actuator choice is the *electrohydraulic* system, which uses an electric actuator to control a hydraulic servomotor or transmission by moving the pilot valve or the swashplate angle of the pump. Such systems combine the power of hydraulics with the advantages of electrical systems. Figure 15 shows a hydraulic motor whose pilot valve motion is caused by an armature-controlled dc motor. The transfer function between the motor voltage and the piston displacement is

$$\frac{X(s)}{V(s)} = \frac{K_1 K_2 C_1}{As^2(\tau s + 1)} \qquad (12)$$

If the rotational inertia of the electric motor is small, then $\tau \approx 0$.

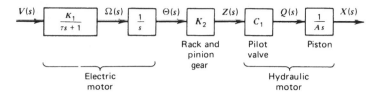

Figure 15   Electrohydraulic system for translation.[1]

4.3   Pneumatic Actuators

Pneumatic actuators are commonly used because they are simple to maintain and use a readily available working medium. Compressed air supplies with the pressures required are commonly available in factories and laboratories. No flammable fluids or electrical sparks are present, so these devices are considered the safest to use with chemical processes. Their power output is less than that of hydraulic systems but greater than that of electric motors.

A device for converting pneumatic pressure into displacement is the bellows shown in Fig. 16. The transfer function for a linearized model of the bellows is of the form

$$\frac{X(s)}{P(s)} = \frac{K}{\tau s + 1} \tag{13}$$

where x and p are deviations of the bellows displacement and input pressure from nominal values.

In many control applications, a device is needed to convert small displacements into relatively large pressure changes. The nozzle–flapper serves this purpose (Fig. 17a). The input displacement y moves the flapper, with little effort required. This changes the opening at the nozzle orifice. For a large enough opening, the nozzle back pressure is approximately the same as atmospheric pressure p_a. At the other extreme position with the flapper completely blocking the orifice, the back pressure equals the supply pressure p_a. This variation is shown in Fig. 17b. Typical supply pressures are between 30 and 100 psia. The orifice diameter is approximately 0.01 in. Flapper displacement is usually less than one orifice diameter.

The nozzle–flapper is operated in the linear portion of the back-pressure curve. The linearized back pressure relation is

$$p = -k_f x \tag{14}$$

where $-K_f$ is the slope of the curve and is a very large number. From the geometry of similar triangles, we have

$$p = -\frac{aK_f}{a + b} y \tag{15}$$

In its operating region, the nozzle–flapper's back pressure is well below the supply pressure.

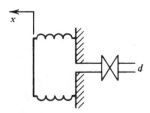

Figure 16   Pneumatic bellows.[1]

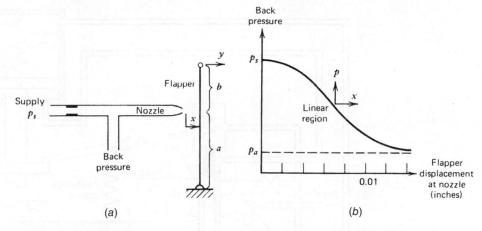

Figure 17 Pneumatic nozzle–flapper amplifier and its characteristic curve.[1]

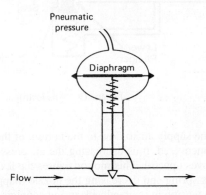

Figure 18 Pneumatic flow control valve.[1]

The output pressure from a pneumatic device can be used to drive a final control element like the pneumatic actuating valve shown in Fig. 18. The pneumatic pressure acts on the upper side of the diaphragm and is opposed by the return spring.

Formerly, many control systems utilized pneumatic devices to implement the control law in analog form. Although the overall, or higher level, control algorithm is now usually implemented in digital form, pneumatic devices are still frequently used for final control corrections at the actuator level, where the control action must eventually be supplied by a mechanical device. An example of this is the electropneumatic valve positioner used in Valtek valves and illustrated in Fig. 19. The heart of the unit is a pilot valve capsule that moves up and down according to the pressure difference across its two supporting diaphragms. The capsule has a plunger at its top and at its bottom. Each plunger has an exhaust seat at one end and a supply seat at the other. When the capsule is in its equilibrium position, no air is supplied to or exhausted from the valve cylinder, so the valve does not move.

The process controller commands a change in the valve stem position by sending the 4–20-mA dc input signal to the positioner. Increasing this signal causes the electromagnetic actuator to rotate the lever counterclockwise about the pivot. This increases the air gap between the nozzle and flapper. This decreases the back pressure on top of the upper diaphragm and causes the capsule to move up. This motion lifts the upper plunger from its supply seat and

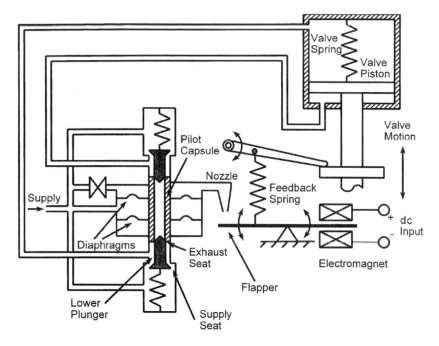

Figure 19   An electropneumatic valve positioner.[1]

allows the supply air to flow to the bottom of the valve cylinder. The lower plunger's exhaust seat is uncovered, thus decreasing the air pressure on top of the valve piston, and the valve stem moves upward. This motion causes the lever arm to rotate, increasing the tension in the feedback spring and decreasing the nozzle–flapper gap. The valve continues to move upward until the tension in the feedback spring counteracts the force produced by the electromagnetic actuator, thus returning the capsule to its equilibrium position. A decrease in the dc input signal causes the opposite actions to occur, and the valve moves downward.

5   CONTROL LAWS

The control logic elements are designed to act on the error signal to produce the control signal. The algorithm that is used for this purpose is called the *control law,* the *control action,* or the *control algorithm.* A nonzero error signal results from either a change in command or a disturbance. The general function of the controller is to keep the controlled variable near its desired value when these occur. More specifically, the control objectives might be stated as follows:

1. Minimize the steady-state error.
2. Minimize the settling time.
3. Achieve other transient specifications, such as minimizing the overshoot.

In practice, the design specifications for a controller are more detailed. For example, the bandwidth might also be specified along with a safety margin for stability. We never know the numerical values of the system's parameters with true certainty, and some controller designs can be more sensitive to such parameter uncertainties than other designs. So a parameter sensitivity specification might also be included. The following control laws form the basis of most control systems.

5.1 Proportional Control

Two-position control is the most familiar type, perhaps because of its use in home thermostats. The control output takes on one of two values. With the *on–off controller,* the controller output is either on or off (e.g., fully open or fully closed). Two-position control is acceptable for many applications in which the requirements are not too severe. However, many situations require finer control.

Consider a liquid-level system in which the input flow rate is controlled by a valve. We might try setting the control valve manually to achieve a flow rate that balances the system at the desired level. We might then add a controller that adjusts this setting in proportion to the deviation of the level from the desired value. This is *proportional control,* the algorithm in which the change in the control signal is proportional to the error. Block diagrams for controllers are often drawn in terms of the deviations from a zero-error equilibrium condition. Applying this convention to the general terminology of Fig. 6, we see that proportional control is described by

$$F(s) = K_P E(s)$$

where $F(s)$ is the deviation in the control signal and K_P is the *proportional gain.* If the total valve displacement is $y(t)$ and the manually created displacement is x, then

$$y(t) = K_P e(t) + x$$

The percent change in error needed to move the valve full scale is the *proportional band.* It is related to the gain as

$$K_P = \frac{100}{\text{band}\%}$$

The zero-error valve displacement x is the *manual reset.*

Proportional Control of a First-Order System

To investigate the behavior of proportional control, consider the speed control system shown in Fig. 20; it is identical to the position controller shown in Fig. 6, except that a tachometer replaces the feedback potentiometer. We can combine the amplifier gains into one, denoted K_P. The system is thus seen to have proportional control. We assume the motor is field controlled and has a negligible electrical time constant. The disturbance is a torque T_d, for example, resulting

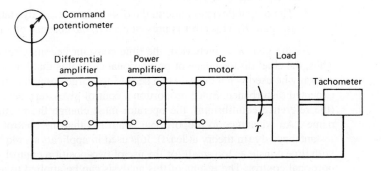

Figure 20   Velocity control system using a dc motor.[1]

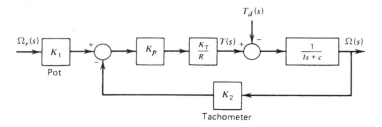

Figure 21   Block diagram of the velocity control system of Fig. 20.[1]

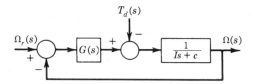

Figure 22   Simplified form of Fig. 21 for the case $K_1 = K_2$.[1]

from friction. Choose the reference equilibrium condition to be $T_d = T = 0$ and $\omega_r = w = 0$. The block diagram is shown in Fig. 21. For a meaningful error signal to be generated, K_1 and K_2 should be chosen to be equal. With this simplification the diagram becomes that shown in Fig. 22, where $G(s) = K = K_1 K_p K_T / R$. A change in desired speed can be simulated by a unit step input for ω_r. For $\Omega_r(s) = 1/s$, the velocity approaches the steady-state value $\omega_{ss} = K/(c + K) < 1$. Thus, the final value is less than the desired value of 1, but it might be close enough if the damping c is small. The time required to reach this value is approximately four time constants, or $4\tau = 4I/(c + K)$. A sudden change in load torque can also be modeled by a unit step function $T_d(s) = 1/s$. The steady-state response due solely to the disturbance is $1/(c + K)$. If $c + K$ is large, this error will be small.

The performance of the proportional control law thus far can be summarized as follows. For a first-order plant with step function inputs:

1. The output never reaches its desired value if damping is present ($c = 0$), although it can be made arbitrarily close by choosing the gain K large enough. This is called offset error.

2. The output approaches its final value without oscillation. The time to reach this value is inversely proportional to K.

3. The output deviation due to the disturbance at steady state is inversely proportional to the gain K. This error is present even in the absence of damping ($c = 0$).

As the gain K is increased, the time constant becomes smaller and the response faster. Thus, the chief disadvantage of proportional control is that it results in steady-state errors and can only be used when the gain can be selected large enough to reduce the effect of the largest expected disturbance. Since proportional control gives zero error only for one load condition (the reference equilibrium), the operator must change the manual reset by hand (hence the name). An advantage to proportional control is that the control signal responds to the error instantaneously (in theory at least). It is used in applications requiring rapid action. Processes with time constants too small for the use of two-position control are likely candidates for proportional control. The results of this analysis can be applied to any type of first-order system (e.g., liquid level, thermal, etc.) having the form in Fig. 22.

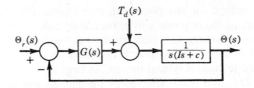

Figure 23   Position servo.[1]

Proportional Control of a Second-Order System

Proportional control of a neutrally stable second-order plant is represented by the position controller of Fig. 6 if the amplifier transfer function is a constant $G_a(s) = K_a$. Let the motor transfer function be $G_m(s) = K_T/R$, as before. The modified block diagram is given in Fig. 23 with $G(s) = K = K_1 K_a K_T/R$. The closed-loop system is stable if I, c, and K are positive. For no damping ($c = 0$), the closed-loop system is neutrally stable. With no disturbance and a unit step command, $\Theta_r(s) = 1/s$, the steady-state output is $\Theta_{ss} = 1$. The offset error is thus zero if the system is stable ($c > 0, K > 0$). The steady-state output deviation due to a unit step disturbance is $-1/K$. This deviation can be reduced by choosing K large. The transient behavior is indicated by the damping ratio, $\zeta = c/2\sqrt{IK}$.

For slight damping, the response to a step input will be very oscillatory and the overshoot large. The situation is aggravated if the gain K is made large to reduce the deviation due to the disturbance. We conclude, therefore, that proportional control of this type of second-order plant is not a good choice unless the damping constant c is large. We will see shortly how to improve the design.

5.2   Integral Control

The offset error that occurs with proportional control is a result of the system reaching an equilibrium in which the control signal no longer changes. This allows a constant error to exist. If the controller is modified to produce an increasing signal as long as the error is nonzero, the offset might be eliminated. This is the principle of *integral control*. In this mode the change in the control signal is proportional to the *integral* of the error. In the terminology of Fig. 7, this gives

$$F(s) = \frac{K_I}{s} E(s) \tag{16}$$

where $F(s)$ is the deviation in the control signal and K_I is the *integral gain*. In the time domain, the relation is

$$f(t) = K_I \int_0^t e(t)dt \tag{17}$$

if $f(0) = 0$. In this form, it can be seen that the integration cannot continue indefinitely because it would theoretically produce an infinite value of $f(t)$ if $e(t)$ does not change sign. This implies that special care must be taken to reinitialize a controller that uses integral action.

Integral Control of a First-Order System

Integral control of the velocity in the system of Fig. 20 has the block diagram shown in Fig. 22, where $G(s) = K/s, K = K_1 K_I K_T/R$. The integrating action of the amplifier is physically obtained by the techniques to be presented in Section 6 or by the digital methods presented in Section 10. The control system is stable if I, c, and K are positive. For a unit step command input, $\omega_{ss} = 1$; so the offset error is zero. For a unit step disturbance, the steady-state deviation is zero if the system is stable. Thus, the steady-state performance using integral control is excellent for this plant with step inputs. The damping ratio is $\zeta = c/2\sqrt{IK}$. For slight damping, the response will be oscillatory rather than exponential as with proportional control. Improved steady-state performance has thus been obtained at the expense of degraded transient performance. The conflict between steady-state and transient specifications is a common theme in control system design. As long as the system is underdamped, the time constant is $\tau = 2I/c$ and is not affected by the gain K, which only influences the oscillation frequency in this case. It might be physically possible to make K small enough so that $\zeta \gg 1$, and the nonoscillatory feature of proportional control recovered, but the response would tend to be sluggish. Transient specifications for fast response generally require that $\zeta < 1$. The difficulty with using $\zeta < 1$ is that τ is fixed by c and I. If c and I are such that $\zeta < 1$, then τ is large if $I \gg c$.

Integral Control of a Second-Order System
Proportional control of the position servomechanism in Fig. 23 gives a nonzero steady-state deviation due to the disturbance. Integral control $[G(s) = K/s]$ applied to this system results in the command transfer function

$$\frac{\Theta(s)}{\Theta_r(s)} = \frac{K}{1s^3 + cs^2 + K} \tag{18}$$

With the Routh criterion, we immediately see that the system is not stable because of the missing s term. Integral control is useful in improving steady-state performance, but in general it does not improve and may even degrade transient performance. Improperly applied, it can produce an unstable control system. It is best used in conjunction with other control modes.

5.3 Proportional-plus-Integral Control

Integral control raised the order of the system by 1 in the preceding examples but did not give a characteristic equation with enough flexibility to achieve acceptable transient behavior. The instantaneous response of proportional control action might introduce enough variability into the coefficients of the characteristic equation to allow both steady-state and transient specifications to be satisfied. This is the basis for using *proportional-plus-integral control* (PI control). The algorithm for this two-mode control is

$$F(s) = K_P E(s) + \frac{K_I}{s} E(s) \tag{19}$$

The integral action provides an automatic, not manual, reset of the controller in the presence of a disturbance. For this reason, it is often called *reset action*.

The algorithm is sometimes expressed as

$$F(s) = K_P \left(1 + \frac{1}{T_I s}\right) E(s) \tag{20}$$

where T_1 is the *reset time*. The reset time is the time required for the integral action signal to equal that of the proportional term if a constant error exists (a hypothetical situation). The reciprocal of reset time is expressed as repeats per minute and is the frequency with which the integral action repeats the proportional correction signal.

The proportional control gain must be reduced when used with integral action. The integral term does not react instantaneously to a zero-error signal but continues to correct, which tends to cause oscillations if the designer does not take this effect into account.

PI Control of a First-Order System
PI action applied to the speed controller of Fig. 20 gives the diagram shown in Fig. 21 with $G(s) = K_P + K_I/s$. The gains K_P and K_I are related to the component gains, as before.

The system is stable for positive values of K_P and K_I. For $\Omega_r(s) = 1/s$, $\omega_{ss} = 1$, and the offset error is zero, as with integral action only. Similarly, the deviation due to a unit step disturbance is zero at steady state. The damping ratio is $\zeta = (c + K_P)/2\sqrt{IK_I}$. The presence of K_P allows the damping ratio to be selected without fixing the value of the dominant time constant. For example, if the system is underdamped ($\zeta < 1$), the time constant is $\tau = 2I/(c + K_P)$. The gain K_P can be picked to obtain the desired time constant, while K_I is used to set the damping ratio. A similar flexibility exists if $\zeta = 1$. Complete description of the transient response requires that the numerator dynamics present in the transfer functions be accounted for.[1,2]

PI Control of a Second-Order System

Integral control for the position servomechanism of Fig. 23 resulted in a third-order system that is unstable. With proportional action, the diagram becomes that of Fig. 22, with $G(s) = K_P + K_I/s$. The steady-state performance is acceptable, as before, if the system is assumed to be stable. This is true if the Routh criterion is satisfied, that is, if I, c, K_P are positive and $cK_P - IK_I > 0$. The difficulty here occurs when the damping is slight. For small c, the gain K_P must be large in order to satisfy the last condition, and this can be difficult to implement physically. Such a condition can also result in an unsatisfactory time constant. The root-locus method of Section 9 provides the tools for analyzing this design further.

5.4 Derivative Control

Integral action tends to produce a control signal even after the error has vanished, which suggests that the controller be made aware that the error is approaching zero. One way to accomplish this is to design the controller to react to the derivative of the error with *derivative control* action, which is

$$F(s) = K_D s E(s) \tag{21}$$

where K_D is the *derivative gain*. This algorithm is also called *rate action*. It is used to damp out oscillations. Since it depends only on the error rate, derivative control should never be used alone. When used with proportional action, the following PD control algorithm results:

$$F(s) = (K_P + K_D s)E(s) = K_P(1 + T_D s)E(s) \tag{22}$$

where T_D is the rate time or derivative time. With integral action included, the proportional-plus-integral-plus-derivative (PID) control law is obtained:

$$F(s) = \left(K_P + \frac{K_I}{s} + K_D s\right) E(s) \tag{23}$$

This is called a three-mode controller.

PD Control of a Second-Order System

The presence of integral action reduces steady-state error but tends to make the system less stable. There are applications of the position servomechanism in which a nonzero derivation resulting from the disturbance can be tolerated but an improvement in transient response over the proportional control result is desired. Integral action would not be required, but rate action can be added to improve the transient response. Application of PD control to this system gives the block diagram of Fig. 23 with $G(s) = K_P + K_D s$.

The system is stable for positive values of K_D and K_P. The presence of rate action does not affect the steady-state response, and the steady-state results are identical to those with proportional control; namely, zero offset error and a deviation of $-1/K_P$, due to the disturbance. The damping ratio is $\zeta = (c + K_D)/2\sqrt{IK_P}$. For proportional control, $\zeta = c/2\sqrt{IK_P}$. Introduction

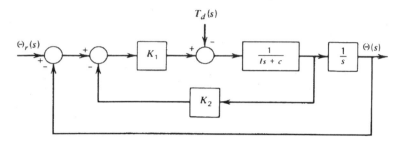

Figure 24   Tachometer feedback arrangement to replace PD control for the position servo.[1]

of rate action allows the proportional gain K_P to be selected large to reduce the steady-state deviation, while K_D can be used to achieve an acceptable damping ratio. The rate action also helps to stabilize the system by adding damping (if $c = 0$, the system with proportional control is not stable).

The equivalent of derivative action can be obtained by using a tachometer to measure the angular velocity of the load. The block diagram is shown in Fig. 24. The gain of the amplifier–motor–potentiometer combination is K_1, and K_2 is the tachometer gain. The advantage of this system is that it does not require signal differentiation, which is difficult to implement if signal noise is present. The gains K_1 and K_2 can be chosen to yield the desired damping ratio and steady-state deviation, as was done with K_P and K_I.

5.5   Proportional–Integral–Derivative Control

The position servomechanism design with PI control is not completely satisfactory because of the difficulties encountered when the damping c is small. This problem can be solved by the use of the full PID control law, as shown in Fig. 23 with $G(s) = K_P + K_D s + K_I/s$.

A stable system results if all gains are positive and if $(c + K_D)K_P - IK_I > 0$. The presence of K_D relaxes somewhat the requirement that K_P be large to achieve stability. The steady-state errors are zero, and the transient response can be improved because three of the coefficients of the characteristic equation can be selected. To make further statements requires the root-locus technique presented in Section 9.

Proportional, integral, and derivative actions and their various combinations are not the only control laws possible, but they are the most common. PID controllers will remain for some time the standard against which any new designs must compete.

The conclusions reached concerning the performance of the various control laws are strictly true only for the plant model forms considered. These are the first-order model without numerator dynamics and the second-order model with a root at $s = 0$ and no numerator zeros. The analysis of a control law for any other linear system follows the preceding pattern. The overall system transfer functions are obtained, and all of the linear system analysis techniques can be applied to predict the system's performance. If the performance is unsatisfactory, a new control law is tried and the process repeated. When this process fails to achieve an acceptable design, more systematic methods of altering the system's structure are needed; they are discussed in later sections. We have used step functions as the test signals because they are the most common and perhaps represent the severest test of system performance. Impulse, ramp, and sinusoidal test signals are also employed. The type to use should be made clear in the design specifications.

6 CONTROLLER HARDWARE

The control law must be implemented by a physical device before the control engineer's task is complete. The earliest devices were purely kinematic and were mechanical elements such as gears, levers, and diaphragms that usually obtained their power from the controlled variable. Most controllers now are analog electronic, hydraulic, pneumatic, or digital electronic devices. We now consider the analog type. Digital controllers are covered starting in Section 10.

6.1 Feedback Compensation and Controller Design

Most controllers that implement versions of the PID algorithm are based on the following feedback principle. Consider the single-loop system shown in Fig. 1. If the open-loop transfer function is large enough that $|G(s)H(s)|$, the closed-loop transfer function is approximately given by

$$T(s) = \frac{G(s)}{1 + G(s)H(s)} \approx \frac{G(s)}{G(s)H(s)} = \frac{1}{H(s)} \tag{24}$$

The principle states that a power unit $G(s)$ can be used with a feedback element $H(s)$ to create a desired transfer function $T(s)$. The power unit must have a gain high enough that $|G(s)H(s)| \gg 1$ and the feedback elements must be selected so that $H(s) = 1/T(s)$. This principle was used in Section 1 to explain the design of a feedback amplifier.

6.2 Electronic Controllers

The *operational amplifier (op amp)* is a high-gain amplifier with a high input impedance. A diagram of an op amp with feedback and input elements with impedances $T(s)$ is shown in Fig. 25. An approximate relation is

$$\frac{E_o(s)}{E_i(s)} = -\frac{T_f(s)}{T_i(s)}$$

The various control modes can be obtained by proper selection of the impedances. A pro-portional controller can be constructed with a multiplier, which uses two resistors, as shown in Fig. 26. An inverter is a multiplier circuit with $R_f = R_i$. It is sometimes needed because of the sign reversal property of the op amp. The multiplier circuit can be modified to act as an adder (Fig. 27).

PI control can be implemented with the circuit of Fig. 28. Figure 29 shows a complete system using op amps for PI control. The inverter is needed to create an error detector. Many industrial controllers provide the operator with a choice of control modes, and the operator can switch from one mode to another when the process characteristics or control objectives change. When a switch occurs, it is necessary to provide any integrators with the proper initial voltages

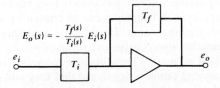

$$E_o(s) = -\frac{T_f(s)}{T_i(s)} E_i(s)$$

Figure 25 Operational amplifier (op amp).[1]

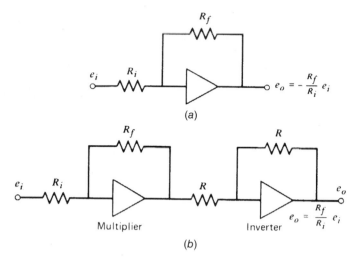

Figure 26   Op-amp implementation of proportional control.[1]

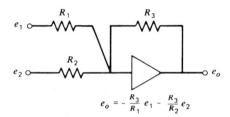

Figure 27   Op-amp adder circuit.[1]

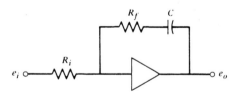

Figure 28   Op-amp implementation of PI control.[1]

or else undesirable transients will occur when the integrator is switched into the system. Commercially available controllers usually have built-in circuits for this purpose.

In theory, a differentiator can be created by interchanging the resistance and capacitance in the integrating op amp. The difficulty with this design is that no electrical signal is "pure." Contamination always exists as a result of voltage spikes, ripple, and other transients generally categorized as "noise." These high-frequency signals have large slopes compared with the more slowly varying primary signal, and thus they will dominate the output of the differentiator. In practice, this problem is solved by filtering out high-frequency signals, either with a low-pass filter inserted in cascade with the differentiator or by using a redesigned differentiator such as the one shown in Fig. 30. For the ideal PD controller, $R_1 = 0$. The attenuation curve for the ideal controller breaks upward at $\omega = 1/R_2C$ with a slope of 20 dB/decade. The curve for

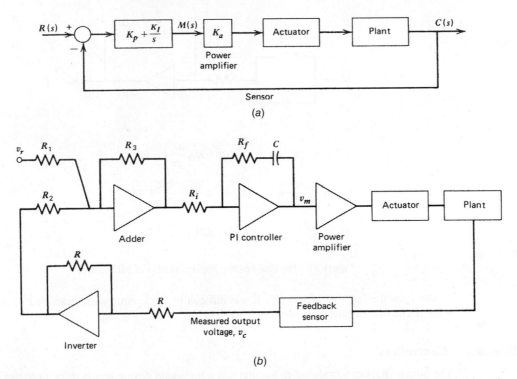

Figure 29   Implementation of a PI controller using op amps. (*a*) Diagram of the system. (*b*) Diagram showing how the op amps are connected.[1]

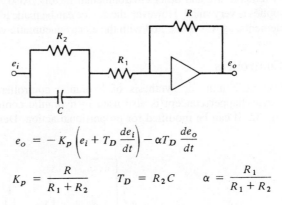

$$e_o = -K_p \left(e_i + T_D \frac{de_i}{dt} \right) - \alpha T_D \frac{de_o}{dt}$$

$$K_p = \frac{R}{R_1 + R_2} \qquad T_D = R_2 C \qquad \alpha = \frac{R_1}{R_1 + R_2}$$

Figure 30   Practical op-amp implementation of PD control.[1]

the practical controller does the same but then becomes flat for $\omega > (R_1 + R_2)/R_1 R_2 C$. This provides the required limiting effect at high frequencies.

PID control can be implemented by joining the PI and PD controllers in parallel, but this is expensive because of the number of op amps and power supplies required. Instead, the usual implementation is that shown in Fig. 31. The circuit limits the effect of frequencies above $\omega = 1/\beta R_1 C_1$. When $R_1 = 0$, ideal PID control results. This is sometimes called the noninteractive algorithm because the effect of each of the three modes is additive, and they do not interfere with one another. The form given for $R_1 \neq 0$ is the *real* or *interactive* algorithm. This name

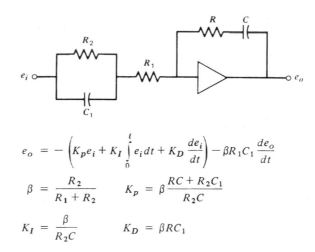

$$e_o = -\left(K_p e_i + K_I \int_0^t e_i \, dt + K_D \frac{de_i}{dt}\right) - \beta R_1 C_1 \frac{de_o}{dt}$$

$$\beta = \frac{R_2}{R_1 + R_2} \qquad K_p = \beta \frac{RC + R_2 C_1}{R_2 C}$$

$$K_I = \frac{\beta}{R_2 C} \qquad K_D = \beta R C_1$$

Figure 31   Practical op-amp implementation of PID control.[2]

results from the fact that historically it was difficult to implement noninteractive PID control with mechanical or pneumatic devices.

6.3   Pneumatic Controllers

The nozzle–flapper introduced in Section 4 is a high-gain device that is difficult to use without modification. The gain K_f is known only imprecisely and is sensitive to changes induced by temperature and other environmental factors. Also, the linear region over which Eq. (14) applies is very small. However, the device can be made useful by compensating it with feedback elements, as was illustrated with the electropneumatic valve positioner shown in Fig. 19.

6.4   Hydraulic Controllers

The basic unit for synthesis of hydraulic controllers is the hydraulic servomotor. The nozzle–flapper concept is also used in hydraulic controllers.[5] A PI controller is shown in Fig. 32. It can be modified for proportional action. Derivative action has not seen much use

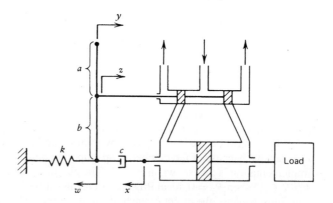

Figure 32   Hydraulic implementation of PI control.[1]

in hydraulic controllers. This action supplies damping to the system, but hydraulic systems are usually highly damped intrinsically because of the viscous working fluid. PI control is the algorithm most commonly implemented with hydraulics.

7 FURTHER CRITERIA FOR GAIN SELECTION

Once the form of the control law has been selected, the gains must be computed in light of the performance specifications. In the examples of the PID family of control laws in Section 5, the damping ratio, dominant time constant, and steady-state error were taken to be the primary indicators of system performance in the interest of simplicity. In practice, the criteria are usually more detailed. For example, the rise time and maximum overshoot, as well as the other transient response specifications of the previous chapter, may be encountered. Requirements can also be stated in terms of frequency response characteristics, such as bandwidth, resonant frequency, and peak amplitude. Whatever specific form they take, a complete set of specifications for control system performance generally should include the following considerations for given forms of the command and disturbance inputs:

1. Equilibrium specifications
   a. Stability
   b. Steady-state error
2. Transient specifications
   a. Speed of response
   b. Form of response
3. Sensitivity specifications
   a. Sensitivity to parameter variations
   b. Sensitivity to model inaccuracies
   c. Noise rejection (bandwidth, etc.)

In addition to these performance stipulations, the usual engineering considerations of initial cost, weight, maintainability, and so on must be taken into account. The considerations are highly specific to the chosen hardware, and it is difficult to deal with such issues in a general way.

Two approaches exist for designing the controller. The proper one depends on the quality of the analytical description of the plant to be controlled. If an accurate model of the plant is easily developed, we can design a specialized controller for the particular application. The range of adjustment of controller gains in this case can usually be made small because the accurate plant model allows the gains to be precomputed with confidence. This technique reduces the cost of the controller and can often be applied to electromechanical systems.

The second approach is used when the plant is relatively difficult to model, which is often the case in process control. A standard controller with several control modes and wide ranges of gains is used, and the proper mode and gain settings are obtained by testing the controller on the process in the field. This approach should be considered when the cost of developing an accurate plant model might exceed the cost of controller tuning in the field. Of course, the plant must be available for testing for this approach to be feasible.

7.1 Performance Indices

The performance criteria encountered thus far require a set of conditions to be specified—for example, one for steady-state error, one for damping ratio, and one for the dominant time constant. If there are many such conditions, and if the system is of high order with several gains to be selected, the design process can get quite complicated because transient and steady-state criteria tend to drive the design in different directions. An alternative approach is to specify the system's desired performance by means of one analytical expression called a *performance index*. Powerful analytical and numerical methods are available that allow the gains to be systematically computed by minimizing (or maximizing) this index.

To be useful, a performance index must be selective. The index must have a sharply defined extremum in the vicinity of the gain values that give the desired performance. If the numerical value of the index does not change very much for large changes in the gains from their optimal values, the index will not be selective.

Any practical choice of a performance index must be easily computed, either analytically, numerically, or experimentally. Four common choices for an index are the following:

$$J = \int_0^\infty |e(t)|dt \quad \text{(IAE index)} \tag{25}$$

$$J = \int_0^\infty t|e(t)|dt \quad \text{(ITAE index)} \tag{26}$$

$$J = \int_0^\infty |e(t)|^2 dt \quad \text{(ISE index)} \tag{27}$$

$$J = \int_0^\infty t|e(t)|^2 dt \quad \text{(ITSE index)} \tag{28}$$

where $e(t)$ is the system error. This error usually is the difference between the desired and the actual values of the output. However, if $e(t)$ does not approach zero as $t \to \infty$, the preceding indices will not have finite values. In this case, $e(t)$ can be defined as $e(t) = c(\infty) - c(t)$, where $c(t)$ is the output variable. If the index is to be computed numerically or experimentally, the infinite upper limit can be replaced by a time t large enough that $e(t)$ is negligible for $t > t_f$.

The *integral absolute-error* (IAE) criterion (25) expresses mathematically that the designer is not concerned with the sign of the error, only its magnitude. In some applications, the IAE criterion describes the fuel consumption of the system. The index says nothing about the relative importance of an error occurring late in the response versus an error occurring early. Because of this, the index is not as selective as the *integral-of-time-multiplied absolute error* (ITAE) criterion (26). Since the multiplier t is small in the early stages of the response, this index weights early errors less heavily than later errors. This makes sense physically. No system can respond instantaneously, and the index is lenient accordingly, while penalizing any design that allows a nonzero error to remain for a long time. Neither criterion allows highly underdamped or highly overdamped systems to be optimum. The ITAE criterion usually results in a system whose step response has a slight overshoot and well-damped oscillations.

The *integral squared-error* (ISE) (27) and *integral-of-time-multiplied squared-error* (ITSE) (28) criteria are analogous to the IAE and ITAE criteria, except that the square of the error is employed for three reasons: (1) in some applications, the squared error represents the system's power consumption; (2) squaring the error weights large errors much more heavily than small errors; and (3) the squared error is much easier to handle analytically. The derivative of a squared term is easier to compute than that of an absolute value and does not have a discontinuity at $e = 0$. These differences are important when the system is of high order with multiple error terms.

The closed-form solution for the response is not required to evaluate a performance index. For a given set of parameter values, the response and the resulting index value can be computed numerically. The optimum solution can be obtained using systematic computer search procedures; this makes this approach suitable for use with nonlinear systems.

7.2   Optimal-Control Methods

Optimal-control theory includes a number of algorithms for systematic design of a control law to minimize a performance index, such as the following generalization of the ISE index, called the *quadratic* index:

$$J = \int_0^\infty (\mathbf{x}^T\mathbf{Q}\mathbf{x} + \mathbf{u}^T\mathbf{R}\mathbf{u})\,dt \tag{29}$$

where **x** and **u** are the deviations of the state and control vectors from the desired reference values. For example, in a servomechanism, the state vector might consist of the position and velocity, and the control vector might be a scalar—the force or torque produced by the actuator. The matrices **Q** and **R** are chosen by the designer to provide relative weighting for the elements of **x** and **u**. If the plant can be described by the linear state-variable model

$$\dot{\mathbf{x}} = \mathbf{A}\mathbf{x} + \mathbf{B}\mathbf{u} \tag{30}$$

$$\mathbf{y} = \mathbf{C}\mathbf{x} + \mathbf{D}\mathbf{u} \tag{31}$$

where **y** is the vector of outputs—for example, position and velocity—then the solution of this *linear-quadratic* control problem is the linear control law:

$$\mathbf{u} = \mathbf{K}\mathbf{y} \tag{32}$$

where **K** is a matrix of gains that can be found by several algorithms.[1,6,7] A valid solution is guaranteed to yield a stable closed-loop system, a major benefit of this method.

Even if it is possible to formulate the control problem in this way, several practical difficulties arise. Some of the terms in (29) might be beyond the influence of the control vector **u**; the system is then *uncontrollable*. Also, there might not be enough information in the output equation (31) to achieve control, and the system is then *unobservable*. Several tests are available to check controllability and observability. Not all of the necessary state variables might be available for feedback or the feedback measurements might be noisy or biased. Algorithms known as *observers, state reconstructors, estimators,* and *digital filters* are available to compensate for the missing information. Another source of error is the uncertainty in the values of the coefficient matrices **A**, **B**, **C**, and **D**. Identification schemes can be used to compare the predicted and the actual system performance and to adjust the coefficient values "on-line."

7.3   The Ziegler–Nichols Rules

The difficulty of obtaining accurate transfer function models for some processes has led to the development of empirically based rules of thumb for computing the optimum gain values for a controller. Commonly used guidelines are the *Ziegler–Nichols rules,* which have proved so helpful that they are still in use 50 years after their development. The rules actually consist of two separate methods. The first method requires the open-loop step response of the plant, while the second uses the results of experiments performed with the controller already installed. While primarily intended for use with systems for which no analytical model is available, the rules are also helpful even when a model can be developed.

Ziegler and Nichols developed their rules from experiments and analysis of various industrial processes. Using the IAE criterion with a unit step response, they found that controllers adjusted according to the following rules usually had a step response that was oscillatory but with enough damping so that the second overshoot was less than 25% of the first (peak) overshoot. This is the *quarter-decay* criterion and is sometimes used as a specification.

The first method is the *process reaction* method and relies on the fact that many processes have an open-loop step response like that shown in Fig. 33. This is the *process signature* and is characterized by two parameters, R and L, where R is the slope of a line tangent to the steepest

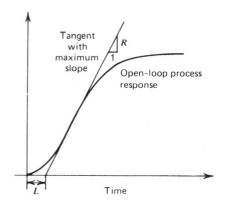

Figure 33 Process signature for a unit step input.[1]

Table 2 The Ziegler–Nichols Rules

Controller transfer function $G(s) = K_P \left(1 + \dfrac{1}{T_I s} + T_D s \right)$

Control Mode	Process Reaction Method	Ultimate-Cycle Method
P Control	$K_P = \dfrac{1}{RL}$	$K_P = 0.5 K_{pu}$
PI Control	$K_P = \dfrac{0.9}{RL}$	$K_P = 0.45 K_{pu}$
	$T_I = 3.3L$	$T_I = 0.83 P_u$
PID Control	$K_P = \dfrac{1.2}{RL}$	$K_P = 0.6 K_{pu}$
	$T_I = 2L$	$T_I = 0.5 P_u$
	$T_D = 0.5L$	$T_D = 0.125 P_u$

part of the response curve and L is the time at which this line intersects the time axis. First- and second-order linear systems do not yield positive values for L, and so the method cannot be applied to such systems. However, third- and higher-order linear systems with sufficient damping do yield such a response. If so, the Ziegler–Nichols rules recommend the controller settings given in Table 2.

The ultimate-cycle method uses experiments with the controller in place. All control modes except proportional are turned off, and the process is started with the proportional gain K_P set at a low value. The gain is slowly increased until the process begins to exhibit sustained oscillations. Denote the period of this oscillation by P_u and the corresponding ultimate gain by K_{Pu}. The Ziegler–Nichols recommendations are given in Table 2 in terms of these parameters. The proportional gain is lower for PI control than for proportional control and is higher for PID control because integral action increases the order of the system and thus tends to destabilize it; thus, a lower gain is needed. On the other hand, derivative action tends to stabilize the system; hence, the proportional gain can be increased without degrading the stability characteristics. Because the rules were developed for a typical case out of many types of processes, final tuning of the gains in the field is usually necessary.

7.4   Nonlinearities and Controller Performance

All physical systems have nonlinear characteristics of some sort, although they can often be modeled as linear systems provided the deviations from the linearization reference condition are not too great. Under certain conditions, however, the nonlinearities have significant effects on the system's performance. One such situation can occur during the start-up of a controller if the initial conditions are much different from the reference condition for linearization. The linearized model is then not accurate, and nonlinearities govern the behavior. If the nonlinearities are mild, there might not be much of a problem. Where the nonlinearities are severe, such as in process control, special consideration must be given to start-up. Usually, in such cases, the control signal sent to the final control elements is manually adjusted until the system variables are within the linear range of the controller. Then the system is switched into automatic mode. Digital computers are often used to replace the manual adjustment process because they can be readily coded to produce complicated functions for the start-up signals. Care must also be taken when switching from manual to automatic. For example, the integrators in electronic controllers must be provided with the proper initial conditions.

7.5   Reset Windup

In practice, all actuators and final control elements have a limited operating range. For example, a motor–amplifier combination can produce a torque proportional to the input voltage over only a limited range. No amplifier can supply an infinite current; there is a maximum current and thus a maximum torque that the system can produce. The final control elements are said to be *overdriven* when they are commanded by the controller to do something they cannot do. Since the limitations of the final control elements are ultimately due to the limited rate at which they can supply energy, it is important that all system performance specifications and controller designs be consistent with the energy delivery capabilities of the elements to be used.

Controllers using integral action can exhibit the phenomenon called *reset windup* or *integrator buildup* when overdriven, if they are not properly designed. For a step change in set point, the proportional term responds instantly and saturates immediately if the set-point change is large enough. On the other hand, the integral term does not respond as fast. It integrates the error signal and saturates some time later if the error remains large for a long enough time. As the error decreases, the proportional term no longer causes saturation. However, the integral term continues to increase as long as the error has not changed sign, and thus the manipulated variable remains saturated. Even though the output is very near its desired value, the manipulated variable remains saturated until after the error has reversed sign. The result can be an undesirable overshoot in the response of the controlled variable.

Limits on the controller prevent the voltages from exceeding the value required to saturate the actuator and thus protect the actuator, but they do not prevent the integral buildup that causes the overshoot. One way to prevent integrator buildup is to select the gains so that saturation will never occur. This requires knowledge of the maximum input magnitude that the system will encounter. General algorithms for doing this are not available; some methods for low-order systems are presented in Ref. 1, Chapter 7; Ref. 2, Chapter 7, and Ref. 4, Chapter 11. Integrator buildup is easier to prevent when using digital control; this is discussed in Section 10.

8   COMPENSATION AND ALTERNATIVE CONTROL STRUCTURES

A common design technique is to insert a *compensator* into the system when the PID control algorithm can be made to satisfy most but not all of the design specifications. A compensator is a device that alters the response of the controller so that the overall system will have satisfactory

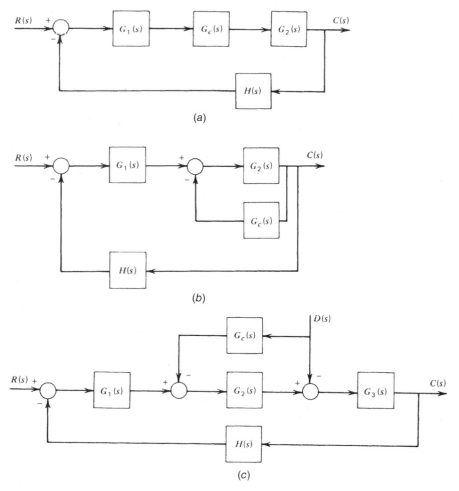

Figure 34 General structures of the three compensation types: (*a*) series, (*b*) parallel (or feedback), and (*c*) feedforward. The compensator transfer function is $G_c(s)$.[1]

performance. The three categories of compensation techniques generally recognized are *series compensation, parallel* (or *feedback*) *compensation,* and *feedforward compensation.*

The three structures are loosely illustrated in Fig. 34, where we assume the final control elements have a unity transfer function. The transfer function of the controller is $G_1(s)$. The feedback elements are represented by $H(s)$, and the compensator by $G_c(s)$. We assume that the plant is unalterable, as is usually the case in control system design. The choice of compensation structure depends on what type of specifications must be satisfied. The physical devices used as compensators are similar to the pneumatic, hydraulic, and electrical devices treated previously. Compensators can be implemented in software for digital control applications.

8.1 Series Compensation

The most commonly used series compensators are the *lead*, the *lag*, and the *lead–lag* compensators. Electrical implementations of these are shown in Fig. 35. Other physical

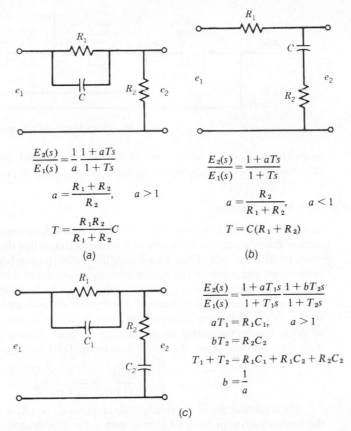

$$\frac{E_2(s)}{E_1(s)} = \frac{1}{a}\frac{1+aTs}{1+Ts}$$

$$a = \frac{R_1+R_2}{R_2}, \qquad a>1$$

$$T = \frac{R_1R_2}{R_1+R_2}C$$

(a)

$$\frac{E_2(s)}{E_1(s)} = \frac{1+aTs}{1+Ts}$$

$$a = \frac{R_2}{R_1+R_2}, \qquad a<1$$

$$T = C(R_1+R_2)$$

(b)

$$\frac{E_2(s)}{E_1(s)} = \frac{1+aT_1s}{1+T_1s}\frac{1+bT_2s}{1+T_2s}$$

$$aT_1 = R_1C_1, \qquad a>1$$

$$bT_2 = R_2C_2$$

$$T_1 + T_2 = R_1C_1 + R_1C_2 + R_2C_2$$

$$b = \frac{1}{a}$$

(c)

Figure 35 Passive electrical compensators: (*a*) lead, (*b*) lag, and (*c*) lead–lag.[1]

implementations are available. Generally, the lead compensator improves the speed of response; the lag compensator decreases the steady-state error; and the lead–lag affects both. Graphical aids, such as the root-locus and frequency response plots, are usually needed to design these compensators (Ref. 1, Chapter 8; Ref. 2, Chapter 9; and Ref. 4, Chapter 11).

8.2 Feedback Compensation and Cascade Control

The use of a tachometer to obtain velocity feedback, as in Fig. 24, is a case of feedback compensation. The feedback compensation principle of Fig. 3 is another. Another form is *cascade control,* in which another controller is inserted within the loop of the original control system (Fig. 36). The new controller can be used to achieve better control of variables within the forward path of the system. Its set point is manipulated by the first controller.

Cascade control is frequently used when the plant cannot be satisfactorily approximated with a model of second order or lower. This is because the difficulty of analysis and control increases rapidly with system order. The characteristic roots of a second-order system can easily be expressed in analytical form. This is not so for third order or higher, and few general design rules are available. When faced with the problem of controlling a high-order system, the designer should first see if the performance requirements can be relaxed so that the system can be approximated with a low-order model. If this is not possible, the designer should attempt

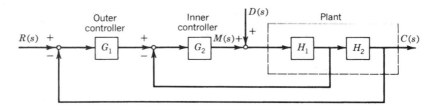

Figure 36   Cascade control structure.

to divide the plant into subsystems, each of which is second order or lower. A controller is then designed for each subsystem. An application using cascade control is given in Section 11.

8.3   Feedforward Compensation

The control algorithms considered thus far have counteracted disturbances by using measurements of the output. One difficulty with this approach is that the effects of the disturbance must show up in the output of the plant before the controller can begin to take action. On the other hand, if we can measure the disturbance, the response of the controller can be improved by using the measurement to augment the control signal sent from the controller to the final control elements. This is the essence of feedforward compensation of the disturbance, as shown in Fig. 34c. Feedforward compensation modified the output of the main controller. Instead of doing this by measuring the disturbance, another form of feedforward compensation utilizes the command input. Figure 37 is an example of this approach. The closed-loop transfer function is

$$\frac{\Omega(s)}{\Omega_r(s)} = \frac{K_{f+}K}{Is + c + K} \tag{33}$$

For a unit step input, the steady-state output is $\omega_{ss} = (K_f + K)/(c + K)$. Thus, if we choose the feedforward gain K_f to be $K_f = c$, then $\omega_{ss} = 1$ as desired, and the error is zero. Note that this form of feedforward compensation does not affect the disturbance response. Its effectiveness depends on how accurately we know the value of c. A digital application of feedforward compensation is presented in Section 11.

8.4   State-Variable Feedback

There are techniques for improving system performance that do not fall entirely into one of the three compensation categories considered previously. In some forms these techniques can be

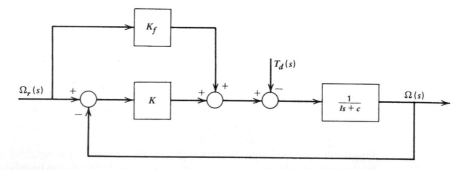

Figure 37   Feedforward compensation of the command input to augment proportional control.[2]

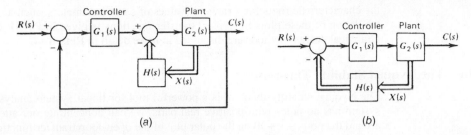

Figure 38   Two forms of state-variable feedback: (*a*) internal compensation of the control signal and (*b*) modification of the actuating signal.[1]

viewed as a type of feedback compensation, while in other forms they constitute a modification of the control law. *State-variable feedback* (SVFB) is a technique that uses information about all the system's state variables to modify either the control signal or the actuating signal. These two forms are illustrated in Fig. 38. Both forms require that the state vector **x** be measurable or at least derivable from other information. Devices or algorithms used to obtain state-variable information other than directly from measurements are variously termed *state reconstructors, estimators, observers,* or *filters* in the literature.

8.5   Pseudoderivative Feedback

Pseudoderivative feedback (PDF) is an extension of the velocity feedback compensation concept of Fig. 24. It uses integral action in the forward path plus an internal feedback loop whose operator $H(s)$ depends on the plant (Fig. 39). For $G(s) = 1/(Is + c)$, $H(s) = K_1$. For $G(s) = 1/Is^2$, $H(s) = K_1 + K_2 s$. The primary advantage of PDF is that it does not need derivative action in the forward path to achieve the desired stability and damping characteristics.

9   GRAPHICAL DESIGN METHODS

Higher order models commonly arise in control systems design. For example, integral action is often used with a second-order plant, and this produces a third-order system to be designed. Although algebraic solutions are available for third- and fourth-order polynomials, these solutions are cumbersome for design purposes. Fortunately, there exist graphical techniques to aid the designer. Frequency response plots of both the open- and closed-loop transfer functions are useful. The *Bode plot* and the *Nyquist plot* present the frequency response information in different forms. Each form has its own advantages. The root-locus plot shows the location of

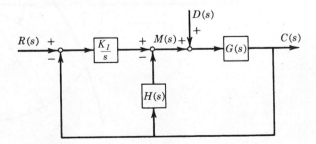

Figure 39   Structure of pseudoderivative feedback.

the characteristic roots for a range of values of some parameters, such as a controller gain. A tabulation of these plots for typical transfer functions is given in Ref. 4, Chapters 11 and 12. Graphical design methods are discussed in more detail in Refs. 1–4.

9.1   The Nyquist Stability Theorem

The Nyquist stability theorem is a powerful tool for linear system analysis. If the open-loop system has no poles with positive real parts, we can concentrate our attention on the region around the point $-1 + i0$ on the polar plot of the open-loop transfer function. Figure 40 shows the polar plot of the open-loop transfer function of an arbitrary system that is assumed to be open-loop stable. The Nyquist stability theorem is stated as follows:

> A system is closed-loop stable if and only if the point $-1 + i0$ lies to the left of the open-loop Nyquist plot relative to an observer traveling along the plot in the direction of increasing frequency ω.

Therefore, the system described by Fig. 39 is closed-loop stable.

The Nyquist theorem provides a convenient measure of the relative stability of a system. A measure of the proximity of the plot to the $-1 + i0$ point is given by the angle between the negative real axis and a line from the origin to the point where the plot crosses the unit circle (see Fig. 39). The frequency corresponding to this intersection is denoted ω_g. This angle is the *phase margin* (PM) and is positive when measured down from the negative real axis. The phase margin is the phase at the frequency ω_g where the magnitude ratio or "gain" of $G(i\omega)H(i\omega)$ is unity, or 0 decibels (dB). The frequency ω_P, the *phase crossover frequency,* is the frequency at which the phase angle is $-180°$. The *gain margin* (GM) is the difference in decibels between the unity gain condition (0 dB) and the value of $|G(\omega_P)H(\omega_P)|$ decibels at the phase crossover frequency ω_P. Thus,

$$\text{Gain margin} = -|G(\omega_P)H(\omega_P)| \quad \text{(dB)} \tag{34}$$

A system is stable only if the phase and gain margins are both positive.

The phase and gain margins can be illustrated on the Bode plots shown in Fig. 41. The phase and gain margins can be stated as safety margins in the design specifications. A typical set of such specifications is as follows:

$$\text{Gain margin} \geq 8 \, \text{dB and phase margin} \geq 30° \tag{35}$$

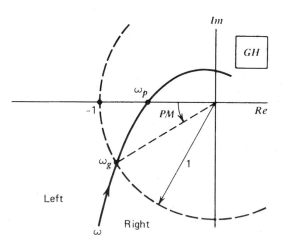

Figure 40   Nyquist plot for a stable system.[1]

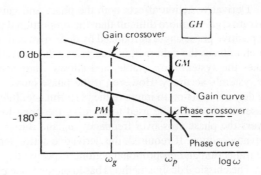

Figure 41 Bode plot showing definitions of phase and gain margin.[1]

In common design situations, only one of these equalities can be met, and the other margin is allowed to be greater than its minimum value. It is not desirable to make the margins too large because this results in a low gain, which might produce sluggish response and a large steady-state error. Another commonly used set of specifications is

$$\text{Gain margin} \geq 6\,\text{dB and phase margin} \geq 40° \tag{36}$$

The 6-dB limit corresponds to the quarter amplitude decay response obtained with the gain settings given by the Ziegler–Nichols ultimate-cycle method (Table 2).

9.2 Systems with Dead-Time Elements

The Nyquist theorem is particularly useful for systems with dead-time elements, especially when the plant is of an order high enough to make the root-locus method cumbersome. A delay D in either the manipulated variable or the measurement will result in an open-loop transfer function of the form

$$G(s)H(s) = e^{-Ds}P(s) \tag{37}$$

Its magnitude and phase angle are

$$|G(i\omega)H(i\omega)| = |P(i\omega)||e^{-i\omega D}| = |P(i\omega)| \tag{38}$$

$$\angle G(i\omega)H(i\omega) = \angle P(i\omega) + \angle P(i\omega) - \omega D \tag{39}$$

Thus, the dead time decreases the phase angle proportionally to the frequency ω, but it does not change the gain curve. This makes the analysis of its effects easier to accomplish with the open-loop frequency response plot.

9.3 Open-Loop Design for PID Control

Some general comments can be made about the effects of proportional, integral, and derivative control actions on the phase and gain margins. Proportional action does not affect the phase curve at all and thus can be used to raise or lower the open-loop gain curve until the specifications for the gain and phase margins are satisfied. If integral action or derivative action is included, the proportional gain is selected last. Therefore, when using this approach to the design, it is best to write the PID algorithm with the proportional gain factored out, as

$$F(s) = K_P\left(1 + \frac{1}{T_I s} + T_D s\right)Es \tag{40}$$

Derivative action affects both the phase and gain curves. Therefore, the selection of the derivative gain is more difficult than the proportional gain. The increase in phase margin due to the positive phase angle introduced by derivative action is partly negated by the derivative gain, which reduces the gain margin. Increasing the derivative gain increases the speed of response, makes the system more stable, and allows a larger proportional gain to be used to improve the system's accuracy. However, if the phase curve is too steep near $-180°$, it is difficult to use derivative action to improve the performance. Integral action also affects both the gain and phase curves. It can be used to increase the open-loop gain at low frequencies. However, it lowers the phase crossover frequency ω_P and thus reduces some of the benefits provided by derivative action. If required, the derivative action term is usually designed first, followed by integral action and proprotional action, respectively.

The classical design methods based on the Bode plots obviously have a large component of trial and error because usually both the phase and gain curves must be manipulated to achieve an acceptable design. Given the same set of specifications, two designers can use these methods and arrive at substantially different designs. Many rules of thumb and ad hoc procedures have been developed, but a general foolproof procedure does not exist. However, an experienced designer can often obtain a good design quickly with these techniques. The use of a computer plotting routine greatly speeds up the design process.

9.4   Design with the Root Locus

The effect of derivative action as a series compensator can be seen with the root locus. The term $1 + T_D s$ in Fig. 32 can be considered as a series compensator to the proportional controller. The derivative action adds an open-loop zero at $s = 1/T_D$. For example, a plant with the transfer function $1/s(s + 1)(s + 2)$, when subjected to proportional control, has the root locus shown in Fig. 42a. If the proportional gain is too high, the system will be unstable. The smallest achievable time constant corresponds to the root $s = -0.42$ and is $\tau = 1/0.42 = 2.4$. If derivative action is used to put an open-loop zero at $s = 1.5$, the resulting root locus is given by Fig. 42b. The derivative action prevents the system from becoming unstable and allows a smaller time constant to be achieved (τ can be made close to $1/0.75 = 1.3$ by using a high proportional gain).

The integral action in PI control can be considered to add an open-loop pole at $s = 0$ and a zero at $s = 1/T_I$. Proportional control of the plant $1/(s + 1)(s + 2)$ gives a root locus like that shown in Fig. 43, with $a = 1$ and $b = 2$. A steady-state error will exist for a step input. With the PI compensator applied to this plant, the root locus is given by Fig. 42b, with $T_I = \frac{2}{3}$. The steady-state error is eliminated, but the response of the system has been slowed because the dominant paths of the root locus of the compensated system lie closer to the imaginary axis than those of the uncompensated system.

As another example, let the plant transfer function be

$$G_P(s) = \frac{1}{s^2 + a_2 s + a_1} \tag{41}$$

where $a_1 > 0$ and $a_2 > 0$. PI control applied to this plant gives the closed-loop command transfer function

$$T_1(s) = \frac{K_P s + K_I}{s^3 + a_2 s^2 + (a_1 + K_P)s + K_I}. \tag{42}$$

Note that the Ziegler–Nichols rules cannot be used to set the gains K_P and K_I. The second-order plant, Eq. (41), does not have the S-shaped signature of Fig. 33, so the process reaction method does not apply. The ultimate-cycle method requires K_I to be set to zero and the ultimate gain K_{P_u} determined. With $K_I = 0$ in Eq. (42) the resulting system is stable for all $K_P > 0$, and thus a positive ultimate gain does not exist.

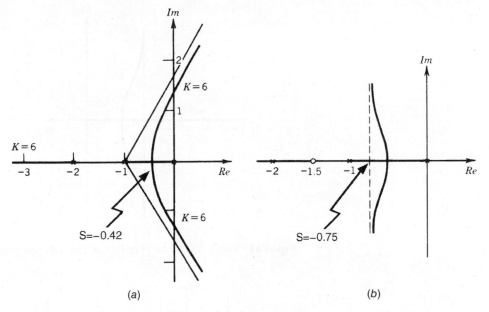

Figure 42 (*a*) Root-locus plot for $s(s+1)(s+2)+K=0$, for $K \geq 0$. (*b*) The effect of PD control with $T_D = \frac{2}{3}$.[1]

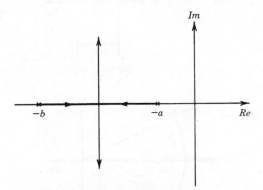

Figure 43 Root-locus plot for $(s+a)(s+b)+K=0$.

Take the form of the PI control law given by Eq. (42) with $T_D = 0$, and assume that the characteristic roots of the plant (Fig. 44) are real values $-r_1$ and $-r_2$ such that $-r_1 < -r_2$. In this case the open-loop transfer function of the control system is

$$G(s)H(s) = \frac{K_P(s+1/T_I)}{s(s+r_1)(s+r_2)} \tag{43}$$

One design approach is to select T_I and plot the locus with K_P as the parameter. If the zero at $s = -1/T_I$ is located to the right of $s = -r_1$, the dominant time constant cannot be made as small as is possible with the zero located between the poles at $s = -r_1$ and $s = -r_2$ (Fig. 44). A large integral gain (small T_I and/or large K_P) is desirable for reducing the overshoot due to a disturbance, but the zero should not be placed to the left of $s = -r_2$ because the dominant

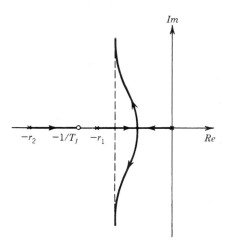

Figure 44 Root-locus plot for PI control of a second-order plant.

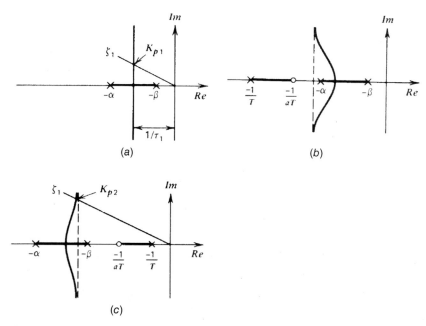

Figure 45 Effects of series lead and lag compensators: (*a*) uncompensated system's root locus, (*b*) root locus with lead compensation, and (*c*) root locus with lag compensation.[1]

time constant will be larger than that obtainable with the placement shown in Fig. 44 for large values of K_P. Sketch the root-locus plots to see this. A similar situation exists if the poles of the plant are complex.

The effects of the lead compensator in terms of time-domain specifications (characteristic roots) can be shown with the root-locus plot. Consider the second-order plant with the real distinct roots $s = -\alpha, s = -\beta$. The root locus for this system with proportional control is shown in Fig. 45*a*. The smallest dominant time constant obtainable is τ_1, marked in the figure.

A lead compensator introduces a pole at $s = -1/T$ and a zero at $s = -1/aT$, and the root locus becomes that shown in Fig. 45b. The pole and zero introduced by the compensator reshape the locus so that a smaller dominant time constant can be obtained. This is done by choosing the proportional gain high enough to place the roots close to the asymptotes.

With reference to the proportional control system whose root locus is shown in Fig. 45a, suppose that the desired damping ratio ζ_1 and desired time constant τ_1 are obtainable with a proportional gain of K_{P1}, but the resulting steady-state error $\alpha\beta/(\alpha\beta + K_{P1})$ due to a step input is too large. We need to increase the gain while preserving the desired damping ratio and time constant. With the lag compensator, the root locus is as shown in Fig. 45c. By considering specific numerical values, one can show that for the compensated system, roots with a damping ratio ζ_1 correspond to a high value of the proportional gain. Call this value K_{P2}. Thus $K_{P2} > K_{P1}$, and the steady-state error will be reduced. If the value of T is chosen large enough, the pole at $s = -1/T$ is approximately canceled by the zero at $s = -1/aT$, and the open-loop transfer function is given approximately by

$$G(s)H(s) = \frac{aK_P}{(s + \alpha)(s + \beta)} \tag{44}$$

Thus, the system's response is governed approximately by the complex roots corresponding to the gain value K_{P2}. By comparing Fig. 45a with 45c, we see that the compensation leaves the time constant relatively unchanged. From Eq. (44) it can be seen that since $a < 1$, K_P can be selected as the larger value K_{P2}. The ratio of K_{P1} to K_{P2} is approximately given by the parameter a.

Design by pole–zero cancellation can be difficult to accomplish because a response pattern of the system is essentially ignored. The pattern corresponds to the behavior generated by the canceled pole and zero, and this response can be shown to be beyond the influence of the controller. In this example, the canceled pole gives a stable response because it lies in the left-hand plane. However, another input not modeled here, such as a disturbance, might excite the response and cause unexpected behavior. The designer should therefore proceed with caution. None of the physical parameters of the system are known exactly, so exact pole–zero cancellation is not possible. A root-locus study of the effects of parameter uncertainty and a simulation study of the response are often advised before the design is accepted as final.

10 PRINCIPLES OF DIGITAL CONTROL

Digital control has several advantages over analog devices. A greater variety of control algorithms is possible, including nonlinear algorithms and ones with time-varying coefficients. Also, greater accuracy is possible with digital systems. However, their additional hardware complexity can result in lower reliability, and their application is limited to signals whose time variation is slow enough to be handled by the samplers and the logic circuitry. This is now less of a problem because of the large increase in the speed of digital systems.

10.1 Digital Controller Structure

Sampling, discrete-time models, the z transform, and pulse transfer functions were outlined in the previous chapter. The basic structure of a single-loop controller is shown in Fig. 46. The computer with its internal clock drives the *digital-to-analog* (D/A) and *analog-to-digital* (A/D) converters. It compares the command signals with the feedback signals and generates the control signals to be sent to the final control elements. These control signals are computed from the control algorithm stored in the memory. Slightly different structures exist, but Fig. 46

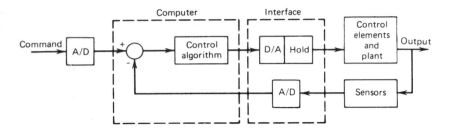

Figure 46   Structure of a digital control system.[1]

shows the important aspects. For example, the comparison between the command and feedback signals can be done with analog elements, and the A/D conversion made on the resulting error signal. The software must also provide for *interrupts,* which are conditions that call for the computer's attention to do something other than computing the control algorithm.

The time required for the control system to complete one loop of the algorithm is the time T, the *sampling time* of the control system. It depends on the time required for the computer to calculate the control algorithm and on the time required for the interfaces to convert data. Modern systems are capable of very high rates, with sample times under 1 μs.

In most digital control applications, the plant is an analog system, but the controller is a discrete-time system. Thus, to design a digital control system, we must either model the controller as an analog system or model the plant as a discrete-time system. Each approach has its own merits, and we will examine both.

If we model the controller as an analog system, we use methods based on *differential* equations to compute the gains. However, a digital control system requires *difference* equations to describe its behavior. Thus, from a strictly mathematical point of view, the gain values we will compute will not give the predicted response exactly. However, if the sampling time is small compared to the smallest time constant in the system, then the digital system will act like an analog system, and our designs will work properly. Because most physical systems of interest have time constants greater than 1 ms and controllers can now achieve sampling times less than 1 μs, controllers designed with analog methods will often be adequate.

10.2   Digital Forms of PID Control

There are a number of ways that PID control can be implemented in software in a digital control system because the integral and derivative terms must be approximated with formulas chosen from a variety of available algorithms. The simplest integral approximation is to replace the integral with a sum of rectangular areas. With this rectangular approximation, the error integral is calculated as

$$\int_0^{(k+1)T} e(t) \approx Te(0) + Te(t_1) + Te(t_2) + \cdots Te(t_k) = T\sum_{i=0}^{k} e(t_i) \tag{45}$$

where $t_k = kT$ and the width of each rectangle is the sampling time $T = t_{i+1} - t_i$. The times t_i are the times at which the computer updates its calculation of the control algorithm after receiving an updated command signal and an updated measurement from the sensor through the A/D interfaces. If the time T is small, then the value of the sum in (45) is close to the value of the integral. After the control algorithm calculation is made, the calculated value of the control signal $f(t_k)$ is sent to the actuator via the output interface. This interface includes a D/A converter and a hold circuit that "holds" or keeps the analog voltage corresponding to the control signal applied to the actuator until the next updated value is passed along from the computer. The simplest digital form of PI control uses (45) for the integral term. It is

$$f(t_k) = K_P e(t_k) + K_I T \sum_{i=0}^{k} e(t_i) \qquad (46)$$

This can be written in a more efficient form by noting that

$$f(t_{k-1}) = K_P e(t_{k-1}) + K_I T \sum_{i=0}^{k-1} e(t_i)$$

and subtracting this from (46) to obtain

$$f(t_k) = f(t_{k-1}) = K_P[e(t_k) - e(t_{k-1})] + K_I Te(t_k) \qquad (47)$$

This form—called the *incremental* or *velocity* algorithm—is well suited for incremental output devices such as stepper motors. Its use also avoids the problem of integrator buildup, the condition in which the actuator saturates but the control algorithm continues to integrate the error.

The simplest approximation to the derivative is the first-order difference approximation

$$\frac{de}{dt} \approx \frac{e(t_k) - e(t_{k-1})}{T} \qquad (48)$$

The corresponding PID approximation using the rectangular integral approximation is

$$f(t_k) = K_P e(t_k) + K_I T \sum_{i=0}^{k} e(t_i) + \frac{K_D}{T}[e(t_k) - e(t_{k-1})] \qquad (49)$$

The accuracy of the integral approximation can be improved by substituting a more sophisticated algorithm, such as the following trapezoidal rule:

$$\int_0^{(k-1)T} e(t)\, dt \approx T \sum_{i=0}^{k} \frac{1}{2}[e(t_i + 1 + e(t_i))] \qquad (50)$$

The accuracy of the derivative approximation can be improved by using values of the sampled error signal at more instants. Using the four-point central-difference method (Refs. 1 and 2), the derivative term is approximated by

$$\frac{de}{dt} \approx \frac{1}{6T}[e(t_k) + 3e(t_{k-1}) - 3e(t_{k-2}) - e(t_{k-3})]$$

The derivative action is sensitive to the resulting rapid change in the error samples that follows a step input. This effect can be eliminated by reformulating the control algorithm as follows (Refs. 1 and 2):

$$ft_k = f(t_{k-1}) + K_p[c(t_{k-1}) - c(t_k)]$$
$$+ K_I T[r(t_k) - c(t_k)]$$
$$+ \frac{K_D}{T}[-c(t_k) + 2c(t_{k-1}) - c(t_{k-2})] \qquad (51)$$

where $r(t_k)$ is the command input and $c(t_k)$ is the variable being controlled. Because the command input $r(t_k)$ appears in this algorithm only in the integral term, we cannot apply this algorithm to PD control; that is, the integral gain K_I must be nonzero.

11   UNIQUELY DIGITAL ALGORITHIMS

Development of analog control algorithms was constrained by the need to design physical devices that could implement the algorithm. However, digital control algorithms simply need to be programmable and are thus less constrained than analog algorithms.

11.1   Digital Feedforward Compensation

Classical control system design methods depend on linear models of the plant. With linearization we can obtain an approximately linear model, which is valid only over a limited operating range. Digital control now allows us to deal with nonlinear models more directly using the concepts of feedforward compensation discussed in Section 8.

Computed Torque Method

Figure 47 illustrates a variation of feedforward compensation of the disturbance called the *computed torque method*. It is used to control the motion of robots. A simple model of a robot arm is the following nonlinear equation:

$$I\ddot{\theta} = T - mgL\sin\theta \tag{52}$$

where θ is the arm angle, I is its inertia, mg is its weight, and L is the distance from its mass center to the arm joint where the motor acts. The motor supplies the torque T. To position the arm at some desired angle θ_r, we can use PID control on the angle error $\theta_r - \theta$. This works well if the arm angle θ is never far from the desired angle θ_r so that we can linearize the plant model about θ_r. However, the controller will work for large-angle excursions if we compute the nonlinear gravity torque term $mgL\sin\theta$ and add it to the PID output. That is, part of the motor torque will be computed specifically to cancel the gravity torque, in effect producing a linear system for the PID algorithm to handle. The nonlinear torque calculations required to control multi-degree-of-freedom robots are very complicated and can be done only with a digital controller.

Feedforward Command Compensation

Computers can store lookup tables, which can be used to control systems that are difficult to model entirely with differential equations and analytical functions. Figure 48 shows a speed control system for an internal combustion engine. The fuel flow rate required to achieve a desired speed depends in a complicated way on many variables not shown in the figure, such

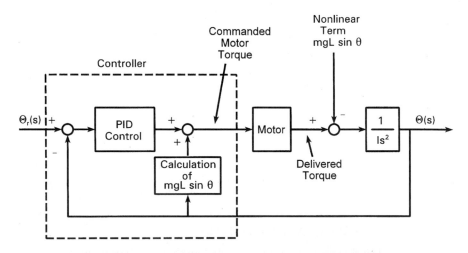

Figure 47   The computed torque method applied to robot arm control.

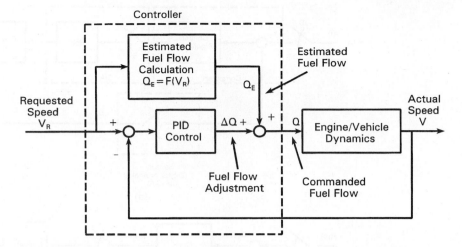

Figure 48 Feedforward compensation applied to engine control.

as temperature, humidity, and so on. This dependence can be summarized in tables stored in the control computer and can be used to estimate the required fuel flow rate. A PID algorithm can be used to adjust the estimate based on the speed error. This application is an example of feedforward compensation of the command input, and it requires a digital computer.

11.2 Control Design in the z Plane

There are two common approaches to designing a digital controller:

1. The performance is specified in terms of the desired continuous-time response, and the controller design is done entirely in the s plane, as with an analog controller. The resulting control law is then converted to discrete-time form, using approximations for the integral and derivative terms. This method can be successfully applied if the sampling time is small. The technique is widely used for two reasons. When existing analog controllers are converted to digital control, the form of the control law and the values of its associated gains are known to have been satisfactory. Therefore, the digital version can use the same control law and gain values. Second, because analog design methods are well established, many engineers prefer to take this route and then convert the design into a discrete-time equivalent.

2. The performance specifications are given in terms of the desired continuous-time response and/or desired root locations in the s plane. From these the corresponding root locations in the z plane are found and a discrete control law is designed. This method avoids the derivative and integral approximation errors that are inherent in the first method and is the preferred method when the sampling time T is large. However, the algebraic manipulations are more cumbersome.

The second approach uses the z transform and pulse transfer functions, which were outlined in the previous chapter. If we have an analog model of the plant, with its transfer function $G(s)$, we can obtain its pulse transfer function $G(z)$ by finding the z transform of the impulse response $g(t) = \mathcal{L}^{-1}[G(s)]$; that is, $G(z) = \mathscr{Z}[g(t)]$. A table of transforms facilitates this

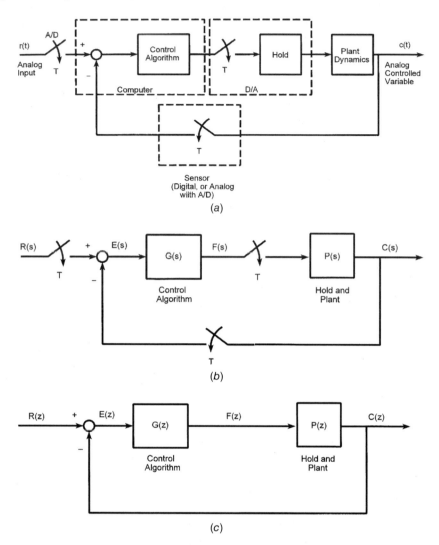

Figure 49   Block diagrams of a typical digital controller. (*a*) Diagram showing the components. (*b*) Diagram of the *s*-plane relations. (*c*) Diagram of the *z*-plane relations.

process; see Refs. 1 and 2. Figure 49*a* shows the basic elements of a digital control system. Figure 49*b* is an equivalent diagram with the analog transfer functions inserted. Figure 49*c* represents the same system in terms of pulse transfer functions. From the diagram we can find the closed-loop pulse transfer function. It is

$$\frac{C(z)}{R(z)} = \frac{G(z)P(z)}{1 + G(z)P(z)} \tag{53}$$

The variable z is related to the Laplace variable s by

$$z = e^{sT} \tag{54}$$

If we know the desired root locations and the sampling time T, we can compute the z roots from this equation.

Digital PI Control Design

For example, the first-order plant $1/(2s + 1)$ with a zero-order hold has the following pulse transfer function (Refs. 1 and 2):

$$P(z) = \frac{1 - e^{-0.5T}}{z - e^{-0.5T}} \tag{55}$$

Suppose we use a control algorithm described by the following pulse transfer function:

$$G(z) = \frac{F(z)}{E(z)} = \frac{K_1 z + K_2}{z - 1} = \frac{K_1 + K_2 z^{-1}}{1 - z^{-1}} \tag{56}$$

The corresponding difference equation that the control computer must implement is

$$f(t_k) = f(t_{k-1}) + K_1 e(t_k) + K_2 e(t_{k-1}) \tag{57}$$

$$\frac{C(z)}{R(z)} = \frac{(1 - b)(K_1 z + K_2)}{z^2 + (K_1 - 1 - b - bK_1)z + b + K_2 - bK_2} \tag{58}$$

where $b = e^{-05T}$.

If the design specifications call for $\tau = 1$ and $\zeta = 1$, then the desired s roots are $s = -1$, -1, and the analog PI gains required to achieve these roots are $K_P = 3$ and $K_I = 2$. Using a sampling time of $T = 0.1$, the z roots must be $z = e^{-0.1}$, $e^{-0.1}$. To achieve these roots, the denominator of the transfer function (58) must be $z^2 - 2e^{-0.1}z + e^{-0.2}$. Thus the control gains must be $K_1 = 2.903$ and $K_2 = 2.717$. These values of K_1 and K_2 correspond to $K_P = 2.72$ and $K_I = 1.86$, which are close to the PI gains computed for an analog controller. If we had used a sampling time smaller than 0.1, say $T = 0.01$, the values of K_P and K_I computed from K_1 and K_2 would be $K_P = 2.97$ and $K_I = 1.98$, which are even closer to the analog gain values. This illustrates the earlier claim that analog design methods can be used when the sampling time is small enough.

Digital Series Compensation

Series compensation can be implemented digitally by applying suitable discrete-time approximations for the derivative and integral to the model represented by the compensator's transfer function $G(s)$. For example, the form of a lead or a lag compensator's transfer function is

$$G_c(s) = \frac{M(s)}{F(s)} = K\frac{s + c}{s + d} \tag{59}$$

where $m(t)$ is the actuator command and $f(t)$ is the control signal produced by the main (PID) controller. The differential equation corresponding to (59) is

$$\dot{m} + dm = K(\dot{f} + cf) \tag{60}$$

Using the simplest approximation for the derivative, Eq. (48), we obtain the following difference equation that the digital compensator must implement:

$$\frac{m(t_k) - m(t_{k-1})}{T} + dm(t_k) = K\left[\frac{f(t_k) - f(t_{k-1})}{T} + cf(t_k)\right]$$

In the z plane, the equation becomes

$$\frac{1 - z^{-1}}{T}M(z) + dM(z) = K\left[\frac{1 - z^{-1}}{T}F(z) + cf(z)\right] \tag{61}$$

The compensator's pulse transfer function is thus seen to be

$$G_c(z) = \frac{M(z)}{F(z)} = \frac{K(1 - z^{-1}) + cT}{1 - z^{-1} + dT}$$

which has the form

$$G_c(z) = K_c \frac{z+a}{z+b} \tag{62}$$

where K_c, a, and b can be expressed in terms of K, c, d, and T if we wish to use analog design methods to design the compensator. When using commercial controllers, the user might be required to enter the values of the gain, the pole, and the zero of the compensator. The user must ascertain whether these values should be entered as s-plane values (i.e., K, c, and d) or as z-plane values (K_c, a, and b).

Note that the digital compensator has the same number of poles and zeros as the analog compensator. This is a result of the simple approximation used for the derivative. Note that Eq. (61) shows that when we use this approximation, we can simply replace s in the analog transfer function with $1 - z^{-1}$. Because the integration operation is the inverse of differentiation, we can replace $1/s$ with $1/(1 - z^{-1})$. when integration is used. [This is equivalent to using the rectangular approximation for the integral, and can be verified by finding the pulse transfer function of the incremental algorithm (47) with $K_P = 0$.]

Some commercial controllers treat the PID algorithm as a series compensator, and the user is expected to enter the controller's values, not as PID gains, but as pole and zero locations in the z plane. The PID transfer function is

$$\frac{F(z)}{E(z)} = K_P + \frac{K_I}{s} + K_D s \tag{63}$$

Making the indicated replacements for the s terms, we obtain

$$\frac{F(z)}{E(z)} = K_P + \frac{K_I}{1 - z^{-1}} + K_D(1 - z^{-1})$$

which has the form

$$\frac{F(z)}{E(z)} = K_c \frac{z^2 - az + b}{z - 1} \tag{64}$$

where K_c, a, and b can be expressed in terms of K_P, K_I, K_D, and T. Note that the algorithm has two zeros and one pole, which is fixed at $z = 1$. Sometimes the algorithm is expressed in the more general form

$$\frac{F(z)}{E(z)} = K_c \frac{z^2 - az + b}{z - c} \tag{65}$$

to allow the user to select the pole as well.

Digital compensator design can be done with frequency response methods or with the root-locus plot applied to the z plane rather than the s plane. However, when better approximations are used for the derivative and integral, the digital series compensator will have more poles and zeros than its analog counterpart. This means that the root-locus plot will have more root paths, and the analysis will be more difficult. This topic is discussed in more detail in Refs. 1–3 and 8.

11.3 Direct Design of Digital Algorithms

Because almost any algorithm can be implemented digitally, we can specify the desired response and work backward to find the required control algorithm. This is the *direct-design* method. If we let $D(z)$ be the desired form of the closed-loop transfer function $C(z)/R(z)$ and solve for the controller transfer function $G(z)$, we obtain

$$G(z) = \frac{D(z)}{P(z)[1 - D(z)]} \tag{66}$$

Where $P(z)$ is the transfer function of the plant.

We can pick $D(z)$ directly or obtain it from the specified input transform $R(z)$ and the desired output transform $C(z)$ because $D(z) = C(z)/R(z)$.

Finite-Settling-Time Algorithm

This method can be used to design a controller to compensate for the effects of process dead time. A plant having such a response can often be approximately described by a first-order model with a dead-time element; that is,

$$G_P(s) = K\frac{e^{-Ds}}{\tau s + 1} \tag{67}$$

where D is the dead time. This model also approximately describes the S-shaped response curve used with the Ziegler–Nichols method (Fig. 33). When combined with a zero-order hold, this plant has the following pulse transfer function:

$$P(z) = Kz^{-n}\frac{1 - a}{z - a} \tag{68}$$

where $a = \exp(-T/\tau)$ and $n = D/T$. If we choose $D(z) = z^{(n+1)}$, then with a step command input, the output $c(k)$ will reach its desired value in $n + 1$ sample times, one more than is in the dead time D. This is the fastest response possible. From (66) the required controller transfer function is

$$G(z) = \frac{1}{K(1 - a)}\frac{1 - az^{-1}}{1 - z^{-(n+1)}} \tag{69}$$

The corresponding difference equation that the control computer must implement is

$$f(t_k) = (ft_{k-n-1}) + \frac{1}{K(1 - a)}[e(t_k) - ae(t_{k-1})] \tag{70}$$

This algorithm is called a *finite-settling-time* algorithm because the response reaches its desired value in a finite, prescribed time. The maximum value of the manipulated variable required by this algorithm occurs at $t = 0$ and is $1/K(1 - a)$ If this value saturates the actuator, this method will not work as predicted. Its success depends also on the accuracy of the plant model.

Dahlin's Algorithm

This sensitivity to plant modeling errors can be reduced by relaxing the minimum response time requirement. For example, choosing $D(z)$ to have the same form as $P(z)$, namely,

$$D(z) = K_d Z^{-n}\frac{1 - a_d}{z - a_d} \tag{71}$$

we obtain from (66) the following controller transfer function:

$$G(z) = \frac{K_d(1 - a_d)}{K(1 - a)}\frac{1 - az^{-1}}{1 - a_d z^{-1} - K_d(1 - a_d)z^{-(n+1)}} \tag{72}$$

This is Dahlin's algorithm.[3] The corresponding difference equation that the control computer must implement is

$$f(t_k) = a_d f(t_{k-1}) + k_d(1 - a_d)f(t_{k-n-1})$$
$$+ \frac{K_d(1 - a_d)}{K(1 - a)}[e(tk) - ae(t_{k-1})] \tag{73}$$

Normally we would first try setting $K_d = K$ and $a_d = a$, but since we might not have good estimates of K and a, we can use K_d and a as tuning parameters to adjust the controller's performance. The constant a_d is related to the time constant τ_d of the desired response: $ad = \exp(-T/\tau d)$. Choosing τ_d smaller gives faster response.

Algorithms such as these are often used for system startup, after which the control mode is switched to PID, which is more capable of handling disturbances.

12   HARDWARE AND SOFTWARE FOR DIGITAL CONTROL

This section provides an overview of the general categories of digital controllers that are commercially available. This is followed by a summary of the software currently available for digital control and for control system design.

12.1   Digital Control Hardware

Commercially available controllers have different capabilities, such as different speeds and operator interfaces, depending on their targeted application.

Programmable Logic Controllers (PLCs)

These are controllers that are programmed with relay ladder logic, which is based on Boolean algebra. Now designed around microprocessors, they are the successors to the large relay panels, mechanical counters, and drum programmers used up to the 1960s for sequencing control and control applications requiring only a finite set of output values (e.g., opening and closing of valves). Some models now have the ability to perform advanced mathematical calculations required for PID control, thus allowing them to be used for modulated control as well as finite-state control. There are numerous manufacturers of PLCs.

Digital Signal Processors (DSPs)

A modern development is the *digital signal processor* (DSP), which has proved useful for feedback control as well as signal processing.[9] This special type of processor chip has separate buses for moving data and instructions and is constructed to perform rapidly the kind of mathematical operations required for digital filtering and signal processing. The separate buses allow the data and the instructions to move in parallel rather than sequentially. Because the PID control algorithm can be written in the form of a digital filter, DSPs can also be used as controllers.

The DSP architecture was developed to handle the types of calculations required for digital filters and discrete Fourier transforms, which form the basis of most signal-processing operations. DSPs usually lack the extensive memory management capabilities of general-purpose computers because they need not store large programs or large amounts of data. Some DSPs contain A/D and D/A converters, serial ports, timers, and other features. They are programmed with specialized software that runs on popular personal computers. Low-cost DSPs are now widely used in consumer electronics and automotive applications, with Texas Instruments being a major supplier.

Motion Controllers

Motion controllers are specialized control systems that provide feedback control for one or more motors. They also provide a convenient operator interface for generating the commanded trajectories. Motion controllers are particularly well suited for applications requiring coordinated motion of two or more axes and for applications where the commanded trajectory is complicated. A higher level host computer might transmit required distance, speed, and acceleration rates to the motion controller, which then constructs and implements the continuous position profile required for each motor. For example, the host computer would supply the required total displacement, the acceleration and deceleration times, and the desired slew speed (the speed during the zero acceleration phase). The motion controller would generate the commanded position versus time for each motor. The motion controller also has the task of providing feedback control for each motor to ensure that the system follows the required position profile.

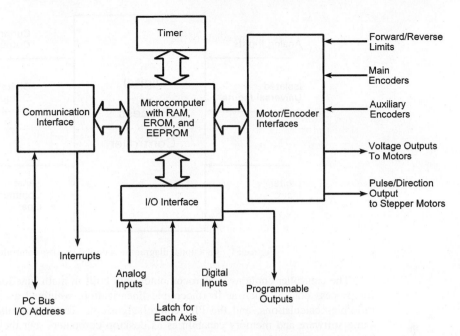

Figure 50   Functional diagram of a motion controller.

Figure 50 shows the functional elements of a typical motion controller, such as those built by Galil Motion Control, Inc. Provision for both analog and digital input signals allows these controllers to perform other control tasks besides motion control. Compared to DSPs, such controllers generally have greater capabilities for motion control and have operator interfaces that are better suited for such applications. Motion controllers are available as plug-in cards for most computer bus types. Some are available as stand-alone units.

Motion controllers use a PID control algorithm to provide feedback control for each motor (some manufacturers call this algorithm a "filter"). The user enters the values of the PID gains (some manufacturers provide preset gain values, which can be changed; others provide tuning software that assists in selecting the proper gain values). Such controllers also have their own language for programming a variety of motion profiles and other applications. For example, they provide for linear and circular interpolation for two-dimensional coordinated motion, motion smoothing (to eliminate jerk), contouring, helical motion, and electronic gearing. The latter is a control mode that emulates mechanical gearing in software in which one motor (the slave) is driven in proportion to the position of another motor (the master) or an encoder.

Process Controllers
Process controllers are designed to handle inputs from sensors, such as thermocouples, and outputs to actuators, such as valve positioners, that are commonly found in process control applications. Figure 51 illustrates the input–output capabilities of a typical process controller such as those manufactured by Honeywell, which is a major supplier of such devices. This device is a stand-alone unit designed to be mounted in an instrumentation panel. The voltage and current ranges of the analog inputs are those normally found with thermocouple-based temperature sensors. The current outputs are designed for devices like valve positioners, which usually require 4–20-mA signals.

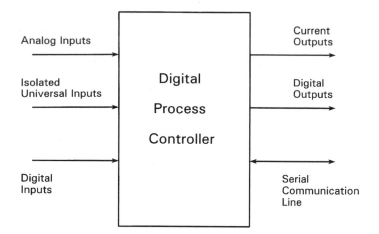

Figure 51 Functional diagram of a digital process controller.

The controller contains a microcomputer with built-in math functions normally required for process control, such as thermocouple linearization, weighted averaging, square roots, ratio/bias calculations, and the PID control algorithm. These controllers do not have the same software and memory capabilities as desktop computers, but they are less expensive. Their operator interface consists of a small keypad with typically fewer than 10 keys, a small graphical display for displaying bargraphs of the set points and the process variables, indicator lights, and an alphanumeric display for programming the controller.

The PID gains are entered by the user. Some units allow multiple sets of gains to be stored; the unit can be programmed to switch between gain settings when certain conditions occur. Some controllers have an adaptive tuning feature that is supposed to adjust the gains to prevent overshoot in startup mode, to adapt to changing process dynamics, and to adapt to disturbances. However, at this time, adaptive tuning cannot claim a 100% success rate, and further research and development in adaptive control is needed.

Some process controllers have more than one PID control loop for controlling several variables. Figure 52 illustrates a boiler feedwater control application for a controller with two PID loops arranged in a cascade control structure. Loop 1 is the main or outer loop controller for maintaining the desired water volume in the boiler. It uses sensing of the steam flow rate to implement feedforward compensation. Loop 2 is the inner loop controller that directly controls the feedwater control valve.

12.2 Software for Digital Control

The software available to the modern control engineer is quite varied and powerful and can be categorized according to the following tasks:

1. Control algorithm design, gain selection, and simulation
2. Tuning
3. Motion programming
4. Instrumentation configuration
5. Real-time control functions

Many analysis and simulation packages now contain algorithms of specific interest to control system designers. MATLAB® is one such package that is widely used. It contains built-in

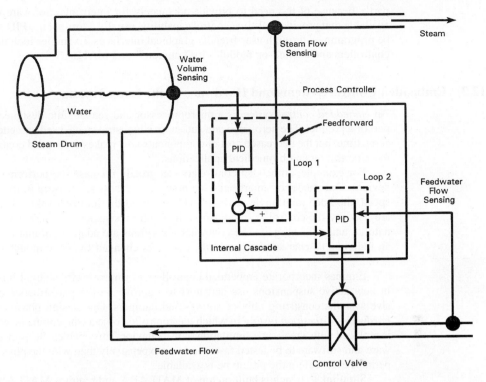

Figure 52   Application of a two-loop process controller for feedwater control.

functions for generating root-locus and frequency response plots, system simulation, digital filtering, calculation of control gains, and data analysis. It can accept model descriptions in the form of transfer functions or as state-variable equations.[1,4,10]

Some manufacturers provide software to assist the engineer in sizing and selecting components. An example is the *Motion Component Selector* (MCS) available from Galil Motion Control, Inc. It assists the engineer in computing the load inertia, including the effects of the mechanical drive, and then selects the proper motor and amplifier based on the user's description of the desired motion profile.

Some hardware manufacturers supply software to assist the engineer in selecting control gains and modifying (*tuning*) them to achieve good response. This might require that the system to be controlled be available for experiments prior to installation. Some controllers, such as some Honeywell process controllers, have an autotuning feature that adjusts the gains in real time to improve performance.

Motion programming software supplied with motion controllers was mentioned previously. Some packages, such as Galil's, allow the user to simulate a multiaxis system having more than one motor and to display the resulting trajectory.

Instrumentation configuration software, such as *LabView*®, provides specialized programming languages for interacting with instruments and for creating graphical real-time displays of instrument outputs.

Until recently, development of real-time digital control software involved tedious programming, often in assembly language. Even when implemented in a higher level language, such as FORTRAN or C++, programming real-time control algorithms can be very challenging, partly because of the need to provide adequately for interrupts. Software packages are now available that provide real-time control capability, usually a form of the PID algorithm, that can be programmed through user-friendly graphical interfaces. Examples include the Galil motion controllers and the add-on modules for Labview® and MATLAB®.

12.3   Embedded Control Systems and Hardware-in-the Loop Testing

An *embedded control system* is a microprocessor and sensor suite designed to be an integral part of a product. The aerospace and automotive industries have used embedded controllers for some time, but the decreased cost of components now makes embedded controllers feasible for more consumer and biomedical applications.

For example, embedded controllers can greatly increase the performance of orthopedic devices. One model of an artificial leg now uses sensors to measure in real time the walking speed, the knee joint angle, and the loading due to the foot and ankle. These measurements are used by the controller to adjust the hydraulic resistance of a piston to produce a stable, natural, and efficient gait. The controller algorithms are adaptive in that they can be tuned to an individual's characteristics and their settings changed to accommodate different physical activities.

Engines incorporate embedded controllers to improve efficiency. Embedded controllers in new active suspensions use actuators to improve on the performance of traditional passive systems consisting only of springs and dampers. One design phase of such systems is *hardware-in-the-loop testing* in which the controlled object (the engine or vehicle suspension) is replaced with a real-time simulation of its behavior. This enables the embedded system hardware and software to be tested faster and less expensively than with the physical prototype and perhaps even before the prototype is available.

Simulink®, which is built on top of MATLAB® and requires MATLAB® to run, is often used to create the simulation model for hardware-in-the-loop testing. Some of the *toolboxes* available for MATLAB®, such as the control systems toolbox, the signal-processing toolbox, and the DSP and fixed-point blocksets, are also useful for such applications.

13   SOFTWARE SUPPORT FOR CONTROL SYSTEM DESIGN

Software packages are available for graphical control system design methods and control system simulation. These greatly reduce the tedious manual computation, plotting, and programming formerly required for control system design and simulation.

13.1   Software for Graphical Design Methods

Several software packages are available to support graphical control system design methods. The most popular of these is MATLAB®, which has extensive capabilities for generation and interactive analysis of root-locus plots and frequency response plots. Some of these capabilities are discussed in Refs. 1 and 4.

13.2   Software for Control Systems Simulation

It is difficult to obtain closed-form expressions for system response when the model contains *dead time* or nonlinear elements that represent realistic control system behavior. Dead time (also called *transport delay*), rate limiters, and actuator saturation are effects that often occur in real control systems, and simulation is often the only way to analyze their response. Several software packages are available to support system simulation. One of the most popular is Simulink.

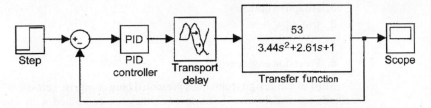

Figure 53 Simulink model of a system with transport delay.

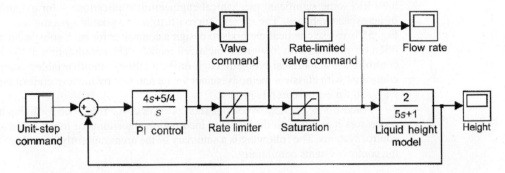

Figure 54 Simulink model of a system with actuator saturation and a rate limiter.

Systems having dead-time elements are easily simulated in Simulink. Figure 53 shows a Simulink model for PID control of the plant $53/(3.44s^2 + 2.61s + 1)$, with a dead time between the output of the controller and the plant. The block implementing the dead-time transfer function e^{-Ds} is called the *transport delay* block. When you run this model, you will see the response in the scope block.

In addition to being limited by saturation, some actuators have limits on how fast they can react. This limitation is independent of the time constant of the actuator and might be due to deliberate restrictions placed on the unit by its manufacturer. An example is a flow control valve whose rate of opening and closing is controlled by a *rate limiter*. Simulink has such a block, and it can be used in series with the saturation block to model the valve behavior. Consider the model of the height h of liquid in a tank whose input is a flow rate q_i. For specific parameter values, such a model has the form $H(s)/Q_i(s) = 2/(5s + 1)$. A Simulink model is shown in Figure 54 for a specific PI controller whose gains are $K_P = 4$ and $K_I = 5/4$. The saturation block models the fact that the valve opening must be between 0 and 100%. The model enables us to experiment with the lower and upper limits of the rate limiter block to see its effect on the system performance.

An introduction to Simulink is given in Refs. 4 and 10. Applications of Simulink to control system simulation are given in Ref. 4.

14 FUTURE TRENDS IN CONTROL SYSTEMS

Microprocessors have rejuvenated the development of controllers for mechanical systems. Currently, there are several applications areas in which new control systems are indispensable to the product's success:

1. Active vibration control
2. Noise cancellation

3. Adaptive optics

4. Robotics

5. Micromachines

6. Precision engineering

Most of the design techniques presented here comprise "classical" control methods. These methods are widely used because when they are combined with some testing and computer simulation, an experienced engineer can rapidly achieve an acceptable design. Modern control algorithms, such as state-variable feedback and the linear–quadratic optimal controller, have had some significant mechanical engineering applications—for example, in the control of aerospace vehicles. The current approach to multivariable systems like the one shown in Fig. 55 is to use classical methods to design a controller for each subsystem because they can often be modeled with low-order linearized models. The coordination of the various low-level controllers is a nonlinear problem. High-order, nonlinear, multivariable systems that cannot be controlled with classical methods cannot yet be handled by modern control theory in a general way, and further research is needed.

In addition to the improvements, such as lower cost, brought on by digital hardware, microprocessors have allowed designers to incorporate algorithms of much greater complexity into control systems. The following is a summary of the areas currently receiving much attention in the control systems community.

14.1   Fuzzy Logic Control

In classical set theory, an object's membership in a set is clearly defined and unambiguous. *Fuzzy logic control* is based on a generalization of classical set theory to allow objects to belong

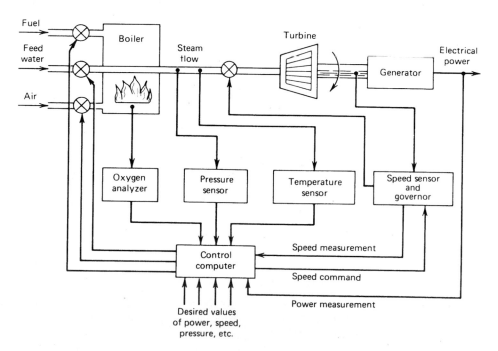

Figure 55   Computer control system for a boiler–generator. Each important variable requires its own controller. The interaction between variables calls for coordinated control of all loops.[1]

to several sets with various degrees of membership. Fuzzy logic can be used to describe processes that defy precise definition or precise measurement, and thus it can be used to model the inexact and subjective aspects of human reasoning. For example, room temperature can be described as cold, cool, just right, warm, or hot. Development of a fuzzy logic temperature controller would require the designer to specify the membership functions that describe "warm" as a function of temperature, and so on. The control logic would then be developed as a linguistic algorithm that models a human operator's decision process (e.g., if the room temperature is "cold," then "greatly" increase the heater output; if the temperature is "cool," then increase the heater output "slightly").

Fuzzy logic controllers have been implemented in a number of applications. Proponents of fuzzy logic control point to its ability to convert a human operator's reasoning process into computer code. Its critics argue that because all the controller's fuzzy calculations must eventually reduce to a specific output that must be given to the actuator (e.g., a specific voltage value or a specific valve position), why not be unambiguous from the start, and define a "cool" temperature to be the range between 65° and 68°, for example? Perhaps the proper role of fuzzy logic is at the human operator interface. Research is active in this area, and the issue is not yet settled.[11,12]

14.2   Nonlinear Control

Most real systems are nonlinear, which means that they must be described by nonlinear differential equations. Control systems designed with the linear control theory described in this chapter depend on a linearized approximation to the original nonlinear model. This linearization can be explicitly performed, or implicitly made, as when we use the small-angle approximation: $\sin \theta \approx \theta$. This approach has been enormously successful because a well-designed controller will keep the system in the operating range where the linearization was done, thus preserving the accuracy of the linear model. However, it is difficult to control some systems accurately in this way because their operating range is too large. Robot arms are a good example.[13,14] Their equations of motion are very nonlinear, due primarily to the fact that their inertia varies greatly as their configuration changes.

Nonlinear systems encompass everything that is "not linear," and thus there is no general theory for nonlinear systems. There have been many nonlinear control methods proposed—too many to summarize here.[15] *Lyapunov's stability theory* and Popov's method play a central role in many such schemes. Adaptive control is a subcase of nonlinear control (see below).

The high speeds of modern digital computers now allow us to implement nonlinear control algorithms not possible with earlier hardware. An example is the computed-torque method for controlling robot arms, which was discussed in Section 11 (see Fig. 47).

14.3   Adaptive Control

The term *adaptive control,* which unfortunately has been loosely used, describes control systems that can change the form of the control algorithm or the values of the control gains in real time, as the controller improves its internal model of the process dynamics or in response to unmodeled disturbances.[16] Constant control gains do not provide adequate response for some systems that exhibit large changes in their dynamics over their entire operating range, and some adaptive controllers use several models of the process, each of which is accurate within a certain operating range. The adaptive controller switches between gain settings that are appropriate for each operating range. Adaptive controllers are difficult to design and are prone to instability. Most existing adaptive controllers change only the gain values, not the form of the control algorithm. Many problems remain to be solved before adaptive control theory becomes widely implemented.

14.4   Optimal Control

A rocket might be required to reach orbit using minimum fuel or it might need to reach a given intercept point in minimum time. These are examples of potential applications of *optimal-control theory*. Optimal-control problems often consist of two subproblems. For the rocket example, these subproblems are (1) the determination of the minimum-fuel (or minimum-time) trajectory and the open-loop control outputs (e.g., rocket thrust as a function of time) required to achieve the trajectory and (2) the design of a feedback controller to keep the system near the optimal trajectory.

Many optimal-control problems are nonlinear, and thus no general theory is available. Two classes of problems that have achieved some practical successes are the *bang-bang control* problem, in which the control variable switches between two fixed values (e.g., on and off or open and closed),[6] and the *linear-quadratic-regulator* (LQG), discussed in Section 7, which has proven useful for high-order systems.[1,6]

Closely related to optimal-control theory are methods based on stochastic process theory,[17] including *stochastic control theory, estimators, Kalman filters,* and *observers.*[1,6,17]

REFERENCES

1. W. J. Palm III, *Modeling, Analysis, and Control of Dynamic Systems,* 2nd ed., Wiley, New York, 2000.
2. W. J. Palm III, *Control Systems Engineering,* Wiley, New York, 1986.
3. D. E. Seborg, T. F. Edgar, F.J. Doyle, and D. A. Mellichamp, *Process Dynamics and Control,* 3rd ed., Wiley, Hoboken, NJ, 2011.
4. W. J. Palm III, *System Dynamics,* 3rd ed., McGraw-Hill, New York, 2013.
5. D. McCloy and H. Martin, *The Control of Fluid Power,* 2nd ed., Halsted, London, 1980.
6. A. E. Bryson and Y. C. Ho, *Applied Optimal Control,* Blaisdell, Waltham, MA, 1969.
7. F. Lewis, D. Vrabie, and V. Syrmos, *Optimal Control,* 3rd ed., Wiley, New York, 2013.
8. K. J. Astrom and B. Wittenmark, *Computer Controlled Systems*, Prentice-Hall, Englewood Cliffs, NJ, 1984.
9. Y. Dote, *Servo Motor and Motion Control Using Digital Signal Processors,* Prentice-Hall, Englewood Cliffs, NJ, 1990.
10. W. J. Palm III, *Introduction to MATLAB for Engineers,* 3rd ed., McGraw-Hill, New York, 2011.
11. G. Klir and B. Yuan, *Fuzzy Sets and Fuzzy Logic*, Prentice-Hall, Englewood Cliffs, NJ, 1995.
12. B. Kosko, *Neural Networks and Fuzzy Systems,* Prentice-Hall, Englewood Cliffs, NJ, 1992.
13. J. Craig, *Introduction to Robotics,* 3rd ed., Addison-Wesley, Reading, MA, 2005.
14. M. W. Spong, S. Hutchinson, and M. Vidyasagar, *Robot Dynamics and Control,* 2nd ed., Wiley, New York, 2006.
15. J. Slotine and W. Li, *Applied Nonlinear Control,* Prentice-Hall, Englewood Cliffs, NJ, 1991.
16. K. J. Astrom, K. J. Wittenmark, *Adaptive Control,* 2nd ed., Addison-Wesley, Reading, MA, 1995.
17. R. Stengel, *Stochastic Optimal Control,* Wiley, New York, 1986.

CHAPTER 22

GENERAL-PURPOSE CONTROL DEVICES

James H. Christensen
Holobloc, Inc.
Cleveland Heights, Ohio

Robert J. Kretschmann
Rockwell Automation
Mayfield Heights, Ohio

Sujeet Chand
Rockwell Automation
Milwaukee, Wisconsin

Kazuhiko Yokoyama
Yaskawa Electric Corporation
Tokyo, Japan

1 CHARACTERISTICS OF GENERAL-PURPOSE CONTROL DEVICES

1.1 Hierarchical Control

As shown in Fig. 1, *general-purpose control devices* (GPCDs) occupy a place in the hierarchy of factory automation above the closed-loop control systems described in Chapters 12–15. The responsibility of the GPCD is the coordinated control of one or more machines or processes.

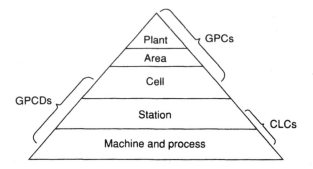

Figure 1   Plant control hierarchy using GPCDs, general-purpose computers (GPCs), and closed-loop controllers (CLCs).

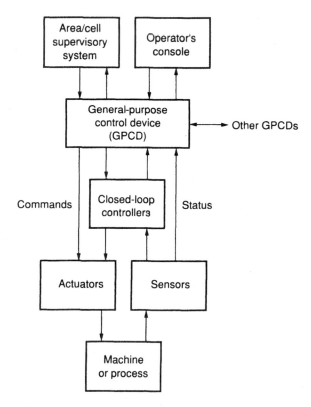

Figure 2   Communication and control paths of a GPCD.

Thus, a GPCD may operate at the "station" level, where it controls part or all of a single machine or process, or at the "cell" level, where it coordinates the operation of multiple stations.

In fulfilling its responsibilities, the GPCD must be capable of performing the following functions, as shown in Fig. 2:

- Issuing commands to and receiving status information from a set of closed-loop controllers that control individual machine and process variables such as velocity, position, and temperature. These closed-loop controllers may be separate devices or integral parts of the GPCD hardware and/or software architecture.

- Issuing commands to and receiving status information from a set of actuators and sensors directly connected to the controlled operation. These actuators and sensors may include signal-processing and transmission elements, as described in Chapter 6. This capability may not be required if all interface to the controlled operation is through the closed-loop controllers described above.
- Receiving commands from and sending information to a control panel or console for the operator of the machine or process.
- Receiving commands from and sending status information to a manual or automated system with the responsibility of supervising the operation of a number of GPCDs within the boundaries of a "cell," for example, a number of coordinated machines or unit operations, or over a wider "area," such as a chemical process or production zone in a factory.
- Interchanging status information with other GPCDs within the same cell or across cell boundaries.

It should be noted that not all of these capabilities are necessarily required for every application of a GPCD. For instance, special closed-loop controllers may not be necessary in an operation requiring only simple on–off or modulating control. Similarly, communication in an automation hierarchy may not be required for simple "stand-alone" applications such as an industrial trash compactor.

However, the capability for expansion into a communicating hierarchy should be inherent in the GPCD architecture if retrofit of stand-alone systems into an integrated production system is considered a future possibility.

1.2 Programmability

To be truly general purpose, a GPCD must be programmable; that is, its operation is controlled by sequences of instructions and data stored in internal memory. The languages used for programming GPCDs are usually *problem oriented*: Programs are expressed in terms directly related to the control to be performed, rather than in a general-purpose programming language such as C + + or BASIC. These languages will be described in appropriate sections for each type of GPCD.

Depending on the application, the responsibility for development and maintenance of GPCD programs may reside with:

- The *original equipment manufacturer* (OEM) of a machine that includes a GPCD as part of its control apparatus
- The *system integrator* who designs and installs an integrated hierarchical control system
- The *end user* who wishes to modify the operation of the installed system

The degree to which the operation of the system can be modified by the end user is a function of:

- The complexity of the system
- The degree to which the end user has been trained in the programming of the system
- The extent to which the operation of the process must be modified over time

For instance, in a high-volume chemical process, only minor modifications of set points may be required over the life of the plant. However, major modifications of the process may be required annually in an automotive assembly plant. In the latter case, complete user programmability of the system is required.

Depending on the complexity of the control program and the degree of reprogrammability required, GPCD programming may be supported by any of several means, including:

- An integral programming panel on the GPCD
- Portable programming and debugging tools

- Minicomputers, personal computers, or engineering workstations that may or may not be connected to the GPCD during control operation
- An online computer system, for example, the area or cell controller shown in Fig. 2.

1.3 Device Architecture

Figure 3 illustrates a GPCD architecture capable of providing the required functional characteristics:

- The *memory* provides storage for the programs and data entered into the system via the *communications processor*.
- The *control processor* performs control actions under the direction of the stored program as well as coordinates the operation of the other functions.
- The *communications processor* provides the means of accepting commands from and providing status information to the supervisory system and operator's console, interchanging status information with other GPCDs, and interacting with program development and configuration tools.
- The *I/O (input/output) processor* provides the means by which the control processor can issue commands to and receive status information from the closed-loop controllers as well as interchange information directly with the controlled operation via the *output* and *input interfaces*.

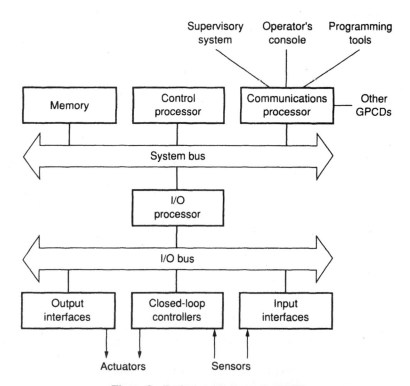

Figure 3 Typical architecture of a GPCD.

- The *system bus* provides for communication among the functional blocks internal to the GPCD, while the *I/O bus* provides for communication between the internal functional blocks and the "outside world" via the closed-loop controllers and I/O interfaces. As an option, the I/O controller may extend the I/O bus functionality to remote locations using data communications methods such as those described in Section 9.

The architecture shown in Fig. 3 is not the only one possible for GPCDs, nor is it necessarily the most desirable for all applications. For instance:

- Large systems may require multiple control processors on the system bus.
- In applications requiring only a few control loops, it may be more economical to perform the closed-loop control function directly in the software of the control processor.
- A separate I/O processor may not be required if the number of separate I/O interface points is less than a few dozen.
- In small systems, the programmer's console may be interfaced directly to the system or I/O bus.

However, if it is anticipated that control system requirements will grow substantially in the future or if total control system requirements are only partially understood, the use of a flexible, extendable GPCD architecture such as that shown in Fig. 3 is recommended.

1.4 Sequential Control

It is obvious that GPCDs must perform complex sequences of control actions when they are applied to the coordination of material handling and machine operation in the fabrication and assembly of discrete parts or in batch and semibatch processes such as blast furnace operation and pharmaceutical manufacture. However, sequential control is also increasing in importance in "continuous" process control, since no process is truly continuous. At the very least, the process must be started up and shut down for maintenance or emergencies by a predetermined sequence of control actions. In large, integrated processes, these sequences are too complicated to be carried out manually and must be performed automatically by GPCDs.

The increasing importance of sequential control, coupled with the increasing complexity of the controlled processes, have generated the need for graphical programming and documentation techniques for the representation of large, complex sequential control plans. These plans must provide a straightforward representation of the relationship between the operation of the control program in the GPCD and the operation of the controlled machine or process as well as the interrelationships between multiple, simultaneous control sequences.

Recognizing this need, the International Electrotechnical Commission (IEC, www.iec.org) has undertaken several efforts to standardize the representation of sequential control plans:

- The IEC 60848 standard[1] defines the GRAFCET specification language for the functional description of the behavior of the sequential part of a control system.
- The IEC 61131-3 standard[2] defines a set of sequential function chart (SFC) constructs, specifically intended to make the IEC 60848 concepts usable for the programming of programmable logic controllers (PLCs).
- The IEC 61512-1 standard[3] defines SFC-like constructs for the description of the sequential operations involved in the control of batch chemical processes.
- The IEC 61499-1 standard[4] defines an execution control chart (ECC) construct which provides for the sequential execution of control algorithms in function blocks for distributed automation systems, similar to the Harel state chart notation of the Unified Modeling Language (UML).[5]

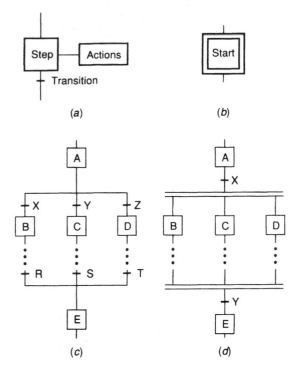

Figure 4 Sequential function chart constructs.

As shown in Fig. 4*a*, an SFC is constructed from three basic types of elements:

- A *step,* representing the current state of the controller and controlled system within the sequential control plan
- A set of associated *actions* at each step
- A *transition condition* that determines when the state of the controller and controlled system is to evolve to another step or set of steps

An SFC consists of a set of independently operating *sequences* of control actions built up out of these basic elements via two mechanisms:

- *Selection* of one of a number of alternate successors to a step based on mutually exclusive transition conditions, as shown in Fig. 4*c*
- *Divergence,* that is, initiation of two or more independently executing sequences based on a transition condition, as shown in Fig. 4*d*

The mechanisms for representing convergence, that is, resumption of a main sequence after step selection or parallel sequence initiation, are also shown in Figs. 4*c* and 4*d*, respectively.

Figure 4*b* illustrates the representation of the *initial step* of each sequence. The operation of control sequences can be visualized by placing a *token* in each initial step upon the initiation of system operation. A step is then said to be *active* while it possesses a token and *reset* when it does not possess a token. The actions associated with the step are performed while the step is active and are not performed when it is reset. The resetting of one or more steps and the activation of one or more successor steps can then be envisioned as the processes of token passing, consumption, and generation, as shown in Figs. 5 and 6. It should be noted that the

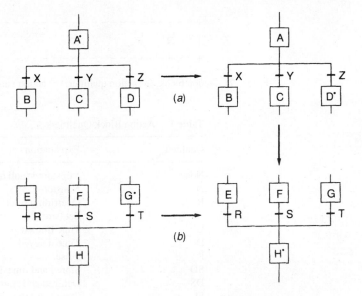

Figure 5 Sequence selection and convergence.

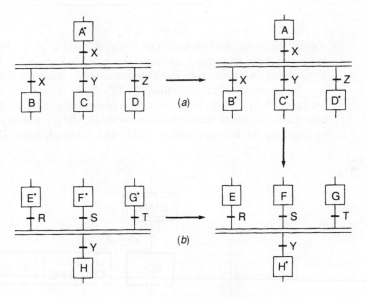

Figure 6 Parallel sequence initiation and convergence.

selection and convergence of alternate paths within a sequence, as shown in Fig. 5, simply involve the passing of a single token. In contrast, divergence to multiple sequences involves the consumption of a single token and the generation of multiple tokens, as shown in Fig. 6a, with the converse operation for termination of multiple sequences shown in Fig. 6b.

When an action associated with a step is a Boolean variable, its association with the step may be expressed in an *action block,* as shown in Fig. 7. The qualifiers which can be used to

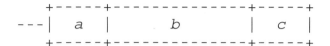

Figure 7   Action block: (*a*) action qualifier; (*b*) action name; (*c*) feedback name.

Table 1   Action Block Qualifiers

Qualifier	Explanation
None	Nonstored (null qualifier)
N	Nonstored
R	Overriding reset
S	Set (stored)
L	Time limited
D	Time delayed
P	Pulse
SD	Stored and time delayed
DS	Delayed and stored
SL	Stored and time limited
P1	Pulse (rising edge)
P0	Pulse (falling edge)

specify the duration of the action are listed in Table 1. More complex actions can be specified via one of the programming languages described in Section 2.3; in this case, the action executes continuously while the associated "action control" shown in Table 1 has the Boolean value 1.

An example of the application of SFCs to the control and monitoring of a single motion, for example, in a robot control system, is given in Fig. 8. Here, the system waits until a motion command is received via the Boolean variable CMD IN. It then initiates the appropriate motion by asserting the Boolean variable CMD. If a feedback signal DONE is not received within a

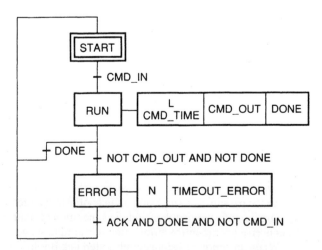

Figure 8   Sequential function chart example.

time interval specified by the variable CMD, the system enters an error step and issues the error message TIMEOUT ERROR. This error condition is cleared when acknowledged by the signal ACK, for example, from an operator's console, combined with the feedback (DONE) that the motion has been accomplished. The system then reenters the initial step and waits for another command.

1.5 Path Control

General-purpose control devices must often perform *path control,* that is, the coordinated control of several variables at once along a continuous path through time. Typical kinds of path control include:

- The path of a metal-cutting tool or a robot manipulator
- The trajectory of a set of continuous process variables such as temperature, pressure, and composition
- The startup of a set of velocity and tension variables in a paper or steel processing line

The numerical control of metal-cutting tools and robot manipulators is discussed in more detail in Sections 3 and 4, respectively.

A typical application to path control of the general-purpose architecture shown in Fig. 3 has the control processor planning the motions to be accomplished and issuing commands to the closed-loop controllers to perform the required motions. Coordination between the closed-loop controllers may be performed by the control processor or by direct interaction among the closed-loop controllers using the I/O bus or special interconnections.

In addition to performing path planning, the control processor also performs sequencing of individual motions and coordination of the motions with other control actions, typically using programming mechanisms such as the SFCs discussed in Section 1.4.

2 PROGRAMMABLE CONTROLLERS

2.1 Principles of Operation

The programmable controller (PLC) is defined by the IEC as[6]:

A digitally operating electronic system, designed for use in an industrial environment, which uses a programmable memory for the internal storage of user-oriented instructions for implementing specific functions such as logic, sequencing, timing, counting, and arithmetic, to control, through digital or analog inputs and outputs, various types of machines or processes. Both the PLC and its associated peripherals are designed to be easily integrable into an industrial control system and easily used in all their intended functions.

The hardware architecture of almost all programmable controllers is the same as that for the GPCD shown in Fig. 3.

As illustrated in Fig. 9, the operation of most programmable controllers consists of a repeated cycle of four major steps:

1. All inputs from interfaces and closed-loop controllers on the I/O bus, and possibly from other GPCDs, are scanned to provide a consistent "image" of the inputs.

2. One "scan" of the user program is performed to derive a new "image" of the desired outputs, as well as internal program variables, from the image of the inputs and the internal and output variables computed during the previous program scan. Typically, the program scan consists of:

   a. Determining the currently active steps of the SFC (see Section 1.4), if any, contained in the program.

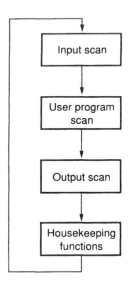

Figure 9 Basic operation cycle of a programmable controller.

b. Scanning the program elements or computing the outputs contained in the active actions of the SFC, if any (if the user program does not contain an SFC, then all program elements are scanned). Scanning of program elements in ladder diagrams or function block diagrams (see Section 2.3) typically proceeds from left to right and from top to bottom. Programming elements are sometimes provided to enable skipping the evaluation of groups of program elements or to force the outputs of a group of elements to zero.

c. Evaluating transition conditions of the SFC (if any) at the end of the program scan, in preparation for step 2a in the next program scan.

3. The data from the updated output image are then transferred to the interfaces and closed-loop controllers on the I/O bus and possibly to other GPCDs as well.

4. Finally, "housekeeping" tasks are performed on a time-available basis. These typically include communication with the operator, a supervisory controller, a programming terminal, or other GPCDs.

After the performance of housekeeping tasks, the cyclic operation of the PLC begins again with the input scan. This may follow immediately upon execution of the housekeeping tasks or may be scheduled to repeat at a fixed execution interval.

Some programmable controller systems with separate I/O and/or communications processors provide for overlapping the scanning of the user program with the scanning of the inputs (step 1) and outputs (step 3) and communication functions (step 4). In these cases, special programming mechanisms may be needed to achieve concurrency and synchronization between the program and I/O scans and between the program and communications processing.

A further feature of some PLCs is the incorporation of a multitasking operating system. As in general computing, the PLC operating system serves to coordinate the multitude of hardware and software resources and capabilities of the PLC. The incorporation of a multitasking operating system serves to allow the PLC to essentially execute multiple instances of the basic operation cycle shown in Fig. 9, rather than a single instance, at the same time. This affords much increased flexibility and capability over a single-tasking operating system PLC.

In PLCs incorporating multitasking operating systems, mechanisms may be needed to achieve appropriate levels of task coordination and synchronization between the multiple instances of the basic operating cycle. As an example, one task's results may affect another task's decisions, especially with regard to I/O.

2.2   Interfaces

The IEC has specified the standard voltage ratings shown in Table 2 for power supplies, digital inputs, and digital outputs of programmable controllers. The IEC standard[7] also defines additional parameters for digital inputs and outputs, shown in Tables 3–5; the parameters specified by the manufacturer should be checked against those defined in the IEC standard in order to assure the suitability of a particular input or output module for its intended use in the control system.

The IEC-specified signal ranges for analog inputs and outputs for programmable controllers are shown in Tables 6 and 7, respectively. The IEC standard lists a number of characteristics whose values are to be provided by the manufacturer, such as input impedance, maximum input error, and conversion time and method, and which must be checked against the requirements of the particular control application.

In addition to simple digital and analog inputs and outputs, closed-loop controllers which can reside on the I/O bus of the programmable controller system may be provided, as illustrated for GPCDs in Fig. 2. In this case, the programming languages for the programmable controller typically provide language elements, in addition to those described in Section 2.3, to support the configuration and supervisory control of these "slave" closed-loop controllers.

Table 2   Rated Values and Operating Ranges for Incoming Power Supplies and Digital I/O Interfaces of Programmable Controllers

	Recommended for	
Rated Voltage	Power Supply	I/O Signals
24 V dc[a]	Yes	Yes
48 V dc[a]	Yes	Yes
120 V rms ac[b]	Yes	Yes
230 V rms ac[b]	Yes	Yes

Note: See the IEC Programmable Controller standard (Ref. 7) for additional notes and rating values.
[a]Voltage tolerance for dc voltage ratings is −15 to +20%.
[b]Voltage tolerance for ac voltage ratings is −15 to +10%.

Table 3   Rated Values for dc and ac Digital Inputs of Programmable Controllers

Rated Voltage	Types
24 V dc	1–3
48 V dc	1–3
120 V ac	1–3
230 V ac	1–3

Note: See the IEC Programmable Controller standard (Ref. 7) for additional notes and ratingvalues.

Table 4 Rated Values for dc Digital Outputs of Programmable Controllers

Rated Current (A)	Maximum Current (A)	Leakage Current (mA)	Input Type Compatibility
0.1	0.12	0.1	1–3
0.25	0.3	0.5	1–3
0.5	0.6	0.5	1–3
1	1.2	1	2, 3
2	2.4	1	2, 3

Note: See the IEC Programmable Controller standard (Ref. 7) for additional notes and rating values.

Table 5 Rated Values for ac Digital Outputs of Programmable Controllers

Rated Current (A)	Maximum Current (A)	Leakage Current (mA)
0.25	0.28	5
0.5	0.55	10
1	1.1	10
2	2.2	10

Note: See the IEC Programmable Controller standard (Ref. 7) for additional notes and ratingvalues.

Table 6 Rated Values for Analog Inputs of Programmable Controllers

Signal Range	Input Impedance
−10–10 V	$\geq 10\,k\Omega$
0–0 V	$\geq 10\,k\Omega$
1–5 V	$\geq 5\,k\Omega$
4–20 mA	$\leq 300\,\Omega$

Note: See the IEC Programmable Controller standard (Ref. 7) for additional notes and rating values.

Table 7 Rated Values for Analog Outputs of Programmable Controllers

Signal Range	Load Impedance
−10–10 V	$\geq 1\,k\Omega$
0–10V	$\geq 1\,k\Omega$
1–5 V	$\geq 500\,\Omega$
4–20 mA	$\leq 600\,\Omega$

Note: See the IEC Programmable Controller standard (Ref. 7) for additional notes and rating values.

Communication interfaces for programmable controllers provide many different combinations of connectors, signal levels, signaling rates, and communication services. The manufacturer's specifications of these characteristics should be checked against applicable standards to assure the achievement of the required levels of system performance and compatibility of all GPCDs in the system.

2.3   Programming

Programmable controllers have historically been used as programmable replacements for relay and solid-state logic control systems. As a result, their programming languages have been oriented around the conventions used to describe the control systems they have replaced, that is, relay ladder logic and function block diagrams. Since these representations are fundamentally graphic in nature, programmable controllers provide one of the first examples of the practical application of graphic programming languages.

Much of the programming of programmable controllers is done in the factory environment while the controlled system is being installed or maintained. Hence, programming was traditionally supported by special-purpose portable programming terminals. In recent years, most of the support for PLC programming has migrated to software packages for personal computers. However, some need still exists for specialized terminals for programming and debugging of programmable controllers in the industrial environment, although most of this functionality can also be supplied by ruggedized, portable personal computers. The selection of support environments is thus an important consideration in the selection and implementation of programmable controller systems.

The standard for PLC programming languages[2] is published by the IEC. This standard specifies a set of mutually compatible programming languages, taking into account the different courses of evolution of programmable controllers in North America, Europe, and Japan and the wide variety of applications of programmable controllers in modern industry. These languages include:

- The SFC elements described in Section 1.4 for sequential control
- Ladder diagrams (LDs) for relay replacement functions
- Function block diagrams (FBDs) for logic, mathematical, and signal-processing functions
- Structured text (ST) for data manipulation
- Instruction list (IL) for assembly-language-level programming

The IL language will not be described in this book; for further details the IEC standard[2] should be consulted.

Figure 10 shows the application of the LD, FBD, and ST languages to implement a simple command execution and monitoring function. In general, a desired functionality can be programmed in any one of the IEC languages. Hence, languages can be chosen depending on their suitability for each particular application.

An exception to this portability is the use of iteration and selection constructs (IF...THEN...ELSIF, CASE, FOR, WHILE, and REPEAT) in the ST language.

The functionality shown in Fig. 10 can be encapsulated into a reusable *function block* by following the declaration process defined in the IEC language standard. An example of the graphical and textual declaration of this functionality is shown in Fig. 11.

In addition to providing mechanisms for the programming of mathematical functions and function blocks, the standard provides a large set of predefined standardized functions and function blocks, as listed in Tables 8 and 9, respectively. The intent is for these to be used as "building blocks" for user programs.

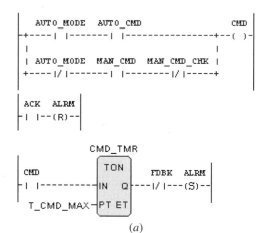

(a)

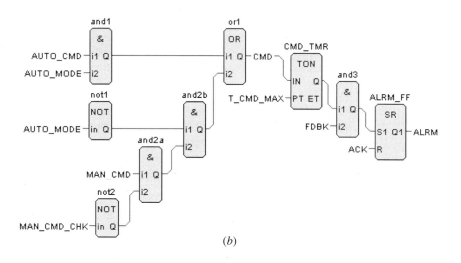

(b)

```
CMD := AUTO_CMD & AUTO_MODE OR MAN_CMD & NOT MAN_CMD_CHECK & NOT
AUTO_MODE;

CMD_TMR(IN := CMD, PT := T_CMD_MAX);

ALRM_FF(S1 := CMD_TMR.Q & NOT FBDK, R := ACK);

ALRM := ALRM_FF.Q1;
```

(c)

Figure 10   Programmable controller programming example.

```
FUNCTION_BLOCK CMD_MONITOR (* Begin definition of FB CMD_MONITOR *)

   (* Definition of external interface *)

   VAR_INPUT

      AUTO_CMD : BOOL;        (* Automatic Command *)

      AUTO_MODE : BOOL;       (* AUTO_CMD Enable *)

      MAN_CMD : BOOL;         (* Manual Command *)

      MAN_CMD_CHK : BOOL;     (* Negated MAN_CMD for debouncing *)

      T_CMD_MAX : TIME;       (* Maximum time from CMD to FDBK *)

      FDBK : BOOL;            (* Confirmation of CMD completion by operative

      unit *)

      ACK : BOOL;             (* Acknowledgement/Cancel ALRM *)

   END_VAR

   VAR_OUTPUT

      CMD : BOOL;             (* Command to operative unit *)

      ALRM : BOOL;            (* T_CMD_MAX expired without FDBK *)

   END_VAR

   (* Definition of internal state variables *)

   VAR

      CMD_TMR : TON;   (* CMD-to-FDBK timer *)

      ALRM_FF : SR;    (* Note over-riding "S1" input,

                        Command must be cancelled before ACK can cancel

      alarm *)

   END_VAR

(* Definition of Function Block Body per
Figure (a), (b) or (c) *)

END_FUNCTION_BLOCK  (* End definition of FB CMD_MONITOR *)
```
                                  (*a*)

Figure 11 Function block encapsulation: (*a*) textual declaration in ST language; (*b*) graphic representation.

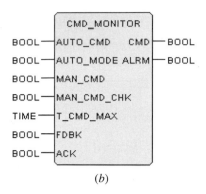

(b)

Figure 11   *(Continued)*

Table 8   IEC Standard Functions for Programmable Controllers

Standard Name	Function	Standard Name	Function
Numeric Functions			
ABS	Absolute value	ACOS	Arc cosine
SQRT	Square root	ATAN	Arc tangent
LN	Natural logarithm	ADD or +	Addition
LOG	Logarithm base 10	SUB or −	Subtraction
EXP	Natural exponential	MUL or *	Multiplication
SIN	Sine	DIV or /	Division
COS	Cosine	MOD	Modulo
TAN	Tangent	EXPT or **	Exponentiation
ASIN	Arc sine	MOVE or : =	Assignment
Bit String Functions			
SHL	Shift left, zero filled	SHR	Shift right, zero filled
ROL	Rotate left circular	ROR	Rotate right circular
AND or &	Bitwise Boolean AND	OR or >= 1	Bitwise Boolean OR
XOR or $= 2k + 1$	Bitwise Boolean exclusive OR	NOT	Bitwise Boolean complement
Selection and Comparison Functions			
SEL	Binary (1 of 2) selection	MUX	Multiplexer (1 of *N*) selection
MIN	Minimum of *N* inputs	MAX	Maximum of *N* inputs
LIM	Hard upper/lower limiter	GT or >	Greater than
GE or > =	Greater than or equal to	EQ or =	Equal to
LE or < =	Less than or equal to	LT or <	Less than
NE or < >	Not equal		
Character String Functions			
CONCAT	Concatenate *N* strings	INSERT	Insert one string into another
DELETE	Delete a portion of a string	REPLACE	Replace a portion of one string with another
NE or < >	Not equal		

Source: From Ref. 2.

Table 9 IEC Standard Function Blocks for Programmable Controllers

Standard Name	Function
Bistable function blocks	
SR	Flip-flop (set dominant)
RS	Flip-flop (reset dominant)
Edge detection function blocks	
R_TRIG	Rising edge detect
R_TRIG	Falling edge detect
Counter function blocks	
CTU	Up counter
CTD	Down counter
Timer function blocks	
TP	One-shot (pulse) timer
TON	On-delay timer
TOF	Off-delay timer
Message transfer and synchronization	
SEND	Messaging requester
RCV	Messaging responder

Source: From Ref. 2.

In addition to being used directly for building functions, function blocks, and programs, the LD, FBD, ST, and IL languages can be used to program the "actions" to be performed under the control of SFCs as described in Section 1.4. These SFCs can then be used to build programs and reusable function blocks using the mechanisms defined in the IEC language standard.[2]

It will be noted in Figure 11 that data types are defined for all variables. The IEC standard provides facilities for strong data typing, with a large set of predefined data types as listed in Table 10. In addition, facilities are provided for user-defined data types as listed in Table 11. The standard allows manufacturers to specify the language features that they support. Users should consult the standard to determine which language features are required by their application and check their language requirements carefully against the manufacturers' specifications when making their choice of a programmable controller system.

2.4 Programmable Controller Standard, IEC 61131

To propagate consistent characteristics and capabilities for PLCs in the marketplace, an international standard describing them has been developed under the auspices of the IEC. It is impossible, in a chapter of this length, to cover all the hardware, software, and programming language characteristics, features, and scope defined in the IEC standard for programmable controllers. The standard comprises seven parts[2,6–11] under the general title "Programmable Controllers," covering various aspects of PLCs.

Since technology is always advancing, some part(s) of the standard, at any given time, is (are) being updated. Copies of draft standards are normally available for review from the appropriate National Committees for the IEC. Additionally, experts are always welcome to participate in the standards generation effort on the recommendation of National Committees. Information about the IEC, its National Committees, and ordering of the various parts of the IEC 61131 standard for PLCs is available at the IEC website.

Table 10   IEC Standard Data Types for Programmable Controllers

Keyword	Data Type	Bits
BOOL	Boolean	1
SINT	Short integer	8
INT	Integer	16
DINT	Double integer	32
LINT	Long integer	64
USINT	Unsigned short integer	8
UINT	Unsigned integer	16
UDINT	Unsigned double integer	32
ULINT	Unsigned long integer	64
REAL	Real number	32
LREAL	Long real	64
TIME	Duration	Implementation dependent
DATE	Date (only)	Implementation dependent
TIME_OF_DAY	Time of day (only)	Implementation dependent
DATE_AND_TIME	Date and time of day	Implementation dependent
STRING	String of 8-bit characters	$8n$
WSTRING	String of 16-bit characters	$16n$
BYTE	Bit string of length 8	8
WORD	Bit string of length 16	16
DWORD	Bit string of length 32	32
LWORD	Bit string of length 64	64

Source: From Ref. 2.

Table 11   Examples of User-Defined Data Types for Programmable Controllers

Direct derivation from elementary types, e.g.:
TYPE RU-REAL : REAL ; END— TYPE
Enumerated data types, e.g.:
TYPE ANALOG_SIGNAL_TYPE : (SINGLE_ENDED, DIFFERENTIAL) ; END_TYPE
Subrange data types, e.g.:
TYPE ANALOG DATA : INT (-4095..4095) ; END_TYPE
Array data types, e.g.:
TYPE ANALOG_16_INPUT_DATA : ARRAY [1..16] OF ANALOG_DATA ; END_TYPE
Structured data types, e.g.:
TYPE
ANALOG_CHANNEL_CONFIGURATION :
STRUCT
RANGE : ANALOG_SIGNAL_RANGE ;
MIN_ SCALE : ANALOG_DATA ;
MAX_ SCALE : ANALOG_DATA ;
END_ STRUCT ;
ANALOG_16_INPUT_CONFIGURATION :
STRUCT
SIGNAL_TYPE : ANALOG_SIGNAL_TYPE ;
FILTER_PARAMETER : SINT (0..99) ;
CHANNEL : ARRAY [1..16] OF ANALOG_CHANNEL_CONFIGURATION ;
END_STRUCT ;
END_TYPE

Source: From Ref. 2.

3 NUMERICAL CONTROLLERS

3.1 Introduction and Applications

The century from 1760 to 1860 saw the development of a large number of machine tools for shaping cylindrical and flat surfaces, threads, grooves, slots, and holes of many shapes and sizes in metals. Some of the machine tools developed were the lathe, the planer, the shaper, the milling machine, drilling machines, and power saws. With increasing applications for metal machining, the cost in terms of manpower and capital equipment grew rapidly. The attempt at automation of the metal removal process gave birth to numerical controllers.

The history of numerical controllers dates back to the late 1940s, when John T. Parsons proposed a method to automatically guide a milling cutter to generate a smooth curve. Parsons proposed that successive coordinates of the tool be punched on cards and fed into the machine. The idea was to move the machine in small incremental steps to achieve a desired path. In 1952, the U.S. Air Force provided funding for a project at the Massachusetts Institute of Technology (MIT) that developed the Whirlwind computer. In a subsequent project, the Servomechanisms Laboratory at MIT developed the concept of the first workable numerical control (NC) system. The NC architecture was designed to exploit the Whirlwind computer with emphasis on five-axis NC for machining complex aircraft parts.

The MIT NC architecture identified three levels of interaction with the numerical controller.[12-14] At the highest level is a machine-independent language, called APT (Automatically Programmed Tools). APT provides a symbolic description of the part geometry, tools, and cutting parameters. The next level, called the cutter location (CL) level, changes the symbolic specification of cutter path and tool control data to numeric data. The CL level is also machine independent. The lowest level, called the G-code level, contains machine-specific commands for the tool and the NC axis motions.

The conversion from APT to CL data involves the computation of cutter offsets and resolution of symbolic constraints. The conversion from CL data to G-code is called postprocessing.[12,14] Postprocessing transforms the tool center line data to machine motion commands, taking into account the various constraints of the machine tool such as machine kinematics and limits on acceleration and speed. The APT-to-CL data conversion and the compilation of CL data to G-code are computationally intensive; these computationally intensive functions were envisioned to be performed by the Whirlwind computer. The numerical controller works with simple G-codes to keep computational requirements low. The G-codes, punched on perforated paper tape, would be the input medium to the numerical controller.

Since their inception in the 1950s, numerical controllers have followed a similar pattern in the evolution of controller technology as the computer industry in the past 30 years. The first numerical controllers were designed with vacuum tube technology. The controllers were bulky and the logic inside the control was hard wired. The hard-wired nature of the controller made it very difficult to change or modify its functionality. Vacuum tubes were replaced by semiconductors in the early 1960s. In the early 1970s, numerical controllers started using microprocessors for control. The first generation of numerical controllers with microprocessor technology were mostly hybrid, with some hard-wired logic and some control functions in software. Today, most NC functionality is in the software. Microprocessor-based NC is also called computer numerical control (CNC).

The concept of distributed numerical control (DNC) was introduced in the 1960s to provide a single point of programming and interface to a large number of numerical controllers. Most NC users agree that the paper tape reader on a numerical controller suffers the most in reliability. DNC can transfer a program to NC through a direct computer link, bypassing the paper tape reader. DNC has two primary functions: (1) computer-assisted programming and storage of NC programs in computer memory and (2) transfer, storage, and display of status and control information from the numerical controllers. DNC can store and transfer programs to as many as

100 numerical controllers. Distributed numerical controllers commonly connect to numerical controllers through a link called Behind the Tape Reader (BTR). The name BTR comes from the fact that the connection between DNC and CNC is made between the paper tape reader and the control unit.

The use of paper tape is no longer the main form of storing or updating in CNC. With the advent of computer technology that is now incorporated into CNC, the CNC now has memory storage capability such that the programs are stored in the CNC as files (similar to a computer). Thus, loading new programs or storing old programs is now done via electronic connection through RS232, Ethernet, or memory storage devices such as flash memory cards. DNC is still used, especially in die-cutting machines where programs are too long and the programs are transferred through a "drip-feed" technique where the CNC runs the program as the program is downloaded from the DNC.

Numerical controllers are widely used in industry today. The predominant application of NC is still for metal-cutting machine tools. Some of the basic operations performed by machine tools in metal machining are turning, boring, drilling, facing, forming, milling, shaping, and planing.[15] Turning is one of the most common operations in metal cutting. Turning is usually accomplished by lathe machines. The part or workpiece is secured in the chuck of a lathe machine and rotated. The tool, held rigidly in a tool post, is moved at a constant speed along the rotational axis of the workpiece, cutting away a layer of metal to form a cylinder or a surface of a more complex profile.

Applications other than metal machining for numerical controllers include flame cutting, water jet cutting, plasma arc cutting, laser beam cutting, spot and arc-welding, and assembly machines.[16]

3.2 Principles of Operation

NC System Components

A block diagram of a NC system is shown in Fig. 12. The three basic components in a NC system are (1) a program input medium; (2) the controller hardware and software, including the feedback transducers and the actuation hardware for moving the tool; and (3) the machine itself.

The controller hardware and software execute programmed commands, compute servo-commands to move the tool along the programmed path, read machine feedback, close the servocontrol loops, and drive the actuation hardware for moving the tool. The actuation hardware consists of servomotors and gearing.

The feedback devices on an NC servosystem provide information about the instantaneous position and velocity of the NC axes. The servofeedback devices can be linear transducers or

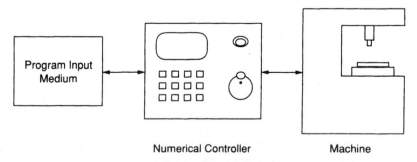

Numerical Controller　　　　　　　　Machine

Figure 12 Components of NC system.

rotary transducers. The two most common rotary transducers are *resolvers* and *encoders*.[17] Resolvers consist of an assembly that resembles a small electric motor with a stator–rotor configuration. As the rotor turns, the phase relationship between the stator and rotor voltages corresponds to the shaft angle such that one electrical degree of phase shift corresponds to one mechanical degree of rotation. An optical encoder produces pulsed output that is generated from a disk containing finely etched lines that rotates between an exciter lamp and one or more photodiodes. The total number of pulses generated in a single revolution is a function of the number of lines etched on the disk. Typical disks contain 2000–10,000 lines. A resolver is an analog device with an analog output signal, whereas an encoder is a numerical device that produces a digital output signal.

Linear transducers may include scales or distance-coded markers. Linear transducers have the advantage that they do not introduce backlash errors that a rotary transducer may not be able to detect depending on its mechanical location.

Traditional input media for entering programs into an NC machine include (1) punched cards, (2) punched tape, (3) magnetic tape, and (4) direct entry of the program into the computer memory of the numerical controller.[18] More modern media include flash memory cards.

Originated by Herman Hollerith in 1887, the punched card as an input medium is almost obsolete. The standard "IBM" card's fixed dimensions are 3.25 in. wide, 7.375 in. long, and 0.007 in. thick. Each card contains 12 rows of hole locations with 80 columns across the card. To edit a part program, cards in the deck are replaced with new cards. With a deck of cards, it is easy to lose sequence or have missing blocks due to the loss of a card. Also, punched cards are a low-density storage medium with an input rate that is slower than most other media. As a result, punched cards are now regarded as obsolete for NC program media.

Punched tape was for many years the most popular input medium for a numerical controller. Although punched tape is mostly obsolete, many numerical controllers still provide a punched tape reader. The specifications of the punched tape are standardized by the EIA (Electronic Industries Association) and the AIA (Aerospace Industries Association). Tapes are made of paper, aluminum–plastic laminates, or other materials. Making editorial changes to the punched tape is difficult; only minor editing is possible by splicing new data into the tape. With the advent of online computer editing techniques, rapid editorial changes to a program can be made on a computer screen. At the end of an editing session, a tape can be automatically punched on command from the keyboard.

Magnetic tape is not used as much as punched tape because of its susceptibility to pollutants in the NC environment. Dust, metal filings, and oil can cause read errors on the tape. Sealed magnetic tapes overcome some of these problems.

Along with flash memory cards, direct entry of a part program into the controller memory is a common input medium for today's NC. The programmer can either type in the NC program from a keyboard and a video display terminal or generate the NC program from an interactive graphics environment.[19] Part programming with the aid of interactive graphics is discussed in more detail in Section 3.5.

Operation of a Numerical Controller: Machine Coordinate System

NC requires a point of reference and a coordinate system to express the coordinates of parts, tools, fixtures, and other components in the workspace of the machine tool. The commonly used coordinates are three orthogonal intersecting axes of a right-handed Cartesian coordinate frame, as shown in Fig. 13. The rotations a, b, and c about the x, y, and z axes are used for NC with more than three axes. In most older numerical controllers, a coordinate frame is marked on the machine and all coordinates are with respect to this fixed frame. In the newer numerical controllers, the machine tool user can program, or "teach," a location for the origin of the reference coordinate system. Since such a reference coordinate system is not permanently attached to the machine, it is sometimes called a *floating coordinate system*.

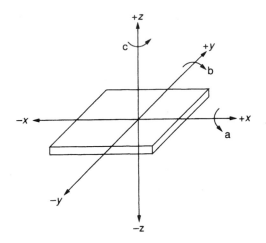

Figure 13   NC coordinate system.

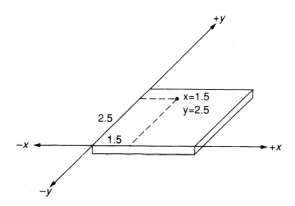

Figure 14   Rectangular plate for drilling.

To illustrate the use of an NC coordinate system, let us consider a simple example of drilling a hole in a rectangular plate with a numerical controller. This example will illustrate the steps in the initial setup and operation of an NC machine. Figure 14 shows the drawing of a simple rectangular plate to be drilled by a drilling machine.

The first step in the programming of any NC operation is getting a drawing of the part. The drawing is usually a blueprint with the dimensions and geometric attributes of the part. Figure 14 shows a rectangular part and the location of the center of the hole to be drilled in the part. Let the lower left-hand corner of the part be the origin of a two-dimensional Cartesian coordinate frame with the x axis and the y axis as shown in Fig. 14. The location of the center of the hole is specified by the coordinates $x = 1.500$ and $y = 2.500$.

The drilling machine must be told the location of the center of the hole. The first step is to establish the location of the NC coordinate system. With the part rigidly held in a fixture, the operator manually moves the machine tool to the lower left-hand corner of the part and presses a button on the machine control panel to teach this point as the origin of the Cartesian coordinate system. Now the machine can locate the center of the hole from the coordinates $x = 1.500$ and $y = 2.500$. In addition to specifying the location of the hole, a programmer can program the rotational speed of the drill, the direction of rotation of the drill (clockwise or counterclockwise), the feed rate, and the depth of cut. The *feed rate* is the distance moved by the tool in an axial direction for each revolution of the workpiece. The *depth of cut* is defined by the thickness of the metal removed from the workpiece, measured in a radial direction.

If the application were milling instead of drilling, the programmer also specifies a cutting speed and a rate of metal removal. The *cutting speed* in a turning or milling operation is the rate at which the uncut surface of the workpiece passes the cutting edge of the tool, usually expressed in millimeters per minute or inches per minute. The *rate of metal removal* is given by the product of the cutting speed, the feed rate, and the depth of cut. The cutting speed and the feed rate are the two most important parameters that a machine operator can adjust to achieve optimum cutting conditions.

3.3   Point-to-Point and Contouring Numerical Controllers

In the example of the preceding section, the drill must be moved to the center of the hole before the drilling operation starts. The drill can start from the origin and traverse to the center of the hole. The machine tool may first move along the x axis by 1.5 in., followed by a movement along the y axis by 2.5 in. It may also simultaneously start moving along the x and y axes. A numerical controller that can position the tool at specified locations without control over the path of the tool between locations is called a *point-to-point* or *positioning controller*.[12] A positioning controller may be able to move along a straight line at 45° by simultaneously driving its axes.

A *contouring controller* provides control over the tool path between positions. For instance, the tool can trace the boundary of a complex part with linear and circular segments in one continuous motion without stopping. For this reason, contouring controllers are also called *continuous-path machines*. The controller can direct the tool along straight lines, circular arcs, and several other geometric curves. The user specifies a desired contour which is typically the boundary or shape of a complex part, and the controller performs the appropriate calculations to continuously drive two or more axes at varying rates to follow this contour.

The specification of a contour in point-to-point NC is tedious because it takes a large number of short connected lines to generate a contour such as a circular arc. The programmer must also provide the appropriate feed rates along these short straight-line segments to control the speed along the contour.

NC Interpolators

In multiaxis contoured NC, multiple independently driven axes are moved in a coordinated manner to direct the tool along a desired path. An interpolator generates the signals to drive the servoloops of multiple actuators along a desired tool path.[20] The interpolator generates a large number of intermediate points along the tool path. The spacing between the intermediate points determines the tool accuracy in tracking the desired path. The closer the points are, the better the accuracy. At each interpolated point along the tool path, the positions of the multiple axes are computed by a kinematic transformation. By successively commanding the NC axes to move from one intermediate point to the next, the tool traces a desired contour with respect to the part.

Older NC systems contained hardware interpolators such as integrators, exponential deceleration circuits, linear interpolators, and circular interpolators. Most of today's numerical controllers use a software interpolator. The software interpolator takes as input the type of curve (linear, circular, parabola, spline, etc.), the start and end points, parameters of the curve such as the radius and the center of a circle, and the speed and acceleration along the curve. The interpolator feeds the input data into a computer program that generates the intermediate target points along the desired curve. At each intermediate target point, the interpolator generates the commands to drive the servoloops of the actuators. The servoloops are closed outside of the interpolator. Several excellent references provide details on closed-loop control of servoactuators.[21,22]

3.4   NC Programming

The two methods of programming numerical controllers are *manual programming* and the use of a high-level, computer-assisted *part programming language*.[23] In both cases, the part programmer uses an engineering drawing or a blueprint of the part to program the geometry of the part. Manual programming typically requires an extensive knowledge of the machining process and the capabilities of the machine tool. In manual programming, the part programmer determines the cutting parameters such as the spindle speed and feed rate from the characteristics of the workpiece, tool material, and machine tool.

The most common programming medium for NC machines was formerly perforated tape. EIA standards RS-273-A and RS-274-B define the formats and codes for punching NC programs on paper tape. The RS-273-A standard is called the interchangeable perforated-tape variable-block format for positioning and straight-cut NC systems, while the RS-274-B is the interchangeable perforated-tape variable-block format for contouring and contouring/positioning NC systems. A typical line on a punched tape according to the RS-273-A standard is as follows:

```
n001 g08 x0.0 y1.0 f225 S100 t6322 m03 (EB)
```

where *n* is the sequential block number; *g* is a preparatory code used to prepare the NC for instructions to follow; *x*, *y* are dimensional words or coordinates; *f* is the feed rate; *s* is the code for the spindle rotation speed; *t* is the tool selection code; *m* denotes a miscellaneous function; and (EB) is the end-of-block character, typically a carriage return. The dimension words *x* and *y* can be expanded for machines with greater than two axes. The order of the dimension words for machines with more than two axes is given as *x, y, z, u, v, w, p, q, r, i, j, k, a, b, c, d, e*.

Punched tape can be created manually or by means of a high-level computer programming system. Manual programming is more suited to simple point-to-point applications. Complex applications are usually programmed by a computer-assisted part programming language. The most common computer-assisted programming system for NCs is the APT system.[23] APT programming is independent of the machine-tool-specific parameters such as tool dimensions. Some of the other programming systems are ADAPT, SPLIT, EXAPT, AUTOSPOT, and COMPACT II.

NC programming today is getting less specialized and less tedious with the help of interactive graphics programming techniques. The part programmer creates the NC program from a display of the part drawing on a high-resolution graphics monitor in a user-friendly, interactive environment. Graphical programming is discussed in Section 3.5.

Manual Programming

Manual programming of a machine tool is performed in units called *blocks*. Each block represents a machining operation, a machine function, or a combination of both. Each block is separated from the succeeding block by an *end-of-block code*.

The standard tape code characters are shown in Table 12. *Preparatory functions,* or G-codes, are used primarily to direct a machine tool through a machining operation. Examples are linear interpolation and circular interpolation. *Miscellaneous* functions, or M-codes, are generally on–off functions such as coolant-on, end-of-program, and program stop. Details on the G-codes and the M-codes and their corresponding functions are given in several excellent tutorials on NC programming.[18,23]

The *feed-rate function f* is followed by a coded number representing the feed rate of the tool. A common coding scheme for the feed-rate number is the inverse-time code, which is

Table 12 Manual Programming Codes

Character Digit or Code	Description
0	Digit 0
1	Digit 1
2	Digit 2
3	Digit 3
4	Digit 4
5	Digit 5
6	Digit 6
7	Digit 7
8	Digit 8
9	Digit 9
a	Angular dimension around x axis
b	Angular dimension around y axis
c	Angular dimension around z axis
d	Angular dimension around special axis or third feed function
e	Angular dimension around special axis or second feed function
f	Feed function
g	Preparatory function
h	Unassigned
i	Distance to arc center parallel to x
j	Distance to arc center parallel to y
k	Distance to arc center parallel to z
m	Miscellaneous function
n	Sequence number
p	Third rapid traverse direction parallel to x
q	Second rapid traverse direction parallel to y
r	First rapid traverse direction parallel to z
s	Spindle speed
t	Tool function
u	Secondary motion dimension parallel to x
v	Secondary motion dimension parallel to y
w	Secondary motion dimension parallel to z
x	Primary x motion dimension
y	Primary y motion dimension
z	Primary z motion dimension
.	Decimal point
,	Unassigned
Check	Unassigned
+	Positive sign
−	Negative sign
Space	Unassigned
Delete	Error delete
Carriage return	End of block
Tab	Tab
Stop Code	Rewind stop
/	Slash code

generated by multiplying by 10 the feed rate along the path divided by the length of the path. If FN denotes the feed-rate number in inverse-time code, FN is given as follows:

```
FN = 10*(feedrate along the path)/(length of the path)
```

For example, if the desired feed rate is 900 mm/min and the length of the path is 180 mm, the feed-rate number is 50 ($f = 0050$). The inverse-time coded feed rate is expressed by a four-digit number ranging from 0001 to 9999. This range of feed-rate numbers corresponds to a minimum interpolation time of 0.06 s and a maximum interpolation time of 10 min.

Spindle speeds are prefixed with the code "s" followed by a coded number denoting the speed. A common code for spindle speed is the *magic-three code*. To illustrate the steps in computing the magic-three code, let us convert a spindle speed of 2345 rpm to magic-three. The first step is to write the spindle speed as 0.2345×10^4. The next step is to round the decimal number to two decimal places and write the spindle speed as 0.23×10^4. The magic-three code can now be derived as 723, where the first digit 7 is given by adding 3 to the power of 10 ($4 + 3$), and the second two digits are the rounded decimal numbers (23). Similarly, a spindle speed of 754 rpm can be written as 0.75×10^3, and the magic-three code is 675.

Manual programming of the machine tool requires the programmer to compute the dimensions of the part from a fixed reference point, called the origin. Since the NC unit controls the path of the tool center point, the programmer must take into account the dimensions of the machine tool before generating the program. For instance, a cylindrical cutting tool in a milling operation will traverse the periphery of the part at a distance equal to its radius. The programmer also takes into account the limits on acceleration and deceleration of the machine tool in generating a program. For example, the feed rate for straight-line milling can be much higher than the feed rate for milling an inside corner of a part. The tool must be slowed down as it approaches the corner to prevent overshoot. Formulas and graphs are often provided to assist the programmer in the above calculations.

To illustrate manual part programming, let us consider an example of milling a simple part shown in Fig. 15. The programmer initially identifies the origin for the Cartesian coordinate system and a starting point for the tool. We will assume that the tool starts from the origin (point PT.S). The programmer builds each block of the program, taking into account the tool diameter and the dynamic constraints of the machine tool. Let us assume that the milling tool in our example is cylindrical with a diameter of 12 mm.

In the first block in the program, the tool starts from rest at point PT.S and moves to point PT.A. Let the desired feed rate be 3000 mm/min. Since the distance from PT.S to PT.A is 100 mm, the inverse-time code for the feed rate is 300. Let the spindle speed be 2000 rpm. The magic-three code for 2000 rpm is 720. We will use a preparatory function code, g 08, for exponential acceleration from rest to the desired feed rate. The first block of the program can now be written as follows:

BLOCK 1 (FROM PT.S TO PT.A)

n	0001	Block 1
g	08	Automatic acceleration
x	0.0	No displacement along the x axis
y	100	A displacement of 100 mm along the y axis
f	300	Feed rate in inverse-time code
s	720	Spindle speed in magic-three code
m	03	Clockwise spindle rotation
(EB)		End of block

Similarly, the remaining blocks for moving the tool center point along the dotted line of Fig. 15 can be derived as follows. Note that the feed rate and spindle speed need not be programmed for each block if there is no change.

BLOCK 2 (FROM PT.A TO PT.B)

n	0002	Block 2
x	406	Displacement of 406 mm along the x axis
y	0.0	No displacement along the y axis
m	08	Turn coolant on
(EB)		End of block

BLOCK 3 (FROM PT.B TO PT.C)

n	0003	Block 3
x	0.0	No displacement along the x axis
y	6	Displacement of 6 mm along the y axis
(EB)		End of block

BLOCK 4 (FROM PT.C TO PT.D)

n	0004	Block 4
g	03	Circular interpolation, counterclockwise
x	106	Displacement of −106 mm along the x axis
y	106	Displacement of 106 mm along the y axis
i	300	The x coordinate of the center of the circle
j	106	The x coordinate of the center of the circle
(EB)		End of block

BLOCK 5 (FROM PT.D TO PT.E)

n	0005	Block 5
x	206	Displacement of −206 mm along the x axis
y	0.0	No displacement along the x axis
(EB)		End of block

BLOCK 6 (FROM PT.E TO PT.F)

n	0006	Block 6
g	09	Automatic deceleration
x	0.0	No displacement along the x axis
y	112	Displacement of −112 mm along the y axis
m	30	Turn off spindle and coolant
(EB)		End of block

These six blocks are entered on a sheet called the *process sheet,* as shown in Fig. 16. The data from the process sheet are punched on a paper tape and input to the NC machine. Notice that each line on the process sheet represents one block of the program. Manual programming as illustrated above can be both tedious and error prone for an inexperienced programmer. In the next section, we illustrate the use of a computer to enter the part program.

Computer-Assisted Programming: Programming in APT

APT was designed to be the common programming language standard for all numerical controllers.[24] APT programs are machine independent. An APT program is a series of English-like statements with a precise set of grammatical rules. Hundreds of keywords embody the huge expanse of NC knowledge into one language. An APT program typically contains process- or part-oriented information. The program does not contain control- and machine-tool-oriented information. APT provides three-dimensional programming for up to five axes.

An APT program containing a description of the part geometry and tool motions is input to a computer program called the *postprocessor.* The postprocessor checks for errors, adds machine-tool-specific control information, transforms the geometric description of the part into tool motion statements, and produces the proper codes for running the machine tool. Each

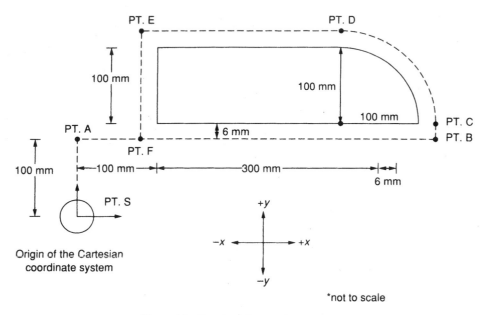

Figure 15   Drawing of a part for machining.

N	G	X	Y	Z	I	J	K	F	S	T	M	COMMENTS
PART NUMBER Sample						TAPE NUMBER 2003A						DATE 6/1/88
PART NAME Example						MACHINE Allen-Bradley 8200						PROGRAMMER S. Chand
001	08	0.0	100					300	720		03	From PT. S to PT. A
002		406	0.0								08	From PT. A to PT. B
003		0.0	6									From PT. B to PT. C
004	0.3	-106	106		300	106						From PT. C to PT. D
005		-206	0.0									From PT. D to PT. E
006	0.9	0.0	-112								30	From PT. E to PT. F

Figure 16   Process sheet for manual programming.

special machine has its own postprocessor. Most statements in APT are composed of two parts separated by a slash. The word to the left of the slash is called the major word or the keyword and the one to the right is called the minor word or the modifier. For example, COOLNT/OFF is an APT statement for turning the coolant off. Comments can be inserted anywhere in an APT program following the keyword REMARK.

An APT program is generated in two steps. The first step is to program the geometric description of the part. The second step is to program the sequence of operations that defines the motion of the tool with respect to the part. The programmer defines the geometry of the part in terms of a few basic geometric shapes such as straight lines and circles. The geometric description comprises a sequence of connected shapes that defines the geometry of the part. No matter how complex a part is, it can always be described in terms of a few basic geometric shapes.

Step 1: Programming Part Geometry in APT. The programmer programs the geometry of a part in terms of points, lines, circles, planes, cylinders, ellipses, hyperbolas, cones, and spheres. APT provides 12 ways of defining a line; 2 of the 12 definitions are (1) a line defined by the intersection of two planes and (2) a line defined by two points. There are 10 ways of defining a circle—for instance, by the coordinates of the center and the radius or by the center and a line to which the circle is tangent. The first part of an APT program typically contains the definitions of points, lines, circles, and other curves on the part. For the part shown in Fig. 15, an APT program may start as shown in Fig. 17. Note that the line numbers in the leftmost column are for reference only and are not a part of the APT program.

Line numbers 3 and 4 define the inside and outside tolerances in millimeters.[12,14] Line 5 defines the tool diameter as 12 mm. Lines 6 and 7 define the start point and point PT1. Lines 8–12 define the lines and the circular arc in the part geometry. Note that a temporary line, LIN2, is introduced to guide the tool to the corner at point PT2 (Fig. 17). Line 13 turns the spindle on with a clockwise rotation. Line 14 specifies the feed rate, and line 15 turns the coolant on.

```
1     PARTNO A345, Revision 2

2     REMARK Part machined in 3/4 inch Aluminum

3     INTOL/0.00005

4     OUTTOL/0.00005

5     CUTTER/12

6     STPT   =   POINT/0,0,0

7     PT1    =   POINT/0,106,0

8     LIN1   =   LINE/(POINT/0, 106, 0), (POINT/400, 106,
      0)

9     CIRC1  =   CIRCLE/CENTER, (POINT/300, 106, 0),
      RADIUS, 100

10    LIN2   =   LINE/(POINT/400, 206, 0), LEFT, TANTO,
      CIRC1

11    LIN3   =   LINE/(POINT/400, 206, 0), (POINT/100, 206, 0)

12    LIN4   =   LINE/(POINT/100, 206, 0), (POINT/100, 106, 0)

13    SPINDL/ON, CLW

14    FEDRAT/300

15    COOLNT/ON
```

Figure 17 APT programming of part for machining.

Step 2: Programming the Tool Motion Statements in APT. In programming the motion of the tool, the programmer usually starts at a point on the part that is closest to the origin. The programmer then traverses the geometry of the part as viewed from the tool, indicating directions of turn such as GOLFT and GORGT. Sometimes the analogy of the programmer "riding" or "straddling" the tool is used to get the sense of direction around the periphery of the part.

An initial sense of direction for the tool is established by the statement INDIRP ("in the direction of a point"). For the part of Fig. 17, let us assume that the tool starts from the point STPT and moves to point PT1. A sense of direction is established by the following statements:

```
FROM/STPT
INDIRP/PT1
```

The statement GO is used in startup of the motion. The termination of the motion is controlled by three statements: TO, ON, and PAST. The difference between the statements GO/TO LIN1, GO/ON LIN1, and GO/PAST LIN1 is illustrated in Fig. 18.

For the example of Fig. 17, the first GO statement is as follows:

```
GO/TO, LIN1
```

The next statement moves the tool to PT2. The sense of direction is set by the previous motion from STPT to PT1. The tool must turn right at PT1 to traverse to PT2. Also, the tool must remain to the right of the line LIN1 as seen from the tool in the direction of motion between PT1 and PT2. This move statement is written as follows:

```
TLRGT, GORGT/LIN1, PAST, LIN2
```

The next step is to move the tool from PT2 to PT3 along the circle CIRC1:

```
TLRGT, GOLFT/CIRC1, PAST, LIN3
```

The next two motion statements to complete the motion around the part are as follows:

```
GOFWD/LIN3, PAST, LIN4
TLRGT, GOLFT/LIN4, PAST, LIN1
```

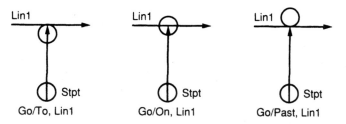

Figure 18   GO statement in APT.

The complete program is shown below:

```
PARTNO A345, Revision 2
REMARK Part machined in 3/4 inch Aluminum
INTOL/0.00005
OUTTOL/0.00005
CUTTER/12
STPT = POINT/0,0,0
PT1 = POINT/0,106,0
LIN1 = LINE/(POINT/0,106,0), (POINT/400, 106, 0)
CIRC1 = CIRCLE/CENTER, (POINT/300, 106, 0), RADIUS, 100
LIN2 = LINE/(POINT/400, 206, 0), LEFT, TANTO, CIRC1
LIN3 = LINE/(POINT/400, 206, 0), (POINT/100, 206, 0)
LIN4 = LINE/(POINT/100, 206, 0), (POINT/100, 106, 0)
SPINDL/ON, CLW
FEDRAT/300
COOLNT/ON
FROM/STPT
INDIRP/PT1
GO/TO, LIN1
TLRGT, GORGT/LIN1, PAST, LIN2
TLRGT, GOLFT/CIRC1, PAST, LIN3
GOFWD/LIN3, PAST, LIN4
TLRGT, GOLFT/LIN4, PAST, LIN1
STOP
FINI
```

3.5   Numerical Controllers and CAD/CAM

A computer-aided design (CAD) system is used for the creation, modification, and analysis of designs. A CAD system typically comprises a graphics display terminal, a computer, and several software packages. The software packages aid the designer in the creation, editing, and analysis of the design. For instance, in the analysis of a part design, the designer can access a library of routines for finite-element analysis, heat transfer study, and dynamic simulation of mechanisms.

Computer-aided manufacturing (CAM) refers to the use of a computer to plan, manage, and control the operations in a factory. CAM typically has two functions: (1) monitoring and control of the data on the factory floor and (2) process planning of operations on the factory floor.

Since the parts manufactured in a factory are usually designed on a CAD system, the design should be integrated with the programming of the machines on the factory floor. The process of integrating design and manufacturing is often referred to as CAD/CAM.

Today, CAD systems are commonly used to design the parts to be machined by NC. The NC programmer uses a CAD drawing of the part in the generation of the part program. Several numerical controllers provide an interactive graphic programming environment that displays the CAD drawing of the part on a graphics terminal. On the screen of the graphics terminal, a programmer can display the cross section of the part, rotate the part in three dimensions, and magnify the part. These operations help the programmer to better visualize the part in three dimensions; a similar visualization of the part from a two-dimensional blueprint is difficult.

In an interactive graphic programming environment, the programmer constructs the tool path from a CAD drawing. In many systems, the tool path is automatically generated by the system following an interactive session with the programmer. The output can be an APT program or a CLFILE, which can be postprocessed to generate the NC punched tape.[19]

There are two basic steps in an interactive graphics programming environment:

1. *Defining the Geometry of the Part.* The geometric definition of the part can be specified during the part design process in a CAD/CAM system. If the geometric definition does not exist, the part programmer must create it on the graphics terminal. This process is interactive, with the part programmer labeling the various edges and surfaces of the part on the graphics screen. After the labeling process is complete, the system automatically generates the APT geometry statements for the part. The definition of the part geometry is much more easily performed on the interactive graphic system than by the time-consuming, error-prone, step-by-step manual process.

2. *Generating the Tool Path.* The programmer starts by defining the starting position of the cutter. The cutter is graphically moved along the geometric surfaces of the part through an interactive environment. Most CAD/CAM systems have built-in software routines for many machining operations, such as surface contouring, profile milling around a part, and point-to-point motion. The programmer can enter feed rates and spindle speeds along each segment of the path. As the program is created, the programmer can visually verify the tool path with respect to the part surface on the graphics screen. The use of color graphics greatly aids in the visual identification of the part, tool, and program parameters.

The advantages of the CAD/CAM approach to NC programming are as follows[25]:

1. The geometric description of the part can be easily derived from the CAD drawing.

2. The programmer can better visualize the part geometry by manipulating the part drawing on the screen, such as rotating the part and viewing cross sections.

3. The interactive approach for the generation of part programs allows the user to visually verify a program on the graphics screen as it is being created.

4. The programmer has access to many machining routines on the CAD system to simplify tool path programming.

5. The CAD environment allows the integration of part design and tool design with programming.

4 ROBOT CONTROLLERS

The robot controller is a device that, via a pretaught software program, controls the operation of the manipulator and peripheral equipment of a robot. This section explains the composition of a typical robot system and the procedure for developing its working program.

4.1 Composition of a Robot System

As shown in Fig. 19, a basic industrial robot system is composed of four components: tools, manipulators, the controller, and the teach pendant. The tool is attached at the end point of the manipulator. The type of tool depends on the work to be done by the manipulator. The manipulator maintains the position and the orientation of the tool. The controller controls the manipulator's motion, tools, and peripherals. The teach pendant is the human–machine interface for the robot controller. The movement of the manipulator and the attached tools are programmed by using the teach pendant.

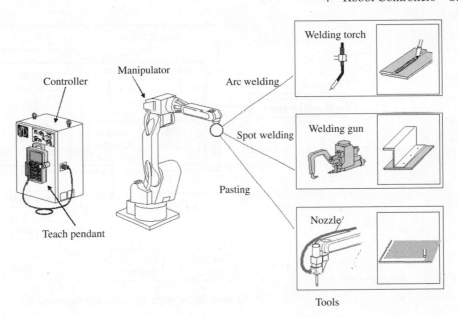

Figure 19 Basic industrial robot system.

Table 13 Relation between Work and Tool

Work	Tool	Peripheral Equipment	Main Controlled Items
Arc welding	Welding torch	Welding power supply, wire supply device	Welding voltage, welding current, arc on/off
Spot welding	Welding gun	Welding power supply	Welding condition, gun on/off, pressurizing power
Pasting/sealing	Nozzle	Paste (seal material) supply device	Gun on/off
Parts transportation	Gripper		Opening and closing
Assembly	Screw driver	Tool exchange device	Tool on /off

Tools

The general industrial robot, which is without a dexterous humanlike hand, works by installing a suitable tool for each job in the arm end. Some applications require the use of peripheral equipment, such as power supply or parts supply for the tools. The robot controller can control the operation of peripheral equipment along with the movement of the manipulator. This subject will be described at length later in this section. Table 13 shows the relation between work and the tool. For multiprocess jobs such as assembly tasks, an automatic tool exchange device is necessary to change tools for each individual process. The robot controller also controls the tool exchange device.

Manipulator

The role of the manipulator is to maintain the tool position and orientation. Most industrial robot arms now are very similar to a human arm in that they have a joint. The joint is driven by an actuator in the form of a servomechanism. Actuators include ac servomotors, dc servomotors,

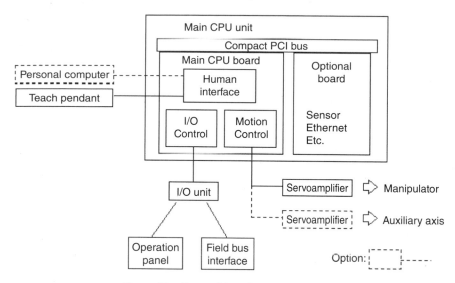

Figure 20 Composition of a robot controller.

and oil pressure motors, for example. In general, industrial robots use six actuators (thereby have six degrees of freedom), but there are some cases where an industrial robot will require less than six. The manipulator needs tools and transfer devices in order to hold and transport payloads.

Controller

As explained earlier, the controller is a device which, using preprogrammed (or "taught") control programs, controls the position and orientation of the manipulator and attached tools. This will be explained in more detail later. A general controller includes the functions to interpret and execute the control program, I/O functions (digital I/O and analog I/O) to control the tools and peripheral equipment, and functions to control servodrives for each actuator.

As shown in Fig. 20, the configuration of a robot controller follows the general scheme of all GPCDs illustrated in Fig. 3, with specialized interfaces as required for the robot control and programming functions. Of particular note is the increasing use of Internet or local-area network (LAN) communication interfaces on the controller, because the robot is frequently working simultaneously with many robots in the factory and also because of the use of remote operation and remote maintenance.

Teach Pendant

The teach pendant is the controller's human–machine interface for making the robot's control program. The screen on which the status of the manipulator and the control program are displayed and the button to operate the manipulator and to teach the position of each joint into the memory of the controller are arranged on the pendant.

4.2 Control Program

The control program is programmed using the teach pendant to make the robot work as described above. The control program is composed of commands, such as those to drive the controller itself or the manipulator. The program saved in the memory of the controller controls the manipulator and executes the software processing, such as external I/O processing or communication processing, by a set of output signals.

Just as with a computer program, very difficult tasks are made possible by control programs that can be changed through either a change in commands or the creation of subroutines via a series of commands. The basic programming method is to teach the position and the posture of the tool according to the work to be done. For jobs such as arc welding, pasting, and sealing, it is necessary to teach the trajectory of the movement of the tool. In such a case, the controller has a function (interpolation function) to automatically generate any trajectory while the control program is executed by teaching only the starting point and ending point for a straight line and/or the starting point, ending point, and intermediate point for any curve, instead of teaching the entire trajectory directly. Through this function the position and orientation of the tool can be moved. When teaching the trajectory, the speed at which the tool will be moved through the trajectory is very important; therefore, the speed of the tool is taught along with the position and orientation. For example, consider the simple arc welding of two parts shown in Fig. 21. The tool is an arc-welding torch, and the peripheral equipment is a welding power supply.

A series of jobs begin with the positioning of the torch at the start point. Next, the arc-welding power supply is turned on. To get the best value from the welding, the torch end matches the parts and welds them together while controlling the welding voltage and the current. The arc-welding power supply is turned off when reaching the position where the welding is finished. To create this program, the manipulator is moved while the welding torch position and orientation are changed by the teach pendant. The teaching is done as follows:

1. Standby position (P1)
2. Welding start point vicinity (P2)
3. Welding start point (P3)
4. Welding end point (P4)
5. Welding end-point vicinity (P5)
6. Standby position (P6)

Two teaching points, the start point (P3) and the end point (P4), are necessary for the arc welding. However, supplementary teaching points should be included in order to assure the manipulator does not interfere with the change in the work object or to assure that there is not interference between the work object and the tool. In an actual work situation, it is necessary to teach more supplementary points according to the work environment. The result of the teaching is shown in Table 14 as commands of a robot language. In step 4, the straight-line trajectory is generated automatically, at 1200 mm/min. In an actual situation commands are added using the teach pendant.

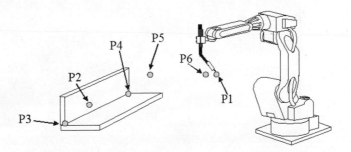

Figure 21 Programming an arc-welding robot.

Table 14 Initial Robot Program

Step commands
1: MOVEP1
2: MOVEP2
3: MOVEP3
4: MOVE LINEAR P4 SPEED = 1200mm/ min
5: MOVE P5
6: MOVE P6

Table 15 Robot Program with Peripheral Equipment Control

Step commands
01: MOVE P1
02: MOVE P2
03: MOVE P3
04: OUT ANALOG PORT = 0 VALUE =< some value >
05: OUTANALOG PORT = 1VALUE =< some value >
06: OUT DIGITAL PORT = 0 VALUE = ON
07: MOVE LINEAR P4 SPEED = 600mm/ sec
08: OUT DIGITAL PORT = 0 VALUE = OFF
09: MOVE P5
10: MOVE P6

Figure 22 Example screen of an off-line programming system.

After the motion of the manipulator is taught, the control of the peripheral equipment is taught. In this case, when moving from P3 to P4, it is necessary to control the arc-welding power supply. Table 15 shows the robot language program.

In the example above, a robot language was used to explain the procedure for creating a robot control program. The teaching method of actually moving the manipulator with the teach pendant is called *online programming*. On the other hand, the method by which the robot is not actually used in making the program but via computer graphics and a robot simulation is called *off-line programming*.

An off-line program can make the control program when an actual manipulator cannot be used, such as the production line design period, and is effective, for example, in the examination of the line arrangement. Online programming and off-line programming are the same as the basic instruction commands shown previously. Figure 22 shows an example screen of a PC-based off-line programming system.

ACKNOWLEDGMENTS

The authors gratefully acknowledge the contributions of Steven Tourangeau of Rockwell Automation to Section 3 as well as the contributions to the previous edition of this chapter by the late Odo J. Struger of Rockwell Automation.

REFERENCES

1. *GRAFCET Specification Language for Sequential Function Charts*, 2nd ed., IEC 60848, International Electrotechnical Commission, Geneva, 2002.
2. *Programmable Controllers—Part 3: Programming Languages*, 2nd ed., IEC 61131-3, International Electrotechnical Commission, Geneva, 2003.
3. *Batch Control—Part 1: Models and Terminology*, 1st ed., IEC 61512-1, International Electrotechnical Commission, Geneva, 1997.
4. *Function Blocks for Industrial-Process Measurement and Control Systems—Part 1: Architecture*, 1st ed., IEC 61499-1, International Electrotechnical Commission, Geneva, 2004.
5. B. P. Douglass, *Real-Time UML*, Addison-Wesley, Reading MA, 1998.
6. *Programmable Controllers—Part 1: General Information*, 2nd ed., IEC 61131-1, International Electrotechnical Commission, Geneva, 2003.
7. *Programmable Controllers—Part 2: Equipment Requirements and Tests*, 2nd ed., IEC 61131-2, International Electrotechnical Commission, Geneva, 2004.
8. *Programmable Controllers—Part 4: User Guidelines*, 2nd ed., IEC 61131-4, International Electrotechnical Commission, Geneva, 2004.
9. *Programmable Controllers—Part 5: Communications*, 1st ed., IEC 61131-5, International Electro-technical Commission, Geneva, 2000.
10. *Programmable Controllers—Part 7: Fuzzy Control Programming*, 1st ed., IEC 61131-7, International Electrotechnical Commission, Geneva, 2000.
11. *Programmable Controllers—Part 8: Guidelines for the Application and Implementation of Programming Languages*, 2nd ed., IEC 61131-8, International Electrotechnical Commission, Geneva, 2003.
12. R. S. Pressman and J. E. Williams, *Numerical Control and Computer-Aided Manufacturing*, Wiley, New York, 1977.
13. W. Leslie, *Numerical Control Users Handbook*, McGraw-Hill, New York, 1970.
14. Y. Koren, *Computer Control of Manufacturing Systems*, McGraw-Hill, New York, 1983.
15. E. M. Trent, *Metal Cutting*, Butterworths, Woburn, MA, 1977.
16. G. Boothroyd, C. Poli, and L. E. Murch, *Automatic Assembly*, Marcel Dekker, New York, 1982.
17. A. Fitzgerald and C. Kingsley, *Electric Machinery*, 2nd ed., McGraw-Hill, New York, 1958.

18. J. Childs, *Numerical Control Part Programming*, Industrial Press, New York, 1973.

19. M. Groover and E. Zimmers, *CAD/CAM Computer-Aided Design and Manufacturing*, Prentice-Hall, Englewood Cliffs, NJ, 1984.

20. Y. Koren, A. Shani, and J. Ben-Uri, "Interpolator for a CNC System," *IEEE Trans. Comput.*, **C-25**(1), 32–37, 1976.

21. *Pulse Width Modulated Servo Drive*, GEK-36203, General Electric, March 1973.

22. J. Beckett and G. Mergler, "Analysis of an Incremental Digital Positioning Servosystem with Digital Rate Feedback," *ASME J. Dyn. Syst. Measur. Control*, 87, March 1965.

23. A. Roberts and R. Prentice, *Programming for Numerical Control Machines*, 2nd ed., McGraw-Hill, New York, 1978.

24. Illinois Institute of Technology Research Institute (IITRI), *APT Part Programming*, McGraw-Hill, New York, 1967.

25. D. Grossman, "Opportunities for Research on Numerical Control Machining," *Commun. ACM*, **29**(6), 515–522, 1986.

BIBLIOGRAPHY

T. A., Hughes, *Programmable Controllers*, 4th ed., ISA, Raleigh, NC, 2004.

R. Lewis, *Modelling Distributed Control Systems Using IEC 61499*, IEE, London, 2001.

R. W. Lewis, *Programming Industrial Control Systems Using IEC 1131-3, rev. ed.*, IEE, London, 1998.

CHAPTER 23

NEURAL NETWORKS IN FEEDBACK CONTROL SYSTEMS

K. G. Vamvoudakis
University of California
Santa Barbara, California

F.L. Lewis
The University of Texas at Arlington
Fort Worth, Texas

Shuzhi Sam Ge
University of Electronic Science and Technology of China
Chengdu, China
and
Engineering
National University of Singapore
Singapore

1   INTRODUCTION

Dynamical systems are ubiquitous in nature and include naturally occurring systems such as the cell and more complex biological organisms, the interactions of populations, and so on as well as man-made systems such as aircraft, satellites, and interacting global economies. A.N. Whitehead and L. von Bertalanffy[1] were among the first to provide a modern theory of systems at the beginning of the twentieth century. Systems are characterized as having outputs that can be measured, inputs that can be manipulated, and internal dynamics. *Feedback control* involves computing suitable control inputs, based on the difference between observed and desired behavior, for a dynamical system such that the observed behavior coincides with a desired behavior prescribed by the user. All biological systems are based on feedback for survival, with even the simplest of cells using chemical diffusion based on feedback to create a potential difference across the membrane to maintain its *homeostasis,* or required equilibrium condition for survival. Volterra was the first to show that feedback is responsible for the balance of two populations of fish in a pond, and Darwin showed that feedback over extended time periods provides the subtle pressures that cause the evolution of species.

There is a large and well-established body of design and analysis techniques for feedback control systems that has been responsible for successes in the industrial revolution, ship and aircraft design, and the space age. Design approaches include classical design methods for linear systems, multivariable control, nonlinear control, optimal control, robust control, H_∞ control, adaptive control, and others. Many systems one desires to control have unknown dynamics, modeling errors, and various sorts of disturbances, uncertainties, and noise. This, coupled with the increasing complexity of today's dynamical systems, creates a need for advanced control design techniques that overcome limitations on traditional feedback control techniques.

In recent years, there has been a great deal of effort to design feedback control systems that mimic the functions of living biological systems. There has been great interest recently in "universal model-free controllers" that do not need a mathematical model of the controlled plant but mimic the functions of biological processes to learn about the systems they are controlling online, so that performance improves automatically. Techniques include fuzzy logic control, which mimics linguistic and reasoning functions, and artificial neural networks, which are based on biological neuronal structures of interconnected nodes, as shown in Fig. 1. By now, the theory and applications of these nonlinear network structures in feedback control have been

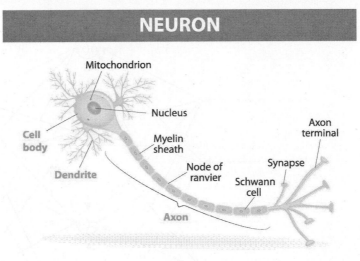

Figure 1   Nervous system cell. Copyright © Designua/Shutterstock.

well documented. It is generally understood that neural networks (NNs) provide an elegant extension of adaptive control techniques to nonlinearly parameterized learning systems.

This chapter shows how NNs fulfill the promise of providing *model-free learning controllers* for a class of nonlinear systems in the sense that a structural or parameterized model of the system dynamics is not needed. The control structures discussed in this chapter are *multiloop controllers* with NNs in some of the loops and an outer tracking unity-gain feedback loop. Throughout, there are repeatable design algorithms and guarantees of system performance including both small tracking errors and bounded NN weights. It is shown that as uncertainty about the controlled system increases or as one desires to consider human user inputs at higher levels of abstraction, the NN controllers acquire more and more structure, eventually acquiring a hierarchical structure that resembles some of the elegant architectures proposed by computer science engineers using high-level design approaches based on cognitive linguistics, reinforcement learning, psychological theories, adaptive critics, or optimal dynamic programming techniques.

Recent results show that approximate dynamic programming (ADP) with critic and actor neural networks,[2] allows the design of adaptive learning controllers that converge online to optimal control solutions,[3,4] while also guaranteeing closed-loop stability. ADP also offers design methods for differential games that can be implemented online in real time.[5] However, obtaining guaranteed performance using ADP requires rigorous design techniques based on mathematical properties of the controlled systems.

Many researchers have contributed to the development of a firm foundation for analysis and design of NNs in control system applications. See Section 8 on historical development and further study.

1.1   Background

Neural Networks

The multilayer NNs is modeled based on the structure of biological nervous systems (see Fig. 1), and provides a nonlinear mapping from an input space $\mathfrak{R}^n$ into an output space $\mathfrak{R}^m$. Its properties include function approximation, learning, generalization, classification, etc. It is known that the two-layer NN has sufficient generality for closed-loop control purposes. The two-layer neural network shown in Fig. 2 consists of two layers of weights and thresholds and

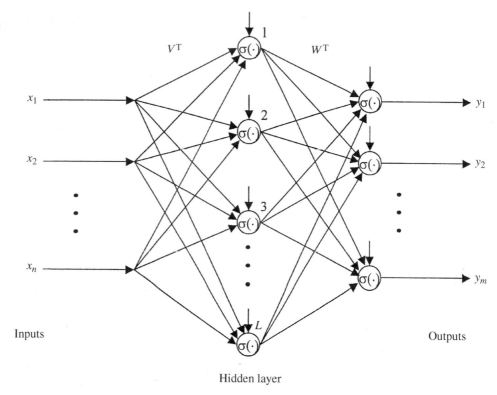

Figure 2 Two-layer neural network (NN).

has a hidden layer and an output layer. The input function $x(t)$ has n components, the hidden layer has L neurons, and the output layer has m neurons.

One may describe the NN mathematically as

$$y = W^T \sigma(V^T x)$$

where V is a matrix of first-layer weights and W is a matrix of second-layer weights. The second-layer thresholds are included as the first column of the matrix W^T by augmenting the vector activation function $\sigma(\cdot)$ by 1 in the first position. Similarly, the first-layer thresholds are included as the first column of matrix V^T by augmenting vector x by 1 in the first position.

The main property of NNs we are concerned with for control and estimation purposes is the *function approximation property*.[6,7] Let $f(x)$ be a smooth function from $\Re^n \to \Re^m$. Then, it can be shown that if the activation functions are suitably selected and x is restricted to a compact set $S \in \Re^n$, then for some sufficiently large number L of hidden-layer neurons, there exist weights and thresholds such that one has

$$f(x) = W^T \sigma(V^T x) + \varepsilon(x)$$

with $\varepsilon(x)$ suitably small. Here $\varepsilon(x)$ is called *the neural network functional approximation error*. In fact, for any choice of a positive number ε_N, one can find an NN of large enough size L such that $\varepsilon(x) \le \varepsilon_N$ for all $x \in S$.

Finding a suitable NN for approximation involves adjusting the parameters V and W to obtain a good fit to $f(x)$. Note that tuning of the weights includes tuning of the thresholds as well. The neural net is *nonlinear in the parameters V*, which makes adjustment of these parameters difficult and was initially one of the major hurdles to be overcome in closed-loop feedback control applications. If the first-layer weights V are fixed, then the NN is linear in the adjustable parameters W [linear in the parameters (LIP)]. It has been shown that, if the first-layer weights V are suitably fixed, then the approximation property can be satisfied by selecting only the output weights W for good approximation. For this to occur, $\sigma(V^T x)$ must provide a *basis*. It is not always straightforward to pick a basis $\sigma(V^T x)$. It has been shown that cerebellar model articulation controller (CMAC),[8] radial basis function (RBF),[9] fuzzy logic,[10] and other structured NN approaches allow one to choose a basis by suitably partitioning the compact set S. However, this can be tedious. If one selects the activation functions suitably, e.g., as sigmoids, then it was shown in Igelnik and Pao[11] that $\sigma(V^T x)$ is almost always a basis if V is selected randomly.

Neural Network Control Topologies

Feedback control involves the measurement of output signals from a dynamical system or *plant* and the use of the *difference* between the measured values and certain prescribed *desired values* to compute system inputs that cause the measured values to follow or *track* the desired values. In feedback control design it is crucial to guarantee by rigorous means both the tracking performance and the internal stability or boundedness of all variables. Failure to do so can cause serious problems in the closed-loop system, including instability and unboundedness of signals that can result in system failure or destruction.

The use of NNs in control systems was first proposed by Werbos[12] and Narendra and Parthasarathy[13] NN control has had two major thrusts: Approximate dynamic programming, which uses NNs to approximately solve the optimal control problem, and NNs in closed-loop feedback control. Many researchers have contributed to the development of these fields. See Section 8 at the end of this chapter.

Several NN feedback control topologies are illustrated in Fig. 3,[14] some of which are derived from standard topologies in adaptive control.[15] Solid lines denote control signal flow

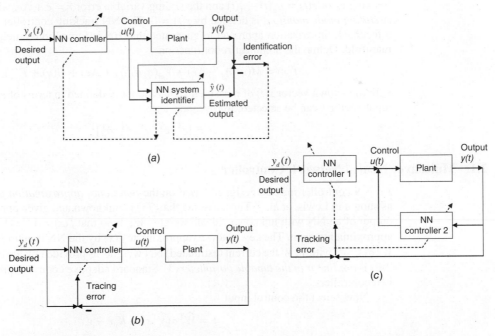

Figure 3   NN control topologies: (*a*) indirect scheme, (*b*) direct scheme, and (*c*) feedback/feedforward scheme.

loops, while dashed lines denote tuning loops. There are basically two sorts of feedback control topologies—indirect techniques and direct techniques. In *indirect* NN control there are two functions; in an identifier block, the NN is tuned to learn the dynamics of the unknown plant, and the controller block then uses this information to control the plant. *Direct* control is more efficient and involves directly tuning the parameters of an adjustable NN controller.

The challenge in using NNs for feedback control purposes is to select a suitable control system structure and then to demonstrate using mathematically acceptable techniques how the NN weights can be tuned so that closed-loop stability and performance are guaranteed. In this chapter, we shall show different methods of NN controller design that yield guaranteed performance for systems of different structure and complexity. Many researchers have participated in the development of the theoretical foundation for NN in control applications. See Section 8.

2   FEEDBACK LINEARIZATION DESIGN OF NEURAL NETWORK TRACKING CONTROLLERS

In this section, the objective is to design an NN feedback controller that causes a robotic system to follow, or track, a prescribed trajectory or path. The dynamics of the robot are unknown, and there are unknown disturbances. The dynamics of an *n*-link robot manipulator may be expressed as[16]

$$M(q)\ddot{q} + V_m(q, \dot{q})\dot{q} + G(q) + F(\dot{q}) + \tau_d = \tau \tag{1}$$

with $q(t) \in R^n$ the joint variable vector, $M(q)$ an inertia matrix, V_m a centripetal/coriolis matrix, $G(q)$ a gravity vector, and $F(\cdot)$ representing friction terms. Bounded unknown disturbances and modeling errors are denoted by τ_d and the control input torque is $\tau(t)$.

The sliding mode control approach of [Slotine and Coetsee[17] and Slotine and Li[18]] can be generalized to NN control systems. Given a desired arm trajectory $q_d(t) \in R^n$ define the tracking error $e(t) = q_d(t) - q(t)$ and the sliding variable error $r = \dot{e} + \Lambda e$, where $\Lambda = \Lambda^T > 0$. A *sliding mode manifold* is defined by $r(t) = 0$. The NN tracking controller is designed using a feedback linearization approach to guarantee that $r(t)$ is forced into a neighborhood of this manifold. Define the nonlinear robot function

$$f(x) = M(q)(\ddot{q}_d + \Lambda \dot{e}) + V_m(q, \dot{q})(\dot{q}_d + \Lambda e) + G(q) + F(\dot{q}) \tag{2}$$

with the known vector $x(t)$ of measured signals suitably defined in terms of $e(t), q_d(t)$. The NN input vector x can be selected, for instance, as

$$x = [e^T \quad \dot{e}^T \quad q_d^T \quad \dot{q}_d^T \quad \ddot{q}_d^T]^T \tag{3}$$

2.1   Multilayer Neural Network Controller

An NN controller may be designed based on the *functional approximation properties* of NNs as shown in Lewis, et al..[19] Thus, assume that $f(x)$ is unknown and given approximately as the output of an NN with unknown "ideal" weights W, V so that $f(x) = W^T\sigma(V^Tx) + \varepsilon$ with ε an approximation error. The key is now to approximate $f(x)$ by the NN functional estimate $\widehat{f}(x) = \widehat{W}^T\sigma(\widehat{V}^Tx)$, with $\widehat{V}, \widehat{W}$ the current (estimated) NN weights as provided by the tuning algorithms. This is *nonlinear in the tunable parameters* $\widehat{V}$. Standard adaptive control approaches only allow LIP controllers.

Now select the control input

$$\tau = \widehat{W}^T\sigma(\widehat{V}^Tx) + K_v r - v \tag{4}$$

with K_v a symmetric positive definite (PD) gain and $v(t)$ a certain robustifying function detailed in the cited reference. This NN control structure is shown in Fig. 4. The outer PD tracking

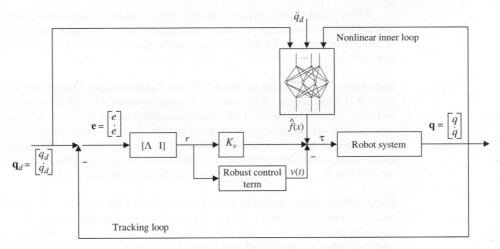

Figure 4 Neural network robot controller.

loop guarantees robust behavior. The inner loop containing the NN is known as a *feedback linearization loop*[20] and the NN effectively learns the unknown dynamics online to cancel the nonlinearities of the system.

Let the estimated sigmoid jacobian be $\widehat{\sigma}' \equiv d\sigma(z)/dz|_{z=\widehat{V}^{T}x}$. Note that this jacobian is *easily computed in terms of the current NN weights*. Then, the next result is representative of the sort of theorems that occur in NN feedback control design. It shows how to tune or train the NN weights to obtain guaranteed closed-loop stability.

Theorem (NN Weight Tuning for Stability). Let the desired trajectory $q_d(t)$ and its derivatives be bounded. Take the control input for (1) as (4). Let NN weight tuning be provided by

$$\dot{\widehat{W}} = F\widehat{\sigma}r^{T} - F\widehat{\sigma}'\widehat{V}^{T}xr^{T} - \kappa F\|r\|\widehat{W}$$

$$\dot{\widehat{V}} = Gx(\widehat{\sigma}'^{T}\widehat{W}r)^{T} - \kappa G\|r\|\widehat{V} \tag{5}$$

with any constant matrices $F = F^{T} > 0, G = G^{T} > 0$, and scalar tuning parameter $\kappa > 0$. Initialize the weight estimates as $\widehat{W} = 0$, $\widehat{V} = $ random. Then the sliding error $r(t)$ and NN weight estimates $\widehat{W}, \widehat{V}$ are uniformly ultimately bounded.

A proof of stability is always needed in control systems design to guarantee performance. Here, the stability is proven using nonlinear stability theory (e.g., an extension of Lyapunov's theorem). A Lyapunov energy function is defined as

$$L = \frac{1}{2}r^{T}M(q)r + \frac{1}{2}\mathrm{tr}\{\widetilde{W}^{T}F^{-1}\widetilde{W}\} + \frac{1}{2}\mathrm{tr}\{\widetilde{V}^{T}F^{-1}\widetilde{V}\}$$

where the weight estimation errors are $\widetilde{V} = V - \widehat{V}$, $\widetilde{W} = W - \widehat{W}$, with $\mathrm{tr}\{\cdot\}$ the trace operator so that the Frobenius norm of the weight errors is used. In the proof, it is shown that the Lyapunov function derivative is negative outside a compact set. This guarantees the boundedness of the sliding variable error $r(t)$ as well as the NN weights. Specific bounds on $r(t)$ and the NN weights are given in Lewis et al.[19] The first terms of (4) are very close to the (continuous-time) backpropagation algorithm.[21] The last terms correspond to Narendra's *e*-modification[22] extended to nonlinear-in-the-parameters adaptive control.

Robust adaptive tuning methods for nonlinear-in-the-parameters NN controllers have been derived based on the adaptive control approaches of *e*-modification, Ioannou's σ-modification, or projection methods. These techniques are compared by Ioannou and Sun[23] for standard adaptive control systems.

Robustness and Passivity of the NN When Tuned Online. Though the NN in Fig. 4 is static, since it is tuned online it becomes a dynamic system with its own internal states (e.g., the weights). It can be shown that the tuning algorithms given in the theorem make the NN *strictly passive* in a certain novel strong sense known as "state-strict passivity," so that the energy in the internal states is bounded above by the power delivered to the system. This makes the closed-loop system *robust* to bounded unknown disturbances. This strict passivity accounts for the fact that no persistence of excitation condition is needed.

Standard adaptive control approaches assume that the unknown function $f(x)$ is linear in the unknown parameters, and a certain regression matrix must be computed. By contrast, the NN design approach allows for nonlinearity in the parameters, and in effect the NN learns its own basis set online to approximate the unknown function $f(x)$. It is not required to find a regression matrix. This is a consequence of the NN universal approximation property.

2.2   Single-layer Neural Network Controller

If the first layer weights V are fixed so that $\widehat{f}(x) = \widehat{W}^{\mathrm{T}}\sigma(V^{\mathrm{T}}x) \equiv \widehat{W}^{\mathrm{T}}\phi(x)$, with $\varphi(x)$ selected as a basis, then one has the simplified tuning algorithm for the output-layer weights given by

$$\dot{\widehat{W}} = F\phi(x)r^{\mathrm{T}} - \kappa F\|r\|\widehat{W}$$

Then, the NN is linear-in-the-parameters and the tuning algorithms resemble those used in adaptive control. However, NN design still offers an advantage in that the NN provides a universal basis for a class of systems, while adaptive control requires one to find a regression matrix, which serves as a basis for each particular system.

2.3   Feedback Linearization of Nonlinear Systems Using Neural Networks

Many systems of interest in industrial, aerospace, and Department of Defense (DoD) applications are in the affine form $\dot{x} = f(x) + g(x)u + d$, with $d(t)$ a bounded unknown disturbance, and nonlinear functions $f(x)$ unknown, and $g(x)$ unknown but bounded below by a known positive value g_b. Using nonlinear stability proof techniques such as those above, one can design a control input of the form

$$u = \frac{-\widehat{f}(x) + v}{\widehat{g}(x)} + u_r \equiv u_c + u_r$$

that has two parts, a *feedback linearization* part $u_c(t)$, plus an *extra robustifying part* $u_r(t)$. Now, *two NNs* are required to manufacture the two estimates $\widehat{f}(x)$, $\widehat{g}(x)$ of the unknown functions. This controller is shown in Fig. 5. The weight updates for the $\widehat{f}(x)$ NN are given exactly as in (5). To tune the $\widehat{g}$ NN, a formula similar to (5) is needed, but it must be modified to ensure that the output $\widehat{g}(x)$ of the second NN is bounded away from zero, to keep the control $u(t)$ finite. It is called a controller singularity problem if $u(t)$ becomes infinity. More advanced control is possible using novel techniques. One good example is the use of integral Lyapunov functions in Ge et al.[24,25].

2.4   Partitioned Neural Networks and Input Preprocessing

In this section we show how NN controller implementation may be streamlined by partitioning the NN into several smaller subnets to obtain more efficient computation. Also discussed in this

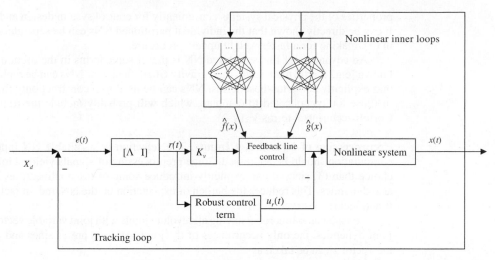

Figure 5   Feedback linearization NN controller.

section is preprocessing of input signals for the NNs to improve the efficiency and accuracy of the approximation.

Partitioned Neural Networks. A major advantage of the NN approach is that it allows one to partition the controller in terms of partitioned NNs or neural subnets. This (i) simplifies the design, (ii) gives added controller structure, and (iii) makes for faster weight tuning algorithms.

The unknown nonlinear robot function (2) can be written as

$$f(x) = M(q)\varsigma_1(x) + V_m(q, \dot{q})\varsigma_2(x) + G(q) + F(\dot{q})$$

with $\varsigma_1(x) = \ddot{q}_d + \Lambda e$, $\varsigma_2(x) = \dot{q}_d + \Lambda e$. Taking the four terms one at a time[26], one can use a small NN to approximate each term as depicted in Fig. 6. This procedure results in four neural subnets, which we term a *structured or partitioned NN*. This approach can also utilize the

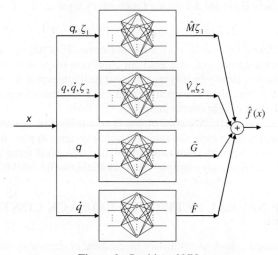

Figure 6   Partitioned NN.

properties of the physical systems conveniently for control system design and implementation. It can be directly shown that the individual partitioned NNs can be separately tuned exactly as in (5), making for a faster weight update procedure.

An advantage of this structured NN is that if some terms in the robot dynamics are well known [e.g., inertia matrix $M(q)$ and gravity $G(q)$], then their NNs can be replaced by equations that explicitly compute these terms. NNs can be used to reconstruct only the unknown terms or those too complicated to compute, which will probably include the friction $F(\dot{q})$ and the Coriolis/centripetal terms $V_m(q, \dot{q})$.

Preprocessing of Neural Net Inputs. The selection of a suitable NN input vector $x(t)$ for computation should be addressed. Some preprocessing of signals yields a more advantageous choice than (3) since it can explicitly introduce some of the nonlinearities inherent to robot arm dynamics. This reduces the burden of expectation on the NN and, in fact, also reduces the functional reconstruction error.

Consider an n-link robot having all revolute joints with joint variable vector $q(t)$. In revolute joint dynamics, the only occurrences of the joint variables are as sines and cosines,[16] so that the vector x can be taken as

$$x = [\varsigma_1^T \ \varsigma_2^T \ (\cos q)^T \ (\sin q)^T \ \dot{q}^T \ sgn(q)^T]^T$$

where the signum function is needed in the friction terms.

3   NEURAL NETWORK CONTROL FOR DISCRETE-TIME SYSTEMS

Most feedback controllers today are implemented on digital computers. This requires the specification of control algorithms in *discrete-time* or digital form.[27] To design such controllers, one may consider the discrete-time dynamics $x(k + 1) = f(x(k)) + g(x(k))u(k)$, with functions $f(\cdot)$ and $g(\cdot)$ unknown. The digital NN controller derived in this situation still has the form of the feedback linearization controller shown in Fig. 4.

One can derive tuning algorithms, for a discrete-time NN controller with N layers, that guarantee system stability and robustness.[19] For the ith layer the weight updates are of the form

$$\widehat{W}_i(k + 1) = \widehat{W}_i(k) - \alpha_i \widehat{\phi}_i(k) \widehat{y}_i^T(k) - \Gamma \| I - \alpha_i \widehat{\varphi}_i(k) \widehat{\varphi}_i^T(k) \| \widehat{W}_i(k)$$

where $\widehat{\varphi}_i(k)$ are the output functions of layer i, $0 < \Gamma < 1$ is a design parameter, and

$$\widehat{y}_i(k) \equiv \widehat{W}_i^T(k) \widehat{\varphi}_i(k) + K_v r(k) \text{ for } i = 1, \ldots, N-1 \quad \text{and} \quad \widehat{y}_N(k) \equiv r(k + 1)$$

for last layer with $r(k)$ a filtered error. This tuning algorithm has two parts: The first two terms correspond to a gradient algorithm often used in the NN literature. The last term is a discrete-time robustifying term that guarantees that the NN weights remain bounded. The latter has been called a "forgetting term" in NN terminology and has been used to avoid the problem of "NN weight overtraining."

Recently, NN control has been successfully extended to systems in the strict-feedback form using a modified tuning law[28] to systems in pure-feedback form using states prediction,[29] to systems with unknown control direction gains using discrete Nussbaum gain[30] as well as to block triangular multiple-input, multiple-output (MIMO) discrete-time systems[31]

4   MULTILOOP NEURAL NETWORK FEEDBACK CONTROL STRUCTURES

Actual industrial or military mechanical systems may have *additional dynamical complications* such as vibratory modes, high-frequency electrical actuator dynamics, compliant couplings or gears, etc. Practical systems may also have *additional performance requirements* such as requirements to exert specific forces or torques as well as perform position trajectory following (e.g., robotic grinding or milling). In such cases, the NN in Fig. 4 still works if it is modified to include *additional inner feedback loops* to deal with the additional plant or performance complexities. Using Lyapunov energy-based techniques, it can be shown that, if each loop is state-strict passive, then the overall multiloop NN controller provides stability, performance, and bounded NN weights. Details appear in Ref. 19.

4.1 Backstepping Neurocontroller for Electrically Driven Robot

Many industrial systems have high-frequency dynamics in addition to the basic system dynamics being controlled. An example of such systems is the *n*-link rigid robot arm with motor electrical dynamics given by

$$M(q)\ddot{q} + V_m(q, \dot{q})\dot{q} + F(\dot{q}) + G(q) + \tau_d = K_T i$$

$$L\dot{i} + R(i, \dot{q}) + \tau_e = u_e$$

with $q(t) \in R^n$ the joint variable, $i(t) \in R^n$ the motor armature currents, $\tau_d(t)$ and $\tau_e(t)$ the mechanical and electrical disturbances, and motor terminal voltage vector $u_e(t) \in R^n$ the control input. This plant has unknown dynamics in both the robot subsystem and the motor subsystem.

The problem with designing a feedback controller for this system is that one desires to control the behavior of the robot joint vector $q(t)$; however, the available control inputs are the motor voltages $u_e(t)$, which only affect the motor torques. As a second-order effect, the torques affect the joint angles.

Backstepping NN Design. The NN tracking controller in Fig. 7 may be designed using the *backstepping* technique.[32] This controller has *two neural networks*, one (NN 1) to estimate the unknown robot dynamics and an additional NN in an inner feedback loop (NN 2) to estimate the unknown motor dynamics. This multiloop controller is typical of control systems designed using rigorous system-theoretic techniques. It can be shown that by selecting suitable weight tuning algorithms for both NN, one can guarantee closed-loop stability as well as tracking performance in spite of the additional high-frequency motor dynamics. Both NN loops are state-strict passive. Proofs are given in terms of a modified Lyapunov approach. The NN tuning algorithms are similar to the ones presented above, but with some extra terms.

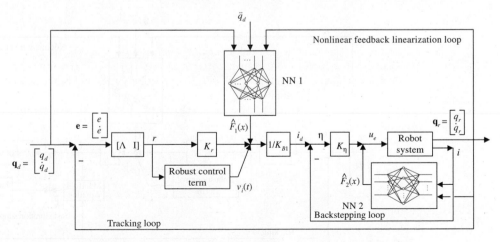

Figure 7 Backstepping NN controller for robot with motor dynamics.

In standard backstepping, one must find several regression matrices, which can be complicated. By contrast, NN backstepping design does not require regression matrices since the NNs provide a universal basis for the unknown functions encountered.

4.2 Compensation of Flexible Modes and High-Frequency Dynamics Using Neural Networks

Actual industrial or military mechanical systems may have additional dynamical complications such as vibratory modes, compliant couplings or gears, etc. Such systems are characterized by having more degrees of freedom than control inputs, which compounds the difficulty of designing feedback controllers with good performance. In such cases, the NN controller in Fig. 4 still works if it is modified to include additional inner feedback loops to deal with the additional plant complexities.

Using the Bernoulli–Euler equation, infinite series expansion, and the assumed mode shapes method, the dynamics of flexible-link robotic systems can be expressed in the form

$$\begin{bmatrix} M_{rr} & M_{rf} \\ M_{fr} & M_{ff} \end{bmatrix} \begin{bmatrix} \ddot{q}_r \\ \ddot{q}_f \end{bmatrix} + \begin{bmatrix} V_{rr} & V_{rf} \\ V_{fr} & V_{ff} \end{bmatrix} \begin{bmatrix} \dot{q}_r \\ \dot{q}_f \end{bmatrix} + \begin{bmatrix} 0 & 0 \\ o & K_{ff} \end{bmatrix} \begin{bmatrix} q_r \\ q_f \end{bmatrix} + \begin{bmatrix} F_r \\ 0 \end{bmatrix} + \begin{bmatrix} G_r \\ 0 \end{bmatrix} = \begin{bmatrix} B_r \\ B_f \end{bmatrix} \tau$$

where $q_r(t)$ is the vector of rigid variables (e.g., joint angles), $q_f(t)$ the vector of flexible mode amplitudes, M an inertia matrix, V a coriolis/centripetal matrix, and matrix partitioning is represented according to subscript r, for the rigid modes, and subscript f, for the flexible modes. Friction F and gravity G apply only for the rigid modes. Stiffness matrix K_{ff} describes the vibratory frequencies of the flexible modes.

The problem in controlling such systems is that the input matrix $B = [B_r^T \ B_f^T]^T$ is not square but has more rows than columns. This means that while one is attempting to control the rigid modes variable $q_r(t)$, one is also affecting $q_f(t)$. This causes undesirable vibrations. Moreover, the zero dynamics of such systems is non-minimum phase, which results in unstable flexible modes if care is not taken in choosing a suitable controller.

Singular Perturbations NN Design. To overcome this problem, an additional *inner feedback loop* based on singular perturbation theory[33] may be designed. The resulting multiloop controller is shown in Fig. 8, where an NN compensates for friction, unknown nonlinearities, and

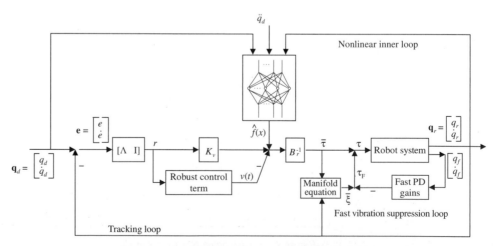

Figure 8 NN controller for flexible-link robotic system.

gravity, and the inner loop manages the flexible modes. The internal dynamics controller in the inner loop may be designed using a variety of techniques including H_∞ robust control and linear quadratic Gaussian/loop transfer recovery (LQG/LTR). Such controllers are capable of compensating for the effects of inexactly known or changing flexible mode frequencies. An observer can be used to avoid strain rate measurements.

In many industrial or aerospace designs, flexibility effects are limited by restricting the speed of motion of the system. This limits performance. By contrast, using the singular perturbations NN controller, *a flexible system can far outperform a rigid system in terms of speed of response*. The key is to use the flexibility effects to speed up the response in much the same manner as the cracking of a whip. That is, the flexibility effects of advanced structures are not merely a debility that must be overcome, but they offer the possibility of *improved performance* over rigid structures, if they are suitably controlled. By exploiting recent advances in materials, such as piezoelectric materials, further improved performance is attainable for the so-called smart materials flexible robots.[25]

4.3 Force Control with Neural Nets

Many practical robot applications require the control of the force exerted by the manipulator normal to a surface along with position control in the plane of the surface. This is the case in milling and grinding, surface finishing, etc. In applications such as microelectromechanical systems (MEMS) assembly, where highly nonlinear forces including van der Waals, surface tension, and electrostatics dominate gravity, advanced control schemes such as NNs are especially required.

In such cases, the NN force/position controller in Fig. 9 can be derived using rigorous Lyapunov-based techniques. It has guaranteed performance in that both the position tracking error $r(t)$ and the force error $\tilde{\lambda}(t)$ are kept small while all the NN weights are kept bounded. The figure has an *additional inner force control loop*. The control input is now given by

$$\tau(t) = \hat{W}^\mathrm{T} \sigma(\hat{V}^\mathrm{T} x) + K_v(Lr) - J^\mathrm{T}(\lambda_d - K_f \tilde{\lambda}) - v$$

where the selection matrix L and Jacobian J are computed based on the decomposition of the joint variable $q(t)$ into two components — the component $q_1(t)$ (e.g., tangential to the given surface) in which position tracking is desired and the component $q_2(t)$ (e.g., normal to the surface)

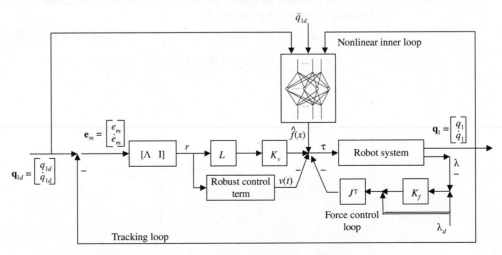

Figure 9 NN force/position controller.

in which force exertion is desired. This is achieved using holonomic constraint techniques based on the prescribed surface that are standard in robotics (e.g., work by McClamroch and Wang[34] and others). The filtered position tracking error in $q_1(t)$ is $r(t)$, that is, $r(t) = q_{1d} - q_1$ with $q_{1d}(t)$ the desired trajectory in the plane of the surface. The desired force is described by $\lambda_d(t)$ and the force exertion error is captured in $\tilde{\lambda}(t) = \lambda(t) - \lambda_d(t)$ with $\lambda(t)$ describing the actual measured force exerted by the manipulator. The position tracking gain is K_v and the force tracking gain is K_f.

5   FEEDFORWARD CONTROL STRUCTURES FOR ACTUATOR COMPENSATION

Industrial, aerospace, DoD, and MEMS assembly systems have actuators that generally contain dead zone, backlash, and hysteresis. Since these actuator nonlinearities appear in the *feedforward* loop, the NN compensator must also appear in the feedforward loop. The design problem for neurocontrollers where the NN appears in the feedforward loop is significantly more complex than for feedback NN controllers. Details are given in Lewis et al.[35]

5.1   Feedforward Neurocontroller for Systems with Unknown Dead Zone

Most industrial and vehicle and aircraft actuators have dead zones. The dead-zone characteristic appears in Fig. 10 and causes motion control problems when the control signal takes on small values or passes through zero, since only values greater than a certain threshold can influence the system.

Feedforward controllers can offset the effects of dead zone if properly designed. It can be shown that an NN dead-zone compensator has the structure shown in Fig. 11. The NN compensator consists of *two* NNs. NN II is in the direct feedforward control loop, and NN I is not directly in the control loop but serves as an observer to estimate the (unmeasured) applied torque $\tau(t)$. The feedback stability and performance of the NN dead-zone compensator have been rigorously proven using nonlinear stability proof techniques.

The two NN were each selected as having one tunable layer, namely the output weights. The activation functions were set as a basis by selecting fixed random values for the first-layer weights.[11] To guarantee stability, the output weights of the inversion NN II and the estimator NN I should be tuned, respectively, as

$$\widehat{W}_i = T\sigma_i(U_i^{\mathrm{T}}w)r^{\mathrm{T}}\widehat{W}^{\mathrm{T}}\sigma'(U^{\mathrm{T}}u)U^{\mathrm{T}} - k_1 T\|r\|\widehat{W}_i - k_2 T\|r\|\,\|\widehat{W}_i\|\widehat{W}_i$$

$$\widehat{W} = -S\sigma'(U^{\mathrm{T}}u)U^{\mathrm{T}}\widehat{W}_i\sigma_i(U_i^{\mathrm{T}}w)r^{\mathrm{T}} - k_1 S\|r\|\widehat{W}$$

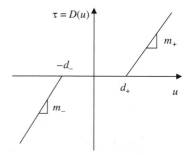

Figure 10   Dead-zone response characteristic.

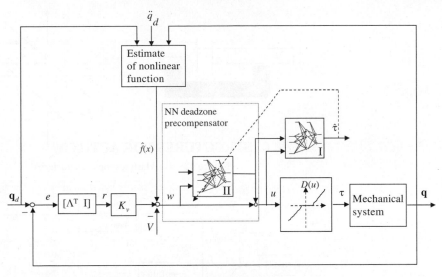

Figure 11   Feedforward NN for dead-zone compensation.

where subscript i denotes weights and sigmoids of the inversion NN II and nonsubscripted variables correspond to NN I. Note that $\sigma\prime$ denotes the Jacobian. Design parameters are the positive definite matrices T and S, and tuning gains k_1, k_2. The form of these tuning laws is intriguing. They form a coupled nonlinear system with each NN helping to tune itself and the other NN. Moreover, signals *are backpropagated through NN I* to tune NN II. That is, the two NNs function as *a single NN with two layers*, first NN II then NN I, but with the second layer not in the direct control path. Note the additional terms, which are a combination of Narendra's e-modification and Ioannou's σ-modification.

Reinforcement Learning Structure. NN I is not in the control path but serves as a *higher level critic* for tuning NN II, the action generating net. The critic NN I actually functions to provide an estimate of the torque supplied to the system in the absence of deadlock, which is a *target torque*. It is intriguing that this use of NN in the feedforward loop (as opposed to the feedback loop) requires such a *reinforcement learning* structure. Reinforcement learning techniques generally have the critic NN outside the main feedback loop, on a higher level of the control hierarchy.

5.2   Dynamic Inversion Neurocontroller for Systems with Backlash

Backlash is a common form of problem in actuators with gearing. The backlash characteristic is shown in Fig. 12 and causes motion control problems when the control signal reverses in direction, often due to dead space between gear teeth.

Dynamic inversion is a popular controller design technique in aircraft control and elsewhere.[36] Dynamic inversion by NN has been used by Calise and co-workers[37] in aircraft control using NN. Using dynamic inversion, an NN controller for systems with backlash is designed in Ref. 38. The neurocontroller appears in the feedforward loop as in Fig. 13 and is a *dynamic or recurrent NN*. In this neurocontroller, a desired torque $\tau_{\text{des}}(t)$ to be applied is determined, then, using a backstepping type of approach,[32] the neurocontroller structure shown in Fig. 13 is derived. An NN is used to approximate certain nonlinear functions appearing in the derivation. Unlike backstepping, dynamic inversion lets the required derivative appear explicitly in

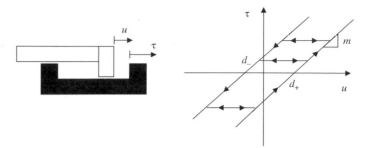

Figure 12 Backlash response characteristic.

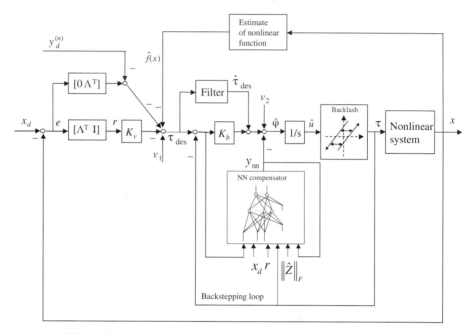

Figure 13 Dynamic inversion NN compensator for system with backlash.

the controller. In the design, a filtered derivative $\xi(t)$ is used to allow implementation in actual systems.

The NN precompensator shown in Fig. 13 effectively adds control energy to invert the dynamical backlash function. The control input into the backlash element is given by

$$\hat{u}(t) = K_b\tilde{\tau} + \xi - y_{nn} + v_2$$

where $\tilde{\tau}(t) = \tau_{\text{des}}(t) - \tau(t)$ is the torque error, $y_{nn}(t)$ is the NN output, and $v_2(t)$ is a certain robust control term detailed in Ref. 35. Weight tuning algorithms given there guarantee closed-loop stability and effective backlash compensation.

5.3 Adaptive Neural Control for Nonlinear Systems with Hysteresis

The most common approach for the control of nonlinear systems preceded by unknow hysteresis nonlinearities is to construct an inverse operator, which, however, has its limits due to the complexity of the hysteresis characteristics. Recently, adaptive neural control strategies have been developed to achieve the stable output tracking performance and mitigation of the effects of hysteresis without constructing the hysteresis inverse, for the classic Prandtl–Ishlinskii (PI) hysteresis.[39]

5.4 Adaptive Neural Control for Nonlinear Systems with Input Constraints

In Chen, et al.[40] adaptive tracking control was proposed for a class of uncertain multi-input and multi-output nonlinear systems with non-symmetric input constraints.

Neural Network Observers for Output Feedback Control

Thus far, we have described NN controllers in the case of full state feedback, where all internal system information is available for feedback. However, in actual industrial and commercial systems, there are usually available only certain restricted measurements of the plant. In this output feedback case one may use an *additional dynamic NN* with its own internal dynamics in the controller. The function of this additional NN is effectively to provide estimates of the unmeasurable plant states, so that the dynamic NN functions as an *observer* in control system theory.

The issues of observer design using NN can be appreciated using the case of rigid robotic systems.[16] For these systems, the dynamics can be written in state-variable form as

$$\dot{x}_1 = x_2$$
$$\dot{x}_2 = M^{-1}(x_1)[-N(x_1, x_2) + \tau]$$

where $x_1 \equiv q$, $x_2 \equiv \dot{q}$ and the nonlinear function $N(x_1, x_2) = V_m(x_1, x_2)x_2 + G(x_1) + F(x_2)$ is assumed to be unknown. It can be shown[41] that the following dynamic NN observer can provide estimates of the entire state $x = [x_1^T \ x_2^T]^T \equiv [q^T \ \dot{q}^T]^T$ given measurements of only $x_1(t) = q(t)$:

$$\dot{\hat{x}}_1 = \hat{x}_2 + k_D x_1$$
$$\dot{\hat{z}}_2 = M^{-1}(x_1)[-\hat{W}_o^T \sigma_o(\hat{x}) + k_P \tilde{x}_1 + \tau]$$
$$\hat{x}_2 = \hat{z}_2 + k_{P2} \tilde{x}_1$$

In this system, the hat denotes estimates and the tilde denotes estimation errors. It is assumed that the inertia matrix $M(q)$ is known, but all other nonlinearities are estimated by the observer NN $\hat{W}_o^T \sigma_o(\hat{x})$, which has output layer weights $\hat{W}_o$ and activation functions $\sigma_o(\cdot)$. Signal $v_o(t)$ is a certain observer robustifying term, and the observer gains k_P, k_D, k_{P2} are positive design constants detailed in Refs. 19 and 41.

The NN output feedback tracking controller shown in Fig. 14 uses the dynamic NN observer to reconstruct the missing measurements $x_2(t) = \dot{q}(t)$, and then employs a second static NN for tracking control, exactly as in Fig. 4. Note that the outer tracking PD loop structure has been retained but an additional dynamic NN loop is needed. In the references, weight tuning algorithms that guarantee stability are given for both the dynamic estimator NN and the static control NN.

6 APPROXIMATE DYNAMIC PROGRAMMING AND ADAPTIVE CRITICS

Approximate dynamic programming is based on the optimal formulation of the feedback control problem. For discrete-time systems, the optimal control problem may be solved using dynamic programming,[42] which is a backwards-in-time procedure and so unsuitable for online implementation. ADP is based on using nonlinear approximators to solve the Hamilton–Jacobi

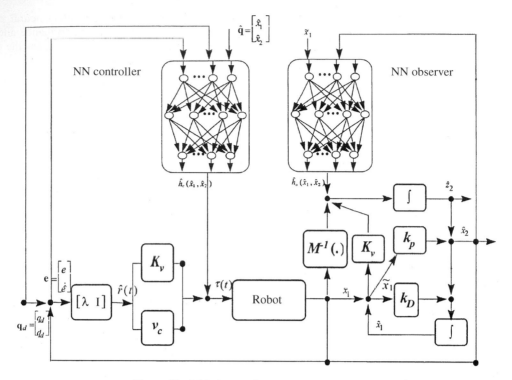

Figure 14   NN observer for output feedback control.

(HJ) equations forward in time and was first suggested by Werbos.[43] See Section 8 for ADP for cited work of major researchers. The current status of work in ADP is given in Si et al.[44]

ADP is a form of reinforcement learning based on an actor/critic structure. Reinforcement learning (RL) is a class of methods used in machine learning to methodically modify the actions of an agent based on observed responses from its environment.[45–48] The RL methods have been developed starting from learning mechanisms observed in mammals. Every decision-making organism interacts with its environment and uses those interactions to improve its own actions in order to maximize the positive effect of its limited available resources; this in turn leads to better survival chances. RL is a means of *learning optimal behaviors by observing the response from the environment to non-optimal control policies*. In engineering terms, RL refers to the learning approach of an actor or agent that modifies its actions, or control policies, based on stimuli received in response to its interaction with its environment. This learning can be extended along two dimensions: (i) nature of interaction (competitive or collaborative) and (ii) the number of decision makers (single or multiagent). The actor/critic structures are RL systems that have two learning structures: A critic network evaluates the performance of a current action policy, and based on that evaluation an actor structure updates the action policy.

6.1   Approximate Dynamic Programming for Discrete-Time Systems

For discrete-time (DT) systems of the form

$$x_{k+1} = f(x_k, u_k)$$

with k the time index, one may select the cost or performance measure

$$V(x_k) = \sum_{i=k}^{\infty} \gamma^{i-k} r(x_i, u_i)$$

with γ a discount factor and $r(x_k, u_k)$ known as the instantaneous utility. A first-difference equivalent to this yields a recursion for the value function given by the Bellman equation

$$V(x_k) = r(x_k, u_k) + \gamma V(x_{k+1}) \tag{6}$$

One may invoke Bellman's principle to find the optimal cost as

$$V^*(x_k) = \min_{u_k}(r(x_k, u_k) + \gamma V^*(x_{k+1})) \tag{7}$$

and the optimal control as

$$u * (x_k) = \arg \min_{u_k}(r(x_k, u_k) + \gamma V^*(x_{k+1}))$$

Determining the optimal controller using these equations requires an iterative procedure known as dynamic programming that progresses backward in time. This is unsuitable for real-time implementation and computationally complex.

The DT Hamiltonian function is

$$H(x_k, h(x_K), \Delta V_k) = r(x_k, h(x_k)) + \gamma V^h(x_{k+1}) - V^h(x_k) \tag{8}$$

where $\Delta V_k = \gamma V_h(x_{k+1}) - V_h(x_k)$ is the forward difference operator. The Hamiltonian function captures the energy content along the trajectories of a system.

The goal of ADP is to provide approximate techniques for evaluating the optimal value and optimal control using techniques that progress forward in time, so that they can be implemented in actual control systems. Howard[47] showed that the following successive iteration scheme, known as *policy iteration*, converges to the optimal solution:

1. *Policy Evaluation*. Find the value for the prescribed policy $u_j(x_k)$:

$$V_j(x_k) = r(x_k, u_j(x_k)) + \gamma V_j(x_{k+1})$$

2. *Policy Improvement*. Update the control action using

$$u_{j+1}(x_k) = \arg \min_{u_k}(r(x_k, u_k) + \gamma V_j(x_{k+1}))$$

This algorithm has an actor/critic structure, with the critic evaluating the policy in step 1 and the actor updating the policy in step 2. Werbos[2] and others (see Section 8) showed how to implement ADP controllers by four basic techniques, HDP, DHP, ADHDP, ADDHP, to be described next.

Heuristic Dynamic Programming (HDP). In HDP, one approximates the value by a *critic* neural network with tunable parameters w_j and the control by an *action generating* neural network[49] with tunable parameters v_j so that

$$\text{Critic NN} \quad V_j(x_k) \approx V(x_k, w_j)$$

$$\text{Action NN} \quad u_j(x_k) \approx u(x_k, v_j)$$

HDP then proceeds as follows:

Critic Update. Find the desired target value using

$$V^D_{k,j+1} = r(x_k, u_j(x_k)) + \gamma V(x_{k+1}, w_j)$$

Update critic weights using *recursive least squares (RLS)*, backprop, etc., e.g.,

$$w_{j+1} = w_j + \alpha_j \frac{\partial V}{\partial w_j}(V^D_{k,j+1} - V(x_k, w_j))$$

Action Update. Find the desired target action using

$$u^D{}_{k,j+1} = \arg\min_{u_k}(r(x_k, u_k) + \gamma V(x_{k+1}, w_{j+1}))$$

Update critic weights using RLS, backprop, etc., e.g.,

$$v_{j+1} = v_j + \alpha_k \frac{\partial u}{\partial v_j}(u^D{}_{k,j+1} - u(x_k, v_j))$$

This procedure is straightforward to implement given today's software, e.g., MATLAB. The value required for the next state x_{k+1} may be found either using the dynamics equation (1) or the next state can be observed from the actual system.

Dual Heuristic Programming (DHP). Noting that the control only depends on the value function gradient [e.g., see (9)], it is advantageous to approximate not the value but its gradient using an NN. This yields a more complex algorithm, but DHP converges faster than HDP. Details are in the Refs. 2, 12, 21, and 43.

Q Learning or Action-Dependent HDP (ADHDP). A function that is more advantageous than the value function for ADP is the Q function, defined by Watkins[50] and Werbos[12] as

$$Q(x_k, u_k) = r(x_k, u_k) + \gamma V(x_{k+1})$$

Note that Q is a function of both x_k and the control action u_k, and that

$$Q_h(x_k, h(x_k)) = V_h(x_k)$$

where subscript h denotes a prescribed control or policy sequence $u_k = h(x_k)$. A recursion for Q is given by

$$Q_h(x_k, u_k) = r(x_k, u_k) + \gamma Q_h(x_{k+1}, h(x_{k+1}))$$

In terms of Q, Bellman's principle is particularly easy to write, in fact, defining the optimal Q value as

$$Q^*(x_k, u_k) = r(x_k, u_k) + \gamma V^*(x_{k+1}))$$

one has the optimal value as

$$V^*(x_k) = \min_{u_k}(Q^*(x_k, u_k))$$

The optimal control policy is given by

$$h * (x_k) = \arg\min_{u_k}(Q^*(x_k, u_k))$$

Watkins[50] showed that the following successive iteration scheme, known as Q *learning*, converges to the optimal solution:

1. Find the Q value for the prescribed policy $h_j(x_k)$:

$$Q_j(x_k, u_k) = r(x_k, u_k) + \gamma Q_j(x_{k+1}, h_j(x_{k+1}))$$

2. Policy improvement

$$h_{j+1}(x_k) = \arg\min_{u_k}(Q_j(x_k, u_k))$$

Using NN to approximate the Q function and the policy, one can write down the ADHDP algorithm in a very straightforward manner. Since the control input action u_k is now explicitly an input to the critic NN, this is known as action-dependent HDP. Q learning converges faster than HDP and can be used in the case of unknown system dynamics.[51]

An action-dependent version of DHP (ADDHP) is also available wherein the gradients of the Q function are approximated using NNs. Note that two NNs are needed since there are two gradients, as Q is a function of both x_k and u_k.

6.2 Integral Reinforcement Learning for Optimal Adaptive Control of Continuous-Time Systems

Reinforcement learning is considerably more difficult for continuous-time (CT) systems than for discrete-time systems, and its development has lagged. See Abu-Khalaf et al.[52] for the development of a PI method for CT systems. Using a method known as *integral reinforcement learning* (IRL)[53] allows the application of RL to formulate online optimal adaptive control methods for CT systems.

Consider the CT nonlinear dynamical system

$$\dot{x} = f(x) + g(x)u \tag{9}$$

with state $x(t) \in R^n$ control input $u(t) \in R^m$, and the usual assumptions required for the existence of unique solutions and an equilibrium point at $x = 0$, e.g., $f(0) = 0$ and $f(x) + g(x)u$ Lipschitz on a set $\Omega \subseteq R^n$ that contains the origin. We assume the system is stabilizable on Ω, that is, there exists a continuous control function $u(t)$ such that the closed-loop system is asymptotically stable on Ω.

Define a performance measure or cost function that has the value associated with the feedback control policy $u = \mu(x)$ given by

$$V^\mu(x(t)) = \int_t^\infty r(x(\tau), u(\tau)) \, d\tau \tag{10}$$

with utility $r(x, u) = Q(x) + u^{\mathrm{T}} R u$, $Q(x)$ positive definite, that is, $Q(x) > 0$ for all x and $x = 0 \Rightarrow Q(x) = 0$, and $R > 0$ a positive definite matrix.

For the CT linear quadratic regulator (LQR) one has

$$\dot{x} = Ax + Bu \tag{11}$$

$$V^\mu(x(t)) = \frac{1}{2}\int_t^\infty (x^{\mathrm{T}} Q x + u^{\mathrm{T}} R u) \, d\tau \tag{12}$$

A policy is called *admissible* if it is continuous, stabilizes the system, and has a finite associated cost. If the cost is smooth, then an infinitesimal equivalent to (10) can be found by differentiation to be the nonlinear equation

$$0 = r(x, \mu(x)) + (\nabla V^\mu)^{\mathrm{T}}(f(x) + g(x)\mu(x)) \quad V^\mu(0) = 0 \tag{13}$$

where ∇V^μ (a column vector) denotes the gradient of the cost function V^μ with respect to x. This is the CT Bellman equation. It is defined based on the CT Hamiltonian function

$$H(x, \mu(x), \nabla V^\mu) = r(x, \mu(x)) + (\nabla V^\mu)^{\mathrm{T}}(f(x) + g(x)\mu(x)) \tag{14}$$

The optimal value satisfies the CT Hamilton–Jacobi–Bellman (HJB) equation

$$0 = \min_\mu H(x, \mu(x), \nabla V^*) \tag{15}$$

and the optimal control satisfies

$$\mu^* = \arg \min_\mu H(x, \mu(x), \nabla V^*) \tag{16}$$

We now see the problem with CT systems immediately. Compare the CT Bellman Hamiltonian (14) to the DT Hamiltonian (8). The former contains the full system dynamics $f(x) + g(x)u$, while the DT Hamiltonian does not. This means the CT Bellman equation (13) cannot be used as a basis for reinforcement learning unless the full dynamics are known.

Reinforcement learning methods based on (13) can be developed.[45,54–56] These have limited use for adaptive control purposes because the system dynamics must be known. In another approach, one can use Euler's method to discretize the CT Bellman equation.[54] Noting that

$$0 = r(x, \mu(x)) + (\nabla V^\mu)^{\mathrm{T}}(f(x) + g(x)\mu(x)) = r(x, \mu(x)) + \dot{V}^\mu \tag{17}$$

one uses Euler's method to discretize this to obtain

$$0 = r(x_k, u_k) + \frac{V^\mu(x_{k+1}) - V^\mu(x_k)}{T} \equiv \frac{r_S(x_k, u_k)}{T} + \frac{V^\mu(x_{k+1}) - V^\mu(x_k)}{T} \tag{18}$$

with sample period T so that $t = kT$. The discrete sampled utility is $r_S(x_k, u_k) = r(x_k, u_k)T$, where it is important to multiply the CT utility by the sample period.

Now note that the discretized CT Bellman equation (18) has the same form as the DT Bellman equation (6). Therefore, all the reinforcement learning methods just described for DT systems can be applied.

However, this is an approximation only. An alternative exact method for CT reinforcement learning was given by Vrabie et al.[53] This is termed *integral reinforcement learning (IRL)*. Note that one may write the cost (10) in the *integral reinforcement form*

$$V^\mu(x(t)) = \int_t^{t+T} r(x(\tau), u(\tau)) \ d\tau + V^\mu(x(t+T)) \tag{19}$$

for any $T > 0$. This is exactly in the form of the DT Bellman equation (6). According to Bellman's principle, the optimal value is given in terms of this construction as

$$V^*(x(t)) = \min_{\overline{u}(t:t+T)} \left(\int_t^{t+T} r(x(\tau), u(\tau)) \ d\tau + V^*(x(t+T)) \right)$$

where $\overline{u}(t:t+T) = \{u(\tau) : t \le \tau < t+T\}$. The optimal control is

$$\mu^*(x(t)) = \arg\min_{\overline{u}(t:t+T)} \left(\int_t^{t+T} r(x(\tau), u(\tau)) \ d\tau + V^*(x(t+T)) \right)$$

It is shown in Vrabie et al.[53] that the nonlinear equation (13) is exactly equivalent to the integral reinforcement form (19). That is, the positive definite solution of both that satisfies $V(0) = 0$ is the value (10) of the policy $u = \mu(x)$. Therefore, integral reinforcement form (19) also serves as a Bellman equation for CT systems and serves a fixed point equation. Thus, one can define the temporal difference error for CT systems as

$$e(t:t+T) = \int_t^{t+T} r(x(\tau), u(\tau)) \ d\tau + V^\mu(x(t+T)) - V^\mu(x(t)) \tag{20}$$

This does not involve the system dynamics.

Now, it is possible to formulate policy iteration and value iteration for CT systems. The following algorithms are termed *IRL* for CT systems.[53] They both give optimal adaptive controllers for CT systems, that is, adaptive control algorithms that converge to optimal control solutions.

IRL Optimal Adaptive Control Using Policy Iteration (PI).
Initialize. Select any admissible control policy $\mu_0(x)$. Do for $j = 0$ until convergence:
Policy Evaluation Step: Solve for $V_{j+1}(x(t))$ using

$$V_{j+1}(x(t)) = \int_t^{t+T} r(x(s), \mu_j(x(s))) \ ds + V_{j+1}(x(t+T)) \quad \text{with}$$

$$V_{j+1}(0) = 0 \tag{21}$$

Policy Improvement Step: Determine an improved policy using

$$\mu_{j+1} = \arg\min_u [H(x, u, \nabla V_{j+1})] \tag{22}$$

which explicitly is

$$\mu_{j+1}(x) = -\frac{1}{2} R^{-1} g^T(x) \nabla V_{j+1} \tag{23}$$

IRL Optimal Adaptive Control Using Value Iteration (VI).
Initialize. Select any control policy $\mu_0(x)$, not necessarily stabilizing. Do for $j = 0$ until convergence:
Policy Evaluation Step: Solve for $V_{j+1}(x(t))$ using

$$V_{j+1}(x(t)) = \int_t^{t+T} r(x(s), \mu_j(x(s))) \ ds + V_j(x(t+T)) \tag{24}$$

Policy Improvement Step: Determine an improved policy using (23).

Note that neither algorithm requires knowledge about the system drift dynamics function $f(x)$. That is, *they work for partially unknown systems.* Convergence of PI is proved in Vrabie et al.[53]

Online Implementation of IRL— A Hybrid Optimal Adaptive Controller. Both of these IRL algorithms may be implemented online by reinforcement learning techniques using value function approximation $V(x) = W^T \varphi(x)$ in a critic approximator network. Using value function approximation (VFA) in the PI algorithm (21) yields

$$W_{j+1}^T \lfloor \varphi(x(t)) - \varphi(x(t+T)) \rfloor = \int_t^{t+T} r(x(s), \mu_j(x(s))) \ ds \tag{25}$$

Using VFA in the VI algorithm (24) yields

$$W_{j+1}^T \varphi(x(t)) = \int_t^{t+T} r(x(s), \mu_j(x(s))) \ ds + W_j^T \varphi(x(t+T)) \tag{26}$$

RLS or batch LS can be used to update the value function parameters in these equations. On convergence of the value parameters, the action is updated using (23). The implementation is shown in Fig. 15. This is an optimal adaptive controller, that is, an adaptive controller that measures data along the system trajectories and converges to optimal control solutions. Note that only the system input coupling dynamics $g(x)$ is needed to implement these algorithms, since it appears in action update (23). The drift dynamics $f(x)$ is not needed.

The time is incremented at each iteration by the reinforcement learning time interval T. This time interval need not be the same at each iteration. T can be changed depending on how long it takes to get meaningful information from the observations. T is *not a sample period* in the standard meaning.

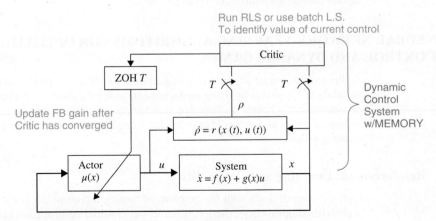

Figure 15 Hybrid optimal adaptive controller.

The measured data at each time increment is $(x(t), x(t + T), \rho(t : t + T))$ where

$$\rho(t : t + T) = \int_t^{t+T} r(x(\tau), u(\tau)) \ d\tau \tag{27}$$

is the *integral reinforcement* measured on each time interval. This can be implemented by introducing an integrator $\dot{\rho} = r(x(t), u(t))$ as shown in Fig. 15. That is, the integral reinforcement $\rho(t)$ is added as an extra CT state. It functions as the memory or controller dynamics. The remainder of the controller is a sampled data controller.

Note that the control policy $\mu(x)$ is updated periodically after the critic weights have converged to the solution of (25) or (26). Therefore, the policy is piecewise constant in time. On the other hand, the control varies continuously with the state between each policy update. It is seen that IRL for CT systems is in fact a hybrid CT/DT adaptive controller that converges to the optimal control solution in real-time without knowing the drift dynamics $f(x)$.

Due to the fact that the policy update (23) for CT systems does not involve the drift dynamics $f(x)$, no actor NN is needed in IRL. Only a critic NN is needed for VFA.

Online Solution of Algebraic Riccati Equation without Full Plant Dynamics.   It can be shown that the integral reinforcement form (19) is equivalent to the nonlinear Lyapunov (13).[53] Thus, the IRL controller solves the Lyapunov equation online without knowing the drift dynamics $f(x)$. Moreover, it converges to the optimal control so that it solves the HJB equation (15).

In the CT LQR case (11) and (12) one has linear state feedback control policies $u = -Kx$. Then, equation (13) is

$$(A - BK)^\mathrm{T} P + P(A - BK) + Q + K^\mathrm{T} RK = 0 \tag{28}$$

which is a Lyapunov equation. The HJB equation (15) becomes the CT alge braic Riccati equation (ARE)

$$A^\mathrm{T} P + PA + Q - PBR^{-1}B^\mathrm{T} P = 0 \tag{29}$$

Thus, IRL solves both the Lyapunov equation and the ARE online in real time, using data measured along the system trajectories, *without knowing the A matrix.*

For the CT LQR, (21) is equivalent to a Lyapunov equation at each step, so that policy iteration is exactly the same as Kleinman's algorithm[57] for solving the CT Riccati equation. This is a Newton method for finding the optimal value. CT value iteration, on the other hand, is a new algorithm that solves the CT ARE based on iterations on certain discrete-time Lyapunov equations that are equivalent to (24).

7   NEURAL NETWORK LEARNING ALGORITHMS FOR OPTIMAL CONTROL AND DYNAMIC GAMES

In this section, online adaptive learning algorithms are developed for optimal control and differential dynamic games by using measurements along the trajectory. These algorithms are based on actor/critic schemes and involve simultaneous tuning of the actor/critic NNs and provide online solutions to complex Hamilton–Jacobi equations, along with convergence and Lyapunov stability proofs.

7.1   Reinforcement Learning and Optimality

Optimal control deals with the problem of finding a control law for a given system such that a prescribed optimality criterion is achieved. Optimal control solutions can be derived using Pontryagin's minimum principle or by solving the HJB equation. Major drawbacks of these solution procedures are that they are offline a priori methods that require full knowledge of the system dynamics. On the other hand, adaptive control allows stabilizing controllers to be learned online in real time for systems with unknown dynamics. Adaptive optimal controllers have been proposed by adding optimality criteria to an adaptive controller, or adding adaptive characteristics to an optimal controller.

In the following sections, online adaptive learning algorithms are developed for optimal control and differential dynamic games by using measurements along the system trajectory. These algorithms are based on actor/critic reinforcement learning schemes. They involve simultaneous tuning of the actor/critic neural networks and provide approximate local smooth solutions to complex HJ equations without explicitly solving them, along with convergence and Lyapunov stability proofs.

Game theory[58] has been very successful in modeling strategic behavior, where the outcome for each player depends on the actions of himself and all the other players. Every player chooses a control to minimize independently from the others his *own* performance objective. None has knowledge of the others' strategy. A lot of applications of optimization theory require the solution of coupled HJ equations.[59,60] In games with N players, each player decides for the Nash equilibrium depending on HJ equations coupled through their quadratic terms.[60,61] Each dynamic game consists of three parts: (i) players, (ii) actions available for each player, and (iii) costs for every player that depend on their actions.

7.2 Optimal Control and the Continuous-Time Hamilton–Jacobi–Bellman Equation

Consider the nonlinear time-invariant affine in the input dynamical system given by

$$\dot{x}(t) = f(x(t)) + g(x(t))\, u(x(t)) \quad x(0) = x_0 \tag{30}$$

with state $x(t) \in \mathbb{R}^n, f(x(t)) \in \mathbb{R}^n, g(x(t)) \in \mathbb{R}^{n \times m}$ and control input $u(t) \in \mathbb{R}^m$. We assume that, $f(0) = 0, f(x) + g(x)u$ is Lipschitz continuous on a set $\Omega \subseteq \mathbb{R}^n$ that contains the origin, and that the system is stabilizable on Ω, i.e., there exists a continuous control function $u(t) \in U$ such that the system is asymptotically stable on Ω. The system dynamics $f(x), \; g(x)$ are assumed known.

Define the infinite horizon integral cost

$$V(x_0) = \int_0^\infty r(x(\tau), u(\tau)) \; d\tau \tag{31}$$

where $r(x, u) = Q(x) + u^T R u$ with $Q(x)$ positive definite, i.e., $\forall x \neq 0, Q(x) > 0$ and $x = 0 \Rightarrow Q(x) = 0$, and $R \in \mathbb{R}^{m \times m}$ a symmetric positive definite matrix.

Definition 1

(Admissible policy) A control policy $\mu(x)$ is defined as admissible with respect to (31) on Ω, denoted by $\mu \in \Psi(\Omega)$, if $\mu(x)$ is continuous on Ω, $\mu(0) = 0$, $u(x) = \mu(x)$ stabilizes (30) on Ω, and $V(x_0)$ is finite $\forall x_0 \in \Omega$.[62]

For any admissible control policy $\mu \in \Psi(\Omega)$, if the associated cost function

$$V^\mu(x_0) = \int_0^\infty r(x(\tau), \mu(x(\tau))) \; d\tau \tag{32}$$

is C^1, then an infinitesimal version of (32) is the Bellman equation

$$0 = r(x, \mu(x)) + (V_x^\mu)^T (f(x) + g(x)\mu(x)), \; V^\mu(0) = 0 \tag{33}$$

where V_x^μ denotes the partial derivative of the value function V^μ with respect to x. (Note that the value function does not depend explicitly on time.)

We define the gradient here as a column vector and use at times the alternative operator notation $\nabla \equiv \partial/\partial x$.

Equation (33) is a Lyapunov equation for nonlinear systems that, given a controller $\mu(x) \in \Psi(\Omega)$, can be solved for the value function $V^{\mu}(x)$ associated with it. Given that $\mu(x)$ is an admissible control policy, if $V^{\mu}(x)$ satisfies (33), with $r(x, \mu(x)) > 0$, then $V^{\mu}(x)$ is a Lyapunov function for the system (30) with control policy $\mu(x)$.

The optimal control problem can now be formulated: Given the CT system (30), the set $\mu \in \Psi(\Omega)$ of admissible control policies and the infinite horizon cost functional (31), find an admissible control policy such that the cost index (31) associated with the system (30) is minimized.

Defining the Hamiltonian of the problem

$$H(x, u, V_x) = r(x(t), u(t)) + V_x^{\mathrm{T}}(f(x(t)) + g(x(t))\mu(t)) \tag{34}$$

the optimal cost function $V^*(x)$ defined by

$$V^*(x_0) = \min_{\mu \in \Psi(\Omega)} \left(\int_0^\infty r(x(\tau), \mu(x(\tau))) \; d\tau \right)$$

with $x_0 = x$ is known as the *value function*, and satisfies the HJB equation

$$0 = \min_{\mu \in \Psi(\Omega)} [H(x, \mu, V_x^*)] \tag{35}$$

Assuming that the minimum on the right-hand side of (35) exists and is unique, then the optimal control function for the given problem is

$$\mu^*(x) = -1/2R^{-1}g^{\mathrm{T}}(x)V_x^*(x) \tag{36}$$

Inserting this optimal control policy in the nonlinear Lyapunov equation, we obtain the formulation of the HJB equation in terms of V_x^*:

$$0 = Q(x) + V_x^{*\mathrm{T}}(x)f(x) - 1/4V_x^{*\mathrm{T}}(x)g(x)R^{-1}g^{\mathrm{T}}(x)V_x^*(x)$$
$$V^*(0) = 0 \tag{37}$$

For the linear system case, considering a quadratic cost functional, the equivalent of this HJB equation is the well-known Riccati equation.

In order to find the optimal control solution for the problem, one only needs to solve the HJB equation (37) for the value function and then substitute the solution in (36) to obtain the optimal control. However, due to the nonlinear nature of the HJB equation, finding its solution is generally difficult or impossible.

Policy Iteration for Optimal Control. The approach of synchronous policy iteration is motivated by policy iteration (PI).[48] Therefore in this section we describe PI.

Policy iteration[48] is an iterative method of reinforcement learning for solving optimal control problems and consists of policy evaluation based on (33) and policy improvement based on (36). Specifically, the PI algorithm consists in solving iteratively the following two equations:

1. Given $\mu^{(i)}(x)$, solve for the value $V^{\mu^{(i)}}(x(t))$ using the Bellman equation:

$$0 = r(x, \mu^{(i)}(x)) + (\nabla V^{\mu^{(i)}})^{\mathrm{T}}(f(x) + g(x)\mu^{(i)}(x))$$
$$V^{\mu^{(i)}}(0) = 0 \tag{38}$$

2. Update the control policy using

$$\mu^{(i+1)} = \arg\min_{u \in \Psi(\Omega)}[H(x, u, \nabla V_x^{(i)})] \tag{39}$$

which explicitly is

$$\mu^{(i+1)}(x) = -\frac{1}{2}R^{-1}g^{\mathrm{T}}(x)\nabla V_x^{(i)} \qquad (40)$$

To ensure convergence of the PI algorithm, an initial admissible policy $\mu^{(0)}(x(t)) \in \Psi(\Omega)$ is required. It is in fact required by the desired completion of the first step in the policy iteration: i.e., finding a value associated with that initial policy (which needs to be admissible to have a finite value and for the nonlinear Lyapunov equation to have a solution). The algorithm then converges to the optimal control policy $\mu^* \in \Psi(\Omega)$ with corresponding cost $V^*(x)$. Proofs of convergence of the PI algorithm have been given in several references.[47,53–56,62–64]

Policy iteration is a Newton method. In the linear time-invariant case, it reduces to the Kleinman algorithm[57] for solution of the Riccati equation, a familiar algorithm in control systems. Then, (38) become a Lyapunov equation.

The PI algorithm, as other reinforcement learning algorithms, can be implemented on an actor/critic structure that consists of two NN structures to approximate the solutions of the two equations (38) and (39) at each step of the iteration.

Online Synchronous Policy Iteration. The critic NN is based on VFA. In the following, it is desired to determine a rigorously justifiable form for the critic NN. One desires approximation in Sobolev norm, that is, approximation of the value $V(x)$ as well as its gradient. It is justified to assume there exist weights W_1 such that the value function $V(x)$ is approximated as

$$V(x) = W_1^{\mathrm{T}}\varphi_1(x) + \varepsilon(x) \qquad (41)$$

Then $\varphi_1(x) : \mathbb{R}^n \to \mathbb{R}^N$ is called the NN activation function vector, N the number of neurons in the hidden layer, and $\varepsilon(x)$ the NN approximation error.

Assumption 1. The solution to (33) is smooth, i.e., $V(x) \in C^1(\Omega)$.

Assumption 2. The solution to (33) is positive definite. This is guaranteed for stabilizable dynamics if the performance functional satisfies zero-state observability,[65] which is guaranteed by the condition that $Q(x) > 0, x \in \Omega - \{0\}; Q(0) = 0$ be positive definite.

The weights of the critic NN, W_1, which provide the best approximate solution for (34), are unknown. Therefore, the output of the critic neural network is

$$\hat{V}(x) = \hat{W}_1^{\mathrm{T}}\varphi_1(x) \qquad (42)$$

where $\hat{W}_1$ are the current estimated values of the ideal critic NN weights W_1. Recall that $\varphi_1(x) : \mathbb{R}^n \to \mathbb{R}^N$ is the vector of activation functions, with N the number of neurons in the hidden layer. The approximate nonlinear Lyapunov equation is then

$$H(x, \hat{W}_1, u_2) = \hat{W}_1^{\mathrm{T}}\nabla\varphi_1(f + gu_2) + Q(x) + u_2^{\mathrm{T}}Ru_2 = e_1 \qquad (43)$$

where the actor NN is given as

$$u_2(x) = -\frac{1}{2}R^{-1}g^{\mathrm{T}}(x)\nabla\varphi_1^{\mathrm{T}}\hat{W}_2 \qquad (44)$$

Define the critic and the actor weight estimation errors as $\tilde{W}_1 = W_1 - \hat{W}_1, \tilde{W}_2 = W_1 - \hat{W}_2$. It is desired to select $\hat{W}_1$ to minimize the squared residual error from (43):

$$E_1 = \frac{1}{2}e_1^{\mathrm{T}}e_1$$

Definition 2

A time signal $\zeta(t)$ is said to be uniformly ultimately bounded (UUB) if there exists a compact set $S \subset \mathbb{R}^n$ so that for all $\zeta(0) \in S$ there exists a bound B and a time $T(B, \zeta(0))$ such that $\|\zeta(t)\| \leq B$ for all $t \geq t_0 + T$.[19]

Persistence of Excitation (PE) Assumption. Let the signal $\overline{\sigma}_{(\cdot)}$ be persistently exciting over the interval $[t, t + T]$, i.e., there exist constants $\beta_1 > 0$, $\beta_2 > 0$, $T > 0$ such that, for all t,

$$\beta_1 I \leq S_0 \equiv \int_t^{t+T} \overline{\sigma}_{(\cdot)}(\tau) \overline{\sigma}_{(\cdot)}{}^T(\tau) \; d\tau \leq \beta_2 I \tag{45}$$

We now present the main theorem, which provides the tuning laws for the actor and critic NNs that guarantee convergence of the synchronous online PI algorithm to the optimal controller, while guaranteeing closed-loop stability.

Theorem 1

Let the dynamics be given by (30), the critic NN be given by (42), and the control input be given by actor NN (44). Let tuning for the critic NN be provided by the normalized gradient descent law

$$\hat{W}_1 = -a_1 \frac{\sigma_2}{(\sigma_2^T \sigma_2 + 1)^2} \; [\sigma_2^T \hat{W}_1 + Q(x) + u_2^T R u_2] \tag{46}$$

where $\sigma_2 = \nabla \varphi_1(f + g u_2)$, and assume that $\overline{\sigma}_2 = \sigma_2/(\sigma_2^T \sigma_2 + 1)$ is persistently exciting.
Let the actor NN be tuned as

$$\hat{W}_2 = -a_2 \{ (F_2 \hat{W}_2 - F_1 \overline{\sigma}_2^T \hat{W}_1) - 1/4 \overline{D}_1(x) \hat{W}_2 m^T(x) \hat{W}_1 \} \tag{47}$$

where

$$\overline{D}_1(x) \equiv \nabla \varphi_1(x) g(x) R^{-1} g^T(x) \nabla \varphi_1{}^T(x) \quad m \equiv \frac{\sigma_2}{(\sigma_2^T \sigma_2 + 1)^2}$$

and $F_1 > 0$ and $F_2 > 0$ are tuning parameters. Let Assumptions 1 and 2 hold, and the tuning parameters be selected appropriately as stated in Vamvoudakis and Lewis.[4] Then there exists an N_0 such that, for the number of hidden layer units $N > N_0$ the closed-loop system state, the critic NN error $\tilde{W}_1$, and the actor NN error $\tilde{W}_2$ are UUB.

Proof:
It is shown in Vamvoudakis and Lewis[4] that under the hypotheses of Theorem 1 the value function approximation (41) approximately satisfies the HJB equation (37) so the optimal control problem is solved.

Simulations of Online Learning of Optimal Control. Here we present a simulation of a nonlinear system to show that the game can be solved *Online* by learning in real time, and we converge to the approximate local smooth solution of HJB without solving it. Consider the following affine in a control input nonlinear system, with a quadratic cost derived as in Nevistic and Primbs.[66]

$$\dot{x} = f(x) + g(x)u, \quad x \in R^2$$

where

$$f(x) = \begin{bmatrix} -x_1 + x_2 \\ -0.5x_1 - 0.5x_2 \left[1 - \left[\cos\left(2x_1\right) + 2 \right]^2 \right] \end{bmatrix}$$

$$g(x) = \begin{bmatrix} 0 \\ \cos\left(2x_1\right) + 2 \end{bmatrix}$$

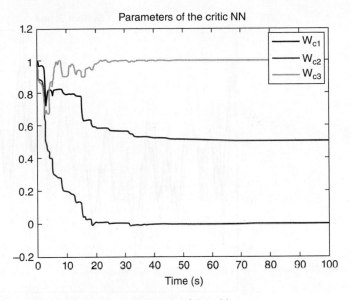

Figure 16   Convergence of the critic parameters.

One selects

$$Q = \begin{bmatrix} 1 & 0 \\ 0 & 1 \end{bmatrix} \qquad R = 1$$

Using the procedure in Nevistic and Primbs[66], the optimal value function is

$$V^*(x) = 1/2x_1^2 + x_2^2$$

and the optimal control signal is

$$u^*(x) = -(\cos(2x_1) + 2)x_2$$

One selects the critic NN vector activation function as

$$\varphi_1(x) = \begin{bmatrix} x_1^2 & x_1 x_2 & x_2^2 \end{bmatrix}^{\mathrm{T}}$$

Figure 16 shows the critic parameters, denoted by

$$\widehat{W}_1 = \begin{bmatrix} W_{c1} & W_{c2} & W_{c3} \end{bmatrix}^{\mathrm{T}}$$

These converge after about 80 s to the correct values of

$$\widehat{W}_1(t_f) = [0.5017 \quad -0.0020 \quad 1.0008]^{\mathrm{T}}$$

The actor parameters after 80 s converge to the values of

$$\widehat{W}_2(t_f) = [0.5017 \quad -0.0020 \quad 1.0008]^{\mathrm{T}}$$

So that the actor NN (44)

$$\widehat{u}_2(x) = -\frac{1}{2}R^{-1} \begin{bmatrix} 0 \\ \cos(2x_1) + 2 \end{bmatrix}^{\mathrm{T}} \begin{bmatrix} 2x_1 & 0 \\ x_2 & x_1 \\ 0 & 2x_2 \end{bmatrix}^{\mathrm{T}} \begin{bmatrix} 0.5017 \\ -0.0020 \\ 1.0008 \end{bmatrix}$$

also converged to the optimal control.

The evolution of the system states is presented in Fig. 17. One can see that after 80 s convergence of the NN weights in both critic and actor has occurred. This shows that the probing noise effectively guaranteed the PE condition. On convergence, the PE condition of the control

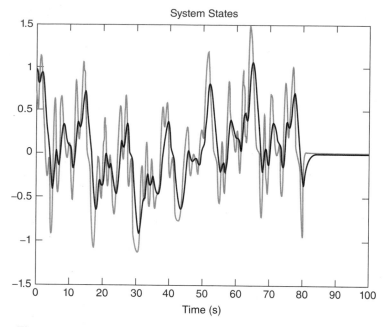

Figure 17   Evolution of the system states for the duration of the experiment.

signal is no longer needed, and the probing signal was turned off. After that, the states remain very close to zero, as required.

Figure 18 shows the three-dimensional (3D) plot of the difference between the approximated value function, by using the online algorithm, and the optimal one. This error is close to zero. Good approximation of the actual value function is being evolved.

7.3   Online Solution of Nonlinear Two-Player Zero-Sum Games and H_∞ Control Using Hamilton–Jacobi–Isaacs Equation

Many systems contain unknown disturbances, and the optimal control approach just given may not be effective.

In this case, one may use the H_∞ design procedure. Consider the nonlinear time-invariant affine in the input dynamical system in Fig. 19 given by

$$\dot{x} = f(x) + g(x)u(x) + k(x)d(x) \tag{48}$$

where state $x(t) \in \mathbb{R}^n$, $f(x(t)) \in \mathbb{R}^n$, $g(x(t)) \in \mathbb{R}^{n \times m}$, control $u(x(t)) \in \mathbb{R}^m$, $k(x(t)) \in \mathbb{R}^{n \times q}$ and disturbance $d(x(t)) \in \mathbb{R}^q$. Assume that $f(x)$ is locally Lipschitz, $f(0) = 0$ so that $x = 0$ is an equilibrium point of the system. Here we take full state feedback $y = x$ and desire to determine the action or control $u(t) = u(x(t))$ such that, under the worst disturbance, one has the L_2 gain bounded by a prescribed γ so that

$$\frac{\int_0^\infty \|z(t)\|^2 dt}{\int_0^\infty \|d(t)\|^2 dt} = \frac{\int_0^\infty (Q(x) + \|u\|^2)\ dt}{\int_0^\infty \|d(t)\|^2 dt} \leq \gamma^2$$

Approximation Error of the Value function

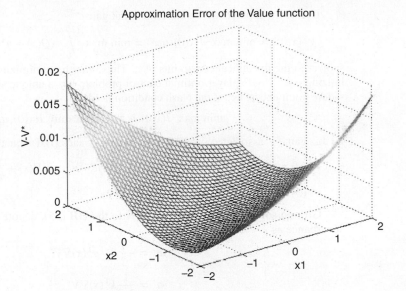

Figure 18 3D plot of the approximation error for the value function.

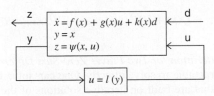

Figure 19 Bounded L_2 gain problem.

Define the performance index[42]

$$J(x(0), u, d) = \int_0^\infty (Q(x) + u^T R u - \gamma^2 \|d\|^2) \ dt \equiv \int_0^\infty r(x, u, d) \ dt \tag{49}$$

for $Q(x) \geq 0$, $R = R^T > 0$, $r(x, u, d) = Q(x) + u^T R u - \gamma^2 \|d\|^2$ and $\gamma \geq \gamma^* \geq 0$, where γ^* is the smallest γ for which the system is stabilized.[65] For feedback policies $u(x)$ and disturbance policies $d(x)$ define the value or cost of the policies as

$$V(x(t), u, d) = \int_t^\infty (Q(x) + u^T R u - \gamma^2 \|d\|^2) \ dt \tag{50}$$

When the value is finite, a differential equivalent to this is the Bellman equation

$$0 = r(x, u, d) + (\nabla V)^T (f(x) + g(x)u(x) + k(x)d(x)) \quad V(0) = 0 \tag{51}$$

where $\nabla V = \partial V / \partial x \in R^n$ is the (transposed) gradient and the Hamiltonian is

$$H(x, \nabla V, u, d) = r(x, u, d) + (\nabla V)^T (f(x) + g(x)u(x) + k(x)d) \tag{52}$$

For feedback policies,[67] a solution $V(x) \geq 0$ to (51) is the value (50) for given feedback policy $u(x)$ and disturbance policy $d(x)$.

Define the 2-player zero-sum differential game[67]

$$V^*(x(0)) = \min_u \max_d \; J(x(0), u, d) = \min_u \max_d \int_0^\infty (Q(x) + u^{\mathrm{T}} R u - \gamma^2 \|d\|^2) \; dt \qquad (53)$$

subject to the dynamical constraints (48). Thus, u is the minimizing player and d is the maximizing one. This 2-player optimal control problem has a unique solution if a game-theoretic saddle point exists, i.e., if the Nash condition holds:

$$\min_u \max_d \; J(x(0), u, d) = \max_d \min_u \; J(x(0), u, d) \qquad (54)$$

To this game is associated the Hamilton–Jacobi–Isaacs (HJI) equation:

$$0 = Q(x) + \nabla V^{\mathrm{T}}(x)f(x) - \frac{1}{4}\nabla V^{\mathrm{T}}(x)g(x)R^{-1}g^{\mathrm{T}}(x)\nabla V(x) + \frac{1}{4\gamma^2}\nabla V^{\mathrm{T}}(x)kk^{\mathrm{T}}\nabla V(x)$$

$$V(0) = 0 \qquad (55)$$

Given a solution $V^*(x) \geq 0 : \mathbb{R}^n \to \mathbb{R}$ to the HJI (55), denote the associated control and disturbance as

$$u^* = -\frac{1}{2}R^{-1}g^{\mathrm{T}}(x)\nabla V^* \qquad (56)$$

$$d^* = \frac{1}{2\gamma^2}k^{\mathrm{T}}(x)\nabla V^* \qquad (57)$$

and write

$$0 = H(x, \nabla V, u^*, d^*)$$

$$= Q(x) + \nabla V^{\mathrm{T}}(x)f(x) - \frac{1}{4}\nabla V^{\mathrm{T}}(x)g(x)R^{-1}g^{\mathrm{T}}(x)\nabla V(x) + \frac{1}{4\gamma^2}\nabla V^{\mathrm{T}}(x)kk^{\mathrm{T}}\nabla V(x) \qquad (58)$$

Policy Iteration for Two Player Zero-Sum Differential Games. The HJI equation (55) is usually intractable to solve directly. One can solve the HJI iteratively using one of several algorithms that are built on iterative solutions of the Lyapunov equation (51). Included are Feng et al.,[68] which uses an inner loop with iterations on the control, and Abu-Khalaf and Lewis,[69,70] Abu-Khalaf et al.,[71] and Van der Schaft,[65] which uses an inner loop with iterations on the disturbance. These are in effect extensions of Kleinman's algorithm[57] to nonlinear 2-player games. The complementarity of these algorithms is shown in Vrabie[65] and Vrabie, et al..[53] Here, we shall use the latter algorithm.[65,69–71].

The structure given in the algorithm below will be used as the basis for approximate online solution techniques in the next section.

Initialization: Start with a stabilizing feedback control policy u_0.

1. For $j = 0, 1, \ldots$ given u_j.

2. For $i = 0, 1, \ldots$ set $d^0 = 0$, solve for $V_j^i(x(t))$, d^{i+1} using

$$0 = Q(x) + \nabla V_j^{i\mathrm{T}}(x)(f + gu_j + kd^i) + u_j^{\mathrm{T}} R u_j - \gamma^2 \|d^i\|^2 \qquad (59)$$

$$d^{i+1} = \arg \max_d [H(x, \nabla V_j^i, u_j, d)] = \frac{1}{2\gamma^2}k^{\mathrm{T}}(x)\nabla V_j^i \qquad (60)$$

On convergence, set $V_{j+1}(x) = V_j^i(x)$.

3. Update the control policy using

$$u_{j+1} = \arg \min_u [H(x, \nabla V_{j+1}), u, d] = -\frac{1}{2}R^{-1}g^{\mathrm{T}}(x)\nabla V_{j+1} \qquad (61)$$

Go to 1.

In practice, the iterations in i and j are continued until some convergence criterion is met, e.g., $\|V_j^{i+1} - V_j^i\|$ or, respectively, $\|V_{j+1} - V_j\|$ is small enough in some suitable norm.

Online Solution for Two-player Zero-Sum Differential Games. The critic NN is based on VFA. Assume there exist NN weights W_1 such that the value function $V(x)$ is approximated as

$$V(x) = W_1{}^{\mathrm{T}}\varphi_1(x) + \varepsilon(x) \tag{62}$$

with $\varphi_1(x) : \mathbb{R}^n \to \mathbb{R}^N$ the NN activation function vector, N the number of neurons in the hidden layer, and $\varepsilon(x)$ the NN approximation error.

The ideal weights of the critic NN, W_1, which provide the best approximate solution for (58), are unknown. Therefore, the output of the critic neural network is

$$\hat{V}(x) = \hat{W}_1^{\mathrm{T}}\varphi_1(x) \tag{63}$$

where $\hat{W}_1$ are the current estimated values of W_1. The approximate nonlinear Lyapunov-like equation is then

$$H(x, \hat{W}_1, \hat{u}, \hat{d}) = \hat{W}_1^{\mathrm{T}}\nabla\varphi_1(f + g\hat{u} + k\hat{d}) + Q(x) + \hat{u}^{\mathrm{T}}R\hat{u} - \gamma^2\|\hat{d}\|^2 = e_1 \tag{64}$$

with the actor and the disturbance NN defined as

$$\hat{u}(x) = -\frac{1}{2}R^{-1}g^{\mathrm{T}}(x)\nabla\varphi_1{}^{\mathrm{T}}\hat{W}_2 \tag{65}$$

$$\hat{d}(x) = \frac{1}{2\gamma^2}k^{\mathrm{T}}(x)\nabla\varphi_1{}^{\mathrm{T}}\hat{W}_3 \tag{66}$$

and e_1 a residual equation error.

Define the critic, actor, and disturbance weight estimation errors

$$\tilde{W}_1 = W_1 - \hat{W}_1, \ \tilde{W}_2 = W_1 - \hat{W}_2, \ \tilde{W}_3 = W_1 - \hat{W}_3$$

where $\hat{W}_2, \hat{W}_3$ denotes the current estimated values of the ideal NN weights W_1.

It is desired to select $\hat{W}_1$ to minimize the squared residual error

$$E_1 = \frac{1}{2}e_1{}^{\mathrm{T}}e_1$$

Select the tuning law for the critic weights as the normalized gradient descent algorithm.

Theorem 2

Let the dynamics be given by (48), the critic NN be given by (63), the control input be given by actor NN (65), and the disturbance input be given by disturbance NN (66). Let tuning for the critic NN be provided by

$$\dot{\hat{W}}_1 = -a_1 \frac{\sigma_2}{(\sigma_2^{\mathrm{T}}\sigma_2 + 1)^2} \ [\sigma_2{}^{\mathrm{T}}\hat{W}_1 + Q(x) - \gamma^2\|\hat{d}\|^2 + \hat{u}^{\mathrm{T}}R\hat{u}] \tag{67}$$

where $\sigma_2 = \nabla\varphi_1(f + g\hat{u} + k\hat{d})$. Let the actor NN be tuned as

$$\dot{\hat{W}}_2 = -a_2\{(F_2\hat{W}_2 - F_1\overline{\sigma}_2^{\mathrm{T}}\hat{W}_1) - 1/4\overline{D}_1(x)\hat{W}_2 m^{\mathrm{T}}(x)\hat{W}_1\} \tag{68}$$

and the disturbance NN be tuned as

$$\dot{\hat{W}}_3 = -a_3\left\{\left(F_4\hat{W}_3 - F_3\overline{\sigma}_2^{\mathrm{T}}\hat{W}_1\right) + \frac{1}{4\gamma^2}\overline{E}_1(x)\hat{W}_3 m^{\mathrm{T}}\hat{W}_1\right\} \tag{69}$$

where $\overline{D}_1(x) \equiv \nabla\varphi_1(x)g(x)R^{-1}g^{\mathrm{T}}(x)\nabla\varphi_1{}^{\mathrm{T}}(x)$, $\overline{E}_1(x) \equiv \nabla\varphi_1(x)kk^{\mathrm{T}}\nabla\varphi_1{}^{\mathrm{T}}(x)$, $m \equiv \sigma_2/(\sigma_2{}^{\mathrm{T}}\sigma_2 + 1)^2$, and $F_1 > 0$, $F_2 > 0$, $F_3 > 0$, $F_4 > 0$ are tuning parameters. Let $Q(x) > 0$.

Suppose that $\bar{\sigma}_2 = \sigma_2/(\sigma_2^T\sigma_2 + 1)$ is persistently exciting. Let the tuning parameters F_1, F_2, F_3, F_4 in (68) and (69) be selected appropriately. Then there exists an N_0 such that, for the number of hidden layer units $N > N_0$ the closed-loop system state, the critic NN error $\widetilde{W}_1$, the actor NN error $\widetilde{W}_2$, and the disturbance NN error $\widetilde{W}_3$ are UUB.

Proof:
It is shown in Vamvoudakis and Lewis[5] that under the hypotheses of Theorem 2 the value function approximation (62) approximately satisfies the HJI equation (58) so the zero sum problem is solved.

Simulation of Online Learning of Zero-Sum Games. Here we present a simulation of a nonlinear system to show that the game can be solved *Online* by learning in real time, and we converge to the approximate local smooth solution of HJI without solving it. Consider the following affine in control input nonlinear system, with a quadratic cost constructed as in Nevistic and Primbs.[66]

$$\dot{x} = f(x) + g(x)u + k(x)d, \quad x \in \mathbb{R}^2$$

where

$$f(x) = \begin{bmatrix} -x_1 + x_2 \\ -x_1^3 - x_2^3 + 0.25x_2\left(\cos\left(2x_1\right) + 2\right)^2 - 0.25x_2\frac{1}{\gamma^2}(\sin(4x_1) + 2)^2 \end{bmatrix}$$

$$g(x) = \begin{bmatrix} 0 \\ \cos\left(2x_1\right) + 2 \end{bmatrix} \quad k(x) = \begin{bmatrix} 0 \\ \left(\sin\left(4x_1\right) + 2\right) \end{bmatrix}$$

One selects

$$Q = \begin{bmatrix} 1 & 0 \\ 0 & 1 \end{bmatrix} \quad R = 1, \quad \gamma = 8$$

Also $a_1 = a_2 = a_3 = 1$, $F_1 = I$, $F_2 = 10I$, $F_3 = I$, $F_4 = 10I$ where I is an identity matrix of appropriate dimensions.

The optimal value function is

$$V^*(x) = 1/4x_1^4 + 1/2x_2^2$$

the optimal control signal is

$$u^*(x) = -\frac{1}{2}[\cos(2x_1) + 2]x_2$$

and

$$d^*(x) = \frac{1}{2\gamma^2}[\sin(4x_1) + 2]x_2$$

One selects the critic NN vector activation function as

$$\phi_1(x) = [x_1^2 \quad x_2^2 \quad x_1^4 \quad x_2^4]$$

Figure 20 shows the critic parameters, denoted by

$$\widehat{W}_1 = \begin{bmatrix} W_{c1} & W_{c2} & W_{c3} & W_{c4} \end{bmatrix}^T$$

by using the synchronous zero-sum game algorithm. After convergence at about 50 s have

$$\widehat{W}_1(t_f) = [-0.0006 \quad 0.4981 \quad 0.2532 \quad 0.0000]^T$$

The actor and disturbance parameters after 80 s converge to the values of

$$\widehat{W}_3(t_f) = \widehat{W}_2(t_f) = \widehat{W}_1(t_f)$$

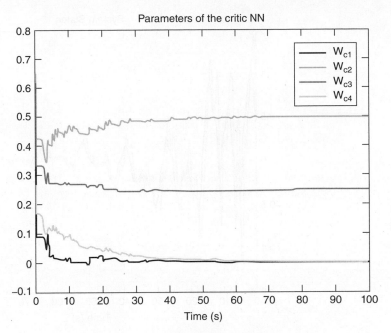

Figure 20   Convergence of the critic parameters.

So that the actor NN

$$\widehat{u}_2(x) = -1/2 R^{-1} \begin{bmatrix} 0 \\ \cos\left(2x_1\right) + 2 \end{bmatrix}^{\mathrm{T}} \begin{bmatrix} 2x_1 & 0 \\ 0 & 2x_2 \\ 4x_1^3 & 0 \\ 0 & 4x_2^3 \end{bmatrix}^{\mathrm{T}} \widehat{W}_2(t_f)$$

also converged to the optimal control, and the disturbance NN

$$\widehat{d}(x) = \frac{1}{2\gamma^2} \begin{bmatrix} 0 \\ \sin\left(4x_1\right) + 2 \end{bmatrix}^{\mathrm{T}} \begin{bmatrix} 2x_1 & 0 \\ 0 & 2x_2 \\ 4x_1^3 & 0 \\ 0 & 4x_2^3 \end{bmatrix}^{\mathrm{T}} \widehat{W}_3(t_f)$$

also converged to the optimal disturbance.

The evolution of the system states is presented in Figure 21.

7.4   Non Zero-Sum Games and Coupled Hamilton–Jacobi Equations

Consider the N-player nonlinear time-invariant differential game on an infinite time horizon

$$\dot{x} = f(x) + \sum_{j=1}^{N} g_j(x) u_j \tag{70}$$

where state $x(t) \in \mathbb{R}^n$, players or controls $u_j(t) \in \mathbb{R}^{m_j}$. Assume that $f(0) = 0$ and $f(x)$, $g_j(x)$ are locally Lipschitz.

The cost functionals associated with each player are

$$J_i(x(0), u_1, u_2, \dots u_N) = \int_0^\infty \left(Q_i(x) + \sum_{j=1}^{N} u_j^{\mathrm{T}} R_{ij} u_j \right) dt \equiv \int_0^\infty r_i(x(t), u_1, u_2, \dots u_N) \, dt \quad i \in N \tag{71}$$

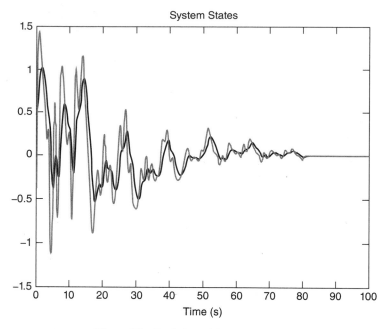

Figure 21 Evolution of the system states.

where function $Q_i(x) \geq 0$ is generally nonlinear, and $R_{ii} > 0, R_{ij} \geq 0$ are symmetric matrices.

We seek optimal controls among the set of feedback control policies with complete state information.

Given admissible feedback policies/strategies $u_i(t) = \mu_i(x)$, the value is

$$V_i(x(0), \mu_1, \mu_2, \ldots \mu_N) = \int_t^\infty \left(Q_i(x) + \sum_{j=1}^N \mu_j^T R_{ij} \mu_j\right) d\tau \equiv \int_t^\infty r_i(x(t), \mu_1, \mu_2, \ldots \mu_N) \, d\tau \quad i \in N$$

(72)

Define the N-player game

$$V_i^*(x(t), \mu_1, \mu_2, \ldots \mu_N) = \min_{\mu_i} \int_t^\infty \left(Q_i(x) + \sum_{j=1}^N \mu_j^T R_{ij} \mu_j\right) d\tau \quad i \in N$$

(73)

By assuming that all the players have the same hierarchical level, we focus on the so-called Nash equilibrium that is given by the following definition.

Definition 3

(Nash equilibrium strategies)[59] An *N-tuple* of strategies $\{\mu_1^*, \mu_2^*, \ldots, \mu_N^*\}$ with $\mu_i^* \in \Omega_i, \; i \in N$ is said to constitute a Nash equilibrium solution for an *N-player* finite game in extensive form, if the following N inequalities are satisfied for all $\mu_i^* \in \Omega_i, \; i \in N$:

$$J_i^* \equiv J_i(\mu_1^*, \mu_2^*, \mu_i^*, \ldots, \mu_N^*) \leq J_i(\mu_1^*, \mu_2^*, \mu_i, \ldots, \mu_N^*), \quad i \in N$$

(74)

The *N*-tuple of quantities $\{J_1^*, J_2^*, \ldots, J_N^*\}$ is known as a Nash equilibrium outcome of the *N*-player game.

Differential equivalents to each value function are given by the following Bellman equations (different form from those given in Basar and Olsder[59]):

$$0 = r(x, u_1, \ldots, u_N) + (\nabla V_i)^T \left(f(x) + \sum_{j=1}^{N} g_j(x)u_j \right) \quad V_i(0) = 0 \quad i \in N \qquad (75)$$

where $\nabla V_i = \partial V_i / \partial x \in \mathbb{R}^{n_i}$ is the gradient vector (e.g., transposed gradient). Then, suitable nonnegative definite solutions to (75) are the values evaluated using the infinite integral (73) along the system trajectories. Define the Hamiltonian functions

$$H_i(x, \nabla V_i, u_1, \ldots, u_N) = r(x, u_1, \ldots, u_N) + (\nabla V_i)^T \left(f(x) + \sum_{j=1}^{N} g_j(x)u_j \right) \quad i \in N \quad (76)$$

According to the stationarity conditions, associated feedback control policies are given by

$$\frac{\partial H_i}{\partial u_i} = 0 \Rightarrow \mu_i(x) = -\frac{1}{2} R_{ii}^{-1} g_i^T(x) \nabla V_i \quad i \in N \qquad (77)$$

Substituting (77) into (76) one obtains the N-coupled Hamilton–Jacobi (HJ) equations

$$0 = \frac{(\nabla V_i)^T \left(f(x) - \frac{1}{2} \sum_{j=1}^{N} g_j(x) R_{jj}^{-1} g_j^T(x) \nabla V_j \right)}{+Q_i(x) + \frac{1}{4} \sum_{j=1}^{N} \nabla V_j^T g_j(x) R_{jj}^{-T} R_{ij} R_{jj}^{-1} g_j^T(x) \nabla V_j,} \quad V_i(0) = 0 \qquad (78)$$

These coupled HJ equations are in "closed-loop" form. The equivalent "open-loop" form is

$$0 = \frac{\nabla V_i^T f(x) + Q_i(x) - \frac{1}{2} \nabla V_i^T \sum_{j=1}^{N} g_j(x) R_{jj}^{-1} g_j^T(x) \nabla V_j}{+\frac{1}{4} \sum_{j=1}^{N} \nabla V_j^T g_j(x) R_{jj}^{-T} R_{ij} R_{jj}^{-1} g_j^T(x) \nabla V_j} \quad V_i(0) = 0 \qquad (79)$$

In linear systems of the form $\dot{x} = Ax + \sum_{j=1}^{N} B_j u_j$, (78) becomes the N-coupled generalized algebraic Riccati equations:

$$0 = P_i A_c + A_c^T P_i + Q_i + \frac{1}{4} \sum_{j=1}^{N} P_j B_j R_{jj}^{-T} R_{ij} R_{jj}^{-1} B_j^T P_j \quad i \in N \qquad (80)$$

where $A_c = A - \frac{1}{2} \sum_{i=1}^{N} B_i R_{ii}^{-1} B_i^T P_i$. It is shown in Basar and Olsder[59] that if there exist solutions to (80) and further satisfying the conditions that for each $i \in N$ the pair

$$A - \frac{1}{2} \sum_{\substack{j \in N \\ j \neq i}} B_j R_{jj}^{-1} B_j^T P_j \quad B_i$$

is stabilizable and the pair

$$A - \frac{1}{2} \sum_{\substack{j \in N \\ j \neq i}} B_j R_{jj}^{-1} B_j^T P_j, Q_i + \frac{1}{4} \sum_{\substack{j \in N \\ j \neq i}} P_j B_j R_{jj}^{-T} R_{ij} R_{jj}^{-1} B_j^T P_j$$

is detectable, then the N-tuple of the stationary feedback policies $\mu_i^*(x) = -K_i x = -\frac{1}{2} R_{ii}^{-1} B_i^{\mathrm{T}} P_i x$, $i \in N$ provides a Nash equilibrium solution for the linear quadratic N-player differential game among feedback policies with full state information. Furthermore the resulting system dynamics, described by $\dot{x} = A_c x$, $x(0) = x_0$ are asymptotically stable.

Policy Iteration for NonZero-Sum Differential Games. Equations (79) are difficult or impossible to solve. An iterative offline solution technique is given by the following policy iteration algorithm. It solves the coupled HJ equations by iterative solution of uncoupled nonlinear Lyapunov equations:

1. Start with stabilizing initial policies $\mu^0_1(x)$, $\dots$, $\mu^0_N(x)$.
2. Given the N-tuple of policies $\mu^k_1(x)$, $\dots$, $\mu^k_N(x)$, solve for the N-tuple of costs $V^k_1(x(t))$, $V^k_2(x(t))$, $\dots$, $V^k_N(x(t))$ using the Bellman equations

$$0 = r(x, \mu^k_1, \dots, \mu^k_N) + (\nabla V_i^k)^{\mathrm{T}} \left(f(x) + \sum_{j=1}^{N} g_j(x) \mu_j^i \right) \quad V^k_i(0) = 0 \quad i \in N \qquad (81)$$

3. Update the N-tuple of control policies using

$$\mu_i^{k+1} = \underset{u_i \in \Psi(\Omega)}{\arg \min} [H_i(x, \nabla V_i, u_1, \dots, u_N)] \quad i \in N \qquad (82)$$

which explicitly is

$$\mu_i^{k+1}(x) = -\frac{1}{2} R_{ii}^{-1} g_i^{\mathrm{T}}(x) \nabla V_i^k \quad i \in N \qquad (83)$$

A linear 2-player version is given in Gajic and Li[61] and can be considered as an extension of Kleiman's algorithm[57] to 2-player games.

In the next section we use PI algorithm 1 to motivate the control structure for an online adaptive N-player game solution algorithm. Then it is proven that an ldquo;optimal adaptive" control algorithm converges online to the solution of coupled HJs (79), while guaranteeing closed-loop stability.

Online Solution for Two-Player Non Zero-Sum Differential Games. In this section we show how to solve the 2-player non-zero sum game online; the approach can easily be extended to more than two players.

Consider the nonlinear time-invariant affine in the input dynamical system given by

$$\dot{x} = f(x) + g(x)u(x) + k(x)d(x) \qquad (84)$$

where state $x(t) \in \mathbb{R}^n$, first control input $u(x) \in \mathbb{R}^m$, and second control input $d(x) \in \mathbb{R}^q$. Assume that $f(0) = 0$ and $f(x)$, $g(x)$, $k(x)$ are locally Lipschitz, $\|f(x)\| < b_f \|x\|$.

This work uses nonlinear approximator structures for VFA[21,72] to solve (79). Therefore, assume there exist constant NN weights W_1 and W_2 such that the value functions $V_1(x)$ and $V_2(x)$ are approximated on a compact set Ω as

$$V_1(x) = W_1^{\mathrm{T}} \varphi_1(x) + \varepsilon_1(x) \qquad (85)$$

$$V_2(x) = W_2^{\mathrm{T}} \varphi_2(x) + \varepsilon_2(x) \qquad (86)$$

with $\varphi_1(x) : \mathbb{R}^n \to \mathbb{R}^K$ and $\varphi_2(x) : \mathbb{R}^n \to \mathbb{R}^K$ the NN activation function basis set vectors, K the number of neurons in the hidden layer, and $\varepsilon_1(x)$ and $\varepsilon_2(x)$ the NN approximation errors. From the approximation literature, the basis functions can be selected as sigmoids, tanh, polynomials, etc.

Assuming current NN weight estimates $\widehat{W}_1$ and $\widehat{W}_2$, the outputs of the two critic NN are given by

$$\widehat{V}_1(x) = \widehat{W}_1^{\mathrm{T}} \varphi_1(x) \tag{87}$$

$$\widehat{V}_2(x) = \widehat{W}_2^{\mathrm{T}} \varphi_2(x) \tag{88}$$

The approximate Lyapunov-like equations are then

$$H_1(x, \widehat{W}_1, u_3, d_4) = Q_1(x) + u_3^{\mathrm{T}} R_{11} u_3 + d_4^{\mathrm{T}} R_{12} d_4 + \widehat{W}_1^{\mathrm{T}} \nabla \varphi_1(f(x) + g(x)u_3 + k(x)d_4) = e_1 \tag{89}$$

$$H_2(x, \widehat{W}_2, u_3, d_4) = Q_2(x) + u_3^{\mathrm{T}} R_{21} u_3 + d_4^{\mathrm{T}} R_{22} d_4 + \widehat{W}_2^{\mathrm{T}} \nabla \varphi_2(f(x) + g(x)u_3 + k(x)d_4) = e_2 \tag{90}$$

It is desired to select $\widehat{W}_1$ and $\widehat{W}_2$ to minimize the square residual error

$$E_1 = \frac{1}{2} e_1^{\mathrm{T}} e_1 + \frac{1}{2} e_2^{\mathrm{T}} e_2$$

Then $\widehat{W}_1(t) \to W_1$, $\widehat{W}_2(t) \to W_2$ and we define the control policies in (89) and (90) in the form of action NNs which compute the control inputs in the structured form:

$$u_3(x) = -\frac{1}{2} R_{11}^{-1} g^{\mathrm{T}}(x) \nabla \varphi_1^{\mathrm{T}} \widehat{W}_3 \tag{91}$$

$$d_4(x) = -\frac{1}{2} R_{22}^{-1} k^{\mathrm{T}}(x) \nabla \varphi_2^{\mathrm{T}} \widehat{W}_4 \tag{92}$$

where $\widehat{W}_3$ and $\widehat{W}_4$ denote the current estimated values of the ideal NN weights W_1 and W_2, respectively. Define the critic and the actor NN estimation errors, respectively, as

$$\widetilde{W}_1 = W_1 - \widehat{W}_1, \quad \widetilde{W}_2 = W_2 - \widehat{W}_2, \quad \widetilde{W}_3 = W_1 - \widehat{W}_3 \quad \widetilde{W}_4 = W_2 - \widehat{W}_4 \tag{93}$$

The main theorem of non zero-sum games is now given, which provide the tuning laws for the actor and critic NNs that guarantee convergence of the 2-player nonzero-sum game algorithm in real-time to the Nash equilibrium solution, while also guaranteeing closed-loop stability.

Theorem 3

Let the dynamics be given by (84), and consider the 2-player non zero-sum game formulation described before. Let the critic NNs be given by (87) and(88) and the control inputs be given (91) and (92). Let tuning for the critic NNs be provided by

$$\dot{\widehat{W}}_1 = -a_1 \frac{\sigma_3}{(\sigma_3^{\mathrm{T}} \sigma_3 + 1)^2} \, [\sigma_3^{\mathrm{T}} \widehat{W}_1 + Q_1(x) + u_3^{\mathrm{T}} R_{11} u_3 + d_4^{\mathrm{T}} R_{12} d_4] \tag{94}$$

$$\dot{\widehat{W}}_2 = -a_2 \frac{\sigma_4}{(\sigma_4^{\mathrm{T}} \sigma_4 + 1)^2} \, [\sigma_4^{\mathrm{T}} \widehat{W}_2 + Q_2(x) + u_3^{\mathrm{T}} R_{21} u_3 + d_4^{\mathrm{T}} R_{22} d_4] \tag{95}$$

where $\sigma_3 = \nabla \varphi_1(f + gu_3 + kd_4)$ and $\sigma_4 = \nabla \varphi_2(f + gu_3 + kd_4)$. Let the first actor NN (first player) be tuned as

$$\dot{\widehat{W}}_3 = -a_3 \{ (F_2 \widehat{W}_3 - F_1 \overline{\sigma}_3^{\mathrm{T}} \widehat{W}_1) - 1/4 (\nabla \varphi_1 g(x) R_{11}^{-\mathrm{T}} R_{21} R_{11}^{-1} g^{\mathrm{T}}(x) \nabla \varphi_1^{\mathrm{T}} \widehat{W}_3 m_2^{\mathrm{T}} \widehat{W}_2$$
$$+ \overline{D}_1(x) \widehat{W}_3 m_1^{\mathrm{T}} \widehat{W}_1) \} \tag{96}$$

and the second actor (second player) NN be tuned as

$$\dot{\widehat{W}}_4 = -a_4 \{ (F_4 \widehat{W}_4 - F_3 \overline{\sigma}_4^{\mathrm{T}} \widehat{W}_2) - 1/4 (\nabla \varphi_2 k(x) R_{22}^{-\mathrm{T}} R_{12} R_{22}^{-1} k^{\mathrm{T}}(x) \nabla \varphi_2^{\mathrm{T}} \widehat{W}_4 m_1^{\mathrm{T}} \widehat{W}_1$$
$$+ \overline{D}_2(x) \widehat{W}_4 m_2^{\mathrm{T}} \widehat{W}_2) \} \tag{97}$$

where

$$\bar{D}_1(x) \equiv \nabla\varphi_1(x)g(x)R_{11}^{-1}g^T(x)\nabla\varphi_1{}^T(x) \quad \bar{D}_2(x) \equiv \nabla\varphi_2(x)kR_{22}^{-1}k^T\nabla\varphi_2{}^T(x)$$

$$m_1 \equiv \frac{\sigma_3}{(\sigma_3{}^T\sigma_3 + 1)^2} \quad m_2 \equiv \frac{\sigma_4}{(\sigma_4{}^T\sigma_4 + 1)^2} \quad F_1 > 0, \ F_2 > 0, \ F_3 > 0, \ F_4 > 0$$

are tuning parameters. Assume that $Q_1(x) > 0$ and $Q_2(x) > 0$. Suppose that $\bar{\sigma}_3 = \sigma_3/(\sigma_3^T\sigma_3 + 1)$ and $\bar{\sigma}_4 = \sigma_4/(\sigma_4^T\sigma_4 + 1)$ are persistently exciting. Let the tuning parameters be selected as detailed in Vamvoudakis and Lewis.[4] Then there exists a K_0 such that, for the number of hidden layer units $K > K_0$ the closed-loop system state, the critic NN errors $\widetilde{W}_1$ and $\widetilde{W}_2$, the first actor NN error $\widetilde{W}_3$ and the second actor NN error $\widetilde{W}_4$ are UUB.

Proof:
See Vamvoudakis and Lewis.[4]

Simulation of Online Learning of Non Zero-Sum Games. Here we present a simulation of a nonlinear system to show that the game can be solved *Online* by learning in real time, using the method described before. PE is needed to guarantee convergence to the Nash solution without solving the coupled HJ equations. In this simulation, exponentially decreasing probing noise is added to the control inputs to ensure PE until convergence is obtained.

Consider the following affine in control input nonlinear system, with a quadratic cost constructed as in Nevistic and Primbs[66]

$$\dot{x} = f(x) + g(x)u + k(x)d, \ x \in \mathbb{R}^2$$

where

$$f(x) = \begin{bmatrix} x_2 \\ -x_2 - \frac{1}{2}x_1 + \frac{1}{4}x_2\left(\cos\left(2x_1\right) + 2\right)^2 + \frac{1}{4}x_2(\sin(4x_1{}^2) + 2)^2 \end{bmatrix}$$

$$g(x) = \begin{bmatrix} 0 \\ \cos\left(2x_1\right) + 2 \end{bmatrix} \quad k(x) = \begin{bmatrix} 0 \\ \sin\left(4x_1{}^2\right) + 2 \end{bmatrix}$$

Select $Q_1 = 2Q_2, R_{11} = 2R_{22}$, and $R_{12} = 2R_{21}$ where Q_2, R_{22}, and R_{21} are identity matrices. Select $a_1 = a_2 = a_3 = a_4 = 1$ and $F_4 = F_3 = F_2 = F_1 = 100I$ for the constants on the tuning laws, where I is an identity matrix of appropriate dimensions.

The optimal value function for the first critic (player 1) is

$$V_1^*(x) = 1/2x_1{}^2 + x_2{}^2$$

and for the second critic (player 2) is

$$V_2^*(x) = 1/4x_1{}^2 + 1/2x_2{}^2$$

The optimal control signal for the first player is

$$u^*(x) = -2(\cos(2x_1) + 2)x_2$$

and the optimal control signal for the second player is

$$d^*(x) = -(\sin(4x_1{}^2) + 2)x_2$$

One selects the NN vector activation function for the critics as $\phi_1(x) = \phi_2(x) \equiv [x_1{}^2 \ x_1x_2 \ x_2^2]$ and uses the control algorithm presented in Theorem 3. Each player maintains two NNs, a critic NN to estimate its current value and an action NN to estimate its current control policy.

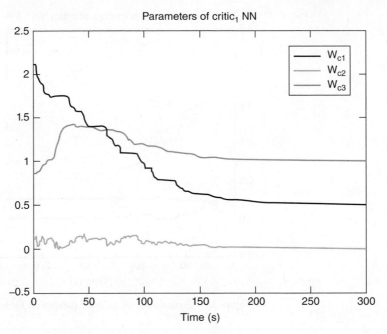

Figure 22   Convergence of the critic parameters for first player.

Figure 22 shows the critic parameters for the first player, denoted by $\widehat{W}_1 = \begin{bmatrix} W_{c1} & W_{c2} & W_{c3} \end{bmatrix}^T$ by using the proposed game algorithm. After convergence at about 150 s one has $\widehat{W}_1(t_f) = [0.5015 \; 0.0007 \; 1.0001]^T$. Figure 23 shows the critic parameters for the second player $\widehat{W}_2 = \begin{bmatrix} W_{2c1} & W_{2c2} & W_{2c3} \end{bmatrix}^T$. After convergence at about 150 s one has

$$\widehat{W}_2(t_f) = [0.2514 \quad 0.0006 \quad 0.5001]^T$$

The actor parameters for the first player after 150 s converge to the values of $\widehat{W}_3(t_f) = [0.5015 \; 0.0007 \; 1.0001]^T$, and the actor parameters for the second player converge to the values of

$$\widehat{W}_4(t_f) = [0.2514 \quad 0.0006 \quad 0.5001]^T$$

Therefore the actor NN for the first player

$$\widehat{u}(x) = -\frac{1}{2} R_{11}^{-1} \begin{bmatrix} 0 \\ \cos\left(2x_1\right) + 2 \end{bmatrix}^T \begin{bmatrix} 2x_1 & 0 \\ x_2 & x_1 \\ 0 & 2x_2 \end{bmatrix}^T \begin{bmatrix} 0.5015 \\ 0.0007 \\ 1.0001 \end{bmatrix}$$

also converged to the optimal control, and similarly for the second player

$$\widehat{d}(x) = -\frac{1}{2} R_{22}^{-1} \begin{bmatrix} 0 \\ \sin\left(4x_1{}^2\right) + 2 \end{bmatrix}^T \begin{bmatrix} 2x_1 & 0 \\ x_2 & x_1 \\ 0 & 2x_2 \end{bmatrix}^T \begin{bmatrix} 0.2514 \\ 0.0006 \\ 0.5001 \end{bmatrix}$$

The evolution of the system states is presented in Fig. 24.

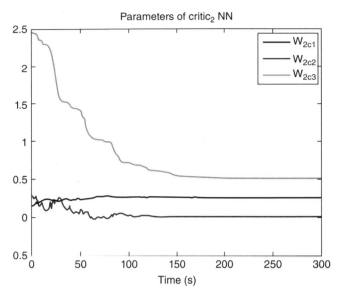

Figure 23 Convergence of the critic parameters for second player.

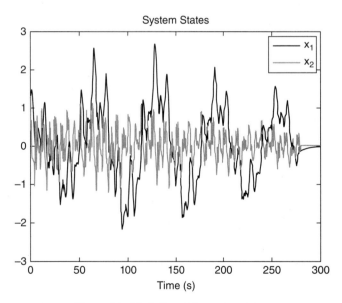

Figure 24 Evolution of the system states.

8 HISTORICAL DEVELOPMENT, REFERENCED WORK, AND FURTHER STUDY

A firm foundation for the use of NNs in feedback control systems has been developed over the years by many researchers. Here is included a historical development and references to the body of work in neurocontrol.

8.1   Neural Network for Feedback Control

The use of NNs in feedback control systems was first proposed by Werbos.[12] Since then, NN control has been studied by many researchers. Recently, NN have entered the mainstream of control theory as a natural extension of adaptive control to systems that are nonlinear in the tunable parameters. The state of NN control is well illustrated by papers in the *Automatica* special issue on NN control.[73] Overviews of the initial work in NN control are provided by Miller et al.[74] and the *Handbook of Intelligent Control*,[75] which highlight a host of difficulties to be addressed for closed-loop control applications. Neural network applications in closed-loop control are fundamentally different from open-loop applications such as classification and image processing. The basic multilayer NN tuning strategy is backpropagation.[21] Basic problems that had to be addressed for closed-loop NN control[2,43] included weight initialization for feedback stability, determining the gradients needed for backpropagation tuning, determining what to backpropagate, obviating the need for preliminary offline tuning, and modifying backpropagation so that it tunes the weights forward through time, providing efficient computer code for implementation. These issues have since been addressed by many approaches.

Initial work was in NNs for system identification and identification-based indirect control. In closed loop control applications, it is necessary to show the stability of the tracking error as well as boundedness of the NN weight estimation errors. Proofs for internal stability, bounded NN weights (e.g., bounded control signals), guaranteed tracking performance, and robustness were absent in early works. Uncertainty as to how to initialize the NN weights led to the necessity for "preliminary offline tuning." Work on offline learning was formalized by Kawato.[76] Offline learning can yield important structural information.

Subsequent work in NN for control addressed closed-loop system structure and stability issues. Work by Sussmann[77] and Albertini and Sontag[72] was important in determining system properties of NN (e.g., minimality and uniqueness of the ideal NN weights, observability of dynamic NN). The seminal work of Narendra and Parthasarathy[13,14] had an emphasis on finding the gradients needed for backpropagation tuning in feedback systems, which, when the plant dynamics are included, become recurrent nets. In recurrent nets, these gradients themselves satisfy difference or differential equations, so they are difficult to find. Sadegh[79] showed that knowing an approximate plant Jacobian is often good enough to guarantee suitable closed-loop performance.

The approximation properties of NN[6,7] are basic to their feedback controls applications. Based on this and analysis of the error dynamics, various modifications to backpropagation were presented that guaranteed closed-loop stability as well as weight error boundedness. These are akin to terms added in adaptive control to make algorithms robust to high-frequency unmodeled dynamics. Sanner and Slotine[9] used radial basis functions in control and showed how to select the NN basis functions, Polycarpou and Ioannou[80,81] used a projection method for weight updates, and Lewis[82] used backprop with an *e*-modification term.[22] This work used NNs that are *linear* in the unknown parameter. In linear NNs, the problem is relegated to determining activation functions that form a *basis set* (e.g., radial basis functions (RBFs),[9] and functional link programmable network (FLPN)[79]). It was shown by Sanner and Slotine[9] how to systematically derive stable NN controllers using approximation theory and basis functions. Barron[83] has shown that using NNs that are linear in the tunable parameters gives a fundamental limitation of the approximation accuracy to the order of $1/L^{2/n}$, where L is the number of hidden-layer neurons and n is the number of inputs. Nonlinear-in-the-parameters NNs overcome this difficulty and were first used by Chen and Khalil[84] who used backpropagation with dead-zone weight tuning, and Lewis et al.[85], who used a Narendra's e-modification term in backpropagation. In nonlinear-in-the-parameters NN, the basis is automatically selected online by tuning the first-layer weights and thresholds. Multilayer NNs were rigorously used for discrete-time control by Jagannathan and Lewis.[86] Polycarpou[87] derived NN controllers that do not assume known bounds on the ideal weights. Dynamic/recurrent NNs were used for control by Rovithakis and Christodoulou,[88] Poznyak et al.,[89] Rovithakis[90] who considered multiplicative disturbances, Zhang and Wang[91], and others.

Most stability results on NN control have been local in nature, and global stability has been treated by Annaswamy, by Kwan et al.,[92] and by others. Recently, NN control has been used in conjunction with other control approaches to extend the class of systems that yields to nonparametric control. Calise and co-workers have used NNs in conjunction with dynamic inversion to control aircraft and missiles. Feedback linearization using NNs has been addressed by Chen and Khalil[84], Yesildirek and Lewis[95], Ge et al.[26] and others. NNs were used for strict-feedback nonlinear systems with backstepping.[19,24,32,96–101] Combining backstepping with an implicit function theorem, input-state stability, NNs were further used for pure feedback nonlinear systems[98,102,103] and nonaffine system.[24,104–106]

NNs have been used in conjunction with the Isidori–Byrnes regulator equations for output tracking control by Wang and Huang.[98] A multimodel NN control approach has been given by Narendra and Balakrishnan.[107] Applications of NN control have been extended to partial differential equation systems by Padhi, et al.[108] NNs have been used for control of stochastic systems by Poznyak and Ljung.[109] Parisini has developed receding horizon controllers based on NN[110] and hybrid discrete event NN controllers.[111]

In practical implementations of NN controllers there remain problems to overcome. A barrier Lyapunov function (BLF) was introduced in Ren et al.[112] to address two open and challenging problems in the neurocontrol area: (i) for any initial compact set, how to determine a priori the compact superset, on which NN approximation is valid; and (ii) how to ensure that the arguments of the unknown functions remain within the specified compact superset. Weight initialization still remains an issue, and one may also find that the NN weights become unbounded despite proofs to the contrary. Practical implementation issues were addressed by Chen and Chang, Gutierrez and Lewis[114], and others. Random initialization of the first-layer NN weights often works in practice, and work by Igelnik and Pao[11] shows that it is theoretically defensible. Computational complexity makes NNs with many hidden-layer neurons difficult to implement. Recently, work has intensified in wavelets, NNs that have localized basis functions, and NNs that are self-organizing in the sense of adding or deleting neurons automatically.[100,115–117]

By now it is understood that NNs offer an elegant extension of adaptive control and other techniques to systems that are nonlinear in the unknown parameters. The universal approximation properties of NNs[6,7] avoid the use of specialized basis sets including regression matrices. Formalized improved proofs avoid the use of assumptions such as certainty equivalence. Robustifying terms avoid the need for persistency of excitation. Recent books on NN feedback control include Refs. 19, 24, 26 35, 41, and 118.

8.2   Reinforcement Learning and Approximate Dynamic Programming

Adaptive critics are reinforcement learning designs that attempt to approximate dynamic programming.[119,120] They approach the optimal solution through forward approximate dynamic programming. Initially, they were proposed by Werbos.[43] Overviews of the initial work in NN control are provided by Miller[74] and the *Handbook of Intelligent Control.*[75] Howard[47] showed the convergence of an algorithm relying on successive policy iteration solution of a nonlinear Lyapunov equation for the cost (value) and an optimizing equation for the control (action). This algorithm relied on perfect knowledge of the system dynamics and is an offline technique. Later, various online dynamic-programming-based reinforcement learning algorithms emerged and were mainly based on Werbos' heuristic dynamic programming (HDP),[2] Sutton's temporal differences (TD) learning methods,[121] and Q learning, which was introduced by Watkins[50] and Werbos[12] (called action-dependent critic schemes there). Critic and action network tuning was provided by RLS, gradient techniques, or the backpropagation algorithm.[21] Early work on dynamic-programming-based reinforcement learning focused on discrete finite state and action spaces. These depended on lookup tables or linear function approximators. Convergence results were shown in this case, such as Dayan.[122]

For continuous state and action spaces, convergence results are more challenging as adaptive critics require the use of nonlinear function approximators. Four schemes for approximate dynamic programming were given in Werbos,[2] the HDP and dual heuristic programming (DHP) algorithms, and their action-dependent versions (ADHDP and ADDHP). The linear quadratic regulation (LQR) problem[42] served as a testbed for much of these studies. Solid convergence results were obtained for various adaptive critic designs for the LQR problem. We mention the work of Bradtke et al.[51] where Q learning was shown to converge when using non-linear function approximators. An important persistence of excitation notion was included. Further work was done by Landelius[123] who studied the four adaptive critic architectures. He demonstrated convergence results for all four cases, in the LQR case and discussed when the design is model free. Hagen and Kröse[124] discussed the effect of model noise and exploration noise when adaptive critic is viewed as a stochastic approximation technique. Prokhorov and Feldkamp[125] look at Lyapunov stability analysis. Other convergence results are due to Balakrishnan and co-workers[108,126] who have also studied optimal control of aircraft and distributed parameter systems governed by partial differential equations (PDEs). Anderson etal.[127] showed convergence and stability for a reinforcement learning scheme. All of these results were done for the discrete-time case.

A thorough treatment of neurodynamic programming is given in the seminal book by Bertsekas and Tsitsiklis.[72] Various successful practical implementations have been reported, including aircraft control examples by Ferrari and Stengel[128], an Auto Lander by Murray, et al.,[129] state estimation using dynamic NNs by Feldkamp and Prokhorov,[130] and neuroobservers by Liu and Balakrishnan.[131] Si and Wang.[132] have provided analysis and applied ADP to aircraft control.[45] An account of the adaptive critic designs is found in Prokhorov and Wunsch.[133]

Applications of adaptive critics in the CT domain were mainly done through discretization, and the application of the well-established DT results..[134] Various CT non dynamic reinforcement learning were discussed by Campos and Lewis[135], and Rovithakis[136], who approximated a Lyapunov function derivative. In Kim and Lewis[41] the HJB equation of dynamic programming is approximated by a Riccati equation, and a suboptimal controller based on NN feedback linearization is implemented with full stability and convergence proofs. Murray, et al.[55] prove convergence of an algorithm that uses system state measurements to find the cost to go. An array of initial conditions is needed. Unknown plant dynamics in the linear case is confronted by estimating a matrix of state derivatives. The cost functional is shown to be a Lyapunov function and approximated using either quadratic functions or an RBF neural network. Saridis and Lee[137] showed the convergence of an offline algorithm relying on successive iteration solution of a nonlinear Lyapunov equation for the cost (value) and an optimizing equation for the control (action). This is the CT equivalent of Howard's work.[47] Beard et al.[63] showed how to actually solve these equations using Galerkin integral approximations, which require much computational effort.

Online adaptive critics algorithms for CT systems have been proposed by Vrabie, et al.[53] with the development of integral reinforcement learning policy iteration that needs partial knowledge of the system dynamics and by Vamvoudakis and Lewis[4] that proposed an adaptive technique for simultaneous tuning of actor and critic NNs.

Q learning is not well posed when sampling times become small, and so is not useful for extension to CT systems. Continuous-time dynamic-programming-based reinforcement learning is reformulated using the so-called advantage learning by Baird[54], who defines a differential increment from the optimal solution and explicitly takes into account the sampling interval Δt. Doya[45] derives results for online updating of the critic using techniques from CT nonlinear optimal control. The advantage function follows naturally from this approach and, in fact, coincides with the CT Hamiltonian function. Doya gives relations with the TD(0) and TD(λ) techniques of Sutton.[121] Tsiotras has used wavelets to find approximate solutions to HJB. Lyshevski[138] has focused on a general parametrized form for the value function and obtained a set of algebraic equations that can be solved for an approximate value function.

ACKNOWLEDGMENTS

The referenced work of Lewis and co-workers was sponsored by NSF grant ECS-1128050, ARO Grant W91NF-05-1-0314, and AFOSR grant FA9550-09-1-0278, and the referenced work of S. S. Ge and co-workers was sponsored by the Basic Research Program of China (973 Program) under Grant 2011CB707005.

REFERENCES

1. Von Bertalanffy, Ludwig, *Organismic Psychology and Systems Theory*, Worchester, MA, Clark University Press, 1968.

2. P. J. Werbos, "Approximate Dynamic Programming for Real-Time Control and Neural Modeling," in D. A. White and D. A. Sofge, (Eds.), *Handbook of Intelligent Control*, Van Nostrand Reinhold, New York, 1992.

3. D. Vrabie, K. Vamvoudakis, and F. Lewis, "Adaptive Optimal Controllers Based on Generalized Policy Iteration in a Continuous-Time Framework," Proc. of the IEEE Mediterranean Conf. on Control and Automation, pp. 1402–1409, 2009.

4. K. G. Vamvoudakis and F. L. Lewis, "Online Actor-Critic Algorithm to Solve the Continuous-Time Infinite Horizon Optimal Control Problem," *Automatica*, **46**(5), 878–888, 2010.

5. K. G. Vamvoudakis and F. L. Lewis, "Multi-Player Non Zero Sum Games: Online Adaptive Learning Solution of Coupled Hamilton-Jacobi Equations," *Automatica*, **47**(8), 1556–1569, 2011.

6. G. Cybenko, "Approximation by Superpositions of a Sigmoidal Function," *Math. Control. Signals Syst.*, **2**(4), 303–314, 1989.

7. J. Park and I. W. Sandberg, "Universal Approximation Using Radial-Basis-Function Networks," *Neural Comp.*, **3**, 246–257, 1991.

8. J. S. Albus, "A New Approach to Manipulator Control: The Cerebellar Model Articulation Controller Equations (CMAC)," *Trans. ASME J. Dynam. Syst. Meas. Control*, **97**, 220–227, 1975.

9. R. M. Sanner and J.-J. E. Slotine, "Gaussian Networks for Direct Adaptive Control," *IEEE Trans. Neural Netw.*, **3**(6), 837–863, November, 1992.

10. L.-X. Wang, *Adaptive Fuzzy Systems and Control: Design and Stability Analysis*, Prentice-Hall, Englewood Cliffs, NJ, 1994.

11. B. Igelnik and Y.-H. Pao, "Stochastic Choice of Basis Functions in Adaptive Function Approximation and Functional-Link Net," *IEEE Trans. Neural Netw.*, **6**(6), 1320–1329, November, 1995.

12. P. J. Werbos, "Neural Networks for Control and System Identification," Proc. IEEE Conf. Decision and Control, Florida, 1989.

13. K. S. Narendra and K. Parthasarathy, "Identification and Control of Dynamical Systems Using Neural Networks," *IEEE Trans. Neural Netw.*, **1**, 4–27, March, 1990.

14. K. S. Narendra and K. Parthasarathy, "Gradient Methods for the Optimization of Dynamical Systems Containing Neural Networks," *IEEE Trans. Neural Netw.*, **2**(2), 252–262, March, 1991.

15. Y. D. Landau, *Adaptive Control*, Marcel Dekker, New York, 1979.

16. F. L. Lewis, D. M. Dawson, and C. T. Abdallah, *Robot Manipulator Control*, Marcel Dekker, New York, 2004.

17. J. J. E. Slotine and J. A. Coetsee, "Adaptive Sliding Controller Synthesis for Nonlinear Systems," *Int. J. Control*, **43**(4), 1631–1651, 1986.

18. J. J. E. Slotine and W. Li, "On the Adaptive Control of Robot Manipulators," *Int. J. Robotics Res.*, **6**(3), 49–59, 1987.

19. F. L. Lewis, S. Jagannathan, and A. Yesildirek, *Neural Network Control of Robot Manipulators and Nonlinear Systems*, Taylor and Francis, London, 1999.

20. L. R. Hunt, R. Su., and G. Meyer, "Global Transformations of Nonlinear Systems," *IEEE Trans. Autom. Control*, **28**, 24–31, 1983.

21. P. J. Werbos, Beyond Regression: New Tools for Prediction and Analysis in the Behavior Sciences, Ph.D. Thesis, Committee on Appl. Math., Harvard Univ., 1974.

22. K. S. Narendra and A. M. Annaswamy, "A New Adaptive Law for Robust Adaptation without Persistent Excitation," *IEEE Trans. Automat. Control*, **AC-32**(2), 134–145, February, 1987.

23. P. Ioannou and J. Sun, *Robust Adaptive Control*, Prentice-Hall, Englewood Cliffs, NJ, 1996. Electronic copy available at http://www-rcf.usc.edu/~ioannou/Robust_Adaptive_Control.htm.

24. S. S. Ge, C. C. Hang, T. H. Lee, and T. Zhang, *Stable Adaptive Neural Network Control*, Kluwer, Boston, 2001.

25. S. S. Ge, T. H. Lee, and Z. P. Wang, "Adaptive Neural Network Control for Smart Materials Robots Using Singular Perturbation Technique," *Asian J. Control*, **3**(2), 143–155, 2001.

26. S. S. Ge, T. H. Lee, and C. J. Harris, *Adaptive Neural Network Control of Robotic Manipulators*, World Scientific, Singapore, 1998.

27. F. L. Lewis, *Applied Optimal Control and Estimation: Digital Design and Implementation*, Prentice-Hall, Englewood Cliffs, NJ, February, 1992.

28. S. S. Ge, G. Y. Li, and T.H. Lee, "Adaptive NN Control for a Class of Strict Feedback Discrete-Time Nonlinear Systems," *Automatica*, **39**, 807–819, 2003.

29. S. S. Ge, C. Yang, and T. H. Lee, "Adaptive Predictive Control Using Neural Network for a Class of Pure-feedback Systems in Discrete-Time," *IEEE Trans. Neural Netw.*, **19**(9), 1599–1614, 2008.

30. C. Yang, S. S. Ge, C. Xiang, T. Chai, and T. H. Lee, "Output Feedback NN Control for two Classes of Discrete-time Systems with Unknown Control Directions in a Unified Approach," *IEEE Trans. Neural Netw.*, 19(11), 1873–1886, November, 2008.

31. Y. Li, C. Yang, S. S. Ge, and T. H. Lee, "Adaptive Output Feedback NN Control of a Class of Discrete-Time MIMO Nonlinear Systems with Unknown Control Directions," *IEEE Trans. Syst., Man Cybern., Part B*, **41**(9), 507–517, 2011.

32. I. Kanellakopoulos, P. V. Kokotovic, and A. S. Morse, "Systematic Design of Adaptive Controllers for Feedback Linearizable Systems," *IEEE Trans. Autom. Control*, **36**, 1241–1253, 1991.

33. P. V. Kokotovic, "Applications of Singular Perturbation Techniques to Control Theory," *SIAM Rev.*, **26**(4), 501–550, 1984.

34. N. H. McClamroch and D. Wang, "Feedback Stabilization and Tracking of Constrained Robots," *IEEE Trans. Automat. Control*, **33**, 419–426, 1988.

35. F. L. Lewis, J. Campos, and R. Selmic, *Neuro-Fuzzy Control of Industrial Systems with Actuator Nonlinearities*, Society of Industrial and Applied Mathematics Press, Philadelphia, 2002.

36. B. L. Stevens and F. L. Lewis, *Aircraft Control and Simulation*, 2nd ed., Wiley, Hoboken, NJ, 2003.

37. A. J. Calise, N. Hovakimyan, and H. Lee, "Adaptive Output Feedback Control of Nonlinear Systems Using Neural Networks," *Automatica*, **37**(8), 1201–1211, August 2001.

38. U.S. Patent No. 6,611,823, R. R. Selmic, F. L. Lewis, A. J. Calise, and M. B. McFarland, "Backlash Compensation Using Neural Network," 2003.

39. B. Ren, S. S. Ge, T. H. Lee, and C.-Y. Su, "Adaptive Neural Control for a Class of Nonlinear Systems with Uncertain Hysteresis Inputs and Time-Varying State Delays," *IEEE Trans. Neural Netw.*, **20**(7), 1148–1164, 2009.

40. M. Chen, S. S. Ge, and B. Ren, "Adaptive Tracking Control of Uncertain MIMO Nonlinear Systems with Input Constraints," *Automatica*, **47**(3), 452–465, 2011.

41. Y. H. Kim and F. L. Lewis, *High-Level Feedback Control with Neural Networks*, World Scientific, Singapore, 1998.

42. F. L. Lewis and V. Syrmos, *Optimal Control*, 2nd ed., Wiley, New York, 1995.

43. P.J. Werbos., "A Menu of Designs for Reinforcement Learning Over Time," in W. T. Miller, R. S. Sutton, P. J. Werbos, (Eds.), *Neural Networks for Control*, pp. 67–95, MIT Press, Cambridge, 1991.

44. J. Si, A. Barto, W. Powell, and D. Wunsch, *Handbook of Learning and Approximate Dynamic Programming*, IEEE Press, West Conshohocken, PA, 2004.

45. K. Doya, "Reinforcement Learning in Continuous Time and Space," *Neural Comput.*, **12**(1), pp. 219-245, MIT Press, 2000.

46. Doya K, Kimura H, Kawato M: Neural mechanisms of learning and control. *IEEE Control Sys.*, **21**, 42–54, (2001).

47. R. Howard, *Dynamic Programming and Markov Processes*, MIT Press, Cambridge, MA, 1960.

48. R. S. Sutton, and A. G. Barto, *Reinforcement Learning – An Introduction*, MIT Press, Cambridge, MA, 1998.

49. A. Al-Tamimi, M. Abu-Khalaf, and F. L. Lewis, "Adaptive Critic Designs for Discrete-Time Zero-Sum Games with Application to H-infinity Control," *IEEE Trans. Sys., Man. and Cyb –B*, **37**(1), 2007.

50. C. Watkins, Learning from Delayed Rewards, Ph.D. Thesis, Cambridge University, Cambridge, England, 1989.

51. S. Bradtke, B. Ydstie, and A. Barto, "Adaptive Linear Quadratic Control Using Policy Iteration," CMPSCI-94-49, University of Massachusetts, June, 1994.

52. M. Abu-Khalaf and F.L. Lewis, "Nearly Optimal Control Laws for Nonlinear Systems with Saturating Actuators Using a Neural Network HJB Approach," *Automatica*, vol. **41**, no. 5, pp. 779-791, 2008.

53. D. Vrabie, O. Pastravanu, F. L. Lewis, and M. Abu-Khalaf, "Adaptive Optimal Control for Continuous-Time Linear Systems Based on Policy Iteration," *Automatica*, **45**(2), 477–484, 2009.

54. L. Baird, "Reinforcement Learning in Continuous Time: Advantage Updating," Proceedings of the International Conference on Neural Networks, Orlando, FL, June, 1994.

55. J. Murray, C. Cox, G. Lendaris, and R. Saeks, "Adaptive Dynamic Programming," *IEEE Trans. Syst., Man, Cybern.*, **32**(2), May, 2002.

56. T. Hanselmann, L. Noakes, and A. Zaknich, "Continuous-Time Adaptive Critics," *IEEE Trans. Neural Netw.*, **18**(3), 631–647, 2007.

57. D. Kleinman, "On an Iterative Technique for Riccati Equation Computations," *IEEE Trans. Autom. Control*, **13**(1), 114–115, 1968.

58. S. Tijs, *Introduction to Game Theory*, Hindustan Book Agency, India, 2003.

59. T. Başar and G. J. Olsder, *Dynamic Noncooperative Game Theory*, 2nd ed., SIAM, Philadelphia, 1999.

60. G. Freiling, G. Jank, and H. Abou-Kandil, "On Global Existence of Solutions to Coupled Matrix Riccati Equations in Closed Loop Nash Games," *IEEE Trans. Autom Control*, **41**(2), 264–269, 2002.

61. Z. Gajic and T-Y Li, "Simulation Results for Two New Algorithms for Solving Coupled Algebraic Riccati Equations," *Third Int. Symp. On Differential Games*, Sophia, Antipolis, France, 1988.

62. M. Abu-Khalaf, F. L. Lewis, and Jie Huang, "Policy Iterations on the Hamilton-Jacobi-Isaacs Equation for H_∞ State Feedback Control with Input Saturation," *IEEE Trans. Autom Control*, **51**(12), pp. 1989–1995, 2005.

63. R. Beard, G. Saridis, and J. Wen, "Approximate Solutions to the Time-Invariant Hamilton-Jacobi-Bellman Equation," *Automatica*, **33**(12), 2159–2177, December 1997.

64. D. Vrabie, Online Adaptive Optimal Control for Continuous Time Systems, Ph.D. Thesis, Dept. of Electrical Engineering, Univ. Texas at Arlington, Arlington, TX, 2009.

65. A. J. Van der Schaft, "$L2$-Gain Analysis of Nonlinear Systems and Nonlinear State Feedback H∞ Control," *IEEE Trans. Autom. Control*, **37**(6), 770–784, 1992.

66. V. Nevistic and J. A. Primbs, Constrained Nonlinear Optimal Control: A Converse HJB Approach, California Institute of Technology, Pasadena, CA 91125, Tech Rep. CIT-CDS **96-021**, 1996.

67. T. Başar and P. Bernard, H_∞ *Optimal Control and Related Minimax Design Problems*, Birkhäuser, Boston, 1995.

68. Y. Feng, B. D. Anderson, and M. Rotkowitz, "A Game Theoretic Algorithm to Compute Local Stabilizing Solutions to HJBI Equations in Nonlinear H∞ Control," *Automatica*, **45**(4), 881–888, 2009.

69. M. Abu-Khalaf and F. L. Lewis, "Neurodynamic Programming and Zero-Sum Games for Constrained Control Systems," *IEEE Trans. Neural Netw.*, **19**(7), 1243–1252, 2008.

70. M. Abu-Khalaf and F. L. Lewis, "Nearly Optimal Control Laws for Nonlinear Systems with Saturating Actuators Using a Neural Network HJB Approach," *Automatica*, **41**(5), 779–791, 2008.

71. M. Abu-Khalaf, F. L. Lewis, and J. Huang, "Policy Iterations and the Hamilton-Jacobi-Isaacs Equation for H-infinity Statefeedback Control with Input Saturation," *IEEE Trans. Automat. Control, http://dx.doi.org/10.1109/TAC.2006.884959*, 1989–1995, December, 2006.

72. D. P. Bertsekas and J. N. Tsitsiklis, *Neuro-Dynamic Programming*, Athena Scientific, MA, 1996.

73. K. S. Narendra and F. L. Lewis, "Special Issue on Neural Network Feedback Control," *Automatica*, **37**(8), August, 2001.

74. W. T. Miller, R. S. Sutton, P. J. Werbos (eds.), *Neural Networks for Control*, MIT Press, Cambridge, 1991.

75. D. A. White and D. A. Sofge (Eds.), *Handbook of Intelligent Control*, Van Nostrand Reinhold, New York, 1992.

76. M. Kawato, "Computational Schemes and Neural Network Models for Formation and Control of Multijoint Arm Trajectory," in W. T. Miller, R. S. Sutton, and P.J. Werbos (Eds.), *Neural Networks for Control*, pp. 197–228, MIT Press, Cambridge, 1991.

77. H. J. Sussmann, "Uniqueness of the Weights for Minimal Feedforward Nets with a Given Input-Output Map," *Neural Netw.*, **5**, 589–593, 1992.

78. F. Albertini and E. D. Sontag, "For Neural Nets, Function Determines Form," Proc. IEEE Conf. Decision and Control, pp. 26–31, December 1992.

79. N. Sadegh, "A Perceptron Network for Functional Identification and Control of Nonlinear Systems," *IEEE Trans. Neural Netw.*, **4**(6), 982–988, November, 1993.

80. M. M. Polycarpou and P. A. Ioannou, "Identification and Control Using Neural Network Models: Design and Stability Analysis," Tech. Report 91-09-01, Dept. Elect. Eng. Sys., Univ. S. Cal., September, 1991.

81. M. M. Polycarpou and P. A. Ioannou, "Neural Networks as On-line Approximators of Nonlinear Systems," Proc. IEEE Conf. Decision and Control, pp. 7–12, Tucson, December, 1992.

82. K. S. Narendra and A. M. Annaswamy, *Stable Adaptive Systems*, New Tork, Courier Dover Publications, 2012.

83. A. R. Barron, "Universal Approximation Bounds for Superpositions of a Sigmoidal Function," *IEEE Trans. Info. Theory*, **39**(3), 930–945, May, 1993.

84. F.-C. Chen and H. K. Khalil, "Adaptive Control of Nonlinear Systems Using Neural Networks," *Int. J. Control*, **55**(6), 1299–1317, 1992.

85. F. L. Lewis, A. Yesildirek, and K. Liu, "Multilayer Neural Net Robot Controller with Guaranteed Tracking Performance," *IEEE Trans. Neural Netw.*, **7**(2), 388–399, March, 1996.

86. S. Jagannathan and F. L. Lewis, "Multilayer Discrete-Time Neural Net Controller with Guaranteed Performance," *IEEE Trans. Neural Netw.*, **7**(1), 107–130, January, 1996.

87. M. M. Polycarpou, "Stable Adaptive Neural Control Scheme for Nonlinear Systems," *IEEE Trans. Automat. Control*, **41**(3), 447–451, March, 1996.

88. G. A. Rovithakis and M. A. Christodoulou, "Adaptive Control of Unknown Plants Using Dynamical Neural Networks," *IEEE Trans. Syst., Man, Cybern.*, **24**(3), 400–412, 1994.

89. A. S. Poznyak, W. Yu, E. N. Sanchez, and J. P. Perez, "Nonlinear Adaptive Trajectory Tracking Using Dynamic Neural Networks," *IEEE Trans. Neural Netw.*, **10**(6), 1402–1411, November, 1999.

90. G. A. Rovithakis, "Performance of a Neural Adaptive Tracking Controller for Multi-input Nonlinear Dynamical Systems," *IEEE Trans. Syst., Man, Cybern. Part A*, **30**(6), 720–730, November, 2000.

91. Y. Zhang and J. Wang, "Recurrent Neural Networks for Nonlinear Output Regulation," *Automatica*, **37**(8), 1161–1173, August, 2001.

92. C. Kwan, D. M. Dawson, and F. L. Lewis, "Robust Adaptive Control of Robots Using Neural Network: Global Stability," *Asian J. Control*, **3**(2), 111–121, June, 2001.

93. J. Leitner, A. J. Calise, and J. V. R. Prasad, "Analysis of Adaptive Neural Networks for Helicopter Flight Control," *J. Guid., Control., Dynam.*, **20**(5), 972–979, September-October, 1997.

94. M. B. McFarland and A. J. Calise, "Adaptive Nonlinear Control of Agile Anti-air Missiles Using Neural Networks," *IEEE Trans. Control Systems Technol.*, **8**(5), 749–756, September, 2000.

95. A. Yesildirek and F. L. Lewis, "Feedback Linearization Using Neural Networks," *Automatica*, **31**(11), 1659–1664, November, 1995.

96. G. Arslan and T. Basar, "Disturbance Attenuating Controller Design for Strict-Feedback Systems with Structurally Unknown Dynamics," *Automatica*, **37**(8), 1175–1188, August 2001.

97. J. Q. Gong and B. Yao, "Neural Network Adaptive Robust Control of Nonlinear Systems in Semistrict Feedback Form," *Automatica*, **37**(8), 1149–1160, August, 2001.

98. D. Wang and J. Huang, "Neural Network Based Adaptive Tracking of Uncertain Nonlinear Systems in Triangular Form," *Automatica*, **38**, 1365–1372, 2002.

99. S. S. Ge and C. Wang, "Adaptive Neural Control of Uncertain MIMO Nonlinear Systems," *IEEE Trans. Neural Netw.*, **15**(3), 674–692, 2004.

100. J. A. Farrell, "Stability and Approximator Convergence in Nonparametric Nonlinear Adaptive Control," *IEEE Trans. Neural Netw.*, **9**(5), 1008–1020, September, 1998.

101.

102. S. S. Ge and C. Wang, "Adaptive NN Control of Uncertain Nonlinear Pure-Feedback Systems," *Automatica*, **38**(4), 671–682, 2002.

103. C. Wang, D. J. Hill, S. S. Ge, and G. R. Chen, "An ISS-Modular Approach for Adaptive Neural Control of Pure-Feedback Systems," *Automatica*, **42**, 723–731, 2006.

104. S. S. Ge, C. C. Hang, and T. Zhang, "Adaptive Neural Network Control of Nonlinear Systems by State and Output Feedback," *IEEE Trans. Syst., Man Cybern. — Part B Cybern.*, **29**(6), 818–828, 1999.

105. S. S. Ge and J. Zhang, "Neural Network Control of Nonaffine Nonlinear Systems with Zero Dynamics by State and Output Feedback," *IEEE Trans. Neural Netw.*, **14**(4), 900–918, July, 2003.

106. S. S. Ge, J. Zhang, and T. H. Lee, "Adaptive MNN Control for a Class of Non-affine NARMAX Systems with Disturbances," *Syst. Control Lett.*, **53**, 1–12, 2004.

107. K. S. Narendra and J. Balakrishnan, "Adaptive Control Using Multiple Models," *IEEE Trans. Autom. Control*, **42**(2), 171–187, February, 1997.

108. R. Padhi, S. N. Balakrishnan, and T. Randolph, "Adaptive-Critic Based Optimal Neuro Control Synthesis for Distributed Parameter Systems," *Automatica*, **37**(8), 1223–1234, August, 2001.

109. A. S. Poznyak and L. Ljung, "On-line Identification and Adaptive Trajectory Tracking for Nonlinear Stochastic Continuous Time Systems Using Differential Neural Networks," *Automatica*, **37**(8), 1257–1268, August, 2001.

110. T. Parisini, M. Sanguineti, and R. Zoppoli, "Nonlinear Stabilization by Receding-Horizon Neural Regulators," *Int. J. Control*, **70**, 341–362, 1998.

111. T. Parisini and S. Sacone, "Stable Hybrid Control Based on Discrete-Event Automata and Receding-Horizon Neural Regulators," *Automatica*, **37**(8), 1279–1292, August, 2001.

112. B. Ren, S. S. Ge, K. P. Tee, and T. H. Lee, "Adaptive Neural Control for Output Feedback Nonlinear Systems Using a Barrier Lyapunov Function," *IEEE Trans. Neural Netw.*, **21**(8), 1339–1345, 2010.

113. F.-C. Chen and C.-H. Chang, "Practical Stability Issues in CMAC neural network control Systems," *IEEE Trans. Control Systems Technol.*, **4**(1), 86–91, January. 1996.

114. L. B. Gutierrez and F. L. Lewis, "Implementation of a Neural Net Tracking Controller for a Single Flexible Link: Comparison with PD and PID Controllers," *IEEE Trans. Ind. Electron.*, **45**(2), 307–318, April, 1998.

115. J. Y. Choi and J. A. Farrell, "Nonlinear Adaptive Control Using Networks of Piecewise Linear Approximators," *IEEE Trans. Neural Netw.*, **11**(2), 390–401, March 5, 2000.

116. R. M. Sanner and J.-J. E. Slotine, "Structurally Dynamic Wavelet Networks for Adaptive Control of Robotic Systems," *Int. J. Control*, **70**(3), 405–421, 1998.

117. Y. Li, N. Sundararajan, and P. Saratchandran, "Neuro-controller Design for Nonlinear Fighter Aircraft Maneuver Using Fully Tuned Neural Networks," *Automatica*, **37**(8), 1293–1301, August, 2001.

118. R. Zbikowski and K. J. Hunt, *Neural Adaptive Control Technology*, World Scientific, Singapore, 1996.

119. A. G. Barto, "Connectionist Learning for Control," in *Neural Networks for Control*, MIT Press, Cambridge, MA, 1991.

120. A. G. Barto and T. G. Dietterich, "Reinforcement Learning and Its Relationship to Supervised Learning," in J. Si, A. Barto, W. Powell, and D. Wunsch, (Eds.) *Handbook of Learning and Approximate Dynamic Programming*, IEEE Press, West Conshohocken, PA, 2004.

121. R. Sutton, "Learning to Predict by the Method of Temporal Differences," *Mach. Learn.*, **3**, 9–44, 1988.

122. P. Dayan, "The Convergence of TD(λ) for General λ," *Mach. Learn.*, **8**(3-4), 341–362, May, 1992.

123. T. Landelius, Reinforcement Learning and Distributed Local Model Synthesis, Ph.D. Dissertation, Linköping University, 1997.

124. S. Hagen and B. Kröse, "Linear Quadratic Regulation Using Reinforcement Learning," in F. Verdenius and W. van den Broek (Eds.), *Proc. of the 8th Belgian-Dutch Conf. on Machine Learning, BENELEARN'98*, pp. 39–46, Wageningen, October, 1998.

125. D. V. Prokhorov and L. A. Feldkamp, "Analyzing for Lyapunov Stability with Adaptive Critics," Proc. Int. Conf. Systems, Man, Cybernetics, pp. 1658–1661, Dearborn, 1998.

126. X. Liu and S. N. Balakrishnan, "Convergence Analysis of Adaptive Critic Based Optimal Control," Proceedings of the American Control Conference, pp. 1929–1933, Chicago, IL, 2000.

127. C. Anderson, R. M. Kretchner, P. M. Young, and D. C. Hittle, "Robust Reinforcement Learning Control with Static and Dynamic Stability," *Int. J. Robust Nonlinear Control*, 11, 2001.

128. S. Ferrari and R. Stengel, "An Adaptive Critic Global Controller," Proceedings of the American Control Conference, pp. 2665–2670, Anchorage, AK, 2002.

129. J. Murray, C. Cox, R. Saeks, and G. Lendaris, "Globally Convergent Approximate Dynamic Programming Applied to an Autolander," Proc. ACC, pp. 2901–2906, Arlington, VA, 2001.

130. L. Feldkamp and D. Prokhorov, "Recurrent Neural Networks for State Estimation," 12th Yale Workshop on Adaptive and Learning Systems, pp. 17–22, New Haven, CT, 2003.

131. X. Liu and S. N. Balakrishnan, "Adaptive Critic Based Neuro-observer," Proceedings of the American Control Conference, pp. 1616–1621, Arlington, VA, 2001.

132. J. Si and Y.-T. Wang, "On-line Control by Association and Reinforcement," *IEEE Trans. Neural Netw.*, **12**(2), 264–276, March, 2001.

133. D. Prokhorov and D. Wunsch, "Adaptive Critic Designs," *IEEE Trans. Neural Netw.*, **8**(5), September, 1997.

134. J. N. Tsitsiklis, "Efficient Algorithms for Globally Optimal Trajectories," *IEEE Trans. Autom. Control*, **40**(9), 1528–1538, September, 1995.

135. J. Campos and F. L. Lewis, "Adaptive Critic Neural Network for Feedforward Compensation," Proceedings of the American Control Conference, San Diego, CA, June, 1999.

136. G. A. Rovithakis, "Stable Adaptive Neuro-control via Lyapunov Function Derivative Estimation," *Automatica*, **37**(8), 1213–1221, August, 2001.

137. G. Saridis and C. S. Lee, "An Approximation Theory of Optimal Control for Trainable Manipulators," *IEEE Trans. Syst., Man, Cybern.*, **9**(3), 152–159, March, 1979.

138. C. Park and P. Tsiotras, "Approximations to Optimal Feedback Control Using a Successive Wavelet-Collocation Algorithm, Proceedings of the American Control Conference, Vol. **3**, pp. 1950-1955, 2003.

139. S. E. Lyshevski, *Control Systems Theory with Engineering Applications*, Birkhauser, Berlin, 2001.

BIBLIOGRAPHY

M. Abu-Khalaf and F. L. Lewis, "Nearly Optimal State Feedback Control of Constrained Nonlinear Systems Using a Neural Networks HJB Approach," *IFAC Ann. Rev. Control*, **28**, 239–251, 2004.

M. Abu-Khalaf, F.L. Lewis, and J. Huang, "Computational Techniques for Constrained Nonlinear State Feedback H-Infinity Optimal Control Using Neural Networks," Proc. Mediterranean Conf. Control and Automation, Paper 1141, Kusadasi, Turkey, June,2004.

L. von Bertalanffy, *General System Theory*, Braziller, New York, 1968.

C. R. Johnson, Jr., *Lectures on Adaptive Parameter Estimation*, Prentice-Hall, Englewood Cliffs, NJ, 1988.

F. L. Lewis, G. Lendaris, and D. Liu, "Special Issue on Approximate Dynamic Programming and Reinforcement Learning for Feedback Control," *IEEE Trans. Syst., Man, Cybern., Part B*, **38**(4), 2008.

F. L. Lewis, K. Liu, and A. Yesildirek, "Neural Net Robot Controller with Guaranteed Tracking Performance," *IEEE Trans. Neural Netw.*, **6**(3), 703–715, 1995.

F. L. Lewis, "Nonlinear Network Structures for Feedback Control," *Asian J. Control*, **1**(4), 205–228, December, 1999.

J .M. Mendel and R. W. MacLaren, "Reinforcement Learning Control and Pattern Recognition Systems," in J. M. Mendel and K. S. Fu (Eds.), *Adaptive, Learning, and Pattern Recognition Systems: Theory and Applications*, pp. 287–318, Academic, New York, 1970.

H. Miyamoto, M. Kawato, T. Setoyama, and R. Suzuki, "Feedback-Error-Learning Neural Network for Trajectory Control of a Robotic Manipulator," *Neural Netw.*, **1**, 251–265, 1988.

K. S. Narendra, "Adaptive Control of Dynamical Systems Using Neural Networks," in D. A. White and D. A. Sofge (Eds.), *Handbook of Intelligent Control*, pp. 141–183, Van Nostrand Reinhold, New York, 1992.

K. S. Narendra and J. Balakrishnan, "Adaptive Control Using Multiple Models," *IEEE Trans. Autom. Control*, **42**(2), 171–187, February, 1997.

R. Selmic, F. L. Lewis, A. J. Calise, and M. B. McFarland, "Backlash Compensation Using Neural Network," U.S. Patent 6,611,823, 26 August, 2003.

K. G. Vamvoudakis and F. L. Lewis, "Online Neural Network Solution of Nonlinear Two-Player Zero-Sum Games Using Synchronous Policy Iteration," *Int. J. Robust Nonlinear Control*.

J. Wang and J. Huang, "Neural Network Enhanced Output Regulation in Nonlinear Systems," *Automatica*, **37**(8), 1189–1200, August, 2001.

T. Zhang, S. S. Ge, and C.C. Hang, "Adaptive Neural Network Control for Strict Feedback Nonlinear Systems Using Backstepping Design," *Automatica*, **36**(12), 1835–1846, 2000.

CHAPTER 24

MECHATRONICS

Shane Farritor and Jeff Hawks
University of Nebraska–Lincoln
Lincoln, Nebraska

Mechatronics is the integration of computers, electronics, and information sciences into mechanical engineering. The advent of cheap and small microcomputers and electronics has lead to the widespread integration into mechanical systems. The goal of this integration is to improve the performance of mechanical systems. Today's mechanical engineers cannot design state-of-the art systems without considering the cross-disciplinary field of mechatronics.

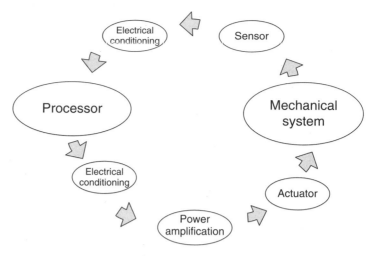

Figure 1 a Mechatronic system.

A typical mechatronic system is pictured in Fig. 1. It shows a mechanical system whose performance is monitored by some type of sensor. The sensor output is conditioned so it can be input into a microcomputer usually embedded in the product. The microcomputer determines an input to the mechanical system that will improve performance. The microcomputer outputs this information, which is converted and amplified to cause an actuator to alter the behavior of the mechanical system. This sense–plan–act cycle is repeated (usually several times per second) to improve the performance of the mechanical system.

Knowledge of mechatronics first and foremost requires a solid understanding of mechanical engineering. Mechanical engineering and control systems are covered throughout this book, so this chapter will give a basic overview of the electrical engineering and computer engineering. Each of these topics is discussed in this chapter.

1 BASIC ANALOG ELECTRONICS

Basic circuit analysis and design is critical in the field of mechatronics. With improving computer technology it is tempting to treat all systems as digital, however, this is an analog world. Analog circuits are often the best choice to condition and modify signals to/and from analog actuators and sensors.

1.1 Definitions

Engineering modeling and analysis is often described as the analysis of the flow of power between various elements. The state of an engineering element can be defined by two variables—the effort variable and the flow variable. The instantaneous power for that element is defined as the product of the effort variable and the flow variable. For example, in traditional linear mechanical systems the power is defined as the product of the velocity of a mass and the net applied force to that mass. In fluid systems this can be written as the product of the volumetric flow rate and the pressure.

These analogies are often helpful to mechanical engineers trying to understand electrical circuits. In electrical systems the power is defined as the product of the electrical current (amperage) and the electrical potential (voltage).

The basic definitions of the variables used in circuit analysis are given below:

1. *Current*. Electrical current is usually designated with the symbol i (or I). It is defined as the time rate of change of the charge (designated q with units of coulombs). The electrical current has the units of the ampere (or amp).

$$i = \text{time rate of change of charge} = \frac{dq}{dt} \quad q = \text{charge (Co)}$$

2. *Voltage*. Voltage is the electromotive force or potential difference that is designated with the symbol V (or e, E, v). It is a measure of the potential of an electric field. The electrical voltage has the units of volts and is generally written

$$V_{AB} = \text{potential of } A \text{ with respect to } B$$

Or more commonly written with a $+$ at point A and a $-$ at point B

3. *Power*. Power is defined as the ability to do work. The flow of power is generally what is described by engineering models. Power has the units of watts and is the product of the voltage and current.

4. *Work*. Work (or energy) has units of joules and is the integral of power over time.

$$W = \int_{t_o}^{t_f} P(t)\, \partial t \qquad (1)$$

1.2 Basic Electrical Elements

Several elements and their constitutive relationships need to be understood to analyze basic electrical circuits. The constitutive relationship for each element describes the relationship between the voltage (effort variable) and the current (flow variable).

1. *Ideal Voltage Source*. The ideal voltage source produces a voltage, or electrical potential, as a function of time. This voltage produced by the source is an input to the system and the voltage source adds power to the system. The current passing through the ideal voltage source is not generally known. This current is determined by the other elements connected to the voltage source. The ideal voltage source is capable of producing infinite current and therefore infinite power. See Fig. 2.

2. *Ideal Current Source*. The ideal current source produces a current as a function of time. This current produced by the source is an input to the system and the current source adds power to the system. The voltage across the ideal current source is not generally known. This voltage is determined by the other elements connected to the current source. The ideal current source is capable of producing infinite voltage and therefore infinite power. See Fig. 3.

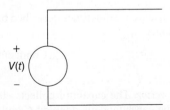

Figure 2 Ideal voltage source.

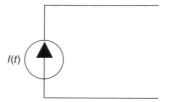

Figure 3   Ideal current source.

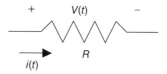

Figure 4   Ideal resistor.

3. *Resistor*. The resistor restricts the flow of current and dissipates power. The magnitude of this resistance is denoted R and described with the units of ohms designated by Ω. The resistor is a passive element in that it does not add energy to the system. The resistor is governed by a passive sign convention, as shown in Fig. 4, in that the current flows across the resistor from the side of higher potential (voltage) to the side of lower potential (voltage). If the current (and the voltage) are negative, this indicates that current flow is in the opposite direction and the potential is higher on the opposite side. In linear resistors the relationship between voltage and current is linear. Linear resistors are governed by Ohm's law given by $V(t) = i(t)R$.

The magnitude of the resistance of a long cylindrical wire is related to the resistivity of the material the wire is made of (copper and gold have low resistivity) and the length of the wire. It is inversely related to the cross sectional area of the wire:

$$R = \frac{\rho \ell}{A} = \frac{\text{(resistivity)(length)}}{\text{(cross-sectional area)}} \tag{2}$$

Series and parallel combinations of resistors can be thought of as an effective resistance. Resistors connected in series (share common current) create an effective resistance equal to the sum of the resistors. Resistors connected in parallel create an effective resistor where the inverse of the effective resistance is related to the sum of the inverse of each of the resistors. See Fig. 5.

If there are two parallel resistors, this expression becomes

$$R_E = \frac{R_1 R_2}{R_1 + R_2} \tag{3}$$

The resistance is sometimes described by the conductance, G, that is defined as the inverse of the resistance:

$$\text{Conductance} = G = \frac{1}{R} \tag{4}$$

4. *Capacitor*. The capacitor collects electrical charge and stores energy. The magnitude of the capacitance is denoted C and described with the units of farads designated by F (or coulombs per volt). The capacitor is an energy storage element. The capacitor,

Series

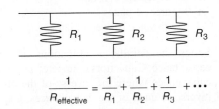

R_1 R_2 R_3

$R_{\text{effective}} = R_1 + R_2 + R_3 + \cdots$

Parallel

$$\frac{1}{R_{\text{effective}}} = \frac{1}{R_1} + \frac{1}{R_2} + \frac{1}{R_3} + \cdots$$

Figure 5   Combinations of resistors.

like the resistor, is governed by a passive sign convention as shown in the figure in that the current flows "across" the capacitor from the side of higher potential (voltage) to the side of lower potential (voltage). If the current (and the voltage) are negative, this indicates that current flow is in the opposite direction and the potential is higher on the opposite side. See Fig. 6.

In linear capacitors, the voltage is linearly related to the time integral of the current. Linear capacitors are governed by the following equation:

$$V(t) = \frac{1}{C} \int_0^t i(t)\, \partial t \tag{5}$$

or

$$i(t) = C\frac{\partial V(t)}{\partial t} \tag{6}$$

The simplest capacitor is a parallel-plate capacitor that consists of two plates separated by a non conducting material called a dielectric. The magnitude of the capacitance for a parallel plate capacitor is proportional to the dielectric constant of the material property between the plates and the area of the plates. It is inversely related to the distance between the plates. Such capacitors are often used as position sensors either by changing the distance between the plates or by changing the area of overlap between the plates. See Fig. 7.

$$C = \frac{KA}{d} \tag{7}$$

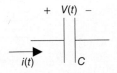

$+$ $V(t)$ $-$

$i(t)$ C

Figure 6   Ideal capacitor.

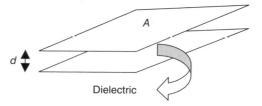

Figure 7 Parallel-plate capacitor.

where K = dielectric constant
A = area of plates
d = distance between plates

Series and parallel combinations of capacitors can be thought of as an effective capacitance. Capacitors connected in series (share common current) create an effective capacitance where the inverse of the effective resistance is related to the sum of the inverse of each of the capacitances. Capacitors connected in parallel create effective capacitance equal to the sum of the capacitances. See Fig. 8.

5. *Inductors.* The inductor stores magnetic energy by creating a magnetic field. The magnitude of the inductor is denoted L and described with the units of henrys designated by H. The inductor is an energy storage element. The inductor, like the resistor, is governed by a passive sign convention as shown in the figure in that the current flows through the inductor from the side of higher potential (voltage) to the side of lower potential (voltage). If the current (and the voltage) are negative, this indicates that current flow is in the opposite direction and the potential is higher on the opposite side. See Fig. 9. In linear inductors, the drop across the inductor, $V(t)$, is linearly related (by L) to the time rate of change of the current passing through the inductor.

Linear inductors are governed by the following equation:

$$V_{(t)} = L\frac{dI_{(t)}}{dt} \tag{8}$$

$$I(t) = \frac{1}{L}\int_0^t V(t)\,\partial t \tag{9}$$

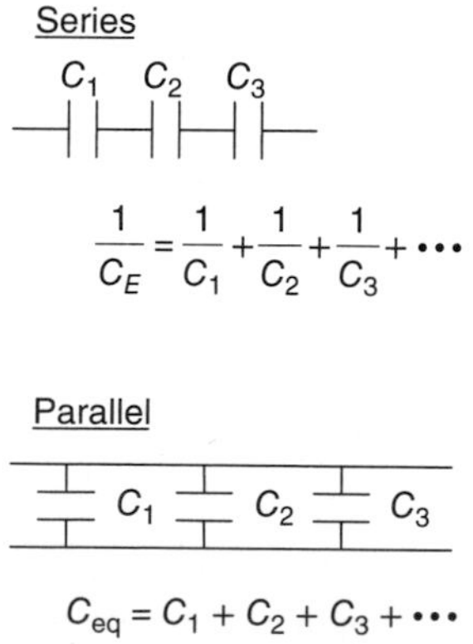

Series

C_1 C_2 C_3

$$\frac{1}{C_E} = \frac{1}{C_1} + \frac{1}{C_2} + \frac{1}{C_3} + \cdots$$

Parallel

C_1 C_2 C_3

$$C_{eq} = C_1 + C_2 + C_3 + \cdots$$

Figure 8 Combinations of capacitors.

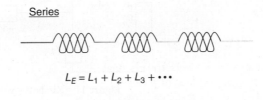

Figure 9   Ideal inductor.

Series

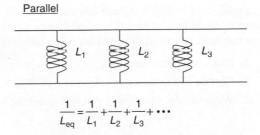

$$L_E = L_1 + L_2 + L_3 + \cdots$$

Parallel

$$\frac{1}{L_{eq}} = \frac{1}{L_1} + \frac{1}{L_2} + \frac{1}{L_3} + \cdots$$

Figure 10   Combinations of inductors.

Series and parallel combinations of inductors can be thought of as an effective inductance. Inductors connected in series (share common current) create effective inductance equal to the sum of the inductors. Inductors connected in parallel create an effective inductance where the inverse of the effective inductance is related to the sum of the inverse of each of the inductors. See Fig. 10.

1.3   Circuit Analysis

Basic circuit analysis is built on two laws (1) Kirchoff's Voltage Law, and (2) Kirchoff's current law. There are several approaches and techniques to circuit analysis, but a straightforward application of these two laws as presented here will be effective in all situations. This straightforward application, however, may lead to some additional algebra. Experience with these problems can lead to shortcuts.

Kirchoff's Voltage Law

$$\sum_{i=1}^{n} V_i = 0 \tag{10}$$

Kirchoff's Voltage Law states that the sum of voltages around a closed loop is zero. To a mechanical engineer this is analogous to the movement of a mass in a potential field (such as gravity). For example, it is clear that if a bowling ball is picked up, rolled down the lane, and then returned to the original position there is no net change in the ball's potential energy as it moved through this closed path.

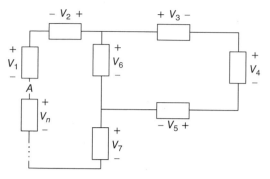

Figure 11   Kirchoff's voltage law.

Similarly, Kirchoff's voltage law states that the sum of the potential changes (voltages) around a closed loop is zero. The closed loops that can be used to formulate these equations are not unique. However, they will not all be independent. See Fig. 11.

1. Pick a start point (e.g., A).
2. Choose a loop direction [clockwise (CW)].
3. Write out the equation as you go around the loop:

$$V_1 + V_2 - V_3 + \cdots + V_n = 0 \tag{11}$$

Kirchoff's Current Law

$$\sum_{i=1}^{N} I_i = 0 \tag{12}$$

Kirchoff's Current Law states that the net sum of the current flowing into a node is zero. To a mechanical engineer this is analogous to fluid flow into and out of a node of a pipe network. If no fluid is stored in the node, then mass conservation requires that the mass flow rates into (positive flow) and out of (negative flow) the node must be zero (i.e. what goes in must come out). See Fig. 12.

1. Assume current directions and draw arrows.
2. Assume current directions must be consistent with assumed voltage drops.
3. Apply law (assume current into node is +):

$$I_1 + I_2 - I_3 = 0 \tag{13}$$

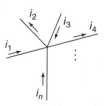

Figure 12   Kirchoff's current law.

1.4   Sources and Meters

Some practical considerations are important when building mechatronic systems. Sources and meters (and all elements) do not behave as the ideal representation described above. Some of the limitations are described here.

1. *Voltage Source.* An ideal voltage source as described above does not have any output resistance (no resistor shown in the above representation) and can supply infinite current. Obviously, a real voltage source does have an output resistance. The output resistance (usually < 1Ω) can be represented as the model given below. Now, as current is output from the source, the output voltage (V_{out}) is no longer the same voltage as the source voltage (V_s). The magnitude of this output resistance (and the corresponding voltage drop) changes with many factors. For example, the output resistance of a rechargeable battery (such as a car battery) increases as the battery ages. The most significant difference between an ideal voltage source and a real source is that the real source can provide only a limited current. This is a limitation on both the instantaneous current that can be produced as well as the total amount (time integral) of current that can be produced. For example, the power supply limitation on instantaneous current is often a constraint for battery-operated devices such as cell phones or robots. See Fig. 13.

2. *Current Source.* An ideal current source as described above does not have any output resistance (no resistor shown in the above representation) and can supply infinite voltage. Obviously, a real current source does have an output resistance. The output resistance (usually > 1 MΩ) can be represented as the model given below. Now, as current is output from the source, the output current (I_{out}) is no longer the same current as the source current (I_s). The magnitude of this output resistance (and the corresponding voltage drop) changes with many factors. The most significant difference between an ideal current source and a real source is that the real source can provide only a limited voltage. This is a limitation on both the instantaneous voltage that can be produced as well as the total amount (time integral) of voltage that can be produced. See Fig. 14.

3. *Voltmeter.* An ideal voltmeter has infinite input resistance and draws no current from the voltage being measured. A real voltmeter has a finite input resistance as modeled below and does draw some current from the source, which can change the voltage being measured. However, the input resistance of most real voltmeters is very large (usually several megaohms), and this makes the voltmeter a very safe device as it does not draw "significant" current. See Fig. 15.

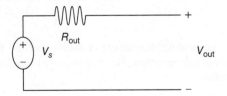

Figure 13   Real voltage source.

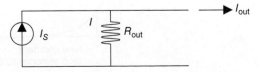

Figure 14   Real current source.

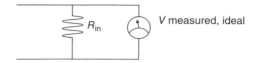

Figure 15 Real voltage meter.

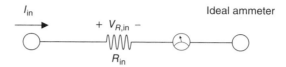

Figure 16 Real current meter.

4. *Ammeter.* An ideal ammeter has zero input resistance and does not produce a voltage drop. A real ammeter has a finite input resistance as modeled below and does have a small voltage drop across the leads of the meter. However, the input resistance of most real ammeters is very small (a few ohms), and this makes the ammeter a device that requires careful consideration before use. If the ammeter leads are placed between two points with a potential difference, a short circuit will occur and very large current will be produced. Therefore, most devices that function as ammeters and voltmeters (multimeters) require that the user remove the leads and place them into new terminals before the devices are used as an ammeter. This requires the user to be very deliberate about using the ammeter. See Fig. 16.

5. *Ohmmeter.* An ohmmeter is used to measure the resistance between the two terminals. The ohmmeter consists of a voltage source and an ammeter. A voltage is applied between the leads, the current is measured, and the resulting resistance is found with Ohm's law.

1.5 *RL* and *RC* Transient Response

To develop a basic understanding of the effects of inductors and capacitors in circuits, it is useful to understand *RL* and *RC* circuits. These circuits contain one resistor and one inductor (*RL*) or it contains one resistor and one capacitor (*RC*). The equations of these circuits (developed with the laws above) are first-order differential equations with constant coefficients of the form

$$X_{(t)} = B\frac{dy}{dt} + Cy \tag{14}$$

If we assume step inputs, a solution of the forms in Figs. 17 and 18 is assumed as the solution to this equation. These first-order responses to step inputs rise or decay according to a time constant *T*.

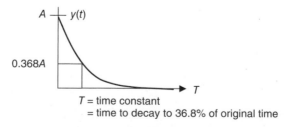

T = time constant
    = time to decay to 36.8% of original time

Figure 17 First-order response.

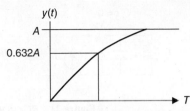

T = time to rise to 63.2% of final value

Figure 18 First-order response.

1. First-order rise:

$$X_{(t)} = Ae - \frac{1}{T}t \qquad (15)$$

2. First-order decay:

$$X_{(t)} = A(1 - e^{\alpha t}) \qquad (16)$$

Consider the following circuits and the resulting equations, solutions, and time responses to step inputs.

1. *RL* circuits:

$$T = \frac{L}{R} \qquad (17)$$

Differential equation:

$$Ri + L\frac{di}{dt} = V \qquad (18a)$$

Solution:

$$i = \frac{V}{R}(1 - e^{-t/T}) \quad \text{(see Fig. 19)} \qquad (18b)$$

Differential equation:

$$Ri + L\frac{di}{dt} = 0 \qquad (19a)$$

Solution:

$$i = \frac{V}{R}(1 - e^{-t/T}) \quad \text{(see Fig. 20)} \qquad (19b)$$

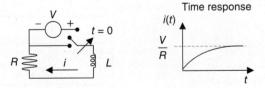

Figure 19 *RL* circuit.

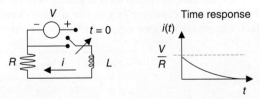

Figure 20 *RL* circuit.

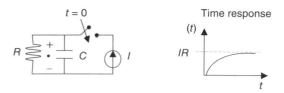

Figure 21   *RC* circuit.

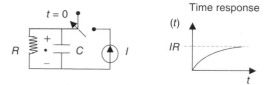

Figure 22   *RC* circuit.

2. *RC* circuits:

$$T = RC \tag{20}$$

Differential equation:

$$\frac{V}{R} + C\frac{dV}{dt} = I \tag{21a}$$

Solution:

$$V = RI(1 - e^{-t/T}) \quad \text{(see Fig. 21)} \tag{21b}$$

Differential Equation

$$\frac{V}{R} + C\frac{dV}{dt} = 0 \tag{22a}$$

Solution:

$$V = RI\,e^{-\frac{t}{T}} \quad \text{(See Fig. 22)} \tag{22b}$$

The resulting observations from these circuits are that the voltage does not change instantly across a capacitor; however, the current can change instantly. Also, the voltage can change instantly across an inductor; however, the current cannot change instantly.

Charge instantaneously?

	V	I
C	NO	YES
L	YES	NO

The instantaneous behavior (*t* is small) steady-state behavior (*t* is large) is also interesting. This shows that a capacitor begins like a short circuit and becomes an open circuit. Also, an inductor begins as an open circuit and becomes a short circuit.

	Capacitor	Inductor
$t = 0$	Short	Open
$t = \infty$	Open	Short

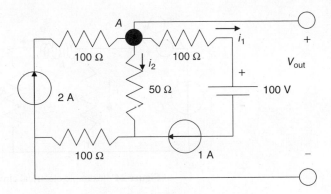

Figure 23 Example electrical circuit.

1.6 Basic Electronics Examples

Consider the circuit in Fig. 23. Applying Kirchoff's current law to node A reveals

$$i_1 + i_2 = 1 \tag{23}$$

Now Kirchoff's voltage law can be written

$$500i_1 + 100i_2 - 100 - 200i_2 = 0 \tag{24}$$

and it is possible to substitute for

$$i_1(i_1 = 1 - i_2) \tag{25}$$

$$500 - 500i_2 - 100i_2 - 100 - 200i_2 = 0 \tag{26}$$

$$800i_2 = 400 \tag{27}$$

$$800i_2 = 400 \tag{28}$$

where $i_2 = \frac{1}{2}$ and $i_1 = 1$

Now Kirchoff's voltage law can be written for the outside loop:

$$V_{out} - 200 - 100i_2 - 100 = 0 \tag{29}$$

Therefore,

$$V_{out} = 350 \ \text{V} \tag{30}$$

Transfer function for an RL circuit

Consider the circuit in Fig. 24.

Kirchoff's current law can be written

$$I_{in} = i_1 + i_2 \tag{31}$$

Kirchoff's voltage law

$$V_{out} + L\frac{di_2}{dt} - R_1 i_1 = 0 \tag{32}$$

and it also can be written as

$$\frac{V_{out}}{R_L} = i_2 \tag{33}$$

now there can be a substitution

$$V_{out} + \frac{L}{R_L}\frac{dV_{out}}{dt} - R_1\left(I_{in} - \frac{V_{out}}{R_L}\right) = 0 \tag{34}$$

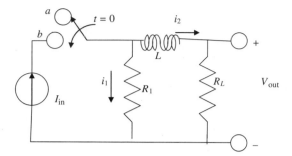

Figure 24 example electrical circuit.

Now the transfer function can be found using the LaPlace transformation:

$$V_{out}(s) + \frac{L}{R_L} s V_{out}(s) - R_1 I_{in}(s) + -\frac{R_1}{R_L} V_{out}(s) = 0 \qquad (35)$$

Rearranging revels

$$\frac{V_{out}(s)}{I_{in}(s)} = \frac{R_1}{(L/R_L)s + (1 + R_1/R_L)} \qquad (36)$$

2 SEMICONDUCTOR ELECTRONICS

Semiconductor electronics, such as diodes and transistors, are used in many circuits and devices for switching and amplification. As the name implies, the materials that make semiconductor devices exhibit neither the low resistance typical of conductors nor the very high resistance associated with insulators. These electronic devices are commonly used as an interface in the operation of real devices, such as motors.

2.1 Diodes

An ideal diode is a passive, non-linear element that allows current to flow in one direction. It does not store, add, or dissipate energy. Diodes are analogous to a hydraulic check valve, which allows fluid to flow in one direction. When current flows in the direction (anode to cathode) indicated in Fig. 25, the resistance, and consequently the voltage, of an ideal diode is zero. The ideal diode behaves as a short circuit (or closed switch). Current does not flow in the

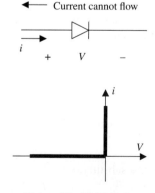

Figure 25 Ideal diode.

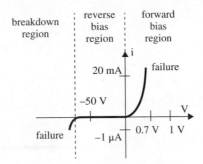

Figure 26 Real diode curve.

direction opposite to what is shown in Fig. 11. In this direction, an ideal diode has infinite resistance. Therefore, current can only be positive and voltage can only be negative as shown in Fig. 25.

A real diode consists of a forward bias region, a reverse bias region, and a breakdown region, shown in Fig. 26. The forward bias region refers to where the anode is at a higher potential than the cathode. Once the voltage exceeds a given threshold, known as the forward voltage (V_f) for the diode, the current flow through the diode increases drastically. A real diode requires about 0.7 V of forward bias to enable significant current flow. Below this voltage, there is very little current flow. In the reverse bias region, only a tiny leakage current (I_l) flows in the diode from cathode to anode until the reverse voltage exceeds a threshold known as the reverse-breakdown voltage (V_r). When this happens, significant currents flow in the opposite direction resulting in a destructive process that should be avoided.

2.2 Analysis of Diode Circuits

The circuit shown in Fig. 27 contains diodes. The procedure for solving diode circuits is as follows:

1. Assume a direction for the currents.
2. Substitute the proper diode state based on the assumed current directions. For example, in Fig. 27 there are four possible options (both shorted, both open, and two variations of one diode open and one closed).
3. Solve the circuit.
4. If solved currents match up with the assumed current directions (signs), the circuit analysis is completed.
5. If the solved currents do not match up with the assumed current directions, then go back to step 1 and repeat the process using different assumptions for the current directions.

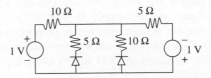

Figure 27 Example diode circuit.

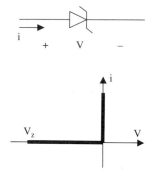

Figure 28 Ideal Zener diode.

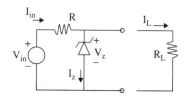

Figure 29 Zener diode voltage regulator.

2.3 Zener Diode

A Zener diode is indicated by the symbol shown in Fig. 28. Zener diodes do not consume power and behave as a non-linear element with three distinct regions shown in Fig. 28. The Zener diode behaves like a normal diode when it is forward biased. For example, when voltage is positive, the resistance is zero and the Zener diode behaves like a regular diode. However, Zener diodes have a set breakdown voltage, or Zener voltage (V_z), that can be much lower than the breakdown voltage for a normal diode. When the voltage is between $-V_z$ and zero, the resistance is infinite and no current flows. Finally, when the voltage is less than $-V_z$ current can flow without destroying itself.

A common use of Zener diodes is to regulate the output voltage in a circuit when the input voltage is variable or unstable. A Zener diode voltage regulator, shown in Fig. 29, produces a smaller, regulated voltage from a variable input voltage. The Zener voltage must be less than the input voltage ($V_z < V_{in}$). This is an effective low-cost solution if the output current load is relatively small and constant. One example is the use of a Zener diode voltage regulator to reduce a 9-V battery supply to 5 V, which is commonly found in integrated circuits.

2.4 Light-Emitting Diode

Another common form of diode is a light-emitting diode (LED). These diodes emit light when forward biased. The voltage drop is typically 1.5 – 2.5 V, which means they can be powered from a digital voltage supply (5 VDC). They can be powered by only a few milliamps, and the amount of light emitted is proportional to the amount of current passing through the LED. Colored plastic casing are used to encase the LED and produce different color LEDs. Fig. 30 shows an LED and its symbol. Notice that the anode (positive terminal) has a longer lead.

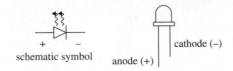

Figure 30   Light-emitting diode (LED).

2.5   Photodiode

A photodiode behaves in a manner opposite to an LED. When light shines on the diode, reverse current (cathode to anode) is allowed to flow. Similarly to the LED, the amount of current is proportional to the amount of light shining on the diode. For this reason, photodiodes are typically used as light sensors.

2.6   Transistor

A transistor is a solid-state switch that opens or closes a circuit. The switching action is non-mechanical and due to the change in the electrical characteristics of the device. Transistors require power, but one advantage is that they can use a small current to control a large current. Current flows in only one direction, and using transistors as a switch in this manner is the base of digital logic circuits. A bipolar junction transistor (BJT), shown in Fig. 31, has three terminals called the base (B), the emitter (E), and the collector (C). V_{CE} is the voltage between the collector and emitter, and V_{BE} is the voltage between the base and emitter. The relationships involving the transistor currents and voltages are as follows:

$$I_E = I_C + I_B$$
$$V_{BE} = V_B - V_E$$
$$V_{CE} = V_C - V_E \tag{37}$$

The BJT has three operational states. When $V_{BE} <\sim 0.6$ V, the transistor is in an off state (non-conducting). In the off state, no current flows between the collector and emitter, so $I_C = 0$. When V_{BE} is $\geq \sim 0.6$ V and $<\sim 0.7$ V and $V_{CE} > 0.2$ V, the transistor is in a linear operation state. In this state, I_C is linearly related to I_B by the following relationship:

$$I_C = \beta(I_B) \tag{38}$$

where β is the current gain, which is often approximately 100, but it could range from 50 to 1000 due to variations in the manufacturing process. There is also a variation with temperature and loading. Finally, when V_{BE} is $\geq \sim 0.7$ V, then transistor is operating in the saturation state. In this state, current flows between the collector and emitter. V_{CE} has value of ~ 0.2 V, and I_C

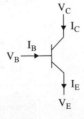

Figure 31   Bipolar junction transistor (BJT).

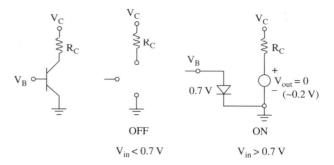

Figure 32 Ideal transistor switch model.

depends on the load of the attached circuit (R_C). The ideal transistor switch model is shown in Fig. 32.

3 OPERATIONAL AMPLIFIERS

Operational amplifiers, or op amps, are extremely useful in mechatronics. They are primarily used to modify an analog signal such as increasing the voltage output of a strain gauge circuit or removing noise from a signal. They are often used to amplify, sum, subtract, integrate, or differentiate analog signals. They are also used to create active filters. They are active devices in that they add energy to the circuit.

Consider the ideal op-amp model shown below. All voltages shown in the figure are referenced to the same ground. As active elements the op amp adds energy to the circuit through the power supplied by $+V_s, -V_s$ (V_s is often, but doesn't have to be 15 V). The supply voltages are often left off Fig. 33 for convenience.

Circuit analysis of op amps requires only two additional observations. For ideal op-amp behavior it must be remembered that

$$\Delta V = (V_+ - V_-) = 0 \tag{39}$$

or

$$V_- = V_+ \tag{40}$$

and

$$i_- = i_+ = 0 \tag{41}$$

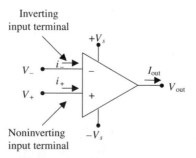

Figure 33 Ideal operational amplifier.

It should also be noted that in general

$$i_{\text{out}} \neq 0 \tag{42}$$

Op amps produce an output voltage (V_{out}) that is related to an input voltage (V_{in}, not necessarily V_+ or V_-). The relationship can frequently be viewed as a gain defined by

$$A_v = \frac{V_{\text{out}}}{V_{\text{in}}} \tag{43}$$

The op amp generally has a very high input impedance (ideally infinite) defined by

$$Z_{\text{in}} = \frac{V_{\text{in}}}{i_{\text{in}}} \tag{44}$$

For real op amps the input impedance is usually > 100 kΩ. Op amps also have a very low output impendence (ideally zero) defined by

$$Z_{\text{out}} = \frac{V_{\text{out}}}{i_{\text{out}}} \tag{45}$$

For real op amps this is usually less than a few ohms.

Two of the most common op-amp circuits are the inverting op amp and the non-inverting op amp, and these two circuits are analyzed in Fig. 34 and 35.

3.1 Inverting Amplifier

Apply op-amp rules:

$$\Delta V = 0 \tag{46}$$

or

$$V_- = V_+ \tag{47}$$

$$i_- = i_+ = 0 \tag{48}$$

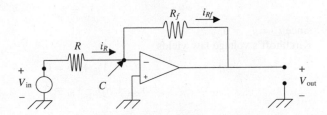

Figure 34 Inverting operational amplifier.

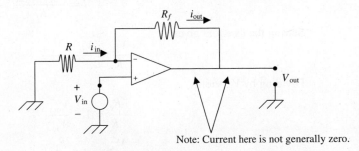

Note: Current here is not generally zero.

Figure 35 NonInverting operational amplifier.

Kirchhoff's current law gives

$$i_R = i_{Rf} = i \tag{49}$$

Kirchhoff's voltage law gives

$$V_{\text{in}} - iR = 0 \tag{50}$$

$$\frac{V_{\text{in}}}{R} = i \tag{51}$$

$$V_{\text{out}} + iR_f = 0 \tag{52}$$

$$-\frac{V_{\text{out}}}{R_f} = i \tag{53}$$

Setting $i_{\text{in}} = i_{\text{out}}$ gives

$$\frac{V_{\text{in}}}{R} = \frac{-V}{R_f} \tag{54}$$

or

$$\frac{V_{\text{out}}}{V_{in}} = -\frac{R_f}{R} \tag{55}$$

Notice for the inverting amplifier the gain is negative, hence the name "inverting." This means that for periodic signals there would be a 180° phase shift between the output voltage and the input voltage. The gain for this circuit is

$$A = -\left(\frac{R_f}{R}\right) \tag{56}$$

$$V_{\text{out}} = A V_{\text{in}} \tag{57}$$

3.2   Noninverting Amplifier

Kirchhoff's current law yields

$$i_{\text{in}} = i_{\text{out}} \tag{58}$$

since $i_+ = i_- = 0$
Kirchhoff's voltage law yields

$$V_{\text{in}} + Ri = 0 \tag{59}$$

$$V_{\text{out}} + R_f i - V_{\text{in}} = 0 \tag{60}$$

$$i = -\frac{V_{\text{in}}}{R} \tag{61}$$

$$i = \frac{V_{\text{in}} - V_{\text{out}}}{R_f} \tag{62}$$

Setting the i's equal gives

$$-\frac{V_{\text{in}}}{R} = \frac{V_{\text{in}} - V_{\text{out}}}{R_f} \tag{63}$$

Dividing by V_{in} yields

$$-\frac{R_f}{R} = 1 - \frac{V_{\text{out}}}{V_{\text{in}}} \tag{64}$$

$$\frac{V_{\text{out}}}{V_{\text{in}}} = 1 + \frac{R_f}{R} \tag{65}$$

Gain ≥ 1

Here V_{out} and V_{in} have the same sign so the amplifier is noninverting. This circuit is often used as a buffer to "isolate" one portion of a circuit from another.

3.3   Other Common Op-Amp Circuits

Some other useful op-amp circuits are described in Fig. 36.

Type	Schematic	Input/Output Equation
Summer	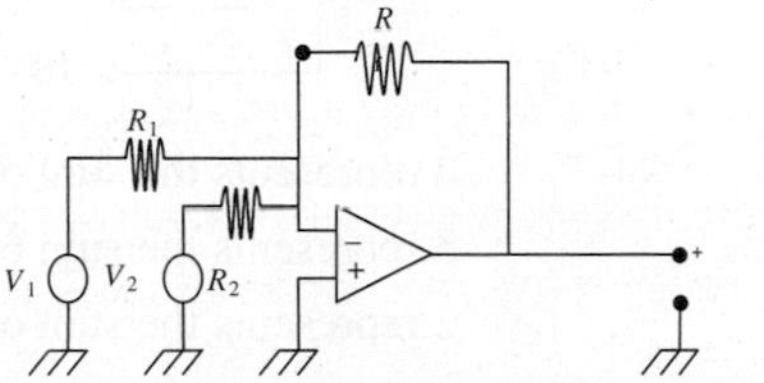	$V_{\text{out}} = -(V_1 + V_2)$
Difference amplifier	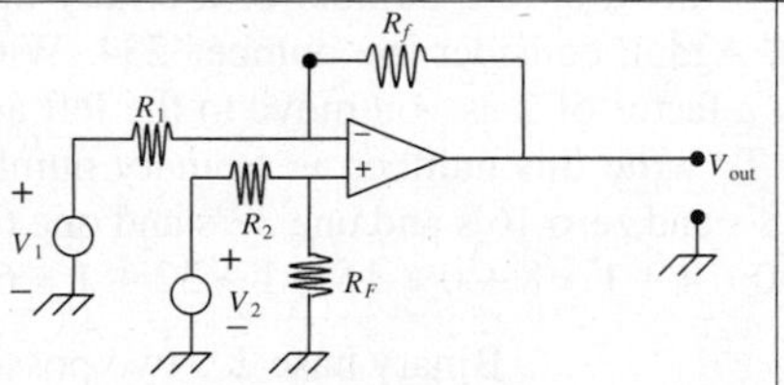	$V_{\text{out}} = \dfrac{R_f}{R}(V_2 - V_1)$
Integrator	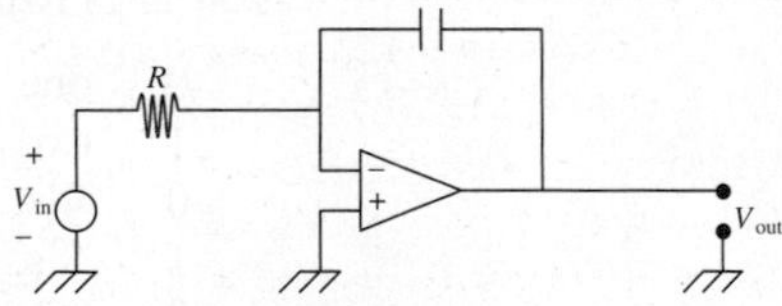	$V_{\text{out}}(t) = -\dfrac{1}{RC}\displaystyle\int_0^t V_{\text{in}}(x)\,dx$
Differentiator	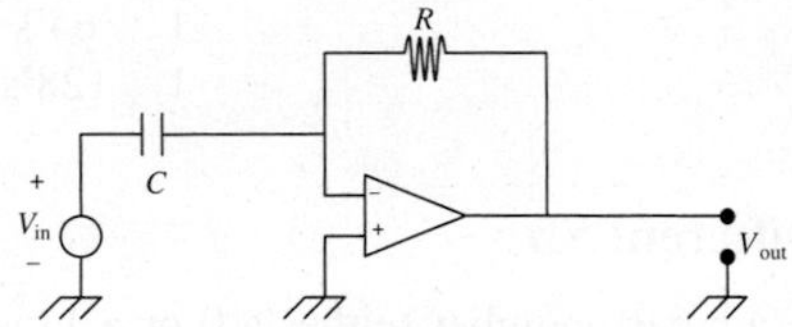	$-RC\left(\dfrac{dV_{\text{in}}}{dt}\right) = V_{\text{out}}(t)$
Gain and shift amplifier	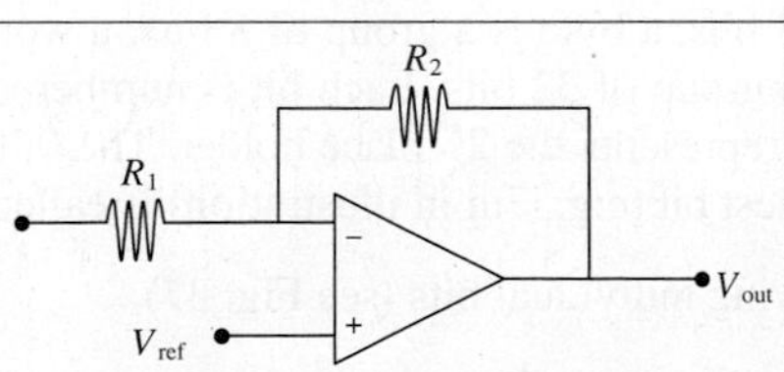	$V_{\text{out}} = -\dfrac{R_2}{R_1}V_1 + \left(1 + \dfrac{R_2}{R_1}\right)V_{\text{ref}}$ if $R_1 = R_2 = R$ $V_{\text{out}} = \dfrac{R_f}{R}(V_2 - V_1)$

Figure 36   Useful op-amp circuits.

4 BINARY NUMBERS

Computers and digital electronics used in mechatronic systems are described by binary arithmetic. Therefore, it is important to understand binary numbers to fully understand the function of computers. First, consider a base 10 number as in standard mathematics. Base 10 numbers have 10 possible digits (0, 1, 2, 3, 4, 5, 6, 7, 8, 9). Consider a number such as 234. Here the 4 represents the 1's digit, the 3 represents the 10's digit, and the 2 represents the 100's digit. The number 234 represents four 1's and three 10's and two 100's (4 * 1 + 3 * 10 + 2 * 100 = 234). The value of each digit increases by a factor of 10 as you move to the left and decreases by a factor of 10 as you move to the right.

Consider a number such as 234.

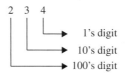

4 represents the sum of nine 1's

3 represents the sum of seven 10's

2 represents the sum of one 100

Now, consider a base-2 number or a binary number. Base-2 numbers have two possible digits (0, 1). Again, consider the number 234. With binary numbers the value of each digit increases by a factor of 2 as you move to the left and decreases by a factor of 2 as you move to the right. To write this number as a binary number requires zero 1's and one 2's and zero 4's and one 8's and zero 16's and one 32's and one 64's and one 128's ($234 \equiv 11101010 = 0 * 1 + 1 * 3 + 0 * 4 + 1 * 8 + 0 * 16 + 1 * 32 + 1 * 64 + 1 * 128$).

Binary base 2 : two possible digits (0, 1)

$$234 \equiv 11101010$$

0	one
1	two
0	4's
1	8's
0	16's
1	32's
1	64's
1	128's

4.1 Binary Numbers of Different Size

Each digit in a binary number (either a 0 or a 1) is called a <u>bit</u> – "binary digit." A nibble is a group of 4 bits, a byte is a group of 8 bits, a word is a group of 16 bits, and a double word (dword) is a group of 32 bits. Each bit is numbered starting with zero and moving to the left. The Kth bit represents the 2^K place holder. The 0th bit is called the least significant bit (LSB) and the highest bit (e.g., 7th in illustration) is called the most significant bit (MSB).

Identifying individual bits (see Fig. 37)

Starting from the right

The Kth bit represents the 2^K slot

7th bit		Kth bit		2nd bit	1st bit	0th bit
1	0	1	1	0	0	1	1

Figure 37   A byte.

Least significant bit (LSB): bit furthest to the right

Most significant bit (MSB): bit furthest to the left

4.2   Hexadecimal Numbers

Hexadecimal Numbers are a version of binary numbers that are easier for humans to work with. Hexadecimal numbers have 16 unique digits (0, 1, 2, 3, 4, 5, 6, 7, 8, 9, A, B, C, D, E, F). Since hexadecimal numbers sometimes use both letters and numbers an "h" is usually placed at the end of the number to distinguish it as a hexadecimal number (e.g., ACFh). Again, the value of each digit increases by a factor of 16 as you move to the left and decreases by a factor of 16 as you move to the right. The advantage of hexadecimal numbers is that each hexadecimal digit represents a 4-bit binary number. This makes it very simple to convert from hexadecimal to binary given in Fig. 38.

Example
34h = 52 :

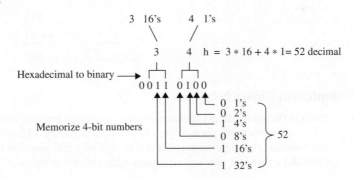

52 decimal = 34h = 0 0 1 1 0 1 0 0

Example
Binary to hexadecimal:

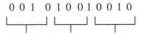

1. Break into 4-bit segments (add two 0's to left).

2. Convert each 4-bit segment into 1 hexadecimal digit.

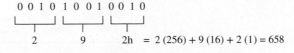

Hexadecimal	Binary
0	0000
1	0001
2	0010
3	0011
4	0100
5	0101
6	0110
7	0111
8	1000
9	1001
A	1010
B	1011
C	1100
D	1101
E	1110
F	1111

Figure 38   Hexadecimal-to-binary conversion.

4.3   Binary Addition

Binary addition is straightforward and similar to decimal addition. The two numbers are first aligned by their respective columns. Then the one's digits are added. If the result is greater than the maximum allowable number for that column (i.e. 1), the remainder is carried into the next column. This process is repeated until all columns are added and no remainder is left.

$$
\begin{array}{cc}
3 & 011 \\
+\ 1 & +\ 001 \\
\hline
4 & 100 \\
\end{array}
$$

4.4   Two's Compliment Binary Numbers

Straight (Regular) Binary notation cannot represent a negative number. A new convention called two's compliment binary numbers is used. Then binary arithmetic can be used to add a positive number to a negative number and in this way it is possible to perform binary subtraction. Consider a three bit two's compliment example shown below:

3 bit example		For N bits
011	3	2^{N-1}
010	2	
001	1	
000	0	0
111	−1	
110	−2	
101	−3	
100	−4	-2^{N-1}

The procedure to convert from a positive number (e.g., 3) to the negative of that number (e.g., −3) is first to invert all bits and then to add one to the result.

To calculate the negative value of a number > 0:

(i) Invert all bits.

(ii) Add 1.

The procedure to convert from a negative number (e.g., −3) to the positive of that number (e.g., 3) is first to subtract 1 (add −1) from the number and then invert all bits.

To calculate the positive value of a number < 0:

(i) Subtract 1.

(ii) Invert all bits.

Consider the following three bit examples.

Example: 3-bit

$$3 \equiv 011$$
$$-3$$

(i) Invert all bits:

$$100$$

(ii) Add 1:

$$100$$
$$\frac{001}{101} = -3$$

Example: 3-bit

$$-2 = 110$$

(i) Subtract 1:

$$110$$
$$\frac{111}{101}$$

(ii) Invert all bits:

$$010 = 2$$

Now two 2's compliment numbers can be added, and this is like binary subtraction. Consider the following example:

$$(-3) + (2) \leftrightarrow 101$$
$$\underline{010} \xleftrightarrow{-1}$$
$$111$$

5 DIGITAL CIRCUITS

A digital signal exists only at a high or low level, and it changes its level in discrete steps. For example, changes occur from a high state to a low state and vice versa. Digital circuits are categorized by their function. The difference between the categories is based on signal timing. For sequential logic devices, the timing of the input plays a role in determining the output. This is not the case with combinatorial logic devices whose outputs depend only on the instantaneous values of the inputs (timing does not matter).

5.1 Combinatorial Logic Operators

The basic operations, schematic symbols, and algebraic expressions for combinatorial logic devices are shown in Fig. 39. These devices are also called gates because they control the flow of signals from the inputs to the output. An inverter is used to invert an input signal.

Operation	Symbol	Expression	Truth Table		
Inverter	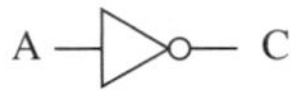 A ─▷o─ C	$C = \overline{A}$	A / C: 0→1, 1→0		
AND	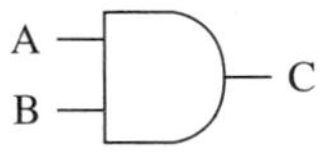 A, B ─ C	$C = A \cdot B$	A B C: 0 0 0; 0 1 0; 1 0 0; 1 1 1		
NAND	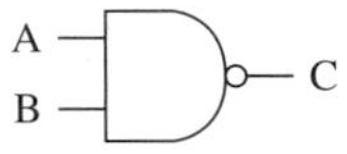 A, B ─ C	$C = \overline{A \cdot B}$	A B C: 0 0 1; 0 1 1; 1 0 1; 1 1 0		
OR	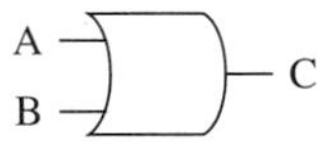 A, B ─ C	$C = A + B$	A B C: 0 0 0; 0 1 1; 1 0 1; 1 1 1		
NOR	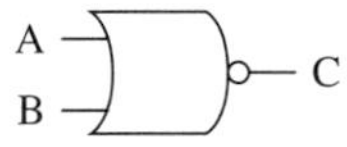 A, B ─ C	$C = \overline{A + B}$	A B C: 0 0 1; 0 1 0; 1 0 0; 1 1 0		
XOR	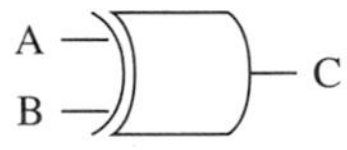 A, B ─ C	$C = A \oplus B$	A B C: 0 0 0; 0 1 1; 1 0 1; 1 1 0		
Buffer	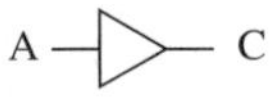 A ─▷─ C	$C = A$	A / C: 0→0, 1→1		

Figure 39 Operations, symbols, algebraic expressions, and truth table for combinatorial logic devices.

For example, a 0 becomes a 1 or a 1 becomes a 0. A small circle at the input or output of a digital device denotes a signal inversion. The truth table for each combinatorial digital device is also shown in Fig. 39. A truth table displays all the possible combinations of inputs and their corresponding outputs. Standard AND, NAND, OR, NOR, and XOR gates only have two inputs, but other forms exist. Truth tables are presented within the data sheets of digital devices, and these should be used when components are not standard.

A buffer is used to increase the output current while retaining the digital state. This is important when driving multiple digital inputs from a single output. Normal digital gates have a limit that defines the number of similar digital inputs that can be driven by the gate's output. The buffer device overcomes these limitations by providing more output current.

5.2 Timing Diagrams

Data sheets also contain timing diagrams for digital devices. Timing diagrams are helpful tools that illustrate the simultaneous levels of inputs and outputs in a device with respect to time. Every possible combination of input values and corresponding outputs will be illustrated within the timing diagram. Fig. 40 illustrates the timing diagram for an AND gate.

5.3 Boolean Algebra Identities

Knowledge of Boolean algebra, which defines the rules binary logic statements, is important in formatting mathematical expressions for logic circuits. Boolean expressions can be used to design logic networks, and Boolean expressions can be derived from truth tables and applied to circuit diagrams. There are many useful Boolean identities, but the basic Boolean laws and identities follow. A bar over a symbol indicates the Boolean operation NOT, which is the same as inverting the signal.

Fundamental Laws.

$$
\begin{array}{ccc}
\text{OR} & \text{AND} & \text{NOT} \\
A + 0 = A & A \cdot 0 = 0 & \\
A + 1 = 1 & A \cdot 1 = A & \overline{\overline{A}} = A \\
A + A = A & A \cdot A = A & \text{(double inversion)} \\
A + \overline{A} = 1 & A \cdot \overline{A} = 0 &
\end{array}
\tag{66}
$$

Commutative Laws.

$$
A + B = B + A
$$
$$
A \cdot B = B \cdot A \tag{67}
$$

Associative Laws.

$$
(A + B) + C = A + (B + C)
$$
$$
(A \cdot B) \cdot C = A \cdot (B \cdot C) \tag{68}
$$

Figure 40 Timing diagram for AND gate.

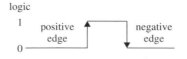

Figure 41   Clock pulse edges.

Distributive Laws.

$$A \cdot (B + C) = (A \cdot B) + (A \cdot C)$$
$$A + (B \cdot C) = (A + B) \cdot (A + C) \tag{69}$$

5.4   Sequential Logic

The timing and sequencing of input variables is important for sequential logic devices. Basic sequential logic devices include flip-flops, counters, and latches. Sequential devices usually respond to inputs when a separate trigger signal transitions from one state to another. The trigger signal is generally referred to as the clock (CK) signal. Fig. 41 illustrates the terminology used to describe the positive (or rising) edge and negative (or falling) edge. Positive-triggered devices respond to the clock's low-to-high (0 to 1) transition, and negative-triggered devices respond to the clock's high-to-low (1 to 0) transition. This topic continues through the discussion of flip-flops.

5.5   Flip-Flops

A flip-flop is the simplest implementation of sequential logic. The flip-flop has two and only two possible output states: 1 (high) and 0 (low). It also has the capability of remaining in a particular output state until the input signals cause it to change state. This is particularly useful in the storing of bits in digital memory devices. Flip-flops perform many of the basic functions of almost all digital devices.

RS Flip-Flop

The RS flip-flop is shown schematically in Fig. 42: S is the set input, R is the reset input, and Q and $\overline{Q}$ are the complementary outputs. Complimentary outputs are common in most flip-flops, where one output is the inverse (NOT) of the other. As long as the inputs (S and R) are both 0, the outputs of the flip-flop remained unchanged. When S is 1 and R is 0, the flip-flop is *set* to Q = 1 and $\overline{Q}$ = 0. When S is 0 and R is 1, the flip-flop is *reset* to Q = 0 and $\overline{Q}$ = 1. It is "not allowed" (NA) to place a 1 on S and R simultaneously. This will cause an unpredictable output. For RS flip-flops, $\overline{Q}$ is the unchanged Q value, or what Q used to be in other words. The truth table is shown in Fig. 42 along with the internal design and timing diagram. The propagation delays Δt_1 and Δt_2 are usually on the order of a nanosecond. All sequential logic devices depend on feedback and propagation delays for their operation.

Edge-Triggered RS Flip-Flop

Many times flip-flops are clocked, and the outputs of different types of clocked flip-flops can change on either a positive edge or a negative edge of a clock pulse. These flip-flops are called edge-triggered flip-flops. The difference between positive and negative edge triggering is indicated schematically as shown in Fig. 43. The edge-triggered RS flip-flop behaves similar to the regular RS flip-flop except that the input states only affect the output states when the clock edge

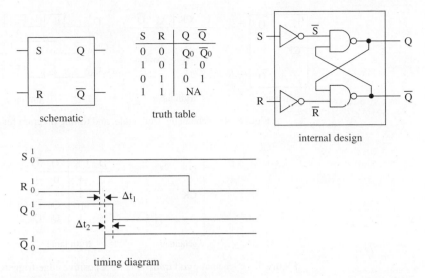

Figure 42 Schematic, truth table, internal design, and timing diagram for RS flip-flop.

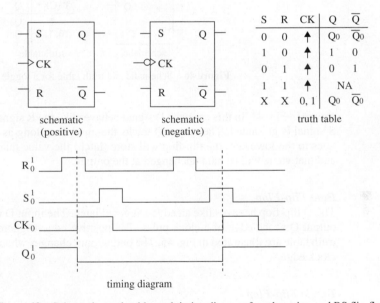

Figure 43 Schematic, truth table, and timing diagram for edge-triggered RS flip-flop.

is encountered. For example, as shown in the truth table for a positive edge-triggered flip-flop in Fig. 43, the output state will always remain unchanged until the clock edge is encountered. As seen in the timing diagram in Fig. 43, when the clock signal rises, the output changes.

Latch
An important device that is not edge triggered is called a latch. Latches can also be called level-triggered flip-flops. The schematic symbol, truth table, and timing diagram for a latch is

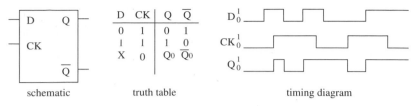

Figure 44   Schematic, truth table, and timing diagram for a latch.

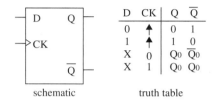

Figure 45   Schematic and truth table for a positive edge-triggered D flip-flop.

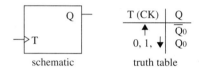

Figure 46   Schematic and truth table for a toggle flip-flop.

shown in Fig. 44. In this case, the D signal behaves like the R signal on the RS flip-flop, and the S signal is grounded. The output Q tracks the input D as long as CK is high. When the clock goes to the low state, the flip-flop will store (latch) the value that D had at the negative edge, and that value will remain unchanged at the output.

Data Flip-Flop
The D flip-flop behaves like an edge-triggered latch. The input D is stored and presented at the output Q at the edge of a clock pulse. The positive edge-triggered D flip-flop schematic and truth table are illustrated in Fig. 45. The output only changes when triggered by the appropriate clock edge.

Toggle Flip-Flop
The toggle flip-flop is perhaps the simplest flip-flop. The input signal T is used as a clock signal. The output signal Q is toggled at the appropriate edge of the T signal pulse. The schematic and truth table for a positive edge-triggered toggle flip-flop is shown in Fig. 46.

5.6   Schmitt Trigger

Digital signal pulses may not exhibit sharp edges in some applications. Many times this is referred to as signal noise. The Schmitt trigger is a device used to reduce noise by filtering an input signal using a high and low threshold. The output goes high when the input exceeds the trigger's high threshold. The output signal remains high until the input signal drops below

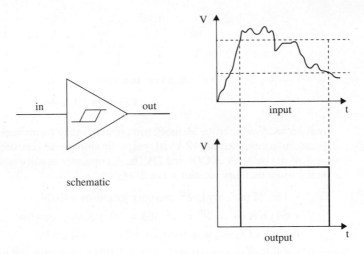

Figure 47 Schematic, input, and output values for a Schmitt trigger.

the trigger's low threshold. Fig. 47 shows the schematic symbol of the Schmitt trigger and an example of the input and output values.

6 DIGITAL COMPUTERS

As mechanical engineers it can be useful to understand some of the basic low-level operation of microcontrollers and computers. This is useful to (1) understand what computers can and cannot do in mechatronic systems, (2) understand how computers are used in mechatronic systems (what their job is), (3) to be able to communicate with electrical and computer engineers on interdisciplinary design teams, and (4) have the ability to select a microcontroller when designing a mechatronic system. With these stated goals only the general concepts are presented here. The details of low-level computer operation are numerous and beyond the scope of this chapter.

6.1 Most Basic Computer

The most basic computer includes a central processing unit (CPU) and some amount of memory. The memory holds numbers while the CPU, sometimes called the microprocessor, holds numbers and processes these numbers. The CPU sends some control information (including addresses) to the memory and the CPU and memory exchange data. The CPU holds and processes data in registers. A register is a set of flip-flops (a digital electronic device with two states—0,1) whose contents are read or written as a group. Registers come in several sizes such as 8, 16, and 32-bits. Consider the 8-bit register shown in Fig. 48 where x can be a 0 or a 1.

Registers are used for temporary storage of numbers and manipulation of numbers (addition, subtraction, etc.).

6.2 Memory

In almost all computers, memory is arranged as a series of bytes. Each memory location can hold a single 8-bit binary number (00000000 to 11111111). More than one memory location is needed to store numbers larger than 255. Memory can be thought of as an 8-bit

(MSB)   (LSB)

Bit #   7   6   5   4   3   2   1   0

| X | X | X | X | X | X | X | X |

where x can be a 0 or a 1.

Figure 48 A register.

register as shown above. Memory can come in many forms such as solid-state electronic chips [random-access memory (RAM)], magnetic storage devices such as floppy and hard disks, and optical devise such as CDs and DVDs. A computer usually has thousands, millions, or many, many more memory locations. For example:

- 1 kb of memory is 2^{10} memory locations = 1024
- 64 kb is $64 \times 2^{10} = 65,563 = 2^{16}$ memory locations
- 1 Mb of memory is 1024 kb = $2^{20} = 1,048,576$
- 1 GB of memory is 1024 Mb = $1,073,741,824 = 2^{30}$ memory locations

One of the most important concepts in understanding computer memory is that memory has an address and contents. The address is a fixed number that is a unique identifier for a specific memory location. The contents (a single byte of data) of a specific address can be changed (see Fig. 49).

To get data from memory, the CPU loads the memory address on a special register and then sends that address to memory. Memory returns the contents of that address to another register on the CPU. The size of the number needed for the address can often be a limitation of the amount of memory a computer can have. For example, most modern desktop computers have 32-bit registers. A single register can hold a number up to 2^{32} or 4 GB or 4,294,967,296. This is the number of memory locations that can be addressed with a single register (although computers often use more than one register for a memory address).

6.3 Storing Information In Memory

Each memory location always contains a number between 0 and 255. This number (along with the contents of other memory locations) can be interpreted in many ways to mean many different things.

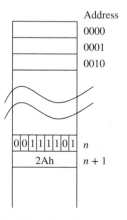

Figure 49 Schematic of computer memory.

Binary Numbers. First, memory can simply store unsigned binary numbers.

Two's Compliment Binary Numbers. Memory can store signed binary numbers in the form of two's compliment binary numbers.

ASCII. ASCII stands for the American Standard Code for Information Interchange. ASCII code is a format for characters such as letters (both upper and lowercase), digits 0 – 9, and punctuation symbols. ASCII only has 128 possible symbols, so only a 7-bit binary number is needed. Since memory comes in 8-bit locations an extra bit is added to the MSB. Consider the example below:

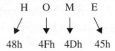

Since ASCII requires only 7 bits, the eigth bit is wasted. For this and several reasons there have been other formats similar to ASCII that encode more information. For example, the American National Standards Institute (ANSI) is another format that includes accented characters, a few Greek letters. Another is UNICODE that uses 16 bits and has 65,536 different symbols.

Binary-Coded Decimal. Binary-coded decimal (BCD) is a format sometimes used in database programs (e.g., business programs). BCD numbers use the first 10 digits of a hexadecimal number, the remaining letter digits (A–F) are not used. Therefore, when the numbers appear in memory in hexadecimal format, the number appears as the original decimal digit. For example, the number 89 will appear as 89h in the memory location. The drawback is that more memory space is required when compared to other formats. For example, the number 255 would require two memory locations (0255h) as a BCD and only one memory location when stored as a binary number (FFh).

BINARY-CODED DECIMAL.

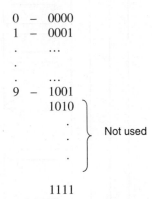

Floating-Point Numbers. Floating-point numbers are stored in several memory locations in scientific notation. For example, 1749 must first be written as 1.749×10^3. There are several different types of floating-point numbers. Single precision floating-point numbers use 32 bits (23 bits for the significant figures 1.749, 1 bit for sign on the these figures, and 8 bits for the exponent and sign). Single-precision numbers can range from 10^{-38} to 10^{38} with only 6–8 digits of accuracy. Double-precision floating-point numbers use 64 bits (53 bits for the significant figures with the sign and 11 bits for the exponent and sign). Double-precision numbers can range from 10^{-308} to 10^{308} with 13–16 digits of accuracy.

6.4 Microcomputers

Microprocessors often refer to the CPU described above. Most microprocessors contain some memory in the form of registers. Microcomputers contain microprocessors connected to other devices by a common data bus. The data bus is a series of conductors that allow information to flow between each subsystem. The other devices can include input/output (I/O) devices such as monitors and keyboards or additional memory devices or I/O devices such as analog-to-digital converters. A microcontroller is just enough microcomputer on a chip to do a specific control job. For example, a microcontroller could be a microprocessor with some digital I/O to control a home security system. See Fig. 50.

7 TRANSFER OF DIGITAL DATA

Digital data can be transferred in several ways. The most common methods are serial and parallel data transfer. In serial communication data bits are sent one after the other along a single conductor. Parallel communication transfers several bits at the same time along separate conductors. There are two types of serial communication—synchronous and asynchronous.

7.1 Parallel Data Transfer

In parallel data transfer all bits occur simultaneously on a set of data lines. Each bit is placed on each data line by the transmitting digital system and then can be read by the receiving

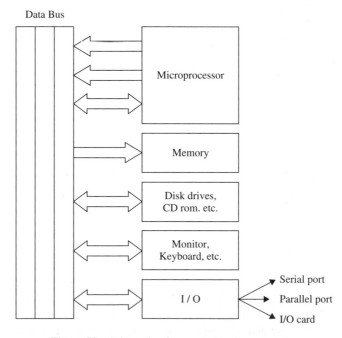

Figure 50   Schematic of an example microcomputer.

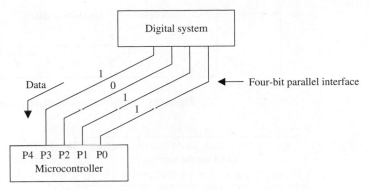

Figure 51 Four-bit parallel transfer of digital data.

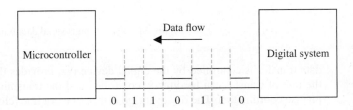

Figure 52 Serial transfer of digital data.

digital system. The advantage of parallel communication is that it is relatively fast (relative to serial communication) because several bits of data can be transferred at the same time. The disadvantage is that several conductors are required. See Fig. 51.

7.2 Serial Data Transfer

In serial data transfer a sequence of bits, or train of pulses, occurs on a single data line as shown in the figure. There are two types of serial communication—synchronous and asynchronous. See Fig. 52.

7.3 Asynchronous Serial Data Transfer

In asynchronous serial communication the data is written to the serial line at a predefined rate. Both the transmitter and receiver must be set for identical timing. This timing rate is called bits per second (bps) or baud rate (e.g., 4800 baud, or 9600 baud). The advantage of asynchronous serial communication is that it only requires one wire (and ground) to communicate the data. This method is used in RS-232 (com ports on PC's) communication. The microprocessor detects the first "edge" of the first data bit and then (because of the baud rate) it knows how long to wait for the next bit. The baud rate is not exactly the rate the data will be transferred because there are some overhead bits required for the transfer (e.g., parity). See Fig. 53.

7.4 Synchronous Serial Data Transfer

In synchronous serial communication the data is written to the serial line and a separate line is used as a clock, or signal, to indicate the data is ready to be transferred. In this case the rate of

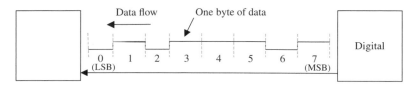

Figure 53 Asynchronous serial data transfer.

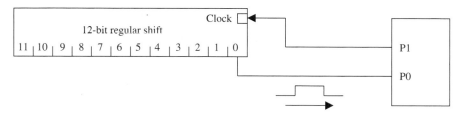

Figure 54 Synchronous serial data transfer.

data transfer is controlled by the digital device that provides the clock. The advantage is that the rate of transfer can be directly controlled and the transmitter and receiver do not require precise coordination. The disadvantage is that an extra line (clock) is required. See Fig. 54.

8 ANALOG-TO-DIGITAL CONVERSION

The world we live in is analog. To interface digital computers to an analog mechanical system an analog-to-digital (A/D) conversion process is required. Converting from an analog signal to a digital number is a two-step process that involves (1) quantization and (2) coding. The process of quantization is where the range of the analog signal is broken into a discrete number of bins. In coding, the bin "location" of the analog signal is converted into a digital number that can be understood by the computer. There are literally hundreds of different types of A/D converters.

Several definitions are required to discuss A/D converters:

Resolution n refers to the number of bits used to digitally approximate the analog value of the input.

Number of possible states $N = 2^n$.

Analog quantization size Q is a measure of the analog change that can be resolved (minimum error):

$$Q = \frac{V_{max} - V_{min}}{N}$$

8.1 Four-Bit A/D Converter

To understand the terms explained above, an example in Fig. 55 will be used. In the example, a tachometer is used to measure the speed of an electric motor. The tachometer produces a voltage between -12 V and $+12$ V that is linearly proportional to the speed of the motor. The analog voltage output of the motor will be read with a 4-bit analog-to-digital converter. Therefore, the A/D converter has a resolution of 4 bits and can have 16 ($= 2^4$) possible states (0000–1111

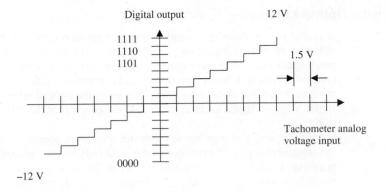

Figure 55 Analog-to-digital conversion.

or 0–15). The analog quantization size will be 1.5 V $\{[12 - (-12)]/16\}$. This means, for example, all voltages between 10.5 and 12 V will be represented by the encoded digital number 1111.

9 DIGITAL-TO-ANALOG CONVERTER

A digital-to-analog converter (DAC) has many of the same issues as an A/D converter. The resolution is given in the number of bits used to create the output analog voltage. For example, a 4-bit DAC can produce 16 different output states. Issues to consider closely when selecting a DAC include the resolution and the amount of current that can be supplied.

10 SENSORS

10.1 Position Sensors

Potentiometers
One of the most basic position sensors is the potentiometer. A potentiometer is a device that generally has a mechanical input in the form of a rotary shaft or a linear slide. The mechanical input moves a wiper across a resistor. The potentiometer also has an electrical input that places a fixed voltage across the resistor. As the position of the mechanical input is changed, the location of the wiper and the voltage of an output electrical lead attached to the wiper changes. The change in voltage is linearly related to the angular change of the mechanical input. Linear potentiometers that use a slide input instead of a shaft are also common.

Potentiometers are traditionally used in many applications such as the volume knob on a radio. See Fig. 56.

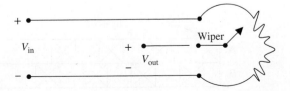

Figure 56 A potentiometer.

10.2 Digital Optical Encoder

A digital optical encoder is another common position sensor (both rotary and linear are available) that provides a digital output. The encoder consists of a disk attached to a shaft that is the mechanical input to be measured (θ). The disk contains several "tracks" of alternating slots. One side of the disk has a light source and the opposite side of the disk has a photo sensor. As the disk rotates the slots alternately either block the light or allow it to pass to the photo sensor. In this way the digital encoder "encodes" the position of the disk as either a 0 (no light) or a 1 (light). See Fig. 57.

In this way it is possible to count the alternating regions of light and dark and determine the angular change in the position of the disk/shaft. Several observations can be made from this principal. First, with only one row (track) of slits it is not possible to determine the direction the disk is rotating. Also, all measurements are made relative to a starting point. Finally, the encoder has a finite number of slits limiting the resolution of the sensor.

These limitations are eliminated in different ways with different types of encoders. A quadrature optical encoder (described in the next section) uses two rows of slits to determine the direction of travel. Other encoders use many tracks so that each position of the shaft is unique, allowing for an absolute measurement of shaft position.

10.3 Quadrature Optical Encoder

A quadrature optical encoder includes two rows of slits that are placed 90° ($^1/_4$ cycle) out of phase. This allows the encoder to measures relative position and direction. Consider the diagram below that represents the two tracks (track A and track B) with the encoder shaft rotating at constant velocity. Fig. 58 represents the alternating regions of light and dark created by the two rows of slits. Since the slits are $^1/_4$ cycle out of phase there are four states created: (1) *A* on, *B* off, (2) *A* on, *B* on, (3) *A* off, *B* on, (4) *A* off, *B* off. Now, the direction of rotation can be determined by the order that these states appear.

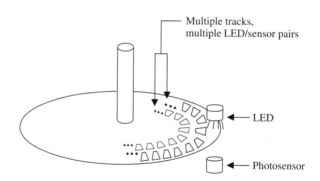

Figure 57   Rotary optical encoder.

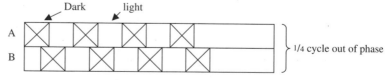

Figure 58   Output of a quadrature optical encoder.

4 States.

$$A\underline{B} \qquad A \text{ on } B \text{ off}$$

$$AB \qquad A \text{ and } B \text{ on}$$

$$\underline{A}B \qquad A \text{ off } B \text{ on}$$

$$\underline{AB} \qquad A \text{ and } B \text{ off}$$

Direction →
Output
$A\underline{B} \quad AB \quad \underline{A}B \quad \underline{AB} \quad A\underline{B} \quad AB \dots$

← Direction
Output
$A\underline{B} \quad \underline{AB} \quad \underline{A}B \quad AB \quad A\underline{B} \quad \underline{AB}$

↗ You can get the direction by noting the change that takes place.

10.4 Linear Variable Differential Transformer

Another common position sensor is the linear variable differential transformer (LVDT). This sensor is based on an electrical transformer depicted in Fig. 59.

The transform consists of a primary coil of wire wrapped around an iron core with a secondary coil also wrapped around the same core. As an alternating current (ac) is passed through the primary core a changing magnetic field is created by the coil. The magnetic field is then transferred through the high permeability iron core. The changing magnetic field induces a current in the secondary coil. The ideal inductor neither adds nor dissipates energy, and power is conserved between the primary and secondary coil ($P_{in} = P_{out}$). The current induced (and hence the voltage) in the secondary coil is related to the current in the primary coil by the ratio of turns of each of the coils ($V_p/n_p = V_s/n_s$ or $n_p i_p = n_s i_s$). This principal is used (in a slightly different arrangement) to produce a position sensor.

The LVDT arrangement shown in Fig. 60 is a schematic that shows how moving the iron core (x direction) will change the coupling between the primary and secondary coil as described above. This arrangement produces a linear region where the magnitude of the ac output voltage ($= V_s$) is linearly related to the position of the movable iron core.

10.5 Force Sensors

Strain Gauge

It is impossible to directly measure a force due to the principle of causality. The effect of a force is often measured and then the force is estimated based on its effect. Strain gauges are frequently used to measure the surface stress on a structure and then estimate the force.

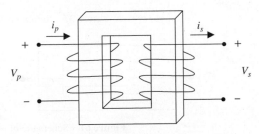

Figure 59 Electrical transformer.

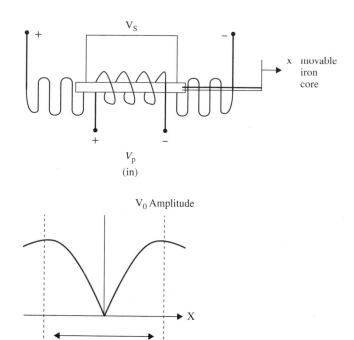

Figure 60   Schematic of an LVDT.

Strain gauges are generally a foil material where resistance changes as the foil is deformed. Strain gauges are generally glued to the surface of a structure and force is measured along a single measuring axis.

Strain gauges can be shown electrically as a variable resistor where the resistance is related to the force. The basic principle is based on the physics of a long square rod as shown in Fig. 61. The electrical resistance of the rod (R) is related to the resistivity of the material, the length of the rod, and inversely related to the cross-sectional area of the rod. As a force is applied along the axis of the rod, its length (and cross sectional area related by Poisson's ratio) will change and thereby change the resistance. Strain gauges are designed to take maximum effect of this basic principle. It should also be noted that resistivity can change with many factors including temperature. Much effort is put into advanced strain gauge applications to make the gauges independent of factors such as temperature variation. See Fig. 62.

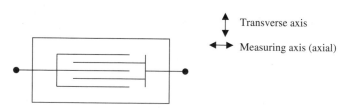

Figure 61   Schematic of a strain gauge.

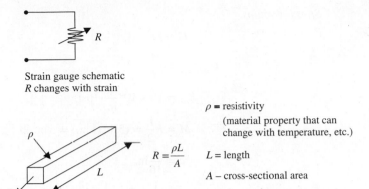

Figure 62 Basic resistance.

Strain gauges are described by the gauge factor. The gauge factor relates the normalized (normalized by the zero load resistance of the gauge) change in resistance to the strain along the axis of the gauge:

$$F = \frac{\Delta R/R}{\varepsilon_{\text{Axial}}} \qquad \text{(often} \approx 2) \tag{70}$$

Consider the cantilevered aluminum beam example shown below. The strain gauge is mounted in the center of the beam 1 in. from the end and 1 in. from the wall. The beam is 2 in. long, 1 in. wide and $\frac{1}{8}$ in. thick. A 5-lb load is applied at the end of the beam and the gauge factor is 2. Figs. 63 and 64 show a mathematical model of the beam and the strain is estimated. This is then translated into the expected change in resistance of the gauge through the gauge factor.

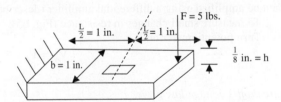

Figure 63 Strain gauge example.

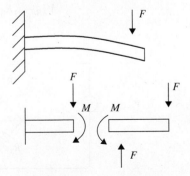

Figure 64 Free-body diagram.

Stress Modulus.

$$G = E\varepsilon \tag{71}$$

$$G_{\text{Bending}} = \frac{M_c}{I} \tag{72}$$

$$E_{A1} = 10 \times 10^6 \frac{\text{lb}}{\text{in.}^2} (\text{psi}) \tag{73}$$

$$M = F\frac{L}{2} = (5\,\text{lb})(1\,\text{in.}) = 5\,\text{in.-lb} \tag{74}$$

$$c = \frac{1}{16}\text{in.} \tag{75}$$

$$I = \frac{1}{12}bh^3 = \frac{1}{12}1\frac{1^3}{8} = 1.6 \times 10^{-4}\ \text{in.}^4 \tag{76}$$

$$\varepsilon = \frac{\frac{(5\,\text{in.-lb})(0.0625\,\text{in.})}{1.6 \times 10^{-4}\ \text{in.}^4}}{10 \times 10^6\ \frac{\text{lb}}{\text{in.}^2}} = 0.000192 = 192\mu\varepsilon \ (\text{unitless}) \tag{77}$$

$$\text{Strain} = \varepsilon = 1.92\mu\varepsilon \tag{78}$$

$$\text{Guage factor} = \frac{\Delta R/R}{\varepsilon_{\text{Axial}}} = Z \tag{79}$$

$$\Delta R = RFE = (120)(2)(0.000192) = 0.046\,\Omega \tag{80}$$

The above change in resistance is very small. This change must be amplified in some way to be able to practically measure the change. If the above strain gauge is used in a voltage divider the change in voltage will be (1) small and (2) the voltage will be centered about a larger unchanged voltage.

For these reasons a Wheatstone bridge is used. It is shown in the analysis below that the change in voltage (V_{AB}) resulting from the change in the gauge is centered about 0 V and can then be amplified using a differential amplifier (described earlier in this chapter).

To measure small changes in resistance (Fig. 65):

Turn potentiometer until

$$V_A = V_B \tag{81}$$

Kirchhoff's voltage law gives.

$$V_{AB} + i_1R_2 - i_2R_1 = 0 \tag{82}$$

$$V_{AB} - i_1R_3 + i_2R_4 = 0 \tag{83}$$

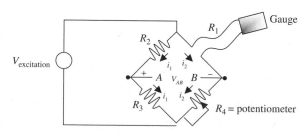

Figure 65 Strain gauge circuit.

and

$$V_{\text{ex}} - i_1 R_2 - i_1 R_3 = 0 \tag{84}$$

$$V_{\text{ex}} - i_1 (R_2 + R_3) = 0 \tag{85}$$

where

$$i_1 = \frac{V_{\text{ex}}}{R_2 + R_3} \tag{86}$$

and

$$V_{\text{ex}} - i_2 R_1 - i_2 R_4 = 0 \tag{87}$$

$$V_{\text{ex}} - i_2 (R_1 + R_4) = 0 \tag{88}$$

where

$$i_2 = \frac{V_{\text{ex}}}{R_1 + R_4} \tag{89}$$

Substituting (86) and (89) into (82) gives

$$V_{AB} + \left(\frac{V_{\text{ex}}}{R_2 + R_3} \right) (R_2) - \left(\frac{V_{\text{ex}}}{R_1 + R_4} \right) R_1 = 0 \tag{90}$$

into (83)

$$V_{AB} - \left(\frac{V_{\text{ex}}}{R_2 + R_3} \right) (R_3) + \left(\frac{V_{\text{ex}}}{R_1 + R_4} \right) R_4 = 0 \tag{91}$$

$$V_{AB} = -V_{\text{ex}} \left(\frac{R_2}{R_2 + R_3} - \frac{R_1}{R_1 + R_4} \right) = V_{\text{ex}} \left(\frac{R_1}{R_1 R_4} - \frac{R_2}{R_2 + R_3} \right) \tag{92}$$

Now we deform the strain gauge (R_1 becomes $R_1 + \Delta R_1$):

$$\frac{V_{AB}}{V_{\text{ex}}} = \left[\frac{R_1 + \Delta R_1}{(R_1 + \Delta R_1) + R_4} - \frac{R_2}{R_2 + R_3} \right] \tag{93}$$

Equation (93) relates

$$V_{AB} \quad \text{to} \quad \Delta R \tag{93a}$$

Which is related to strain (ε).

Shuffle (93)

$$\Delta R_1 = R_1 \left[\frac{\frac{R_4}{R_1} \left(\frac{V_{AB}}{V_{\text{ex}}} + \frac{R_2}{R_2 + R_3} \right)}{\left(1 - \frac{V_{AB}}{V_{\text{ex}}} - \frac{R_2}{R_2 + R_3} \right)} - 1 \right] \tag{93b}$$

Use (93b) and

$$F = \frac{\Delta R / R}{\varepsilon_{\text{Axial}}} \tag{94}$$

Use these equations to design strain sensors.

11 ELECTRO MECHANICAL MODELING EXAMPLE

Consider the following mechatronic system shown in Fig. 66. Here a permanent magnet direct current motor is used to drive a one degree of freedom robotic arm in a gravity field. The motor is connected to the arm with a pair of spur gears. The arm is used to actuate a spring (that could be used to represent interaction with the environment, i.e. applying a force).

The motor can be modeled electrically as an inductor, a resistor, and a voltage source connected in parallel; see Fig. 67. If the gears are assumed to be ideal, the following assumptions may be appropriate:

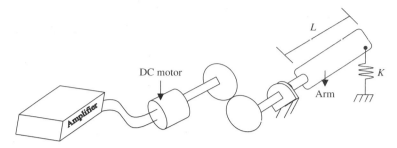

Figure 66 Example mechatronic system.

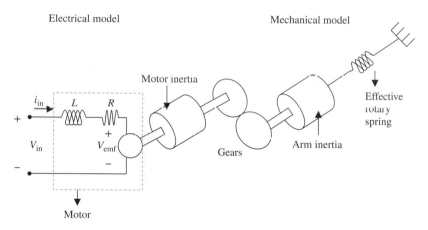

Figure 67 Model of the mechatronic system.

Assumptions.

1. No friction
2. Motor rotor and arm are only significant masses
3. Small motions

Now the constitutive relationships for the motor and gears can be written and Kirchhoff's voltage law can be applied to the electrical circuit. Then, free-body diagrams can be created for each of the mechanical elements; see Fig. 68.

Motor Relations.

$$\tau = K_t \, i_{\text{in}} \tag{95}$$

where K_t is the torque constant.

$$V_{\text{emf}} = \omega K_e = K_e \, \ddot{\theta}_{\text{motor}} \tag{96}$$

where K_e is the electromotive force (emf) constant.

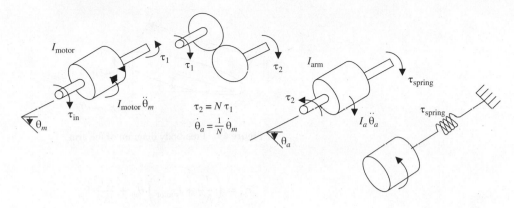

Figure 68   Free-body diagrams.

$$\tau_{\text{spring}} = K_{\text{effective}}\, \theta_a$$

Electrical Equations.

$$V_{\text{in}} = L\frac{di_{\text{in}}}{dt} + R_{\text{in}} + K_e\,\theta_{\text{motor}} \tag{97}$$

$$\tau_{\text{in}} - \tau_1 = I_{\text{motor}}\theta_{\text{motor}} \tag{98}$$

$$\tau_2 - \tau_{\text{spring}} = Ia\,\ddot{\theta}_a \tag{99}$$

where $\tau_1 = 1/N\tau_2$ and $\tau_2 = \tau_{\text{spring}} + I_a\,\ddot{\theta}_a$. Substituting into (98) yields

$$\tau_{\text{in}} - \frac{1}{N}(\tau_{\text{spring}} + I_a\,\ddot{\theta}_a) = I_{\text{motor}}\,\ddot{\theta}_{\text{motor}}$$

$$\ddot{\theta}_a = \frac{1}{N}\ddot{\theta}_m \tag{100}$$

The arm can be modeled with an effective spring constan:

$$\tau_{\text{in}} - \frac{\tau_{\text{spring}}}{N} - \frac{I_a}{N}\left(\frac{1}{N}\ddot{\theta}_m\right) = I_{\text{motor}}\ddot{\theta}_m \tag{101}$$

To find τ_{spring}:

$$\tau_{\text{spring}} = K_{\text{effective}}\,\theta_a \tag{102}$$

$$\tau_{\text{spring}} = LF = KL^2\theta_a \quad \text{see Fig. 69} \tag{103}$$

$$\underbrace{\qquad\qquad}_{K_{\text{effective}}}$$

Substitute.

$$\tau_{\text{in}} - \frac{KL^2}{N^2}\theta_{am} - \frac{I_a}{N^2}\ddot{\theta}_m = I_{\text{motor}}\,\theta_{\text{motor}} \tag{104}$$

$$\underbrace{\text{Inertial forces}}_{} \qquad \underbrace{\text{Spring force reflected to motor axis}}_{}$$

$$\underbrace{\qquad\qquad\qquad}_{} \qquad \underbrace{\qquad\qquad}_{} \tag{105}$$

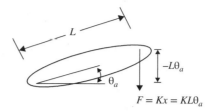

Figure 69 Free-body diagram of the arm.

$$\tau_{\text{in}} = \left(\frac{I_a}{N^2} + I_{\text{motor}} \right) \ddot{\theta}_m + \frac{KL^2}{N^2} \theta_m$$

Arm inertia
reflected to
motor axis

Motor inertia

$$\tau_{\text{in}} = K_t \, i_{in} \tag{106}$$

$$K_t \, i_{in} = \left(\frac{I_a}{N^2} + I_m \right) \ddot{\theta}_m + \frac{KL^2}{N^2} \theta_m \tag{107}$$

substitute (106) into (97)

$$V_{\text{in}} = L \frac{d}{dt} \left[\frac{1}{K_t} \left\{ \left(\frac{I_a}{N^2} + I_m \right) \ddot{\theta}_m + \frac{KL^2}{N^2} \theta_m \right\} \right]$$

$$+ R \left[\frac{1}{K_t} \left\{ \left(\frac{I_a}{N^2} + I_m \right) \ddot{\theta}_m + \frac{KL^2}{N^2} \theta_m \right\} \right] + K_e \theta_{\text{motor}} \tag{108}$$

This is one differential equation that relates θ_m to V_{in}; you can take the Laplace transformation to obtain the transfer function:

$$\frac{\text{Output}}{\text{Input}} = \frac{\theta_m(s)}{V_{\text{in}}(s)} \tag{109}$$

You can also substitute $\theta_a = (1/N)\theta_m$ and obtain $T_{(s)} = \theta_a(s)/V_{\text{in}}(s)$.

BIBLIOGRAPHY

The following references can provide further information useful in the study of mechatronics.

D. M. Asulander, and C. J. Kempf, *Mechatronics: Mechanical System Interfacing*, Prentice Hall, Englewood Cliffs, NJ, 1996.

J. E. Carryer, R. M. Ohline, and T. W. Kenny, *Introduction to Mechatronic Design*, Pearson Education, Upper Saddle River, NJ, 2011.

J. R. Cogdell, *Foundations of Electrical Engineering*, 2nd ed., Prentice-Hall, Englewood Cliffs, NJ, 1996.

R. E. Fraser, *Process Measurement and Control: Introduction to Sensors, Communication, Adjustment, and Control*, Prentice Hall, Upper Saddle River, NJ, 2001.

M. B. Histand and D. G. Alciatore, *Introduction to Mechatronics and Measurement Systems*, 2nd ed., WCB/McGraw-Hill, New York, 2003.

P. Horowitz and W. Hill, *The Art of Electronics*, 2nd ed., University of Cambridge, Cambridge, UK, 1989.

M. Jouaneh, *Fundamentals of Mechatronics*, Cengage Learning, Boston, MA, 2013.

D. Necsulescu, *Mechatronics*, Prentice Hall, Upper Saddle River, NJ, 2002.

C. R. Paul, S. A. Nasar, and L. E. Unnewehr, *Introduction to Electrical Engineering*, 2nd ed., McGraw-Hill, New York, 1992.

G. Rizzoni, *Principles and Applications of Electrical Engineering*, 3rd ed., McGraw-Hill, New York, 2000.

G. Rizzoni, *Principles and Applications of Electrical Engineering*, 4th ed., McGraw-Hill, New York, 2003.

M. Sargent and R. Shocmaker, *The Personal Computer from the Inside Out*, 3rd ed., Addison Wesley, Reading, MA, 1995.

D. Shetty, and R. A. Kolk, *Mechatronics System Design*, PWS Publishing Company, Boston, 1997.

CHAPTER 25

INTRODUCTION TO MICROELECTROMECHANICAL SYSTEMS (MEMS): DESIGN AND APPLICATION

M. E. Zaghloul
George Washington University
Washington, D.C.

1 INTRODUCTION

In general, microelectromechanical systems have features in the micrometer- and, increasingly, nanometer-size range. Often, they are miniaturized systems that combine sensors and actuators with high-performance embedded processors on a single integrated chip. The word *electromechanical* implies the transfer of technology from mechanical to electrical and vice versa. Those devices embedded in functional systems are sometimes referred to as microsystems. This field is increasingly leading to devices and material systems whose size is on the order of a nanometer, that is, the size of molecules. Microsystems and nanotechnology enable the building of very complex systems with high performance at a fraction of the cost and size of ordinary systems. As such, these systems are the enabling technology for today's explosive growth in computer, biomedical, communication, magnetic storage, transportation, and many other technologies and industries. Microsystems and nanotechnology challenges range from the deeply intellectual to the explicitly commercial. This field is by its very nature a link between academic research and commercial applications in the aforementioned and other disciplines. Indeed, these disciplines span a very broad range of industries that are at the forefront of current technological growth.

The integration of microelectronics and micromechanics is a historic advance in the technology of small-scale systems and is very challenging for designers and producers of MEMS. The addition of micromachined parts to microelectronics opens up a large and very important parameter space to technological development and exploitation.

The MEMS structures and devices result from the sequence of design, simulation, fabrication, packaging, and testing. There are varieties of devices that can be classified as MEMS. There are passive devices, that is, nonmoving structures. There are devices that involve sensors and devices that involve actuators, which have micromechanical components. These are conceptually reciprocal in that sensors respond to the world and provide information and actuators use information to influence something in the world. Another class includes systems that integrate both sensors and actuators to provide some useful function. This classification, like most, is imperfect. For example, some devices that are dominantly sensors have actuators built into them for self-testing. Airbag triggers are an example. However, the framework provides a simple but quite comprehensive framework for considering MEMS devices.

In this chapter we will discuss some aspects of the design of these devices and introduce the reader to the technology used. In addition, we will discuss the structure of some of those devices.

2   MICROFABRICATION PROCEDURES

The fact that the field of MEMS largely grew out of the integrated circuit (IC) industry has been noted often. There is no doubt that the use of fabrication processes and associated equipment that were developed initially for the semiconductor industry has given the MEMS industry the impetus it needed to overcome the massive infrastructure requirements. However, it is noted that the field of MEMS has gone far beyond the materials and processes used for IC production. The situation is indicated schematically in Fig. 1. About a half dozen materials, notably silicon and its oxide and nitride, and standard microfabrication processes, such as lithography and ion implantation, oxidation, deposition, and etching, have generally been employed to make ICs. The set of materials used in IC devices is expanding to include, for example, low-dielectric-constant materials, polymers, and other nonconventional IC materials.

Many MEMS can be made with the same set of materials and processes as used for microelectronics. However, one of the hallmarks of the emerging MEMS industry is the use of numerous other materials and processes. Most basically, substrates other than silicon are being employed for MEMS. Silicon carbide has been demonstrated to be a good basis for many mechanisms that can stand higher temperature service than silicon. Diverse materials can be used within MEMS devices. While aluminum and, recently, copper are the metals used in IC devices, micromachining of many other metals and alloys has been demonstrated. Magnetic materials have been incorporated into some MEMS devices. Piezoelectric materials are especially attractive for MEMS because of their electrical–mechanical reciprocity. That is, application of a voltage to a piezoelectric material deforms it, and application of a strain produces a voltage. Zinc oxide and lead zirconium titinate (PZT) are important piezoelectric materials for MEMS. Many other examples of materials employed for MEMS could be given. However, the point is clear. Micromechanics are made of many more kinds of materials than microelectronics.

Because of the varieties of materials used in MEMS fabrication, the processes for producing and modifying them widened far beyond those found in the IC industry. However,

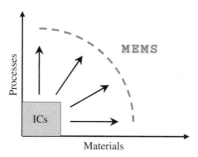

Figure 1   The number of materials and processes employed to make MEMS greatly exceeds those used to manufacture integrated circuits.

something more fundamental is at work when it comes to processes for making MEMS. Integrated circuits are monolithic and, despite up to 30 layers in some cases, are made by largely two-dimensional thin-film processes that yield what some call 2.5-dimensional structures. By contrast, micromechanical devices must have space between their parts so they can move, and the dimension perpendicular to the substrate is often very fundamentally necessary for their performance. Development of processes to make micrometer-scale parts that can move relative to each other was the breakthrough that enabled MEMS. Such micromachining processes fall into three major categories, which will now be reviewed briefly.

Surface micromachining involves the buildup of micromechanical structures on the surface of a substrate by deposition, patterning, and etching processes. The key step is the etching away of an earlier deposited and patterned sacrificial layer in order to free the mechanism. Figure 2 shows the steps needed for such processing.[1,2]

This process was first demonstrated about 35 years ago, when a metal–oxide–semiconductor field effect transistor (MOSFET) with a cantilever mechanical gate was produced.[3] The most common sacrificial material now is silicon dioxide, which is conveniently dissolved from under a movable part using hydrofluoric acid. Surface micromachining has been used to produce an amazing variety of micromechanical devices, some of which are now in large-scale production. Microaccelerometers and MEMS angle rate sensors are examples.

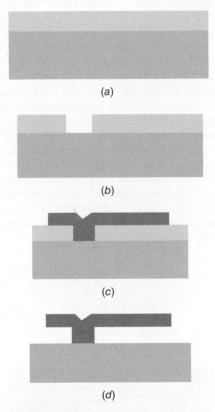

Figure 2 Surface micromachining steps: (*a*) Step 1: Deposit sacrificial layer. (*b*) Step 2: Pattern layers. (*c*) Step 3: Deposit structure layer. (*d*) Step 4: Etch sacrificial layer.

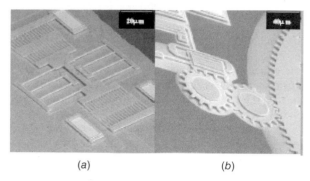

(a)   (b)

Figure 3   Examples of surface micromachining: (*a*) simple sensors and actuators; (*b*) gear train.

Figure 3 shows examples of mechanical structures built by surface micromachining.

Bulk micromachining, as the name implies, involves etching into the substrate to produce structures of interest. It can be done with either wet or "dry," that is, plasma, processes, either of which can attack the substrate in any direction (isotropically) or in preferred directions (anisotropically). Bulk micromachining has two primary variants. The first depends on the remarkable property of some wet chemical etches to attack single-crystal silicon as much as 600 times faster along some crystallographic directions compared to others. This anisotropic process is called orientation-dependent etching (ODE). It was known long before the emergence of MEMS technologies and has become a mainstay of the industry. ODE is especially useful for producing thin membranes that serve as the sensitive element in micropressure sensors. It is employed for production of these and other commercial MEMS devices. The second approach to bulk micromachining is to use plasma-based etching processes that attack the substrate, usually silicon, in preferential directions. Deep reactive ion etching (DRIE) is a plasma process that is used increasingly to make MEMS. It can produce structures that are over 10 times as deep as they are wide. Bulk micromachining steps are shown in Fig. 4*a*. Examples of devices developed using bulk micromachining are shown in Fig. 4*b*.

The third general class of micromachining processes is a collection of the numerous and varied techniques that can produce structures and mechanisms on the micrometer scale. Laser-induced etching and deposition of materials, electroetching and electroplating, ultrasonic and electron discharge milling, ink jetting, molding, and embossing are all available to the MEMS designer.

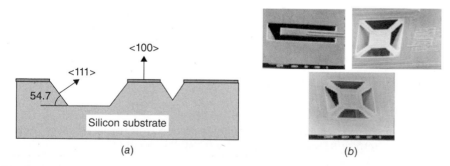

(a)   (b)

Figure 4   (*a*) Bulk micromachining anisotropic etch. (*b*) Examples of bulk micromachining devices.

Similar to ICs, MEMS devices are made using creative combinations of the materials and processes noted above. Some remarkable micromechanisms have been demonstrated, largely in academic fabrication facilities, and commercialized using diverse foundries.

3   DESIGN AND SIMULATIONS

To verify that the devices function, the designer has to model the MEMS device. The modeling involves writing the equation of motion or physical modeling of the performance of the device. Finite-element techniques are used to solve these modeling equations. There are a variety of computer-aided design (CAD) tools to aid the designer in the simulation and modeling of the device. In a very fundamental way, these tools are more complicated than the software for design of either solely ICs or solely mechanical devices. This is due to the close coupling of both electrical and mechanical effects within many MEMS. Consider a microcantilever that is pulled down by electrostatic forces. Its simulation has to take into account both the flow of electrical charge and mechanical elasticity in an iterative and self-consistent fashion.

Thermal, optical, magnetic, fluidic, and other mechanisms are also active in some MEMS and have to be handled self-consistently in the simulation phase.

Two basic approaches have been taken in the past decade to the need for specialized software for the design and simulation of MEMS. In the first approach, CAD design, tools, and available software from electronic design were modified to accommodate the requirement for MEMS design. In the second approach finite-element modeling was applied to MEMS. Software from the Tanner Tools very large scale integrated (VLSI) design suite were used for MEMS, for example, MEMS-PRO (www.tanner.com), which was recently acquired by MEMSCAP (www.memscap.com), as was the popular mechanical engineering software from ANSYS (www.ansys.com). Recently, new suites of software specifically developed for MEMS were marketed. Most of them include electronic, mechanical, and thermal simulation, and some have other physical mechanisms as well as processing simulation tools. Such software is available from CFD Research Corporation, Coventor (formerly called CRONOS Technologies), IntelliSense Corporation, and Integrated Systems Engineering. These tools vary widely in the mechanisms and material parameters that they include, the details of design and simulation of devices, and the fabrication facilities with which they interface. The choice of suitable software to use for MEMS design is still challenging. MEMSCAP is based on CADENCE (which is the most popular IC design tool, www.CADENCE.com). It consists of a set of tools which enable the design flow either bottom up or top down. It incorporates the MEMS design environment into existing and well-known environments with easy intellectual property (IP) cells or design reuse and the ability to exchange data between multidisplinary teams. The MEMSCAP simulator is based on the CADENCE environment, so the designer can simulate MEMS devices with the IC schematics and simulation. Models can be generated from the ANSYS finite-element model or from written analytical equations. Behavior models/scalable symbolic view can be generated. The Verilog-A model can also be generated. The generated model can be used to perform optimization simulations inside the environment or to realize a system simulation. In addition, emulators are available for etching cross section projection of the different material layers.

The design of MEMS devices involves knowledge of the sequence of materials to be used to realize the device. The sequence of materials used could be the standard sequence, in which case a standard technology process may be used in conjunction with other processing steps, for example, postprocessing. The sequence of materials to be used could be custom designed by the designer, which requires knowledge of the materials and their thin-film properties. Designers usually design the device and identify the material to be used and then use CAD tools to verify the performance. Iteration procedures are part of the design until the required performance is reached. After satisfactory simulation performance, the device is sent to fabrication foundries.

4 FABRICATION FOUNDRIES

After designing and simulating MEMS and deciding on the materials they will contain and the processes needed to make them, the next concern is which fabrication facility to employ. Sometimes, a standard IC facility can be used with postprocessing steps. Postprocessing involves adding or removing materials from the standard fabricated device. For example, using a standard complementary metal–oxide–semiconductor (CMOS) fabrication facility, we could realize the suspended-plate structure shown in Fig. 5. In this case the CMOS is removed from the bulk substrate to create the suspended structure on top of the etched pit. For example, the micro-hot-plate shown in Fig. 5 was realized in CMOS technology in a standard foundry with a postprocessing step of bulk micromachining to produce the suspended thin film with a resistive heater. The small mass of the heated element permits temperature changes of 300°C in a few milliseconds.

Many MEMS structures and devices have been produced by such postprocessing of CMOS chips.[4-6] Techniques which are not compatible with CMOS have also been used, in which case surface micromachining techniques produce mechanical structures on top of the substrate. A micromirror fabricated using surface micromachining is an example of such a device.

Figure 6 is a schematic of two pixels of the Digital Mirror Device manufactured by Texas Instruments. The torsion hinges are 5×1 μm in area and about 100 nm thick. The individual mirrors are 16 μm square. Over 500,000 of them are found in a single device, making this the system with the most moving parts produced in the history of mankind. The inventor, Larry Hornbeck, and the company received Emmy Awards in 1998 for outstanding achievement in engineering development.

There are now several foundries specifically for the production of MEMS. The fact that design rules in MEMS are roughly two generations behind those in ICs is significant. This enables MEMS foundries to buy used equipment from the microelectronics industry. Mass production of many MEMS now is done using 100- and 150-mm wafers. Several companies

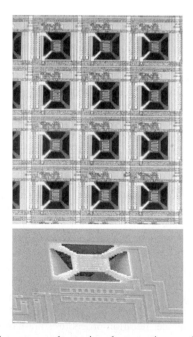

Figure 5 Micro-hot-plate array and scanning electron micrograph of one of the elements.

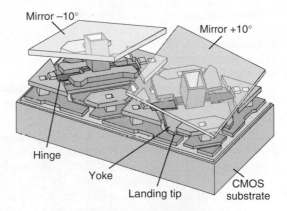

Figure 6   Micromirror.

and organizations in the United States and abroad offer fabrication services for MEMS somewhat analogous to those in IC foundries. They include BFGoodrich, Advanced MicroMachines, CMP (France), Institute of Microelectronics (Singapore), IntelliSense, ISSYS, Surface Technology Systems (U.K.), and many more.

Most of these foundries have all the facilities in-house to produce complete MEMS devices. However, the wide variety of materials and processes that can be designed into MEMS means that it is not always possible to find all the needed tools under one roof. Hence, the Defense Advanced Research Projects Agency instituted a new type of foundry service several years ago. It is called the MEMS-Exchange (www.MEMS_EXCHANGE.com). This organization contracts with diverse industrial and academic fabrication facilities for a wide range of services. The MEMS designer can draw from any of them. A completed design is sent to MEMS-Exchange, which handles scheduling, production, billing, and other factors, such as the protection of proprietary designs.

5   EXAMPLES OF MEMS DEVICES AND THEIR APPLICATIONS

There are varieties of applications of MEMS. This section gives a brief overview of MEMS applications with reference to commercial devices. There are many fields in which MEMS devices have been introduced. Table 1 shows examples of MEMS applications.

Table 1   Examples of MEMS Applications

Pressure Sensors	Inertia Sensors	Optical Devices	Data Storage	RF-MEMS	Acoustic MEMS	MicroFluidic	Chemical Sensors
Aeronautical	Air bag accelerometer	Optical beam steering	Hard-disk component	Miniature antenna	Microphone	Micropumps	Polymer gas sensors
Blood pressure	Motion control sensors	Microlasers	Miniature read/write	RF switch	Acoustic vibratos	Microvalves, micro-channels	Tin oxide gas sensors
Auto tire	Automobile suspension	Optical switch	Magnetic devices	Filters and resonators		Lab on chip	Preconcen-tarors
Touch pressure	Vibration sensor	Micromirrors	Optical storage	Inductors and capacitors		Ink bubble jet nozzle	Smart gas sensors

Table 1 summarizes some of the applications of MEMS and shows the air bag accelerometer developed by Analog Devices in which the structure of the sensors is based on a variable-capacitor device. Figure 7 shows the surface micromachining of the Analog Devices accelerometer. Mechanical structures were studied to develop miroresonators, such as the fixed–fixed beam of Fig. 8, and circular resonators.

Researchers are using MEMS techniques to produce an array of nanoresonators that can be integrated with other components. Figure 8 shows a working radial contour-mode disk

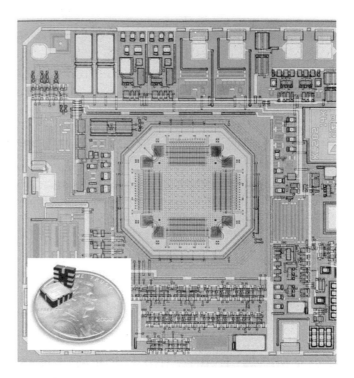

Figure 7   Photograph of exterior of new two-axis microaccelerometer in leadless packages on a penny and micrograph of chip from Analog Devices. In this device, the microelectronic and micromechanical components are tightly integrated on the silicon substrate.

Figure 8   Circular resonator.

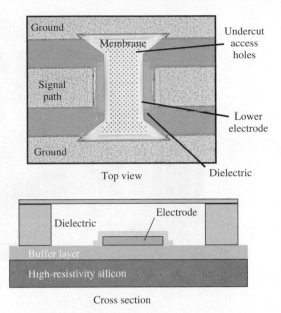

Figure 9 Microwave switch.

resonator with 10 μm radius and quality factor $Q = 1595$ at atmospheric pressure.[7] Work is aimed at coupling such resonators together to make large arrays. These devices were used in the design of electric filters for high-frequency communication systems. The benefit of using MEMS device is the high quality factor, which implies a high-efficiency circuit. Figure 9 shows an example of a mechanical switch which is an electrostatic switch. It is used in high-frequency circuits as a small-loss switch. The radio-frequency switch is small, on the order of 50 μm × 50 μm.

Figure 9 shows a micrograph (top) and schematic of the Raytheon MEMS microwave switch. The electrode under the flexible membrane is the actuator. The capacitance of the switch varies from near zero (open) to 3.4 pF (closed). The signal path is about 50 μm wide.

Figure 10 shows a gas sensor developed using CMOS technology and the associated circuits that make it a smart gas sensor.[8]

The above examples illustrate the variety of applications of MEMS devices as well as the variety of materials used. Other examples can be found in Ref. 9.

The last letter in MEMS stands for systems. This is due partly to the fact that a MEMS device is quite complex. However, MEMS devices are "only" components which are used in larger and more complex systems. That is, individual sensors or actuators can be used as components and incorporated into subsystems or systems in order to perform some useful function. The accelerometer in the air bag subsystem of an automobile and the DMD in a projection system in a theater are examples. However, it is also possible to closely couple both MEMS sensors and actuators into miniature systems all on one substrate. These are called "systems on a chip." Microfluidics with all the needed functionality on a substrate, including pumps and valves, as well as channels, mixers, separators, and detectors, are under development for compact analyzers. These will be relatively evolutionary advances over current microfluidic chips. High-density data storage systems with both actuation and sensing functions represent a more revolutionary example of integrated microsystems.

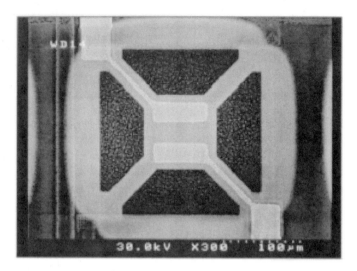

Figure 10   Scanning electron micrograph of CMOS gas sensor.

The variety of commercially available MEMS and their applications have both increased dramatically in recent years. The production of MEMS is now more than a $20 billion industry worldwide, with about 100 million devices marketed annually. While this industry grew out of the microelectronics industry, it is more complex in many important ways. Most fundamentally, it requires the integration of both microelectronics and micromechanics. Many MEMS involve several closely coupled mechanisms, some of which behave differently on the micrometer spatial scale than on familiar macroscopic scales. This complicates both the design and simulation of MEMS. So also does the much wider variety of materials and processes used to make MEMS compared to microelectronics. Because many MEMS have to be open to the atmosphere, their packaging, calibration, and testing are complex. Questions about the long-term reliability of MEMS are being answered as MEMS devices spend more years in use by consumers and industries.

6   CONCLUSIONS

We have discussed the advances in the microfabrication structures that allowed the realization of three-dimensional structures at the micrometer scale. As listed above, the materials and microfabrication processes used are unlimited to realize MEMS devices. Many MEMS involve several mechanical structures which behave differently on the micrometer spatial scale. Several CAD tools were developed to aid in the design of such devices. The design of such miniature devices is challenging and complicated as compared to devices at the macroscale level. Despite such engineering challenges, MEMS offer high performance and are small, have low power, and are relatively inexpensive. They both improve on some existing applications and enable entirely new systems. Some applications are targeted from the outset of design, but others are opportunistic. That is, the large number of MEMS components on the market makes them available to design engineers for a very wide variety of uses.

APPENDIX: BOOKS ON MEMS

A great deal of information on the design, simulation, fabrication, packaging, testing, and application of MEMS is available. This information is presented as journal articles, conference proceedings, and books as well as being available from the World Wide Web. In the past several years many books on MEMS have been published:

1995: *Integrated Optics, Microstructures and Sensors*, M. Tabib-Azar (Ed.), Kluwer Academic.

1996: *Micromachines* (*A New Era in Mechanical Engineering*), I. Fujimasa, Oxford University Press.

1997: *Micromechanics and MEMS* (*Classical and Seminal Papers to 1990*), W. S. Trimmer (Ed.), IEEE Press, New York.

1997: *Fundamentals of Microfabrication*, M. Madou, CRC Press, Boca Raton, FL.

1997: *Handbook of Microlothography, Micromachining and Microfabrication*, Vol. 2: *Micromachining and Microfabrication*, P. Rai-Choudhury (Ed.), SPIE Press.

1998: *Micromachined Transducers Sourcebook*, G. T. A. Kovacs, WCB McGraw-Hill, New York.

1998: *Microactuators*, M. Tabib-Azar, Kluwer Academic.

1998: *Modern Inertial Technology*, 2nd ed., A. Lawrence, Springer.

1998: *Methodology for Modeling and Simulation of Microsystems*, B. F. Romanowicz, Kluwer Academic.

1999: *Selected Papers on Optical MEMS*, V. M. Bright and B. J. Thompson (Eds.), SPIE Milestone Series, Vol. MS 153, SPIE Press.

1999: *Microsystem Technology in Chemistry and Life Sciences*, A. Manz and H. Becker (Eds.), Springer.

2000: *An Introduction to Microelectromechanical Systems Engineering*, N. Maluf, Artech House.

2000: *MEMS and MOEMS Technology and Applications*, Vol. PM85, P. Rai-Choudhury (Ed.), SPIE Press.

2000: *Electromechanical Systems, Electric Machines and Applied Mechatronics*, S. E. Lyshevski, CRC Press, Boca Raton, FL.

2000: *Handbook of Micro/Nano Tribology*, 2nd ed., B. Bhushan, CRC Press, Boca Raton, FL.

2001: *MEMS Handbook*, Mohamed Gad-El-Hak (Editor in Chief), CRC Press, Boca Raton, FL.

2001: *Microsystem Design*, S. D. Senturia, Kluwer Academic.

2001: *MEMS and Microsystems: Design and Manufacture*, T.-R. Hsu, McGraw-Hill College Division, New York.

2001: *Mechanical Microsensors*, M. Elwenspoek and R. Wiegerink, Springer.

2001: *Nano- and Microelectromechanical Systems*, S. E. Lyshevski, CRC Press, Boca Raton, FL.

2001: *Microflows: Fundamentals and Simulations*, 2nd ed., G. E. Karniadakis and A. Berskok, Springer Verlag.

2001: *Microsensors, MEMS and Smart Devices*, J. W. Gardner, V. K. Varadan, and O. O. Awadelkarim, Wiley, New York.

2001: *Microstereolithography and Other Fabrication Techniques for 3D MEMS*, V. K. Varadan, X. Jiang, and V. V. Varadan, Wiley, New York.

2002: *Fundamentals of Microfabrication* (*The Science of Miniaturization*), 2nd ed., M. Madou, CRC Press, Boca Raton, FL.

2002: *MEMS and NEMS: Systems, Devices and Structures*, S. E. Lyshevsky, CRC Press, Boca Raton, FL.

2002: *Microfluidic Technology and Applications*, M. Koch, A. Evans, and A. Brunnschweiler, Research Studies Press.

2002: *Fundamentals and Applications of Microfluidics*, N.-T. Nguyen and S. T. Wereley, Artech House.

2002: *Microelectrofluidic Systems Modeling and Simulation*, T. Zhang, K. Chakrabarty, R. B. Fair, and S. E. Lyshevsky, CRC Press, Boca Raton, FL.

2002: *Modeling MEMS and NEMS*, J. A. Pelesko and D. H. Bernstein, CRC Press, Boca Raton, FL.

2002: *Nanoelectromechanics in Engineering and Biology*, M. P. Hughes, CRC Press, Boca Raton, FL.

2002: *Optical Microscanners and Microspectrometers Using Thermal Bimorph Actuators*, G. Lammel, S. Schweizer, and P. Renaud, Kluwer.
2003: *RF MEMS Theory, Design and Technology*, G. M. Rebeiz, Wiley-Interscience, New York.
2003: *MEMS and Their Applications*, V. K. Varadan, K. J. Vinoy, and K. A. Jose, Wiley, New York.

REFERENCES

1. M. Madou, *Fundamental of Microfabrication,* CRC, Boca Raton, FL, 1779.

2. M. Madou, *Fundamental of Microfabrication, The Science of Miniaturization,* CRC Press, Boca Raton, FL, 2002.

3. H. C. Nathanson et al., "The Resonant Gate Transistor," *IEEE Trans. Electron Devices*, **ED-14**(3), 117–133, 1967.

4. V. Milanovic, M. Gaitan, E. Bowen, N. Tea, and M. E. Zaghloul, "Thermoelectric Power Sensor for Microwave Applications by Commercial CMOS Fabrication," *Trans IEEE Electron Device Lett.*, **18**(9), 450–452, 1997.

5. V. Milanovic, M. Gaitan, E. Bowen, and M. E. Zaghloul, "Micromachining Microwave Transmission Lines in CMOS Technology," *IEEE Trans. Microwave Theory Techniques*, **45**(5), 630–635, 1997.

6. M. Ozgur, M. E. Zaghloul, and M. Gaitan, "High Q Backside Micromachined CMOS Inductors," paper presented at the IEEE International Symposium on Circuits and Systems, Orlando, FL, May 1999, pp. II-577–II-580.

7. J. Wang, Z. Ren, and C. T. C. Nguyen, "1.14 GHz Self Aligned Vibrating Micromechanical Disk Resonator," paper presented at the IEEE Radio Frequency Integrated Circuits (RFIC) Symposium, June 2003, pp. 335–338.

8. M. Afridi, J. S. Suehle, M. E. Zaghloul, D. W. Berning, A. R. Hefner, R. E. Cavicchi, S. Semacik, C. B. Montgomery, and C. J. Taylor, "A Monolithic CMOS Microhotplate-Based Gas Sensor System," *IEEE SENSORS Journal* **2**(6), 644–655 (2002).

9. *Proceedings of the IEEE Special Issue on Integrated Sensors, Microactuators, and Microsystems [MEMS]*, IEEE, New York, August 1998.

Index

Fundamental attribution error, 647
Fundamental laws of Boolean algebra, 921
Fuzzy logic control, 802–803, 847

G

Gain:
 characteristic types of, 550–551
 of controlled sources, 474
Gain and shift amplifiers, 915
Gain function, 549, 552
Gain margin (GM), 782
Gain selection criteria, 773–777
 nonlinearities and control performance, 777
 optimal-control methods, 775
 performance indices, 773–774
 reset windup, 777
 Ziegler–Nichols rules, 775–776
Galil Motion Control, Inc., 797, 799, 800
Gamma distribution, intensity function for, 272
GAMS (General Algebraic Modeling System), 120
Gantt charts, 142
Gap analysis, 141
Gas sensors, 951, 952
Gating functions, 497–498
Gauge factor, 935
Gaussian distribution, 727
Gaussian frequency distribution, 572
GB (gigabyte), 926
G-codes (preparatory functions), 828
G-code level, 823
Gel permeation chromatography, 331
General Algebraic Modeling System (GAMS), 120
General distribution (hazard rate model), 152
General Electric, 303, 307
Generalized reduced gradient (GRG) method, 118–119
General Motors, 179
General-purpose control devices (GPCDs), 805–841
 characteristics of, 805–813
 device architecture of, 808–809
 hierarchical control of, 805–807
 numerical, 823–836
 path control of, 813
 programmability of, 807–808
 programmable, 813–822
 robot, 836–840
 sequential control of, 809–813
General sinusoid, 501
Genetic algorithms, 114
Geological engineering, virtual reality applied to, 409, 410

Geometric distributions, 260
Geometric elements, definition of, 30–31
Geometric features, contaminant trapping, 323
Geometric modeling, 9–11
Geometric programming problems, 114
Gigabyte (GB), 926
Glass transition temperature, 333–334
Global climate change, 182, 183
Global O2 Network, 208
Global Reporting Initiative (GRI), 179
Global Sullivan Principles, 178
Global truncation error, 722
GM (gain margin), 782
Goal constraints, 109
Goal programming, 99, 109–110
Government organizations, 79
GPCDs, *see* General-purpose control devices
GPU, *see* Graphics processing unit
Gradient-based methods, 115–116
GRAFCET specification language, 809
GRANTA, 88
Graphical design methods, 781–787
 dead-time elements, systems with, 783
 Nyquist stability theorem, 782–783
 open-loop for PID control, 783–784
 with root locus, 784–787
 software for, 800
Graphical representation of image data, 38–39
Graphics cards, 15
Graphics processing unit (GPU), 9, 15, 16
Green design, 210
Green Design Institute, 229
Green engineering, 210
Green Seal product standards, 179
GRG (generalized reduced gradient) method, 118–119
GRI (Global Reporting Initiative), 179
Ground:
 definition of, 679
 for electric circuits, 443
Group meetings, 639–640
Guard filter, 741
Guideline/checklist documents, 86–87
Gyrating transducers, 675
Gyrational resistance, 675
Gyration ratio, 675

H

Habit, 647
Half-power frequency, 550
Hamilton–Jacobi–Bellman equation, 867–872
Hamilton–Jacobi–Isaacs equation, 872–877